Handbook of
Metallurgical Process Design

MATERIALS ENGINEERING

1. Modern Ceramic Engineering: Properties, Processing, and Use in Design. Second Edition, Revised and Expanded, *David W. Richerson*
2. Introduction to Engineering Materials: Behavior, Properties, and Selection, *G. T. Murray*
3. Rapidly Solidified Alloys: Processes • Structures • Applications, *edited by Howard H. Liebermann*
4. Fiber and Whisker Reinforced Ceramics for Structural Applications, *David Belitskus*
5. Thermal Analysis of Ceramics, *Robert F. Speyer*
6. Friction and Wear of Ceramics, *edited by Said Jahanmir*
7. Mechanical Properties of Metallic Composites, *edited by Shojiro Ochiai*
8. Chemical Processing of Ceramics, *edited by Burtrand I. Lee and Edward J. A. Pope*
9. Handbook of Advanced Materials Testing, *edited by Nicholas P. Cheremisinoff and Paul N. Cheremisinoff*
10. Ceramic Processing and Sintering, *M. N. Rahaman*
11. Composites Engineering Handbook, *edited by P. K. Mallick*
12. Porosity of Ceramics, *Roy W. Rice*
13. Intermetallic and Ceramic Coatings, *edited by Narendra B. Dahotre and T. S. Sudarshan*
14. Adhesion Promotion Techniques: Technological Applications, *edited by K. L. Mittal and A. Pizzi*
15. Impurities in Engineering Materials: Impact, Reliability, and Control, *edited by Clyde L. Briant*
16. Ferroelectric Devices, *Kenji Uchino*
17. Mechanical Properties of Ceramics and Composites: Grain and Particle Effects, *Roy W. Rice*
18. Solid Lubrication Fundamentals and Applications, *Kazuhisa Miyoshi*
19. Modeling for Casting and Solidification Processing, *edited by Kuang-O (Oscar) Yu*
20. Ceramic Fabrication Technology, *Roy W. Rice*
21. Coatings of Polymers and Plastics, *edited by Rose A. Ryntz and Philip V. Yaneff*
22. Micromechatronics, *Kenji Uchino and Jayne R. Giniewicz*
23. Ceramic Processing and Sintering: Second Edition, *M. N. Rahaman*
24. Handbook of Metallurgical Process Design, *edited by George E. Totten, Kiyoshi Funatani, and Lin Xie*

Additional Volumes in Preparation

Handbook of
Metallurgical Process Design

edited by

George E. Totten
G. E. Totten & Associates, LLC
Seattle, Washington, U.S.A.

Kiyoshi Funatani
IMST Institute
Nagoya, Japan

Lin Xie
SolidWorks Corporation
Concord, Massachusetts, U.S.A.

MARCEL DEKKER, INC.   NEW YORK · BASEL

Although great care has been taken to provide accurate and current information, neither the author(s) nor the publisher, nor anyone else associated with this publication, shall be liable for any loss, damage, or liability directly or indirectly caused or alleged to be caused by this book. The material contained herein is not intended to provide specific advice or recommendations for any specific situation.

Trademark notice: Product or corporate names may be trademarks or registered trademarks and are used only for identification and explanation without intent to infringe.

Library of Congress Cataloging-in-Publication Data
A catalog record for this book is available from the Library of Congress.

ISBN: 0-8247-4106-4

This book is printed on acid-free paper.

Headquarters
Marcel Dekker, Inc., 270 Madison Avenue, New York, NY 10016, U.S.A.
tel: 212-696-9000; fax: 212-685-4540

Distribution and Customer Service
Marcel Dekker, Inc., Cimarron Road, Monticello, New York 12701, U.S.A.
tel: 800-228-1160; fax: 845-796-1772

Eastern Hemisphere Distribution
Marcel Dekker AG, Hutgasse 4, Postfach 812, CH-4001 Basel, Switzerland
tel: 41-61-260-6300; fax: 41-61-260-6333

World Wide Web
http://www.dekker.com

The publisher offers discounts on this book when ordered in bulk quantities. For more information, write to Special Sales/ Professional Marketing at the headquarters address above.

Preface

In addition to material selection and component design there are other equally important considerations that must be addressed in the overall process of design selection. One of these is process design, which not only affects cost and ease of production, but may also impact the final microstructure and mechanical properties of the component being produced. While there are various texts which address a particular process design such as forging, casting, and rolling, there is a need for a single text that will provide an overview of these processes as they relate to metallurgical component design. The objective of this text is to provide a thorough overview of the more important processes from the standpoint of the effect of design.

There are an extensive array of process designs discussed in this book. In Part One, Chapters 1 and 2 provide an overview of hot and cold forming process design, which includes forging process design. Chapter 3 details the effect of steel rolling process on microstructure and properties. Chapter 4 provides the most thorough and current overview on aluminum rolling process design available anywhere. Chapter 5 discusses semisolid metal-forming design. Chapter 6 provides a rigorous overview of the principles of aluminum extrusion process design and Chapter 7 is a comprehensive review of superplastic forming design.

Part Two focuses on casting process design for steel and aluminum, including continuous process designs in addition to a summary of various foundry casting process designs. Extensive guidelines for die casting process design are also included.

Various heat treatment practices are conducted to achieve the desired microstructural and mechanical properties of a particular material. Proper design is vital to the end-use properties of the component being produced. Part Three deals with various heat-treatment topics including: an overview of the effect of heat-treatment process design on hardening, tempering, annealing and other properties carburizing and carbonitriding, nitriding, induction heating, and laser hardening. Chapter 17 discusses the use of quench factor analysis for selection of appropriate quench media for aluminum processing. Chapter 18 covers the use of intensive quenching methodology to provide superior compressive stresses and fatigue properties and/or the replacement of more expensive steel alloys with less expensive plain-carbon steels.

Part Four deals with a topic of ever-increasing importance—surface engineering. This section includes topics on ion implantation, physical vapor deposition (PVD), chemical vapor deposition (CVD), and thermal spray process design. Coating process design for surface endurance is also discussed.

In Part Five, Chapter 22 provides information on designing for machining processes, which is a key topic in metallurgical process design.

This book is an invaluable reference for persons involved in any aspect of product design including metallurgists, material scientists, product and process engineers, and component designers. It is also appropriate for use in an advanced undergraduate or graduate class on material design.

We are indebted to the persistence and thorough work of the contributors to this book. We are also especially grateful for the patience and invaluable assistance provided by the staff at Marcel Dekker, Inc. throughout the preparation of this text.

George E. Totten
Kiyoshi Funatani
Lin Xie

Contents

Preface   *iii*

Contributors   *vii*

Part One   Hot and Cold Forming

1.   Design of Forming Processes: Bulk Forming   1
     Chester J. Van Tyne

2.   Design of Forming Processes: Sheet Metal Forming   23
     T. Wanheim

3.   Design of Microstructures and Properties of Steel by Hot and Cold Rolling   47
     Rafael Colás, Roumen Petrov, and Yvan Houbaert

4.   Design of Aluminum Rolling Processes for Foil, Sheet, and Plate   69
     Julian H. Driver and Olaf Engler

5.   Design of Semisolid Metal-Forming Processes   115
     Manabu Kiuchi

6.   Extrusion   137
     Sigurd Støren and Per Thomas Moe

7.   Superplastic Materials and Superplastic Metal Forming   205
     Namas Chandra

Part Two   Casting

8.   The Design of Continuous Casting Processes for Steel   251
     Roderick I.L. Guthrie and Mihaiela Isac

v

9.   Continuous Casting Design by the Stepanov Method   295
     Stanislav Prochorovich Nikanorov and Vsevolod Vladimirovich Peller

10. Production and Inspection of Quality Aluminum and Iron Sand Castings   349
     William D. Scott, Hanjun Li, John Griffin, and Charles E. Bates

11. Die Casting Process Design   401
     Frank E. Goodwin

Part Three   Heat Treatment

12. Heat-Treating Process Design   453
     Lauralice Campos Franeschini Canale, George E. Totten, and David Pye

13. Design of Carburizing and Carbonitriding Processes   507
     Małgorzata Przyłecka, Wojciech Gęstwa, Kiyoshi Funatani, George E. Totten, and David Pye

14. Design of Nitrided and Nitocarburized Materials   545
     Michel J. Korwin, Witold K. Liliental, Christopher D. Morawski, and George J. Tymowski

15. Design Principles for Induction Heating and Hardening   591
     Valentin S. Nemkov and Robert C. Goldstein

16. Laser Surface Hardening   641
     Janez Grum

17. Design of Steel-Intensive Quench Processes   733
     Nikolai I. Kobasko, Boris K. Ushakov, and Wytal S. Morhuniuk

18. Design of Quench Systems for Aluminum Heat Treating   765
     D. Scott MacKenzie

Part Four   Surface Engineering

19. Surface Engineering Methods   791
     Paul K. Chu, Xiubo Tian, and Liuhe Li

20. Design of Thermal Spray Processes   833
     Bernhard Wielage, Johannes Wilden, and Andreas Wank

21. Designing a Surface for Endurance: Coating Deposition Technologies   857
     Joaquin Lira-Olivares

Part Five   Machining

22. Designing for Machining: Machinability and Machining Performance Considerations   919
     I. S. Jawahir

Index   959

Contributors

Charles E. Bates, Ph.D.   University of Alabama at Birmingham, Birmingham, Alabama, U.S.A.

Lauralice Campos Franeschini Canale, Ph.D.   University of São Paulo, São Carlos, Brazil

Namas Chandra, Ph.D.   Florida State University, Tallahassee, Florida, U.S.A.

Paul K. Chu, Ph.D.   City University of Hong Kong, Kowloon, Hong Kong

Rafael Colás, B.Eng. (Met.), M. Met. Ph.D.   Universidad Autónoma de Nuevo León, San Nicolás de los Garza, N.L., Mexico

Julian H. Driver, B.Sc., Ph.D.   Ecole des Mines de Saint Etienne, St. Etienne, France

Olaf Engler, Ph.D.   Hydro Aluminium Deutschland, Bonn, Germany

Kiyoshi Funatani, Ph.D.   IMST Institute, Nagoya, Japan

Wojciech Gęstwa, Ph.D.   Poznań University of Technology, Poznań, Poland

Robert C. Goldstein, B.S.Ch.E.   Centre for Induction Technology, Inc., Auburn Hills, Michigan, U.S.A.

Frank E. Goodwin, Sc.D.   International Lead Zinc Research Organization, Inc., Research Triangle Park, North Carolina, U.S.A.

John Griffin   University of Alabama at Birmingham, Birmingham, Alabama, U.S.A.

Janez Grum, Ph.D.   University of Ljubljana, Ljubljana, Slovenia

Roderick I.L. Guthrie, A.R.S.M., Ph.D. D.I.C. F.R.S.C., F.A.E., F.C.I.M.Eng.   McGill University, Montreal, Quebec, Canada

Yvan Houbaert, Dr.Ir.   Ghent University, Ghent, Belgium

Mihaiela Isac, B.Sc.Eng., M.Eng., Ph.D.   McGill University, Montreal, Quebec, Canada

I. S. Jawahir   University of Kentucky, Lexington, Kentucky, U.S.A.

Manabu Kiuchi, Ph.D.   Kiuchi Laboratory, Tokyo, Japan

Nikolai I. Kobasko, Ph.D.   Intensive Technologies Ltd., Kiev, Ukraine

Michel J. Korwin   Nitrex Metal, Inc., St. Laurent, Quebec, Canada

Hanjun Li, Ph.D.   City University of Hong Kong, Kowloon, Hong Kong

Liuhe Li, Ph.D.   City University of Hong Kong, Kowloon, Hong Kong

Wiltold K. Liliental   Nitrex Metal Technologies, Inc., Burlington, Ontario, Canada

Joaquin Lira-Olivares, Ph.D.   Simon Bolívar University, Caracas, Venezuela

D. Scott MacKenzie, Ph.D.   Houghton International Inc., Valley Forge, Pennsylvania, U.S.A.

Per Thomas Moe, M.Sc.-Eng.   Norwegian University of Science and Technology, Trondheim, Norway

Christopher D. Morawski   Nitrex Metal, Inc., St. Laurent, Quebec, Canada

Wytal S. Morhuniuk, Ph.D.   Intensive Technologies Ltd., Kiev, Ukraine

Valentin S. Nemkov, Ph.D.   Centre for Induction Technology, Inc., Auburn Hills, Michigan, U.S.A.

Stanislav Prochorovich Nikanorov, Dr.Sc.   A.F. Ioffe Physical Technical Institute of Russian Academy of Sciences, Saint Petersburg, Russia

Vsevolod Vladimirovich Peller, M.D.   A.F. Ioffe Physical Technical Institute of Russian Academy of Sciences, Saint Petersburg, Russia

Roumen Petrov, Ph.D.   Ghent University, Ghent, Belgium

Małgorzata Przyłęcka, D.Sc.   Poznań University of Technology, Poznań, Poland

David Pye   Pye Metallurgical Consulting, Inc., Meadville, Pennsylvania, U.S.A.

William D. Scott, P.E.   AAA Alchemy, Birmingham, Alabama, U.S.A.

Sigurd Støren, Ph.D.   Norwegian University of Science and Technology, Trondheim, Norway

Xiubo Tian, Ph.D.   City University of Hong Kong, Kowloon, Hong Kong

George E. Totten, Ph.D., F.A.S.M.   G. E. Totten & Associates, LLC, Seattle, Washington, U.S.A.

George J. Tymowski   Nitrex Metal, Inc., St. Laurent, Quebec, Canada

Boris K. Ushakov, Ph.D.   Moscow State Evening Metallurgical Institute, Moscow, Russia

Chester J. Van Tyne, Ph.D.   Colorado School of Mines, Golden, Colorado, U.S.A.

T. Wanheim   Technical University of Denmark, Lyngby, Denmark

Andreas Wank   Chemnitz University of Technology, Chemnitz, Germany

Bernhard Wielage   Chemnitz University of Technology, Chemnitz, Germany

Johannes Wilden   Technical University Ilmenau, Ilmenau, Germany

Handbook of
Metallurgical Process Design

1

Design of Forming Processes: Bulk Forming

Chester J. Van Tyne
Colorado School of Mines, Golden, Colorado, U.S.A.

I. BULK DEFORMATION

Bulk deformation is a metal-forming process where the deformation is three-dimensional in nature. The primary use of the term *bulk deformation* is to distinguish it from sheet-forming processes. In sheet-forming operations, the deformation stresses are usually in the plane of the sheet metal, whereas in bulk deformation, the deformation stresses possess components in all three coordinate directions. Bulk deformation includes metal working processes such as forging, extrusion, rolling, and drawing.

II. CLASSIFICATION OF DEFORMATION PROCESSES

The classification of deformation processes can be done in one of several ways. The more common classification schemes are based on temperature, flow behavior, and stress state. The temperature of the deformation process is under direct control of the operator and has a profound effect on the viability of the process and the resulting shape and microstructure of the finished product. The flow behavior and the stress state differ from temperature in that they are a result of the actual deformation process that one chooses.

A. Temperature Classification

The temperature classification scheme is normally divided into two primary regions—cold working and hot working. Cold working occurs at relatively low temperatures relative to the melting point of the metal. Hot working occurs at temperatures above the recrystallization temperature of the metal. There is a third temperature range, warm working, which is being critically examined due to energy savings and is, in some cases, used by industries.

1. Cold Working Temperatures

Cold working usually refers to metal deformation that is carried out at room temperature. The phenomenon associated with cold work occurs when the metal is deformed at temperatures that are about 30% or less of its melting temperature on an absolute temperature scale. During cold work, the metal experiences an increased number of dislocations and entanglement of these dislocations, causing strain hardening. With strain hardening, the strength of the metal increases with deformation. To recrystallize the metal, a thermal treatment, called an anneal, is often needed. During annealing, the strength of the metal can be drastically reduced with a significant increase in ductility. The ductility increase often allows further deformation to occur before fracture. The final surface finish and dimensional tolerances can be well controlled in a cold work process.

2. Hot Working Temperatures

Hot working occurs at temperatures of 60% or above of the melting temperature of the metal on an absolute scale. At elevated temperatures, the metal has decreased strength, hence the forces needed for deformation are reduced. Recrystallization occurs readily, causing new grains to continually form during deformation. The continual formation of new grains causes the ductility of the metal to remain high, allowing large amounts of deformation to be imparted without fracture. Control of final dimensions is more difficult in a hot-worked metal due to scale formation and volumetrical changes in the part during subsequent cooling.

3. Warm Working Temperatures

Warm working occurs between hot working and cold working. It occurs in the approximate temperature range of 30–60% of the melting temperature of the metal on an absolute scale. The forces required to deform metal in the warm working regime are higher than during hot working. The final finish and dimensional tolerances are better than hot working but not nearly as good as a cold working process. Although warm work seems to have drawbacks, the primary driver for warm working is economic. There is significant cost in heating a metal up to hot working temperatures. If the working temperature is lowered, there can be major cost savings in the process.

B. Flow Behavior Classification

The flow behavior of a metal or alloy during bulk deformation processes falls into one of two categories— continuous flow or quasi-static. The easiest way to distinguish between these two types of flow is to imagine a movie being made of the deformation region during processing. If the shape of the deformation region changes during each frame of the movie, the process is a continuous-flow process. If in each frame of the movie the shape of the deformation region remains the same, even though a different material is in the region, it is a quasi-static-flow process. The bulk deformation process of forging is an example of a continuous-flow process. As the metal is being shaped in the forging die cavity, the deforming region, which is often the entire amount of metal, is continuously undergoing change. Processes such as rolling, wire drawing, and extrusion are examples of quasi-static flow. For example, in rolling, the deformation region is the metal being squeezed between two rolls. The shape of the deformation region does not vary, aside from initial startup and final finish, although different material flows into and out of the region.

The classification based on flow is useful in determining what type of modeling scheme can be used to simulate the bulk deformation process. For a quasi-static-flow process, the deformation region can often be handled as a single region and a steady-state type of analysis can be applied. For a continuous-flow process, a more complex analysis needs to be used to simulate the process accurately. The complex analysis needs to account for the continually changing shape of the deformation region.

C. Stress State Classification

In all bulk deformation processes, the primary deformation stress is compressive in nature. This is in contrast to sheet metal forming where tensile stresses are often used. Stress state classification consists of two categories for bulk deformation—direct compression and indirect compression. In direct compression, the tools or dies directly squeeze the workpiece. Forging, extrusion, and rolling are examples of direct compression processes. In indirect compression, the deformation region of the workpiece is in a compressive stress state but the application of these compressive stresses occurs by indirect means. Wire drawing is an example of an indirect compression process, where the wire is pulled through a die. The workpiece contacts the converging surfaces of the dies, creating high forces normal to the die surface. The dies react to these forces by pushing back on the workpiece, causing a compressive stress state to exist in the deforming region of the metal. Thus although the equipment action is of a tensile (pulling) nature, the plastic deforming region is being squeezed.

It should be noted that although the stress state for bulk deformation is compressive, there are situations where tensile stress components may be present within the workpiece and fracture may occur. The metal-forming engineer needs to be aware of these types of situations and to properly design the process to avoid the potential fracturing that can occur on the workpiece due to the tensile stress components. For example, in the forging of a right circular cylinder between two flat dies in the axial direction, if friction on the top and bottom surfaces is high, the sides of the cylinder will bulge and some tensile hoop stress may occur on the outside surface of the workpiece. A more insidious example is an extrusion process where a small reduction is performed through a die with a high die angle. For this situation, the deformation region may be limited to the surface region of the workpiece, causing some internal tensile stress components along the centerline of the workpiece. If the internal tensile stress components become excessively high, they can cause an internal fracture in the workpiece. This fracture is referred to as central burst. The worst aspect of central burst is that it cannot be detected via visual methods.

III. TYPES OF BULK DEFORMATION PROCESSES

A. Forging

Forging is a metalworking process where a workpiece is shaped by compressive forces using various dies and tools. The forging process produces discrete parts. Some finishing operations are usually required. Similarly shaped parts can often be produced by casting or powder metallurgy operations, but the mechanical properties of a forged component are usually superior compared to other processing methods. Forging can be done hot or cold. Warm forging is a process that is

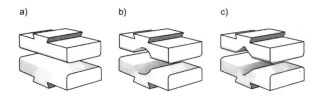

Figure 2 Open die forging tools: (a) flat dies; (b) U-shaped dies; and (c) V-shaped dies. (From Ref. 2.)

growing in popularity due primarily to thermal energy costs. Typical forged parts are shown in Fig. 1.

Open die forging consists of dies with very simple geometry. The dies are usually flat, U-shaped, or V-shaped, as seen in Fig. 2. The shaping of the metal occurs through manipulation of the workpiece and skill of the operator. It is a process that is useful in producing a small number of pieces. It is difficult to hold to close tolerance in this type of forging. Open die forging between two flat dies is often called upsetting. Cogging or drawing out is an open die forging process where the thickness of the workpiece is reduced by successive small strikes along the length of the metal. Open die forging is closely related to blacksmithing.

Closed die or impression die forging consists of a die set with a machined impression, as shown in Fig. 3. There is good use of metal in this operation as compared to open die forging. Excess metal beyond the size needed for forging is used and flows into the gutter portion of the die set to produce flash. The excess metal helps to insure that the cavities are completely filled at the end of the press stroke. Good tolerances and accuracy of the final forging are attainable. The die costs for closed die forging are fairly high due to their property requirements and machining costs.

Closed die forging often occurs in a sequence of steps. Each step of the operation usually has its own impression in the die block. The first step distributes metals into regions where extra volume is required in the final component. This step often involves edging, where extra metal is gathered, or fullering, where metal is moved away from the local region. In hot forging, the first step is referred to as busting because the scale on the surface of the workpiece is busted off. The second step is blocking, where the part is formed into a rough shape. The third step is finishing, where the final shape of the component is imparted to the workpiece. The fourth step is trimming, where the excess metal in the flash region is trimmed from the component. Figure 3 illustrates these various steps.

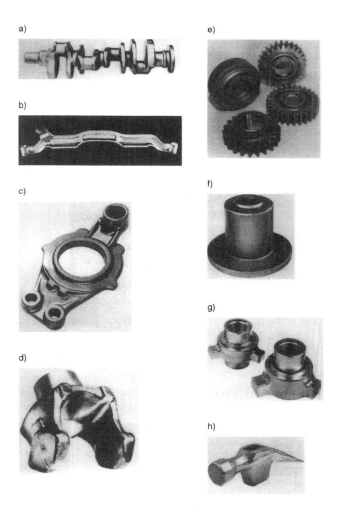

Figure 1 Examples of forged parts: (a) automotive crankshaft; (b) truck axle; (c) truck bracket; (d) universal joint; (e) automotive gears; (f) truck assembly part; (g) coupling fittings; and (h) hammer head. (From Ref. 1.)

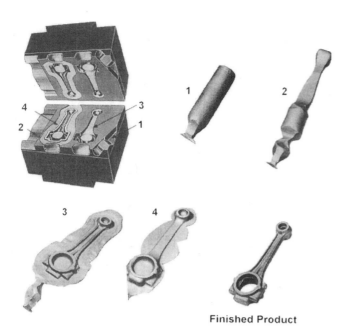

Finished Product

Figure 3 Impression forging dies with forging sequence. (From Ref. 3.)

The machines used for forging are hammers and presses. Hammers are energy-limited equipment and can be a simple gravity drop machine where a free-falling ram strikes the workpiece. Augmentation of the energy supplied to the hammer can be done in the form of pressured air, steam or hydraulic fluid. In a hot closed die operation, multiple blows are usually needed during each step, especially the blocking and finishing steps, when using a hammer to forge metal. Table 1 provides some numerical details about hammers for a typical gear blank forging.

For forging, there are three types of presses used—mechanical press, hydraulic press, and screw press.

Mechanical presses are stroke-limited equipment with a large flywheel powered by an electrical motor. The up-and-down motion of the ram is handled via a connecting rod attached to a crank shaft. The travel distance during each press stroke is controlled by machine design and operation. Hydraulic presses are load-limited equipment where the press will stop once its load capacity is reached. The power comes from pressurized hydraulic fluid. Screw presses, similar to hammers, are energy-limited equipment. A large flywheel transmits power through a vertical screw, which causes the ram to move. The ram movement stops when all the energy from the flywheel has dissipated. Table 2 provides some numerical details about hydraulic presses to produce the same gear blank as in Table 1.

In order to be successful in forging a metal, the formability of the metal needs to be understood, especially with regard to temperature and speed. The impression die shape needs to be carefully designed and machined to allow a good flow of metal without seams or laps developing. The die material needs to be carefully chosen to match the metal being shaped and the temperature of the operation.

B. Extrusion

Extrusion is a bulk deformation process where a billet, generally cylindrical, is placed in a chamber and forced through a die. The die opening can be round to produce a cylindrical product, or the opening can have a variety of shapes. Typical extrusion products are shown in Fig. 4. Because of the large reductions imparted during the extrusion process, most extrusion processes are performed hot in order to reduce the flow strength of the metal. Cold extrusion can occur but it is usually one step in a multistep cold forging operation.

Forward or direct extrusion is where the billet is pushed from the backside and the front side flows

Table 1 Characteristics of Hammers for Forging a 4.45-lb Steel Gear Blank

Hammer size	Process time (sec)	Minimum part temperature (°F)	Maximum part temperature (°F)	Die temperature (°F)	Load (tons)
4000 lb, 1 blow	0.003	2143	2359	502	850
2500 lb, 3 blows	2	2110	2219	418	874
1500 lb, 6 blows	5	2031	2158	506	818
1000 lb, 12 blows	11	1970	2117	553	389

Temperature buildup in dies is lower than press systems.
A 4000-lb hammer had 40% of initial energy available.
Good uniformity of temperature in part.
Source: Ref. 4.

Table 2 Characteristics of Hydraulic Presses for Forging a 4.45-lb Steel Gear Blank

Press size	Process time (sec)	Minimum part temperature (°F)	Maximum part temperature (°F)	Die temperature (°F)	Load (tons)
250 tons, slow	1.6	1458	2159	1233	250
500 tons	0.75	1533	2181	1164	500
1000 tons	0.33	1639	2194	1072	676
2000 tons, fast	0.18	1721	2198	996	705

Two-hundred-fifty-ton press stalled and left underfilled on outer diameter.
Fast 2000-ton press is similar to mechanical or screw press.
Smaller presses resulted in increased die temperature.
Source: Ref. 4.

through the die. Indirect or inverse or backward extrusion is where the die, which imparts shape, moves into the billet. The equipment used to perform an indirect extrusion is more complex than for a forward extrusion. To overcome the significant friction resistance between the billet and the chamber in a forward extrusion, hydrostatic extrusion has been developed. In hydrostatic extrusion, the billet is smaller than the chamber and is surrounded by hydraulic fluid. The hydraulic fluid is pressurized, which squeezes the billet through the die opening. Caution with both the sealing of the fluid and at the end of the process, where the final part of the billet could become a high-velocity projectile, needs to be exerted. Impact extrusion is similar to indirect extrusion and is often performed cold. The tooling, usually a solid punch, moves rapidly into the workpiece, causing it to flow backward and around the face of the punch. This produces a tubular-shaped type of product. These types of extrusions are schematically shown in Fig. 5.

The equipment for extrusion is normally a horizontal hydraulic press. A large shape change is imparted to the billet during a single stroke of the press. The shape change causes significant distortion in the metal during the deformation.

For success in extrusion, the temperature and speed of the process need to be determined based on the formability of the metal being deformed. Excessive temperature, speed, or friction can cause surface cracks to propagate along grain boundaries, which are referred to as fir tree cracking, due to hot shortness of the metal. Improper geometrical configuration of the tooling can cause central bursts if the angle of the die opening is too large, or the reduction is too small. Piping or cavitation at the end of the extrusion can be minimized by reducing the severity of the distortion in the product, or by reducing friction.

C. Rolling

Rolling is a direct compression deformation process, which reduces the thickness or changes the cross section of a long workpiece. The process occurs through a set of rolls, which supply the compressive forces needed to plastically deform the metal. Flat rolled products are classified as plate, sheet, or foil, depending on the thickness of the product. A plate has thickness greater than 6 mm, whereas a foil has thickness less than 0.1 mm. A sheet has thickness between that of the plate and the foil. Rolling can be done hot or cold. In many products, initial reductions are performed hot, where the metal can experience large shape changes without fracturing, and the final reductions are performed cold, so that better surface finish and tolerances can be achieved.

Flat rolling reduces the thickness of the metal, producing a product with flat upper and lower surfaces. Shape rolling can also reduce the thickness of the metal but, more importantly, it imparts a more complex cross-section shape. Shape rolling can be used to produce bars, rods, I-beams, channels, rails, etc. Ring rolling can be used to produce a seamless product by reducing the wall thickness of a ring through the action of two rolls. Seamless pipes can be produced and sized by specialized rolling operations such as rotary tube pierc-

Figure 4 Examples of extruded parts. (From Ref. 5.)

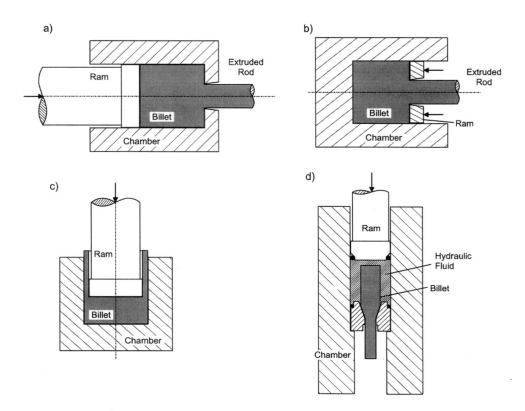

Figure 5 Schematics of extrusion processes: (a) direct or forward extrusion; (b) indirect or reverse extrusion; (c) impact extrusion; and (d) hydrostatic extrusion. (From Ref. 6.)

ing, tube rolling, and pilgering. A variety of rolling processes for steel are schematically shown in Fig. 6.

Although large, the rolling mill equipment is relatively simple. A two-high mill consists of two rolls, and a three-high mill consists of three rolls, which also allows reduction to occur on reverse directional flow of the metal. A four-high mill consists of two work rolls in contact with the metal and two back up rolls. A six-high mill is like a four-high mill, but has two additional rolls between the work roll and backup roll called intermediate rolls, which allow in essence some control over the crown and camber of the work rolls. Cluster mills exist usually for the production of thin foil products. A cluster mill will have a pair of small-diameter work rolls and a series of intermediate and backup rolls to support the work rolls. A tandem rolling mill will have a series of rolling stands where each stand imparts a specific amount of reduction. The operation of a tandem mill is challenging due to coupling effects between the stands.

Defects can be present in sheet and plate products if the rolling operation is not performed correctly. Wavy edges, waves along the centerline, zipper cracks along the centerline, or edge cracks can occur if the reduction is not uniform across the width of the metal. Crowned rolls, six-high mills, and sleeved rolls can be used to correct these types of defects by properly controlling the amount of roll bending that occurs. Small amount of waviness in a sheet product can be eliminated by a postdeformation leveling operation, where the sheet passes over a series of rollers while under tension. Alligatoring or fish tails can occur at the front end or back end of the workpiece. Proper alignment of the feed stock into roll gap, proper balancing of the friction between the top and bottom rolls, and proper choice of roll size for reduction can be used to minimize or to eliminate these two types of defects.

D. Drawing

Drawing of a round rod or wire is an indirect compression process where the cross-sectional area of the metal is reduced by pulling it though a converging die. A schematic illustration of wire drawing is seen in Fig. 7. The process is normally done at ambient temperatures. The major factors that need to be controlled include: reduction, die angle, friction at the die–workpiece in-

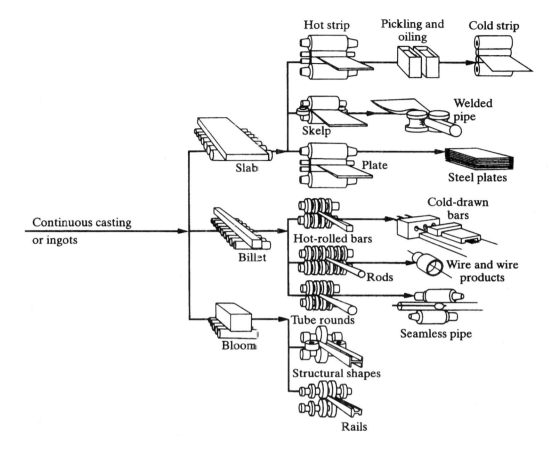

Figure 6 Schematics of various rolling processes for steel. (From Ref. 7.)

terface, and drawing speed. Tubes can also be drawn in a similar process. To control the interior diameter of a tube, a mandrel, which can be fixed, moving, or floating, is used. Because the metal is pulled through the die, the final product, which has the reduced cross section, is subjected to tensile stresses. If these tensile stresses become excessive, then the wire would fracture in a mode similar to a tensile test. The limit on the value of the tensile stress that can be supported limits the amount of reduction that can be achieved in one pass. Multiple reduction passes with multiple dies are needed to achieve large reductions in cross-sectional areas. The approach is analogous to a tandem rolling mill with multiple stands. The theoretical maximum reduction for a frictionless, perfectly plastic material is 63%. In production processing, the reduction that is used is often limited to 35% or 40%. The ironing process, which is used to reduce the wall thickness of a sheet metal, is also a drawing-type operation.

The configuration of the opening in the final die will control the configuration of the product produced. Although a cylindrical shape is the most common, other shapes can be imparted to the wire in the process.

The metal is cold-worked during the wire drawing process and intermediate anneals may be needed to increase its ductility to sufficient levels in order to reach the final reduction desired. Internal fractures, called

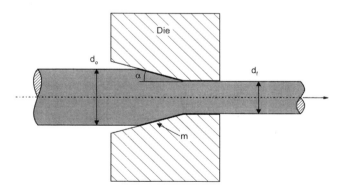

Figure 7 Schematic of a wire drawing process.

central busts, can occur if the die angle is too large, or the reduction is too small. For rods, tubular products, or high-strength wires, postdeformation straightening may be required.

IV. PROCESSING ASPECTS

A. Temperature

In bulk working operations, thermal energy is often supplied to the workpiece to increase its temperature. There are a number of methods used to heat up metal workpieces. Heating in a gas-fired furnace, induction heating, and electrical resistance heating are the most common methods that are used in industries. The operation and control of the heating process are critical features in controlling the deformation process. The workpiece needs to be at the proper working temperature in order to achieve the desired shape change and to have the proper microstructure for deformation.

The deformation in the workpiece is produced by mechanical work. Most of the mechanical work imparted into the workpiece during deformation is converted into heat. The heat causes the workpiece to increase in temperature. The maximum possible increase in temperature is often referred to as adiabatic heating and is calculated by assuming that the entire amount of mechanical work is converted in the temperature rise. The adiabatic temperature rise for a bulk deformation process can be calculated by:

$$\Delta T = \frac{W}{\rho C_{\mathrm{P}}} \tag{1}$$

where W is the mechanical work per unit volume for the deformation process, ρ is the density of the workpiece, and C_{P} is the heat capacity for the workpiece.

B. Strain

During bulk plastic deformation, a shape change is imposed on the workpiece. Strain is the normal measure to quantify the amount of deformation. In operations such as rolling, extrusion, and wire drawing, the cross-sectional area A of the workpiece normally decreases as the length L increases. In forging, the opposite usually occurs where the cross-sectional area increases and the height h of the workpiece decreases.

In most forming operations, the volume of the workpiece remains constant. The constancy of volume is expressed as:

$$A_0 L_0 = A_1 L_1 \tag{2}$$

Plastic deformation is often measured by the engineering strain:

$$e = \frac{L_1 - L_0}{L_0} = \frac{A_0 - A_1}{A_1} \tag{3}$$

or by the true strain:

$$\varepsilon = \ln\left(\frac{L_1}{L_0}\right) = \ln\left(\frac{A_0}{A_1}\right) = \ln(e + 1) \tag{4}$$

Often the measure of deformation for bulk deformation processes is expressed by the reduction in area:

$$R = \frac{A_0 - A_1}{A_0} \tag{5}$$

For forging, the equations will be similar:

$$A_0 h_0 = A_1 h_1 \tag{6}$$

$$e = \frac{h_0 - h_1}{h_0} = \frac{A_1 - A_0}{A_1} \tag{7}$$

$$\varepsilon = \ln\left(\frac{h_0}{h_1}\right) = \ln\left(\frac{A_1}{A_0}\right) = \ln(e + 1) \tag{8}$$

It should be noted that these equations are simplified measures for strain during the process. In bulk deformation, the strain in the workpiece will usually vary from point to point, and for a continuous-flow process, the strain will also vary at each time instant in the process. In its true form, strain is a second-order tensor, which, during deformation, has six unique components—three normal components and three shear components. In deformation operations, strain is often expressed by its three principal components ε_1, ε_2, and ε_3. For deformation processes, which have undergone proportional loading, the effective strain at a point in the workpiece is often given by the Mises equivalent strain:

$$\bar{\varepsilon} = \sqrt{\frac{2}{3}\left(\varepsilon_1^2 + \varepsilon_2^2 + \varepsilon_3^2\right)} \tag{9}$$

C. Strain Rate

During deformation processes, the speed of the operation is usually measured by strain rate. Strain rate $\dot{\varepsilon}$ is the time rate of the change of strain:

$$\dot{\varepsilon} = \frac{d\varepsilon}{dt} = \frac{1}{L}\frac{dL}{dt} = \frac{v}{L} \tag{10}$$

where v is the velocity.

Strain rate is an important variable because the strength and microstructural response of many metals is dependent on the strain rate. Like strain, strain rate in its true form is also a second-order tensor. The effective strain rate at a point in the workpiece can be expressed as:

$$\dot{\bar{\varepsilon}} = \sqrt{\frac{2}{3}(\dot{\varepsilon}_1^2 + \dot{\varepsilon}_2^2 + \dot{\varepsilon}_3^2)} \qquad (11)$$

where $\dot{\varepsilon}_1$, $\dot{\varepsilon}_2$, and $\dot{\varepsilon}_3$ are the principal strain rate components of the strain rate tensor.

D. Stress

In bulk deformation operations, stress has two meanings. The first meaning of stress is related to the equipment used to deform the workpiece. It is a measure of the load requirements necessary to get the workpiece to plastically deform. This is an important aspect that needs to be considered because the sizing of the equipment for bulk deformation is fundamentally dependent on the load requirements for plastic flow.

The second meaning of stress is related to the workpiece. During deformation, each point in the workpiece has a stress state, which is a measure of the materials' internal resistance to the externally supplied forces. These two meanings are interrelated.

In bulk metalworking operations, the external loads supplied are often compressive in nature. Wire drawing is an exception, where the supplied load is a tensile force. For compressive deformation processes, the pressure required for deformation usually describes the external stress. The pressure can vary from point to point along the tool–workpiece interface, often due to the friction resistance present. An average pressure for deformation to occur is:

$$p_{AVG} = \frac{F}{A} \qquad (12)$$

where F is the force or load supplied by the equipment, and A is the area over which the load is being supplied. For wire drawing, a similar equation can be used, but it determines the average drawing stress on the wire being pulled through the die:

$$\sigma_{AVG} = \frac{F}{A} \qquad (13)$$

The internal resistance within the workpiece to these external loads varies from point to point. The measure of this resistance is the internal stress that exists in the workpiece. If the specific point in the workpiece undergoes plastic deformation, then the internal stress is equal to the flow strength of the material at that point.

Internal stress, such as strain and strain rate, is a second-order tensor. This second-order tensor has six components—three normal components and three shear component. The stress tensor is often expressed in terms of the three principal components σ_1, σ_2, and σ_3.

The effective stress at a point within the workpiece is given by:

$$\bar{\sigma} = \sqrt{\frac{1}{2}\left((\sigma_1 - \sigma_2)^2 + (\sigma_2 - \sigma_3)^2 + (\sigma_3 - \sigma_1)^2\right)} \qquad (14)$$

If the effective stress at a point within the workpiece has reached the value of the flow strength of the material at that point, then plastic flow will occur.

If the effective stress and effective strain are known for the deformation process, then the work per unit volume of material for deformation W can be determined by

$$W = \int \bar{\sigma} d\bar{\varepsilon} \qquad (15)$$

Another important stress measure is the mean stress component or hydrostatic stress component:

$$\sigma_M = \frac{1}{3}(\sigma_1 + \sigma_2 + \sigma_3) \qquad (16)$$

For deformation processes, the stress components must be of a sufficient deviation from the hydrostatic stress to cause plastic flow to occur. A pure hydrostatic stress cannot cause plastic flow to occur within a normal material.

E. Friction

During bulk deformation processes, frictional resistance to sliding occurs at the interface between the workpiece and the tooling. The frictional resistance is due to the surface asperities that are present at the microscale on both the tools and the workpiece. These asperities impede the sliding motion that can occur during contact under pressure. Figure 8 schematically shows how the asperities interact to impede motion.

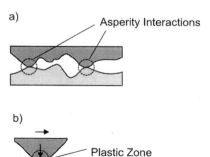

Figure 8 Schematic of frictional resistance and wear on sliding metal surfaces: (a) interactions of asperities; and (b) localized plastic deformation. (From Ref. 8.)

Friction causes the required deformation loads to increase. Friction causes the flow of the material to be less homogeneous. High levels of friction can result in surface damage to the workpiece, or seizing of the workpiece to the tooling.

Frictional resistance is usually described by a shear stress component τ_F. There are two basic models that are used describe the frictional stress component that occurs during metalworking operations. Both of these models are highly simplified and only capture the major aspect of the very complex interaction that occurs at the tool–workpiece interface.

The first model is referred to as Coulomb's law. The frictional stress component is directly proportional to the pressure that exists between the tool and the workpiece at the point of interest, or:

$$\tau_F = \mu p \tag{17}$$

where μ is the coefficient of friction. The value of μ can vary from 0 to $1/\sqrt{3}$ (i.e., 0.577). At low-pressure levels, this equation is a good description of the frictional stress component.

The second model is a better description at higher pressures at the interface. It is referred to as the constant friction factor equation. It assumes that the frictional stress component is some fraction of the flow strength σ_o of the workpiece:

$$\tau_F = m \frac{\sigma_o}{\sqrt{3}} \tag{18}$$

where m is the constant friction factor. The value of m can vary from 0.0 for an ideal frictionless interface to 1.0 for an interface where full sticking between the workpiece and tool occurs.

Friction is controlled through lubrication. The role of the lubricant in metalworking is important in reducing frictional resistance. Lubrication can also play a vital role in cooling the tooling, preventing heat flow from a hot workpiece into the tooling and protecting the new surfaces created during the deformation from oxidation or chemical reactions.

F. Yield Criteria

The ease with which a metal flows plastically is an important factor in deformation processes. The dominant factors that influence the flow (or yield) strength of a metal are the temperature and the amount of prior cold work. Yield criterion is the relationship between the stress state and the strength of the metal. When the criterion is met, then plastic deformation occurs. In uniaxial tensile tests, the yield criteria predict that flow will occur when the uniaxial tensile stress reaches the metals' yield strength. For bulk deformation processes, the stress state is not a simple uniaxial state, hence the criteria for yielding are more complex relationships.

The Tresca yield criterion or maximum shear stress criterion indicates that plastic flow will occur when:

$$\tau_{max} = \frac{1}{2}(\sigma_1 - \sigma_3) = \sigma_o \tag{19}$$

where σ_1 is the largest principal component of the stress state, σ_3 is the smallest principal component of the stress state, and σ_o is the flow strength of the metal. If Eq. (19) is satisfied, then plastic deformation will occur.

A more generally applicable criterion is the Mises criterion or maximum distortion energy criterion, which is:

$$\sqrt{\frac{1}{2}\left((\sigma_1 - \sigma_2)^2 + (\sigma_2 - \sigma_3)^2 + (\sigma_3 - \sigma_1)^2\right)} = \sigma_o \tag{20}$$

Other criteria for the relationship between the applied stress state and the flow strength of the metal, which can cause plastic deformation, do exist, but the two equations given here are the ones most often used to describe bulk deformation processes.

In three-dimensional principal stress space, both yield criteria will plot as surfaces. Thus the yield criteria are often called the yield surface for the metal. The surface for the Tresca yield criterion is a hexagonal-shaped prism, whereas the surface for the Mises yield criterion is cylindrical. If $\sigma_3 = 0$, then the yield surface reduces to yield loci curves in the two-dimensional $\sigma_1 - \sigma_2$ space. Figure 9 shows the relationship between the Tresca and Mises yield criteria in this reduced two-dimensional space.

G. Hardening

During cold work, the metal increases in strength with increased deformation. This phenomenon is referred to

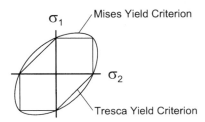

Figure 9 Comparison of Tresca and Mises yield criteria in reduced principal component stress space.

as hardening. Plastic hardening in metals is often reasonably well characterized by a power law equation, where the strength is dependent on the amount of plastic strain imposed:

$$\sigma_0 = K\bar{\varepsilon}^n \qquad (21)$$

where K is a strength coefficient for the hardening behavior and n is the strain hardening exponent. These two material parameters are usually obtained via a tensile or a compression test. Equation (21) indicates that the metal strengthens as the strain increases, which is isotropic hardening. In isotropic hardening, the yield surface is continually expanding with strain. If the strain path imposed on the metal during deformation is changed (e.g., if it is reversed), the yield strength on reversal may be different than expected for the strain imposed before the change. This difference is a manifestation of kinematical hardening, where the center point of the yield surface moves with strain. Figure 10 shows the difference between the yield surface changes that occur for isotropic hardening as compared to kinematical hardening.

V. DESIGN ISSUES TO PREVENT FAILURES

A. Geometrical and Mechanics Issues

The shape of the tooling and the initial shape of the workpiece are important geometrical factors for bulk deformation processes. Incorrect choices of these geometrical factors can lead to problems during deforma-

tion, or lead to process-induced defects in the final product being produced.

In extrusion, rolling and drawing the size and shape of the deformation zone have a strong influence on a variety of forming parameters, such as friction work, redundant work, and deformation loads, as well as properties in the formed part, such as internal porosity, internal cracking, distortion, homogeneity of strength, and residual stresses. A common single parameter measure of the deformation zone geometry is the Δ parameter. The Δ parameter is defined as the ratio of the average thickness or diameter h of the deformation region to the contact length L between the tooling and the workpiece, or:

$$\Delta = \frac{h}{L} \qquad (22)$$

It has been found that deformation under conditions of high Δ parameters can lead to microporosity along the center line of the workpiece, or, in extreme cases, can lead to internal cracks. Caution needs to be used when $\Delta > 2$ because it is this condition that can lead to problems. Figure 11 shows data from an extrusion process that exhibit both sound flow behavior and central burst.

Flow localization can occur in the workpiece during deformation. The common cause of flow localization is a dead metal zone between the workpiece and the tooling. Poor lubrication in forging can cause sticking friction between the die and the workpiece, and in the

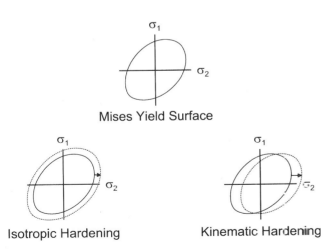

Figure 10 Comparison of isotropic hardening to kinematical hardening for a Mises material in reduced principal component stress space.

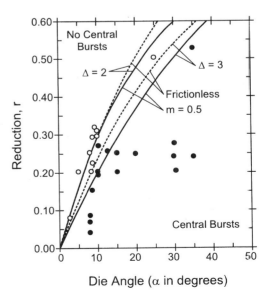

Figure 11 Criteria of the prevention of central burst in extrusions. (From Ref. 9.)

sticking region, a dead metal zone can occur. Forging dies, which are cooler than the workpiece, can extract heat from the metal, causing localized cooling. The metal at a lower temperature has higher flow strength and is more resistant to plastic deformation, which can lead to a dead region in the workpiece. In extrusion, dead metal zones can occur due to very large die angles and the metal will shear over itself, leaving a dead metal region adjacent to the die.

In closed die forging, the width and thickness of the land region are very important parameters. The land region is the choke point for metal flowing into the flash region of the forging. As multiple parts are forged, the land will wear away. The small thickness and large width of the land opening provide restrictive flow into the flash and cause increased pressure to occur in the die cavity. The increased pressure in the cavity allows for better filling of the impression, but at the cost of higher load requirements. If the flow of the metal inside of a cavity during forging is not properly controlled, a lap, a flow-through defect, or a suck-in defect may occur. A lap is where the metal folds back on itself. A flow-through defect occurs when the metal is forced to flow across a recess in the die that is already filled. A suck-in defect occurs when there is too much metal flow into a centrally located rib region. These types of defects can be avoided or minimized by proper redesign of the die cavity.

B. Metallurgical and Microstructure Issues

The common failure modes that occur in cold work deformation processes include: free surface cracking, shear bands, shear cracks, central bursts, and galling. In hot work processes, the common failures are hot shortness, central bursts, triple-point cracks, grain boundary cavities, and shear bands. Metallurgical aspects and microstructure features can have a strong influence on the tendency of the workpiece to experience one of these failure modes.

Because of the segregation and cast microstructure in ingots, these types of workpieces need to be hot-worked. Due to chemical segregation and microstructural inhomogeneities, the properties of an ingot are not constant from one location to another. Care must be taken to provide enough deformation to break down the cast structure. Low melting point phases may also be present and can lead to hot shortness if the temperature during deformation is not carefully controlled.

Hot working can lead to creep-type fractures, especially at slower working speeds in metals with low workability. It is also important not to let the workpiece be locally chilled during hot working processes. Chilling can lead to strength variations in the metal and cause the promotion of shear banding.

Cold working causes the strength of the workpiece to increase during deformation. Thus regions where significant cold work has been imparted to the metal are regions of higher strength. These strength variations can lead to internal shear banding. The grain size of the workpiece also can have an influence on the final product produced. Working of large grained metals can lead to a surface roughening phenomenon called orange peel, which is usually undesirable.

VI. WORKABILITY AND TESTING METHODS

A. Definition

Workability is a characteristic that is usually attributed to the metal or alloy. It is a relative measure of how easily the metal can be plastically deformed without fracture. It should be noted that workability depends not only on the metal itself, but also on other external processing factors. The temperature and stress state imposed by the processing conditions will strongly influence workability. Most metals have high higher workabilities at higher temperatures. Workability is usually higher under compressive states of stress as compared to tensile states. Terms such as formability, forgeability, extrudability, and drawability are often used to describe the workability within a specific metal-forming process.

B. Tests

A number of different mechanical tests are used to assess the workability of a metal or alloy. The best test is the one that most closely mimics the actual stress state that would exist in the metal during the bulk deformation operation. Unfortunately, the optimum is often not the easiest one to perform on the amount of material available, or is constrained by the type of laboratory testing equipment available for use.

1. Tensile Tests

The tensile test is the most common test used to evaluate the mechanical properties of a metal or alloy. The tensile test can be set at a variety of speeds to study strain rate effects and a variety of temperatures to study the properties of the metal as a function of temperature.

In a tensile test, a specimen of known initial geometry is placed in testing apparatus and pulled until fracture.

The pulling load and the tensile elongation are measured throughout the test with a strip chart or computerized data acquisition. Load and elongation are converted into engineering stress–strain data. From the engineering stress–strain curve elastic modulus, the yield strength, ultimate tensile strength, fracture stress, and tensile elongation can be determined. Figure 12 shows a typical engineering stress–strain curve for a metal. After the test specimen is removed from the testing apparatus, the final cross-sectional area in the fracture region can be measured and the reduction in area can be calculated. The reduction in area and the tensile elongation are the two primary measures for the ductility of the metal. The ductility determined from a tensile test is for the tensile stress state, temperature, and strain rate imposed on the specimen during the testing.

The engineering stress–strain curve can be transformed into a true stress–true strain curve for the metal. The transformation is valid between the yield point and the ultimate point, where uniaxial plastic deformation occurs and localized necking has not occurred. The data from a true stress–true strain curve can be plotted on a log–log scale. From such a plot, the slope is the strain hardening exponent n and the intercept is the logarithm of the strength coefficient K.

2. Torsion Tests

The torsion test is a fairly straightforward process. The specimen is held fixed on one end and the other end is twisted at a constant angular velocity. The torque needed to twist the sample and the angle of twist are the measured parameters. The deformation is caused by

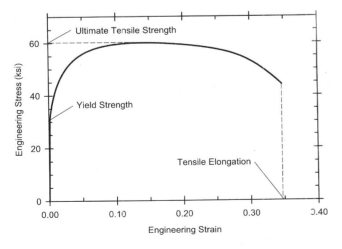

Figure 12 Engineering stress–strain curve from a uniaxial tensile test with material properties indicated.

pure shear and large strains can be achieved without flow localization and necking, which occurs in a tensile test, or barreling, which occurs during a compression test. The test is suitable in providing flow stress and ductility data for materials as a function of strain, strain rate, temperature, and prior processing. The test is frequently used to determine these material properties under hot working conditions. Because the strain rate imposed on the material is proportional to the rotational speed of the test, high strain rates (up to 10^3 sec^{-1}) are obtainable in a torsion test.

Because a torque is being applied to the specimen during the torsion test, the stress state in the material will vary from the centerline to the surface of the specimen. The variation in stress state in a torsion-tested specimen is in contrast to the tensile and compression tests where the stress state in the deforming region of the specimen is relatively uniform. The analysis of the torque twist data to produce stress–strain curves for the material needs to be done carefully, with an understanding of the test itself.

3. Compression Tests

Because most bulk deformation processes involve compressive states of stress, a compression test is often more desirable in assessing the workability of a metal that will be deformed by such a process. In theory, the compressive force imposed on the metal during a compression test creates a uniaxial stress state within the metal. If this were the case, then the analysis of the experiment would be handled in a manner similar to the data acquired via a tensile test. Unfortunately, the existence of a uniaxial stress state in a compression sample is not achieved because the specimen is compressed between two flat platens. The compression causes the cross-sectional area to increase and the friction that exists at the top and bottom surfaces, where the specimen is in contact with the platens, causes nonuniform flow. The unconstrained sides of the sample will show the nonuniform flow by bulging. A bulged sample is a clear indication that the stress state was not uniaxial.

To overcome this difficulty with friction, a variety of specimen geometries have been used, as shown in Fig. 13. Each specimen is compressed and the compressive strain in the axial direction and the diametrical strain are measured. Measurement is usually performed by imposing a grid onto the side surface of the specimen and periodically stopping the test to measure the change in dimensions of the grid pattern. When a cylindrical specimen is compressed, the strain path that it follows can be different, as shown in Fig. 14. The specimens are

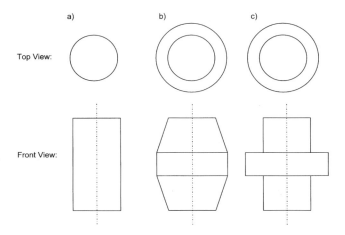

Figure 13 Schematics of compression test specimen geometries: (a) cylindrical sample; (b) tapered sample; and (c) flanged sample. (From Ref. 10.)

compressed until fracture occurs to assess the metals' workability during compression and produce a forming limit curve. Typical fracture curves (or forming limit diagrams) for 1020 steel, 303 stainless steel, and 2024-T351 aluminum are shown in Fig. 15.

4. Friction Tests

The most common method used to determine the friction factor for a forging process is the ring compression test. The test can be conducted at varying temperature and speed, and with the lubricant and workpiece mate-

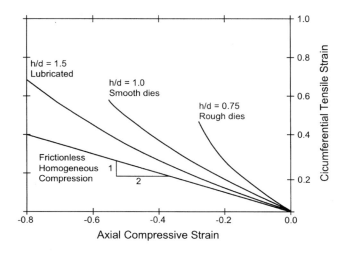

Figure 14 Strain paths for compression tests of cylindrical specimens with various height (*h*)-to-diameter (*d*) ratios and various lubrication conditions. (From Ref. 11.)

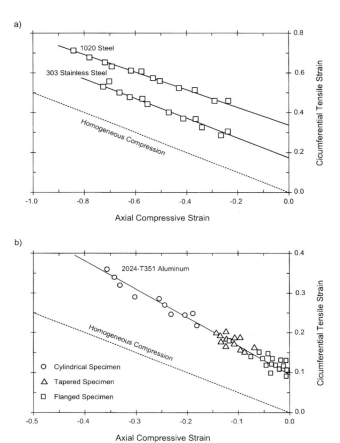

Figure 15 Cold upset compression failure criteria: (a) 1020 steel and 303 stainless steel; and (b) 2024-T351 aluminum. (From Ref. 11.)

rial of interest. The workpiece material is machined into a ring with dimensions usually in a 6:3:2 ratio of the outer diameter to the inner diameter to the thickness. The ring is compressed in the thickness direction to a given level of deformation and the new inside diameter is measured. Friction calibration curves can be used to determine the friction factor from the amount of deformation imparted to the ring and the change in inner diameter (Fig. 16). Rings of other dimensions can be used but the appropriate calibration curves must be used for the specific starting geometry.

VII. DEFORMATION MODELING METHODS

A diagram illustrating the input and output as well as the constraints, which must be considered when trying to model a bulk deformation process, is shown in Fig. 17. The input parameters fall into three major catego-

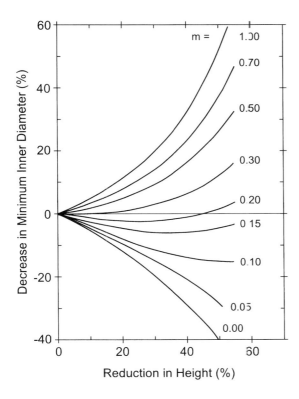

Figure 16 Ring test calibration curve for the determination of constant friction factor for rings with a 6:3:2 geometrical ratio of outer diameter/inner diameter/height. (From Ref. 12.)

ries—geometrical parameters, process parameters, and material parameters. Constraints imposed by either the product requirements or by the equipment should also be considered and incorporated into the model. Often models flag situations where one of the constraints is exceeded, rather than directly imposing the constraints.

The result of the modeling effort is the determination of process geometry and process performance conditions. Models, especially if they are complex and account for the fine details of the process, can take a long time to run and often the results cannot be determined in "real time." The models are normally used to provide a more detailed understanding of the process, rather than in a control scheme. For control of a specific bulk deformation process, empirical models based on historical operating data are often best suited for the task.

What occurs within the core of a model is shown in Fig. 18. In essence, the model must adhere to the laws of deformation mechanics. The relationships between stress and strain both within the deforming metal as well as within the tooling and at the interface between the workpiece and the tooling must be obeyed.

The stresses that are generated within the workpiece and the tooling must satisfy the equilibrium equations, yield criteria, metal flow properties, and stress boundary conditions. Likewise, the strains generated from these stresses must satisfy compatibility equations as well as incompressibility requirements and any imposed displacement boundary conditions.

For a model to be exact and complete, all of the requirements in Fig. 18 must be met for a given set of input parameters. The complete and exact solution, except in very simple cases, cannot be obtained. Often it is necessary to simplify the model by allowing some of the deformation mechanics requirements to be relaxed. Although this simplification does not give an exact solution, the solution obtained is often quite reliable for many processing situations. Simplifications are often necessary to obtain solutions. The amount of time and effort one is willing to invest is often directly proportional to the closeness of the solution to the exact solution. To get extremely close, a large investment of time, personnel, and funds is often needed.

To describe each of the individual techniques, a specific example will be used. The sample problem will be the open die compression forging of a right circular cylinder between two flat parallel platens (Fig. 19). This simple example is used primarily for illustrative purposes. It is equivalent to the initial breakdown (or pancaking) of an ingot or bar in an open die press or forge. This problem will be examined via the slab equilibrium, slip line, upper bound, and finite element method (FEM) techniques. The methods describe herein can be applied to other bulk deformation processes.

A. Slab Equilibrium

In the slab equilibrium technique, a small element (or slab) is extracted from the deforming workpiece (Fig. 20). A force balance is performed on this small slab. This balance of forces leads to a differential equation, which relates the stresses in the workpiece to the geometrical variables of the process. With the use of a yield criterion, an assumption of the principal stress directions, and some knowledge of the boundary conditions, a solution to the differential equation can be obtained. For simple geometrical shapes, an analytical solution is often achieved. For more complex shapes, the solution can only be obtained by numerically solving the differential equation. The solution relates the actual values of the pressure needed for deformation to the geometry, friction, and material properties.

For the forging of a cylindrical disk, an analytical solution can be obtained for pressure as a function of

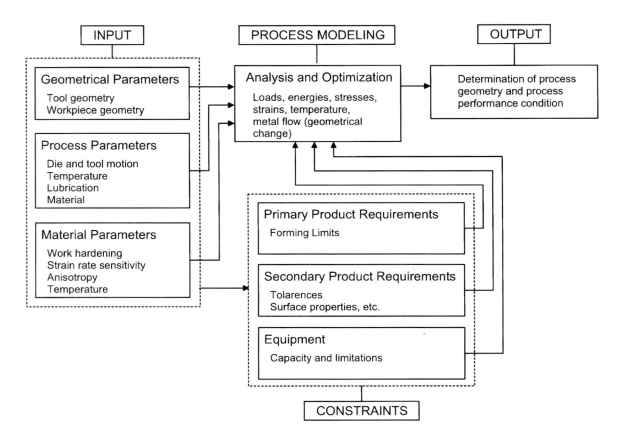

Figure 17   Factors involved in modeling of bulk deformation processes. (From Ref. 13.)

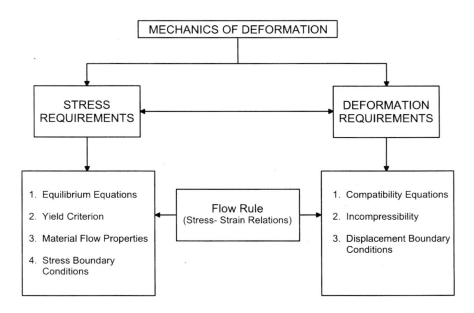

Figure 18   Fundamental mechanics involved in the core of the modeling of metalworking processes. (From Ref. 14.)

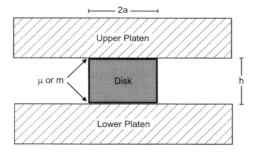

Figure 19   Schematic of open die disk forging process.

Table 3   Properties and Dimensions for Open Die Disk Forging Example

Variable	Description	Value	
a	Radius of disk	0.50 in.	
h	Height of disk	1.00 in.	
m	Coefficient of friction	0.25	For slab and FEM
m	Constant friction factor	0.50	For upper bound
s_o	Flow strength of metal	10.0 ksi	

the radial position along the disk. The solution is as follows:

$$p = \sigma_o e^{\frac{2\mu}{h}(a-r)} \tag{23}$$

$$p_{AVG} = \frac{1}{2}\left(\frac{h}{\mu a}\right)^2 \sigma_o \left[e^{\frac{2\mu a}{h}} - \frac{2\mu}{a} - 1\right] \tag{24}$$

$$F = p_{AVG}\pi a^2 \tag{25}$$

where p is the pressure at any point, σ_o is the material flow strength, μ is the coefficient of friction, a is the radius of the disk, r is the radial position, h is the thickness of the disk, p_{AVG} is the average pressure, and F is the load.

The slab equilibrium provides a solution at a discrete point in time. To determine how the load varies with displacement, an assumption of how the metal changes shape as a function of time must be used. If a uniform shape change is assumed (i.e., the disk remains as a right circular cylinder during the deformation—no bulge or foldover), then a load–displacement curved can be determined.

For an initial disk with the values for the parameters listed in Table 3, the load–displacement curve, up to a 75% reduction in thickness, is shown in Fig. 21. The

pressure distribution across the top of the disk can also be obtained from this method by using Eq. (23). Figure 22 illustrates this distribution for three different reductions—25%, 50%, and 75%. The large increase in the center of the disk is due to friction and this shape is usually called the friction hill.

B.   Slip Line Method

The slip line method is a classical approach to the analysis of deforming bodies. The term *slip line* is misleading to many metallurgists because they have a specific definition for the term. In mechanics, the slip line method probably should be called "maximum shear stress plane" technique.

In slip line method, a network of maximum shear stress planes is superimposed onto the deforming body. There are a variety of restrictions on the generation of such a network. The network must adhere to specific shape requirements and boundary conditions, and provide a realistic flow field for the deforming material. The method is only valid for plane–strain conditions. Because the open die compression of a right circular

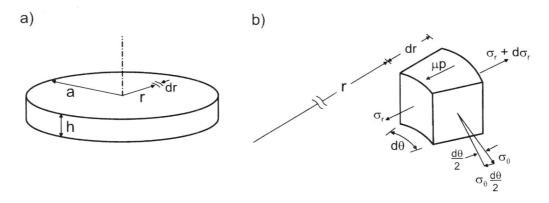

Figure 20   Schematic of slab equilibrium analysis for disk forging: (a) general geometry; and (b) slab element used for analysis.

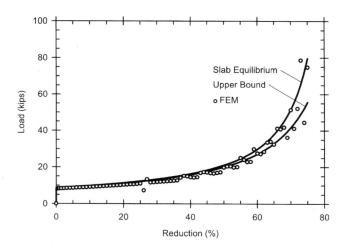

Figure 21 Comparison of load vs. reduction curves for the modeling of disk forging via several methods.

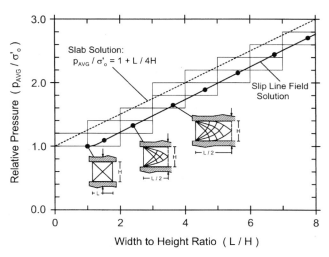

Figure 23 Comparison of the plane–strain forging analysis by slab equilibrium method and slip line field method. (From Ref. 15.)

cylinder is axisymmetrical and not plane strain, the analysis of this problem cannot be performed with the slip line technique.

Figure 23 shows a plane–strain open die forging, which has been solved by the slip line method. The figure also contains the relative averaged pressure for the deformation as predicted by the slab equilibrium technique. The plane–strain flow strength of the metal σ_o' is $2/\sqrt{3}$ times greater than the uniaxial flow strength σ_o. The inserted diagrams show the network of maximum shear stress planes, which is used for each point in the solution. The slip line method predicts a forging load,

which is lower than the load predicted by the slab equilibrium method.

The slip line technique imposes a velocity field on the deforming material through the positioning and orientation of the maximum shear stress network. Hence the velocity field is an implicit assumption within the method.

C. Upper-Bound Models

The upper-bound technique is an energy method where the energy per unit time needed by the workpiece to undergo deformation is set equal to the externally supplied energy per unit time. The primary power (energy per time) terms that must be calculated for the workpiece include: the internal power of deformation, the power to overcome friction, and the shear power. The internal power is determined from the assumed velocity field and is calculated from the strain rate field. The frictional power term is the power needed to overcome any tool–workpiece frictional interaction. The constant friction factor model is usually assumed for this type of analysis. The shear power is determined by calculating the energy per unit time associated with the internal shear that occurs over any assumed internal surfaces of velocity discontinuity.

For the open die forging of a right circular cylinder, the upper-bound solution is given as:

$$p_{\mathrm{AVG}} = \sigma_o \left(1 + \frac{2}{3} \frac{ma}{\sqrt{3}h} \right) \qquad (26)$$

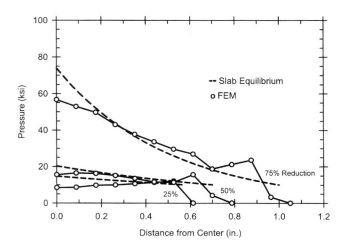

Figure 22 Comparison of pressure distribution over the top of the disk during forging via two different modeling methods.

where m is the constant friction factor. The first term inside the parentheses in Eq. (26) is the internal power term and the second term is the frictional term. For the simple forging process being examined here, there are no shear power losses.

The upper bound, such as the slab equilibrium method, only determines a solution at a discrete instance in time. Because the velocity is assumed, the solution at other time increments is readily available as long as the flow does not change the shape of the workpiece to one for which the solution is invalid. Figure 21 shows the load–displacement curve for the forging of a right circular cylinder with the same properties assumed for the slab equilibrium solution. A constant friction factor of 0.50 was assumed, rather than the value for a coefficient of friction.

The upper-bound solution does not provide a stress field, hence a plot similar to Fig. 22 for the upper-bound approach cannot be determined.

One of the advantages of the upper-bound technique is that it determines a value for the deformation load, which is greater or equal to the actual load. Hence with the use of this method, there is a built-in safety factor for specifying the size of the equipment to be used.

A major use of the upper-bound method is to predict conditions where a process-induced defect may form within the workpiece. Because it is an energy technique, a comparison between the energy needed for sound flow can be made to the energy needed for defect flow. The flow field, which requires the least amount of energy, is the one most likely to occur. For example, this method has been successful in developing criteria for the prevention of central bursts in wire drawing and extrusions, central bursts in double hub forging, central bursts in rolling, side surface cracking in forging with double action presses, cavitation in impact extrusion, fishskin defects in impact extrusion, and the beginning of the piping defect in extrusions.

D. Finite Element Analysis

The finite element method (FEM) is the technique that has received the most research effort during the last several decades. It is the one that produces an overwhelming amount of information about the process that is being modeled. The technique was developed in the 1960s for the analysis of elastic deformation in large complex structures (e.g., aircraft, bridges, buildings, etc.), which have a variety of constraints and loading conditions. The technique was extended in the 1970s and 1980s to the plastic deformation of metals.

In a FEM analysis, the workpiece and tools are discretized into a number of points, called nodes. The more points in the model present, the more accurate is the solution, but the more time it takes for the computer to calculate a solution. The nodes are linked to one another by elements, which obey specific deformation laws. The workpiece is given specific constraints, loads, and displacements, and an equilibrium solution is sought. If the displacements and loads are given as a function of time, the solution can be obtained as a function of time. The solution consists of the stresses and strains that exist at every node within the body and the tooling. Various interpolation methods are used to calculate values between the nodes. The solution to metal deformation problems requires the use of a computer and a skilled operator to interpret the results properly.

For the forging of a right circular cylinder with the properties given in Table 3, the load–displacement curve is shown in Fig. 21. The pressure across the top surface of the disk at reductions of 25%, 50%, and 75% is shown in Fig. 22. In both of these figures, the FEM solution is compared to other solutions. A mesh for this quarter disk was a grid of 20×20 square elements with a width of 0.025 in. The tooling was meshed with 16×7 rectangular elements 0.0714×0.0875 in. The original mesh and the deformed mesh at 75% reduction are given in Fig. 24.

In contrast to the other techniques, the velocity field is not assumed by the FEM analysis but is generated within the analysis itself. This forging of a right circular cylinder at 75% reduction exhibits both foldover and bulge (Fig. 24). Foldover is when the side surface of the disk comes in contact with the tooling surface. Bulge is when the center region of the free surface moves outward at a greater rate than the regions closer to the platens. Because the FEM is a numerical method, which produces a solution at a discrete number of points, the curves shown in Figs. 21 and 22 for the FEM analysis are not smooth.

Finite element method analysis can provide a large amount of information about the process. For example, the effective strain contours that exist within the forging at 75% reduction are shown in Fig. 25. The maximum strains occur in the center of the disk and at the original corners of the disk. The material directly beneath the platens in the center of the disk undergoes the least amount of strain. This type of information is useful for the prediction of possible shear banding. In addition, if the final properties of the product are dependent on the amount of strain, an indication of property gradients within the workpiece might be obtained from such a figure.

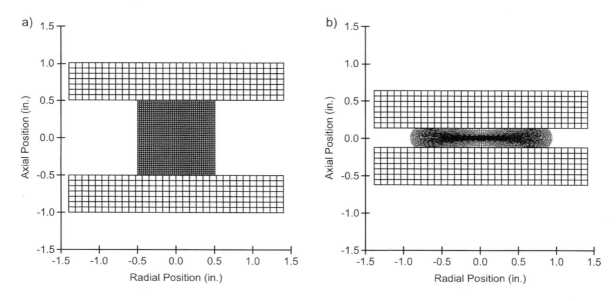

Figure 24 Finite element mesh for open die disk forging: (a) before deformation; and (b) after 75% reduction in height. (From Ref. 16.)

One of the advantages of the FEM technique is that realistic material properties can be assumed for the deforming workpiece and the tooling. All the other analysis methods normally are performed with idealized mechanical properties for the workpiece and the tools.

E. Modeling Limitations

Although modeling of bulk deformation processes is a very powerful and useful tool, there are several limitations that exist in all of the techniques. The first is an adequate description of the constitutive behavior of the deforming workpiece. In almost all cases, some simplification of the actual material flow behavior is assumed. To be accurate, the flow behavior should be known and mathematically characterized as a function of strain, strain rate, and temperature. If a good mathematical description for the material behavior exists,

then FEM analysis could use it. Unfortunately, these descriptions, even for common metals and alloys, are not often available.

The second limitation for all of these methods is in the modeling of the frictional interfaces between the tooling and the workpiece. The two friction models, which are used in these modeling methods, are simplifications for the complex interactions that occur at the tool–workpiece interface.

A third limitation is the specification of boundary conditions. The boundary conditions used for the analysis have a direct and profound effect on the results that are calculated. Poor choice of the boundary conditions, or choosing conditions that make the analysis easier rather than reflective of the real operation can result in misleading or erroneous results. The boundary conditions must be chosen with caution and care to ensure that the results validly reflect the reality of the process.

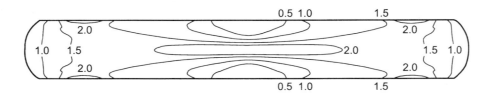

Figure 25 True strain contours predicted by FEM for the open die disk forging after 75% reduction in height. (From Ref. 16.)

FURTHER READING

Altan, T.; Boulger, F.W.; Becker, J.R.; Akgerman, N.; Henning, H.J. *Forging Equipment, Materials and Practices*; MCIC-HB-03: Battelle, Columbus, OH, USA, 1973.

Altan, T.; Gegel, H.L.; Oh, S.I. *Metal Forming—Fundamentals and Applications*; ASM: Metals Park, OH, USA, 1983.

Avitzur, B. *Metal Forming: Processes and Analysis*; McGraw-Hill: New York, NY, USA, 1968.

Avitzur, B. *Metal Forming: The Application of Limit Analysis*; Marcel Dekker: New York, NY, USA, 1980.

Avitzur, B. *Handbook of Metal Forming Processes*; John Wiley: New York, NY, USA, 1983.

Beddoes, J.; Bibby, M.J. *Principles of Metal Manufacturing Processes*; Arnold: London, England, 1999.

Blazynski, T.Z., Ed.; *Plasticity and Modern Metal-Forming Technology*; Elsevier: London, England, 1989.

Boër, C.R.; Rebelo, N.; Rystad, H.; Schröder, G. *Process Modelling of Metal Forming and Thermomechanical Treatments*; Springer-Verlag: Berlin, Germany, 1986.

Byrer, T.G., Semiatin, S.L., Vollmer, D.C., Eds.; *Forging Handbook*; Forging Industry Association: Cleveland, OH, USA, 1985.

Calladine, C.R. *Plasticity for Engineers—Theory and Applications*; Horwood: Chichester, England, 2000.

DeGarmo, E.P.; Black, J.T.; Kohser, R.A. *Materials and Processes in Manufacturing*; 8th Ed.; Prentice-Hall: Upper Saddle River, NJ, USA, 1997.

Dieter, G.E., Ed.; *Workability Testing Techniques*; ASM: Metals Park, OH, USA, 1984.

Dieter, G.E. *Mechanical Metallurgy*: 3rd Ed.; McGraw-Hill: New York, NY, USA, 1986.

Everhart, J.L. *Impact and Cold Extrusion of Metals*; Chemical Publishing: New York, NY, 1964.

Ginzburg, V.B.; Ballas, R. *Flat Rolling Fundamentals*; Marcel Dekker: New York, NY, USA, 2000.

Hartley, P., Pillinger, I., Sturgess, C., Eds. *Numerical Modelling of Material Deformation Process—Research, Development and Applications*; Springer-Verlag: Berlin, Germany, 1992.

Hill, R. *The Mathematical Theory of Plasticity*; Oxford University Press: Oxford, England, 1950.

Hosford, W.F.; Caddell, R.M. *Metal Forming—Mechanics and Metallurgy*; 2nd Ed.; Prentice-Hall: Englewood Cliffs, NJ, USA, 1993.

Johnson, W.; Mellor, P.B. *Engineering Plasticity*; Van Nostrand: London, England, 1973.

Johnson, W.; Sowerby, R.; Haddow, J.B. *Plane–Strain Slip Line Fields: Theory and Bibliography*; Elsevier: London, England, 1970.

Kalpakjian, S.; Schmid, S.R. *Manufacturing Engineering and Technology*; 4th Ed.; Prentice-Hall: Upper Saddle River, NJ, USA, 2001.

Kobayashi, S.; Oh, S.I.; Altan, T. *Metal Forming and the Finite-Element Method*; Oxford University Press: Oxford, England, 1989.

Larke, E.C. *The Rolling of Strip, Sheet and Plate*; 2nd Ed.; Chapman and Hall: London, England, 1963.

Laue, K.; Stenger, H. *Extrusion-Processes, Machinery, Tooling*; ASM: Metals Park, OH, USA, 1981.

Metals Handbook: Forming and Forging; 9th Ed.; ASM International: Metals Park: OH, USA, 1988; Vol. 14.

Mielnik, E.M. *Metalworking Science and Engineering*; McGraw-Hill: New York, NY, USA, 1991.

Open Die Forging Technology; Forging Industry Association: Cleveland, OH, USA, 1993.

Roberts, W.L. *Cold Rolling of Steel*; Marcel Dekker: New York, NY, USA, 1978.

Schey, J.A. *Tribology in Metalworking—Friction, Lubrication and Wear*; ASM: Metals Park, OH, USA, 1983.

Schrader, G.F.; Elshennawy, A.K. *Manufacturing-Processes and Materials*; SME: Dearborn, MI, USA, 2000.

Slater, R.A.C. *Engineering Plasticity—Theory and Application to Metal Forming Processes*; John Wiley: New York, NY, USA, 1977.

Spencer, G.C. *Introduction to Plasticity*; Chapman and Hall: London, England, 1968.

Talbert, S.H.; Avitzur, B. *Elementary Mechanics of Plastic Flow in Metal Forming*; John Wiley: New York, NY, USA, 1996.

Thomsen, E.G.; Yang, C.T.; Kobayashi, S. *Mechanics of Deformation in Metal Processing*; Macmillan: New York, NY, USA, 1965.

Tlusty, G. *Manufacturing Processes and Equipment*; Prentice-Hall: Upper Saddle River, NJ, USA, 2000.

Wagoner, R.H.; Chenot, J.L. *Fundamentals of Metal Forming*; John Wiley: New York, NY, USA, 1997.

Wagoner, R.H.; Chenot, J.L. *Metal Forming Analysis*; Cambridge University Press: Cambridge, England, 2001.

REFERENCES

1. Byrer, T.G., Semiatin, S.L., Vollmer, D.C., Eds.; *Forging Handbook*; Forging Industry Association: Cleveland, OH, USA, 1985; pp. 16–18.
2. *Metals Handbook: Forming and Forging*; 9th Ed.; Vol. 14. ASM International: Metals Park, OH, 1988; pp. 44.
3. DeGarmo, E.P.; Black, J.T.; Kohser, R.A. *Materials and Processes in Manufacturing*; 8th Ed.; Prentice-Hall: Upper Saddle River, NJ, USA, 1997; p. 476.
4. Walters, J. *Scientific Forming Technologies*; Columbus, OH, USA, 2000, personal communication
5. DeGarmo, E.P.; Black, J.T.; Kohser, R.A. *Materials and Processes in Manufacturing*; 8th Ed.; Prentice-Hall: Upper Saddle River, NJ, USA, 1997; p. 486.
6. Avitzur, B. *Handbook of Metal Forming Processes*; John Wiley: New York, NY, USA, 1983; p. 150.

7. Kalpakjian, S.; Schmid, S.R. *Manufacturing Engineering and Technology*; 4th Ed.; Prentice-Hall: Upper Saddle River, NJ, USA, 2001; p. 321.

8. Kalpakjian, S.; Schmid, S.R. *Manufacturing Engineering and Technology*; 4th Ed.; Prentice-Hall: Upper Saddle River, NJ, USA, 2001; p. 888.

9. Zimerman, Z.; Avitzur, B. Analysis of the effect of strain hardening on central bursting of strain hardening in drawing and extrusion. Trans. ASME J. Eng. Ind. 1970, *92*, 135–145.

10. Lee, P.W.; Kuhn, H.A. Cold upset testing. In *Workability Testing Techniques*; Dieter, G.E., Ed.; ASM: Metals Park, OH, USA, 1984; pp. 37–50.

11. Kuhn, H.A.; Lee, P.W.; Ertuk, T. A fracture criteria for cold forging. Trans. ASME J. Eng. Mater. Technol. 1973, *95*, 213–218.

12. Schey, J.A. *Tribology in Metalworking—Friction, Lubrication and Wear*; ASM: Metals Park, OH, USA, 1983; p. 451.

13. Kobayashi, S. Metalworking Process Modelling and the Finite Element Method. Proceedings of NAMRC IX; pp. 16–21.

14. Boër, C.R.; Rebelo, N.; Rystad, H.; Schröder, G. *Process Modelling of Metal Forming and Thermomechanical Treatments*; Springer-Verlag: Berlin, Germany, 1986; p. 20.

15. Hosford, W.F.; Caddell, R.M. *Metal Forming—Mechanics and Metallurgy*; 2nd Ed.; Prentice-Hall: Englewood Cliffs, NJ, USA, 1993; p. 203.

16. Van Tyne, C.J. Modeling of Open Die Forging Processes. *Mechanical Working and Steel Processing Conference Proceedings*; 1989; pp. 209–218.

2

Design of Forming Processes: Sheet Metal Forming

T. Wanheim
Technical University of Denmark, Lyngby, Denmark

I. SHEET METAL DEFORMATION

This chapter, dealing with sheet metal deformation processes, is closely connected with the chapter on bulk deformation. Much of the necessary basic information has been given there, so for many aspects a reference to the text in this part will be sufficient. The method of description of the area of sheet metal forming will follow the layout of the description of bulk metal forming (Chap. 1) to avoid confusion.

Therefore it is recommended that the reader goes through Chap. 1 before reading the present part.

As already mentioned in Chap. 1, sheet metal forming is, from a practical and industrial point of view, considered as a parallel to bulk metal forming but in many cases completely different. The plane stress situation normally present in sheet forming compared to the three-dimensional stresses in bulk forming has been a classical criterion for defining sheet metal forming, but also the differences in tooling and machinery have lead to this distinction. Bulk forming machines are heavy things, often designed for large forces and product weights. Sheet forming machinery is often designed for very bulky products and the press tables are often large, forces can be low in comparison to bulk forming, and tool loads and stresses are relatively light in comparison. Much bulk forming belongs to the group of primary processes, but almost all sheet metal forming is in the group of secondary processing.

However, the distinction between bulk forming and sheet forming, understandable as it is, leads to uncomfortable contradictions. Bending of a wide sheet is a plane strain process, and bending of a narrow bar is very

close to a plane stress process. Bending of sheet is considered as sheet forming, and bending of a narrow bar is bulk forming, even if the continuum mechanical treatment of the two is highly interrelated.

A rolling mill producing hot slabs or profiles in steel is considered a bulk forming machine, which seems right, but what about a rolling mill producing sheet for car industry or even producing aluminum foil for domestic use?

Deep drawing of a 1–2 mm sheet into a cup is clearly a sheet process even if the strain and stress situation is extremely complicated with large strains in the thickness direction. If the work piece is thick with a small diameter, e.g., for forming the first step in a sequence for shell manufacturing, the process is considered as a bulk forming process.

The manufacture of coins is normally considered a bulk process, but changing the lower tool into a leather cushion to produce bracteates with only one hard tool as it was carried out in the middle ages for lower denominations changes the process into a sheet process.

The classic division of metal forming processes in bulk and sheet forming is certainly not satisfactory from a mechanistic point of view, but is nevertheless very difficult to avoid from an industrial point of view.

II. CLASSIFICATION OF DEFORMATION PROCESSES

As stated in the chapter on bulk forming, classification can be performed in several ways as a general principle. The conventional way of basing the classification schemes on temperature, flow behavior, and stress state holds, of course, but some expansion and modification has to be performed.

The classification schemes based on temperature, flow behavior, and stress state have already been discussed in the bulk forming chapter and will not be repeated here.

The most rational way to classify forming processes and especially sheet metal forming processes is to follow the German DIN system for process classification, which is basically a stress state system. It should be noted that in the German system, there are three main groups: forming, separation and joining, also called mass preserving processes, mass reducing processes, and mass increasing processes.

Only sheet material processes belonging to the main group forming will be described below.

The main subdivisions within this group are processes with predominantly compressive stresses, processes with combined compressive and tensile stresses, processes with predominantly tensile stresses, processes with bending, and processes with shear.

This will be treated in some detail below.

A. Processes with Compressive Stresses

To this group belong processes where the flow in the deformation zone mainly is established by a compressive load from the surroundings. The group is dominated by processes from the bulk forming area, as, e.g., extrusion, forging, and plate rolling. However, some sheet forming processes can also be included in this group. Examples are:

Stretching
Dome forming
Cold heading of cups
Flanging
Coining
Necking-in
Edge rolling
Bulging

B. Processes with Compressive and Tensile Stresses

To this group belong processes where the flow in the deformation zone is established by a combination of tensile and compressive stresses. A large part of the sheet metal forming processes belong to this group. Examples are

Deep drawing
Rubber forming

Hydromechanical deep drawing
Spinning
Ironing

C. Processes with Tensile Stresses

To this group belong processes where the flow in the deformation zone is mainly caused by tensile stresses from the surroundings. Also, this group contains many sheet forming processes. Examples are:

Stretcher Leveling
Stretch forming
Bulging
Expanding
Beading
High energy rate processes

D. Processes with Bending

To this group belong processes where the flow in the deformation zone is caused by a bending moment, created either by external compressive or tensile loads or by external moments. Examples are:

Air bending
Die bending
Folding
Rollbending
Rollforming
Roller flanging

E. Processes with Shear

To this group belong processes where the flow in the deformation zone is caused by shear forces or moments from the surrounding tool. This group is not of any significance in sheet forming, and will not be further treated here. It must be mentioned that cutting, punching, blanking, nibbling, etc., are processes that are based on shear stresses, but because they belong to the main group ("separation"), they will not be treated here.

III. TYPES OF SHEET METAL FORMING PROCESSES

A. Processes with Compressive Stresses

1. Stretching

In this process, a sheet metal workpiece is shaped by stretching one of its sides by means of hammer blows or by notching with a wedge-shaped tool (Fig. 1a,b). The process, which is very flexible, can be carried out as well by hand as in special machinery.

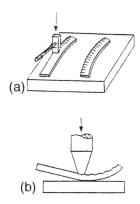

Figure 1 (a) Plane stretching; (b) three-dimensional stretching.

2. Dome Forming

In this process, a dome is formed from a sheet metal blank by local stretching made by multiple blows from a rounded hammer-shaped tool (Fig. 2). The process, which is very flexible, can be carried out as well by hand as in special machinery.

3. Cold Heading of Cups

In this process, the rim of a cup is reduced in length and increased in thickness with the aid of two counteracting tools (Fig. 3).

4. Flanging

In this process, a flange is formed on a hollow component by means of a punch and a die (Fig. 4).

5. Coining

The manufacture of coins, tokens, etc., can be considered as well a sheet metal forming process as a bulk process. The process takes place in special presses and highly loaded tools, often with a ring around the upper and lower coining tool, to increase the hydrostatic

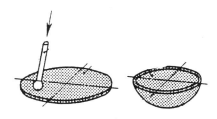

Figure 2 Dome forming.

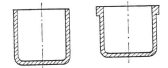

Figure 3 Cold heading of cups.

pressure in the process and to improve the coining precision (Fig. 5).

6. Necking In

In this process, the open-end diameter of a hollow component is reduced (Fig. 6).

7. Edge Rolling

In this process, the edge of a sheet or a sheet component is rolled back upon itself, usually by means of a punch and guide to give rigidity or a better appearance to an article (Fig. 7).

8. Bulging

In this process, an outward channel is formed in a hollow component by the application of axial pressure to the rim of the wall (Fig. 8).

B. Processes with Compressive and Tensile Stresses

1. Deep Drawing

This process, or rather group of processes, is one of the dominating manufacturing methods in sheet metal forming. Several types can be identified.

Deep Drawing Without a Blank Holder

This simple version of deep drawing uses only a die and a punch (Fig. 9). The maximum drawing ratio β (or

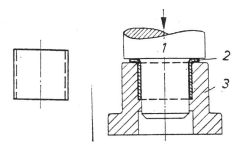

Figure 4 Flanging.

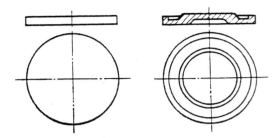

Figure 5 Coining.

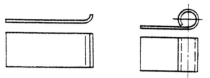

Figure 7 Edge rolling.

Limiting Drawing Ratio, LDR), defined as the ratio of the largest diameter of the blank that can be drawn without failure to the diameter of the punch ($\beta = D_{blank}/d_{punch}$), will be around 1.3, and the process limitation will be wrinkles in the flange.

It is often favorable to shape the contour of the die as a tractrix. The rim of the flange is in this case very stable against tangential compressive stresses. If the ratio between the punch diameter and the sheet thickness is below 40, drawing ratios up to 2.8 are possible. Other benefits are:

Simple press system (no blank holder force, only punch movement).
Easy to combine with ironing steps in the same punch movement.
Good surface quality.
Less lubrication troubles.

The drawbacks are:

Long punch travel.
Somewhat complicated manufacturing of the die.
Almost exclusively limited to axisymmetric geometries.
Large thickness variations in cup wall.
Drawing ratio must be above a certain value.

Conventional Deep Drawing

In conventional deep drawing, the tool system consists of punch, die, and blank holder. Figure 10 shows

the principal layout of a conventional deep drawing. The blank can be divided into three zones, as shown. The outer annular zone X consists of material in contact with die and blank holder. The annular zone Y is at the beginning of the process neither in contact with the punch nor the drawing die. Zone Z is in contact with the punch nose.

In the drawing process, zone X is pulled in toward the profile radius of the drawing die by the radial tensile stresses. The blank holder has to keep the sheet from wrinkling caused by the circumferential compressive stresses during the decrease of the blank diameter. These stresses also result in increase of the thickness of the outer part of the blank.

When the material in zone X passes over the die radius, it will become thinner because of the simultaneous stretching/bending and stretching/unbending. The inner part of zone X will undergo a further thinning because of the tensile stresses between punch and die.

Zone Y will undergo bending, unbending, and sliding under tension over the die radius, pure stretching between punch and die, and bending and sliding under tension over the punch nose radius.

Finally, zone Z undergoes stretching and sliding over the punch nose. The thinning is dependent on the geometry, the friction, and of course the tensile load from the drawing process.

Thus in zone Y, there is a narrow band under pure tension between two bands of combined bending and stretching. This will result in a thicker band between two circular bands with necking A and B (Fig. 11), showing

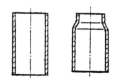

Figure 6 Necking in.

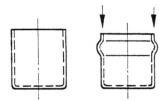

Figure 8 Bulging.

Figure 9 Deep drawing without blank holder.

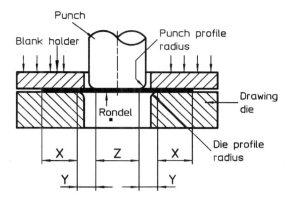

Figure 10 Principal layout of a conventional deep drawing.

the thickness variations (exaggerated) in a cylindrical cup with flat and hemispherical head, respectively.

As a simple approximation, one can assume that there is no net thickness change from the blank to the finished cup, and simple geometrical considerations then give for the height to diameter ratio, h/d, of a finished flat-bottomed cup:

$$h/d = 1/4(\beta^2 - 1) \tag{1}$$

Several different processes in the blank can be identified during the transformation from blank to cup:

Pure radial drawing and sliding against tools on outer annular area.
Pure radial drawing between die and blank holder.
Bending when entering die radius.
Sliding against die radius combined with continued radial drawing.
Unbending in exit of die radius.
Stretching between die and punch.
Bending over punch profile radius.
Sliding over punch profile radius.
Stretching and sliding over punch nose.

DEEP DRAWING OF CIRCULAR BLANKS. (1) Calculation of maximum drawing force. The maximum drawing force $P_{z,max}$ can be calculated by the following formula, which is based on the slab method used on the separate elements of the process described above.

$$P_{z,max} = \pi d_m t \cdot [B(A + C) + D] \tag{2}$$

where

$B = e^{\mu\pi/2}$	is the contribution from friction in the die radius
$A = 1.1\sigma_{0,1} \ln(d_{P,max}/d_m)$	is the contribution from the ideal plastic work in flange
$C = 2\mu P_{BH}/\pi d_{P,max} t$	is the contribution from the blank-holder friction
$D = t\sigma_{0,2}/2\rho_{die}$	is the contribution from the bending/unbending at the die radius

In these expressions, the following symbols are used:

D_0	is the original blank diameter
$d_m = d_1 + t$	is the middle cup diameter
d_1	is the punch diameter
t	is the sheet thickness
μ	is the coefficient of friction at die radius and at outer rim of blank
$d_{P,max}$	is the diameter of the blank when maximum drawing force is reached. This is dependent on the strain-hardening properties of the sheet. Thus for a drawing ratio of 2, $n = 0.1$ gives $d_{P,max}/D_0 = 0.94$ and $n = 0.5$ gives $d_{P,max}/D_0 = 0.79$.
$\sigma_{0,1}$	is the mean yield stress in the flange between rim and start of die radius
$\sigma_{0,2}$	is the mean yield stress in the die radius curvature
P_{BH}	is the blank-holder force
ρ_{die}	is the die radius

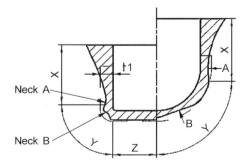

Figure 11 Thickness variations (exaggerated) in deep-drawn cylindrical cup with flat and hemispherical head.

(2) Calculation of blank holder pressure. The blank holder can be of one of three types:

Constant gap blank holder for single-action presses.
Spring-loaded blank holder for single-action presses.
Pneumatic or hydraulic blank holder for single-action presses.
Outer ram activated blank holder for double-action presses.

The most common type for serious production is the outer ram activated blank holder.

The thickness increase described above means that only an outer annulus of the blank will carry the load from the blank holder, and the friction from this will result in a back-pull acting against the tensile stresses originating from the drawing process. Therefore the main part of the blank is not in contact with the blank holder.

The right pressure is found as the minimum pressure necessary for preventing wrinkles in the blank. As a rule of thumb, the blank-holder force is about 1/3 of the drawing force.

The blank-holder pressure p_{BH} and the blank-holder force P_{BH} are given by the following two equations:

$$p_{BH} = (2 - 3) \times 10^3 [(\beta - 1)^3 + 0.5 \times 10^{-2} \qquad (3)$$
$$\times D_0/t] \sigma_{UTS}$$

where

$$P_{BH} = p_{BH} A_{BH} \qquad (4)$$

A_{BH} is calculated as the area of the sheet covered by the blank holder, even if the actual area of contact as mentioned is much smaller and limited to the rim of the blank.

(3) Tooling geometry. If the profile radius of the drawing die is larger than 10 times the sheet thickness, the influence on the drawing force is rather small. The drawing force will increase for decreasing radii as a consequence of the increased bending and unbending strains. Radii substantially larger than 10 times will increase the tendency to wrinkle.

The profile radius of the punch is important as the fracture always occurs at the lower neck (B in Fig. 11). For a constant blank diameter, a larger radius gives a smoother increase in the drawing force and a longer punch travel, but the maximum drawing force is unchanged. A larger profile radius will also give an increased thinning of the bottom of the cup.

The clearance between the punch and the die is normally chosen to 10–30% in addition to the sheet thickness, as the outer zone of the blank has an increased thickness. If the clearance is too small, the blank may be sheared or pierced by the punch.

(4) Material and lubrication. The stress–strain curve of the sheet material is important. A high strain-hardening exponent results in that the maximum drawing force is reached later in the process than for a low exponent. A low exponent will mean lower load at the punch nose and consequently that the thinning of the bottom of the cup decreases; in other words, a high exponent will give better drawing ratios. Anisotropy, treated later in "Isotropy and Anisotropy," is important for deep drawing, as explained there. Lubrication is important for the process, and if possible the blank should be lubricated where it slides against the die and against the blank holder, and not where it touches the punch, where no sliding is wanted.

Deep Drawing of Square and Oval Components

When deep-drawing circular components, the deformation is evenly distributed around the axis. This is not the case when deep-drawing square and oval components. Here the highest strains are located at the corners, and lower strains are found along the side walls. It is possible to apply the rules for drawing of circular components to the drawing of noncircular parts by replacing the punch and the blank areas by circles of equal size and calculating their respective equivalent diameters $d_{punch,eq.}$ and $D_{blank,eq.}$

$$d_{punch,eq.} = 2(A_{punch}/\pi)^{1/2} \qquad (5)$$

and

$$D_{blank,eq.} = 2(A_{blank}/\pi)^{1/2} \qquad (6)$$

Drawbeads are often used to control the flow of the blank into the die cavity. A draw bead is typically a small oblong obstacle shaped like a small half cylinder, built into the blank holder or the die surface to increase the local stresses needed to pull the sheet into the die cavity. It can also be shaped as a step in the common contour between die surface and blank holder. Beads restrict the flow of the sheet metal by bending and unbending it during drawing.

Draw beads also help to reduce the required blank-holder force, because the beaded sheet has a higher stiffness and hence a lower tendency to wrinkle.

Redrawing

A typical maximum drawing ratio can be $\beta = 2$, which means that a blank with a diameter of 200 mm can be drawn into a cup with a diameter of 100 mm and a height of around 75 mm. To produce a cup with a larger ratio of height to diameter, one or more redraws are required.

The methods for redrawing falls into two main groups: direct redrawing and reverse redrawing.

DIRECT REDRAWING. The tooling for direct redrawing is different from the first drawing tool (Fig. 12a,b). In Fig. 12a, the blank has to go through two sequences of right angle bending and unbending. This is not so severe in Fig. 12b, where the die is made conical with an angle of 30–45°, and the blank holder, which fills out the drawn cup, consequently has a conical nose. The redrawing punch slides in the blank holder in the process.

Several redrawing steps can follow each other, but the drawing ratio in these steps must decrease because of the strain hardening. The resulting drawing ratio can be written as:

$$\beta_{tot} = \beta_1 \cdot \beta_2 \cdot \beta_3 \cdots \beta_n. \tag{7}$$

Resulting drawing ratios up to about 6 can be reached without annealing for deep-drawing qualities of steel sheet.

For a component having to undergo many redraws, a low strain-hardening exponent is advantageous unless interstage annealing is introduced. This is contrary to the situation where only one draw is necessary, where a high strain hardening helps to prevent excessive thinning of the bottom of the cup.

REVERSE REDRAWING. In reverse drawing, the original inner side of the first drawn cup becomes the outer side after the reverse draw. If the tool has the geometry shown in Fig. 12c, only one bending and unbending is necessary. The two draws can be carried out in the same operation, as the punch from the first draw acts as die ring for the reverse draw, as shown in Fig. 12d. This of course gives reduced tool costs.

2. Rubber Forming I (Guerin Process)

In this process, a rubber pad constitutes the one half of the usual die set (mostly the female part) and plays a universal role in that it adapts to different shapes of the counterpart (Fig. 13). The advantage is the inexpensive tooling; the disadvantages are high press forces and the limited life of the rubber pad, which, depending on the severity of the operation, may not endure beyond about 20,000 pieces. This makes the process interesting for small batch production. Polyurethanes are normally used because of their resistance to abrasion, their resistance against being damaged by sharp edges or burrs, and their long fatigue life. The rubber pad is either a

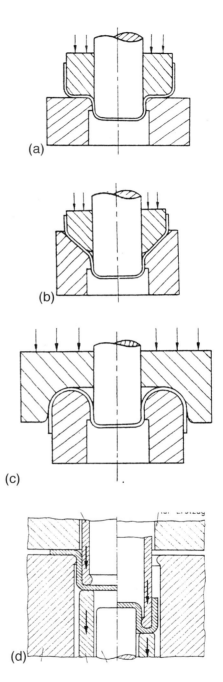

Figure 12 (a) Direct redrawing, right-angle tool; (b) direct redrawing, conical tool; (c) reverse redrawing, separate tool; (d) reverse redrawing, combined tool.

solid block or is consisting of several slabs cemented together, and it is contained in a steel housing. The advantage of cementing rubber slabs together is that different hardness can be used in different layers, normally the hardest closest to the forming surface. A wear pad is often used, placed on the blank before the

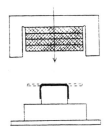

Figure 13 Rubber forming, Guerin process.

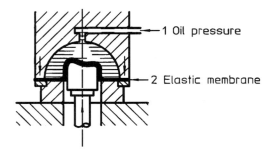

Figure 15 Rubber forming, hydroform process.

pressing. The thickness of the total pad must be about 30% greater than the height of the form block.

3. Rubber Forming II (Marform Process)

In the Marform process, deep-recessed parts with either vertical or sloped walls can be formed. The blank is held against the rubber pad by a blank holder, through which a punch is acting as in conventional deep drawing (Fig. 14). A controlled pressure is applied on the blank holder during the process, whereby a pressure from the rubber is exerted between the partly drawn cup and the punch. The resulting friction allows deeper draws to be carried out, and the blank-holder pressure will give a component free from wrinkles.

4. Rubber Forming III (Hydroform Process)

In the hydroform or fluid form process, a controlled hydraulic pressure is acting on a rubber diaphragm covering the blank (Fig. 15). In this way, a more uniform pressure can be obtained and more severe draws can be achieved than by the other two methods.

5. Hydromechanical Deep Drawing

In hydromechanical deep drawing, a hydraulic pressure squeezes the partly drawn cup directly around the punch (Fig. 16). This results again in an unloading of the tensile stresses around the punch nose and thereby a

considerable increase in the possible drawing ratio. The process can be used for very complicated geometries.

6. Spinning

The group of spinning processes involves the forming of axisymmetric parts over a mandrel with tools or rollers. There are three basic types of spinning processes: conventional spinning, shear spinning, and tube spinning. The spinning machines are from a kinematic point of view similar to lathes, but the static and dynamic properties are very different, the main difference being that in a lathe the tangential (normally vertical) force is an order of magnitude larger than the radial force, this being reversed for a spinning machine.

Conventional Spinning

In conventional spinning, a circular blank of flat or preformed sheet material is held against a mandrel and rotated while a rigid tool deforms and shapes the material over a mandrel (Fig. 17). The tools may be ac-

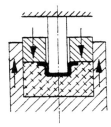

Figure 14 Rubber forming, Marform process.

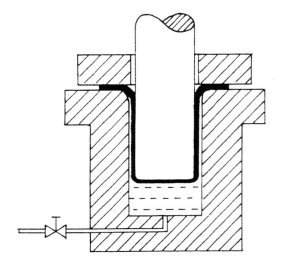

Figure 16 Hydromechanical deep drawing.

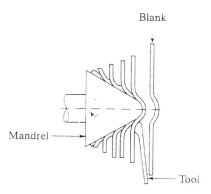

Figure 17 Conventional spinning.

tivated manually or by a computer-controlled hydraulic mechanism. The process involves a sequence of passes and requires considerable skill. The process is particularly suited for conical or curvilinear shapes in relatively small batches. As the particles of the blank are drawn into smaller radii, the wrinkling problem is serious. The process is normally carried out at room temperature, but spinning at elevated temperatures is used, e.g., for low-ductility or high-strength materials. Very large parts, with diameters up to 6 m, are formed this way.

Shear Spinning

In shear spinning, axisymmetric conical or curvilinear components are formed by reducing the thickness and maintaining the diameter of the blank (Fig. 18). If the geometry is correct, this means that no radial displacements are present, and consequently no circumferential stresses are introduced. From this follows, for instance, that a conical shape can be spun on a square blank without dimensional changes or distortions in the plane of the blank. Components for space technology

and military use are produced this way. Parts up to 3 m in diameter can be produced.

Tube Spinning

In tube spinning, the thickness of a cylindrical part is reduced by spinning it on a cylindrical mandrel using rollers (Fig. 19). The process can be carried out with the rollers working internally as well as externally. Superficially, the process resembles an ironing or an extrusion process, but it should be noted that the repeated deformations after each revolution, where the roller has moved horizontally because of the feed, is closer to a stepwise open-die forging. Tube spinning is used to make pressure vessels, automotive components, jet engine, rocket, and missile parts.

7. Ironing

A drawn cup will normally have a nonuniform thickness distribution in the wall. To create a uniform wall thickness and to increase the height of the cup, the cup is normally pushed through a die using a punch in an ironing process (Fig. 20). The clearance between punch and die determines the wall thickness of the cup. The cup can go through several ironing processes, sometimes in the same stroke. Ironing can also be directly carried out in the drawing process by controlling the clearance.

As there is practically no diameter change, the ironing process can often be regarded as a plane strain process in an analysis, e.g., using the slab method. The ironing process resembles the plane strain drawing of a flat strip, and the same approach is used, just remembering that the friction in strip drawing is acting backward on both sides of the strip. In an ironing process, the friction against the die is still acting backward. but as the cup elongates and the punch remains undeformed, the friction between punch and cup is opposite to that between cup and die. If the friction on the inside is higher than the friction at the outside, friction can

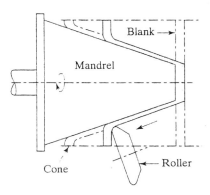

Figure 18 Shear spinning.

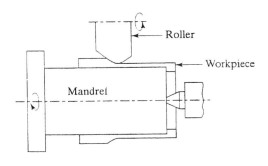

Figure 19 Tube spinning.

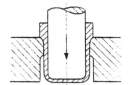

Figure 20 Ironing.

actually help in obtaining high reductions in the ironing process.

C. Processes with Tensile Stresses

1. Stretcher Leveling

In stretcher leveling, the sheet or strip is straightened with tensile forces between jaws (Fig. 21). For discussion, see the paragraph on tensile testing of strips.

2. Stretch Forming

In the stretch forming process, a metal sheet is formed by the application of tensile loads to the material to produce the required shape. The sheet is clamped in jaws at both ends and stretched into plastic condition over a forming tool as shown in Fig. 22a,b. Two versions of the process are used: stretch forming and stretch-wrap forming. In the stretch forming process, the tool moves into the clamped sheet as shown; in stretch-wrap forming, the sheet is first stretched beyond the plastic limit and then wrapped over the form block as shown. The process is used to produce body parts for aircrafts, trucks, cars, and rocket motor housings, where the geometry has rather gentle shape and is free of sharp bends.

The advantages of stretch forming are:

Only one die is needed.
The die can be made of inexpensive material.
Contours with compound contours are possible.
Spring back is almost absent.

The stretching and resulting sliding at the interface between die and sheet causes friction, and consequently the tension in the sheet will vary from a maximum at the jaws and gradually decreasing inward against the sliding direction.

Figure 21 Stretcher leveling.

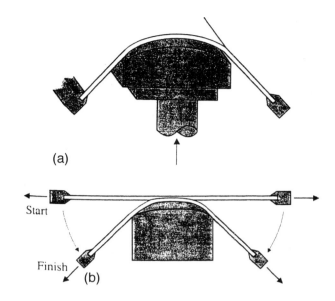

Figure 22 (a) Stretch forming; (b) stretch-wrap forming.

If we let

R = local radius of curvature at element in consideration
$d\theta$ = center angle covering element
ds = length of the sheet–die interface for the element
T = varying tensile force in sheet for unit width
T_0 = tensile force at A for unit width
T_B = tensile force at B for unit width
p = normal pressure in interface
μ = friction coefficient in interface

the force equilibrium in the radial direction gives (Fig. 23):

$$p = T/R \tag{8}$$

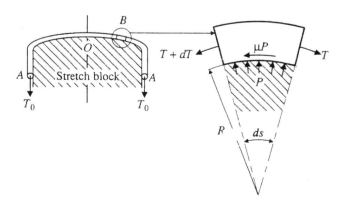

Figure 23 Element in stretch-forming process.

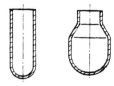

Figure 24 Bulging.

the force equilibrium in the tangential direction gives:

$$dT/T = -(\mu/R)ds \qquad (9)$$

and combining and integrating gives:

$$\ln(T_B/T_0) = -\mu \int_A^B (1/R)ds \qquad (10)$$

The right-hand side of the above equation can be numerically evaluated for any given die contour.

As shown later in the discussion on bending, spring back is a problem for pure bending. If tensile stresses are superimposed, the neutral line will move toward the concave side of the bend. If those tensile stresses become so large that the neutral line disappears, i.e., the outer side as well as the inner side of the bend are deformed in tension, the spring-back problem is almost eliminated, which is an important advantage for stretch forming processes.

3. Bulging

In the bulging process, the internal diameter of a hollow component is increased by the use of a fluid or elastic medium. The main process limitations here are instability or local thinning, described in "Instability in Tension" (Fig. 24). Bulging can also be performed by spinning, described under processes with combined tensile and compressive stresses. The mentioned process limitations originating from the tensile stress situation do not apply there.

4. Expanding

In this process, the diameter of the open end or another part of a hollow component is increased (Fig. 25).

5. Beading

In this process, an outward (beading) or inward (necking in) channel in the wall of a hollow component by the operation of shaped matching rolls is produced (Fig. 26).

6. High-Energy Rate Processes

A small group of nonconventional processes attracted much interest in the 1960s and 1970s where explosives, electrical discharges, and magnetic fields were used to shape domes in sheet blanks and bulges on tubes, often using water as a energy-transmitting medium. They are playing insignificant roles in modern production technology.

D. Processes with Bending

1. Bending of Bar and Sheet with Pure Moment

It is necessary to distinguish from the very beginning the difference between bending of a narrow bar and bending of a wide sheet. Figure 27 shows an element of a bar subjected to a moment M_z. The bar will be deformed into a curved shape, where the outer fibers necessarily have to be stretched and the inner fibers will be compressed. Somewhere in the middle, there must then be a layer that is neither stretched nor compressed. This layer is called the neutral line, and with this line as border, the tensile forces and the compressive forces are in equilibrium. In a bar, there will not be any restraint for sideways deformation, and therefore one can compare the elements above the neutral line with tensile specimens and the elements below with elements subjected to compression. The material originally above the neutral line will as a whole be stretched and thinned, and below the neutral line it will undergo longitudinal compression and sideways expansion. Therefore the situation in bending a narrow bar is with good approximation one of plane stress. Obviously, this situation will lead to a piling up of material below the neutral line and a thinning of material above this line. Therefore the sides of the bar will start to tilt inward toward the "stretch"

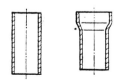

Figure 25 Expanding.

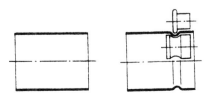

Figure 26 Beading.

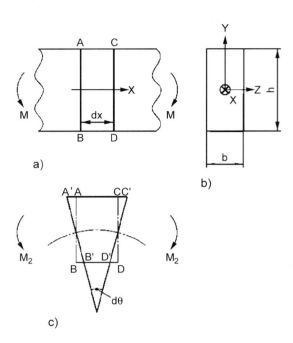

a)

b)

c)

Figure 27 Element of a bar subjected to plastic bending by pure moment (a) seen from the side; (b) seen from the end; (c) deformed shape.

side, and as a consequence a curvature at right angles to the bending curvature will appear on the upper and lower surface of the bar. This curvature, concave at the top and convex at the bottom, is called the "anticlastic curvature."

As there always must be static equilibrium in longitudinal direction, this piling up of material on the inner side of the bend and thinning of material on the outer side means that the neutral line has to move downward as the bending proceeds.

Bending of a wide sheet is somewhat different. The piling up of material below the neutral line has to happen without relief in sideways flow because of the width of the sheet, so only a thickening of the compressive zone can happen, except, of course, for very narrow zones close to the edges, where some lateral deformation is possible. The neutral zone will have to move downward all the same. The main thing to remember is that the bending of a wide sheet basically is a plane strain process, and that the bending of a narrow beam is a plane stress process.

2. Strains, Stresses, and Moments in Sheet Bending

The strains in the length direction x of the sheet will be proportional to the distance from the neutral line, and

as value for the engineering strain in the x direction one obtains in the distance y from the neutral line:

$$e_x((R + y)d\theta - R\,d\theta)/Rd\theta = y/R \qquad (11)$$

or, for the logarithmic strain:

$$\epsilon_x = \ln(1 + (y/R)) \qquad (12)$$

where R is the radius of the neutral line and y is the distance of the element from the neutral line.

If the sheet is totally in the elastic state, the stresses in the x direction are expressed by

$$\sigma_x = E(y/R) \qquad (13)$$

as shown in Fig. 28a. For increasing moment, the outer fibers will start to yield first, and if we assume a nonhardening material for simplicity, the plastic zone will gradually spread inward as shown in Fig. 28b,c.

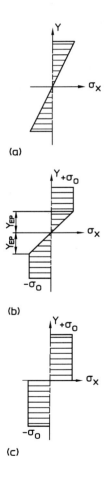

Figure 28 Stress distribution in bending of a sheet, elastic-nonhardening plastic material. (a) Elastic state. (b) Elastic-plastic state. (c) Fully plastic state.

From this, it is easy to show that the necessary moment for elastic bending is:

$$M_E = (1/6)bh^2\sigma_x \qquad (14)$$

and the moment necessary for full plastic condition is:

$$M_P = (1/4)bh^2\sigma_0' \qquad (15)$$

σ_0' being the yield stress in plane strain, $\sigma_0' = (2/\sqrt{3})\sigma_0 = 1.15\sigma_0$.

3. Residual Stresses and Spring Back

As shown above, the necessary external moment to load the sheet to full plastic condition can be expressed by:

$$M_P = (1/4)bh^2\sigma_0' \qquad (16)$$

If this moment is released, the sheet will straighten somewhat. This will happen as an elastic unloading, i.e., equivalent to a stress linearly changing through the thickness of the sheet (Fig. 29). Assume that the associated, imaginary, elastic stress distribution reaches the value σ_A in the outer fibers of the sheet.

This moment of elastic unloading M_A must be equal to the plastic moment of maximum load, which means:

$$M_A = M_P \qquad (17)$$

or

$$(1/6)bh^2\sigma_A = (1/4)bh^2\sigma_0' \qquad (18)$$

giving:

$$\sigma_A = (3/2)\sigma_0' \qquad (19)$$

The residual stresses are found as the difference between the elastic and the plastic stress distribution. The outer

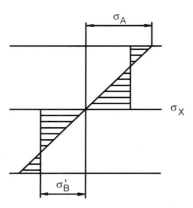

Figure 29 Unloading of bent sheet. (A) Assumed elastic unloading stress distribution. (B) Resulting residual stress distribution.

sixth on the top of the sheet will end up with compressive residual stresses with the maximum value:

$$\sigma_{\text{res,max,top}} = -1/2\sigma_0' \qquad (20)$$

and the lower part of the sheet has tensile residual stresses with the maximum value:

$$\sigma_{\text{res,max,bottom}} = +1/2\sigma_0' \qquad (21)$$

The release of the moment that was used in the bending process and the resulting elastic unloading will result in an opening of the bent angle. This is called the elastic spring back.

The derivation is somewhat lengthy, but if we set

α = bending angle before spring back
α' = bending angle after spring back
α = 0 before bending
R = neutral bending radius before spring back
R' = neutral bending radius after spring back
$\beta = 2\sqrt{3}(1 - v^2)(\sigma_0/E)(R'/t)$

we can write

$$\alpha = \alpha'(1 + \beta) \qquad (22)$$
$$R = R'/(1 + \beta) \qquad (23)$$

4. Position of the Neutral Line

As earlier stated, the neutral line has to move inward, toward the concave side of the bend. For a sheet, where the width can be assumed constant from convex to concave side and plane strain can be assumed, the slab method used on a curved sheet will give

$$R = (a/b)^{1/2} \qquad (24)$$

or

$$R/a = (1 + t/a)^{1/2} \qquad (25)$$

where

R = neutral radius
a = inner radius
b = outer radius
t = sheet thickness

5. Bending Allowance

The position of the neutral line is not only of academic interest. It is deciding for a very important parameter in bending jobs: the bending allowance.

If the convex (male) part of a bending tool has the radius of a, it is clear that if the finished component has to satisfy tolerances in its final shape, one must know where the neutral line is positioned in order to design the

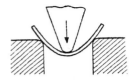

Figure 30 Air bending.

process and the size of the original blank properly. From the drawing, one can read the tool dimensions, but a sheet with a thickness t will bend along a curvature that is determined of the position of the neutral line where the length of the bent part does not change. One simple suggestion is of course just to add half of the sheet thickness to the radius a, and calculate the necessary blank size from this. But as we have seen, the neutral line does not stay in the middle of the original sheet thickness, it moves inward. This simple suggestion will then give to large blanks for the final component.

Several suggestions have been forwarded, but the problem is rather complicated, among other things because there will be a theoretical discontinuity between the unbent part of the component, where the neutral line of course stays nicely in the middle of the sheet thickness, and the curved part, where the neutral line has moved from this position. This geometrical discontinuity does not exist in reality, but is evened out in different ways, among other things depending on the strain-hardening characteristics of the material.

The given expression in Eq. (24) is a good choice for the calculation of the neutral radius.

6. Air Bending

In this process, a tool, typical the upper tool in a press brake, is causing a bend of a sheet without any forming support from the lower tool (Fig. 30).

7. Die Bending

In this process, a female die is used together with the male die to give better tolerances to the bent profile. The

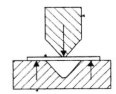

Figure 31 Die bending.

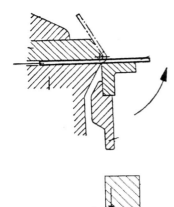

Figure 32 Folding.

final stage of the process is often a calibrating force exerted between the dies (Fig. 31).

8. Folding and Wiping

In these bending operations, one part of the sheet is clamped and the free part is moved in a folding motion by a forming tool (Fig. 32a,b).

9. Roll Bending

In this process, a cylindrical component is produced from a sheet by passing it between three staggered rolls (Fig. 33).

10. Roll Forming

This process is the shaping of a (often complicated) long profile by passing a strip material from a coil through a number of forming rolls (Fig. 34).

11. Roller Flanging

This is the production of a flange on a hollow component by means of two shaped opposing rolls (Fig. 35).

Figure 33 Roll bending.

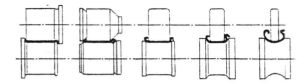

Figure 34 Roll forming.

IV. PROCESSING ASPECTS

A. Temperature

The basic considerations of temperature are given in Chap. 1. The differences between sheet forming and bulk forming are determined by the large differences in the surface/volume ratio between the two types of processes, meaning that as bulk forming processes as a rule can be considered as adiabatic, but in the case of sheet metal forming one has to be more careful. A deep drawing of a thin cup in a brass alloy could be considered as isothermal, but the same process carried out with a titanium alloy (low density, high strength) could be close to adiabatic.

Some processes can be improved by a temperature increase. For instance, in a deep-drawing process, the drawing ratio can be increased by heating the flange of the blank through the die and blank holder and keeping the punch cold.

Spinning processes can be carried out hot. Difficult alloys can be formed in heated tools, in the so-called isothermal processes.

B. Strain

The basic considerations have already been given in the bulk forming chapter. One important aspect will be mentioned here, namely the relatively uncomplicated method of measuring strains in sheet metal forming processes. Because of the almost uniform strain distribution through the thickness of the sheet in most sheet forming processes, strain can be experimentally determined with the aid of a grid on the surface of the specimen put on by, e.g., photographic, serigraphic, or etching techniques. If the strain is not uniform, as, e.g., in bending processes, strain distribution through the thickness can normally be easily calculated on the basis of the surface strains. The grid is applied on the flat sheet before the process, and measured on the deformed workpiece, either after the full process or after selected steps.

The normal grid pattern is a system of circles with small diameters, touching each other or slightly over-

lapping. The circles will deform into ellipses, and the magnitude and the directions of the principal strains can be found by identifying the major and minor axes of the ellipses.

If D_0 indicates the original diameter of the circle and D_{max} and D_{min} are measured as the maximum and minimum diameters of the ellipse after forming, their directions will give principal strain direction, and the values $\ln(D_{max}/D_0)$ and $\ln(D_{min}/D_0)$ will give the two principal strains in the surface of the sheet. The thickness strain can then be calculated from volume constancy.

C. Strain Rate

This has been covered in the bulk forming chapter.

D. Stress

This has been covered in the bulk forming chapter.

E. Friction

This area has been described in the bulk forming chapter. Because of the comparatively low interface pressures between tool and sheet specimen, the Coulomb friction law is most often used. Experimental methods for determining friction coefficients will be discussed in Sec. VI.B.1.3.

F. Yield Criteria—Isotropic and Anisotropic

For simple analyses, the von Mises or Tresca criterion discussed in Chap. 1 can be and have been used. These criteria describe isotropic materials. For some approaches in sheet forming, a compromise between the Tresca and the von Mises criterion is used. This happens, e.g., in the classical derivation of maximum drawing force in deep drawing, where the yield stress in the flange is set to $1.1\sigma_0$.

As explained later, the rolling process, which is the basis for sheet metal production, creates anisotropy,

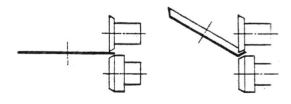

Figure 35 Roller flanging.

i.e., different properties in different directions. The accepted criterion assumes three mutually orthogonal planes of symmetry throughout the material; the lines of intersection of these planes being called the principal axes of symmetry. The rolling direction, the transverse direction, and the thickness direction are as a rule considered principal axes of symmetry.

The definition and experimental determination of anisotropy will be discussed later in the text. If it is necessary to bring anisotropy into plasticity calculations, the more complicated Hill anisotropic yield criterion should be used replacing the von Mises criterion shown in Eq. (20) of Chap. 1.

The Hill criterion is written as

$$
\sigma_0 = [(3/2)/(F + G + H)]^{1/2}[F(\sigma_2 - \sigma_3)^2 \\
+ G(\sigma_3 - \sigma_1)^2 + H(\sigma_1 - \sigma_2)^2]^{1/2}
\tag{26}
$$

reducing to the von Mises criterion for $F = G = H$.

Determination of the constants is discussed later in this chapter.

G. Hardening

This has been covered in Chap. 1.

V. DESIGN ISSUES TO PREVENT FAILURES

A. Geometrical and Mechanical Issues

The description of this in Chap. 1 is mainly relevant for three-dimensional flow in bulk specimens under compressive stresses, but is applicable in general terms to sheet metal forming. However, some additional remarks are necessary.

The main problem with sheet metal forming processes is that neither too large tensile stresses nor too large compressive stresses in the plane of the sheet are acceptable. This seems a paradox when one considers the multitude of process alternatives in sheet forming, but probably the large number of different processes is a consequence of the rather narrow windows in stresses and strains that are at hand in this process group.

Tensile stress and strain leads to local thinning and eventually cracks, and compressive stresses in the plane of the sheet can lead to buckling or wrinkling. They are both phenomena constituting serious process limitations that should be avoided by clever tool-and process design.

Because of the large surface-to-volume ratio in a typical blank, lubrication plays an important role in sheet metal forming, and may significantly change the flow. However, this can be used in a constructive way, using local application of lubricant, as discussed in the paragraphs on deep drawing.

Sliding lengths are often large, and therefore lubricant breakdown and resulting surface defects and local pick up can appear. This is often unacceptable because of the cosmetic requirements on, e.g., car parts. On the other hand, if the lubrication is too efficient or the surface is free, an orange-peel topography may appear, which is equally unwelcome. This is because of the unrestrained deformation of the individual grains according to their varying orientation and available slip planes.

Residual stresses are important, mainly because they are the origin of spring back, which is a large problem in sheet forming, necessary to cope with in the process design stage.

B. Metallurgical and Microstructure Issues

Most of the issues regarding metallurgy and microstructure are determined when the material goes through the primary processing in the casting, hot rolling, and cold rolling stages. The main considerations for sheet metal forming processes are given in the following paragraphs.

Anisotropy, discussed later in this chapter, gives rise to formability properties in drawing and stretching processes. Large plane anisotropy gives rise to the formation of ears in deep-drawing processes and consequently to uneven wall thickness distribution in the circumference of the drawn cup and material waste in the final trimming. On the other hand, a high value of the normal anisotropy is normally beneficial.

Strain hardening must be considered in each process. Normally, a high strain hardening will delay the moment of local thinning and instability and is thus beneficial, but in special cases, as redrawing without interstage annealing, a low strain hardening ratio could be preferable.

The upper and lower yield point in steel can give rise to easy-glide bands that appear as stretcher strains or Lüder's bands. They can be avoided by eliminating the upper yield point, by a slight reduction (2%) in temper rolling shortly before the forming process. Too long delay will give the carbon and nitrogen in the steel possibility to diffuse and thereby to recreate the upper yield point.

VI. FORMABILITY AND TESTING METHODS

A. Definition

For bulk forming, the concept of workability has been discussed. Because of high tool stresses, large and inhomogeneous flow, risk for catastrophic shear, and internal and external cracks, a set of tests and workability definitions have been developed. The situation for sheet metal processes is different.

The tool stresses are typically lower than in bulk forming processes; internal catastrophic shear and internal cracking are normally exceptions and an external crack will normally be preceded by a local thinning. The prevalent plane stress situation in sheet metal forming processes means in broad terms that the process limitations are either caused by excessive compressive stresses or excessive tensile stresses in the plane of the sheet.

B. Tests

Testing of sheet material for sheet forming processes can broadly be divided into:

materials testing
simulative testing
friction tests

1. Materials Testing

Uniaxial Tensile Testing

The uniaxial testing of a metal strip is mechanically carried out in the same way as the testing of a round specimen, the gripping jaws of course being modified. However, some very important differences have to be noted.

ISOTROPY AND ANISOTROPY. If a material is isotropic, it has the same value of its properties in all direction. An anisotropic material obviously then has different properties in different directions. One has to define which properties that are under consideration, a certain material could, e.g., be anisotropic from the point of view of strength, but isotropic if one considers thermal conductivity.

Because of the rolling processes, which are the basis for sheet metal production, anisotropy is a phenomenon that must be taken seriously in sheet forming. In the sheet rolling process, the material is deformed more or less under plane strain conditions, resulting in mainly thickness reduction and elongation. For calculation purposes of forces in wide sheet rolling, the influence of the width increase is normally neglected as a process phenomenon, and some average of the contact area is chosen.

For sheet material, a very well-defined sort of isotropy/anisotropy is defined with great importance for metal forming. This definition of anisotropy deals only with strains.

We will consider this phenomenon with a tensile test of a metal strip as reference.

The yield stress is:

$$\sigma_o = P/(wt) \tag{27}$$

and the principal strains are given by:

$$\epsilon_l = \ln(l/l_1) \tag{28}$$
$$\epsilon_w = \ln(w/w_1) \tag{29}$$
$$\epsilon_t = \ln(t/t_1) \tag{30}$$

where

P = tensile force read on the tensile testing machine
l = momentary length of the considered element
w = momentary width of the test strip
t = momentary thickness of the test strip
l_1 = original length of the considered element
w_1 = original width of the test strip
t_1 = original thickness of the test strip

For an isotropic material, the reduction in thickness is the same as the reduction in width, so $\epsilon_t = \epsilon_w$ and thus $\epsilon_l = -2\epsilon_t = -2\epsilon_w$ according to the equation of volume constancy:

$$\epsilon_t + \epsilon_w + \epsilon_l = 0$$

For an anisotropic material with different deformatoric properties in different directions, ϵ_t and ϵ_w will not be equal. The anisotropy of a sheet metal is normally determined by a tensile test and is calculated as:

$$R = \epsilon_w/\epsilon_t \qquad 0 < R < \infty \tag{31}$$

If the tensile specimen keeps its width and all deformation happens as thinning, we obtain $R = \infty$, and if the specimen keeps its thickness and only width is decreasing, $R = 0$.

To describe the anisotropy of a sheet material, tensile tests in the directions $0°$, $45°$, and $90°$ to the rolling direction are carried out. $R = R(\alpha)$ normally has a maximum or a minimum for these angles. From this data, two values of anisotropy are defined, called normal anisotropy $\bar{R}$ and the plane anisotropy ΔR.

$\bar{R}$ is determined as a simple average of the R values in all eight 45° directions relative to the rolling direction. One obtains the following expression:

$$\bar{R} = (R_0 + R_{90} + 2R_{45})/4 \tag{32}$$

expressing the difference in ease of deformation between the average values in the plane of the sheet and the value normal to the sheet plane.

The plane anisotropy ΔR expresses the variation of R in the plane of the sheet. This is calculated taking the values in all eight 45° directions. One obtains the following expression:

$$\Delta R = (R_0 + R_{90} - 2R_{45})/2 \tag{33}$$

The practical importance of anisotropy is rather large, so a close control of a certain amount of anisotropy is beneficial. In deep drawing for instance, the anisotropy in the plane of the sheet, ΔR, will give ears on the drawn cup, resulting in waste of material and uneven wall thickness. One tries to reduce this anisotropy for this reason. The normal anisotropy, on the contrary, is beneficial, as it allows drawing of higher cups.

The constants in the Hill anisotropic yield criterion can be determined using tensile testing in three directions, σ_{01}, σ_{02}, σ_{03} determined in the rolling direction, transverse rolling direction, and normal to the sheet plane. This last determination can be performed by one of the compression methods described later.

One obtains:

$$2F = \sigma_{02}^{-2} + \sigma_{03}^{-2} - \sigma_{01}^{-2} \tag{34a}$$

$$2G = \sigma_{03}^{-2} + \sigma_{01}^{-2} - \sigma_{02}^{-2} \tag{34b}$$

$$2H = \sigma_{01}^{-2} + \sigma_{02}^{-2} - \sigma_{03}^{-2} \tag{34c}$$

Biaxial Tensile Testing

Biaxial testing or bulge testing is carried out to investigate the behavior of a sheet material subjected to tensile stresses in two perpendicular directions. It is normally carried out by clamping a circular disk of the sheet metal—a membrane—around its rim and subject it to an oil pressure on one side. The membrane will bulge and the strain can be measured the central region.

BIAXIAL TENSILE TESTING, EQUAL STRAINING. If the membrane is circular, the test method is often called "balanced biaxial testing." The strains and also the stresses in the central region can be determined if the strain, the bulge radius, and the oil pressure are measured during the bulging. The strain is measured with an extensometer, the curvature with a spherometer, and the oil pressure with a pressure gage. Extensometer and spherometer should be in one integrated unit. The relevant equations are:

$$\sigma_0 = p\rho/2t \tag{35}$$

$$\varepsilon = 2\ln D/D_0 \tag{36}$$

where

σ_0 = yield stress
ε = equivalent strain
p = fluid pressure
ρ = radius of curvature
t = sheet thickness
D = momentary distance between the extensometer contact points
D_0 = original distance between the extensometer contact points

The bulge will be unstable and start local thinning when $\varepsilon = (4/11)(2n + 1)$; that is, for a considerably higher strain than the uniaxial tensile test. For an n value of 0.25, representative for steel, the uniaxial tensile test will become unstable for $\epsilon = 0.25$, but the balanced biaxial test will give useful data up to $\epsilon = 0.55$.

BIAXIAL TENSILE TESTING, UNEQUAL STRAINING. If the circular die is substituted with an elliptic die, the strains will not longer be equal in the longitudinal and the transverse directions. The strain in the transverse direction will be larger than the strain in the longitudinal. With a circle grid etched or photographed on the surface, this can be used for formability measurements, as discussed later in *Formability Testing*.

Compression Test

Two compressive tests for sheet material should be mentioned—the multiple layer upsetting test and the plane strain upsetting test.

MULTIPLE LAYER UPSETTING TEST. The multiple layer upsetting test can be used as a substitute for the biaxial balanced tensile test, if the objective is to obtain stress–strain curves, but useless for experimental determination of instability. However, this condition can be calculated in the actual situation from the experimentally determined value of n.

The test is useful when the sheet pieces are too small for a tensile test or a balanced biaxial test. It can give stress–strain curves up to considerably higher strains than the tensile tests.

The test is carried out by taking a number of small blanks possibly with a small hole punched in the center of the blanks and stacking them to a cylindrical slug of dimensions $1 < h/D < 1.5$. In a press or testing machine, the stack is then upset as a normal solid slug for bulk forming upsetting with measurement of force and displacement. The slugs are held concentric by a pin in the lower tool and the upper tool has a corresponding hole in it to accept the pin as the upsetting proceeds.

The test relies on that the strain condition in compressing a flat blank between frictionless tools is exactly the same as the condition in a balanced biaxial test. From Eqs. (14) and (20) in Chap. 1, it is seen that adding or subtracting the hydrostatic stress component σ_M [Eq. (16), Chap. 1] in the von Mises yield criterion does not influence the value of the yield stress.

In the uniaxial tensile test σ_M has the value $(\sigma_0 + 0 + 0)/3 = \sigma_0/3$; in the balanced biaxial test, the value is $(\sigma_0 + \sigma_0 + 0)/3 = (2/3)\sigma_0$.; in the multiple layer upsetting test, the value of σ_M is $(-\sigma_0 + 0 + 0)/3 = -\sigma_0/3$.

This difference in hydrostatic stress component of $[(2/3)\sigma_0 - (-\sigma_0/3)] = \sigma_0$ will only have negligible influence on the stress–strain curve.

PLANE STRAIN COMPRESSION TEST. The plane strain compression test (the Watts and Ford compression test) was originally developed to establish stress–strain curves for the rolling process.

Even if rolling is normally considered a bulk forming process, the test is developed for establishing the stress–strain curves for sheet material, and therefore should be mentioned here.

Two flat tools are pressed into a sheet strip that has a width w at least six times as large as the tool contact breadth a in order to secure a plane strain deformation (Fig. 36). The proportion between the breadth of the tool a and the thickness of the sheet h should satisfy the inequality $2 < (a/h) < 4$. The reason for this is the following:

If $a = h$, the upper bound field is a simple cross under $45°$, similar to the slip line field shown in Chap. 1, Fig.

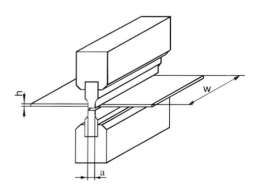

Figure 36 Plane strain compression test.

23. The necessary pressure at this moment is $p/\sigma_0' = 1$ or 1.15 times the uniaxial yield stress.

The value of the yield stress in plane strain σ_0' is often written $2k$.

When the compression starts, this "forging cross" will distort, and an upper bound model of the situation will just show a gradually flattening cross, as shown for the situation $(a/h) = 1.41$, when the upper bound model predicts a switch to two narrow crosses. The dimensionless pressure at this moment is $p/\sigma_0' = 1.06$.

From here, the dimensionless pressure will decrease until the situation can be described with two right-angle crosses for $a = 2h$ and again giving $p/\sigma_0' = 1$. The two crosses are gradually flattened until they, for the (a/h) value of 2.44, flip over from two wide to three narrow crosses. At this moment, $p/\sigma_0' = 1.02$. The three narrow crosses go through the right angle geometry at $a = 3h$ and here again $p/\sigma_0' = 1$.

The mechanism of flattening crosses and flip over to one cross more is repeated, and this shift from three to four occurs at $(a/h) = 3.46$ with a $p/\sigma_0' = 1.01$.

To avoid the first pressure maximum, it is recommended not to start the test for $a = h$, but for $a = 2h$.

When a reaches $4h$, friction starts to play a role. As shown in Chap. 1 in the description of the slab method, the parameter constellation of $\mu(a/h)$ plays an important role in the determination of the contribution of friction to the measured pressure [see, e.g., Eq. (23) in Chap. 1]. In this case, forging between flat anvils in plane strain the dimensionless mean pressure can be written as:

$$p/\sigma_0' = 1 + (\mu/2)(a/h) \tag{37}$$

A coefficient of 0.01, which is rather low and very difficult to obtain, and requires polished tools and careful lubrication will thus give an error of 2% for an $(a/h) = 4$.

In this way, the recommended window of $2 < (a/h) < 4$ is determined.

In the testing process, one starts with a tool twice as broad as the thickness of the strip. This tool pair is used for the first reduction in compression to half thickness. Now the tool is exchanged to a second tool with half the breadth of the first tool, and compression goes on to 1/4 of the original sheet thickness. Now a new tool with yet half breadth can be inserted and so on. Thus a plane strain compression test can be carried out keeping (a/h) between the recommended limits.

It must be remembered that the equivalent strain in this test must be calculated as

$$\epsilon = (2/\sqrt{3})\ln(h/h_1) \tag{38}$$

and the uniaxial yield stress must be calculated as

$$\sigma_0 = (1/2\sqrt{3})(P/ab) \tag{39}$$

Here h_1 is the original thickness of the sheet and b is the width of the sheet.

Instability in Tension

When treating sheet metal processes, where stresses are predominantly tensile, it is necessary to go a bit deeper into what happens at the point of necking in the tensile test than when treating bulk forming processes.

INSTABILITY IN A ROUND TENSILE SPECIMEN. For simplicity, we will start with reviewing the tensile testing of a cylindrical rod.

For this rod being tested in tension, the yield stress σ_0 increases but the cross-sectional area A decreases. The tensile force F can be written as

$$F = \sigma_0 \cdot A \tag{40}$$

At some point of the testing, the decrease of area wins over the strength increase, and at this moment the engineering stress–strain curve shows a maximum, (Fig. 12, Chap. 1).

Here we have

$$dF = d(\sigma_0 \cdot A) = 0 \tag{41}$$

Constancy of volume V gives

$$dV = d(l \cdot A) = 0 \tag{42}$$

l being some defined length on the specimen.

At the maximum of the engineering stress–strain curve, we thus can write the following two equations:

$$\sigma_0 \cdot dA + A d\sigma_0 = 0 \tag{43}$$

$$l \cdot dA + A \cdot dl = 0 \tag{44}$$

which easily give

$$\sigma_0/l = d\sigma_0/dl \tag{45}$$

or

$$d\sigma_0/d\epsilon = \sigma_0 \tag{46}$$

The importance of this equation becomes clear when it is combined with the power law hardening expression given in Eq. (21) in Chap. 1. One obtains:

$$d(K\epsilon^n)/d\epsilon = K\epsilon^n \tag{47}$$

or

$$\epsilon_{\text{instab}} = n \tag{48}$$

The value of the strain hardening exponent n lies between 0 for a non-strain-hardening material, where

the instability in a tensile test sets in immediately and 0.4/0.5 for very fast strain hardening materials. Steel and aluminum lie somewhere in the middle of this range, annealed qualities having higher n values than work-hardened.

The fact that necking starts when the longitudinal strain reaches the value of n is a serious limitation for the use of tensile testing as a material test for forming operations. In the area of bulk forming, curves for strain hardening at higher strains can be obtained by upsetting of circular cylinders and measuring deformation force plus the axial deformation. The friction problems mentioned in Chap. 1 can be overcome by special measures, as, e.g., by the Rastegayev method, where a shallow depression in the end surfaces of the specimen are filled with, e.g., paraffin or stearin.

Another approach using tensile testing is to repeatedly use the wire drawing process and after each reduction cut a suitable length of the wire and subject this it to a tensile test. Arranging the resulting stress–strain curves on a common strain axis, where the starting point for each curve is the accumulated strain up to that point, possibly with the addition of redundant strain, can give reliable stress–strain curves.

The same approach can be used in sheet metal testing using the rolling process + tensile testing of rolled strips instead of wire drawing. It has to be remembered that the rolling is carried out under plain strain conditions and the subsequent tensile testing under uniaxial conditions.

INSTABILITY IN A STRIP. As in the case of the rod, the homogeneous deformation of a strip in tensile testing will only be possible as long as the strain hardening can compensate for the decreasing area. The decrease of area will invariably win this tug-o-war, and the deformation will change from homogeneous into necking with local deformation. The deformation will be concentrated in the neck, the deformation force will decrease and consequently the plastic deformation in the remainder of the test piece will stop. The necking will take the shape of a straight shallow ditch with the borders along a direction where no deformation exists. If the width was constant ($R = \infty$), this would be a straight line along 45°. However, for an isotropic material, the decrease in width will be equal to the decrease the thickness. Calculations show that this direction will lie under the angle $\varphi = 35.3°$ from horizontal if the tensile direction is vertical.

Obviously, the value of this angle will be dependent on the value of R.

It can be shown that:

$$tg\phi = (R/(1 + R))^{1/2} \tag{49}$$

which gives the above angle for $R = 1$.

INSTABILITY OF A THIN-WALLED SPHERE. A hollow thin-walled sphere with wall thickness t and radius r is loaded with an internal pressure p. Assume that the stress σ_r in the thickness direction is compressive and equal to $-p$. Along the shell surface, the stress is everywhere tensile an has the value $\sigma_\theta = \sigma_\varphi$. As no shear stresses exist, σ_θ, σ_φ, and σ_r must be principal stresses, θ and φ being two orthogonal directions in the shell surface.

Equilibrium of a small circular part of the shell surface gives:

$$p\pi r^2 = 2\pi r t\sigma_\theta \tag{50}$$

or

$$p = 2(t/r)\sigma_\theta \tag{51}$$

or

$$\sigma_r = -2(t/r)\sigma_\theta \tag{52}$$

If t/r is small, say 0.01–0.02, the influence of σ_r in the yield criterion [Eq. (14) in Chap. 1] is negligible and we can safely write:

$$\sigma_\theta = \sigma_\phi = \sigma_o, \sigma_r = 0 \tag{53}$$

We will assume that the criterion for instability can be considered as $dp/d\epsilon = 0$.

This means that the sphere becomes unstable and keeps expanding when a certain inner pressure is reached, even if this is not further increased. Calculations will give:

$$d\sigma_0/d\epsilon = (3/2)\sigma_0 \tag{54}$$

or, using the power hardening law:

$$\epsilon_{instab} = (2/3)n \tag{55}$$

Testing Conditions

Care should be taken to carry out the testing in conditions as close to the process conditions as possible. This is not only relevant for strain rate but also for temperature, even if a process as well as tensile testing is carried out at room temperature. In studying the deformation of a titanium alloy sheet for a heat exchanger, it turned out that tensile testing in free air gave much too low yield stresses to be reliable in the process analysis. The reason was that the close contact between the thin titanium and the very large heat sink created by the pressing tool gave almost isothermal conditions, but the testing of the titanium strip in air resulted in a very hot specimen. Testing the strip submerged in a water container solved the problem.

Formability Testing

A very important way of describing the deformatoric properties of a sheet metal is developed in the so called formability diagrams or forming limit diagrams (FLD). These diagrams are based on a plane coordinate system, where the abscissa shows the minimum strain ϵ_{min} and the ordinate is ϵ_{max} (Fig. 37). Below the lower full line, no fracture will occur. Above the upper full line, fracture is certain. The first quadrant in the diagram describes conditions where ϵ_{min} and ϵ_{max} are both positive: There are tensile stresses everywhere in the plane of the sheet. This area is called the stretch forming area. Thus the balanced biaxial tensile test will be represented by a straight line going upward from origo under 45° (slope $+1$) dividing the first quadrant in two equal sectors.

In the fourth quadrant, the strains are of opposite sign, and this is called the draw forming area. Thus a pure isotropic tensile test will be depicted in a straight line through origo under 26.6° (slope -2).

The vertical axis $\epsilon_{min} = 0$ describes forming conditions (stretching) at right angles over a long ridge in the forming tool.

The strains are measured by a circle grid applied on the surface of the sheet to be tested, as described in Sec. IV.B. When a crack appears, the strains are measured and the coordinates marked in the diagram. Thus the

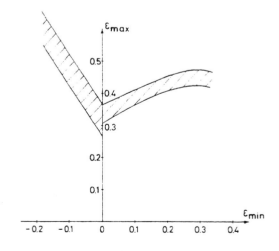

Figure 37 Formability diagram.

points for the balanced biaxial test and the uniaxial test are easily found.

For intermediate points in the FLD curve, the elliptical biaxial test discussed in "Biaxial Tensile Testing, Unequal Straining" can be used.

A modification of the Erichsen test described in below (see "Erichsen Test") can also be used. The conventional Erichsen test will give balanced biaxial condition on the top of the sphere, but if rectangular sheet specimens with smaller width than the ball diameter are used, condition between balanced biaxial state and uniaxial condition can be obtained. The ball diameter should be larger than in the standard Erichsen test.

2. Simulative Tests

The philosophy behind the simulative tests is that they in a simplified way are supposed to simulate different sheet metal forming processes and thereby facilitate the selection of the right material. The simulative tests are named after their initiators: Erichsen, Olsen, Fukui, Swift, Sachs, etc. The two most well-known tests are the Erichsen test and the Swift test.

Erichsen Test

The Erichsen test, schematically shown in Fig. 38, is a stretch forming test. A circular blank from the material that is going to be tested is clamped between a die and a blank holder with a force of 10 kN. A spherical punch is pressed down into the blank and the travel length of the punch before fracture occurs is measured. This travel length, measured in millimeter, is a characteristic property of the sheet metal: the Erichsen number. A high Erichsen number indicates a good "stretchability."

Swift Test

The Swift test, schematically shown in Fig. 39, is a deep-drawing test. A series of blanks with increasing diameters are cut out of the sheet metal in question. Blanks with steadily increasing diameters are deep-drawn, and at one point a diameter is reached, where

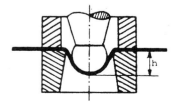

Figure 38 Erichsen test.

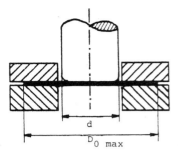

Figure 39 Swift test.

the punch penetrates the not yet completely drawn cup. In this way, the maximum drawing ratio β_{max} is found. β_{max} is defined as the ratio between the largest blank diameter, $D_{0,max}$, that can be drawn without fracture and the punch diameter d_1. A high Swift number indicates a good "drawability."

3. Friction Tests

The ring test described in Chap. 1, Sec. VI.B.4, can in principle be used for sheet metal forming, but as the thickness of the sheet normally is rather small, say 1–3 mm, the dimensions of the ring become very small. Furthermore, interface pressures between sheet and tool in sheet metal forming are normally below the yield stress because of the tensile stresses that as a rule exist in the plane of the sheet, and sliding lengths in sheet forming operations are often large compared with the sliding lengths in the ring test. This test is consequently avoided in sheet metal friction testing.

Here two types of test are predominant, the plane strip tests without deformation and strip tests with bending under tension.

Strip tests without plastic deformation have been developed with varying geometry (Fig. 40). To avoid misleading results caused by the lubricant being scraped off by a sharp edge or caused by misalignment of the two sliding shoes, hinged versions have been used. If a back tension is applied, the friction coefficient changes.

For tests with bending under tension, two versions have been used: the die ring simulating test and the draw bead simulating test.

The die ring simulating test is shown in Fig. 41a,b. If the radius is small, a considerable amount of the measured drawing force is caused by the bending and unbending of the strip. This can be measured and subtracted by using a free rotating roll as die in a second experiment (Fig. 41b).

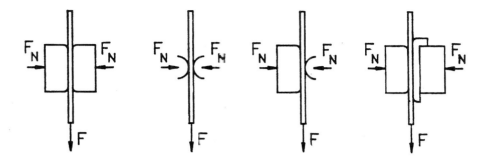

Figure 40 Friction simulation tests without plastic deformation.

The draw bead simulating test is schematically shown in Fig. 42. It is one of the most used simulative tests in sheet metal forming. In the test, a strip of the sheet is pulled through the draw bead simulator, and is going through several bending and unbending steps. This force has to be measured and separated by exchanging the fixed draw beads with free rotating rolls.

The drawing force and the tool separating force C has to be measured to determine the friction coefficient.

VII. DEFORMATION MODELING METHODS

In Chap. 1, dealing with bulk forming processes, the main modeling methods have been described and there is no need to repeat that here. Because of the great variety of process configurations in sheet metal forming, all of them, the slab equilibrium, the slip line method, the upper bound method, and the finite element simulation, have been used. Early literature was much based on the slab equilibrium method because of the rather homogenous deformation in many of the processes, but slip lines and load bounding were also used. Interesting work has been performed using the slip line method to design the shape of the blanks for noncircular deep-drawing processes.

These "classic" approaches should not be neglected, as the feeling for the mechanics of the processes in a very high degree relies on them. The basic understanding created by these methods is extremely valuable whether the task is to select a process, design it, establish the diagnosis and the cure for failures in it, or it comes to designing experiments or appraising the results of a simulation.

At present, almost all process simulation and modeling is carried out using different FEM software packages, an area that in sheet metal forming really has been developing fast. Especially the development of, e.g., high-strength sheet materials, new nonferrous alloys, study of nonlinear strain paths, and the use of tailored

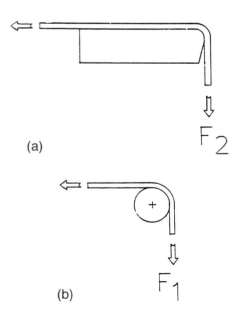

Figure 41 (a) Die ring simulative test. (b) Profile radius replaced by roller.

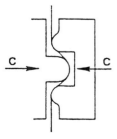

Figure 42 Draw bead simulative test.

blanks (laser-welded blanks with different material qualities, different thicknesses, etc.) have put very large demands on new software, and much work has been spent on the development of this.

The forming of high-precision miniature components has also stimulated this development.

SELECT REFERENCES

Avitzur, B. *Handbook of Metal Forming Processes*; John Wiley: New York, NY, USA, 1983.

CIRP. *Dictionary of Production Engineering*. Sheet Metal Forming: Girardet, Essen, Germany, 1965; Vol. 3.

Hill, R. *The Mathematical Theory of Plasticity*; Oxford University Press: Oxford, UK, 1950.

Hoffman, O.; Sachs, G. *Introduction to the Theory of Plasticity for Engineers*; McGraw-Hill: New York, USA, 1953.

Johnson, W.; Mellor, P.B. *Engineering Plasticity*; Van Nostrand: London UK, 1973.

Kalpakjian, S.; Schmid, S.R. *Manufacturing Engineering and Technology*, 4th ed.; Prentice Hall: Upper Saddle River, NJ, USA, 2001.

Kienzle, O.; Dohmen, H.G. *Mechanische Umformtechnik. Plastizitätstheorie, Werkstoffmechanik, Fertigung*; Springer-Verlag: Berlin, 1968.

Lange, K. *Lehrbuch der Umformtechnik, I–III*; Springer Verlag: Berlin, 1975.

Lange, K. *Handbook of Metal Forming*; McGraw-Hill: New York, NY, USA, 1985; (Translation of "Lehrbuch der Umformtechnik).

Lange, K. *Umformtechnik, Handbuch für Industrie und Wissenschaft, I–IV*; Springer-Verlag: Berlin, 1990.

Mielnik, E.M. *Metal Working Science and Engineering*; McGraw-Hill: New York, NY, USA, 1991.

Sachs, G. *Principles and Methods of Sheet Metal Fabricating*; Reinhold Publishing Corporation: New York, 1966.

Shey, J.A. *Metal Deformation Processes, Friction and Lubrication*; Marcel Dekker Inc: New York, 1970.

Schey, J.A. *Tribology in Metal Working—Friction, Lubrication and Wear*; ASM: Metals Park OH, USA, 1983.

Tlusty, G. *Manufacturing Processes and Equipment*; Prentice Hall: Upper Saddle River, NJ, USA, 2000.

3

Design of Microstructures and Properties of Steel by Hot and Cold Rolling

Rafael Colás
Universidad Autónoma de Nuevo León, San Nicolás de los Garza, N.L., Mexico

Roumen Petrov and Yvan Houbaert
Ghent University, Ghent, Belgium

I. INTRODUCTION

Rolling as a metalworking process dates from just over 500 years, the time at which it may have been developed from mills used to extract sugar from sugarcane. The earlier recorded designs of rolling mills were made by Leonardo da Vinci; these sketches are thought to have been made between 1480 and 1490, but there is no evidence that they were actually built. The drawings can be found in the *Codex Atlanticus* and *Manuscript B*, kept at the *Biblioteca Ambrosiana* in Milan and in the *Institute de France* in Paris. Two different designs appear in this last document; a model of one of them, a two-high mill for the production of 300-mm-wide tin strip, is exhibited in the *Museo Leonardiano di Vinci* at Vinci, Italy. The second design is that of a four-high mill.

Not a single mention to rolling appears in *De Re Metallica*, first published by Agricola [1] in 1556, although the reference of hammering to produce flat plates and strips is made, and this may be indicative of the novelty of this process. Rolling became more common around mid- and late 16th century as different European states used it to produce gold and silver strips of even profile and thickness, which were used for coining and minting [2]. The need for long iron products, such as nails and bolts, during the Industrial Revolution provided for the development of slitting mills, in which a piece of flat-hammered iron passed between a pair of rotating disc cutters. It was normal practice to feed the strip by means of rotating rolls [3,4]; a plate of such a mill is illustrated in the *Encyclopédie* [5].

The energy required to rotate the rolls and deform the metal in earlier mills was supplied either by the operators or by horse gins [5]. The Industrial Revolution required great amounts of iron in several shapes and sizes, and this promoted the installation of rolling mills in sites formerly occupied by wheat mills (therefore the name used up to now) to take advantage of the water wheels available in rapid flowing rivers. Steam machines were also used for rolling; at the beginning, they were limited to pump water from reservoirs located at different heights, making it possible to install rolling mills in sites far away from rivers, and, later on, they were used to impulse the mills [2–4]. An example of such a case is the mill installed at Teplitz, Czech Republic, commissioned in 1892, and it may be considered the first modern hot strip steel mill. It had two steam machines, one used to drive two roughing mills, whereas the second machine was used to drive five finishing stands. Unfortunately, this mill was phased out in 1907, as the machines were not able to roll fast enough to avoid chilling of the stock [6,7].

II. EQUIPMENT

Deformation is carried out in rolling by a series of rotating cylindrical or conical tools called rolls. The tools that contact the workpiece are called work rolls, whereas cylinders that support or provide flexural strength to them are called backup rolls. Rolling mills can be classified either by the number or type of rolls being used, by the shape of the final product, by the way or temperature at which the process is carried out, etc. Full description of different mill types, configurations, and facilities, as well as other ancillary equipment, can be found elsewhere [3,6–12], so a brief description will follow.

A. Mills

It was mentioned before that the term mill, used to describe either the individual stands or whole continuous lines, was derived from wheat or sugarcane mills, as the new process was installed in sites occupied by the formers to use the power of available water wheels. A mill stand is the individual piece of equipment capable of carrying out the deformation. It may comprise work and backup rolls, machines or engines that provide the power, mechanical parts that provide for rigidity, and any other ancillary equipment needed to obtain the desired shape, profile, or thickness.

A continuous mill is obtained when a series of stands are put together to deform the workpiece in a sequential manner. Figure 1 shows a schematic diagram of a six-stand continuous mill used to roll steel strip at temperatures above 850°C (hot rolling). The individual stands are of the four-high type, as each of them contains two work and two backup rolls; each pair of work rolls is

impulsed by an electric motor. The temperature of the workpiece is recorded by pyrometers located towards the entry and exit sides; the descaler located after the entry pyrometer removes the high-temperature oxide crust that has been formed during reheating or in between previous rolling passes. A characteristic of rolling, as will be seen later, is that the speed of the workpiece increases as its cross-sectional area is reduced to maintain a constant mass flow, but since the width of the strip does not change while being processed (due to plane strain conditions), the reduction in thickness imparted at each pass produces an equivalent increase in speed. Therefore multistand continuous mills require a series of electro-mechanical devices, known as loopers, that have the function to control the rotating velocity of the work rolls and maintain the tension between stands within narrow limits. Every rolling stand has a series of headers that are used to control and maintain the temperature of the work rolls within certain limits; this is done to reduce the risk of damaging them and to avoid distortion of the roll gap.

A characteristic of the mills such as the one shown in Fig. 1 is that the material being deformed flows in only one direction. Reversible mills, on the other hand, are those in which the flow of material is reversed after one pass is completed. Figure 2 shows an example of a single reversible mill used for cold rolling of flat steel. The stand is of the four-high type; two down-coilers maintain the tension required to deform the material. The headers shown in the diagram are not only for controlling the work rolls temperature, but also for providing the means for lubrication at the roll gap.

Figure 3 shows the diagram of a cross-country mill used to obtain nonflat hot-rolled products. All the work rolls in this mill are impulsed by one motor, so it is not

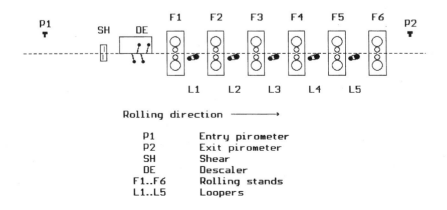

Rolling direction ⟶

P1	Entry pirometer
P2	Exit pirometer
SH	Shear
DE	Descaler
F1..F6	Rolling stands
L1..L5	Loopers

Figure 1 Example of a six-stand continuous mill.

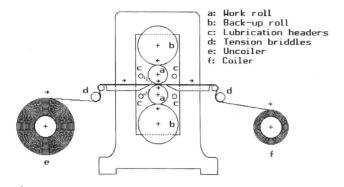

a: Work roll
b: Back-up roll
c: Lubrication headers
d: Tension briddles
e: Uncoiler
f: Coiler

Figure 2 Diagram of a reversible cold-rolling mill.

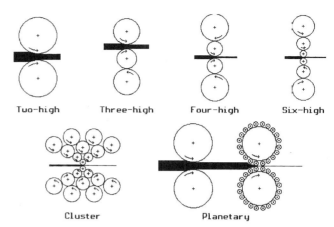

Two-high   Three-high   Four-high   Six-high

Cluster   Planetary

Figure 4 Different configurations of stands used for rolling flat products.

possible to change or control the rotating speed of each pair of work rolls; therefore the changes in the length of the bar as it exits each stand are accommodated by looping. The individual stands of the mill shown are of the three- and two-high type, as they comprise three or two work rolls; the flow of material is reversed in three-high mills by deforming the stockpiece between different pairs of work rolls.

Figure 4 shows a series of common configurations of stands used for rolling flat products. Most of these designs have only two work rolls, the exception being the three-high and the planetary mill; the number of backup roll, as well as the complexity of the design, increase as the requirements for thinner gages and dimensional quality augment. Most of the stands shown in this figure can be found being used in continuous or reversible operations. Exceptions are the three-high mill, which is always reversible (the workpiece is rolled in one direction between the central and one of the other work rolls

and in the other direction between the central and the opposing roll), and the planetary mill, which is used in tandem with a two-high stand, that feeds the stock into the mill. Another characteristic of the work rolls in the planetary mill is that they rotate against the direction of rolling [13].

B. Rolls

The rolls are the principal tools used to deform the metal. They can be made of different types of materials, ranging from highly alloyed steels to cast irons; the steel rolls can be either cast or forged. The work rolls used in high-speed and wear conditions, such as the ones encountered when steel reinforcing bars are being produced, or in which dimensional quality is of paramount importance, can have inserts made of sintered carbides [14]. Casting of rolls can be static or centrifugal; pouring can be carried out with a single or multiple melts. All the rolls are heat-treated either to reduce the residual stresses generated during their processing or to obtain a given hardness profile [15,16]. The operational life of work rolls is short, as their required values of hardness are limited to a narrow superficial layer or skin, dropping towards the center of the roll. All the rolls have to be machined or rectified after a rolling cycle or campaign. This is done either to eliminate any damage or residual stresses that may have been generated during processing or to restore their profile, which was lost as the surface wears out [17].

Hot rolling of steel is carried out at temperatures above 700°C, conditions that subject the work rolls to thermal shock and to mechanical and thermal fatigue

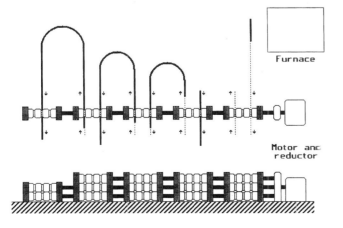

Furnace

Motor and reductor

Figure 3 Schematic diagram of a cross-country mill.

and promote the formation of the so-called fire cracks, which may propagate to the interior of the roll producing its breakage or spalling [18,19]. Contact with the workpiece also promotes the oxidation of the work roll surface [17–21], and, if cracks are present, oxidation may proceed within the cracks, promoting high wear rates [22].

Another problem encountered at these working temperatures is that caused by the oxide layer formed on the steel surface, which is hard and brittle and promotes wear of work rolls [23]. Cold-rolling work rolls are not subjected to these high wear rates, as it is normal practice to remove the oxide layer before processing the steel and to use lubricants during rolling; however, they are subjected to thermal and mechanical cycling [7].

C. Ancillary Equipment

The equipment needed for rolling is more than the stands or the work rolls; for instance, production of thin gage strip, characteristic of cold rolling, requires tension, and this is provided by coilers and bridles in individual mills or by controlling the speeds of different stands in continuous mills. Tension can also be applied during hot rolling in a continuous mill to reduce the separation forces in a given stand, although this is not a common practice due to the possibility of reducing the width of the strip. The forces and tensions developed during rolling are recorded by different types of load cells [12].

Water is used in hot rolling of steel to control the temperature at which the strip is coiled at the end of the run-out table; the water from the cooling headers may be sprayed at high pressure or may establish a laminar flow [24,25]. Water headers are also installed in modern hot rolling mills to provide for interstand cooling and control the rolling and finishing temperatures [26].

Most of the ancillary equipment installed in strip mills have the function to assure the dimensional quality of the rolled product. Shape of the strip can be controlled by machining diverse profiles into the work rolls; by use of roll bending or roll shifting systems, control of the actuators is achieved in real time with the aid of mechanical or optical shape meters. Changes in width of the strip can be recorded by optical means, and such measurements can be used to feedback edging mills or presses. Variation of the thickness, either at a fixed point or along the width of the strip, can be recorded either with x-ray or gamma-ray sources; these measurements are also used to feedback automatic gage control systems that are based on the elastic characteristics of the stand and assembly [12,27].

III. ROLLING MECHANICS

As was mentioned before, rolling is carried out by direct contact between rotating working tools and the stock or workpiece. Two different conditions can be considered. The first one is that in which the width of the stock does not change, which is normally assumed to be the case of flat products, and the second one is that of long products, in which longitudinal changes are accompanied by lateral spreading.

A. Flat Products

Figure 5 shows a schematic diagram of the geometric conditions existing during flat rolling [6–8,10–12,23,27, 28]. The initial thickness of the strip h_b is reduced to h_a by the pressure exerted by two work rolls of radius R, rotating at a constant angular speed (ω). The reduction of height or draft is identified as Δh, the rolling or contact angle is identified by α, and the arc of contact (L) is given by:

$$L = \left(R\Delta h - \frac{\Delta h^2}{4} \right)^{0.5} \tag{1}$$

but as Δh is small, the second term within the parenthesis is normally neglected. The average thickness ($\bar{h}$) during rolling can be calculated by:

$$\bar{h} = (h_b h_a)^{0.5} \tag{2}$$

The width (w) of the strip remains constant during rolling, as deformation proceeds in plane strain con-

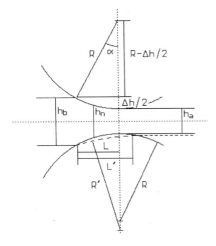

Figure 5 Schematic diagram of the geometry of the roll bite.

ditions, causing a continuous increment on the strip speed during rolling, as the mass flow has to remain constant:

$$v_a w h_a = R\omega w h_n = v_b w h_b \tag{3}$$

where v_a and v_b are the delivery and the entry speed of the strip, respectively. h_n is the thickness at the neutral plane that corresponds to the position at which the velocities of the strip and that of the work rolls are equal.

The force per unit length (P/w) required to roll a material can be calculated by:

$$\frac{P}{w} = \bar{\sigma}(R\Delta h)^{0.5} Q_p \tag{4}$$

where Q_p is a geometrical term and $\bar{\sigma}$ is the average strength of the material that is defined by:

$$\bar{\sigma} = \int_{\varepsilon_0}^{\varepsilon_f} \sigma d\varepsilon \tag{5}$$

where the integral represents the area under the corresponding stress–strain curve. Evaluation of the magnitude and dependence of Q_p with respect to various rolling and geometrical parameter is the aim of different rolling theories [6–8,10–12,27,28]. Figure 6 shows, as an example, the variation of Q_p with respect to the pass reduction for different ratios of the radius of the deformed work rolls (R') over the exit thickness of the strip assuming adhesive friction at the work roll–strip interface [29].

The pressure (p_r) exerted by the material on the work rolls causes the elastic deformation of the work rolls, which translates as an increment of projected length of contact from L to L' (see Fig. 5). The apparent radius of

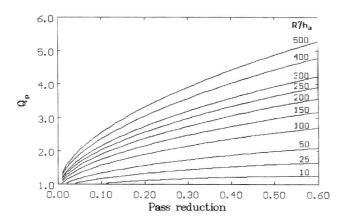

Figure 6 Variation of Q_p for hot rolling of steel assuming adhesive frictional conditions. (From Ref. 29.)

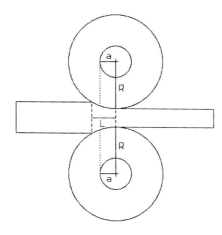

Figure 7 Diagram showing the method used to calculate the torque during rolling.

the work roll R' can be calculated assuming that the deformed arc of contact maintains an elliptical shape [30]:

$$R' = R\left(1 + \frac{C}{\Delta h}\frac{P}{w}\right) \tag{6}$$

where C depends on the elastic properties of the material used as work roll:

$$C = \frac{16(1 - v^2)}{\pi E} \tag{7}$$

where E is the Young's modulus of elasticity and v is the Poisson's ratio. The value of C for steelwork rolls is of around 2×10^{-5} mm^2/N.

The torque (Γ) required to drive the work rolls and deform the material can be calculated assuming that the separation force (P) is concentrated at a level arm (a) which is a fraction of the projected arc of contact length [6–8,10–12,27] (Fig. 7):

$$\Gamma = 2Pa \tag{8}$$

where the number 2 comes from having two work rolls. The fraction (a/L) of the projected arc of contact length can also be associated with geometrical parameters [31–33] (Fig. 8).

B. Long Products

A series of products such as rail, beams, bars, and wire rods are manufactured by rolling them through a series of work rolls in which different shapes are machined into them. All these products are rolled from an initial round or square bloom or billet into the desired shape

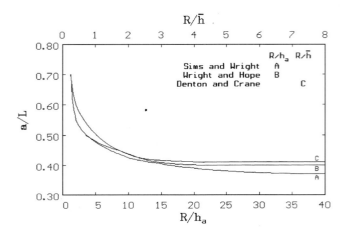

Figure 8 Variation of the fraction of the projected arc of contact length (a/L) as a function of the ratio of the work roll radius over either the final thickness or the average thickness of the strip. (From Refs. 31–33.)

by passing them through calibrated work rolls. The sequences of passes required to obtain a given product are traditionally established by empirical trials [3,8,9], but the development of new mathematical formulations coupled with cheap and powerful computers has opened this field to improvement [34–36].

The main differences between rolling flat and long products are the lateral spreading involved in the changes of shape and the area of contact between the work rolls and stockpiece. Figure 9 shows an example of the change in geometry involved in an oval to round pass. The total reduction involved in the pass has to be expressed in terms of the cross-sectional areas of the start (oval) and final (round) products, as the pass involves a reduction in thickness, Δt, and a positive spreading, Δw.

It is worth noticing that the area of contact between work rolls and the piece follows an elliptical shape (Fig. 9c), as the oval only touches one point in the periphery of the work roll as it gets in contact with them; but, as deformation progresses, different regions of the formerly oval will touch the work rolls. This contact conditions imply that the radius of the work roll will change from R_c, at the central point, to R_e, at the edges.

Roll pass design is critical in rolling long products, as it is normal practice to invert the direction of reduction and spreading after each pass. Such sequence of successive spreading and reductions may cause superficial longitudinal defects if the material is allowed to spread beyond the cavities machined into the work rolls in a given pass. In such a case, a pair of protrusions may be

generated, and, when compressed in the following pass, they will fold against the surface of the product leaving a couple of seams.

C. Frictional Effects

Friction is required in rolling to feed the stockpiece into the roll bite. Figure 10 shows a diagram in which the horizontal components of the frictional (F) and radial (P_r) forces are drawn. Rolling can only succeed if the horizontal component of friction is higher than the opposing radial force component that results from the resistance to deformation:

$$F\cos(\alpha) > P_r \sin(\alpha) \qquad (9a)$$

or

$$\frac{F}{P_r} > \tan(\alpha) \qquad (9b)$$

where α is the roll bite angle (see Fig. 5).

The maximum draft (Δh) that can be achieved in a given rolling pass depends on the radius of the work roll and on the frictional conditions at the interface, and it can be deduced from Eq. (9b):

$$\Delta h = \mu^2 R \qquad (10)$$

where μ is the friction coefficient at the work roll–stockpiece interphase. The minimum exit thickness (h_{min}) that can be obtained in a strip depends on the average strength of the material ($\bar{\sigma}$), defined by Eq. (5), the magnitude of the friction coefficient, and the elastic properties of both work roll and stockpiece [37,38]. An expression to calculate h_{min} is [37]:

$$h_{min} = \frac{A\mu R}{E_r}\left(1 - v_r^2\right)(\bar{\sigma} - \sigma_t) \qquad (11)$$

where A is a coefficient that takes a value between 7 and 8, E_r and v_r are the values of the Young's modulus and Poisson's ratio of the work rolls, respectively, and σ_t is the tensile stress (if any) applied to the rolled piece during its processing. Equation (11) implies that thin strip can only be rolled in mills with work rolls of very small diameter, made of alloys that posses a high Young's modulus.

The direct effect of an increase of the friction coefficient is that of increasing the separation forces registered and the power requirements during rolling. Figure 11 shows the pressure distribution along the arc of contact calculated for a 20% cold reduction of low-carbon steel strip of 1-m width. The formulation used in the

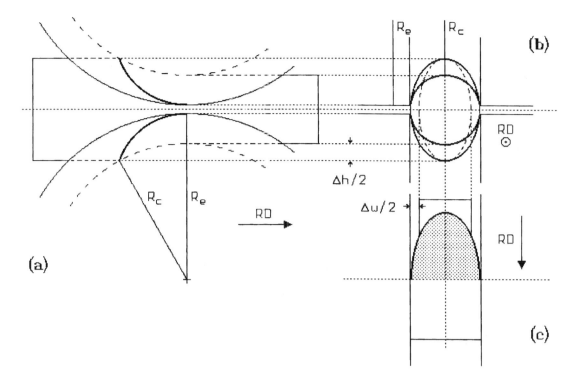

Figure 9 Schematic diagram of the change in shape that takes place in an oval to round pass. The traces of the contact area are drawn with thicker lines.

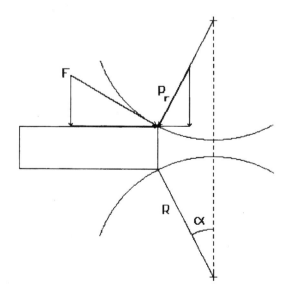

Figure 10 Horizontal components of the frictional and radial forces encountered during rolling.

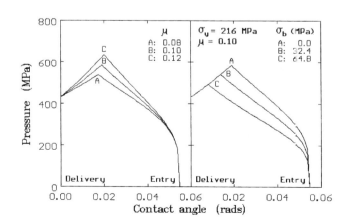

Figure 11 Pressure distribution along the arc of contact computed for cold rolling of steel. (From Ref. 39.)

computations was that proposed by Bland and Ford [39]. The radius of the work rolls was set at 170 mm, the entry (h_b) and exit (h_a) thickness were 2.54 and 2.03 mm, respectively, and the contact angle (α) was 0.0547 rad (3.14°). The constitutive equation for the steel was that given by a power relationship:

$$\sigma = k\varepsilon^n \qquad (12)$$

with k and n equal to 530 MPa and 0.26 [40], respectively, and the yield strength (σ_y) was calculated to be equal to 216 MPa. In all cases, the delivery or exit side was set at the origin, with the contact angle increasing in a clockwise sense taking Fig. 5 as reference. The left-hand side diagram of Fig. 11 shows how the pressure distribution increases, and its peak (that corresponds to the position of the neutral plane) moves towards the entry position as the friction coefficient augments, whereas the diagram on the right-hand side shows how the friction hill is reduced, while the neutral plane moves towards delivery, as the tension applied at the entry increases. Further reductions of the radial pressure, coupled with the displacement of the neutral plane towards the entry side, can be achieved with the increase in tension on delivery.

The friction coefficient can be reduced during rolling by using liquid emulsions of different formulations and by increasing the rolling speed [7,12,23], but care has to be taken with geometrical changes that may take place at the roll gap due to the reduction in the separation force, which may propitiate shape defects [7]. It can be assumed that friction during hot rolling of steel is sticking [10–12,23,29], although it can be reduced with lubricants [12,23]. Recent work carried out in modern hot rolling mills designed to impart large reductions at high speed indicates that the friction conditions change from sticking to sliding as a result of oxidation or oxide transfer to the work roll surfaces or to pulverization of the oxide crust grown in the steel strip surface [41,42].

D. Dimensional Control

The most important quality attributes for commercial grade steel strip are those related to surface appearance and dimensional consistency. Most of the superficial defects of rolled strips can be traced back to teeming or casting incidents, or the inadequate removal of high-temperature oxides during hot rolling (descaling), and will be dealt with in further sections. Dimensional control refers not only to the consistency in gage (thickness) and width along the length of the strip, but also on how the cross-sectional profile varies with rolling to avoid shape defects.

The longitudinal variation of the strip thickness can be recorded with x-ray or gamma radiation sources located at different points within the rolling line. The gage can be measured at a fixed position, normally at mid-width of the strip, or at different positions across the width. Changes in gage across the width (profile) can be recorded by mounting both source and detector on a "C" shape device that traverse the width of the rolling line when a strip is being produced. The gage readings are used to feedback closed-loop automated gage control systems. Periodic variation in gage along the strip can be attributed to eccentricities of rolling bearings or rolls. Gage changes in cold-rolled strips can be due to variation of rolling speeds, whereas those in hot-rolled bands can be attributed to variation in temperature [7,10–12,27]. Figure 12 shows, as an example, the variation in thickness (specified by the deviation from the expected one) recorded in a low-carbon strip that resulted from the temperature fall from front to back end and from the cold spots produced by the skids installed in the reheating furnace used.

Strategies and practices to assure the width of strips have changed with the advent of continuous casting. Universal or slabbing mills were used to break the cast structure in ingots and provide for hot rolling slabs. These mills were necessary due to the inconvenience of having more than three or four different sizes of ingot molds which cast the steel. The slabs produced were left to cool down to room temperature, inspected for any imperfection, so any superficial defect was removed before the slab was charged into a reheating furnace. The

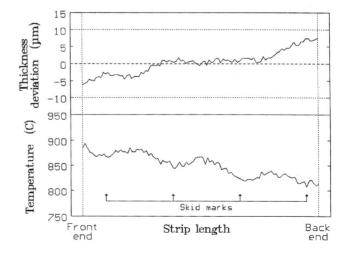

Figure 12 Variation in thickness along the length of a strip that resulted from thermal gradients.

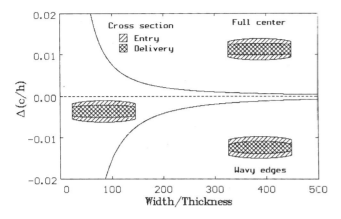

Figure 13 Graphical representation of the shape criterion represented by Eq. (13).

rolling mills had a series of edging stands that were used only to assure that the width was within specifications [12,27,43].

Continuously cast slabs have replaced those obtained from ingots in most rolling operations. This was due not only to the higher yields that can be obtained in continuous casting, but also due to their higher homogeneity and surface quality [9]. The difficulties that arise when changing the width of slabs during pouring had resulted in the installation of universal roughing and edging mills or sizing presses within rolling lines. The slabs fed into reheating furnaces are more or less of constant width, so that the required width is produced by edging mills or sizing presses; width meters installed in rolling mills are used to feedback and control these devices [12,27].

Production of strip by rolling is carried out in plane strain, i.e., the width of the strip remains constant, although lateral spreading may occur depending on geometrical conditions [8,10–12,27,44–46]. Shape defects, identified by the presence of waves or buckles along the width, may develop during rolling when the in-going profile of the strip is inconsistent with the geometry of the roll gap. In such a case, some regions of the strip are deformed to a greater extent than in others, producing local elongations that transform into waves. This problem is more accentuated in thinner gages, as the force required to bend the strip is proportional to the momentum of area of the cross section, and thicker gages are able to spread more easily [8,10–12,27].

It is normal practice to refer to the profile by its crowning (c), which is defined as the difference in thickness at mid-width and at a point close to one of the edges. Problems in shape may develop in a given pass when the reduction in crowning is not consistent with

the reduction in thickness. A shape criterion was developed in terms of the change in the in-going and outgoing crown over thickness ratios [47,48]:

$$-80\left(\frac{h_a}{w}\right)^{1.86} < \Delta\left(\frac{c}{h}\right) < 40\left(\frac{h_a}{w}\right)^{1.86} \tag{13}$$

where h_a is the exit thickness (see Fig. 5) and w is the width of the strip. The change in the crown-over-thickness ratio, $\Delta(c/h)$, is given by:

$$\Delta\left(\frac{c}{h}\right) = \left(\frac{c_b}{h_b} - \frac{c_a}{h_a}\right) \tag{14}$$

where the subindexes a and b refer to the outgoing and in-going profile, respectively. Figure 13 shows the graphic representation of the shape criterion expressed by Eq. (13).

IV. METALLURGICAL PHENOMENA

Steels constitute a versatile group of engineering alloys with more applications and uses than any other type of materials. The reason behind this is the various types of microstructures that can be obtained by alloying or processing. The main constituent in steels is iron, which crystallizes in two types of structures: body-centered cubic (BCC), stable either at temperatures above 1486°C or below 910°C, and face-centered cubic (FCC), stable at the intermediate range, causing a series of allotropic solid-state transformations. Figure 14 shows the calculated change in length of a piece of iron as it is heated up [49–51].

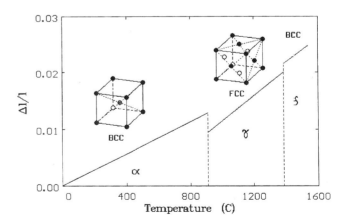

Figure 14 Crystallographic structures and changes in elongation as a function of temperature for pure iron. (From Ref. 51.)

A. Casting

Almost all steel destined to be rolled into flat products, and a big proportion of long products, is cast in a continuous way. Up to the advent of this later process, steel was poured into open top or bottom ingot molds that were reduced in high-lift or slabbing mills to obtain flat products. Most of the commercial grades were of the rimming type, so-called due to the carbon monoxide evolution concurrent with solidification that forms an almost pure iron rim around the ingot surface. Production yield of this material was close to 90%, as most of the contraction due to solidification was counteracted by trapping carbon monoxide into blowholes. Steels with carbon contents higher than 0.15% or alloyed with strong carbide or oxide-forming elements (such as silicon, aluminum, niobium, titanium, etc.) were deoxidized and cast into hot top molds. Yield of these materials varied from 80% to 85% due to contraction and crop losses [9].

Continuous casting has displaced ingot mold casting not only due to higher yielding, close to 98%, but also due to the higher homogeneity of the product, as mold casting is also accompanied with segregation from bottom to top, where most of the impurities and inclusions remain. It should also be mentioned that heterogeneity in rimming products was encountered also across width and thickness. Another point in favor of continuous casting is that of surface quality, as blowholes in rimming steel were susceptible to oxidation during reheating leaving behind different types of inclusions. Another source of inclusions in ingot cast steels was the lack of protection against oxidation of the liquid steel stream while pouring [9,52].

Inclusions in continuous cast steels can be avoided with an adequate secondary metallurgy and subsequent protection of the steel surface by synthetic slag and inert gases. Inclusions that may have been formed in the liquid stage can be trapped into the slag by installing flow waffles within the casting tundish and with the use of submerged nozzles. Most superficial inclusions in continuous cast products can be associated when lubrication powders, used to protect the mold, are trapped by the solidification front, which reflect poor operational practices [9,53,54].

B. Hot Rolling

It was normal practice to reheat ingots in soaking pits before rolling them in slabbing mills. The pieces so produced were left to cool down to room temperature, inspected for surface defects, to be eliminated before charging into a reheating furnace. Continuously cast slabs can also be inspected at room temperature before being charged for hot rolling, but improvement in steelmaking and casting practices has made it possible to establish direct rolling procedures that reduce the energy required for reheating and shorten processing time [11,12]. Special mention has to be made to thin slab casting that is able to produce pieces of 50 to 100 mm in thickness. These slabs are reheated in holding furnaces as soon as they exit the casting line and are hot-rolled directly without having to cool down to room temperature [42,53,54].

Oxidation of the steel surface is always present at the temperatures involved in hot rolling. Shallow surface defects can be removed by light oxidation during reheating, but any discontinuity, such as cracks or pinholes, can promote internal oxidation that results in the formation of inclusions [52]. The oxide layer has to be removed before the steel is deformed to avoid the imprint and the consequent damage of the steel surface. It is normal practice to use high-pressure water jets (descaling) to break and remove the oxide crust formed at different stages. Different mechanisms are thought to be responsible for oxide removal. Impingement by water produces a localized chilling of the surface that generates stresses at the steel–oxide interphase, which aids in oxide removal by the water sprayed [57].

The oxide layer at temperatures above $570°C$ is made of different species: FeO, around 90%, Fe_3O_4, around 8%, and Fe_2O_3, the remaining 1–2%. Elements present in the steel affect the oxidation behavior; of them, the cases of silicon, added to steels aimed for electrical applications, and copper, added in some weathering steel grades, but mostly present as trap element, undergo difficult oxide removal during descaling. Silicon forms a hybrid oxide, fayalite, that strongly adheres to the oxide surface, increasing the difficulty for its removal. Copper tends to concentrate at the surface at a given temperature range and also increases the adhesion of the oxide crust [58,59].

Deformation in the hot-working range promotes the occurrence of thermally activated restoration mechanisms that modify the microstructure of the steel. Restoration can be associated with ordering or elimination of punctual and linear defects (recovery) or with the reconstruction of the microstructure by a sweeping high-angle boundary (recrystallization). These mechanisms can be concurrent with deformation or can take place after the material was strained, being called dynamic or static in either case. The special case in which restoration starts during deformation, but continues after straining has finished, is named metadynamic. High-

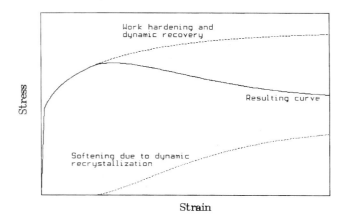

Figure 15 Stress–strain curve for a steel deformed at high temperature assuming different hardening and softening mechanisms. (From Ref. 65.)

angle boundaries increase the internal energy of the material; therefore most steels are subjected to grain growth at the end of recrystallization to minimize its internal energy [10–12,49–51,60–63].

Deformation at high temperature results in the balance of the increment in the density of dislocations, required to produce deformation [64], and their elimination by different restoration mechanisms [61,62]. Figure 15 shows an example of a stress–strain or flow curve for a steel being deformed at high temperature that was computed based on the assumption of a curve in which only hardening and dynamic recovery mechanisms take place, to which softening due to dynamic recrystallization is subtracted [65].

Steels that are able to recrystallize during hot rolling do not present a strong texture [66]. Addition of strong carbide, nitride, or carbonitride formers, such as Nb, Ti, or V, to steel is common to suppress recrystallization and promote a fine-grained structure at room temperature, which may be able to inherit the texture developed during hot rolling in the austenitic range [63,66–68]. Description of the texture components developed by hot rolling, as well as those inherited to the various transformation products, has been reviewed elsewhere [69].

The room temperature microstructure of hot-rolled steel depends on the amount and conditions at which deformation was imparted, as well as on the cooling rate after straining. Combination of such effects constitutes the basis for thermomechanical processing such as controlled rolling and accelerated cooling, which are reviewed elsewhere [62,63,70–72] and will be discussed in the following sections.

C. Cold Rolling

Cold rolling is normally applied to obtain final products with narrow dimensional tolerances and better surface appearance in comparison to those of hot-rolled products. Most of hot rolling is carried out from the viewpoint of optimizing installations and facilities. It is normal practice to have continuous mills in tandem with pickling lines that are used to remove the high-temperature oxide crusts formed during hot rolling. The speeds and reductions that can impart such mills are limited by factors such as the thickness and speed of the incoming hot-rolled strip, as personnel in charge of pickling lines prefer to deliver thicker sections to reduce the surface area per volume of steel processed, whereas the speed will depend on the characteristics and the capacity of the pickling line. The total reduction of a given mill depends on factors such as width and strength of the strip and the total number of individual stands available. The total reduction imparted in reversible mills can be more flexible, but the number of passes is normally minimized to increase productivity. In some cases, due to layout, the number of passes has to be either odd or even [6–9,11,12,27].

Deformation by cold rolling accumulates in the material producing the increment in strength (Fig. 16). The grains elongate along the rolling direction, but their size is not affected, as recrystallization does not occur. As a result of this, cold-rolling materials present strong and sharp textures that are reviewed elsewhere [73–75].

Cold-rolled material in process for galvanizing by immersion is normally delivered in its as-rolled condition since the steel will be annealed prior to immersion.

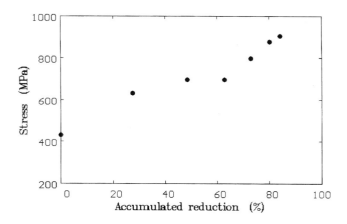

Figure 16 Increase in yield strength of low-carbon steel as result of cold rolling.

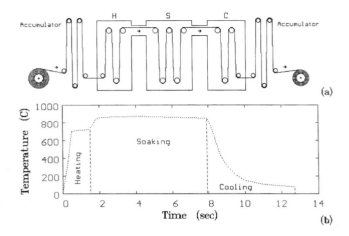

(a)

(b)

Figure 17 Schematic diagram of a continuous annealing line (a) and the corresponding heating cycle (b).

All other products are annealed and, after this, subjected to small reductions, normally less than 3%, to eliminate strain aging effects [49–51]. This later process is called tempering and can be used to control the surface aspect of the strip by engraving a certain roughness on the surface of the work rolls [7,11,12].

D. Annealing

Annealing is used to restore the values of strength and ductility that were modified by the plastic deformation imparted by cold rolling. It is common to associate the restoration of such properties by recrystallization of the highly distorted microstructure, although some processes rely only on the arrangements of dislocations by recovery mechanisms [76].

Cold-rolled steel strips can be processed in continuous lines or in batch furnaces. In the first case, the strip goes through installations that subject the material to heating, soaking, and cooling cycles (Fig. 17a). A direct advantage of such processing is the high productivity due to the short times involved, as the low thermal inertia of the strip allows for rapid heating and cooling (Fig. 17b). The atmosphere within the furnace is controlled to avoid oxidation. Annealing cycles and other parameters can be modified by changing the speed at which the strip moves within the furnace [7,9,77,78].

Batch annealing is normally conducted on tight wound coils arranged within a protective cover that avoids oxidation. Figure 18a shows a schematic diagram of a load consisting in four coils of cold-rolled strips, and Fig. 18b shows the thermal evolution re-

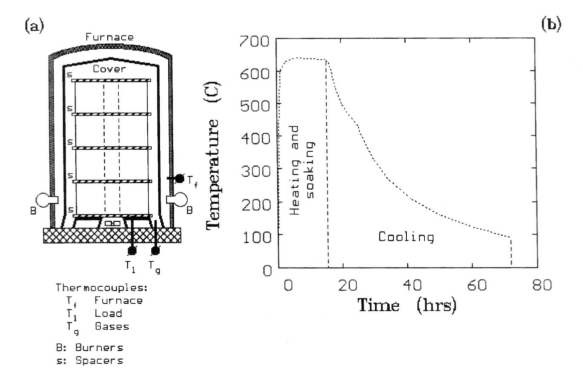

Figure 18 Schematic diagram of a continuous annealing line (a) and the thermal cycle recorded by the thermocouple close to the load (b).

corded by the thermocouple located in contact with the coil located at the bottom; the periods involved in heating, soaking, and cooling are indicated. The most important advantages of this type of process, when compared with the continuous one, are the flexibility of its operation, the low initial investment costs required, and the possibility for expanding production capacity by adding extra basses and furnaces. The high thermal inertia of the wound coils implies strong thermal gradients within the material, which, when not controlled, may result in the variation of mechanical properties along the length of the coiled strip [7,9,79,80].

Comparison of Figs. 17b and 18b shows that the metallurgical conditions are different for either type of process, as restoration in continuous annealing involves the transformation to austenite in heating and to ferrite in cooling, whereas that in batch annealing is only by recrystallization [7,9,49–51]. The texture present in cold-rolled and annealed strips [75] affects the forming characteristic of the material in the way described in the following section.

V. SPECIAL PRODUCTS

The great variety of uses that steels can provide is due to the possibility of achieving different properties by tailoring their microstructure to specific applications. Changes in size, shape, and distribution of the different microstructural components in steels can be done by varying the processing route as well as the chemical composition.

A. High-Strength Low-Alloy Steels

Steels for structural applications require for both high strength and toughness a combination that can only be achieved by the reduction or refining of ferritic grain size [49–51]. These steels are processed following either controlled rolling [63,70] or accelerated cooling [63,71,72] practices. The first one (Fig. 19) relies on the progressive refining of the austenite grain size by recrystallization between, or within, rolling passes. Further reduction of the ferritic grain size can be achieved by the addition of elements able to precipitate during processing. Such precipitation inhibits recrystallization and allows for the accumulation of strain within the austenite, which is reflected by the increase of the surface-to-volume ratio (S_v) of elongated grains. Further increase in S_v occurs when deformation bands start to appear. Refining of ferrite results from the enhanced nucleation rate propitiated by the distorted microstructure [62,63,70].

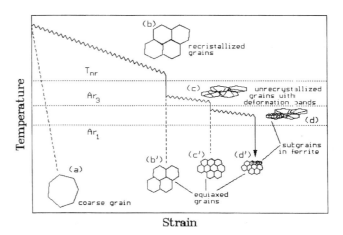

Figure 19 Schematic diagram of controlled rolling.

Normal controlled rolling is carried out at temperatures at which recrystallization does not proceed, called T_{nr}, but above that for the start of transformation to austenite (Ar$_3$). Further reduction in ferritic grain size can be obtained by rolling within the intercritical range (between Ar$_3$ and Ar$_1$), where deformation proceeds in a mixture of ferrite and austenite, although some authors are worried about the texture-related effects arising from the deformed ferrite [63,69,70,81,82].

The temperature at which recrystallization stops (T_{nr}) depends on processing, and higher reductions imparted in a given rolling pass, coupled with higher interpass times, will result in the increase of such temperature [62,63,70,83,84]. Microalloying affects this temperature to a different degree depending on the amount and the type of element added. Figure 20 [85] shows the temperature at which recrystallization stops in a carbon–manganese steel modified by the addition of various elements. The difference in behavior shown in this last figure is not only due to the solubilities of the particles able to precipitate (Fig. 21) [86], but also on the crystallographic structure of the precipitates [63,87]. Figure 22 shows the microstructures of a Nb–V microalloyed steel (0.08 C, 1.54 Mn, 0.35 Si, 0.055 Nb, 0.078 V, wt.%) after roughing (a) and controlled rolling (b) stages in which the differences in grain size can be appreciated.

Further reduction in the ferritic grain size can be achieved by the depression of the temperature at which the austenite to ferrite starts (Ar$_3$), as the rate for nucleation of new ferrite grains will be increased [63,71,72,88–90]. Figure 23 [71] shows the effect that the addition of Ni and the increase on cooling rate exert on the depression of Ar$_3$ and the consequent reduction of the grain size of ferrite. Intensive use of water to reduce the temperature during or at the end of rolling has resulted

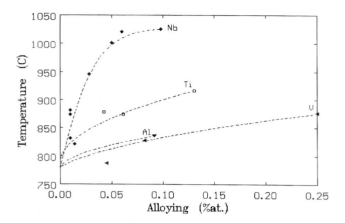

Figure 20 Temperature at which recrystallization stops in a C–Mn steel. (From Ref. 85.)

in accelerated cooling [63,71,72]. Such process can be used in combination with controlled rolling to reduce the time involved in cooling from the high-temperature regime, at which austenite recrystallizes, to those at which this phenomenon does not take place (see Fig. 19) or at the end of the rolling process [91]. Variation in toughness and strength of the steel can be achieved by accelerated cooling. Figure 24 shows the way in which the yield strength and the temperature at which the ductile to brittle transition takes place as a function of the cooling rate are achieved and the temperatures at which accelerated cooling starts and finishes [63,71].

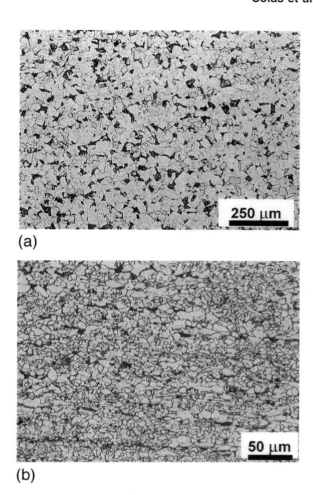

(a)

(b)

Figure 22 Microstructures, at different magnification, of a Nb–V microalloyed steel after the roughing (a) and controlled rolling (b) stages.

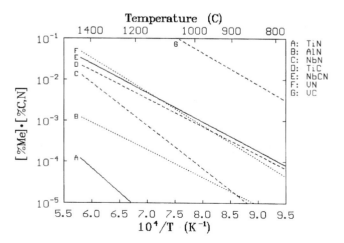

Figure 21 Solubility products for various carbides and nitrides in austenite. (From Ref. 86.)

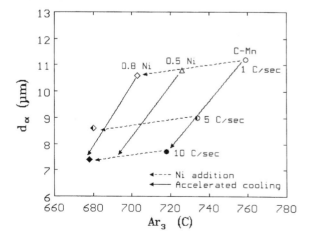

Figure 23 Variation of Ar_3 and ferritic grain size due to the increase in Ni content and cooling rate. (From Ref. 71.)

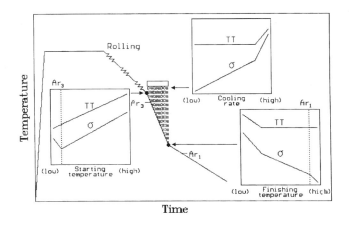

Figure 24 Schematic diagram of accelerated cooling showing the changes in mechanical properties as result of the variation of various processing parameters.

Thermomechanical processing was developed to obtain steels with a good combination of high strength and toughness as result of the fine ferritic grain size obtained [49–51,62,63]. Further increase in strength can be obtained by other mechanisms, such as subgrain formation or precipitation [63,92–94].

B. Multiphase Steels

The oil crisis of the mid-1970s provided for the driving force to search for new materials with higher strengths that will be able to contribute in reducing the weight of automobiles. One of the most promising materials were dual-phase steels, so-called because they consist of a mixture of finely dispersed martensite within a ferritic matrix. These steels exhibit a superior combination of high strength while retaining good ductility (Fig. 25) [96–103]. Another point of favor for these materials is that they do not show the abrupt yield point characteristic of annealed steels that results in stretch marks in deformed pieces [49]. The reason behind the mechanical behavior of such steels has been explained in terms of the combination of a hard and strong dispersed second phase, martensite, that strengthens the soft and ductile matrix, ferrite. The absence of yield point is attributed to the high density of dislocations caused by the transformation to martensite [98–100].

The dual-phase structure of these steels can be obtained by the rapid cooling of a material held at the intercritical range (between A_1 and A_3). It is normal practice to alloy these steels with silicon and manganese to expand the dual-phase region and retard the diffusive transformation of austenite [95–100]. Two different processing routes can be attempted. The first one (Fig. 26a) consists in heating the material at temperatures above A_3 for the time required to fully transform the structure into austenite; the material is then allowed to cool down to a temperature within the intercritical range to promote the formation of ferrite. The temperature and the time required to obtain the mixture of structures depend on the chemical composition of the

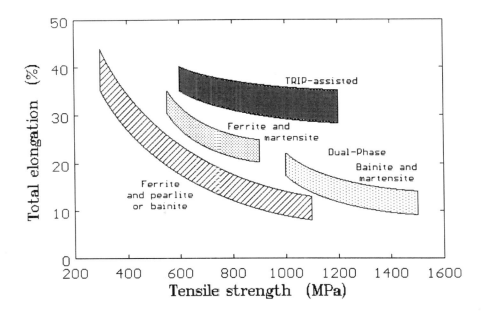

Figure 25 Typical values of strength and ductility of different types of steels.

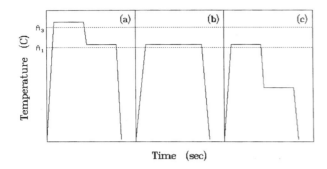

Figure 26 Thermal cycles that result in dual-phase (a) and (b) or TRIP (c) steels.

material. The steel is quenched after this intercritical treatment to transform the austenite. The second processing route (Fig. 26b) consists in holding the material within the intercritical range; the rate of transformation of carbides and ferrite, as well as the enrichment of carbon to the austenite, will depend on the chemical composition and the temperature at which the treatment is carried out. As in the former case, the mixture of ferrite and martensite is achieved by quenching the steel. Figure 27 shows the microstructure of a dual-phase steel (0.11 C, 1.26 Si, 1.53 Mn, wt.%) treated at 800°C for 10 min before quenching it in water.

Evolution of dual-phase steels resulted in a new type called TRIP due to the transformation-induced plasticity effect that occurs when retained austenite transforms into martensite during deformation. These effects allow for high values of ductility and strength when compared

with other steels (Fig. 25). The enhancement in ductility in these steels is attributed to the increase in the work-hardening rate that results from the transformation to martensite, which will delay the onset necking, or other type of plastic instabilities [101–103].

The microstructure required to obtain the TRIP effect in steels is attained by following the cycle shown in Fig. 26c, which consists in holding the material at a temperature between A_1 and A_3, followed by holding at a temperature that allows partial transformation of the austenite into bainite. This partial transformation allows for the retention of austenite once the steel is cooled down to room temperature, as the best properties are obtained in steels containing 7% to 11% of retained austenite [102,103]. Figure 28 shows a sample of a carbon steel (0.13 C, 0.78 Si, 1.66 Mn, wt.%) treated for 5 min at 780°C and for 30 sec at 410°C to develop the TRIP microstructure.

C. Steel for Electrical Uses

Steels are used in electrical devices due to the ferromagnetic characteristics of iron. Various types are used in transformers or in electrical motors; in general, low-carbon steels are used in fractional motors and low-power transformers, whereas silicon-bearing steels are used in more powerful devices. Steels for high-power applications can be processed to develop a strong <001> texture to take advantage of the easy magnetization along this particular crystallographic direction (Fig. 29) [49,104–106]. Steels for electrical uses are rated in terms of the power lost when subjected to magnetization in the

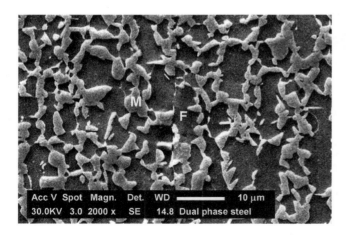

Figure 27 Scanning electron microscope image of a steel treated to develop a dual-phase structure (M: martensite, F: ferrite).

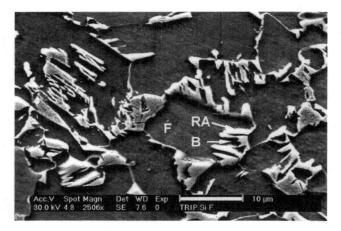

Figure 28 Scanning electron microscope image of a steel treated to develop a TRIP structure (RA: retained austenite, F: ferrite, B: bainite).

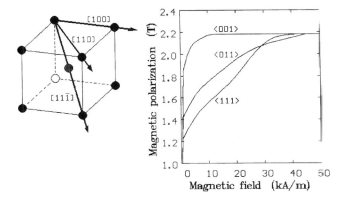

Figure 29 Magnetization in iron as a function of different crystal orientations. (From Ref. 104.)

field produced by alternate current [107,108]. The steel in these applications is used as thin strips or shapes and laminations; their surface can be either coated or oxidized to reduce power losses due to eddy currents [106–108].

A ferromagnetic material consists of domains of different sizes that can align along the direction of magnetization. The domains in a nonmagnetized material will be distributed in a random manner, so each cancels the effect of others. Magnetization occurs when a magnetic field is applied; it is possible for either the domains to rotate and align in the direction of magnetization or for the better-oriented domains to grow at expenses of those that are not oriented to the field. The borders between neighboring domains are around 50 nm; therefore precipitates and grain boundaries constitute obstacles to the domains during magnetization [49,104–106].

The material used for laminations in fractional motors and low-power transformers is cast with low contents of carbon and nitrogen; addition of elements that may form precipitates are avoided, as these particles reduce the mobility of magnetic domains. Hot rolling is carried out following normal practices; care is taken to finish rolling and coil the strip at high temperatures to promote grain growth. Two cold-rolling reductions, with an intermediate annealing, are applied. The first one, used to obtain a close to final thickness, can be in the order of 50% to 80%, depending on the thickness of the hot-rolled band. The second reduction is limited to 5% to 8%; its intention is to provide for enough distortion that will allow for grain coarsening during the last stage of annealing. The intermediate annealing can be conducted either in tight wound or in open coils; the latter is used in conjunction with a decarburizing atmosphere. The steel so produced is normally sent to the manufacturer of the electrical devices, and it is called as semiprocessed. The manufacturer punches out the laminations and anneals them in a controlled atmosphere that will cause decarburization of the steel and the formation of a thin isolating layer of oxide [106–111].

Steels for higher-power applications are alloyed with silicon to increase their electrical resistivity and, with it, reduce electrical losses due to eddy currents. Iron can dissolve up to 6.3% of silicon in weight, but, due to brittleness caused by formation of various ordered structures, it is normally kept below 4.5% [106,112]. Figure 29 shows that the iron can be easily magnetized along the [001] direction; therefore oriented steels are processed to obtain such texture. The first successful case to obtain a close texture was that of Goss [113], who found that steels containing 3.5% Si were able to develop a strong [001](011), cube on edge, or Goss texture by annealing thin samples within a 1050°C to 1200°C temperature range. The ideal [001](001), cube on face, texture was developed by conducting a similar annealing in a controlled atmosphere [114].

D. High Formability Steels

Formability in sheet metal is related to the capacity that the material has to extend rather than thin down. This capacity is expressed by the parameter r that is evaluated in a tensile sample by measuring the ratio of the logarithmic strain measured on the width of the specimen (ε_w):

$$\varepsilon_w = \ln\left(\frac{w_o}{w_i}\right) \tag{15a}$$

where w_o and w_i are the initial and instantaneous width of the sample, respectively, over the logarithmic strain as a function of the thickness of the specimen (ε_t):

$$\varepsilon_t = \ln\left(\frac{t_o}{t_i}\right) \tag{15b}$$

where t_o and t_i are the initial and instantaneous thickness on the gage length, respectively. The value of r is calculated by:

$$r = \frac{\varepsilon_w}{\varepsilon_t} \tag{16}$$

An average value of this parameter $\bar{r}$ is obtained by testing samples cut along 0°, 45°, and 90° with respect to the rolling direction:

$$\bar{r} = \left(\frac{r_0 + 2r_{45} + r_{90}}{4}\right) \tag{17}$$

that is normally called the average strain ratio. Another parameter used to evaluate the formability of sheet metals is the planar anisotropy (Δr) defined as:

$$\Delta r = \left(\frac{r_0 + r_{90} - 2r_{45}}{2} \right) \quad (18)$$

that is associated with the tendency for ear formation during deep drawing. Ears along 0° and 90°, with respect to the rolling direction, are found when Δr is greater than 0° and at 45° when negative [50,73–76].

It has been shown that high values of $\bar{r}$ are displayed by materials that have a sharp and well-defined {111} texture, i.e., they have a high proportion of grains with their {111} planes parallel to the sheet plane, whereas components such as the {001} have been found to reduce the drawing capacity, being possible to relate $\bar{r}$ with the integrated intensity ratio of the components of texture for {111} over {001} (Fig. 30) [50,73–76,115]. The formation of texture in steels is affected by alloy chemistry and by processing conditions for hot and cold rolling, as well as cooling and annealing practices [50, 73–76].

Table 1 [116] summarizes the texture components found in cold-rolled and annealed low-carbon steel together with the average strain ration and planar anisotropy. Examination of the values presented in Table 1 yields to the conclusion that the best combination, from the formability point of view (maximum $\bar{r}$, coupled with a minimum $|\Delta r|$), is obtained with a mixed texture of {111} $\langle 011 \rangle$ and {111} $\langle 112 \rangle$ [75,116–118].

Commercial sheet steels can be produced with controlled compositions that are processed to maximize the

Table 1 Major Texture Components in Cold-Rolled and Annealed Low-Carbon Steels, and Their Effect on Formability

Component	$\bar{r}$	Δr
{001} <011>	0.4	−0.8
{112} <011>	2.1	−2.7
{111} <011>	2.1	0
{111} <012>	2.1	0
{554} <225>	2.1	1.1
{011} <001>	5.6	8.9

Source: Ref. 116.

average strain ratio. The maximum values of $\bar{r}$ displayed in rimming steels are close to 1.3 when batch-annealed, whereas aluminum-killed batch-annealed steels can easily reach the 1.7 value. Both types of steels exhibit a dramatic decrease to around 1.1 when annealed in continuous lines [50,73–76]. The reduction in the value of $\bar{r}$ resulted in the formulation of new types of steels susceptible to retain their good formability while being annealed in a continuous way. These new types of steels are normally referred to as ultralow carbon (ULC) or interstitial free (IF) steels and are able to exhibit values of $\bar{r}$ above 2.2. These steels are produced in modern degassing steel-making facilities and are added with strong carbide and nitride formers, such as Ti and Nb, to put C and N out of solution [75,77,78,119–121].

Different parameters affect $\bar{r}$; for instance, reduction in the temperature at which aluminum-killed hot-rolled strip is coiled or decrease in the heating rate during annealing tends to increase such parameters, and these features are associated with the permanence in solution of aluminum nitrate, which will be able to precipitate when annealed with a low heating rate. Neither the coiling temperature nor the heating rate affects to a high extent ULC or IF steels [50,73–76].

Conventional low-carbon steels are cold-rolling to a reduction of around 70% to maximize $\bar{r}$, although it seems that the reduction at which $\bar{r}$ is maximum tends to increase with the reduction of carbon content [50, 73–76]. A maximum value of $\bar{r}$ seems to be achieved at around 10 ppm of either carbon or nitrogen [122,123].

VI. CONCLUDING REMARKS

Steels constitute the most abundant and varied group of engineering alloys due not only to the possibility of alloying iron with various other elements, but also due to the capacity that such other elements have to affect

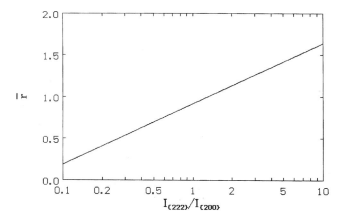

Figure 30 Relationship between the integrated intensities for the texture components for {111} over {001} and $\bar{r}$. (From Ref. 115.)

the behavior of steels under controlled and adequate processing conditions.

Control in chemical composition and processing allows for tailoring the shape, size, and distribution of different microstructural components of steels. Recent developments in texture measurement and control have put in evidence the importance of knowing, and controlling, the crystallographic distribution of the microstructural aggregates. This knowledge allows for tailoring the properties required by the material.

Care has to be taken with respect to surface quality and dimensional control of steel products, as they may be degraded if either their surface or shape does not comply with specifications. These aspects are not only affected by mechanical processing, but are also influenced by steelmaking practices and environmental conditions.

ACKNOWLEDGMENTS

Rafael Colás would like to thank the support by *Consejo Nacional de Ciencia y Tecnología*, CONACYT, Mexico.

REFERENCES

1. Agricola, G. *De Re Metallica*; Hoover, H.C., Hoover, L.H., Eds.; Dover Pub. Inc.: New York, 1950.
2. Cooper, D. *Coins and Minting*; Shire Pubs.: Aylesbury, 1983.
3. Roll Pass Design, British Steel Corp., Sheffield, 1960.
4. Gale, W.K.V. *Ironworking*; Shire Pubs: Aylesbury, 1985.
5. d'Alambert, J.leR.; Diderot, D. *L'Encyclopé die ou dictionnaire raisonné des sciences, des arts et des métiers, par une societé de gens de lettres*, http://encyclopedie.inalf.fr.
6. Larke, E.C. *The Rolling of Strip, Sheet and Plate*; 2nd Ed; Chapman and Hall: London, 1963.
7. Roberts, W. *Cold Rolling of Steel*; M. Dekker, Inc: New York, 1978.
8. Wuzatowski, Z. *Fundamentals of Rolling*; Pergamon Press: Oxford, 1969.
9. McGannon, H.E., Ed.; *The Making, Shaping and Treating of Steel*; 9th Ed.; United States Steel: Pittsburgh, 1971.
10. Roberts, W. *Hot Rolling of Steel*; M. Dekker, Inc.: New York, 1982.
11. Roberts, W. *Flat Processing of Steel*; M. Dekker, Inc.: New York, 1982.
12. Ginzburg, W. *Steel Rolling: Theory and Practice*; M. Dekker, Inc.: New York, 1989.
13. Giles, J.L.; Gutteridge, C. J. Iron Steel Inst. 1973, *211*, 9.
14. Simâo, J.; Aspinwall, D.K.; El-Menshawy, F.; Meadows, K. 41st Mechanical Working and Steel Processing Conf. Proc., 37, ISS, Warrendale, 1999; 567 pp.
15. Adams, T.A.; Collins, D.B. 40th Mechanical Working and Steel Processing Conf. Proc., 36, ISS, Warrendale, 1999; 427 pp.
16. Gaspard, C.; Bataille, S.; Batazzi, D.; Thonus, P. 42nd Mechanical Working and Steel Processing Conf. Proc., 38, ISS, Warrendale, 2000; 655 pp.
17. de Carvalho, M.A.; Xavier, R.R.; Pontes Filho, C da S.; Morane, C.; Bocallini, M. Jr.; Sinatora, A. 42nd Mechanical Working and Steel Processing Conf. Proc., 38, ISS. Warrendale, 2000; 697 pp.
18. Schröder, K.H. 42nd Mechanical Working and Steel Processing Conf. Proc., 38, ISS, Warrendale, 2000; 697 pp.
19. Farber, D.W. 39th Mechanical Working and Steel Processing Conf. Proc., 35, ISS, Warrendale, 1998; 451 pp.
20. Kapadia, B.M.; Marsden, K.W. 39th Mechanical Working and Steel Processing Conf. Proc., 35, ISS, Warrendale, 1998; 349 pp.
21. Ohkomori, Y.; Sakae, C.; Murakani, Y. 42nd Mechanical Working and Steel Processing Conf. Proc., 38, ISS, Warrendale, 2000; 723 pp.
22. Colás, R.; Ramírez, J.; Sandoval, I.; Morales, J.C.; Leduc, L.A. Wear 1999, *230*, 56.
23. Schey, J.A. *Tribology in Metalworking: Friction, Lubrication and Wear*; ASM: Metals Park, 1983.
24. Colás, R.; Sellars, C.M. *Accelerated Cooling of Rolled Steel*; Ruddle, G.E., Crawley, A.F., Eds.; Pergamon Press: New York, 1988; 121 pp.
25. Colás, R. *Manufacturing Science and Engineering*; Alzheimer, W.E., Ed.; ASME (PED-Vol. 68-2): New York, 1994; Vol. 2, 611.
26. Colás, R.; Elizondo, L.; Leduc, L.A. 2nd Int. Conf. Modelling Metal Rolling Processes; Beynon, J.H. Ingham, P., Teichert, H., Watson, K., Eds.; Institute of Materials: London, 1996; 12 pp.
27. Ginzburg, V.B. *High Quality Steel Rolling*; M. Dekker, Inc.: New York, 1993.
28. Rowe, G.W. *Principles of Industrial Metalworking Processes*; E. Arnold: London, 1986.
29. Sims, R.B. Proc. Inst. Mech. Eng. 1954, *168*, 191.
30. Hitchock, J.H. *Roll Neck Bearing, Appendix I, ASME Report of Special Research Committee*; ASME: New York, 1953.
31. Sims, R.B.; Wright, H. J. Iron Steel Inst. 1963, *201*, 261.
32. Denton, B.K.; Crane, F.A.A. J. Iron Steel Inst. 1972, *210*, 606.
33. Wright, H.; Hope, T. Met. Technol. 1975, *2*, 565.
34. Mori, K.; Osakada, K. Int. J. Numer. Methods Eng. 1990, *30*, 1431.
35. Hensel, A.; Wehage, H. Neue Hutte 1990, *9*, 35.
36. Kim, N.; Lee, S.M.; Shin, W.; Shivpuri, R. J. Eng. Ind. 1992, *114*, 329.

37. Stone, M.D. AISE Yearly Proceedings, AISE, Pittsburgh, 1953; 115 pp.

38. Ford, H.; Alexander, D.M. J. Inst. Met. 1959, *88*, 193.

39. Bland, D.R.; Ford, H. Proc. Inst. Mech. Eng. 1948, *159*, 144.

40. Dieter, G.E. *Mechanical Metallurgy*; McGraw-Hill: London, 1988.

41. Reiner, C.; Hriman, R.L. Ironmak. Steelmak. 1993, *20*, 275.

42. Morales, J.; Sandoval, I.; Murillo, G. AISE Steel Technol. Nov. 1999, *25* (11), 46.

43. Keefe, J.M.; Earnshaw, I.; Schofield, P.A. Ironmak. Steelmak. 1979, *4*, 156.

44. El-Kalay, A.K.E.H.A.; Sparling, L.G.M. J. Iron Steel Inst. 1968, *206*, 152.

45. Helmi, A.; Alexander, J.M. J. Iron Steel Inst. 1968, *206*, 1110.

46. Beese, J.G. J. Iron Steel Inst. 1972, *210*, 433.

47. Shohet, K.N.; Towsend, N.A. J. Iron Steel Inst. 1968, *205*, 1088.

48. Shohet, K.N.; Towsend, N.A. J. Iron Steel Inst. 1971, *209*, 765.

49. Leslie, W.C. *The Physical Metallurgy of Steels*; McGraw Hill: New York, 1982.

50. Pickering, F.B. *Physical Metallurgy and the Design of Steels*; Appl. Sc. Pub.: London, 1983.

51. Honeycombe, R.W.K.; Bhadeshia, H.K.D.H. *Steels: Microstructure and Properties*; 2a Ed; Arnold: London, 1995.

52. Sims, C.E. Trans. AIME 1959, *215*, 367.

53. Pike, T.J. Ironmak. Steelmak. 1989, *16*, 168.

54. English, T.H.; Dyson, D.J.; Walker, K.D.; Pike, T.J. Ironmak. Steelmak. 1993, *20*, 97.

55. Flemming, G.; Hennig, W.; Hofmann, F.; Pleschiutschingg, F.-P.; Rosental, D.; Schwellenbach, J. Metall. Plant Technol. Int. 1997, *16* (3), 64.

56. Leduc-Lezama, L.A.; Muñoz-Baca, J. 39th Mech. Worl. Steel Proc. Conf. ISS-AIME: Warrendale, 1997; Vol. 35, 89.

57. Browne, K.W. *Recent Advances in Heat Transfer and Micro-Structure Modelling for Metal Processing*; Guo, R.-M., Too, J.J., Eds.; ASME (MD-Vol. 67): New York, 1995; 187 pp.

58. Llewellyn, D.T. Ironmak. Steelmak. 1995, *22*, 25.

59. Fukagawa, T.; Okada, H.; Maehara, Y. ISIJ Int. 1994, *34*, 906.

60. Sellars, C.M.; Tegart, W.J.McG. Int. Metall. Rev. 1972, *17*, 1.

61. Jonas, J.J.; McQueen, H.J. Plastic Deformation of Materials. In *Treatise on Materials Science and Technology*; Arsenault, R.J., Ed.; Academic Press: New York, 1975; Vol. 6, 216.

62. Sellars, C.M. *Hot Working and Forming Processes*; Sellars, C.M., Davies, G.J., Eds.; Metals Soc.: London, 1980, 3 pp.

63. Tamura, I.; Ouchi, C.; Tanaka, T.; Sekine, H. *Thermomechanical Processing of High Strength Low Alloy Steels*; Butterworths: London, 1988.

64. Kocks, U.F.; Argon, A.S.; Ashby, M.F. *Thermodynamics and Kinetics of Slip, Progress in Materials Science*; Pergamon Press: Oxford, 1975; Vol. 19, 135.

65. Colás, R. J. Mater. Process. Technol. 1996, *62*, 180.

66. Kallend, J.S.; Morris, P.P.; Davies, G.J. Acta Metall. 1976, *24*, 361.

67. Nakamura, T.; Sakaki, T.; Roe, Y.; Fukushima, E.; Inagaki, H. Trans. Iron Steel Inst. Jpn. 1975, *15*, 561.

68. Inagaki, H. Z. Met.kd. 1983, *74*, 716.

69. Ray, R.K.; Jonas, J.J. Int. Met. Rev. 1990, *35*, 1.

70. Tanaka, T. Int. Met. Rev. 1981, *26*, 185.

71. Kozasu, I. *Accelerated Cooling of Steel*; Southwick, P.D., Ed.; T.M.S.-A.I.M.E: Warrendale, 1986; 15 pp.

72. DeArdo, A.J. *Accelerated Cooling of Steel*; Southwick, P.D., Ed.; TMS-AIME: Warrendale, 1986; 97 pp.

73. Mishra, S.; Därmann, C. Int. Met. Rev. 1982, *27*, 307.

74. Hutchinson, W.B. Int. Met. Rev. 1984, *29*, 25.

75. Ray, R.K.; Jonas, J.J.; Hook, R.E. Int. Mater. Rev. 1994, *39*, 129.

76. Humphreys, F.J.; Hatherly, M. *Recrystallization and Related Annealing Phenomena*; Pergamon Press: Oxford, 1995.

77. Matsudo, T.; Osawa, K.; Kurihara, K. *Technology of Continuous Annealed Cold-Rolled Sheet Steels*; Pradham, R., Ed.; TMS-AIME: Warrendale, 1985; 3 pp.

78. Pradham, R.; Battisti, J.J. *Hot- and Cold-Rolled Sheet Steels*; Pradham, R. Ludkovsky, G., Eds.; TMS-AIME: Warrendale, 1988; 41 pp.

79. Rovito, A. Iron Steel Eng. April 1991, *68* (4), 31.

80. Liesch, J.; Blum, F.; Hubert, R.; Christophe, J. Stahl Eisen 1992, *112*, 91.

81. Inagaki, H. Trans. ISIJ 1979, *17*, 166.

82. Iino, M.; Mimura, H.; Namura, N. Trans. ISIJ 1978, *18*, 33.

83. Bai, D.Q.; Yue, S.; Sun, W.P.; Jonas, J.J. Metall. Trans. A 1993, *24A*, 2151.

84. Abad, R.; Fernández, A.I.; López, B.; Rodríguez-Ibabe, J.M. ISIJ Int. 2001, *41*, 1373.

85. Cuddy, L.J. *Thermomechanical Processing of Microalloyed Austenite*; DeArdo, A.J., Ratz, G.A., Wray, P.J., Eds.; TMS-AIME: Warrendale, 1982; 129 pp.

86. Obehauser, F.M.; Listhuber, F.E.; Wallner, F. *Microalloying '75*; Union Carbide Co.: New York, 1977; 665 pp.

87. Villars, P., Ed.; *Pearson's Handbook on Crystallographic Data for Intermetallic Phases*; 2nd Ed; ASM International: Materials Park, 1977.

88. Christian, J.W. *The Theory of Transformation in Metals and Alloys*; Pergamon Press: Oxford, 1975.

89. Aaronson, H.I. Metall. Trans. A 1993, *24A*, 14.

90. Morales, J.C.; García, C.; Colás, R.; Leduc, L.A. *Hot Workability of Steels and Light Alloys-Composites*; McQueen, H.J., Konopleva, E.V., Ryan, N.D., Eds.; Can. Inst. Mining Metall.: Montréal, 1996; 365 pp.

91. Colás, R. *Advances in Hot Deformation Textures and Microstructures*; Jonas, J.J., Bieler, T.R., Bowman, K.J., Eds.; TMS-AIME: Warrendale, 1994; 63 pp.

92. Gladman, T.; Dulieu, D.; McIvor, I.D. *Microalloying'75*; Union Carbide Co.: New York, 1977; 32 pp.

93. Rashid, M.S. *SAE Preprint 760206*; Soc. Aut. Eng.: Detroit, 1976.

94. Baker, T.N.; McPherson, N.A. Met. Sci. 1979, *13*, 611.

95. He, K.J.; Baker. T.N. Mater. Sci. Eng. A 1993, *A169*, 53.

96. Davies, R.G. Metall. Trans. A 1978, *9A*, 41.

97. Owen, W.S. Met. Technol. 1980, *7*, 1.

98. Kim, N.J.; Thomas, G. Scripta Metall. 1984, *18*, 817.

99. Llewellyn, D.T.; Hillis, D.J. Ironmak. Steelmak. 1996, *23*, 471.

100. García-Navarro, L.G.; Rodríguez, P.; Pérez-Unzueta, A.J.; Colás, R.; Lizcano, C.J.; Alvarez, I.; Thomas, G. *1st Int. Automotive Heat Treatment Conf.*; Colás, R., Funatani, K., Stickels, C.A., Eds.; ASM International: Materials Park, OH, 1999; 456 pp.

101. Zackay, V.F.; Porter, E.R.; Fahr, D.; Bush, R. Trans. Am. Soc. Met. 1967, *60*, 252.

102. De Meyer, M.; Vanderschueren, D.; De Cooman, B.C. ISIJ Int. 1999, *39*, 813.

103. Petrov, R.; Kestens, L.; Houbaert, Y. ISIJ Int. 2001, *41*, 883.

104. Bozorth, R.M. *Ferromagnetism*; Van Nostrand: New York, 1951.

105. Staley, J.K. *Electrical and Magnetical Properties of Metals*; Am. Soc. Met.: Metals Park, 1963.

106. Chen, C.-W. *Magnetism and Metallurgy of Soft Magnetic Materials*; Dover Pub.: New York, 1986.

107. Werner, F.E. *Energy Efficient Electrical Steels*; Marder, A.R., Stephenson, E.T., Eds.; TMS-AIME: Warrendale, 1981; 1 pp.

108. Werner, F.E. J. Mater. Eng. Perform. 1992, *1*, 227.

109. Arato, P.; Boc, I.; Gret, T. J. Magn. Magn. Mater. 1984, *41*, 53.

110. Liao, K.C. Metall. Trans. A 1986, *17A*, 1259.

111. Ueno, K.; Tachino, I.; Kubota, T. *Metallurgy of Vacuum-Degassed Carbon-Steel Products*; Pradham, P., Ed.; TMS-AIME: Warrendale, 1990; 347 pp.

112. Reviprasad, K.; Aoki, K.; Chattopadhyay, K. Mater. Sci. Eng. A 1993, *A172*, 125.

113. Goss, N.P. Trans. Am. Soc. Met. 1935, *23*, 515.

114. Wiener, G.; Albert, P.A.; Trapp, R.H.; Littmann, M.F. J. Appl. Phys. 1958, *29*, 366.

115. Held, J.F. *Mech. Working Steel. Proc. IV*; Edgecombe, D.A., Ed.; AIME: New York, 1965; 3 pp.

116. Daniel, D.; Jonas, J.J. Metall. Trans. A 1990, *21A*, 331.

117. Daniel, D.; Sakata, K.; Jonas, J.J. ISIJ. Int. 1991, *31*, 696.

118. Ray, R.K.; Jonas, J.J.; Butrón-Guillén, M.P.; Savoie, J. ISIJ Int. 1994, *34*, 927.

119. Takahashi, N.; Abe, M.; Akisue, O.; Katoh, H. *Metallurgy of Continuous-Annealed Sheet Steel*; Bramfitt, B.L., Mangonon, P.L. Jr., Eds.; TMS-AIME: Warrendale, 1982; 51 pp.

120. Ono, S.; Nozoe, O.; Shimomura, T.; Matsudo, K. *Metallurgy of Continuous-Annealed Sheet Steel*; Bramfitt, B.L., Mangonon, P.L. Jr., Eds.; TMS-AIME: Warrendale, 1982; 99 pp.

121. Obara, T.; Satoh, S.; Nishida, M.; Irie, T. Scand. J. Metal. 1984, *13*, 201.

122. Takahashi, M.; Okamoto, A. Trans. ISIJ 1979, *19*, 391.

123. Hutchinson, W.B.; Nilsson, K.-I.; Hirsch, J. *Metallurgy of Vacuum-Degassed Steel Products*; Pradhma, R., Ed.; TMS-AIME: Warrendale, 1990; 109 pp.

4

Design of Aluminum Rolling Processes for Foil, Sheet, and Plate

Julian H. Driver
Ecole des Mines de Saint Etienne, St. Etienne, France

Olaf Engler
Hydro Aluminium Deutschland, Bonn, Germany

I. INTRODUCTION

A large proportion (~50%) of all aluminum alloys are used as rolled products in the form of sheet, foil, or plate for an increasingly wide variety of applications. Over the last few decades, there has been a strong development of aluminum sheet and foil for the packaging industry, typically for beverage cans, foil containers, and foil wrapping. The building industry is now also a major user of sheet for roofing and siding, whereas the transport industry, one of the first to use aluminum alloys, is an expanding market for strong, light alloys; both sheet and plate are used for aircraft construction, ships, high-speed trains, and military vehicles. In this context, the high-volume production of lighter automobile frames and body parts in aluminum is now expected to become a major feature of the drive to improve fuel consumption and to reduce gas emissions. Finally, a small but significant part of the foil material is used in specialized "niche" markets such as electrical equipment, heat exchangers, and lithographical plates.

The alloys are used essentially because of their overall combination of lightness, conductivity (electrical and thermal), corrosion resistance, machinability, and a wide range of mechanical strengths. Other important properties such as formability and weldability vary significantly with the composition and the processing method. The reader will find a number of useful descriptions of alloys [1], metallurgy [2], and processing routes [3,4].

In virtually all practical cases, the required properties are tailored to a particular application by a close control of their microstructures via their composition and the thermal and thermomechanical treatments. Rolling, as one of the most important solid state processing routes, is obviously a thermomechanical treatment that offers considerable potential for microstructure control. If, traditionally, rolling has been used to transform the material shape, for example, from ingot to sheet, then it has also become a cost-effective process for microstructure control and property improvement. A well-known example is the breakup of the as-cast structure and the refinement of grain size.

In more detail, the microstructural variables for a given composition are grain size, shape, and orientation (or macroscopical texture); the deformation substructures are subgrain size and dislocation densities, and, of course, the second-phase particle distributions (size, shape, and spatial distributions). These variables influence the mechanical strength, ductility, and anisotropy of the product (i.e., most of the critical properties for important industrial applications). They depend on the composition of the alloy and the thermomechanical treatments, sometimes in a rather complicated way.

The aim of this chapter is, first, to describe the basic relations between composition, rolling process, microstructures, and properties, and, second, to illustrate how rolling processes can be designed to achieve optimal properties for specific applications.

II. TYPICAL ALLOYS AND PROPERTIES

Pure aluminum is a light (relative density 2.7 g/cm^3) but very soft metal only used for specific electronic applications. It is usually alloyed (with Mg, Mn, Si, Zn, Cu, etc.) to develop mechanical strengths (UTS) ranging from 50 to 650 MPa. The most common types of alloys that undergo rolling are given in Table 1.

More specific examples of the compositions and properties of some well-known alloys are given in Table 2.

For a given alloy, the properties obviously depend on the metallurgical state of the material as classified by the characters O, $H(x)$, and $T(x)$. It is recalled that O refers to the annealed (i.e., soft and ductile) condition, H is for a work-hardened state, typically cold-rolled, and T denotes the tempered condition (i.e., solutionized and aged for precipitation hardening). The reader is referred to Ref. 1 for further details of the nomenclature concerning alloys and their treatments.

Very schematically, foil material is made from the 1xxx (and 8xxx) series for packaging; most sheets are from the work-hardened alloys (3xxx, 5xxx; for can stock and vehicle bodies) and most plates are from the high-strength 2xxx and 7xxx age-hardened alloys for aircraft, aerospace, etc.

III. ROLLING SCHEDULES

Most aluminum alloy ingots are produced by direct chill (DC) semicontinuous casting with typical dimensions of 0.4–0.6 m thickness, 2 m width, and up to 9 m length (weight, 20–30 tons). These ingots are destined to be rolled down to plate, sheet, etc., by a rolling schedule comprising hot rolling, cold rolling, and often intermediate anneals, as detailed below.

A. Homogenization

The ingot is first homogenized (i.e., heated to a temperature in the range 500–600°C) for relatively long time (at least a few hours) to reduce segregation and to remove nonequilibrium, low-melting point eutectics. This facilitates subsequent hot working and improves homogeneity. The homogenization treatment can also have a further effect on the final microstructures of some alloys in that precipitation reactions can occur during the treatment. For example, the high-strength 7xxx alloys containing Mn or Zr tend to precipitate out submicron particles (dispersoids) of Al_6Mn or Al_3Zr during the heating up stage. These dispersoids are very stable and can strongly influence the recrystallization behavior, and thereby the grain size and structure.

B. Hot Rolling

After homogenization, the ingot is usually hot-rolled down to a 30- to 10-mm-thick strip in a reversible (breakdown) rolling mill (Fig. 1). The number of passes varies from 9 to 25. The strip from the single-stand breakdown mill is either coiled to await cold rolling or, in modern processing lines, further hot-rolled in a multiple stand tandem mill. Figure 2 illustrates a conventional reversible three stand tandem mill set up. In current practice, the tandem mills have between two and six stands.

Table 1 Generic Aluminum Alloys, Properties, and Applications (Minor Elements in Parentheses)

Series	Major alloying elements	UTS [MPa]	Major applications
1xxx	<1% (Fe + Si)	50–150	Cooking utensils, heat exchangers, packaging foil, electrical conductors
2xxx	2–6% Cu (+ Si, Mg)	300–480	Sheet and plate for aerospace, armaments, general mechanical engineering
3xxx	0.5–1.5% Mn (+ Mg, Cu)	100–240	Packaging, can stock, cooking utensils, car radiators, construction
5xxx	0.5–1.5% Mg (+ Mn, Cu)	100–340	Construction, ship building, cars, trucks (sheet panels), can ends, wire
6xxx	0.5–1.5% Mg + 1–2% Si (Cu, Cr)	200–400	Mostly extrusions but rolled sheet for transport
7xxx	5–7% Zn + 1–2% Mg (Cu, Zr)	320–650	Aerospace, mechanical construction, sport and leisure equipment
8xxx	>1% (Fe + Si)	130–190	Packaging foil, heat exchangers

Table 2 Some Specific Alloys, Treatments, and Properties (Typically 2–4 mm Sheets)

Alloy	Composition [wt.%]	State	YS [MPa]	UTS [MPa]	Elongation [%]
1050A	0.2Si, 0.4Fe	O	35	90	40
		H14	115	125	20
2024	3.8–4.9Cu, 1.2–1.8Mg (Mn, Si, Fe)	O	140	220	13
		T4	275	425	14
		T8	400	460	5
3104	0.8–1.4Mn, 0.8–1.3Mg (Fe, Si)	O	60	155	15
		H14	180	220	2
		H18	230	260	1
5052	2.2–2.8Mg (0.25Si, 0.4Fe)	O	65	170	15
		H14	180	230	4
		H18	240	270	2
5182	4.0–5.0Mg (Si, Fe)	O	110	255	12
		H19	320	380	1
6082	0.7–1.3Si, 0.6–1.2Mg	O	85	150	16
		T6	260	310	10
7075	5.1–6.1Zn, 2.1–2.9Mg, 1.2–2Cu	O	145	275	10
		T6	470	540	7

Cold rolling is usually carried out in a reversible cold mill between two coilers, as shown in Fig. 3. When correctly set up, this equipment can be used to roll down the "softer" alloys to a thickness of 15–20 μm. To obtain a very thin packaging foil of about 6 μm in thickness, the foils are then doubled up and rerolled. Intermediate annealing is frequently necessary to achieve large cold rolling reductions.

It should be noted that an increasing proportion of the less strongly alloyed sheet products are now produced by continuous strip casting methods. As shown in Fig. 4, the hot metal is poured between rotating cylin-

ders to produce a "thick" sheet (10–20 mm), which is then immediately rolled in a tandem mill.

C. Rolling Conditions

Table 3 indicates some typical dimensions and temperatures for the different rolling processes. Figure 5 also shows typical temperature/time schedules for the production of (a) can stock and (b) foil. Close microstructure control requires knowledge of the average and local strains, strain rates, and temperatures as a function of time, together with the interpass times for a given pass.

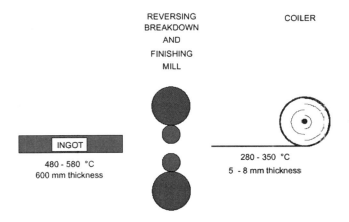

Figure 1 Schematic of reversible breakdown rolling mill. (From Ref. 3.)

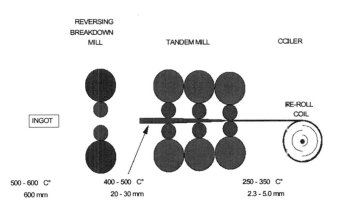

Figure 2 Schematic of reversible and (three-stand) tandem mill setup. (From Ref. 3.)

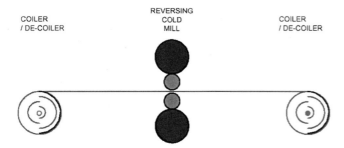

Figure 3 Schematic of cold rolling mill. (From Ref. 3.)

The values of these parameters should then be evaluated throughout the entire rolling schedule.

The average deformation per pass is given by the thickness reduction $h_0 \rightarrow h$ as $\varepsilon = \ln h/h_0$, and the corresponding average strain rate $\dot{\varepsilon}$ is obtained by dividing by the pass time. It is now becoming standard to use the more general description of strain (and strain rate), that is, the von Mises equivalent strain defined as $\bar{\varepsilon} = \sqrt{2/3 \varepsilon_{ij} \varepsilon_{ij}}$ (or strain rate $\bar{\dot{\varepsilon}} = \sqrt{2/3 \dot{\varepsilon}_{ij} \dot{\varepsilon}_{ij}}$), adopting the repeated indices summation convention. To a very rough approximation for macroscopical sheets, rolling deformation is plane strain compression for which the transverse strain is zero and the averaged shear components are taken as zero so that, in terms of strain rate components:

$$\dot{\bar{\varepsilon}} = \sqrt{\frac{2}{3} \left(\dot{\varepsilon}_{11}^2 + \dot{\varepsilon}_{22}^2 + \dot{\varepsilon}_{33}^2 + 2\dot{\varepsilon}_{12}^2 + 2\dot{\varepsilon}_{13}^2 + 2\dot{\varepsilon}_{23}^2 \right)}$$
$$= \frac{2}{\sqrt{3}} \dot{\varepsilon}_{11} \qquad (1)$$

The real instantaneous strain rates depend on the geometry of the rolls, and the work piece and the friction conditions between them. They vary through the roll bite and also from the surface to the center. For cold rolling thin sheet, the deformation can be considered close to plane strain compression as assumed above. However, for hot rolling of thick plates, there are significant redundant shear deformations through a large part of the thickness. The metal roll contact zone for this situation is illustrated schematically in Fig. 6; only at the neutral point is the local metal velocity exactly equal to that of the rolls. Before the neutral point, the metal advances more slowly than the rolls and, after the neutral point, it exits more quickly; the difference gives rise to additional shear components whose sign changes during the pass and which can be

large near the surface. The exact evaluation of the local deformations during the process has been the subject of much research. Approximate solutions are obtained by slip line theory, but most of the recent analyses adopt finite element methods (FEMs).

Figure 7 gives some finite element results for the cumulated shear and compression strain components through the roll gap of a breakdown mill, at two depths in the sample and for two passes (strain rate and temperature fields calculated for a 3104 alloy in the stationary regime [5]). In the first passes, the (negative and positive) shear terms are roughly equivalent so they balance out, whereas during the later passes, the small initial negative shear is more than compensated for by the large positive shear at the exit. Note also that there are temperature variations in the metal (on the order of 10–20°C) between the surface and the hotter center.

IV. MICROSTRUCTURE-PROPERTY EVOLUTION

A. Work Hardening

As seen above, many alloys are used directly in the work-hardened state, so their work hardening characteristics are a major issue. Work hardening is also important in a more indirect way because of its influence on both formability and subsequent softening processes such as recrystallization. The role of work hardening during forming is well known because, according to the Considère criterion, the limiting strain for homogeneous deformation is given by the work hardening coefficient n; this is a major advantage of the Al–Mg alloys that possess relatively high n values.

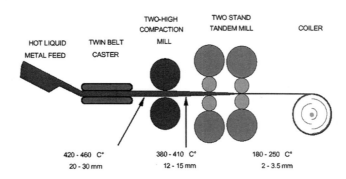

Figure 4 Schematic of continuous strip casting. (From Ref. 3.)

Table 3 Some Typical Rolling Conditions

	Reversible	Tandem	Cold rolling
Start temperature [°C]	500–600	400–500	20
Finish temperature [°C]	400–500	250–350	100
Number of passes	9–25	2–5	2–10
Initial thickness [mm]	400–600	45–15	2–6
Final thickness [mm]	45–15	2–9	0.01–1
Strain per pass	0.1–0.5	0.7	0.3–0.7
Total strain	3.5	3	<5
Strain rates [sec^{-1}]	1–10	10–100	>50
Inter-stand times [sec]	10–300	<5	

The role of work hardening on microstructure evolution is critical for subsequent softening by annealing to enable recovery or recrystallization; in many practical cases, this occurs during cooling after hot rolling, so the hardening and softening phenomena during hot working strongly influence the final structure. Finally, they also control the loads required during the rolling process for metal deformation. Concurrent softening during high-temperature working is, of course, the fundamental reason for carrying out hot rolling.

The work hardening of a metal is usually represented by the flow stress σ and its evolution during deformation (i.e., $d\sigma/d\varepsilon$ or $d\sigma/dt$). A general stress is described by the Cauchy stress tensor $[\sigma]$, which is often reduced to a scalar equivalent stress or von Mises stress $\bar{\sigma}$, defined in a similar way to the equivalent strain (Eq. (1)). During plastic strain, this equivalent flow stress is identical to the flow stress of a material as measured during a standard, uniaxial, tensile test.

1. Cold Deformation

Work hardening will be described initially for the temperature ranges where thermally activated processes do not play a significant role (typically for homologous temperatures below $0.4T_m$ or ~100°C for Al). At these temperatures, material viscosity effects are negligible so that, for example, plastic behavior is independent of the strain rate. One can then define a unique yield stress σ_y, the stress at which plastic flow begins. Figure 8 illustrates some typical room temperature stress–strain curves for Al–Mg alloys. Clearly, these alloys exhibit high work hardening rates with flow stresses evolving from ~100 to ~400 MPa (and even higher during cold rolling).

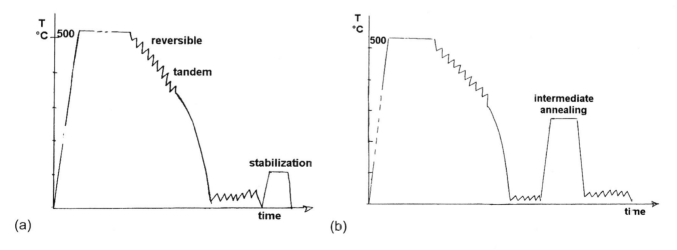

(a) (b)

Figure 5 Two typical temperature/time schedules for the production of (a) can stock and (b) foil.

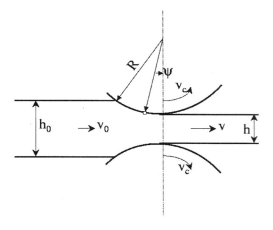

Figure 6 Schematic of metal roll contact zone during hot rolling of thick plate. The neutral point is situated at an angle ψ to the vertical.

Macroscopical Laws

Stress–strain curves are sometimes represented by the simple Ludwik power law:

$$\sigma = \sigma_y + k_1(\bar{\varepsilon})^n \tag{2}$$

where k_1 is the flow stress increase at $\bar{\varepsilon} = 1$ and n is the work hardening coefficient ($\partial \ln \sigma / \partial \ln \bar{\varepsilon}$), which for most metals takes values between 0.05 and 0.5.

If the initial yield stress is low compared with the work hardening, the above relation can be reduced to the Hollomon law:

$$\sigma = k_2(\bar{\varepsilon})^{n'} \tag{3}$$

These power laws apply reasonably well to the parabolical part of the stress–strain curve, typically for strains ≤ 0.5. At higher strains, the hardening rate decreases so that the flow stress tends to vary nearly

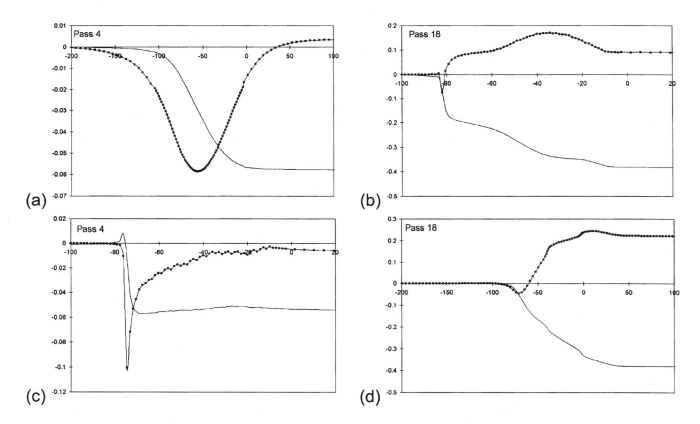

Figure 7 The cumulated shear and compression strain components through the roll gap of a breakdown mill, at two depths (a, b, near center; c, d, near surface), and for two passes (strain rate and temperature fields calculated by finite element methods for a 3104 alloy in the stationary regime). (From Ref. 5.)

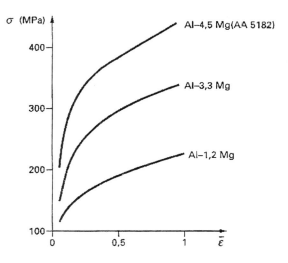

Figure 8 Typical stress–strain curves of Al–Mg alloys. (From Ref. 6.)

linearly with the strain. Two or more pairs of (k,n) values are then required to describe the stress–strain curve over a wide range of deformations. Alternatively, and preferably for large strains, one can use the fact that the flow stress tends to saturate at a value denoted σ_s. In this case, the constitutive law can be approximately described by an exponential relation initially proposed by Voce [7]:

$$\sigma = \sigma_s - \left(\sigma_s - \sigma_y\right)\exp(-\alpha\bar{\varepsilon}) \tag{4}$$

where α is a dimensionless constant characteristic of the work hardening behavior. In fact, as can be seen in Fig. 9, the saturation stress of aluminum alloys de-

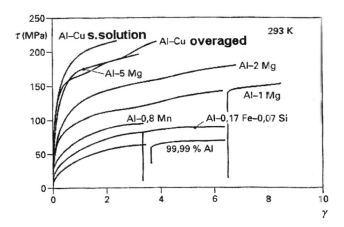

Figure 9 Torsion stress–strain curves of several aluminum alloys. (From Ref. 8.)

formed at room temperature is not defined with any great accuracy. Other variants of the Voce law have therefore been proposed of which the most usual, initially due to Hockett and Sherby [9], is:

$$\sigma = \sigma_s - \left(\sigma_s - \sigma_y\right)\exp(-\alpha\bar{\varepsilon}^p) \tag{5}$$

where the exponent p takes values ~0.5 (see Lloyd and Kenny [6,10]).

Deriving the Voce law gives the current work hardening rate $\bar{\theta}$ as a linear function of the flow stress:

$$\bar{\theta} = \frac{d\sigma}{d\varepsilon} = \bar{\theta}_0\left(1 - \frac{\sigma_0}{\sigma_s}\right) \tag{6}$$

where $\bar{\theta}_0$ is the initial work hardening rate. The experimental values of this law are identified by plotting the work hardening rate $\bar{\theta}$ as a function of the flow stress σ. Both parameters are often normalized by the shear modulus μ [i.e., $\theta/\mu = f(\sigma/\mu)$]. Figure 10 illustrates this type of plot at different temperatures ($< 0.4T_m$). Clearly,

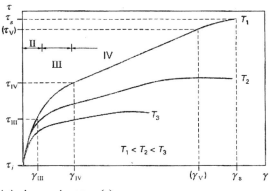

(a) shear stress $\tau(\gamma)$

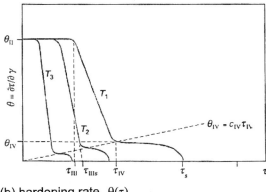

(b) hardening rate $\theta(\tau)$

Figure 10 Schematic of work hardening stages and rates as a function of flow stress at three temperatures.

the work hardening rate, at a given flow stress, decreases with temperature. In addition, it is apparent that at high stresses (and therefore large strains), the work hardening rate does not go to zero; there is a roughly constant residual $\bar{\theta} \approx 10^{-4}\,\mu$, independent of temperature.

Microscopical Mechanisms

The plastic flow of crystalline materials takes place by the movement of dislocations along crystal planes under the influence of an applied stress. Work hardening is a consequence of the fact that the stress required for dislocation movement usually increases during plastic flow as the dislocations become increasingly hindered by microstructural obstacles. Of the order of increasing size, these obstacles are solute atoms, dislocations, precipitates, and grain boundaries. The most important variation in obstacle density is due to the dislocations themselves.

Dislocations slip in crystallographical directions on planes of easy glide under the influence of a critical applied shear stress τ_c. The emergence of each dislocation creates a displacement along the slip direction of quantity b (the magnitude of the Burgers vector). The amount of shear strain due to the emergence of length l of dislocation in a small volume element is $d\gamma = \rho b l$, where ρ is the dislocation density (in m/m^3 or m^{-2}). In terms of the shear strain rate, this is written as $\dot{\gamma} = \rho b v$, where v is the average speed of the mobile dislocations. According to the von Mises law, the plastic deformation of a crystal requires the operation of as many slip systems as imposed strain rate components (≤ 5), so usually several systems (combinations of slip planes and directions) operate and interact in each grain. The shear stress for plastic flow then evolves with strain because the interaction of a moving dislocation with other dislocations usually constitutes a barrier to further movement. This requires an increase in flow stress for the dislocation to continue moving either by circumventing the barrier, or more often by creating a new segment of dislocation, which is then activated in the vicinity. The flow stress therefore increases with dislocation density and the relation between them is usually written:

$$\tau_c = \tau_0 + \alpha\mu b\sqrt{\rho} \tag{7}$$

where τ_0 is the glide stress at zero ρ and α is a material constant (0.2–0.5). For aluminum at 20°C, $\mu = 26$ GPa, $b = 0.286$ nm, and $\alpha \approx 0.3$, so that $\Delta\tau = \tau - \tau_0 \approx 2.23\sqrt{\rho}$ Pa (ρ is in m^{-2}). For dislocation densities typical of a softened, recrystallized metal $\rho \approx 10^{10}\,m^{-2}$ and so $\Delta\tau \approx 0.2$ MPa, which is negligible. For $\rho \approx 10^{15}\,m^{-2}$, as in heavily cold-worked alloys, $\Delta\tau \approx 70$, MPa which is

obviously significant. Note that the applied macroscopical tensile stress would be about three times τ_c (Section IV.A.1, "Microscopical–Macroscopical Relations") i.e., $\Delta\sigma \approx 210$ MPa, which is of the order of the difference in flow stress between the annealed (O-temper) and work-hardened (H) states (Table 2).

This elementary order of magnitude analysis of work hardening can be developed further to characterize the evolution of the dislocation density with strain, alloy content, microstructure, and, at high temperatures, with T and $\dot{\varepsilon}$. The details and the physical laws can be complicated (and controversial) but the essential features are schematized in Fig. 10 in terms of four stages [11], as follows:

Stage I: The dislocations are mostly confined to their slip planes and the interactions occur principally between the dislocation pileups and the grain boundaries. The resistance of the grain boundaries to dislocation movement gives rise to the well-known Hall–Petch grain size (d) hardening relation:

$$\sigma(d) = \sigma_\infty + \frac{k_3}{\sqrt{d}} \tag{8}$$

where σ_∞ is the flow stress of the material at a very large grain size. In the latter case, the dislocations can, initially, move freely with little interaction and the flow stress remains low, but in the general case of polycrystals, Stage I is very short and often neglected, except for the above Hall–Petch relation.

Stage II: The dislocation interactions on different slip systems give rise to a rapid multiplication of the dislocations and the development of dislocation tangles, which often adopt a cellular pattern. This Stage II, which extends up to strains of 0.05–0.2 (or even more at low temperature), is associated with a high work hardening rate $\sim 30 \times 10^{-4}$ to $50 \times 10^{-4}\,\mu$.

Stage III: Subsequently and up to strains of about unity, the work hardening rate tends to decrease progressively to values of $\sim 1 \times 10^{-4}\,\mu$ as the dislocation multiplication processes are counterbalanced by local annihilations (dynamic recovery due to localized climb and/or annihilation of segments of opposite sign). The microstructure evolves toward a well-defined cell substructure (Fig. 11) composed of dislocation cell walls, which delimit cell interiors of low dislocation density. The cells have dimensions that decrease during deformation, typically from a few microns to some tenths of microns. Simultaneously, the misorientation between adjacent cells increases from about 1–3° to 5° (and often higher for heterogeneous deformations).

Stage IV: At higher strains ≥ 1 typical of many rolling and extrusion processes, many grains break up into

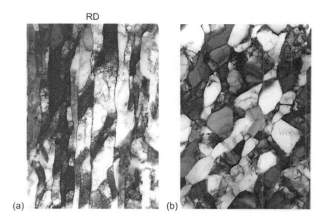

Figure 11 Substructure of aluminum alloy deformed at room temperature (Al–1.3% Mn). (From Ref. 48.) (a) Microband structure (as-deformed); (b) subgrain structure (after recovery).

bands of different orientations, separated by transition zones and grain boundaries (whose total surface per unit volume increases significantly by large plastic deformations). This is characteristic of fibrous microstructures. As noted above, the work hardening rate in Stage IV, although low, does not go to zero but retains a near-constant value of $\sim 1 \times 10^{-4} \mu$.

The crystal defects, particularly the dislocations, generated by plastic deformation possess high elastic energies, which are stored in the deformed material essentially in the stress fields around the dislocations. The energy per unit length of dislocation is:

$$E_l = \frac{\mu b^2}{K} \cdot \ln\left(\frac{R}{R_0}\right) \qquad (9)$$

(with $K = 4\pi$ for screws, or $4\pi(1-v)$ for edge-type dislocations). R is the outer cutoff radius of the dislocation stress field ($\approx$ the distance between dislocations, or $\rho^{-1/2}$) and R_0 is the core radius ($\approx 2b$). For typical densities of 10^{-14} to 10^{-15} m^{-2}, $E_l \approx 0.5\mu b^2$ and the stored energy/unit volume $E_v \approx 0.5\rho\mu b^2$. This value varies from 50 kJ/m^3 for lightly deformed aluminum to 2×10^4 kJ/m^3 for heavily deformed steel.

Influence of Alloying Elements

The strong influence of alloying elements on work hardening behavior is related to the metallurgical state of the material and depends particularly on whether the elements are in solid solution, or in a dispersion of second-phase particles.

In solid solutions, certain elements such as Mg and Cu segregate to the dislocation cores and significantly

reduce their mobility. This tends to confine the dislocations to slip planes, reduce their capacity to recover dynamically by local climb and cross slip, inhibit the formation of "clean" cell structures, and thereby increase the dislocation density for a given deformation. In general, Stage II work hardening is prolonged and Stages III and IV are retarded so the alloy work hardens extensively (see the example of Al–Mg; Fig. 8). At moderate temperatures (between room temperature and $\sim 300\,°$C), the reduced dislocation mobility by solute segregation tends to favor heterogeneous deformation by Lüders bands and Portevin LeChatelier serrations in the stress–strain curves [12] (Fig. 12).

Elements in the form of second-phase particles usually harden the material by requiring the dislocations to expend additional energy by either cutting the particle (fine particles of radius $r_p \leq 10$ or 20 nm), or by looping around them ($r_p \geq 20$ nm). The effect of these processes on the yield strength is well documented because they control, in part, the final properties (Fig. 13).

The effect of second-phase particles on work hardening (i.e., after yield) depends critically on the particle size but in a rather different way. The work hardening rate tends to be rather low for very fine particles (including the peak hardness distributions) because the dislocations, once they have cut through one particle, can cut through entire fields of particles at roughly the same stress. This leads to the formation of shear bands in which plastic deformation is heavily concentrated (e.g., certain Al–Cu alloys). On the other hand, if the particles have dimensions ≥ 1 μm, the dislocations loop around them and, on continued straining, build up

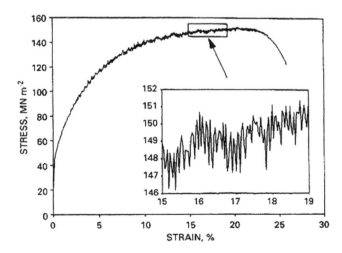

Figure 12 Stress–strain curves showing serrated flow and the PLC effect in Al–2% Mg. (From Ref. 12.)

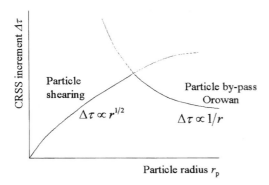

Figure 13 The influence of particle size on yield strength.

high local dislocation densities near the particles and therefore relatively high work hardening rates. In fact, a considerable fraction of the dislocation population is required to accommodate the difference in plastic strain between the hard particles and the soft matrix; they are termed geometrically necessary dislocations (GNBs) [13]. The accumulation of dislocations in these particle deformation zones creates high local stresses if the particles are sufficiently large to withstand them (≥ 1 μm). This process creates a strongly dislocated substructure with a dimension characteristic of the interarticle distance and is particularly pronounced if the particles are nonspherical [e.g., discs (Al–Cu) or fibers].

Microscopical Hardening Laws

The macroscopical hardening laws of Section IV.A.1, "Macroscopical Laws") are essentially empirical descriptions of some parts of the flow curves; they are incapable of describing a complete through-process material behavior during rolling schedules. It is now recognized that microscopical-based models of work hardening using internal variables such as the dislocation density are required for an accurate modeling of plastic flow. The particular microscopical laws are often complex so the relatively simple analysis given below in terms of the total dislocation density can be considered as an illustration of the principles.

The microscopical work hardening rate $d\tau_c/d\gamma$ can be written in terms of the variation of the dislocation density as:

$$\frac{d\tau}{d\gamma} = \frac{d\tau}{d\rho} \cdot \frac{d\rho}{d\gamma} \tag{10}$$

Using the standard $\tau_c(\rho)$ relation (Eq. (7)), the variation of the critical shear stress with dislocation density is:

$$\frac{d\tau}{d\rho} = \frac{\alpha\mu b}{2\sqrt{\rho}} \tag{11}$$

According to the analysis of Kocks [14] and Mecking [15], the rate of creation of dislocations ($d\rho^+$) during a small strain increment $d\gamma$ is inversely proportional to the mean dislocation slip distance λ so that $d\rho^+/d\gamma = 1/b\lambda$ and the mean slip distance $\lambda = C_1/\sqrt{\rho}$, where C_1 is the average number of obstacles that the dislocation meets before stopping:

$$\frac{d\rho^+}{d\gamma} = \frac{\sqrt{\rho}}{C_1 b} \tag{12}$$

It can be noted that this relation between the rate of creation of dislocations and $\sqrt{\rho}$ is not accepted by all authors, some of whom (e.g., Laasraoui and Jonas [16]) prefer a constant value.

During Stage II hardening, where dislocation annihilation is small and can be neglected, the above model gives a constant hardening rate:

$$\frac{d\tau^{II}}{d\gamma} = \frac{\alpha\mu b}{2\sqrt{\rho}} \cdot \frac{\sqrt{\rho}}{C_1 b} = \frac{\alpha\mu}{2C_1} \tag{13}$$

If the experimental value of this rate is 5×10^{-3} μ, C_1 is about 50 μm^{-1}, and assuming that $\rho \approx 10^{14}$ m^{-2}, the average slip length would be about 5 μm.

During Stage III, the hardening rate decreases continuously as some of the dislocations are annihilated by dynamic recovery at a rate written as $d\rho^-/d\gamma$. The exact mechanisms of annihilation (cross slip, climb) are the subject of current research, but one can write that the rate of annihilation is proportional to the current density ρ and a probability of elimination $P(T)$, which is strongly temperature-dependent:

$$\frac{d\rho^{III}}{d\gamma} = \frac{d\rho^+}{d\gamma} - \frac{d\rho^-}{d\gamma} = \frac{\sqrt{\rho}}{bC_1} - C_2 P(T)\rho \tag{14}$$

consequently, Stage III work hardening rate becomes linear in τ_c:

$$\frac{d\tau_c^{III}}{d\gamma} = \frac{\alpha\mu}{2C_1} - \frac{C_2 P(T)}{2} \cdot \tau_c \tag{15}$$

This law can be expressed in the following form for the macroscopical stress: $d\sigma^{III}/d\varepsilon = \theta_0(1-\sigma/\sigma_s)$, which is the hardening rate in the Voce law (Eq. (6)).

At large strains and at low temperatures, dislocation annihilation is insufficient to completely balance the rate of creation. This results in the low, but nonzero, work hardening rate of Stage IV; there is no general agreement on the basic physical causes of this stage.

Finally, it should be noted that the above analysis assumes only one internal variable, the total dislocation density. Several recent two-parameter or three-parameter models take account of the different dislocation distributions (cell walls, cell interiors, etc.). The added complexity usually enables one to better define transition behavior (varying T, $\dot{\varepsilon}$, etc.) [17,18].

Microscopical–Macroscopical Relations

The plastic flow properties (e.g., the work hardening behavior) of all crystalline solids are described in terms of basic mechanisms by the microscopical quantities (τ_c, $\dot{\gamma}$...) but are usually measured in terms of the macroscopical parameters (σ, $\dot{\varepsilon}$...). The section below briefly describes the relations between them and the way they can be used to understand and to analyze the important properties of textures and anisotropy.

The shear stress on a slip system (defined by its plane normal **n** and slip direction **b**) is related to the macroscopically applied stress σ by the standard relation:

$$\tau^s = b_i^s \sigma_{ij} n_j^s \tag{16}$$

where b_i^s and n_j^s are the direction cosines of the slip direction **b** and the slip plane normal **n** (with respect to the reference coordinate system). The same geometrical factors enter into the relation linking the components of the strain rate tensor $\dot{\varepsilon}$ with the slip rates $\dot{\gamma}$ of the glide systems:

$$\dot{\varepsilon}_{ij} = \sum_s \frac{1}{2}\left(b_i^s n_j^s + b_j^s n_i^s\right)\dot{\gamma}^s \tag{17}$$

The sum of the slip rates $\Sigma\dot{\gamma}^s$ in a grain g is used to define the Taylor factor of the grain $M(g)$. This factor relates the microscopical and macroscopical flow properties by the plastic work rate $\dot{W}(g)$:

$$\dot{W}(g) = \sum_s \dot{\gamma}^s \tau_c^s = \sigma(g)\dot{\varepsilon}(g) \tag{18}$$

If one assumes that the critical shear stresses of the different slip systems are identical (termed isotropic latent hardening and probably quite reasonable for aluminum), then:

$$M(g) = \frac{\dot{W}(g)}{\tau_c \cdot \dot{\varepsilon}} = \frac{\sum \dot{\gamma}^s(g)}{\dot{\varepsilon}} = \frac{\sigma(g)}{\tau_c} \tag{19}$$

For a polycrystalline aggregate, the average Taylor factor over all grains $\langle M \rangle$ can be calculated for a given distribution of orientations. In the case of the tensile deformation of an fcc metal with a random orientation distribution, $\langle M \rangle = 3.06$. However, nonrandom distributions as in textured polycrystals lead to other $\langle M \rangle$ values; typically, they lie in the range 2–4. The Taylor factor is also important for the work hardening behavior. To first order for small strains $\sigma\langle M \rangle\tau_c = \langle M \rangle/d\varepsilon$ and therefore $d\sigma/d\varepsilon \approx \langle M \rangle(d\tau/d\varepsilon)$ so that if $d\varepsilon = \Sigma d\gamma/\langle M \rangle$, then $d\sigma/d\varepsilon \approx \langle M \rangle^2(d\tau/\Sigma d\gamma)$ (i.e., the microscopical and macroscopical work hardening rates are related by the square of the Taylor factor).

During plastic deformation, the individual grains undergo lattice rotations toward certain orientations, leading to the formation of deformation textures (Section IV.C.1). During this process, the grains rotate at rates that are functions of the above geometrical parameters b_i^s and n_j^s. In general, the rotation rate of the crystal lattice during plastic deformation is $\dot{r} = \dot{d} - \sum_s b_i^s n_j^s \dot{\gamma}^s$, where $\dot{d}$ is the velocity gradient tensor imposed on the material by the deformation process and the second term is the velocity gradient tensor of the grain due to slip on the activated systems.

2. Hot Deformation

Flow Stresses

At homologous temperatures above $0.4T_m$, plastic deformation is strongly influenced by thermally activated processes so that the flow stress becomes temperature-dependent and strain rate-dependent (viscoplastic). The processes involved are mostly controlled by local atomic diffusion and give rise to a strong dynamic recovery of the dislocation substructure and reduced flow stresses. The latter tend to saturate at a value function of T, $\dot{\varepsilon}$, and alloy content. It is generally recognized that the T and $\dot{\varepsilon}$ terms can be regrouped into a temperature-corrected strain rate known as the Zener–Hollomon parameter $Z = \dot{\varepsilon}\exp(Q/RT)$, where R is the gas constant and Q is an apparent activation energy for plastic flow. For many aluminum alloys, $Q \approx 156$ kJ/mol

At low Z values (typically $\ln Z \leq 26$ for Al), plastic flow is purely viscous after an initial transition strain of about 0.1. At higher Z ($26 \leq \ln Z \leq 50$), the flow stress also depends on the applied strain [i.e., $\sigma = f(T,\dot{\varepsilon},\varepsilon)$]. Note that in typical hot rolling schedules, $\ln Z$ varies from about 22 in the first passes to about 33 near the end.

In the low Z regime, many studies have shown that the saturation flow stress σ_s can be related to the Zener–Hollomon parameter by a power law:

$$\sigma_s = A \cdot Z^m \tag{20}$$

where m is the strain rate sensitivity $\partial \ln \sigma / \partial \ln \dot{\varepsilon}$, typically 0.1–0.3.

Blum [19] has shown that the saturation flow stresses of different materials can be correlated by normalizing with respect to the self-diffusion coefficient D and the shear modulus μ (Fig. 14):

$$\left(\frac{\sigma_s}{\mu}\right) = \left(\frac{\dot{\varepsilon}k_B T}{D\mu b}\right)^m \tag{21}$$

In the higher Z regime typical of most hot rolling schedules, the relations $\sigma(T,\varepsilon,\dot{\varepsilon})$ are more complex and

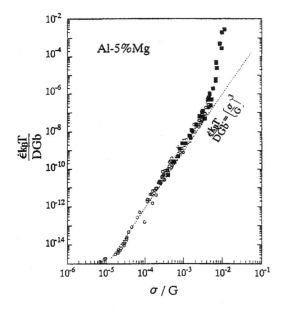

Figure 14 The temperature and strain rate dependence of saturation flow stress for Al–5% Mg. (From Ref. 19.) Note the power law relation at low σ (low Z) and the power law departure at higher σ (high Z).

often represented by an empirical law due to Sellars and Tegart [20]:

$$Z = A_2(\sinh \varpi \sigma_s)^{n'} \qquad (22)$$

where A_2, ω, and n' are parameters to be evaluated experimentally. At low Z values, this law is similar to the above power law. At high Z, the sinh law can be reasonably well approximated by an exponential function:

$$Z \cong A_3 \exp [\beta \sigma(\varepsilon)] \qquad (23)$$

where $\sigma(\varepsilon)$ is the flow stress at a given value of the applied strain and $\beta \approx \omega n'$.

To describe an entire work hardening curve in this regime, there are two possible approaches: (a) purely empirical, using a variant of the Voce law where the stresses are functions of Z (e.g., the law used recently by Shi et al. [21]):

$$\sigma = \sigma_e + (\sigma_s - \sigma_e)\left[1 - \exp\left(\frac{-\varepsilon}{\varepsilon_r}\right)\right]^m \qquad (24)$$

where ε_r is a transition deformation and $m \sim 0.5$. An example of this type of representation for the stress–strain curves of an Al–1% Mn alloy at different T and $\dot{\varepsilon}$ is given in Fig. 15; and (b) modified Kocks–Mecking

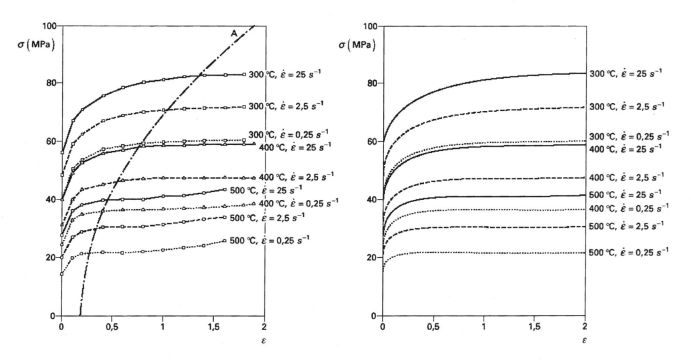

Figure 15 Experimental and calculated plane strain compression stress–strain curves of Al–1% Mn at different temperatures and strain rates. (From Ref. 21.)

approach, which allows for the effect of different microscopical variables on the rates of accumulation and annihilation of the dislocation populations (usually more than one). The reader is referred to Mughrabi [17], Nes [18], and the reviews of Gil Sevillano [22] and Argon [23].

Hot Deformation Microstructures

The hot deformation microstructures of most aluminum alloys are essentially composed of dislocation subgrains within each of which there is a relatively low density of free dislocations. The processes of dynamic recovery (cross-slip, dislocation annihilation by local climb, etc.) limit the accumulation of free dislocations so the flow stress saturates rapidly at a level controlled by the subgrain size (δ) and the dislocation contents of the subgrain walls and interiors. These microstructural parameters depend essentially on the T, $\dot{\varepsilon}$, and solute atom dislocation interactions. During plastic flow, the subgrain boundaries (unlike the grain boundaries) are continually annihilated and replaced by new ones by a dynamic polygonization of the free dislocations, so that their average size (δ) remains constant over very large strains. Their average misorientation tends to saturate at strains about unity but may also continue to increase for some alloys and deformation modes.

A frequently used relation between subgrain size and flow stress is of the form:

$$\sigma = \sigma_0 + \frac{k \cdot \mu b}{\delta^c} \qquad (25)$$

where the constant $k \sim 8$ and the exponent c is between 0.75 and 1.5 according to the alloy and deformation conditions; to the first order, it is often taken as 1.

A more sophisticated version of the above is taken from Nes [18] and allows for the influence of free dislocations within the cells ρ_i, that is,

$$\sigma = \sigma_0 + \alpha_1 \mu b \sqrt{\rho_i} + \alpha_2 \frac{\mu b}{\delta} \qquad (26)$$

B. Softening by Recovery and Recrystallization

1. Phenomenology

During deformation, the strength of a metallic material increases substantially, whereas its residual deformability decreases. As described above, this work hardening is caused by the formation and the storage of dislocations on deformation. Although these dislocations are thermodynamically unstable, at room temperature, processes to rebuild the resulting dislocation structure

are too slow to enable significant changes. On a subsequent annealing treatment, however, thermally activated processes give rise to a release in dislocation energy accompanied by a decrease in material strength. As a result, the material may be subjected to further deformations. For instance, thin aluminum converter foil is rolled from a 600-mm cast slab down to a 6.5-μm final gauge (see Section V.G). Such a total strain of as much as 99.999% thickness reduction ($\varepsilon > 10$) can only be achieved with the help of repeated softening processes that occur along the entire thermomechanical processing of the foil. As an example, Fig. 16 shows the evolution of Vickers microhardness with annealing time for different annealing temperatures. The hardness curves imply that the softening behavior may roughly be subdivided into three steps, which correspond to the mechanisms of recovery/nucleation, (primary) recrystallization, and grain growth (e.g., Refs. 24 and 25). Figure 17 schematically summarizes the microstructural processes during these steps.

During the early stages of annealing, the material gradually softens, which is caused by recovery reactions. The term "recovery" combines all reactions leading to changes in the as-deformed microstructure through dislocation reactions, including dislocation, annihilation, and rearrangement of dislocations into more stable cell or subgrain structures (Fig. 17a and b). Under certain circumstances, recovery is so extended that essentially the entire excess dislocation density is re-

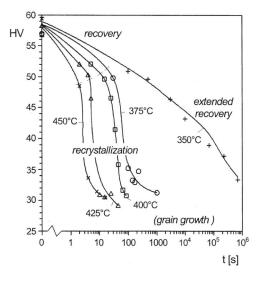

Figure 16 Evolution of Vickers microhardness H_V with annealing time t as a function of annealing temperature (Al–1.3% Mn, 97% cold-rolled). (From Ref. 48.)

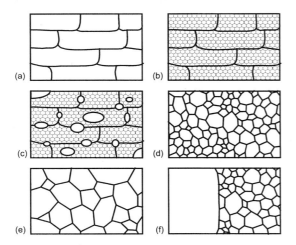

Figure 17 Schematic diagram of the processes that may occur during the annealing of a deformed metallic material. (a) Deformed grain structure; (b) recovered subgrain structure; (c) partially recrystallized structure; (d) recrystallized structure; (e) continuous grain growth; and (f) discontinuous grain growth. (From Ref. 24.)

moved through recovery (see, for instance, the curve for 350°C in Fig. 16, Section IV.B.5).

In most cases, by contrast, recovery reactions give rise to the formation of recrystallization nuclei, which, after a certain incubation period, appear at distinct locations in the as-deformed microstructure (Fig. 17c, Section IV.B.6). During further annealing, these nuclei grow by consumption of the deformed matrix until they impinge with other growing grains. These processes lead to a complete transformation of the microstructure, which, originally, gave rise to the term "recrystallization." The driving force for this so-called primary recrystallization is provided by the reduction in dislocation density. Recrystallization is characterized by the formation of new grain boundaries during nucleation and their subsequent motion into the as-deformed microstructure. Accordingly, recrystallization is accompanied by characteristic texture changes (Section IV.C.2), which may be used to monitor the progress of recrystallization and to elucidate the underlying recrystallization mechanisms. In terms of the mechanical properties, the progress of recrystallization is manifested by a rapid hardness decrease (Fig. 16).

Following this primary recrystallization, the recrystallized grains may coarsen further. Here, the driving force is provided by the reduction in the energy stored in the grain boundaries. Grain growth processes may take place either continuously (Fig. 17e), or discontinuously

(secondary recrystallization; Fig. 17f), with the latter being favored in the presence of second-phase particles, or in materials with a pronounced texture. In most cases, grain growth processes stop when the grain size reaches the smallest specimen dimension (e.g., the thickness of a rolled sheet). In some cases, especially in very thin sheets, however, a few grains may continue to grow discontinuously. Depending on the annealing atmosphere, these processes may be accelerated, suppressed, or even reversed. To differentiate this from secondary recrystallization, the latter process—driven by differences in the anisotropy of the specific energy of different crystal surfaces—is labeled tertiary recrystallization.

2. Driving Force

In contrast to the exact atomistic processes acting on recrystallization, the energetical causes of recrystallization are quite well understood. The driving force for primary recrystallization is provided by the reduction in elastic energy due to the dislocations stored in the as-deformed state. A brief estimate of the driving force p_D yields:

$$p_D = \Gamma \cdot \Delta\rho = \frac{\mu \cdot b^2}{2}(\rho_{\text{def}} - \rho_{\text{RX}}) \approx \frac{\mu \cdot b^2}{2} \cdot \rho_{\text{def}} \quad (27)$$

(where Γ is the line energy and ρ is the density of the dislocations). For heavily deformed aluminum alloys with a dislocation density of as much as $\rho = 10^{16}$ m^{-2}, this results in a driving force on the order of 10^7 N/m^2 = 10 MPa.

As noted above, recovery reactions that take place dynamically during the deformation or statically during the first seconds of a subsequent recrystallization annealing will transform the dislocations into a more or less well-defined cell or subgrain structure (e.g., Refs. 11 and 26; Section IV.A.1, "Microscopical Mechanisms"). In that case, the driving force p_D is comprised of two contributions, the energy stored in the cell or subgrain boundaries, and the energy of the free dislocations in the cell interior:

$$p_D = \frac{\alpha\gamma_{\text{SB}}}{\delta} + \frac{1}{2}\mu b^2 \cdot \rho_i \quad (28)$$

(where α is the geometrical constant on the order of 3, and γ_{SB} is the specific subgrain boundary energy). The dislocation density within the subgrains ρ_i and the average subgrain size δ have been found to be linked through the relation $\sqrt{\rho_i} = C/\delta$, where C is an alloy-dependent constant on the order of 2. Both the energy of a subgrain or low-angle grain boundary γ_{SB} and the energy of an ordinary high-angle grain boundary γ_{GB}

can be expressed in terms of the well-known Read–Shockley relation, so that the driving force p_D can be rewritten as:

$$p_D = \frac{\alpha \gamma_{GB} \theta}{\delta \theta_c} \ln\left(\frac{e\theta_c}{\theta}\right) + \frac{C^2 \mu b^2}{2\delta^2} \qquad (29)$$

θ and θ_c, respectively, denote the average and the maximum angle between neighboring subgrains. Whereas θ_c is usually estimated to be 15°, θ varies with strain ε to values from 3° to 5°.

For typical data of heavily cold-rolled aluminum ($\gamma_{GB} = 0.25$ J/m², $\theta = 3°$, and $\delta = 0.3$ μm), the contribution of the free dislocations is negligible compared with the energy stored in the subgrain boundaries. The driving force becomes $p_D \approx 10^6$ N/m² $= 1$ MPa, which is about one order of magnitude smaller than the value derived for a homogeneous distribution of dislocations (Eq. (27)). In hot-rolled aluminum ($\theta = 5°$ and $\delta = 1.5$ μm), the driving force is even smaller, $p_D \approx 3 \times 10^5$ N/m² $= 0.3$ MPa.

The driving force for grain growth processes p_{GG} is provided by the reduction in grain boundary area and, thus, in (absolute) grain boundary energy:

$$p_{GG} = \frac{2\gamma_{GB}}{R}; \quad R \approx 5, \ldots, 10d \qquad (30)$$

(where R is the curvature of the grain boundary and d is the grain size).

3. Kinetics

From a technological point of view, recrystallization kinetics is of outmost importance. As an example, Fig. 18 shows the evolution of the recrystallized volume fraction X as a function of the annealing time t for different tensile-strained Al samples [27]. The data of the evolution of the recrystallized volume fraction with time $X(t)$ can be determined metallographically, or derived from the changes in mechanical properties (Fig. 16). A promising technique to determine recrystallization kinetics is provided by calorimetrical measurements of the energies released during recovery and recrystallization [28], although it is difficult to interpret DSC data for aluminum alloys that undergo precipitation reactions.

Primary recrystallization takes place by nucleation and subsequent growth. Thus, for a quantitative description of recrystallization, the thermal dependency of both processes must be taken into account:

$$\dot{N} = \dot{N}_0 \cdot \exp[-Q_N(kT)]; \quad v = v_0 \cdot \exp[-Q_v(kT)] \qquad (31)$$

(where k is the Boltzmann constant and T is the temperature). The nucleation rate $\dot{N}$ is the number of newly formed grains per volume and time in the nonrecrystal-

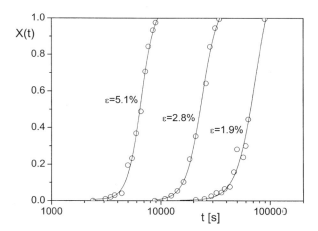

Figure 18 Recrystallized fraction X as a function of annealing time t in tensile-strained aluminum (annealing temperature 350°C). (From Ref. 27.)

lized volume $1 - X$; v is the growth velocity of the new grains. Q_N and Q_v are the activation energies of thermally activated nucleation and growth, respectively.

At variance with the classical nucleation processes of solidification and phase transformations, the reduction in dislocation density during recrystallization is an irreversible process whose driving force is essentially independent of temperature (except for the minor temperature dependency of Burgers vector and shear modulus; Eqs. (27) and (29)). Nonetheless, the progress of recrystallization is often characterized by a recrystallization temperature T_{RX}, at which recrystallization is completed in a given time, say 1 hr. Because of the thermal activation of both nucleation and growth processes, a minor change in temperature leads to large changes in recrystallization time; in contrast, changes in the prescribed time necessary to complete recrystallization to, say, 0.5 or 2 hr give rise to only marginally changed recrystallization temperatures.

For a quantitative description of the characteristic S-shaped kinetics curves (Fig. 18), the Johnson–Mehl–Avrami–Kolmogorov (JMAK) equation is generally utilized [29–31]:

$$X(t) = 1 - \exp\left(-\left(\frac{t}{t_{RX}}\right)^q\right) \qquad (32)$$

(where q is the so-called Avrami exponent and t_{RX} is the recrystallization time.) The JMAK equations allow one to determine parameters of technological interest, including recrystallization time t_{RX} and recrystallized grain size d_{RX}. Under certain conditions, $\dot{N}$ and v can be derived and the progress of recrystallization can be quantified. The assumption of a constant nucleation rate and isotropic growth yields an Avrami exponent of $q = 4$, and it follows that:

$$t_{RX} = \sqrt[q]{\frac{3}{\pi \dot{N} v^3}}; \qquad d_{RX} = 2v t_{RX} = 2\sqrt[q]{\frac{3v}{\pi \dot{N}}} \qquad (33)$$

If nucleation prevails at the early stages of recrystallization ("site-saturated nucleation"), $q = 3$. Experimentally, the Avrami exponent can be determined from a plot of the expression $\ln\{\ln[1/(1-X)]\}$ versus $\ln t$.

Based on the JMAK equation, the kinetics and microstructure evolution accompanying recrystallization can be modeled (e.g., Refs. 32–35). Nuclei are assumed to form in a given simulation volume with a certain rate. Textural effects may be incorporated through a distinct spatial or orientational correlation of the nuclei, and/or through orientation-dependent nucleation rates. Then, the nuclei grow in accordance with the JMAK equation until impingement. Again, orientation-dependent growth rates may be considered. The resulting microstructure (grain size and shape) is revealed by two-dimensional sections through the three-dimensional simulation space.

4. Recovery

From Fig. 16, it is evident that during the early stages of annealing, the material gradually softens, which is attributed to recovery reactions. The term "recovery" combines all reactions leading to changes in the as-deformed microstructure through dislocation reactions. Recovery can take place dynamically (i.e., during the deformation) or statically (i.e., after deformation has been completed, typically during a subsequent annealing treatment). Because of the high stacking fault energy and the high homologous temperature of aluminum recovery, reactions are typically quite pronounced. At elevated temperatures, dislocations may cross-slip and, at higher temperatures, may climb so that blocked dislocations are remobilized. Thus, dislocations can annihilate, which reduces the overall dislocation density. Furthermore, the as-deformed substructure characterized by microbands (Fig. 11a) may be rearranged into energetically more stable, lower-energy configurations such as cell structures and, on further recovery, into a well-defined subgrain structure (Fig. 11b). During further annealing, the subgrains coarsen by the reduction in subboundary energy. Subgrain coarsening may occur through the mechanism of subgrain coalescence as originally proposed by Li [36] and Hu [37]. In most cases, however, subgrain coarsening will take place

through subgrain boundary motion (i.e., analogous to ordinary grain growth) [38,39]. During this subgrain growth, the misorientations between neighboring subgrains may slightly increase as, for example, recently studied in detail in commercial-purity aluminum by Furu et al. [40]. Despite the obvious microstructural changes during recovery, the textures remain virtually unaffected, except for a slight sharpening of the texture (e.g., Refs. 41 and 42).

5. Extended Recovery and Continuous Recrystallization

In principle, the dislocations brought into the material during deformation can be released solely by recovery. For instance, on deformation of high-purity metals, or in samples deformed at high temperatures, dynamic recovery can reduce the dislocation energy so efficiently that no recrystallization is initiated [43]. In Al alloys, this extended recovery is of particular importance when solution-treated alloys are annealed at low temperatures such that precipitation reactions can interfere with the recrystallization process [44,45]. Under such circumstances, the subgrain boundaries may be pinned by finely dispersed precipitates and, therewith, nucleation of recrystallization is strongly retarded or even completely suppressed (Fig. 19a). Thus, provided the subgrain boundaries cannot break free from the particles, the substructure evolution is controlled by subgrain growth governed by the coarsening of the particles [45–47] (Fig. 19b). Under extreme circumstances, especially in highly strained two-phase alloys with large local orientation perturbations, these processes can eventually lead to the formation of high-angle grain boundaries and microstructures that strongly resemble those of materials that have undergone discontinuous recrystallization [49,50] (Fig. 19c). This gave rise to the terms "continuous recrystallization" or "recrystallization in situ," although the underlying mechanisms are inherently different from genuine, discontinuous recrystallization, which will now be discussed.

6. Nucleation

In most cases, during the annealing of a deformed microstructure, genuine recrystallization occurs through the formation of recrystallization nuclei at some distinct places in the microstructure followed by their growth under consumption of the as-deformed neighborhood. According to classical nucleation theory, supercritical nuclei that will be able to grow form through thermal fluctuations. At critical sizes, the reduction in volume free enthalpy $\Delta g_V \cdot dV$ associated with the forma-

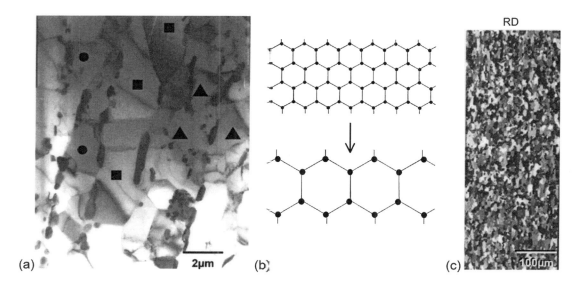

Figure 19 Extended recovery as caused by the coarsening of finely dispersed second-phase particles. (a) Heavily pinned subgrain structure in cold-rolled, supersaturated Al–1.3% Mn (annealed for ~10^6 sec at 350°C; see Fig. 16). (From Ref. 48.) (b) Schematic sketch of subgrain growth controlled by particle coarsening. (From Ref. 2.) (c) Microstructure after continuous recrystallization in Al–1.8% Fe–1.1% Si (96% cold-rolled, annealed for 10^4 sec at 300°C). (From Ref. 50.)

tion of a nucleus of volume V just balances the energy increase associated with the creation of a new grain boundary of area A. Thus, for the critical nucleus size r_{crit}, it holds:

$$r_{crit} = \frac{2\gamma_{GB}}{p_D} \qquad (34)$$

This estimate yields a critical nucleus size on the order of $0.1,\ldots,0.5$ μm and, correspondingly, a critical nucleation energy on the order of $\Delta G_C \approx 2 \times 10^{-13}$ J/mol $\approx 10^5$ eV. This results in a nucleation probability on the order of $N/N_0 \approx \exp(-10^7)$, which is so small that thermal nucleation of recrystallization can completely be ruled out.

The above argument clearly proves that the "recrystallization nuclei" cannot form by thermal fluctuations, but must be provided by other means. In materials with a high stacking fault energy such as aluminum, the formation of new orientations by recrystallization twinning does not play an important role, so that the orientations of the recrystallization nuclei must already preexist in the deformed microstructure (preformed nuclei; e.g., Refs. 51–53). Nonetheless, because of the strong analogies between the microstructural processes acting during recrystallization and nucleation and growth during thermodynamic phase transformations, the same terms are commonly used to describe recrystallization.

Considering the substructure of (recovered) deformed aluminum and the critical nucleus size on the order of a few tenths of a micron (Eq. (34)), this implies that, in principle, all dislocation cells or subgrains may be regarded as potential nuclei. This, however, raises the question of why so few subgrains actually act as nuclei. For an average subgrain size of 1 μm after recovery and a typical final grain size of 50 μm, only one out of 125,000 subgrains becomes a recrystallized grain.

The reason for this extremely low proportion of successful nuclei is attributed to the necessity that the boundary between a nucleus and the surrounding matrix must principally be able to move (i.e., the nucleus must possess a high-angle grain boundary). This implies that during the nucleation period, local misorientations in excess of about 15° have to be formed, whereas recovery in the deformed matrix only provides misorientations that scarcely exceed 10° [40]. Consequently, nucleation is generally restricted to sites in the vicinity of deformation inhomogeneities, where substantially larger local misorientations are built up already during the preceding deformation.

The requirements of steep orientation gradients with high local misorientations and, therewith, sufficiently high grain boundary mobility are naturally given at the preexisting grain boundaries between the deformed grains. Recrystallization nuclei may form by bulging of a given grain boundary into a grain with a lower dislocation density (strain-induced boundary migration, or SIBM). In Al alloys, which usually comprise a recovered subgrain structure prior to the onset of recrystallization (e.g., Fig. 11b), nucleation at a grain boundary can proceed by the growth of a given subgrain on one side of the boundary into the deformed structure on the other side of the boundary (Fig. 20a). Accordingly, the resulting recrystallized grains will have orientations of the former deformation texture [54], including the characteristic R-orientation [55]. Furthermore, in regions close to the grain boundaries, generally higher dislocation densities and stronger orientation gradients exist due to the activation of additional slip systems to reduce strain incompatibilities at the grain boundaries during deformation. Therefore, favorable conditions for recovery and, consequently, for successful nucleation events prevail in regions close to the grain boundaries (Fig. 20b).

Similarly, the disturbed zones close to large particles can act as nucleation sites. According to Humphreys [56], these so-called deformation zones are caused by the deformation incompatibilities at the matrix/particle interface, where very high dislocation densities and, consequently, strong lattice rotations are being built up during rolling. During recovery, the dislocations recover rapidly into subgrains that are substantially smaller than the subgrains in the matrix well away from the particles. In the next stages of annealing, subgrain growth takes place in this very fine local subgrain

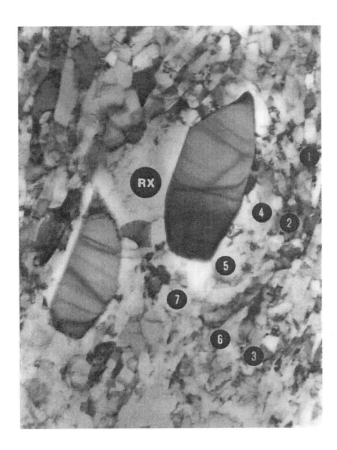

Figure 21 Example of PSN in two-phase Al–1.3% Mn containing large Al_6Mn precipitates (80% rolled, annealed for 30 sec at 350°C, transmission electron microscopy or TEM).

structure, until finally one or a few large subgrains have consumed the deformation zone and have formed high-angle grain boundaries to the surroundings (Fig. 21). Finally, if their size exceeds the critical diameter of ~ 1 μm, some of the enlarged subgrains may be able to start growing into the deformed matrix. The textures of samples where recrystallization is dominated by this so-called particle-stimulated nucleation (PSN) are typically quite weak with a high fraction of randomly oriented grains. It is worth mentioning here that the formation of the deformation zones depends on the strain rate and particularly on the deformation temperature. At high deformation temperatures and/or low strain rates, dynamic recovery counteracts the formation of the deformation zones and, hence, prevents the occurrence of PSN, as discussed in detail elsewhere [57].

Furthermore, other deformation heterogeneities, including transition bands and shear bands, may provide

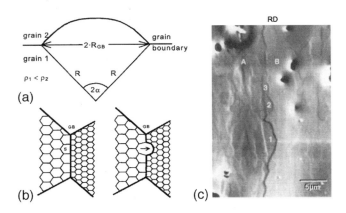

Figure 20 Nucleation of recrystallization at a preexisting grain boundary. (a) Nucleation through strain-induced boundary migration; (b) nucleation at a grain boundary in a subgrain structure; and (c) example of nucleation at a grain boundary in commercial purity aluminum AA1145 (86% rolled, annealed for 1000 sec at 250°C, scanning electron microscopy).

recrystallization nuclei. Because of the highly localized strain, they comprise an elevated dislocation density, which, during recovery, leads to rapid subgrain growth and, accordingly, to favorable nucleation conditions. When the critical nucleus size is exceeded, the growth of the nuclei can readily take place due to the high local lattice rotations.

Shear bands are characteristic bandlike deformation heterogeneities that typically form at an angle of approximately 35° to the RD in several Al alloys (Fig. 22) (e.g., in the presence of fine shearable particles [58,59], or in alloys with a high concentration of Mg atoms [60,61]). Transition bands form when an unstable crystal orientation tends to split off into two different orientations on deformation. In Al alloys, cube-oriented grains are often assumed to nucleate in transition bands, which form through a mechanism originally proposed by Dillamore and Katoh [62]. By Taylor-type model calculations (see Section IV.A.1, "Microscopical–Macroscopical Relations"), it can be shown that during plane strain deformation, all orientations close to the cube orientation in the starting texture rotate about the RD toward the rolling texture orientations. This rotation is divergent, which means that a grain with an orientation along the cube RD rotation path can split during rolling and that the two parts of the grain rotate apart toward two symmetrically equivalent rolling texture orientations. These rotation paths are shown schematically in Fig. 23. The transition band then represents the bandlike remaining part of the splitting grain that still comprises its original orientation.

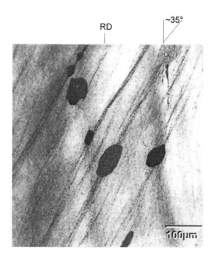

Figure 22 Nucleation at shear bands in a rolled single crystal with {112}⟨111⟩ C orientation (Al–1.8% Cu, 80% cold-rolled, annealed for 100 sec at 300°C).

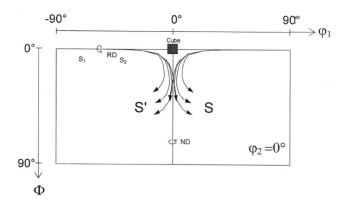

Figure 23 Scheme of the orientation changes during the formation of transition bands according to the Dillamore–Katoh [62] mechanism, represented in the ODF $\varphi_2 = 0°$ section.

7. Growth Processes

The critical size of a recrystallization nucleus in heavily deformed aluminum is on the order of 0.5–1 μm (Section IV.B.6), whereas typical recrystallized structures reveal grain sizes of 10–50 μm. This size difference underlines the importance of growth effects after the nucleation stage on microstructure and texture of recrystallized materials. Growth takes place through the movement of grain boundaries into the as-deformed microstructure. Hence, growth is controlled by the mobility of the grain boundaries involved, which, in turn, depends on the structure of the grain boundary plus external factors such as annealing temperature, driving force, and alloy composition, including impurities, etc. [63–66].

Thermally activated grain boundary motion takes place through the transfer of a given atom from the lattice of the shrinking grain, through the boundary and onto the lattice of the growing grain (Fig. 24a) [67–69]. When a grain boundary moves under a driving forced p_D, this elementary step is associated with an energy gain of $p_D b^3$ (b^3 is the atomic volume) (Fig. 24b). The grain boundary velocity v is given by the difference in thermally activated diffusion jumps Γ_i between the two grains:

$$v = \Delta x \cdot (\Gamma_1 - \Gamma_2)$$

$$= b v_0 c_v^{GB} \left[\exp \frac{-\Delta G}{kT} - \exp \left(-\frac{\Delta G + p_D b^3}{kT} \right) \right] \quad (35)$$

(v_0 is the Debye frequency, $\sim 10^{13}$ sec^{-1}; $\Delta x \approx b$ is the jump distance; ΔG is the activation enthalpy for the diffusion jump; and c_v^{GD} is the vacancy concentration

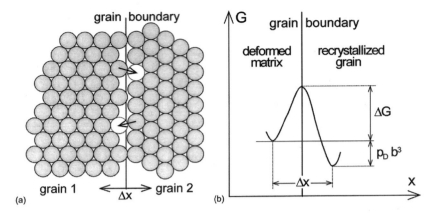

(a)   (b)

Figure 24 On grain boundary motion during the growth of recrystallization nuclei. (a) Schematic sketch of the grain boundary structure; and (b) evolution of the free enthalpy G across a grain boundary under a driving force p_D.

in the boundary). For all recrystallization processes $p_D \ll b^3 kT$, so that:

$$v = b^4 v_0 c_v^{GB} \frac{1}{kT} \exp \frac{-\Delta G}{kT} \cdot p_D = m \cdot p_D \quad (36)$$

Thus, the grain boundary velocity v is proportional to the driving force p_D; the proportionality factor is denoted by mobility m. It also follows from Eq. (36) that the mobility reveals an Arrhenius-type temperature dependency $m = m_0 \cdot \exp(-Q/kT)$. For an accurate analysis of grain boundary mobility, dedicated experiments with specially prepared bicrystals are best suited. Such experiments have given good evidence of the proportionality between grain boundary velocity and driving force, which, in turn, enables one to derive the mobility as a function of temperature, alloy composition, misorientation, etc. [70].

Besides temperature, the grain boundary mobility also depends on the misorientation between the two grains. In general, low-angle grain boundaries as well as (coherent) twin boundaries display very low mobility. In contrast, the so-called "special" grain boundaries may have higher mobility than average (i.e., "random" boundaries). This behavior is commonly explained in terms of the so-called "coincidence site lattice" (CSL) theory. The CSL boundaries are thought to be densely packed such that they will absorb less solutes that lower the mobility of random boundaries (see below). However, the grain boundary mobility not only depends on the misorientation, but may also depend on the exact position of the grain boundary plane. Figure 25 reproduces the well-known results of Liebmann et al. on the grain boundary velocity as a function of the misorien-

tation angle for $\langle 111 \rangle$ tilt boundaries in aluminum [71]. Evidently, grains with a $40°\langle 111 \rangle$ orientation relation depict maximum mobility, which will become important with a view to the interpretation of recrystallization textures where this special orientation relationship is of great importance.

Alloying elements, both in solid solution and in the form of precipitates, may strongly retard recrystalliza-

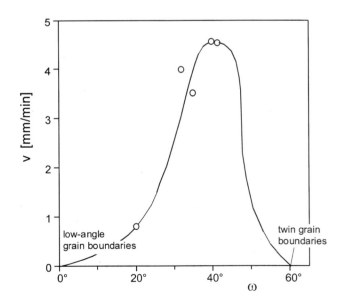

Figure 25 Grain boundary velocity as a function of the misorientation angle for $\langle 111 \rangle$ tilt grain boundaries in aluminum. (From Ref. 71.)

tion by exerting a retarding force on the moving grain boundaries. Most works on the influence of solutes on grain boundary motion are based on a model developed simultaneously by Cahn [72] and Lücke and Stüwe [73]. This Cahn–Lücke–Stüwe theory is based on the idea that, due to local lattice distortions, solutes close to the grain boundaries have a lower energy than in the undistorted matrix in the grain interior. This difference in energy results in an attractive force dU/dx between the grain boundary and the solute (Fig. 26a) and, therefore, an increased concentration of solutes in the vicinity of the grain boundaries:

$$c_{GB} = C_0 \cdot \exp \frac{U(x)}{kT} \tag{37}$$

(where c_0 is the average solute concentration and c_{GB} is the concentration of solutes in the vicinity of the grain boundary).

If the grain boundary moves, the solutes have to follow the grain boundary by diffusion processes so as to retain their low-energy position (Fig. 26a, dashed line). This results in a retarding force p_F that reduces the driving force and, hence, the grain boundary velocity ("impurity drag"):

$$v = m \cdot (p_D - p_F) \tag{38}$$

At low velocity, the grain boundaries are "loaded" with solutes and the impurity drag increases with grain boundary velocity. At high velocity, the grain boundaries will break away from the solutes and then move more or less freely. In the intermediate regime, there is transition between the behavior of loaded and free boundaries, where grain boundary velocity and driving force may be nonproportional (Fig. 26b).

Second-phase particles likewise exert a drag on the moving grain boundary. This so-called Zener drag [74] is caused by the interaction of grain boundary and particles so as to reduce grain boundary area (Fig. 27). The resulting retarding force p_Z is given by:

$$p_Z = \frac{-3\gamma_{GB}f}{2r_P} \tag{39}$$

(where f is the volume fraction and r_P is the radius of the second-phase particles). Thus, the Zener drag is largest in the presence of small, finely dispersed particles.

C. Textures

The large plastic strains encountered in rolling operations develop directional structures, or textures, which can strongly influence the resulting mechanical (or

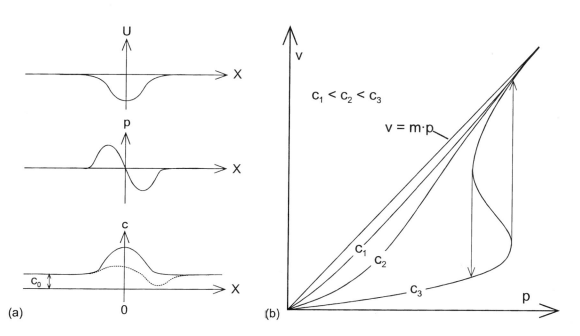

Figure 26 Solute drag. (a) Development of potential U, interaction force p, and concentration of solutes c along distance x across a grain boundary. The dashed line indicates the changes in the concentration profile $c(x)$ for a grain boundary that moves from right to left; and (b) grain boundary velocity v as a function of the driving force p for different concentrations of solutes c.

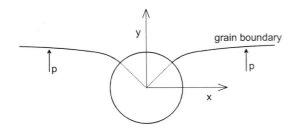

Figure 27 Interaction between a particle and a grain boundary moving under a driving force p (Zener drag).

physicochemical) properties of the material. The most well-known effect is the plastic anisotropy of sheet products, an example of which is illustrated in Fig. 28 [75]; the properties in the sheet plane [e.g. σ_y, at an angle α to RD ($\alpha \approx 55°$)] are significantly lower than along RD. This type of anisotropy can also strongly influence formability (e.g., the product shape during a deep drawing operation), as can occur for beverage cans (Section V.D). The deformation textures developed during rolling can, of course, be modified by annealing to generate new recrystallization textures.

Crystallographical textures in polycrystalline aggregates are characterized by the volume fractions of a particular orientation (g) compared with its theoretical value in a perfectly random set of orientations. Formally, they are represented by an orientation distribu-

tion function (ODF) defined by the volume fraction of a material (dv/V), which possesses an orientation within a small angular range dg [i.e., $f(g) = (dv/V)/dg$]. In practice, textured metals are usually composed of grains with orientations close to one or more "ideal" texture components and other grains oriented at random well away from the latter (often termed random orientations).

An individual orientation can be defined in several ways; the parameters most used by the texture community are the three Euler angles (φ_1, Φ, φ_2), which, by three successive rotations, bring the orientation (g) to some common reference orientation. In cubic materials, the latter would be the three $\langle 100 \rangle$ crystallographical axes parallel to the principal axes of the deformation process (i.e., RD, TD and ND in rolling). The metallurgical community often prefers the Miller indices $\{hkl\}\langle uvw \rangle$, giving the crystal planes and directions of the major texture components aligned with the principal straining directions [i.e., $\{hkl\}$ for the sheet plane and $\langle uvw \rangle$ for the rolling direction]. The correspondences of Miller indices and Euler angles for some typical sheet metal orientations are given in Table 4.

For symmetry reasons, the possible ranges of the Euler angles can be restricted to specific values (e.g., $\varphi_1 \leq 90°$ for orthotropic symmetry typical of rolled sheet) so that each orientation is defined by its (φ_1, Φ, φ_2) coordinates in a well-defined 3D Euler space. For practical visualization, this 3D Euler space is sectioned into 2D sections of constant φ_2 in which the ideal texture components are easily located. The reader is referred to Bunge [76], Kocks et al. [77], and Randle and Engler [78]

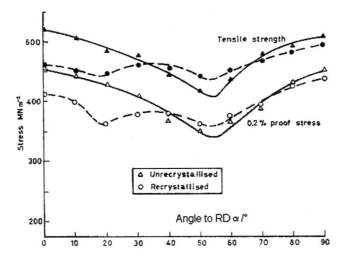

Figure 28 Yield stress anisotropy in a high-strength Al–Li alloy. (From Ref. 75.)

Table 4 Miller Indices and Euler Angles of the Most Important Orientations of Al and Al Alloys After Rolling and After Recrystallization (Approximated)

Component	Miller indices $\{hkl\}\langle uvw \rangle$	Euler angles (φ_1, Φ, φ_2)
Cu	$\{112\}\langle 111 \rangle$	90°, 30°, 45°
S	$\{123\}\langle 634 \rangle$	59°, 34°, 65°
Bs	$\{011\}\langle 211 \rangle$	35°, 45°, 0/90°
Goss	$\{011\}\langle 100 \rangle$	0°, 45°, 0°/90°
Cube	$\{001\}\langle 100 \rangle$	0°, 0°, 0°/90°
Cube$_{RD}$	$\{013\}\langle 100 \rangle$	22°, 0°, 0°/90°
Cube$_{ND}$	$\{001\}\langle 310 \rangle$	0°, 22°, 0°/90°
P	$\{011\}\langle 122 \rangle$	70°, 45°, 0°
Q	$\{013\}\langle 231 \rangle$	45°, 15°, 10°
R	$\{124\}\langle 211 \rangle$	53°, 36°, 60°

for further details and particularly the diffraction techniques used to determine textures and ODFs.

1. Rolling Textures

Sheet rolling textures of fcc metals are usually characterized by continuous spreads of orientations along one of two fibers, denoted as the α and β fibers. The β fiber orientations, usually developed in rolled aluminum, include "Cu" {112}⟨111⟩, "S" {123}⟨634⟩, and "Bs" {011}⟨211⟩. The α fiber components extend from the same "Bs" to "Goss" {011}⟨100⟩, (i.e., possess an {011} crystal plane parallel to the sheet plane). Figure 29 gives an example of the {111} and {100} pole figures of rolled aluminium sheet, and Fig. 30 gives the corresponding ODF sections.

In cold-rolled aluminum sheets, the strongest components at the sheet center are usually the S and Cu components, together with small but significant amounts of Bs and Goss, in this order. However, the rolling conditions, the alloy content, and the initial texture prior to rolling can influence this order. Cold rolling is always very close to plane strain compression, so the deformation textures tend to be very similar if the same deformation mechanism operates. An important case, treated in detail in Section V.D, is cold rolling of cube textured sheet for beverage can applications; the

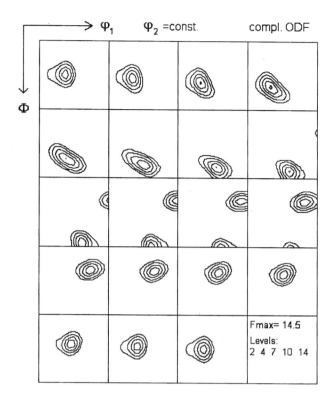

Figure 30 The ODF of the pole figures shown in Fig. 29 (92% cold-rolled Al alloy AA5182).

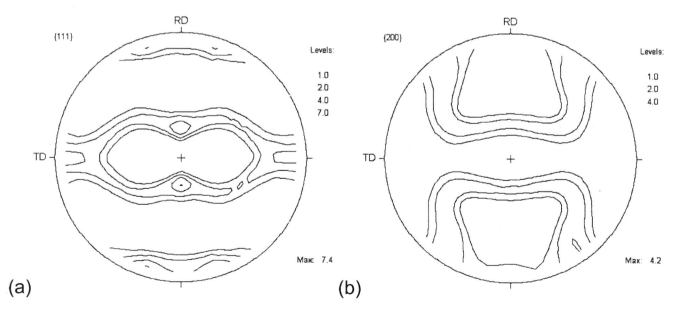

(a)

(b)

Figure 29 (a) {111} and (b) {200} pole figures of cold-rolled Al (AA5182, 92% rolled).

cube texture gradually decreases in intensity to be replaced by the β fiber components at high strains.

The rolling textures can be modified only if the strain path and/or the deformation mechanisms change. This occurs at room temperature if the usual slip processes are replaced at high strains by pronounced shear banding, as is observed in some particle-strengthened alloys containing fine shearable precipitates (Al–Li and Al–Cu). In this case, the intense localized shears along the bands tends to strengthen the Bs and Goss components (see Ref. 58). Mechanical twinning has a similar effect but is usually not encountered in current aluminum alloys.

During high-temperature rolling, the deformation textures can be significantly different because there are fewer constraints on both the strain path and the slip systems. The strain path varies because the hot rolling of an ingot by reversible rolling and then tandem rolling (Section III.B) superimposes significant shear components on the plane strain compression mode; the shears vary from zero at the center to high values near the surface. Also at temperatures $> 0.5T_m$, slip in the aluminum is no longer restricted to the {111} planes but can occur on the {110} and {100} planes (and possibly others) (see Maurice and Driver [79] and Perocheau and Driver [80]). An example of the texture variation through the thickness of a reversibly rolled AA3103 alloy is represented in Fig. 31 by {111} pole figures. Near the center ($L = 0.2$), there is a typical hot rolling β fiber texture albeit with a stronger Bs component than usually found after cold rolling. Midway between the center and the surface ($L = 0.5$), the texture is quite different and, near the surface ($L = 0.9$), tends to a unique {001}⟨110⟩ shear component. This develops as a consequence of the reversed shear deformation that material elements undergo both within a pass and from

pass to pass. On subsequent tandem rolling, the shear is reduced in amplitude and rolling is unidirectional. The deformation mode during tandem rolling then transforms the {001}⟨110⟩ shear component into two symmetrical {112}⟨111⟩ Cu components, which form part of the usual β fiber texture. The final hot-rolled plate from the tandem mill has, therefore, a β fiber texture in which the Bs component tends to be strongest in the center and the Cu component tends to be stronger near the surface [81].

An important point with regard to subsequent recrystallization texture development is the amount of cube texture left in the as-hot-rolled material. During cold rolling, the cube component is unstable so cube grains gradually break up by deformation banding before ultimately (at $\varepsilon \geq 2$) rotating to the β fiber (near-S) components. However, even after large strains, some fragments of the cube grains are left over and can become sites for recrystallization nuclei. During high-temperature rolling, many studies have now shown that the cube grains are stable (as a result of the change of slip systems) up to very large strains and therefore survive the entire hot rolling process as elongated "cube bands."

The evolution of these rolling textures can be modeled quite successfully by the crystal plasticity models described briefly in Section IV.A.1, "Microscopical–Macroscopical Relations." An example of a cold rolling texture simulation assuming {111}⟨110⟩ and plane strain compression is given in Fig. 32a. This shows strong S and Cu components with some Bs. Hot rolling textures and their gradients have also been modeled more recently using FE data for the strains along different material flow lines as shown in Fig. 32b and c. At the center, there is a classical β fiber with strong Bs, S, and Cu components, together with some Bs and Goss. Toward the surface, the alternating shears in the breakdown mill lead to a well-developed {001}⟨110⟩ shear texture and subsequent (virtual) deformation in the tandem, then destroy most of this shear components to reform a β fiber texture.

2. Recrystallization Textures

The most common recrystallization textures of rolled and annealed aluminum alloys are the cube {001}⟨100⟩, the R (or retained S) {124}⟨211⟩ and, under certain circumstances, the P {011}⟨122⟩ or Q {013}⟨231⟩ components. Texture evolution during recrystallization is a complex (and often controversial) subject, so the following will attempt to describe the essential features of the problem, at the risk of oversimplification. The tex-

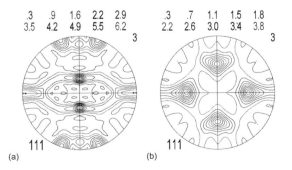

.3	.9	1.6	2.2	2.9	.3	.7	1.1	1.5	1.8
3.5	4.2	4.9	5.5	6.2	2.2	2.6	3.0	3.4	3.8

(a)   (b)

Figure 31 Examples of texture variations through the thickness of hot-rolled AA3103. (a) Center; and (b) near surface.

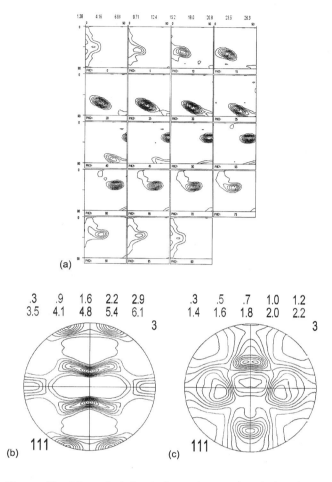

Figure 32 Numerical simulations of texture evolution during the rolling of aluminum alloys. (a) Cold rolling; (b) reversible hot rolling along center line; and (c) reversible hot rolling near the surface. (From Ref. 5.)

Lücke [82] have shown that cold-rolled Al–0.007% Fe, when annealed at 360°C, leads to a R + cube texture, probably due to the influence of the precipitation of Fe on the migrating boundaries. When both Fe and Si are present (as in the commercial $1xxx$ alloys), they form Al–Fe–Si phases during casting, which act as nucleation sites for PSN. Subsequent rolling and recrystallization then promote a degree of texture randomization by PSN (Fig. 33). Note that the cube texture still dominates but f_{max} ~10 due to PSN randomization. Because most commercial alloys contain typically ~1% (Fe + Si), clearly PSN will be a major factor controlling recrystallization textures, at least after cold rolling. Recrystallization after hot rolling, as pointed out in Section IV.B.6, is less influenced by PSN because the deformation zones around the particles are reduced by dynamic recovery [57].

In the more highly alloyed $2xxx$, $5xxx$, and $7xxx$ series, which contain either fine shearable precipitates or high Mg contents, shear banding can occur (Section IV.A.1, "Influence of Alloying Elements"). These deformation inhomogeneities are also nucleation sites for recrystallization (Section IV.B.6) and have been extensively studied by Engler [83]. It appears that shear bands

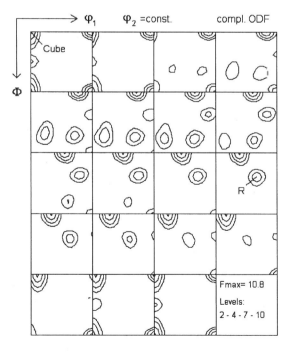

Figure 33 Recrystallization texture evolution in commercial purity aluminum due to the presence of (Fe + Si) particles.

tures depend very sensitively on the details of the nucleation and growth mechanisms outlined in Section IV.B (i.e., are functions of the deformation texture, the microstructure, and, particularly, the solute and precipitate distributions, together with the annealing conditions)

In sheets of pure fcc metals such as aluminum, large rolling reductions and high annealing temperatures strongly favor the cube recrystallization texture as a consequence of rapid nucleation of cube grains from the cube fragments and their growth into the adjacent, highly dislocated matrix (cube with $f_{max} > 20$). Recrystallization after hot rolling can produce even stronger cube textures. However, even small quantities of Fe in solid solution, as occurs in most commercial aluminum alloys, can modify this texture. For example, Hirsch and tend to favor the Goss {011}⟨100⟩ component and, at very high strains, the Q orientation. If the same alloys are deformed without shear banding, then the other mechanisms such as PSN, SIBM, etc., operate.

Clearly, recrystallization texture development is sensitive to the relative rates of nucleation and subsequent growth from a wide variety of sites in the deformed metal (cube bands, PSN, SIBM, and shear bands) and which obviously depend on the deformation microstructure (i.e., the deformation conditions and the alloy). Theoretical prevision of recrystallization texture development is not easy. However, there is a general tendency for the recrystallization texture to be controlled by the first stages of nucleation and growth, often termed microgrowth, of small micron-sized elements out of the deformation structure and which are often related by 30–40° ⟨111⟩ rotations with respect to their local environment. It follows that recrystallization textures can sometimes be modeled by simply taking the deformation texture and by imposing 40°⟨111⟩ rotations to each component of the deformation texture [50,84]. As shown in Fig. 34, for random nucleation, which is approximately given when PSN prevails, this transformation texture can give quite a reasonable

approximation of the recrystallization texture. A somewhat more sophisticated model has been developed by the Trondheim group [85] for the case of hot deformed alloys (where shear banding does not intervene). The analysis relies on models for the rates of nucleation and the growth of the PSN, SIBM, and cube band sites as a function of the deformation structure and texture. It can also be extended to predict recrystallization grain sizes.

In a more generalized context, it has been demonstrated that the recrystallization textures of most rolled Al sheets can be modeled by the multiplication of a function representing the probability of the nucleation of the new grains with their growth probability function, with the latter being identical to the 40°⟨111⟩ transformation texture [86,87]. The probability of nucleation is given by the distribution of potential nucleus orientations that form at the corresponding nucleation sites, which means that it can be determined experimentally.

More recently, Engler and Vatne have combined these two approaches in that the Trondheim model is utilized to determine the contribution of nucleation at the various characteristic nucleation sites acting in rolled Al alloys (cube bands, grain boundaries, and large particles) to the distribution of nucleus orientations. This approach has been successfully applied to simulate the recrystallization textures of a variety of Al alloys with different microstructural characteristics processed under different processing parameters, viz. strain, strain rate, and deformation temperature [88].

3. Anisotropy

The crystallographical textures developed by rolling deformation and eventually subsequent recrystallization lead to anisotropic mechanical properties. An example is the variation, with direction in the sheet plane, of the yield strength of an Al–Li alloy as illustrated in Fig. 28. A second case concerns the plastic anisotropy during a forming operation such as deep drawing. A circular blank stamped out of a sheet is drawn down by punching through a die into a cup shape, but the resulting "cup" develops irregular wall heights commonly denoted "ears" (Fig. 35). The ears are situated at certain angles to the sheet directions according to the plastic anisotropy, or the texture, of the sheet (see Section V.D).

Plastic anisotropy of sheet materials is measured experimentally by the contraction ratios of flat tensile samples taken at angles α to the rolling direction. The ratio of the width (plastic) strain ε_w to the through-

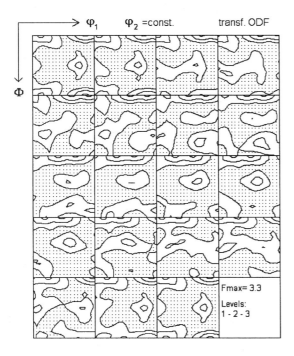

Figure 34 Rolling-to-recrystallization transformation texture obtained by 40°⟨111⟩ rotations (Al–Fe–Si, 90% cold-rolled; regions with intensity <1 are dotted).

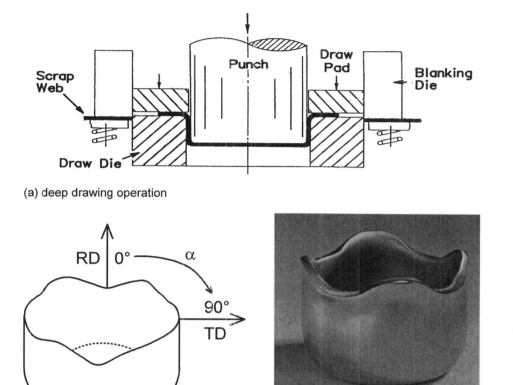

(a) deep drawing operation

(b) cup with uneven rim

(c) example showing pronounced earing

Figure 35 (a) Deep drawing of cylindrical cups and (b, c) typical earing profile with ears under 0° and 90° to the rolling direction developed during cupping of a circular aluminum sheet with strong cube texture (AA3104 hot band). (Courtesy of Dr. J. Hirsch, Hydro Aluminium Deutschland GmbH.)

thickness strain ε_t is denoted by the Lankford coefficient or the R value:

$$R = \frac{\varepsilon_w}{\varepsilon_t} \left(\text{or } r = \frac{\dot{\varepsilon}_w}{\dot{\varepsilon}_t} \right) \qquad (40)$$

In theory, R varies from 0 (no widening) to ∞ (no thickening) and equals 1 for isotropic materials. Note that some authors use the contraction ratio $q = -\dot{\varepsilon}_w/\dot{\varepsilon}_l$ measured in terms of the strain rates along the sample width and length (i.e., $q = r/1 + r$ and varies from 0 to 1)

To characterize the overall resistance to thinning of a sheet, an average of the R values at three alpha angles is usually determined:

$$\overline{R} = \frac{(R_0 + R_{90} + 2R_{45})}{4} \qquad (41)$$

and to measure the amplitude of the in-plane anisotropy (related to the ear heights), the difference value is used:

$$\Delta R = \frac{(R_0 + R_{90} - 2R_{45})}{2} \qquad (42)$$

An $\overline{R}$ value greater than 1 means that during elongation, the sheet thins less than it contracts and often indicates good formability. Average $\overline{R}$ values greater than 1 are not common in most aluminum alloys. Only in materials that have been produced to include pronounced shear textures have $\overline{R}$ values of about unity been obtained [89].

A more general description of plastic anisotropy is contained in the yield surface. In theory, these are five dimensional spaces that describe the current yield stress as a function of any 5D stress direction. In general practice, they are measured in the sheet plane by applying different combinations of in-plane loads so that experimental yield surfaces are essentially 2D. An example of the sheet plane yield surface of an aluminum alloy is given in Fig. 36.

Anisotropy is frequently modeled using the crystal plasticity analysis outlined in Section IV.A.1, "Microscopical–Macroscopical Relations." For the extreme case of a single crystal, the basic equation is the generalized Schmid criterion for slip on a system under a general stress state [σ]:

$$\tau_c^s = \tau^s = b_i^s \sigma_{ij} n_j^s \tag{43}$$

Assuming that the stress normal to the slip plane has no influence, this can be written in terms of the five-component stress deviator [s] and defines a set of hyperplanes in 5D space. Because, according to the above equation, the glide stress cannot exceed the value τ_c^s, the surface made up of the inner set of hyperplanes describes the stress states for plastic flow; this is termed the (lower bound) single crystal yield surface, or yield locus. For a textured polycrystalline aggregate, the condition of plastic flow is determined both by the Schmid criterion and by the requirements of strain compatibility between grains. An upper bound yield locus is given by the Taylor, Bishop, and Hill analysis by calculating, for all possible strain increments, the asso-

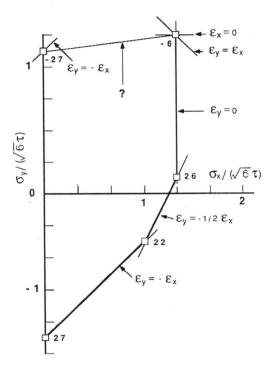

Figure 37 Upper-bound single crystal yield surface, {110} ⟨112⟩ Bs orientation. (From Ref. 90.)

ciated stress states (i.e., the stress corresponding to the maximum plastic work rate of the material). An example of an upper-bound single crystal yield surface is given in Fig. 37.

For a textured polycrystalline aggregate, the stress state for each strain increment is obtained by an averaging procedure over a representative population of grains, usually based on the orientation distribution function. This then gives the plastic work rate and the average Taylor factor ⟨M⟩. The methods are outlined by Hosford [90] and Kocks et al. [77]. The yield loci of textured polycrystals tend to be intermediate between the single crystal case of Fig. 37 and the (isotropic) von Mises ellipse (see Fig. 36). All yield loci obey the normality rule according to which the strain rate vector, for a particular stress state on the yield surface, is normal to the yield surface at that point. One of the important features of these textured yield loci are the vertices, which, for a single grain, correspond to a discontinuous change in slip systems for a small change in stress state. According to the normality rule, they imply large changes in strain path with two important consequences: (a) the Lankford coefficient can be very dependent on texture and straining conditions; and (b)

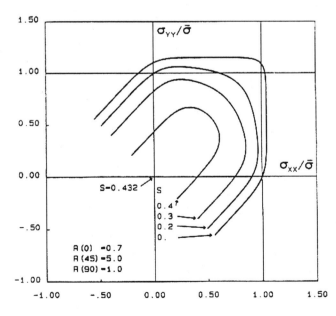

Figure 36 Anisotropic 2D sheet plane yield surface of textured Al from Ref. 90. The stresses are normalized to the x-direction yield strength. S indicates the normalized level of τ_{xy}.

pronounced discontinuities of the normal to the yield surface often imply plastic instabilities, which can be more or less localized.

V. PROPERTIES AND SPECIFIC APPLICATIONS

A. Introduction

Rolled products (i.e., sheets, plates, and foils) constitute almost 50% of all Al alloys used. In North America and Western Europe, the packaging industry consumes the majority of the sheet and foil for making beverage cans, foil containers, and foil wrapping. Sheet is used extensively in buildings for roofing and siding; in transport for airframes, road, and rail vehicles; in marine applications, including offshore platforms; and superstructures and hulls of boats. Although, to date, only relatively little Al is used in the manufacture of high-volume production automobiles, it is expected that the next decade will see a substantial increase of aluminium sheet used for body panels. Plates are used for airframes, military vehicles and bridges, ship superstructures, cryogenic and chemical vessels, and as tooling plates for the production of plastic products. Foil applications outside packaging include electrical equipment, insulation for buildings, lithographical plate, and foil for heat exchangers.

In this chapter, several typical examples of the use of aluminum-rolled products will be presented in order to illustrate the impact of the various steps during the processing of rolled Al products on the resulting product properties. In the sequence of thickness variations during rolling, the examples start with the use of thick hot-rolled transfer slab for aeronautical applications. Hot bands with thickness of 4–8 mm may be used for undercarriage and wheels in automotive applications. Examples of cold-rolled strip include hard (i.e., unrecrystallized) can stock for beverage cans as well as soft (i.e., recrystallized) clad brazing sheets for heat exchangers. Furthermore, solution-treated sheets of age-hardenable Al alloys for applications in car body-in-white applications will be addressed. Finally, we will discuss the production of Al foil that is used in numerous applications in packaging.

These examples were chosen to cover a wide range of use of Al alloys in applications of technological importance. In each example, we will discuss the main property requirements as well as the important processing parameters that need to be controlled so as to achieve these requirements. It is worth noting that most of the new products have been developed by a combination of engineering ingenuity and improved scientific knowledge of the physical mechanisms.

B. Use of Thick Al Plate in Aeronautical Applications

The development of modern air traffic was virtually impossible without the use of aluminum. The combination of good corrosion resistance, high formability, and high strength, and, in particular, the low specific weight, makes Al the most versatile material for aircraft applications. For instance, the Boeing 747 "jumbo jet" contains 75,000 kg of aluminum. Despite the development of new composite materials, Al remains the primary aircraft material, comprising about 70% of a modern aircraft's weight. The alloys used most widely for aircraft applications are the heat-treatable alloys of the Al–Cu system ($2xxx$ series) and, for components requiring higher strength, the Al–Zn–Mg system ($7xxx$ series). For many years, alloys 2024 and 7075 dominated in aircraft applications. By limiting the contents of accompanying elements, especially Fe and Si, and by optimizing the thermomechanical processing, damage-tolerant variants of these alloys have been established (e.g., Refs. 91–93).

The fuselage of commercial aircraft consists of a skin with longitudinal stiffeners (stringers) and regularly spaced frames in the circumferential direction. An important design driver is damage tolerance of the structure, associated, for example, with the tensile stresses introduced in the skin by the pressurization cycle. However, other loadings lead to additional requirements, such as stability in the compression of the lower part of the fuselage. Currently, there is a major development toward welded structures as opposed to riveted ones so as to reduce productions costs.

Wings of commercial aircraft consist of upper and lower wing covers reinforced with riveted or integrally machined longitudinal stringers. Longitudinal spars and transverse ribs are used to stiffen the wing box. Auxiliary structures include leading and trailing edges, control surfaces (ailerons, flaps, slats, etc.), and engine beams. Upper wing components are compression-dominated structures, whereas bottom wing structures are designed for damage tolerance requirements. Intermediate spar and rib components must achieve the best compromise between the technical needs of the upper and lower wings.

The mechanical structure of the horizontal and vertical stabilizers of commercial aircraft can be compared with the structure of the wings, albeit with smaller dimensions, with the fin box and the tailplane outerbox, and the center box and auxiliary structures (leading and trailing edges, elevator, etc.). Loads on the horizontal stabilizers are typically inverted compared with those on a wing, with the need of compression-resistant solutions for the lower parts and damage-tolerant alloys for the upper structures. As far as the vertical stabilizers are concerned, complex loads involving overall bending and local gust loads may occur.

Sheets are typically used in an aircraft for fuselage panels, skins of auxiliary structures (leading/trailing edges), and control surfaces, as well as for bent stiffeners or frames. Tempers vary from annealed "O" to quenched "T" or as-fabricated "F," depending on the forming operation of the component and the alloy formability. In addition, for outside panels, corrosion protection through cladding with commercial purity Al is often required.

Plates are increasingly used in the structure of commercial aircraft to substitute for assemblies of sheet or small gauge plate and stiffeners by so-called integral or monolithic structures. A conventional assembly usually consists of a number of formed sheets and extruded, forged, and/or machined parts, which are riveted together or joined otherwise. An integral structure, on the other hand, integrates the function of all these individual parts into one large structure with the advantage of weight reduction and production cost savings due to the lower number of components. Manufacturing of the integrated structure itself is cost-efficient due to high-speed milling machines, with up to 25,000 rpm and high feed rates, which are available today for machining of high-strength Al alloys.

The major challenge for the material suppliers in view of this substitution process is that the property combination of the raw material should not limit the performance of the integral structure when compared with a classical assembly. For that reason, a number of limitations of conventional Al plates had to be overcome. The maximum thickness of conventional Al plates had to be increased in order to allow the machining of very heavy sections. The residual stress level of the plate material had to be controlled to an extremely low level to guarantee distortion-free machinability even with the large dimensions and the high aspect ratios involved. Residual stresses are relieved by a controlled stretching operation wherein the plates are stretched by a few percent in their long dimension. The development of new alloys with lower quench sensitivity such as 7010 and 7050 led to improved through-hardening properties of thick plates. Additionally, several key properties of the plate material such as corrosion and fatigue resistance, toughness, and ductility had to be improved in order to compete with properties of other product forms.

In general, the fracture toughness of 7xxx series alloys depends on yield stress, chemistry, and on the distribution of second-phase particles [94–98]. Furthermore, to improve toughness and resistance against stress corrosion cracking, it is necessary to suppress recrystallization. Figure 38 shows the microstructure of alloy 7010 after laboratory deformation through plane strain compression tests [42], displaying a fine recovered subgrain structure plus a few strongly enlarged recrystallized grains. In order to suppress recrystallization, elements such as Cr and/or Zr are usually added. In addition, the thermomechanical processing must be controlled so as to minimize postdynamic recrystallization.

Thus, the optimization of the chemical composition, together with an appropriate thermomechanical processing, led to the development of aircraft alloys that combine high strength, high corrosion resistance, and high fracture toughness. These alloys are used in the most recent generation of airplanes such as the Boeing 777, or the planned new Airbus A380 (Fig. 39).

C. Hot Band for Welded Aluminum Wheels

Light metal wheels made from Al alloys have acquired a large market. However, such Al wheels, which are produced by casting or forging, are used mainly because of their attractive styling, but with weights of up to 10 kg, they do not offer a large potential for saving weight compared with conventional steel wheels. Therefore, there is an increasing interest in welded wheels made of aluminum hot band (e.g., Ref. 99). The deep-drawn disk is joined to the welded rim by MIG welding. With a weight of 5.6 kg, these wheels are about 30% lighter than their steel counterparts (Fig. 40). Thus, although Al wheels are more expensive than steel wheels, they offer the potential of reducing unsprung mass of the chassis, resulting in better ground contact and, there-

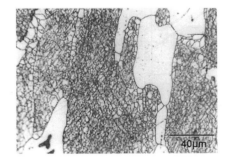

Figure 38 Optical micrograph of hot deformed, partially recrystallized AA7010 (etched in chromium acid).

Figure 39 Airbus A380 (computer simulation; courtesy of EADS).

with, improved safety, handling, and driving comfort, as well as improved fuel efficiency.

For such applications, and for other components of the chassis, naturally hard alloys of the Al–Mg–Mn system (5*xxx* series) are well suitable [100]. In the soft state, 5*xxx* series alloys exhibit very good formability as

required for producing the deep-drawn disks. The strength of Al–Mg alloys strongly increases with strain (Fig. 8), with the entire strength level depending on the Mg concentration. The corrosion behavior of Al–Mg alloys is generally excellent, such that many automotive parts may be used without a protective layer. However, alloys with Mg contents in excess of 3% may suffer from intergranular corrosion when they are exposed for a long time to temperatures in the range 60–200°C. Figure 41 shows the influence of Mg concentration on

Figure 40 Light metal wheel made from 8.6-mm-thick hot band of alloy AA5454. (Courtesy of Hydro Aluminium Deutschland GmbH.)

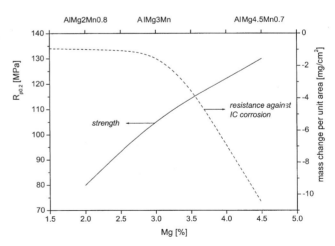

Figure 41 Yield strength $R_{p0.2}$ and resistance against intergranular corrosion of 5*xxx* series Al–Mg alloys as a function of the Mg content. (Courtesy of Dr. D. Wieser, Hydro Aluminium Deutschland GmbH.)

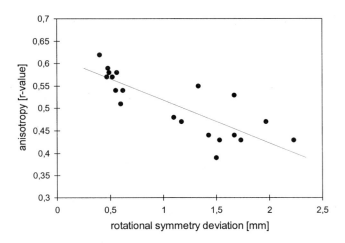

Figure 42 Correlation between the *r*-value and the deviation in rotational symmetry in a deep-drawn wheel disk. (Courtesy of Dr. J. Hasenclever, Hydro Aluminium Deutschland GmbH.)

strength and on the resistance against intergranular corrosion, the latter being expressed as weight changes during annealing for 17 hr at 130°C (ASTM G67). Thus, 5*xxx* alloys with 2–3%Mg—including the alloys AA5049 (EN AW–AlMg2Mn0.8), AA5454 (AlMg3Mn), and AA5918 (AlMg3.5Mn)—offer a good compromise between strength and resistance against intergranular corrosion. Age-hardenable 6*xxx* alloys would allow for further weight savings through downgauging, yet the production must be tailored so as to allow for the age hardening effect to be exploited.

The deep-drawn disk is fabricated from hot strip with a thickness of about 8 mm. Here, by controlled partial recrystallization, some of the hot rolling texture is preserved to effectively balance the opposite anisotropy of the cube recrystallization texture (see Section V.D). Thus, an almost even cup rim profile can be produced, which is required because strong anisotropy effects can lead to severe deviations from the rotational symmetry. The necessary condition of deviations below 1 mm correlate well with the deformation anisotropy measured in form of the *R*-value (Fig. 42).

D. Production of Can Body Stock for Beverage Cans

One of the probably best-known examples of products from rolled Al sheet is the aluminum beverage can, recognized as one of the most innovative packaging containers in the world. In 1999, more than 112 billion aluminum beverage cans were produced in North America (i.e., about 300 cans per person). As illustrated in Fig. 43, the production of the standard two-piece beverage can is carried out in a sequence of drawing and ironing operations. In a press operation, 12–14 circular blanks are stamped out in parallel from highly cold-rolled sheet with thickness of 0.25–0.30 mm, followed by deep drawing into a cup. In a second operation, the "bodymaker," the cup walls are further ironed down to about 0.1 mm thickness with concurrent forming of the can base. The can top is then trimmed to length. After chemical cleaning, the cans are given their final design by a sequence of internal and external coatings and printing operations. Finally, the cans are necked-in at the top to reduce can diameter and flanged so that the top of the can (Fig. 43) can be attached and sealed after filling. The can end is produced separately in a sequence of high-speed stamping and forming operations, with up to 28 blanks stamped out in parallel from coil-coated Al. The so-called shells are fed through a high-precision press where rivet making, scoring, and tabbing occur in progressive operations. After the cans and ends are inspected, they are ready for shipping and filling.

The common Al alloys used for the production of can bodies are Al–Mn–Mg alloys of the 3*xxx* series (AA3004 and AA3104), which meet best the requirements for can strength and formability. Continuous efforts to increase strength for further materials savings by downgauging led to the addition of small amounts of Cu (up to maximum of 0.25%). For the can end, higher strength is achieved through a higher Mg content, ranging from 4% to 5%, as in alloy AA5182. Can tabs used to be produced from an intermediate-strength variant, such as AA5042, but for several years, AA5182 has been used almost exclusively. Typical alloy compositions are listed in Table 5.

The conventional fabrication route for can body sheet consists of DC casting of large ingots, a two-step preannealing (up to 600°C), breakdown hot rolling (at about 500°C) to 20–40 mm transfer slab gauge, followed by tandem hot rolling to about 3-mm hot band with exit (coiling) temperatures of about 300–350°C (Section III). Modern multistand hot rolling lines that are optimized for maximum mass output are characterized by a combination of high speed, short interstand times, and large reduction steps, which prevents interstand recrystallization and, accordingly, gives rise to a fairly high-stored energy. Thus, for sufficiently high coiling temperatures, recrystallization will take place in the coiled hot strip, which is referred to as "self-annealing." The final cold rolling, usually without interannealing, yields a finish gauge sheet with a thickness below 0.3 mm in a high-strength state (condition "H19").

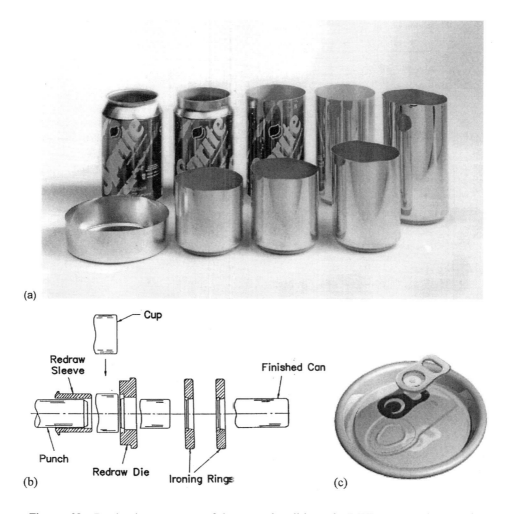

(a)

(b)

Cup

Redraw
Sleeve

Punch

Redraw Die

Ironing Rings

Finished Can

(c)

Figure 43 Production sequence of drawn and wall-ironed "DWI" cans and can ends.

The major requirements for AA3104 can body stock include sufficiently high strength and good formability. High strength is needed to avoid the buckling of the can base under high internal pressure (dome reversal) and to achieve sufficient structural stability (column strength), which is of particular importance for the very thin can wall thickness below 0.1 mm. The increase in yield strength is achieved by cold rolling. Figure 44 shows the evolution of yield strength $R_{p0,2}$ as a function of (laboratory) cold rolling level, starting from two differently recrystallized hot bands.

Evidently, the properties of the finish gauge sheet are largely controlled by the hot rolling parameters. The hot strip thickness dictates the reduction by cold rolling required to achieve the final sheet gauge and, therewith, the final strength (Fig. 44). Furthermore, the hot band determines the formability and anisotropy of the final

sheet, which is important because of the heavy forming operations accompanying can making (Fig. 43) [101–105]. Plastic anisotropy due to the crystallographical texture of the sheet gives rise to the formation of an uneven rim of the can during deep drawing and ironing operations (Fig. 35, Section IV.C.3). When cups are

Table 5 Chemical Composition of Alloys Used for Can Production (wt.%)

AA designation	Si	Fe	Mn	Mg	Cu	Cr
AA3004	0.3	0.7	1.0–1.5	0.8–1.3	0.25	
AA3104	0.6	0.8	0.8–1.4	0.8–1.3	0.05–0.25	
AA5042	0.2	0.35	0.2–0.4	3.0–4.0	0.15	0.1
AA5182	0.2	0.35	0.2–0.5	4.0–5.0	0.15	0.1

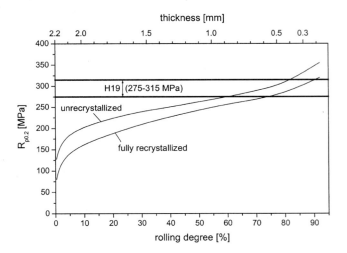

Figure 44 Effect of the level of cold rolling on yield strength $R_{p0.2}$. (From Ref. 106.)

deep-drawn from circular blanks cut from a textured sheet, the plastic flow properties vary with the angle α around the sheet. Therefore, the flow of metal will be uneven and gives rise to the formation of an undulating rim with a number of high points, known as "ears," and an equal number of low points, known as "troughs." Highly uneven cup rims are detrimental for the transport of the can bodies and affect the whole process when ears are stretched and clipped off during ironing, leading to machine downtime reducing efficiency. Thus, the amount of earing must be minimized in can body stock production to ensure the optimum performance of the can-making processes.

The characteristic cube texture of a recrystallized hot band generates four ears at 0° and 90° to the former rolling direction, whereas a cold-rolled sheet with the typical β fiber rolling texture will develop four ears at ±45°. Figure 45 illustrates how earing in can stock can be minimized by controlling the texture such that both texture components are mixed. As already mentioned earlier, multistand hot rolling produces a recrystallized hot strip with a strong cube texture, which, during deep drawing, generates four 0°/90° ears. The quite slow rotation of cube-oriented grains toward the stable rolling texture β fiber and the resultant mixture of both texture components efficiently balances these opposite effects.

Incomplete recrystallization at too low coiling temperatures leads to a hot band with a less pronounced cube texture but stronger rolling texture components (Fig. 45a and b). Consequently, there is a tendency to form 45° ears in the hot band (Fig. 45d), which is further

enhanced during subsequent cold rolling (Fig. 45e). Complete self-annealing is achieved only at coiling temperatures in excess of 300°C. Otherwise, additional hot band annealing or interannealing treatments must be applied. On the other hand, at too high rolling temperatures, recrystallization may occur between rolling passes, which also reduces cube texture intensity [106]. Thus, for maximum cube texture—as necessary for optimum earing properties at final gauge—an optimum combination of hot rolling process parameters is required, which is best achieved in multistand hot rolling lines [107].

By cold rolling, the final material properties are reached. As illustrated in Fig. 44, for a 2.2-mm hot strip with a fully recrystallized microstructure, the required final yield strength in excess of 275 MPa (H19) is reached only at cold rolling reductions of as much as 80–90% and a finished gauge thickness below 0.5 mm. For a partially recrystallized or unrecrystallized hot strip, this level is reached sooner, but as outlined earlier, because of the less pronounced cube texture in the hot strip, the acceptable level of 45° earing in the cold-rolled sheet is reached at too low rolling degrees (Fig. 45).

Another factor affecting finish gauge strength is the cold rolling temperature, which, under industrial rolling conditions, may rise up to as much as 150°C, significantly reducing cold rolling strength by recovery reactions. Likewise, any further heat treatment will give rise to recovery and/or recrystallization and, hence, will affect final material properties. For instance, in subsequent paint bake annealing operations, a certain decrease in strength is implied due to the recovery processes involved. Thus, such processes must be integrated in a complete description of final strength effective in the product application (e.g., Ref. 107).

In conclusion, the production of such an apparently simple object as the aluminum beverage can requires a very detailed knowledge of the deformation and recrystallization mechanisms of the corresponding materials as well as a sophisticated control of the main processing parameters.

E. Brazing Sheet

Up to the 1970s, radiators and other heat exchangers for vehicles were manufactured from copper and copper alloys. In order to utilize the lower weight of aluminum alloys, brazing techniques suitable for aluminum have since been developed. Today, aluminum alloys are most commonly used in heat exchanger applications, including radiators, heater cores, oil coolers, as well as evaporators and condensers in air-conditioning devices

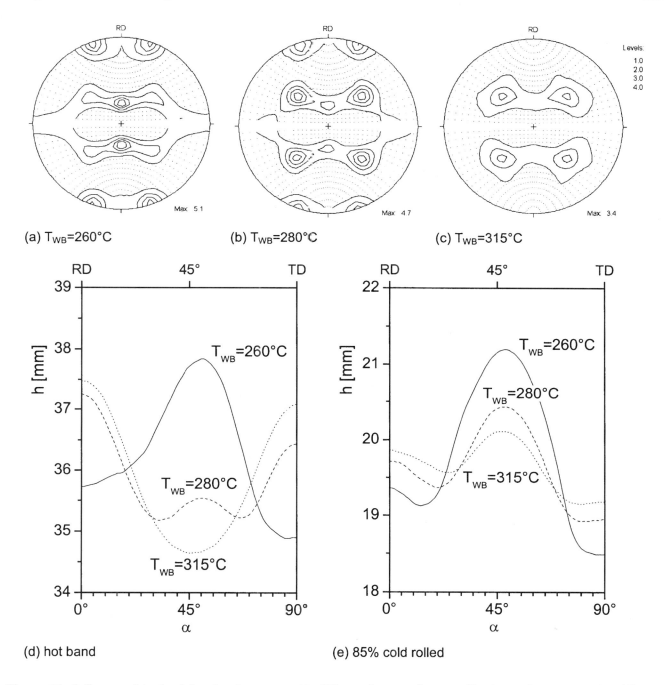

(a) T_{WB}=260°C   (b) T_{WB}=280°C   (c) T_{WB}=315°C

(d) hot band   (e) 85% cold rolled

Figure 45 Influence of (a, b, c) hot band textures with different degrees of recrystallization and, consequently, different fractions of the cube texture ({111} pole figures) on anisotropy in (d) hot strip and (e) final cold-rolled can body stock (symmetrized earing profiles). A decrease in hot band temperature results in incomplete recrystallization with less pronounced cube recrystallization texture. Accordingly, the 45° ears in the hot strip are increased at the expense of the desired 0°/90° ears, which, on cold rolling, results in too strong 45° earing profiles. (Courtesy of Dr. P. Wagner, Hydro Aluminium Deutschland GmbH.)

[108–111]. Figure 46a shows a typical modern car radiator with a brazed core made of different aluminum alloys in tubes, end plates, side plates and fins, and a plastic tank on each side. The plastic tank is mechanically fastened to the core by bending the outer rim of the end plate. New development is focused toward the development of fully recyclable all-aluminum heat exchangers (e.g., Fig. 46b). This design requires both good formability and high mechanical strength of the end plate material. The brazing of aluminum is most often performed using sandwich material in which the core alloy—usually a 3*xxx* Al–Mn alloy or, less frequently, a 6*xxx* Al–Mg–Si alloy—is clad with a thin layer of a braze with a lower melting point, typically a near-eutectic Al–Si alloy of the AA4*xxx* series containing 7.5–12.5% Si (Fig. 47a). The cladding alloy melts during brazing and accumulates at inward bends in order to minimize the surface tension (Fig. 47b).

Semifinished products of brazing alloys are produced through conventional hot and cold rolling processes. The core alloy is clad with a layer of 5–10% thickness of the core either on one or on two sides with the filler. The sandwich of as-cast ingot plus—mechanically joined or welded—clad is preheated, with the temperatures involved depending on the demands regarding final grain size. Heating of the ingots to about 500°C gives a coarse grain size because alloying elements such as Mn will essentially remain in solid solution. Thus, there is a large supersaturation of Mn, which, during the following steps of thermomechanical processing, will give rise to finely dispersed precipitates that interfere strongly with the progress of recrystallization. Homogenization at a temperature of about 600°C for 10 hr and hot rolling at 500°C will give rise to fairly coarse precipitates and, consequently, result in a material with a much smaller grain size. The heated sandwich is introduced into the rolling mill, where the connection between core and clad

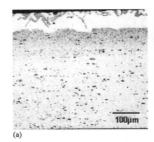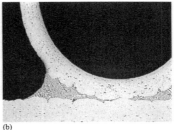

Figure 47 Section through a brazing material. (a) Core material, AA3005, one-side plated with a 10% layer of AA 4045 after a simulated brazing cycle. (Courtesy of Dr. R. Sicking, Hydro Aluminium Deutschland GmbH). (b) Cross-section of the joint between a tube and a fin in a brazed car radiator. (Courtesy of Dr. H. -E. Eckström, Sapa Technology.)

material is achieved by roll bonding during the first hot rolling passes. The material is hot-rolled to a thickness of 3–7 mm and then cold-rolled to final gauge. The material may be delivered soft-annealed (temper O), or back-annealed (temper H22, H24, or H26). Interannealing may be performed at a suitable sheet gauge for materials to be delivered in the cold-rolled tempers: H12 (= 1/4 hard), H14 (= 1/2 hard), or H16 (= 3/4 hard).

Besides the technological factors affecting brazeability, a necessary metallurgical condition for good brazeability is that the core material is recrystallized, preferably with a reasonably coarse—or rather strongly elongated—grain size. When the as-received material is not yet recrystallized, it must be ensured that the core recrystallizes during the brazing cycle, well before the braze reaches its melting temperature (550–590°C). A coarse-grained structure is required to prevent the silicon in the cladding from diffusing along the grain boundaries into the core material (Fig. 48) and to improve sagging resistance. Loss of silicon due to diffusion would give rise to an increase in the melting temperature of the filler material, together with a reduction in its ability to flow and to form joints.

All parts of the heat exchanger must have good corrosion resistance in the environment where it is used. A radiator in a car, for example, should sustain the salt and wet atmosphere on the roads in northern countries without any extra corrosion protection. Special alloys have been developed for that reason. In general, silicon penetration into the core is also detrimental to the corrosion resistance, which means that the grain size should be large.

Another important consideration is mechanical strength of the brazed parts. Saving vehicle weight by

Figure 46 (a) A typical modern car radiator with a brazed core made of different aluminium alloys in tubes, end plates, side plates and fins, and a plastic tank on each side. (Courtesy of AKG Autokühler, Germany.) (b) Prototype of an all-aluminum car heat exchanger. (Courtesy of Behr, Germany.)

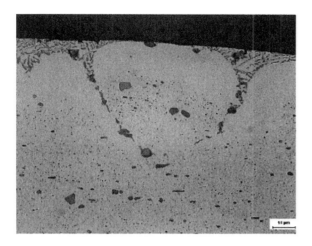

Figure 48 Penetration of silicon from the cladding into the core material. (Courtesy of Dr. H. -E. Eckström, Sapa Technology.)

downgauging is an important issue for car manufacturers, resulting in a demand for alloys with higher mechanical strength. However, most of the usual strengthening methods of common Al alloys cannot be used, either because the alloying elements would interfere with the brazing process, or because of the temperatures imposed during the brazing cycle. For instance, the solidus temperature of the core material must be at least 10°C higher than that of the filler (i.e., typically higher than 600°C). With regard to age-hardenable alloys, the cooling down from brazing temperature is too slow to allow for good age hardenability. As mentioned earlier, the grain size of brazing alloys must be large; hence, the only means to increase mechanical strength are solute and, particularly, dispersion hardening. Therefore, the common brazing alloys of the AA3xxx series alloys contain up to 0.5% Fe and 1–1.5% Mn (e.g., AA3103 and AA3005). The most modern method for brazing, controlled atmosphere brazing (CAB), uses a flux that is only capable of operating for alloys with maximum Mg concentrations of 0.3%. Alloys with higher Mg contents, including AA6xxx series alloys such as AA6063, have to be brazed under vacuum or with—very expensive—special Cs-containing fluxes. Additions of Cu and Zn are limited because they lower the melting point of the core material. However, with concentrations of a few tenths of a percent, both elements can be used to adjust the corrosion potential of the various components of a heat exchanger.

In order to enable one to produce the rather complex shaped parts of a heat exchanger, the material must have good formability. Although forming of parts is usually made before brazing, some ductility is needed when the brazed parts are joined to other parts of a radiator using mechanical joining. To obtain good formability, the grain size should be small, whereas high amounts of alloying elements added to increase mechanical strength tend to decrease the ductility. This means that a compromise must frequently be made among optimal brazeability, formability, and mechanical strength after brazing.

F. Aluminum Sheet for Car Body Panels

The demands for weight reduction in automotive construction led to increasing interest in sheets made from aluminum alloys for autobody applications so as to increase fuel efficiency and reduce vehicle emissions (e.g., Refs. 111–118). Two types of Al alloys are being used, the nonheat-treatable Al–Mg (or Al–Mg–Cu) alloys of the 5xxx series and the heat-treatable alloys of the 2xxx (Al–Cu) and 6xxx (Al–Mg–Si) series. The main requirement for automotive sheets is to have a good formability such that complex panels can be stamped at economical rates. The 5xxx series alloys stand out by very good formability, but have relatively poor strength, and they suffer from strain markings during forming (e.g., Ref. 100). Therefore, they are generally used for interior structural applications. The heat-treatable alloys, on the other hand, achieve their final strength only during the automotive paint bake cycle. Thus, they combine the good formability of the solution-treated state (T4 temper) with the increased service strength of the age-hardened state (T6 or T8 temper). Therefore, these alloys are superior for automotive skins where high dent resistance is required. Originally, the 2xxx series (e.g., alloys AA2008, AA2010, and AA2036) were most frequently used, for instance, as inner and outer hood panels for the Lincoln Towncar. However, these alloys suffer from poor corrosion resistance and merely offer a limited hardenability at the usual paint bake temperatures, so that their strengthening potential cannot fully be exploited.

For that reason, in the last years, the interest in 6xxx series alloys has increased markedly. The 6xxx series alloys contain magnesium and silicon, both with and without additions of copper. Compared with 5xxx and 2xxx alloys, sheets made from 6xxx series alloys stand out by good formability, good corrosion resistance, and sufficient strengthening potential during standard paint bake cycles. The 1991 Honda NSX all-aluminum sports

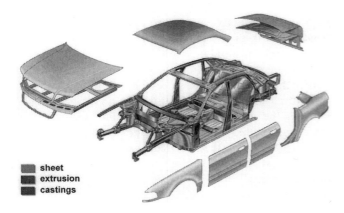

Figure 49 Parts made from Al alloys in the Audi A8 with space frame technology. (Courtesy of Audi, Germany.)

car was built from a combination of sheets made from 5*xxx* series alloys for the body skeleton and the inner parts of the doors, hood, and trunk, and 6*xxx* sheets for the outer panels. In 1994, the Audi A8 was introduced as the first series vehicle with an aluminum space frame (ASF) technology, giving it a weight reduction of 40% compared with a corresponding steel body (Fig. 49). The outer panels are made from the alloy AA6016 supplied by Alusuisse, Switzerland (now Alcan International). Since 2000, the new compact-size Audi A2 has been available, which relies on a more advanced second-generation space frame.

The potential 6*xxx* alloys that may be used for autobody sheets include AA6009, AA6010, AA6016, and AA6111; recently, alloy AA6181A was introduced for recycling aspects (Table 6). These alloys are formed in the soft, naturally aged T4 condition and then harden during paint curing. In the United States, AA6111 is often used for outer panels in gauges of 0.9–1.0 mm. Alloy 6111 combines high strength and, hence, high dent resistance with good formability. However, due to the high copper content, 6111 may show susceptibility to filiform corrosion, although a study of recovered hang-

on parts showed good performance after many years of service [119]. In Europe, 6016 is typically selected in gauges of around 1–1.2 mm. Alloy 6016 gives superior formability to 6111, has better filiform corrosion resistance, and allows flat hems even on parts with local predeformations of more than 10%. However, the bake-hardened strength of 6016 is significantly lower than that of 6111.

In commercial production of AA6*xxx* sheets, the material goes through a specific thermomechanical treatment before reaching the final gauge. Property control depends on most of these process steps individually as well as in rather complex interactions. In preparation for the hot rolling, the DC-cast ingots are scalped and then preheated to a temperature between 480°C and 580°C, in a cycle which may last up to 48 hr. During the preheating, the material is homogenized, short-range segregations are removed, and soluble phases are dissolved. The hot ingots are then transferred to the rolling line, which typically consists of a reversible breakdown mill where the ingots are rolled in a number of passes to a transfer gauge of 25–40 mm, followed by a high-speed multistand tandem mill where the thickness of the slab is reduced in three or four steps to a strip with a thickness between 3 and 6 mm at a defined exit temperature. The hot band is coiled and allowed to cool before it is cold-rolled to its final gauge of around 0.9–1.2 mm. The cold-rolled sheet is then unwound from the coil and passed single-stranded through a continuous annealing line. In this line, the material is rapidly heated to temperatures between 500°C and 570°C to dissolve the hardening phases and then quenched to retain the major alloying additions in solid solution; concurrently, the as-deformed material recrystallizes. Finally, the annealed sheets are precoated and/or prelubricated before being supplied for blanking and stamping.

The final in-service strength of the manufactured parts is only achieved after the forming operations through age hardening, preferably during the final automotive paint baking. This is at variance to the non-age hardening 5*xxx* series alloys that lose part of their

Table 6 Chemical Composition (wt.%) of the Most Important AA6*xxx* Alloys Used for Automotive Sheets

Alloy	Mg	Si	Cu	Fe	Mn	Zn	Ti
AA 6009	0.4–0.8	0.6–1.0	0.15–0.60	<0.5	0.2–0.8	<0.25	<0.1
AA 6010	0.6–1.0	0.8–1.2	0.15–0.60	<0.5	0.2–0.8	<0.25	<0.1
AA 6016	0.3–0.6	1.0–1.5	<0.2	<0.5	<0.2	–	–
AA 6111	0.5–1.0	0.6–1.1	0.5–0.9	<0.4	<0.4	–	<0.1
AA6181A	0.6–1.0	0.7–1.1	<0.25	0.15–0.5	<0.4	<0.3	<0.25

strength during paint baking due to recovery reactions. The complex temperature/time cycle during a typical paint bake cure may be simulated through annealing for 20–30 min at about 180°C. At this temperature, the standard 6xxx car body sheet alloys would reach peak strength only after several hours, so finished body panels would be employed in a heavily underaged condition (Fig. 50). Therefore, the bodies of the first generation Audi A8 were subjected to a separate bake-hardening annealing for 30 min at 205°C to achieve extra strength. Conversely, this means that the aging response of the car body sheet alloys had to be tailored so as to give optimum results during the paint bake cycle. High levels of magnesium, copper, and silicon increase both T4 strength and, in particular, the bake-hardening response of 6xxx series alloys [117]. Prolonged solutionizing times, increased temperature, and accelerated quenching rates all result in better age hardenability, yet at the cost of formability [115]. The bake-hardening response can be improved considerably using a preaging treatment, which may consist of an interrupted (two-step) quenching, extra preannealing, or retrogression [120,121]. A slight predeformation of a few percent prior to warm aging also accelerates the age-hardening response (Fig. 50). However, a full scientific understanding of the underlying processes is still lacking due to the complexity of the precipitation reactions in the Al–Mg–Si(–Cu) system, and this is an area of considerable current interest [116,122].

An important consideration for sheets that are to be used as outer panels for car body applications is surface appearance after sheet processing and forming operations. Effects such as orange peel, strain markings in sheets of 5xxx series alloys, and the appearance of so-called paint brushes lines in 6xxx series alloys may prevent the use of such parts for exterior automotive applications. Orange peel can be avoided by controlling final grain size at levels below around 50 μm. Strain markings in 5xxx alloys can be reduced by keeping the grain size above a certain limit, by quenching the material after the final recrystallization anneal (O-temper), by prestraining, or, in theory, by changing the temperature and strain rate during forming. The appearance of paint brushes lines in 6xxx alloys—commonly referred to as ridging or roping—has been linked to the presence of bands of similar orientation in the sheets [123,124]. Figure 51a shows an optical micrograph of AA6016 in condition T4 after anodical oxidation, displaying strong texture banding. The texture bands result in surface roughening, which, in severe cases, can be unacceptable for paint finish and may degrade forming limits [125]. In Fig. 51b, a roughness plot after 20% tensile deformation perpendicular to the rolling direction is displayed, which reveals the roping tendency. Bands of similar crystallographical orientation tend to deform collectively under the formation of elevated or depressed bandlike regions. Thus, in order to avoid roping, the texture sharpness of the sheet in T4 temper needs to be weakened using appropriate processing schedules. In conventional processing lines, cooling of the hot band is accompanied by heavy precipitation of fine dispersoids, mostly Mg_2Si precipitates (Fig. 52). These dispersoids may strongly interfere with the progress of recrystallization during the final solution annealing [126]. Accordingly, control of the thermomechanical hot rolling is an essential tool in optimizing the texture and, ultimately, the surface appearance of 6xxx sheet material.

Thus, age-hardening Al alloys of the 6xxx series offer a great potential for substituting steels in automotive body panels. With a detailed understanding of the microstructural development and the resulting properties during sheet production, the thermomechanical processing can be optimized to produce sheets of high and stable quality as required for the high demands in sheet metal forming operations.

G. Aluminum Foil Products for Packaging

The intrinsic properties of aluminum, in particular its impermeability and barrier properties, together with its high level of corrosion resistance and high formability, make Al foil an excellent material for packaging appli-

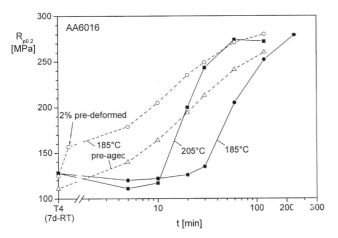

Figure 50 Warm age-hardening response of alloy AA6016 as a function of the aging time t for different aging temperatures and pretreatments. (Courtesy of Dr. E. Brünger, Hydro Aluminium Deutschland GmbH.)

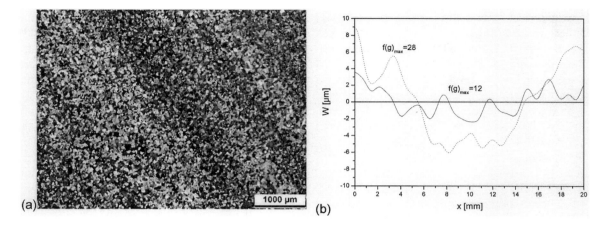

Figure 51 On the appearance of ridging in conventionally processed AA6016 sheet (T4 temper). (a) Optical micrograph showing strong texture banding; (b) waviness W (smoothed roughness profile) after 20% tensile deformation perpendicular to the rolling direction, for two textures with different sharpness $f(g)_{max}$ illustrating the influence of texture on ridging.

cations. Al foil offers excellent barrier properties to light, ultraviolet rays, water vapor, oils and fats, oxygen, and microorganisms. When used to package sensitive products such as pharmaceuticals or food, aluminium is hygienic, nontoxic, nontainting, and retains the product's flavor. A very thin layer of Al foil—in some cases only 6.35 μm—can render a packaging laminate completely light-proof and liquid-tight. Al foil has been

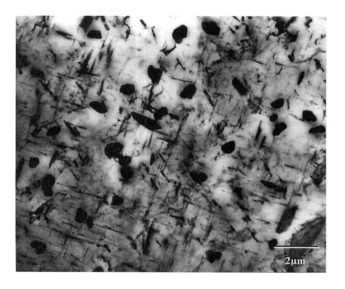

Figure 52 Transmission electron microscopy micrograph of the coiled hot band, showing several dispersoids of size 0.1–0.5 μm plus numerous finely dispersed platelike Mg_2Si precipitates.

in use in the packaging of foodstuffs for some 70–80 years since it was first introduced to replace tin foil in the linings of tea chests and to wrap chocolate bars. Recent statistics published by the European Aluminium Foil Association (EAFA) have shown that in 2000, the demand for aluminium foil continued to increase to 682,000 tons, representing a growth rate of more than 4%.

Foil products include a wide range of final gauges and applications. In general, foil is defined as rolled and annealed Al product with thickness below 150 or 200 μm. Foils for food containers are about 40–100 μm thick, lidding foils are between about 30 and 40 μm, with household wrapping and cooking foil between 10 and 20 μm. The thinnest foil, used to wrap chocolates and cigarettes or in packaging of liquids, may be only 6–7 μm thick. This so-called "converter foil" is used in a laminated form with paper, plastics, and/or lacquers, with the lamination process being termed as "conversion."

By far, the largest demand for Al foil comes from the packaging industry. Packaging applications may include a wide spectrum of products, such as "ready meals" and catering, dehydrated foods (soups, sauces, pasta, etc.), preserved foods and other sterilized products (including pet foods), confectionery products, liquid products (long-life drinks, single-use portions, etc.), dairy products (milk, yogurt, butter, cream, cheese), and bottle necking and decorative labels (see Fig. 53a). Besides food packaging, aluminium foil is used extensively in the packing of tobacco products, cosmetics, and pharmaceutical products, either unsupported or in multilayered combinations ("blister foil"; see Fig. 53b).

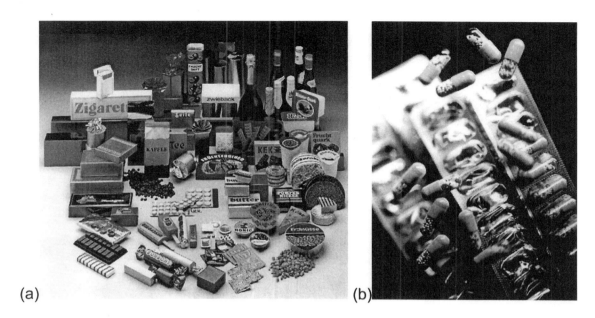

(a)
(b)

Figure 53 Use of Al foil for packaging of (a) food products and (b) pharmaceuticals products. (Photographs from Hydro Aluminium Deutschland GmbH.)

The remaining quantities of Al foil—on the order of 25%—goes into heat exchangers for cars (Section V.E), air conditioning, insulation, cable wrap, and other technical applications.

For most foil applications, commercial purity aluminum is used, which consists of more than 99% Al with additions of up to 0.8% Fe and about 0.25–0.3% Si. In the fully annealed condition (O-temper), such commercial purity Al of the AA1xxx series, including the well-known variants AA1050 and AA1200, have rather low strength. Higher strength—to allow for downgauging, for example—may be achieved through increased levels of Fe, for instance in the 8xxx series alloys such as AA8079. Al foil is produced in two different ways, either starting from a conventional DC cast ingot, which is hot-rolled to a strip of about 2–5 mm, or from a 6- to 7-mm-thick sheet produced by continuous casting routes (twin roll casting or belt casting) [127]. This strip is then cold-rolled to an intermediate (foilstock) gauge of about 0.4–1 mm, followed by annealing, typically in the temperature range 350–400°C. Then, the material is cold-rolled to final foil thickness. To achieve thinner foil gauges below, say, 100 μm, closed gap rolling must be used, which is performed in a dedicated foil mill. In order to increase productivity and to reduce the risk of damaging the extremely thin foils, in the last rolling step, typically below 40 μm, the foil is doubled so that two layers of foil are passed through the rollers together.

That is why these foils have a brightly polished surface, which is imparted by the rolling cylinders on one side, whereas the other side has a matte finish. Most Al foil products will be used in the O-temper, which means that the foil must finally be soft-annealed. In addition to the recrystallization, the final gauge annealing is required to remove rolling lubricants from the foil [128].

The ratio of ductility and strength at the final gauge largely depends on the solution/precipitation state of the alloying elements Fe and Si. For processing reasons, the solute level must be kept low. This means that there is a limitation to the amount of strengthening through solid solution hardening, but the dominant strengthening mechanism is dispersion hardening with some contribution from solutes [128–130]. Hence, the volume fraction and the size distribution of the dispersoids are an important factor in dispersion strengthening. Grain size has a relatively small influence on strength, but largely impacts ductility, with finer grain size giving higher ductility. Thus, to achieve a good balance of strength and formability in the final gauge product, recrystallization to a fine, uniform grain structure is required.

The recrystallization mechanism at final gauge is dependent on the level of cold reduction [55,127]. At low to moderate strains, recrystallization occurs in a discontinuous manner (i.e., by nucleation and growth; (Section IV.B.6). In such cases, the final grain structure will be governed by particle-stimulated nucleation. At the high rolling reductions of thin gauge foils, however, continuous recrystallization may occur during annealing (Section IV.B.5), promoting a fine grain structure [47,127].

The size distribution of dispersoids and the levels of elements in solid solution are also important in controlling the development of grain structure during final annealing. Dispersoids will tend to refine the grain structure by inhibiting boundary movement during recrystallization. Solute elements will inhibit recovery, which generates recrystallization nuclei, thus tending to increase grain size.

For many packaging applications, burst strength is a requirement (i.e., the product must be resistant to splitting and puncture when in use, for example, as a wrapping or as a container lid). Burst strength is dependent on both the tensile strength and ductility, as well as the gauge [131]. For wrapping applications, low springback is necessary to impart the so-called "dead-fold" characteristics, and this is achieved by using the alloy in the relatively soft, fully O-temper.

For converter foil, the mechanical requirements are related not to the end application in the laminated form, but rather to the lamination process itself. Again, tensile strength and ductility are important to minimize the number of strip breaks occurring through tearing during the conversion process and to allow the laminator to operate at higher tensions to avoid wrinkles in the laminate.

Obviously, the fortuitous combination of properties—especially its high formability and sufficient strength, together with excellent barrier properties—renders aluminum a very suitable material for packaging applications. With a variety of alloys and tempers plus a range of thicknesses varying from as little as 6.5 μm up to an (defined) upper limit of 0.2 mm, Al foil may be used in numerous ways to package goods such as food, dairy products, pharmaceuticals, or industrial products.

REFERENCES

1. Davis, J.R., Ed.; *Aluminum and Aluminum Alloys*; ASM Specialty Handbook; ASM International: Ohio, 1993.
2. Polmear, I.J. *Light Alloys: Metallurgy of the Light Metals;* 3rd Ed; Butterworth-Heinemann: Oxford, 1995.
3. Woodward, R. The rolling of aluminum, the process and the product. TALAT Lecture 1301, E.A.A., 1994.
4. Singh, R.V. *Aluminum-Rolling (Process, Principles and Applications)*; TMS: Warrendale, PA, 2000.
5. Perocheau, F. Hot deformation textures of aluminum alloys. PhD dissertation. Ecole des Mines: St.-Étienne, 1999.
6. Lloyd, D.J.; Kenny, D. The large strain deformation of some aluminum alloys. Metall. Trans. 1982, *13A*, 1445–1452.
7. Voce, E. The relationship between stress and strain for homogeneous deformation. J. Inst. Met. 1948, *74*, 537–562.
8. Rollett, A.D.; Kocks, U.F. A review of the stages of work hardening. Solid State Phenom. 1994, *35–36*, 1–18.
9. Hockett, J.E.; Sherby, O.D. Large strain deformation of polycrystalline metals at low homologous temperatures. J. Mech. Phys. Solids 1975, *23*, 87–98.
10. Lloyd, D.J.; Kenny, D. The structure and properties of some heavily cold worked aluminum alloys. Acta Metall. 1980, *28*, 639–649.
11. Gil Sevillano, J.; van Houtte, P.; Aernoudt, E. Large strain work hardening and textures. Prog. Mater. Sci. 1980, *25*, 69–412.
12. Robinson, J.M. Serrated flow in Al–base alloys. Int. Mater. Rev. 1994, *39*, 217–227.
13. Ashby, M.F. The deformation of plastically non-homogeneous materials. Philos. Mag. 1970, *21*, 399–424.
14. Kocks, U.F. Laws for work-hardening and low temperature creep. J. Eng. Mater. Technol. (ASME-H) 1976, *98*, 76–85.
15. Mecking, H.; Kocks, U.F. Kinetics of flow and strain hardening. Acta Metall. 1981, *29*, 1865–1875.
16. Laasraoui, A.; Jonas, J.J. Prediction of steel flow stresses at high temperatures and strain rates. Metall. Trans. 1991, *22A*, 1545–1558.
17. Mughrabi, H. A two-parameter description of heterogeneous dislocation distributions in deformed metal crystals. Mater. Sci. Eng. 1987, *85*, 15–31.
18. Nes, E. Modeling work hardening and stress saturation in fcc metals. Prog. Mater. Sci. 1998, *41*, 129–193.
19. Blum, W. Creep of aluminum and aluminum alloys. *Hot Deformation of Aluminum alloys*; Langdon, T.G., Ed.; The Minerals, Metals and Materials Society: Warrendale, PA, 1991; 181–209.
20. Sellars, C.M.; Tegart, W.J. McG. La relation entre la résistance et la structure dans la déformation à chaud. Mem. Sci. Rev. Metall. 1966, *63*, 731–746.
21. Shi, H.; Mclaren, A.J.; Sellars, C.M.; Shahani, R.; Bolingbroke, R. Constitutive equations for high temperature flow stress of aluminium alloys. Mater. Sci. Technol. 1997, *13*, 210–216.
22. Gil Sevillano, J. Flow stress and work hardening. In *Materials Science and Technology; A Comprehensive Treatment*; Cahn, R.W., Haasen, P., Kramer, E.J., Eds.; VCH: Weinheim, 1993; Vol. 6, 19–88.
23. Argon, A.S. Mechanical properties of single phase crystalline media. In *Physical Metallurgy*; Cahn, R.W.,

Haasen, P., Eds.; 4th Ed.; North-Holland, 1996; 1875–1955.

24. Humphreys, F.J.; Hatherly, M. *Recrystallization and Related Phenomena*; Pergamon: Oxford, 1995.

25. Cahn, R.W. Recovery and recrystallization. In *Physical Metallurgy*; Cahn, R.W., Haasen, P., Eds.; *Elsevier: Amsterdam, 1996; 2399–2500.*

26. Nes, E.; Sæter, J.A. Recovery, modelling and experiments. In Proceedings of the 16th Risø International Symposium on Materials Science; Hansen, N., et al., Ed.; Risø Nat. Lab: Roskilde, Denmark, 1995; 169–192.

27. Anderson, W.A.; Mehl, R.F. Recrystallization of aluminum in terms of rate of nucleation and rate of growth. Trans. AIME 1945, *161*, 140–172.

28. Haeßner, F. The study of recrystallization by calorimetric methods. In *Proceedings of Recrystallization '90*; Chandra, T., Ed.; TMS: Warrendale, PA, 1990; 511–516.

29. Avrami, M. Kinetics of phase changes. I. General theory. J. Chem. Phys. 1939, *7*, 1103–1112. Avrami, M. Kinetics of phase changes: II. Transformation–time relations for random distribution of nuclei. J. Chem. Phys. 1940, *8*, 212–224.

30. Johnson, W.A.; Mehl, R.F. Reaction kinetics in processes of nucleation and growth. Trans. AIME 1939, *135*, 416–458.

31. Kolmogorov, A.N. On the statistical theory of metal crystallization. Izv. Akad. Nauk. USSR, Ser. Mat. 1937, *1*(3), 355–359.

32. Mahin, K.W.; Hanson, K.; Morris, J.W. Jr Comparative analysis of the cellular and Johnson–Mehl microstructures through computer simulation. Acta Metall. 1980, *28*, 443–453.

33. Marthinsen, K.; Lohne, O.; Nes, E. The development of recrystallization microstructures studied experimentally and by computer simulation. Acta Metall. 1989, *37*, 135–145.

34. Rollett, A.D.; Srolovitz, D.J.; Doherty, R.D.; Anderson, M.P. Computer simulation of recrystallization in non-uniformly deformed metals. Acta Metall. 1989, *37*, 627–639.

35. Hesselbarth, H.W.; Kaps, L.; Haeßner, F. Two dimensional simulation of the recrystallization kinetics in the case of inhomogeneous stored energy. Mater. Sci. Forum 1993, *113–115*, 317–322.

36. Li, J.C.M. Possibility of subgrain rotation during recrystallization. J. Appl. Phys. 1962, *33*, 2958–2965.

37. Hu, H. Annealing of silicon–iron single crystals. In Recovery and Recrystallization of Meals; Himmel, L., Ed.; Interscience Publ.: New York, 1962; 311–378.

38. Gleiter, H. The migration of small angle boundaries. Philos. Mag. 1969, *20*, 821–830.

39. Sandström, R.; Lehtinen, B.; Hedman, E.; Groza, I.; Karlsson, S. Subgrain growth in Al and Al–1% Mn during annealing. J. Mater. Sci. 1978, *13*, 1229–1242.

40. Furu, T.; Ørsund, R.; Nes, E. Subgrain growth in heav-

41. Engler, O.; Heckelmann, I.; Rickert, T.; Hirsch, J.; Lücke, K. Effect of pretreatment and texture on recovery and recrystallization in Al–4.5Mg–0.7Mn alloy. Mater. Sci. Technol. 1994, *10*, 771–781.

42. Engler, O.; Sachot, E.; Ehrström, J.C.; Reeves, A.; Shahani, R. Recrystallisation and texture in hot deformed aluminium alloy 7010 thick plates. Mater. Sci. Technol. 1996, *12*, 717–729.

43. Hjelen, J.; Ørsund, R.; Nes, E. On the origin of recrystallization textures in aluminum. Acta Metall. Mater. 1991, *39*, 1377–1404.

44. Engler, O.; Hirsch, J.; Lücke, K. Texture development in Al–1.8% Cu depending on the precipitation state: Part II. Recrystallization textures. Acta Metall. Mater. 1995, *43*, 121–138.

45. Ahlborn, H.; Hornbogen, E.; Köster, U. Recrystallization mechanism and annealing texture in aluminium–copper alloys. J. Mater. Sci. 1969, *4*, 944–950.

46. Humphreys, F.J.; Chan, H.M. Discontinuous and continuous annealing phenomena in aluminum–nickel alloy. Mater. Sci. Technol. 1996, *12*, 143–148.

47. Davies, R.K.; Randle, V.; Marshall, G.J. Continuous recrystallization-related phenomena in a commercial Al–Fe–Si alloy. Acta Mater. 1998, *46*, 6021–6032.

48. Engler, O.; Yang, P. Progress of continuous recrystallization within individual rolling texture orientations in supersaturated Al–1.3% Mn. Proceedings of the 16th Risø International Symposium on Materials Science; Hansen, N., et al. Eds.; Risø Nat. Lab: Roskilde, Denmark, 1995; 335–342.

49. Oscarsson, A.; Ekström, H.-E.; Hutchinson W.B. Transition from discontinuous to continuous recrystallisation in strip-cast aluminium alloys. Mater. Sci. Forum 1993, *113–115*, 177–182.

50. Lücke, K.; Engler, O. Recrystallization textures in non-heat-treatable and heat-treatable Al-alloys. In Proceedings of the 3rd International Conference on Al-Alloys (ICAA3); Arnberg, L., et al., Eds.; The Norwegian Institute of Technology: Trondheim, Norway, 1992; Vol. III, 439–452.

51. Beck, P.A. Annealing of cold worked metals. Adv. Phys. 1954, *3*, 245–324.

52. Hutchinson, W.B. Nucleation of recrystallization. Scr. Metall. Mater. 1992, *27*, 1471–1475.

53. Doherty, R.D. Recrystallization and texture Prog. Mater. Sci. 1997, *42*, 39–58.

54. Bellier, S.P.; Doherty, R.D. Structure of deformed aluminum and its recrystallization: investigations with transmission Kossel diffraction. Acta Metall. 1977, *25*, 521–538.

55. Engler, O. On the origin of the *R*-orientation in the recrystallization textures of aluminum alloys. Metall. Mater. Trans. 1999, *30A*, 1517–1527.

56. Humphreys, F.J. Nucleation of recrystallization at second phase particles in deformed aluminium. Acta Metall. 1977, 25, 1323–1344.

57. Humphreys, F.J.; Kalu, P.N. Dislocation-particle interaction during high-temperature deformation of two-phase aluminium alloys. Acta Metall. 1987, 35, 2815–2829.

58. Engler, O.; Hirsch, J.; Lücke, K. Texture development in Al–1.8% Cu depending on the precipitation state: Part I. rolling textures. Acta Metall. 1989, 37, 2743–2753.

59. Liu, J.; Mato, M.; Doherty, R.D. Shear banding in rolled dispersion hardened Al–Mg₂Si alloys. Scr. Metall. 1989, 23, 1811–1816.

60. Lloyd, D.J.; Butryn, E.F.; Ryvola, M. Deformation morphology in cold rolled Al–Mg alloys. Microstruct. Sci. 1982, 10, 373–384.

61. Koken, E.; Embury, J.D.; Ramachandran, T.R.; Malis, T. Recrystallization at shear bands in Al–Mg. Scr. Metall. 1988, 22, 99–103.

62. Dillamore, I.L.; Katoh, H. The mechanisms of recrystallization in cubic metals with particular reference to their orientation-dependence. Met. Sci. 1974, 8, 73–83.

63. Gordon, P.; Vandermeer, R.A. Grain boundary migration. In Recrystallization, Grain Growth and Textures; Margolin, H., Ed.; ASM: Metals Park, OH, 1966; 205–266.

64. Lücke, K. Recrystallization. Mater. Sci. Eng 1976, 25, 153–158.

65. Gottstein, G.; Molodov, D.A.; Czubayko, U.; Shvindlerman, L.S. High-angle grain boundary migration in aluminium bicrystals. J. Phys. IV, Coll. C3, Suppl./J. Phys. III 1995, 5, 89–106.

66. Dimitrov, O.; Fromageau, R.; Dimitrov, C. Effects of trace impurities on recrystallization phenomena. In Haeßner, F., Ed.; Recrystallization of Metallic Materials; Riederer-Verlag: Stuttgart, 1978; 137–157.

67. Turnbull, D. Theory of grain boundary migration rates. Trans. AIME 1951, 191, 661–665.

68. Gleiter, H. Theory of grain boundary migration rate. Acta Metall. 1969, 17, 853–862.

69. Haeßner, F.; Hofmann, S. Über die Mechanismen bei der thermisch aktivierten Korngrenzenbewegung. Z. Met.kd. 1971, 62, 807–810.

70. Molodov, D.A.; Czubayko, U.; Gottstein, G.; Shvindlerman, L.S. On the effect of purity and orientation on grain boundary motion. Acta Mater. 1998, 46, 553–564.

71. Liebmann, B.; Lücke, K.; Masing, G. Untersuchungen über die Wachstumsgeschwindigkeit bei der primären Rekristallisation von Aluminium-Einkristallen. Z. Met.kd. 1956, 47, 57–63.

72. Cahn, J.W. The impurity drag effect in grain boundary motion. Acta Metall. 1962, 10, 789–798.

73. Lücke, K.; Stüwe, H.P. On the theory of impurity controlled grain boundary motion. Acta Metall. 1971, 19, 1087–1099.

74. Smith, C.S. Grains, phases and interfaces; an interpretation of microstructures. Trans. Met. Soc. AIME 1948, 175, 15–51.

75. Bowen, A.W. Texture development in high strength aluminum alloys. Mater. Sci. Technol. 1990, 6, 1058–1071.

76. Bunge, H.J. Texture Analysis in Materials Science; Mathematical Methods; Butterworth: London, 1982.

77. Kocks, U.F.; Tomé, C.N.; Wenk, H.-R. Texture and Anisotropy, Preferred Orientations in Polycrystals and Their Effect on Materials Properties; Cambridge University Press: Cambridge, 1998.

78. Randle, V.; Engler, O. Texture Analysis, Macrotexture, Microtexture and Orientation Mapping; Gordon and Breach: Amsterdam, 2000.

79. Maurice, C.; Driver, J.H. High temperature plane strain compression of cube-oriented aluminum crystals. Acta Metall. Mater. 1993, 41, 1644–1653.

80. Perocheau, F.; Driver, J.H. Slip system rheology of Al–1% Mn crystals deformed by hot plane strain compression. Int. J. Plast. 2002, 18, 185–202.

81. Driver, J.H.; Perocheau, F.; Maurice, C. Modeling hot deformation and textures of aluminum alloys. Mater. Sci. Forum 2000, 331–337, 43–56.

82. Hirsch, J.; Lücke, K. The application of quantitative texture analysis for investigating continuous and discontinuous recrystallization processes of Al–0.01Fe. Acta Metall. 1985, 33, 1927–1938.

83. Engler, O. Nucleation and growth during recrystallization of aluminum alloys investigated by local texture analysis. Mater. Sci. Technol. 1996, 12, 859–872.

84. Pospiech, J.; Lücke, K. Comparison between the ODFs of the recrystallization texture and the 40°⟨111⟩ transformation rolling texture of Cu–20% Zn. Z. Met.kd. 1979, 70, 567–572.

85. Vatne, H.E.; Furu, T.; Orsund, R.; Nes, E. Modeling recrystallization after hot deformation of aluminum. Acta Mater. 1996, 44, 4463–4473.

86. Engler, O. Simulation of the recrystallization textures of Al-alloys on the basis of nucleation and growth probability of the various textures components. Textures Microstruct. 1997, 28, 197–209.

87. Engler, O. A simulation of recrystallization textures of Al-alloys with consideration of the probabilities of nucleation and growth. Textures Microstruct. 1999, 32, 197–219.

88. Engler, O.; Vatne, H.E. Modeling the recrystallization textures of aluminum alloys after hot deformation. JOM 1998, 50 (6), 23–27.

89. Engler, O.; Kim, H.C.; Huh, M.Y. Formation of a {111} fibre texture in recrystallised aluminium sheet. Mater. Sci. Technol. 2001, 17, 75–86.

90. Hosford, W.F. The Mechanics of Crystals and Textured Polycrystals; Oxford Science Publ.: Oxford, 1993.

91. Rendigs, K.-H. Aluminium structures used in aerospace—status and prospects. Mater. Sci. Forum 1997, 242, 11–24.

92. Saintfort, P.; Sigli, C.; Raynaud, G.M.; Gomiero, Ph.

Structure and property control of aerospace alloys. Mater. Sci. Forum 1997, *242*, 25–32.

93. Liu, J.; Kulak, M. A new paradigm in the design of aluminum alloys for aerospace applications. Mater. Sci. Forum 2000, *331–337*, 127–140.

94. Santner, J.S. A study of fracture in high purity 7075 aluminum alloys. Metall. Trans. 1978, *9A*, 769–779.

95. Suzuki, H.; Kanno, M.; Saito, H. Differences in effects produced by Zr and Cr additions on recrystallization of hot-rolled Al–Zn–Mg–Cu alloys. J. Jpn. Inst. Light Met. 1986, *36*, 22–28.

96. Dorward, R.C.; Beerntsen, D.J. Grain structure and quench-rate effects on strength and toughness of AA7050 Al–Zn–Mg–Cu–Zr alloy plate. Metall. Mater. Trans. 1995, *26A*, 2481–2484.

97. Deshpande, N.U.; Gokhale, A.M.; Denzer, D.K.; Liu, J. Relationship between fracture toughness, fracture path, and microstructure of 7050 aluminum alloy: Part I. Quantitative characterization. Metall. Mater. Trans. 1998, *29A*, 1191–1201.

98. Morere, B.; Maurice, C.; Shahani, R.; Driver, J. The influence of Al₃Zr dispersoids on the recrystallization of hot-deformed AA 7010 alloys. Metall. Mater. Trans. 2001, *32A*, 625–632.

99. Ostermann, F. *Anwendungstechnologie Aluminium*; Springer-Verlag: Berlin, 1998.

100. Wieser, D.; Brünger, E. Maßgeschneiderte Aluminium-Blechwerkstoffe für den Automobilbau—vom Fahrwerk bis zur Karosserie. In Proceedings of the 10th Aachen Colloquium Automobile and Engine Technology; Pischinger, S., Wallentowitz, H., Eds.; fka Forschungsgesellschaft: Aachen, 2001; Vol. 1, 495–507.

101. Doherty, R.D.; Fricke, W.G., Jr.; Rollett, A.D. Investigations into the origin of cube texture recrystallization in aluminium alloys. In *Aluminium Technology 1987;* Sheppard, T., Ed.; The Institute of Metals: London, 1987; 289–302.

102. Hutchinson, W.B.; Oscarsson, A.; Karlsson, Å. Control of microstructure and earing behaviour in aluminium alloy AA3004 hot bands. Mater. Sci. Technol. 1989, *5*, 1118–1127.

103. Hirsch, J.; Hasenclever, J. Cube texture and earing in Al sheet. In Proceedings of the 3rd International Conference on Al-Alloys (ICAA3); Arnberg, L., et al., Eds.; The Norwegian Institute of Technology: Trondheim, Norway, 1992; Vol. II, 305–310.

104. Marshall, G.J. Simulation of commercial hot rolling by laboratory plane strain compression and its application to aluminium industry challenges. In *Hot Deformation of Al Alloys II;* Bieler, T.R., Lalli, L.A., MacEwen, S.R., Eds.; (1998). TMS: Warrendale, PA, 1998; 367–382.

105. Courbon, J. Mechanical metallurgy of aluminium alloys for the beverage can. Mater. Sci. Forum 2000, *331–337*, 17–30.

106. Hirsch, J.; Karhausen, K.; Kopp, R. Microstructure simulation during hot rolling of Al–Mg alloys. In Proceedings of the 4th International Conference on Al-Alloys (ICAA4); Sanders, T.H., Starke, E.A., Eds.; The Georgia Institute of Technology: Atlanta, GA, 1994; Vol. 1, 476–483.

107. Hirsch, J.; Karhausen, K.; Wagner, P. Practical application of modeling in the industrial sheet production. Mater. Sci. Forum 2000, *331–337*, 421–430.

108. Schultze, W.; Schoer, H. Fluxless brazing of aluminum. Welding J. 1973, *52*, 644–651.

109. Marshall, G.J.; Bolingbroke, R.K.; Gray, A. Microstructure control in an aluminum core alloy for brazing sheet applications. Metall. Trans. 1993, *24A*, 1935–1942.

110. Senaneuch, J.; Nylén, M.; Hutchinson, B. Metallurgy of brazed aluminium heat-exchangers. Aluminium *77*, 896–899, 1008–1011.

111. Miller, W.S.; Zhuang, L.; Bottema, J.; Wittebrood, A.J.; De Smet, P.; Haszler, A.; Vieregge, A. Recent development in aluminium alloys for the automotive industry. Mater. Sci. Eng., A Struct. Mater.: Prop. Microstruct. Process. 2000, *280*, 37–49.

112. Muraoka, Y.; Miyaoka, H. Development of an all-aluminum automotive body. J. Mater. Process. Technol. 1993, *38*, 655–674.

113. Cole, G.S.; Sherman, A.M. Lightweight materials for automotive applications. Mater. Charact. 1995, *35*, 3–9.

114. Burger, G.B.; Gupta, A.K.; Jeffrey, P.W.; Lloyd, D.J. Microstructural control of aluminum sheet used in automotive applications. Mater. Charact. 1995, *35*, 23–39.

115. Hirsch, J. Aluminium alloys for automotive applications. Mater. Sci. Forum 1997, *242*, 33–50.

116. Gupta, A.K.; Burger, G.B.; Jeffrey, P.W.; Lloyd, D.J. The development of microstructure in 6000 series aluminum sheet for automotive outer body panel applications. In Proceedings of the 4th International Conference on Al-Alloys (ICAA4); Sanders, T.H., Starke, E.A., Eds.; The Georgia Institute of Technology: Atlanta, GA, 1994; Vol. III, 177–186.

117. Bloeck, M.; Timm, J. Aluminium-Karosseriebleche der Legierungsfamilie AlMgSi. Aluminium 1994, *70*, 87–92.

118. Hirsch, J.; Dumont, C.; Engler, O. *Eigenschaften aushärtbarer AlMgSi-Legierungen für den Karosseriebau. VDI-Berichte 1151*; VDI Verlag: Düsseldorf. 1995; 469–476.

119. Brown, K.R.; Woods, R.A.; Springer, W.J.; Fujikura, C.; Nabae, M.; Bekki, Y.; Mace, R.; Ehrström, J.C.; Warner, T. *The Corrosion Performance of Aluminum Automotive Body Panels in Service. SAE Paper 980460*; Society of Automotive Engineers: Warrendale, PA, 1998.

120. Pashley, D.W.; Jacobs, M.H.; Vietz, J.T. The basic processes affecting two-step ageing in an Al–Mg–Si alloy. Philos. Mag. 1967, *16*, 51–76.

121. Bryant, J.D. The effect of preaging treatments on aging kinetics and mechanical properties in AA6111 aluminum autobody sheet. Metall. Mater. Trans. 1999, *30A*, 1999–2006.

122. Huppert, G.; Hornbogen, E. The effect of Mg-additions on precipitation behaviour of Al–Si alloys. In Proceedings of the 4th International Conference on Al-Alloys (ICAA4); Sanders, T.H., Starke, E.A., Eds.; The Georgia Institute of Technology: Atlanta, GA, 1994; Vol. I, 628–635.

123. Beaudoin, A.J.; Bryant, J.D.; Korzekwa, D.A. Analysis of ridging in aluminum auto body sheet metal. Metall. Mater. Trans. 1998, *29A*, 2323–2332.

124. Baczynski, G.J.; Guzzo, R.; Ball, M.D.; Lloyd, D.J. Development of roping in an aluminum automotive alloy AA6111. Acta Mater. 2000, *48*, 3361–3376.

125. Bate, P.S. Texture inhomogeneity and limit strains in aluminium sheet. Scr. Metall. Mater. 1992, *27*, 515–520.

126. Engler, O.; Hirsch, J. Recrystallization textures and plastic anisotropy in Al–Mg–Si sheet alloys. Mater. Sci. Forum 1996, *217–222*, 479–486.

127. Ekström, H.-E.; Oscarsson, A.; Charlier, P.; Ben Harrath, F. Influence of the microstructure on the mechanical properties of aluminium foil. In Proceedings of the 4th International Conference on Al-Alloys (ICAA4); Sanders, T.H., Starke, E.A., Eds.; The Georgia Institute of Technology: Atlanta, GA, 1994; Vol. I, 297–304.

128. Mahon, G.J.; Marshall, G.J. Microstructure–property relationship in *O*-temper foil alloys. JOM June 1996, *48*, 39–42.

129. Raynaud, G.M.; Grange, B.; Sigli, C. Structure and property control in non heat treatable alloys. In Proceedings of the 3rd International Conference on Al-Alloys (ICAA3); Arnberg, L., et al., Eds.; The Norwegian Institute of Technology: Trondheim, Norway, 1992; Vol. III, 169–213.

130. Hasenclever, J. Behaviour and effect of dissolved iron during the production of lithographic sheet and foil of Al–Fe–Si alloys. In Proceedings of the 3rd International Conference on Al-Alloys (ICAA3); Arnberg, L., et al., Eds.; The Norwegian Institute of Technology: Trondheim, Norway, 1992; Vol. II, 251–256.

131. Ekström, H.-E.; Charlier, P. Strip cast aluminium foil. In *Aluminum Alloys for Packaging II;* Morris, J.G., Dash, S.K., Goodrich, H.S., Eds.; (1996). TMS: Warrendale, PA, 1996; 245–251.

5

Design of Semisolid Metal-Forming Processes

Manabu Kiuchi
Kiuchi Laboratory, Tokyo, Japan

I. INTRODUCTION

Hereinafter, the word "metal" is used to represent a metal and/or its alloy. The metal becomes soft and its flow stress decreases as its temperature rises. This characteristic of the metal at high temperature range is widely utilized in various hot metal-forming processes. The question arises of what does happen, however, when the metal is heated to a temperature higher than the temperature of its solidus line. In this chapter, the morphology and mechanical properties of the metal in the temperature range from its solidus line to its liquidus line, where it consists of solid component and liquid component, respectively, will be explained. In addition, the utilization of those properties of the metal for various forming and shaping will be discussed.

The background of the explanation and discussion in this chapter is as follows. In recent years, various shaping and forming processes of metals in the temperature range from the solidus line to the liquidus line are becoming more and more popular. These processes are widely conducted for manufacturing mechanical parts and structural components of passenger cars as well as electric and electronic appliances. For instance, frames, pads, shoes, discs, wheels, and other mechanical parts for braking and steering systems, high-pressure accumulative cylinders and valves for fuel supply systems, frame joints for car body, piston heads, connecting rods, and other parts for combustion engines are manufactured of aluminum alloys by several types of two-phase (solid/liquid) die-casting processes. Casings of portable computers and mobile communicating devices, wheels and frames of sport utility bicycles, structures of handheld electric tooling machines, and components of furniture and housings are made of magnesium alloys by utilizing a type of injection molding technology.

However, customer's requirements on products are becoming more and more severe. Higher dimensional accuracy, better surface quality, uniformity of mechanical property, superior strength, thinner wall thickness, smaller weight, geometrical complexity, and functional flexibility are required. In addition, environmental consciousness as well as excellent productivity and drastic cost reduction are strongly requested.

To respond to such requirements, extensive technological improvements of those processes, machines, dies, and tools are needed. To do so, better understanding of fundamental characteristics of the metal consisting of solid and liquid component is necessary. To obtain the required technological knowledge and data, development of the comprehensive theory of flow and deformation of the two-phase (solid/liquid) metal is indispensable. Such theory makes it possible to simulate the flow and deformation of the two-phase metal in shaping and/or forming processes and control product quality. In this chapter, mathematical modeling for characterizing the two-phase metal will be explained.

II. MORPHOLOGY OF MUSHY/SEMISOLID METAL

Before starting the discussion, a brief explanation of the terminologies such as "mushy" and "semisolid" used in this chapter is necessary.

When the solid metal is heated to a temperature higher than its solidus temperature, it starts to melt and the liquid component appears in it. This state of the metal is called "mushy state" and the metal is called "mush metal."

When the molten metal is cooled down to a temperature lower than its liquidus temperature, some solidified grains appear in it. This state of the metal is described as "semisolid state" and the metal is called "semisolid metal" (see Fig. 1).

The mushy state and the semisolid state are similar to each other in the meaning that the metal includes both solid and liquid components in both states. Therefore, many researchers and engineers use only the word "semisolid" to express both states. However, the metal that is starting to melt has quite different mechanical properties from the metal that is starting to solidify. The difference is too big to express by a common word, as will be explained later.

Of course, the difference between the mushy metal and the semisolid metal becomes small when they are in the close or same-temperature range. As the temperature rises, the liquid component in the mushy metal increases; on the other hand, the solid component increases in the semisolid metal as the temperature drops. Consequently, when both metals include a similar proportion of the solid or liquid component, as naturally expected, they have fundamentally similar mechanical properties.

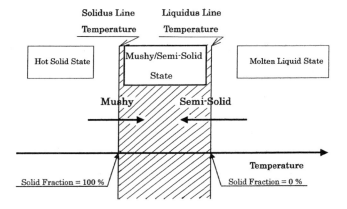

Figure 1 Temperature ranges of mushy state and semisolid state.

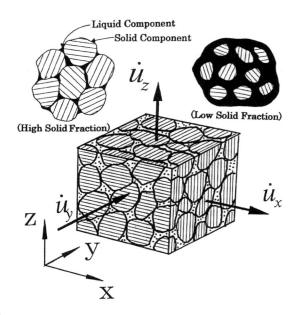

Figure 2 Solid and liquid components in mushy/semisolid metal.

Generally, the mushy metal includes a high rate of the solid component and deforms like a cray and a sherbet. The semisolid metal includes a high proportion of the liquid component and flows like a non-Newtonian fluid. Because of this difference, in this chapter the words mushy and semisolid are used appropriately following the situation concerned.

The liquid component in the mushy metal is generated through partial melting of the solid metal. The partial melting of the solid metal takes place on its grain boundaries and the liquid component is usually present there. The solid component, consisting of remaining solid grains, and the liquid component are separated from each other by their boundaries (see Fig. 2).

The solid component in the semisolid metal is generated through partial solidification of the molten metal. When the molten metal is cooled, the solidification starts on the wall of the crucible and the dendritic structure grows on it. If mechanical or electromagnetic stirring is applied to this partially solidified metal, the dendritic structure is broken into solid grains and dispersed in the remaining molten metal. This metal consisting of the remaining molten (liquid) metal and the dispersed solid grains is called semisolid metal.

Because of the coexistence of the solid and liquid components, the mushy metal has properties quite different from those of the solid metal as follows:

1.   Because of the liquid component existing on the grain boundaries, the connecting force between

neighboring solid grains is very weak and sometimes almost zero. In accordance with the low connecting force, the relative slip between the grains and the rotation of each individual grain occurs very easily in the mushy metal. This implies that the deformation and flow of the mushy metal take place under very low working forces.

2. The solid fraction, defined as the weight in percent of the solid component and denoted by φ, is a very important parameter that represents the state of the mushy metal. When the solid fraction φ is in the range roughly from 60% to 95%, the mushy metal deforms or flows like a cray or a sherbet. When φ is approximately less than 60%, the mushy metal flows like a slurry even under gravitative force. When φ is higher than 95%, the mushy metal deforms like a solid metal.

3. The mushy metal of which solid fraction φ is lower than 90% can be stirred and mixed with other materials such as ceramic particles, ceramic fibers, and graphite fibers (see Fig. 3). It follows that various kinds of mushy mixtures of metal and other materials can be made by utilizing such characteristic of the mushy metal.

4. The mushy metal can easily be separated into particles because of the weak intergrain connecting force. If the mushy metal is stirred and cooled simultaneously, the grains are separated by the stirring force and solidify without connecting with each other. Thus the mushy metal is made into particles and/or powder.

5. On the other hand, two kinds of mushy metals can be joined together by utilizing the peculiar characteristics of the mushy metal. When their surfaces are pressed together, the liquid com-

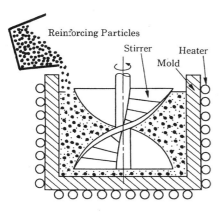

Figure 3 Mushy stirring and mixing.

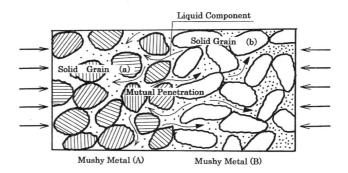

Figure 4 Joining of mushy metals based on mutual penetration and solidification of liquid components.

ponents of both of the mushy metals penetrate and diffuse through the interface and solidify together (see Fig. 4). Owing to the mutual penetration of the liquid component and the unified solidification, the desirable joining between the mushy metals is accomplished. The liquid components of both mushy metals act as effective bonding substances.

6. The semisolid metal is a viscous fluid. Its viscosity is higher than that of the molten metal because of the existence of the solid grains dispersed in it. The viscosity of the semisolid metal increases as the solid fraction φ increases.

7. The viscosity of the semisolid metal is influenced by the number and size of the dispersed solid grains, which are determined by the cooling rate and the shear rate associated with the stirring.

8. The size of the solid grain becomes smaller as the shear rate increases with the stirring. On the other hand, during the stirring, the grain grows rather fast. Therefore, temperature control is important to get the required solid grain size and the desirable viscosity of the semisolid metal.

9. It is very important for the semisolid metal to obtain low viscosity and high solid fraction. Low viscosity provides good fluidity, which is indispensable for casting. The high solid fraction is effective for preventing various defects and obtaining fine internal structure and excellent quality of cast products.

III. DEFORMATION CHARACTERISTICS OF MUSHY METAL

Generally, the macroscopic deformation of the solid metal consists of microscopic deformation and rotation of individual grains, and relative slip among neighboring grains. In the solid metal, the mechanical restraints among neighboring grains strictly work against each other and restrict the deformation, slip, and rotation of the grains. The grains in the solid metal, therefore, are not able to deform, rotate, and displace freely. If mechanical interactions and restraints are released according to the melting of the grain boundaries, the rotation and the relative slip of the remaining solid grains can take place quite freely, and consequently the macroscopic deformation of the mushy metal can occur under a small external working force (see Fig. 5).

It is for this reason that the flow stress of the mushy metal is very small. As the partial melting of the grain boundary proceeds, the mechanism of deformation of the mushy metal changes drastically from that of the solid metal. The restraints among the solid grains are lifted rapidly and the mechanical strength of the multi-grain structure disappears fast. When the volume of the liquid component increases adequately for the solid grains to rotate, slip, and displace independently, the mushy metal flows like a slurry.

The solid fraction φ gives dominant influences on the flow stress of the mushy metal. From a viewpoint of actual forming and shaping processes in the mushy state, it is strongly requested that the flow stress of the mushy metal be known in relation to its temperature. It is indispensable to know the change in the solid fraction φ corresponding to the change in the temperature. However, in many cases, the solid fraction φ of the mushy metal cannot be known directly from its temperature. Therefore, some appropriate method should be developed to know the direct relationship between the temperature and the solid fraction φ of the mushy metal.

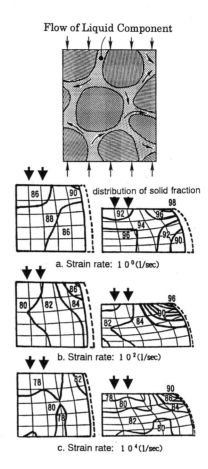

Figure 6 Flow of liquid component and distribution of solid fraction in mushy metal.

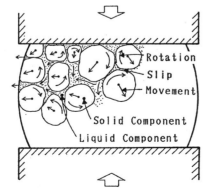

Figure 5 Rotation, slipping, and movement of solid grains in mushy metal.

When the mushy metal, which has a rather high solid fraction φ deforms, the liquid component usually flows through gaps or channels among the solid grains separately from the deformation, rotation, and movement of the solid grains. This is similar to the flow of a viscous liquid through a labyrinth. As the solid fraction φ decreases, the liquid component flows more and more smoothly through the intergranular gaps. In special cases, the liquid component flows out of the mushy metal through the gaps or channels that lead the liquid component from its inner portion to its free surface (see Fig. 6).

However, when the solid fraction φ is much higher than this level, the liquid component is apt to be trapped in the closed gaps among the solid grains. In this state, the restraint among the solid grains is not lifted so much; therefore, the flow stress of the mushy metal does not drop so remarkably. It should be noticed that even though the solid grains restrict their rotation and relative slip, the connecting force between them is weakened to a very low level because of the presence of the liquid component on the grain boundaries. This means that the mushy metal, even if its solid fraction φ is high, usually has very low ability of elongation.

IV. STRESS–STRAIN CURVE OF MUSHY METAL

In this section, some results of the uniaxial compression test of the mushy metal will be explained. Typical Al alloys and Cu alloys were subjected to the uniaxial compression test in the mushy state and their stress-strain curves were measured. From the results, the general features of dependence of the flow stress of the mushy metal upon the solid fraction φ are known.

The method for measuring the flow stress of the mushy metal is as follows (see Fig. 7). A cylindrical specimen with 16 mm diameter and 18–22 mm height is set in a subpress unit, which consists of a flat-top punch, a flat bottom die, and heavy wall container. The subpress unit is used not only to upset the specimen but also to protect it from cooling. The assembly of the specimen and the subpress unit is heated in an electric induction furnace up to the scheduled temperature, which corresponds to the required solid fraction φ, and is kept at the same temperature for a while to ensure the uniformity of the temperature distribution in the mushy specimen. Then the subpress unit that contains the specimen is

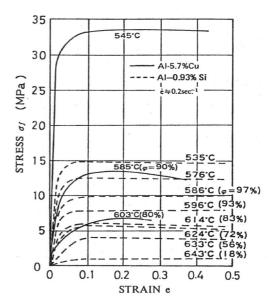

Figure 8 Measured stress–strain curves of mushy metals.

transferred to a cam plastometer and is subjected to the uniaxial compression test under a constant strain-rate.

The solid fraction φ of the specimen is, in some cases, calculated from its temperature by using its phase diagram. When the mushy specimen consists of two chemical compositions, its phase diagram shows directly the relationship between the temperature t and the solid fraction φ. But when the tested specimen includes many chemical compositions, its phase diagram fails to show the relationship between the solid fraction and the temperature. In such a case, a reasonable and effective method should be developed to obtain the solid fraction φ of the mushy specimen.

Some examples of the measured stress-strain curves of the mushy specimens, whose solid fraction ω are either known from their phase diagrams or estimated by a newly developed calibration curve (see Fig. 10), are shown in Fig. 8. It follows from the figure that the flow stress of the tested mushy metal drops remarkably as the solid fraction φ decreases from 100% to, for instance, 80%. Similar stress–strain relationships, especially the drop of the flow stress, are observed with respect to every mushy metal.

Figure 9 shows the relationship between the flow stress σ of the mushy metal and its temperature t more directly, concerning Al and Cu alloys. The abscissa gives the test temperature in both the solid state and the mushy state. The ordinate shows the flow stress of the specimen at the stage of 4% compressive strain. Here, it

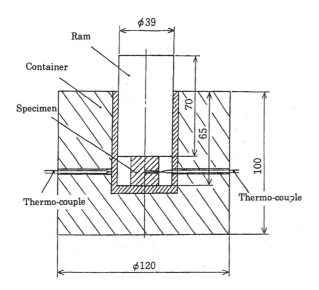

Figure 7 Installation of a subpress unit and a specimen for uniaxial compression test.

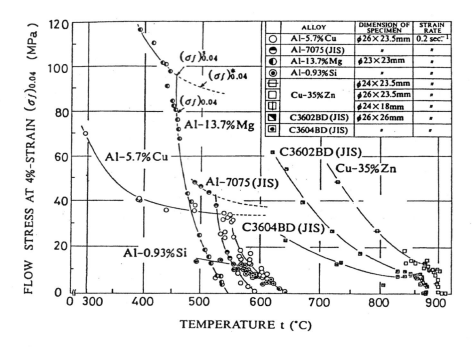

Figure 9 Relationship between flow stress and temperature of metals.

is again clearly observed that the flow stress of the mushy metal drops rapidly when its temperature exceeds its solidus temperature.

V. CORE RELATIONSHIP BETWEEN FLOW STRESS AND SOLID FRACTION OF MUSHY MEAL

To understand the effect of the liquid component of the mushy metal on its flow stress quantitatively, the measured flow stresses are rearranged from a point of view of the relationship between the normalized flow stress σ_n and the solid fraction φ. The result is shown in Fig. 10. The abscissa gives the solid fraction φ of the tested mushy metal and the ordinate shows the normalized flow stress σ_n. The normalized flow stress σ_n is defined as the ratio of the flow stress of the mushy metal at the tested temperature to its flow stress at the solidus temperature (which corresponds to the solid fraction $\varphi = 100\%$).

An important fact emerges from Fig. 10, namely, that the relationship between the normalized flow stress σ_n and the solid fraction φ can be expressed by one characteristic curve concerning all of the tested mushy metals. This means that the drop rate of the flow stress of the mushy metal is generally dominated by the solid fraction φ, in other words, the relative

volumetric ratio of the liquid component included in the mushy metal.

The chemical composition of the mushy metal gives only a slight influence on the rate of drop in its flow stress. This is quite understandable because the difference between the flow stress of solid component and the viscosity of liquid component—in the same mushy metal—is always much larger than the difference between the flow stresses of solid components present in different mushy metals. Furthermore, it is also larger than the difference between the viscosities of liquid components in the different mushy metals. Thus, the effect of the amount of the liquid component, which corresponds to the solid fraction φ, is dominant in the flow and deformation of the mushy metal.

The initial internal structure of the solid metal, on the other hand, gives some kind of influence on its flow stress in the mushy state. When the grain size is extraordinarily large or the grain shape is extraordinarily long, the drop in the flow stress of the mushy metal caused by the decrease in the solid fraction φ delays, compared with the drop in the flow stress of the mushy metal, including round and small solid grains. In such cases, the relaxation of the restraint between the long or large solid grains does not proceed as the relaxation among the round and small solid grains when the partial melting on the grain boundary occurs. In some cases, the rate of decrease in the flow stress of such mushy

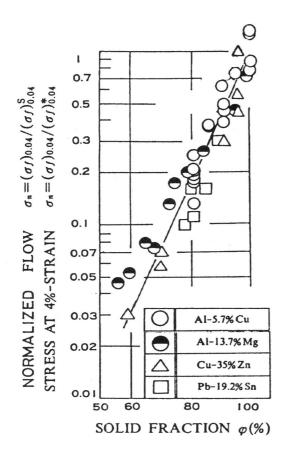

Figure 10 Relationship between normalized flow stress and solid fraction of mushy metal.

metal, and the obtained data are systematized and utilized for design and operation of various casting processes. The value of viscosity is indispensable for control of flow of the molten metal in casting processes. The primary objective of every casting process is to fill up the cavity of the mold with the molten metal, and to cope with the task, it is absolutely necessary to know the flow characteristics of the molten metal. The flow behavior of the molten metal is dominantly determined by its viscosity. This is the reason why the viscosity of the molten metal has been the primary target of the investigation into casting.

Generally, the viscosity of the molten metal is measured by the rotational coaxial double-cylinder method (see Fig. 11). In this method, coaxial double cylinders are installed and the gap between the outer cylinder and the inner cylinder is filled by an appropriate amount of the molten metal concerned. Then the inner cylinder is rotated at the required speed and the driving torque G is measured. The viscosity η is calculated from G by the following equation

$$\eta = \frac{(r_2^2 - r_1^2)G}{4\pi r_1^2 r_2^2 \omega L} \tag{1}$$

Here,

G = driving torque
r_1 = radius of inner cylinder
r_2 = radius of outer cylinder
ω = angle velocity of rotation
L = contact length between inner cylinder and molten metal

By this method the viscosity of the semisolid metal was investigated from various viewpoints. Effects of the

metal is far smaller than that of the mushy metal consisting of round and small solid grains.

The influence of the strain rate on the flow stress of the mushy metal is generally observed. It should be discussed as well when the forming and shaping of the mushy metal is designed and conducted. At present, however, the available data are not enough; therefore, the quantitative explanation about the strain rate effect on the flow stress is not done here. One important characteristic can be pointed out as follows. The strain rate effects on the deformation of the mushy metal are not stronger than the strain rate effects on the hot solid metal. A similar rate of effects is observed on both metals.

VI. VISCOSITY OF SEMISOLID METAL

The viscosity of the molten metal has hardly been investigated in the long history of casting. So many researchers tried to measure the viscosity of the molten

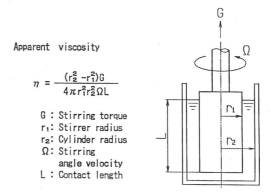

Figure 11 Illustration of coaxial rotating double-cylinder method.

process variables on the viscosity were widely measured and analyzed, such as the solid fraction φ, shear rate that is calculated from the rotational speed and the gap between two cylinders, and chemical composition and size of solid grains in the remaining molten metal. Through the investigations, the fundamental characteristics of the semisolid metal were systematically clarified.

Figure 12(a)–(c) shows some examples of the measured relationships between the (apparent) viscosity and the solid fraction φ of the semisolid metal. It follows from the results shown in the figures that the viscosity of the semisolid metal rises up very rapidly after its solid fraction φ exceeds a critical value. The increasing rate of the viscosity in this range is remarkable. The flow mechanism of the semisolid metal is considered to change fundamentally as well.

The critical value of the solid fraction φ becomes higher as the shear rate increases. However, the upper limit of the critical value may exist in the range from 45% to 50%.

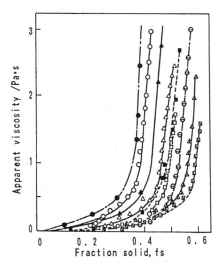

SHIBUTANI et. al.		
ALLOY	$\dot{r}$	C.R.
▱	364	0.63
△ Sn-5%Pb	256	0.86
⊖	158	0.84
▢	377	16.1
△ Sn-15%Pb	266	0.53
○	161	1.42

JOLY et. al.		
ALLOY	$\dot{r}$	C.R.
■	350	0.33
▲ Sn-15%Pb	230	"
●	110	"

$\dot{r}$: SHEAR RATE/s^{-1}
C.R. : COOLING RATE/K·min^{-1}

Figure 12 Relationship between apparent viscosity and solid fraction.

VII. FORMULA TO PREDICT VISCOSITY OF SEMISOLID METAL

The measured values of the viscosity of the semisolid metal were analyzed statistically and summarized in several mathematical formulas with aims to systematically and conveniently predict the viscosity. Each formula includes the process parameters such as chemical composition of semisolid metal, solid fraction φ, solid grain size, temperature, and shear rate.

The resulting formula proposed by M. Hirai is shown as follows.

$$\eta = \eta_L \left\{ 1 + \frac{a}{2(1/\varphi - 1/\varphi_{cr})} \right\} \quad (2)$$

Here,

$$a = \alpha \rho_m C^{1/3} \dot{\gamma}^{-4/3}$$
$$f_{scr} = 0.72 - \beta C^{-1/3}$$
$$\alpha = 2.03 \times 10^2 (X/100)^{1/3}$$
$$\beta = 19.0 (X/100)^{1/3}$$

where

C = solidification rate $d\varphi/dt$ (sec^{-1}), ($\varphi = 0$–0.4)
φ = solid fraction
φ_{cr} = critical value of solid fraction (upper limit to flow)
$\dot{\gamma}$ = shear rate (sec^{-1})
X = primary alloying component (mass%)
ρ_m = density of molten metal at liquidus temperature
η_L = apparent viscosity of liquid component

This formula can cover the wide range of the solid fraction, chemical composition, temperature, and solid grain size of the semisolid metal. It provides useful values of the viscosity for the engineering of the semisolid metal processing and forming (see Fig. 13).

VIII. YIELD CRITERION AND CONSTITUTIVE EQUATION OF MUSHY METAL

A. Characterization of Low Shear Strength

Several uniaxial compression tests concerning mushy Al alloys and Cu alloys clarify the fundamental features of flow and deformation of the mushy metal. The observed results indicate the procedure to obtain a reasonable formulation of mathematical model of the deformation of the mushy metal.

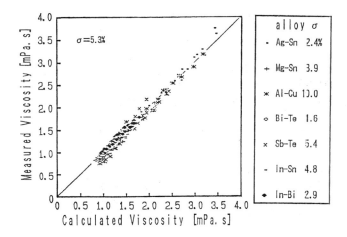

Figure 13 Summarized relationship between viscosity and solid fraction presented by M. Hirai.

The solid component of the mushy metal consists of the gathering of solid grains. This gathering of solid grains is expressed as "solid skeleton." Even though the billet is heated up to the mushy state, as long as the solid grains restrict each other and the solid skeleton is stiff enough, its flow stress does not drop so rapidly. When the billet's temperature rises much higher and its solid fraction decreases to still lower level, then the solid skeleton is apt to collapse and the flow stress of the billet drops notably [see Fig. 14(a)–(c)].

Through the investigation into the dropping feature of the flow stress, it was found that the solid skeleton, that is, the gathering of solid grains, of the mushy metal has especially low shear strength. This is considered to be the most important characteristic of the flow stress of the mushy metal. It should be taken into the mathematical expression of the yield criterion of the mushy metal.

Hill's theory of plasticity indicates that the shear strength of the solid metal is half of its normal strength. This theory can be used and is effective, for instance, for the analysis of deformation of the solid skeleton of sintered porous metal, because it has the metallurgically unified solid skeleton, not the gathering of solid grains (see Fig. 15).

However, as for the solid skeleton of the mushy metal, this relationship between the normal strength and the shear strength is not acceptable. This is because the solid skeleton of the mushy metal is composed of the independent solid grains. They contact with each other, but are not metallurgically connected (see Fig. 16).

Figure 14(a)–(c) shows some examples of the measured relationships between the flow stress at 4% uniaxial compressive strain and its temperature. From these measurements, the following results are obtained.

1. As mentioned repeatedly, as the billet's temperature exceeds the solidus temperature, its flow stress drops remarkably.
2. However, the feature of dropping of the flow stress of the mushy billet is not always consistent. The flow stress of one mushy billet drops very rapidly, but the flow stress of another mushy billet drops slowly.

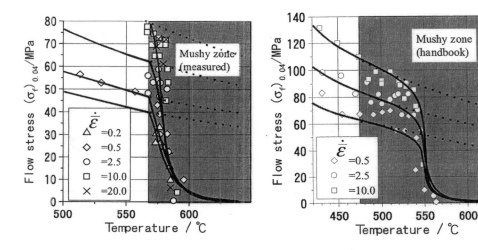

Figure 14 Drop of flow stress of mushy metal.

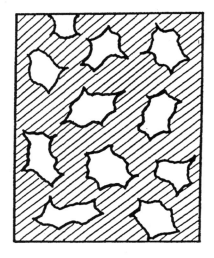

Figure 15 Unified solid skeleton of sintered porous metal.

Thus, the shear strength of the solid skeleton of the mushy metal is lower than that of the solid skeleton of the sintered metal. This is the reason why the yield criteria of the sintered porous metal and the constitutive equations derived from them cannot give satisfactory analytical results concerning the flow and deformation of the mushy metal. The mushy metal needs a different criterion that can deal with the peculiar drop of the shear strength of its solid skeleton, i.e., the very low shear strength of its solid component.

B. Yield Criterion

It should be noted that the solid grains in the mushy metal, no matter whether the solid fraction is high or low, are not metallurgically joined to others, because the grain boundaries melt in the mushy state. Therefore, the

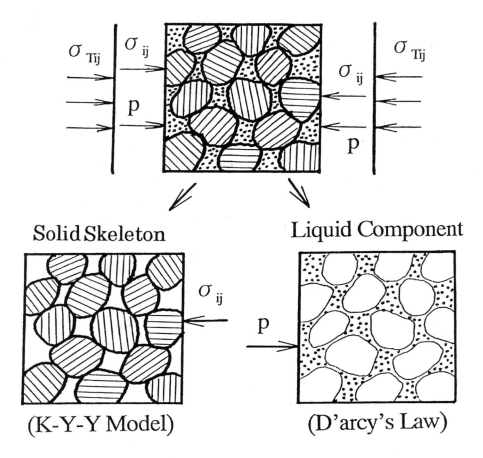

Figure 16 Solid skeleton, gathering of solid grains, of mushy metal.

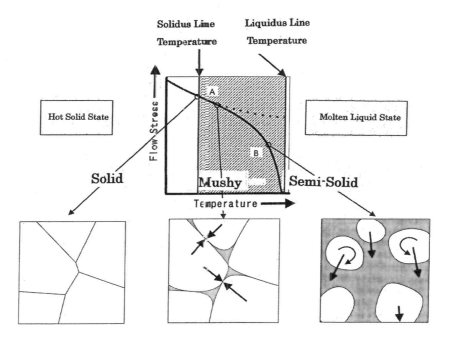

Figure 17 Effect of mutual restraint among solid grains on flow stress.

solid grains can slip on each other under a very small working force. The slipping occurs more and more easily as the solid fraction φ decreases (see Fig. 17).

As the result of the careful investigation into the remarkable drop of the shear strength of the solid component of the mushy metal and the consistent consideration on the conformity with Hill's mathematical theory of plasticity, Kiuchi et al. proposed the following formula of the yield criterion of the mushy metal (K-Y-Y model of yield criterion)

$$\left(\frac{1}{f_0}\right)\left[f_2\left(3J_2 - f_3\sqrt[3]{J_3^2}\right) + \left(\frac{\sigma_m}{f_1}\right)^2\right]^{1/2} = Y_0$$

$$f_1 = a(1 - f_s)^{-m}, \quad f_0 = f_s^n \tag{3}$$

$$0 < f_3 < \left(9\sqrt[3]{2}/2\right), \quad f_2 = \left[1 - f_3\left(2/9\sqrt[3]{2}\right)\right]^{-1}$$

f_s = solid fraction of mushy metal
f_0 = (apparent average stress working to mushy metal)/(effective stress working to solid skeleton)
Y_0 = normal yield stress of mushy metal
σ_m = mean normal stress (= hydrostatic pressure)
J_2, J_3 = invariants of deviatoric stress tensor
a, m, n = material's constants
f_1, f_2, f_3 = effect coefficients

Figure 18 shows the profile of the yield curve based upon the K-Y-Y Model plotted on a π plane. As shown in the figure, the yielding takes place under very low stress when the mushy metal is subjected to a pure shear deformation. Here, the value of $\gamma_{min}/\gamma_{max}$ shown in the figure depends on the value of an effect coefficient f_3. On the other hand, when the mushy metal is subjected to a

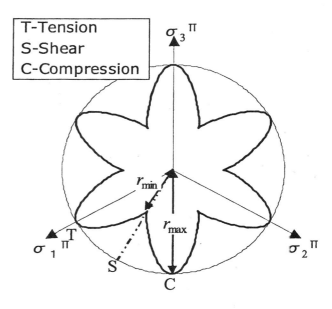

Figure 18 Yield curve based on the K-Y-Y model.

normal (uniaxial) deformation, the yield stress is relatively high compared with the shear yield stress.

Here it should be noticed that the K-Y-Y model is formulated for the solid fraction range higher than 60%.

C. Constitutive Equations

From the K-Y-Y model, the constitutive equations can be derived as follows.

$$\dot{\varepsilon}_{ij} = \dot{\lambda}\left\{\sigma'_{ij} + \frac{2\sigma_m\delta_{ij}}{9f_1^2f_2} - \frac{2}{9}\frac{f_3}{\sqrt[3]{J_3}}\left(\sigma'_{ik}\sigma'_{kj} - \frac{2}{3}J_2\delta_{ij}\right)\right\} \quad (4)$$

Here

$$\dot{\lambda} = \left[3f_2f_s/2(f_0)^2\right](\dot{\bar{\varepsilon}}\bar{\sigma}), \quad \sigma_m = (1/3)\sum\sigma_{ii}$$

Figure 19 shows the calculated load-reduction relationship in a combined forward and backward squeezing process. Two punches are squeezed simultaneously into a cylindrical solid billet.

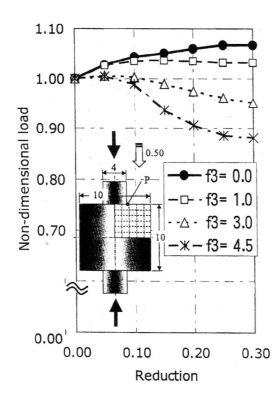

Figure 19 Analysis of forward and backward squeezing based on the K-Y-Y model.

In this case, because of the circumstance, the billet is not in the mushy state but in the solid state; however, it is assumed that the billet has a especially low shear yield stress compared with ordinary Mises materials and the yield criterion and the constitutive equations based on the K-Y-Y model can be used for the analysis.

From the calculated load-reduction curves, the following are noted.

1. When the effect coefficient f_3 is larger than 2.0, the squeezing force becomes lower than the initial value, that is, a kind of strain softening is observed.
2. When $f_3 = 4.5$, the drop of 20% of the squeezing force is observed at the stage of 30% reduction.
3. Even for the solid metal, the low shear yield stress reduces the force necessary to deform the billet.
4. If the yield criterion and the constitutive equations derived from K-Y-Y model are used to analyze the deformation of the solid skeleton of the mushy metal, it can be expected to simulate the remarkable drop of the force necessary to deform the mushy billet.

Figure 20 shows the results of the uniaxial compression test of the mushy billets (Fig. 10). The figure also shows the results of the simulations based on the S-O model and the K-Y-Y model. The strain rate is $0.5\,\mathrm{sec^{-1}}$. The abscissa shows the initial solid fraction φ of the billet before starting the compression. The ordinate shows the normalized flow stress σ_n of the billet at 4% uniaxial strain.

The figure indicates some important features of the analysis as follows.

1. When the grain size of the mushy billet is in the same range, the rate of decrease in the normalized flow stress σn is almost the same as with other billets.
2. When the grain size of the mushy billet is extraordinarily large, the rate of decrease in the normalized flow stress σn is small. In Fig. 20, Al–0.93%Si billet corresponds to such a case.
3. The FEM simulation based on the S-O model, which was originally proposed for sintered porous metals consisting of metallurgically unified solid skeleton, cannot predict the drastic drop of the normalized flow stress σn of the mushy metal caused by the drop of its solid fraction φ. But the result of the simulation shows good agreement with the measurements con-

Table 1: Value f_3 used for calculation.

Solid Fraction "f_S"	90%	80%	70%	60%
Parameter "f_3"	3.0	5.0	5.2	5.2

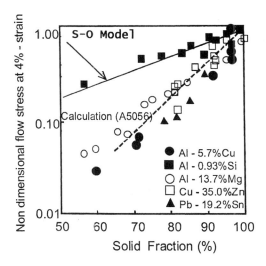

Figure 20 Relationship between normalized flow stress and solid fraction, experimental results and analytical results.

cerning Al–0.93%Si billet. In this case, Al–0.93%Si billet had an extraordinarily large grain size, and thus. the shear yield stress of its solid skeleton was higher than the shear yield stresses of others.

4. The broken line shows an expected normalized flow stress σn–solid fraction φ relationship, which may be obtained by the K-Y-Y model. For the numerical analysis, however, the value of effect coefficient f_3 should be adjusted corresponding to the value of the solid fraction φ as follows.

Solid fraction f_s	90%	80%	70%	60%
Coefficient f_3	3.0	5.0	5.2	5.2

IX. FLOW STRESS IN WHOLE MUSHY/ SEMISOLID RANGE

In the whole mushy/semisolid temperature range, which corresponds to the range from the solidus temperature to the liquidus temperature, the flow stress of the metal changes much more drastically than the change just in the mushy state or in the semisolid state. Figure 21

shows the overall change in the flow stress of the mushy and/or semisolid metal relating to the change in the solid fraction φ from 100% to 0%.

When the solid fraction φ is roughly lower than 40%, the semisolid metal flows like a non-Newtonian fluid. In this solid fraction range, owing to the fluidity of the semisolid metal, die casting and injection molding are widely conducted for manufacturing various parts and components of cars and electronic devices.

In the range where the solid fraction φ is higher than 60%, the mushy metal has excellent deformability and it deforms like clay. The deformability and the low flow stress of the mushy metal are utilized to conduct die forging, die molding, rolling, and extrusion.

When the solid fraction changes from 40% to 60%, the flow stress changes drastically. The flow of the semisolid metal changes to the deformation of the mushy metal; in other words, the fundamental mechanical characteristic changes from the fluidity to the deformability.

The flow stress of the semisolid metal in the range of low solid fraction shown in Fig. 21 are the flow stress converted from the viscosity of the semisolid metal measured by the rotating double-cylinders method (see Fig. 22). In the method, as mentioned before, the inner cylinder is rotated and the necessary driving torque G and the rotating angle velocity ω are measured. From the measured results, the viscosity is defined (Fig. 11).

If the distribution of the shear stress in the semisolid metal at the gap between two cylinders is assumed to be

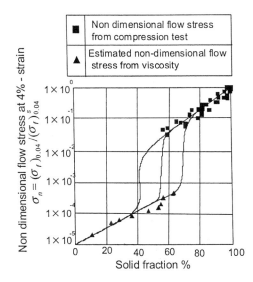

Figure 21 Flow stress in whole temperature range of mushy/semisolid state.

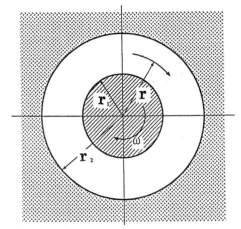

Co-axial Double Cylinder System

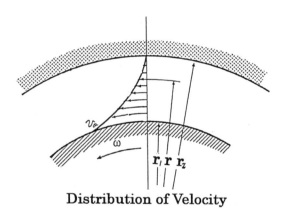

Distribution of Velocity

Figure 22 Transformation of viscosity to shear flow stress.

as shown in Fig. 22, the shear flow stress τ of the semisolid metal is derived from the measured torque G by using the next equation.

$$\tau = \frac{G}{2\pi\gamma_1^2 L} \qquad (5)$$

Here,

γ_1 = radius of inner cylinder
L = contact length between inner cylinder and semisolid metal

From the calculated shear flow stress τ, the normal flow stress σ is derived on the assumption of the ratio of τ to σ. The exact value of the ratio is not yet clear, but at present, the following equation is tentatively introduced.

$$\sigma = \alpha\tau, \qquad \alpha \div \sqrt{3} \qquad (6)$$

Then, the relationship between the normalized flow stress σ_n and the solid fraction φ of the semisolid metal in the low solid fraction range can be expressed as follows.

$$\sigma_n = A\varphi^n \qquad (n = -6.0-12.0, \qquad \varphi \leq 40\%) \qquad (7)$$

This is one of the fundamental mechanical characteristics of the semisolid metal. The value of n depends on the shear rate; however, its effect is not so serious, but rather slight.

The relationship between the normalized flow stress σ_n and the solid fraction φ of the mushy metal in the range of $\varphi > 70\%$ can be expressed as follows.

$$\sigma_n = B\varphi^m \qquad (m = 4.0-6.0, \qquad \varphi \geq 70\%) \qquad (8)$$

The value of m depends upon the grain size, grain shape, and the strain rate as well. In the case shown in Fig. 21, the value of m is roughly 6.0.

The relationship between the normalized flow stress σ_n and the solid fraction φ in the intermediate range of φ is not clarified yet. More detailed investigations into the shift of the deformation mechanism and more extended measurements of the flow stress and the viscosity are required.

X. MANUFACTURING OF SEMISOLID METAL

The semisolid metal is made by simultaneous cooling and stirring of the molten metal. This approach is well known as the "rheocasting" process (Fig. 23). In this process, the molten metal in a crucible is cooled while stirring conducted by a mechanical, electromagnetic, or gravitative method. Due to the cooling, the molten metal starts to solidify on the wall surface of the crucible and the dendritic structure grows there. However, the dendritic structure is soon broken into particles by the stirring and the particles are dispersed back into the molten metal. Through the cooling and stirring, the growing and the breaking of the dendritic structure are repeated and as a result, a mixture of solidified particles and molten metal, i.e., the semisolid metal, is made. The solid fraction of the molten metal depends on the temperature, the stirring speed, and the thermal equilibrium of the semisolid metal undergoing the stirring operation. Thus the solid fraction and the particle size can be changed by controlling the stirring speed and the cooling rate.

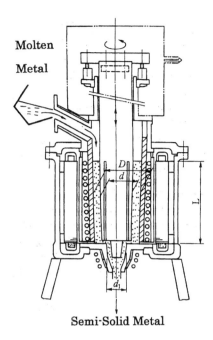

Molten

Metal

Semi-Solid Metal

Figure 23 Illustration of stirring and cooling to make semisolid metal.

At the end of the 1960s, rheocasting was proposed with the aim of innovating the quality of cast products (see Fig. 12). In this process, a molten metal is made into a semisolid metal. The semisolid metal is cast into a mold and solidified to form the product. The semisolid metal is effective to obtain the fine grains in the cast product because the solid particles in the semisolid metal become kernels of the fine multigrain structure. In addition, the semisolid metal is good for improving the life of the mold because the temperature of the semisolid metal is much lower than that of the molten metal for the conventional casting.

After rheocasting was proposed, many investigations were performed to obtain the necessary data relating to the viscosity of the semisolid metal. The semisolid metal has much higher viscosity than the molten metal, and this means that the control of its flow and the complete filling up of the cavity of the mold are more difficult than those in the conventional casting process. To overcome such difficulties, accurate data concerning the viscosity of the semisolid metal were strongly requested. This is the reason why so many investigations were done in this field.

In the original rheocasting process, the semisolid metal obtained is expected to be directly subjected to casting and molding and made into either semifinished or near-net-shape products. Conventional die forging and other forming processes may be used to finish the

rheocast semifinished products to final products. Important advantages of rheocasting are as follows.

1. The internal structure of the product is fine. The grain size is very much smaller than that obtained by any conventional casting process.
2. So-called segregation in the semisolid metal and the cast product is effectively prevented by stirring and high level of homogeneity of the product's internal structure is obtained. Because of this, various internal defects of the products, such as porosity, which occurs very often in conventional cast products, are eliminated, and the finished products display good mechanical and metallurgical properties and surface quality.
3. The temperature of the semisolid metal is much lower than that of the molten metal and, therefore, the heat shock imposed on the dies and other tools is lessened in the casting process. This is highly desirable because the life-span of the dies and other tools is considerably improved.
4. Utilizing the semisolid metal, the productivity of casting and forming of metal products can be improved. Moreover, the energy saving and the material can be attained.

Studies on rheocasting have been conducted worldwide, but they are concerned with the basic aspects of the effects of the stirring and the cooling on the semisolid metal. The effects of the stirring speed, shear rate, and cooling rate on the grain size, internal structure, and strength of the cast product have been mainly investigated. Studies of the application of the original rheocasting process are less frequent.

There exist a number of technological problems. For instance, how can a homogeneous metal slurry be made continuously? How can it be held or transported? How can the slurry be cast or shaped to semi- or near-net-shaped products of homogeneous structure? What tools and materials are available for semisolid processing at high temperatures?

Notwithstanding these unclarified technological difficulties as yet, the rheocasting process is expected to provide new economical manufacturing of metallic products and to play a significant role in the innovative forming technology.

XI. SEMISOLID DIE CASTING AND SEMISOLID INJECTION MOLDING

In actual cases, it is not so easy to transfer the semisolid metal, which is made by the stirring and cooling method

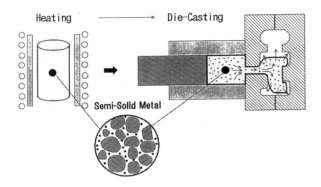

Figure 24 Illustration of semisolid die-casting.

and has a rather low solid friction, directly to a forming or shaping machine. Its solid fraction should be kept constant and uniform during the transference. Therefore, the semisolid metal is usually solidified once into solid bars by the continuous casting method. After the complete cooling, they are cut into solid billets with the required geometry. Then the billets are heated again to the necessary semisolid state and subjected to the required forming and shaping, through die casting and injection molding.

In these cases, the solid billets are heated and partial melting is induced. In this sense, the billets can be called "mushy billet," and the processes "mushy die casting" and "mushy injection molding." However, the solid fraction of the billet is very low compared to the semisolid metal treated in the rheocasting process, therefore the expression "semisolid" is usually used instead of "mushy."

Semisolid die-casting is now used widely and many kinds of mechanical and structural parts of automobiles, bicycles, electronic appliances, mobile telecommunication devices, and handheld electric tools are manufactured from Al alloys and Mg alloys. The working conditions necessary to obtain high quality of products are investigated and the characterization of the process is promoted (see Fig. 24).

In some cases, the numerical simulation of the process is carried out where the flow of the semisolid metal in the die cavity is clarified. Using the results of the detailed flow analysis, the die design and the process design are conducted. Thus the utilization of semisolid die casting is becoming more and more popular and the variety of products is extending.

Semisolid injection molding, or "thixoforming," is mainly used to manufacture products of Mg alloys. In this process, chips and flakes of a Mg alloy are supplied to an injection molding machine. They are heated to the semisolid state and stirred to obtain homogeneity in the machine, then the semisolid Mg alloy is injected into the die cavity, cooled, solidified, and made into the required product (see Fig. 25). The semisolid injection molding is an advantageous process for Mg alloys. It is a safe, clean, productive, convenient, easy, and energy saving process. Because of this, the application is now rapidly extending.

XII. MANUFACTURING OF MUSHY METAL

Mushy metal is made by heating the solid metal. Billets, blanks, particles, and powders are heated up to the

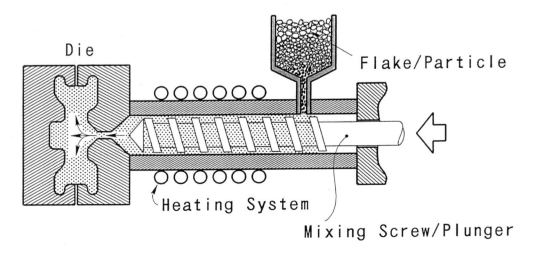

Figure 25 Illustration of semisolid injection molding (thixoforming).

scheduled temperature and made into required mushy metals. In the heating process, some portions of solid grains in the mushy metal are left from partial melting. It follows, therefore, that the shape and size of each unmelted solid grain is affected by the initial internal structure of the solid metal.

The solid grains partially melt in the heating process; on the other hand, they grow at the same time. Therefore, if the heating process is not appropriate, for instance, if it takes an excessively long time, grain growth becomes too fast and the size of the solid grains becomes too large in the mushy metal. Such coarsening of the solid grains results in reducing the quality of the mushy metal and products made of them.

To get fine solid grains in the mushy metal, it is effective to impose a high rate of strain upon the solid metal before the heating. The deformation energy accumulated in the solid metal induces recrystallization of the solid grains in the heating process; as a result, the remaining unmelted solid grains in the mushy metal becomes small. This type of preprocessing of the solid metal is carried out according to the situation required.

XIII. MUSHY EXTRUSION

Appreciating the mechanical properties of the mushy metal, especially remarkably low flow stress, the extrusion of wires, bars, and tubes from mushy billets is conducted.

The outline of the mushy extrusion is as follows (see Fig. 26). The billet is heated to the required temperature in an appropriate electric induction furnace. While the rate of liquid component is less than 20–30% by weight or the rate of solid fraction is higher than 70–80%, the billet is able to retain its initial shape and dimension. However, when the rate of liquid component exceeds

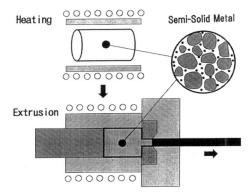

Figure 26 Illustration of mushy extrusion.

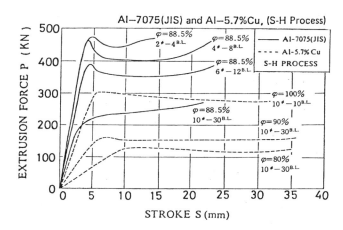

Figure 27 Relationship between force and stroke in mushy extrusion.

this range, the billet may deform under the gravitative force, and therefore a container is necessary to hold it. After the desired mushy state of the billet is attained, the billet is transferred to an extrusion press by a manipulator. The billet is inserted into a container and then extruded. During this operation the billet should be protected from cooling and be kept in the scheduled mushy state. To accomplish this, the transference path for the billet from the furnace to the extrusion press needs to be heated to afford the necessary thermal protection. The container, die, and punch must be preheated. The preheating temperature should be decided taking into consideration the heat capacity of the billet, container, die, ram, and other tooling.

Figure 27 shows the relationship between extrusion force and ram stroke in the mushy extrusion process. The diameter of the billet is 40 mm. The diameter of the extruded wires and bars, as well as the solid fraction φ of the heated mushy billet are shown in the figure. These curves are very similar to those obtained in conventional hot or cold extrusion processes. However, the extrusion force necessary for the mushy-state operation is very low compared with that for the hot extrusion process.

Figure 28 shows the effect of the solid fraction φ on the extrusion pressure p required in the mushy extrusion of Al alloys. The working conditions adopted for the extrusion are stipulated in the figure. The solid fraction $\varphi = 100\%$ means that the billet is maintained at the solidus temperature before the extrusion. That temperature is, of course, higher than the temperature adopted for the conventional hot extrusion. It should be noted here that the extrusion pressure p drops remarkably as the solid fraction φ of the billet decreases. When the solid fraction φ lies between 70% and 80%, the extru-

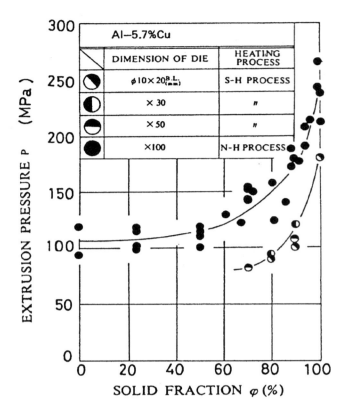

Figure 28 Relationship between extrusion pressure and solid fraction.

sion pressure p is about one quarter of that required for $\varphi = 100\%$.

Figure 29 illustrates another interesting result concerning the mushy extrusion of bars. The working conditions are indicated in the figure where the abscissa shows the extrusion ratio λ and the ordinate shows the normalized extrusion pressure p_n; the latter is defined as the ratio of the mushy extrusion pressure to the flow stress of the billet at the solidus temperature. The broken line in the figure represents the theoretically calculated normalized extrusion pressure p_n needed for the extrusion of the solid billet at the solidus temperature assuming that the friction is zero on the die surface. It follows that in those cases when the solid fraction φ is 70% or less, the measured normalized extrusion pressure p_n is almost equal to the calculated value under the above-mentioned assumption. This implies that the liquid component of the mushy billet acts as a very effective lubricant.

From these results, the characteristics of the mushy extrusion can be summed-up as follows:

1. Because of the pronounced low flow stress of the billet, the extrusion pressure needed in the mushy

extrusion is very low compared with that for the conventional hot extrusion. The extrusion pressure is about one fourth to one fifth of that for the hot extrusion.

2. Because of the low flow stress and the low extrusion pressure, very high reduction or extrusion ratio is attainable in a single pass.

3. The liquid component of the billet acts as an effective lubricant on the surfaces of the container and die.

4. Because of the low extrusion pressure and the lubricating effect of the liquid component, high-quality products with complicated cross sections can be extruded.

XIV. MUSHY ROLLING

The rolling of metal sheets in the mushy state is a rather new but potential process. The fundamental features and characteristics of the mushy rolling are summarized as follows.

1. In the mushy rolling, the deformation zone of the mushy sheet is very narrow corresponding to the contact length between the sheet and roll. On the other hand, the mushy sheet has wide free surfaces, which locate near by the deformation zone. Because of this geometrical condition, it is difficult to hold the deformation zone under a stable high hydrostatic pressure. The liquid component, which presents in the deformation zone at the roll gap, is pressed and likely to flow out from the narrow deformation zone to other parts of the mushy sheet. This means, in turn, that the solid and liquid components do not always flow together.

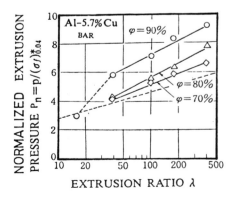

Figure 29 Relationship between extrusion pressure and extrusion ratio.

2. The presence of the liquid component is linked with the poor ability of the sheet to elongate. This is because the intergranular connection between the solid grains is very weak because of the partial melting of the grain boundaries. Each grain is easily separated from others even by a small tensile stress. Under the hydrostatic pressure, the separation of grains and the incidence of cracks on grain boundaries are prevented. Therefore the hydrostatic pressure is essential for the elongation of the mushy sheet. However, as already pointed out, it is not easy to maintain a sufficiently high hydrostatic pressure in the mushy sheet at the roll-gap.

3. In spite of such difficulty, mushy rolling is considered to be a potential process to manufacture various types of metal sheets, which have different internal structures compared with those made by conventional hot or cold rolling processes.

Figure 30 shows a schematic illustration of the mushy rolling. Before the rolling operation, the sheet is heated to the mushy temperature range by using an induction furnace. At the entrance to the roll-gap, the liquid component is likely to flow to the top and bottom surfaces of the mushy sheet under the hydrostatic pressure induced in the deformation zone. However, the liquid component does not flow out of the mushy sheet; it is cooled by the roll, solidifies, and is then drawn into the roll-gap by the friction force. The solid component, consisting of grains, is compressed, deformed, and then drawn into the roll-gap. In the roll-gap, each grain is deformed and elongated. Before the mushy sheet reaches the exit of the roll-gap, all of the liquid component usually completes its solidification. In some cases, the rolled sheet may include some amount of liquid because of incompleteness of solidification, but the amount of the residue should be less than a few weight percent.

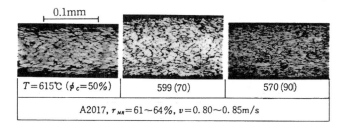

$T = 615°C$ ($\phi_c = 50\%$)   599 (70)   570 (90)

A2017, $r_{MR} = 61 \sim 64\%$, $v = 0.80 \sim 0.85 m/s$

Figure 31 Internal structure of sheet rolled in mushy state.

Figure 31 shows a few examples of the internal structure of the Al-alloy sheets manufactured by mushy rolling. The solid fraction before rolling, thickness reduction, rolling speed, and other working conditions are shown in the figure. The roll surface temperature is kept at ambient before rolling and so the mushy sheet is rolled and cooled simultaneously at a rather high cooling rate as well.

The following results can be deduced from the figure.

1. When the mushy sheet includes an adequate amount of the liquid component before rolling and the thickness reduction in rolling is not so high, the solid component does not deform at the roll-gap. Almost all of the solid grains keep their initial geometry during the rolling. They appear to be wafting in the liquid component and flow through the roll-gap like a slurry—without deformation. Furthermore, in such a case, the liquid component does not complete the solidification during passage through the roll-gap.

2. When the thickness reduction is high and/or the solid fraction of the mushy sheet is high enough, the solid grains are not able to pass through the roll-gap without colliding with each other and without changing their shapes. The solid grains restrict each other's moving and slipping, and therefore they are forced to deform and elongate at the roll gap. Each solid grain is elongated by the rolls and the assortment of fiber structure is formed in the mushy sheet. As the thickness reduction and the solid fraction increase, the elongation of the solid grains is further promoted and a typical fiber structure is observed in the rolled sheet.

3. Another result to be noted is that the rolled sheet does not always have a homogeneous internal structure. When the mushy sheet includes an adequate amount of the liquid component before rolling, the liquid tends to solidify in the surface layer. On the contrary, the

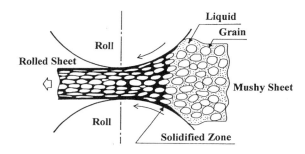

Figure 30 Illustration of mushy rolling.

solid grains gather at the inner zone. As a result, the rolled sheet displays two different structures—one in the surface layer and the other within the bulk.

This type of inhomogeneity is responsible for the zoning mechanical properties, such as hardness and yield strength, across the thickness of the rolled sheet. To attain the homogeneity of the rolled sheet, those dual structures should be prevented by choosing appropriate mushy rolling conditions. However, if such distribution of the mechanical properties is desirable, mushy rolling can manufacture multilayer sheet products.

XV. MUSHY FORGING

By utilizing unique characteristics of the mushy metal, new processes such as mushy mixing, mushy powdering, and mushy forging are available. These processes can produce a variety of metal–ceramic composites, which in turn can be formed into bars, wires, tubes, sheets, and machine parts. In this section, the manufacture of so-called particle-reinforced metals (PRMs) by mushy mixing and mushy forging is explained.

A schematic diagram of the framework of applicable mushy mixing and forging processes is shown in Fig. 32. The fundamental functions and characteristics of these processes are as follows.

1. The metal matrix is heated up to the mushy-state where it contains the required weight percentage of the solid and liquid components. As mentioned before, when the solid fraction is less than a critical value, the mushy metal matrix becomes very soft and therefore can be stirred. During the stirring, the reinforcing particles, such as particles of Al_2O_3 and SiC, are put into the mushy metal matrix. Thus a mushy mixture of metal matrix and reinforcing particles is obtained.

2. The mushy mixture, which is a raw material for an aimed composite, can be transferred directly to the extrusion press, forging press or rolling mill. Using a conventional extrusion press or a forging press the mushy raw mixture is formed into bars, wires, tubes, and machine parts. In the forging process, the mushy mixture is formed and cooled simultaneously and the liquid component in the mushy metal matrix solidifies. Bonding between the metal matrix and the reinforcing particles is attained mechanically and effectively by a high pressure imposed on the mixture in the process.

3. Even though hot/cold forging and extrusion of the solid mixture of metal matrix and reinforcing particles is attempted, an extremely large force and high pressure are necessary to conduct the required operation. This is because the compo-

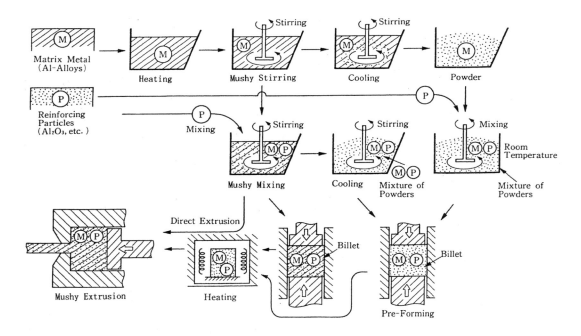

Figure 32 Illustration of framework of processes to manufacture composites by utilizing mushy forging.

site mixture, which includes ceramic particles in a rather high volumetric ratio, is very hard and has poor deformability. The forming of such a composite mixture is sometimes practically not possible.

4. However, when an appropriate mushy forging is employed, the mushy matrix is very soft and its liquid component acts once again as an effective lubricant between the solid grains of the metal matrix and the reinforcing particle. Consequently, the forging can be conducted without serious difficulties under a rather small force and a low pressure.

5. When the mushy metal matrix is cooled gradually during the stirring, the solid grains in it are separated into particles and then solidify. Thus, the mushy metal matrix is made into particles and/or powder. The reinforcing particles can be put into it; thus the mixture of the metal matrix powder and the reinforcing particles is obtained in the same way.

6. The mixture of the metal matrix powder and the reinforcing particles is usually preformed into billets and blanks. These are heated up to the mushy-state and are subjected to various die-forging processes. This procedure constitutes an alternative to the above-mentioned direct forging of the mushy mixture.

7. The process, shown in Fig. 32, has two important aspects. One is the process to make the raw mixture of metal matrix and reinforcing particles; the other is the forging process of the raw composite into products. Both are dependent on the mechanical properties of the mushy metal and both can be conducted only in the mushy-state.

The particle-reinforced metal products usually have very high hardness and excellent antiwear resistance. These characteristics depend, naturally, on the volumetric percentage and on the size of the reinforcing particles.

The higher the volumetric percentage and the smaller the particle size, the higher are the hardness and the wear resistance. On the other hand, the ductility, deformability, and impact strength of the particle-reinforced metal decrease with the increase in the volumetric percentage of the reinforcement. Such disadvantages are not due to the mushy-state forging, but they are usually observed as for the metals reinforced by ceramic particles. When the volumetric percentages of the mixed reinforcing particles are in the same range, the small-sized reinforcing particles will have larger integrated surface area than the large-sized reinforcing particles. Thus, the metal reinforced by the small-size particles needs larger energy dissipation in the forging process than the metal reinforced by large-sized particles, and the former has higher hardness and larger flow stress than the latter.

The mechanism of bonding between the metal matrix and the reinforcing particles is mechanical and not diffusive. In the mushy forging, the liquid component included in the mushy metal matrix fills up all the cavities between the solid grains and the reinforcing particles. Then the liquid component solidifies under very high hydrostatic pressure generated by the motion of the dies. The location of an individual reinforcing particle in the metal matrix is thus firmly established. This can be regarded as the principal mechanism of bonding that takes place in the mushy-state forging. Other bonding mechanisms, such as diffusion, can also be expected when the pretreatment of surfaces of the reinforced particles is conducted. However, in the mushy forging, the mechanical bonding is dominant.

Studies on the application of mushy forging to manufacture metal–ceramic composites are now widely carried out. In this field, the mushy forging can provide a considerable possibility of development of new types of composites.

REFERENCES

1. Mehrabian, R., et al. Metal Trans. 1974, *5*, 1899.
2. Fleming, M.C., et al. *Japan–US Joint Seminar on Solidification of Metals and Alloys,* 1977; 1 pp.
3. Kiuchi, M., et al. *Proc. 20th Int. Machine Tool Design and Research Conf,* 1979; 71–79 pp.
4. Kiuchi, M., et al. J. Jpn. Soc. Technol. Plast. 1979, *20-224*, 826.
5. Kiuchi, M., et al. Mechanical Behaviour of Materials 1984, *IV*, 1013.
6. Kiuchi, M., et al. *Proc. 14th North American Manufacturing Research Conf.*; 1986; 359 pp.
7. Toyoshima, S. ISIJ Int. 1991, *31-6*, 577.
8. Hirai, M., et al. ISIJ Int. 1993, *33*, 2.
9. Kiuchi, M., et al. Proc. JSTP Spring Conf.; 2000; 277 pp.
10. Kiuchi, M., et al. CIRP Ann. 2000, *50*(1), 157.

6

Extrusion

Sigurd Støren and Per Thomas Moe
Norwegian University of Science and Technology, Trondheim, Norway

I. INTRODUCTION

This chapter is devoted to extrusion of aluminum alloys and divided into three main sections. Section II covers the basic parameters of extrusion needed for designing an aluminum section and a die, for understanding the processing steps, and for optimizing productivity, cost, and product quality. A specific section shape is used to illustrate the interaction between these parameters. Section III is focused on the commercial application aspects of extruded sections, life cycle aspects, alloy selection, and section design guidelines. Section IV covers the extrusion process in some detail, focusing on the basics of quantitative modeling of metal flow in the container and through the die. In the final section, some of the outstanding research challenges in the theory of extrusion of thin-walled aluminum sections are discussed: (1) 3D-modeling of thin-walled extrusion, (2) the bearing channel friction in interaction with die deflections and section surface formation, (3) stability of flow, and (4) limits of extrudability.

The intention is that the chapter should give the reader an overview of the practical aspects of extrusion as well as an understanding of the present state of the theoretical work and some challenges in this branch of metal-forming science and technology. However, the study of extrusion as a process is both relatively complex and multidisciplinary, and this chapter can hardly give the answer to all problems that may be encountered. Thus before making detailed section design and alloy decisions, the reader is advised to contact an extrusion plant. Although theoretical and experimental work has managed to explain a number of relevant phenomena, the quality of an extruded profile and naturally also of a complete product based on extrusions is still mainly dependent on the experience of personnel close to or at the extrusion plant. One may also confer with more general works on extrusion [1,2].

II. BASIC PARAMETERS OF EXTRUSION

A. The Process

The most common method for producing aluminum profiles is that of direct extrusion (Fig. 1). Here, the ram is moving into the container at one end and pushes the billet through the opening of the die at the other end. The temperature of the deforming aluminum alloy is in the range of 450°C to 600°C during the process cycle. In contrast to the extrusion of steel, aluminum extrusion is taking place in the absence of any lubrication of the die. Hence the material sticks to the container and the die, giving a highly inhomogeneous flow with large degree of viscoplastic shear flow (see Section IV). The material faraway from the wall is flowing easier than that closer to it, with the surface of the billet remaining in the container. The billet and the container are normally circular cylindrical but can, in special cases, be rectangular with rounded corners.

A special feature in extrusion of aluminum alloys is the production of hollow sections (Fig. 2). In this case the metal flows into the opening between the die and the mandrel. The mandrel is kept in position by bridges.

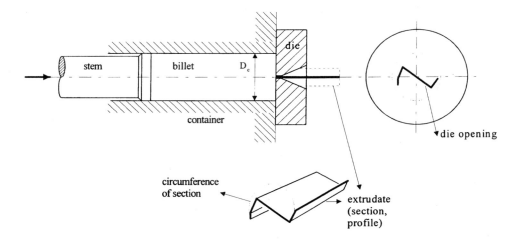

Figure 1   Direct extrusion of an open section.

The billet material is forced, by the movement of the ram, into the portholes in the bridge die, called the feeder ports. Under the bridges, adjoining metal streams meet and are forge-welded together in the weld chamber, before flowing through the bearing channel, i.e., the opening between the die and the mandrel.

Besides direct extrusion, two other special extrusion methods are used, indirect extrusion, and continuous extrusion, the Conform method [3]. In indirect extrusion (Fig. 3) the die is pushed into the container, whereas the extrudate is flowing in opposite direction through the hollow stem. In the continuous extrusion (Fig. 4) a continuous feedstock is fed into a groove in a rotating wheel. Pressure is built up by friction between the groove walls and the feedstock in the gap between the wheel groove, the feeder plate, and the abutment. The metal is then forced to the die opening in a continuous flow. Both open and hollow sections can be produced.

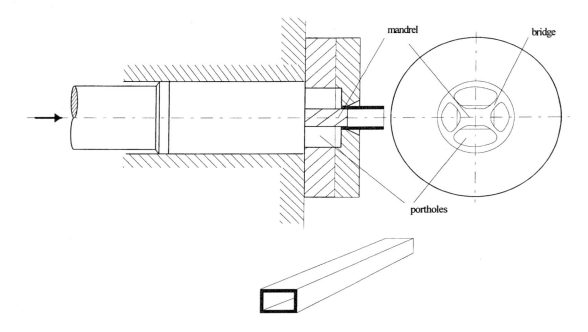

Figure 2   Direct extrusion of a hollow section.

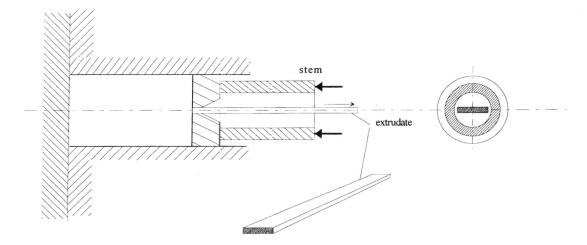

Figure 3 Indirect extrusion.

Extrusion in rectangular containers, indirect extrusion, and continuous extrusion are used for special products in limited quantities. Therefore in the rest of this chapter the direct extrusion of open and hollow sections are dealt with.

The main parameters of the billet, the container, and the extruded section are (Fig. 1):

- Diameter of the container: D_c [m]
- Cross-sectional area of the container: $A_{container} = A_c = \pi/4 D_c^2 [m^2]$
- Billet weight: $W_b = \rho\pi/4 D_b^2 L_b [kg]$
- Billet diameter: D_b [m]
- Billet length: L_b [m]
- Density of aluminum: $\rho = 2700$ [kg/m^3]
- The circumscribed diameter of the section: d [m]

- Section thickness: t [m]
- Cross-sectional area of the section: $A_{section} = A_s$ [m^2]
- Weight of section per meter length: $w_s = A_s \rho$ [kg/m]
- Reduction ratio: $R = A_c / A_s$

The most common values for the diameter of the container are 0.178 and 0.208 m. The billet diameter is usually 5 to 10 mm less than the container diameter, allowing the billet to enter the container easily. The circumscribed diameter of the profile is usually less than 0.9 times the diameter of the container, but specially designed dies with a so-called expansion chamber may actually allow for $d > D_c$. The section thickness often varies over the cross section of the profile. The reduction

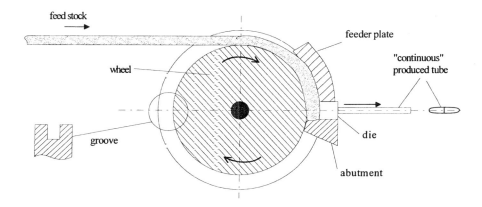

Figure 4 Continuous extrusion.

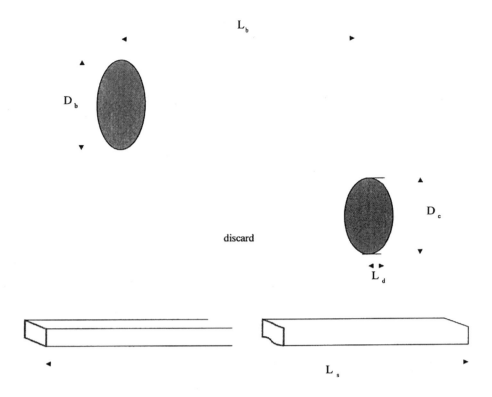

Figure 5 A multihole die.

ratio is normally in the range of 20 to 80. If R is very high ($R > 70$) and the section is of a proper shape, the die is usually designed with more than one die opening (Fig. 5). In this case, the reduction ratio is:

$$R = \frac{A_c}{A_s n} \qquad (n = \text{number of die openings})$$

When an extrusion press cycle is carried out (see Section IV for details), a small part of the billet is left in the container, the *discard* (Fig. 6). The length of the discard is normally around 10–20 mm.

- Discard length: L_d
- Discard weight: $W_d = \rho \pi / 4 D_c^2 L_d$ [kg]

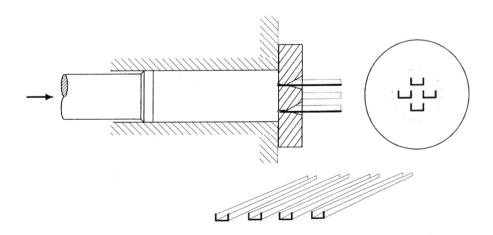

Figure 6 Billet, discard, and the extruded section.

- The weight of the extruded section: $W_s = W_t - W_d$ [kg]
- Length of the extruded section: $L_s = W_s/w_s$ [m]

B. The Die

The tooling package is to perform the deformation of the aluminum and must naturally withstand very large forces. Tools are generally made of high-strength steels such as H11 and H13, and the surface in direct contact with the flowing material is hardened through nitriding prior to any use. Furthermore, the complete tooling package will be composed of a great number of parts which are all meant to support the die when pressure is applied by the stem. The complete tooling package will be designed differently for the extrusion of hollow or open profiles. In any case, however, a bolster will be situated directly behind the die and provide the main

support. The die and bolster will then be placed in a horseshoe clamp, which is firmly attached to the press structure.

In the case of extrusion of open sections one die design does not differ significantly from another although the bolster may provide varying degrees of support. Various die designs have, however, been developed for the extrusion of hollow profiles. The names of the most commonly used die types are porthole, spider, and bridge, and for the extrusion of 6XXX alloys porthole dies have traditionally been most popular, partly because of the ease with which they can be cleaned after extrusion.

The design of a porthole die is displayed in Fig. 7. The outer contour of the section is formed by the die plate (Fig. 7a). The tongue will be less stiff and weaker than the rest of the plate because it supports the pressure from the deforming material on the tongue only along

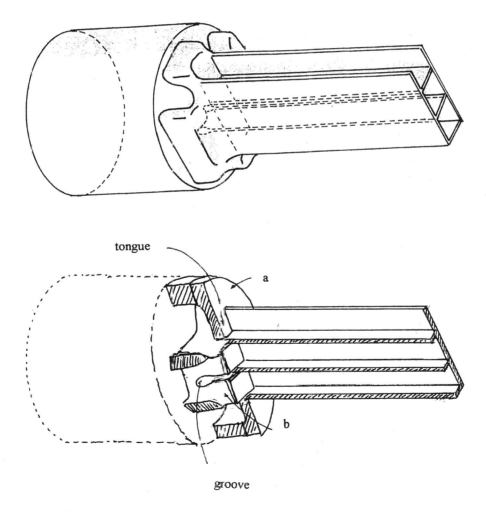

Figure 7   Billet, die, and extruded section in the process of extrusion.

one edge. The inner circumference of the section is formed by the mandrel (Fig. 7b). The mandrel is an integrated part of the porthole die, connected to the rest of the die by webs, or bridges. In the mandrel a groove is machined out. This groove enables the internal rib in the hollow section to be formed.

The deforming alloy is flowing over the bridges and down into the feeder ports. Under each bridge, in the weld chamber, the two neighboring metal streams are forge-welded together. In this process the temperature of the material will not exceed that of melting, but welding will take place because of high pressures and diffusion rates. The alloy is also flowing into the groove in the mandrel from two sides, and in the center of the groove the two streams of metal are forge-welded, before the material flows into the bearing channel. All such welds are denoted seam welds. If pressures are not high enough in the weld zones, insufficient welding will take place. Furthermore, if material flows in an uncontrolled manner, one will not be able to predict the exact position of the weld. All these phenomena are highly unwanted and hence detailed studies of such can be found in the literature [4].

When designing mandrels one has to keep the following in mind:

- The stiffness and strength of the bridges should be optimized. The feeder ports should at the same time be as large as possible in order to reduce the load on the mandrel and allow for higher extrusion speed. This will, however, result in a weak bridge construction with unwanted flexibility and an increased risk of die deflection.
- Controlled flow out of the bearing channel should be sought. The die and the bearing channel should be designed so that the section leaves the bearings at a uniform speed and without generating excessive tensile or compressive stresses. Of special importance is the control of metal flow and die welding of the inner rib, because this cannot be inspected from outside during the press cycle.
- The surface of the section should be homogeneous and leave the die without streaks and stripes at the highest acceptable speed.

Clearly, there is a complex, but a very fascinating design-optimizing challenge here. Today, die-design competence exists mainly as practical knowledge by highly skilled die designers, die producers, and die correctors in the die shops. As will be pointed out in Sections IV and V, however, the development of 3D computer simulation of hot extrusion processes is approaching such a level of precision that it can be used as a tool for die design. It must, however, be done in close cooperation with skilled and experienced die specialists.

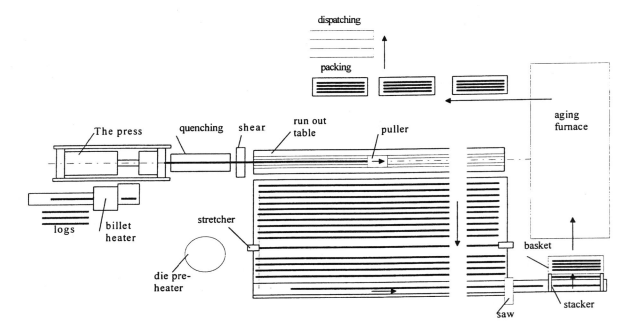

Figure 8 Layout of an extrusion plant.

C. The Manufacturing System

Satisfactory control of the material flow may be viewed as the key element in a successful production of aluminum profiles. In this context the last assertion has two alternative interpretations, and both are in fact equally correct. In order to produce extrusions with the desired quality at an optimum pace, one has to establish some sort of understanding of the mechanisms of plastic flow of material in the container and die. However, if an enterprise is to succeed economically in the extrusion business, it is as important that it masters the logistics, i.e., the control of the material flow in and around the production facilities. The extrusion process is carried out in an extrusion plant, which often has a layout similar to that presented in Fig. 8. Although heat treatment in general is the most time consuming part of the production system, other process steps may in fact constitute the actual bottlenecks. The pressing of profiles is one such step, as it is noncontinuous and as considerable time is spent on changing dies, reloading new material into the container, and performing maintenance tasks. Procedures are made even more complicated as new production orders for profiles often may necessitate several trial runs on the press. If the material is not transported effectively, down times may easily be long, and the most important parameter of all, productivity, will, consequently, be low.

As is seen in Fig. 8 the extrusion process is composed of a great number of steps. One of the most important, however, is the production of raw material for the process, and this usually does not take place in the plant. Feed stock for the process is logs, normally in lengths 6–7 m. They are supplied from the cast house of primary aluminum smelter or a secondary (recycled) aluminum cast house. The logs are produced as visualized in Fig. 9.

The liquid metal at temperature above 700°C is cleaned, added alloying elements and grain refiner before entering the casting table. By passing the casting molds with direct water cooling, the liquid aluminum alloy solidifies into a log. After casting, the log is homogenized in a temperature cycle that secures the best possible extrudability by establishing a homogeneous distribution of alloying elements and by dissolving phases with low melting points, typically Mg_2Si [5,6]. The logs are then transported to the extrusion plant.

In the plant a number of distinct processing steps take place (Fig. 8). The logs are first taken one by one from the log stacker and transported to the induction

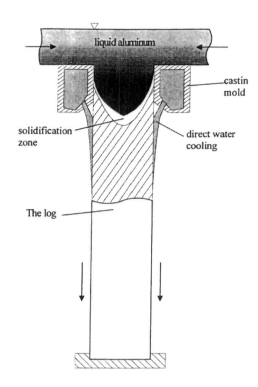

Figure 9 Direct chilling casting (DC-casting) of logs.

heater. Here a certain temperature profile is imposed on the log, and it is then cut into billets of a prescribed weight. In some plants the logs are cut prior to any heating.

The billet is then loaded into the extrusion press, where the ram pushes it into the container. The end of the billet surface in contact with the ram has been given a coating so that it does not stick to the *dummy block* between the ram and the billet. Because the billet has a smaller diameter than the container bore, it is given an upsetting in order to fill the container. In this phase there is a risk of entrapping air in the container, and thus the ram stops after upsetting, unloads, and moves a small distance backward to let the possible entrapment leave. This is called the *burb cycle*.

Thereafter, the extrusion process commences. The ram pushes the billet through the die opening. The load capacity of the press with a container diameter of 0.178 m is normally 16 MN, which corresponds to a specific pressure of 643 MPa. If the container diameter is 0.208 m, the load capacity is normally 22 MN and the specific pressure 647 MPa. The temperature of the billet prior to extrusion is in the range 450–470°C. In the induction heater, the billet may have been given varying temperature along its length in order to compensate for the heat generation caused by the shearing along the container walls when it is pushed through the container (see also Section IV). This is called *tapering*, and the highest temperature is usually in the front end of the billet. The temperature of the section leaving the die is in the range of 550–600 °C. The taper should be given in such a way that the runout temperature is constant as this will result in minimum variation of dimensions and properties during the press cycle.

As the section front leaves the die, it is gripped by a *puller*, which guides the section out on the *runout table*. The profile is then quenched and further cooled down when moving sideways along the table. The lengths of the profiles upon leaving the die may be from 20 to 50 m, depending on the length of the billet and the reduction ratio. Normally, a number of charges (billets) are performed with the same die in a production set up. In this case, one may weld the profile from the new charge directly to the one produced in the foregoing charge, creating a so-called *charge weld*. This procedure simplifies production but necessitates cutting of the profile during extrusion. On the cooling table the section is given a plastic deformation of 0.5–2% elongation in order to eliminate internal stresses due to uneven cooling over the cross section of the profile and straighten up possible bends and twists before going into the cutting saw. The extruded section is finally cut into prescribed lengths, normally 6 m. The process of cutting may vary somewhat from one plant to another.

The cut sections are stacked in bins and transported through the aging oven where they spend 3–6 hr at temperature in the range of 170–190 °C. After aging, the sections are inspected and packed before they are delivered to the customer for further fabrication and surface treatment, followed by joining and assembling into the finished component or product.

With a generic aluminum section (Fig. 10) some important features and characteristics of die design and productivity for aluminum extrusions will be demonstrated.

An order of 200 sections and 6 m of alloy AA6060 (Al-MgSi0.5) shall be produced in a 16-MN press with a container diameter of 0.178 m and runout table length of 42 m. The following typical process parameters can be calculated and controlled:

- The cross-sectional area of the container is:

$$A_c = \frac{\pi}{4} \cdot 0.178^2 = 24.9 \cdot 10^{-3} \ [\mathrm{m^2}]$$

- The cross-sectional area of the extruded section is:

$$A_s = 0.084 \cdot 0.02 - 2 \cdot 0.028 \cdot 0.016$$
$$+ \left(0.003^2 - \pi \cdot 0.0015^2\right) \frac{10}{4}$$
$$= 0.437 \cdot 10^{-3} \ [\mathrm{m^2}]$$

- The reduction ratio can thus be calculated to:

$$R = \frac{A_c}{A_s} = \frac{24.885}{0.437} = 57$$

This is a reduction ratio within the acceptable range for a one-hole die.

- The circumscribed diameter of the section is:

$$d = \sqrt{0.02^2 + 0.084^2} = 0.086 \ [\mathrm{m}]$$

This is well bellow the maximum recommended diameter of $0.9 \cdot 0.178 = 0.16 \ [\mathrm{m}]$

- Special features of the section shape that should be noticed are a hollow rectangular section with constant wall thickness, an outer tongue, and an inner rib. Furthermore, the profile is symmetric about the horizontal axis.

D. Productivity and Cost

A number of aspects are important to consider for a customer who is to choose between the many different suppliers of aluminum profiles. As a great number of sections ought to and have to be designed and manufactured for only one product type, customer service stands out as particularly important. Furthermore, the supplier must of course be able to deliver the section requested within an agreed time limit and to the specified quality. If the profile geometry is fairly complicated or a very high strength alloy is chosen, some suppliers may fall out of the race, but for most profiles one may not be able to differ on these grounds alone. And in the

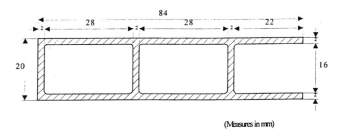

(Measures in mm)

Figure 10 A "generic" extruded aluminum section.

end, thus, all usually comes down to money. The basic parameters in the extrusion business are the prices per meter or per kilogram extruded section. These measures are dependent on the choice of alloy and the geometry of the section, and one has to contact different suppliers in order to determine the exact prices. These should not differ too much as there is an active market mechanism working. This mechanism will, however, also pressure the suppliers to continuously seek to increase productivity and cut costs. It is in the creative negotiations between the customer and the supplier that the right price is agreed upon as a consequence of a section design with the right balance between requirements for functionality and the cost efficiency in the extrusion plant. Important parameters that determine the productivity and cost of the extruded section are:

- Length produced per press cycle
- Length of end cuts that have to be scrapped
- Number of cut lengths per billet
- The discard weight per billet
- Number of billets produced, i.e., gross weight delivered to the press
- Net weight ordered
- The dead cycle, i.e., the time between each press cycle
- The ram speed
- The acceleration time, i.e., the time to reach the full ram speed
- Time for die change
- The price of billet delivered at the press
- Die cost
- Production cost.
- Unpredicted press stop.
- Unpredicted quality scrap, i.e., the number of sections produced, which are not conforming to the required quality.

The production cost may be measured as cost per minute extrusion time spent. This measure contains all direct costs and man-hour costs in the plant, divided by the estimated availability of the press in minutes.

The following example is meant to illustrate a typical calculation of the cost per meter and cost per kilogram extruded profile. The calculations are meant to refer to the section in Fig. 10, and data from the example of the previous section are used.

- The gross material mass (weight) is first calculated.

The sections are cut to lengths of 6 m, of which six may be produced from each billet. In addition the first and the last meter of the total section are assumed to be of inferior quality and therefore scrapped. The total section length produced in the press cycle is then:

$$6 \cdot 6 + 2 = 38 \ [\text{m}]$$

The mass per meter section can be calculated from the cross-section area and the density:

$$0.437 \cdot 10^{-3} \cdot 2700 = 1.18 \ [\text{kg/m}]$$

The previously calculated data can be used to find the runout mass per billet:

$$1.18 \cdot 38 = 44.84 \ [\text{kg}]$$

The length of the billet is chosen so that the discard length is 0.02 m. As the container diameter is 0.178 m, the mass of the discard is:

$$\pi \cdot (0.178/2)^2 \cdot 0.02 \cdot 2700 = 1.34 \ [\text{kg}]$$

The total mass of each billet may be calculated to be:

$$44.84 + 1.34 = 46.18 \ [\text{m}]$$

In order to compensate for possible quality scrap of 6%, 12 more lengths than ordered are produced. The total number of cut lengths is then 212, and the corresponding number of billets is:

$$212/6 = 35.3$$

Hence, 36 billets must be ordered and the gross material mass will consequently be:

$$36 \cdot 44.18 = 1614 \ [\text{kg}]$$

- The net mass of the sections delivered to the customer is, however, only:

$$200 \cdot 6 \cdot 1.18 = 1416 \ [\text{kg}]$$

- The yield, which is the gross mass of the material divided by the delivered mass, is then:

$$\frac{1416}{1614} = 0.877 = 87.7 \ [\%]$$

- The total production time should then be calculated.

The runout speed of the press for this particular alloy and geometry is found to be 36 m/min or 0.6 m/sec. The additional time spent per charge on reaching the desired runout speed, the acceleration time, is found to be 7 sec. Because the length of the billet (38m) is known, the total time of a press cycle can be calculated to be:

$$38/0.6 + 7 = 70 \ [\text{sec}]$$

The dead-cycle time then has to be assessed. The time spent on cutting of the discard after extrusion, inserting a new billet, and performing the burb cycle is found to be 15 sec. If one expects no additional unexpected stops, one only has to add the time spent on changing the die prior to extrusion. For this particular press this is found to be 180 sec. The total production time without any unpredicted stops and delays is then:

$$(70 + 15) \cdot 36 + 180 = 3240 \; [\text{sec}] = 54 \; [\text{min}]$$
$$= 0.9 [\text{hr}]$$

- The productivity is viewed as the net mass delivered divided by the total extrusion time and can optimistically be calculated to be:

$$1416/0.9 = 1573 \; [\text{kg/hr}]$$

- Finally, a calculation of cost has to be performed.

The material cost of the billet is set to \$1.5/kg. Production cost in this example is found to be \$50/min and the die cost for the order is \$2000. The total cost respectively without and with the die cost is then:

$$1.5 \cdot 1614 + 50 \cdot 54 = 5121 \; [\$]$$
$$5121 + 2000 = 7121 \; [\$]$$

The corresponding costs per meter section delivered can be calculated:

$$5121/1200 = 4.27 [\$/\text{m}]$$
$$7121/1200 = 5.93 \; [\$/\text{m}]$$

Finally, the cost per kilogram section delivered is:

$$5121/1416 = 3.62 [\$/\text{kg}]$$
$$7121/1416 = 5.03 [\$/\text{kg}]$$

The same die can often be used in several production orders. If all die cost is placed on the first order, one can produce the next orders without any die cost. Maintenance cost of the die is included in the production costs.

E. Measures of Section Quality

1. Process Variability

The aluminum extrusion process is unique in the sense that it offers the possibility to produce almost ready-to-use profiles of high quality and with large functional freedom at a relatively low cost. The product quality, which in the very end will be judged by the product's ability to satisfy customer demands, relies heavily on the restrictions imposed on design by the process itself. Purely geometrical considerations indicate that profile dimensions necessarily will be limited by the press capacity and size, and it is known that material flow also puts restrictions on both wall thickness and changes in such. However, product quality is probably to the largest extent affected by the mere variability in chemical composition, microstructure, geometrical dimensions, mechanical properties, and surface finish over the length and width of the profile. Such variations are connected to the transient nature of the extrusion process and to the difficulties in establishing a system of measurement and control of important parameters during the press cycle. Changes in temperature, deformation history, and material composition of the extruded profile will be encountered both during the course of a press cycle and from one billet to another. Furthermore, both production parameters and dies have to be changed in order to extrude different aluminum alloys because both their metallurgical and thermomechanical properties may differ considerably. If attention is not paid to controlling the material flow in the factory, the production of profiles with uneven and thereby also inferior properties will ultimately lead to either the distribution of products of poor quality or to low productivity due to excessive scrapping. Therefore most producers of aluminum profiles stress the use of housekeeping and have established routines for production of profiles of different alloys based on experience. However, in order to make proper use of such routines sufficiently reliable and consistent, measurements of production parameters such as temperatures and pressures have to be obtained. This task is not easily performed due to the noise inherent in the process.

2. Dimensional Variability

In order to make a direct assembly of extruded profiles possible, the characteristic dimensions of the product such as straightness, thickness, height, width, length, and angles have to be made within sufficiently narrow tolerances. Dimensional variability is, to a certain extent, always existent and often in the order of 0.25 mm on thin-walled profiles. Open or partly open profiles tend to experience larger variation than closed ones, whose die construction is more robust. Based on experience with when the process can be expected to be under control, tolerance on wall thickness is often set to be around 10% of the nominal measure. Measurement of profile dimensions is usually implemented as a standard procedure at extrusion plants.

Table 1, which is taken from the German standard DIN 17615, gives an indication of within which tolerances the open profile in Fig. 11a can be delivered. Thickness variations may be caused by the changing deflection and temperature of the die during a press cycle. Another cause of thickness variations is wear. In the case of large production series, dies are often bought from die manufacturers with so narrow bearing surfaces as to compensate for future wear. Furthermore, dies are also produced within certain tolerances, although somewhat narrower than those of the profiles. As a result of both these factors, die changes may cause variation in profile geometry. A last reason for variation in wall thickness is the deformation of dies through fracture when extruding hollow profiles. This is caused by uneven loading, which very often is a result of flow imbalance and is most common when extruding alloys of higher flow resistance such as, for instance, the 7XXX series. Fracture need not always immediately be fatal, and the presence of a crack may very well lead to a gradual reduction of the die strength and increasing deviation from nominal thickness. The presence of a crack combined with flow imbalance will also often result in thickness variations along a wall.

Variation in height and width of the profile may be due to variation in manufactured die dimensions or wear. However, larger deviations will be measured if profile walls are curved. In this case, tolerances are set with regard to the maximum curvature that can be accepted. The curving of profile walls is often a result of varying flow velocity across the profile, which again is due to variation in wall thickness and friction conditions in different parts of the outlet of the die. Table 2 shows tolerances with regard to curvature of walls in the profile given in Fig. 11b.

Prior to stretching, operation profiles very often have a certain curvature, a warping in the direction of extrusion (Fig. 11c). This is often a result also of variations in flow velocity in the profile but can be caused by the uneven cooling rates of walls with different thicknesses. DIN 17615 gives tolerances on the deviation from straightness as a function of length as given in Table 3 and Fig. 11c. Flow imbalance may also lead to twisting of the profile along the extrusion direction. Figure 11d and Table 4 show that this deformation is often measured as a distance, v, which can be taken to be a function of both the length and the width of the profile.

Although the height, width, and thickness of a profile may be within tolerances, assembly may be hindered by deviations in angular measures as shown in Fig. 11e. Direct measurement of angles on the profile is most easily done with the help of electronic equipment but is relatively time consuming in comparison with simpler mechanical methods. DIN 17615 proposes the use of the length w as a measure of angular deviation and this can be taken as a function of the profile width as seen in Table 5.

3. *Variability in Surface Properties*

High surface quality is usually obtainable when extruding aluminum profiles, and the combination of a very even surface, outstanding optical properties, and large corrosion resistance makes the use of aluminum preferable to, for instance, steel in many applications. However, surface quality is extremely dependent on die design, billet quality, and extrusion practices in general, and a series of surface defects may develop if proper attention is not paid to controlling the process [7]. Given the excessive noise in extrusion equipment, it is not always possible to sort out the causes for defects, and very often several different error mechanisms may be operating simultaneously.

Corrosion properties are related to the content of different alloying elements, in particular Cu, and are therefore mostly dependent on the properties of the billet. However, by extruding with the wrong parameters, changes in chemical composition, grain sizes, and surface roughness may be unfavorable to corrosion properties.

Die lines and pick ups are common surface defects shown in Fig. 12a and b and give an indication that the profile surface is formed in the presence of hard and uneven attributes or particles in or around the bearing channel. Rough and worn-out dies or abrupt changes in bearing lengths may be the cause, but it may just as well be that oxide build-ups behind bearings or hard par-

Table 1 Tolerances on Section Thickness After DIN 17 615

Measures of thickness, s, from [mm]	Measures of thickness to [mm]	Allowed deviation [mm]
—	1.5	±0.15
1.5	3	±0.20
3	6	±0.25
6	10	±0.30
10	15	±0.40
15	20	±0.50
20	30	±0.60
30	40	±0.70

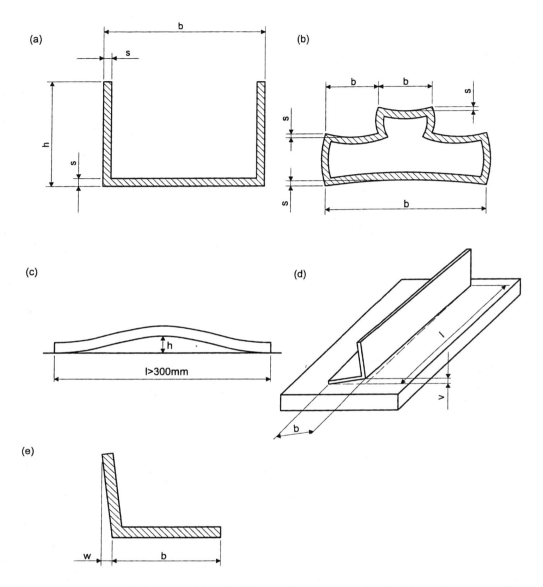

Figure 11 Measures of geometrical deviation: (a) wall thickness, (b) curvature of walls, (c) profile warping, (d) profile twisting, (e) angular deviation.

ticles from the billet cause the defects. In the first case dies should either be polished and nitrided [8,9] or scrapped. Oxide build-ups behind bearings are often the result of too small a relief area (Fig. 13), and this phenomenon can be reduced by either increasing the relief or by using nitrogen shrouding behind the bearings in order to avoid the oxidation. Hard particles in the extruded sections often originate from oxidized and even dirty material on the billet surface even when performing direct extrusion. Inflow of material from

billet surface will take place if the container temperature is too high, causing a low yield resistance and easy flow of surface material [10]. Inflow to the bearing channel may also take place if too large a profile is extruded or the container is misaligned so that surface material may flow directly out of the bearing channel. Hard particles may also exist in the interior of the billet if it is not properly homogenized, or, in the case of 6061 and 6063 alloys, the iron content is so high that AlFeSi particles are formed. And to complicate matters further, research

Table 2 Tolerances on Transversal Curvature After DIN 17 615

Measure of width of wall, b, from [mm]	Measure of width of wall, b, to [mm]	Tolerance on straightness, e, [mm]
—	40	0.20
40	60	0.30
60	90	0.40
90	120	0.45
120	150	0.55
150	180	0.65
180	210	0.70
210	240	0.75
240	270	0.80
270	300	0.90

Table 4 Tolerances on Profile Twisting After DIN 17 615

Width, b		Length, l					
From	To	0–1000	1000–2000	2000–3000	3000–4000	4000–5000	5000–6000
—	25	1.0	1.5	1.5	2.0	2.0	2.0
25	50	1.0	1.2	1.5	1.8	2.0	2.0
50	75	1.0	1.2	1.2	1.5	2.0	2.0
75	100	1.0	1.2	1.5	2.0	2.2	2.5
100	125	1.0	1.5	1.8	2.2	2.5	3.0
125	150	1.2	1.5	1.8	2.2	2.5	3.0
150	200	1.5	1.8	2.2	2.6	3.0	3.5
200	300	1.8	2.5	3.0	3.5	4.0	4.5

All Measures in mm.

has shown that the presence of die lines will be heavily dependent on the size of the bearing angles when defects are caused by AlFeSi particles.

The profile surface may contain blisters as shown in Fig. 12c, which may be inclusions of air, oxides, or even oil. Such blisters degrade surface appearance and mechanical properties, and when the profile is cooled, blisters may also crack, creating a distinct sound and causing open holes in the profile surface. They may be a result of inflow of material from the surface of the billet or inclusions that are already contained in the interior of the billet prior to extrusion. However, blisters containing air are often caused by air trapped in the container due to too fast upsetting or air trapped in hollow porthole dies prior to extrusion. Hence they may very often be found close to the forward part of the profile or charge welds. When extruding light metals, lubricants are not used. Contamination of oil must therefore be avoided. Lamination is also a result of contamination and should be avoided in order to preserve both mechanical strength and, in some cases, also appearance. This phenomenon may be a result of inflow of oxidized and contaminated material from the surface or from the charge weld.

Sometimes, when studying an extruded aluminum profile, areas of different shades of gray color may be detected. This surface defect is called structural streaking and is a result of varying reflective properties across the surface. While streaking seldom will be a reason for scrapping the material, efforts are often made to extrude under conditions that give even optical properties. This is especially the case if surface treatments such as etching and/or anodizing are to be employed because these processes tend to accentuate streaking. Streaking is a result of variations in grain size and grain orientation of the finished product, and three somewhat different types have been identified. These are streaks caused by variations in bearing surfaces, variation in temperature, amount of hot work, and recrystallization. However, as streaking in general is a result of the whole thermomechanical and metallurgical treatment of the material, it is not always possible to differ between the different types. So-called bearing streaks are often due to uneven bearing surfaces, having created depressions in the profile surface. Hence light is reflected in different planes and streaks are only visible when viewed from specific directions. A type of combined bearing and grain size streaking is caused by sudden shifts in bearing lengths due to varying profile thickness. Apart from leading to problems connected to filling, such a die

Table 3 Tolerances on Longitudinal Curvature on Profile After DIN 17 615

Length to l [mm]	1000	2000	3000	4000	5000	6000	Above
Tolerance h [mm]	0.7	1.3	1.8	2.2	2.6	3.0	3.5

Table 5 Tolerances on Deviation from Straightness of Angles

Width, b, from [mm]	Width, b, to [mm]	Tolerance deformation, w
—	40	0.3
40	100	0.6
100	300	0.8

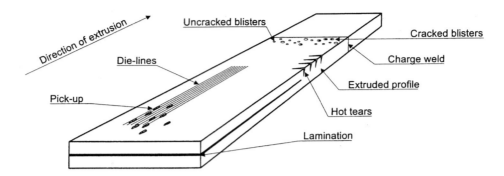

Figure 12   Visualization of various surface defects.

design will give abrupt changes in grain sizes and orientations and also reflective properties over short distances in the profile. Furthermore, as the amount of heating is varying, the temperatures and the degree of recrystallization may also be expected to change. Oxide streaks can be recognized as dark streaking areas of varying width and intensity on etched or anodized parts.

The presence of such streaks is linked to the inflow of oxidized material from the billet surface to the bearing channel. Oxide streaks will be avoided if inflow is hindered. In general, structural streaking will be less of a problem if material in all parts of the profile cross-section undergoes much of the same thermomechanical loading. Furthermore, experience has shown that a minimum degree of choke on bearings should be sought in order to reduce variation in reflective properties.

Cracking or tearing of the extruded profile is experienced when process control is lacking. If the temperature of the extruded metal on the bearings is too high as a result of preheating of billet or high extrusion rate, partial melting of Mg_2Si particles will take place [11].

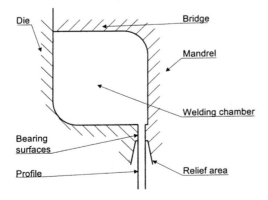

Figure 13   Detail from a porthole die: An area of relief may be found behind the bearings.

Thus cracking of profile surface or so-called hot tearing may be the result. If cracks develop in the die or pick-ups are created through insufficient clearance behind bearings, edges may be torn.

Surface quality is also often reduced through scratches and gouges asserted in material-handling processes on the run-out table, lift-overs, walking beams, saw tables, or in stations for packaging and stacking. Efforts must be made so that profiles are not damaged in transport. Great emphasis has been laid on establishing best practice in handling materials and on securing that the equipment used can only cause minimal damage. However, because of the large variability in profile geometry and the great demands on material handling, many operations must be performed manually at high costs.

III.   APPLICATIONS, ALLOY SELECTION, AND DESIGN CONSIDERATIONS

By developing products out of materials found in nature mankind has managed to differentiate itself from the animals. The fact that historians have tended to denote periods by the names of the materials found in the tools and other equipment of the times indicates that one assesses the use of materials to be of prime importance to human life. Today, one recognizes the stone, bronze, and iron age, but not surprisingly, when dealing with the present, the determination of the material with the greatest importance turns out to be harder. Some emphasize that relatively new materials such as plastics are omnipresent and have brought enormous changes to human life over the last decades while others maintain that the increasing use of computers marks the entrance to the silicon age. However, the fact still remains that iron in the form of steel even today by far is the most favored material for most applications. The strongest contender of the metals is aluminum, but as for volumes produced, steel is about 20 times larger. So, despite the several industrial revolutions that have taken place in the last century, it may still be claimed that the contemporary period is the one that was allegedly introduced by the Hittite development of iron around 1300 B.C. [12].

A. Product Development

Innovative thinking is maybe the one most important virtue of a designer, but if ideas are to be transformed into innovations, proper use of knowledge of and experience with both design principles, processes, and materials is mandatory. In fact, modern product developers often stress that focus must be placed not only on the mere functional properties of the product and the ability of the product to satisfy consumer's demands but also on the chain of processes from raw material to components, joining, surface treatment, and assembly. Obviously, a quality product will not only be satisfactory to the user but also to the producer in that it creates the possibility to generate a surplus.

Three aspects are of equally great importance when creating a new component that satisfies the user's notion of quality. In Fig. 14 product development is given as a combination of function, production, and material. The functional side is linked to the transforming of new concepts of, for instance, physical or structural origin to products that are of lasting value to the user. The goal for the production system is to establish processes and routines so that new products can be manufactured within calculated costs and time limits. When developing a new product one should attempt to make use of possibilities offered by manufacturing systems to forward functionality rather than letting the system impose restrictions or additional costs. Obviously, upon having decided on a concept, the most optimal production process should always be sought. However, one should also bear in mind that manufacturing processes automatically offer new degrees of functionality to the product. Therefore the alternative candidates of production methods should be evaluated at a very early stage in the design process.

Another fundamental aspect of product development is the use of materials. Traditional use of materials is limited to choosing on the basis of only certain tabulated data such as yield limit or tensile strength and, in some cases, corrosion properties, and often because of limited knowledge of alternative materials, steel is the most favored material for most constructional applications. However, materials should not only be chosen but also designed so that both functional demands are satisfied and so that production is simplified. For a long time, more advanced users of materials, such as the aircraft industry, have realized that optimal solutions can only be reached by making use of the whole specter of new materials and by obtaining knowledge about and manipulating the microstructure. In this way, the design process has been brought all the way down to atomic level, thus spanning more than 10 orders of magnitude (Fig. 14). Over the last decade it

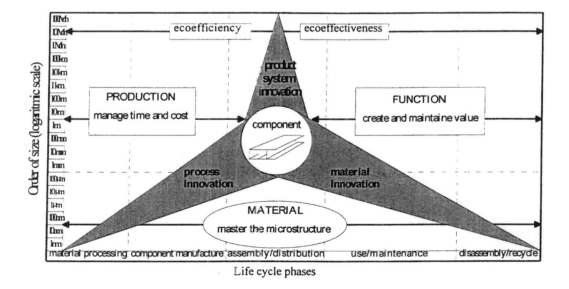

Figure 14 The life cycle view (LCV) of a component.

has become apparent that thorough knowledge of material properties is a prerequisite in almost all design work and that the traditional practice is in accordance neither with customers' demands nor with international standards.

As shown in Fig. 14 product innovation is taking place in parallel to innovation connected to the properties of process and material. New or improved materials or processes will increase the possibilities to create products with new functions or improve older products. As shown, a large number of fields of research will in one way or another contribute to the development of new and better products. As the possibility to improve products increases, so do consumer demands to quality. While factories earlier could produce and sell enormous numbers of standardized goods with often inferior quality, today's consumers demand products with

which they can identify themselves and which are virtually perfect. An example of enterprises that meet such demands each day are of course those of the automotive industry.

Furthermore, one has over the last couple of decades witnessed an increasing consciousness of environmental protection and sustainability, and in the very recent years, this emphasis has not only focused attention to cleaner production, health, and safety within the individual production plant but also on the product life cycle and loop closing of products, components, and materials. A consequence of the interest in including industrial ecology and ecodesign into industrial practice is that important new concepts and methods are under development [13]. They are life cycle assessment (LCA), life cycle cost (LCC), eco-efficiency, and eco-effectiveness. It is not possible to go into these concepts here, but

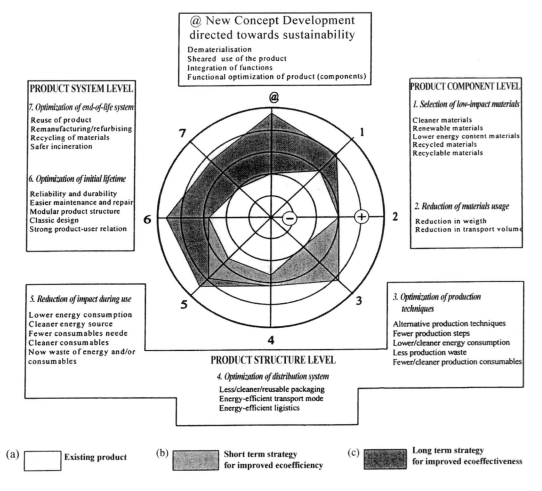

Figure 15 The ecodesign strategy wheel (ESW) with (a) reference product profile, (b) short-term development strategy, (c) long-term development strategy.

the combined ecologic and economic life cycle performance of products and processes will probably be among the more important features of successful products and processes in the years ahead. Thus it seems natural that the designer must think more in terms of establishing life cycle systems than isolated products. Innovations for exploiting the life cycle merits of aluminum alloys should be the hallmark of aluminum components and structures. Figure 14 shows a life cycle view (LCV) of a component, placed in a life cycle time scale (abscissa) and an order of size scale (ordinate). Here the interactions between the details and the whole are visualized, so that critical success parameters can be identified more easily. Very often one sees that successful products are a result of combined innovations of process, material, product, and production.

In the UNEP-manual [14] van Hamel has designed a so-called "ecodesign strategy wheel" (ESW) as a tool for formulating strategies for improvements of economic and ecological performance of products, processes, and practice, both in a short-term and in a long-term perspective (Fig. 15). An existing product is used as a present time reference of improvement (improved eco-efficiency) and a scenario of a possible future sustainable society as the goal to strive for (measure of eco-effectiveness). It may be useful to see the connection between Figs. 14 and 15. The LCV can be transformed into the ESW by bending the LCV into a cylinder and view it from above (Fig. 16).

As can be seen clearly, the work of product developers is both getting increasingly difficult and challenging, and greater demands are placed on their ability to master all aspects of the product development process. Innovative thinking and a well-documented understanding of customer's needs will of course still be at the center of attention, but the model discussed may be an important tool when establishing an integrated method of product development, which in any way is bound to take function, production, and material aspects into account.

Table 6 gives an overview of extrusion-based systems and components and the generic alloy selection for the different applications areas. With a generic alloy selection one here understands an alloy, generally used for the specific application, which will be available in most markets and with properties best documented. It should then be the alloy first selected by the designer and only be changed if special combinations of properties, not found in the generic alloys, are needed. It is then important to contact potential suppliers of the section to ensure that the alternative alloy is available to acceptable price and with properties documented. A short description of some typical products in these main application areas are given below. It is recommended, however, to actively collect information brochures and inspiration material from aluminum producers, extrusion plants, and final-product manufacturers as basis for understanding the rich diversity of design solutions based on extruded aluminum rods, tube, and sections that is possible [15].

B. Designing with Aluminum Sections

Although one may find that the ideal material to be used for a product is aluminum and that manufacturing should take place in the form of extrusion of profiles, there is still great freedom with regard to the functions that the product could fulfill [16]. Aluminum profiles are used in a number of applications, and the following five groups may be identified:

- Buildings, architecture, and furniture
- Structural
- Transport
- Heat exchangers and electrical conductors
- Durables and mechatronics

A short listing of some typical products in these main application areas is given in Table 6. Although profiles are produced as long beams and, thus, originally only contain variations in two dimensions, forming or cutting processes may be used to modify the product so that

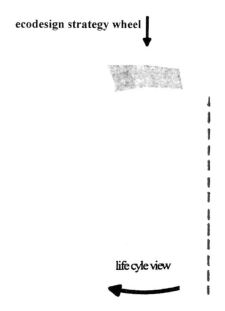

ecodesign strategy wheel

life cyle view

Figure 16 Two aspects of life cycle thinking.

Table 6 Applications of Extruded Aluminum Rod, Tube, and Sections

Main process	Systems components	Generic alloys	Selected applications				
			Building architectural furniture	Welded structural	Transport	Heat exchangers, electrical conductors	Durables, mechatronics
HOT EXTRUSION OF ALUMINUM * ROD * OPEN SECTION * TUBE * HOLLOW SECTION,	systems of section	AA6060 AA6063	windows and doors shelves shower cabinets		utility vehicles	conductors cables multiport tubes	housings
	structural sections	AA6082, AA6061	building structures scaffolding	Bridges	utility vehicles		ladders stairs
	bended and locally formed sections	AA6060, AA6063, AA6082	furniture ship windows frames	Space frames	bumpers seating spaceframe	bended tubing in heat exchangers	
	components manufactured by shearing, drilling, milling, etc.	AA6060, AA6063	hinges building elements			cooling elements	tooling hosing multifunctional components
	cold drawn tubes	AA3103			radiators air condition systems fuel lines	air condition systems heating equipment	
	coldforged components from rod, tube, and section	AA6351			steering column		
	hot forged components from rod	AA7075			air craft components		

parts with variations in all three directions may be obtained. Furthermore, as a large number of alloys with a specter of properties have been developed, the choice of aluminum puts very few limitations on design. It is recommended to actively collect information brochures and inspiration material from aluminum producers, extrusion plants, and final-product manufacturers as basis for understanding the rich diversity of design solutions based on extruded aluminum rods, tube and sections that is possible.

Aluminum is, at present, undoubtedly one of the most popular engineering materials. Its popularity is mirrored by the fact that the metal, in terms of volumes produced, is second only to steel and as large as all other nonferrous metals combined although it was first introduced only about a hundred years ago. The reason for the material's enormous success lies in its thermal, mechanical, and electrical properties, or, in short, in its microstructure. Aluminum is light, but some of its alloys are able to compete with comparable steel alloys in terms of both yield limit and tensile strength. At the same time much less effort is needed in hot forming of aluminum than of steel with the same yield limit at room temperature. This is because aluminum has a rather low melting point and that hardening processes, such as aging, are taking place. Furthermore, although Young's modulus of aluminum is only 1/3 of the modulus of steel, aluminum beams can be stiffer than steel beams of the same weight because aluminum is lighter than steel. The ratio of Young's modulus to weight is about the same for the two materials, but such a comparison is unfair to aluminum because the lower density offers the opportunity to place material at farther from the neutral axis, thus creating larger momentum. Aluminum is also a very efficient conductor of both heat and electricity. Because an undamaged and untreated surface reflects almost 90% of visible incoming light, aluminum exposed to direct illumination will be extremely resistant to heating. This makes the material useful for structural purposes. If, however, aluminum attains higher temperatures elongation will be larger although thermal stresses will be smaller, a result of the aluminum's higher thermal coefficient of elongation but lower elastic modulus. One of the appealing characteristics of many aluminum alloys is a high corrosion resistance.

Over the last 30 years questions have been raised as to whether the production of aluminum is in accordance with the principles of sustainable development and related thinking. The major concern has been that although aluminum is the most common metal on earth, so-called high-grade deposits of bauxite are, to a certain extent, limited. Another objection has been related to the large amounts of energy needed to produce primary aluminum. The aluminum industry has responded by developing LCAs and by modifying and improving the Bayer process so that it accepts bauxite that was formerly assessed as low-grade. However, it turns out that recycling of aluminum is a result of market mechanisms and the properties of aluminum and not of state legislation as is the case for plastic materials and, to a certain extent, steel [17]. The fact that a much larger amount of energy is needed to produce primary aluminum from raw materials than secondary metal through remelting makes efforts invested in recycling highly profitable. One is today talking about an aluminum bank, which exchanges metal with the market by selling finished products and buying scrap. A closed loop has long ago been established. At present, about 30% of the material going into new components is of secondary kind. The driving force in the material bank is of course energy, usually supplied by hydroelectric power plants. Had it not been for human interference aluminum in the form of pure metal would be nonexistent in nature as the spontaneous process of oxidation through the times has degraded all material, leaving only Al_2O_3. Large amounts of energy are needed to extract metal from its oxide. Hence aluminum metal might be looked upon as an energy investment. As questions are raised regarding the soundness of using energy producing metal one has to assess the alternatives, i.e., merely comparing energy investments. Much smaller energy investments are done when producing steel, but one might say that the oxide layer of aluminum represents a more secure bank than steel does. However, by employing light metals in, for instance, the transport sector one can obtain reductions

in fuel consumption, which is the equivalent of being paid back on the initial energy investment. In structural applications the payback on the energy investment cannot be so easily detected, but if the need for maintenance or use of materials can be reduced, energy is eventually saved. This analysis should be an integrated part of the LCA and product development process, and the designer should always ask whether energy can be gained by applying aluminum in a construction. In the aircraft industry a conclusion on this question was reached, if not formally, but at least intuitively, already in the 1920s by the construction of the world's first aluminum aircraft Ju-7. Today, airframes consist about 70% to 80% by weight of aluminum, of which a large part is in the form of profiles. In other parts of the transport industry demands for lightweight have until recently not been that strict, but as environmental issues are pressed and competition is increasing both with regard to prices and velocity, new materials and concepts are brought forward. Examples of these are the employment of aluminum in high-speed trains such as the French TGV duplex, the German Maglev system, and the Japanese superconducting Maglev system. Changes are even taking place within the shipping industry, which traditionally has been viewed as a notoriously conservative enterprise. Passenger transport has already for some time been carried out by aluminum-intensive fast ferries, and so-called high-speed surface effect ships for transport of goods are also under development. As aluminum frames are being employed to greater extent in both buses and commercial vehicles, the next breakthrough is expected to take place in the automotive industry. However, although LCAs are performed in this industry, consumers still tend to look more at the initial cost than costs related to use. Hence the lightweight solutions are not sought as vigorously as in the other branches of the transport industry. New cars contain about 70 kg of aluminum parts, and smaller production series of more expensive cars have been made with so-called space frames of aluminum. Such frames comprise the structural elements of the car.

In connection with material forming, one should notice that aluminum is unique in the sense that it can be extruded to a beam with a cross section of almost any form, open or hollow. Because of the large forces generated when extruding steel, geometrical forms must be kept simple and only smaller reductions in sizes of cross sections can be obtained. Another light metal, of which some use has been made in, for instance, the automotive industry, magnesium, is less extrudable because of its hexagonal close-packed structure.

A world of new opportunities with regard to functionality and form arises when designing with aluminum profiles. In all there are very few limitations to the forms that can be produced by extrusion of aluminum, the largest problem being that variations in geometrical features are only two-dimensional (2D) in nature. However, beams may be given different lengths and profiles can also be altered by the processes of bending and hydroforming. Therefore fully 3D structures as the space frame of a car or a window frame may be designed. The case in the following chapter gives an example of a simpler but successful design that managed to not only satisfy the original demands imposed on functionality but also incorporate other useful functions. This is a result of the freedom that aluminum, as a light metal, and extrusion, as a versatile process, offer. The most serious restrictions encountered in the design process are caused by designer's experience with steel constructions. Steel products comparable to extruded profiles are manufactured in standardized forms and dimensions, and a steel design will often be an assembly of a series of such parts. Machining operations must be applied in order to impose modifications. Although aluminum profiles of standard shapes are sold, it is usually better to think in terms of new profile designs better suited for the application. Extrusion tools are relatively cheap to produce. The prices of open dies can be expected to range from $1000 to 2000, while hollow dies will be more expensive, from $1500 to $4000. Tools for complicated profiles will naturally be manufactured at higher costs. However, if profiles can be made so that machining, welding, and assembly operations can be avoided, investments in dies may be worthwhile even for smaller production series. Hence the development of tailor-made products may prove to be cheaper than mass production of standardized products with a simpler shape. This fact indicates that the extrusion process is well adapted to consumers' demands and their notion of quality.

As for functions that can be integrated in a profile, limitations are only imposed by the human imagination. An example of a function that is in common use is the locking mechanism, which simplifies assembly and reduces the number of parts needed in the design. Generally, such a solution also reduces the cost of the product and is often necessary to secure competitiveness. Figure 17 displays examples of various locking devices. The basis for all these is aluminum with relatively low modulus of elasticity.

Cast and extruded aluminum parts complement each other in most applications, but if both processing

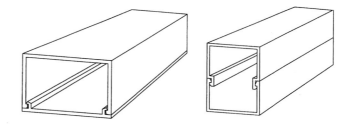

Figure 17 Joining of sections by the use of snap solutions.

methods can be used to produce a part, extruded products are usually preferred. Tools needed for casting operations are more expensive than dies, and the production rates of the two processes are not comparable. This indicates that wherever possible, solutions including profiles should be sought. Another reason for choosing profiles to cast products is that wrought alloys usually have better mechanical properties than cast alloys. This is partly because of defects created during casting and partly because cast alloys contain large amounts of silicon and copper, which cause a heterogeneous structure with brittle secondary faces. In general, cast alloys have lower elongation and strength, especially in fatigue. The progress in improving casting alloys and controlling the casting process in recent years, however, has been impressive, both for aluminum and magnesium alloys, as well as for steel. The designer should therefore take care to make process selection based on the present state of the art.

Alloy development is the subject of continuous research. By systematically varying the content of different alloying elements improvements in properties such as tensile strength, ductility, fracture strength, fatigue strength, corrosion resistance, and formability are sought. At present about 350 wrought alloys are commercially available, but not all of these are interesting from a designer's point of view. While the aluminum industry continuously must seek alloys with improved properties, the designer should concentrate on a group of so-called generic alloys. In Table 6 an overview of extrusion-based systems, components, and the generic alloy selection for different application areas is given. A generic alloy selection is taken to be an alloy, which is generally used for the specific application, is available in most markets, and has properties that can be expected to be thoroughly documented. It should be the alloy preferred by the designer, and the choice should only be changed if special combinations of properties, not found in the generic alloys, are needed. It is important to contact potential suppliers of the section to ensure that the alternative alloy is available to acceptable price and with properties documented. When choosing particular alloys one must also remember that an alloy, which is optimal with regard to all properties, does not exist. Alloys with high yield and tensile strengths are usually harder to extrude, thus resulting in lower production rates (Fig. 18), higher prices, and limitations on product geometry. Thin-walled sections of high-strength material are for instance not extruded easily. Some alloys are also in possession of relatively low corrosion resistance although mechanical strength may be high. The group of generic alloys should contain elements that can be used for most applications.

Profiles containing precipitation hardening alloys are relatively easily formed but gain high strength after heat treatment. This explains why about 80% of all extruded products are made of the 6XXX series of alloys. Members of this group can gain from medium to high strengths. High extrusion speeds and very high productivity can be obtained when extruding the alloys with medium strength. 6XXX alloys are generally relatively corrosion resistant, but this property is both dependent on the chemical composition and on the thermal treatment the material has undergone. While the alloy 6060 contains limited amounts of magnesium and silicon and is only of medium tensile strength the high alloy metal 6082 has a tensile strength of about 340 MPa at room temperature in the T6 condition.

For sections that are not carrying loads the 6XXX series is the natural choice. In such cases even the 3XXX

and 1XXX alloys may be applied as these alloys are in possession of superior corrosion and conduction properties. If a product is designed to carry loads, different 6XXX alloys may still represent alternatives, but one should always evaluate whether it is profitable to reduce weight by using alloys of higher strength. It has been found that both in structural applications and in many areas of transport designing with the 6XXX series is most rational because of the low cost of extrusion. In some branches of transportation such as aviation and of course space flight the use of material with the highest strengths is mandatory. Production costs connected to extrusion will be substantial, but by decreasing weight or increasing the load capacity these expenses are soon covered.

The strengthening mechanism of the 7XXX series is also precipitation hardening, and as for load capacity and strength these alloys represent the next step on the ladder. In aerospace construction 7075 is a preferred alloy. The automotive industry has also made use of several of the 7XXX alloys. However, the fact that dies will experience more wear and that extrusion rates will be lower for higher strength metals applies also to this group and is limiting its usefulness in most applications. The low extrusion rates will not only be a result of the material's higher flow resistance but also of a somewhat lower melting point of some phases. Besides, if the quench rate after extrusion is low, corrosion resistance will generally be unsatisfactory. Larger contents of copper are causing the high strength/low extrudability as well as the degraded corrosion properties.

The 2XXX series has also been traditionally used in aerospace construction and shows extremely good damage tolerance. It has both high fracture toughness and high resistance to fatigue crack propagation. As the 2XXX alloys contain much copper, they tend to show low corrosion resistance. Important alloys are the 2X24 [18] and the 2X19. The 2020 and 2090 alloys are so-called lithium alloys. For every weight percent of lithium added, the elastic modulus of the material is increased by 6% and the density lowered by 3%. Hence very stiff and light aluminum constructions can be developed by the application of lithium alloys. Use of such materials has been made both in fighter aircrafts and space shuttles, but only to a limited extent in commercial aircrafts. Another important property of the 2XXX series is that high strength can be obtained at relatively high temperatures. This, however, complicates extrusion, and products made of 2XXX alloys are today mainly manufactured in other forming processes.

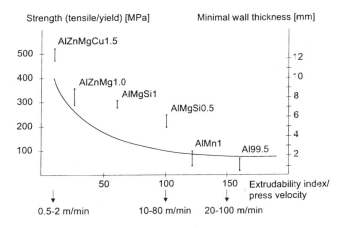

Figure 18 Relation between material strength, minimum wall thickness, and extrudability/press velocity. The results apply only to a section with a specific geometry, but similar curve may be established for all profiles.

C. Limitations on Section Design

Although one should focus on possibilities when designing with extruded profiles, one is sooner or later bound to encounter the limitations that the process imposes. Evidently these limits are dependent on both the process equipment and practice and on the choice of material. If the process is not properly controlled or the design and choice of material is not in accordance with the choice of process, poor product quality will unavoidably be the final result. In the last part of the second section of this chapter some of the symptoms of low product quality were discussed, and their cause has been and will be further discussed. Suppliers of extruded profiles have established general design rules, which can be used to secure that a design is in harmony with the process. Some of these are general in character, while others are referring to specific dimensions and are necessarily dependent both on material and process equipment. However, apart from restrictions on the size of the largest sections produced, there are seldom any absolute limitations. Very few profiles may prove to be impossible to manufacture, but there is always the danger that the quality or the price of the product may be unsatisfactory to the consumer. In order to develop a quality product, the designer must try to reach production-friendly solutions through discussions with experienced people at an extrusion plant.

If full freedom in designing a functional product is to be obtained, no limitations should be placed on form. However, as will be understood from a study of flow patterns, a key word in relation to the process of extrusion is symmetry. Asymmetric profiles cause flow imbalance and necessitate complicated die design. Flow velocity in the cross section must be controlled, and this often leads to low extrusion rates. Furthermore, as dies may experience uneven loading, there is an increased danger of fracture and unstable tools, especially when extruding higher strength alloys. Asymmetric profiles may also cause thermal gradients in both die and profile during cooling. Hence not all parts of the profile will be given the same thermomechanical treatment, and the result of this loss of control of metallurgical processes is a product with a large variation in both microstructure and properties. A last problem connected to asymmetric profiles is that possible bending and stretching operations may be more complicated to perform. Examples of deviations from symmetry are given in Fig. 19. The mass distribution over the profile cross section should not be uneven. Large ratios between the thickest and thinnest walls in a profile may also be difficult to handle, and large eccentric hollows also cause an unwanted flow pattern.

Naturally strict limitations exist with regard to the size of the profile, and specific numerical values must be sought from the producer. If the profile has too large a circumference circle diameter, i.e., the smallest circle surrounding the profile, problems connected to inflow of material from the billet surface may arise. Furthermore, extrusion of large profiles is often synonymous with very open die designs, which usually are weakly supported, and in the case of hollow profiles, larger

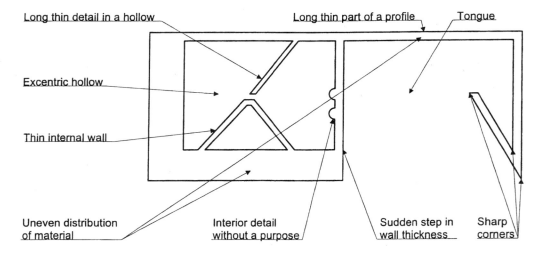

Figure 19 An extruded profile with a complex geometry.

forces on the mandrel of the die are generated. The result will be larger dimensional variations and also poorer surface quality because of either die lines or streaking. On the other hand, profiles with too small dimensions give larger press ratios. In this case the press may not be able to supply the needed force to generate the profile. The solution is then often to press several strands simultaneously in order to increase productivity (Fig. 5).

Simplicity is another key word in almost all areas of production. This certainly also applies to the extrusion process although its largest virtue maybe is the complex profiles it offers. Both hollow profiles and profiles with large tongues tend to increase the complexity of the production and thus should in fact be avoided if possible. Dies that are made for such profiles are more complex and necessarily also more expensive. However, the largest problem is that they tend to generate larger forces because of the restrictions they impose on flow, and that their weaker design at the same time gives them a lower load capacity. Die breakage and large deflections are usually the results. Dimensional variability and poor surface quality can usually be expected in such cases. The solution may often be that tongues are made with smaller length-to-width ratios so that they can be properly supported and that hollow profiles are made as two open sections, which later can be assembled.

Many special forms and features are usually included in the extruded profile so that the product is able to perform a large number of functions. However, the addition of even small attributes may lead to large changes in both productivity and product quality, and one should always assess whether a feature is necessary and whether it could be made in a more process-friendly manner. Examples of features that are difficult to handle include the following:

- Sharp corners may, in some cases, cause incomplete filling and in others tearing (Fig. 19). Besides, crack initiation and growth usually take place in the parts of the die where sharp angles exist. Hence corners should always be rounded. The smallest radiuses that should be used are around 0.4–1.0 mm. Only for special purposes may a radius of 0.2 mm be applied.
- Sudden steps in wall thickness and thus also the bearing lengths of the die should be avoided (Fig 19). Two related types of flow problems will be the result of such changes. On the one hand, complete filling of the sharp corners around steps in wall thickness will be hard to perform if bearings are not carefully designed. On the other hand,

streaking may be the result of microstructural differences caused by the abrupt change in the conditions of deformation and cooling from areas of large to small thickness. The solution to this problem is a gradual change from thicker to thinner regions. If this proves impossible, the streaking can be made less appalling simply by constructing a notch that may mark a natural border between thicker and thinner regions. However, this does nothing to decrease the problems connected to varying mechanical properties and may in turn cause a larger flow resistance.

- At a certain point extrusion of thin-walled parts will always represent a problem. For such parts friction forces are high, and flow speed is hard to control. The result may be incomplete filling of some parts of the cross section and of course a need for greater extrusion force. If internal walls are too thin (Fig. 19), complete filling is even harder to obtain, and pressures in the welding chamber may not be sufficiently high to secure a proper weld. Weak or nonexistent seam welds are both disastrous to profile quality and hard to detect. Hence internal thin walls should be avoided. Often it may be better to use larger wall thickness to secure filling although material cost increases. Another problem with internal walls is that the whole die concept becomes weaker and that there is a greater danger of die breakage.
- Details generally cause problems connected to flow balance and friction heating, and as a result, production rates are limited (Fig. 19). Furthermore, tearing of the surface and incomplete filling may be additional problems. Therefore details that do not fill any functional purpose should not be added to a profile. Obviously, one should try to design details sufficiently thick and not too long. At the same time corners must be rounded. One also ought to bear in mind that internal details are usually more problematic than external.

Apart from these limitations on form connected to the process, there are many others that are related to the use of the product. One such limitation may, for instance, be that corrosion should be prevented by avoiding geometry that can lead to the gathering of water in the profile. Whereas designers who are not accustomed to working with aluminum profiles easily may overlook one of the many restrictions imposed by both process and material aspects, the experienced ones will develop a product which integrates most of the aspects previously mentioned. However, to all designers, the establishment of a method of design that systematically incorporates the treatment of all aspects of importance is a necessity when applying aluminum for constructional purposes.

D. Case: Helicopter Landing Deck on Offshore Platform

Usually aluminum profiles are most competitive in applications where they must be designed to fulfil many functions at the same time. Helicopter decks on offshore platforms are such an application. On platforms, such as the tension leg platform, Snorre, in the North Sea, weight aspects are often critical. Furthermore, the corrosion properties of the material must be outstanding, as weather conditions are often very harsh. Originally, helicopter decks were made out of steel plates that were welded together and supported by traditional steel beams, but the solution was by far optimal. The construction has gradually been modified, and the aluminum deck in use today is about 60% lighter than the original one. By using profiles, large design flexibility has also been obtained, and many more functions have been implemented in the construction. It was expected that the surface construction of the helicopter deck should perform the following functions:

- Carry structural loads
- Carry concentrated loads
- Carry torsion loads
- Be simple to assemble
- Prevent slipping
- Lead away petrol and rain water
- Allow circulation of air in order to prevent crevice corrosion
- Lead fire extinguisher fluid
- Lead deicing cables

Demands were that the helicopter deck had to be designed so that it easily could be fitted to the platform, and that it was constructed in accordance with regulations and standards. Evaluation criteria would be related to the weight and the strength of the construction and to corrosion properties. Furthermore, as almost always the product had to be evaluated on the basis of life cycle cost.

The original steel construction did not have the desired functionality, but it was able to carry the specified loads. Hence when the first modifications in aluminum were made, the design was not altered but dimensions were changed to suit the properties of aluminum better. Of course, modifications to dimensions can be done in a number of ways. If the length of the rung is kept constant the thickness of the profile may

be increased about three times, resulting in a profile of the same weight and only marginally increased stiffness. A simple calculation can be made to show the effect of a uniform thickening of the whole profile. The geometry is shown in Fig. 20. The moment of inertia can be calculated to be:

$$I_x = \frac{1}{12}\left(\left(8 \cdot h_1^3 + 12 \cdot h_1^2 h_2 + 6 \cdot h_1 h_2^2\right)\right.$$
$$\left. \times w_1 - h_2^3 w_2\right) \tag{1}$$

For the steel beam both h_1 and w_2 can be set to 2, and h_2 and w_1 to 1. This gives a moment of inertia equal to:

$$I_x^S = \frac{1}{12}\left(5tl^3 + 12t^2l^2 + 8t^3l\right) \tag{2}$$

Because the density of aluminum is about 1/3 than that of steel the dimensions h_1 and w_2 may be set to $3t$. This gives a moment of inertia of:

$$I_x^A = \frac{1}{12}\left(15tl^3 + 108t^2l^2 + 216t^3l\right) \tag{3}$$

The stiffness obtained for steel and aluminum is respectively:

$$W_x^S = \frac{E}{12}\left(15tl^3 + 36t^2l^2 + 24t^3l\right) \tag{4}$$

$$W_x^A = \frac{E}{12}\left(15tl^3 + 108t^2l^2 + 216t^3l\right) \tag{5}$$

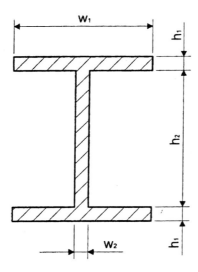

Figure 20 Simple I-beam originally in use in helicopter decks and manufactured through the processes of rolling and welding.

E represents the elastic modulus of aluminum. As l is much greater than t the first term will be dominant, and it can be seen that no large gains can be achieved by using aluminum in this way. However, a much larger increase in stiffness can be obtained if the rungs are made slimmer and taller and the flanges are designed thicker. If for instance h_2 and b_1 are set to $3l$ as the thickness of both flanges, and if the rungs keep the thickness t, the profile can gain a stiffness of:

$$W_x^A = \frac{E}{12} \left(135tl^3 - 108t^2l^2 + 24t^3l \right) \tag{6}$$

Clearly, by intelligent design bending stiffness may be increased enormously. However, the calculations just made are extreme cases. Problems would arise if the last cross section were to be used, both because the flanges could not carry the possible concentrated forces and because rotations due to torsion of such a section would be large. The torsion momentum would depend on the thickness of the section in the third power. The torsion stiffness of the original steel section will be about one third of that of the extreme aluminum section. The aluminum profile, however, will be more severely loaded because of the long flanges and rungs. The optimal cross section of the type given in Fig. 20 could be reached by maximizing both bending and torsion stiffness with respect to the different measures. Such an analysis reveals that the use of aluminum profiles generally is preferable to the use of steel. However, when constructing with profiles even larger gains in torsion stiffness can be made by applying hollow profiles. By turning to extruded profiles large freedom in functionality can be obtained, and a positive side effect of this is that the profile may be given such a form so that all the desired functions mentioned above can be fulfilled. Different stages in the development process are shown in Fig. 21. The same figure shows the aluminum profile that is currently in use in platform decks in the North Sea.

IV. THE EXTRUSION PROCESS

A. Describing the Conditions of Flow

Traditionally, advances in extrusion technology have come as a result of experimenting. In fact, as the process has until recently been viewed as too complex to be understood in its entirety, the key to success in the extrusion business has been the establishment of a system of best practice based on experience gained through trial and error. Experiments are still expected

to be an important element in establishing knowledge of fundamental aspects, and because of the large amount of noise in the data received from extrusion presses, one can hardly expect to avoid errors, from which insight should be gained. However, the impetus for discovering the improving methods for predicting results a priori is large. Experiments are usually expensive and time-consuming, and practices such as die correction, which even to this day to a large extent is based on the experience of the workman, will most probably not satisfy demands to either productivity or quality in terms of, for instance, dimensional variability. Hence research is today focused on the development of mathematical methods that are able to describe and thereby also predict both the macroscopic and microscopic changes taking place during deformation. In this way the manufacture of dies and the control of process parameters can be performed in such a manner that profiles will be produced within the narrowest tolerances and to a low cost. As extrusion is a process taking place in the presence of large deformations and at relatively high temperatures an establishment of such a mathematical fundament is both a complicated and to a large extent a multidisciplinary task. A unified approach must link knowledge of metallurgical processes at high temperatures and strain rates with the continuum theories of rheology [19] or plasticity. Furthermore, as analytical results then hardly can be obtained, the development of effective numerical procedures will be of greatest importance. Naturally, a thorough understanding and experience with all aspects of the process will be a prerequisite for establishing a model.

1. Experimental Studies of Flow

Hot extrusion of aluminum is performed at temperatures from about 450°C to something above 600°C, depending on the melting point of the alloy. At these temperatures the material has a relatively low resistance against dislocation movement, and shear deformation will therefore be initiated when the extrusion force reaches a certain limit. The material then starts flowing out of the die and will permanently change shape. If homogenous deformation had taken place, the longitudinal logarithmic strain would be equal to $\ln(R)$. Strains of magnitude $\varepsilon_z = 4$ are therefore not unusual as profiles quite commonly are extruded with reduction ratios of 50 and above. In fact, strains may locally be much larger than this value as the deformation during extrusion is extremely inhomogeneous due to extensive shearing. During extrusion aluminum has the characteristics of a viscoplastic fluid, which start to flow when the stress

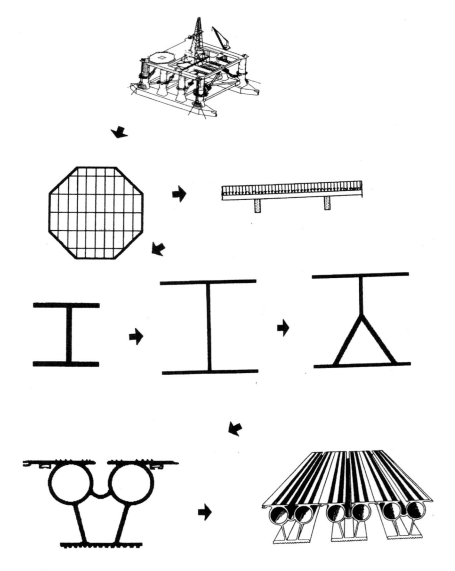

Figure 21   The profile development process and the profile in use today.

reaches the yield limit, and the extrusion process itself will, in principle, be a forced unsteady flow through a reduction. Macroscopically, the flow field will be characterized by local velocities, temperatures, and stresses, for which values can be measured at the boundary between aluminum and container/die.

The velocity field describes the particle velocities at all points in the container and bearing channel and therefore also the flow at all times. The rates of strain and rotation of particles may be of larger interest in the study of changes to microstructure during extrusion, but these quantities can be derived directly from the velocity field. However, because of the high pressure and temperature in the container, flow rates are extremely difficult to measure. Conventional flow meters are generally not constructed for the relevant conditions. Furthermore, the velocity field may be expected to be relatively complex and inhomogeneous, especially if complicated profiles are extruded. A few measuring points at the boundary would therefore not reveal all the characteristics of the flow. Only when the material leaves the die can particle velocities be measured directly and easily. The flow velocity ought then to be approximately uniform across the profile cross section, directed normal to the opening, and of a magnitude equal to the reduction rate times the velocity of the stem.

As hot aluminum behaves viscoplastically, the material will stop flowing when the extrusion force is relaxed so that stresses fall below yield. This property is important because information about the total deformation of the material will be saved in its structure after extrusion. This is not the case for perfect fluids such as for instance water, for which the measurement of total strains is both of minimal interest and impossible as the material deforms also when forces are removed. In order to establish the deformation history during extrusion, one performs a number of tests which may be interrupted at various stroke lengths. This will provide information on the total deformation at different stages. If the velocity of the stem is known, the approximate flow rate and strain rate at each point may also be calculated. There are a number of variations of this technique which bears the name viscoplasticity. Model materials such as wax, clay, plasticine, or even lead have earlier been used extensively when simulating extrusion of aluminum [20]. The billets are first parted, and a rectangular grid is applied on the surfaces of each half. The two parts are thereafter extruded together, and the distortion of the grid is in the end studied. The changes in geometry can be used directly to calculate strains and rotations. The use of model material is advantageous in

that extrusion can be performed with a low force and that both the equipment and the model material are relatively cheap. There are, however, also a number of shortcomings connected to such a use. It is always difficult to be certain that the material models the aluminum correctly, especially as temperature effects, which are totally neglected when using model materials, are known to be of large importance to the flow characteristics of aluminum. Furthermore, model materials are also susceptible to plastic deformation during post-extrusion treatment.

The interest in model materials has over the last years fallen, as modeling of extrusion process has increasingly become the realm of finite element programs. However, the method described above may also be applied to the extrusion of aluminum as shown in Fig. 22. The unmodified version of the technique works well for reduction ratios up to about 3, but at this point the material is so deformed that the grid may be erased locally, especially in shear zones. Valberg [21–24] has developed an alternative method, which can be applied when extruding at much larger ratios. Some alloys of aluminum share mechanical properties although their composition may differ. This is the case for a number of AlMgSi and AlCu alloys, the first group remaining gray the second turning black when etched. Thus the AlCu alloy may therefore work as a so-called indicator or marker

(a) (b)

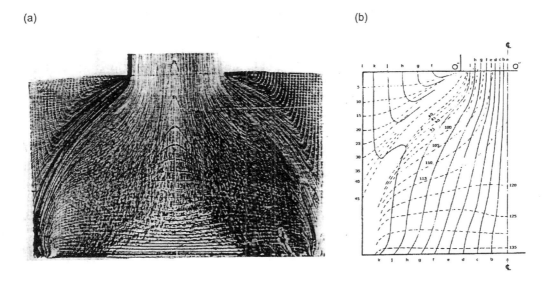

Figure 22 The split-billet technique applied on the extrusion of aluminum. The partial extrusion of a billet at a reduction ratio of about $R = 2$ is shown.

material. By drilling evenly separated longitudinal and transversal holes in different parts of a plane of symmetry and inserting pins of indicator material, a grid is formed (Fig. 23). A certain portion of the billet is then extruded, and the aluminum is carefully removed from the container and die. By splitting the billet and profile along the axis of symmetry, grinding, and etching, the deformation of the material is made visible (Fig. 24). A grid pattern may then be reproduced. By this technique metal flow may be investigated up to logarithmic strains of about $\varepsilon = 10$.

The marker material technique has been applied in the study of porthole die extrusion [25], two-hole die extrusion [26], flow adjacent to bearing walls [27,28], and with a 3D-version [29] also in the study of more general flow patterns. However, the most easily analyzable results are provided by simple axisymmetric extrusion, which has also been the standard test case for earlier techniques. General theory connected to direct extrusion describes four categories of flow in the container (Fig. 25a). These differ because of the varying degrees of friction between metal flow and container

walls. Only flow type B is of interest in the study of aluminum extrusion, the reason being that the other types either underestimate the influence of friction or assume inhomogeneous material behavior. After in-depth study of flow patterns during both direct and indirect extrusion, Valberg [30] has proposed two new general flow patterns more in accordance with observations, flow types A_1 and B_1. Figure 25b shows a typical grid on a partially extruded billet at various stroke lengths. When studying such grids one must bear in mind that the extrusion process is transient in its nature and that the deformation paths will change during extrusion. Whereas the initial deformation may be relatively homogeneous, deformation will in later periods be characterized by localized shearing. In the very end the direction of flow will even change as the material starts flowing in the radial direction toward the die opening. Figure 24b represents material flow at an early part of an intermediate period of almost steady state and reveals that the flow pattern will be strongly influenced by shearing toward the container walls. Hence deformation can be viewed as inhomogeneous, and a number of distinct regions may be identified. The one closest to both the centerline and die opening is called the primary zone of deformation, and in this the material necessarily has to undergo relatively large deformations as it enters the bearing channel. In the secondary zone of deformation only a relatively small distortion of the grid may be observed, and friction between the stem and the aluminum will in fact even prevent deformation in the uppermost part, creating a zone of minimal deformation, a dead zone [31]. The zone of intense shear will stretch from the die opening and to the stem. This deformation mode is caused by the condition of full sticking of material particles to the container wall. In the corner close to the die surface the sticking condition will immobilize the material, and another dead zone, bordering to the area of shear, will be formed.

Experiments reveal that the flow pattern in the container actually will depend on temperature conditions. If the container is relatively hot, the material in the outer part of the billet will be more mobile because of lower material resistance to flow. As a result the aluminum may no longer stick perfectly to the container wall and inflow of surface material may take place either along the surface of the stem or directly through the shear zone and into the die opening. This is an unwanted effect as the surface material usually contains impurities and defects caused by rough handling of billets or inverse segregation. The flow pattern will also change because of geometrical variations. A small reduction ratio will for instance yield a smaller dead zone close to

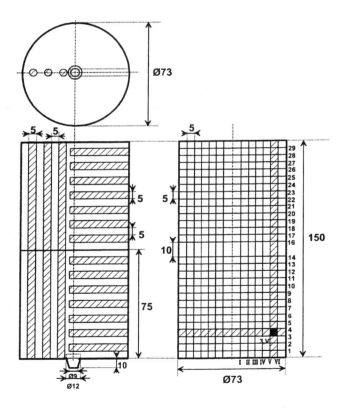

Figure 23 The preparation of gridded billets in accordance with Valberg's technique.

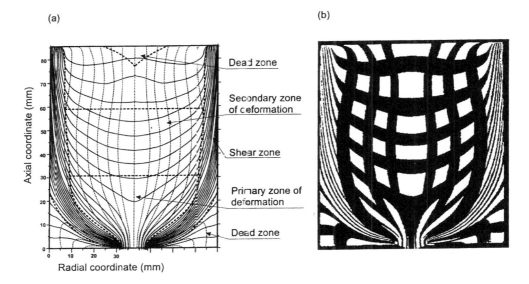

Figure 24 Determination pattern in a partially extruded billet. (a) Reconstructed deformation field. (b) The original patterns on the partially extruded billet.

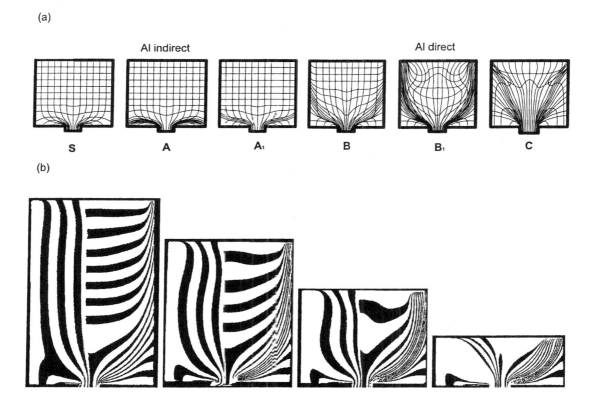

Figure 25 (a) Classes of flow patterns during axisymmetric extrusion according to Pearson, Dürrschnabel, and Valberg. (b) Experimentally determined flow pattern for direct extrusion of aluminum at various stroke lengths.

the die, and the result may also then be that material flows directly from the billet surface and out into the die opening.

A study of the deformed grid in the profile after a complete extrusion charge reveals that extruded products are far from homogeneous with respect to the deformation undergone. Figure 26 is a representation of such a grid where the longitudinal axis is scaled down. The first material leaving the die opening will have undergone a relatively small degree of deformation, but as extrusion proceeds, the material in the profile will have been deformed while passing through at least a part of the primary deformation zone. Region 2 is characterized by a grid that is very homogeneous, indicating that material particles originally coming from the secondary deformation zone have undergone almost the same deformation history. Hence when extruding this material the process will resemble one of steady state. Region 3, however, is constituted of the

material that was hindered from deforming by the stem during the extrusion charge, and the degree of deformation can therefore be expected to be much smaller. A layer of heavily shear-deformed material will exist closest to the profile surface [32]. The growth of this layer toward the end of the profile indicates that the shear zone in the billet gradually will flow out of the container as the stem is brought closer to the die. In order to find how the container is emptied during extrusion, Valberg has developed so-called emptying diagrams. These consist of lines, on which all particles will need the same amount of time to reach the exit. Hence the lines are denoted iso-residence-time lines. They are constructed by placing a grid over the extruded profile, finding the crossing points with the deformed grid of the marking material and then tracing these points back to the corresponding undeformed grid in the billet. In Fig. 27c, a number of such lines are shown. As expected, the bulk of the dead zone close to the die will not flow into the die

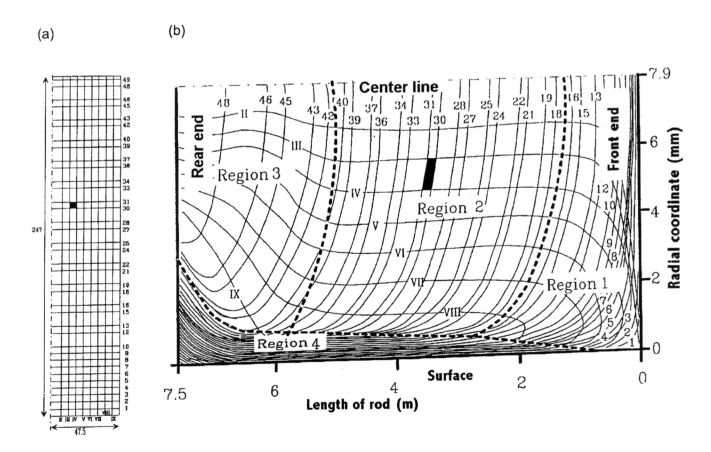

Figure 26 (a) Original grid on an axisymmetric billet. (b) Deformed grid on an axisymmetric profile (Valberg plot). The longitudinal axis is scaled down.

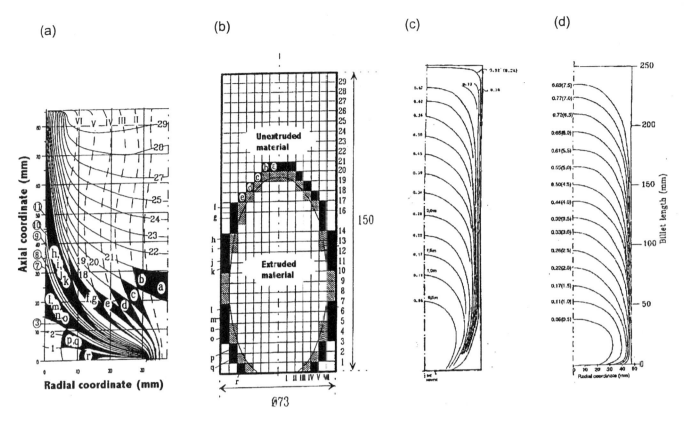

Figure 27 Emptying diagrams for axisymmetric extrusion. (a,b) Determination of one iso-residence-time curve. (c,d) Complete diagrams for direct and indirect extrusion, respectively.

opening until the very end of the charge. During the whole extrusion process, however, parts of the shear zone will be transported out into the profile, and the dead zone will gradually be reduced in size. Figure 27a and b also reveals that the surface layer of the profile will be generated from the material in the shear zone of the billet.

The observation that the size of the dead zone varies during an extrusion charge is extremely important to the understanding of the extrusion process as a whole. Energetically, the process of extrusion through flat dies seems favorable to that through tapered dies, as the metal is allowed to find the most optimal flow path by varying the inclination of the shear zone through the charge. However, as the flow pattern is unsteady, the properties of the product will necessarily also vary along its length. One of the objectives for using feeders such as that shown in Fig. 28 is to stabilize the conditions at the inlet to the bearing channel. In this way deformation paths for particles in the back and in the front of the billet will be more equal. Other reasons for applying

feeders, however, are usually viewed as more important. As the metal in the feeder is not removed when the butt of the billet is cut, one may weld metal from two charges together and almost extrude continuously. The advantage of applying a feeder then is that production rates will be increased. The drawback is that the charge weld formed most probably will contain inclusions of oxides and therefore will have inferior mechanical and corrosion properties. As charge welds under the influence of

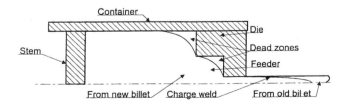

Figure 28 A schematic example of direct extrusion with a feeder.

varying material velocity over the profile cross section will become a parabolic line extending often many meters along its length, low weld quality may lead to extensive scrapping.

2. Temperature and Metallurgy

Variations in temperatures during extrusion seem to influence flow behavior in a number of ways. As indicated, flow patterns may be changed considerably by rendering the temperature distribution in the container. Furthermore, it is known that the extrusion pressure may be lowered if either the temperature of the billet or the velocity of the stem is increased, and that there are certain limitations, because the material starts melting or cracking if it leaves the die with too high a temperature. The observations indicate that there is a strong coupling between what would be regarded as thermal and mechanical effects [33]. In order to describe the temperature field during extrusion, a system for measuring temperature at surface between the flowing alloy and tooling has been developed by Lefstad [34,35]. The method has its limitations because it does not reveal the complete temperature field, but it is well suited for studying the temperature of the flowing metal on the bearing surfaces close to the outlet. Because of frictional forces the temperature can be expected to be highest in this part of the flow.

In principle, all the effects witnessed on a macroscopic level could be explained with an atomistic perspective. The close coupling between the temperature field and the mechanical forces is simply a result of the exchange of kinetic and potential energy of the atoms. Today some phenomena such as frictional behavior and dislocation movement are explained partly with such a perspective [36], but no complete atomistic description of the extrusion process can be found in literature. There are presently no ways of performing satisfactory measurements on flow, and the modeling methods are not able to handle the complexity of the problem. Hence simplifications are sought. As will be explained in the next section, modeling is today primarily performed with the thermomechanical continuum theory. This is, however, merely a mathematical tool, which can be used to quantify states of deformation and pressures, and it will not reveal any new fundamental mechanisms. The problems of extrusion, related to speed limits, flow resistance, evolution of microstructure, stability of flow, surface quality, and so forth are generally of metallurgical and microstructural origins. Furthermore, as the flow and temperature fields are influenced by the flow resistance of the hot metal, which is strongly dependent on the alloy constituents, microstructure and metallurgical state of the deforming material, there is a strong interaction between flow, temperature, and microstructural evolution. If the continuum theory shall be able to model and quantify such phenomena properly, one must be able to relate the parameters of the model to the thermodynamics and kinetics of the material. An example of such is the description of recrystallization, of whose degree is influenced by such factors as temperature, time, and deformation [37,38]. When extruded under similar conditions, different alloys need not experience the same amount of recrystallization. If quantitative description of the degree of deformation is to be given by a continuum model, the rate of recrystallization must be related to the values of deformation provided by a mathematical calculation as well as the constitution of the alloy. In the present state of the art of extrusion technology, the continuum description of thin-walled extrusion has not yet reached the point where the metallurgical phenomena can be explained quantitatively based on the continuum-mechanical description of the flow and temperature field.

An important fact that is often forgotten is that the complete thermomechanical history of the material should be known if one is to assess the microstructure. Focus should not only be placed on the mere extrusion charge but also on all the process steps. A study of the influence of Mg_2Si particles on the maximal allowed extrusion speed gives an example of this [39,40]. The example applies probably to most 6XXX alloys and in fact also to some of the 7XXX series. The billet is prepared for extrusion after casting by homogenizing, controlled cooling after homogenizing and preheating before extrusion. Based on the chemical constituents of the alloy (Fig. 29a), given here for an AA6060 alloy (also showing AA6082 for comparison) with equilibrium diagram (Fig. 29b), the initial temperature distribution in the billet at the starting point of extrusion should be such that the magnesium and silicon are completely in solid solution before the alloy leaves the die. At the same time the extrusion speed should be selected in order to give the shortest possible press-cycle time without causing overheating or unacceptable risk for production stops and/or quality problems. If there is eutectic left with melting point at 585°C (Fig. 29d), then this will be the maximum temperature without overheating. Otherwise, the solidus line will give the upper temperature limit. Clearly, by preparing the billet in an optimal manner, high gains in productivity can be achieved.

Studies of microstructure in profiles often reveal a relatively large degree of inhomogeneity over the cross

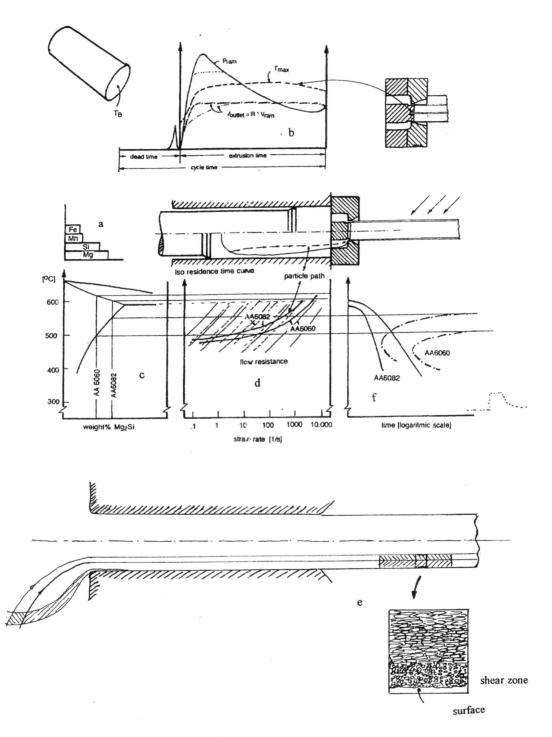

Figure 29 Relation between metallurgy and process parameters.

section. This applies to the size and form of crystals and to the dislocation density. Mechanical testing usually gives results in accordance with such observations. As Valberg's experiments explain, this is to be expected because the deformation of the material during extrusion is extremely inhomogeneous, and the same will naturally also be the case for the temperature history. In the early studies of microstructure evolution during extrusion insufficient attention was paid to these facts, and very often only average values of deformation and temperature were assessed. In this way the limits of the extrusion process can hardly be studied and the variability of the process only poorly understood. Figure 29 also explains the importance of knowing the exact temperature-, strain-, and strain-rate history for each particle in the deforming alloy and especially for those that undergo the largest strains. The largest differences will be experienced when the material particles enter the bearing channel. Here the material close to the die will be exposed to shear strain rates up to about 10,000 [l/sec], whereas the strain rate in the center of the bearing channel is in the range of 0.1 [l/sec]. As shown in the micrograph of Fig. 29e, there is a considerable degree of variation in microstructure in an extruded section. Partial recrystallization over the cross section may also occur here, and this gives rise to unsatisfactory large variability of properties and surface appearance. Because every material element in the billet goes through different strain-rate-, temperature-, and strain histories as shown in Fig. 29d, it is important that the alloy is not very sensitive to these variations with respect to flow resistance and final microstructure after extrusion and quenching and aging (Fig. 29f).

B. Predicting the Conditions of Flow

1. Mathematical Approaches

The traditional method of treating macroscopic problems within both solid and fluid mechanics has since the 18th century been through the principles of continuum mechanics [41]. The fundamental equations obtained within the frames of this perspective are those of motion and conservation of heat, expressed incrementally. Furthermore, the assumption is that processes on a microscopic scale will average out so that the material properties will be continuous functions in space except in the case of discrete discontinuities. Material behavior will, however, be determined by processes taking place at levels from atomic to grain-size. In mechanics, constitutive equations can be utilized to characterize the relationship between measures such as deforma-

tion and stress. Purely elastic atomic lattice deformation may be represented by Hooke's law, which is a linear relationship between stress and strain. As elastic behavior generates no permanent deformations and in a thermodynamic sense may be looked upon as ideally reversible, it may also mathematically be characterized by functions of states. From a computational point of view this makes the material model attractive [42]. Deformations taking place during extrusion, however, contain only a minor, and in fact in many cases, a negligible elastic component. As the profile is generated from a billet, the material has to undergo permanent deformation. On a microscopic scale this deformation is caused by the sliding of dislocations through the grains [43]. As long as force is applied, energy will be dissipated, and therefore the process is thermodynamically irreversible. Consequently, the material behavior should be presented mathematically by history-dependent functionals [44]. A further complication is that plastic material behavior usually can be taken to be nonlinear. Hence a theory of plasticity is bound to be of another dimension of complexity compared to purely elastic theory.

2. Different Perspectives

As extrusion basically is a nonsteady flow problem, principles of fluid mechanics may be applied to evaluate both deformation and stresses. The viscosity will in this case be a function of the strain rate. As the material will not flow below a certain yield limit, a constitutive model such as that of Bingham may be appropriate if the necessary corrections are made for temperature effects on yield stress. The fluid mechanical or rheological approach, which addresses strain rates or time increments of strain, is favorable for many reasons.

Firstly, stresses generated through plastic deformation at high temperatures are found to be almost independent of the total strain but extremely dependent on the strain rate. Secondly, flow is viewed in an Eulerian sense, by which is meant that components of flow velocity and stress are related to spatial coordinates, and that deformation is assessed in an incremental manner, the reference state at all times being that of the last time step.

Hence an Eulerian formulation of the constitutive relation will therefore never violate the principle of material objectivity, which states that the constitutive relation should be independent of the choice of reference frame [45]. A rotation of the element should not, for instance, cause changes in stresses as long as the strains are held constant. Incremental deformations will, to a close approximation, always be in accordance with this principle.

Another favorable aspect of an Eulerian description is that computer code can be made very efficient because velocities and stresses can be evaluated in a mesh that does not deform. However, as the Eulerian approach only assesses increments, information about each particle's total stress/strain history is lost, and elastic deformation can only, with some difficulty, be described. When evaluating the flow pattern, elastic strains are generally small, and it can be assumed that they are only of minor influence. However, elastic stresses may be of importance for instance to the surface quality of the product as friction in the bearing channel is affected by pressure due to elastic strains. A last problem connected to the Eulerian description in connection to extrusion is that the extruded profile is free to move as a rigid body in any direction as it leaves the bearing channel. Furthermore, deformation is then not plastic, and the movement will not be confined within certain limits but should be calculated for the rigid body from the laws of motion. Furthermore, the boundary conditions of traction caused by a puller may not easily be described in the Eulerian system.

The Lagrangian description has traditionally been most popular when describing the movement and deformation of a rigid body. This approach assesses the deformation of particles in relation to a fixed reference state in space, thus making use of so-called material coordinates. This view is favorable in that the total strain history is kept, and therefore that both elastic and plastic deformation may be included in calculations. Furthermore, an arbitrary movement of particles in space can be described when boundary conditions are given. This makes it possible to predict deformations also when the material moves out of the bearing channel. A problem, however, is that deformation and rotations tend to be large during extrusion. A Lagrangian description of finite deformation is not automatically in accordance with the principle of objectivity, the cause being that the material derivative of stress is not an objective measure although stress is. Again this is a minor problem when evaluating only plastic strains as it is done on an incremental basis. However, when adding elastic deformation components the principle of material objectivity is satisfied only by making use of an alternative material derivative of stress that compensates for any rotations of particles. In plasticity theory the Jaumann derivative is in common use. An analytical treatment of finite elastoplastic deformation is by no means trivial. Lagrangian numerical calculations, however, are widely performed, and these make use of an

element net that moves and deforms with the material particles. Calculation times tend to be higher for Lagrangian codes than for Eulerian even in the case when elastic deformations are disregarded, the reason being that meshes have to be regenerated during simulations because of large distortions and loss of numerical accuracy. A promising method for the evaluation of the state of deformation during extrusion seems to be the combination of the two descriptions in an arbitrary Lagrangian–Eulerian (ALE) code.

3. *Plasticity and Extrusion*

Whereas the theory of rheology is focused on solving problems where stresses are given as functions of temperature and strain rates, the classical theory of plasticity was originally established to handle isothermal solid-state problems where the stress after having exceeded the yield point still was a function of strain [46–48]. An example of such a problem is that of the low-temperature uniaxial tensile test where a plot of true stress to true strain will usually give a monotonically increasing curve. Plastic deformation can be assumed to be initiated when the yield limit is reached. An increase in yield stress as a result of further deformation is called hardening. Several simplified material models have been established so that calculations can be made easier. When applying a perfectly plastic model it is assumed that plastic deformations are dominating and that the material does not strain harden. An elastic/plastic model can be utilized when the elastic strain component is of a certain value (Fig. 30). Strain hardening models such as that of Ramberg and Osgood also simplify matters in that hardening behavior is characterized by one parameter, the hardening exponent.

In principle, one is not to expect that the material shows any strain hardening behavior during extrusion. Such is traditionally connected to the pileup of dislocations, but at characteristic temperatures from 450°C to 600°C recovery and recrystallization mechanisms will contribute to the reduction of dislocation tangles during deformation. At a given stress, continuous deformation should therefore be possible. However, the strain rate will be of importance to the yield stress as the rates of creation and destruction of dislocation tangles will determine the equilibrium dislocation density. As for temperatures, the yield stress may be expected to follow the same Arrhenius relationship as metallurgical processes generally do. Hence the yield stress can generally be expressed as a function:

$$\overline{\sigma} = f(\overline{\varepsilon}, \dot{\overline{\varepsilon}}, T) \qquad (7)$$

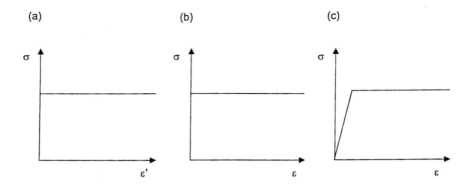

Figure 30 Simple constitutive relations: (a) Bingham fluid without strain rate hardening, (b) rigid plastic material without strain hardening, (c) elastic plastic material without strain rate hardening.

The equation is given in terms of equivalent stress and plastic strain:

$$\overline{\sigma} = \sqrt{\frac{3}{2}(s_{ij}s_{ij})}$$

$$= \sqrt{\frac{3}{2}\left[(\sigma_x - \sigma_o)^2 + (\sigma_y - \sigma_o)^2 + (\sigma_z - \sigma_o)^2 + 2\tau_{xy} + 2\tau_{yz} + 2\tau_{zx}\right]}$$

$$= \frac{1}{\sqrt{2}}\sqrt{(\sigma_x - \sigma_y)^2 + (\sigma_y - \sigma_z)^2 + (\sigma_z - \sigma_x)^2 + 6\tau_{xy} + 6\tau_{yz} + 6\tau_{zx}}$$

$$\tag{8}$$

$$\dot{\overline{\varepsilon}} = \sqrt{\frac{2}{3}(\dot{\varepsilon}_{ij}\dot{\varepsilon}_{ij})} = \sqrt{\frac{1}{3}\left(2\dot{\varepsilon}_x^2 + 2\dot{\varepsilon}_y^2 + 2\dot{\varepsilon}_z^2 + \dot{\gamma}_{xy}^2 + \dot{\gamma}_{yz}^2 + \dot{\gamma}_{zx}^2\right)} \tag{9}$$

The equivalent measures of stress and strain are of interest when the state of stress is multiaxial. It can be shown that dissipation in general can be given as $\dot{\omega} = \sigma_{ij}\dot{\varepsilon}_{ij} = \overline{\sigma}\dot{\overline{\varepsilon}}$, and for a purely tensile test the equivalent stresses and strains will reduce to σ_z and ε_z. In simulation of extrusion extensive use is made of the Norton–Hoff relation:

$$\overline{\sigma} = K\overline{\varepsilon}^n \dot{\overline{\varepsilon}}^m e^{\frac{\beta}{T}} \tag{10}$$

where $n = 0$ can be used to remove any strain dependence. n and m are material constants. β is a parameter often set equal to Q/R, where Q is the activation energy and R is the universal gas constant. Another frequently used relation in the study of extrusion is that of Zener and Hollomon [49] which is given by:

$$\overline{\sigma} = \frac{1}{\alpha}\arcsin h\left(\frac{Z}{A}\right) = \frac{1}{\alpha}\arcsin h\left(\frac{\dot{\overline{\varepsilon}} \cdot e^{\frac{Q}{RT}}}{A}\right) \tag{11}$$

In this case the yield stress is taken to be independent of the total strain. α and A are parameters which can be applied in curve fitting. Experiments indicate that the Zener–Hollomon relation may be satisfactory both

when performing torsion tests on aluminum and during extrusion. Improvements to this material law have, however, been suggested. The main difference between torsion testing and extrusion is that the former assumes steady state while the strain rates may vary considerably in the latter.

Typical stress–strain curves obtained through torsion, compression, or tension tests at temperatures around that of extrusion are shown in Fig. 31. As changes in strain rates and temperatures tend to influence yield stress more than the total strain, rheological modeling seems preferable to modeling through plasticity theory. However, if a perfectly plastic or a plastic material model with low strain hardening is adopted, effects of strain rate and temperature can be implemented by calculating the yield stress on the basis of relations such as Norton–Hoff or Zener–Hollomon. The important point is that a deformation mechanism based on dislocation glide is one of shear and, hence, should be modeled macroscopically as one that relates

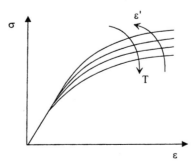

Figure 31 More general stress–strain curves' dependence on temperature and strain rates.

increments of strains to the deviatoric stress state. Both theories of rheology and plasticity have this perspective and have for a long time been widely used to describe the extrusion process. Early analytical and semi-analytical work has not managed, however, to solve the complex problem of extrusion in its entirety. Rheological approaches have been based on simple material models and have only assessed very simplified flow fields. Techniques related to plasticity theory have traditionally been popular when calculating very rough estimates of, for instance, the necessary extrusion force. By applying the so-called slip-line theory one has been able to find closed-form solutions for particle paths and stress states for simple geometries, but material models are generally limited to perfectly plastic, and temperature effects and material anisotropy have been neglected. However, modern numerical programs offer the user the possibility to implement constitutive models based on microstructural considerations and are able to perform coupled thermomechanical analysis for complex geometries. The last being of greatest importance because of the large variation in temperature in the container and bearing channel. Furthermore, by obtaining the temperature and strain history of the individual particles numerical programs should be able to describe changes in microstructure and the consequences of these. Both continuum plasticity theory and rheology are the building bricks for such programs.

4. Yield Criteria

For tensile tests, yielding is initiated when the axial stress component reaches the yield limit, $F(\sigma_z) = 0$. In the general case it can be expected that all components of stress have an influence on the point of yielding, and in a 6D space a yield surface $F(\sigma_{ij}) = 0$ may be defined. The assumption is made that deformation is purely elastic in the case where $F(\sigma_{ij}) < 0$. In classical plasticity theory one then usually resorts to two simplifications. The first is that the material is isotropic, thus implying that the yield surface is determined only by the principal stresses, $F(\sigma_1, \sigma_2, \sigma_3) = 0$, and therefore also the stress invariants, $F(J_1, J_2, J_3) = 0$. The second is that the hydrostatic pressure has no effect on yielding. Then the yield criterion can be written in terms of the deviatoric invariants:

$$F(J_2', J_3') = 0 \tag{12}$$

where

$$J_1' = s_{ii} = 0 \qquad J_2' = \frac{1}{2} s_{ij} s_{ij} \qquad J_3' = \det(s_{ij})$$

and s_{ij} is the deviatoric stress, $s_{ij} = \sigma_{ij} - \sigma_0 \delta_{ij}$. σ_0 is the hydrostatic stress. The assumptions made are more or less applicable to most engineering materials. However, alternative anisotropic yield criterions have been developed so that materials with strong anisotropy can be handled.

When applying the assumption of isotropy, graphical representation of the yield criterion can be made in the three-dimensional $\sigma_1, \sigma_2, \sigma_3$-space. By assuming that hydrostatic pressures do not influence yielding, one is able to reduce the yield surface to a yield locus in the deviatoric plane, which is defined by the normal $(\sigma_1, \sigma_2, \sigma_3) = (1, 1, 1)$. If it is assumed that the material is isotropic and that yielding will take place at the same stress level both in tension and compression, the yield locus must be symmetric about the six axes defined by the projections of the σ_1-, σ_2-, and σ_3-axes into the deviatoric plane. The last assumption is, however, not always correct because of the Bauschinger effect, which will be explained later.

The most commonly applied yield criteria are those of von Mises and Tresca. The von Mises criterion assumes that F only is a function of the second invariant of deviatoric stress. For many materials, experiments often indicate that yielding only to a very limited extent is dependent on the third invariant and thus confirms the hypothesis of von Mises. Mathematically, the representation is:

$$\begin{aligned} J_2' - \kappa_M^2 &= \frac{1}{6}\Big[(\sigma_1 - \sigma_2)^2 + (\sigma_2 - \sigma_3)^2 \\ &\quad + (\sigma_3 - \sigma_1)^2\Big] - \kappa_M^2 = 0 \end{aligned} \tag{13}$$

Hence the von Mises criterion has a nice property in that it appears as a circle in the deviatoric plane. Furthermore, the second invariant of deviatoric stress is simply the square of the expression for equivalent stress. This indicates that the von Mises criterion bases the criterion of yielding on the deformation energy absorbed by the material. The natural alternative to this is the Tresca criterion, which states that yielding will take place when the maximum shear stress reaches a critical value. This may be expressed as:

$$\tau_{\max} = \max\left[\frac{1}{2}|\sigma_1 - \sigma_2|, \frac{1}{2}|\sigma_2 - \sigma_3|, \frac{1}{2}|\sigma_3 - \sigma_1|\right] = \kappa_T \tag{14}$$

The yield locus will in this case be a hexagon in the deviatoric plane. Also, the Tresca criterion may be expressed in terms of the second and third invariants. In

Fig. 32 both the von Mises and the Tresca yield loci are presented. Which one of the criteria that are the most conservative depends on the choice of κ_M and κ_T. If it is assumed that yielding first is taking place at a characteristic yield stress for pure tension, σ_Y, one would expect that:

$$\kappa_M = \frac{\sigma_Y}{\sqrt{3}} \qquad \kappa_T = \frac{\sigma_Y}{2} \qquad (15)$$

The result is that the locus of the von Mises criterion will circumscribe that of Tresca. However, if shear stress is the reference for yielding then both are given as $\kappa_M = \kappa_T = \kappa_Y$. In this case the Tresca criterion is the least conservative except in the point of maximal shear.

An interesting property of the yield locus is that singular points may exist, such is the case for the Tresca criterion. The singularity should, from a macroscopic perspective, be explained as a result of the sudden change of plane of maximum shear stress when loading from one state of stress to another. Consequently, the plane of shear deformation is also changed. However, the Tresca yield locus can also be explained on the basis of knowledge about glide systems in fcc-crystals. As a simplification, a state of plane stress is assumed. In this case shear deformation through dislocation movement may take place in three separate groups of planes (Fig. 33). The principal stresses are directed along the axes of the coordinate system. The same axes also define the (1 0 0) [1,0,0] texture of the fcc-lattice, which is taken to be homogeneous, so that all the grains have oriented the {1 0 0} plane normal to the x-direction. If it is assumed that deformation is only taking place on the glide plane with the largest shear stress, the ratios of the stresses $\sigma_x/\sigma_y/\sigma_z$ will determine the deformation of the grains. If σ_x is the largest stress and $\sigma_z = 0$ the smallest, one can expect that the first glide system will be operative. Subsequently, if σ_y is the largest stress and $\sigma_z = 0$ the smallest, deformation will be of the second type. The last alternative is that σ_x and σ_y are the extreme principal stresses in which case the third modus of deformation is relevant. With the help of Schmidt's law a yield locus can be constructed in the $\sigma_x\sigma_y$-plane. Figure 33 shows that such an analysis renders the Tresca yield locus. The singular points are the results of a limited amount of glide systems. While single-crystal materials might follow a yield criterion such as that of Tresca, most materials contain grains with all sorts of orientations and glide systems. In this case deformation will take place in a series of directions yielding a number of straight lines defining a yield locus that resembles that of von Mises. A locus without singularities is often denoted regular.

5. Strain Hardening and Plastic Flow

Material behavior at stresses above the yield limit will depend on both the hardening behavior and boundary conditions. Whereas flow of a perfectly plastic material will be completely restricted by outer constraints, deformation of a strain hardening material can only take place if the stresses are increased gradually. Furthermore, in order to calculate deformation a constitutive relation between stresses and strains has to be established. An interesting observation in relation to the preceding determination of the yield locus on the basis of microplasticity is that any increment of plastic strain will be in a direction normal to the yield surface. Thus an indication is given that both stresses and strains can be determined if the form of the yield surface is known during deformation.

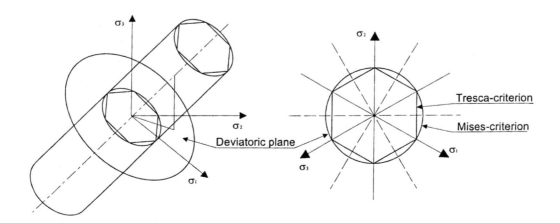

Figure 32   The yield surface, the yield locus, and the deviatoric plane.

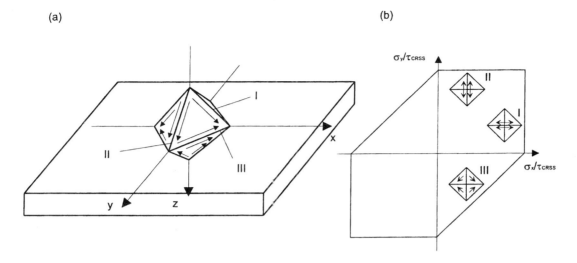

Figure 33 (a) An ideal (1 0 0) [1 0 0] texture. (b) The yield locus for ideal (1 0 0) [1 0 0] texture material.

A uniaxial tensile test can be performed to determine the increase of yield strength during plastic deformation. However, as such a test alone is not able to describe the form of the yield surface during triaxial loading, firm knowledge of changes caused by yielding is even harder to obtain. The most attractive hardening principle mathematically is the isotropic (Fig. 34a). When applying this principle it is assumed that plastic deformation causes the yield surface to expand uniformly. Hence upon yielding the expression

$$F(J_2', J_3', \bar{\sigma}) = f(J_2', J_3') - \bar{\sigma} = 0 \qquad (16)$$

applies. The equivalent stress is taken to be a function either of the energy spent during plastic deformation or the total plastic strain. These are denoted the work hardening and the strain hardening hypothesis, respectively. Such functions can be obtained through tensile and compressive testing. The main problem with an assumption of isotropic hardening is that it is not in accordance with the Bauschinger effect, which can be witnessed during cyclic loading. After the material has been deformed plastically in tension, yielding will take place at lower compressive stresses than that of the expected yield point in compression. Kinematic hard-

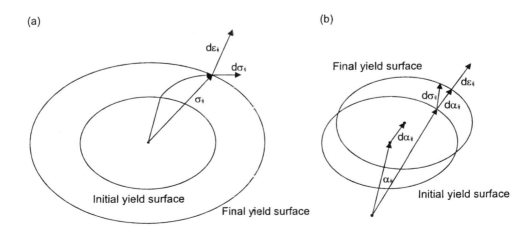

Figure 34 Strain hardening rules: (a) isotropic, (b) kinematical.

ening rules describe a movement of the yield surface in space so that the ratio of the yield limits in different directions is altered. A mathematical expression that can be used to describe such a behavior is:

$$F(s_{ij}, \alpha_{ij}) = f(s_{ij} - \alpha_{ij}) - \kappa = 0 \qquad (17)$$

where α_{ij} describes the movement of the yield surface and is known as the back ratio. κ will, for a strict kinematic rule, be a constant. However, newer models often let both α_{ij} and κ vary so that the proposed law can be better fitted to data.

In the mathematical treatment of plastic strains generated by loading it is rational to define a plastic potential, g, which is set to the ratios of the plastic strain increments. One can expect that this function, as the yield function, originally is one of all stress components. If such a relation is known, the plastic stresses can be derived from the mere definition of a potential function, namely, that the increment of strain shall be normal to the potential surface:

$$d\varepsilon_{ij}^{p} = \frac{\partial g}{\partial \sigma_{ij}} d\lambda \qquad (18)$$

$d\lambda$ is simply a constant. The previous microplastical considerations have shown that the yield surface, at least in the case of yield by the Tresca criterion, also can be expected to be a potential surface. If g is set equal to f, the associated flow rule is adopted. Such an assumption can, in most cases, be expected to yield satisfactory results because many of the characteristics of g are equal to those of f. Whereas the assumption that yielding is unaffected by hydrostatic stress limits the evaluation of yield to the deviatoric plane, a choice of an incompressible material model leads to the conclusion that the hydrostatic stress line in a $\sigma_1\sigma_2\sigma_3$-system nowhere can be perpendicular to a potential surface. Hence a potential surface will be fully described by its locus in the deviatoric plane. As plastic deformation is one of shear, an assumption about incompressibility leads only to marginal errors. Mathematically, such a relation is written:

$$d\varepsilon_x^p + d\varepsilon_y^p d\varepsilon_z^p = \begin{bmatrix} 1 \\ 1 \\ 1 \end{bmatrix} \cdot \begin{bmatrix} d\varepsilon_1^p & d\varepsilon_2^p & d\varepsilon_3^p \end{bmatrix} = 0 \qquad (19)$$

In the $\sigma_1\sigma_2\sigma_3$-system an incremental change in strain will generate the stress component $2G(d\varepsilon_1^p, d\varepsilon_2^p, d\varepsilon_3^p)$, where G is a constant of proportionality. It can be observed that this component has a direction perpendicular to the line $(\sigma_1, \sigma_2, \sigma_3) = (1,1,1)$ and therefore lies

in the deviatoric plane. The locus described in such a plane by a potential surface also has to be symmetrical with respect to the stress axes because of assumptions of isotropy and independence of sign reversals. When one discusses states of stresses and the corresponding increment of strains, the incompressibility assumption allows one to view these states as vectors in the deviatoric plane (Fig. 35).

Drucker has proved that unless the material shows strain softening behavior such as that of, for instance, soils and rocks, the flow rule will be associated and the normality rule will hold. Drucker then has defined a stabile material as one for which the inequality

$$W = \int_{C_0} \Delta\sigma_{ij} d\varepsilon_{ij} = \int_{C_0} \left(\sigma_{ij} - \sigma_{ij}^0\right) d\varepsilon_{ij} \geq 0 \qquad (20)$$

holds during a complete load cycle where the original stress state, σ^0, may or may not be one of yielding. A material on which an external agency does positive work during an elastic–plastic stress cycle is considered strain hardening. As the inequality will not be satisfied if the material is strain softening, the postulate is often referred to as Drucker's strain hardening postulate. To a close approximation the net plastic work done over the loading part of the cycle can be expressed as:

$$\left(\sigma_{ij} - \sigma_{ij}^0\right) d\varepsilon_{ij}^p + \frac{1}{2} d\sigma_{ij} d\varepsilon_{ij}^p > 0 \qquad (21)$$

Drucker's first and second postulate follows from this inequality. If it is assumed that the original state of stress is one of yielding, the first term cancels. In accordance with the inequality, the first postulate states that the plastic work done by an external agency during the application of additional stress is positive for a work hardening and zero for a nonhardening material. If the last part of the postulate is to be true, vectors representing the increments of stress and plastic strain have to be

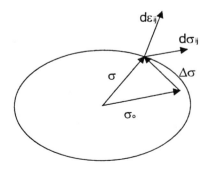

Figure 35 The yield locus in relation to Drucker's hardening postulates.

perpendicular. In a nonhardening material loading is assumed to be neutral, i.e., all load paths are tangential to the yield surface. Therefore the increment of strain is directed normally to it, thus resulting in the normality law. If the material is hardening, it must be assumed that the increment of stress is a linear combination of a plastic and an elastic part, and that the last is directed tangentially to the yield surface. As the elastic part of the stress increment does no irreversible work, the increment of plastic strain is also in this case directed normal to the yield surface.

Drucker's second postulate is connected to the first part of the inequality above and states that the net work done by an external agency during a cycle of addition and removal of stresses is nonnegative. If it is assumed that the original state of stress is not one of yielding, the last term in the inequality can be neglected. The result is:

$$\left(\sigma_{ij} - \sigma_{ij}^{\mathrm{o}} \right) \mathrm{d}\varepsilon_{ij}^{\mathrm{p}} > 0 \tag{22}$$

This inequality also bears the name the maximum work theorem. From Drucker's second postulate it can be seen that the angle between the vectors $\sigma - \sigma^{\mathrm{o}}$ and ε^{p} cannot be obtuse. Hence as the vector ε^{p} is always normal to the yield surface and σ^{o} is located inside the yield surface, this surface has to be convex at all points.

If it is assumed that yielding is initiated according to the von Mises criterion and that the flow rule is associative, the plastic strains can be calculated to be:

$$\mathrm{d}\varepsilon_{ij}^{\mathrm{p}} = \frac{\partial f}{\partial \sigma_{ij}} \mathrm{d}\lambda = \frac{3}{2\overline{\sigma}} \left(\sigma_{ij} - \sigma_{ij}^{\mathrm{o}} \delta_{ij} \right) \mathrm{d}\lambda = \frac{3s_{ij}}{2\overline{\sigma}} \mathrm{d}\lambda \tag{23}$$

This rule can also be written in the form:

$$\frac{\mathrm{d}\varepsilon_x^{\mathrm{p}}}{s_x} = \frac{\mathrm{d}\varepsilon_y^{\mathrm{p}}}{s_y} = \frac{\mathrm{d}\varepsilon_z^{\mathrm{p}}}{s_z} = \frac{\mathrm{d}\gamma_{xy}^{\mathrm{p}}}{\tau_{xy}} = \frac{\mathrm{d}\gamma_{yz}^{\mathrm{p}}}{\tau_{yz}} = \frac{\mathrm{d}\gamma_{xz}^{\mathrm{p}}}{\tau_{xz}} = \mathrm{d}\lambda \tag{24}$$

indicating that the increments of strain are proportional to the deviatoric stress components. If the material is assumed to be rigid/plastic, the names of Lévy and von Mises are connected to the equations. If the material model also contains an elastic strain component the above relations are those of Prandtl and Reuss.

The constant of proportionality $\mathrm{d}\lambda$ is expected to be related to the increment of equivalent strain, the reason being that the increment of plastic strain is determined by the strain hardening behavior of the material. By applying the measures of equivalent stress and strain one can relate the hardening behavior of a uniaxial test to that of the multiaxial actual load case. From the definition of equivalent stress and strain it is known that:

$$\dot{\omega} = \overline{\sigma}\dot{\overline{\varepsilon}} = \sigma_{ij}\dot{\varepsilon}_{ij} = s_{ij}\dot{\varepsilon}_{ij} \tag{25}$$

If only plastic components of strain are evaluated, the flow rule can be applied in order to eliminate the increment or in this case the material derivative of plastic strain:

$$\overline{\sigma}\dot{\overline{\varepsilon}} = s_{ij} \left(\frac{3}{2\overline{\sigma}} s_{ij} \right) \dot{\lambda} = \overline{\sigma}^2 \frac{\dot{\lambda}}{\overline{\sigma}} = \overline{\sigma}\dot{\lambda} \tag{26}$$

This gives the result $\dot{\lambda} = \dot{\overline{\varepsilon}}$ or $\mathrm{d}\lambda = \mathrm{d}\overline{\varepsilon}$ and a final flow rule of the form:

$$\mathrm{d}\varepsilon_{ij}^{\mathrm{p}} = \frac{3\mathrm{d}\overline{\varepsilon}^{\mathrm{p}}}{2\overline{\sigma}} \left(\sigma_{ij} - \sigma_{ij}^{\mathrm{o}}\delta_{ij} \right) = \frac{3\mathrm{d}\overline{\varepsilon}^{\mathrm{p}}}{2\overline{\sigma}} = s_{ij} \tag{27}$$

The total strain increment will generally be the sum of both an elastic and a plastic part. The elastic component can be obtained by the use of Hooke's law on an incremental form, and from this equation it can be seen that the ratio of the components of elastic strain will be determined by the increment of stress and not the full deviatoric stress as the case is for the plastic component. A particular formulation of stress–strain relationship developed by Hencky assumes that also the ratios of plastic strains are given by those of deviatoric stress. This can only be expected to hold if the loading is proportional, i.e., if the stress components experience a proportional increase during loading. As initially indicated the load path is generally history dependent and must be calculated incrementally.

6. The Uniqueness Theorem

The Lévy–Mises equations reveal both the strengths and the weaknesses connected to the description of extrusion through the theory of plasticity. Deformation is described as one that is initiated by deviatoric stresses and the components of deformation will be determined by their ratio. This is in accordance with the notion that plastic deformation is connected to the glide of dislocations. However, the material is expected to experience strain hardening defined by a curve relating the equivalent stress and strain. At the relevant temperatures little hardening is experienced and therefore the derivative of a stress–strain curve will be almost zero. This means that it is not possible to calculate the total strain from the stress alone. Thus the stress state in the material during extrusion will not be determined by the strain but by the geometry of the tooling and by temperatures and strain rates.

An important property that applies to hardening and nonhardening materials alike is that the state of stress is unique when certain tractions, T_j, and velocities, v_j, are defined on separate parts, S_F and S_v, of the boundary. The proof of this is as follows for a nonhardening material strains. One may assume that two consistent solutions, $(\sigma, \mathbf{v})$ and $(\sigma^{\circ}, \mathbf{v}^{\circ})$, for the stress and associated velocity distribution, which corresponds to the same boundary conditions, exist. The principle of virtual work is then applied to the differences in the field quantities over the volume of interest:

$$\int \left(T_j - T_j^{\circ} \right) \left(v_j - v_j^{\circ} \right) \mathrm{d}S$$

$$= \int \left(\sigma_{ij} - \sigma_{ij}^{\circ} \right) \left(\dot{\varepsilon}_{ij} - \dot{\varepsilon}_{ij}^{\circ} \right) \mathrm{d}V + \int (k - \tau^{\circ}) \qquad (28)$$

$$\times [v]\mathrm{d}S_{\mathrm{D}} + \int (k - \tau)[v^{\circ}]\mathrm{d}S_{\mathrm{D}}^{\circ}$$

The two last terms are connected to the virtual energy dissipated due to velocity discontinuities $[v]$ and $[v^{\circ}]$ on surfaces S and S° in the volumes with the solutions $(\sigma, \mathbf{v})$ and $(\sigma^{\circ}, \mathbf{v}^{\circ})$, respectively. k is the shear stress on the surfaces of discontinuity, and it is know that neither τ nor τ° may be larger than k. Hence the two last integrals on the right-hand side will yield solutions greater than or equal to zero. The integral on the left side must be equal to zero as it is assumed that the same boundary conditions are applied in the two cases. The maximum work inequality gives:

$$\left(\sigma_{ij} - \sigma_{ij}^{\circ} \right) \left(\dot{\varepsilon}_{ij} - \dot{\varepsilon}_{ij}^{\circ} \right) = \left(\sigma_{ij} - \sigma_{ij}^{\circ} \right) \dot{\varepsilon}_{ij}$$

$$+ \left(\sigma_{ij}^{\circ} - \sigma_{ij} \right) \dot{\varepsilon}_{ij}^{\circ} \geq 0 \qquad (29)$$

It follows that both terms on the right-hand side are positive and that they must separately vanish. This can only take place if $\sigma = \sigma^{\circ}$. The two states of stress are equal. In the case of extrusion the value of the yield strength will not be a constant but a function of temperature and strain rate and therefore of spatial coordinates. The above analysis, however, seems to apply also in this case because the volume of material can always be made so small that spatial variations can be overlooked.

In relation to extrusion the uniqueness theorem states that if the load on the tools can be determined uniquely one may, in theory, also expect to find a unique stress distribution in the flowing metal. However, the matters are complicated by the fact that in reality boundary conditions on the tools are not prescribed but determined by a friction rule. Tresca friction and assumption of full sticking will not cause any problems as it is characterized by a tangential shear stress with a value of yield, but some simplifications have to be made so that the uniqueness theorem may hold also for Coloumb friction.

The determination of the unique state of stress constitutes no small problem although such a state is known to exist. In the early parts of the development of the theory of plasticity the extrusion process represented a natural test case for analytical work. Therefore several solutions based on larger and smaller simplifications have been developed. Usually one assumes the material model to be perfectly plastic, and the yield stress to be independent of both strain rate and temperature. The geometry is taken to be either plane or axially symmetric. A state of plane stain corresponds to the extrusion of an infinitely thick plate while axial symmetry only yields cylinders.

7. Upper and Lower Bound Solutions

The plasticity theory has the advantage that it facilitates a simple procedure for calculation of an upper and a lower estimate for the boundary forces needed to cause plastic deformation. Such estimates can be obtained for a range of material models and in the presence of both small and large deformation. If it is assumed that the yield behavior is perfectly plastic the simplest estimates can be reached. In order to obtain a lower bound on applied forces one assumes a stress field, σ°, that satisfies the equilibrium equations and boundary conditions without violating the yield condition. As the stresses need not be in accordance with a constitutive relation, such a field will generally not be the actual field and is instead denoted statically admissible. The actual stress and associated strain are σ and ε, respectively. The principle of virtual work can in this case be written as:

$$\int \left(T_j - T_j^{\circ} \right) v_j \mathrm{d}S = \int \left(\sigma_{ij} - \sigma_{ij}^{\circ} \right) \dot{\varepsilon}_{ij} \mathrm{d}V$$

$$+ \int (k - \tau^{\circ})[v]\mathrm{d}S_{\mathrm{D}} \qquad (30)$$

where S_{D} are the surfaces of velocity discontinuity for the actual solution, and k is the shear yield stress. $[v]$ is the actual velocity discontinuity. As the admissible shear stress on the actual discontinuity, τ°, will not be larger than k, the last integral on the right-hand side is not negative. The first integral is nonnegative by the maximum work inequality. In this case the traction cause by the actual stresses and statically admissible ones need not be the same. If the velocity normal to the boundaries where forces are prescribed, S_{F}, are assumed to be zero, the following will hold on the rest of the boundary, S_{v}:

$$\int T_j v_j \mathrm{d}S_{\mathrm{v}} \geq \int T_j^{\circ} v_j \mathrm{d}S_{\mathrm{v}} = \int l_i \sigma_{ij}^{\circ} v_j \mathrm{d}S_{\mathrm{v}} \qquad (31)$$

The lower bound theorem thus states that the rate of work done by actual surface tractions on S_v is greater than or equal to that done by surface tractions in a statically admissible stress field. If the v_j is uniform on S_v, one will find that the load of the statically admissible field will give a lower bound to the actual one. The principle can be directly applied in extrusion. The velocities normal to the container wall and the die will be zero. S_v can be assumed to consist of two parts, the surface between the billet and the stem and surface defined by the die opening. The tractions on the last part of the surface can be assumed to be approximately zero. Hence as the velocity on the surface of the stem is uniform, an analysis will yield a lower estimate of the extrusion force at a certain time during extrusion.

Figure 36a gives an example of a statically admissible flow field. The extrusion is assumed to be frictionless. Lines of stress discontinuity are drawn. The normal stress components to such lines as well as the shear stress have to be continuous if equilibrium is to be achieved. The normal component of stress along the line, however, may differ from one side of the discontinuity to the other as shown in Fig. 37. In Fig. 36b the Mohr circles for all the stress states are drawn. Two interesting observations can be made. Firstly, the yield criterion will nowhere be violated, and, secondly, a stress discontinuity only amounts to a rotation of the Mohr circle about a vertical line going through the point (σ, τ). Geometrical considerations limit the use of the proposed field to cases where the reduction ratio is equal to or larger than 3. The necessary pressure applied to the billet in order to cause yielding has been calculated to be $p = 5k(R-1)/R$ where k is the shear yield stress.

In order to obtain an upper estimate on the extrusion force the upper bound theorem is applied. One can assume that $\mathbf{v}^o$ is a continuous velocity field satisfying the incompressibility condition. Then the material derivative of ε^o may be calculated and the corresponding stress σ^o is finally obtained from the normality rule. The actual strain rate is denoted σ. The virtual work principle can be used to obtain the following relation:

$$\int T_j v_j^o \,dS = \int \sigma_{ij} \dot\varepsilon_{ij}^o \,dV + \int \tau[\mathbf{v}^o]\,dS_D^o \tag{32}$$

The assumed *velocity field* may contain surfaces of discontinuity $[\mathbf{v}^o]$. It is known that the actual shear stress, τ, on the virtual surface of discontinuity cannot be larger than k. By the maximum work inequality, $\sigma_{ij} \dot\varepsilon_{ij}^o \le \sigma_{ij}^o \dot\varepsilon_{ij}^o$ as σ^o is on the yield surface while σ may either be inside or on. If the virtual *velocity field*, $\mathbf{v}^o$, is regarded to be kinematically admissible, it also satisfies the boundary conditions on the part of the surface where the velocity is prescribed, S_v. The last equation is then rendered, resulting in the inequality:

$$\int T_j v_j^o \,dS_v \le \int \sigma_{ij}^o \dot\varepsilon_{ij}^o \,dV + \int k[\mathbf{v}^o]\,dS_D^o - \int T_j v_j^o \,dS_F$$
$$= \int T_{ij}^o v_j^o \,dS_v \tag{33}$$

The upper boundary theorem thus states that the rate of work done by the unknown surface tractions on S_v is less than or equal to the rate of internal energy dissipated in any kinematically admissible velocity field. The last equation to the right of the inequality sign may be used to obtain an expression for an upper limit on the extrusion force applied. This limit is denoted T_j^o.

In Fig. 38a a kinematically admissible flow field for axisymmetric extrusion analog to that proposed by Avitzur [50–52] is shown. The material is assumed to

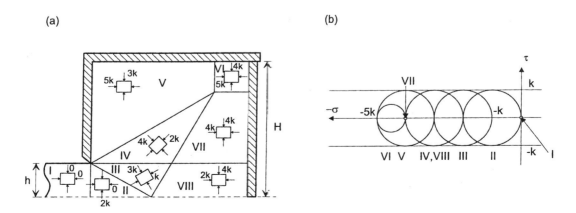

Figure 36 (a) Statically admissible flow field during extrusion. (b) The corresponding states of stress in a Mohr diagram.

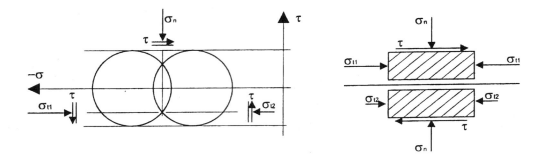

Figure 37 Possible stress discontinuity in the deforming material.

flow like rigid body with speed of the stem, v_0, in the leftmost part of the container. At the surface where the material enters the deformation zone there is a discontinuity in velocity of $v = v_0 \cdot \cos \theta$ tangential to surface marking the inlet to the primary deformation zone (Fig. 38b). Then the material velocity increases according to the relation $v = v_0 \cdot (r_0/r)^2 \cdot \cos \theta$. At the end of the zone a new discontinuity is encountered. In the dead zone surrounding the deformation zone the material is assumed to be rigid. The velocity field described is kinematically admissible, but it need not be the actual field. The experiments performed by Valberg and presented earlier show that the dead zone will communicate with the flow and that no absolute velocity discontinuity will exist between the flow and the dead zone. Instead, the flow velocity is expected to gradually decrease toward the container wall. At all times a boundary layer will exist along the whole container wall and one cannot expect the leftmost zone to be perfectly rigid. Furthermore, the form of the velocity field will change as the stem is brought closer to the die. In the limiting case, the

material will flow in a radial direction toward the bearing channel and the whole dead zone will disappear. As extrusion proves to be a transient process, the solution obtained will only be applicable at a certain stage. However, as long as all the relevant effects are taken into consideration, the proposed velocity field will produce an upper bound on the force needed to initiate deformation. Better upper bounds may be found in the literature [53–56], but for these calculations may be more complicated.

According to the last equation and Fig. 38a the work that the stem does on the metal has to compensate for the dissipation of energy in the deformation zone, on the surfaces of velocity discontinuity, and, in connection with shearing of material, on the discontinuity close to the dead zone. In addition, the extrusion force has to be increased because of friction in the container and on the yield surface. Fortunately, in an upper boundary analysis each effect can be treated separately. The force needed to deform the material in the two velocity discontinuities can be shown to be equal. The change in

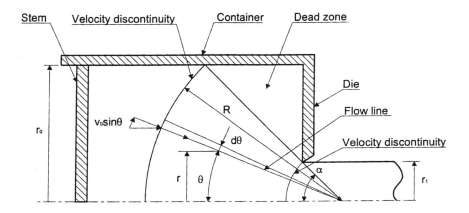

Figure 38 Kinematically admissible flow field similar to that proposed by Avitzur.

velocity over the first discontinuity is, according to Fig. 38b, $[\mathbf{v}] = v_0 \sin \theta$. Integration has to be performed then over the whole surface of discontinuity, which is assumed to be spherical:

$$F_D \cdot v_0 = 2 \int_A k[v]dA = \int_0^\alpha kv_0 \sin \theta \cdot 2\pi R^2 \sin \theta \, d\theta$$

$$= 2k \cdot \pi r_0^2 \left(\frac{\alpha}{\sin^2 \alpha} - \cot \alpha \right) \cdot v_0 \tag{34}$$

As v_0 appears on both sides and can be dropped, an expression for the force F_D is obtained.

An expression for the force F_F caused by the friction between the dead zone and the flowing material can be obtained if the third term on the right side of the upper bound inequality is integrated over a conical surface:

$$F_F \cdot v_0 = \int_{A_2} \tau_j v_j dA = k \int_{A_1} v_0 \left(\frac{r_0}{r} \right)^2 \cos \alpha \, dA$$

$$= kv_0 r_0^2 \cos \alpha \int_{\frac{r_1}{\sin\alpha}}^{\frac{r_0}{\sin\alpha}} \frac{1}{r^2} 2\pi r \sin \alpha \, dr \tag{35}$$

$$= \pi r_0^2 \sin \alpha \cos \alpha \ln \left(\frac{r_0}{r_1} \right)^2 \cdot v_0$$

Again an expression for the force is obtained by dropping the term v_0, which appears on both sides of the equality.

The force needed to deform the material in the primary deformation zone can be estimated by using the first term on the right-hand side of the upper bound inequality. Given the velocity field v, the components of strain may be expressed as:

$$\dot{\varepsilon}_r = -2\dot{\varepsilon}_\theta = -2\dot{\varepsilon}_\varphi = -2v_0 \frac{r_0^2}{r^3} \cos \theta$$

$$\dot{\varepsilon}_{r\theta} = -\frac{1}{2} v_0 \frac{r_0^2}{r^3} \sin \theta \qquad \dot{\varepsilon}_{r\varphi} = \dot{\varepsilon}_{\theta\varphi} = 0 \tag{36}$$

The equivalent strain rate is expressed in the same way in the $r\theta\varphi$-coordinate system as in one of xyz, and as a result the equivalent strain will be given as $(v_0/3^{1/2}) \cdot (r_0^2/r^3) \cdot (11 \cos^2 \theta - 1)$. A calculation of the given conical volume integral then results in the following expression:

$$F_V \cdot v_0 = \int_V \overline{\sigma} \dot{\overline{\varepsilon}} dV = \overline{\sigma} \int_{r_1}^{r_0} \int_0^\alpha \dot{\overline{\varepsilon}} \cdot 2\pi r \cdot r d\theta dr$$

$$= \frac{\overline{\sigma}}{\sqrt{3}} \pi r_0^2 \int_0^\alpha \sin \theta \sqrt{11\cos^2 \theta + 1} \, d\theta \ln \left(\frac{r_0}{r_1} \right)^2 \cdot v_0 \tag{37}$$

As expected all components of force are dependent on the yield stress given in terms of shear or equivalent

stress. Furthermore, the angle α, which defines the dead zone, also has an influence on the extrusion force. Most importantly, however, the expressions indicate that the force is a function of the reduction ratio. If all terms are added and the reduction ratio, R, is the only parameter of interest the force equation may be written:

$$F = a + b \ln R \tag{38}$$

Experiments confirm this relationship for various metals over a range of extrusion speeds. The equation may also be expected to hold for various profile geometries, but more complex profiles will generally necessitate a higher extrusion force than simpler ones with the same R-ratio as the friction surface to total volume ratio can be expected to be larger [57,58]. Therefore the constants a and b will be dependent on profile geometry. The value of a and F will also vary over the press cycle as shown in Fig. 39. In the case of direct extrusion the largest variations are due to the fact that both the surface area and friction force between the billet and the container decrease over time. However, other variations may be expected as the movement of the stem alters the geometry of the flow field. In the last phase of the extrusion charge, the press force will experience a sharp increase as the material from the dead zone moves toward the bearing channel. Another reason for variations in press force is an increase or a decrease in yield stress caused by variations in strain rates or temperatures. The degree of hardening will be highest in the last part of the charge as the metal then experiences very high strain rates. A high degree of plastic deformation causes dissipation of energy and potentially also temperature variations in time and space. If the billet is not properly preheated, temperatures particularly in the central deformation zone, close to the container wall and in the bearing

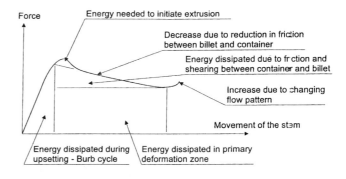

Figure 39 General relation between extrusion force and stroke length.

channel will increase gradually, resulting in lower flow resistance.

8. Slip-Line Theory

As shown, the calculation of an upper and a lower boundary estimate may be carried out relatively effortlessly, but such estimates only yield a certain indication of one parameter of interest, the extrusion force. Whereas the lower boundary estimate reveals nothing in connection to the velocity field, the upper boundary calculation gives no indication of the stresses present in the flowing metal. Furthermore, the proposed velocity and stress fields are only kinematically and statically admissible, respectively, and will only in the limiting case be equal to the actual fields. An exact solution has only for a few problems been indicated by the equivalence of the upper and lower bounds. Generally, a principle of maximizing or minimizing of energy cannot be expected to yield the exact solution as the proposed form of the field cannot be expected to be correct in the very beginning. Therefore the most interesting parameters to the quality of the finished products such as temperature, strain, and strain-rate history of individual material elements cannot be obtained through the presented limit analysis.

The theory of slip-line fields assumes that either a state of plane strain or axial symmetry exists. Furthermore, when elastic strains are neglected, calculations may be made for hardening as well as nonhardening materials. An assumption of a constant yield stress, however, leads to the simplest results. The method will at least provide an estimate of both the stress and velocity field. The slip-line method is also one that is based on an initial proposal of a kinematically admissible velocity field. However, by combining the kinematical evaluation with the use of equilibrium equations, constitutive relations, and stress boundary conditions, the proposed field will yield stresses that are in equilibrium and actually also satisfy boundary conditions. However, it is not guaranteed that the stresses in the assumed rigid region do not violate the yield condition and that they are in equilibrium. Therefore a proposed slip-line field may be viewed as an upper bound, which in the limit will be an exact solution. But as this solution is obtained in a more rigorous manner through a relatively systematic procedure of the slip-line theory than through a standard upper bound analysis, a firmer knowledge of the strain and strain-rate history for a particle can be gained. Furthermore, by obtaining knowledge of the flow line of the material, one will also be able to calculate adiabatic changes in temperature.

If it is assumed that the material responds perfectly plastic to loading, the constitutive relation will be that of Lévy and von Mises. Plane strain is assumed and the z-direction is taken to be perpendicular to the plane. Because the material is incompressible, the coordinate strain increments, $d\varepsilon_x$ and $d\varepsilon_y$, will be equal in magnitude but opposite in sign. Hence the deformation will be one of pure shear. As $d\varepsilon_z$ is set to zero, the Lévy–Mises equation for the z-direction can be applied in order to obtain:

$$\sigma_z = \frac{1}{2}(\sigma_x + \sigma_y) \tag{39}$$

Equation (39) shows that the largest and smallest stresses at all times may be found in the xy-plane, which is in accordance with the fact that deformation is one of shear in this plane. Furthermore, the stress σ_z is equal to the hydrostatic stress. Because extrusion is a process in which the hydrostatic stress may be assumed to be negative, the negative sign convention is applied, $p = -\sigma_z$. If it is assumed that the material follows the von Mises criterion for yielding, the last equation can be applied in Eq. (13). The result is:

$$(\sigma_x - \sigma_y)^2 + 4\tau_{xy}^2 = 4k^2 \tag{40}$$

where k is the shear yield stress. It can be observed that if relations for the normal stress, σ, and the shear stress, τ, on a surface with an inclination of ϕ to the x-axis are calculated, squared, and added, the Mohr circle emerges as:

$$\left(\sigma - \frac{1}{2}(\sigma_x + \sigma_y)\right)^2 + \tau^2 = \frac{1}{4}(\sigma_x - \sigma_y)^2 + \tau_{xy} \tag{41}$$

Hence in a Mohr diagram all stress states in the plastic region will be described as circles with radiuses of magnitude k. From the typical Mohr circle in Fig. 40b it can be seen that the state of stress at each point in the material can be described merely by the hydrostatic pressure, p, and the orientation, ϕ, of the plane with the largest shear stress:

$$\sigma_x = -p - k\sin 2\phi \qquad \sigma_y = -p + k\sin 2\phi$$
$$\tau_{xy} = k\cos 2\phi \tag{42}$$

A new coordinate system $\alpha\beta$ can then be defined so that the shear stress has its maximum value along the axes α and β. A convention is then that the line of action of the algebraically greatest principal stress makes a counterclockwise angle of $\pi/4$ with the α-direction (Fig. 40). As the plane of greatest shear changes from point to point in the material, the coordinate system will have to be curved. However, the α- and β-lines will still be orthog-

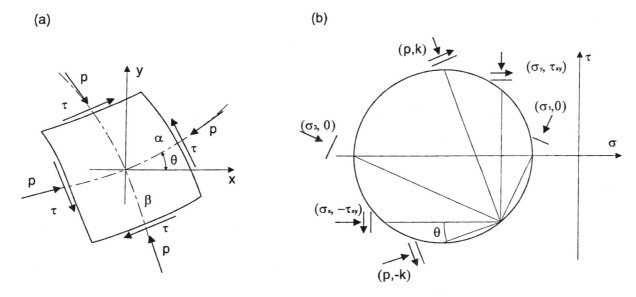

Figure 40 (a) The slip lines in relation to the Cartesian coordinate system. (b) State of stress in a material particle.

onal at each point. Because the deformation is expected to be one of shear, the α- and β-lines are called slip lines or shearlines.

As the deformation is assumed to be quasi-static and body forces are neglected, the equations of equilibrium reduce to:

$$\frac{\partial \sigma_x}{\partial x} + \frac{\partial \tau_{xy}}{\partial y} = 0 \qquad \frac{\partial \tau_{xy}}{\partial x} + \frac{\partial \sigma_y}{\partial y} = 0 \qquad (43)$$

These are the differential equations that have to be solved if the state of stress in the material during extrusion is to be obtained. In order to reduce the number of unknowns, σ_x, σ_y, and σ_{xy} are substituted with p and ϕ in accordance with Eq. (42):

$$\frac{\partial p}{\partial x} + 2k \left(\cos 2\phi \frac{\partial \phi}{\partial x} + \sin 2\phi \frac{\partial \phi}{\partial y} \right) = 0$$

$$\frac{\partial p}{\partial y} + 2k \left(\sin 2\phi \frac{\partial \phi}{\partial x} - \cos 2\phi \frac{\partial \phi}{\partial y} \right) = 0 \qquad (44)$$

Equation (44) may be described as hyperbolic. In such a case, solutions for p and ϕ can be obtained in parts of the xy-plane merely by solving a simplified version of the equations of interest along certain curves that cross a line, along which a boundary condition is prescribed. The lines of interest are called characteristics and are defined as curves in the xy-plane, across which the derivatives of p and ϕ may be discontinuous. It turns out that for this problem there are two distinct and

perpendicular characteristics going through each point, having slopes $\tan \phi$ and $-\cot \phi$ to the x-axis. Thus in the slip-line analysis, the α- and β-lines are actually the characteristics of interest. If the state of stress is given at a certain boundary, it will also be uniquely defined in all points of the xy-plane which share both its characteristics with the boundary line (Fig. 41a). Thus some peculiarities exist. It turns out the state of stress will not be influenced by conditions on other parts of the boundary of the material. Furthermore, by the definition of characteristics, the state of stress or any other field may change abruptly when crossing the α- or β-lines. These observations, however, are actually in accordance with the understanding of plastic deformation.

The velocity distribution in the flowing metal may be calculated only if isotropy and incompressibility are assumed:

$$\frac{2\dot{\gamma}_{xy}}{\dot{\varepsilon}_x - \dot{\varepsilon}_y} = \frac{\tau_{xy}}{\sigma_x - \sigma_y} \qquad \frac{\partial v_x}{\partial x} + \frac{\partial v_y}{\partial y} = 0 \qquad (45)$$

By introducing the Eq. (42) in the isotropy condition, the following relation may be obtained:

$$\cos 2\phi \frac{\partial v_x}{\partial x} + \sin 2\phi \frac{\partial v_x}{\partial y} + \sin 2\phi \frac{\partial v_y}{\partial x}$$

$$- \cos 2\phi \frac{\partial v_y}{\partial y} = 0 \qquad (46)$$

The velocity equations are also hyperbolic, and the characteristics of stress and strain are found to coincide

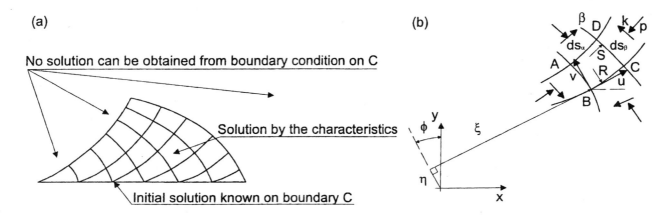

Figure 41 (a) Area with uniquely defined slip-line solution. (b) Mikhlin coordinates, ξ, $\bar{x}$ and η, $\bar{y}$ in relation to slip-line grid.

at all points because of isotropy. Thus the characteristic directions are those of both maximum shear stress and strain.

Both equations of stress and of velocity should be solved in the coordinate system defined by the slip lines. If it is assumed that the x- and y-directions at a particular point are oriented along the tangents to the α- and β-lines, the angle ϕ may be set to zero, rendering Eq. (44):

$$\frac{\partial p}{\partial x} + 2k\frac{\partial \phi}{\partial x} = 0 \qquad \frac{\partial p}{\partial y} - 2k\frac{\partial \phi}{\partial y} = 0 \qquad (47)$$

Because the original point chosen was arbitrary, the relations will hold at any point. If the equations in Eq. (47) are integrated along α- and β-lines, respectively, the relation between the hydrostatic pressure p and the orientation, ϕ, of the slip-line field relative to a rigid x-axis is:

Constant along an α – line: Constant along a β – line:

$$p + 2k\phi = C_1 \qquad\qquad p - 2k\phi = C_2 \qquad (48)$$

These are the Hencky equations, which simply represent equilibrium along a slip line. It can be observed that the hydrostatic pressure will not change along a straight line. Generally, p will vary along a curved line, the result being that the center of the Mohr circle is translated along the σ-axis.

If the velocity equations are viewed in a coordinate system with axes tangential to the slip lines, the rate of extension vanishes along the slip lines:

$$\left(\frac{\partial v_x}{\partial x}\right)_{\phi=0} = \left(\frac{\partial v_y}{\partial y}\right)_{\phi=0} = 0 \qquad (49)$$

If u and v are the velocity components along the slip lines, the velocity components along the x- and y-coordinate axes can be generally expressed as:

$$v_x = u\cos\phi - v\sin\phi$$
$$v_y = u\sin\phi + v\cos\phi \qquad (50)$$

By substituting Eq. (50) into Eq. (49) and setting $\phi = 0$, differential relations along the slip lines are obtained:

Along an α – line: Along a β – line:

$$du - v\,d\phi = 0 \qquad\qquad dv + u\,d\phi = 0 \qquad (51)$$

These are the Geiringer equations [59]. Although incompressibility is assumed, the velocity component tangential to a slip line may change. Such a variation is introduced by the curving of the slip lines. A constant tangential velocity along a slip line can be expected if the slip line is straight or if the velocity component normal to the slip line is zero. The last is the case for curved slip lines marking the boundary between flowing metal and dead zones.

Velocity discontinuities can be expected to be present when the hyperbolic equations are solved. As mass has to be conserved the velocity component normal to a line of discontinuity cannot alter when passing through it. The tangential component, however, may change (Fig. 41b). Hence the discontinuity line may be looked upon as one along which the rate of shear is infinite. The discontinuity line will then also be a slip line. Along such a line the change in tangential velocity will be constant.

Two theorems that are of practical interest when a slip-line field is to be drawn are Hencky's first and second theorem. These follow directly from the Hencky equations. The first states that the following relations can be taken to hold for the values of ϕ and p in the points A, B, C, and D in Fig. 41b:

$$\phi_C - \phi_D = \phi_A - \phi_B \qquad p_C - p_D = p_A - p_B \qquad (52)$$

Thus when going from one slip line to another in the same family the angle is turned and the pressure change will always be the same.

Hencky's second theorem is based on the first and states that the curvature of lines of the other family decreases in proportion to the distance traveled along a slip line. If R and S are the curvatures along the α- and β-lines, respectively, and s_α and s_β the corresponding coordinates, the theorem yields:

$$\frac{\partial R}{\partial s_\beta} = -1 \; \frac{\partial S}{\partial s_\alpha} = -1 \qquad \Rightarrow$$

Along an α – line:	Along a β – line:	
$dS + R d\phi = 0$	$dR - S d\phi = 0$	(53)

The last is a result of the mere definition of curvature. As can be seen, the curvature will decrease steadily as one moves to the concave side of the slip line. If the plastic zone extends sufficiently far, the radius of curvature finally vanishes. Discontinuities may, however, exist. The differential equations [Eq. (53)] are of the same form as Eq. (51). Furthermore, by moving along the slip lines, the Mikhlin coordinates shown in Fig. 41b will change according to an equation of the same form:

Along an α – line:	Along a β – line:	
$d\bar{y} + \bar{x} d\phi = 0$	$d\bar{x} - \bar{y} d\phi = 0$	(54)

The equations of the slip-line theory can either be solved analytically, numerically, or graphically. Complete analytical solutions are available for only a few problems. An example that will be provided is the process of frictionless plane strain extrusion at $R = 3$. If an area is determined uniquely by the stress boundary conditions, however, the pressures and slip-line directions at a point (m,n) can be determined directly from known values at neighboring points by the Hencky equations:

$$p(m,n) = p(m,n-1) + p(m-1,n)$$
$$- p(m-1,n-1) \qquad (55)$$

$$\phi(m,n) = \phi(m,n-1) + \phi(m-1,n)$$
$$- \phi(m-1,n-1) \qquad (56)$$

The velocity field and the geometry of the slip-line field have to be calculated with the help of Eqs. (51), (53), and (54) within the area that is defined uniquely by the boundary solutions. If the slip lines are curved at all

points a closed-form solution may be obtained by combining each pair of equations to the equation of telegraphy:

$$\frac{\partial^2 f}{\partial \alpha \partial \beta} = f \qquad (57)$$

f may in this case be either the velocities, curvatures, or Mikhlin variables. α and β are the coordinates along the slip lines and are related to ϕ as $\phi = \phi_0 - \alpha + \beta$, where ϕ_0 is a reference. Depending on the curvatures of the slip lines, a solution giving the form of the field and the velocities will be provided by either the modified Bessel function of the first kind or the Bessel function of the first kind:

$$f(\alpha, \beta) = \sum_{n=0}^{\infty} [a_n f_n(\alpha, \beta) + c_n f_{n+1}(\beta, \alpha)]$$

$$f_n(\alpha, \beta) = \left(\frac{\alpha}{\beta}\right)^{n/2} I_n\left(2\sqrt{\alpha\beta}\right) \qquad (58)$$

$$f(\alpha, \beta) = \sum_{n=0}^{\infty} [a_n g_n(\alpha, \beta) + c_n g_{n+1}(\beta, \alpha)]$$

$$g_n(\alpha, \beta) = \left(\frac{\alpha}{\beta}\right)^{n/2} J_n\left(2\sqrt{\alpha\beta}\right) \qquad (59)$$

The constants will then be determined from the boundary conditions.

As the analytical solution may be hard to perform, the geometry of the slip-line field may be calculated numerically by the discretization of Eq. (54). A constant angular distance in both α- and β-direction is chosen. The values of the Mikhlin coordinates in a point (m,n) are then calculated as those in neighboring points $(m,n-1)$ and $(m-1,n)$ are already known.

$$\bar{x}(m,n) - \bar{x}(m,n-1) = \frac{1}{2}[y(m,n)$$
$$+ \bar{y}(m,n-1)]\mu\Delta\phi \qquad (60)$$

$$\bar{y}(m,n) - \bar{y}(m-1,n) = -\frac{1}{2}[\bar{x}(m,n)$$
$$+ \bar{x}(m-1,n)]\lambda\Delta\phi \qquad (61)$$

μ and λ are either -1 or 1 depending on whether $\Delta\phi$ decreases or increases when going from the neighboring points. The velocity field can be found in the same manner.

A complete geometrical slip-line solution is provided by Prager's method [60] (Fig. 42). As three aspects are of interest, the stress state, the velocity distribution, and the geometry of the slip-line field, it seems natural to generate three geometrical representations, the stress plane,

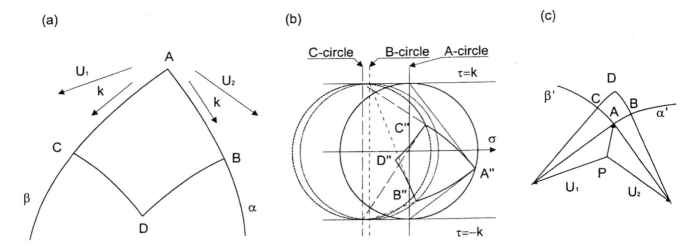

Figure 42 Slip-line solution by Prager's method: (a) the physical plane, (b) the stress plane, (c) the hodograph.

the hodograph, and the physical plane. The stress plane consists of a number of Mohr's circles, representing the state of stress at all points in the material. As the material is assumed to yield, all states provide circles with the same radius, k. The position of the pole will then be of primary interest as it will characterize the state of stress. When one moves along a curved slip line, the stresses are observed to change and consequently also the position of the pole. Lines showing its movement can then be drawn in the stress plane. The lines corresponding to the movement along an α- and β-lines are the cycloids that would be generated if the Mohr circle had rolled without slipping on the lines $\tau = k$ and $\tau = -k$. More importantly, it can be shown that their form corresponds to the form of the slip lines in the physical plane.

The hodograph is simply a representation of the velocity vectors at each point in the physical plane. The vectors are constructed from the same point so that the changes in velocity between two points will be given by the vector drawn between their arrow heads. An important property is that while the slope of the α-line is $\tan \phi$, the relation dv_y/dv_x will be $-\cot \phi$ along the line. Hence the vector giving the change in velocity at a point will be directed normal to the slip lines. This is in accordance with the fact that no elongation will be expected along the slip lines.

9.   Frictionless Extrusion with $R=3$

A simple analytical slip-line solution may be obtained if it is assumed that extrusion of $R=3$ with no friction

forces along the container wall and die is performed as shown in Fig. 43a. The problem is regarded as one of steady state. A velocity is prescribed for the piston. Other boundary conditions are assumed to be expressed in terms of stress. Neither a bearing channel nor a puller is included in the analysis. Therefore the material will be free from tractions at the die opening, $\sigma_x = 0$ and $\tau_{xy} = 0$. It may then be seen that some kind of discontinuity must exist as equilibrium does not allow the stress component along the x-axis to be zero for the material close to the die. The solution is the construction of a fan field as this allows an abrupt change in the hydrostatic pressure along a given line.

If it is assumed that the state of stress in zone I is one of yielding, it can be determined uniquely by the boundary conditions. The Mohr circle reveals that the stress component $\sigma_y = -2k$. As a result the lines of maximum shear must be inclined at an angle of $\pi/4$ and $-\pi/4$ to the x-axis. Hencky's first theorem indicates that both the angle ϕ and the hydrostatic stress, $p = k$, will be uniform in zone I, and therefore it will have the form of a triangle with straight edges.

In zone II the α-lines are taken to be radial and β-lines tangential. As the radial lines are straight, the hydrostatic pressure can be expected to be uniform along these lines. A singularity will then exist at the point where the radial lines run together, the presence of a sharp edge being the cause. Along the β-lines the expression $p - 2k\phi$ is constant. If ϕ is chosen to be $-\pi/4$ at the boundary between zone I and zone II where $p = k$, the pressure along an arbitrary β-line will be $p = k + k\pi/2 + 2k\phi$. The boundary to zone III will be a straight line at $\phi = \pi/4$.

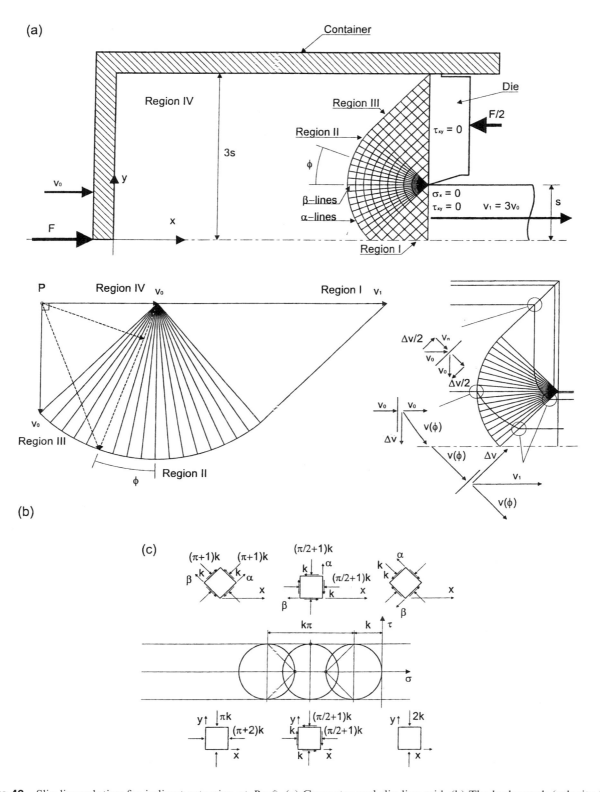

Figure 43 Slip-line solution for indirect extrusion at $R = 3$. (a) Geometry and slip-line grid. (b) The hodograph (velocity field) and the flow lines. (c) The state of stress.

The reason is that the characteristics also in this region must be straight and inclined at an angle $\pi/4$ and $-\pi/4$ because of the boundary condition $\tau_{xy} = 0$ at the die. Hence the hydrostatic pressure increases from k to $k(1 + \pi)$ from regions I to III. This corresponds to a movement of the center of the Mohr circle a distance $k\pi$ to the left. The coordinate stresses in region III may then be calculated to be:

$$\sigma_x = -p - k \sin 2\phi = -k(1 + \pi) - k \sin 2\frac{\pi}{4}$$
$$= -2k\left(1 + \frac{\pi}{2}\right) \tag{62}$$

$$\sigma_y = -p + k \sin 2\phi = -k(1 + \pi) + k \sin 2\frac{\pi}{4} = -k\pi \tag{63}$$

$\tau_{xy} = 0$ in accordance with the boundary condition and Hencky's first theorem. These results can also be obtained graphically through Fig. 43c. The stress component σ_x constitutes the pressure experienced by the die.

The velocity field may then be calculated. The material to the left of the plastic zone is assumed to be rigid and is therefore expected to move with the same velocity, V, as the stem. As the incompressibility condition can be expected to hold, the velocity of the extruded profile will be $R \cdot V$. The straight characteristics of region I indicate a uniform velocity of the same magnitude in the x-direction. In region III the material must flow along the die surface. As the characteristics are straight also in this region, one must expect that the velocity is uniform. The line that separates region III from the rest of the container must then be a velocity discontinuity. As expected, this is also a slip line. The change in flow velocity will be directed tangentially to the discontinuity line and be $\sqrt{2}V$ (Fig. 43b). As can be seen, the flow velocity in region III will also be V.

The remaining question is then how velocity changes take place in the fan. The simplest way to answer this is by constructing the hodograph (Fig. 50b). The velocity vectors representing regions I, III, and IV are drawn first. The vector representing the discontinuity between regions III and IV is then added, confirming assumptions made earlier. One should notice, however, that the magnitude of a velocity discontinuity is constant along a characteristic. Therefore the velocity discontinuity at each point along the outer β-characteristic may be drawn as a radial line with a length $\sqrt{2}V$ starting from the end of the velocity vector of region IV. The circle sector generated in the hodograph will span the same angle, $\pi/2$, as the sector in the physical plane. Thus there must also be a velocity discontinuity of the same magnitude on the border between the fan and region I. As

expected, the geometry of the hodograph resembles that of the characteristics in the physical plane.

The main objective for using the slip-line theory is to obtain a simple mathematical description of the flow paths and deformation history. The uniform flow velocity in regions I, II, and IV may easily be described by constant components in the x- and y-direction. In order to obtain expressions for the velocity in region II one has to apply the Geiringer equations. In polar coordinates the criterion of no elongation along stream lines may be written as:

$$\dot{\varepsilon}_r = \frac{\partial v_r}{\partial r} = 0 \qquad \dot{\varepsilon}_\phi = \frac{\partial v_\phi}{r \partial \phi} + \frac{v_r}{r} = 0 \tag{64}$$

The first equation confirms that the radial velocity only will be a function of the angle ϕ. As no radial discontinuity in velocity can be allowed for a material particle going from the outer field and into the fan, v_r must be a cosine function which yields $-V$ for $\phi = 0$. The second equation of Eq. (64) may then be solved with the boundary condition $v_\phi (\pi/4) = -\sqrt{2}V$. The velocity components in the fan are then:

$$v_r = -V \cos \phi \qquad v_\phi = V \sin \phi - \sqrt{2}V \tag{65}$$

The shear–strain rate and equivalent strain rate may be calculated from the above expressions:

$$\dot{\gamma}_{r\phi} = \frac{1}{r} \frac{\partial v_r}{\partial \phi} + \frac{\partial v_\phi}{\partial r} - \frac{v_\phi}{r} = \sqrt{2}\frac{V}{r}$$
$$\dot{\bar{\varepsilon}} = \frac{\dot{\gamma}_{r\phi}}{\sqrt{3}} = \sqrt{\frac{2}{3}}\frac{V}{r} \tag{66}$$

With the help of the definition of radial and tangential velocity, the flowline through the fan for a particle starting at (r_0, ϕ_0) on the boundary may be calculated:

$$\frac{v_r}{v_\phi} = \frac{dr}{r d\phi} = \frac{-V \cos \phi}{V \sin \phi - \sqrt{2}V} = \frac{\frac{1}{\sqrt{2}} \cos \phi}{1 - \frac{1}{\sqrt{2}} \sin \phi} \tag{67}$$

Integration and use of the known starting point (r_0, ϕ_0) at an edge of the fan provide the equation for the flowline:

$$r = r_0 \frac{\sqrt{2} - \sin \phi_0}{\sqrt{2} - \sin \phi} \tag{68}$$

Some of the possible flowlines are shown in Fig. 43b. If such a line enters the fan at r_0 on the characteristic $\phi = \pi/4$, the ratio r/r_0 will be independent of the value of r_0. This is because of the lack of variation in particle velocity along an α-characteristic. As a result, all such flowlines in the fan have the same form. By inserting data one finds that $r < r_0$ for all ϕ except when $r = r_0 = 0$. This is in accordance with the fact that the material is compressed during extrusion.

The study of deformation of rectangular grid patterns constructed on the billet proves valuable as it opens for a comparison of the experimental and analytical results. Furthermore, emptying diagrams, which can be derived numerically, may also provide information about the origin of material particles in the profile. Thus at least qualitative data on the deformation of material particles and the danger of inferior surface quality due to the inflow of particles from the billet surface may be obtained through the use of slip-line theory. If one is to relate the position of a particle in the billet to that in the extruded profile, one has to calculate both the coordinates of the path line and time spent on traveling along the line. As material particles in all but the fan region are expected to run along straight lines with a prescribed velocity, most of the necessary calculations are trivial. In the fan, Eq. (68) describes the flowlines. The corresponding expression for the time needed for a particle to travel from one point on the

flowline to another must be derived from the definition of tangential velocity, $v_\phi = r\, d\phi/dt$. An integration then results in:

$$
\begin{aligned}
t &= \int_{\phi_0}^{\phi} \frac{r}{v_\phi}\,d\phi = -\frac{r_0}{V}\left(\sqrt{2} - \sin\phi_0\right) \\
&\quad \times \int_{\phi_0}^{\phi} \frac{d\phi}{\left(\sqrt{2} - \sin\phi\right)^2} \\
&= \frac{r_0}{V}\left(\sqrt{2} - \sin\phi_0\right)\left[\frac{\cos\phi}{\sqrt{2} - \sin\phi}\right. \\
&\quad \left. - 2\sqrt{2}\,\arctan\left(\sqrt{2}\tan\frac{\phi}{2} - 1\right)\right]_{\phi_0}^{\phi}
\end{aligned}
\tag{69}
$$

where ϕ_0 and ϕ are the angles marking the starting point and the endpoint of the flowline, respectively. In order to calculate the time spent on going through the whole fan, one must set $\phi = -\pi/4$ and (r_0, ϕ_0) equal to either $(r_0 < R, -\pi/4)$ or $(R, -\pi/4 < \phi_0 < \pi/4)$ depending on the boundary section of the fan crossed.

Figure 44a reveals the geometry of the calculation model. The deformation of a straight line starting at

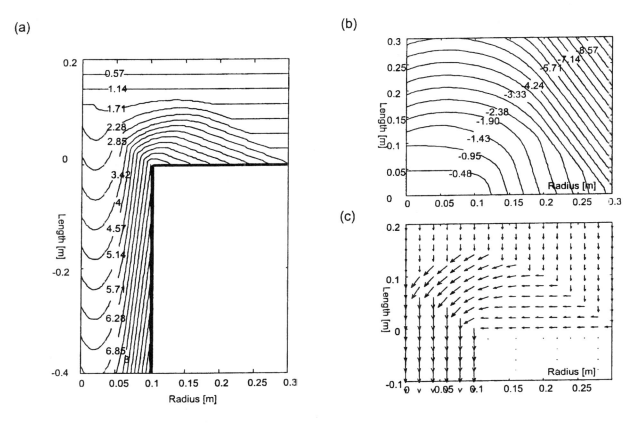

Figure 44 Material deformation in extrusion with $R = 3$ and $v_0 = 0.05$ m/sec. All times in seconds. (a) The deformation of a line which is straight at $x = 2L = 0.2$ m and $t = 0$ sec. (b) The emptying diagram. (c) Velocity vectors.

$x = 2 \cdot L$ is then studied by calculating the time taken for a particle to flow from the line to various points and then drawing iso-time curves. In the region denoted IV, the material will move uniformly and the line will remain straight until a part of it touches the fan. The material in the fan will then accelerate, and because of the geometry of the fan field part of the line at a short distance from the centerline will move forward relative to line segments both on the left and the right side. Particles on the centerline will experience a sudden increase in velocity when entering region I. However, it turns out that some particles that have been accelerated through the fan on the average will move the fastest and thus constitute the most forward part of the iso-time curve in the profile. The particles close to the container walls will be delayed because of the long distance they have to travel. Because zero friction conditions are assumed, however, material from the container surface will eventually reach the profile surface and potentially cause poor surface quality. Dead zones will not be present. The last observation is confirmed by Fig. 44b, which is the emptying diagram for the model used. The diagram is obtained by calculating the time each particle in the container will need to reach the die opening of the container. Iso-residence-time curves are then drawn. The two diagrams presented essentially give the same description of the extrusion process.

If it is assumed that extrusion is performed with a high flow rate, heat transfer will mainly be convective and the temperature of a particle will at any point be proportional to the amount of heat received. Conditions are then assumed to be adiabatic. As a material is deformed, the energy spent on deforming it will be either stored in the microstructure through a high dislocation density or dissipated as heat. In the case of cold working the first part will not amount to more than about 5%. As both work hardening and recovery/recrystallization take place during extrusion the dislocation density does not necessarily increase and therefore the part of the energy stored in the microstructure may be neglected. As a result one may roughly state that the increase in thermal energy of a particle will be equivalent to the heat provided by dissipation:

$$\rho \cdot c \cdot \dot{T} = \sigma_{ij}\dot{\varepsilon}_{ij} = \overline{\sigma}\dot{\overline{\varepsilon}} \tag{70}$$

where ρ is the density and c the heat capacity. The temperature will increase through the fan and when crossing a velocity discontinuity. In the areas where the material moves with a uniform velocity, no strain rates are expected and therefore no increase in temperature either. An analytic expression for the rise in temperature through the fan is obtained by performing an integra-

tion of Eq. (70) with respect to the time spent on traveling from one point, (r_0, ϕ_0), to another, (r, ϕ). Equation (66) is an expression for the equivalent strain, and Eqs. (68) and (69) must be applied as they describe the path followed and relate the time increment to an increment in the angle, ϕ. If k is taken to be constant, an expression for the difference between the temperatures at two points in the fan is:

$$T - T_0 = \int_{t_u}^{t} \frac{\overline{\sigma}}{\rho \cdot c}\dot{\overline{\varepsilon}}dt = \sqrt{\frac{2}{3}}\frac{\overline{\sigma}}{\rho \cdot c}\int_{t_0}^{t} \frac{V}{r_f}dt$$

$$= \sqrt{\frac{2}{3}}\frac{\overline{\sigma}}{\rho \cdot c}\int_{\phi_0}^{\phi} \frac{1 - \sqrt{2}\sin\phi}{\sqrt{2} - \sin\phi} - 2 \tag{71}$$

$$\times \frac{\sec^2\phi\left(\sqrt{2} - \sin\phi\right)}{1 + \left(\sqrt{2}\tan\frac{\phi}{2} - 1\right)^2}dt$$

If $\overline{\sigma} = \sqrt{3}\,k$ and integration is performed the result is:

$$T - T_0 = 2\sqrt{2}\frac{k}{\rho \cdot c}\left(\arctan\left(\sqrt{2}\tan\frac{\phi_0}{2} - 1\right)\right.$$

$$\left. - \arctan\left(\sqrt{2}\tan\frac{\phi}{2} - 1\right)\right) \tag{72}$$

The adiabatic temperature change through the fan is only dependent on the point where the material enters and on the total angle turned. In other words, an increment of temperature change will only depend on the increment of change in angle. The material that enters the fan at the angle $\phi = \pi/4$ will experience a temperature rise that is independent of r_0, and therefore the material closest to the surface will leave with a uniform temperature. Although the value of the strain rate goes to infinity for small radiuses, temperature changes will be limited as the time spent in the fan approaches zero. If, however, a material particle enters the fan at an angle $\phi_0 < \pi/4$, the increase in temperature through the fan will be smaller, and the material in the center of the profile can be expected to be colder than that on the surface.

Velocity discontinuities represent areas of concentrated shear. If it is assumed that the discontinuity has a certain thickness, δ, the strain rate experienced by a particle going through a discontinuity will be $\dot{\gamma} = \Delta v/\delta$, where Δv is the sudden change in velocity tangentially to the discontinuity. The time spent on passing the discontinuity will be $\Delta t = \delta/v_n$, where v_n is the component of the velocity normal to the discontinuity line. The total straining of a material particle going through the discontinuity will then be $\gamma = \dot{\gamma}\cdot\Delta t = \Delta v/v_n$. On the discontinuity between regions III and IV, v_n can be expressed as:

$$\Delta \gamma = \frac{\Delta v}{v_n} = \frac{V\sqrt{2}}{V/\sqrt{2}} = 2 \qquad (73)$$

In the same way the discontinuities between the regions I and II and II and IV can be shown to give the total strain of 2/3 and $\sqrt{2}/\cos \phi$, respectively. The rise in temperature due to a number of velocity discontinuities can be calculated to be:

$$\Delta T = \frac{k \sum \Delta \gamma}{\rho \cdot c} \qquad (74)$$

If it is assumed that the billet originally has a uniform temperature of 450°C, the adiabatic temperature at a certain point may be calculated simply by adding the contributions from all areas of shear the particle at the point has passed. The result is shown in Fig. 45a. In reality, both the friction at the container wall and at bearing surfaces will contribute to even higher temperatures close to the surface of the profile. The adiabatic assumption may also often prove to be not entirely correct, and as a result, the temperatures at least in the profile leaving the container may be lower and more uniform than expected.

10. Alternative Slip-Line Fields

The motivation for choosing a case with an R-ratio of 3 and frictionless conditions is mathematical simplicity and not the model's coherence with reality. Extrusion usually takes place at R-ratios that can be many tenfolds larger than 3, and because of the presence of friction, the velocity fields differ significantly from the one proposed.

However, as the extrusion process traditionally has been a popular test case for the application of theory, a large number of analytical and semianalytical slip-line solutions have been found. Books on the classical slip-line theory [61,62] present quite a few problems with different assumptions connected to friction and reductions.

Figure 46 gives an example of a slip-line field in the case where there is sticking friction between the flowing metal and the tooling. If the friction model is to be that of Tresca, the shear stress between the container wall and the flowing metal is equal to the shear yield stress of the metal. As this is also the maximum shear stress possible, the slip lines that interfere with the walls have to be either normal or tangential to it except at singularities. The slip lines of both families also have to meet the center axis at an angle of $\pi/4$ because of the symmetry condition. The slip-line field may then be calculated both analytically and numerically as earlier described.

When applying numerical methods the slip lines in the grid should be separated by a constant increment of the angle, $\Delta\phi$. The geometry of the container, particularly the R-ratio, will uniquely determine the angle between the outermost radial lines of the fan. If R reaches about 12.5, the analysis breaksdown as the upper radial line then is tangential to the surface of the die. When the slip-line field is determined, the stresses follow directly from Hencky's equations. The coordinates in an xy-system may be calculated by starting at the fan and working outward under the assumption that the line segment between two nodes is approximately straight.

In order to obtain the velocity field, the boundary conditions have to be determined. The material to the right of the slip-line field is assumed to move uniformly

(a)   **(b)**

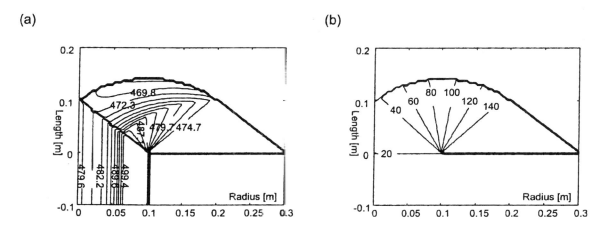

Figure 45 (a) Temperatures given under adiabatic conditions with billet preheated to 450°C. (b) Hydrostatic pressure during extrusion (MPa).

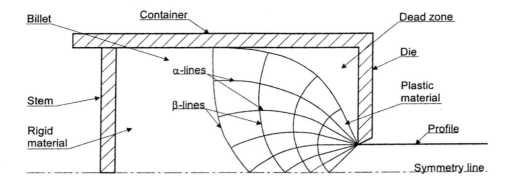

Figure 46 Slip-line field for the direct extrusion of aluminum under the assumption of full sticking at the walls of the container and the die.

at the pace of the piston. This can be used to calculate the velocity components along the velocity field in the rightmost part of the slip-line field. In the dead zone the material velocity is expected to be zero, and velocity components normal to the closest slip line vanish. Consequently, the velocity tangential to this slip line is constant and equal to the velocity of the piston, which is then also the magnitude of the velocity discontinuity along the line. A last boundary equation is that the components of velocity along the symmetry line must be equal as it is a flowline. In order to obtain a numerical solution to the problem a discrete form of the Geiringer equations has to be applied. The values of the components of velocity in a node are then found from the corresponding values in the neighboring nodes. By a method of trial and error, the velocity components at all points on the symmetry lines are varied until all boundary conditions are satisfied simultaneously.

11. The Slab Method Applied in the Study of Friction on Bearing Surfaces

The strip or slab method, which is part of the elementary plasticity theory, is based on a simplified view of the state of stress in the material. Usually, only conditions of plane strain are considered. The work piece is directed along an axis, say the x-axis, and then divided into a number of infinitesimally thin strips, each with a thickness dx. These are studied individually. In order to make this division worthwhile one has to neglect all other velocity components than that along the x-axis. At the same time one assumes that the principal directions of stress and strain are along the coordinate axes. Differential relations describing both the velocity field and the stress field may then be obtained simply by asserting

that the equations of conservation of volume and motion in the x-direction have to be satisfied. The relation between different components of stress is obtained through the Tresca yield criterion.

The slab method has been applied on a range of problems connected to materials forming, such as drawing, rolling, forging, and extrusion. The results obtained are usually the forming forces, which must be taken strictly to be the lower bounds as the method is based on equilibrium considerations. However, the slab methods will at best only give a simplified view of the stresses and deformation in the metal and at worst provide totally erroneous results. This is especially the case when the slab method is applied in the study of the relatively complex flow in the container during extrusion, as the assumption regarding the principal directions of stress need not be correct. Another problem is that the slab method does not accept large changes in geometry along the x-axis because the analysis then will be inconsistent. The extrusion force may be calculated to be in accordance with the relation $F = a + b \ln(R)$ also with the slab method, but one must be aware of the fact that the method of calculation only guarantees the result in case of small values of the reduction ratio.

In relation to the extrusion process, the slab method may probably most effectively be applied in the study of the pressure build-up through the bearing channel. Usually, analytical solutions neglect the presence of such a channel altogether, but in practical extrusion bearing surfaces of zero length are neither possible nor desirable. The pressure build-up caused by the friction between the flowing material and the bearing surfaces may actually be utilized to control the flow and therefore also the profile quality. The aim will then always be that the velocity in the cross section of the profile leaving the die should be as uniform as possible and that no internal stresses should be generated. The general rule is that a material particle will flow in the direction of the lowest pressure gradient. If the material flows too fast over a certain part of the cross section as a result of low flow resistance, one must attempt to force the material flow in other directions by increasing the length of the bearing channel and thereby also the total friction force. As less material then is expected to enter the region, the flow speed is reduced and hopefully made more equal to that of the neighboring parts of the cross section.

When bearing channels are not properly adjusted, as often is the case in connection with complex die geometry, the result will be unbalanced and uncontrolled flow but not necessarily totally unusable products. If the material flows faster in parts of the cross section, the profile can be expected to bend as it leaves the die. If velocity gradients perpendicular to the flow direction also exist during further extrusion, the profile walls may buckle or twist in areas with too great speed and experience thinning or even tearing where the velocity is too low. However, as more of the profile is extruded the material will usually leave the die at an almost uniform speed. This is because velocity gradients will cause shear stresses, which in turn will contribute to the reduction of such gradients. The result is that residual compressive stresses can be found in parts of the profile that experienced the largest flow velocity in the bearing channel. Tensile stresses will be generated in slow flowing parts so that a force equilibrium will eventually exist. This self-stabilizing effect is present in a number of metal-forming processes and explains why relatively satisfactory product quality can be obtained without total process control. However, complete reliance on such an effect is not desirable as the control of flow velocity in products with variable wall thickness and especially in thin-walled section is inefficient. Furthermore, residual stresses have to be properly removed by a stretching operation. This is particularly important if further operations such as, for instance, bending are to be performed on the profile.

The lengths of the bearing surfaces may be from a couple of millimeters to about a centimeter, and the surface area constitutes only a very small percentage of the total area of interaction between the flowing metal and the die. Furthermore, short bearing surfaces are generally preferred as they generate less friction and therefore reduce the need for a large extrusion force. However, even relatively short bearing lengths generate large increases in the needed extrusion pressure. And most importantly, the interaction between the flowing metal and the bearing surfaces is maybe the most complex and variable element of the extrusion process and thus can be expected to be of prime importance to product quality. As will be explained in the next section, the study of interaction between flow stability, die deflection, and friction in the bearing channel is one of the most interesting fields of research.

A simple slab analysis will be performed in order to calculate a rough mean value for the pressure rise through the bearing channel because of the presence of friction. As will be explained later, one may assume that the flowing material sticks to the bearing surface over the first 4 mm and that the shear stress in the sticking region is constant and equal to the shear yield stress k. The friction close to the outlet will be of Coulomb type, and it is assumed that the shear stress in the slipping region decreases linearly from the slip point and toward the outlet. The last assumption is probably only correct if the normal stress to the bearing can also be expected to decrease toward the opening. The bearing surfaces are taken to be parallel at all points. Usually, bearing surfaces are either converging or diverging and are said to be designed with a choke or a release, respectively. The establishment of the position of the slip point where material starts gliding along the bearings and a surface may be said to be created represents the main problem. Although experiments have confirmed the presence of such a point, and its position has been determined for various choke angles, bearing lengths, alloys, and extrusion rates, no rigorous method of estimation for this parameter has yet been developed. Hence when performing both analytical and numerical calculations, assumptions are usually made regarding the position of the transition region. This introduces an uncertainty into the analysis. In the following slab calculations (Fig. 47) the friction against the bearings will be described by the equations:

$$0 < x < 4 \text{ [mm]}: \quad \tau = 30 \text{ [MPa]}$$

$$4 < x < 8 \text{ [mm]}: \quad \tau = 55 - \frac{25}{4}x \text{ [MPa]} \tag{75}$$

The slip point is thus assumed to be at $x = 4$ mm, and the shear stress at the opening is set to 5 MPa. Finally, the principal stress directions are chosen to be parallel to the coordinate directions. This is actually only correct near the x-axis, along which the coordinate shear stresses may be expected to vanish because of symmetry. Close to the bearing surfaces the presence of the boundary condition makes it evident that the assumption regarding principal directions must be incorrect. However, if it is assumed that the inconsistencies may be

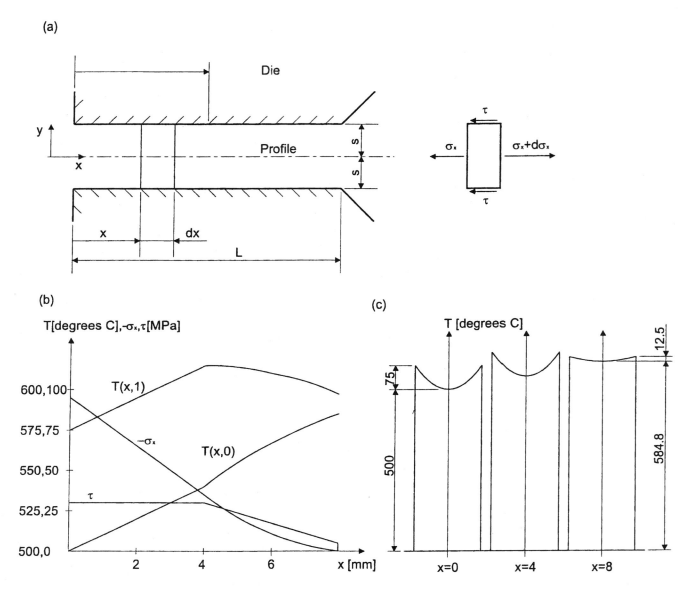

Figure 47 Slab model of the material flow in the bearing channel. (a) The geometry of the model. (b) Variation in stresses and temperatures throughout the bearing channel ($L = 8$ mm). (c) Development of the temperature field in the bearing channel.

neglected, a force equilibrium will yield a differential equation for the x-component of stress:

$$\left(\sigma_x + \frac{d\sigma_x}{dx}dx\right) \cdot 2s - \sigma_x \cdot 2s - 2\tau \cdot dx = 0$$

$$d\sigma_x = \frac{\tau}{s}dx \tag{76}$$

Integration from $x = 8$ to $x = 0$ mm then provides the stress along the bearing channel. First, the integration is performed from $x = L = 8$ mm to $x = x_s = 4$ mm with the assumption that $\sigma_x(L) = 0$. Then, $\sigma_x(x_s) = -70$ MPa is

found and the data are used in the integration from $x = x_s$ to $x = 0$. The result is:

$$0 < x < 4 \text{ [mm]}: \quad \sigma_x(x) = \sigma_x(x_s) - \int_x^{x_s} 30 d\xi = -190 + 30x$$

$$4 < x < 8 \text{ [mm]}: \quad \sigma_x(x) = \sigma_x(L) - \int_x^L \left(55 - \frac{25}{4}\xi\right)d\xi$$

$$= -\frac{25}{8}x^2 + 55x - 240 \tag{77}$$

From the criterion of full sticking and the above equations it may be seen that the stress component normal to the bearing surface will decrease linearly in the sticking region, which is in accordance with the assumptions made. At the inlet to the bearing channel the x-component of normal stress is then calculated to be -190 MPa. This result may be used as a rough estimate in a slip-line analysis.

The slab method may also yield a rough estimate of the temperature increase through the bearing channel. The y-axis is assumed to be directed normal to the bearing surfaces. As heat will be generated on the surface between the flowing metal and the die it is natural to expect that the temperatures of greatest magnitude may be found in this area. At the same time, the temperature will increase steadily toward the outlet of the bearing channel. A simple approximate temperature field can be assumed to satisfy the relation:

$$T = T_s + \Delta T \left(\frac{y}{s}\right)^2 \tag{78}$$

where $2s$ is the width of the bearing channel, and the parameters T_s and ΔT are only dependent on x. It is then assumed that all heat generated on the boundary between the die and the profile will contribute to an increased temperature in the profile and that heat conduction will only take place in the y-direction. Thus

$$q_y = -\lambda \left(\frac{dT}{dy}\right)_{y=s} = -\frac{2\lambda \cdot \Delta T}{s} = -\tau \cdot v$$

$$\Delta T = \frac{v \cdot s}{2\lambda} \tau \tag{79}$$

where v is the extrusion speed and τ is the shear stress on the boundary. ΔT will then be a function of τ and therefore also of x. In this example the following material parameters are used: $v = 1$ m/sec, $\lambda = 200$ W/mK, $\rho = 2700$ kg/m^3 and $c = 1100$ W/mK. For $x < 4$ mm, ΔT will be constant and equal to the original value of $75\,°$C as the shear stress is constant. From $x = 4$ to $x = 8$ mm the value of ΔT will decrease linearly to $12.5\,°$C at the outlet because of decreasing τ. As the heat generation connected to friction will be spent on increasing the temperature, T_s must be expected to change along x. As ΔT will not be altered for $x < 4$ mm, all heat added will contribute to the increase in T_s. The heat balance for an element is then:

$$-q_y \cdot dx \cdot dt = \rho \cdot c \cdot dT_s \cdot s \cdot dx \tag{80}$$

$$T_s(x) = -\frac{q_y \cdot x}{\rho \cdot c \cdot s \cdot v} = \frac{\tau \cdot x}{\rho \cdot c \cdot s} = 10.1x$$

x is given in millimeters. As only changes in temperature are to be assessed, $T_s(0)$ is set to 0, and the value of T_s at $x = 4$ mm can be calculated to be $40.4\,°$C. A similar

equation may be derived from the heat balance in the case where ΔT varies along x:

$$-q_y \cdot dx \cdot dt = \rho \cdot c \cdot \left(dT_s + \frac{1}{3} d\Delta T\right) \cdot s \cdot dx \tag{81}$$

The term $1/3$ is due to the parabolic ΔT-distribution. Integration and substitution of $q = -\tau \cdot v$ then yield:

$$T_s(x) = T_s(x_s) + \frac{1}{3}\Delta T(x_s) - \frac{1}{3}\Delta T(x)$$

$$+ \frac{1}{\rho \cdot c \cdot s} \int_{x_s}^{x} \tau(\xi) d\xi \tag{82}$$

In the present example a closed-form solution can be reached if the functions $\Delta T(x)$ and $\tau(x)$ are inserted. x is to be given in millimeters.

$$T_s(x) = -37.67 + 23.73x - 1.05x^2 \tag{83}$$

The temperature in the middle of the profile at the outlet ($x = 8$ mm) may thereafter be calculated to be approximately $84.8\,°$C. As expected the temperature in the midsection of the profile has risen through the bearing channel, but the transversal temperature gradients are reduced because of decreasing amount of dissipation through friction in the slipping zone. If x and y are still taken to be in millimeters, the total temperature function may be written:

$$T(x, y) = 10.1x + 75y^2$$
$$T(x, y) = -37.67 + 23.73x - 1.05x^2$$
$$+ 2.5\left(55 - \frac{25}{4}x\right)y^2 \tag{84}$$

The curves for $T(x, 0)$ and $T(x, 1)$ are given in Fig. 47. Here it is assumed that $T_s(0) = 500\,°$C and not 0.

12. Numerical Analysis

A complete analytical description of the extrusion process is of interest because it simplifies parameter studies, and because numerical results can easily be obtained and therefore also applied in online process control. However, such solutions are rarely found for more complex geometries with multiaxial stress and strain states and will only, under the simplifying assumption of adiabatic conditions, yield a proper estimate on both the strain rate and temperature history. Today, it seems the most satisfactory alternative or complement to closed-form solutions has been provided by the finite element method (FEM) [63–65].

A FEM solution for a plastically deforming material is, however, nothing but an upper bound solution, as the method utilizes the calculation of work done on the volume of the elements during a virtual deformation. The main difference between a numerical and an analytical approach is that fewer elements usually are applied in the latter, and therefore that the corresponding kinematically admissible velocity field provides a higher upper bound. If one assumes that the material model and boundary conditions give an accurate description of reality, a numerical method will, in the very limit of infinitesimal elements, yield a solution that is correct and continuous. Such is evidently unachievable as the number of elements and therefore also the computation time would be infinite. Presently, the numerical codes for axisymmetric and plane strain solutions work satisfactorily, and some approaches have been made in order to solve problems of the full three dimensions.

Undoubtedly, the finite element method constitutes a valuable tool when one is to simulate both the continuum thermomechanical and metallurgical aspects of the extrusion process [66,67]. The approach, however, has to be indirect as the FEM method does not address problems at the micromechanical level. Firstly, constitutive relations such as that of Zener–Hollomon may be determined experimentally, and one should therefore be able theoretically to predict material behavior with the help of the principles of metallurgy. Furthermore, material anisotropy may be determined from metallurgical studies, and such information may in theory, although still not in practice, be implemented in a numerical code. Shorter routines for calculation of changes in the microstructure of the material after and during extrusion may also be added to the FEM program. The elongation, shearing, and rotation of grains can then be calculated through a Taylor analysis [68], and although the theoretical fundament of this approach is far from flawless, experiments tend to give results in fair accordance with calculations. By obtaining information about the stain, strain-rate, and temperature history of each particle one should also be able to assess the degree of recrystallization and changes in dislocation density.

Numerical modeling of extrusion also has other advantages over analytical calculations. As conduction of heat may be simulated, one need not limit the analysis to one of large extrusion rates and adiabatic conditions [69]. Furthermore, the equilibrium and energy equations may be solved simultaneously, the result being that one manages to capture the strong two-way thermomechanical coupling inherent in the equations. Whereas simpler calculations may be performed so that the temperature field is affected by mechanical dissipation,

one will hardly be able to model the temperature's influence on the constitutive relation. Consequently, information on the softening effect on material during extrusion is lost in analytical calculations. One may, however, argue that calculations with one-way coupling and a perfectly plastic material model will give results not far away from those provided by a more complete model. The reason for this is that larger strain rates and the higher temperature caused by increased dissipation affect the shear yield stress in opposite directions. Furthermore, as shear deformation preferentially takes place on planes with the lowest shear resistance, there is some kind of self-regulating mechanism, which establishes a state of quasi-equilibrium between the deformation and temperature field at all times. Although it has not been proven, one may expect that the shear stress on the planes of deformation will approximately be a constant, a yield stress.

Numerical simulations are also advantageous in that they have the potential for capturing the thermomechanical interaction between the flowing metal and the tooling. Until now only two aspects of this coupling have been properly exploited, the description of friction on the bearings and on the container wall and the description of heat conduction between the tools and the aluminum. A solution of the complete heat conduction problem can only be found if the geometry of the stem, die, and container is prescribed, the mode of heat exchange between the flow and the tools is determined and the temperature field is calculated by FEM. Various FEM packages perform this calculation in a satisfying manner. Few numerical codes, however, provide solutions for deformation of and stresses in the die, container, and stem, and as most programs only manage to describe plane or axisymmetric geometry, the solutions that exist yield insufficient information for use, for instance, in the construction of dies [70]. Full thermomechanical description of this coupling for 3D geometry would be valuable as it would increase the understanding of how the deformation of the die influences the aspects of profile quality such as dimension and surface finish, and because it could be used as a tool for designing new dies and profiles. A study of such is today hindered by the extremely long calculation times for 3D codes and incomplete understanding of friction phenomena especially in the bearing channel.

$$\rho \left[\frac{\partial u_i}{\partial t} + \left(\mathbf{u}_j - \mathbf{u}_{0j} \right) \frac{\partial u_i}{\partial x_j} \right] = \frac{\partial \sigma_{ij}}{\partial x_j} \tag{85}$$

$$\rho \cdot c \left[\frac{\partial T}{\partial t} + \left(\mathbf{u}_i - \mathbf{u}_{0i} \right) \frac{\partial T}{\partial x_i} \right] = -\frac{\partial q_i}{\partial x_i} + s_{ij} \dot{\varepsilon}_{ij} \tag{86}$$

Equations (85) and (86) are the equations of motion and energy, respectively. ρ is the density of the material, $\mathbf{u}_0$ is simply a reference velocity. The total stress σ_{ij} is composed of a deviatoric and a hydrostatic part, s_{ij} and $-p\delta_{ij}$, respectively. Equation (85) reveals that a numerical procedure takes the acceleration terms into account. This stands in sharp contrast to analytical solutions, which usually assume steady-state conditions. Such an assumption is, however, only satisfactory in the mid-part of the extrusion charge. The extrusion process is transient in nature, and especially in the first and last parts of the extrusion run will a steady-state assumption lead to numerical errors of some magnitude.

The description of the problem is, however, not complete as the constitutive relations are not defined. The usual assumption is that conduction is determined by Fourier's law, and a Zener–Hollomon relation may be applied in the mechanical equation. The last defines only a relation between shear stresses and strains, and the last equation of interest is that of incompressibility. Elastic effects may also be simulated, but as earlier explained calculations then tend to be more complicated.

An element formulation may then be reached by multiplying Eq. (85) by a virtual velocity, Eq. (86) by a virtual temperature, and the incompressibility equation by a virtual pressure, applying the constitutive relations and then integrating over the complete volume of the element. A system of equations which yields the change in velocity, pressure, and temperature over a time increment is then reached. If the temperature and velocity fields are to be solved simultaneously, calculation times may be very high. Instead, equations of temperatures and velocities/pressures are often uncoupled and calculated separately by an iterative technique. The reference list provides an example of a system of equations for one element, which is solved in the numerical program ALMA2π [71]:

$$\begin{bmatrix} \mathbf{k}_{uu} & \mathbf{k}_{up} \\ \mathbf{k}_{up}^{\mathrm{T}} & 0 \end{bmatrix} \begin{bmatrix} \Delta\mathbf{u} \\ \Delta\mathbf{p} \end{bmatrix} = \begin{bmatrix} \mathbf{S}_{nod} \\ 0 \end{bmatrix} - \begin{bmatrix} \mathbf{S}_{sig} + \mathbf{S}_{acc} \\ \mathbf{S}_{inc} \end{bmatrix} \quad (87)$$

$$\mathbf{m}\Delta\dot{\mathbf{T}} + (\mathbf{k}_{con} + \mathbf{k}_{dif})\Delta\mathbf{T} = \mathbf{S}_{nod} - \mathbf{m}\dot{\mathbf{T}}$$
$$- (\mathbf{k}_{con} + \mathbf{k}_{dif})\mathbf{T} + \mathbf{S}_{heat} \quad (88)$$

The different $\mathbf{k}$'s represent "stiffness" matrices and the S's are the "loads." In the energy equation [Eq. (88)] the temperature state of the last time step as well as the dissipative heat will be the loads. Besides the traditional loads on the nodes, the acceleration term will be regarded as a load in the mechanical equation [Eq. (87)].

A number of codes, which can be utilized in the study of extrusion, have been developed. Some of these programs are made with somewhat more general perspectives but function quite well in the study of extrusion although they are not ideally suited. One such program is FIDAP, which is addressing more general fluid mechanical problems with an Eulerian perspective, and others include Forge2, Autoforge, and Deform, which are Lagrangian programs meant to handle various problems in the field of materials forming. The main weakness of all these codes is that they originally were not meant to handle the special geometry of the extrusion process. Problems arise when material is entering the bearing channel and undergoes extreme deformation. In Lagrangian programs elements are severely deformed and will not yield proper results unless remeshing is performed continuously. In Eulerian codes a constant velocity at the outlet cannot be specified, and therefore one will not be able to simulate the self-stabilizing effect that usually takes place in the bearing channel. Special programs that handle most case-sensitive aspects have been developed. One such program is ALMA2π, which has been developed at SINTEF/ Norwegian University of Science and Technology supported by Hydro Aluminium.

Experiments performed with the split-billet technique indicate that material deformation during extrusion is localized to very narrow shear zones as the one extending from the outlet of the container toward and along the container wall. This is typical for plastic deformation as the underlying equations in the case of constant yield shear strength will be hyperbolic and in fact allow distinct velocity discontinuities. Infinitely large spatial changes in velocity cannot, however, be found for results provided by FEM, as the underlying equations are no longer hyperbolic, and as the solution itself is not given as a continuous function in space. A perfectly plastic material behavior for the whole billet is in fact impossible to simulate with methods known today, as the stiffness matrix will be singular and the velocity field indeterminate. Furthermore, if the material shows very low strain-rate hardening, calculation times tend to be very long because a great number of iterations are needed to determine the states of deformation and stress. If sufficiently small elements are applied and the strain and strain-rate hardening exponents in a Norton–Hoff relation are taken to be sufficiently low, however, results in satisfactory accordance with as well analytical as experimental result can be obtained. Figure 48 provides a comparison of a deformed network found experimentally and numerically. When studying the figure, one should remember that

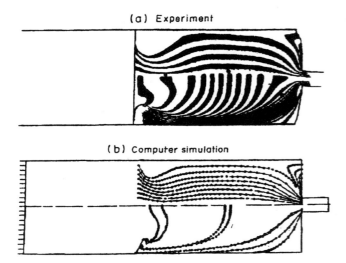

(a) Experiment

(b) Computer simulation

Figure 48 Comparison of numerically calculated (a) and experimentally found flow field (b) for a reduction ratio of about 6 and under the assumption of full sticking at walls.

the flow in the container is more easily simulated than that in the bearing channel, and that numerical results need not be that satisfactory in regions with large strain rates.

V. RESEARCH TOPICS

The use of aluminum sections in buildings, architecture, furniture, transport, electronic equipment, heat exchangers, and in mechanical design generally is well established. The thin-walled complex, multifunctional shape, with its low cost dies, with its availability, flexibility in shape, ease of fabrication, and attractive surface, has made it a favorite for the creative designer. Many successful products have been created. Still there are considerable innovative potentials in exploiting the extrusion technology, its downstream processes, and the aluminum alloys. Particular challenges are within the following areas:

- Reducing the variability of dimensions, shape, properties, and surface appearances without cost increase.
- Reducing wall thickness with narrow tolerances and increasing the strength of a section at reduced cost.
- Combining generally available sections to large sections by stir welding, instead of using large presses for large sections with limited availability and high cost.

- Combining extrusion, bending, and hydroforming for cost-effective production of complex 3D shapes with narrow tolerances and properties.

As will be and has been shown in this chapter, the hot extrusion process is characterized by:

- A strong interaction between mechanical, thermal, and metallurgical parameters during a press cycle.
- A continuous and transient variation of temperature distribution and metal flow field in container and die during the press cycle. Each material element goes through a different thermomechanical history.
- Sharp gradients in strain rate and temperature, both spatial and temporal, when the deforming material flows into and through the bearing channel.
- An interaction between the displacements of the bearing channel walls, bearing channel friction, formation of the section surface, and the stability of flow.
- An absence of adequate inline sensors, predictors, and actuators for controlling the variations in dimensions and shape of the extruded sections.
- An absence of analytical models of the extrusion process being able to "catch" the basic feature of thin-walled extrusion; the self-stabilization of the process.
- An absence of 3D numerical codes that are at a stage of development where studies of the self-stabilization–phenomenon can be studied.

Some selected research topics will be discussed that now seem ripe for "attack." These topics are considered as fundamental, precompetitive problems that will form the base for the scientific theory of thin-walled extrusion. According to Støren [72], by the theory of extrusion we understand a theory that is able to make predictions about:

- The flow pattern, the distribution of temperature and stresses, and the evolution of the microstructure of the deforming material in the container and the whole process.
- The properties of the extruded, heat-treated, and fabricated section as a function of chemical composition, initial microstructure, shape of the section, and the parameters of the processes that the section passes through from raw material to finished product.
- The sensitivity to variations in the die and section design and processing history on the material

properties, surface, dimensional tolerances, and optimal processing speed with a specified shape, alloy, and production setup.

As pointed out by Bishop [73] already in 1957 one has to apply the continuum thermomechanics of extrusion to quantify phenomena that are primarily of metallurgical origin, such as speed limit phenomena, flow resistance, the evolution of microstructure, and the properties of the extruded section. A fully coupled theory of mechanical, thermal, and metallurgical parameters has to be developed therefore. This then may give the basis for the synthesis of alloy development, process innovation, and production optimization that is needed to release the potential of extrusion-based components and products with respect to the improvements in quality properties, economic efficiency, and ecological effectiveness (Figs. 14 and 15). In the following, a possible "research strategy" to achieve this is outlined.

A. Numerical 3D Simulation and Laboratory Extrusion Experiment Validation

The development of software for 3D thermoelastc-viscoplastic flow of metals in interaction with thermc-elastic deflections of the tooling and the die is now approaching a level of precision and speed that the basic problem of the thin-walled extrusion process can be attacked [74], namely, the phenomena of self-stabilization. With self-stabilization one understands the ability of the process to react to variations in the flow-, temperature-, and flow-stress field of the material approaching the bearing channel in such a way that the variation in dimension, shape, microstructure, and surface properties is kept within the required limits during a press cycle. In order to study this basic phenomenon in a systematic and quantitative way, the following set of systematic experiments is proposed (Fig. 49):

- Thin strip extrusion
- Thin-walled tube extrusion
- Rectangular hollow thin-walled section

For the thin strip the width is kept constant, whereas the thickness is gradually reduced, giving an increase in reduction ratio, until instability in the form of buckling of the section is reached. In the tube, the reduction ratio is kept constant, whereas the tube diameter is increased and the wall thickness is decreased until the limit of satisfactory extruded section is reached. In the rectangular hollow section, the cross-sectional area of the mandrel and the thickness of the section are kept constant, whereas the width/height ratio is increased

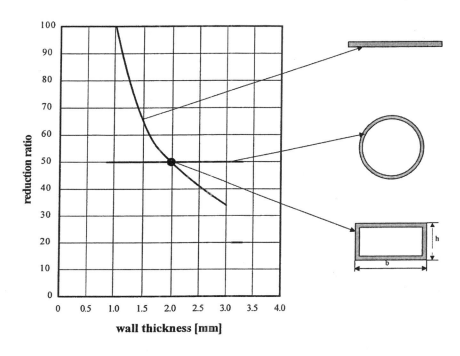

Figure 49 Generic sections for 3D modeling.

until the limit of satisfactory extruded section is reached. In the numerical simulation of these processes one will see that bearing channel phenomena will have a major influence on the simulation results.

B. Die Deflections, Friction, and Surface Formation in the Bearing Channel

A proper understanding of the interaction between the aluminum and the bearing surfaces has been viewed as a key to controlling dimensional variability and surface quality of the extruded profiles, and intensive research has, over the last few years, been undertaken in this field [75]. Most of the interest has been connected to the study of choked dies as only these are of commercial interest. If the die is designed with a release, friction forces will naturally be lower. However, the surface quality will usually be unsatisfactory as tearing, die lines, or streaking may be caused by pickup deposited directly behind the main area of contact between the die and the extruded metal.

Because of the existence of friction forces and a nonlinear material model, the flow through the bearing channel is a plug flow driven by the change in pressure from the inlet to the outlet. However, as the bearings are relatively narrow, a steady state can hardly be obtained. As a result, the velocity profile may be expected to change almost continuously through the bearing channel. In the inlet, velocity components normal to the bearings will exist as the material flow enters the channel from the container. At the outlet the velocity field ought to be uniform.

Figure 50 shows some of the phenomena observed in an experimental setup of extrusion of a thin strip with a split die, advised by Abtahi [76–78] and further studied by Tverlid [79] and Aukrust et al. [80]. At the inlet to the bearing channel, the flowing material is sticking to the die to high contact pressures. By this, one understands that there will be no relative velocity between particles on the boundary between the metal and the tool and therefore no distinct surface of the extruded metal either. The shear stress will be given by the constitutive

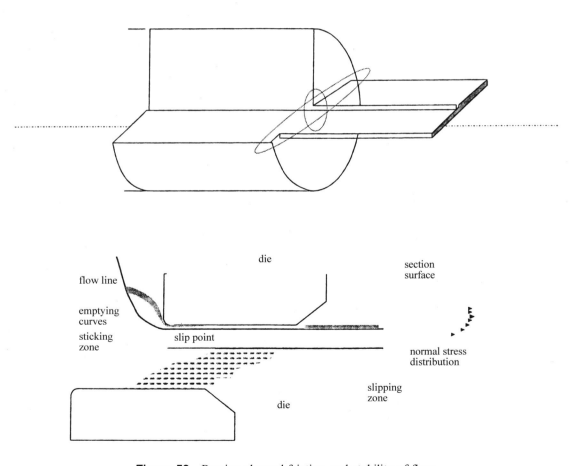

Figure 50 Bearing channel friction and stability of flow.

equation of the deforming metal and may be relatively high because of high strain rates, 1000 to 10,000 [1/sec]. As the pressure normal to the bearing surface decreases toward the outlet, however, one is to expect that the sticking friction at a certain point, the slip point, must be replaced by sliding. The sliding friction is regarded to be of Coulomb type as there will only be partial contact between deforming material and the bearings, and the magnitude of shear stress is found to be dependent on the pressure normal to the bearing surface. Over a relative short distance denoted the transition region the slipping speed is increased from zero to full outlet extrusion speed. A positive gradient in the slipping speed means that new surface is forming. In the same region material particles will also move in a direction normal to the bearings. The position of the slip point will be determined partly by the normal stresses in the end of the bearing channel, the angle of choke, which affects the pressure build-up, and the inlet radius to the bearing channel. Only in the limiting case of very large angles should one expect the slip point to be close to the outlet.

The surface structure and topology of an extruded section is created and modified in the transition and slipping regions. Furthermore, the presence of a layer of oxidized metal is thought to be important to both the friction conditions and surface generation. While such a layer is broken up and removed from the sticking region of the bearing surfaces, it remains attached to the bearings and simplifies gliding in the region of slipping. In the case of extrusion with the 6XXX series of alloys, it has been found that microdie lines present on the profile surface can be related to hard particles existing in the adhesive layer of the slipping region. The regions of slip and stick may easily be identified on the bearings of a die after extrusion as a relatively thick adhesive or oxidized layer is a witness to the presence of the former. Between the areas of slipping and sticking, a transition region of a certain length is also found to exist.

During the press cycle, the forces on the die and the temperature distribution in the die will vary, giving variations in die deflections. This then, depending on the tooling and die design, may cause variations in the choke angle and the bearing channel opening during a press cycle. These parameters, the normal stress at the outlet and the deflection of the bearing surfaces of the die, influence the friction and the local reduction ratio and thus the pressure build-up in the bearing channel. Physical understanding and quantification of these effects constitute a major scientific challenge to the study of thin-walled, dry friction, hot extrusion of aluminum.

C. The Stability of Flow

When the material in some parts of a thin section tends to flow faster than the rest of the section, compressive stresses are set up in those parts and tensile stresses in others (Fig. 50). The variations in normal stresses over the cross section of the extruded profile leaving the bearing channel will influence the position of the slip point. Under the influence of compressive stresses the slip point moves toward the outlet, and a higher friction and thus also a higher pressure build-up in the bearing channel is experienced. Tensile stresses have the opposite effect. The slip point is moved away from the outlet, and friction is reduced in the bearing channel, causing a reduced pressure build up in the inlet. This conclusion can be drawn directly from the simple slab analysis in the previous chapter. The change in pressure distribution, depending of the material response to stress variations, will promote a more balanced outlet flow.

If the change in outlet normal stress distribution causes buckling in the compressive region or thinning and even cracking in the tensile stress region, or if the pressure build-up in the inlet to the bearing channel causes redistribution of loads on the die that give die defections that act against the self-stabilization effect or gives unacceptable variations in section thickness, then the limits of extrudability are reached.

From the description it is clear that a detailed thermomechanical quantitative model of these effects is critical if successful 3D computer codes describing the extrusion process are to be developed. Detailed physical models of material response, of the conditions in the bearing channel, of surface formation, and of stick-slip mechanisms cover several levels of magnitude, from atomistic to continuum level.

D. Alloy Development, Process Innovations, and the Limits of Extrudability

Based on the progress of 3D simulation and experimental validation of flow in thin-walled extrusion, combined with the

- Computation of the tool and die deflection
- Detailed quantitative description of the response of the alloy with respect to microstructural evolution and flow resistance
- Understanding and modeling of the bearing channel friction and surface formation phenomena
- Prediction of the limits of self-stabilization of flow

the limits of extrudabilty can be studied in a systematic and quantitative way. Combined with the knowledge and creativity of experienced extrusion metallurgists, production and die experts as well as the product designers, this new scientific-based knowledge of extrusion will give way to alloy systems based on recycled aluminum, new innovative principles of die, tooling and press design, section handling, and downstream processes with quality properties, economic efficiency, and ecological effectiveness that mark the sustainable products of tomorrow.

Symbols (used in Section IV)

x, y, z	Cartesian coordinates
r, ϕ, z	cylindrical coordinates
r, θ, φ	spherical coordinates
I_x	moment of inertia about the x-axis
W_x	stiffness about the x-axis
$\bar{\sigma}$	equivalent stress
$\bar{\varepsilon}$	equivalent strain
$\dot{\bar{\varepsilon}}$	equivalent strain rate
σ_{ij}	coordinate stresses
	in Cartesian coordinates: $\sigma_x, \sigma_y, \sigma_z, \tau_{xy}, \tau_{yz}, \tau_{xz}$
	in cylindrical coordinates: $\sigma_r, \sigma_\phi, \sigma_z, \tau_{r\phi}, \tau_{\phi z}, \tau_{rz}$
	in spherical coordinates: $\sigma_r, \sigma_\theta, \sigma_\varphi, \tau_{r\theta}, \tau_{\theta\varphi}, \tau_{r\varphi}$
s_{ij}	deviatoric coordinate stresses
	in Cartesian coordinates: $s_x, s_y, s_z, s_{xy}, s_{yz}, s_{xz}$
	in cylindrical coordinates: $s_r, s_\phi, s_z, s_{r\phi}, s_{\phi z}, s_{rz}$
	in spherical coordinates: $s_r, s_\theta, s_\varphi, s_{r\theta}, s_{\theta\varphi}, s_{r\varphi}$
u_j	velocity field
ε_{ij}	coordinate strains
	in Cartesian coordinates: $\varepsilon_x, \varepsilon_y, \varepsilon_z, \gamma_{xy}/2, \gamma_{yz}/2, \gamma_{xz}/2$
	in cylindrical coordinates: $\varepsilon_r, \varepsilon_\phi, \varepsilon_z, \gamma_{r\phi}/2, \gamma_{\phi z}/2, \gamma_{rz}/2$
	in spherical coordinates: $\varepsilon_r, \varepsilon_\theta, \varepsilon_\varphi, \gamma_{r\theta}/2, \gamma_{\theta\varphi}/2, \gamma_{r\varphi}/2$
σ_o	hydrostatic stress, $\sigma_o = -p$
σ_i	principal stresses ($\sigma_1, \sigma_2, \sigma_3$)
J_i	stress invariants (J_1, J_2, J_3)
J_i^{I}	deviatoric stress invariants ($J_1^{\mathrm{I}}, J_2^{\mathrm{I}}, J_3^{\mathrm{I}}$)
ω	dissipative work
K, n, m	parameters in the Norton–Hoff relation
A, n, α	parameters in the Zener–Hollomon relation
Z	Zener parameter
Q	activation energy
R	universal gas constant
β	Q/R
δ_{ij}	Kronecker's delta
K	shear yield stress ($k = \kappa = \tau_Y$: κ_M by von Mises, κ_T by Tresca)
σ_Y	tensile yield stress
$F(\sigma_{ij})$	yield limit ($= f(\sigma_{ij}) - \kappa$)
$g(\sigma_{ij})$	plastic potential
α_{ij}	back stress movement of the yield locus
σ_{ij}^o	arbitrary stress state at or bellow tension, in some instances either
	a statically admissible stress state or
	a stress state corresponding to a kinematically admissible velocity field
$d\lambda$	constant of proportionality
S_F	parts of outer boundary where tractions are prescribed
S_v	parts of outer boundary where velocities are prescribed
T_j	tractions at boundary S_F (T_j^o corresponds to σ_{ij}^o)
v_j	velocities at boundary S_v (v_j^o corresponds to σ_{ij}^o)
R	reduction ratio
v_o	velocity of the stem ($= V$)
a, b	Constants of the force-R equation
$\bar{x}, \bar{y}$	Mikhlin coordinates ($\bar{x} = \xi, \bar{y} = \eta$)
R, S	curvatures of α- and β-lines, respectively
α, β	coordinates along α- and β-lines, respectively
u, v	components of velocity along α- and β-lines, respectively
v_n	velocity normal to a line of discontinuity
Δv	velocity change parallel to a line of discontinuity
δ	thickness of the velocity discontinuity
τ	shear stress along a specific line (τ^o corresponds to σ_{ij}^o)
t	time
ρ	density
c	specific heat capacity
q_i	heat flux (in Cartesian coordinates $q_x, q_y,$ and q_z)
T	temperature
s	thickness of profile
$\mathbf{k}$	stiffness matrix
$\mathbf{S}$	load vector

REFERENCES

1. Laue, K.; Stenger, H. *Extrusion: Process, Machinery, Tooling*; American Society for Metals: Metals Park, OH, 1981.

2. Bello, L. *Die Corrections for Changing Flow Characteristics*; Proc. 2th Int. Al. Extr. Techn. Sem., Chicago, Illinois, 1980; 89–115 pp.

3. Dawson, J.R. *Development of Conform Technology for the Manufacture of Multi-Port Tube*; Proc. 6th Int. Al. Extr. Techn. Sem., Chicago, Illinois, 1996; Vol. II, 435–439.

4. Valberg, H. *Extrusion Welding in Porthole Die Extrusion*; Proc. 6th Int. Al. Extr. Techn. Sem., Chicago, Illinois, 1996; Vol. II, 213–224.

5. Reiso, O., et al. *The Effect of Cooling Rate After Homogenization and Billet Preheating Practice on Extrudability*; Proc. 6th Int. Al. Extr. Techn. Sem., Chicago, Illinois, 1996; Vol. I, 1–10.

6. Reiso, O., et al. *The Effect of Cooling Rate After Homogenization and Billet Preheating Practice on Extrud-*

ability; Proc. 6th Int. Al. Extr. Techn. Sem., Chicago, Illinois, 1996; Vol. I, 141–148.

7. Parson, N.C., et al. *Surface Defects on 6XXX Alloy Extrusions*; Proc. 6th Int. Al. Extr. Techn. Sem., Chicago, Illinois, 1996; Vol. I, 57–67.

8. Totten, G.E.; Howes, M.A.H. *Steel Heat Treatment Handbook*; Marcel Dekker 1997; 692-715 pp.

9. Pye, D. *A Review of Surface Modification Techniques for Pre-Heat-Treated H 13 Extrusion Dies and Emerging Technologies*; Proc. 6th Int. Al. Extr. Techn. Sem., Chicago, Illinois, 1996; Vol. II, 197–200.

10. Hanssen, L.; Lindviksmoen, P.; Rystad, S. *Effect of Blend and Extrusion Parameters on Material Flow*; Proc. 6th Int. Al. Extr. Techn. Sem., Chicago, Illinois, 1996; Vol. II, 83–87.

11. Lefstad, M.; Reiso, O. *Metallurgical Speed Limitations During the Extrusion of AlMgSi-Alloys*; Proc. 6th Int. Al. Extr. Techn. Sem., Chicago, Illinois 1996; Vol. I, 11–21

12. Micropaedia. *Encyclopedia Britannica*; 15th Ed.; (Hittite), Chicago, 1992; Vol. 5, 951 pp.

13. Graedel, T.E.; Allenby, B.R. *Industrial Ecology*; Prentice-Hall: Englewood Cliffs, NJ, 1995.

14. Brezet, H.; van Hemel, C. *Ecodesign: A Promising Approach to Sustainable Production and Consumption, Modul B: Optimization of the End-of -Life System*; United Nations Environment Programme: Paris, 1997.

15. *Alulib—a source of inspiration*. Interactive CD-rom. http://www.alulib.com. Skanaluminium. Oslo, 1997.

16. Bralla, J.G., et al. *Handbook of Product Design for Manufacturing*; McGraw-Hill: New York, 1986 (Ch. 3.2 Metal Extrusions).

17. Altenpohl, D.G. *Aluminum: Technology, Applications, and Environment*; 6th Ed.; The Minerals, Metals and Materials Society, 1998.

18. Yao, C.; Mueller, K.B. *Metal Flow and Temperature Developed During Direct Extrusion of AA2024*; Proc. 6th Int. Al. Extr. Techn. Sem., Chicago, Illinois, 1996; Vol. II, 141–146.

19. Barnes, H.A.; Hutton, J.F.; Walters, K. *An Introduction to Rheology*; Elsevier: Amsterdam, 1989.

20. Lange, K., et al. *Handbook of Metal Forming*; McGraw-Hill: New York, 1985.

21. Valberg, H. *Metal Flow in Direct Axisymmetric Extrusion*; Proc. Int. Conf. Dev. Form. Techn., Lisbon, Portugal, 1990; Vol. 2, 1.11–1.38.

22. Valberg, H. Metal flow in the direct axisymmetric extrusion of aluminium. J. Mat. Proc. Techn. 1992, *31*, 39–55.

23. Valberg, H.; Groenseth, R.A. *Metal Flow in Direct, Indirect and Porthold Die Extrusion*; Proc. 5th Int. Al. Extr. Techn. Sem., Chicago, Illinois, 1992; Vol. I, 337–357.

24. Valberg, H. Forming of Metals by Forging, Rolling, Extrusion, Drawing and Sheet-Metalforming, Department of Machine Design and Materials Technology, The Norwegian University of Science and Technology, 7491 Trondheim, Norway, unpublished work (in Norwegian) p 125–134.

25. Valberg, H.; Groenseth, R.A. *Deformation and Metal Flow when Extruding Hollow Aluminium Profiles*; The 2nd. East–West Symp. Mat. Proc.; (also published in: Int. J. Mat. Prod. Techn., 1993; *8* (1), 1-22.

26. Valberg, H.; Coenen, F.P.; Kopp, R. *Metal Flow in Two-Hole Extrusion*; Proc. 6th Int. Al. Extr. Techn. Sem., Chicago, Illinois, 1996; Vol. II, 113–124.

27. Valberg, H.; Malvik, T. An experimental investigation of the material flow inside the bearing channel in aluminium extrusion. Int. J. Mat. Prod. Techn. 1994, *9* (4/5/6), 428–463.

28. Valberg, H.; Malvik, T. *Metal Flow in Die Channels of Extrusion*; Proc. 6th Int. Al. Extr. Techn. Sem., Chicago, Illinois, 1996; Vol. II, 17–28.

29. Valberg, H.; Hansen, A.W.; Kovacs, R. *Deformation in Hot Extrusion Investigated by Means of a 3-D Grid Pattern Technique*; Proc. 4th Int. Conf. Techn. Plasticity, Beijing, China, 1993; Vol. I, 637–645.

30. Valberg, H. *A Modified Classification System for Metal Flow Adapted to Unlubricated Hot Extrusion of Aluminum and Aluminum Alloys*; Proc. 6th Int. Al. Extr. Techn. Sem., Chicago, Illinois, 1996; Vol. II, 95–100.

31. Kialka, J.; Misiolek, W.Z. *Studies of Dead Metal Zone Formation in Aluminum Extrusion*; Proc. 6th Int. Al. Extr. Techn. Sem., Chicago, Illinois, 1996; Vol. II, 107–111.

32. Valberg, H. The profile surface formation during the extrusion of metals, Doctoral Thesis NTH, 1988.

33. Saha, P.K. *Influence of Plastic Strain and Strain Rate on Temperature Rise in Aluminum Extrusion*; Proc. 6th Int. Al. Extr. Techn. Sem., Chicago, Illinois 1996; Vol. I, 355–359.

34. Lefstad, M. *Metallurgical Speed Limitations in Extrusion of AlMgSi-alloys*; Doctoral Thesis, University of Trondheim, 1993.

35. Hansen, A.W.; Valberg, H. *Accurate Measurements Inside the Tool in Hot Working of Metals*; Proc. 6th Int. Al. Extr. Techn. Sem., Chicago, Illinois, 1996; Vol. II, 11–15.

36. Dieter, G.E. *Mechanical Metallurgy*; McGraw-Hill: New York, 1986.

37. Hufnagel, W., et al. *Aluminium Taschenbuch*; Aluminium-Verlag: Dusseldorf, 1983; 266-296.

38. Sheppard, T. *Development of Structure, Recrystallization Kinetics and Prediction of Recrystallized Layer Thickness in Some Al-alloys*; Proc. 6th Int. Al. Extr. Techn. Sem., Chicago, Illinois, 1996; Vol. I, 163–170.

39. Jackson, A.; Sheppard, T. *Observations on Production and Limit Diagrams for the Extrusion Process*; Proc. 6th Int. Al. Extr. Techn. Sem., Chicago, Illinois, 1996; Vol. I, 209–216.

40. Lefstad, M.; Reiso, O. *Metallurgical Speed Limitations During the Extrusion of AlMgSi-Alloys*; Proc. 6th Int. Al. Extr. Sem., Chicago, Illinois, 1996; Vol. I, 11–21.

41. Malvern, L.E. *Introduction to the Mechanics of a Continuous Medium*; Prentice-Hall: New Jersey, 1969.

42. Timoshenko, S.P; Goodier, J.N. *Theory of Elasticity* 3rd Ed.; McGraw-Hill: New York, 1970.

43. Yang, W.; Lee, W.B. *Mesoplasticity and Its Applications*; Springer Verlag; 1993.

44. Lemaitre, J.; Chaboche, J.L.*Mechanics of Solid Materials*; Cambridge University Press, 1990.

45. Khan, A.S.; Huang, S. *Continuum Theory of Plasticity*; John Wiley & Sons: New York, 1995.

46. Chakrabarty, J. *Theory of Plasticity*; McGraw-Hill: New York, 1987.

47. Hill, R. *The Mathematical Theory of Plasticity*; Oxford Clarendon Press: Oxford, 1950.

48. Druyanov, B.A.; Nepershin, R.I. *Problems of Technological Plasticity*; Elsevier: Amsterdam, 1994.

49. Zener, C.; Hollomon, J.H. Effect of strain rate upon plastic flow of steels. J. Appl. Phys. 1944, *15*, 22.

50. Avitzur, B. *Metal Forming: Process and Analysis*; McGraw-Hill: New York, 1968.

51. Avitzur, B.; Pachla, W. The upper bound approach to plane strain problems using linear and rotational velocity fields: Part I. Basic concepts. J. Eng. Ind. Nov. 1986, *108*, 295–306.

52. Avitzur, B.; Pachla, W. The upper bound approach to plane strain problems using linear and rotational velocity fields: Part I. applications. J. Eng. Ind. Nov. 1986, *108*, 307–316.

53. Grasmo, G. Friction and Flow Behaviour in Aluminium Extrusion, Doctoral Thesis 1995:37 NTH, 1995.

54. Jia, Z., et al. *Application of Upper Bound Element Technique (UBET) for Aluminium Extrusion*; Proc. 6th Int. Al. Extr. Techn. Sem., Chicago, Illinois, 1996; Vol. II, 247–252.

55. Tibbets, B.; Wen, J. *Control Framework and Deformation Modeling of Extrusion Processes: An Upper Bound Approach*; Proc. 6th Int. Al. Extr. Techn. Sem., Chicago, Illinois, 1996; Vol. I, 375–385.

56. Kakinoki, T.; Katoh, K.; Kiuchi, M. *Application of Upper Bound Method to Extrusion Die Design*; Proc. 6th Int. Al. Extr. Techn. Sem., Chicago, Illinois, 1996; Vol. II, 5–9.

57. Akeret, R. Influence of cross-sectional shape and die-design in the extrusion of aluminium: Part I. Processes in the deformation zone. Alum. Engl. 1983, *59* (9), 276–280.

58. Akeret, R. *Influence of Shape and Die Design on Metal Flow, Extrusion Load and Speed*; Proc. Extr.-Sci. Tech. Dev. Deutsche Gesellschaft fur Metallkunde E.V.Obenwisel, 1981; 191–204.

59. Geiringer, H. Fundéments mathématiques de la théorie des corps plastiques isotrops. Mém. Sci. Math. 1937, *86*.

60. Prager, W. A geometrical discussion of the slip line field in plane plastic flow. Trans. R. Inst. Technol., Stockholm, Sweden 1953, *65*.

61. Johnson, W.; Sowerby, R.; Haddow, J.B. *Plane-Strain Slip-line Fields: Theory and Bibliography*; Edward Arnold: London, 1970.

62. Johnson, W.; Sowerby, R.; Venter, R.D. *Plane Slip Line Fields for Metal Deformation Processes*; Pergamon Press: Oxford, 1982.

63. Cook, R.D.; Malkhus, D.S.; Plesha, M.E. *Concepts and Applications of Finite Element Analysis*, 3rd Ed. John Wiley & Sons: New York, 1989.

64. Zienkiewicz, O.C.; Taylor, R.L. *The Finite Element Method*; 4th Ed.;McGraw-Hill: London, 1991; Vol. 1.

65. Zienkiewicz, O.C.; Taylor, R.L. *The Finite Element Method*; 4th Ed.; McGraw-Hill: London, 1991; Vol.2.

66. Kopp, R. *Zur Simulation Strangpressens, Symposium Strangpressen*; Garmisch-Partenkirchen: Germany, 1997; 69–83.

67. Kusiak, J., et al. *Application of the Finite-Element Technique to the Simulation of the Aluminium Extrusion Process*; Proc. 6th Int. Al. Extr. Sem., Chicago, Illinois, 1996; Vol. I, 361–367.

68. Furu, T.; Pedersen, K.; Abtahi, S. *Microstructurally Based Modeling Applied to Cold Extrusion of Aluminium*; Proc. 6th Int. Al. Extr. Techn. Sem., Chicago, Illinois, 1996; Vol. I, 341–347.

69. Holthe, K.; Tjøtta, S. *The Heat Balance During Multiple Press Cycles*; Proc. 6th Int. Al. Extr. Techn. Sem., Chicago, Illinois, 1996; Vol. I, 387–392.

70. Skauvik, I., et al. *Numerical Simulation in Extrusion Die Design*; Proc. 6th Int. Al. Extr. Sem., Chicago, Illinois, 1996; Vol. II, 79–82.

71. Holthe, K.; Støren, S.; Hansen, L. *Numerical Simulation of the Aluminium Extrusion Process in a Series of Press Cycles*; Proc. 4th Int. Conf. Num. Meth. Ind. Form. Proc., Valbonne, France,1992; 611–618.

72. Støren, S. The theory of extrusion—Advances and challenges. Int. J. Mech. Sci. 1993, 35 (12), 1007–1020.

73. Bishop, J.F.W. Metall. Rev. 1957, *2*, 361.

74. Støren, S.; Grasmo, G. *High Velocity Extrusion of Thin-Walled Aluminium Sections*; Proc. 5th Int. Al. Extr. Techn. Sem., Chicago, Illinois, 1992; 353–357.

75. Akeret, R. Influence of cross-sectional shape and die-design in the extrusion of aluminium: Part II. Friction in the die land, no. 10, 1983,Vol. 59, 355–360.

76. Abtahi, S.; Friction and Interface Reactions on the Die Land in Thin-walled Extrusion, Doctoral Thesis 1995:42 NTH, 1995.

77. Abtahi, S.; Welo, T.; Støren, S. *Interface Mechanisms on the Bearing Surface in Extrusion*; Proc. 6th Int. Al. Extr. Techn. Sem., Chicago, Illinois, 1996; Vol. II, 125–131.

78. Welo, T.; Abtahi, S.; Skauvik, I. *An Experimental and Numerical Investigation of the Thermo-Mechanical Conditions on the Bearing Surface of Extrusion Dies*; Proc. 6th Int. Al. Extr. Techn. Sem., Chicago, Illinois, 1996; Vol. II, 101–106.

79. Tverlid, S. Modelling of Friction in the Bearing Channel of Dies for Extrusion of Aluminium Sections, Doctoral Thesis 1997:147 NTH, 1997.

80. Aukrust, T., et al. *Texture and Grain Structure in Aluminium Sections*; Proc. 6th Int. Al. Extr. Techn. Sem., Chicago, Illinois, 1996; Vol. I, 171–177.

7

Superplastic Materials and Superplastic Metal Forming

Namas Chandra
Florida State University, Tallahassee, Florida, U.S.A.

I. HISTORICAL NOTE OF SUPERPLASTIC MATERIALS

A. Historical Milestones

After considerable debate and deliberations at the International Conference on Superplasticity (ICSAM-91) in Osaka, Japan, superplasticity (SP) was defined as the ability of polycrystalline materials to exhibit, in a generally isotropic manner, very high tensile elongations prior to failure [1]. Figure 1 shows a quick snap shot of historical events that have marked the progress in SP. Figure 1a is a speculative application of superplasticity in ancient swords, whereas the dramatic elongation of 1800% in Pb–Sn alloy (Fig. 1c) marked the real beginning of its modern history. An 8000% elongation in Cu–Al alloy (see Fig. 1b) recorded in the *Guiness Book of World Record* invites the attention of scientists and engineers alike to this interesting field. Finally, Fig. 1d shows the unusual 120% elongation obtained in ceramics that usually shows an elongation of lesser than 2%. As much as we struggle to define superplasticity, we find it difficult to trace the earliest commercial applications of superplastic forming (SPF). Sherby et al. [2] speculate that the ancient bronzes (copper–arsenic) as far back as in 2500 BC and Damascus steels (ultra-high carbon) in 350 BC were probably the first applications of SPF. Novel applications use SP materials ranging from aluminum-based, titanium-based, iron-based, and zinc-based metals and alloys to a host of other materials in a wide range of operating strain rates, as seen in Fig. 2. Manufacturing

methods include single sheets, multiple sheets with and without diffusion bonding, and isothermal bulk-forming processes. As SPF method competes with other sheet and bulk-forming processes right from the early stages of design, the unique processing conditions of SPF require that this manufacturing method be computer-simulated to yield the final product shape and thickness distribution. This need has led to the development of SPF process models, usually based on finite element formulations and solution procedures. Although superplastic forming can be applied to both bulk and sheet forms, the present technology is concentrated on sheet forms, using mainly aluminum-based and titanium-based materials. In addition, superplastic forming application is focussed on aerospace industry. Although the market share based on cost heavily favors aeronautical and space industries, based on the number of products, architectural applications lead the way. Since the discovery of superplasticity in hard-to-form materials (nanocrystalline and ceramic materials), metal matrix composites, and high strain rate superplastic materials, the interest in superplasticity has been increasing rapidly. However, it is important to understand the market forces before a new technology (primary and secondary processing of materials and products) becomes a success story.

B. Why Superplasticity?

Superplastic forming has become a viable industrial manufacturing process for specific applications. How-

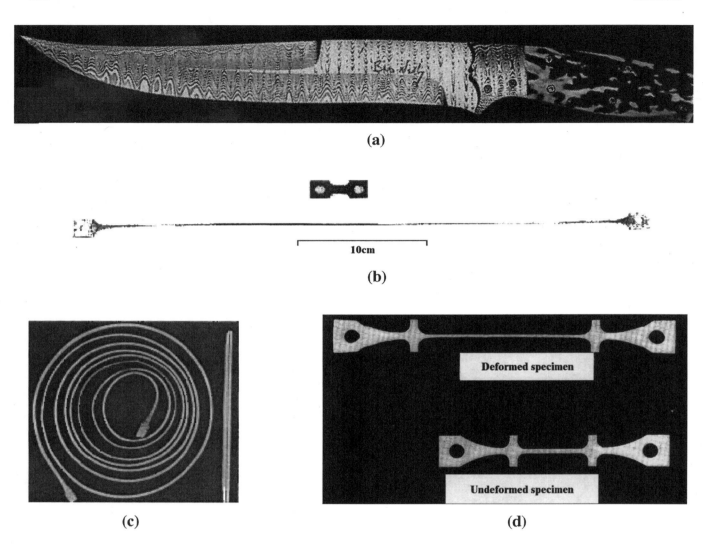

(a)

(b)

(c)

(d)

Figure 1 Historical milestones in superplasticity.

ever, SPF has not become as widespread as expected because of its prohibitive cost. Attempts over the years have been made to reduce the overall cost in a number of ways. Metallurgists and material scientists have been working on reducing the cost of basic materials and sheet production processes. Manufacturing engineers are attempting to form a given superplastic material at the fastest strain rate and lowest temperature, and are maximizing the number of parts in one forming operation. The important question is: "How can one retain the desirable superplastic properties without heating to the typical superplastic temperature while forming at the fastest rate?" In other words, what is the lowest possible temperature and the highest possible strain rate where acceptable superplasticity can be found? This question has spurred significant volumes of research in the past few decades [3].

Before examining the topic of reducing the temperature at which useful superplasticity is possible, it is worthwhile to understand some of the key scientific and technological issues. The key cost and weight savings in using SPF are clearly demonstrated in Fig. 3. The advantages of SPF over conventional forming methods are as follows:

1. Very complex and intricate shapes can be produced in one operation.
2. Integrated structures are lighter and stronger, and can replace cumbersome assemblies.
3. It eliminates the need for fabrication and assembly.
4. It reduces the amount of material required.

The above advantages result in lower cost to manufacture a component [4]. Because the forming tempera-

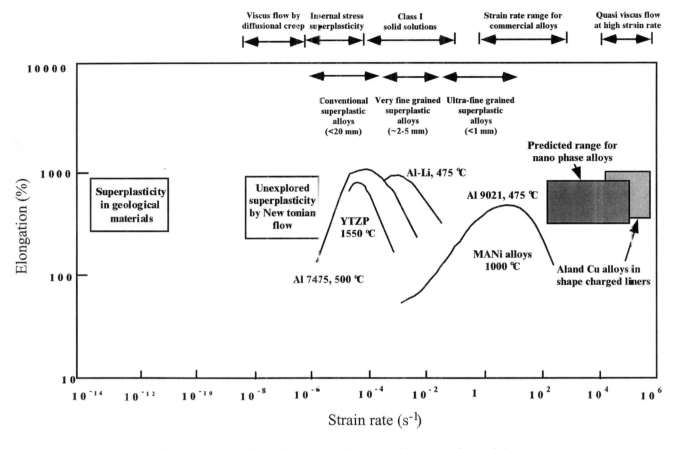

Figure 2 Overview of superplasticity in a wide range of materials.

ture is above the annealing temperature of the material, SPF components are dimensionally stable and free from processing-induced residual stresses [5]. When SPF replaces the conventional forming process, it leads to a reduction on the order of 30% in weight as well as a staggering 40% in cost (Fig. 3). SPF process has been most effectively applied in the aerospace, sports health, and automotive industries.

There are also certain disadvantages to the SPF process that restrict its widespread use. They are as follows:

1. Processing at high temperatures
2. Extended forming times compared to other forming methods
3. Narrow range of strain rate for superplastic behavior
4. Undesirable thickness distribution in the final component.

Because superplastic forming is a high-temperature process, the die material must be dimensionally as well as structurally stable at those temperatures for extended periods of time. Compared to existing hot and cold

forming methods (rolling, stamping, etc.), superplastic forming is slow and, therefore, the production quantity for SPF is small to medium in size. SP is extremely sensitive to the rate of deformation; usually, there is a very narrow range of strain rate at which optimum superplastic conditions exist. This requires lengthy trial and error procedures, or very accurate process models. These factors contribute to the high cost of the SPF process, which can only be justified by a few industries, or highly complex parts. Hence, the lowest possible forming temperature and highest reachable strain rate are constantly being sought to ensure a more widespread use of superplastic forming.

C. Expanding Literature Base on Superplasticity

Superplasticity has been one of the well-researched and documented subjects in modern times. The subject happens to be at the confluence of interest to physicists; chemists; metallurgists; mechanical, chemical, industrial, and manufacturing engineers; and industrialists.

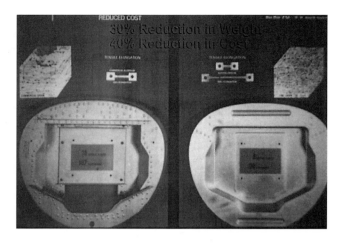

Figure 3 Cost and weight savings in superplastic forming vs. conventional forming.

The total number of publications on the subject has been constantly expanding. There have been numerous symposiums, conferences, and workshops held all over the world. There is an international committee on superplasticity, which sponsors a very well-attended meeting every 3 years starting with San Diego, CA, USA (1982); then Grenoble, France (1985); Blaine, WA, USA (1988) [6]; Osaka, Japan (1991) [1]; Moscow, Russia (1994) [7]; Bangalore, India (1997) [8]; and Orlando, FL, USA (2000) [9]; the next one in 2003 is slated to be held in Oxford, UK. The last few of these conference proceedings have appeared as a publication in the *Journal of Materials Science Forum* and the readers are referred to these valuable sources of information. A number of monographs on the subject have also appeared recently [10–12].

II. SUPERPLASTIC MATERIALS

It has been extensively observed and experimentally verified that very fine, equiaxed, and stable grain size is a prerequirement for SP in materials. In this context, we focus our attention on one specific type of superplastic behavior observed in fine-grained materials. Although all the SP materials are made up of fine grains, the converse is not true. All fine-grained materials do not exhibit superplasticity, and that has frustrated many experimentalists. There is a constant effort to identify the range of temperature and the rate of deformation at which superplasticity is exhibited in a given material, be it be in metals, alloys, glassy polymers, nanocrystals, or composites thereof. It is also fully realized that super-

plasticity may never be found in a given material. The property of fine stable, equiaxed grain size has determined many of the processing methods that have gone into the making of these materials, mainly through thermomechanical processing of precursor materials. The requirement of stable grain size is met by using either two different phases (duplex or microduplex), or small particulates that can participate in the grain refinement and stability (static or dynamic recrystallization). There is yet another class of superplasticity called *internal stress superplasticity* in which the very large elongations are achieved by the thermal cycling of polycrystalline materials through a phase change. Because this internal stress type has very limited practical applications, we do not pursue that here.

A. Metals

Although the number of material systems in which superplasticity occurs has been growing, the commercial exploitation of materials has been limited to titanium-based, aluminum-based, iron-based, and magnesium-based alloys. Grain refinement and grain size stability in each of these cases are achieved quite differently based on the material constitutents and their high-temperature behavior. In duplex or microduplex systems (e.g., titanium and Cu–bronze) at a temperature when superplasticity occurs, the microstructure consists essentially of equal volumes of the two phases. Typically, one of these phases has a higher structural rigidity (Young's modulus and yield strength) compared to the other. The rigid (harder) phase essentially remains undeformed, whereas the softer phase flows around to maintain material continuity, inhibiting the formation of microvoids and cavities. A large and important class of superplastic materials belong to this category. They include titanium-based alloys (α/β titanium), copper-based alloys (α/β copper), stainless steel (α/τ phases), ultra-high carbon steels (ferrites/carbides), and other eutectics such as Pb–Sn alloys. Table 1 lists some of the industrially useful titanium materials. *m* in Table 1 refers to the strain sensitivity parameter as given in Eq. (10).

The basic requirement of equiaxed and fine-grain structure can also be obtained in pseudo-single phase alloys in which the precipitates are essentially used both to refine and to stabilize the grains. A series of warm working and heat treatment cycles can be effectively used to refine the microstructure. During the warm working process, heavy mechanical deformation leads to a high density of dislocations, forming walls that eventually become subgrain boundaries. Submicron particles prevent the recovery process. During subse-

Table 1 Titanium-Based Superplastic Materials

Alloy	Temperature (°C)	Strain rate (1/sec)	Stress range (MPa)	Elongation (%)	m
Ti–6Al–4V	790–940	1×10^{-4}–1×10^{-3}	5–30	700–1400	0.8
Ti–6Al–5V	850	8×10^{-4}		700–1100	0.7
Ti–6Al–2Sn–4Zr–2Mo	900	8×10^{-5}–8×10^{-4}		>500	0.6–0.7
Ti–6Al–4V–2Co	815–950	8×10^{-5}–8×10^{-4}	26	670	0.75–0.85
Ti–6Al–4V–2Ni	815–950	8×10^{-5}–8×10^{-4}	13	720	0.7–0.95
Ti–6Al–4V–2Fe	815	2×10^{-4}	19	650	0.54
Ti–14% Al–20% Nb–3% V–2% Mo (Super Alpha 2)	940–980	1×10^{-4}–5×10^{-4}		1350	
Ti–4.5Al–5Mo–1Cr	840–870	2×10^{-4}		>500	0.8
Ti–5Al–2.5Sn	900–1100	2×10^{-4}		>400	0.5
Ti–4Al–4Mo–2Sn–0.5Si	880–930	8×10^{-5}–8×10^{-4}		500–1200	0.48–0.65
Ti–4% Al–4% Mo–2% Sn–0.5% Si (IMI550)	810–930	1×10^{-4}–1×10^{-3}		1600	
Ti–4.5% Al–3% V–2% Fe–2% Mo (SP 700)	750–830	1×10^{-4}–2×10^{-3}		700	
Ti–5.8% Al–4% Sn–3.5% Zr–0.5% Mo–0.3% Si–0.05% C	950–990	1×10^{-4}–1×10^{-3}		400	0.82
Ti–15V–3Cr–3Sn–3Al	760–850	2×10^{-4}–8×10^{-4}		200	0.5
Ti–8Mn	750			150	0.43

quent heat treatment, these high-energy sites act as nucleation sites, leading to a very-fine-grain microstructure. Major alloy systems that exhibit superplasticity in this manner include Al–Cu–Zr (2000 series including SUPRAL, a patented material system), Al–Mg–Zr (5000 series), Al–Zn–Mg–Cr (7000 series), and Al–Cu–Li (8000 series). A list of industrially important aluminum alloys with their mechanical properties is shown in Table 2.

B. Ceramics

Superplasticity studies and applications in ceramics are relatively recent phenomena that started in the 1980s. Typically, a grain size of 1 μm and lower is necessary for superplastic ceramics compared to that of 10 μm commonly observed in metals. Wakai is credited as being the first to observe superplasticity in a Yttria-stabilized Tetragonal Zirconia Polycrystal (YTZP) in 1986. Al-though many large elongations have been observed in ceramics during that period, a clear distinction should be made that superplasticity in a compressive state not only does not qualify as superplasticity in the traditional sense, but also does not translate to a similar behavior in tension. A number of polycrystalline ceramic materials and their composites show superplastic behavior including YTZP, Y_2O_3-doped or MgO-doped Al_2O_3, hydroxyapatite, glass ceramic, Al_2O_3-reinforced YTZP, SiC-reinforced Si_3N_4, and iron/iron carbide composites. A detailed description of the different types of monolithic ceramics and ceramic composites that show superplastic behavior is given in Ref. 12.

C. Composites

Many discontinuously reinforced metal matrix composites show superplasticity, viz., aluminum-based, magnesium-based, and zinc-based composites. Super-

Table 2 Superplastic Aluminum-Based Alloys

Alloy	Temperature (°C)	Strain rate (1/sec)	Stress range (MPa)	Elongation (%)	m
Al–33Cu	400–520	8×10^{-4}		400–1200	0.5–0.8
Al–5Ca–5Zn	450–525	8×10^{-3}		600	0.4
SUPRAL 100	400–480	5×10^{-4}– 1×10^{-2}	5–35	1800	0.4–0.55
SUPRAL 220	500	4.5×10^{-3}		1060	
SUPRAL 150	470	4.5×10^{-3}		890	
SUPRAL 5000	500	4.5×10^{-3}		230	
FORMALL 700	510	1×10^{-3}		240	
FORMALL 545	530	1×10^{-3}		290	
FORMALL 548	530	1×10^{-3}		430	
Al–6% Cu–0.4% Zr–0.3% Mg–0.2Si–0.1Ge	460	8×10^{-4}		>1800	0.65
Al–5Mg–0.6Cu–0.7Mn–0.15Cr	480–530	8×10^{-4}		700	0.45–0.7
Al–6Mg–0.4Zr	520	2×10^{-4}		885	0.6
Al–6Zn–3Mg	320–360	8×10^{-4}– 8×10^{-3}		200–400	0.3–0.35
Al–3Cu–2Li–1Mg–0.2Zr	500	1.3×10^{-3}		878	0.4
Al–5.5% Zn–2.0% Mg–1.5% Cu–0.2% Cr (7475)	510–530	2×10^{-4}– 1×10^{-3}	1–20	1400	0.5–0.8
Al–6.2Zn–2.5Mg–1.7Cu (7010)	520	5×10^{-4}		>350	0.65
Al–2.7% Cu–2% Li–0.7% Mg–0.12% Zr (2090)	510–530	1×10^{-4}– 1×10^{-2}	2–20	800	
Al–2.5Li–1.2Cu–0.6Mg–0.1Zr (8090)	500–540				
Al–4.8% Cu–1.3% Li–0.4% Mg–0.14% Zr (Weldalite)	470–530	2×10^{-4}– 8×10^{-4}	10–20	1000	
Al–3Li–0.5Zr	425–575	2×10^{-3}– 5×10^{-3}		500–1000	0.45
Al–4Cu–3Li–0.5Zr	450	2×10^{-3}– 2×10^{-2}		500–800	0.45
Al–4.7Mg–0.65Mn (5038 Al)	475–560	5×10^{-4}– 1×10^{-2}	0.8–20	250–450	

plasticity has been primarily observed in metal matrix composites (MMCs) that have been produced by powder metallurgy routes.

D. Metallic Glasses, Nanocrystalline, and Other Materials

When the size of the grains is below a few hundreds of a micrometer, then the materials fall under the category of nanocrystalline materials. Because obtaining as fine a grain size as possible has been one of the primary mantras of SP, nanocrystalline materials automatically come under its scope. Many grain refining methods, originally developed for superplasticity, have been quickly adapted for producing nanocrystaline materials. For example, an equal channel angular extrusion method originally developed for superplastic metals has become a manufacturing method of choice for nanomaterials.

Nanomaterials have been produced using either the powder consolidation method or severe plastic deformation route. Samples of Ti–6Al–4V, Ni3Al, and Al–5% Mg–2% Li–0.1% Zr have been produced using severe plastic deformation methods with grain sizes less than 100 nm. Although the superplastic tendencies have been observed in these materials, exact mechanisms and possible industrial use remain unclear.

E. High Strain Rate Superplasticity

The above discussions were based on the type of materials. There is an important class of superplastic materials that exhibits superplasticity at rates of deformation much higher than 10^{-4} sec^{-1} strain rate typically observed in the classical titanium-based and aluminum-based systems. Although strain rates with positive exponents (higher than 10^{0}) have been observed in many materials, there is very little practical use for such high rates. This contradiction arises from the fact that the time consumed in heating the part to superplastic temperature will far outweigh the forming time requirement, and hence does not lead to any reduction in the total time and the manufacturing cost. Some of the familiar high strain rate materials with their properties are listed in Table 3.

III. DEFORMATION CHARACTERISTICS OF SUPERPLASTIC MATERIALS

A. Mechanical Behavior

Thermomechanical constitutive equations of superplastic materials describe the relationship between flow stress, strain, strain rate, temperature, and other microstructural quantities, usually the grain size. The mathematical relationship employs a number of model/material parameters (coefficients) and material constants. Lattice and bulk diffusion coefficients, shear, bulk and Young's moduli, and universal gas constant are examples of material constants, whereas a number of fitting parameters (e.g., K, m, n, p) are model/material parameters. There are other microscopic variables (e.g., density of dislocation, density and orientation of cavi-

ties, and texture) that are not only difficult to measure but much more difficult to be used in practical computations. Although hundreds, if not thousands, of materials (metals, alloys, intermetallics, ceramics, and composites) exhibit superplasticity in a narrow range of temperature and strain rate (see Fig. 2), it is the desire of constitutive modelers to build a general mathematical framework to describe their mechanical behavior. This is a very challenging proposition.

As theoretical and computational mechanics, they all seek the Holy Grail of an ideal constitutive equation that can describe the material behavior at all ranges of thermomechanical loading conditions, knowing fully well that the microstructure strongly influences the inelastic behavior and that the microstructure is a product of the initial chemistry and the history of the processing conditions beginning with the melt. In the case of superplastic material, we seek such a relation involving $\sigma - \dot{\epsilon} - \epsilon$, taking into account the effects of:

Temperature
Strain hardening/softening
Grain growth (static and dynamic)
Cavitation (initiation, growth, and coalescence)
Deterioration in postdeformed thermomechanical properties.

The model should be flexible enough to allow minor changes in the chemical composition of the ingot, as well as small changes in the thermomechanical primary and secondary processing of the melt into the product form, and allow for variations in the deformation history

Table 3 High Strain Rate Superplastic Materials

Alloy	Temperature (°C)	Strain rate (1/sec)	Stress range (MPa)	Elongation (%)	m
2124-Zr	475	3×10^{-1}	35	490	0.5
7475-Zr	520	10^{-1}	11	600–900	0.6
Al–Mg–Zr	500	10^{-1}	21	570	0.3
Al–Cu–Zr	470	10^{-1}	25	480	0.3
SiCp/1100	630	10^{-1}	7.5	200	0.5
SiCw/2124	525	3×10^{-1}	10	300	0.33
Si3N4w/6061	545	2×10^{-1}	7	600	0.5
Si3N4p/6061	560	2	5.2	620	0.3–0.5
Si3N4w/7064	545	5×10^{-1}	14	240	0.5
Si3N4p/7064	545	1	10	330	0.45
Al–Ni–Mm–Zr 600	600	1	15	650	0.5
IN9021	550	50	18	1250	0.5
SiCp/IN9021	550	5	5	600	0.5
Al–Mg–Li–Zr	350	10^{-2}	85	1180	0.5
Zn–Al	200	3×10^{-2}	27	1970	0.6

(temporal and spatial variations of process parameters) during superplastic forming. Additionally, the constitutive equation should be *simple enough but not too simple*. It should facilitate the experimental evaluation of material constants and model parameters with the least number of tests and high degree of accuracy and reliability. Finally, the constitutive equation should be numerically simple to be implemented in a computational model.

The ability of constitutive equations to accurately model the deformation conditions is critical in superplasticity. For example, the temperature has a profound effect on the behavior (e.g., transition from plastic to superplastic behavior), as shown in Fig. 4. This figure shows the deformation behavior of Al 5083 (Al–4.7% Mg–0.65% Mn) under uniaxial loading condition at a constant true strain rate of 1×10^{-4} sec^{-1}, with the material showing optimum superplasticity at 540°C. As one can see, the deformation characteristics change significantly within a range of ±10°C. In addition, note that the material exhibits elastoplastic behavior below 200°C.

A properly formulated and validated constitutive equation helps material scientists and metallurgists to modify composition and processing conditions to induce superplasticity, if not present, or to obtain optimum superplastic conditions. There is an ever-increasing need for the material scientist to lower the superplastic temperature, increase the superplastic strain rate, extend the superplastic range (e.g., higher m over a wider range of temperature and strain rate), and, finally, enhance the postformed mechanical properties, by reducing cavitation. Mechanical/manufacturing engineers, on the other hand, conceive the constitutive equation as the mathematical representation of the material during its entire

forming operation at all ranges of strain rate, strain, and temperature to obtain the optimal formed shape with the least deterioration in the postformed mechanical properties. These engineers seek a form of constitutive equation that is simple for numerical implementation, and hope that the equation (or set of equations) is valid for a range of processing conditions (multiaxial state of stress, strain, and strain rate) during the entire duration of the forming process. Thus, there is a clear dichotomy of requirements in the accurate and efficient formulation of constitutive equations between material scientists and mechanical engineers. Not surprisingly, this dichotomy leads to the divergence in the formulation, validation, and application of constitutive equations among the two groups.

Constitutive equations can be broadly classified into *phenomenological* and *(micro)mechanical* types. Phenomenological equations are formulated based on a given set of experimental data and then finding a function that best *fits* the measured data with a specified number of coefficients. If those equations were to be considered as *material laws*, they should possess a general character with a predictive capability, and with a field of validity outside the range of fitted data. On the other hand, the mechanics-based approach attempts to model the deformation and fracture by using continuum mechanics and thermodynamics by the process of averaging field variables at a scale of volume elements. Typically, this approach uses mechanical analogs of springs, dashpots, and viscous and frictional elements in some combinations to represent the mechanical behavior. This approach is very appealing to process modelers because the equations can easily be cast into three-dimensional (3D) forms and made available as matrices. The mechanics approach naturally lends itself

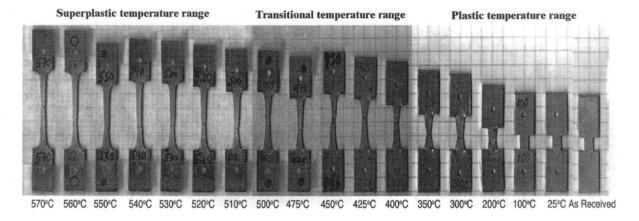

Figure 4 Effect of temperature on the transition from plastic to superplastic behavior.

to modeling yielding, and strain and strain rate harden-ing (isotropic, kinematical, or combined) through a combination of elastic, viscoelastic, and viscoplastic formulations. It will be shown later that these formula-tions—although very relevant to room temperature elastoplasticity, high-temperature viscoplasticity, and creeping solids—do not adequately capture the unique nature of superplastic mechanical behavior.

B. Microstructural Observations

It is very well established from the study of surface markers and from the study of deformation-induced textures that the dominant strain-producing mechanism in superplasticity is grain boundary sliding (GBS). It should be realized that because the total strain achieved in superplasticity is several hundred percent, there should be significant levels of grain boundary sliding. Although the geometrical details of the grain boundary sliding are not well understood, there is a general agreement as to the microstructural features associated with the process. Strain in a given direction is accumu-lated by the motion of individual grains or clusters of grains relative to each other by sliding and rolling. Grains are observed to change their neighbors and are seen to emerge at the free surface from the interior. During deformation, the grains remain equiaxed or become equiaxed. Texture, an indicator of the volume of grains in a given spatial orientation, decreases in intensity with superplastic flow. This is in direct contrast to the plastic flow, where the texture increases along the principal direction of plastic strain. The motion of an individual grain is dependent on both the normal and shear strains acting on the grain boundary and therefore depends on the mechanical and geometrical character-istics of the boundary surface. It also depends on the thermomechanical properties and deformation charac-teristics of the interior of the grains. Translation and rotation being stochastic in nature, they occur in differ-ent directions to different extents at different locations. Thus, the overall deformation is highly inhomogeneous at the scale of the grains.

Because crystalline materials are made up of contig-uous grains in three dimensions separated only by grain boundaries, and if grains were to rearrange during deformation, there has to be some accommodation process permitting the geometrical rearrangement of the grains. If grain boundary sliding were to occur as a completely rigid system, then voids would develop in the microstructure. The creation of such cavities is omnipresent barring a few cases, and, in fact, cavitation is the leading cause of failure. Many superplastic mate-rials do not cavitate, and even in cavitating materials, the cavities are far from homogeneous, noticeable only at larger strain. Hence, cavities do not accommodate grain boundary sliding. Diffusion, dislocation, and/or grain boundary mobility can act as accommodation mechanisms. Even partial melting has been suggested as a possible source in some cases. The exact nature of the accommodation process and the rate at which the process occurs have been the subjects of intense research for a very long time, and is still not understood properly.

Figure 5 shows the evolution of microstructure and texture with increasing levels of plastic strain. The orientation of grains that are initially random (or with minor texture) rotate along the maximum principal direction, keeping intact the same neighbors throughout the deformation. With increased levels of plastic strain, the number of crystals oriented along the principal direction continues to increase as shown in the pole diagrams. The rotations are achieved primarily by dislocation motions, with the dislocation density in-creasing with plastic strain. The material becomes in-creasingly anisotropic with strengthening along the rolling (principal strain) direction in relation to other directions. Voids nucleate at the particle/grain and grain boundary interfaces (at all locations with sharp stiff-ness/strength gradients), and quickly lead to material instabilities and eventual failure.

On the other hand, in superplasticity, original grain shape and size remain essentially identical for most of the accumulated strain. The randomness of the crystal orientation increases, thus weakening the texture. As shown in Fig. 6, the neighbors continue to change. The inelastic strain is produced primarily through grain boundary sliding accommodated by some mechanism. It is believed that the rate of deformation is controlled by the rate with which accommodation can occur. Dislocation activity is restricted to the role of one of the many possible accommodation mechanisms; inter-estingly, there is no increase in dislocation density with superplastic deformation. Some possible accommoda-tion mechanisms include grain boundary mobility (grain boundary traversing normal to the boundary surface), dislocation in the interior of the grain and in the boundary, diffusion in the grain and the boundary, partial melting/recovery, and void creation and con-sumption. The debate even today has focused on the questions as to which of them (any, many, or all) contributes to maintaining the material continuity, to what extent, and at what rate. Lack of accommodation causes cavities to originate especially at grain boundary ledges, triple points, and particles located on grain boundaries. These minute cavities continue to grow,

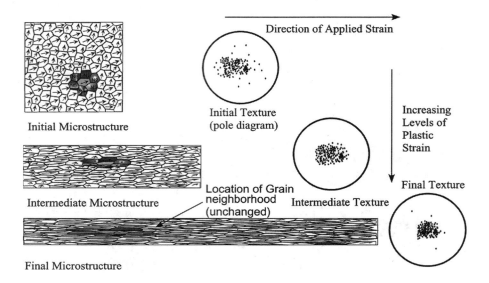

Figure 5 Evolution of microstructure and texture during plastic deformation.

coalesce, and result in material discontinuity and final failure. The shape of the cavities is much more rounded compared to that of elastoplastic deformation. Material does not cause work hardening (i.e., the flow stress essentially remains constant). Some work hardening experienced by a few materials is essentially due to thermally (static) or strain (dynamic)-induced grain growth. Postformed SP properties show a remarkable level of isotropy indicating that the grain orientations are random and that there is no preferential strengthening along any direction.

It is clear from the foregoing discussions that the microstructural features of superplasticity are very different from that of other inelastic processes. It is also evident that superplasticity is not an isolated anomalous behavior, but occurs in a large class of materials. At various stages of deformation, grains (without change in shape and size) switch neighbors and rearrange in

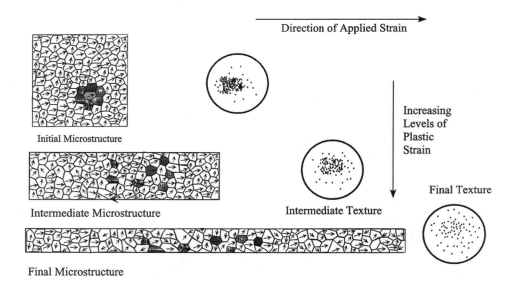

Figure 6 Evolution of microstructure and texture during superplastic deformation.

the direction of principal strains. Thus, it is possible to produce very large levels of strain provided the switching process is achieved without creating any material discontinuity (cavity). Thus, the primary criterion is that during superplastic deformation, internal surfaces (cavities) are not created. Even if present, the cavities are subcritical in that they do not grow but are subsumed.

The other feature of superplasticity is the ability of the material to resist local necking. It is well known that geometrical/material imperfections during deformation feed onto itself, creating a localized necking process leading to geometrical discontinuity and eventual failure. In a simple sense, a region with a reduced cross-section experiences a higher level of stress ($\Delta\sigma$) compared to the rest of the section, causing further reduction in area and consequent additional increase in stress. This quickly leads to an unstable situation where the failure occurs. Superplasticity reduces the tendency to local necking by reducing the differential stress $\Delta\sigma$, delaying the process of localization. This effect is related to the mechanics of deformation and is not material-specific.

C. Cavitation Damage and Failure

Superplastic materials under deformation (SPD) eventually fail under two different modes: one as a result of excessive localized plastic flow, and the other as a consequence of growth and coalescence of cavities. In the latter case, the cavities may either preexist (prior to SPD) or nucleate concurrent with the deformation; here, the failure occurs as a result of cavitation damage and the full level of superplastic strain cannot be achieved. It is well established that grain boundary sliding and rotation are the main strain-producing mechanisms; some kind of accommodation mechanisms should be continually and amply available if material separation were to be avoided. However, as it happens in a number of superplastic material systems including aluminum-based, copper-based, iron-based, and zinc-based systems, microcavities that limit the achievable levels of superplasticity in those materials are formed during SPD. Typically, cavities originate at the grain boundary triple points or precipitates/impurities along the grain/phase boundaries, when those areas are under a state of tensile stress. Interestingly, cavitation is not an essential feature of superplasticity when one of the phases is soft enough to accommodate the deformation process, as in the case of titanium-based two-phase superplastic systems.

Although externally visible levels of cavitation damage may not be reached in many practical parts, any level of cavitation in the material leads to reduced mechanical properties especially at elevated temperatures and is a cause for concern. Hence, cavitation in superplastically formable materials should be reduced as much as possible, if not eliminated. Fortunately, there is a simple solution to this problem. Because tensile stress is a prerequirement for the origin of cavities, superimposition of hydrostatic stress during the forming process practically eliminates cavitation. However, such imposition becomes impractical in some cases because the overall load levels required by the forming press become very prohibitive, and many industries prefer to form without the back pressure, sacrificing some loss in mechanical properties. That brings the problem back a full circle. One needs to understand the origin and growth of cavities, and their effect on mechanical properties and modes of failure. Cavitation has been extensively studied and a vast literature base is available.

One of the fundamental questions to be answered is: When do cavities nucleate, in the material preprocessing stages (preexisting type) or during superplastic deformation (concurrent type)? A very recent work by Chen [13] indicates that in Al 5083, material cavities do preexist (see Figs. 7 and 8). The distribution of these cavities that are isolated in a row (stringers) depends on the rolling direction during the sheet manufacturing process. The work also finds that cavitation growth depends on the state of stress, viz., uniaxial, biaxial, or triaxial. Typically, one studies uniaxial tests and extends the results to multiaxial cases based on some principle of equivalence. One widely used criterion is based on distortional energy, postulated by von Mises. Such extensions based on the von Mises criterion have been

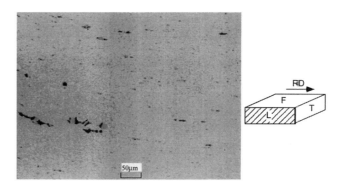

Figure 7 Preexisting cavities in as-received Al 5083.

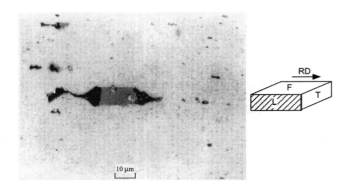

Figure 8 Preexisting cavities at higher magnifications.

broadly used in the literature but are found to be inaccurate. If one uses the logarithmic thickness strain as the basis for the extension, a better fit to the experimental data can be made. Figure 9 clearly shows the increase of the number and the size of the cavities in a biaxial cone test conducted on Al 5083 materials.

The next question of the influence of cavitation on the mode of failure has been well studied by Watts [3]. The work of Watts (see Fig. 10) clearly shows that the level of cavitation significantly increases as one moves from either end of a tensile specimen previously subjected to SPD. As the level of cavitation increases, all the room temperature mechanical properties (yield strength, Young's modulus, and percent elongation) decrease. However, the decrease is a stepwise function of temperature coinciding with the plastic flow, transitional flow, and superplastic flow. The slope in each temperature regime is different. Although this behavior is reproducible in Al 5083, it is expected that this behavior will be similar to other cavitating materials.

IV. PHENOMENOLOGICAL MODELS

A. Plasticity and Superplasticity

In order to develop a basic understanding of the superplastic behavior of materials, it is worthwhile to compare superplasticity with other inelastic processes, especially elastic–plastic and elevated temperature creep processes. Table 4 outlines some of the basic differences between plastic and superplastic behavior. Figure 11 shows that the stress–strain response of the popular Ti–6Al–4V alloy at the superplastic temperature of 927°C (1800°F) occurs at various levels of strain rates. It can be noted that the other popular aluminum alloy, Al

7475 (Al–5.7% Zn–2.2Mg–1.6% Cu–0.2Zr), exhibits superplasticity at a temperature of 550°C (1022°F). We should note here that although Al 7475 requires special processing to achieve the fine equiaxed grain structure, Ti–GA1–4V exhibits superplasticity as commercially processed. The plot shows that the stress is a strong function of strain rate, and in a log–log scale, the two show linearity in some range (10^{-4}–10^{-3}) and the curve shifts lower with temperature.

In general, the elastic–plastic behavior can be represented by:

$$\sigma = \sigma(\varepsilon^{\text{elastic}}, \alpha_{ij}) \tag{1}$$

where α_{ij} is the tensorial form of some internal state variable. Although in theory α_{ij} can either be scalar or tensor of any order and may even be more than one in number ($\alpha_{ij}^1, \alpha_{ij}^2, \ldots$), for discussion purposes, it is assumed that α_{ij} is a unique tensor given by the evolution law:

$$\alpha_{ij} = \int_{-\infty}^{t} \dot{\alpha}_{ij}(\varepsilon, \tau) d\tau \tag{2}$$

representing the history dependence of the variable, and in the case of elastoplastic material, it represents the growth law of the equivalent plastic strain.

For superplastic material, the behavior is represented by:

$$\sigma = \sigma(\dot{\varepsilon}, \beta_{ij}) \tag{3}$$

where the stress (or the flow stress) is a function of strain rate and some history-dependent internal state variables. Thus superplastic behavior is more akin to fluid (Newtonian or non-Newtonian viscous) behavior rather than a solid behavior where the state of stress is purely governed by the elastic strain within the body.

Consider a basic form of an elastoplastic behavior represented by:

$$\sigma = \begin{cases} E_\epsilon & \text{if } \epsilon \leq \epsilon_Y \left(= \frac{\sigma_Y}{E}\right) \\ \sigma_Y \left(\frac{E_\epsilon}{\sigma_Y}\right)^n & \text{if } \epsilon > \epsilon_Y \end{cases} \tag{4}$$

with a tensile yield stress σ_Y and a strain hardening exponent n. Note that $n = \infty$ corresponds to an elastic, perfectly plastic solid. In a simplified form, the stress is a function of strain as in $\sigma = A\varepsilon^n$. On the other hand, superplastic models are typically written as:

$$\sigma = k\dot{\varepsilon}^m \tag{5}$$

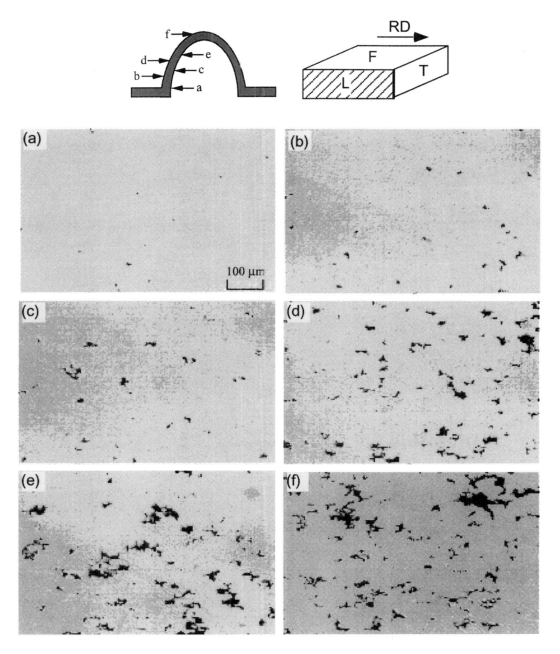

Figure 9 Cavity distribution in a biaxial cone testing, thickness strain: (a) 0.288, (b) 0.416, (c) 0.562, (d) 0.713, (e) 0.904, and (f) 0.879.

where the slope $m = \partial(\ln \sigma)/\partial(\ln \varepsilon)$ and represents the strain rate sensitivity parameter. The higher the value of m is, the higher are the superplastic properties (large elongation). It can be theoretically shown that m represents the resistance to necking and provides more diffused necking during deformation, prolonging the stretching process. There is a good experimental cor-

relation between m and the total elongation for a wide range of materials. The trouble, of course, is that the model is simple but then too simple to capture many important material and process variations.

Based on the phenomenological form of superplastic behavior, the uniaxial flow stress σ is seen to be a strong function of inelastic strain rate $\dot{\varepsilon}^P$ and a weak function of

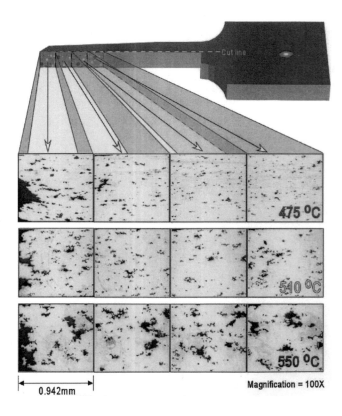

Figure 10 Cavity distribution in a tensile specimen.

strain ϵ and grain size d. The material is assumed to be purely inelastic and incompressible. A functional form of the constitutive relationship is given by:

$$\sigma = f(\epsilon, \dot{\epsilon}, d) \qquad (6)$$

Alternatively, the function can also be expressed in terms of logarithmic quantities as:

$$\ln \sigma = F(\ln \epsilon, \ln \dot{\epsilon}, \ln d) \qquad (7)$$

By expanding the above form in terms of Taylor's series at a given equilibrium state and by neglecting the higher-order terms, we can write Eq. (7) as:

$$\sigma = K_1 \dot{\epsilon}^m \epsilon^n d^p \qquad (8)$$

where:

$$m = \frac{\partial (\ln \sigma)}{\partial (\ln \dot{\epsilon})} \qquad n = \frac{\partial (\ln \sigma)}{\partial (\ln \epsilon)} \qquad p = \frac{\partial (\ln \sigma)}{\partial (\ln d)}$$

and K_1, m, n, and p are material constants. It can be recognized that m is the strain rate sensitivity, n is the

strain hardening exponent, and p is the grain growth exponent. The above equation can further be simplified by neglecting the effect of grain growth, leading to extended power law model given by Eq. (9):

$$\sigma = K \dot{\epsilon}^m \epsilon^n \qquad (9)$$

The above equation can further be simplified as:

$$\sigma = K \dot{\epsilon}^m \qquad (10)$$

B. Equations Based on Deformation Mechanisms

From the very early stages of superplastic material development, various researchers have proposed constitutive relationship based on *postulated* deformation mechanisms. It is very widely believed that the major contributor of superplastic strain is grain boundary sliding. Because it is geometrically impossible for grains to slide past each other in three dimensions without changing shape or creating voids, certain accommodation processes should accompany GBS. Traditionally, diffusional and dislocation activities (and probably partial melting in the case of high strain rate superplastic materials) are believed to be some possible accommodation mechanisms. Based on such assumptions, many forms of constitutive relations have been proposed. Table 5 shows a few of the models developed over the years. In that table, K_1–K_8 are material constants, σ_0 is the threshold stress, T is the absolute temperature, d is the grain size, b is Burger's vector, E is Young's modulus, Q is the activation energy, k is the Boltzmann constant, and D_{gb}, D_L, D_{IPB}, and D_{eff} are grain boundary, lattice, interphase boundary, and effective diffusion coefficient, respectively. It can also be seen from Table 5 that all the equations have $m = 0.5$, indicating that it is a fixed value and the grain size has an inverse proportionality with the flow stress (i.e., $p < 0$; refer to Eq. (8)). None of the models clearly addresses the inhomogeneity in the deformations at the grain level due to a distribution of grain size, orientation, and particles, and different types of grain boundary structure, energy, and misorientations.

C. Law Based on Grain Growth

Hamilton et al. [14] formulated an equation incorporating the effects of grain growth and back stress. In this

Table 4 Comparison of Plasticity and Superplasticity

Superplasticity	Plasticity
Superplasticity represents an inelastic behavior with high strain rate sensitivity.	Plasticity represents inelastic behavior with no rate dependence.
The effect of strain hardening is secondary.	Flow stress primarily increased due to strain hardening.
Grain switching and grain boundary sliding are the primary mechanisms.	Grain neighbors remain as such at all times.
Texture decreases (grain orientation becomes random increasing with strain).	Texture decreases (preferred orientation in the principle plastic strain direction).
Deformation is primarily due to GBS, with diffusion and dislocation as the accommodating mechanisms.	Deformation is primarily due to dislocational activities.
Deformation reduces the initial anisotropy.	Deformation induces a strong anisotropy.
Failure is due to cavity initiation, growth, and, finally, geometrical instability.	Failure is due to material and geometrical instability.

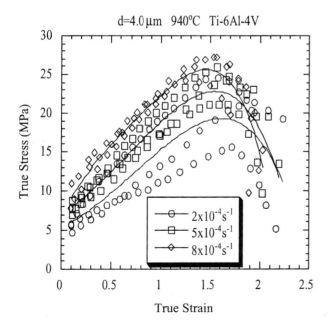

(a) Stress-strain behavior of Ti-6Al-4V in the superplastic range

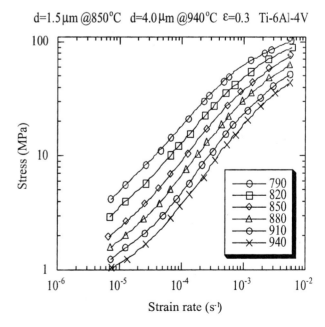

(b) Replot of the data in (a) showing the stress-strain-rate response

Figure 11 Thermomechanical behavior of Ti–6Al–4V alloy.

Table 5 Superplastic Models Developed Over the Years

Name	Year	Equation	Comments
Ball-Hutchison	1969	$\dot{\varepsilon} = K_1\left(\dfrac{b}{d}\right)^2 D_{gb}\left(\dfrac{\sigma}{E}\right)^2$	GBS accommodated by dislocation climb
Langdon	1970	$\dot{\varepsilon} = K_2\left(\dfrac{b}{d}\right)^1 D_L\left(\dfrac{\sigma}{E}\right)^2$	Movement of dislocations adjacent to GBs
Gifkins	1976	$\dot{\varepsilon} = K_3\left(\dfrac{b}{d}\right)^2 D_{gb}\left(\dfrac{\sigma}{E}\right)^2$	Pile-up at triple points (core–mantle)
Gittus	1977	$\dot{\varepsilon} = K_4\left(\dfrac{b}{d}\right)^2 D_{IPB}\left(\dfrac{\sigma - \sigma_0}{E}\right)^2$	Pile-up at interphase boundary
Arieli and Mukherjee	1980	$\dot{\varepsilon} = K_5\left(\dfrac{b}{d}\right)^2 D_{gb}\left(\dfrac{\sigma}{E}\right)^2$	Climb of individual dislocations near GBs
Ruano and Sherby	1984	$\dot{\varepsilon} = 6.4 \times 10^9 \left(\dfrac{b}{d}\right)^2 \dfrac{D_L}{b^2}\left(\dfrac{\sigma}{E}\right)^2$	Phenomenological $(T = (0.4 - 0.6)T_m)$
Wadsworth and White	1984	$\dot{\varepsilon} = 5.6 \times 10^8 \left(\dfrac{b}{d}\right)^3 \dfrac{D_{gb}}{b^2}\left(\dfrac{\sigma}{E}\right)^2$	Phenomenological $(T > 0.6T_m)$
Kaibyshev et al.	1985	$\dot{\varepsilon} = \dfrac{K_6}{kT}\left(\dfrac{b}{d}\right)^2 D_0 \exp\left(\dfrac{-Q}{kT}\right)\left(\dfrac{\sigma - \sigma_0}{E}\right)^2$	Hardening and recovery of dislocations at GBs
Ashby-Verrall	1973	$\dot{\varepsilon} = K_7\left(\dfrac{b}{d}\right)^2 D_{eff}\left(\dfrac{\sigma - \sigma_0}{E}\right)$ $D_{eff} = D_L\left[1 + \left(\dfrac{3.3w}{d}\right)\left(\dfrac{D_{gb}}{D_L}\right)\right]$	Rate-controlling diffusional accommodation
Padmanabhan	1980	$\dot{\varepsilon} = K_8\left(\dfrac{b}{d}\right)^2 D\left(\dfrac{\sigma}{E}\right)^2$	Nonrate-controlling diffusional accommodation D may differ from D_L and D_{gb}

model, the strain hardening is embedded as a grain growth effect and is represented by:

$$\bar{\varepsilon} = \frac{K_{II}(\bar{\sigma} - \sigma_0)^{\frac{1}{m}}}{d^p} + K_{III}\bar{\sigma}^n \tag{11}$$

with the grain growth law given by:

$$d = \left[d_0^q + Bt\right]^{\frac{1}{q}} + \int_{t_0}^t \lambda d\dot{\varepsilon}dt \tag{12}$$

where K_{II}, K_{III}, m, p, n, q, B, and λ are all constants.

D. Polynomial Forms

It is clear from the foregoing discussions that there are a number of constitutive equations that attempt to mimic the mechanical behavior of materials. There are a number of material parameters that need to be determined

before any of the equations can be used. The number of such parameters can vary anywhere between 2 and 15, and the determination of these parameters from experimental data is not a trivial task. Once determined, these equations are supposed to *predict* the behavior in a wider range of process and operating conditions.

A generalized polynomial form of the power law equation has been proposed by Chandra et al. purely from a curve-fitting exercise of fitting the experimental data. The proposed form:

$$\sigma = \sum_{i=0}^{N} A_i(\ln \dot{\varepsilon})^i \tag{13}$$

where a value of $N = 7$ with eight constants A_0, A_1,..., A_7 seems to fit the data very well. Eq. (13) has been applied to a specific material and the eight constants were evaluated. We should note that these constants do

Table 6 Parameters for Power Law Model for Al 5083

Temperature (K)	k (MPa)	m	R (%)
783	112.4	0.374	99.46
803	102.0	0.407	99.02
823	133.6	0.469	98.60
843	125.0	0.509	98.48

not have any physical meaning, except that it fits the data much better those of the historical forms.

E. Determination of Material Parameters for Al 5083

We examine the constitutive equations for Al 5083 alloy using two types of models given by Eqs. (10) and (13). Al 5083 SPF is a recently developed SPF-grade aluminum alloy. It has become a very strong candidate for the automotive industry, and is being manufactured in different parts of the world. There are three known major manufacturers, Alusuisse (Formall 545) from Switzerland, Alcan (5083-SPF) from the UK and United States, and Sky Aluminum from Japan. Specific thermomechanical treatments are given to obtain superplasticity, and the structure shows equiaxed grain structure with an average grain size of 17 μm.

The flow stress–strain rate behavior of the material (manufactured by Sky Aluminum) has been reported by Iwasaki et al. [15]. The experimental data are then used to fit three different models: (a) power law $\sigma = k\dot{\epsilon}^m$ (refer to Eq. (10) with two parameters); and (b) logarithmic polynomial proposed in this work: $\sigma = \sum_{i=0}^{7} A_i (\ln \dot{\epsilon}^i)$ (refer to Eq. (13) with eight parameters).

Table 6 gives the best fit parameters for Al 5083, and is different for different temperatures. Figure 12a shows the experimental data and the matching equation predicted by this data is shown in the table. R denotes the correlation coefficient between the experimental values and the fitted data.

Next, the same set of data was fitted with the logarithmic form, and the set of material parameters obtained is shown in Table 7 and the fitting is shown in Fig. 12b. It can be seen from this figure that the logarithmic fitting is much better than the simple power law model not only in region II, but for the entire range of strain rate. It is important to realize that during SPF, although a small portion of the sheet is being formed at the optimal value, a significant portion of the sheet is being formed at lower strain rate and hence the need for accurate equation.

F. Multiaxial Form of the Quantities

All the above equations are developed in uniaxial form. In order to use them in multiaxial form required by 3D process models, some hypotheses need to be postu-

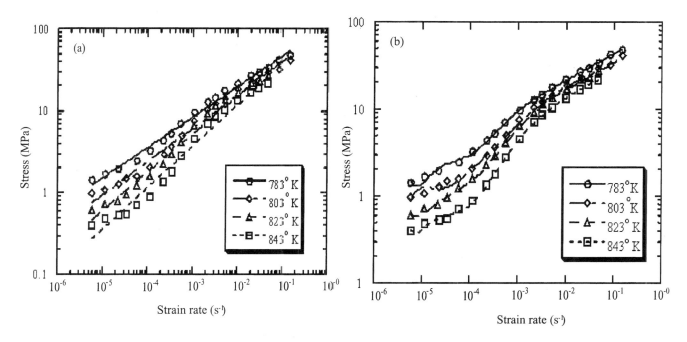

Figure 12 Power law and logarithmic fit of experimental data for Al 5083.

Table 7 Parameters for Logarithmic Polynomial Model for Al 5083

Temperature (K)	A_0 (MPa)	A_1	A_2	A_3	A_4	A_5	A_6	A_7	R (%)
783	23.64	−154.6	−267.1	−193.6	−75.60	−16.46	−1.875	−0.08706	99.93
803	159.7	321.7	341.6	201.4	68.00	13.16	1.363	0.05871	99.96
823	192.1	361.2	332.9	168.6	48.65	7.957	0.6852	0.02402	99.97
843	93.14	135.9	110.8	51.11	12.78	1.613	0.08069	—	99.99

lated. One of the well-known concepts is based on von Mises distortional energy criterion. In essence, the yielding (and hence flow stress) is affected only by the deviatoric component of stress and is independent of hydrostatic stress. Using that principle, the multiaxial kinetic and kinematical quantities can be written in terms of equivalent quantities. For example, Eq. (8) can be written as:

$$\bar{\sigma} = K_1 \bar{\epsilon}^n \dot{\bar{\epsilon}}^m d^p \tag{14}$$

with

$$\bar{\epsilon} = \sqrt{\frac{2}{3}\epsilon_{ij}\epsilon_{ij}} \quad \dot{\bar{\epsilon}} = \sqrt{\frac{2}{3}\dot{\epsilon}_{ij}\dot{\epsilon}_{ij}} \quad \bar{\sigma} = \sqrt{\frac{3}{2}\hat{\sigma}_{ij}\hat{\sigma}_{ij}}$$

where $\hat{\sigma}_{ij}$ is the deviatoric component of stress.

V. SUPERPLASTIC METAL FORMING

The continued interest in superplasticity hinges on its ability to manufacture complex components in a cost-effective manner. In this respect, superplastic metal forming should compete successfully with other bulk and sheet metal forming processes. Although SPF has been used to form both bulk materials and sheet metals, the predominant quantity has been in the latter category and will be discussed in detail here. Superplastic forming of sheet metals can be broadly classified into one with (SPF/DB) and without concurrent diffusion bonding. Most of the forming methods exploit the fact that superplastic materials exhibit large tensile elongations at relatively low flow stresses. The concomitant high-temperature requirement and low rates of deformation demand that the sheet be housed in a furnace for extended periods of time. Low flow stresses (between 1 and 10 MPa) permit the use of inert gas pressure in the range of 10–100 MPa to act as the loading medium. Figure 13

shows various stages in the sequence of formation of a square deep pan.

A. Female Forming

This is one of the most common and simple methods used to form sheet metals. In this method shown in Fig. 14, the die and the sheet are nominally maintained at the forming temperature and the gas pressure is imposed over the sheet, causing it to form into the lower die. The gas within the lower die chamber is either vented to the atmosphere, or subjected to vacuum or positive back pressure. The pressure (female) forming process has the advantage of no moving parts and does not require mating die members. Production rates can also be increased by forming multiple components in a single cycle.

When the deformation proceeds, some portions of the sheet contact the die surface and suffer very little thinning after the contact. Only the sheet metal that is not in contact continues to deform, and this leads to the fact that the formed component has different thick-

Figure 13 Typical stages in deep dish forming using SPF.

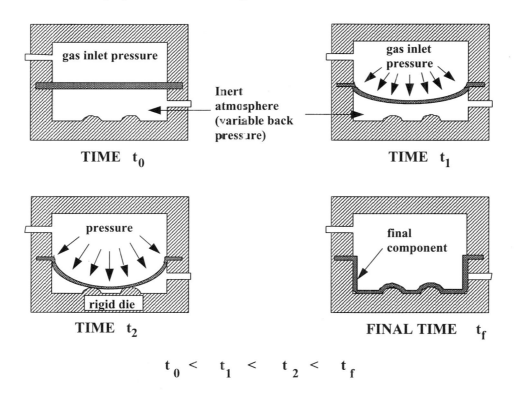

Figure 14 Schematic diagram of superplastic forming process.

nesses at different locations of the die, depending on the time it took for the initial contact to occur. The segment of the sheet that contacts last is the thinnest and need not necessarily coincide with the design requirements based on service conditions. The severity of relative thinning profile increases with the complexity of the part, especially one in which there is a substantial time differential between the time of the first contact and the final forming time. The female forming method works very well for shallow pans, even with complex features, as long as the draft angles are such as to permit the part withdrawal, subsequent to forming.

Important processing information required for a successful and economic fabrication of components are as follows:

The pressure–time cycle to maintain optimum strain rate

The resulting thickness distribution and location of maximum thinning.

B. Drape Forming

Other forming methods attempt to address the short-coming of differential thinning problem described in the

female forming method. In the drape forming, a male tool is introduced such that the sheet–die contact is made at the critical zone where the thinning is limited (see Fig. 15). The male tool can either be stationary or moving, depending on the design of the component. A number of variations in stationary/moving male inserts are being used in practice to obtain the required thickness profile. In such cases, experience with specific material, forming temperature, and forming rate plays a crucial role in the die design. As will be discussed later, computer simulation models based on numerical method [with computer-aided design/computer-aided manufacturing (CAD/CAM)] may go a long way in solving the important problem of thinning especially when the design is complex and also while working with material and process range with limited experience.

C. Reverse Forming

Because a free forming superplastic dome has thickness variation from that of the pole (apex) to the edge (die), this thickness variation can be judiciously brought to bear in reverse forming, as shown in Fig. 16. In the first step, the dome is formed as shown in (b), and the sheet is

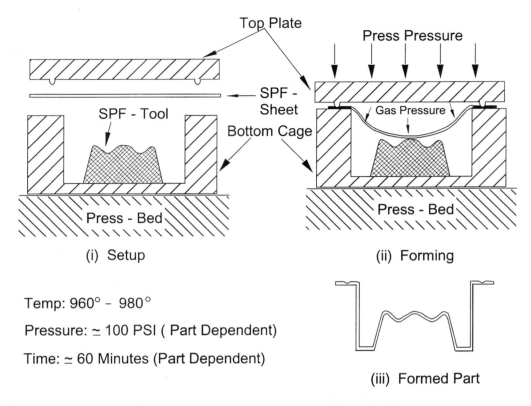

Temp: 960° – 980°

Pressure: ≃ 100 PSI (Part Dependent)

Time: ≃ 60 Minutes (Part Dependent)

Figure 15 Drape forming process with a male die.

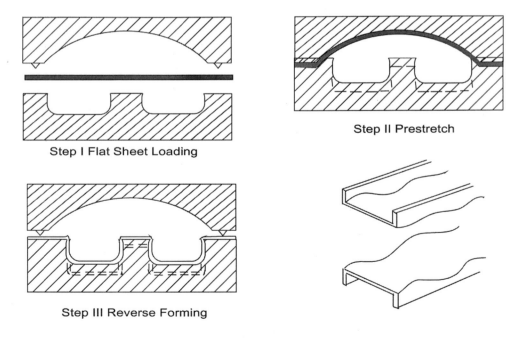

Figure 16 Obtaining uniform thinning using reverse forming.

reversed into the die. The corner region (where, in the traditional female forming, thinning is the maximum) has an initial higher thickness, making the final formed part more uniform.

D. Chemical Milling and Female Forming

The concept of thickness variation in the initial configuration can be alternatively achieved through the use of differential thickness reduction using chemical milling. The general idea is to have an initial thickness distribution of the sheet inversely proportional to the anticipated time of forming (hence, thinning) of a given section. As can be imagined, this requires either a trial-and-error operation or a good model, which can take into account the effect of varying gage thickness on the forming characteristics.

E. Diaphragm Forming

The ability of the SPF process to form complex shapes has been innovatively used in the forming of hard-to-form materials that include some ceramics and polymeric matrix composites. In this method, one or two *slave* superplastic sheets either juxtapose or sandwich the composite and form in a conventional SPF press. Because the composite is essentially confined, the formed shape conforms to that of the (sacrificial) SPF sheets and the die.

F. SPF with Diffusion Bonding

When certain materials are brought into intimate contact under elevated temperature and pressure, the process results in a strong mechanical bonding due to metal-to-metal diffusion at the interface. Fortunately, titanium alloy exhibits strong tendency to diffuse and hence can be formed superplastically with concurrent diffusion bonding. In the diffusion bonding process, when the two mating surfaces are brought together, at first, there is only a partial penetration of the surfaces creating a line with intermittent bonding. With exposure time and pressure, atomic diffusion takes place in the cavity such as regions and eventually results in a solid surface with no trace of initial interface. The process variable for diffusion bonding includes the surface finish, temperature, contact pressure, and time of exposure. Unfortunately, materials that quickly oxidize and form an impenetrable oxide layer are not suitable for diffusion bonding (e.g., aluminum-based materials). Industries, especially the aerospace industry, have used this feature very widely in the design of SPF/DB of multiple titanium sheets, through two-sheet, three-sheet, and four-sheet processes.

1. Two-Sheet SPF/DB Process

In this process, two sheets of titanium are bonded along the circumference (for a spherical dome) or along the length (for a long rectangular closed box). A stopoff

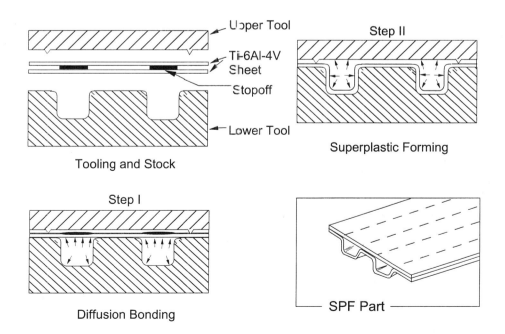

Figure 17 Two-sheet SPF/DB process.

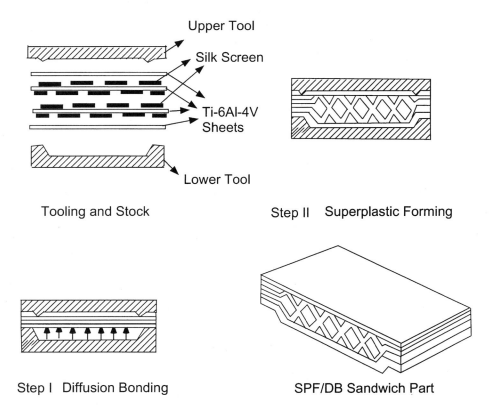

Figure 18 Three-sheet SPF/DB process.

piece or agent is placed in areas where diffusion bonding is not desired, before the application of temperature and pressure. Once bonded at a selected region, the stopoff agent is burned. An internal pressure corresponding to the desired strain rate is applied in the cavity and the sheets are allowed to deform in either direction until complete contact is made with the die surfaces. A perfect dome with sheet metals can be formed in this manner, a shape very difficult to be obtained by any other method. The schematic of the two sheet process is shown in Fig. 17.

2. Three-Sheet and Four-Sheet SPF/DB Process

The same concept of selectively bonding predetermined areas and then blowing the bonded pieces into the die has been used in the manufacture of sandwich beams that are structurally very stable and almost impossible to be fabricated by other means. In the three-sheet process illustrated in Fig. 18, there are two outside (face) sheets on either side of a core sheet. By providing the stopoff at alternate regions, an integrally stiffened structure can be fabricated.

The four-sheet process uses two core sheets that are spot-welded, as shown in Fig. 19. All the four sheets are

made into a gas pack and placed in the press. The face sheets, upon pressurizing, form the top and the bottom surfaces, whereas the core sheets bond to form the cell walls.

The SPF/DB, in general, has produced quite a few product lines in a number of defense and commercial aircrafts. But the specific design of the parts and the methods to manufacture them still depend on experience rather than sound engineering design principles. Methods to maintain uniform temperature, the selection of gas inlet/outlet locations and flow rates, the number and location of intrusive thermocouples, the control of oxygen-rich layers, and techniques to avoid hydrogen pick-up during final cleaning are all relevant issues that need to addressed before the choice of SPF/DB as a viable process can be made.

VI. PROCESS MODELING

As discussed in the previous sections, in order to manufacture cost-effective SPF parts, it is important to utilize the ability of superplastic materials to form complex shapes with very limited loading (press capacity) requirements. Superplastic forming competes with a

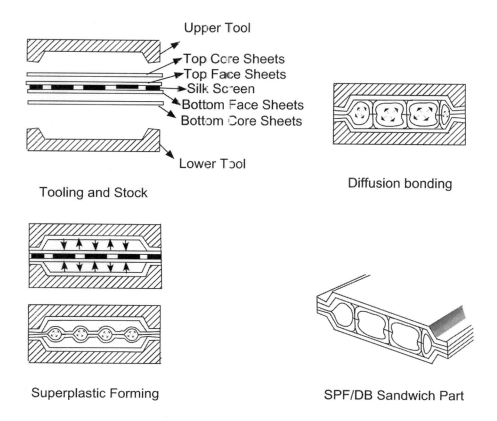

Tooling and Stock

Diffusion bonding

Superplastic Forming

SPF/DB Sandwich Part

Figure 19 Four-sheet SPF/DB process.

number of casting (die casting, investment casting, and sand casting) and sheet metal (stamping, stretch forming, hydroforming, folding/bending/rolling, and weldments) for a final place in the selection process. Most of these processes are well established and have been effectively incorporated in a CAD/CAM environment such that a designer not only conceives the shape of the part, but has the ability to visualize its detailed shape and manufacturing feasibility. In order for SPF to be a viable manufacturing method, it is imperative that computer models that stimulate the actual fabrication process be available during the design evolution process. Thus, the industry requires process models:

To examine whether a conventionally formed part can be replaced by superplastic forming

To study the effect of material and process parameters such as changes in m, n, and ϵ, and applied pressure–time loading details, original sheet thickness, etc., on the formed component

To interactively study the formed part for minimum thickness and maximum stress regions

To determine the process parameters and incorporate the models in a CAD/CAM environment.

A. Processing Modeling Methods

Process models are mathematical models in which geometry, equilibrium, and constitutive laws for the material are all satisfied at each incremental time step and provide a fairly accurate description of the process. Such models are then incorporated into a computation-

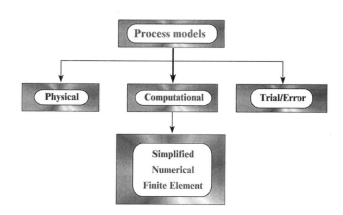

Figure 20 Types of SPF process models.

al procedure and a user-friendly visualization tool. Such computational process models enable the designer and the manufacturer to not only evaluate the feasibility of fabricating a given component using SPF, but help one modify the geometrical details to achieve a cost-effective product. There are essentially two kinds of models, computational and physical, as shown in Fig. 20, apart from the trial-and-error method practiced in many shop floors.

1. Simplified Methods

In simplified methods, pressure–time loading is determined by assuming the shape of formation and also uniform thinning. This approach is applied to the superplastic formation of long plane strain rectangular box (Fig. 21), cones, deep cups, and domes (Fig. 22). Acceptable results can be obtained through this method for long rectangular box formation and in other cases where m is very high. This method cannot be used for practical material (m between 0.3 and 0.7) for the formation of cups, cones, and domes.

2. Numerical Methods

In the numerical modeling method, pressure–time loading and thickness distribution are determined by making certain kinetic and kinematical assumptions. It is to be recognized that except in plane strain box case, there will be thickness variation because each point is subjected to a different state of stress (different equivalent stress) and hence different rate of thinning. Hence, any assumption of uniform thinning cannot represent the forming process accurately. These numerical methods invariably involve additional equations, which assume a different thickness distribution and/or force equilibrium with applied pressure load. In general, the solution procedure will involve some iterative schemes. Numerical models including the membrane element method, which uses the concept of finite elements in space and finite difference in time, have been developed. These

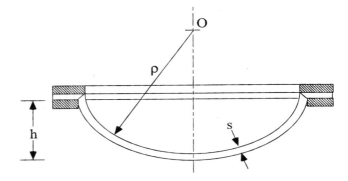

Figure 22 Axisymmetrical forming with uniform thinning.

methods can be applied to complex shapes and also to a variety of constitutive equations through innovative methods.

3. Finite Element Methods

For complex shapes, material behavior, and consideration of friction at the tool–die interface, the use of nonlinear finite element method (FEM) offers a potentially viable approach. In the finite element method, the undeformed sheet is divided into a number of continuum or structural elements depending on the planes of symmetry. Because of the complexity of material, geometry, loading, and other nonlinearities, the finite element method is more appropriate for its use. The analysis using finite element solution method involves high computational effort, computer software, and personnel, and may be uneconomical for a general iterative design process.

4. Physical Modeling Methods

In the physical model, forming process is simulated in general to establish similarity in die and workpiece geometry, mechanics, and frictional loading conditions. In practice, physical models cannot always achieve similarities in all the processing conditions; however, they will still be applicable in simulating specific physical conditions. Physical models are very inexpensive and, if properly designed, can provide valuable information on the kinematics of deformation of complex geometry. Further conducting experiments on actual superplastic materials can be difficult because: (a) deformation at an intermediate stage of formation cannot be studied; (b) material properties and forming parameters can greatly influence the results and mask the fundamental mechanisms of sheet metal deformation; (c) it is

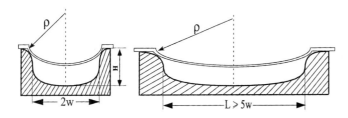

Figure 21 Geometry of long rectangular box.

very expensive (in terms of material, time, and labor); and (d) such facility will not be easily available in a university laboratory or medium-sized manufacturing plant.

B. Detailed Discussion of the Models

In the following sections, we will develop specific equations that can be used in the process modeling of SPF parts using simplified, numerical, and finite element methods.

1. Simplified Methods

Long Plane Strain Rectangular Box

In this analysis, the length of the box is assumed to be much greater than the width. In practice, however, this condition is satisfied when $l \geq 5 \times w$, where l is the length of the box and w is the semiwidth. A condition of plane strain exists when the longitudinal strain and strain rates $\bar{\epsilon}_2$ and $\dot{\bar{\epsilon}}$ vanish, leaving $\dot{\bar{\epsilon}}_1 = -\dot{\bar{\epsilon}}_3$ with $\dot{\bar{\epsilon}} = -(2/\sqrt{3})\dot{\epsilon}_3$. It is further assumed in this simplified analysis that the corner radius is zero and frictional effects are negligible.

If the design strain rate is $\dot{\bar{\epsilon}}$, then the following equations apply before the sheet metal contacts the bottom or side, depending on whether the box is shallow or deep. For a given radius ρ, the time of formation is given by:

$$t = \frac{2}{\sqrt{3}\dot{\bar{\epsilon}}}\ln\left[\frac{\rho}{w}\sin^{-1}\frac{w}{\rho}\right] \quad (15)$$

where w is the semiwidth and the pressure p is given by:

$$p = \frac{2}{3}\frac{s_0}{\rho}e^{-\frac{\sqrt{3}}{2}\dot{\bar{\epsilon}}t}\bar{\sigma} \quad (16)$$

where s_0 is the original thickness and $\bar{\sigma}$ can be determined using any of the relevant constitutive equations (Eqs. (9) and (10)) for a specified design strain rate. From the above two equations, the p–t relationship can be determined for any ρ until contact is made in the bottom. The thickness at any time is assumed to be constant and given by:

$$s = s_0 e^{-\frac{\sqrt{3}}{2}\dot{\bar{\epsilon}}t}$$

After contact is made on the side or the bottom, depending on the ratio of w/h where h is the depth of the pan, the radius ρ is given by:

$$\rho = \frac{y^2 + x^2}{2y}$$

where y and x are the unsupported distance along the side and bottom die surface, respectively. The time of formation can be determined as:

$$t_{i+1} = t_i + \frac{2}{\sqrt{3}\dot{\bar{\epsilon}}}\ln\left\{\frac{\rho_{i+1}\Phi_{i+1} + \frac{\Delta y}{2} + \frac{\Delta x}{2}}{\rho_i\Phi_i - \frac{\Delta y}{2} - \frac{\Delta x}{2}}\right\} \quad (17)$$

where the subscript is the time increment index and Δx and Δy are the incremental distance of contact in x and y directions, respectively. The thickness formulation remains the same as in the first stage.

Hemispherical Dome

The principal assumptions in this analysis are as follows: (a) the thickness is uniform over the domain; (b) the shape of the deformation is spherical; (c) the stress component in the thickness direction is negligible compared to membrane stresses; (d) the state of stress is balanced biaxially over the entire domain; (e) the deformation follows thin shell formulations; and (f) the material is isotropic, incompressible, and purely inelastic. The fourth assumption does not satisfy the boundary condition of the problem. It can be easily seen that balanced biaxial stress exists at the pole and varies with plane strain at the edge with the above assumptions. However:

$$\sigma_1 = \sigma_2 = \bar{\sigma} \quad \sigma_3 = 0 \quad (18)$$

$$\dot{\epsilon}_1 = \dot{\epsilon}_2 = \frac{\dot{\bar{\epsilon}}}{2} \quad \dot{\epsilon}_3 = -\dot{\bar{\epsilon}} \quad (19)$$

In terms of dimensionless height $H = h/a$, pressure p can be determined from:

$$\bar{\sigma} = \frac{a}{4s_0}\frac{(1 + H^2)^2}{H}p \quad (20)$$

The time of deformation is given by:

$$t = \frac{1}{\dot{\bar{\epsilon}}}\left[\ln\left(\frac{1}{1 + H^2}\right)\right] \quad (21)$$

If the constitutive equation of the material is known in functional form $\bar{\sigma} = \bar{\sigma}(\dot{\bar{\epsilon}})$ or $\bar{\sigma} = \bar{\sigma}(\bar{\epsilon}, \dot{\bar{\epsilon}})$, then $\bar{\sigma}$ can be found from a specified design strain rate $\dot{\bar{\epsilon}}$ and total maximum $\bar{\epsilon} = \dot{\bar{\epsilon}}t$. Thickness can be determined using:

$$\frac{s}{s_0} = \frac{1}{1 + H^2} \quad (22)$$

Thus, for any given pole height h (or H), the relationships between the radius of curvature ρ, thickness s, time

t, and pressure p can be evaluated using the above equations. By evaluating at different heights, pressure-vs.-time schedule, thickness-vs.-time, and height-vs.-time graphs can be generated.

Deep Stretched Cup

For the analysis of deformation of sheet, the following assumptions are made: (a) the volume of the material remains constant; (b) the free surface deforms as a part of thin spherical surface; (c) the thickness of the unsupported spherical cap is uniform; (d) once the material comes in contact with the rigid surface, it does not deform further; and (e) the change in the thickness does not change the radius of cylinder.

The formation of the deep-drawn cup can be divided into three stages as shown in Fig. 23, based on the geometry of the deformation. In stage I, the sheet metal deforms in the shape of spherical surface with its radius of curvature changing continuously until it is formed into a hemisphere. In the next stage (stage II) of deformation, the formed hemisphere moves down inside

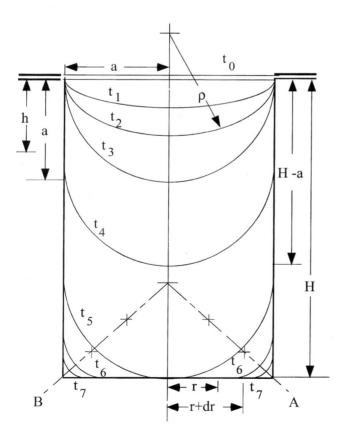

Figure 23 Various stages in the deformation in deep cup forming.

the die without further change in the radius of curvature and sheet metal lays over the cylindrical surface.

The third stage (stage III) of deformation starts when the hemisphere just touches the flat bottom surface of the die. In this stage, the material is overlaid on cylindrical surface as well as on bottom surface as the deformation proceeds. The free forming surface is a section of a toroid and its radius, and the center of curvature changes with deformation.

The thickness variation during the various stages of deep stretching of superplastic sheet will be given by the following equations:

Stage I $[0 \leq h \leq a]$

$$s = \frac{a^2 s_0}{h^2 + a^2} \tag{23}$$

Stage II $[a \leq h \leq H - a]$

$$s = \frac{s_0}{2} e^{-\left(\frac{h-a}{a}\right)} \tag{24}$$

Stage III $[0 \leq r \leq a]$

$$\ln\left(\frac{S'}{s}\right) = \frac{1 + a_3}{a_3} \ln\left[\frac{a_1 + a_2 r + a_3 r^2}{a_1}\right] \tag{25}$$

$$+ \frac{1}{q}\left[(2a + a_2) - (1 + a_3)\frac{a_2}{a_3}\right]$$

$$\times \ln\left[\frac{2a_3 r + a_2 - q}{2a_3 r + a_2 - q} \frac{a_2 + q}{a_2 - q}\right] \tag{26}$$

where:

$$S' = \frac{s_0}{2} e^{-\left(\frac{H-a}{a}\right)}, \quad q = \sqrt{a_2^2 - 4a_1 a_3}, \quad a_1 = 2a^2 \tag{27}$$

$$a_2 = a(\pi - 4) \quad a_3 = (2 - \pi) \tag{28}$$

2. Numerical Methods

In this method, we are primarily concerned with the numerical formulation that considers the thickness variations in the sheet. One type of method assumes a thickness distribution prompted by experimental observation, and the other is a purely numerical iterative scheme such as the *membrane element method*.

Axisymmetrical Dome Forming

The main assumptions of this analysis are: (1) the shape of the deformation is spherical; (2) the stress in the thickness direction is negligible (i.e., the entire domain is in a state of plane stress); (3) the state of stress is balanced biaxial at the pole and the plane strain at the edge; (4) the engineering thickness strain varies parabolically from the pole to the edge, which results in a logarithmic variation of thinning from the pole to the edge; and (5) the force in the membrane direction 1 of the sheet metal remains constant. Assumption (3) is exact if plane stress conditions throughout the domain exist. The assumption of a parabolic variation of thickness strain is based on experimental results from the dome formation of titanium sheet metal.

The constitutive relationship is assumed to be $\bar{\sigma} = k\dot{\bar{\epsilon}}^m$ in which $\bar{\sigma}$ can be calculated for a specified $\dot{\bar{\epsilon}}$ because k and m are known material parameters. However, for any other generic functional relationship (e.g., $\bar{\sigma} = \bar{\sigma}(\dot{\bar{\epsilon}})$, can still be found if $\dot{\bar{\epsilon}}$ is specified.

At the pole (invoking the condition for balanced biaxial stress state):

$$\dot{\epsilon}_1 = \dot{\epsilon}_2 \quad \dot{\epsilon}_3 = -2 \ \dot{\epsilon}_1 = -\dot{\bar{\epsilon}}$$

$$\epsilon_1 = \int \dot{\epsilon}_1 dt = \dot{\epsilon}_1 t$$

$$\sigma_1 = \sigma_2 = \bar{\sigma} = \bar{\sigma}(\dot{\epsilon}) \quad \sigma_3 = 0$$

From force equilibrium:

$$\sigma_{1,p} s_p = \sigma_{1,e} s_e$$

where subscripts p and e refer to pole and edge, respectively.

Using the kinematical quantities at the pole and at the edge, the total length of the arc of deformation can be determined for any specified strain distribution. If a parabolic distribution of engineering strain $e_\epsilon^{(x)}$ at radius x is assumed to be a function of e^{ϵ_p} and e^{ϵ_p}, then (Fig. 24):

$$f(x) = e^{\epsilon(x)} = A + Bx^2, \text{ with boundary conditions}$$

$$f(0) = e^{\epsilon_p} \quad f(a) = e^{\epsilon_e}$$

It can be shown that:

$$2\rho \sin^{-1}\left(\frac{a}{\rho}\right) = \frac{2a}{3}[2e^{\epsilon_{1,p}} + e^{\epsilon_{1,e}}]$$

which is a nonlinear equation in ρ. Because all other quantities are already known, once the radius of curvature ρ is known, all other process parameters can be

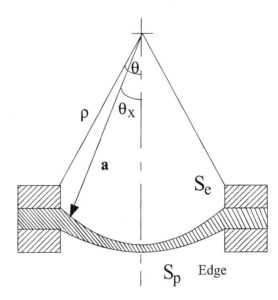

Figure 24 Axisymmetrical forming with parabolic strain distribution.

found using the above equations. The explicit relationship is then given by:

$$p = \frac{2\sigma_{1,p} s_p}{\rho}$$

$$\epsilon_{3,x} = \ln\frac{s_x}{s_0} \quad s_x = s_0 e^{\epsilon_{3,\epsilon^x}}$$

By incrementing ρ at time t, the new process parameters can be solved at time $t + \Delta t$ and proceeded further. Thus, pressure–time data and thickness distribution can be found for the assumed variation. Similar expressions can be derived by assuming distributions other than the parabolic of engineering meridional strain.

Dome Forming Using Membrane Element Analysis

The theory of membrane element presented here is valid for axisymmetrical and nonaxisymmetrical formations. The sheet metal is divided into a number of linear segments in the case of plane strain and axisymmetrical problems. When complex geometries are to be modeled, the sheet is divided into either triangles or rectangles. The one-dimensional (1D) line and the two-dimensional (2D) plane stress elements are called *membrane elements* (Fig. 25). The deformation at any given time is first approximated based on the kinetic and the kinematical quantities in the earlier time step. Force equilibrium is imposed in each of the elements and the resulting stress quantities are used to predict a new approximation. The numerical iterative scheme is carried out until the known boundary conditions at the pole and at the edges are satisfied.

The assumptions made in the analysis are: (1) the shape of the deformed unsupported segment is spherical; experimental evidence shows that this is a valid assumption; (2) the sheet is in a state of plane stress; (3) the deformation at the pole (geometrical center) is balanced biaxial in axisymmetrical parts, and is a function of edge dimensions (length and width) in nonaxisymmetrical parts; and (4) the edges are tightly held, and are thus in a state of plane strain.

The sheet metal is made up of materials with constitutive equations of the form $\bar{\sigma} = \sigma(\dot{\bar{\epsilon}}, \epsilon)$ with $\bar{\sigma} = \sqrt{(3/2)\hat{\sigma}_{ij}\hat{\sigma}_{ij}}$ and $\dot{\bar{\epsilon}} = \sqrt{(2/3)\dot{\bar{\epsilon}}_{ij}\dot{\bar{\epsilon}}_{ij}}$, where $\bar{\sigma}$ and $\dot{\bar{\epsilon}}$ are the equivalent quantities. A hat denotes that the quantity is deviatoric.

The constitutive equation can be written in Lévy–Mises form as $\hat{\sigma}_{ij} = (2/3)(\bar{\sigma}/\epsilon)\epsilon_{ij}$, with $\sigma_{ij} = \hat{\sigma}_{ij} + (\sigma_{kk}/3)$ and $\dot{\bar{\epsilon}}_{kk} = 0$ with the boundary condition at the pole:

$$\bar{\sigma} = \sigma_1 = \sigma_2 = \frac{p\rho}{2s} \quad \sigma_3 = 0$$

$$\dot{\epsilon}_1 = \dot{\epsilon}_2 = 2\dot{\bar{\epsilon}} \quad \dot{\bar{\epsilon}} = -\dot{\epsilon}_3$$

and at the edge:

$$\dot{\epsilon}_2 = 0 \quad \dot{\epsilon}_1 = -\dot{\epsilon}_3 \quad \dot{\bar{\epsilon}} = \frac{2}{\sqrt{3}}\dot{\epsilon}_1 \quad \sigma_1 = 2\sigma_2 \quad \bar{\sigma} = \sqrt{3}\sigma_2$$

The problem can be formulated as an incompressible nonlinear viscous material subjected to a time-dependent pressure loading with a set of boundary conditions. The loading is to be determined for a specified strain rate

$\dot{\bar{\epsilon}}^*$ at *the critical location*. The critical location is at the pole of a freely forming axisymmetrical part. Once the pole contacts the die, the critical location shifts to the geometrical center of the largest undeformed segment.

The sheet metal is divided into N elements with node numbers 0 to n. Assume that the equilibrium configuration, stress, and strain rate quantities are all known at time t. It is required to determine all these quantities at time $t + \Delta t$, where Δt is the chosen time increment. Using the incompressibility conditions and the definition of thickness strain, the change in surface area δA_I of element I can be written as:

$$\delta A_I = -\frac{{}^t\dot{\bar{\epsilon}}}{2{}^t\bar{\sigma}}({}^t\sigma_1 + {}^t\sigma_2)({}^tA_I)\Delta t$$

where the left superscript indicates the time of reference. In the above equation, all the data refer to quantities at time t and the latest iteration.

The total change in surface area ΔA of the sheet metal in the incremental time Δt can be determined by summing δA_I over all the elements 1 to N. The global quantities such as radius of curvature ρ, angle α, and height H can be calculated from the new area. The pressure P is calculated from the equivalent stress at the pole using $\bar{\sigma} = P_\rho/2S_p$. Because $\dot{\bar{\epsilon}}^*$ is specified at the pole, at the pole can be evaluated using the functional form of the constitutive equation. The thickness s_I is calculated based on a linear variation of thickness within an element. From the force equilibrium of a spherical segment subjected to uniform pressure, we have:

$$\sigma_{1,p}s_p = \sigma_{1,I}s_I$$

where the p refers to the pole and 1 refers to the meridional direction. σ_1 in each element can be evaluated

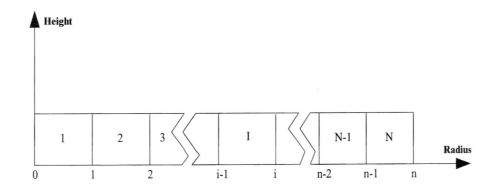

Figure 25 Membrane element discretization of sheet metal.

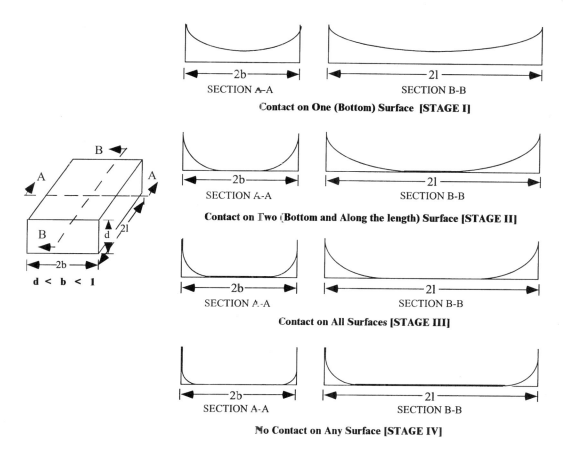

Figure 26 Stages of formation of a box with finite dimensions.

from the above equation. From the geometry of deformation, thickness at each node, and force equilibrium, the new stress and strain rate quantities are calculated in each element. The updated values of stress and strain rate quantities are used to recalculate the area and other geometrical quantities until the boundary conditions at the pole and the edge are satisfied. This procedure is repeated until the dome height reaches any specified value.

Nonaxisymmetrical Box Formation

The membrane element concepts described previously can be extended to model *nonaxisymmetrical* problems. As an illustrative example, the formation of a square box as shown in Fig. 26 is considered.

In this case, it is made sure that every line segment passing through the geometrical center deforms as a part of a sphere with maximum height at the pole and with the radius of curvature corresponding to half the edge-to-edge distance. An additional geometrical constraint is that the vertical deflection and the thickness at

the geometrical center are identical to each of the domes. Thus, every material point can be considered as a part of a radial line from the geometrical center through the point under consideration. The deformation at the point is determined based on the history of stress and the depth at the geometrical center. The contact of the sheet with any of the surfaces is determined from the geometry of each node and the equation of the surface. Thus, a square box (or box with finite dimension) is formed in four stages with free forming in the first stage and is shifted to the pole of unsupported sheet in the subsequent stages.

C. Finite Element Methods

A goal in the process design of any manufacturing process is to determine the critical process parameters and to optimize these factors in order to produce the final part within the required specifications. Over the years, several process models based on analytical and simplified numerical methods have been developed to

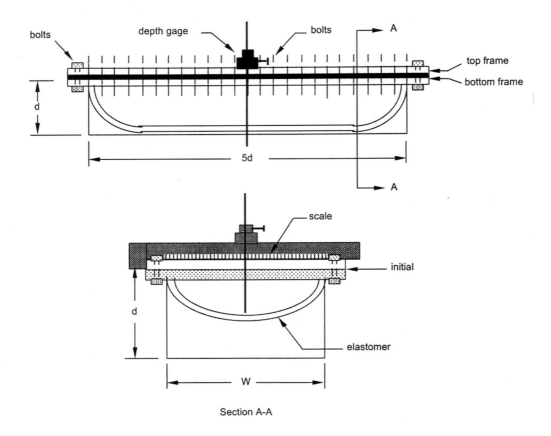

Figure 27 Model of a long rectangular pan.

gain an understanding of the process mechanics, to predict forming loads and overall geometry changes, and to determine optimum process conditions. Such methods are usually restricted to modeling axisymmetrical components such as cups and cones, or the formation of a long rectangular box section with complex cross-sectional details (Fig. 27). In the analytical and some numerical modeling methods, the shape of deformation is assumed a priori to be spherical or cylindrical, and the thickness is assumed to be either constant or to vary linearly or parabolically. The constitutive equation of the material is assumed to be the simple power law model ($\bar{\sigma} = K\dot{\bar{\epsilon}}^m$), enabling the development of closed form solutions [16–18].

Despite the ease of use and computational efficiency, simplified analytical and numerical process models are restricted to two-dimensional plane strain and simple axisymmetrical geometries. To model 3D geometry and/or complex 2D shapes considering sheet–die frictional effects, it is necessary to resort to finite element methods. Additionally, FEM can consider material, geometry, loading, and other nonlinearities with no

added difficulty or major assumptions. As the power of the computers continues to increase and as the software (code) is bundled with easy-to-use graphical user interfaces, the impact of FEM is on the increase. In the next few sections, we provide an overview of the available commercial and noncommercial codes for SPF process modeling.

Since the early 1980s, several research groups have proposed various FEM techniques for modeling SPF process. In general, to analyze sheet metal forming, the solid formulation and the flow formulation can be used with either a quasi-static implicit or a dynamic explicit solution procedures. Additionally, in SPF of single sheets, contact occurs between a rigid die and moving sheet; however, in addition, the SPF/DB process encounters contact between two moving and deformable sheets. These contact conditions are implemented differently through different approaches. In contrast to sheet metal stamping, which is a displacement-driven process, SPF is a force/pressure-controlled process. The optimum pressure–time loading cycle for a successful SPF is to be determined by FEM, whereas in a general

FEM, displacement/force is the input. This requirement is achieved differently in different codes. Finally, the ease of use and the integration with CAD tools are other aspects that discriminate one code from the other.

1. Solid and Flow Formulations

In the solid approach, when posing the boundary value problem, it is assumed that at a given stage, the shape of the body and the internal distribution of stresses are known. The boundary conditions are specified in terms of displacement and traction, or in terms of the velocity and traction rate if the equilibrium equations are presented in the rate form. The distribution of displacements and/or stresses is usually the solution to the problem. The formulation uses a total Lagrangian description with an elastic–plastic material model developed in an incremental rate form of Prandtl–Reuss equations. Hence, the Jaumann rate of Kirchhoff stress and the Euler–Lagrangian strain rate are used in the constitutive equations. The solid formulation has been popular in modeling sheet stamping processes and has been adapted by many commercial explicit dynamic codes such as LS-DYNA, OPTRIS, and PAM-STAMP, and also by quasi-static implicit codes such as MARC [19] and ABAQUS.

Within the confines of the flow formulation, the metal is modeled as a rate-dependent rigid plastic, rigid viscoplastic, or purely viscous material. The velocity vector is prescribed on a part of the surface with traction on rest of the surface. Solutions to this problem are the velocity and/or stress distributions that satisfy the governing equations and the boundary conditions. Most sheet metal forming processes including hydrostatic bulging, punch stretching, and deep drawing were solved using a rigid plastic material approach. Onate and Zienkiewicz [20] were one of the first to use thin shell theory and viscous flow formulation to model sheet metal deformation for two-dimensional and three-dimensional configurations. The majority of FEM process models for SPF use the flow formulation.

2. Formulations for SPF/FEM Process Models

Argyris and Doltsinis [21] used a natural formulation to solve plane strain and axisymmetrical problems of superplastic forming. Zhang et al. [22] used the viscous flow formulation to model the free forming of a superplastic dome (Fig. 28). Chandra [23] used a continuum finite element formulation to model plane strain and axisymmetrical cases of superplastic forming process, together with a penalty approach for the enforcement of the incompressibility of the material. Because continuum elements are used across thickness, this model was able to handle both bending and transverse shear effects, which may be important especially at die entry radii and at die corners. All the works mentioned above involve the modeling of superplastic sheet forming for two-

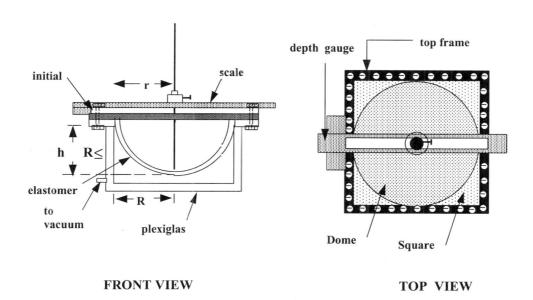

FRONT VIEW   **TOP VIEW**

Figure 28 Physical model of a dome and a square pan.

dimensional geometries. Work on sheet forming processes for three-dimensional die geometries has been performed since late the 1980s. The updated Lagrangian finite element formulation by Nakamachi [24] used a membrane shell theory and an elastic–plastic material model to simulate punch drawing and stretch forming problems.

The same membrane approach was used by Bellet and Chenot [25] who used eight-node isoparametric quadrilateral elements to model the sheet as a membrane with the use of convective coordinate system. The convected coordinate system enables the representation of the 3D geometry into 2D by embedding a coordinate system in the plane of the sheet. The authors verified the model by simulating many simple configurations and a few complex shapes with a fair amount of accuracy. A semi-implicit time integration scheme was used for geometry, and velocity updates are involved but are unconditionally stable. Bonet and Wood [26] used the same membrane approach but used triangular constant stress element based on a convective coordinate system, which simplified the formulation to a great extent when compared to the approach of Bellet and Chenot [25]. This formulation was based on an explicit time stepping scheme, which was later dropped in favor of the semi-implicit scheme [27]. To capture the effects of bending, Kirchhoff shell kinematics was later implemented by the authors in a simplified form with the use of a constant moment bending element (Morley element) superimposed over the membrane triangular element. Bonet et al. [27] devised a pressure control scheme that was integrated into the formulation as a constraint equation, which is later discussed in this paper.

Following the two-dimensional continuum formulation [23] SPASM2D (Superplastic Analysis of Sheet Metals—2-Dimensional), Chandra and Rama [28] extended the studies to include the effects of die radius and sheet–die friction on SPF. They proposed a pressure prediction–correction scheme [29], which proved to be independent of geometry and the type of formulation. Subsequently, they suggested the use of a dual-strain rate criterion for strain rate control for the SPF of dynamically recrystallizing materials [29] (e.g., Al–Li alloys). Later, a new three-dimensional thin shell formulation was developed and named SPASM-3D [30,31]. A summary of the details of this formulation is presented below.

3. SPASM-3D Formulation

Because superplastic sheet forming is predominantly a membrane deformation process, a three-dimensional finite element formulation using thin shell kinematics with Kirchhoff assumptions to model the sheet deformation has been developed. The shell mechanics is based on a convective coordinate system, which is embedded on the midsurface of the deforming sheet. The in-plane reference coordinates (ξ^1, ξ^2) remain tangent to the midsurface of the deforming thin shell, and the out-of-plane coordinate ξ^3 always remains normal to the midsurface. Describing $\mathbf{e}_\alpha$ and $\mathbf{e}_\beta$ as the covariant base vectors with respect to the ξ^i coordinate system, the covariant convective components of the rate of deformation tensor $\mathbf{D}$ can be derived as:

$$D_{\alpha\beta} = \frac{1}{2}\dot{g}\alpha\beta = D_{\alpha\beta}^{M} + \frac{h}{2}\xi^3 D_{\alpha\beta}^{B} \quad \text{where} \quad \alpha, \beta = 1, 2 \tag{29}$$

where $g_{\alpha\beta} = \mathbf{e}_\alpha\mathbf{e}_\beta$ and $\dot{g}_{\alpha\beta}$ describe the time derivative.

Using the equivalent stress and equivalent strain rate relationships at constant equivalent strain, the membrane and stress resultants can be derived as:

$$\sigma_M^{\alpha\beta} = 2\overset{M}{\mu} h\, G^{\alpha\beta\gamma\delta}D_{\gamma\delta}^{M} \quad \sigma_B^{\alpha\beta} = 2\overset{B}{\mu} \frac{h^3}{12} G^{\alpha\beta\gamma\delta}D_{\gamma\delta}^{B} \tag{30}$$

where:

$$G^{\alpha\beta\gamma\delta} = g^{\alpha\beta}g^{\gamma\delta} + g^{\alpha\gamma}g^{\beta\delta}$$

and μ^M and μ^B are the membrane and bending viscosities, respectively. The equilibrium equations presented in terms of virtual work can be written as:

$$\delta\dot{W}_{\text{int}} = \int_{m\Gamma}\sigma_M^{\alpha\beta}\delta D_{\alpha\beta}^{M}\,\mathrm{d}\bar{A} + \int_{m\Gamma}\sigma_B^{\alpha\beta}D_{\alpha\beta}^{B}\,\mathrm{d}\bar{A} \tag{31}$$

$$\delta\dot{W}_{\text{ext}} = \int_{m\Gamma}p^i\delta\bar{v}^i\,\mathrm{d}\bar{A}$$

A computationally efficient six-noded constant stress constant moment triangular element was used to discretize the sheet. The element has three corner nodes with three translational degrees of freedom and three midside nodes with only rotational degrees of freedom about the side.

The tangent membrane stiffness and bending stiffness matrices are computed explicitly because the finite element considered has constant stress and moment. To keep the formulation simple, only symmetrical matrices are considered. A Newton–Raphson iterative scheme is used to solve for the Δv^i, incremental velocities to compute the final velocity vector (the solution variable) at the end of every time step in this quasi-static approach. The convergence is decided from the magnitude of two different norms: one based on the residual forces and the other based on the incremental velocity field. The details of the contact algorithm and the pressure–prediction scheme used by the authors and others are discussed later.

4. Contact and Friction

This is one of the critical areas in terms of numerical accuracy and stability. Because the shape of the die or tool determines the final shape of the superplastic component, it is crucial that the die be accurately represented. In two-dimensional formulations of axisymmetrical and plane strain problems, an analytical function can be directly used to describe the die geometry. However, the complexity of general three-dimensional dies renders this approach impractical. Hence, the process modelers of metal forming, in general, have unanimously chosen to represent the die with finite elements, which has clear advantages in formulation in the context of finite element simulations. Defining contact between a deforming sheet and an arbitrarily shaped die has posed a challenge for a long time, and only in the last decade, several methods have emerged to handle it effectively in most commercial finite element codes that perform metal forming simulations. The first challenge in the process identifying contact between the deforming sheet and the die is to devise a strategy to identify the intersection between a line derived from the path of a node on the deforming sheet and the finite element that represents the rigid die.

Bellet and Chenot [25] use a simple technique to identify contact between two surfaces. They use triangular 3D elements to represent the die so that it is possible to evaluate the relative positions of sheet points (nodes) at each deformation increment. In the contacting regions, the tangential friction stress τ is given by a Coulomb model $\tau = -\mu p(1/|U|)U$, where p is the applied forming pressure and μ is the friction coefficient. Bellet and Chenot [25] successfully used their contact and friction algorithm in performing complex 3D shapes and also used the same technique in 2D to model SPF with diffusion bonding. Wood et al. [32] used triangular elements to define the die. During the initial phases of their formulation, they considered only sticking contact with the assumption that no subsequent separation may take place. They solved the intersection problem between the node path and the die surface through an unrelated method that was originally devised for the solution of geometrical searching problems. They effectively used the *alternating digital tree* technique based on the use of binary tree structures.

Chandra et al. [33] and Rama [31] used a novel technique called the *pseudo-equilibrium* method to handle contact. In this method, the die was represented as a line segment in 2D models and a three-noded triangular element for the 3D models. This research work uses *primary* followed by *secondary* search processes. The primary process is used to only identify if the "coordinate limits" of the node path (a line) coincides with the "coordinate limits" of the die element. Once a list of nodes that could potentially be in contact with the die is established through the first process, the second process is used to find if the node path and the die element really intersect. Once a node penetrates the die surface, the coordinates of the contact point along with the depth of penetration are determined. This deformation stage at which die penetration is detected is described as *pseudo-equilibrium* stage. Upon evaluating the depth of penetration, the *compatibility load step* is executed, pushing the penetrated sheet one to the die surface, as shown in Fig. 29.

Thus, force equilibrium without violating the contact compatibility conditions is established. This method is more comprehensive and complete than other methods, which physically relocate the penetrated node either to the point of contact on the die surface, or, in some cases,

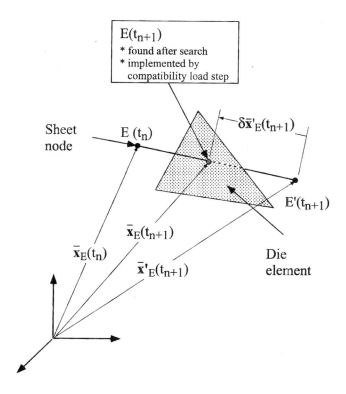

Figure 29   Schematic of a contact algorithm.

where the node is fully constrained once it penetrates the die. When dealing with intricate shapes, such procedures violate the force equilibrium of the deforming body and could potentially miscalculate the pressure requirements for maintaining superplastic conditions. It should be noted that the compatibility load step is not included as part of the deformation stage and is only executed once to achieve a force equilibrium of the deforming sheet after penetration. The sheet–die friction is handled through the estimation of normal and tangential forces to be applied on the node in contact with the die. Coulomb's model is used for this purpose using both the static and dynamic friction coefficients, depending upon whether the contacted node was stationery or moving in the previous load step. A local coordinate system for the die element with the contacted node is established to make it convenient for implementation.

In the commercial code MARC [19], whenever a penetration occurs, the ratio of the incremental displacement is divided based on the contact point on the die (either a line, a regular surface, or an analytical NURB surface). Three different strategies are implemented: (a) retracing the deformation to the earlier iteration; (b) applying a force to return back to the contact point from the deformed configuration; and (c) specifying an incremental deformation from the earlier iteration. The user can select the method based on the requirement and experience. Although the search techniques to identify contact point are slightly different in all cases, the basic principle is very similar. However, the friction is handled quite differently in each of the formulations.

5. Pressure Prediction Algorithms

Maintaining optimum conditions for superplastic deformation is critical and it can only be achieved by altering the deforming pressure as deformation progresses. Bellet and Chenot [25] achieved pressure control by trial and error using a posteriori adjustment technique, at the end of each time step. Bonet and Wood [9] treated the unknown pressure parameter as an additional variable to be determined from an additional equation constraining the maximum strain rate or a method based on the rate of energy dissipated over the volume of material. They seem to observe less fluctuations in the pressure cycles than those obtained using the posteriori adjustment method. This method of pressure control is not easily transferable to other formulations without a major effort.

Rama and Chandra [29] developed a very simple and flexible method of pressure control. They assumed that

the free forming sheet deforms like a thin membrane and hence establishes a relationship between the stress in the sheet and the pressure load. Considering that the maximum strain rate will only occur in the free forming regions and also that the sheet tends to deform similar to the free expansion of a cylindrical membrane, new pressure loads for subsequent load steps were explicitly predicted along with an iterative correction procedure working to determine the required pressure in that time step. With the knowledge of pressure at time step t, the initial pressure for time step $t + \Delta t$ is:

$$p^{t+\Delta t} = (1-v)\left(\frac{\dot{\bar{\epsilon}}_{opt}}{\dot{\bar{\epsilon}}^t}\right)^m p^t + v\left(\frac{\dot{\bar{\epsilon}}_{opt}}{\dot{\bar{\epsilon}}^{t-\Delta t}}\right)^m p^{t-\Delta t} \quad (32)$$

The optimum pressure is determined iteratively within each time step before the initial pressure for the next time step is predicted using the above equation. The proposed method is quite flexible and efficient, and can be easily incorporated in any existing formulation with-

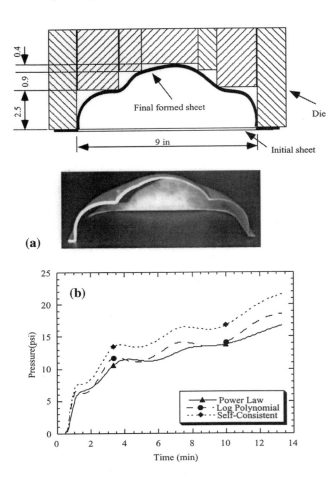

(a)

(b)

Figure 30 SPF of an Al 5083 pan: (a) cross-section, and (b) pressure–time used in the forming.

out any difficulty. A FEM code SPASM3D based on the formulation was developed and verified by Rama, through the SPF simulation of several superplastic parts including a complex aerospace component.

MARC uses a similar concept for the pressure prediction. However, instead of the maximum strain $\bar{\epsilon}^t$, the prediction is based on an average $\dot{\epsilon}$ computed over all the elements, after neglecting all the elements with strain rate below a user-specified cutoff $\dot{\epsilon}_{cutoff}$.

D. Process Modeling of an Al 5083 Deep Pan

Using the constitutive models developed earlier, finite element analysis was carried out using the commercial code MARC [19]. The actual details of the configuration and the cross-section of the pan are shown in Figure 30. Figure 30 shows the experimental thickness distribution obtained and that predicted by the three models. Although it is found that all the three models do a reasonable job, logarithmic Eq. (13), followed closely by the micromechanical model, does an excellent match compared to the simple power law (Eq. (10)). This is to be expected because the constitutive Eq. (10) did not closely follow the experimental data (see Fig. 12).

VII. INDUSTRIAL APPLICATIONS

Applications for superplasticity and superplastic forming (with and without diffusion bonding) are wide-ranging (Fig. 31). Uses can be found in areas ranging from sports health to aerospace applications. Anytime there is a complex shape or form, requiring lightness,

Table 8 Superplastic Material Cost

Material	$/lb	Normalized per unit volume	Normalized for equal stiffness
Steel coil	0.2	1	1
Aluminum strip	1.15	1.4	2
SP 5083 sheet	3.5	4.3	6.2
SP 2004	5.0	6.5	9.4
SP 7475	11.0	14	20.2
SP 8090	16.0	19	27
SP MMC 2024	30+	37	47

accuracy, and minimal pieces for assembly, superplasticity is the prime candidate for the forming process. Cost is invariably the deciding factor in the selection of any production process. Both energy costs (in terms of heat energy and time of forming) and tooling need to be considered in the total cost. Increases in forming temperature and time must be studied closely in order to justify the feasibility of superplasticity in a production environment [34].

A. Energy/Cost Considerations

1. Material Cost

The foremost cost consideration for production is material cost. Engineering will determine the material properties required for the integrity and safety of a product and then seek the most suitable (cheapest and most cost-effective) material that will fulfill the requirements. Al 5083, although not as cheap as steel, is very affordable in sheet form when compared with the other SP alloys listed in Table 8. When the normalized stiffness is compared with steel, SP 5083 sheet is much higher at 6.2. Superplastic materials can be rather expensive; but, in the case of Al 5083, they can still be reasonably justified when cost and performance criteria are considered.

2. Forming Time

The key to streamlining production costs is minimizing the amount of time taken to form that part. The amount of time it takes to form a part superplastically is dependent on the strain rate at which the forming material can be deformed. Maximizing the strain rate during forming will minimize the amount of time taken and hence lower production cost. If the material is deformed too fast, the integrity of the finished product will be compromised. Superplasticity is observed in Al 5083 at strain rates ranging from 10^{-1} to 10^{-5} sec^{-1}. So

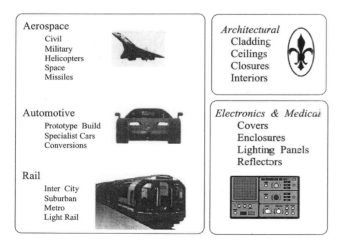

Figure 31 Application of SPF in different industries.

Aerospace
Civil
Military
Helicopters
Space
Missiles

Automotive
Prototype Build
Specialist Cars
Conversions

Rail
Inter City
Suburban
Metro
Light Rail

Architectural
Cladding
Ceilings
Closures
Interiors

Electronics & Medical
Covers
Enclosures
Lighting Panels
Reflectors

far, in low-volume-production niche market applications, Al 5083 is being formed at strain rates of about 10^{-2} sec^{-1} and strains of up to 300%. For high-volume mass production, a target strain rate of 10^{-1} sec^{-1} is necessary to meet necessary cost and time constraints. Lower forming times mean, of course, increased production, but there are drawbacks as well. As forming time reduces, gas management time increases with strain rate and gas pressures. This means that other factors need close consideration when optimizing the strain rate for production.

3. Role of Temperature in SPF

The role of temperature in superplastic forming is manyfold. The primary role that temperature plays is to introduce the superplastic deformation mechanisms and to allow the material to deform, such as "silly putty" or chewing gum. Because the superplastic state of a material is reached above the annealing temperature, any memory of the forming process is erased, and a residual stress-free part results. The higher the temperature, the easier the material deforms, requiring less gas pressure. The forming time can also be reduced as the strain rate can be increased to a point without any undue side effects to the finished product. Temperature also plays an important economic role in SPF. Higher forming temperatures require increased energy to homogeneously heat the SPF material and surrounding furnace internals before any forming can take place. Also a major consideration at high forming temperatures is the structural and dimensional integrity not only of the die material, but also of the furnace itself. If the furnace's structural/dimensional integrity is compromised, the whole forming process can be compromised. Material and engineering costs for extremely high-temperature furnaces can be astronomical.

4. Tooling

Fundamental to the success of most component manufacturing processes is tooling. Superplastic forming is no exception; in fact, it is at the heart of the economic viability for commercial superplastic aluminum forming.

Tooling needs to be [34]:

Durable—able to continuously operate at elevated temperature without degrading and safely contains the gas pressures and mechanical forces applied

Accurate—to consistently produce forming to the required dimensions and surface quality

Productive—yielding maximum production output commensurate with the superplastic strain rate limits of the alloy being formed and the pressure containment capacity of the forming equipment (forming presses) being used

Cost—effective-manufactured at a minimum cost for "optimized forming conditions."

As the forming temperature rises, not only is more energy required in the superplastic manufacturing process, but higher cost of the tooling is needed as well. For example, when forming at or near 500°C (Al 5083), dies made of steel have adequate creep resistance as well as corrosion resistance and will last almost indefinitely for a finalized part. If a more exotic material is formed such that the forming temperature is high (927°C for Ti 6Al–4V or titanium β-21s), special tooling requirements will drive the initial investment to astronomical heights. Higher temperatures dictate the use of exotic materials such as molybdenum or high-nickel superalloys for the construction of dies along with key elements of the forming press itself. Machining for these materials, due to extreme hardness, requires specialized tooling. This more expensive, specialized tooling will tend to experience wear at an accelerated rate due to working conditions, and thus have to be replaced more often than conventional tooling for conventional materials. If the material of the die is hard to form, then the tooling method (e.g., electric discharge machining (EDM)) becomes even more expensive. Exacerbating the cost of high-temperature forming would be the inevitable degradation of the die itself. The die would have a finite lifetime and would require refurbishing and/or remanufacturing periodically. Extensive use of ceramics is a key to lowering these manufacturing costs, but is still a less viable alternative to a low-temperature steel alloy die.

B. Engineering Applications

There are many current applications for superplastic forming in production environments. In comparison to traditional plastic forming methods, SPF has many benefits which can justify its use economically.

1. Aerospace

Performance is the first objective when developing components for the aerospace industry. Critical to the performance of an aircraft or spacecraft is minimum weight and maximum strength. Also critical is packaging concerns, which can often call for very unorthodox,

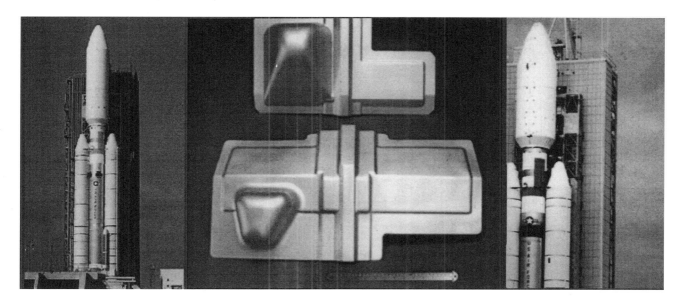

Figure 32 Titan IV heavy payload rocket.

complex geometries. These aerospace parts usually require multiple components riveted, bolted, or welded to form the final part. The added fasteners and/or welding may add weight and possibly compromise the integrity of the final product. Distortion is also of major concern. Uneven heating/expansion due to welding often causes deformation and induces residual stresses to the point where the final product will not fit within the required tolerance range for their prospective application. When this happens, the part is then scrapped and the time spent in assembling, forming, etc., is wasted.

Superplastic forming is the one-step process that uses minimum material to form complex geometries with minimum weight and unparalleled repeatability. The key to the repeatability lies in the fact that the material is formed at a temperature above its annealing temperature. This means that there is little or no residual stress in the final product, and therefore no springback to compensate for in the initial die manufacturing. The repeatability of SPF has procured a nearly 0% rejection rate for parts that once had a nearly 40% rate rejection after inspection using previous production methods.

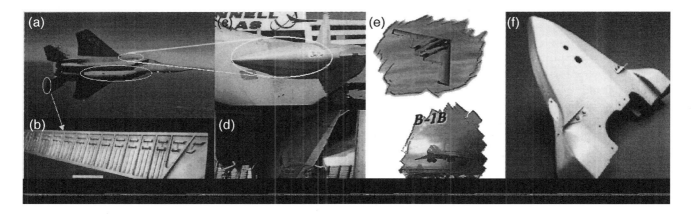

Figure 33 (a) F-15 Eagle, (b) wing root, (c) reconnaissance pad, (d) horizontal stabilizer, (e) B1-B and B-2 bombers, and (f) ejection seats.

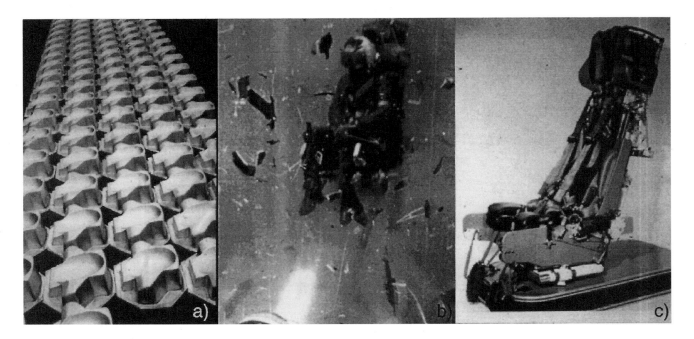

Figure 34 (a) Ejection seat and head rest, (b) seat in action, and (c) SPF parts in the seat.

Even the relatively slow strain rate of SPF can be justified when one part can be formed to eliminate several manufacturing processes and the excess assembly time of several parts into a singular and accurate forming process. Examples of aerospace applications include the Titan IV heavy payload rocket (Fig. 32), various parts for the B1-B and B2 bombers, F-15 Eagle (Fig. 33), and ejection seat components (Fig. 34). The Titan IV rocket (Fig. 32) uses several SPF parts to address critical weight issues and packaging concerns within a heavy lift launch vehicle. The F-15 has many nice examples of SPF-manufactured Al 5083 components. Components successfully manufactured using SPF on the F-15 range from the complex structure of the horizontal stabilizers to the deep, curvaceous structures of the wing root leading edge and the reconnaissance pod mounted on the belly of the airplane. Of the ejection seat components manufactured with the SPF process, the Al 5083 alloy head rests are good examples of deep and complex parts formed in one SPF process, eliminating assembly time and the manufacture of several cold-stamped, riveted, and welded components [34,35].

2. General/Public Transportation

Weight is also a major performance consideration for other industries. Vehicles that carry heavy loads regu-

larly can benefit from the weight reduction SPF components can generate. The less initial weight a vehicle has, the more revenue-generating load capacity it can safely sustain while also reducing the maintenance required for the said vehicle, which may not operate at its maximum capacity for a majority of its service life. Parts that may be spared due to the decreased loading would be wheel bearings, breaks, suspension components, and the railways themselves (if applicable). Public transportation in London has found use for SPF processing in its underground railways, as seen in Fig. 35. The London Underground has incorporated SPF-formed window panes into its newer underground trains. Their light weight adds efficiency to the rail system by trimming unnecessary weight from the train with no degradation of structural integrity.

3. Automotive

In the automotive industry, minimum forming time is critical for the feasibility of a part to be produced. The finished shape should contain a number distinct features, which only superplasticity can reproduce repeatedly and reliably. In Fig. 36 are presented some typical applications for SPF's viability in the automotive industry with parts that would otherwise be stamped when cold-formed. These are semipractical applications which, alone, cannot justify the expense for the tooling

Figure 35 Superplastic forming door panels in London underground trains.

Figure 36 Al 5083 autoparts: (a) rear door, and (b) vehicle housing.

Courtesy: A.J. Barnes - Superform USA

Figure 37 Panoz AIV featuring SPF door panels.

and/or the slow forming time. The key to their viability lies in a single forming process for several parts from a single sheet, and the accuracy of the finished part. Some specific examples of automobiles with SPF parts in current production include the Dodge Viper, with SPF-formed rocker panels, and the Oldsmobile Aurora, with SPF-formed trunklid. The company that produces the vehicles which are the most SPF-intensive automobiles in production today is Panoz. The Panoz AIV roadster and the brand new Esperante feature full SPF-formed body panels.

The performance of the Panoz Esperante roadster (Fig. 37) is among the world's best. Initial tests show that with its 320-hp, 32-valve aluminum V8, the Esperante will outrun a Ferrari F355 Spyder to 60 miles/hr. It also stops shorter than an Acura NSX and has more lateral grip than a Porsche 911 Cabriolet. The Esperante's lightweight aluminum body is made using Superplastic batch forming, a process developed for the

aerospace industry and pioneered in the autoindustry by Panoz. Its modular extruded aluminum chassis is an exercise in technology and as rigid as those in most high-performance coupes. The Panoz Esperante's superplastically formed Al 5083 inner door panel is a classic example of a piece that can only be formed using SPF. Its complex contours and ultra-deep profile contain strains approaching 300%. Forming temperatures remain less than 500°C for every SPF part on this sports car, including the inner door panel. The body panels are all SPF-formed as well. Not as complex as the inner door panel, the body shows clean, smooth lines with residual stress-free SPF Al 5083 panels. The panels and other components SPF-formed for this automobile are batch-formed (several panels simultaneously) to streamline production using SPF forming.

Both Chrysler and General Motors have utilized SPF to form select body panels for the Dodge Viper and the

Figure 38 Superplastic forming body panes in Dodge viper (left) and Oldsmobile Aurora (right).

Oldsmobile Aurora (Fig. 38). The Viper features rocker panels made of SPF-formed Al 5083 and represent a small production run of SPF parts. The Oldsmobile Aurora features an SPF-formed trunk lid. The Oldsmobile production represents a medium to large production volume with about 80,000 units produced per year. Both automakers have demonstrated and proven the feasibility of superplastic forming in mass-produced automobiles.

4. Architectural

Although weight is not as much a concern in architecture (unless designing sky scrapers and the like), appearance can be. A boring box of a building can be

Figure 39 Superplasticity-formed exterior panels/windows.

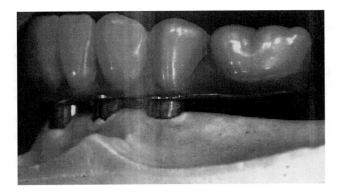

Figure 40 Superplasticity-formed dental implants.

transformed from just another drab building to a tasteful design that is pleasing to the eye. Artful design of architectural structures is a viable use for SPF. The fact that complex forms can be recreated precisely using SPF can be exploited in architectural components that serve a dual role of a functional and aesthetically pleasing design. An SPF window pane can be seen in Fig. 39.

5. Dental Health

Traditionally, superstructures for dental implants are cast in gold alloy onto premachined gold alloy cylinders. Castings are quite bulky and 24% do not fit correctly. This requires sectioning and soldering for a fit that is clinically acceptable and would not affect the integrity of the commercially pure titanium implants osseointegrated with the bone. Here we have a superplastically formed dental implant (see Fig. 40). Measurements of flow stress were used in conjunction with an analytical model to superplastically form a fixed-bridge dental implant superstructure using titanium alloy sheet. Using these tools, dental implant superstructures can be superplastically formed from Ti–6Al–4V alloy sheet to produce accurately fitting prostheses that are strong in thin section. Material costs are significantly reduced compared to materials used in traditional casting techniques.

VIII. CRITICAL ISSUES AND FUTURE CHALLENGES

Although economics drives the final decision, it is a result of supply and demand. Although SPF and SPF/DB are extremely powerful manufacturing methods for forming near-net shapes with the least number of joints, their full potential has not been realized. Superplastic forming is resorted to only when all other conventional

manufacturing methods fail to meet the demand. Because SP components yield structurally stable and reliable light weight components, they deserve better attention and extensive use. We anticipate that the future designs will be carried out in virtual environments with inputs from concept, design, analysis, engineering, and manufacture modules. Such an integrated product and process design (IPPD) should include superplastic forming as a module, as shown in Fig. 41. Superplastic module represents the process model that accurately predicts the design and process parameters. The module also contains information to determine the geometry envelope (material, shape, size, and details), where superplasticity is not economically viable and not worth the consideration. The process models are essentially mathematical models that predict the pressure–time cycle to maintain optimum forming conditions, and the resulting thickness distribution and maximum thinning.

A. Critical Issues

Superplastic forming components are now in use for the past two decades. However, it has not captured the market on the same scale as many other sheet metal or bulk forming methods such as brake forming, deep drawing, or extrusion. Lack of knowledge on the part of designer, which was one of the major concerns, is also slowly disappearing as more and more designers are at least aware of the process. It is clear that there are certain technological and economic issues shrouding the process, hindering its use. Some of them can be solved if the research community pays attention to the mundane but important issues as briefly addressed below.

B. Thickness Control

Because SPF deforms by stretching rather than by bending, the segment that forms last thins the most, and many times the thinning levels are not acceptable from postformed strength considerations. Thinning profile has an impact on the geometry of the part that can be formed using SPF, rather than the hundreds of percent of elongation that SP is famous for. Typically, the thickness can be controlled in a number of ways, and are listed below:

(1) Premachining: If the initial thickness can be made nonuniform to compensate for thinning, then final shape and thickness distribution can be controlled. What is required is an inverse design that can predict the initial thickness,

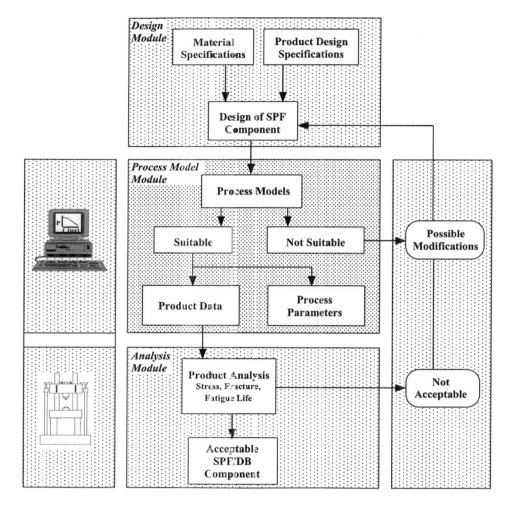

Figure 41 Integrated process design of SPF components.

given the final desired thickness, in addition to a new pressure-time forming cycle. The cost of machining (chem milling) the part should be accounted for in the analysis.

(2) Two-stage forming: The initial thickness variation can possibly be obtained by forming the sheet in the reverse direction first to obtain a nonuniform thinning, and then by using the sheet to obtain the final forming. New process models need to be developed to optimize the process. Additionally, this procedure requires two separate dies and a much longer time. These factors should be considered in the cost analysis.

(3) Male forming: Because, in SPF, the sheet does not thin (very little, if at all) once it contacts the die, a movable die can be placed in critical areas

to avoid further thinning. Male forming is a specific case where thinning is controlled all along the male surface. The design of the die is complicated and increases the cost.

(4) Diffusion bonding: If the material allows diffusion bonding (e.g., Ti–6Al–4V), this leads to an effective way to reinforce regions where thickness needs to be increased. Judicious choice of the thickness of the bonded sheet and the location needs to be made at the design stage.

C. Die and Lubrication Material

Die design plays a critical role in determining the overall cost of a part. Although it is true that SPF requires only one die and the sheet does not spring back, the prorated cost of the die depends on the life of the die (total number of parts to be made), the thermal cycle (peak temperature and heating/cooling excursions), and the geometrical complexity (die entry radius and lateral angles). There are a number of choices of ceramic and metallic (carbon, alloy, and maraging steels) dies with varying capabilities and manufacturing costs. There are no clear guidelines on the selection criteria and the effect of the die material on the surface finish of the part. In combination with the lubrication used, the surface of the die gets affected differently, leading to poor product quality. A systematic study on the effect of desire quality of the part on the die material, design, and choice of lubricant is required.

D. Issues Specific to Ti–6Al–4V

Ti–6Al–4V is the work horse SP material and proves its worth when used with diffusion bonding. However, when SPF/DB is used in three-sheet or four-sheet configurations, the web segment of the honeycomb structure is difficult to control. This is especially true when the configuration is not parallel (i.e., top and bottom sheets are not parallel). This causes many design and manufacturing issues. Also due to the high-temperature forming, a coating of α is formed, which needs to be cleaned using a chemical process that is a health hazard as well as a cost issue. Both these problems need to be addressed.

IX. SUMMARY

With the continued use of superplastic forming process in the manufacture of various aerospace and automotive components, there is an imminent need to accurately predict the process parameters for complex 3D configurations. New SPF products are finding their way into architectural, rail, heavy industry, biomedical, sports, and textile sectors apart from the traditional aeroindustry and autoindustry. High-fidelity process models, in combination with other concurrent design tools (CAD, stress analysis, and cost analysis), should be able to provide the platform necessary for the wider use of superplastic products across a range of industries. The objective of this chapter is to first introduce the various industrial applications of SPF products and the design bottlenecks hindering their wider use. Practical considerations in the development of appropriate material constitutive models for superplastic materials and their relevance to numerical simulations are then

discussed. Three different models, including a micromechanics-based model, are used to describe the constitutive behavior of a selected SP material (Al 5083). The performance of the models is then compared with respect to experimental results. The effect of the actual material parameters on the simulated results, and hence the design, are outlined. Some of the commercial and other finite element codes used in SPF process modeling are enumerated, and the features of each of the codes with respect to the type of elements, contact/friction algorithms, and pressure–prediction schemes are compared. In closing, other practical considerations in the manufacture of SPF products (e.g., die materials, optimal SPF/DB design, α-casing, product thickness control) are also discussed.

REFERENCES

1. Superplasticity in Advanced Materials (ICSAM-91); Tokzane, M., Hori, S., Furushiro, N., Eds. JSRS: Osaka, Japan, 1991.
2. Sherby, O.D.; Nieh, T.G.; Wadsworth, J. Overview on superplasticity research on small grained materials. Mater. Sci. Forum 1994, *170–172*, 13–22.
3. Watts, J.D. *Temperature Effect on the Transition from Plastic to Superplastic Behavior of Al 5083. M.S. Thesis.* Florida State University: Tallahassee, FL, 2000.
4. Pellerin, C.J. Air Force Perspective on spf and spf/db Titanium Technology. Proceedings of the Symposium on Superplastic Forming, Los Angeles, California, USA, 1984.
5. Khalleel, M.A.; Johnson, K.I.; Smith, M.T. Process Simulation for Optimizing Superplastic Forming of Sheet Metal. Proceedings of the MARC International Users Conference, 1994.
6. In *Superlasticity and Superplastic Forming (ICSAM-88)*; Hamilotn, C.H., Paton, N.E., Eds. Minerals, Metals and Materials Society: USA, 1989.
7. In *Superplasticity in Advanced Materials (ICSAM-94)*. Materials Science Forum; Langdon, T.G., Eds. Trans Tech Publications: Switzerland, 1994; vol. 170–172.
8. In *Superplasticity in Advanced Materials (ICSAM-97)*. Materials Science Forum; Chokshi, A., Eds. Trans Tech Publications: Switzerland, 1994; Vol. 243–245.
9. In *Superplasticity in Advanced Materials (ICSAM-2000)*. Materials Science Forum; Chandra, N., eds. Trans Tech Publications: Switzerland, 2001; Vol. 357–359.
10. Enikeev, F.U.; Padmanabhan, K.A.; Vasin, R.A. *Superplastic Flow: Phenomenology and Mechanics*; Springer-Verlag: New York, 2001.
11. Pilling, J.; Ridley, N. *Superplasticity in Crystalline Solids*; The Institute of Metals: London, UK, 1989.
12. Wadsworth, J.; Nieh, T.G.; Sherby, O.D. *Superplasticity in Metals and Ceramics*; Cambridge University Press, 1997.

13. Chen, Z. Experimental and Numerical analysis of Cavitation Damage in Superplastic Aluminum Alloys. In *Ph.D. Dissertation*. Florida State University: Tallahassee, FL, 2002.

14. Hamilton, C.H.: Ash, B.A.; Sherwood, D.; Heikkinen, H.C. Effect of microstructural dynamics on superplasticity in al alloys. Heikkinen, H.C. McNelley, T.R., Eds.; In Superplasticity in Aerospace; TMS Publishers: Pennsylvania, 1988; 29–50 pp.

15. Iwasaki, H.; Mori, T.; Tagata, T.; Higashi, K. Experimental evaluation for superplastic properties and analysis of deformation mechanisms in commercial 5033 alloy. Trans. Jpn. Soc. Mech. Eng. 1998, *64*, 1390–1396.

16. Cornfield, G.C.; Johnson, R.H. The forming of superplastic sheet metal. Int. J. Mech. Sci. 1970, *12*, 479–490.

17. Holt, D.L. Analysis of the bulging of a superplastic sheet by lateral pressure. Int. J. Mech. Sci. 1970, *12*, 491–497.

18. Jovane, F. An appropriate analysis of the superplastic forming of a thin circular diaphragm: theory and experiments. Int. J. Mech. Sci. 1968, *10*, 403–427.

19. *MARC Finite Element Code;* Version k7 Ed.; MARC Analysis Research Corporation, Pal. Alto: CA, 94306 USA, 1997.

20. Onate, E.; Zienkiewicz, O.C. A viscous shell formulation for the analysis of thin shell metal forming. Int. J. Solids Struct. 1983, *25*, 305–335.

21. Argyris, J.H.; Doltsinis, J.S.L. A primer on superplasticity in natural formulation. Comput. Methods Appl. Mech. Eng. 1984, *45*, 83–131.

22. Zhang, W.C.; Wood, R.D.; Zienkiewicz, O.C. Superplastic Forming Analysis Using a Finite Element Viscous Flow Formulation. Proceedings of Aluminum Technology '86, London, 1986; 11.1–11.6.

23. Chandra, N. Analysis of superplastic metal forming by a finite element method. Int. J. Numer. Methods Eng. 1988, *26*, 1925–1944.

24. Nakamachi, E. A finite element simulation of the sheet metal forming process. Int. J. Numer. Methods Eng. 1988, *25*, 283–292.

25. Bellet, M.; Chenot, J.L. Numerical modeling of thin sheet superplastic forming. In NUMIFORM '89; Thompson,

et al., Eds.; Balkema Publishers: Rotterdam, USA, 1989; 401–406 pp.

26. Bonet, J.; Wood, R.D. Solution procedures for the finite element analysis of superplastic forming of thin sheet. Proceedings of the International Conference of Computational Plasticity—Models, Software and Applications, 1987; Part 2, 875–899.

27. Bonet, J.; Wood, R.D.; Wargadipura, A.H.S. Simulation of the superplastic forming of thin sheet components using the finite element method. In *NUMIFORM '89*. Thompson, E.G., et al., Ed. Balkema Publishers: Rotterdam, 1989; 85–93 pp.

28. Chandra, N.; Rama, S.C. Application of finite element method to the design of superplastic forming process. ASME J. Eng. Ind. 1992, *114* (4), 452–459.

29. Rama, S.C.; Chandra, N. Development of a pressure prediction method for superplastic forming processes. Int. J. Non-Linear Mech. 1991, *26* (5), 711–725.

30. Chandra, N.; Rama, S.C.; Rama, J. Computational modeling of 3-D superplastic components. Mater. Sci. Forum 1994, *170–172*, 577–582.

31. Rama, S.C. *Finite Element Analysis and Design of 3-Dimensional Superplastic Sheet Forming Processes. Ph.D. Dissertation.* Texas A&M University: College Station, TX, 1992.

32. Wood, R.D.; Bonet, J.; Wargadipura, A.H.S. Numerical simulation of the superplastic forming of thin sheet components using the finite element method. Int. J. Numer. Methods Eng. 1990, *30*, 1719–1737.

33. Chandra, N.; Haisler, W.E.; Goforth, R.E. A finite element solution method for contact problems with friction. Int. J. Numer. Methods Eng. 1987, *24*, 477–495.

34. Barnes, A.J. Industrial applications of superplastic forming—trends and prospects. In: Chandra, N., Ed. Superplasticity in Advanced Materials, ICSAM 2000; Trans Tech Publishers, Switzerland, 2000; pp. 3–17.

35. Sanders, D.G. The current state of the art and the future in airframe manufacturing using superplastic forming technologies. In *Superplasticity in Advanced Materials, ICSAM 2000*; Chandra, N., Ed.; Trans Tech Publishers, Switzerland, 2001.

8

The Design of Continuous Casting Processes for Steel

Roderick I.L. Guthrie and Mihaiela Isac
McGill University, Montreal, Quebec, Canada

I. INTRODUCTION

The world's current output of steel is approximately 0.8 billion tonnes per annum. To put this figure in perspective, this represents some 97% of the world's supply of all metals on a tonnage basis. Aluminum runs at 3% on a global basis (or 8% on a volume basis), while the remaining metals represent fractions of a percent. Most of the molten steel now produced is continuously cast, rather than cast into ingot molds as was the case up until the mid-1970s. Figure 1 illustrates in schematic form the main processes of ironmaking, steelmaking, continuous casting, and rolling, together with the main products of the steel industry. In the case of steel sheet production, for instance, most automobiles use steel sheet for side panels, etc. One can confidently expect that steel sheets will remain an important automobile body material for years to come, and that the continuing development of higher-strength steels with high-formability steel will achieve the environmentally desirable specific weight reductions needed in the transportation sector. Similarly, current markets for steel long products (bars, tubes, rods, rails, wire, rounds, I-beams, H-beams, etc.) are unlikely to be significantly usurped by other materials, given the increasingly competitive strength/weight/cost ratios possible with steel products.

Nonetheless, Fig. 2 shows that many unit-processing operations are still needed to convert cold iron ore to molten iron, to then transform this hot metal into "raw" liquid steel, and to refine this raw steel prior to casting. Thirty years ago, the refined steel was cast into solid ingots for subsequent rolling into steel sheet products. However, given the high capital costs of blast furnaces, melt shops, and hot and cold rolling mill complexes, major research and development efforts have been made within the industry with the objective of drastically reducing the number of process steps needed to produce a final product. The ultimate goal for a future steel plant would be to transform iron oxide to a final steel product in just one continuous operation. Realistically, one can imagine reducing the number of major processing steps, as we now know them, down to two: direct steelmaking and direct, net shape casting [1–3].

In direct steelmaking (step 1), the aim is to feed cheaper coal (rather than coke), together with iron ore pellets and lime flux, into an autogenous reactor to produce the various liquid steel compositions and melt quality needed for casting.

In direct casting (step 2), the aim is to develop technologies suitable for directly casting steel sheets to thicknesses of 0.5 to 10 mm (0.02 to 0.4 in.), at tonnage rates of 100–200 tonnes/hr/m width (35 to 70 tons/hr/ft width). Such performance characteristics would match those of the big slab casters of the present day, producing 2–5 million tonnes/annum, but, by eliminating reheat furnaces and hot-rolling mills and possibly cold-rolling mills, would have a dramatic impact on the capital and operating costs of future steel sheet producing plants.

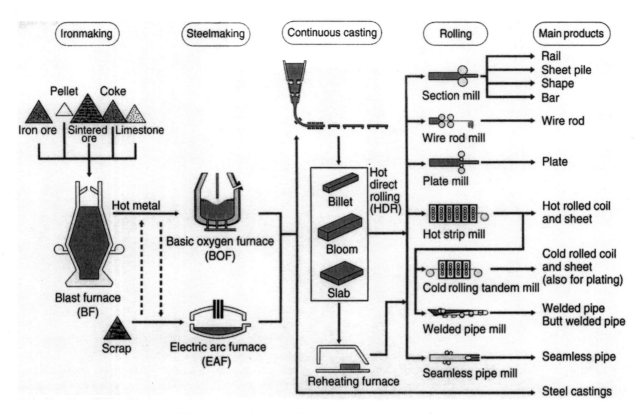

Figure 1 Manufacturing processes for iron and steel.

Today's integrated steelmaking plants currently include the following generic-type processing operations:

1. Blast furnaces to produce molten iron from hematite ores (Fe_2O_3) by smelting with coke and limestone.
2. Hot metal (molten iron) pretreatment for further removal of sulfur, and sometimes phosphorus.
3. Oxygen steelmaking furnaces to decarburize the hot metal and to dephosphorize the molten steel, giving a raw steel containing some 400 ppm O, and 700 ppm C.
4. Teeming ladles, for adjusting the chemistry of the raw steel during "tapping" or pouring of the steel from the furnace into the ladle. These ladle additions include ferroalloys (Fe–Mn, Fe–V, Fe–Nb, etc.) and steel deoxidizers (Al and/or Fe–Si) needed for steel chemistry and steel castability

and secondary refining operations, illustrated in Fig. 3, which include,

5. RH- and VOD-type vacuum degassers to remove the gas-forming dissolved metalloids, oxygen, nitrogen, and hydrogen from the liquid steel and

the further elimination of carbon as carbon monoxide, to 20 ppm C for the ultralow carbon steel grades. The alternative approach for degassing is the AOD vessel where varying argon/oxygen ratios are used during decarburization of stainless steels (generally) so as to lower oxygen partial pressures toward those achieved by vacuum operations.

6. Ladle furnaces for further alloying, for temperature control (heating), for final steel desulfurization (S) with reducing fluxes, and for further removal of inclusions into the slag using inert gas bubbling, and sometimes lime powder injection.

prior to continuous casting.

Minimill operations, which in North America now account for some 50% of steel production, bypass steps 1 and 2 by remelting steel scrap and prereduced pellets in electric arc furnaces, rather than pneumatic oxygen processing vessels (BOF, OBM, Q-BOP, etc.) They then face equivalent or greater challenges in preparing the molten steel for chemistry and quality before casting. These additional challenges are associated with residual contaminating elements (copper and tin) in the scrap

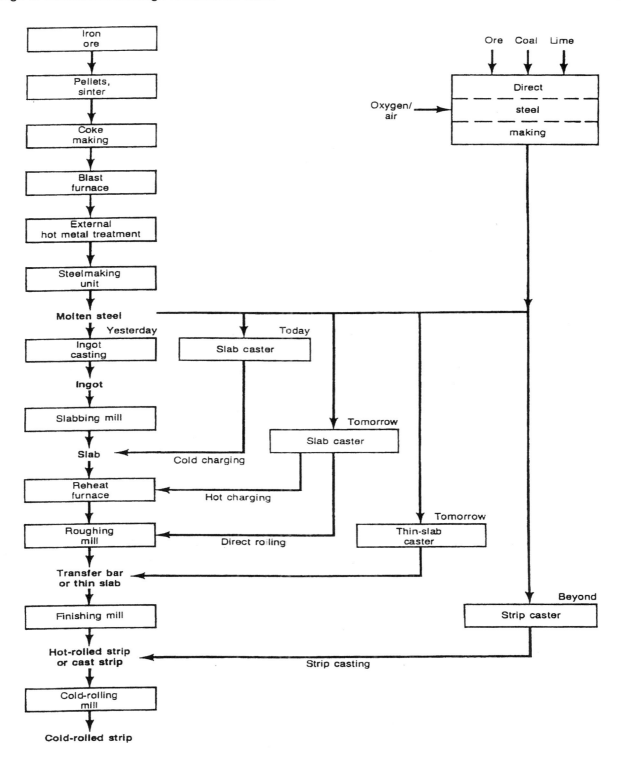

Figure 2 Past, present, and future steel-processing operations.

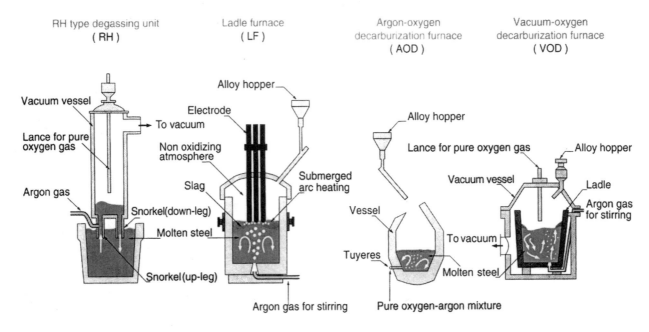

Figure 3 Typical facilities for secondary refining.

recycle process, and the more likely blockage of smaller metering nozzles by alumina inclusions vis-a-vis slab caster operations for aluminum vs. silicon-killed grades of steel.

II. CONTINUOUS CASTING PROCESSES FOR STEEL

A. Conventional Slab Casting

Before the advent of North America's first commercially successful vertical mold continuous casting machine, in 1954, for the production of stainless steel slabs at the Welland Plant of Atlas Steels in Ontario, all liquid steel was teemed from teeming ladles and cast into stationary molds in the ingot aisle. Groups of these ingots were then placed in an ingot-soaking pit, reheated, and then withdrawn at temperatures appropriate for hot rolling (~1100°C or 2700°F). They then passed through a slabbing mill where the "fish-ends" of the resulting slab had to be cropped off at a cost of some 10% in yield. The resulting 6- to 10-in.-thick slabs were inspected, "scarfed" (ground down) if necessary for major surface defects, and then either reheated in a slab reheating furnace, or in some plants such as DOFASCO, following passage through the high-water-pressure "descaler" operation, rolled directly in a 7-stand hot strip mill.

Today, ingot casting has been largely eliminated worldwide, and steel yields thereby increased by about 10%, thanks to the successful development of the slag-lubricated, oscillating, open-ended mold. This represents the heart of the various fixed-mold continuous casting machines. A key role in the development of this process was played by Iain Halliday at Barrow Steel Works, who perfected the Junghans–Rossi oscillation mode [4] (i.e., three quarters of the cycle down and one quarter up) by imposing a relative velocity during the downstroke—with the mold going slightly faster than the strand—which he termed "negative strip." This invention helped resolve sticking of the growing frozen steel shell (see Fig. 16) to the copper mold, and allowed him to reach casting speeds of 14.5 m/min in 1958. This is still a world record [5,6]. These machines, nearing process perfection in recent years, allow partially solidified slabs, blooms, or billets to be continuously withdrawn through their bottom, at the productivity levels associated with the ingot aisles of yesteryear.

As shown in Fig. 4, conventional continuous casting operations comprise a "tundish," a water-cooled, chromium-plated, copper alloy mold, a mold oscillator, a group of cast strand supporting rolls, rolls for bending and straightening the strand, rolls to pinch and withdraw the cast strands, groups of water spray nozzles to extract heat from the strand, a torch cutter for cutting the cast strand, and a dummy bar. This dummy bar fits in the bottom of the open mold at start-up, providing a starting base on which the strand can freeze. The dummy bar then extracts the cast strand, and finally moves away

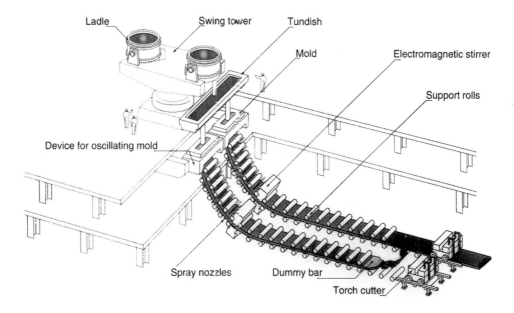

Figure 4 Two-strand continuous slab caster.

to "storage," once the forming strand has approached the torch cutter. The overall setup is generic, in that a "ladle" sits on a rotating "turret" or "swing tower," and empties liquid steel into a tundish set below via a protective ladle shroud (or tube). This tundish is used to distribute metal to one, two, four, six, or even eight, oscillating molds set below. In the case of slab casting operations, it is common to have twin slab production. In the diagram, a submerged entry nozzle (SEN) is used to protect the streams entering the mold from reoxidation. Once the ladle has been emptied of steel, the tundish is allowed to drain while the turret rotates a new filled ladle into position. In this way, sequence, or continuous, continuous casting can be achieved, with uninterrupted runs on the order of 1000 ladle changes being possible, using the flying tundish technique for rapid replacement of a spent one.

Figure 5 shows the continuously cast slabs from such a curved mold caster entering the unbending section of the machine, where rolls straighten the slabs, before passage to a walking beam slab-reheating furnace, for reheating (including scale formation), before hot rolling.

Another currently popular type of fixed-mold caster is the vertical and bending-type caster, in which the mold and support rolls are arranged vertically so as to maximize the chances for inclusions to float out of the deep sump of molten steel before being trapped in the solidifying shell. The strand is then bent and straightened to deliver the strand horizontally and continuously. In comparison, the first "stick," or vertical mold casters, froze single, vertical strands in the same way as today's direct chill (DC) casters for the aluminum industry. The fact that curved mold and low-head curved mold machines tend to capture a band of inclusions about mid-distance from the top surface of the strand and its central axial plane has led to the popularity of the vertical-bending machine class of casters.

In technically advanced steel mills, where surface quality is good (freedom from scabs, slivers, and deep oscillation marks) these slabs, blooms, and billets can be hot charged directly to a reheat furnace without surface inspection before further size reduction. A certain amount of scale (iron oxide) forms on the slabs during their time in the reheating furnace. This can be useful in moderation since surface blemishes can be removed, following passage through the high-powered water jets of the descaler. In the case of slabs for sheet products, a roughing mill then rolls the slab down to a "transfer bar" some 25–40 mm (1–1.5 in.) thick.

In most current commercial operations for sheet production in integrated steel plants, the transfer bar is then fed into a 6- or 7-stand hot strip mill, where steel sheet, 1–4 mm thick, is produced at speeds approaching 15 m/sec (3000 ft/min). Strip temperatures, alloy compositions, and rates of compression are all carefully controlled, so as to manipulate the solid-state phase transformations and thereby optimize steel microstructures and attendant physical properties.

Figure 5 Continuous caster operation—unbending of slab.

B. Thin Slab Casting

In the last 15 years, progress toward the goal of process rationalization (Fig. 2) as a result of thin slab casting developments based on the fixed, oscillating mold, or conventional approach have proved successful. This development has allowed scrap based-EAF minimills to enter the flat products business. Thin slab casting technology eliminates the need for a hot-rolling mill as such. As a result, an integrated line comprising an advanced EAF thin slab caster can produce up to about 800,000 tonnes/year with much lower investment in facilities and much less energy consumption as compared with an integrated mill. Figure 6 illustrates the layout of such an SMS machine, and the funnel shaped, tapered type of mold that allows for the placement of a SEN between its narrow faces. An alternative approach by Demag/Arvedi uses a straight, tapered, oscillating fixed mold. In both cases, casting velocities ~5 m/min are needed to compete with the highly productive conventional casters typically running at 1.2–1.6 m/min.

In the 1980s, alternate twin-belt moving-mold technologies were researched in Japan and the United States. Work by Sumitomo in collaboration with the Hazelett Strip Casting Corporation of the United States led to the successful casting of 60-mm-thick, thin slab. However, the control of liquid steel delivery proved to be very challenging, as did problems with transverse microcracks and poor upper surfaces owing to entrapped and floated inclusions.

C. Strip Casting

In 2000 A.D., Nippon Steel and Mitsubishi Heavy Industries announced the world's first commercial operation of a twin-roll caster for the production of stainless steel sheet 2–5 mm thick, 0.76–1.3 m wide, at their Hikari Works in Southern Japan. A schematic of their twin-drum, or twin-roll caster, is shown in Fig. 7 [7]. As seen, a conventional ladle and single-strand tundish feeds steel into a liquid sump of steel created by the twin, counterrotating rolls and two side dams. The roll diameters are about 1.2 m, and peripheral roll speeds are on the order of 1 m/sec. A pinch roll gathers the strip and, following minimal hot working, the stainless steel strip is cooled and coiled directly into finished product.

In other ground-breaking news, Castrip (NUCOR/ BHP/IHI) recently announced (*ForeCast*, Vol. II, issue 2, a publication of Castrip, July 2002) the production of

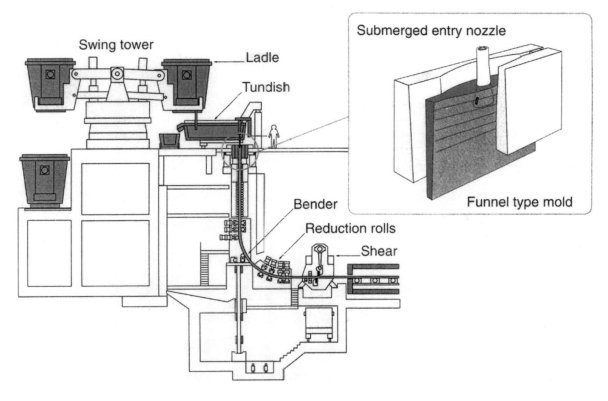

Figure 6 Thin slab caster operation.

the world's first commercial strip cast coils of low-carbon steel, 1.7–1.9 mm thick (Fig. 8). This follows the construction and start-up of a twin-roll strip caster (TRC) for the production of low-carbon sheet products in NUCOR's plant in Crawfordsville, Indiana, U.S.A. Table 1 lists the machine's operational characteristics, and Table 2 lists the Castrip process parameters. As seen, its rolls are only 0.5 m in diameter, running at speeds of 1–2 m/sec. While steel makers still await the promise of direct steelmaking, they may have finally realized Bessemer's visionary intuition, and achieved the long sought-after rationalization of downstream processing of steel.

However, given a fivefold mismatch in the productivity of a twin-roll caster vs. current slab casting operations (e.g., 500,000 vs. 2,500,000 tonnes/year), illustrated in Fig. 7. and thermomechanical considerations to be discussed later, alternative moving-mold processes such as the Hazelett single- and twin-belt casters deserve further attention, given their much higher potential productivities. The higher productivity of such moving-mold machines arises from the arbitrarily long lengths of water-cooled belts that can be incor-

porated in such machine designs vis-a-vis twin-roll casters. A typical Hazelett twin-belt caster used for the continuous casting of thin slabs of aluminum, copper, and zinc alloys is depicted in Fig. 9.

Exploratory experimental work at Clausthal University, Germany, and MEFOS, Sweden, together with physical and mathematical models of novel metal delivery systems at McGill University, all point to the validity of a single-belt machine for the casting of steel at 100–200 tonnes/hr/m width, at strand thicknesses 7–12 mm, and speeds of 1–2 m/sec [8].

The philosophy for developing such a process is somewhat different from that which led to the twin-roll process. First, the process is primarily intended for casting common steel grades (carbon steels). Many of these steels require some hot rolling prior to cold rolling for closing internal porosity. Therefore, the strand thickness is about 10 mm so as to allow for in-line hot rolling with 60% to 80% reduction to attain the normal range of hot band thicknesses (1–3 mm). Second, the process is designed to reach high productivities, which is also in accordance with the requirements for mass steel production.

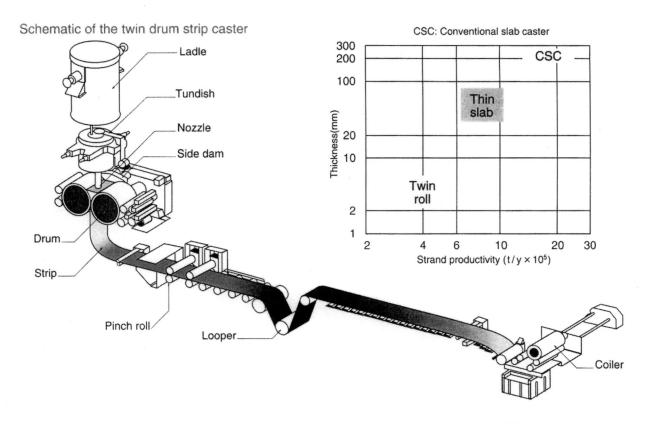

Figure 7 Schematic of the twin-drum strip casting of stainless steel at NSC.

Thus, such a machine could equal, or exceed, the productivity of today's slab casting machines and, by producing 10-mm-thick strip, provide the necessary downstream thermomechanical processing needed to break down the large grains of austenite ($\sim 400\,\mu$m) into 10- to 20-μm-diameter grains of ferrite. This could then match today's steel properties in terms of ultimate tensile strength (UTS) and elongation. Such a pilot-scale Hazelett machine, with a cooled belt length of 2.6 m and a belt width of 400 mm, originally at BHP Australia and now located at the McGill Metals Processing Centre, is shown in Fig. 10, producing 7-mm-thick steel strands at 0.4 m/sec.

D. Summary of Machine Characteristics

Figure 11 presents an overview of the various types of new steel casting machines being developed in terms of their productivity capabilities. To maintain equivalent volumetric flows of steel, one sees that a 2-mm-thick product needs to be running at a casting speed of 70 m/min for 50 tonnes/hr/m width, or 140 m/min for 100 tonnes/hr/m width. At the other extreme, a slab caster

producing slab 200 mm thick at 100 tonnes/hr/m width would only need to run at 1% of that speed, i.e., 1.4 m/min. Since the maximum casting speeds associated with fixed-mold machines are estimated to be on the order of 10 m/min, it is clear that the goal of net shape casting will require a friction-free, moving-mold machine. Table 2 presents a summary of CC products and casting parameters, while Fig. 12 presents an economic assessment of the relative specific investment costs for the installation of thin slab, twin-roll (5–8 mm), twin-roll (1–2 mm), and a direct strip cast (DSC) single horizontal belt caster, producing 5- to 10-mm-thick strands. As seen, beyond a production rate of about 0.9 million tonnes/year, the DSC caster becomes the most economic option of all scenarios.

III. CONTROL OF STEEL QUALITY; LADLE, TUNDISH, AND MOLD METALLURGY

A. Ladle and Tundish Metallurgy

Liquid steel in the ladle, following alloying, deoxidation, and secondary refining, is at its cleanest. The chal-

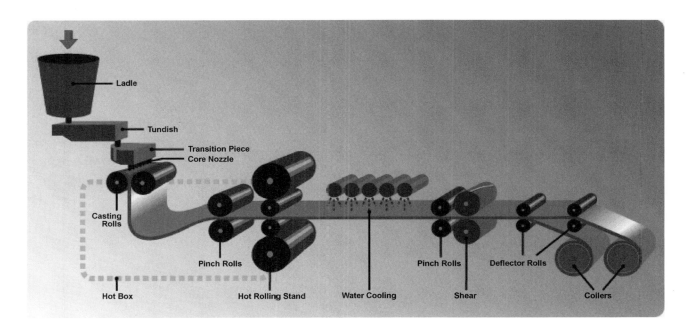

Figure 8 The "Castrip" process.

Table 1 Characteristics of NUCOR's Twin-Roll Strip Casting Facility

No.	Characteristic	Specification
1	Ladle turret caster	Twin arm turret
2	Heat size	110 tonnes
3	Mode of operation	4 Heat sequence casting
4	Caster type	Twin-roll, 500-mm diameter
5	Casting speed	80 m/min (typical), 150 m/min (maximum)
6	Product strip thickness	0.7 to 2.1 mm
7	Product strip width	2000 mm maximum
8	Coil mass	25 tonnes
9	Ladle size	110 tonnes
10	Tundish	23 tonnes
11	Tundish flow control	Slide gate
12	In line reduction mill	
13	Mill stand	Single stand—4 high with hydraulic AGC
14	Work roll dimensions	475 × 2050 mm
15	Backup roll dimensions	1550 × 2050 mm
16	Rolling force	30 MN maximum
17	Main drive	3500 kW
18	Product capability	300–1000 ktonnes/year

All dimensions and tons are metric.

lenge for the steelmaker is to maintain this quality, in terms of chemistry, inclusion levels, and temperature uniformity, during the emptying of the steel through the tundish and into the mold. Figure 13 illustrates many of the potential sources of contamination of purified steel while passing from the ladle into the tundish and thence into the mold. As illustrated, while bubbling of gas within a ladle using submerged gas injection helps to remove the indigenous inclusions generated by steel deoxidation procedures into the upper slag layer, excessive bubbling can disrupt the

Table 2 Continuous Casting Products and Process Parameters

Parameters	Thick slab	Thin slab	Strip casting
Cast thickness (mm)	240	60	2
Casting speed (m/min)	1 to 2	5	80
Average mold heat flux (MW/m^2)	1.3	2.5	15
Total solidification time (sec)	1070	45	0.15
Average shell cooling rate (°C/sec)	12	50	700

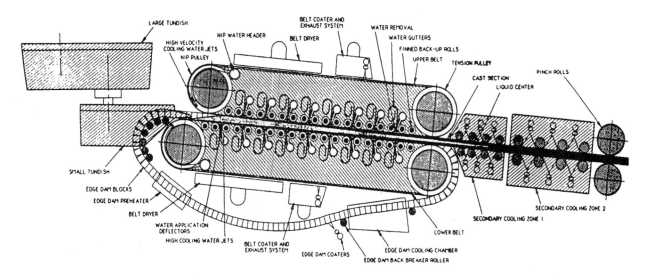

Figure 9 Twin-belt caster of Hazelett Strip Casting Corporation.

Figure 10 McGill–BHP single-belt Hazelett caster.

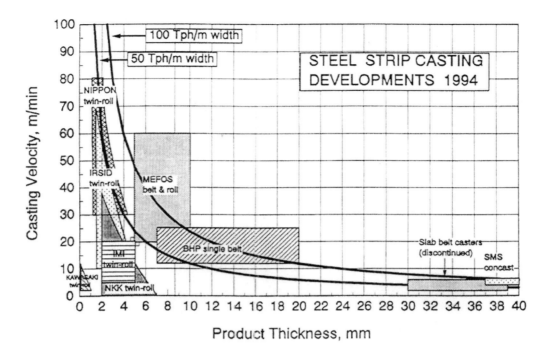

Figure 11 Steel strip casting developments, 1994.

upper slag phase, parts of which can become entrained as microdroplets. These microdroplets of slags can then track through the tundish and enter the final solidified product. In transferring the steel from the ladle into the tundish, a sliding gate nozzle meters steel from the bottom of the teeming ladle via a protective ladle shroud down into a tundish. Given a head of liquid steel ~1.4 m above the level of steel in the tundish, the static pressure in the liquid steel exiting the ladle is practically zero. In the absence of a perfect seal, inhalation of air into the exiting stream of liquid steel requires that the top of the ladle shroud be protected with a positive flow of argon. If not, air infiltration leads to reoxidation and often renitrogenation of the steel with the consequent generation of inclusions.

The tundish, in addition to acting as a metal distributor to two or more casters, serves as a cleansing unit for inclusion removal. Therefore, current practices often use dam and weir combinations so as to modify the flow of steel within the tundish in order to enhance inclusion separation. This has led to a trend toward tundishes with larger volumes and thus residence times for a given throughput (e.g., a 60-tonne tundish with a 7-min residence time is typical for a 320-tonne ladle full of steel). Flow modifiers have no influence on very small inclusions, but they can help clean the steel of midsize inclu-

sions in the 50- to 200-μm (2 to 8 mils) range. For larger inclusions with Stokes rising velocity greater than 5 mm/sec (0.2 in./sec), these flow controls are not needed for the set of operating conditions noted.

Tundishes are normally fitted with insulating covers to conserve heat. For highly deoxidized steels, they are protected with an argon gas cover to reduce reoxidation and inclusion formation. An artificial slag can also be added to absorb those inclusions that are floating out. Large inclusions can have a deleterious effect on the surface quality, paintability, and zinc-coating characteristics of steel sheet. Similarly, as such inclusions (of alumina or manganese silicates, and so on) are rolled out into long stringers, the transverse properties of steel sheet or plate, such as percent elongation and ultimate tensile strength, are severely compromised, as is metal formability. Consequently, the modification of these inclusions into calcium aluminate inclusions, which are refractory at rolling temperatures and retain their original spherical shape following rolling, is much preferred.

For other critical applications, the presence of inclusions with a diameter greater than about 50 μm (2 mils) needs to be prevented. Figure 14 illustrates a break in a steel cord wire used in the fabrication of steel belted tires. The break was caused by the presence of a 50-μm inclusion within the steel. In recent developments by Heraeus-

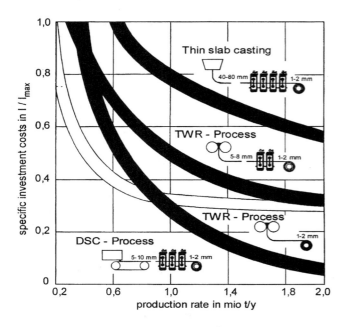

Figure 12 Specific investment costs for various continuous casting processes.

Electronite, the frequency and size distributions of such inclusions can now be monitored in situ using an inclusion sensing probe, illustrated in Fig. 15.

B. Mold Metallurgy

The last opportunity for inclusions to be removed is in the mold. Metal enters the mold of a conventional continuous caster through a submerged entry nozzle (Fig. 16), in which the ports of the nozzle are often angled so as to direct some of the flow from the exiting jets of metal up toward the steel surface. There, a layer of lubricating slag from fused mold powder further assimilates inclusions while it simultaneously protects the steel from reoxidation and provides lubrication between the forming shell of the steel and the surfaces of the oscillating mold.

It is preferred that the final structure of the solid steel be equiaxed rather than columnar, so that cracking of the billet, slab, or bloom during unbending operations is less likely. Precise control of the metal superheat temperature is needed to prevent dendrite tips, broken from the advancing columnar freezing front of steel, from remelting, because these dendrite tips are needed to act as nuclei for grain growth within the remaining melt. Electromagnetic stirring is also used to enhance uniformity of chemistry and structure, and to eliminate centerline segregation of solute-rich material. The cast steel is

then cut, as previously noted, with traveling oxy-torches into slabs, billets, or blooms of appropriate length for further processing. The slabs are about 4 m (13 ft) long, 1 m (3.3 ft) wide, and 100–200 mm (4–8 in.) thick. These slabs are inspected and then charged to a slab-reheating furnace for subsequent hot-rolling operations. Alternatively, in plants with advanced steelmaking practices where slab surface quality is guaranteed to be acceptable (i.e., no scarfing is required), the slabs can be directly charged into the slab-reheat furnace.

IV. FLUID FLOWS, SOLIDIFICATION, AND HEAT TRANSFER IN CONVENTIONAL FIXED-MOLD MACHINES

A. Heat Transfer

The mold and spray chamber of a continuous casting machine must fulfill important thermal requirements if steel is to be cast efficiently with a minimum of internal or external defects. The mold must be capable of extracting sufficient heat from the incoming metal to solidify a shell, which can support the liquid pool at the mold exit, and thereby minimize breakouts. It is also important that the mold be able to remove the heat uniformly so as to avoid the formation of locally thin regions in the shell, which could rupture and lead also to breakouts or surface cracks. To achieve a uniform shell with sufficient thickness for a given set of casting conditions, several mold parameters have to be regulated. The most important of these are the working length, the taper, and the corner radius.

Continuing the heat extraction process begun in the mold (Fig. 16a), the sprays must remove sufficient heat from the steel to virtually complete solidification of the cast section. The rate at which the heat extraction proceeds is critical to the smooth operation of the process, because undercooling can result in excessively long liquid pools and overcooling can lead to the formation of cracks at the bending or straightening rolls. The heat extraction in the sprays must also be arranged so that there is a smooth transition of the surface temperature, with a minimum of reheating, as the steel passes from the mold to the sprays and from the sprays to the radiation cooling zones. Excessive reheating has been shown to be a primary cause of halfway cracks in strand-cast steel. There are two major variables, which can be adjusted to optimize the performance of the sprays: water flux distribution and length of the spray zone. From an operational standpoint, the optimum heat extraction can be achieved by a suitable combination of spray nozzle types, water pressures, and nozzle arrangements.

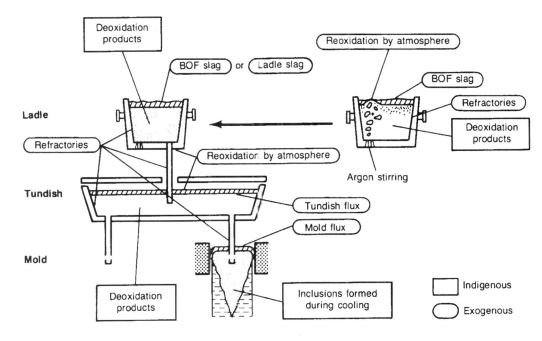

Figure 13   Potential sources of contamination in the continuous casting of steel.

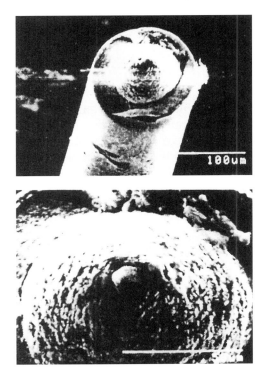

Figure 14   Scanning electron micrograph of inclusion at a break in a steel cord wire.

It is important to realize that in addition to the thermal requirements outlined above, a continuous casting machine has vital mechanical functions. The oscillation of the mold, the support and guidance of the strand with pinch rolls, bending and straightening rolls and, in some cases, cutting of the strand with mechanical shears, must all be carefully designed and controlled if trouble-free casting is to be achieved.

The extraction of heat from the surface of the strand proceeds by different mechanisms in each of the three cooling zones. Of these, the heat extraction processes are most complex in the mold and spray regions. In the mold, heat transfer from the steel surface is influenced markedly by the formation of a gap between the solid shell and slag film and the mold face as illustrated in Fig. 16b. As the width of the gap appears to be very small (~1–2 mm), considerable heat can flow from the shell surface to the mold by conduction through gas in the gap as well as by radiation. A complication in describing the rate of heat flow arises, however, because the gap width does not appear to be constant, but varies in both vertical and horizontal directions over the mold face. Unfortunately, the magnitude of this variation and the resultant effect on the rate of heat extraction are unknown. In the absence of mechanistic data of this kind, efforts have been made to obtain useful heat-flow data

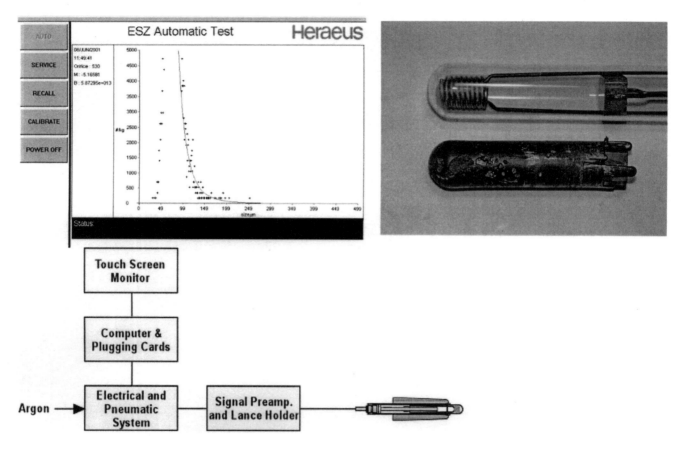

Figure 15 Schematic of the inclusion sensor system showing internal construction probe before and after a sampling sequence and display of results, by Heraeus-Electronite.

in the mold from a heat balance on the mold cooling water. By this method, the average mold heat flux, $\bar{q}_m$, can be obtained for a given set of casting conditions, the most important parameter of which appears to be the mold dwell time, t_m. They are related by the following expression:

$$\bar{q}_m = 64 - 5.3\sqrt{t_m} \quad (\text{cal cm}^{-2}\text{sec}^{-1}\text{sec}) \quad (1)$$

for a wide range of casting conditions [9], as shown in Fig. 17. The solid line in the figure is a plot of this equation, which can be seen to pass through the data obtained from several different casting machines. The instantaneous heat flux, q_m, which varies with time [equivalent to distance/casting speed ($=z/u$)] below the meniscus, is then deduced to be given by:

$$q_m = 64 - 8\sqrt{t} \quad (\text{cal cm}^{-2}\text{sec}^{-1}\text{sec}) \quad (2)$$

It is this expression, which first was measured by Savage and Pritchard [10], that can be used to describe mathematically the rate of heat extraction in the mold.

It is important to note that according to this equation, q_m is a function only of t or distance below meniscus, whereas, as suggested above, it probably should also vary with horizontal position relative to the corners, as well as steel chemistry. To determine such information, this refinement requires that measurements be made of the heat flow distribution over the entire mold face using an array of thermocouples inserted in the mold wall. This approach has led to the use of instrumented molds in many casters that can be used in control strategies to eliminate breakouts. The simplest of these is to slow casting rates if and when the mold surface temperature further down a mold exceeds the mold face temperature immediately above it.

B. Role of Steel Chemistry on Heat Fluxes

Singh and Blazek [11] studied the influence of steel chemistry on steel shell formation through a series of controlled breakout experiments in a bloom caster, and

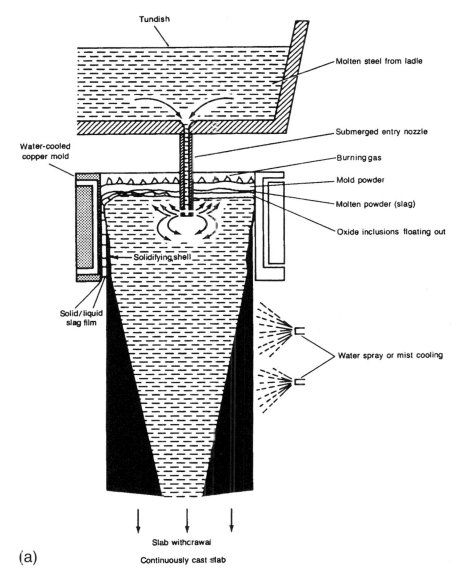

Figure 16 (a) Typical continuous casting operation. (b) Schematic showing role of lubricating film of molten slag, frozen slag shell attached to mold, and formation of air gap lower in the mold.

found the shell profile to be dependent on the steel's carbon content. Figure 18 shows their data, in which the thickness of the forming shells are plotted vs. distance below the meniscus for a 0.1%C and 0.9%C steel, respectively. As seen, the freezing shell forms in an erratic manner, in the case of the low-carbon steel. These observations correlate with wavy wrinkled outer surfaces in the low-carbon steels, which decrease at higher carbon levels, as illustrated in Fig. 19. There the shot-blasted surface of the 0.1%C steel is far more wavy than that for the 0.4%C steel. The reason for

this can be explained in terms of the peritectic reaction, in which the initial shell of steel to form at low carbon levels is in the delta (BCC) phase region. This high-temperature form of iron must then transform into gamma (FCC) iron, which is slightly more dense (0.38 g/cm^3). The consequent contraction leads to buckling of the forming shell away from the mold surface, creating the surface waviness observed. Heat flux measurements by Singh and Blazek, shown in Fig. 20, reveal a drastic drop in mean heat transfer rates in the 0.08–0.14%C range. This drop suggests a large increase in

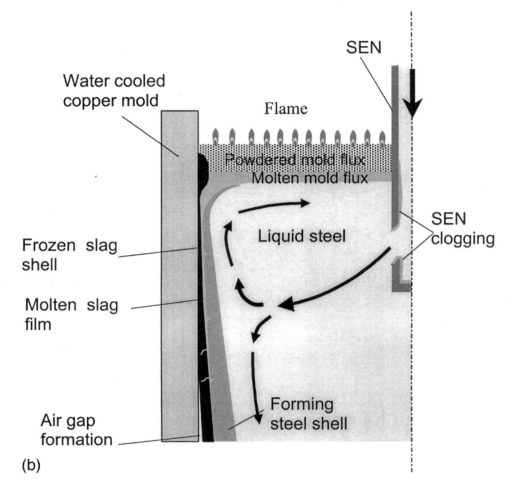

Figure 16 Continued.

interfacial thermal resistance. This would be expected from the poorer contact of the forming steel shell with the surface. As a final comment, peritectic steel grades are well known for their poor castability. A recent review of early work by Dr Keith Brimacombe and coworkers on these topics is presented by I. Samarasekera in the Brimacombe Memorial Symposium [12].

C. Role of Cooling Water Flow Rate and Mold Lubricants

The consistency in the measurements of overall heat transfer rates, such as those shown in Fig. 17 for a wide range of casters, point to the fact that the overriding controlling factor on heat extraction from the solidifying steel is the interface boundary between the steel and the hot face of the copper mold. As such, flow rates

of water through the mold are of secondary influence. Figure 21 shows the effect of various mold lubricants on (1) heat flux distribution down the mold of a billet caster and (2) on hot face copper temperatures. One may note that rapeseed oil led to higher heat fluxes vs. mold powders, particularly in the meniscus region, no doubt on account of the generation of hydrogen during thermal decomposition of the oil.

D. Spray Chamber

In the spray chamber, heat is removed from the strand by fast-moving water droplets, which, having been ejected by the spray nozzle, penetrate the steam film adjacent to the steel surface and evaporate. Like the mold, this process is too complex to be treated theoretically. Instead, the rate of heat extraction, q_s, can best be

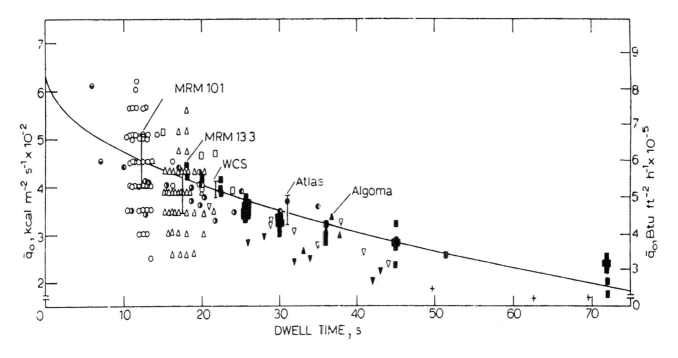

Figure 17 Average mold heat flux as a function of dwell time.

characterized mathematically by a spray heat-transfer coefficient, h_s, as follows:

$$q_s = h_s(T_0 - T_s) \tag{3}$$

where T_0 and T_s are the steel surface temperature and spray water temperature, respectively. Values of h_s must be determined experimentally for the spray conditions—water flux, spray pressure, nozzle type, and steel surface temperature—common to continuous casting. In secondary cooling operations using water alone, primary atomization is used to create a wide-angle spray (up to 120°) at each inter-roll space (spray zone). In water–air mist-spray cooling systems, the cooling water is mixed with compressed air in a mixing chamber ahead of the nozzle, and the mixture emerges as a finely atomized, high-impulse, wide-angled spray. Such sprays are particularly suitable for high-grade steels that are susceptible to cracking. They also offer a very wide volume flow rate control, ~1:12 vs. 1:4 for the water-only type spray nozzles.

E. Coupled Turbulent Flows and Solidification for Continuous Casting of Steel Slabs

In the recent past, many mathematical models of fluid flows within the molds have been reported in the literature, thanks to the advent of computational fluid dynamics (CFD), the ever-increasing processing power of computer systems, and the availability of commercial software systems. Thus, mathematical modeling, coupled with physical models and experiments on operating casters, play an increasingly important role in the improvement of our understanding and control of various aspects of this process. Figure 22a, for instance, illustrates an interesting example of CFD computations carried out by Nippon Steel engineers during the development of directly cast clad steels [13]. By applying a level d.c. magnetic field (LMF) between liquid stainless steel fed to the upper part of a slab caster mold, and less expensive low-carbon steel fed to the lower portion of the mold via two SENs, almost perfect delineation of the two pools is possible; that is, the mixed region of liquid stainless steel and low-carbon steel is constrained to a small region with the LMF on (right side of the diagram), but is badly mixed (see left diagram) without the magnetic field. The first part of the shell to freeze is thus stainless steel, while the inside is an ultra-low-carbon steel, as illustrated in Fig. 22b.

As a simpler example of the use of such mathematical models, a comparison of the spatial distribution of inclusions measured in a full-scale water model are compared with the computational predictions of Fig. 23A in Fig. 23B. One sees that good agreement is achieved,

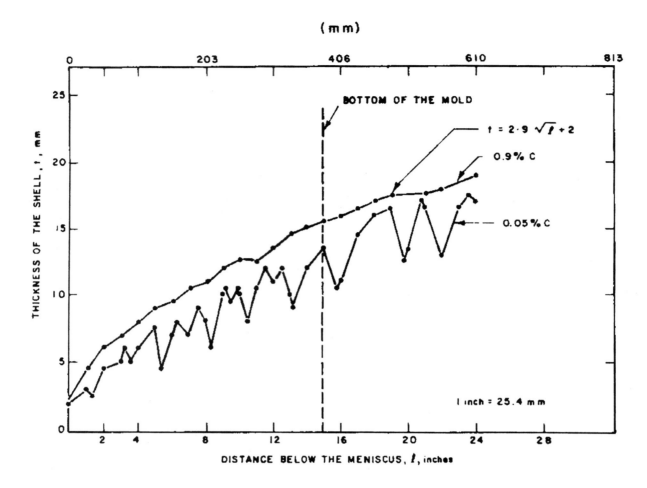

THICKNESS OF THE SHELL IN THE MOLD BELOW THE MENISCUS

Figure 18   Steel shell profiles for 0.05% C and 0.90% C steels.

confirming that strand quality 4 m below the meniscus can vary spatially within a forming slab by as much as 25% for 100-μm inclusions, given a casting speed of 1 m/min. The reason for this relates to the downflow of steel nearer the narrow faces, rendering the float-out of buoyant inclusions more difficult vs. the central regions of the sump, where there is a general upflow of melt back toward the SEN, aiding inclusion float-out.

F. Thermomechanical Behavior of Molds and Slabs

Many of the early problems associated with continuous casting was the problem of crack formation. Unless casting machines are carefully designed and set up with the correct thermal and mechanical management fea-

tures, they are capable of producing material with a rich range of defects! In contrast to the pouring of steel into ingots, where thermal cracking was never an issue, continuous casters can be very efficient crack producers. The variety of cracks in billet casters, for example, range from halfway or midway cracks, longitudinal corner cracks, transverse cracks, and depressions—both mid-face and corner, off-corner internal cracks, pinch roll cracks, diagonal cracks, on through to centerline cracks. Similarly, rhomboid vs. square of billets is an outcome of poor thermal management of the mold cooling water. Figure 24 illustrates the various defects in slabs and billets that can be attributed to poor water-spray management. Early mathematical models by Brimacombe and colleagues [14] helped greatly in explaining the appearance of the midway cracks illustrated in Fig.

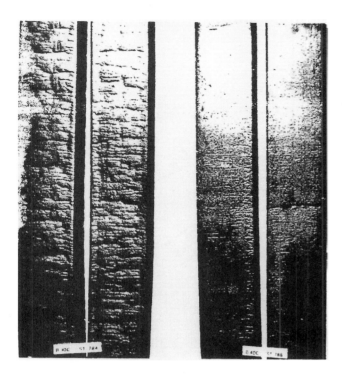

Figure 19 As-cast surface of 0.10% C and 0.40% C steels that have been shot blasted.

24 in slab and billet casting operations. Their appearance within a casting can be ascribed to excessive reheating of the outer surfaces of the billet or slab as it exits the mold into the spray chamber. This reheating leads to a transient expansion of the frozen outer shell of the slab, bloom, or billet, which can place the hot soft steel adjacent to the sump of liquid steel into tension. This tensile force can lead to hot tearing of the steel and result in the generation of midway cracks. Reasons for the various casting defects are now well understood, and can be found in various surveys of the International Iron and Steel Institute (IISI) [15].

V. FLUID FLOWS, SOLIDIFICATION, AND HEAT TRANSFER IN MOVING MOLD MACHINES

A. Twin-Roll Casting

1. Fluid Flows in the Design of Metal Delivery Systems

While process details of metal flow delivery systems for commercial twin-roll casters remain confidential, mathematical and physical modeling of twin-roll casting processes has raised a number of interesting issues as

to how the liquid metal should best be introduced so as to ensure the even development of the solidification fronts across the widths of the two rolls. It has also revealed inherent problems associated with rapid grain growth of austenite as the strip leaves the nip of the rolls. This can lead to relatively poor ferrite/pearlite microstructures with coarse grains.

Figure 25, depicting a half section of the sump of liquid steel and one-roll surface, illustrates the predicted flow patterns that would be developed within a twin-roll caster when a "conventional" bifurcated submerged entry nozzle used in slab casters delivers liquid steel to the sump. As seen, two hot jets of liquid steel pass across the sump of cooler liquid steel to either side dam, losing heat as they travel and slowly mix in with the bulk liquid. They then split up to recirculate back toward the surfaces of the rotating rolls, losing superheat all the way. The net result in terms of solidification is that the shells of steel forming on the rolls will tend to be thicker near the midsection of the rolls, giving rise to the situation shown in Fig. 26, where the fractional amount of solidification is seen to be greater in the central region [16]. This will lead to uneven solidification of the forming shells, giving unstable operations, uneven microstructures and associated variations in strip properties.

Computations have confirmed that the optimum way to effect solidification in an open metal delivery system to a twin-roll caster [17] is to arrange for a low velocity, horizontally directed flow of incoming steel from the entry nozzle to be directed toward the two major menisci regions (or triple points) where the rotating roll surfaces first meet the pool of liquid steel. The use of a vertically slotted nozzle the full width of the rolls represents a reasonable alternative, although surface turbulence and freezing of the meniscus are predicted to be

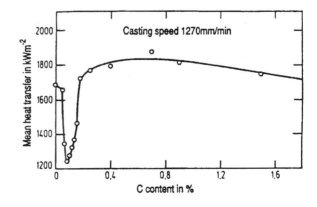

Figure 20 Average mold heat flux as a function of carbon content.

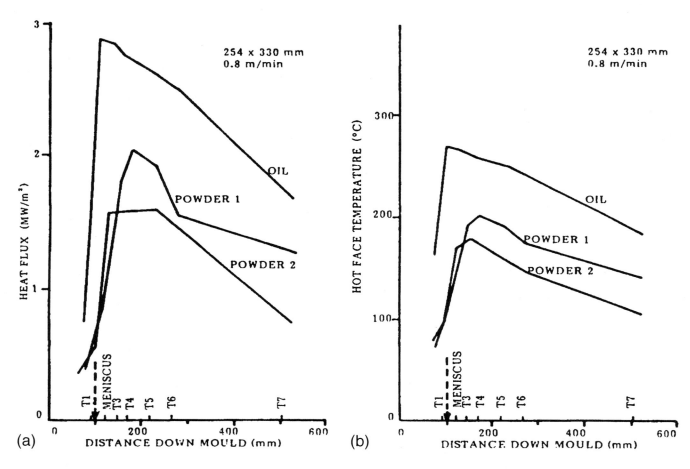

Figure 21   Effect of lubricant on (a) heat fluxes and (b) mold hot face temperature.

more problematic, together with a certain amount of shell remelting lower in the sump (i.e., wedge-shaped pool of liquid steel between the rolls). Figure 27A illustrates the patterns of solidification predicted using the twin-roll version of the CFD Metflo Code. One sees that the uniformity of shell growth is badly compromised by using less wide (and less expensive) vertical slot nozzles (Fig. 27B). A submerged cylindrical vertical nozzle would be even worse, the hot vertical jet of liquid steel remelting the dendritically growing shells on the central regions of the roll surfaces. These modeling efforts are helpful in understanding the finer details of appropriate ways to effect a uniform casting.

2.   Twin-Roll Casting—Heat Transfer

Introduction

The heat transfer literature on twin-roll casting processes is relatively sparse. Chen et al. [18] studied interfacial heat transfer behavior when free-falling streams of liquid metal (wood's alloy) or of hot water impinge at

0.65 m/sec on to a horizontal plate of steel or aluminum, moving at 0.5 m/sec. The authors observed a rapid rise in heat flux to a peak value followed by a sharp drop during solidification and then a slight rise in interfacial heat flux, ascribed to the release of the latent heat of solidification. By contrast, the hot water jet exhibited a maximum heat flux at the point of stagnation or stream impingement, and this then dropped off exponentially with distance from the point of impact.

Strezov and Herbertson [19] have reported on instantaneous heat flux measurements made in a simulator of a TRC in order to study the initial stages of solidification (0–100 msec) of stainless steel onto variously textured cold copper substrates. These copper slabs were set within a sturdy instrumented immersion paddle. This was rapidly plunged into an inductively stirred bath of stainless steel, and then quickly withdrawn for subsequent thermal monitoring and microstructural analysis of the thin shells of steel so produced.

Only one set of thermal data by Tavares et al. [16,20] is currently reported in the literature concerning the

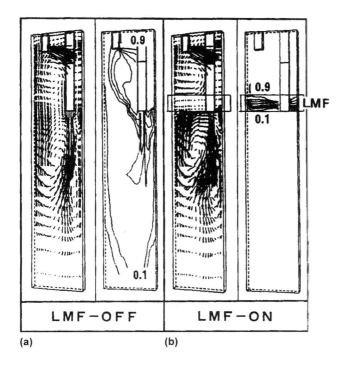

Figure 22 Flow and solute concentration distributions as predicted by MHD model.

from the integrated forms of their instantaneous measurements of $h(t)$ (kW/m^2 K), presented later, fit reasonably well with the data of others, which all roughly fit on the correlation

$$\overline{h_s} = 17.3 V_s^{0.65} \qquad (4)$$

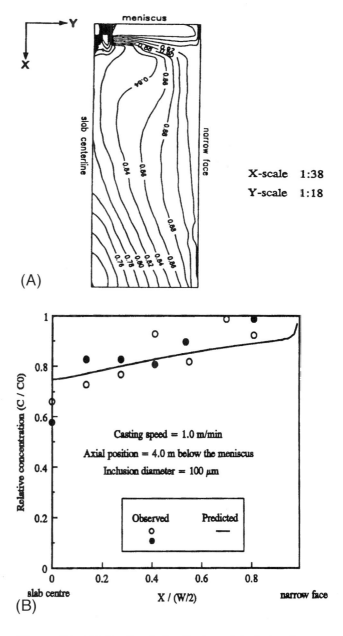

(A)

(B)

Figure 23 (A) Predicted distribution of inclusion isoconcentrations in the upper 4 m of a slab caster. (B) Relative concentration ratios of inclusions-predicted and measured, using full-scale water model of slab caster mold and ESZ probe.

direct measurement of instantaneous heat fluxes to the rolls of a twin-roll caster. This is not surprising, given that such measurements are rather challenging, particularly at high rpm. On the other hand, overall values of heat fluxes and heat transfer coefficients are relatively easily measured by measuring temperature rises in mold cooling water, together with exit strip temperatures from the roll nip.

Average Heat Fluxes to Twin-Roll Caster Systems

Wang and Matthy's recent literature review [21,22] report on various average heat transfer coefficients for thin-strip casting. These are plotted in Fig. 28 as a function of casting speed. Although there is some scatter around the line, one sees that moving-mold processes lead to increasing heat fluxes and heat transfer coefficients, as cooling substrate velocities increase. The scatter around the curve for various machines presumably relate to second-order effects, such as substrate texture, mold material (i.e., copper alloy), protective coatings (e.g., plasma-coated layers of alumina or zirconia, etc.), gas environment, etc.

It is to be noted that the average heat transfer coefficients obtained by Tavares et al. [16,20] determined

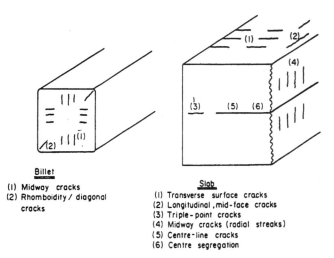

Figure 24 Schematic drawing of spray-related defects found in billets and slabs.

proposed by Wang and Matthys [22]. In this relationship, V_s represents the rolls casting speed (m/sec).

These results show that the productivity of a strip casting machine can be increased the higher the rolling speed, V_s, and suggest that Eq. (2) can give an adequate engineering estimate of the average heat transfer coefficient in near-net-shape casting operations, in the same way as the correlations proposed by Blazek and by Brimacombe for fixed-mold casting operations. However, any accurate evaluations of local cooling rates and temperatures across the strip during the solidification event itself require a knowledge of instantaneous values of interfacial heat fluxes and heat transfer coefficients during the time of contact of metal with the cooling substrate. Also, while the exponent 0.65 in Eq. (2) is similar to the theoretical exponent of 0.5 for perfect thermal contact, it is to be emphasized that perfect thermal contact would have resulted in much higher values, as explained in the following analysis, based on instantaneous measurements of heat fluxes.

Instantaneous Heat Flux Measurements for Twin-Roll Casters

Using embedded thermocouples, and an inverse heat transfer analysis using the methods described by Beck, temperature–time data were used to deduce instantaneous heat fluxes to the roll surfaces during the production of steel strips. Steel strip compositions are listed in Table 3, while the experimental setup is shown in Fig. 29. There a tubular SEN with two horizontal ports in the direction of the side dams was used to deliver liquid steel into the twin-roll caster. In all experiments, the superheat in the molten steel was approximately 25°C.

Figure 30 shows the cyclical responses of the external and internal thermocouples during rotation of the copper rolls. As seen, peak temperatures approaching 160°C were measured by the outermost thermocouple located just beneath the roll surface, and up to 100°C at the inner thermocouple, 12 mm or so below the roll surface. During periods of noncontact with the liquid steel melt, temperatures returned close to cooling water temperatures of some 20°C.

The variation in heat flux with contact time is presented in more detail in Fig. 31. There, following thermal radiation and liquid metal contact, a rapidly increasing heat flux to the roll surface was recorded. A maximum value was reached, usually after one half to two thirds of the metal sump-roll contact time. One will appreciate from Fig. 31 that actual heat fluxes were quite different from those expected on the basis of quasi-perfect thermal contact, for which a maximum heat flux would have been registered at the first moment of contact between the liquid metal and the roll surface.

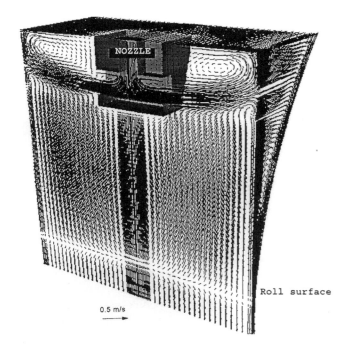

Figure 25 Predicted velocity field of liquid steel within the sump of a twin-roll caster using a conventional tubular nozzle with horizontal ports; casting speed: 0.188m/sec, strip thickness: 4 mm, nozzle penetration: 5 cm.

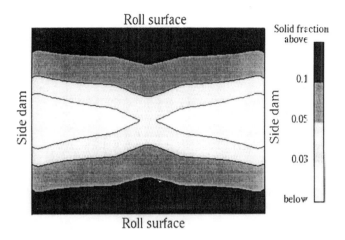

Figure 26 Solid fraction contours 0.1 m below the top surface; superheat: 50°C.

Figure 32 illustrates cyclic variations in heat fuxes during one of the "thin-strip, higher-speed" experiments, corresponding to 4-mm strips and roll speeds of 8 m/min. In general, the peak heat fluxes were higher than those obtained in the experiments of "thick-strip/low-speed" experiments, corresponding to 4 m/min and 6- to 7.5-mm-thick strip. The higher peak heat fluxes were presumably because of the higher casting speeds. Figure 33 presents the transient heat fluxes during one period of roll contact with the freezing shell. Up until about 70% of the contact time, the heat flux variations are similar to those shown in Fig. 31. The mechanism proposed for the experiments of slow-speed group should also apply here. However, for most of the higher-speed experiments, a second peak in the heat flux was also observed during the last 30% of the contact time between the strip and the roll.

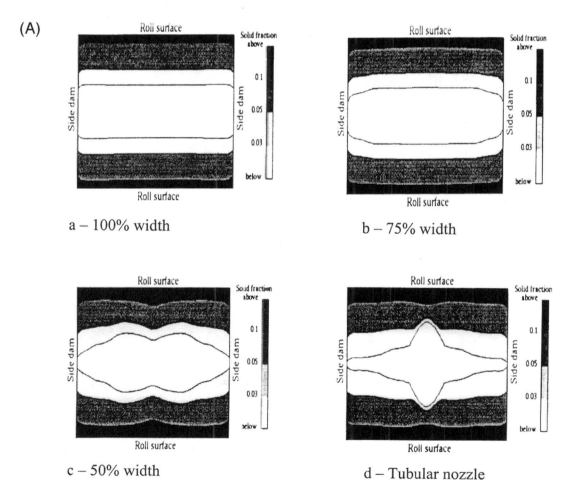

a – 100% width　　　　　　　　b – 75% width

c – 50% width　　　　　　　　d – Tubular nozzle

Figure 27 (A) Solid fraction contours at mid-depth in the sump for different widths of vertical slot nozzles. (B) Velocity fields associated with various slot nozzles for TRC systems.

(B)

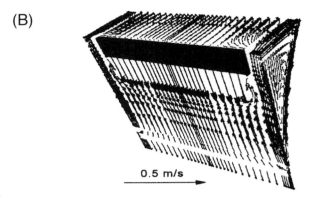

0.5 m/s

e - 75 % of roll width.

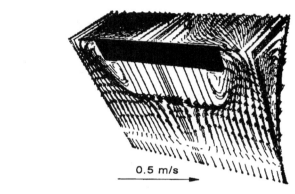

0.5 m/s

f - 50 % of roll width.

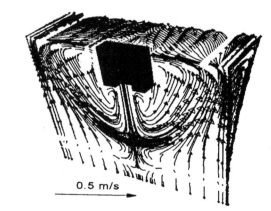

0.5 m/s

g - Tubular nozzle

Figure 27 Continued.

The second peak in the heat flux was probably associated with a more intense interaction of the two dendritic mushy zones formed between the rolls as the solid fraction at the center of the caster continuously increases. At some critical solid fraction, the dendrites form a cohesive network that begins to develop some

strength and to exert pressure against the thin solidified shell, as the distance between the surfaces of the rolls continuously diminishes toward the roll nip. It is believed that the relatively thinner solid shell in the second set of experiments was not able to withstand this internal pressure and was pushed against the surface of the roll, increasing the contact pressure and thereby the heat flux, as illustrated in Fig. 34. Similar to what happens in rolling, the contact pressure then reduces after reaching a maximum.

The fact that a second peak was not detected in the "thick-strip" experiments is probably associated with the lower cooling rate during solidification because of the lower heat fluxes. Thus, in the experiments of low-speed group, the solidified shells are thicker and did not deform to the same extent as the thinner shells associated with the thin-strip set, once the rolling zone was reached. The latent heat of the mushy zone was then absorbed by the solid shells and not transferred to the working rolls. Finally, it should be noted that a constant force of 20 kN was applied to the 400-mm-wide rolls, independent of strip thickness being cast. It is therefore possible that the thinner shells were deformed such that better thermal contact was reestablished with the rolls.

Theoretical Heat Fluxes Based on Perfect and Imperfect Contact

While measured, interfacial heat transfer coefficients were initially low at the melt meniscus, and only gradually increased to a maximum as the freezing shell moved down into the sump and better contact was es-

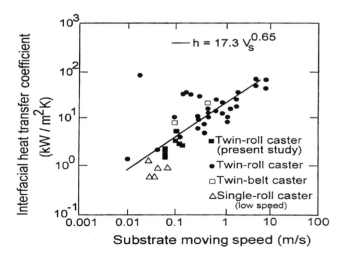

Figure 28 Average interfacial heat transfer coefficient vs. casting speed for twin-roll, single-roll, and twin-belt, casters.

Table 3 Chemical Analyses of the Low-Carbon Steels Used in McGill/EMI Experiments

Element/Steel	C	Mn	Si	P	S	Al	N
A	0.11	0.42	0.087	0.009	0.009	0.0004	0.0056
B	0.127	0.42	0.120	0.012	0.012	0.0030	0.0065
C	0.146	0.38	0.10	0.008	0.005	0.002	0.010

tablished, our time-averaged values coincided well with the mean values obtained by other workers, as illustrated in Fig. 28. However, if perfect thermal contact between the melt and the smooth roll surfaces were to be achieved in twin-roll casting practices, heat fluxes would be given by:

$$q = -k \frac{\partial T}{\partial x}\Big|_{x=0} = \frac{k(T_M - T_S)}{\sqrt{\pi \alpha t}} \tag{5}$$

where q is the interfacial heat flux, T_M the melt temperature, T_S the substrate temperature, and t the contact time between the melt moving down in contact with the rolls. In assessing the relative thermal resistances offered by the thin shell of freezing steel and the copper substrate, it is clear that a significant interfacial thermal resistance must have been responsible for the low heat fluxes.

On the basis of much evidentiary work in the literature, these low heat fluxes can be ascribed to the presence of a gas film. Within a thin gas film, the heat is transferred by conduction because radiation is of secondary importance, i.e., $q = h\Delta T \cong (k/\Delta x)\Delta T$. The

air gap thickness, Δx, is related to the conductivity of the air and the heat transfer coefficient according to:

$$\Delta x = \frac{k_{air}}{h} \tag{6}$$

while the heat flux between the melt and the substrate, in the presence of an interfacial resistance of negligible thermal capacity, becomes:

$$q = \frac{(T_M - T_S)}{(M/k') + (1/h_g) + \sqrt{\pi t/\rho C k}} \tag{7}$$

Accepting Eq. 2 for the mean interfacial heat transfer coefficient ($h_s = 17.3 V_s^{0.65}$) and ascribing this resistance primarily to heat conduction over an air gap of thickness, δ, one can write:

$$h_g = \frac{k_g}{\delta_g} = 17.3 \times 10^3 V_s^{0.65} \quad (\text{W/m}^2\text{K}) \tag{8}$$

or

$$\Delta_g \approx \frac{k_g}{17{,}300 \times V_s^{0.65}} \quad (\text{m}) \tag{9}$$

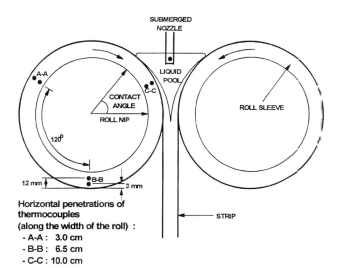

Figure 29 Schematic of an instrumented roll on the IMI twin-roll pilot-scale caster.

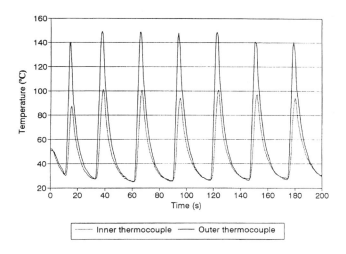

Figure 30 Typical variation of the corrected temperatures during casting of thicker strip ($u = 4$ m/min, strip 6–7.5 mm thick).

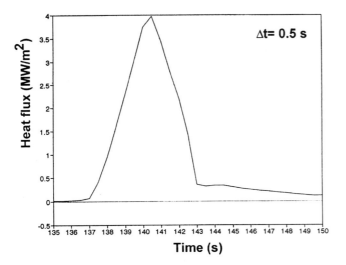

Figure 31 Typical variation in heat fluxes during contact of roll surface with steel in sump of twin-roll caster for thicker strip experiments.

Taking $k_g \sim 0.08$ W m^{-1} K^{-1} at 1000°C, and $V_s \sim 0.1$ m/sec, the mean air (or nitrogen) gap thickness would be 20 μm, but this would be significantly higher at the meniscus, where heat fluxes are much lower than average, according to the present data.

3. Solidification and Strip Microstructures

In conventional continuous casting operations for slabs, blooms, and billets, the as-cast microstructure is completely or almost completely changed during slab

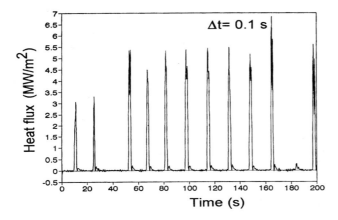

Figure 32 Typical variation in heat fluxes at the roll–melt interface for thinner strip experiments (u = 7.5 m/min, strip 3.5 mm).

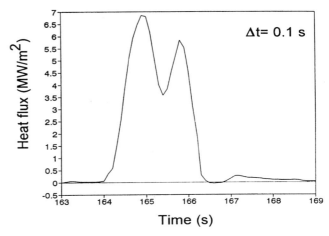

Figure 33 Variation in heat fluxes during the contact time—thinner strip.

reheating operations, where the structure transforms to FCC austenite prior to hot-rolling operations. However, in strip casting processes, the as-cast structure can be of major significance in influencing the nature of the final microstructures and associated properties of the steel. As such, final microstructures will be affected by instantaneous heat fluxes at the roll–strip interfaces. As with heat fluxes, little quantitative work has been published in the open literature regarding the as-cast structure of low-carbon steels strips produced by twin-roll casting. The investigation of Shiang and Wray [23] was mainly devoted to an analysis of how strip microstructures respond to different thermal processing operations, while Ueshima et al. [24] studied the effect of MnS precipitation during the strip's solidification, on ferrite grain size distribution. None of these works correlated the characteristics of the microstructures, particularly grain sizes, to the corresponding rates of heat extraction and to caster conditions.

Figure 35 shows a transverse section of the dendritic structure of 6-mm-thick strip, cut from a hypoeutectoid steel, cast at slow speed, 4 m/min, in the Bessemer twin-roll caster at IMI Boucherville. The primary solidification, or dendritic, structure was revealed using Oberhoffer's reagent. Basically, this etching solution reveals the different levels of microsegregation. The areas richer in iron (dendrites) appear black, while regions with higher concentrations of impurities (interdendritic spaces) appear white. The structure is essentially columnar, with the radial growth of columnar dendrites meeting approximately at the center of the strip, and angled at about 15° in the direction of casting. Signs of interaction of the columnar dendrites

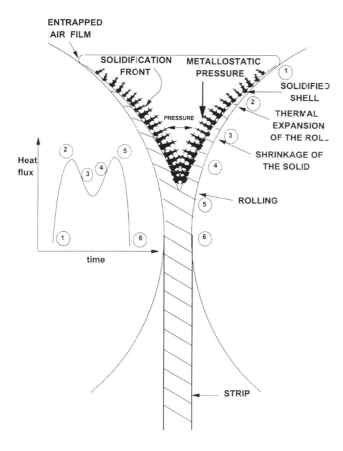

Figure 34 Mechanism to explain transient heat flux variations in thinner-strip experiments associated with higher casting speeds.

coming from the two surfaces of the strip can be seen; various broken and bent dendrites are apparent, as indicated by the arrows. This is more likely to occur in situations of relatively large interdendritic spaces, associated with the lower cooling rates pertaining to the thick-strip experiments.

A transverse section of the dendritic structure of a strip of 3.5-mm-thick, thin strip produced at the higher casting speeds, ~7.5 m/min, is illustrated in Fig. 36. Again, columnar dendrites are formed close to the surfaces of the strip. The primary dendrite spacings here are much smaller, in agreement with the higher cooling rates (higher heat fluxes) for the higher-speed experiments. It is also seen, in contrast to Fig. 35, that the columnar dendrites do not reach the center of the strip. Previous investigations [25,26] also reported on an equiaxed zone in the center of twin-roll cast strips This was attributed to incomplete solidification of the strip at the roll gap. In the work of Tavares et al., simulations of heat transfer and solidification [27], using average heat fluxes determined experimentally, revealed that the strips were usually fully solid just above the roll gap. It is also seen in Fig. 36 that the noncolumnar zone is darker than the surrounding areas. Considering that the Oberhoffer's reagent makes the regions richer in impurities appear white, this suggests that the noncolumnar zone was depleted of these impurities, possibly because of squeezing of the interdendritic liquid back into the sump. This evidence is consistent with the reverse flows of steel that are predicted to occur as the two shells of

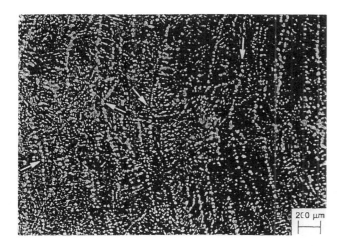

Figure 35 Typical dendritic structure revealed in the transverse section of a strip (6 mm thick, casting speed 4 m/min), using Oberhoffer's etchant.

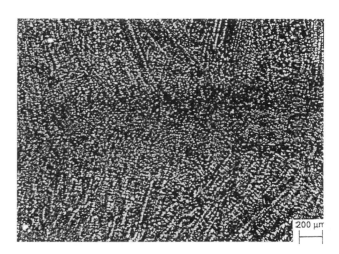

Figure 36 Typical dendritic structure in the transverse section of a thinner strip (3.5 mm thick, casting speed 7.5 m/min), using Oberhoffer's etchant.

steel converge as they approach the roll nip. Excess entrained steel in the roll boundary layers are returned back up toward the surface of the steel (see Figs. 25 and 27B). This suggests that a more intense interaction of dendrites, together with a certain amount of hot deformation of the strip before leaving the rolls, were responsible for the equiaxed zone.

Figure 37 shows the as-cast structure across the thickness of one strip produced in a thick-strip experiment as revealed by a nital etching solution. This section was taken from a region located approximately halfway between the center and the edge of the strip (along its width). As seen, the room-temperature structure was mainly composed of ferrite and pearlite. The morphology of the ferrite in this sample is usually called a Widmanstätten structure [28,29], which is a microstructure produced when the prior austenite grains are large and the cooling rates during transformation from austenite to ferrite and pearlite are relatively high.

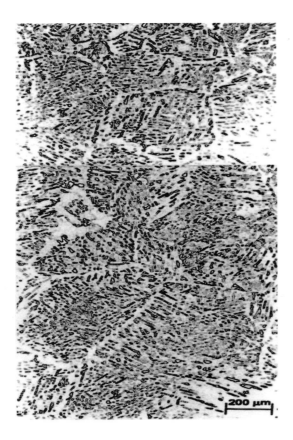

Figure 37 As-cast structure in a transverse section of a strip produced (6 mm thick, casting speed 4 m/min). Nital etch reveals grain boundaries of prior austenite and Widmanstatten type microstructure.

The Widmanstätten plates nucleate at the austenite grain boundaries and grow along well-defined matrix planes [29]. They grow either from the grain boundaries or from preexisting ferrite allotriomorphs. Nucleation can also occur inside the grains, at the austenite twin planes. This kind of nucleation pattern, mainly at the grain boundaries, allows one to approximately identify the sizes and shapes of austenite grains prior to their transformation. These are delineated by the white network of ferrite. This feature was used to estimate the prior austenite grain sizes, which can be identified through inspection of Fig. 37. The prior austenite grains are seen to be large, especially near the strip's surface (bottom of photograph), where they show a tendency to columnar shapes.

Mathematical simulations of heat transfer, solidification, and subsequent grain growth across the strip thickness [30] indicate that the surface region experiences a significant reheating after leaving the roll gap, in much the same way as conventional casting (bloom back). This reheating also favors a more pronounced grain growth of austenite in the surface regions of the strip.

Figure 38 presents predictions for the variation of austenite grain size, for the thick-strip/slow-speed group, as a function of distance below meniscus, for three positions across the strip's thickness. As seen, three different equations were used for calculations of austenite grain growth. Equation (10), obtained from the solidification studies of Yasumoto et al. [31], predicts a more pronounced and faster grain growth than those given by the equation proposed by Maehara et al. [32], and by the equation proposed by Ikawa et al. [33].

$$D_Y^3 - D_{Yo}^3 = 3.434 \times 10^{34} \sum_{i=1}^{n} \Delta t_i \exp\left(\frac{-135767.5}{T_i}\right) \quad (10)$$

To verify these predictions, a few approximate measurements of austenite grain sizes were made. The prior austenite grains presented very irregular shapes. Two dimensions were then used to specify their sizes (Fig. 39), a major and a minor length, and the following average values were obtained:

Major length: 780 μm (standard deviation: 328 μm, $n = 10$)

Minor length: 453 μm (standard deviation: 117 μm, $n = 10$)

These results are mainly for the surface region. In the center of the strip, the austenite grains that were identified present a more equiaxed shape, with sizes ranging from 130 to 360 μm.

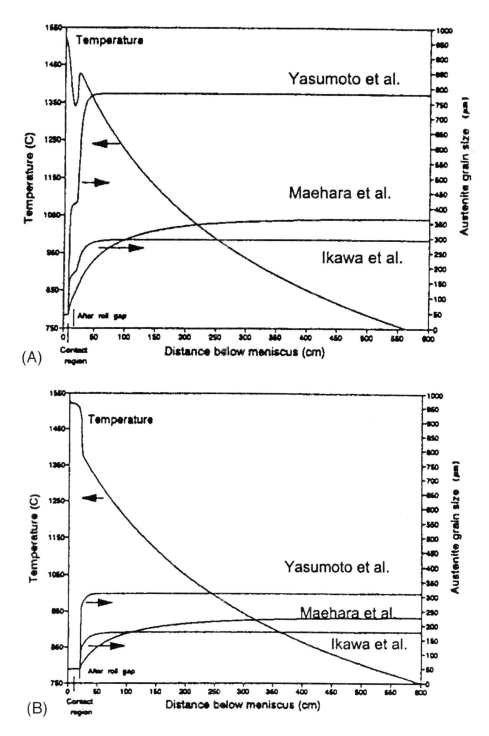

Figure 38 Predictions of austenite grains sizes and strip temperatures as a function of distance below meniscus of IMI TRC, for 6-mm-thick strip. (A) Strip surface, (B) strip center.

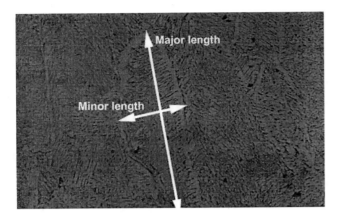

Figure 39 Dimensions used to define the prior austenite grain sizes.

Considering that the austenite grains form initially surrounding a primary dendrite arm, it seems reasonable to associate the major length of the grain to the length of the primary dendrite arm. The austenite grain then grows perpendicularly to this primary arm. This would lead to more reduction in the specific area of grain boundaries, which is the main driving force for grain growth in the solid state. Therefore, the predictions of the models for austenite grain size should be compared to the minor length, which is basically the direction of growth. In this case, the equation proposed by Maehara et al. used in conjunction with thermal history predictions following the strip's exit from the rolls gives results in best agreement with measurements.

The predictions for prior-austenite grain size for the sample strips of the thin-strip group are presented in Fig. 40. Grain sizes are significantly smaller than those of Fig. 38, presumably on account of faster cooling rates (thinner strip and higher casting speed) giving less time at high temperature for gamma grain growth as the strip leaves the nip. The measurements that were made close to the surface gave the following results:

Major length: 205 μm (standard deviation: 36 μm, $n = 5$)
Minor length: 168 μm (standard deviation: 27 μm, $n = 5$)

In the center of the sample, the equiaxed grains had a size of approximately 120 μm. Again the equation proposed by Maehara et al. gave the best predictions for grain size.

The variation in grain size across the strip's thickness can be analyzed in terms of previous work [31,33] that demonstrate that the austenite grains grow only in the temperature range where austenite is the single phase present. The surface of the strip reaches this temperature range before the center does. As a consequence, the grains there have more time to grow. As such, if cast strips are to be hot rolled directly, without cooling to ambient temperature for subsequent reheating, it seems important to reduce the austenite grain growth during the casting operation and to adjust the postprocessing route in order to obtain products with mechanical properties within specifications. If not, typical UTS and YS values of 450 and 310 MPa for hot-strip products can be predicted to increase to 540 and 370 MPa, and maximum elongations of 40% to drop to elongations of no more than 20%, on the basis of these simulations.

B. Twin-Belt Casters

Finally, let us turn our attention to belt casting operations, and their potential for development as machines for high-productivity, near-net-shape steel casting operations. Some 15 years ago, Hazelett Strip Casting Corporation provided twin-belt machines to study the production of thin slab at Sumitomo Metals Inc. and Gary works USX in connection with a USX–Bethlehem Steel–DOE research partnership. Metal delivery in both cases proved to be exceptionally difficult. A double-tundish control system was successfully developed at SMI in order to pour steel at a closely controlled rate into the gap between the (off-set) belts of an inclined belt caster, while a large slot nozzle was used at USX. Problems with freezing steel in the gaps between the nozzle and the cold belts in the USX case led to the development of frozen skulls attached to the nozzle's tips, which quickly led to their destruction, as well as unacceptable strip. For Sumitomo, it was found that transverse hairline cracks could be minimized by running the pinch rolls at slightly slower speed than the belts to reduce tensile forces on the growing shells of steel. These casters were abandoned in favor of the SMS thin-slab, caster-type designs. Other problems arose on account of inclusion float out, compromising near top surface quality.

Nonetheless, the productivity potential for such belt casters for thick-strip production still remains, in that the length of the belt needed for cooling the steel can presumably be extended practically without limit, in comparison to the massive roll diameter scenario for twin-roll casters producing 10-mm-thick strip. It is for these reasons that researchers at MEFOS, Clausthal University, and McGill University have all favored the single horizontal belt caster for the production of steel strip. Based on the most recent work at MEFOS [34,35],

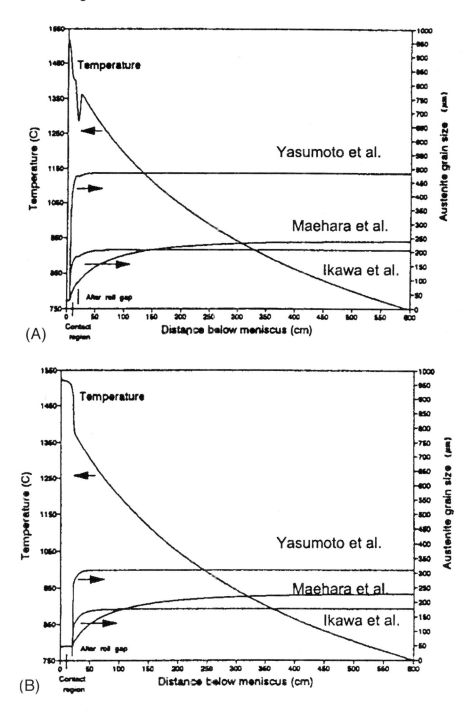

Figure 40 Predictions of austenite grains sizes as a function of distance below meniscus in IMI twin-roll caster, for a 3 5-mm-thick strip. (A) Strip surface, (B) Strip center.

single-belt casting still remains a promising technology, suitable for high-productivity demands in the mass production of low-carbon steels.

C. Single Horizontal Belt Casting Processes

1. Introduction

Thus, in the single-belt process [also called the direct-strip-casting (DSC) process] the liquid steel is fed from the supply system onto a moving belt ~1 mm thick, made from steel or copper sheet, which is cooled from below by water emerging from nozzles located in the cooling chamber. Thick strip is produced with thicknesses between 7 and 15 mm.

Compared to the twin-roll process, developments have not yet led to production-scale installations. Table 4 gives data on the known casters. The most advanced one is that operated at the Clausthal Technical University in cooperation with Salzgitter AG, SMS Demag AG, and Thyssen Krupp Stahl AG. This caster (Clausthal II in Table 4) is connected to a rolling stand for in-line hot rolling. The sketch of the installation is shown in Fig. 41 [36,37]. It has been demonstrated in Clausthal that casting and hot rolling can be coupled without major difficulties [38]. The MEFOS caster is somewhat larger, and heat sizes considerably greater, allowing wider strip to be cast. The McGill caster originally operated at BHP, Newcastle, is a research machine used for exploring high-speed, thin-strip casting of both ferrous and nonferrous strip material.

2. Fluid Flows; Design of Metal Delivery Systems

An obvious requirement of the flow control system to such casters is that liquid steel must emerge from a dispenser at the same mass rate with which it is withdrawn by the traveling belt. At the same time, uniform solidification requires that lateral flows be minimal. Ideally, the melt should be delivered to the belt in the form of a flat stream, the same width as that of the strand. Also ideally, the metal stream should emerge from the feeding nozzle with a bulk velocity as close as possible to that of the belt (isokinetically) and with a thickness as close as possible to that of the final strand, i.e., no hydraulic jumps. Unfortunately, it is very difficult to fully realize these additional requirements at low belt velocities. For such cases, the effective head, or height difference Δh between the level of steel in the dispenser and its level in the strand level has to be made very small. Furthermore, this low head must be kept constant if the exit velocity of steel from the delivery system v is to equal the casting velocity v_c. Since this has not yet been properly resolved in practice, the effective head, Δh, has been larger than that leading to isokinetic conditions. Consequently, in the Clausthal experiments, v has been larger than v_c, and the flat stream at the nozzle exit thinner than the strand thickness and faster than the belt speed. This mismatch has caused hydraulic jumps with strong turbulence where the steel stream impinges on the belt. Another problem in designing an appropriate metal delivery system is that thin flat streams with an upper free surface do not have shape or form stability; rather, they tend to contract as a result of surface tension forces to form a narrower, thicker stream with surface waves [36]. If the thin stream is constrained to remain flat (and wide) by using a closed channel with upper and lower confinement (slit nozzle), local freezing may occur easily in the narrow slit gap and cause clogging.

The general equation relating the streaming velocity v of the metal at the nozzle end to the height difference

Table 4 Data on Available Single-Belt Casting Installations

	Clausthal I	Clausthal II (pilot)	MEFOS	McGill
Cooled belt length, mm	1500	3200	3800	2600
Belt width, mm	300	500	1050	400
Belt material	Low-C steel	Low-C steel	Copper, low-C steel	Copper, low-C steel
Strand thickness, mm	10 to 15	10 to 15	10 to 12	10 to 15
Cast strand width, mm	150	300	450, 870	200
Realized velocity, cm/sec (at 10 mm)	8	18	29	15
Furnace capacity, kg	120	1300	4000	250
In-line rolling	No	Yes	No	No
Roll diameter, mm	—	300	—	—
Roll width, mm	—	450	—	—

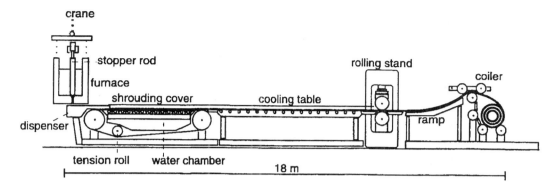

Figure 41 Pilot plant installation for the DSC single-belt casting/hot-rolling process at Clausthal Technical University.

Δh between the meniscus in the dispenser and the strand level is

$$v = \sqrt{\frac{2}{1+C}\left(g\Delta h + \frac{\Delta p}{\delta}\right)} \qquad (11)$$

where g is the gravitational constant, Δp the pressure difference between interior and exit of the dispenser, ρ the density of the liquid metal, and C an effective friction factor. Furthermore, the continuity equation is valid in the form

$$v_c A_c = vA \qquad (12)$$

where A_c and A are the cross-sectional areas of the strand and of the feeding stream, respectively.

Many types of feeding system have been proposed and explored experimentally [17,39–42]. Two methods developed at MEFOS and at Clausthal subdivide the feeding stream into several streams so as to make the stream area A smaller than the strand area A_c. This allows one to use a larger exit velocity v from the dispenser, with a concomitant increase in Δh to a more practical level. This general principle has been applied in the so-called "zig-zag nozzle" and the "multihole nozzle." Unfortunately, such subdivisions of the flow into several streams leads to nonuniform solidification, for similar reasons to those for TRC using standard SENs. Similarly, it can lead to feeding, or meniscus marks on the lower strand surface and/or stripes of decreased strand thickness, or even nozzle clogging.

Two dispenser types involving single flat streams ($A_c = A$) are depicted in Figs. 42 and 43. In the kind used at Clausthal Technical University (Fig. 42), the metal stream flows over a refractory weir substrate with two steps. Two so-called argon rakes are used to prevent contraction of the open stream on the weir and distribute it laterally on to the belt so that the edges are filled. The

argon rakes consist of a series of argon streams emerging from bores in the gas distributors. The distributors extend over the width of the strip, are water cooled, and are made of copper.

In the dispenser proposed by the McGill group (Fig. 43), the feeding occurs through a reticulated medium (e.g., open-pore zirconia filter) at the bottom of the dispenser chamber. The high Δh values (~80 mm) are isokinetically matched to belt speeds ~1.2 m/sec. At lower metal production rates and belt speeds associated with pilot-scale casters producing 7-mm-thick steel strip, flow constrictions are required to drop the effective head, Δh, in these extended metal delivery systems.

While the head loss due to such flow modifiers are minimal, their role in promoting an even, pressurized, vertical flow of slowly moving liquid steel over an extended region onto the chilled moving belt below is

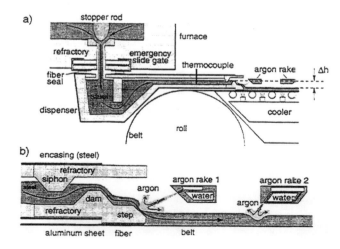

Figure 42 Feeding system with argon rakes used on the Clausthal caster II.

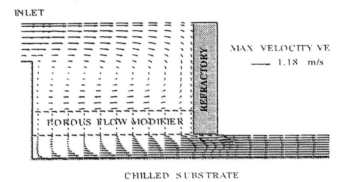

INLET

MAX VELOCITY VE

———— 1.18 m/s

REFRACTORY

POROUS FLOW MODIFIER

CHILLED SUBSTRATE

Figure 43 Predicted velocity fields in an extended pool of liquid steel placed over a water-cooled, horizontal belt casting machine using a porous flow modifier. Belt speed 1.18 m/sec, porous filter 35 ppi.

illustrated by considering their absence. Figure 44 shows the short circuiting of metal flow to the exit slot without flow control. The extended nozzle configuration is thought to provide the optimum means for delivering steel under pressure without significant turbulence in a chemically controlled environment to the moving mold below, provided it can be engineered to avoid freezing at the back wall. The need for a flow modifier is evident when comparing Fig. 43 with Fig. 44. The "no-flow modifier" case leads to recirculation, regions of high turbulence, freezing at the back wall, as well as remelting of the frozen moving shell due to short circuiting of the

steel flow to the exit. Computations show that belt lengths on the order of 10–15 m are required to produce 10-mm-thick strip at belt speeds of 1 to 2 m/sec. This advanced delivery system has yet to be tested on a belt caster. One prerequisite for the proper functioning of such a delivery system would be very clean steel, or steels with liquid inclusions (e.g., silicon-killed grades), such that clogging of the flow modifiers as a result of a build-up in solid rafts of inclusions does not occur.

Another way to enable a larger height difference Δh to be achieved is by the application of underpressure. This is achieved in the dispenser designed by Mannesmann Demag (SMS Demag). The dispenser has two chambers and a vacuum installation. In one chamber, which is open, the steel is added, and from the other chamber, which is closed and connected to the vacuum pump, the steel is fed via an inclined channel to the belt. The idea is to compensate the high Δh between the meniscus in the second chamber and the strand that is necessary for constructional reasons by the underpressure Δp (Eq. 11). This system operated well in water modeling trials but proved, in practice, to be very difficult, to start up.

3. Heat Withdrawal

At the top surface of the strand, cooling is by radiation into the free gaseous space and by natural convection, and at the bottom by heat transfer through the belt to the cooling water. The heat withdrawal at the top surface is considerably slower than at the bottom sur-

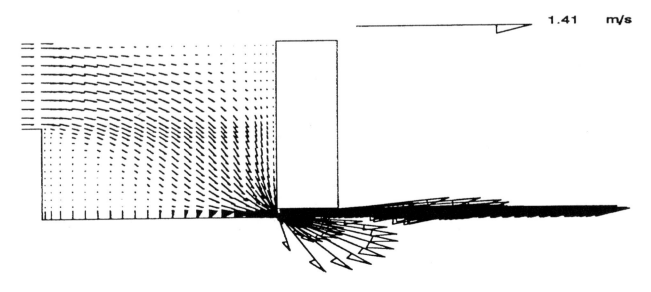

1.41   m/s

Figure 44 Predicted velocity fields in an extended pool of liquid steel placed over a water-cooled horizontal belt casting machine with no flow modification, exit slot, $d = 4$ mm, meniscus level in extended nozzle $= 80$ mm.

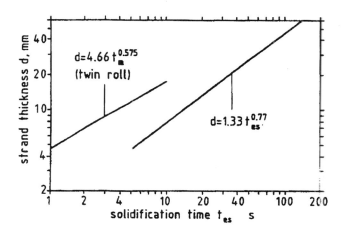

Figure 45 Relationship between strand thickness and total solidification time for the single-belt process. Steel grade ST 37.

face. Because of this difference, the temperature field in the strand is asymmetrical with respect to its central horizontal plane. On the water side of the belt, temperatures must be below the burnout point ($\sim 100\,^\circ$C). These requirements are met with a minimum heat transfer coefficient, h, of 50 kW m^{-2} K^{-1} for the water. Radiation cooling at the top surface depends strongly on the state of the surface.

Data for the heat flux density and heat transfer coefficient were determined in model experiments using a laboratory setup. The relationship between strand thickness d and total solidification time t_{es} was computed from these data and is presented in Fig. 45 plotted in the usual manner as d vs. t_{es}. The function is given as:

$$d = 1.33 t_{es}^{0.77} \tag{13}$$

where d is in millimeters and t_{es}, in seconds. The corresponding relationship for the twin-roll process is in-

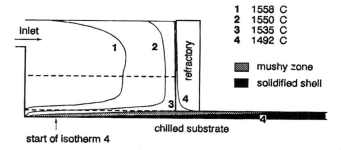

Figure 47 Isotherms for exit gap, $d = 4$ mm, with flow modifier.

cluded in Fig. 45 for comparison. As seen, the time for complete solidification in the single-belt process is longer than in the twin-roll process at any given strand thickness. This is the result of the slow heat transfer at the upper free surface and the generally lower rate of heat transfer to the belt, because of the generally lower contact pressures in the single-belt process than at the rolls in the twin-roll process. Figures 46 and 47 illustrate the predicted operation of such a caster using the extended nozzle concept. It demonstrates how the initial line of solidification can be separated from the back wall of the delivery system, provided the flows, velocities, and superheats of the entering steel are correctly optimized.

VI. THE PROPERTIES AND MICROSTRUCTURES OF STEEL SHEET PRODUCTS

A. Introduction

There are four methods for increasing the hardness of steel materials: (1) solid-solution hardening, (2) precipitation hardening, (3) work hardening, and (4) hardening by grain refinement. Solid solution hardening, precipitation hardening, and work hardening all decrease elongation and toughness, although strength is increased. By contrast, hardening by grain refinement does not have this undesirable effect. Modern sheet steel products aim at producing fine grains, which correlate with strength according to the following relationship, attributed to Hall–Petch:

$$\sigma_{0.2} = 0.15 + 0.017 d^{-0.5} \tag{14}$$

for the stainless steel sheet product shown in Fig. 48. As seen, such correlations hold down to grain sizes in the

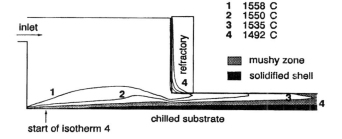

Figure 46 Predicted isotherms in extended delivery system using an h_{tf} varying with position along chilled substrate.

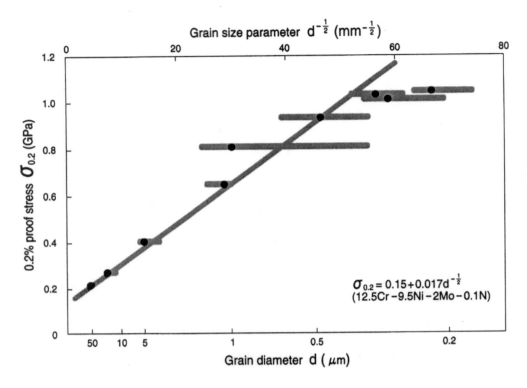

Figure 48 Hardening by grain refinement—Hall–Petch relationship.

order of 1 μm, as shown by the scale on the lower axis of the graph, for grain diameter. The increase in strength then lowers off because subgrains with low-angle grain boundaries contribute less to strengthening than do normal grains.

The grain sizes in conventional hot-rolled structural steels are typically about 20 μm. These can be reduced by controlling hot-rolling conditions and water-cooling conditions after hot rolling, and the grains can be further refined to 5–1 μm by quench-tempering procedures. While strength is the most important factor for a structural steel material, allowing less material to be used, formability is equally important for bend and press forming and deep drawing applications. Similarly, good fatigue properties and resistance to corrosion can also be critical factors in determining the choice of specific steels for a given application.

Figure 49 shows the formability–strength relationship for conventional steels, where the formability, plotted on the ordinate, is plotted vs. strength in megapascals. As seen, for parts needing good formability, such as the inner and outer panels of a car body, only low-strength sheets of low-carbon steel and ultra-low-carbon steels can be used, because there is an inverse

correlation between strength and elongation (and formability), as seen from the graph. For parts not requiring high formability, high-tensile-strength steels are an appropriate choice. The theoretical maximum strength of steel is about 22,000 MPa in contrast to structural steels, which vary from 400 to 800 MPa. Iron whiskers approach this UTS, reaching ~12,000 MPa. The development of materials that have both high strength and high formability has contributed substantially to the reductions in car body weights that have been made possible thereby.

B. Comparison of Conventional and Strip Cast Products and Properties

1. Twin-Roll Cast Strip

A comparison of the typical properties of a silicon-killed, low-carbon steel strip obtained from strip casting and an aluminum-killed steel from the conventional hot strip mill route [43] is presented in Fig. 50. Although the strength levels are slightly higher and elongation levels lower for the TRC product, it is claimed that overall material properties of the as-cast TRC strip compare favorably with steel strip produced via the hot strip mill

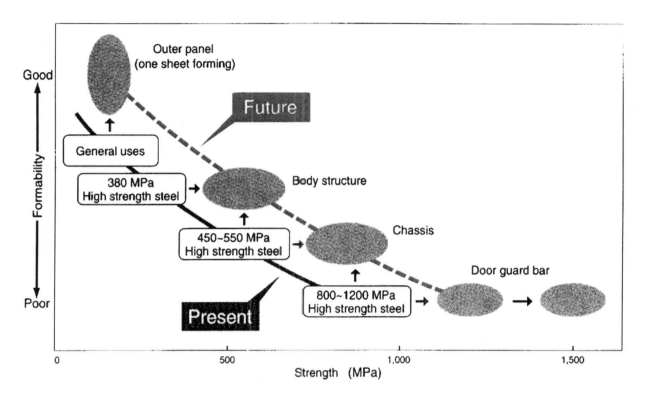

Figure 49 Strengthening of steel for reducing weight of automobile.

route. Nonetheless, the authors note that there are significant differences in the final microstructures of the strip produced by the two routes as illustrated in Fig. 51. Key differences between the two microstructures are summarized in Table 5. Thus, conventional hot strip mill products exhibit a fine equiaxed ferrite microstructure while twin-roll cast (TRC) strip microstructures

are predominantly a mixture of coarse polygonal and Widmanstatten/acicular ferrite, confirming our previous analysis [27]. The evolution of microstructures in the HSM casting routes are well understood: These products are an outcome of large reductions that break up the original cast structure resulting in significant refinement of the austenite grains, which upon further

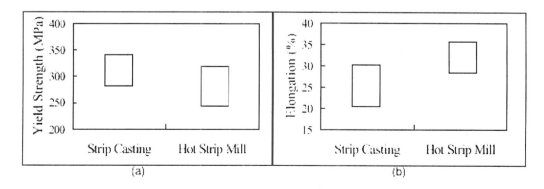

Figure 50 Comparison of mechanical properties of low-carbon steel strip produced via strip casting and hot strip mill routes. (a) Yield strength, (b) elongation.

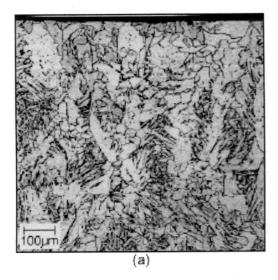

Figure 51 Comparison of the final strip microstructures produced (a) by Castrip, (b) in a conventional hot strip mill.

transformation produce a fine equiaxed ferrite grain structure. In this respect, product metallurgy knowledge developed for the HSM route is closely applicable to DC (single horizontal belt) strip cast material, but not to TRC strip cast material.

In the case of TRC, the ability, or not, to control ferrite microstructure is crucial as it determines the mechanical properties of the final product. Figure 52 shows the dramatic impact various cooling rates have on TRC strip microstructures [44].

Changes in the solidification parameters such as melt chemistry, substrate texture, and casting speed can also strongly influence the final microstructure of the as-cast strip.

Scale Formation

For all near-net-shape casters, the problem of excessive oxidation of the strip surfaces at high temperatures, owing to their much larger surface/volume ratios vs. conventional slab products, require that a protective atmosphere, such as carbon dioxide, be used. The same applies to the liquid and solidifying upper surface of steel in the DSC caster. Abuluwefa and Guthrie include a report on the kinetics of steel oxidation in various gaseous mixtures of oxygen, nitrogen, water vapor, and carbon dioxide at high temperatures [45].

In-Line Hot Rolling of Thin-Roll Cast Strip

It is reported that in-line hot rolling is effective in refining the as-cast microstructure of TRC strip. A minimum 15% hot reduction was needed to start altering microstructures generated during the casting process. For similar hot reduction levels, rolling temperatures had a profound impact on the microstructure refinement. Fine ferrite microstructures were observed when rolling at around 860°C. Ultrafine ferrite grains in the range of 2 to 4 μm were observed near the surface and this region extended up to 30% of the thickness. The rest of the sample consisted of fine polygonal ferrite grains in the range of 10 to 20 μm. It is claimed that material with high yield strength (400 MPa) and a total elongation in excess of 30% can be produced with this

Table 5 Key Differences Between Conventional and Castrip Microstructures

	Strip casting	Hot strip mill
Prior austenite grain morphology, size	Columnar shape 100 to 250 μm wide, 300 to 700 μm long	Equiaxed 25 μm
Final microstructure constituents	30% to 60% polygonal ferrite, 70% to 40% Widmanstatten, acicular ferrite	100% equiaxed ferrite
Ferrite grain morphology, size	Polygonal 10 to 50 μm wide, 50 to 250 μm long	Equiaxed 10 μm

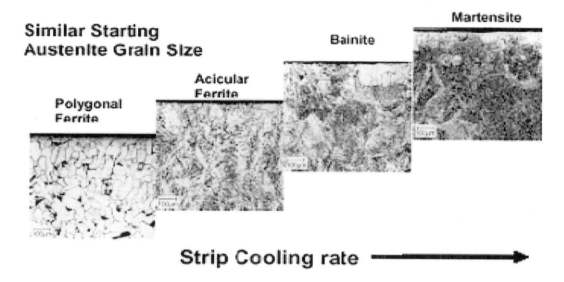

Figure 52 Effect of cooling rates on final microstructures during the transformation of austenite. Respective cooling rates: 0.1, 13, 25, and 100°C/sec.

microstructure. Hot-rolling temperatures of around 1050°C on the other hand produced microstructures with polygonal ferrite content estimated to be more than 80%, with grains in the size range of 10 to 40 μm.

Thus, a wide range of strip microstructures can be produced through adequate control, or not, of the degree of hot reduction and rolling temperature.

2. Product Properties of Direct Chill (Single Horizontal Belt) Strip Material

Gauge Characteristics

In reviewing the properties of the as-cast strand and hot-rolled material from horizontal belt casters, these again are reported to be very satisfactory. One of the major concerns re the application of single-belt technology has been whether strip of uniform gauge can be produced by such a (unconstrained) process. Figure 53 demonstrates the shape profiles across the width of the sheet that could be achieved on the Clausthal II caster, before and after hot reduction.

To avoid the problems associated with belt distortion created by thermal stresses generated when the molten steel first contacts the belt, which can lead to buckling, the cooling chamber of both the Clausthal and MEFOS machines were operated under partial vacuum. This represents one of the techniques that can be used for maintaining a flat belt under thermal loads. Figure 53 shows thickness profiles of the strand in the as-cast

state, after one in-line hot-rolling pass and after three additional off-line hot-rolling passes. The various profiles shown in the diagram for the as-cast strand were measured at distances of 20 cm apart along the strand.

The variations of about +1 mm at 15-mm thickness are typical of those experienced when casting peritectic grades of steel on a conventional slab caster. These variations proved to be stochastic, and practically invisible following in-line pass rolling. As such, they proved harmless.

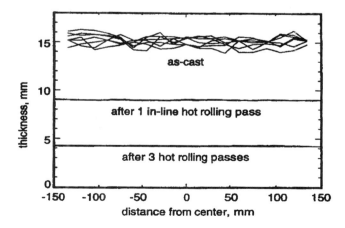

Figure 53 Profiles of DSC strip in the as-cast state and after hot rolling.

Internal Quality of Direct Strip Cast Strip

It was found that columnar dendritic zones extended from both the lower and upper surfaces. In between there may be an equiaxed zone that may exhibit porosity in carbon steels.

Figure 54 shows the data on secondary dendrite arm spacing and on microsegregation of manganese in carbon steel ST37 (I). It is confirmed by these measurements that the dendritic structure is finer compared to that in the interior of conventionally cast slabs. Similarly microsegregation is in the same range as in conventionally cast material. Macrosegregation was found to be very small in carbon (0.2%C) steel and type 304 stainless steel.

The internal porosity in DSC cast material is in the form of shrinkage voids and disappears on sufficient hot reduction. Figure 55 shows that for a 48% thickness reduction (cp = 0.66) almost all the internal voids are closed, and are eliminated at 66% (*p* = 1.08).

Internal porosity has also been observed in twin-roll casting of carbon steels.

Mechanical Properties

Owing to the different cooling rates at the lower and upper surfaces of the DC cast strand, the dendritic structure and microsegregation are asymmetrical with respect to the center plane, as already demonstrated in Fig. 54.

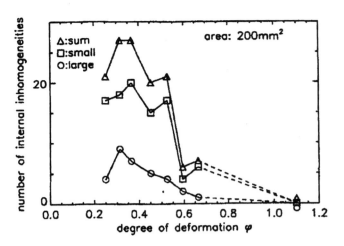

Figure 55 Decrease in internal porosity by hot rolling. Grade ST44.

Nonetheless, tests carried out on samples taken from the upper and lower halves of the material did not reveal significant differences in mechanical properties after hot rolling, in terms of yield and tensile strengths, elongation, and notch impact energies.

VII. CONCLUSIONS

To conclude, as a result of the promised profitability of net-shape casting of steel sheet products, there ap-

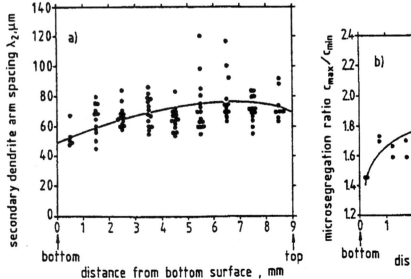

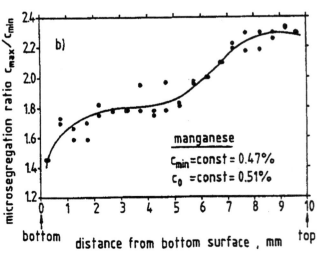

Figure 54 Data on secondary dendrite arm spacing and microsegregation in strand of carbon steel ST37, cast on a single-belt caster. (a) Secondary arm spacing. (b) c_{max}/c_{min} ratio for manganese.

pears to be an inexorable move worldwide toward the development of moving-mold technologies. As with NUCOR's successful commercialization of Mannesmann's thin slab casting technology, the final charge toward commercialization for low-carbon sheet material, is again being driven by NUCOR, this time in collaboration with BHP and IHI.

While the evolution of microstructure in twin-roll casting strip casting is fundamentally different from strip products produced via conventional hot strip mills, a variety of ferrite microstructures can be produced through the control of solidification and strip cooling (during the austenite–ferrite/pearlite transformation), as well as with in-line hot rolling and residuals. The combinations of material strength and elongation levels that can be potentially obtained with strip casting using a single, low-carbon steel chemistry, are summarized in Fig. 56.

In the conventional strip production processes, significant chemistry changes are necessary to produce a broad range of properties. Figure 57 summarizes a typical range of steel chemistries used to produce a variety of hot-rolled structural products. Generally, higher C, Mn, and microalloys (Ti, Nb) are necessary to produce high-strength material. Comparison of the strength–elongation curves from the two processes are very similar, the TRC curve being slightly lower. However, one should note that the ferrite grain sizes for TRC (Table 5) are considerably larger, signifying that comparisons of formability should be considered where appropriate.

Thus, many commercial issues need clarification. For TRC, operating costs in terms of roll wear, life of sidewall dams, refractory costs for the metal delivery

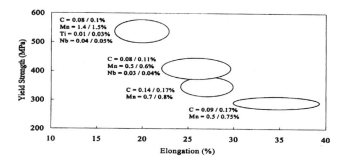

Figure 57 Chemistry requirements in conventional hot strip mill route to produce range of hot-rolled structural products.

system, strip porosity, consistency and properties of products, etc. are currently being established to determine the commercial viability of the process and products.

While major economies are promised through the process rationalization that near-net shape casting can bring, low-carbon steels, because of their metallurgical characteristics, are such that a good deal of thermomechanical working is conventionally applied to produce acceptable grain sizes and associated mechanical properties. This suggests that hot strips cast in the order of 10-mm thickness will be required to closely reproduce the properties and characteristics of conventional products. This issue favors the DSC process, or an equivalent, over the TRC process, particularly for high-tonnage operations (2–3 million tonnes/year). Similarly, most hot strip is sold in the 2- to 3-mm-thick range, which again favors the DSC route. On the other hand, the new TRC steel products may be able to penetrate the market for cold-rolled strip, provided the roll caster can operate at 0.7-mm-thick strip for extended periods of time.

For a single-belt caster for the mass production of steel, many questions remain unanswered. First, the DSC machine must be able to function over many hours at belt speeds of 1 m/sec or more. This is five times faster than any of the three pilot casters engaged in producing 10-mm-thick strand. Similarly, the effective belt length for a commercial machine must be ~12 m and must stay planar. For this, the mechanical and thermal loads acting on the belt and its guidance system will be severe. Similarly, the gasket between the moving belt and the side walls of the water chamber must be sealed perfectly for tightness, but must experience sufficiently low friction to avoid damage of the belt at its sides.

For commercial production, the width of the strip may be up to 2 m or more, so that an even delivery of

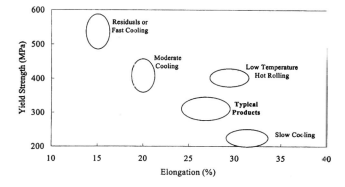

Figure 56 Product properties (opportunities) for a TRC strip casting machine, based on a single, low-carbon-steel chemistry.

metal across the width of the caster will be crucial, as in the case of a twin-roll caster. In compensation, the feeding of steel onto belts moving at high speeds is much less difficult than feeding metal onto the slow-moving belts (~0.2 m/sec) associated with the pilot casters. For these, metal flows have had to be throttled or the effective metal head values kept exceedingly small to achieve isokinetic conditions with a slow-moving belt during the freezing of 10-mm-thick strand.

Other technical issues for DSC casters, such as melt containment (side wall problems), strip edge profiles, and metal feeding, have all been successfully demonstrated on the large pilot-scale caster at MEFOS, Lulea, Sweden, as discussed in articles by Carlsson et al. [34] and Nystrom et al. [35].

Finally, casting with the single-belt DSC process will have to be coupled with slow-speed, in-line hot rolling. Mathematical modeling of these issues suggest that this is feasible, but a full-scale demonstration plant is needed for the next step, and the responsibility must rest with an integrated steelmaker.

As a final conclusion therefore, the fixed-mold continuous casting of slabs, blooms, and billets have neared process perfection. One can expect that these casters will be the mainstay of the world's continuous casting operations for sheet production for many years to come, or until an integrated plant matches the resolve of the minimill operators, and builds a demonstration plant for the direct strip casting machine.

REFERENCES

1. Guthrie, R.I.L.; Isac, M. Steel strip in the context of near net shape casting production. Steel Res. 1999, 70 (8/9), 343–348.
2. Guthrie, R.I.L.; Isac, M. Steel casting in the third millenium; strip casting interfacial heat fluxes and microstructures. The Brimacombe Symposium Proceedings, Met Soc., 2000; 209–243 pp.
3. Guthrie, R.I.L.; Jonas, J.J. *Steel processing technology, Metals Handbook*, 10th Ed.; ASM International: Ohio, 1989; Vol. 1, 107–125 pp.
4. Halliday, I.M.D. Continuous casting at barrow. J. Iron Steel Inst. 1959, *191*, 121–163.
5. Wolf, M.M. On the art of continuous casting. BHM 2000, *145* (1), 35–45.
6. Wolf, M.M. History of continuous casting in steelmaking in the 20th century. ISS, 75th Steelmaking Conference, April, Toronto, 1992; 49 pp.
7. An Introduction to Iron and Steel Processing. Kawasaki Steel 21st Century Foundation, 1997.
8. Schwerdtfeger, K. Benefits, challenges and limits in new routes for hot strip production. Yukawa Memorial Lecture, 1997; 31–46.
9. Brimacombe, J.K. Design of continuous casting machines based on a heat-flow analysis: state-of-the-art review. Continuous Casting, Heat Flow, Solidification and Crack Formation. ISS of AIME, 1984; 9–17.
10. Savage, J.; Pritchard, W.H. J. Iron Steel Inst 1954, *178*, 269–277.
11. Singh, S.N.; Blazek, K.E. Open Hearth Proceedings, AIME, 1977; 60 pp.
12. Samarasekera, I.V. The cornerstone of research in continuous casting of steel billets. The Brimacombe Symposium Proceedings, Met Soc., 2000; 399–419.
13. Zeze, M.; Tanaka, H.; Takeuchi, E.; Mizoguchi, S. Continuous casting of clad steel slab with level magnetic field brake. Proceedings of 79th Steelmaking Conference, ISS-AIME, Pittsburgh, 1996.
14. Brimacombe, J.K.; Agarwal, P.K.; Hibbins, S.; Prabhaker, B.; Baptista, L.A. Spray cooling in the continuous casting of steel. Continuous Casting Proceedings; ISS of AIME, 1984; Vol. 2.
15. Irving, W.R. Continuous Casting of Steel. 1993.
16. Tavares, R.P. Vertical twin-roll caster: metal-mould heat transfer, solidification and product characterization, Ph.D. thesis; McGill University, 1998; 234–268.
17. Herbertson, J.; Guthrie, R.I.L. Continuous Casting of Thin Metal Strip, Canadian Patent 536533, March 3, 1992, and U.S. Patent 4,928,748, May 29, 1990.
18. Chen, S.J.; Ren, R.C.; Tseng, A.A. Interface heat transfer in metal casting on a moving substrate. J. Mater. Process. Manuf. Sci. 1995, *3*, 373–386.
19. Strezov, L.; Herbertson, J. Mechanisms of initial melt/substrate heat transfer pertinent to strip casting. The Belton Memorial Symposium Proceedings, Australia, 2000; 289–299.
20. Tavares, R.P.; Isac, M.; Hamel, F.G.; Guthrie, R.I.L. Instantaneous interfacial heat fluxes during the 4 to 8 m/min casting of carbon steels in a twin roll caster. Metall. Mater. Trans. B February 2001, *32*, 55–67.
21. Wang, G.-X.; Matthys, E.F. Trans. ASME-J. Heat Transfer. 1996, *118*, 157–163.
22. Wang, G.-X.; Matthys, F. On the heat transfer at the interface between a solidifying metal and a solid substrate. Melt Spinning, Strip Casting and Slab Casting; Matthys, E.F., Truckner, W.G., Eds.; The Minerals, Metals & Materials Society, 1996; 205–236.
23. Shiang, L.-T.; Wray, P.J. Metall. Trans. A, 1989, *20A*, 1191.
24. Ueshima, Y.; Sawai, T.; Mizoguchi, T.; Mizoguchi, S. Proceedings of the 6th International Iron and Steel Congress, 1990; 642.
25. Mizoguchi, T.; Miyazawa, K. ISIJ Int. 1995, *35* (6), 771.
26. Mizoguchi, T.; Miyazawa, K. Adv. Mater. Process. 1990, *1*, 93.
27. Tavares, R.P.; Isac, M.; Guthrie, R.I.L. Roll-strip interfacial heat fluxes in twin-roll casting of low carbon

steels and their effects on strip microstructure. ISIJ Int. 1998, *38*, 1353–1361.

28. Shewmon, P.G. *Transformations in Metals*; McGraw Hill Book Co.: New York, 1969; 364 pp.

29. Honeycombe, R.W.K.; Bradeshia, H.K.D.H. *Steels, Microstructures and Properties*; John & Sons, 1995; 324 pp.

30. Guthrie, R.I.L.; Isac, M.; Kim, J.S.; Tavares, R.P. Measurements, simulations, and analyses of instantaneous heat fluxes from solidifying steels to the surface of two roll casters and of aluminium to plasma-coated metal substrates. Metall. Mater. Trans. B October 2000, *31*, 1031–1047.

31. Yasumoto, K.; Nagamichi, T.; Maehara, Y.; Gurji, K. Effects of alloying elements and cooling rate on austenite grain growth in solidification and the subsequent cooling processes of low alloy steels. Tetsu-to-Hagane 1987, *73*, 1738–1745.

32. Maehara, Y.; Yasumoto, K.; Sugitani, Y.; Gunji, K. Effect of carbon on hot ductility of as-cast low alloy steels. Trans. ISIJ 1985, *25*, 1045–1052.

33. Ikawa, H.; Shin, S.; Osihige, H.; Mekuchi, Y. Austenite grain growth of steels during thermal cycles. Trans. Jpn. Weld. Soc. 1977, *8* (2), 46–51.

34. Carlsson, G.; Nystrom, R.; Sandberg, H.; Reichelt, W.; Urlau, U. Single belt casting—a promising technique for the production of steel strip. Iron and Steel—Today, Yesterday and Tomorrow, Stockholm, Sweden 1997, Vol. 1, 161–174.

35. Nyström, R.; Burström, E.; Reichelt, W.; Urlau, V. DSC—a high productivity concept for strip production of steel. Metec 94, 2nd European Conference on Continuous Casting, Düsseldorf, 1994.

36. Schwerdtfeger, K. Belt casting for steel—a critical review. The Brimacombe Symposium Proceedings. Met Soc. 2000; 613–629.

37. Schwerdtfeger, K.; Spitzer, K.-H.; Reichelt, W.; Voss-Spilker, P. Stahl Eisen 1991, *91* (6), 37–43.

38. Schwerdtfeger, K. ISIJ Int. 1998, *38* (8), 852–861

39. Reichelt, W.; Schwerdtfeger, K.; Voss-Spitzer, P. Vorrichtung zum kontinuierlichen Gieben von Metallschmelze, isobesondere von Stahlschmelze. German Patent No DE 3423834 C2, 28 June 1984.

40. Jefferies, C.; Hasan, M.; Guthrie, R.I.L. A coupled fluid flow and heat transfer study for planar thin strip steel casting processes. Proceedings 10th Process Technology Conference, 2nd International Symposium on Modelling in the Iron & Steel Industry, ISS of AIME, Toronto, 1992; 355–363.

41. Jefferies, C. Modelling a novel, thin strip, continuous steel caster delivery system. Ph.D. thesis, McGill University, 1995.

42. Guthrie, R.I.L.; Herbertson, J. A novel concept for metal delivery to a thin strip caster. Proceedings of Casting of Near Net Shape Products, Honolulu, Hawaii; Sahai, Y., Bethles, G.E., Carbonara, R.S., Mobley, C.E., Eds.; November 1998; 335–349.

43. Mukunthan, K.; Strezov, L.; Mahapatra, R.; Blejde, W. Evolution of microstructures and product opportunities in low carbon steel strip casting. The Brimacombe Symposium Proceedings, Met Soc., 2000; 421–439.

44. Campbell, P.; Wechsler, R. The CASTRIP process; a revolutionary casting technology, an exciting opportunity for unique steel products or a new model for steel micro-mills?. Proceedings of Innovative Technologies for Steel and other Materials, COM 2001, Met Soc, CIM; Guerard, J., Essadiqi, E., Eds.; 2001; 201–215.

45. Abuluwefa, H.; Guthrie, R.I.L.; Ajersch, F. Oxidation of low carbon steel in multicomponent gases: Part II. Reaction mechanisms during reheating. Metall. Trans. A *28*, 1643–1651.

9

Continuous Casting Design by the Stepanov Method

Stanislav Prochorovich Nikanorov and Vsevolod Vladimirovich Peller
A.F. Ioffe Physical Technical Institute of Russian Academy of Sciences, Saint Petersburg, Russia

I. HISTORY AND CONCEPT OF STEPANOV'S METHOD

Casting is the least expensive shaping process. About half of the machine parts that are produced are by a casting process. However, the production of high-quality, long, thin-walled castings especially, with a composite cross-sectional shape, is a challenging design and manufacturing task. These parts are typically produced by continuous (or semicontinuous) casting processes that possess greater stability and more homogeneous properties at a reasonable cost.

Several continuous casting processes were proposed in the 19th century [1,2]. Laing (1843) suggested the production of tubular billets by an uninterrupted casting of molten metal into a water-cooled cylindrical crystallizer. The metal solidified at the inner work surface of the crystallizer. The resulting rind is pulled from the crystallizer. Bessemer (1858) proposed casting of molten metal into the gap between rotating rollers to produce a foil or sheet. Continuous casting processes derived from these two concepts include the use of a rotating crystallizer instead of a stationary unit. In this case the solidifying rind is forced against the surface of the crystallizer by hydrostatic pressure which produced structure defects. However, continuous casting into a crystallizer produces a billet that is not part of the desired shape.

In 1902 Brines proposed the production of a rod or tube by freezing the molten metal on the end of the bar pulled from the melt, and tube shape was produced by blowing air. The pulled rod was rolled and the inner surface of the pulled tube was smoothed with a special device. This design was the first step in the development of continuous casting without a crystallizer. In 1917 Czochralski [3] pulled crystal ingots from the melt. More recently, this process was used to obtain wire or thin rods. Gomperz (1922) pulled a wire of Pb, Zn, Sn, Al, Cd, and Be from the molten metal through the opening of mica plate situated on the melt surface [4]. The objective was to obtain smooth-surfaced wire or rod. The mica plate with the opening on the surface of the melt according to Gomperz was necessary to eliminate the growth of defects (intakings) on the surface of the wire. Mark and coworkers (1923) grew metal wire by a similar way [5]. Kapitza (1928) used the same technique for growing bismuth rods [6]. However, the casting process itself was not an object of these studies.

In 1938 Stepanov [7] proposed a new concept of shaping solids where the process would be conducted in the liquid state so that the solidifying metal near the liquid–solid interface does not touch the walls of the crucible and shaper. In Stepanov's method (SM), the liquid is shaped without any vessel by applying a high-frequency electromagnetic field, under action of ultrasound, utilizing a hydrodynamic effect. The simplest method of shaping the liquid is to utilize the forces of surface tension using a shaper that is placed on the surface of the melt. In the simplest case, the shaper is a refractory plate with an orifice. A seed is introduced into the orifice which is coupled with the melt. As the seed moves up, it draws the liquid column of the melt using surface tension forces. The melt crystallizes in the area of lower temperatures keeping the shape given in the liquid state. The molten material at the solid–liquid interface is not touched by the shaper and the crucible. Therefore the orifice of the shaper is not a die. Not only the shaper geometry but also the shape of the meniscus bounding the liquid column, the position and form of the liquid–solid interface, and the shape of a seed determine the shape of the article crystallized. The shape of the meniscus and the position of the interface are connected with the pressure in the melt.

The material of a shaper can be wetted or nonwetted by a melt. In the first case (the wetting angle $\theta < 90°$) the melt can rise along vertical walls of the shaper orifice due to forces surface tension. If the material of a shaper is nonwetted ($\theta > 90°$) by the melt it is necessary to create a surplus hydrostatic pressure in the liquid by the pressure of the shaper. Stepanov's method initially was used to grow crystalline articles in the shape of rods, tubes, and plates with shapers nonwetted and wetted by melt [7].

Figure 1 shows some versions of the Stepanov's method. Versions A and B correspond to the nonwetted shaper. In version A the lower boundary of the liquid column is fixed on the internal wall because of the constancy of the wetting angle. In the second case, coupling the melt with the shaper takes place at the internal edge of the shaper. These versions can transform each other when the pressure in the melt is changed. These variants of the Stepanov method are general for growing articles of light metals and their alloys as well as in experiments with cast iron (Table 1).

Some versions of the SM with wetted shaper were used early in the history of the technique. Using this technique, sheets of ionic crystals of LiF, CsI from Pt crucible, AgCl from Mo crucible, and tubes of KNO_3 from brass crucible [20] were grown. Figure 2 shows one of the variants of SM used for growing tubes of Ge [21].

In 1967 LaBelle and Mlavsky [22] described the growth of a sapphire filament from the melt which was used as the reinforcement of metal matrix of composites, using a wetted die. Figure 3 illustrates this edge-defined film-fed growth (EFG) technique. This is a very stable method and is effective for growing not only filament but also bulk-shaped sapphire crystals. The capillary design enables the production of the melt from the crucible during crystallizing. Sapphire is a material exhibiting excellent optical, mechanical, and chemical properties. This method has permitted the mass production of shaped crystals of sapphire. Additional capillary shaping methods have been developed to grow shaped crystals of different semiconductors and dielectrics for electronics, sun energetic, medicine, and other engineering applications [7,23]. However, articles of light metals and high-carbon iron are grown by the classical Stepanov method with a nonwetted shaper, and process selection depends on the chemical activity of metal and alloy melts with respect to wetted materials of shapers.

To grow crystalline parts with a constant cross-section size, it is necessary to maintain the shape of the meniscus of the liquid column and the level of the

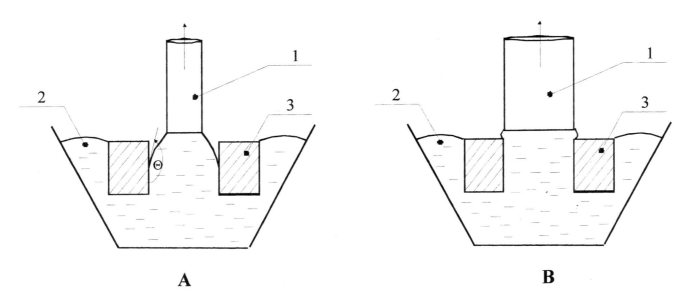

A   **B**

Figure 1 Scheme of growing shaped metal crystalline articles by Stepanov's method for coupling the low border of liquid column with smooth internal surface of the non-wetted shaper (A) and with the upper edge of shaping orifice (B).

Table 1 The First Investigations on Growing Shaped Articles of Metals and Alloys by the Stepanov Method

Reference	Material	Profile	Material of shaper	Rate 10^{-3} cm sec^{-1}
7,8	Al	sheets, rods, tubes	cast iron	56–280
9,10	Zn, Al, duralumin, brass 68	plates	steatite	220
11	Al–Cu, Al–Si alloys	sheets, tubes	cast iron	140–330
12	Cu, tin bronze, aluminum bronze,	rods	superchromium steel, SiC	110–770
13	AL–Mn, Al–Mg alloys, aviation Al alloy	tubes in the sheet	cast iron	125–194
14	Hg	rods of circle and rectangular cross section	glass	6.2
15	High carbon alloy of iron	tubes and circle rods	BNC	1.7–17
16	Composition: Al matrix reinforced uninterrupted steel wires	rods of rectangular cross section	cast iron	100–117
17	Mg-based alloy	tubes, plates	steel	33–300
18	Mg-Me (Me–Al, Zn, Mn, Zr, Y, Nd)	lattes, circle rods	steel	33–170
19	Composition: Al matrix reinforced by fibers of steel, Ni–Cr, W, Mo, SiO$_2$, Al$_2$O$_3$, B, SiC	plates, circle rods, tubes	cast iron	28–280

liquid–solid interface surface. These conditions demand temperature stabilization of the melt and the pulling rate in addition to a constant hydrostatic pressure in the melt. These conditions are also dependent on the decrease of the melt volume in a crucible during processing. A theoretical solution to this problem with experimental validation was possible using microscopic theory of the stable growth of shaped crystals using Stepanov's method [24,25].

II. APPLICABILITY OF SM TO VARIOUS METAL ALLOYS

Although the SM permits the growth of shaped crystalline articles of various materials (metals, dielectrics, semiconductors), only its use for metallic systems will be considered in the remaining discussion. At the Ioffe Physical Technical Institute, Saint-Petersburg, Russia, shaped polycrystalline articles of Al, Mg, Cu, Fe, Bi,

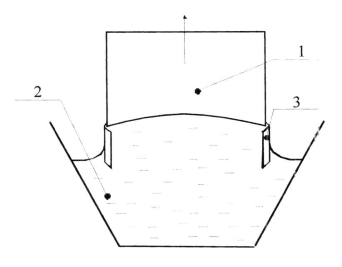

Figure 2 One of variants of growing the shaped crystal by SM with the wetted shaper.

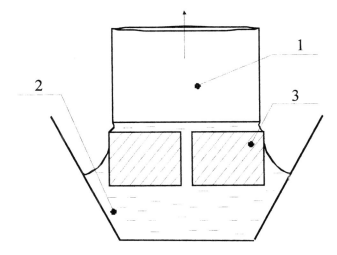

Figure 3 Scheme of EFG-method.

Sn, Zn, and their alloys, single crystals and also eutectic and reinforced Al matrix materials were formed [7,9]. The growth process will be discussed here, which will include a general discussion of equipment, technology, and quality parameters that are applicable to all metal systems.

III. EQUIPMENT

Of primary importance is the equipment required for pulling the liquid column of the crystallized melt. The general operational and design principle is based on the Czohralski method [26] and vertical continuous casting technology [2]. The distinction of the SM is that the process is dependent on metal alloy, melting point, and other characteristics of the crystallized material, and on the size of the pulled article. A schematic of the generalized scheme is provided in Fig. 4.

The metal is melted and fed from a crucible placed in a furnace unit (smelter) (1). The melt of the desired composition is prepared in a similar smelter (1). Usually, resistance furnaces or inductive furnaces are used. Shaping of melt column is performed using a shaping device (SD) (2), which is part of a changeable tool. Pulling up of shaped profile is performed by the pulling device (3), which may be a sliding carriage along vertical guide bars or by a roller mechanism. The drive mechanism is usually electric or hydraulic. It is important to maintain a constant level of the melt in a crucible with respect to the SD (4), which may be done by sinking the SD into the crucible when the volume of the melt decreases, by lifting the crucible or by adding additional melt to the crucible.

Additional components include a water cooler (5) with different units including the SD situated near the melt. Control instrumentation (6) is used to regulate the processing parameters. A power unit delivers the required electric energy. There is also an air cooler for the part as it is being pulled out from the open mirror of the melt. Water–air mixtures or other cooling agents may also be used for cooling. Generally, it is preferred for the system components to be operationally independent from each other.

Although SM can be used as a continuous process, more often it is used in a semicontinuous process. As a rule, the systems are designed to be used for as broad a product mix of different shapes and sizes as possible.

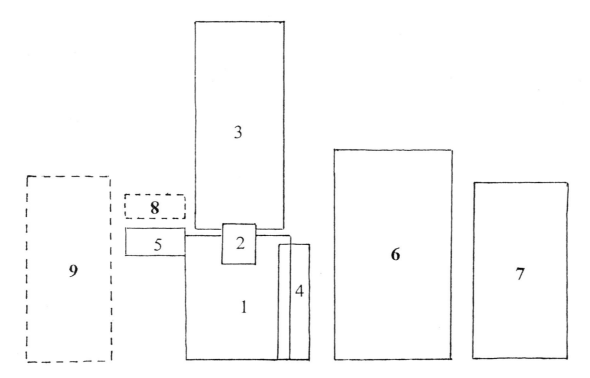

Figure 4 Scheme of installation of pulling out shaped crystalline articles by SM and making up the set. 1. Furnace unit with melting—working crucible; 2. Shaping device; 3. Pulling device with the drive; 4. System controlling the melt level relative to a shaper; 5. System of water cooling; 6. Control desk; 7. Power unit; 8. Cooler of profile pulled; 9. Vacuum system. Blocks that can be absent at installations for pulling individual materials are shown by dotted line.

It is critically important that there are no vibrations in the shaping zone and when the part is being pulled. Therefore the furnace unit is set on its own bulky frame and foundation. Dampers are used for drivers and reducers.

IV. TOOLS

The set of pulling out is made up of a changeable technological tool named by heat technological zone [7,26]. The main part of the tool is SD. It includes the shaper realizing capillary shaping liquid column of the melt of the assigned cross section. The shaper restricts the work piece of the melt that the growth of a profile is carrying from. It determines the temperature regime in the melt near the liquid–solid interface and the additional pressure in the melt.

For growing metallic profiles the nonwetted shapers are used as a rule. The material of the shaper must be highly chemically stable, so as not to interact with the melt and atmosphere in the work zone. It must have high heat conductivity, high density, and mechanical strength for a long period of service. In the case of obtaining shaped profiles with small content of impurities and perfect structure for electronics (in particular, for single crystals) the high purity and chemical inertness of the shaper material are very important [26]. Accessory elements of the SD are a seed and a seed holder. The seed is plunged into the melt and pulls up the liquid column in the beginning of processing. The seed holder gives mechanical coupling of the pulling device and the profile.

One of the main parts of the sets with the open melt mirror is the cooler.

The pulling device is developed separately for every profile or for geometrically similar profiles. A tool includes also, apart from interchangeable SDs, crucibles, the system of heating, heat-insulating elements (screens), dampers, and other parts.

V. TECHNOLOGICAL PARAMETERS

The geometry and quality of profiles pulled up by SM depend on both construction and sizes of the shaper as well as on technological parameters [7]. The main parameters are the temperature of the melt, the pulling rate and the pressure in the melt determined by the level of the melt relative to the shaper, the intensity of cooling. The size of a cross section of the pulled profile increases when the pressure in the melt and cooling intensity rise. The cross section increases also when the

temperature of the melt and the pulling rate decrease. These parameters have influence on the structure and quality of the profile pulled. They together determine the regime of the process. They must be found for each individual alloy. At the same time, sizes, and structure of articles depend on outer factors: the vibration, the oscillation of a voltage, the motion of surrounding air. They are controlled with difficulty. Therefore it is necessary to take precautions when attempting to cancel the influence of outer factors.

In the case of growing crystals of high purity it is necessary to use the tool of high purity material and to take precautions against the penetration of impurities from an atmosphere.

VI. CHEMICAL COMPOSITION ALONG THE LENGTH AND OVER THE CROSS SECTION

The composition of an article crystallized differs from that of the melt, which is due to the different solubility of its components in the solid and liquid phases [27,28]. Impurity solidification from the liquid phase is dependent on the effective distribution coefficient K, which is the ratio of concentrations in solid to liquid phases:

$$K = C_S/C_L$$

The value of K depends on the crystallization rate, convection intensity, and diffusion coefficient. The character of impurity distribution in a liquid for stationary, conservative processes of directed crystallization and flat liquid–solid interface can be determined by the Tiller equation [28,7]:

$$C_L = C_o + C_o[(1 - K)/K]e^{-Rx/D} \qquad (1)$$

where C_L is the impurity concentration in the liquid as a function of the distance of x from the interface, C_o is the initial concentration of the impurity in the liquid, K is the distribution coefficient of the impurity, R is the motion rate of the solid–liquid interface, and D is the diffusion coefficient of the impurity in the melt. At $x = 0$, the impurity concentration limit is $C_L = C_o/K$. The dependence of C_L from x is significantly dependent on the motion rate of the solid–liquid interface. This equation permits the evaluation of the behavior of alloy components for pulling out parts of various materials by SM.

It is possible to consider SM as a conservative directed process of crystallization. To obtain metal polycrystalline profiles at a given pulling rate (m hr^{-1}) it is necessary that the impurities forced back during part growth be small to provide a constant composition of the part along its length. But the impurities forced back may be substantial for growing metallic single crystals with the growth rate in the range from 0.5 to 5 cm/min. This is the same problem encountered for the growth of single crystals of semiconductors. However, variation of the chemical composition over the cross section of a part is typically small and it occurs because of the vacancies or vacancy cluster segregation and their association with impurities on the part's surface and the volatility of some components.

VII. QUALITY CHARACTERISTICS OF SHAPED SINGLE CRYSTALS AND ARTICLES

The quality of parts produced by SM is determined by various factors including crystal structure (perfection) and the material composition, which is dependent on the conditions and regime of the crystallization process. They include melt purity and atmosphere, chemical and mechanical stability of operating equipment and tooling, operational parameters and their control.

Quality characterization of the metal can be performed by determination of part geometry, surface quality, micro- and macrostructure, mechanical behavior, physical properties, chemical purity, electrochemical and corrosion parameters, and in-service properties. The role and importance of separate parameters depend on specific application.

A. Part Geometry

The Stepanov method is capable of producing parts of complex geometric shapes, which are often superior to alternative production methods. Geometric size tolerances along the length and over the cross section of each part and for a group of parts can be very tight. However, there are some specific characteristics of parts pulled by SM such as the small-rounded angles of the cross-section shape. Profiles of parts pulled out of unclosed shaper slots typically possess more size variation relative to those produced with a closed contour. It is necessary to take additional care to obtain constant sizes of the many parts grown by a multiple part growth process.

The greatest influence on profile geometry is the precision of the tool and pulling device, stability of operational process parameters, accounting for shrinkage factor, and the capillary characteristics of the melt.

B. Surface Quality

Perfection (quality) of a surface is characterized by roughness and surface defects. Roughness can be determined by micro- and macroroughness. Surface defects are transverse waviness, longitudinal furrows or floating, local small freezing, dents, warping and cracks, parasitic crystals exhibited as tubercles or irregularities.

The essence of SM is melt solidification without any contact with a crystallizer as the part is shaped in the liquid state. This results in the absence of surface pores and shrinkage shells. The surface quality is substantially higher than for conventionally cast parts.

The greatest surface perfection is obtained for shaped single crystals and articles of alloys of eutectic composition. In this case, the surface quality improves with decreasing temperature interval of crystallization (solidus–liquidus temperature interval for the given composition of the alloy).

Surface quality requirements depend largely on the specific application of the part such as the stringent requirements of decorative articles or parts for critical applications such as wave guides, high heat-radiating or heat-absorbing surfaces, or superior corrosion resistance.

C. Structure

Polycrystalline articles manufactured by SM and other methods of direct crystallization typically exhibit dendritic structure with the texture oriented along the crystal growth direction. More frequently, columnar or fan structures are observed. But under some growth conditions described later, equiaxial structure can be obtained. Microstructures are characterized by concentration and dispersion of different inclusions (structure elements) and specific interphase surfaces.

Alloy composition and growth conditions determine the structure of the article. The Stepanov method can influence the structure by controlling the thermal parameters of crystallization to obtain optimum structure, which may be required by the end-use application of the part being produced.

D. Chemical Purity

Chemical purity is dependent on raw material purity and chemical interactions of the melt with a crucible, shaping device, and atmosphere. Therefore single crystals with a perfect structure are pulled out of a melt in vacuum or under an inert gas. High purity is required for polycrystalline metal parts used in the food industry or electrical engineering.

E. Properties

Property requirements are application dependent. Properties are controlled by SM by varying the operating conditions as it is possible to vary the growth profiles by varying a wide range of parameters including temperature, pulling rate, and cooling, which results in a design improvement. It is also possible to improve the mechanical properties by thermomechanical treatment during the growth process. Process parameter variations that may be used include enlarging wall thickness, alloy selection, additional heat treatment, manufacturing specific profile constructions giving strengthening along certain direction for separate loading at individual application and increasing stiffness which can only be obtained by SM, specific surface strengthening by film covering and reinforcing the profile by high strength fibers or layers. Furthermore, the SM process provides parts without pores. Therefore the density of the metal is higher than for alternative methods.

The Stepanov method may be used to provide for metal parts with improved corrosion resistance. Parts produced by some alloys exhibit very high cathodic protection of metals because such protection is dependent on the purity as well as on the quality of the structure and of the surface.

The Stepanov method also enables to grow articles of high damper alloys [29]. Such articles can be used for parts subjected to high vibration conditions. Recently, SM was used to grow parts with a shape memory effect, which is especially useful because it was possible to achieve single-crystal growth at a convenient temperature interval of thermoelastic transition and stable shape memory effect.

VIII. STEPANOV METHOD PROCESS CLASSIFICATIONS

The Stepanov method is used for growing parts of single crystal and polycrystalline structure of various materials. These parts can be obtained using various conditions: at moderate and high temperature, in vacuum, inert gas or air, with forced cooling or without. There are two SM process classifications. The first process utilizes an open mirror of a melt (having some protection or without it). The second is a vacuum (germetic) process. Experimental and small-production size process will be discussed in this section.

A. Processing with Open Mirror of the Melt

It is possible to pull polycrystalline articles of some alloys out of a melt in the air with an open mirror of the melt when the melting point is moderate and interaction between the melt and atmosphere is small. Articles of Al-based alloys are grown with the open mirror of melt. In the case of Mg-based alloys, growth of shaped single crystal can be performed only with vacuum or gas atmosphere chamber for protection because of the high purity requirements of the process.

1. Crystallization of Al-Based Alloys Without Melt Protection

When articles of Al-based alloys are grown out of the melt with open mirror of the melt without protection from air, the surface of the melt of aluminum and the most of its alloys is covered with thin density film of oxide, which prevents the further oxidation and the penetration of impurities into the pulling article. Rather low melting points (the melting point of Al is $660°C$, for many alloys it is lower) decrease the chance of impurity penetration out of the crucible and the environment. Besides the requirement of chemical purity (which depends on microimpurities) most of the applications of Al-based alloy polycrystalline articles are not particularly rigid. Aluminum content in room air when pulling of articles of Al-based alloy is lower than the permissible limit because of the relatively rapid surface oxidation of melt and the low volatility of its oxide.

System design with open mirror of a melt is simpler and cheaper than when vacuum is used as they are relatively large and simple and permit growth of long articles of large cross sections.

Design Features

Shaped articles are growing with continuous and semicontinuous acting sets. Semicontinuous acting sets possess simpler design, smaller dimensions, less disturbance in the zone of crystallization than for continuous sets. For semicontinuous sets, an article is pulled continuously during one cycle at which time the cycle is interrupted because the height of the set may exceed the room height.

Continuous set processing is used for pulling out rather small cross sections of long articles of low small rigidity, especially for articles that are sufficiently flexible along the length when demands for a perfect geometry are not high. Continuous sets provide greater productivity because there are only relatively small amounts of ancillary auxiliary operations. Minimum seeding time, lower raw material expense, and lower total electrical energy production costs provide a significant overall production cost savings.

A typical scheme of a continuous set is shown in Fig. 4. Vertical-pulling motion of the article is created by the pulling device. Sets of semicontinuous pulling possess vertical guides with a carrier moving with them. The motion of the carrier is provided by an electrical drive and reducer through winding roller, block, and steel line. Some sets utilize a screw or hydraulic mechanism to provide motion. A more optimal design of the pulling device for the continuous set is to use a roll mechanism. Continuous production of very large lengths can be performed by winging on a roller. It is possible also to use sets of the carriage type. In this case, there is cutting on-the-go and with an interrupted mechanism.

Melting of aluminum alloys in a crucible is usually performed using a nichrom resistance furnace or with a special alloy heater. In principle, it is possible to use an induction heater. The furnace has metallic carcass and platform for movement. For semicontinuous sets, melting and pulling are performed in the same furnace. The supplemental addition to the melt in the crucible is accomplished periodically. But for semicontinuous sets and especially for continuous sets it is advisable to use a separate melting and working furnaces which will increase manufacturing productivity and process stability and improve the quality of the product.

It is very important to provide a constant level of the melt surface relative to the shaper. It is very difficult, or impossible, to raise the crucible containing a large quantity of melted metal. The simplest design is one where a movable table relative to the direction of pulling is used. The shaping device (SD) is fixed on it. The table is drawn using an electrodrive and reducer with the rate equal to the rate of lowering the melt level in the crucible. When the melting and working crucibles are independent, they can be coupled by a special channel for supporting constant melt level in the working crucible.

An important design feature of the sets is the air-cooling system. This system cools the zone near the liquid–solid interface and the crystalline article pulled. It includes a compressor, tube-line, and distributing chamber. Heat removal from SD and parts of mechanisms closed to the melt is carried out by the water-cooling system. It consists of a water distributing unit, cooler, and water collector.

The control panel contains start-control apparatus, control-measuring instrument, indicators, and other electrical units.

All principal parts of the set with the exception of the furnace and control desk are arranged on one mount.

The semicontinuous operation includes the following operations: set preparation, starting, pulling of an article, stopping process, and extracting the pulled article. During the preparation of the set for the start, metal is charged into the crucible and melted. The shaping device is assembled and the seed is fixed in a holder. All parts of the set and the parameters of the cooling system are checked for the start. The start involves cleaning the oxide on the surface of the melt, plunging the shaper into the melt to the required depth, flowing the melt into a slot of shaper, dropping a seed into the slot of the shaper, switching on electrical motors of the pulling mechanism and moving table, turning on air-cooling system, and fixing the necessary flow. Pulling out of an article is performed according to an optimum regime with respect to the profile and the alloy selected. Stopping is achieved by turning the air-cooling system, the drive of the table, and the pulling device. After it new portion of the metal is added into the crucible as needed and the preparation of the following operations is begun. During the preparation of the set and its work, it is necessary to control the state of the crucible, SD, and the slot of a shaper in addition to centering different parts of the set.

Besides the general demands of vibration suppression to all sets discussed above, for Al-based alloys profiles of high geometric purity it is necessary to: guarantee smooth steady motion of the pulling mechanism for the crystallizing article with side displacement of less than 0.001 mm per 1 mm of a run; deviation of a guide column from vertical direction must be less than 0.1 mm per 1 m of a height; the parts of the pulling device, motion table, and SD must be accurately centered and uniaxial; and the center of gravity must be conjoined with pulling axis.

The sizes of the set are determined by the dimensions of articles pulled out. For semicontinuous sets, the width of the cross section of articles is limited by the furnace and crucible, permeable internal sizes of the motion table, and the distance between the guides of the pulling carriage. The length of a profile is determined by the height of the carriage guides. The clearance of the continuous set depends also on the design of the receiver of the profile.

A number of sets of semicontinuous and continuous acting have been developed at Ioffe Physico-Technical Institute, St. Petersburg, Russia. One set permitted the production of 1-m-wide panels. The shape of these panels was analogous to "many tubes in the sheet." The maximum length of articles pulled with the semicontinuous sets was 4.5 m and possibly up to 6 m. These sets illustrated that it was possible to develop equipment for larger cross-section sizes and to obtain long profiles when using continuous sets.

Figure 5 illustrates the operation of semicontinuous sets: Fig. 5a is an experimental-industrial set ASP-1. Fig. 5b is a scheme of the group of sets assembled on unit carcass. The set for continuous operation is shown in Fig. 6.

For manufacturing shaped aluminum products by SM, it is necessary to use accessory equipment for etching shapers in water alkaline solution, water washing, drying and painting, transverse cutting and control, assembly, dismantling and repair of shapers and SD. It is desirable to have a centralized preparation of a melt of raw alloys, which includes melting and purification.

Tool Peculiarities

SHAPING DEVICE. The shaping device [7] predetermines the shape of the cross section of the liquid column of the melt and the article pulled if the liquid–solid interface has no contact with the walls of SD and the crucible. Therefore SD is a contactless crystallizer. The shaping device together with automatic control system regulates the temperature regime of crystallization and cooling of a profile and isolates the warm shaper from a cooler [30]. The shaping device should be simple, relatively light but of rigid construction. At the same time it must provide a sufficient zone of shaping for article and exhibit an acceptable lifetime.

The main parts of SD are the shaper and the cooler. Accessory parts are the seed holder with seeds and supporting carcass. The scheme of SD for obtaining a tube is shown in Fig. 7.

a. Shaper. A Shaper performs the capillary shaping of the melt column. Shapers for growing aluminum articles are constructed of a material that is not wetted by the melt. Therefore the shape of the melt column is created by coupling with the internal edge of a profiling slot. The melt fills the profiling slot when the shaper is plunged into it. The shape of the slot is similar to the cross-section shape of the article being pulled [30]. Linear sizes of the slot differ slightly from necessary sizes of the article. Solidification of the liquid column occurs until about 1 mm, above the slot where the sizes of its cross section differ from the sizes of the slot. It is also necessary to account for crystallization shrinkage [31].

The shaper for profiles of small sizes without internal hollows at its cross section is constructed with a plate and a slot. But the shaper for larger articles and for articles with interval hollows must possess a more complex shape and is usually a compound construction.

The shaper for obtaining a tube shown in Fig. 7 possesses a body (3), outside (1), and inside (2) insets.

The shaper must not exhibit any deformation during heating to 750–800°C and during thermal cycling for tuning in and out. In partial cases it must resist enlargement during recrystallization. The body of a shaper may possess a special shape, e.g., box form or Π-form with high rigidity.

It is very important that the material of the shaper does not dissolve in the aluminum melt. Shapers for pulling out of Al-based alloys can be constructed from high-density special cast iron, ceramics, metal ceramics, to increase the resistance to interaction with the melt covers, and paints for shapers are used.

b. Coolers. The cooler removes heat during crystallization from different parts of the article being pulled. Thus the cooler, together with SD, supports the liquid–solid interface on the given level for obtaining an article. At the same time the cooler is used to increase the pulling rate of a profile with the cross section of necessary sizes [7,30].

The cooler is shown in Fig. 7. It is an air (gas or vapor) receiver–distributor system. The cooler contains a two-chamber section with the channel for water cooling. One chamber is the receiver, the other is the distributor. Outlets of distributing chamber have a form of orifices for the stream blowing or slots for wide-band blowing made directly in the body of the cooler or on the surface. The cooler of wide-band blowing gives high uniformity of the air flow along a perimeter of crystallizing zone. It is recommended to use such cooler for cooling simple profiles of constant thickness over the perimeter. The cooler of the stream blowing provides for good localized cooling. They are used for cooling articles of complex shape with variable wall thickness at the cross section. It is advisable to sectionalize a distributing chamber to intensify the blowing action on the thick-walled parts of a profile with two- or multirow orifices arrangement and to decrease the step between orifices.

The necessary amount of cooling for an article being pulled can be done by controlling the flow rate of the air and its distribution along the perimeter of the article. It is also important to compensate for temperature fluctuations near the crystallizing zone by supplying the cool medium to the desired surface location [32]

c. Shaping Device Accessories. Extracting the first portion of the melt out from profiling orifice at the beginning of processing is performed by seeding, which couples the pulling device with the article being pulled out. The seed is wetted by the melt of aluminum alloys. Therefore usually copper or nickel or other metals are

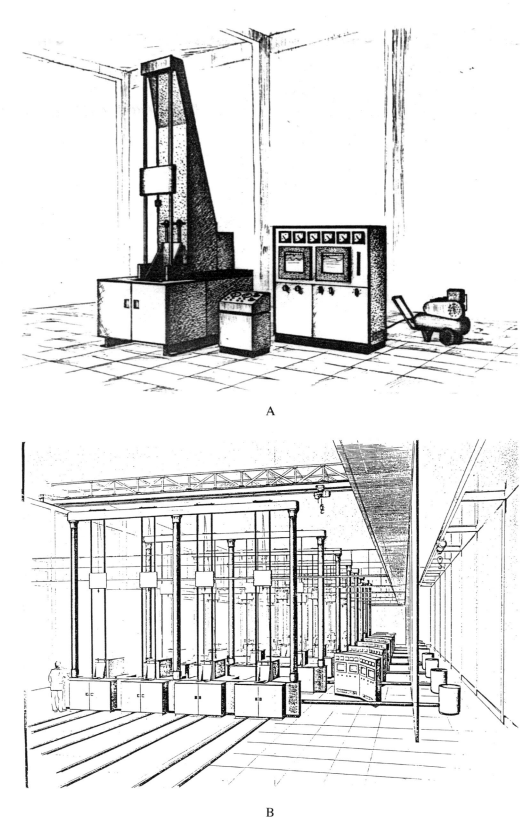

A

B

Figure 5 Sets of semicontinuous operating: A—set ASP-1; B—scheme of group of sets assembled on one carcase.

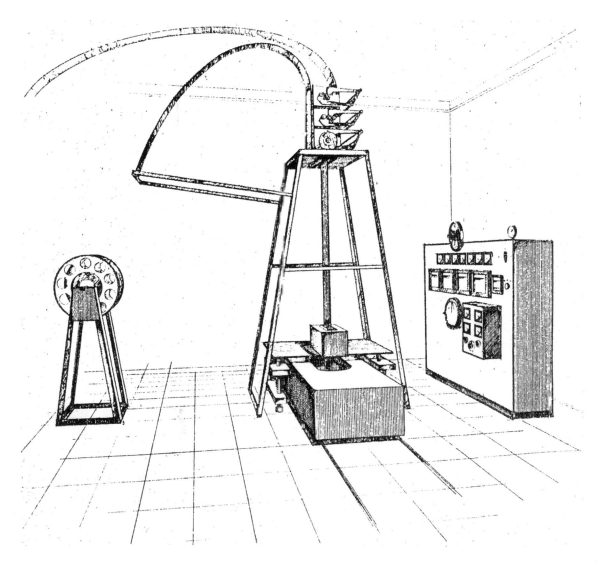

Figure 6 Set of continuous operating.

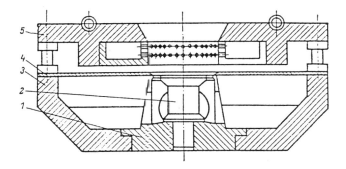

Figure 7 Shaper of pulling a tube. 1. Outside bush; 2. insert bush; 3. body of shaper; 4. heat shield; 5. cooler (outside).

used. The ribbon of such metal is bent accordingly to the shape of the profiling orifice and it is fixed in the seed holder.

The accessory parts of SD include bearing carcass for connecting a shaper and a cooler and fixing them on the motion table. They also include supply tubes of water and air and various coupling pins.

d. Unification of Shaping Device. It is very important to unify the SD [30] to include multiple profiles for simplification and to reduce production and maintenance costs. It is possible to combine construction of various sizes of SD and the simplest approach is to couple separate parts of SD and SD with the set (except for coupling seed holder with seed). Unification of the main parts of SD as a whole (shaper and cooler) and their elements is beneficial for profiles of rather small cross sections and simple shapes, especially for a wide variety of profiles. Shapers and coolers can have unified carcasses with specific inserts (for shapers) and nozzles (for coolers) for articles of like or similar shapes and sizes.

e. Other Tools. Crucibles of sets are changed periodically because of wear and when converting to pulling articles of another aluminum alloy. The set of a continuous action of roller-type is constructed by changeable pulling rollers depending on the shape and sizes of profiles being pulled out.

Processing Parameters

Maintenance of SM pulling out sets is possible only with knowledge of the operation of the installations with SD and the various profiles being manufactured. The profiles must have an assigned level of deflection to maintain their geometry within specification limits with respect to cross section and length, taking into account the necessity to maintain a constant height of the liquid–solid interface for different parts during the production process. Therefore an optimum combination of processing parameters must be established. For an aluminum alloy, in addition to melt temperature, melt pressure, and pulling rate, it is necessary to design for cooling intensity, temperature, moisture, and use of cooling air or another cooling agent [7]. In the above discussion, these parameters were considered separately and in combination with varying cross-section sizes. Every profile must be grown using optimum processing conditions, which are dependent on the aluminum alloys used, and they are established in advance.

Processing begins at start regime and transition to the main work regime. Although the optimum pulling rate and other operating parameters may vary with crystallization of individual alloys, the control parameters must be constant throughout the entire process. This is especially important for SM capillary shaping which is processing-parameter sensitive. Parameter stability is less sensitive for continuous processes. For semicontinuous acting sets, it is necessary to design for varying temperature conditions due to loss of the melt in the crucible.

If the processing parameters are not held constant, varying geometry and defects will result. Examples include: excessive pulling out rates, high melt temperature, or insufficient interfacial cooling, which will produce decreasing cross-section sizes and poor control. At lower pulling rates or melt temperatures, the cross-section increases. If the interface has sunk very low, the article begins to freeze into the profiling orifice of the shaper.

Processing conditions can affect oxide film formation on the surface of the melt at 670–720°C. Although this film protects the bulk of metal from further oxidation, it also changes the wettability, surface tension, and reciprocal viscosity of the melt making it difficult to fill the profile orifice by the melt and subsequent seeding. In some cases, vibration of a melt was used to decrease this phenomenon [7].

It is very important to remain within the processing tolerances when size variation and surface quality affect tolerances. The resulting design tolerances must reflect the combined influence of all parameters on the overall quality. It is also necessary to determine the highest pulling out rate for any given cooling system and cooling agent. Generally, it is more difficult to control the temperature in the shaping zone of an article.

Continuous pulling process can be automated. For several sets with an automated control system, the melt temperature deviations from rated temperature in the shaping zone must be less than $\pm 1.5°C$; the pulling rate deviation must be less than $\pm(1–1.5)\%$ of the rated value. Complex automatization can be made according to two variants. The first is stiff automatization. In that case the system gives stability of main parameters with given precision. This variant is simpler and for many cases permits the necessary process control precision. The second is flexible automatization, when constant geometric sizes of article are created by controlling some operational parameters by feedback with respect to geometric size deviations. Such a system is useful when very close size and quality tolerances are demanded. In this case, the greatest difficulty is the precision of size measurement.

Thermal Conditions and Processing Parameters

Thermal conditions and crystallization parameters determine the crystallization rate, shape, and structure of the shaped articles [7,33]. The pulling rate is determined by the crystallization rate, which depends on the heat removal rate. The heat exchange for growing shaped articles can be described by accounting for heat flows in the melt, in the crystalline article during growth, and within the production environment itself.

TEMPERATURE DISTRIBUTION IN DIFFERENT ZONES. Thermal conditions of crystallizing by SM are similar to conditions for the Czochralski method or vertical continuous casting. There is an immediate connection of the pulling article with the melt. Therefore the heat conditions for growth are more stable than for conventional casting.

The scheme of cooling of a part being pulled and then cooled by air, water, or air–water sprays is shown in Fig. 8 [7]. In the liquid part of the profile, called "the melt column," the heat is conducted to the liquid–solid

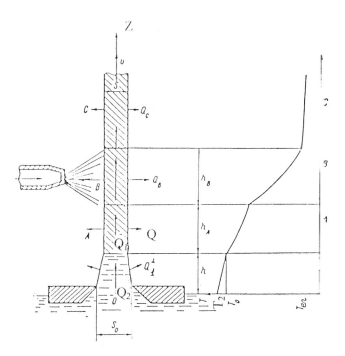

Figure 8 Heat scheme. From the left there is cooler; A is buffer zone; B is cooled zone; C is part cooled by free convection, T_{en}, T_o, and T_1 are the temperature of medium, crystallization, and melt, respectively; h and Q are height and heat flows for zones, respectively; S_0 is the width of shaping orifice, S is the thickness of article. Arrows show heat flow.

interface due to the thermal conductivity and convection (the flow Q_2), and above it due to the heat conductivity (Q_1). At the liquid–solid interface the equation of continuity of heat flow is:

$$Q_1 = Q_2 + \rho L v. \tag{2}$$

Here Q is a heat flow, index 1 corresponds to the solid and index 2 to the liquid, L is the crystallization heat (latent heat) of an alloy, ρ is the density, v is the crystallizing rate (the pulling rate). From here

$$\rho L v = \lambda_1 G_1 - \lambda_2 G_2, \tag{3}$$

where λ is the heat conductivity, G is the temperature gradient, indexes 1 and 2 correspond to the solid and the liquid, respectively.

The crystallization heat is removed only by the heat conductivity. The heat flow from the melt column to the surroundings can be neglected because the height of the melt column is small (0.5–5.0 mm). In this case, the liquid–solid interface is considered to be flat. In the buffer zone A and zone C the profile is cooled because of convection, B is the zone of force cooling [7].

In the case of alloy crystallization, the existence of the crystallization zone must be considered. The length of this zone from the liquidus to the solidus depends on the alloy composition, profile geometry, and pulling regimes.

The alloy composition is specified as percent by weight. For ribbon profiles of Al–2% Mg–0.4% Mn alloy with the temperature interval of crystallization ΔT_{cr} equal to 25°C, the zone length is changed from 2 to 4 mm. For the same profiles of Al–6% Zn–2.2% Mg–1.7% Ca–0.4% Mn–0.2% Cr alloy with $\Delta T_{cr} = 161$°C, the length of the zone of crystallization is 15–40 mm. But for 29-mm-diameter circle tubes of the latter alloy, for one-sided cooling of the outer surface the length is changed from 60 to 110 mm.

According to Eq. (3), the crystallization rate increases when the temperature gradient in the melt decreases. The temperature gradient depends on the contribution of the heat conductivity and the (mixing) convection during heat transfer. The change of the temperature gradient in the melt results in variation of the pulling rate to provide the shape and the geometry of the profile grown [33].

The change in the temperature of the melt along the pulling axes Z near the interface is described by the following empirical equation [34]:

$$T = T_2 - (T_2 - T_L)\exp(Kz). \tag{4}$$

Here T_2 is the temperature of the melt in a crucible, T_L is the temperature of the liquidus, K is a coefficient.

In the case of pulling of a ribbon of commercially pure aluminum the temperature of the melt in the crucible is higher than at the interface by about 40°C.

The pulling rate v is determined by the thickness of the profile and cooling intensity. Therefore it is necessary to know the heat emission coefficient α for a specific condition. This can be estimated based on an empirical equation [7]: Thus for the profile without internal hollows, α can be calculated from Newton law [35]:

$$\alpha = (\rho L v / \Delta T)(F_1 / F_2), \tag{5}$$

where ΔT is the difference in temperatures of crystallization and of environment, F_1 is the area of the cross section of a profile, F_2 is the surface of air cooling. This equation for stable growth profile is based on the condition that at least the crystallization heat is removed by frontal blowing off.

For tubes pulled under the force cooling of the outer side only, the heat emission coefficient is as follows [35]:

$$\alpha = \rho L v F_1 / (\Delta T F_2) - \alpha_1 F \Delta T_1 / (F_2 \Delta T), \tag{6}$$

where α_1 is the heat emission coefficient for the free convection heat exchange on the internal surface of the tube, F is the internal surface of the piece of the tube conformable to outer blowing zone, ΔT_1 is the difference in crystallization temperature and temperature of the atmosphere inside the tube.

The values of α calculated by Eqs. (5) and (6) are in fact effective heat emission coefficients. The zone of most intensive air-cooling is usually higher than the interface. The height of the cooling zone above a shaper and the intensity of blowing off are selected to exclude perturbation of the melt column by airflow. The buffer zone is lower than the blowing zone. In the buffer zone the blowing off is practically absent and the cooling is due to the free convection [7].

It was shown [36] that the temperature as a function of the article length (the cooling curve) can be described by means of some linear segments. The rate of cooling m can be determined for every section:

$$m = \left[\ln(T - T_{en}) - \ln\left(T' - T'_{en}\right)\right] / (t' - t) \tag{7}$$

where T and T_{en} are the temperatures of article and environment, respectively, at time t; T', and T'_{en} are the same at time t'.

The crystalline structure of an article depends most of all, relative to other thermal parameters, on the cooling rate in the crystallization interval called the "crystallization index" [37]:

$$\chi_{cr} = \Delta T_{cr} / \Delta t_{cr} = v G_{cr} \tag{8}$$

where ΔT_{cr} is the temperature interval of crystallization, Δt_{cr} is the time of passing by the article of the crystallization interval, G_{cr}, is the temperature gradient in the crystallizing interval.

The crystallizing rate determines the weight of a profile pulled at the unit time which is proportional to their heat capacity and dependent on the shape and cross-section sizes [38] of the profile. It is possible to say that v determines the metal capacity of an article (the weight of the unit length). Practically, it is possible to consider that G_{cr} is equal to $(T_2 - T_0)/h$, where T_0 is the crystallization temperature, h is the height of the melt column [39]. It was shown [38] that in the case of forced cooling the temperature gradient distribution along the length of an article exhibits a maximum in the zone of most intense cooling (Fig. 9).

The thermal conditions of pulling 20-mm-diameter commercially pure aluminum rod of 2 m length were studied based on a typical approach utilized for continuous casting [17]. In contrast to casting, the pulling direction for the Stepanov method is opposite to the force of gravity, and a contact of pulled metal near the interface with surrounding walls is absent. As is known, the primary characteristic feature of continuous casting is a small crater. In the case of SM crystallizing, it is possible to consider in the crater capacity, melt volume limited by the interface and the horizontal plane going through the upper surface of a shaper.

Figure 10a shows the distribution of temperature in a melt at the crystallizing zone and near it. The cooling

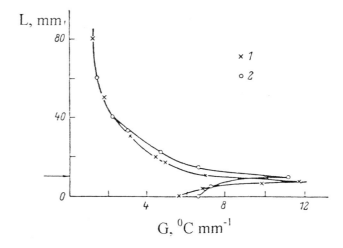

Figure 9 Temperature gradient along the length of the Al-3.1%Mg-0.5%Mn-0.7%Si alloy ribbon pulled with various rates (1 is 71 mm min^{-1}, 2 is 127 mm min^{-1}) under two-side air blowing. The arrow shows the cooled zone.

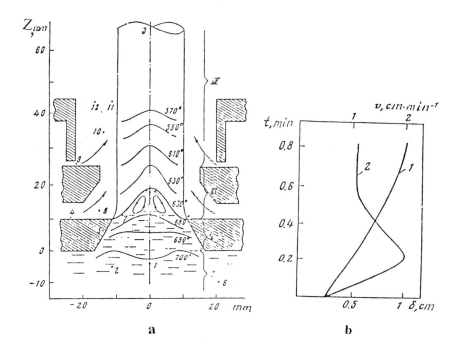

Figure 10 Heat conditions of growing 20 mm diameter aluminum rod. a—Isotherms at crystallization zone: 1–12 are numbers of thermocouples (1–3 thermocouples move together with pulled crystal; the rest are fixed); I–IV are cooled zones. Arrows show cooling air flows. b—Length δ of crystallized part of the rod (1) and the rate of radial growth v (2) as a function of the time of a motion of the rod t. (From Ref. 17.)

rate of metal in zones shown in Fig. 10a was determined by the equation: $v_{cool} = (t_s - t_f)/\tau$, where t_s and t_f are the temperatures in the beginning and in the end of the zone, τ is the duration of passing the zone. The cooling rate at the zones is given in Table 2. The dependencies of a size of crystallized rod piece and of the radial crystallization rate on time are shown in Fig. 10b. The solidification of metal takes place in zone III (the zone of air ejection) where the cooling rate (Table 2) is maximum (0.75–0.85 K sec^{-1}).

It is seen that crystallization of the entire cross section of the rod occurs within 0.8 min. The maximum crystallization rate of 2 cm min^{-1} corresponds to about 0.2 min from the beginning of the crystallization. This can be explained by the relatively high value of α of liquid metal and by considering the linear temperature

distribution in the oxide film of the metal which produces the relative cooling rates of the metal in zone III (0.85 K/sec) and the solidification rate (about 1.5 cm/min). These are close to the corresponding rates for continuous casting.

The pulled article is cooled by forced air and natural convective heat exchange. It is clear that in this zone the maximum contribution is due to forced air cooling. After leaving this zone, an article is cooled mainly by convective heat exchange. The temperature and temperature gradient along the length of the article smoothly decrease outside the crystallizing zone [34].

Thin-walled profiles (the thickness of a wall of about 1.2 mm and less) have rather large specific surface in comparison with the thick-walled examples leading to increased heat removal by air cooling and to increased

Table 2 Cooling Rate in Zones (in units K s^{-1}).

The number of thermocouples according to Fig. 10a	Zone I	Zone II	Zone III	Zone IV	Zone V
1	0.06	0	0.75	0.37	0.09
2	0.04	0.04	0.85	0.35	0.09

pulling rates to maintain the optimum axis temperature gradient [40]. It has been shown that uniformity of the growth of ribs on ribbed tubes depends both on the distribution of airflow over the perimeter and on the temperature field in the melt [41]. To obtain flat tubes under one-side outer cooling, the height of the interface surface of the tube is higher than the outer surface (for 3-mm wall thickness the difference of the height is about 0.3–0.5 mm).

The heating zone determines the thermal conditions. It provides a given vertical temperature gradient in the profiling orifice and in the pulled article near the interface. The heating zone controls the change of temperature (T) within necessary limits at the time of start regime, the stability of T during the time of pulling, and the reproducibility of thermal conditions from one process to another [33].

SHAPER INFLUENCE THERMAL CONDITIONS. The shaper plays an important role in forming the temperature distribution near the liquid–solid interface [7]. The temperature of different parts of a shaper in a melt is diverse. The coolest part is the upper open surface [42]. The heat emission from the open surface reduces overheating of the melt in the volume enclosed by the shaper. The melt within the volume of the shaper is supplemented uninterruptedly when the article is being pulled. The intensity of heat removal from the melt within the shaper is estimated by the value $W = F_{cool}/F_{heat}$. Here F_{cool} is the open cooled surface of the shaper, and F_{heat} is the surface heated by the melt. F_{heat} is determined as the area on the horizontal upper surface of the shaper limited by the counter of the shaping orifice and of the envelop remote from the counter of the orifice for the distance equal to the depth of the shaping orifice [42]. The heat removal from overheated melt that is in the volume inside the shaper goes through the body of the shaper and through the liquid melt. But the intensity of the heat removal from the melt inside the shaper depends on shaper design. Control of the heat removal is achieved by changing the proportion of the heat dispersed from the open surface of the shaper.

Shapers can be of three types according to the intensity of heat removal. A "very hot" shaper has relative heat removal W of less than 0.5. Dividing walls are washed by the melt both from the inner and outer sides, which restricts the shaping orifice. The upper cooling surface of the shaper, nonwetted by the melt, has an area per unit length of the orifice smaller than the surface of the shaping orifice for the same unit length. A "semi-hot" shaper has W in the range from 0.5 to 2.0.

The shaping slot is restricted by flat or relief cooled surfaces of the shaper. The area per unit length of the orifice is larger than the surface of the shaping orifice for this piece. "Very cold" shaper has $W > 2$. The shaper is heated by the melt only into its volume. The upper and side walls are open. All shapers permit pulling profiles for certain minimum overheating of the melt. This minimum overheating decreases when the heat removal decreases.

A semiempirical method for estimating the thermal characteristics of shapers is offered in Ref. 43. It was based on an exponential distribution of temperature along the pulling axes. The conclusion was made that the exponent is the characteristic of the thermal resistance of the shaper. It depends on geometric sizes and thermal properties of the shaper as well as on thermal properties of the melt.

COOLING OF ARTICLE. Forced cooling provides faster pulling rates and improves the structure and properties of aluminum articles. Typically, forced air or air–water sprays are used. In many cases it is necessary to change the thickness of a profile in different places of its cross section, which may lead to added expense. A multisection cooler provides controlled delivery of the cooling agent to specific surfaces of the article and regulates the thickness along the perimeter of its cross section [32]. Two-sided cooling helps to level the liquid–solid interface on the wall thickness and to increase the pulling rate and achieve a consistent structure and shape along the length of the profile for complex geometries. The article may be cooled from the inside and outside, which is possible if the design of the article permits it, such as an article with many hollow sections [32]. The cooling of additional zones using an air–water spray improves the structure and process productivity [44,45].

Part Geometry

Consistent geometric control, including shape, varying cross-section sizes, deflection along the length, with very tight tolerances is the design objective for the SM. For a curvilinear longitudinal profiles, it is also necessary to maintain consistent size control of longitudinal sections. This is accomplished by controlling the shape of the melt column.

There are some general points related to geometry and related operational parameters:

1. In many applications, SM profiles without calibration and finishing operations. For demanding dimensional control, some machining may be necessary.

2. Cross-sectional sizes of an article depend on several factors but the most important are profiling orifice size and operational parameters such as v and overheating of the melt ΔT. For example, in general case [46], there is a following dependence for the wall thickness S:

$$S = f_1(v, \Delta T, \alpha, h, h_A, h_B, c) \qquad (9)$$

and

$$h = f_2(a, S_0, S) \qquad (10)$$

where h is the height of the interface, h_A and h_B are the sizes according to Fig. 8, a is the capillary constant of the molten alloy, S_0 is the width of the slot of profiling orifice of the shaper, $\Delta T = T_2 - T_0$ (see the caption for Fig. 8), c is a constant for a given alloy. The capillary constant $a = (2\gamma/\rho g)^{1/2}$, where γ is the surface tension, ρ is the density, and g is the gravity acceleration.

There are empirical and semiempirical formulas and graphic dependencies for general sizes of cross sections and wall thickness. For example, the radius r of a circular rod can be found by empirical equation [47]:

$$r = r_0 - 0.18(1 - 0.1h)h^3 \text{ mm}, \qquad (11)$$

where r_0 is the radius of the profiling orifice of a shaper. This equation can be also used for extended parts of complex profiles with small curvature.

By analogy, the thickness of a ribbon S can be found by empirical formula:

$$S = S_0 - 0.18(1 - 0.1h)h^3 \text{ mm}. \qquad (12)$$

It is worthwhile to note that the height of the interface h is a resulting regime parameter dependent on v, overheating of the melt, the pressure in the melt, and the cooling intensity.

Graphic dependencies of the outer and internal diameters and the wall thickness of tubes on the pulling rate are given in Ref. 47 for different outer and internal diameters of a slot.

These equations and their dependencies can be used both for more complex cross-section shapes and for structural components. But the greater the interface the greater the deviation of the shape of a complex profile from the shape of the slot. Smaller interfaces (2 mm and lower) lead to smaller geometric sensitivity of an article to changing pulling rates.

3. The shaper design must account for linear shrinkage ε for the article of the given alloy. For conventional cast $\varepsilon = (l_f - l_0)/l_0$, where l_f is the linear size of an element in a cast mold, and l_0 is the linear size of an article. In the case of SM, crystal growth is used without the casting shape, and by convention ε is considered as linear shrinkage and l_f is the size of the element of the profiling orifice [31]. Profiles obtained by SM have no pores and blisters. Therefore the shrinkage is created by the difference in thermal expansion coefficients of the crystallizing alloy and the material of a shaper for the given temperature range. This difference is rather large in the case of a shaper made from cast iron. But this difference increases still further for ceramic shapers. It is also necessary to account for the difference in sizes (diameter) of a profiling orifice and the interface for the shaping capillary. This capillary shrinkage increases with enlarging height of the interface. It is necessary to have experimental data for consideration of two mechanisms of the shrinkage to construct the shaper well with allowance for conditional ε.

It was shown that the conditional shrinkage of large profiles, especially of closed sections with partitions, is from 1.5 to 2 times smaller than for profiles of average sizes when linear sizes of a cross section are not more than 200–250 mm [31]. For different Al-based alloy profiles by SM with different shapes and sizes, ε is in the range from 0.8% to 2.5%.

4. Some peculiarities of the cross-section shape of Al-based alloy articles were noted in Sect. 7. It was shown that the longitudinal deflection of long articles depends on the axes stability during pulling connected with surface tension of a melt column. Estimates of maximum mechanical disturbances were made in Ref. 48.

The dynamic pressure of cooling air has a strong influence on the stability of processing [49]. Its influence is different for thick-walled ($S > 5$ mm) and thin-walled ($S < 1$ mm) profiles. In the case of thin-walled profiles it is very important to support an interface height [40].

5. Many investigations were devoted to the constancy of the shape and sizes of profiles [7,30,31,47,50]. The constancy of geometry depends on the stability of operational parameters. The constancy of the cross section of an article depends on the constancy of the width of the profiling orifice and temperature field. The operational regimes must be kept constant during pulling whether it is for one article or groups of articles.

It is necessary to distinguish the constancy of outer and internal sizes of profiles, their outer and internal elements and walls. A different thickness of walls can be both undesirable and dictated by design.

A great number of measurements of sizes of profiles were made. Profiles with an outer size of more than 20 mm and wall thickness of more than 1.2 mm were acceptable clearances. An allowance for cross section can be ±1%; for wall thickness it must be lower, ±10%; for longitudinal deflection, per unit length should not to be higher than 0.5–1.0 mm/m [7]. Such values correspond practically to pressed profiles or better. At the same time pressed profiles meet demands of most customers.

Similar to pressed profiles, allowances of sizes of articles obtained by SM depend on absolute nominal values of the sizes. Thus for profiles of outer sizes ranging from 5 to 20 mm and profiles of wall thickness less than 1.0–1.2 mm, allowances are greater than the values shown above. However, for large-size articles, deviations of cross-section sizes are significantly smaller ranging from ±0.2% to ±0.8% [31].

A scatter of sizes of wall thickness for thick-walled profiles is given in Table 3. The data of this table provide the following conclusion [49]:

a) Scatter of wall thickness of thick-walled square beams along their length is less than for thick-walled circular tubes.

b) In the case of square profiles the outer surface has smaller scatter of sizes than internal surfaces and for circular profiles. It is possible that four right angles in the cross section of a square beam fixes the outer surface of the profile and stabilizes both its profile and the wall thickness.

c) Scatter of sizes of 36-rib tubes is less than for smooth circular tubes. It can be explained by fixing action of ribs. As the scatter of the outer diameter of ribbed tube is significantly less than for the internal diameter.

d) Scatter of diameter sizes of 36-rib tubes depends mainly on the precision of the cooling agent delivered in the space between all ribs of the pulled profile.

It was shown in Ref. 51 that the relative scatter of ribbon thickness increases with enlarging the value $(S_0 - S)$ or with increasing the height of the interface.

Table 3 Different Thickness of Thick-Walled Profiles

Type of profile	$(S_{max} - S_{min})/S_{max}$
1. High quality beams of square cross section 32·32 mm, $S = 6$ mm	0.012–0.058
2. Circle tubes, internal diameter is 64 mm, $S = 5$ mm	0.12–0.55
3. Thirty-six-rib tubes, internal diameter is 60 mm, diameter together with ribs is 95–96 mm, $S = 4$–6 mm	0.12–0.18

Most deviations of sizes from those assigned occur at the start and at processing transitions. They can be lower at pulling profiles with a continuous acting set [7].

Stirring of a melt near the shaping zone can decrease thickness size scatter [32]. Decreasing thickness size scatter is also facilitated by using a precision shaper. It was shown in Refs. 31 and 50 that the scatter of sizes of the thickness of a box profile can be decreased about 2 times to about ±6% when accounting for the influence of changing the width of a shaper slot.

The root-mean-square deviation of wall thickness and the thickness variation coefficient were determined for various shapes (tubes, blades with partitions, cylindrical two-walled profiles with partitions, complex profiles of air heater–adsorber, and others) [31,32]. For the best samples of wall thickness in the range from 1.5 to 2 mm, obtained at optimum conditions, the root-mean-square deviation ranges from 0.18 to 0.22 mm, the variation coefficient ranges from 0.08 to 0.13. In the case of obtaining smooth wall tubes the rotation can decrease the variation coefficient down to 0.03–0.06.

When producing different cross-section thickness profiles, it is necessary to construct a shaper that provides different interface heights in places of different thickness. It is also necessary to cool different places of a cross section with an intensity proportional to the thickness of crystallized element using a multisection cooler.

Surface and Structure of Articles

SURFACE QUALITY. Surface quality is determined by topography, structure, and specific physical-chemical properties. Chemical composition and crystallization interval exhibit the greatest influence on the surface quality. Other factors influencing surface quality include melt purity, interaction of the melt with a crucible and environment, and stability of processing parameters. Corrosion, heat radiation and absorption, strength, and friction coefficient are dependent on surface quality. In some applications, surface color, which depends on oxide properties of the crystallized alloys, is important. The surface of aluminum profiles is usually smooth and shiny, duralumin surface is bluish, Al–Mg-based alloys exhibit a dead-blue surface [9].

The microroughness of articles of Al-based alloys as determined by profilometer measurements is typically less than 14 μm and optical measurements have shown even lower roughness. Macroroughness is on the same order as exhibited by pressed articles [7,52]. Estimations of macroroughness for millimeter or centimeter scale (the distance between asperities along certain direction) show highest purity class except that some samples have centimeter roughness somewhat higher. Measurements of the macroroughness by means of universal measuring microscope revealed transverse waves of 0.05-mm height (in some cases up to 0.1 mm) spaced from 20 to 40 mm.

It is shown also that the quality of the inside surface of tubes is better than the outside surface because the inside surface is more protected from oxidation and airflow disturbances [13]. The most perfect surface, nearly a glassy surface, is observed for articles with the column structure [39].

Practically, it is possible to consider that the quality of articles produced by SM is similar to articles obtained by pressing and much better than those produced by conventional methods. In many cases the quality of the surface of SM grown articles meets the demands of most applications. However, particularly for alloys with a wide temperature interval of crystallization, some machining and subsequent coverage of the surface by a thin film are required. A very high quality surface is observed for the eutectic Al–12.6% Si alloy, which is typically sufficient for decorative uses.

STRUCTURE. 1. A dendritic structure is the most typical form of the growth for Al-based alloy articles produced by the SM. In this case cellular, twinned, and equiaxial dendrites form columnar, fan, and equiaxial structures, respectively, which are dependent on the morphological stability of dendrite.

Columnar macrostructure consists of coarse, drawn grains of developed cellular substructure [53]. There are some differences in the sizes of grains and cells as a result of the composition of an alloy, the geometry of profiles, and the pulling rate. Thus in the case of 2-mm-thick ribbons of commercially pure Al, grown at pulling rates from 3 to 10 m hr^{-1}, the grains exhibited a length from 50 to 500 mm; transverse sizes grains varied from 0.1 to 40 mm. For tubes of the diameter from 3.6 to 5.0 mm and wall thickness from 0.2 to 0.8 mm, grown at pulling rates ranging from 6 to 8 m hr^{-1}, the grain length changed from 8 to 16 mm, grain width ranged from 1.5 to 3.5 mm [39]. Al–1% Mn ribbons of 0.3–0.4-mm thickness processed at $T = 680°C$ initially possessed a columnar structure with grains drawn along the pulling direction. When the pulling rate increased from 5.5 to 13 m hr^{-1}, the grains decreased from 20×3 mm^2 to 6×1 mm^2. In transverse and longitudinal sections of grains the chains of the MnAl$_6$ precipitates along the pulling direction were observed [51].

Macrostructure of Al–(0.8–1.6%) Mg tubes, obtained at a pulling rate of 9.5 m hr^{-1} in the first part of processing, was also columnar. A β-phase of Al$_5$Mg$_8$ in the form of broken chains between axes of dendrites was obtained [7]. When duralumin was pulled from tubes at a pulling rate greater than 3.4 m hr^{-1} for $G_2 = $ (10–12) K mm^{-1} and $G_1 = $ (1–2) K mm^{-1}, the typical structure was columnar which was created by cellular dendrites with long shafts drawn along the side surface of the tube [54]. For pulling rates equal to 3.4 m hr^{-1}, the grains were about 2–2.5 mm. In a longitudinal section, the grain consists of parallel primary shafts of cellular dendrites. In transverse sections, the grains are created by 45-μm dendritic cells. Grain boundaries consist of the eutectic of α-solid solution and intermetallic compound of Al$_2$Cu. The grain boundary thickness is about 5 μm.

Fiber structure is formed in the manner of columnar structures by cellular dendrites. This formation occurs at the same pulling rate as for columnar structure but for higher melt temperature. In the case of commercially pure Al, grain sizes ranged from 4 to 50 mm [39]. The fiber structure of 29-mm-diameter duralumin tubes formed at the high overheat (the temperature of the melt was higher than 700°C) and the pulling rate was higher than 3.8 m hr^{-1} [55]. The grains were 5 to 10 mm in length, drawn along the pulling direction and lightly rotated. They may also exhibit cellular structure where the cell length ranged from 200 to 300 μm and the width was about 40 μm. The cell size was reduced and became more equiaxial when the pulling rate increased [55].

Fan structure is formed particularly when Al–1% Mn and Al–(0.8–1.6)% Mg profiles are pulled at usual conditions. The latter alloy transforms to the fan structure from columnar one [7]. The transformation to the fan structure may occur when the impurity concentration near the liquid–solid interface reaches a critical value and the constitutional supercooling begins to influence on crystallization. This is corroborated by the columnar structure along all the length of pure Al article. The fan structure of Al–1% Mn ribbons consists of colonies of decades of fan-shaped divergent crystals, which are flat twin dendrites of 100–200-mm thickness.

Fan structure of 29-mm-diameter duralumin tubes that were pulled from gently overheated melt (660–680°C) is formed in a wide range of pulling rates (from 2.1 to 3.8 m hr^{-1}) [54,55]. The structure consisted of twinned dendrites, which grew in the twin plate creating long lamellae along the pulling direction. Dendritic cells are about 44 μm. In the area of secondary branches and between dendrites, liquation is observed which is typical for impurities of $K < 1$. The concentration of Cu in the dendritic boundary is about 2.3%, in the area of secondary dendritic branches it is about 3.4%, in the interlayer between dendrites it is about 7% [55]. It was shown that a critical pulling rate exists. For rates higher than the critical rate, twin dendrites of the fan structure transform into the columnar or fiber structure. The critical rate decreases when the heat removal from the overheated melt in the volume of a shaper is reduced.

Equiaxial structure arises when the overheating of the melt is very small and is also due to a negative temperature gradient for a specific design of the shaper as in the case when only an upper zone of water cooling was used. [55].

2.   The structure of the profiles twisted during their pulling and stirring of the melt at the same time was studied [54]. The process was investigated for duralumin tubes under the following conditions: a) pulling rate was $2.9\,m\,hr^{-1}$, twist rate was 0.1 and 0.4 rpm; b) pulling rate was 2.9 and $1.45\,m\,hr^{-1}$, twist rate was 0.4 rpm.

Tubes obtained at the pulling rate of $2.9\,m\,hr^{-1}$ and the angular velocity of 0.1 rpm exhibited a columnar structure with about 65-μm-diameter cells. Increasing the angular velocity up to 0.4 rpm leads to the decrease from 1.5 to 1 mm of macrograins created by cellular dendrites. Decreasing the pulling rate to $1.45\,m\,hr^{-1}$ for the angular velocity 0.4 rpm does not change the type of a structure but it does make it coarser (the cell diameter is about 85 μm).

3.   The structure of articles was estimated quantitatively using stereometric optical metallographic methods. The quantity of inclusions (structure elements) per unit area and the extent of their orientation were measured for areas in some transverse and longitudinal sections. For this purpose, the quantity of inclusions per some known area (for ribbed twisted tubes of Al–1% Mn alloy) or the quantity of intersections of inclusions by the secant of a given value (for twisted flat-walled tubes of duralumin) was calculated [32] with an optical microscope of magnification to 320×. Microstructures

(a)   (b)

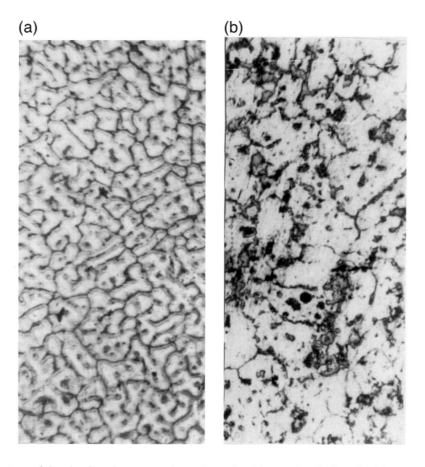

Figure 11   Microstructure of duralumin tubes grown from the melt with rotating (b,d) and without rotating (a,c); (a) and (b) are transverse sections, (c) and (d) are longitudinal ones (200×). (From Ref. 32.)

(c)   (d)

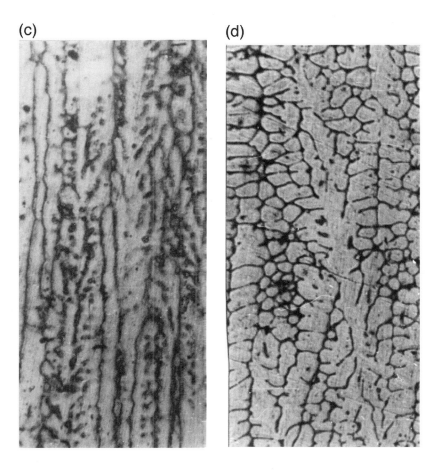

Figure 11 Continued.

of latter articles and of articles obtained without rotation are shown in Fig. 11. The extent of linear orientation of inclusions α_1 in article volume was determined by the following equations [32]:

$$\alpha_1 = (X_\perp - X_\parallel)/(X_\perp + X_\parallel) \text{ for Al} - 1\%\text{Mn,} \quad (13)$$

$$\alpha_1' = (m_\perp - m_\parallel)/m_\perp \text{ for duralumin} \quad (14)$$

Here $X_\perp$ and $X_\parallel$ are averages of inclusions per 1 mm^2 for transverse and longitudinal sections; $m_\perp$ and $m_\parallel$ are averages of intersections of inclusions by 1 mm secant normal or parallel to the growing direction for longitudinal section. Results of the measurements are given in Table 4. Some conclusions can be made.

a) Al–1% Mn alloy:

- For the tube part of the article the linear orientation α_1 does not depend on rotation. $X_\perp$ and $X_\parallel$ decrease by about a factor of two, microstructure becomes coarser.

- For twisted ribs α_1 is lower than for tube part. It is probably due to more intensive stirring of the melt in the range of their formatting. This may also be related to specific force two-side cooling. X for ribs is larger than for tube part. This can be explained both by stirring effect and by an increase in the resulting pulling rate. Microstructure maintains its orientation parallel to the direction of preferred heat removal. This direction coincides with the resulting direction of rib growth.

b) Duralumin:

Twisting of tubes coarsens the microstructure of the transverse section (Fig. 11b as compared with Fig. 11a) and at the same time decreases $m_\perp$. Twisting decreases also α_1' and the length of microstructure elements of the longitudinal section (Fig. 11d in comparison with Fig. 11b). This may be related to a change of appearance of inclusions and creating additional dendritic

Table 4 Parameters of Microstructure of Article as Functions of Rotation Rate ω

Parameter	Three-rib tubes of Al–1% Mn			Smooth duralumin tube	
	Tube		Ribs		
ω, rpm	0	0.34	0.34	0	0.42
$X_\perp$, 10^3 per mm^2	15.0 ± 1.0	8.4 ± 0.5	11.1 ± 0.6	–	–
$X_\parallel$, 10^3 per mm^2	2.3 ± 0.2	1.2 ± 0.05	2.4 ± 0.1	–	–
$m_\perp$, per mm	–	–	–	23.9 ± 0.6	18.6 ± 0.5
$m_\parallel$, per mm	–	–	–	10.0 ± 0.6	12.4 ± 0.5
α_1	0.73 ± 0.07	0.74 ± 0.07	0.65 ± 0.05	–	–
α_1'	–	–	–	0.58 ± 0.04	0.33 ± 0.02

Diameter is 30 mm, $v = 48$ mm min^{-1}.

branches under twisting. It is significant for duralumin, which is a multicomponent alloy with wide crystallizing interval [32].

4. Useful quantity characteristic of a dispersity of microstructure is an average quantity of dendritic cells N per the unit area of the section. It was shown that [56]:

$$N = kS_n v \qquad (15)$$

Here $S_n = F/L$, F is the area of a cross section, L is the perimeter of the cross section, $S_n v$ is the processing productivity, k is the constant for the material.

Average area of a transverse section of a cell is $f = 1/N$. Therefore

$$fvS_n = 1/k \qquad (16)$$

Here the dispersion is determined by dendrite boundaries because they are much larger than the boundaries of grains and blocks.

It is seen from Eq. (16) that f may be constant in the case of changing S_n and v to opposite directions.

$S_n v$ dependence of N for Al–1% Mn alloy cylindrical rods and ribbons (curve 1) and for 29-mm-diameter duralumin tubes (straight line 2) is shown in Fig. 12. $S_n v$ dependencies of N for Al–1% Mn cylindrical and ribbon samples coincide with each other (straight line 1, tangent of slope to an abscissa $k = 0.255 \cdot 10^8$ cm^{-4} s). This means that the reduction of the thickness of the cross section of an article S_n influences the dispersity of a structure more intensively than the cross-section shape of the article and average temperature gradient within the solidification interval. The straight line slope to coordinate axis depends on alloy properties. Thus for duralumin tubes $k = 0.0862 \cdot 10^8$ cm^{-4} s (line 2), for eutectic silumin ribbons [57] $k = 0.023 \cdot 10^8$ cm^{-4} s (in this

case the dendrites of α solid solution are oriented along the pulling axis and fill 35.7% of the alloy volume).

Emperical dependencies (15) and (16) allow the control of microstructure dispersion for a given thickness of article by means of v. It is possible also to calculate N if sizes of article and v are known [56].

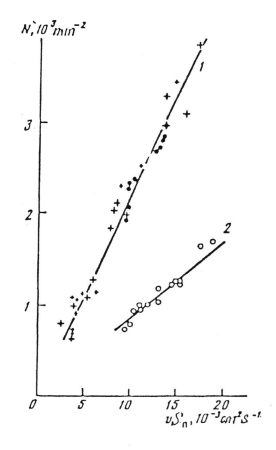

Figure 12 Dispersion N of structure as a function of physical productivity of processing $v \cdot S_n$: 1—Al-1%Mn alloy cylindrical rod and ribbon; 2—duraluminum tube.

Surface and Structure Defects

Aluminum profiles obtained by different methods, especially by conventional casting, exhibit various surface and structural defects including: shells, shrinkage cavities, pores, fins, steps, liquations, and others [29]. These are related to contact of solidifying metal with crystallizer walls. There are also many surface defects on extruded profiles such as: fissures, deflections, and others. Those faults are corrected by added deformation operations [29].

Profiles obtained by SM exhibit no pores and shrinkage cavities but they can exhibit some surface and structural defects, especially during operational transitions and when encountering raw material quality problems.

Localized freezing, erosion, or floating are frequently due to interruptions in the manufacturing process. These defects are eliminated when the process problem is corrected [7]. Fluctuation of the relative level of the melt can induce a growth of the profile with periodic thickness variations [9]. When the melt splashes out onto the edges of the shaping orifice, floating or thicker parts in fillets along the article can be created. Contamination of a melt with scale or other solid particles in the shaping orifice may induce longitudinal channels or slots [7,9]. Slots can also arise due to the weak wetting of a seed by the melt and drastic change of cooling regime [9]. Some transverse waviness and hatching, ring-shaping, shears, and breaks can be produced by disturbance of the normal operation of mechanisms of the set (irregularity of moving pulling device, oscillations, and others) [7]. Formation of waves on surface of thick-wall profiles (wall thickness larger than 5 mm) can be evoked by high dynamic pressure of cooling air. Such a mechanism has been described in Ref. 49 and is shown in Fig. 13. The melt column couples the upper edge of the shaper (Fig. 13a). Dynamic pressure of cooling air deflects the surface of the melt column on the cooler side. The cross section decreases (AA′→BB′). The liquid–solid interface lowers to the level BB′ in connection with conditions of process stability. The thickness of crystallized article decreases to BB′. When constant operational parameters are restored, the thickness is restored slowly and the interface rises to the level of AA′. This results in small periodic dents of small depth on the cooler side (Fig. 13b). But the dynamic pressure of cooling air can also create deep dents at thick-walled articles. This occurs for thick shaping plate (thickness to 10 mm). The dynamic pressure must be sufficiently high so that liquid column deflects. The place of coupling is not controlled. The melt column is interrupted from the upper edge of a shaper [49]. The shaper ceases to take away the heat from the melt and the melt temperature increases. The decrease in the cross section of the liquid column results in the decrease in the wall thickness of the article and therefore in the heat removal from the melt to the solid through the interface decreases. The stable position of the interface goes to a higher level. The pulled article is crystallized with the dent along the entire length. In some cases the origin of the cross section is reduced slowly. There is a heat contact of the liquid and the plate of the shaper. Such situation is quasi-stable. An accidental increase in air pressure can initiate formation of defects. In some cases the wall is very thin and an interruption of air pressure may occur. Then the chink is created through the article [49].

The growth of oxide film can create sticking and distort the article shape. Such defects arise at lower temperatures of a melt and at low pulling rate [7]. They can be corrected if the melt is refined or by other procedures.

A serious problem for a number of alloys is warping and creating "hot" cracks especially for large-size profiles. Irregular shrink stresses inducing such defects are due to an incompatibility of local thermal deformations in the case of high temperature gradients on the cross section and along the length of the article [7]. It is necessary to improve the temperature regime. Such defects can be minimized if the content of Fe is more than Si in an alloy.

In some cases the structure and surface defects arise simultaneously. Thus thickenings of different length or backstitch structure arises on the surface of shaped articles of fan structure when twin planes of dendritic colony have the slope to the surface about 30–45°. Such surface defects induced by fault of structure under breaking thermal regime were observed particularly for ribbons of Al–1% Mn and Al–13% Zn alloys [58]. They are removed by a choice of the optimum seed

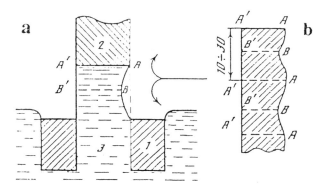

Figure 13 Effect of dynamical pressure of cooling air on the melt column shaping the article [49]: (a) is deflection of the liquid column engaged with upper edge of the shaper. 1 is the shaper, 2 is the crystallized alloy, 3 is the melt. (b) is longitudinal section of the article. (From Ref. 56.)

orientation, by decreasing the length of a two-phase zone for enlarging G_2 as well as by decreasing turbulent air-cooling flow.

Defects called by tubercular roughs arose on the article surface of Al-based alloys of wide temperature interval of crystallization (Al–3.5% Mg–0.5% Mn, duralumin and others) [7]. They appeared as a narrow ridge with one top extending along the longitudinal direction. These surface defects were accompanied by structural orientation faults. Therefore they can be considered also as structure defects. Their origin is related to constitutional supercooling near the liquidus. Supercooling induces the growth of dendrites of irregular orientation. Such dendrites originate because of stiff oxide film and induce tubercular roughs [48]. If such dendrites extend beyond the oxide film, they grow into a regular structure and form a structure defect. The tubercular roughs do not influence the mechanical properties of articles but they spoil the appearance. Such defects are absent when the oxide film is removed. These defects can also be created due to minimal heat removal through a shaper. In that case, the liquidus decreases to shaper and the stiff oxide film that originates will be taken by the solid phase in the crystallization zone above solidus, which is interrupted from the edge of the shaper and it is carried away by the article as it is pulled up. If the oxide film is absent the tubercular roughs do not arise [48].

It can be concluded that the defects described above are due to the control of the manufacturing operation and related equipment and not to the SM itself. Because they are equipment or process related, they can be corrected and eliminated.

Properties of Articles

Al-based alloys and articles constructed from them exhibit excellent properties including low density, high specific strength and stiffness, plasticity, high heat and electrical conductivity, and corrosion resistance, which makes them useful in various environments. They are also weldable and machinable. Properties of Al-based alloy articles obtained by SM depend on the growth regime. Property comparison illustrates the advantages of the SM compared to alternative production methods.

MECHANICAL PROPERTIES. Mechanical properties including tensile strength σ_B and the specific elongation δ of both deformable alloys and casting alloys were compared. It was shown that increasing impurities of Fe and Si in 0.5–2.0-mm thickness of commercially pure Al ribbon increases σ_B and decreases δ [59]. The data in Table 5 show that plasticity of transverse samples improves with low levels of impurities; however, increasing impurities decreases δ and results in two times smaller specific elongation than for longitudinal samples. Yield stress, $\sigma_{0.2}$, increases monotonically with increasing impurities. Ti, 0.1% to 0.2%, yields dendritic structure and it was shown that the dendritic structure causes an (0.3–0.8%) increase in σ_B and an increase in δ from 23% to 43% for longitudinal samples. This also explains increasing δ from 20% to 24% for transverse sample [59]. Similar results were obtained for 0.3–4.0-mm thickness Al–1.16% Mn ribbons with impurities of 0.64% Fe, 0.33% Si, 0.1% Cu (Table 6) [51].

Most of the information on the mechanical behavior was obtained for Al–(0.9–1.2)% Mn alloy as this alloy is often used for processing by the SM. Also, this widely used alloy exhibits acceptable surface quality, good weldability, high heat conductivity, and is corrosion resistant.

One reason for mechanical property variation is that different sizes exhibit different orientation of dendritic branches for ramified dendritic structure. The ramified structure of the $MnAl_6$ phase results in high plasticity [51]. Increasing Fe and Si in that alloy increases

Table 5 Influence of Impurities and Orientation of Samples on Characteristics of Al Ribbons

Impurity content %			σ, MPa			
Fe	Si	Orientation of samples	Average	Interval	δ, % (average)	$\sigma_{0.2}$, MPa
0.10	0.02	longitudinal	40	38–42	35	22
		transverse	40	38–44	52	–
0.14	0.17	longitudinal	55	54–57	22	–
		transverse	53	51–56	35	31
0.29	0.32	longitudinal	69	69–70	22	41
		transverse	57	52–61	12	–
0.87	0.32	longitudinal	89	81–94	19	46
		transverse	81	73–92	8	–

Table 6 Mechanical Properties of Al–1.16% Mn Alloy Ribbon

Structure	Orientation of samples	σ_B, MPa (average)	δ, % (average)
columnar	longitudinal	146	13
	transverse	123	8
dendritic	longitudinal	146	23
	transverse	139	16

the content of MnAl$_6$ because the solubility of Mn in Al decreases.

The mechanical behavior of Al–1% Mn alloy tubes of 203×102 mm^2 rectangular cross section with $S = 3$ mm and with internal ribs and without them was studied in detail [50]. It was found out that σ_B is in the range from 120 to 130 MPa, δ ranges from 7% to 29%, relative narrowing ψ changes from 16% to 32%. It was revealed that the Young's modulus E was in the range from $7.28 \cdot 10^4$ to $7.34 \cdot 10^4$ MPa, $\sigma_{0.05}$ was 12–30 MPa, the yield stress $\sigma_{0.2}$ ranged from 36.5 to 39.1 MPa, real tensile stress was 150–185 MPa, shock viscosity ranged from $3.8 \cdot 10^5$ to $6.2 \cdot 10^5$ J m^{-2}. Most of those characteristics are analogous to annealed pressed profiles. There is a small scatter of σ_B and E for different samples, but δ and ψ vary over a wide range. The profile grown at the beginning of the process exhibited lower δ and even lower ψ where there is the copper from a seed. Lower values of σ_B were for the profiles obtained with one-side air-cooling. It was found that there was very high shock viscosity. Fracture of that sample under shock impact had plastic character. Technological 180° bending showed no cracks. Lower values of $\sigma_{0.2}$ may have a positive role. The ratio of σ_B to $\sigma_{0.2}$ ranged from 3.0 to 3.5. But it is known that for $\sigma_B/\sigma_{0.2} \geq 1.4$, a material is strengthened for cyclic loading [60]. Therefore Al–Mn alloy profiles must be strengthened at the beginning of the operation. For optimal control during pulling of Al–Mn alloys, hollow profiles with outer ribbing had σ_B in the range from 130 to 150 MPa and δ about 30% with small scatter.

Al–(0.8–1.6)% Mg alloy hollow profiles of dendritic structure had δ higher than for columnar structure: 20–35% for longitudinal direction vs. 7–14% [7]. Mechanical properties of Al–Mg alloy profiles were compared with annealed profiles obtained by pressing. As cold hardening increases the strength two times and decreases the plasticity three or four times due to formation of brittle β-phase of Al$_3$Mg$_2$.

Al–Mg alloy profiles obtained by SM exhibit the same or better values of σ_B and δ than for pressed profiles as annealed [7]. Al–0.9% Mg–0.9% Si–0.7% Cu–0.15% Mn alloy tubes obtained by SM after quenching and aging had average values of $\sigma_B = 336$ MPa and $\delta = 9.7\%$, which were somewhat higher than standard values for pressed profiles after the heat treatment. A hydraulic test of those tubes showed high mechanical characteristics: tensile cracks arose in longitudinal ribbed tubes of 17.5-mm internal diameter and 1.5-mm wall thickness that were 200 at κGcm^{-2}. Longitudinal ribbed tubes of 8-mm internal diameter and 1-mm wall thickness had no cracks up to 300 at.; there was no water flowing [9]. Al–6% Zn–2.3% Mg–1.7% Cu–0.4% Mn–0.15% Cr high strength alloy 29-mm-diameter tubes, obtained by SM, after quenching and heat treatment had σ_B in the range from 460 to 480 MPa, which approached to σ_B for pressed profiles after analogous heat treatment [7].

The influence of operational factors on profile strength obtained by SM was also studied. It was shown that σ_B of 3.6–5.0-mm diameter and 0.2–0.8-mm wall thickness of commercially pure aluminum tubes increased under increasing intensity of cooling for constant v [39]. Two-zone blowing of pulled duralumin tubes exhibited increased σ_B to 290 MPa from 250 MPa. Compared to one-zone blowing, the value of δ increased from 3–4% to 8–9% [55].

For studying casting alloys, the eutectic silumin (Al–Si) was used. Samples obtained by SM had $\sigma_B = 205$ MPa, $\delta = 11\%$. Samples obtained by casting into earth had $\sigma_B = 147$ MPa, $\delta = 4\%$, for chill casting $\sigma_B = 156$ MPa, $\delta = 2\%$ [7]. Improving mechanical characteristic for pulled profiles can be explained only by low porosity because Si forms no strengthened compounds with Al. Comparative investigation of circular rods of commercially pure aluminum for SM exhibited $\sigma_B = 56$ MPa, for conventional casting $\sigma_B = 50$ MPa [39].

Experimental dependencies of mechanical properties on crystallization parameters and structure were studied for 1.5-mm-thickness Al–1% Mn alloy ribbon for four values of v and air flow [56]. It was revealed that decreasing average grain size (or increasing the number of grains n over a width of a sample) for every pulling regime resulted in an increase of δ from 6–7% to 8–11% and does not influence σ_B. The reason δ is increasing is that grain size and δ decrease when the dispersion N

increases or, in other words, when physical productivity increases. As a result, the average value of δ changes moderately for transferring regime (about 8–9%). But physical productivity changes nearly five times. It is seen in Fig. 14a that δ does not change by increasing the dispersion of structure N (10^3 mm^{-2}). As opposed to δ, the average value of σ_B increases with increasing N.

Indirect verification that σ_B exhibits a direct relationship with internal grain structure but not with respect to grain size is analogous to the dependence of microhardness H on N (Fig. 14b). H was measured for a grain. If the reason for increased σ_B was increased specific grain boundary then H would be constant similar to pure metal. These results suggest that the dispersion of microstructure N determines the strength of pulled Al-based alloy articles. The change of v or reduced thickness of an article S_n controls the structural dispersion [15] and its strength [56].

The strength of articles pulled by SM is sufficient for many fields of application. But there are limits to the increase possible, which were described in Section VII [29].

CORROSION RESISTANCE. Corrosion resistance was tested for profiles of Al–Mn and Al–Mg–Mn alloys and for commercially pure duralumin and eutectic casting

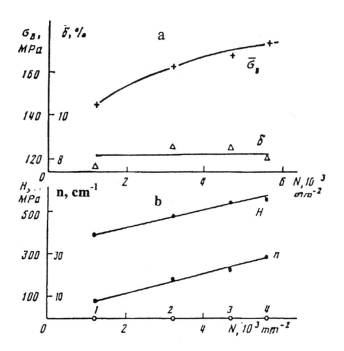

Figure 14 σ_B and δ (a), microhardness H and number of grains along width n (b) of Al-Mn ribbons versus dispersion of structure N. (From Ref. 56.)

silumin (Al–Si). Tests were performed in sodium chloride solution, in water of water supply line, in water of the Caspian Sea [61,62]. It was revealed that the pitting is the most typical kind of corrosion of the profiles. Pitting is usually oriented along the direction of pulling. The speed of a corrosion of technical aluminum is greater than for Al–Mn, Al–Mg–Mn alloys as opposed to profiles obtained by deformation methods [61]. This can be explained by a specific structure. An arrangement of pitting does not depend on geometry of samples in contrast to profiles of the same alloys obtained by deformation methods, but pitting is related to the absence of internal stresses at positions of changing shape or when joining of separate parts of the article [61].

Corrosion stability of pulled articles is greater than for conventional castings because there are no pores. Particularly, commercially pure aluminum and Al–Mn alloy articles obtained by SM have no stratifying corrosion. Sea water or water line testing showed more intensive pitting corrosion for internal surfaces of tubes than for the outer surfaces because of crystallization of the thinner Al_2O_3 film on internal surfaces without forced cooling [62].

Duralumin exhibited the greatest corrosion rates because of formation of intermetallic compounds of $CuAl_2$, which act as a cathode. The best corrosion-resisting profiles are of eutectic silumin (Al–Si). Corrosion fracture of a surface for this material is uniform, pitting is absent or it penetrates to only small depths which is probably due to the absence of intermetallic compounds in Al–Si, because the solid solution of Si in Al and Al in Si are in equilibrium [62].

Special covers can improve corrosion resistance as design, lifetime, and other qualities are dependant on the use of a cover. Therefore surface adsorption is important. Articles obtained by SM can be covered without any preliminary mechanical treatment. It has been shown that Al–Mn alloy profiles that are anodized and nickel-plated are good for this purpose [7].

Comparison of Alternative Methods—Perspective Kinds of Articles

The Stepanov method may be compared with two other methods that provide long-walled profiles: continuous casting and deformation technology (DT) based on extrusion [29]. Conventional casting into a form and mechanical treatment methods yield articles of lower quality and that are more expensive. However, continuous casting may be used to produce articles of the simplest shapes: rod, sheet, plate, smooth-walled tube. Therefore it is of interest to compare SM and DT with respect to engineering characteristics, quality, and cost.

ENGINEERING CHARACTERISTICS. The Stepanov method is differentiated from DT in that it is a one-operation method producing an article that is ready for use and uses a simpler and more space-saving equipment. Because it requires relatively limited production space, is quick and inexpensive to construct, and will utilize industrial power. Process transitions from obtaining one type of article to another are shorter. The Stepanov method is adaptable to a variety of alloys. It is suitable for use with wrought and casting alloys, and with alloys of high brittleness and toughness. Deformation technology processes are used only for wrought alloys. The efficiency of raw alloy with SM is higher than for DT and 85–90% efficiencies are possible.

For DT processes, power is used not only for melting an alloy but also for deformation. The Stepanov method is a single-operation technique. Therefore unit-specific expenses for obtaining profiles by SM are much lower. The Stepanov method produces less scrap, is a power-saving technology, and is an ecologically friendly process.

Process productivity is determined by the rate of metal flow for DT and by the pulling rate for SM. The latter is much lower than for DT but productivity must also account for complexity and amount of equipment. The output per man-hour must be attributed to the cost of all the production equipment. With respect to multi-process operation, DT and SM are of approximately equal value. In the case of hard alloys, complex shape, and large cross-section articles, the productivity of SM is equal to or higher than for DT. There are also additional resources enlarging productivity for SM connected with shortening accessory operation time and pulling a group of articles at the same time.

The Stepanov method is usually used for small-scale production. Specific labor expenses are higher in comparison with high effective equipment of broad-scale production by DT; however, labor costs are less for large-scale production by SM.

Articles of Al alloys are produced by SM with the open mirror of the melt. It permits the production of parts with larger cross sections than DT. The length of articles is approximately equal for both methods.

The Stepanov method, like DT, allows the production of a wide range of configurations including complex-shaped articles. But there are shapes that can be obtained only with one of these methods. Thus DT is used for thick-walled bulk profiles and SM is used to produce thin-walled profiles, articles with a curvilinear

longitudinal profile, and profiles of casting alloys. The Stepanov method is the most effective technique to manufacture polyhedrons, tubes with outer and inner ribs, partitions, different members, and spiral ribs. Articles with channels, twisted channels, hollow, and multichannel hollow products as well as other peculiar complex unique shapes can be obtained by SM. Many of these shapes cannot be produced by DT. It is possible to consider SM as a progressive shaping design without deformation. Figure 15 shows some articles obtained by SM.

QUALITY CHARACTERISTICS OF ARTICLES. Cross-section size variation and the wall thickness of articles as extruded are higher than for SM. But after a number of operations of a treatment the allowances are lower compared with SM or they are the same. The longitudinal curvature (deflection) of profiles for SM is less than for DT.

The surface quality of profiles made by SM depends essentially on alloy composition. In the case of DT the quality of surface of products as extruded is inferior relative to SM. Surface quality of DT and SM articles may be improved by subsequent treatments.

The structure of DT profiles is determined by the manufacturing process and it is usually fine-grained. Continuous process of pulling for SM yields mostly columnar or dendritic structure, although in some alloys and processing of temperatures the structure of an article can approach to equiaxial.

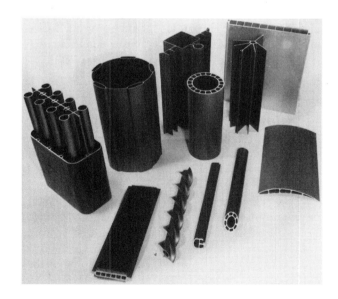

Figure 15 Some shaped articles pulled by SM. (From Ref. 29.)

The Stepanov method's articles possess no pores and therefore exhibit the greatest material density after processing.

Mechanical properties of SM shaped articles agree closely with the values of annealed samples obtained by DT from the same alloys. Cold-hardened DT samples exhibit higher strength but the strength of SM articles made from heat treatable alloys can be increased after heat treatment. The section "Properties of Articles" describes methods for increasing strength properties of SM articles, and by using strengthening structural partitions it is possible to provide greater hardness by SM. In all cases, the strength of SM articles of casting alloys is higher than for the conventional cast of the same alloys. However, some applications of profiles have no need of high strength.

The Stepanov method provides articles that may be produced from a wider range of Al alloys, which may be further varied by alloy modifications. This provides the possibility to obtain profiles with special structures and necessary mechanical, electrochemical, acoustic, electric properties, thus resulting in a wider potential range of applications.

ECONOMICS OF THE STEPANOV METHOD. Specific cost of simpler and single-type equipment (and amortization deductions) for SM is well below those for bulky, complex, and multiple-type equipment for DT. Tool cost is also much less for SM because of small force impact during working and simpler design. Estimations show that the net cost of profiles for both technologies is about identical. The following trends can be noted: DT is more economical for small-sized, simpler-shape product, SM is more economical for large-sized, complex-shape profiles.

NEW ARTICLE TECHNOLOGIES. Recent engineering advances have created the need for complex-shaped articles and also miniaturized designs [48]. Large-scale, effective production with DT (extrusion and other) provides wide assortment of profiles but in many instances it is advantageous to use profiles obtained by SM. Of special interest are cases when DT articles do not meet shape or size cross-section requirements, a combination of mechanical and physical properties, or in the case of some alloys, articles cannot be produced by DT. Below is a summary of the most perspective kinds of profiles for obtaining with SM. Some of these profiles are shown in Fig. 16.

THIN-WALLED PROFILES. The thickness of the wall of such profiles is in the range from 1 to 1.2 mm or less

(Fig. 16d). It is possible to produce profiles of similar sizes by extrusion but only for wrought alloys with rather low strength (hardness) and small temperature interval of crystallization, e.g., Al–Mn, Al–Mg–Si. The Stepanov method is more effective for these alloys in comparison with DT and especially with conventional continuous cast alloys. Thus SM may be used to produce articles (e.g., 150-mm-diameter tubes and ribbons) of Al–Mn alloy with the thickness of wall up to 0.3 mm. Minimum thickness of profiles of higher strength alloys (magnalium, duralumin, Al–Zn–Mg–Cu alloys, and other) for extrusion is about 2–3 mm or more.

MULTICHANNEL HOLLOW PROFILES. Such articles are cylindrical profiles with partitions between tubes (Fig. 16c). Profiles of such types with additional outer ribs or with internal ribs along the inner tube are used for heat exchangers. It is impossible or very hard to obtain such articles, especially of thin walls for small cells with another technique. In some cases, such profiles are fabricated from components; however, this decreases the thermal effectiveness of the articles.

Articles with cross section with triangular framework (Fig. 16k) with relatively small sizes of cross section about 80–120 mm, and outer contours such as, broken, curves are profiles of such type. They are used for exhibition stands and advertisement panels.

LARGE-SIZED SHAPED ARTICLES. Large-sized shaped articles are likely to be a future growth area for SM. Articles with sizes of cross sections greater than those that can be readily produced by extrusion are of greatest interest, such as articles where the maximum diameter (D_{max}) of extrusion tubes is equal to or less than 400–500 mm for wall thickness (S) of about 5 mm. At the same time there are the following trends: as D_{max} increases, S_{min} decreases, hardness increases; as D_{max} increases, S_{min} decreases.

The Stepanov method permits the production of tubes of greater diameters with less wall thickness including alloys of high hardness. Increasing D_{max} therewith does not demand the corresponding increase in S_{min}.

Circular two-walled tubes with partitions between them can be formed by extrusion when D_{max} is in the range from 100 to 150 mm for a minimum wall thickness of about 3.0 mm. At the same time SM provides tubes with 36 partitions for D of about 370 mm. Figure 16c shows these tubes. Please note that there are no principal restrictions for increasing D.

An important priority field of SM is producing different panels of higher than usual width (B): panels

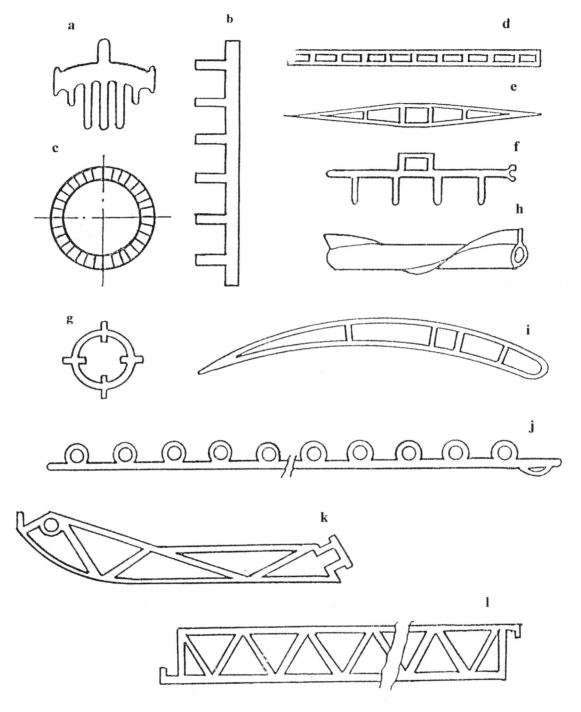

Figure 16 Cross-sections of perspective types of articles grown by SM. (From Ref. 48.)

of solid cross section with ribs (Fig. 16b) and with tubes on solid framework (Fig. 16j) as well as hollow panels (Fig. 16d,l). Some regularity is observed when such panels are obtained by extrusion. B_{max} can rise with the decrease of the thickness of panel (h), with the increase of the wall thickness of tubes, with the increase of the step (t) between ribs, tubes or partitions, with lowering hardness of Al alloy used. Therefore DT in that case is of limited usefulness.

Such limitations are absent for SM. There are possibilities to obtain panels of more B and h, less t and S, not to lower the hardness of alloys. Thus the panels of solid cross section with ribs (Fig. 16b) of moderate hardness alloys can be obtained by an extrusion with the following maximum parameters: B is in the range from 1700 to 1800 mm for h of about 40 mm, t is in the range from 50 to 150 mm, S is about 4 mm. (If S changes from 7 to 8 mm then B can be risen to 2000 mm.) Here we keep in mind obtaining panels by means of cutting and correcting extrusion tubes.

In the case of SM circular tubes D is more than for extrusion. Therefore panels also can have more B. Maximum width of panels with tubes on the solid framework (Fig. 16j) obtained by extrusion changes from 600 to 800 mm (for about 100–170 mm, S is approximately equal to 2.5–3.0 mm). The Stepanov method enabled to produce panels of B equal to 1000 mm with 19 tubes. Maximum width of hollow panels obtained by extrusion (Fig. 16d,l) is in the range from 600 to 800 mm (for h_{max} of about 60–80 mm). The Stepanov method has no principal limitations for the width of hollow panels. At the same time it is possible to increase h and to decrease t and S. The use of SM technology for obtaining large-sized aluminum shaped articles is more economical than DT.

CONVENTIONAL PROFILES WITH CROSS-SECTION GEOMETRIC SIZES UNAVAILABLE TO EXTRUSION. Forming metal in solid state during extrusion has a limitation on ratios of some sizes of a cross section of profiles. For example, these limitations take place for densely ribbed heat exchange tubes of about 60-mm diameter of tube part and of 100-mm diameter with account for ribs. Tubes of 5–6-mm wall thickness and more can be obtained by extrusion. Such tubes are heavy in excess. Such tubes of 2–3-mm thickness wall and less can be produced by SM only.

SHAPED ARTICLES OF CURVILINEAR LONGITUDINAL PROFILE. The Stepanov method has much wider facilities for producing such articles than the extrusion method. It opens a prospect to use such profiles in various applications. They can be made without closed hollows as with closed hollows. Thus tubes with ribs

twisted around the central axis (Fig. 16h) and nonsymmetrical blades twisted around a certain axis (Fig. 16i) were obtained as well as panels with zigzag ribs and rectilinear channel (Fig. 16f).

PROFILES OF SOME HIGHER CHARACTERISTICS. In some cases profiles obtained by SM have the scatter of geometric sizes of a cross section and of wall thickness as well as a sag less than after extrusion. This extends the possibility of the application of shaped articles.

The Stepanov method is perspective at producing stores, e.g., circle tubes (against casting and following extrusion) for a treatment by rolling and dragging up to given required properties.

OTHER PERSPECTIVE PROFILES.

a. Profiles of Casting Aluminum Alloys of High Quality Characteristics. Deformation technology cannot produce the profiles of casting alloys. Obtaining profiles of casting alloys of high-quality characteristic by conventional cast is impossible or very expensive and complex. The Stepanov method enables to obtain profiles of casting alloys with high characteristics. The possibility of using SM for Al–Si system of high resistance to wear is also shown. The Stepanov method can be used also for Al–Cu system. Some of these alloys have high refractory (to 450°C) and can be used for plungers of internal-combustion engine as well as for cylinders of airplane engines.

b. Composition Profiles. Research has shown that the outlook for aluminum profiles may be reinforced during crystallization by SM. Wire, fibers, heater, or design features as reinforcements may be used to obtain shaped articles of high strength or required electrical properties. Thus SM possesses its own field of applications among perspective types of articles (based on shape, composition, and properties) among alternative technologies.

Application of Shaped Articles *

A wide assortment of thin-walled light profiles from miniature to large-sized designs, of very complex cross sections of heat exchangers and other equipment for use in the energy industry may be manufactured by SM.

NONTRADITIONAL POLLUTION-FREE POWER PLANT.

a. Solar Power. Solar power plants offer potential savings in conventional fuel, low thermal pollution,

---

* This section was written with the help of O.B. Ioffe.

and reduced impact on the environment. A number of devices, obtained by SM, have been used for the construction of water heater, air heater, combined water–air heaters; adsorber, condensers, and evaporators for a solar refrigerator; agricultural drying-plants, solar energy concentrators of paraboloid type; parts used for heaters of communal-general service and for protection from excess solar radiation of "solar houses." Other devices that have been manufactured by SM include solar heat exchangers for transportation of viscous materials, low temperature heat tubes for cooling of GaAs photoelectric converter plant with concentrators of solar energy, and heat exchangers for increasing the efficiency of Si solar cells.

The Stepanov method facilitates the manufacture of desired articles without further fabrication. These include such complex designs as panel tubes, combined air–water heaters with longitudinally bent (zigzag) ribs (turbulators); warm feeder heat exchangers in the form of circular tube with several ribs; low temperature flat multichannel heat tubes with dense outer ribbing. Many of these profiles cannot be obtained by any other method.

b. Wind Power Technology. The Stepanov method can be used to produce aluminum blades of optimum aerodynamics as rectilinear along longitudinal direction or they may be twisted. Such articles can create the base for manufacturing effective wind motors of various power and their supporting structures.

Fans of Aggregates of Air Cooling and Graduators; Jalousie

Recent trends are directed toward the conversion of water-cooling by various gas and liquid materials with waste water disposal by highly effective aggregates of air cooling. Cooling of gas and liquid in a heat exchanger is performed by using blowers and ventilators. The thermal efficiency of such air coolers depends on the aerodynamic perfection of the blades of a fan. The Stepanov method can be used to produce hollow, thin-walled, wing-shaped blades with inner partitions of programmed variable angle of twisting along the longitudinal axis according to optimum aerodynamic flow. Hollow jalousie shutters of ideal aerodynamic shape are produced by SM as well. They guarantee close shutting off of the air cooler as it is very important if there is a need to heat gas not to cool it. Hollow profiles can change the direction of flow motion in air cooler. The Stepanov method is used to manufacture blade wheels to graduators for ventilators, which are used to control circulating water supply.

Flat-Tubular Heat Exchange Profiles and Lamellar-Ribbed Heat Exchangers

Heat exchangers that are based on thin-walled flat-tubular profiles (panels) with inner longitudinal partitions (ribs) for increasing heat transfer and creating turbulence on the outer surface of the heat exchanger covered with the transversal widespread ribbing may also be manufactured by the SM [40]. There are three possible methods of manufacture: planning of a thin layer of the outer surface, building up of the transverse outer ribbing from aluminum corrugated foil, or using a transverse knurl.

Such complex profiles, obtained by SM, are used for assembling compact light heat exchangers for removal of heat from heat-liberating members to the surrounding, to supply heat to members of a design or to heat transfer agents. There are a number of important applications of such flat-tubular heat exchangers: heat exchange equipment for thermal siphon cooling of semiconductor sets (evaporators, condensers), thyristors for all kinds of transport; engine radiators, heater of taxis, and automotive air conditioners, coolers for gas and oil compressors, general-purpose and gas-pumping plants; coolers and heaters of heat transfer fluids used in the metallurgical, chemical, and food industry; and water coolers and steam condensers used in power plants.

Lamellar-ribbed heat exchangers (LRHE) manufactured by the SM have excellent characteristics, which make them superior to widely used tubular-ribbed heat exchangers, including the following compactness, which is about 100–200 m^2 of the total surface of a heat exchange per 1 m^3 volume of the exchanger; specific weight is changed from 500 to 800 kg/m^3; thermal effectiveness is in the range from 400 to 700 kW m^{-2} K^{-1}; provide excellent safety, and only minimal soldering is used (more like hermetic sealing).

Round-Tube Exchangers

The Stepanov method can be used to manufacture round-tube exchangers with tubes of outer, inner, or two-sided longitudinal ribbing. In some cases outer ribbing is transverse. It is possible also to produce bimetallic exchangers, with inner part of stainless steel, for example. Longitudinal ribs can be made with turbulizing members to intensify heat and mass exchange and to trap products normally discharged to the atmosphere. Although most applications of round-tube exchangers are the same as for plane-tube exchangers, it is possible to obtain clusters of longitudinally ribbed tubes. Such clusters are jointed by thin films for use as gas separator heat exchangers.

Cooling and Freezing Equipment

It was shown that aluminum profiles of high heat exchange capacity could be obtained by SM including evaporators, condensers, and other members of coolers and freezers that exhibit superior gas and vapor compression, heat utilization, and thermoelectric properties. Reduction of natural gas from 8–9 to 0.2–0.3 at. for cooling is an important development. (It is also possible to generate heat.) In large volume cooling chambers (up to 600–800 m), bracket and ceiling heat exchangers with hollow panels with inner partitions are used, which may be manufactured by SM.

The Stepanov method is also used for the manufacture of equipment used in ice making such as the cylindrical round double-walled profiles with diameters of 100–500 mm of an ice generator. Ammonia, freon, or other refrigerant is among walls of tubes and partitions of these generators where ice grows on the inner and outer surface of the ice generator and is cut with a special knife. Similar multichannel ice generators can be used for the purification of wastewater by successive freezing and melting sediments. Precipitate colloids are destroyed during freezing.

Flexible Heating Unit

The Stepanov method is used to produce flexible heater units with wire conductors (e.g., nichrome) within high temperature electric insulator covered by aluminum shell. Such compact units may be of any length, sizes, and cross-sectional shape. They may have different numbers of conductor wires and the outer surface may have longitudinal or transverse ribbing. Such flexible heating units are highly effective for heating petroleum in pipelines and chemical equipment and in boilers. They can be used successfully as air heaters (thermal ventilator) and gas heaters.

Autonomous, Electric Heat and Cold Sources

Autonomous motion power plants require effective engine heat removal. The Stepanov method is used for profiles for compact heat exchangers having, for example, flat-tube panels with internal partitions.

Heat and Ventilator Systems

Ventilator systems are more efficient when recycling heat removed from the room as it may save about 50% of the heating energy. Such economy is especially important for cold climatic areas. Similarly, for coolers, heat recuperators may also provide for more economic heating. Many industrial ventilator systems are based

on axis and centrifugal fans of various designs in conjunction with recuperative heat exchangers to facilitate heating and cooling. They also have air heaters and air coolers with the surface of effective heat exchange and possibly heat conditioners. All the heat exchange equipment described can be produced by means of SM. The heat exchanger, obtained by SM, can have a number of important preferences: (a) higher effectiveness and compactness or smaller clearance and mass; (b) higher vibration-stability and safety; (c) lighter, simpler, and more compact construction and more heat exchanger design options.

Decorative Design and Other Purposes

A wide assortment of shaped articles, obtained by SM, can be used for the fabrication of articles of constructional, decorative, and other functions including hollow panels with partitions for heaters of buildings; hollow panels, trusses, and other designs for advertising screens; various frameworks, cases, guides; sail boat masts; "green house" construction; sport inventory; carriers for aviation, ship building, space, and others; for railroad trucks as facing wall panels with air-distributing channels and ducts of electric wiring. Perspective articles can be made as combinations of aluminum profiles and units of another material.

Large-diameter fan blades with force beam (axis) made of aluminum and aerodynamic fiberglass covers are an example of a "complex unit" that is light, strong, and inexpensive.

B. Shaped Polycrystalline Articles Obtaining with Melt Protection

Dense, oxide film on an aluminum (or aluminum alloy) melt surface permits pull-up of shaped articles by sets using open mirror of a melt without any of its protection. Advantages of such sets are described in Section VIII.A.1. Specific protection of a melt is required for use with sets for growing articles of some metals including magnesium and its alloys of moderate melting temperature (600–650°C), and Cu-based alloy (bronze, brass) of melting temperature in the range from 800°C to 1200°C.

1. Magnesium and Its Alloy Article

Processing Requirements

As opposed to aluminum, liquid magnesium and its alloys are readily combustible in air. Because the magnesium oxide film is friable and does not protect the metal from further oxygen penetration. Magnesium film also exhibits low heat conductivity leading to overheating under the film and subsequently fire [17]. In experiments of pulling profiles of different Mg-based alloys without any protection of the melt mirror, fire was observed in the shaping orifice after pulling 10–40 cm of the article [17], and it was necessary to stop the process. Least flammability was observed during crystallizing of thin-walled articles (their wall thickness was about 1 mm). Attempts to alleviate this problem by an inflow of argon or by sulfur had no effect.

Precautions had been taken for fire suppression. Operational parameters of crystallizing were kept constant. The melt mirror was cleaned with great care from oxide film before pulling and was covered by flux. Seeding was performed quickly. The article was pulled flux-shielded for small height of the liquid–solid interface and airflow was kept from the interface [17,63].

Regimes of processing Mg-based alloy are similar to the ones for Al-based alloys because these metals have a slight difference in such physical characteristics as the capillary constant, density, melting point, viscosity. The specific heat capacity of magnesium is two times lower than for aluminum, which permits the growth of magnesium-based alloy articles with greater productivity than for aluminum alloys. Pulling rate dependencies of profile thickness for Al and Mg–Al–Zn–Mn alloy are shown in Fig. 17, where it is seen that for the same wall thickness, the Mg-based alloy v is significantly higher [64].

Components and Impurities in Mg-Based Alloys

Chemical and spectrum analyses of the alloy and pulled articles were performed. The central and peripheral parts of circular rod cross sections as well as raw alloys were analyzed for content of Ni, Si, Mn, Cu, Al, and Fe. The following impurities are contained in rods of Mg-based alloys [63,64]: 0.0002–0.001% Ni, 0.001–0.011% Si, 0.003–0.016% Mn, 0.0003–0.01% Cu, 0.0001–0.012% Al, 0.001–0.046% Fe, 0.001–0.002% Ti, and 0.001–0.009% Cl⁻. Impurities of Fe can decrease the corrosion resistance of Mg-based alloys. Separate samples had higher content of Fe in comparison with charging crucible. This may be explained by the use of cast-iron shapers. For Mg–Mn alloys, the Fe content in samples is significantly lower than for charging, which can be related to binding of Fe into αFe–Mn solid solution. Complete solution of Mn in Mg has a long time lag.

The content of Si, Cu, and Fe in peripheral parts of a cross section of samples is usually greater than in the center. Thus Cu in the central part of samples ranged

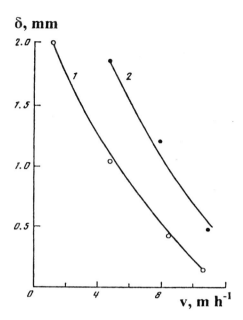

Figure 17 Thickness of a wall δ of Al (1) and Mg-Al-Zn-Mn alloy (2) profiles against the pulling rate v. (From Ref. 64.)

from 0.0004% to 0.006%, while in the peripheral part it was about 0.01%, which can be explained by the high volatility of Mg and enrichment of the sample surface by nonvolatile components.

It is shown that the content of Fe, Si, and Cu usually increases from the beginning to the end of the sample. The increase of Fe to the end of samples is 0.001–0.002% for its content in the beginning of the sample ranged from 0.001 to 0.008%, the increase of Cu is 0.0001–0.0013% for its content in the beginning part of the sample ranged from 0.0004% to 0.0029%, the increase of Si is 0.001% for its content in the beginning part of the sample ranged from 0.001% to 0.005%.

Estimations in Section VI and tests of aluminum articles showed that for usual pulling rates of polycrystalline aluminum articles by SM there was no practical force back. Therefore the above distribution of components of alloy and impurities along the length of magnesium-based alloy profiles is probably due to highly volatile Mg. This must be accounted for in the process design.

Structure and Mechanical Properties

The structure of various Mg-based alloy profiles was studied [17,18,63,64]. Most of these studies involved binary systems. Binary alloys were studied over wide range of composition. Profiles of some multicomponent alloys of industrial application were also investigated.

Their phase diagrams exhibit a eutectic point for temperatures much lower than their melting points. There are areas of Mg solid solution for extended temperature interval. These analyses were based on phase diagrams with regard to the processing regime for the articles of interest. Processing is influenced by chemical interaction between components of alloys, particularly by the interaction of components of the alloy with magnesium.

The microstructure of Mg–Al system was investigated in the range from 1% to 35% Al [18,63]. Polycrystalline Mg had coarse 1-mm grains. These grains consist of cells and are drawn along the pulling direction. The structure of solid solution of Al in Mg profiles exhibits no differences from that of pure Mg. Starting from 2.9% Al, isolated intermetallic $Mg_{17}Al_{12}$ inclusions are observed in the alloy. They are drawn in the form of chains along the pulling direction. In the structure of Mg–5.2% Al alloy article grains of δ-solid solution of Al in Mg and intermetallic nets of $Mg_{17}Al_{12}$ eutectic at grain boundaries are observed. Increasing the Al content beyond 6% leads to an increase in the quantity of $Mg_{17}Al_{12}$ phase crystals and, consequently, the density of chains. Inclusions become dendritic as is seen particularly for Mg–8% Al alloy in Fig. 18a [63]. The orientation of inclusions of $Mg_{17}Al_{12}$ disappears after 12% Al. The structure of alloy is created by uniform arrangement of dendrites. The structure of the alloy as it approaches a eutectic composition is characterized by eutectic colonies drawn along crystallization direction (Fig. 18b).

The microstructure of Mg–Zn alloys is similar to the Mg–Al system. Similar microstructures are observed for alloys of Mg–Nb and Mg–Y systems despite higher eutectic transformation temperatures.

The phase diagram of the Mg–Mn system exhibits a small crystallization interval with no eutectic transformation. There are no chains of intermetallic compound and no dendrites (Fig. 18c for Mg–1.5% Mn alloy) [18]. Mg–Zr alloys have fine grains due to the modifying effect of Zr. The same structure is in multicomponent Mg–Zr–Y–Zn–Cd alloy (Fig. 18d).

It is found that the influence of regime of processing on a structure of shaped articles is small for alloys without eutectic transformation and with a small area of magnesium solid solution (Mg–Mn, Mg–Zr alloys). Conversely, the structure of alloys with eutectic transformation (Mg–Al, Mg–Zn) exhibits high sensitivity to processing parameters (high pulling rate gives fine structure).

Mg-based alloys have a number of valuable properties. They have low density and high specific strength. They have high resistance to shock impact loading. Mg-based alloys have very high damping and high heat absorption. Their thermal conductivity is higher than for most construction materials. These advantages enable the use of magnesium-based alloys for all fields of machine building. The mechanical properties of some magnesium-based alloy shaped samples, obtained by SM, as a function of their geometry and chemical composition are shown in Table 7 [18,63].

The strength for magnesium samples, obtained by SM, is the same as for conventional casting. But specific elongation δ for SM is much higher than for casting. The strength of Mg-based alloys increases with increasing content of Al. Only for Mg–20% Al alloy is the strength a little bit lower. Increasing content of Al decreases δ considerably. The alloy of the composition approaching the eutectic was so brittle that it was impossible to make test samples.

For circular rod samples, a maximum strength was obtained for 9–12% Al. This Al content approaches the maximum solubility of Al in solid Mg.

Maximum σ_B of alloys for samples produced by SM was higher than for casting in earth (175–190 MPa), which can be explained by smaller shrinkage porosity of article obtained by SM [18].

The structure and mechanical characteristics of profiles are determined to a large extent by the phase diagram of the alloy. This relationship was studied for Mg–Al system that is the basis of many industrial alloys [65]. It was suggested that for real pulling rates, nonequilibrium crystallization occurs to some extent. In that case, it is necessary to analyze results on the basis of nonequilibrium phase diagram and to account for the influence of eutectic network of intermetallic compound $Mg_{17}Al_{12}$ on alloy structure and properties.

Comparison of the structure and mechanical properties of circular rods of diameters ranging from 8 to 10 mm, obtained by SM and conventional casting (in earth, in chill mold, under pressure), was performed. According to the nonequilibrium phase diagram for cooling rates commonly used for SM, eutectic net corresponds to 4.5–5.0% Al in alloy. This coincides with the results of structure investigations described above. For pressure-die casting, the eutectic net forms at about 1% Al, for chill casting at about 3% Al, for sand casting at an Al content of more than 5% [66].

Dependencies of σ_B and δ on the composition of Mg–Al alloy for different methods for obtaining articles are shown in Fig. 19. It is seen that the most significant break of the curve of δ for SM corresponds to about 5% Al. This concentration coincides with one for the eutec-

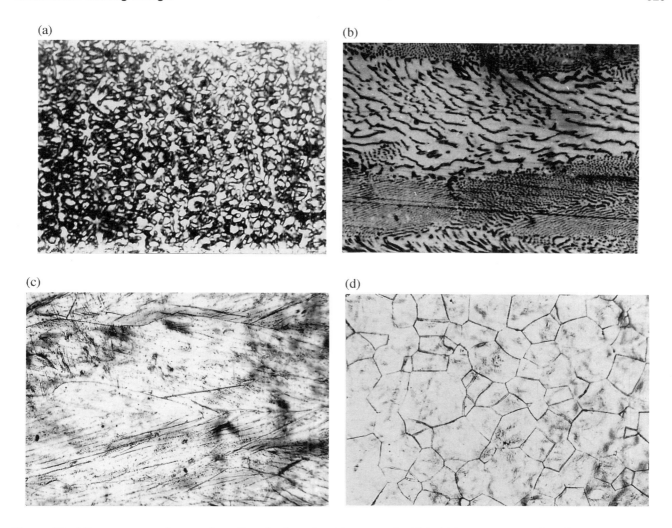

Figure 18 Microstructure of magnesium alloys in shaped articles grown by SM. (a) is transverse section; (b), (c), and (d) are longitudinal sections: (a) for Mg-8,0%Al, 100×; (b) for Mg-32%Al, 540×; (c) for Mg-1,5%Mn, 70×; (d) for Mg-Zr-Y-Zn-Cd, 70×. (From Refs. 18 and 63.)

tic net. The break of the curve of σ_B is displaced to about 9% Al. It can be explained by the effect of maximum solubility of Al in solid Mg, although it is also possible to consider that critical points of nonequilibrium phase diagram correspond to extreme values of the curves' "mechanical properties composition" [65].

Table 7 shows that Mn exhibits small influence on the strength characteristics of the alloy but lowers its plasticity. Zn increases the σ_B and $\sigma_{0.2}$ of the Mg-based alloy. The same influence of Nd and Y is observed. The strength of these alloys, such as Mg-Al alloys in shaped articles obtained by SM, is somewhat higher than for conventional casting.

It should be noted that the pulling of Mg-4% Zn rod at a rate of 6 m hr^{-1} against 4.3 m hr^{-1} for Mg-0.5% Zn alloy resulted in some increase in strength and a drastic

increase in δ (up to 17.8%). This is probably due to a change of structure. For this reason, ribbons pulled with higher rates than for the circular rod have, as a rule, higher values of σ_B and δ.

The structure and mechanical properties of Mg-based alloys with eutectic transformation (Mg-Al, Mg-Zn, Mg-Nd, Mg-Y) are sensitive to processing (increasing v increases strength). For alloys without eutectic transformation and with the area of magnesium solid solution spread along the concentration axis of the phase diagram (Mg-Mn, Mg-Zr), the influence of processing on mechanical properties is small [63].

It is seen from Table 7 that there is a correlation between the yield stress $\sigma_{0.2}$ and Young's modulus E. Gradually increasing Al in the alloy yields a small increase in E and a significant increase in $\sigma_{0.2}$. The density

Table 7   Mechanical Properties of Shaped Articles of Mg-Based Alloys

Composition	Sizes of cross section, mm	σ_B, MPa	δ, %%	E, 10^{10} N/m²	ρ, G/cm³	$\sigma_{0.2}$, MPa
Mg (99.9%)	≈10 × 2	113	26.5	17.92	1.77	9
Mg (99.9%)	⌀8–10	99	16.2	–	–	17
Mg–1.1% Al	≈10 × 2	222	10.3	18.8	1,778	35
Mg–2.9% Al	9.5 × 2.3	230	8.9	18.7	1.795	35
Mg–5% Al	⌀8–10	243	5.1	–	–	118
Mg–5.2% Al	10.5 × 2.4	255	9.0	19.175	1.873	110
Mg–8% Al	≈10 × 2	287	4.5	18.7	1,879	26
Mg–12.2% Al	10.7 × 2.4	349	6.5	18.77	1.825	155
Mg–20% Al	⌀8–10	192	0.1	–	–	–
Mg–27% Al	≈10 × 2	–	–	21.415	1.917	–
Mg–34.6% Al	≈10 × 2	–	–	23.93	1.998	–
Mg–0.5% Zn	⌀8–10	220	2.3	–	–	45
Mg–3.3% Zn	⌀10 × 2	168	20.0	18.14	1.814	–
Mg–4% Zn	⌀8–10	234	17.8	–	–	58
Mg–0.5% Mn	⌀8–10	116	3.0	–	–	42
Mg–2% Mn	⌀8–10	110	6.8	–	–	33
Mg–0.1% Zr	⌀8–10	172	1.9	–	–	52
Mg–0.9% Zr	⌀8–10	183	7.5	–	–	67
Mg–Zr–Y–Zn–Cd (MZY)	⌀11.5	123	27.1	–	–	–
Mg–Zr–Y–Zn–Cd (MZY)	⌀12.0	116	25.2	–	–	–
Mg–Zr–Y–Zn–Cd (MZY)	19.0 × 3.6	169	18.0	–	–	–
Mg–Zr–Y–Zn–Cd (MZY)	17.7 × 3.3	152	22.0	–	–	–

of Mg was measured by the hydrostatic method and was found to be 1.77 g cm⁻³, which is higher than reported from reference sources, which can be explained by the microstructure of Mg in the samples obtained by SM which contains no pores, hollows, and cracks. Experimental values of the density of Mg–Al alloys up to Al equal to 12.2% are also higher than rated values [63].

The mechanical properties of Mg–Zr–Y–Zn–Cd alloy (MZY) were investigated [63,64,67]. Every component of this alloy exhibits a high damping function [67]. Zr assists in the formation of small nucleates of crystallization and fine grains in the alloy, which is due to the values of Zr and Mg lattice parameters. Zr improves mechanical and casting properties and does not increase hot-shortness. Cd has limitless solubility in Mg, does not change mechanical properties of the alloy, but improves its casting properties considerably. Y improves grinding properties and improves strength.

Pressure treatment significantly reduces the damping capacity of this multicomponent Mg-based alloy. Decrement-free torsion oscillations for the strain amplitude 3.3·10⁻⁴ in the case of shaped casting were 20%; after pressing with the shrinkage of about 60% decrement was 4–5%. The Stepanov method provided articles of 2 m

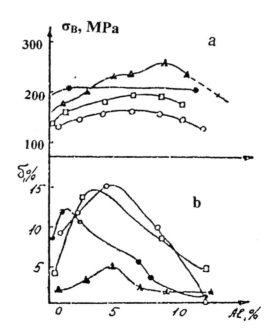

Figure 19   The influence of the composition of Mg-Al alloy obtained of various methods on σ_B (a) and δ (b): open circle is sand casting, open square is chill casting, black circle is pressure–die casting, and black triangles are SM. (From Ref. 65.)

length as 12-mm-diameter rods or 18×3 mm² bands. Mean $\sigma_{0.2}$ was 35 MPa, σ_B and δ as shown in Table 7. The damping capacity has record values: the value for torsion oscillations is 3.5% at shear amplitude $3 \cdot 10^{-4}$ [63,67]. High damping is explained by microplastic deformation in a large volume because of fine grain in that alloy [67]. The alloy provides increased margin of safety design when used in construction by decreasing chatter and noise, sensitivity to resonance. The alloy provides average strength and can be argon-arc welded. This alloy has high cutting workability and high corrosion resistance. Special processing was developed to grow from the melt profiles for frame assemblies and housings. [67].

The average scatter of σ_B for different profiles of this multicomponent alloy was ±10.9%; for δ, ±13%. For ribbons, σ_B along the growing direction was 120 MPa, δ was 11.75%, transverse to the direction σ_B was equal to 11.9 kG/mm², δ was equal to 14.7%. The anisotropy is small because of fine equiaxial structure [63].

A sensitivity to mechanical properties of shaped articles with respect to alloy content was observed. Thus 0.4% increase in Zr and 0.15% decrease in Zn resulted in a significant change of mechanical characteristics ($\sigma_B = 191$ MPa, $\sigma_{0.2} = 109$ MPa, $\delta = 8\%$) [67].

Electrochemical Properties

The electrochemical properties [64,68–70] of magnesium-based alloy profiles are very important because of their potential for use as corrosion protection of metal (steel or aluminum) in sea (salt) water, in chemical operations, or for underground constructions. The protection is based on the shift of corrosion potential from its stationary value to protective one, which is achieved by contact with other more electronegative metals called "the protector" (sacrificial metal) [71]. Materials protectors must exhibit low anodic polarizability, be stable in time with a negative potential, and exhibit no changes for different degrees of anodic polarization. They must also exhibit satisfactory mechanical and use properties. Reducing nonproductive metal waste and increasing protector life can be provided only by high and stable real current efficiency Q_r measured in A hr/kG:

$$Q_r = Q_t - Q_w,$$
$$\dot{\eta} = Q_r/Q_t \cdot 100\%. \tag{17}$$

where Q_t is the theoretical current efficiency, Q_w is the waste current efficiency consumable on self-solution as a result of working micropores, $\dot{\eta}$ is efficiency.

Protectors derived from Mg-based alloys are used because they exhibit long and stable current efficiency Q_r. They possess sufficiently high electronegative potential both at stationary conditions and toward anodic polarization, which provides good protection for various metals, and they are potentially useful for seawater salt contents up to 10% and in soil of low electric conductivity [71]. Standard Mg-based alloys possess low efficiency, about 30–40%. Therefore specific protector alloys of 50–60% and some higher efficiency were developed. Increasing efficiency to 70–80% is very important and is an ongoing engineering problem. Protectors can be produced by casting or pressing. Mg-based alloy profiles obtained by SM for use as protectors exhibit a specific structure [18].

The electrochemical characteristics of circular rods of a number of experimental and multicomponent industrial binary Mg-based alloys with diameters from 10 to 12 mm and 12 × 12-mm blades obtained by SM from the melt were investigated. The study of protector properties was performed in water from the Caspian Sea with a total salt content of 12.8%. The following electrochemical characteristics were determined: real current efficiency Q_r, efficiency η, stationary potential φ_{st}, and polarization potential φ_p. Theoretical current efficiency Q_t was calculated for the given alloy content. Tests were performed to determine constant current density, 5 A m^{-2}. The test duration was 7 days. Results of the tests for a number of alloys are shown in Table 8.

Table 8 shows that electrochemical characteristics depend on the chemical composition of the alloy. Commercially pure Mg exhibits a low Q_r and efficiency η. This is explained by the high speed of its self-dissolution [71]. Binary Mg–Al alloys were investigated over a wide range of Al content [69]. Al has better electrochemical properties than Mg ($Q_t = 2980$ A hr kG^{-1} as compared with 2200 A hr kG^{-1} for Mg), which suggests that for a constant efficiency η of Mg–Al alloys, Q_r had to increase by 2.6 A hr kG^{-1} per 1% of Al. However, Q_r of Mg–Al alloy decreases compared with Mg when the content of Al increases. (The same behavior is observed for η.) This is due to the formation of the intermetallic compound FeAl$_3$. The intermetallic compound is the effective cathode owing to more positive electrode potential, -34 V, in seawater, which results in an increase in self-dissolution of the alloy. Increasing Al in the alloy decreases negative potentials φ_{st} and φ_p because of the formation of FeAl$_3$ and more positive electrode potential of Al in comparison with Mg in seawater [69]. Table 8 also shows that the electrochemical characteristics of alloys depend not only on their composition but also on the sample shape. Thus the efficiency and positive φ_{st}

Table 8 Electrochemical Characteristics of Mg-Based Alloys

Composition (in charge/in profile, %%) or type of alloy	Sizes of cross section, mm	Q_t, A h kG^{-1}	Q_r, A h kG^{-1}	Efficiency, %%	$-\varphi_{st}$, mV	$-\varphi_p$, mV	Remarks
Mg	⌀10	2203	767	34.8	1351	1310	
Mg–5.0 Al	⌀10	2240	585	26.1	1265	1232	
Mg–5.0 Al	11×2	2240	598	26.7	1252	1215	
Mg–12.2 Al	⌀9	2297	405	17.6	1230	1174	
Mg–12.2 Al	11×2.5	2297	454	19.8	1213	1160	
Mg–0.5/0.33 (Zn)	⌀10	2196	1126	51.4	1345	1320	
Mg–0.5/0.33 (Zn)	12×3	2196	1271	57.9	1340	1312	
Mg–4.0/3.3 (Zn)	11×2.5	2148	1046	48.6	1320	1300	
Mg–0.5/0.06 (Mn)	⌀12	2202	1092	49.6	1350	1322	
Mg–1.0/0.13 (Mn)	⌀11	2201	1150	51.9	1340	1310	
Mg–1.0/0.13 (Mn)	⌀11	2201	1661	75.4	1300	1260	thermotreatment 1
Mg–1.0/0.13 (Mn)	⌀11	2201	1761	79.55	1288	1258	thermotreatment 2
Mg–2.0/1.11 (Mn)	12×3	2189	1021	46.6	1335	1306	
Mg–1.0 Ga	⌀10	2193	1143	52.1	1355	1316	
Mg–1.0 In		2188	1199	54.8	1385	1348	
Mg–5.0 In		2128	1219	57.3	1402	1385	
Mg–0.5 Y		2190	865	39.5	1312	1278	
Mg–6.0 Y		2125	431	20.3	1305	1285	
Mg–0.1 Zr		2202	1467	66.6	1402	1348	
Mg–0.1 Zr		2202	1641	74.5	1256	1232	thermotreatment 1
Mg–0.9 Zr		2194	1454	66.3	1396	1354	
ML5 (commercial)		2257	1478	65.5	1280	1235	
ML5 (commercial)		2257	1885	83.5	1240	1210	thermotreatment 1
MZY (commercial pure alloy)		2190	1349	62.8	1520	1436	
ML4 (high purity)		2212	1380	62.4	1290	1242	conventional casting
MP1		2210	1390	62.9	1236	1215	conventional casting

Composition of commercially pure alloys (%): 1) ML5: Mg–(7.5–8.5) Al–(0.5–0.6) Zn–0.2 Mn; 2) MZY: Mg–Zr–Y–Zn–Cd; 3) ML4 (high purity): Mg–(5–7) Al–(2–3) Zn–(0.15–0.5) Mn; 4) MP1: Mg–(5–7) Al–(2–4) Zn–(0.002–0.5) Mn–0.01 Ti.

and φ_p for plates are higher than for circular rods. Probably smaller grain sizes in plates result in the decrease of a portion of the self-dissolution resulting in a smaller increase in Q_r. The Fe content in the peripheral parts of Mg-based alloy samples obtained by SM was higher than in the central part due to the high volatility of Mg [63]. This may be the reason for a more positive φ for plates because of the larger surface of the evaporation of Mg for plates than for rods. The influence of FeAl$_3$ on electrochemical properties of plates is stronger than for rods, although the influence of FeAl$_3$ on the electromechanical characteristics of binary system Mg–Al is unsatisfactory in all.

Interesting results were obtained for Mg–Zn alloys with high solubility of Zn in Mg. Stationary potentials of Mg and Zn in seawater are −1.40 and −0.82 V, re-

spectively, which is a very large difference between them suggesting that alloying of Mg is meaningless. However, Zn forms binary compounds with Cu and Ni and ternary intermetallic compounds with Mg. It is known that impurities of Fe, Ni, Cu in Mg increase the speed of its self-dissolution and decrease Q_r and efficiency [71], which explains increasing Q_r and the efficiency of Mg–Zn alloys in comparison with Mg (Table 8).

Some reduction of Q_r with increasing Zn in Mg to 3.3% is related to the formation and growth of intermetallic compounds of the elements with cathode potentials of −0.64V relative to the solid solution. Consideration of binary Mg–Mn alloys shows that the difference in φ_{st} for Mg and Mn is equal to −1.18 V, but in seawater it reduces to −0.44 V [71]. Therefore Mn is the active cathode relative to Mg which can induce its intensive selective etching. Q_t of Mg is up to 2.2 times larger than for Mn [71], and despite alloying by Mn, as seen from Table 8, Q_r and the efficiency of Mg increase. Q_r and the efficiency increase with increasing Mn in charge to 1%. Then for 2% of Mn there is some decrease. The increase in Q_r and the efficiency can be explained by the formation of the solid solution of Mn in Mg, resulting in a decrease of the solubility of Fe in Mg [71]. The addition of Mn to Mg leads to the formation of the α-Fe, Mn solid solution. Harmful action of the impurities of Fe, which is the reason for the increase of the Mg self-dissolution rate and the decrease in Q_r and efficiency, is partially neutralized, which is corroborated by measurements of the Fe content in the crystallized samples. In Mg–0.5% Mn charge, the Fe content was 0.04%, after crystallization the Fe content was 0.03%. The content of Fe was reduced from 0.04% to 0.02% after crystallization of Mg–2.0% Mn alloy. This phenomenon is confirmed by the difference in Mn content for the charge and for crystallized sample. The positive influence of Mn on the decrease of Fe content in the alloy as grown is finished for 2% of Mn. An excess of Mg induces a nonuniformity of alloys, the formation of internal stresses, and the increase of the effect of micropores.

From Table 8 it is seen that Q_r and $\acute{\eta}$ of Mg–Ga and Mg–In alloys are rather high, which is important for a number of industrial applications. It is especially important for Mg–1.0% In alloy, which exhibits a high plasticity [64].

Very high Q_r, efficiency, and φ of Mg–Zr alloys, in conjunction with high mechanical characteristics above, appear to have considerable promise for protector applications [64].

High electrochemical characteristics are also obtained for multicomponent industrial alloy samples obtained by SM. They are usually higher than obtained by conventional casting ML4 (high purity) alloy and special protection alloy MP1 [64]. The alloys ML5 and MZY exhibit record high electrochemical potentials for sufficiently high efficiency.

Stepwise thermal treatment in the liquid and solid state exhibits a very positive influence on the efficiency. Sharp efficiency increases were observed for all thermal treated alloys (Mg–Mn, Mg–Zr, ML5). The decrease of φ after thermal treatment was small except for the Mg–0.1% Zr alloy [64].

The structure-specific increase of electrochemical characteristics of all considered alloys can be connected with the combined effect of a number of factors: chemical composition and impurities for articles obtained by SM [64]. It is necessary to note especially fine

structure for Mg–Zr system including MZY alloy and decreasing or eliminating grain boundary reduction for electrochemical characteristics by a self-dissolution Mg [64]. It is possible to consider that shaped articles obtained by SM from Mg-based alloys may be useful as protective coatings.

Finally, it is necessary to note that mechanical, physical, and electrochemical characteristics of shaped articles facilitate their use in engineering applications such as light constructions, dampers, and protectors.

2. Copper-Based Alloys

Cu and Cu-based alloy shaped articles are used because of their high electric and thermal conductivity, high corrosion resistivity, and other properties [7,8,10,12,72]. For example, brass radiators and condenser tubes, bronze bushes, and gears can be applied for machine building or as copper circuit, and anodes or brass and other alloys can be used for cryogen engineering [7].

Shaped Cu-based articles can be grown by means of SM with vacuum sets without forced cooling. The pulling rate will not be small owing to high heat conductivity and radiation heat removal. However, the technique for obtaining the articles by means of sets with open mirror of the melt permitting to grow long and large cross-section articles for high pulling rate has been developed [8,10,12,72].

The physical properties of Cu and selected alloys are compared with alloys of Al and Mg in Table 9 along with properties of Fe. The higher melting point of Cu-based alloys than for Al and Mg required an improved smelting furnace for the sets used earlier for growing Al- and Mg-based articles. Melting Cu-based alloys in graphite–chamotte crucible was performed using a high-frequency inductor, although a resistance furnace using a special alloy heater could be used as well.

Other properties of Cu and its alloys (high heat conductivity, low latent crystallization temperature, high capillary constant) are favorable for using sets with the open melt mirror. It was necessary only to choose the material for a shaper. Different steels, graphite materials, corundum, and silicon carbide were tested in addition to carbon, chromium–nickel, and chromium steels containing up to 13% Cr and special steels used for extrusion dissolved in copper melt. High-chromium steel (Cr contents of 23–27%) was stable. Graphite shapers did not react with the melt but burned out rapidly, and they could be used only for inert atmosphere. With an air atmosphere, a shaper with graphite–chamotte floats provided an extended working time. Corundum was stable in the melt but was very sen-

Table 9 Physical Properties of Some Metals and Cu-Based Alloys

Material	Density 10^3 kG m^{-3}	Melting point, °C	Latent melt heat, 10^3 J/kG	Heat conductivity 10^3 J/(kG K)	Surface tension 10^{-3} J/m^2	Capillary constant 10^{-2} m
Cu	8.95	1083	213.5	1398	1220	0.53
Brass (68%Cu)	8.6	940	159.1			
Al bronze (Cu–10% Al–4.4% Fe–3.3% Ni)	7.5	1040				
Al	2.7	660	401.9	728.5	914	0.84
Mg	1.75	651	343.3	554.8	553	1.9
Fe	7.87	1040	268.0	224.4	1460	0.63

sitive to drastic changes of temperature. Shapers of SiC were stable in the melt and insensitive to drastic changes in temperature, but they were difficult to machine and could not be used for profiles of complex geometry [12].

Rods of diameters ranging from 2 to 8 mm were obtained from copper, aluminum bronze (Cu–10% Al–4.4% Fe–3.3% Ni), stannous bronze (Cu–6% Sn–0.6% Al–0.7% Ni–0.4% Mn), stannous brass (Cu–39% Zn–1.0% Sn), and siliceous brass (Cu–14% Zn–3.6% Si). The overheat of the melt was changed from 40°C to 120°C; the pulling rate was from 4 to 28 m hr^{-1} [12]. Seeds were made of steel or sheet iron.

Experiments were also performed for pulling aluminum bronze with 3-mm-diameter rods without air cooling. In that case the shaper was water cooled. The pulling rate was changed from 1.5 to 2.5 m hr^{-1} depending on the melt temperature. It was critically important that the copper be protected from oxidation [12].

Rods pulled from the cooper melt without any protection exhibited a strongly oxidized rough surface. Attempts at using a cover of commercially pure nitrogen were unsuccessful. Good results were obtained by the introduction of copper phosphide into the melt. Pulling from the melt covered with liquid oxidation of phosphorus provided a better surface. Mechanical properties of the cooper rods were excellent ($\sigma_B = 180$ MPa, $\delta = 40\%$).

Oxide film (mainly of aluminum oxide) is created on the surface of aluminum bronze in processing. This film protects the alloy but the film is rough and stiff leading to poor surface quality with deep longitudinal marks (to 0.3 mm) on the surface. Surface quality is somewhat better by increasing v, lowering the shaper a little deeper into the melt, and by vibrationally separating the oxide

film. Before separating the oxide film, the rod is pulled out from the film as "from a stocking."

The surface of the article was good when the profile was pulled under a reducing atmosphere (nitrogen or argon transmitted through incandescent coil) [72]. Drawn samples exhibited coarse grain columnar microstructure. The length of the grains was up to 300 mm, the width was 15–30 mm for overheating at 20–25°C, and the cooling rate was about 5–10 K sec^{-1}. When overheated and the rate of cooling is increased, the grains become smaller, many exhibiting a needle shape, and the rectilinearity of grain boundaries is fractured. For both low (4–5 K sec^{-1}) and high (55–60 K sec^{-1}) cooling rates, the microstructure exhibits a complex character: there are dendrites and almost equiaxial micrograins. The microstructure became finer with increasing cooling rate. σ_B was equal to 540 MPa, δ was 15% [72]. Note that the aluminum bronze profiles were obtained in the form of 4–14 -mm-diameter rods and in the form of ribbons. The ribbon had a width from 30 to 80 mm, the thickness ranged from 0.3 to 2 mm. Tubes of 10–25-mm diameter and 0.5–2-mm wall thickness were obtained also [72].

In the case of stannous bronze the oxide film was thinner and more friable than for aluminum bronze, but surface marks were less deep owing to less stiff oxide film. For stannous bronze rods σ_B was 320 MPa and δ was 28% [12].

Pulling stannous and siliceous brass articles was possible only for small (to 50°C) overheating melt because of the high volatility of Zn. Despite the thick oxide film, the pulling was not complex. Probably the melt surface in the profiling orifice of the shaper had no time for strong oxidation for rather high pulling rate (15–28 m hr^{-1}) of the brass rods. The surface of brass rods was better than for bronze rods, but protecting is advisable not only for improving the article surface but also for preventing zinc burnout.

At equivalent pulling rates, processing productivity for the Cu-based alloy profiles is greater than for Al-based alloy because of the three times larger density. Also, the higher heat conductivity and lower crystallization heat permit growth of Cu-based alloy profiles at a higher rate. The difference in weight productivity in that case will be greater relative to Al-based alloys.

IX. METAL ARTICLES GROWN WITH HERMETIC (VACUUM) SETS

The interaction of the melt of refractory metals and alloys with air atmosphere is very intensive for alloys based on iron, nickel, cobalt, and other metals. In these cases, vacuum sets are required especially when single-crystal articles are the objective.

A. Polycrystalline Articles of Fe-Based Alloys

Fe-based alloys are used extensively in engineering [7,15]. But many of them cannot be machined. Therefore obtaining articles of a given shape from the melt directly by SM is very desirable. The experience for growing articles of Fe-based alloys can be applied to cobalt and nickel because many properties of these metals are similar to iron. Samples of widespread brittle iron carbide alloys with high content of carbon (Fe–3.4% C–2.34% Mo–1.93% Si–0.19% P–0.07% V–0.06% Ti) were pulled with a vacuum set of inductor, heating at a frequency of 5.3 MHz in vacuum 10^{-4} mm Hg [15]. The rod coupling with the melt column pulled was an integral part of the shaper-crucible system shaping the sample. The shaping rod was made from boron carbonitride or other materials wetted and nonwetted by the melt. Tubes with a diameter from 6 to 12 mm and wall thickness ranging from 0.2 to 2 mm were grown with such shaping system and 3-mm-diameter rods were also grown. In the case of coupling of shaping rod edge by the melt column, the diameter of the rod pulled or the internal diameter of the tube pulled was less than the diameter of the central shaping rod. When the melt wet the shaping rod the internal diameter of the tube pulled was larger than the diameter of the central shaping rod; the pulling rate was changed from 1 to 10 mm min^{-1}. Forced cooling was not used. Changing of the pulling rate and power of heater as

well as of mutual arrangement of the inductor heater and the crucible influences the diameter and wall thickness of the tube.

The chemical composition of shaped articles was constant along the length. In some cases a small change in molybdenum impurity was observed, which was due to the solution of the molybdenum seed. The surface of the rods and tubes was perfect with structure and properties depending on the pulling rate. The structure and properties of tubes for v equal to 1.5 mm min^{-1} were the same as for gray iron with the Vickers' hardness equal to 196. Increasing v did not whiten iron completely but increased the hardness because the metal component was quenched. At $v = 3$ mm min^{-1}, the hardness was 480 with a martensitic microstructure with coarse inclusions of cementite. At $v = 10$ mm min^{-1} the hardness was equal to 640 with a dendritic structure with an increased cementite content [15].

X-ray measurements revealed axial structure for all samples. The pulling direction was $\langle 100 \rangle$. This can be used for regulating some properties, such as magnetic properties, during processing.

B. Single Crystals of Shape Memory Effect Cu-Based Alloys

The shape memory effect or spontaneous recovery of a shape is observed for different materials with thermo-martensite transition [73–75]. A typical example of the shape memory effect is recovery of the original shape of a deformed metal article under heating when martensite crystals disappear. But the effect can occur with isothermal cooling also [73]. The reversible shape memory effect reveals itself in direct shape dependence on temperature as if the article shape follows the course of the temperature. For some conditions, thermoelastic unstable equilibrium is achieved between the parent phase and martensite variants. This results in superelasticity when the loaded sample strain, which is essentially plastic deformation, disappears entirely after unloading [75]. The superelasticity is the basis of the reverse memory effect.

A number of shape memory effect materials are coming into use in different applications including space, aeronautics, and machine engineering (thermoreactive elements and drives, coupling constructions, bellows, force elements, and others); electrical and radio engineering (contact breakers, starters, quasi-elastic antennas, and others); systems of automatization and robot engineering (transducers, drives); medical engineering (traumatology and surgery-distracter for automatic traction of bones and scoliosis treatment, apparatus for fracture treatment, vasodilators; stomatology).

Currently, only polycrystalline TiNi applications have been developed, but the range of thermoelastic transitions and some characteristics of that material do not meet demands of a number of applications. Improving some characteristics of the shape memory effect and changing the range of thermoelastic transition can be accomplished with single-crystal Cu-based alloys. Ternary compounds Cu–Al–Ni and Cu–Zn–Al exhibit the best characteristics. Cu–Al–Ni alloys are more preferable, especially as single crystals. Cu–Al–Ni crystals grow more homogenous because the volatility of Ni is less than Zn.

1. Obtaining Shaped Single Crystals

Shaped single crystal of Cu–Al–Ni alloy was pulled by set with resistance furnace [7,76,77]. The pulling rate ranged from 2 to 5 mm min^{-1}. Processing was performed in vacuum (10^{-3}–10^{-5} mm Hg) or atmosphere of spectro-pure argon. The temperature was maintained with an error less than ± 0.1 K. The shaper was made from spectro-pure graphite nonwetted by the melt. The liquid–solid interface was about 1–2 mm higher than the upper edge of the shaper.

Single-crystal rods of 2–4-mm diameter and tubes (internal diameter was 2–6 mm, wall thickness was 1–2 mm), ribbons (1–2-mm thickness) were grown. Single crystals exhibited a perfect structure with practically no low-angle boundaries. The dislocation density was $4 \cdot 10^3$–$7 \cdot 10^4$ cm^{-2} and rods had no visible defects such as floating. At the optimum pulling rate banding is absent too. Using an electric drive, deviations of the diameter of a rod varied from 0.2 to 0.5 mm. This defect was absent for pneumatically driven systems. Pulled ribbons possessed defects in the form of strips parallel to the liquid–solid interface. These striations were drastically decreased with convective stirring of the melt. Pulled crystals were oriented along directions $\langle 100 \rangle$, $\langle 110 \rangle$, and $\langle 111 \rangle$. Yield stress of single crystals of $\langle 110 \rangle$ orientation was 5% of those values for the $\langle 100 \rangle$ orientation. Crystals of $\langle 111 \rangle$ orientation were destroyed under stresses in the area of pure elastic strains. Therefore $\langle 100 \rangle$ rods are the most promising. Samples before investigation and use were heat treated (annealing at high temperature of about 1000°C and quenching into room temperature water).

2. Properties

To obtain the shape memory effect, it is necessary to induce reactions of martensite transformation by heating the crystal to a temperature higher than the M_s temperature and then cooling under mechanical stress to the temperature lower than M_s. Phase transformation takes place within in the crystal. High temperature β-phase transforms into β$_1$-phase of the annealed [74]. In some conditions, other phases may form. In Cu–Al–Ni crystals, thermoelastic reverse phase transformation β$_1$↔γ$_1'$ is observed when cooled to temperatures below the M_s and when heating to a temperature higher than the beginning of the reverse transformation A_s. Under deformation in isothermal condition at a temperature higher than the temperature of the end of the reverse transformation A_f, the elastic reversible β$_1$↔β$_1'$ transformation is obtained. The structure of single crystals, their composition, and heat treatment drastically influence the phase transformations and the shape memory effect.

Single-crystal samples of Cu–13.0% Al–4.0% Ni alloy were tested by cyclically changing temperature through the martensitic transformation range for two constant mechanical stresses equal to 12 and 36 MPa [76]. The maximum temperature dependence of temperature derivative of strain (dε/dT) for 12-MPa stress is observed at 435 K. For 36 MPa, the complex character of the strain recovery occurs. Two maxima of dε/dT dependence on T take place. The first maxima is for 437 K and the second for 443 K. The second maximum is higher than the first one. The ε dependence on T exhibits a break. Such regularities can be explained by a martensitic transformation created by two types of martensitic phases [76].

The influence of the chemical composition of Cu–Al–4% Ni single crystals on the character of the martensitic transformation and on the shape memory effect has been shown [77]. The alloy with 4% Ni and 12.5–14.5% Al was investigated. Figure 20 (curve 1) shows that concentration dependence of strain ε for cooling through the range of direct martensitic transformation under a stress of 30 MPa has a maximum equal to 3% at 13.5% Al. At this strain the type of the martensite is changed. The maximum corresponds to M_s about 370 K (M_s decreases with increasing content of Al, straight line 2 of Fig. 20).

The effect of heat treatment regime on martensitic transformation for Al contents was also shown [77]. Crystal sample of the alloy with 14.0% Al quenched into water from 1223 K exhibited the following values of the start and finish temperature of martensitic and austenitic transformation: $M_s = 250$ K, $M_f = 210$ K, $A_s = 290$ K, $A_f = 340$ K. The transformation temperatures of the alloy with 14.5% Al after the same heat treatment were reduced by 60–120 K. Annealing at 673 K after quenching of samples with 14% Al increases the temperature of

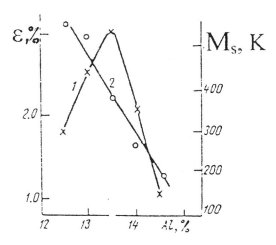

Figure 20 The effect of the Al content in Cu–Al–4%Ni single crystal on stain ε accumulated during cooling under stress 30 MPa through the interval of direct martensitic transformation (1) and on the temperature of beginning of direct martensitic transformation M_s (2). (From Ref. 77.)

reversible transformation: $M_s = 360$ K, $M_f = 300$ K, $A_s = 400$ K, $A_f = 470$ K, which is probably related to diffusion separating into layers [77]. The strain depended on the number of cycles for cyclic temperature variation through the range of reversible martensitic transformation until some stable value is obtained. Alloys with low M_s exhibited no effect of multiple reversible memory. High-temperature martensitic transformation samples always exhibited a reversible shape memory effect. Thus the reverse strain was observed for even a small number of cycles under a stress of 10 MPa for the alloy with 13.5% Al ($M_s = 370$ K).

The nature and regularities of various types of martensitic transformations of Cu–Al–Ni single crystals were investigated [78,79]. The shape memory effect materials deform under cooling at direct transformation along the direction of the force applied. Plasticity of a transformation is obtained [73].

The transformation plasticity was studied for 2-mm-diameter single-crystal rods of Cu–13.4% Al–4.0% Ni alloy with the following characteristic temperatures: $M_s = 355$ K, $M_f = 338$ K, $A_s = 347$ K, $A_f = 373$ K at thermal circling under a load of 15 MPa [79]. The rod was oriented in austenitic state along ⟨100⟩. Samples were annealed at 1225 K about 15 min and quenched into water. After heat treatment they exhibited a β_1' martensitic structure. At 400 K (it is higher than A_f), the sample was under tension under a load equal to 15 MPa and its temperature was circle changed. The temperature change rate for cooling and heating was 5 K min^{-1}.

In the first nine cycles, for direct martensitic transformation with cooling, the alloy dilated and upon heating its shape is recovered totally. The typical changing of the strain as a function of temperature for the first circle is shown in Fig. 21a. Accumulation of a strain at cooling was smooth and monotonic. The recovery of the strain at heating had an analogous character; temperature hysteresis was 15–20 K. It is characteristic for $\beta_1 - \beta_1'$ transformation. The temperature dependence of ε for 2–9 circles was similar to that for the first circle. The strain for the first circle was 6.0%, for the ninth one it was 6.4%. It is seen from Fig. 21b that the sharp change in the direction of deformation before the completion of the martensitic transformation takes place after the 10th circle, and the partial recovery of an accumulated strain or in other words the shape memory effect occurs (curve 1). The value of the recovery strain is 1.1%. Deformation during subsequent cooling is absent. Heating under mechanical stress is accompanied by a two-stage restoring shape (curve 2) and is completed by total recovery of the deformation. Temperature hysteresis for the first stage of the martensitic transformation is 20 K, for the second one it is 40 K.

Optical structural investigations showed that after cyclic changing temperature under stress not only thermoelastic but explosive martensitic transformation can be observed [79]. Before 350 K the surface of the single crystal did not change. But at 352 K two morphological types of martensitic relief were exhibited: needle-shapes in the form of thin straight lines and zigzag type. At 370 K at boundaries of two morphological modifications the lamellar relief in the form of dark strips appeared. The strips expanded gradually with increasing temperature (thermoelastic martensitic transformation) and about 380-K explosive change provided the final sizes. Two types of the martensite (β_1' and γ_1') were revealed in the alloy. It is known that the needle-shaped relief is connected with the β_1' martensite, the lamellar relief with γ_1' martensite. It is possible to suggest that γ_1' martensite had explosive kinetics during creation because of diffusion stratification and decomposition of the solid solution during multiple thermal cycling under a stress.

The reason for the various behavior of Cu–Al–Ni single crystal at cooling is different channels of reversible shape changes. The spectrum of motion forces can contribute in accumulating and recovery of strain by inducing shape changes in one range and breaking in others. Material reaction is determined by combination of all microdeformations in the volume [79].

In connection with the complex regularities of the behavior of Cu–Al–Ni alloy under a load, it is necessary

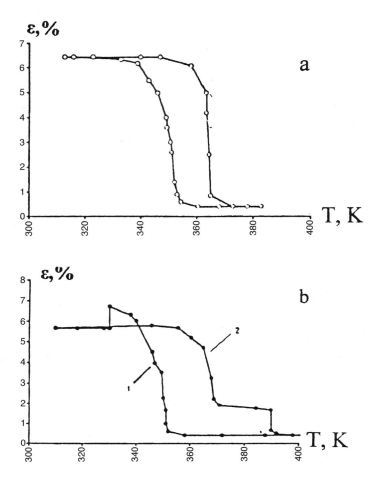

Figure 21 The temperature dependence of strain $\varepsilon\%$ for heat circling under constant load 15 MPa. (a) for the first circle (b) for the tenth circle. (From Ref. 79.)

for maximum effectiveness to take the following precautions when growing single crystals by SM: (1) choose the growth regime to obtain single crystals of perfect structure with minimum possible density of defects; (2) choose the most favorable chemical composition of the alloy for a given application; (3) choose the regime and stages of thermal treatment for the best shape memory effect and superelasticity; and (4) use the possibility for improving properties due to dispersion strengthening by microparticles of oxides or carbides arising during internal oxidation or carbonization of alloying components [80].

The following preferences of Cu–Al–Ni alloys in comparison with TiNi are exhibited [81]: higher heat conductivity and electric conductivity; higher stability properties during working; absence of irreversible plastic deformations and arising stresses during circle

changing of temperature; and lower stress of deformation in martensitic state.

X. SHAPED COMPOSITE ARTICLES WITH ALUMINUM MATRIX

Aluminum matrix composites are advanced engineering materials and the SM may be used to obtain directly from liquid phase both composite materials and shaped articles of given geometry from composite materials. Therefore SM provides the possibility to obtain shaped composite materials directly. This is a new technology at the interface between crystallization from the melt and material science. Various investigations of the growth by SM, structure and properties of the shaped composite construction with aluminum matrix have been performed. The basics of the process of growing shaped eutectic compositions and shaped reinforced articles have been developed.

A. Shaped Eutectic Compositions

The Stepanov method is a direct crystallization method that permits the growth of eutectics of oriented structures including fiber, lamellar, or more complex types. In a eutectic composite, a high-strength component is used to strengthen a matrix of suitable structure and physico-chemical properties. The usual problem of compatibility of different composite materials is avoided. Coupling the matrix and shaping is nearly ideal. At the same time, there is the possibility to control phase morphology, phase dispersion, and orientation [82]. However, there are some problems with forming a regular structure of eutectic composites by direct crystallization. Low rate of crystal growth and moderate cross-section sizes are among the most important deficiencies.

The Bridgmen and Czochralski methods can be used to obtain eutectic single crystals as with single crystals of any other material. There is increasing interest in machining articles from ingots; however, defects are created during the machining process. With the Bridgmen method, stresses within the growing ingot arise because of the difference in the heat dilatation of the crystallized material and the crucible. The Stepanov method does not possess these disadvantages. Processing by SM is also more stable than for the Bridgmen and Czochralski methods [82].

Eutectic compositions of two types were studied: lamellar composition of Al–CuAl$_2$ and the composition of fiber structure of Al–Al$_3$Ni [82–85]. These eutectics were grown with a set of inductive heating in vacuum about $5 \cdot 10^{-5}$ mm Hg or in argon with 0.008% N$_2$, 0.001% O$_2$, and 0.01 g m^{-3} of moisture. A nonwetted crucible and a shaper were fabricated of high-purity graphite. High-purity aluminum, copper of 99.99% purity, and carbonyl nickel were used for the charge [83,84].

Samples of square cross section from 4 to 49 mm^2 and 2·3 and 4·5 mm^2 rectangular and circle cross section with a diameter ranging from 1.5 to 6 mm and ribbons of thickness from 1 to 3 mm were grown. A copper seed was used. At first the sample exhibited a polycrystalline structure, then owing to concurrent growth only one grain (from three) remained. The sample possessed a regular lamellar structure and was used to make one-grain seed for subsequent growing.

Al–Al$_2$Cu of eutectic temperature of 548°C exhibited a regular lamellar microstructure for the pulling rate in the range from 0.55 to 2.5 mm min^{-1}. Its microstructure was columnar as the pulling rate was about 4.5 mm min^{-1} [84]. The spacing between intermetallic plates of 2-mm thickness was from 2 to 3.7 μm depending on v [83]. It was shown that the interplate spacing was proportional to $v^{-1/2}$. The strength of Al–Al$_2$Cu eutectic was in the range from 300 to 500 MPa; the strength of aluminum matrix was about 44 MPa [83]. In the case of one-grain seed of Al–Al$_2$Cu eutectic the effect of the inheritance of the orientation of microstructure was observed [84,85]. At the same time no effect was observed for the eutectic when the Czochralski method was used. This can be explained by the difference in temperature gradients: for SM it was up to 120 K cm^{-1}, for the Czochralski method it was about 15 K cm^{-1}. In the case of high-temperature gradient close to the melt–crystal interface there was no time for the diffusion leveling of concentration in the post-crystallization period [85].

The microstructure of Al–Al$_3$Ni eutectic (the eutectic temperature is 640°C) changed from fiber to plate-type for v in the range from 0.55 to 2.5 mm min^{-1} and higher [84]. As a rule the samples were one-grain for fiber structure. The ratio of the length to the diameter of the fiber was in the range from 10^3 to 10^4. The diameter of Al$_3$Ni fiber was about 0.5 μm. The strength of samples of Al–Al$_3$Ni eutectic changed from 350 to 420 MPa [83].

B. Longitudinal Reinforced Composites

It is necessary to combine pulling of shaped aluminum matrix with its reinforcement by strength elements for obtaining shaped composition articles by SM. Such process has to meet a number of demands [82]. The adhesion of a matrix and reinforcing elements must be high. It can be performed for good wettability of reinforcing elements by the melt of a matrix. The original properties of reinforcing element must be maintained after pulling through the melt of a matrix. The matrix must keep its plasticity in a composite, which requires minimization of the chemical interaction between them and intermetallic compounds being formed. An added complexity is that a uniform arrangement or special distribution of reinforcement must be included, which can be achieved only with specific construction of the set and a shaper. Functions of the shaper in that case are expanded.

Solution to the problem is possible only for the compatibility of material components of the aluminum matrix [82]. The Stepanov method possesses a number of advantages compared to other methods. For example, SM is better than deformation methods because long and large cross-section shaped composite articles under lower force acting on components with continuous or semicontinuous processing are possible. Additional advantages are provided in the section "Comparison Alternative Methods. Perspective Kinds of Articles."

1. Model Compositions

The general regularities for obtaining shaped reinforced composite articles were studied with a matrix of different aluminum alloys reinforced by individual unit fibers, plates, or other components of various materials [16,86–88]. These regularities could be used thereafter for real composites of high volume share of reinforcement achieved to tens of percents [82].

It was shown that stainless steel wire of 300–400-μm diameter disrupts the regular cellular microstructure of aluminum matrix and at about 100–120 μm, the cells increase and extend. By reinforcing by several steel fibers, the structure of the matrix depends on the arrangement of the wires. If the distance between fibers is equal to or less than 500–600 μm, the structure of all interfiber space is changed; in some places regular cellular structure is lost. For 700–800 μm and higher interfiber spacings, the change of the structure is only around every individual fiber such as unit fibers. That means that there is the critical ratio of the diameter of steel wire to the distance between them. If real ratio is higher than the critical one all microstructure between reinforcing fibers will be disrupted.

If steel wire is exposed to aluminum melt for about 4–10 sec, then intermetallic compounds are created. Maximum thickness of such layer ranges from 20 to 25 μm. For "aluminum–200-μm-diameter quartz fiber" composite, the microstructure exhibits no change. A cellular-like structure of shaped aluminum articles is obtained by SM without any reinforcement [87]. In this work, peculiarities of the modification of matrix substructure are explained by local change of thermal condition near the reinforcement. An approximate equation is suggested for estimating the modification. The main factor of a disturbance of heat field near the reinforcement is radial components of temperature gradient near the liquid–crystal interface. They were calculated for various reinforcements of Al matrix

when model composite had been obtained for the same condition [87]. The ratio of radial temperature gradient of various materials to that for quartz fiber in aluminum is shown in Table 10. This ratio explains the disturbance of the substructure of matrix by different reinforcements.

The modification of the substructure of aluminum matrix by various fibers observed in experiments corresponds to the theoretical effect only qualitatively. It is necessary to take into account the effect of the Al_2O_3 oxidation film and the chemical compounds forming at the reinforcement–matrix boundary [87] and the wettability of the reinforcement by the melt. In particular, it is shown [86] that tungsten fiber in model composite possesses strength equal to the initial condition. It is possible that the chemical interaction of W with aluminum melt is small.

During crystallization of composite by SM, the acceleration of a nucleation for lower supercooling is due to the reinforcing fibers which allow an increase in the pulling rate resulting in fine-grained structure and increased lateral strength [88].

2. Real Fiber Composites

The regularities revealed for model composites led to the development of commercial composites for construction applications. B-Al composites have received the most consideration [7,82,89–93]. They combine high specific strength and high rigidity of fibers with technological effectiveness and high construction safety of aluminum alloys. Boron–aluminum has high resistivity to cyclic loading and creep and to loading at high temperature. Electric and heat conductivity are also rather high. These properties and low density are the reason for the wide interest in that material [90,94].

The mechanics of fiber composites is determined by three factors. The first is the relative conserved fiber strength, which is the ratio of the strength of fiber in composite to the original strength. The second is the ratio of the shearing strength of the composite to the cohesion strength of the matrix. The third is the relative conserved plasticity of the matrix. That is, the ratio of the plasticity of the matrix in the composite to its original plasticity [94]. The product of these factors

Table 10 Ratio of Radial Temperature Gradient at the Liquid–Solid Interface in Al-Based Composite for Various Reinforcing Fibers to the Gradient for Quartz Fiber

Fiber	SiO_2	B	B + BN	SiC	Ni–Cr	Al_2O_3	Steel	C	W	Mo
Ratio	1	1.1	1.3	14	15	20	56	59	85	111

can be used as a rate of compatibility of composite components. Boron fibers have the best compatibility with aluminum matrix. The product is about 0.7.

Currently, most of the B-Al composites are made using semifinished products by one of the following methods: diffusion welding single layers obtained by plasma spraying Al on B fibers or diffusion welding multilayer packets of aluminum foils and B fibers specially stacked. The Stepanov method can provide a number of advantages relative to diffusion techniques: small force acting on components and obtaining long and large cross-section composites and lower cost at continuous or semicontinuous automated process. However, it is necessary to minimize the interaction of the reinforcement and the melt of a matrix.

Shaped B-Al fiber composites were produced by SM with the sets used for obtaining aluminum alloy articles. Shaping devices had some differences from those used for pulling aluminum alloy articles. Those devices were used to shape an article pulled from a melt and to distribute fibers over the cross section of the article. Also, the shaping device had to permit different pulling rates. The set had to supply fibers to the shaping device. The temperature of a melt is the same as that used for crystallization of metal, but the pulling rate is higher owing to the smaller weight of an aluminum alloy per the unit cross section crystallized and the heterogeneity of crystallization. For simplification of processing, the reinforcing elements of B fibers group separated of thin Al or Al-based alloy layer were used against unit fiber. As a rule, the fibers possessed safety covers of refractory compounds.

3. Matrix Structure

The macro- and microstructure of a matrix of B-Al composite obtained by SM is very sensitive to the matrix composition, distribution of fibers, position of reinforcing zones, and to thermal properties of the shaper [91–93].

The matrix of Al–1% Mn and Al–6% Mg–0.6% Mn alloys (with microcomponents of Ti and Be) exhibits a different macrostructure in the reinforcing part and in the part without reinforcement. The matrix of a part without reinforcement contains grains drawn in the pulling direction as observed for articles of such alloys. The length of grain is in centimeters, the thickness is in millimeters. In the reinforcing zone, grains are smaller and the transition from directional to equiaxial structure is observed. The average size of grains of the alloy with 6% Mg is about 0.5 mm. The reason for the change is lower heat conductivity of B fibers than of the matrix.

It results in the increase of the length of the crystallization zone and the decrease of the temperature gradient in the crystallization zone.

The length of the crystallization zone (the crystallization interval is 11°C) is 1 mm for the part of 2-mm thickness B-Al–1% Mn ribbon without reinforcement. The crystallization zone increases to 5–10 mm as a function of volume share of fibers. For alloy with 6% Mg (the crystallization interval is 62°C) the length of the crystallization zone in the reinforced part of the ribbon under the same crystallization condition is a few centimeters. A significant change in the microstructure occurs near the fibers. In the case of aluminum (about 99% Al) B fibers have no influence on the macrostructure of the matrix up to the area about 0.1-mm radius from the fiber. It can be explained by the absence of the crystallization interval for Al.

Boron fibers of composite of Al–6% Mg–0.6% Mn (microcomponents of Ti and Be) matrix are surrounded by dendrites normal to the fiber surface. The length of the dendrites is about 0.2 mm. Their sizes are larger near the fiber. The change in the microstructure is observed also in the matrix of Al–1% Mn. But the change is less than 0.1 mm. The change in the matrix microstructure of both alloys is observed both near the fiber and near the reinforcing zone. The growth of dendrites (for Al–1% Mn) normal to the fiber surface can be explained by nucleation on the reinforcement below the interface. Heat removal from the dendrite can be both into the melt and into the fiber. No change in the microstructure of 99% Al matrix near fibers was observed as for the macrostructure.

4. Strength

(1) The first major problem for obtaining B-Al fiber composites is conserving the original high strength of boron fibers (original σ_B ranges from 2000 to 4000 MPa). The ratio σ_c/σ_o is a measure of the conserved strength. Here σ_o is the original strength of fiber, σ_c is the strength of the fiber extracted from the composite (after being in the melt) [94]. The value of the ratio depends on the chemical interaction of the fiber with the melted matrix and oxidation of fibers. Boron interacts intensively with melted aluminum with forming reaction zone consisting of borides AlB_2, AlB_{12}, and Al layer saturated by B [90,95]. The strength of the fiber will be high if the intermetallic layer on the fiber is not more than 0.5–1.0 μm. The mechanical properties of the composite become worse as a result of recrystallization of fibers, an embrittlement of fibers, and a matrix and lowering fiber surface quality.

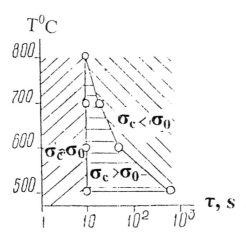

Figure 22 Diagram of the strength of boron fibers depending on the air temperature and the contact time τ for heat treatment of monolayer B-Al ribbon. (From Ref. 89.)

The effect of the regime of pulling B-Al composite profile on the strength of various types of B fibers and semifinished fiber products was investigated [89–92]. Boron fibers were extracted from composite material or from semifinished products by etching 10% water solution of KOH (or NaOH) or 70% water solution of HCl. Tensile strength was measured for many of the fibers [90].

It was known that boron fibers without cover are weakened not only under acting Al melt but also under air heat treatment during 1 min and longer for temperature higher than 500°C because of intensive oxidation. A thermal stability cover (BC, BN, SiC) was used to prevent the boron fibers and melted Al from oxidation and helped to conserve their strength. The strength of boron fibers with BN cover after being exposed to the Al melt during 1 sec decreased from 3000 to 2400 MPa. Acting melt on fibers longer than 1 sec resulted in small increasing σ_B; for 10 sec, σ_B was equal to 2700 MPa [89]. In that case, a contact of the group of boron fibers with the melt during 1 sec induces weakening of fibers by 22–40%. It was revealed that the reason for the effect is connected with the hot shock. It is more effective at reinforcing with packed group of boron fibers divided of thin Al layers. But it was found that even for optimum temperature of the melt because of heat shock the conserved strength of fibers is about 0.7–0.75. The spread of strength after contact with the melt is higher, the variation coefficient increases from 9% to 22% [90–92]. Detailed investigation results in finding the regime of preliminary heat treatment for improving the conserved strength. It was shown that there were ranges of conserving the original strength of fibers ($\sigma_c = \sigma_o$ for air

heat treatment during 10 sec and shorter), strengthening ($\sigma_c > \sigma_o$), and weakening ($\sigma_c < \sigma_o$) of fibers (Fig. 22). Different result of heat treatment is due to two contrary factors. Weakening is connected with an oxidation of fibers. Strengthening is explained by the redistribution or decrease of internal stresses in a fiber. The strengthening may also be due to plasticizing fibers by adjoining aluminum matrix. The conserved strength after heat treatment is determined by these factors. In most cases heat treatment of reinforcing elements resulted in decreasing the spread of the conserved strength from 1.5 to 2 times.

Boron fibers with thermal stable cover (B_4C and other) were joined in groups, then packets were made from the groups (in the kind of bunches or ribbons). Aluminum film surrounding individual fibers in packets decreases thermal shock in the time of bringing into the melt. The air heat treatment in that case is not obligatory. Figure 23 shows the conservation of boron fibers strength (σ_c/σ_o) vs. the stand time in Al–1% Mn melt and on the melt temperature for the packet of 10 bunches; every bunch consists of seven individual fibers joined with aluminum matrix (99% Al) [90,91]. In Fig. 23 numbering 1–4 denotes areas of strengthening (shaded area) and weakening of fibers as projection on horizontal plane between axes of temperature and time. Area 1 is characterized by strengthening of fibers. This strengthening arises in the case of relatively slow heating

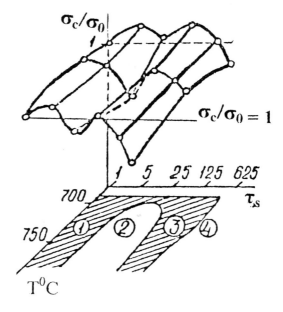

Figure 23 Conservation strength of fiber (σ_{Bc}/σ_0) depending on the contact time with the melt τ and on the melt temperature T. (From Refs. 91 and 92.)

of the bunch. The heat shock is absent and internal stresses in fibers are low. Area 2 is characterized by weakening of fibers. It is explained by disjoining fibers when temperature will be equal to the melting point of primary aluminum matrix joined by individual fibers in bunches and by thermal acting due to the temperature difference between the melting point of aluminum (99% Al) and Al–1% Mn (the secondary matrix). Strengthening in area 3 is caused by a relaxation of internal stresses that arose in area 2 because of standing at constant temperature in the melt. Weakening in area 4 for a time of about 2–4 min is connected with destroying ceramic cover of fibers and chemical interaction of fibers with the melt.

Weakening of packets of boron fibers covered by B_4C can be minimized for optimum regimes of obtaining boron–aluminum fiber composites by SM. Conservation of the strength of fibers σ_c/σ_o under the melt or a heat treatment was lower when their original strength was higher. It is suggested [90] that some surface defects of boron fibers can be a gutter for internal stresses.

(2) The second important problem of obtaining B–Al composites by SM is achievement of high adhesion of boron fibers and aluminum matrix, which determines the shear strength of the composites. Temperature dependencies of shear strength τ_s of B–Al fiber composite ribbons with 0.3–0.4 volume share of fibers and the same samples without reinforcement were studied [92]. τ_s was measured as a ratio of fracture load to shear area. Temperature dependencies of shear strength in the range from 20°C to 350°C are shown in Fig. 24. It is seen that reinforcing of aluminum alloys, in particular Al–1% Mn, can increase the composite longevity for elevated temperature. Curves 1 and 3 of Fig. 24 show that τ_s of composites of Al–1% Mn alloy and Al–6% Mg–0.6% Mn alloy as well as with microcomponent Ti and Be is much higher for all investigated temperature range than for composites of aluminum matrix (99% Al). It is the result of stronger adhesion of fibers and matrix of Al-based alloys than with aluminum matrix. No deposition of matrix of the alloys from fibers is observed for such composites. This is related to the longer time traversing of reinforcement through the melt of Al-based alloys than boron fibers through the aluminum melt.

Primary aluminum matrix jointed by boron fibers in reinforcing packet was rather porous. It was replaced totally by a secondary matrix of Al-based alloy melt. In the case of composites of aluminum matrix the time of contact of boron fibers with aluminum melt should be small, about 1.4 sec, because the absence of the thermal stability cover. Primary porous matrix was not exchanged. Moreover, the wettability of boron

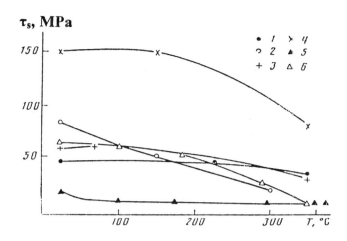

Figure 24 The shear stress as a function of test temperature T: 1—B + B4C in Al–1%Mn matrix; 2—Al–1%Mn alloy; 3—B + B4C in Al–6%Mg–0.6%Mn with impurity of Ti and Be; 4—Al–6%Mg–0.6%Mn with micro impurities of Ti and Be; 5—B in Al(99%) matrix; 6—Al(99%) matrix.

fibers by liquid Al-based alloys is higher than for molten aluminum.

(3) The investigation was also used for developing the process of obtaining shaped B–Al fiber composites. Boron fibers without special cover or with thermal stability covers of B_4C, BN, and SiC are arranged in a shaped profile both over all cross section and certain cross-section pieces loaded in the time of application. The second arrangement was called "local reinforcement." Regimes of pulling for shaped composite profiles without any degradation of reinforcement were developed. Composites of different Al-based alloy matrix of length up to 2 m and different cross sections were obtained. It was shown that longer profiles with complex shape of cross section could be obtained.

Volume content of fibers in reinforced zones achieved 50%. For many cases the original strength of fibers was conserved and σ_B of shaped fiber composites was about 800 MPa. It is close to the σ_B of samples obtained by liquid-phase methods and approaches to the strength of composites obtained by solid-phase methods [89,90,92].

5. Other Kinds of Longitudinal Reinforcing Composites

Development of process of obtaining shaped layer metallic composites was performed [82]. For example, bimetallic aluminum profiles with an internal steel layer were pulled from the aluminum melt. It was shown that a complex shape of the surface of such articles could be obtained. Both high strength and high corrosion resistance are attained.

The process of obtaining a flexible heat element (FHE) in aluminum cover was developed. It can be called a "thermal electric composite" article. Flexible heat element uses Ni–Cr wires in SiO_2 insulation covered by aluminum shell. Tests showed the effectiveness for heating pipelines, e.g., for oil pipeline, for fire hazard places as well as for explosion hazard areas. The operating temperature of FHE can be about 240–250°C.

XI. CONCLUSION

The method of capillary shaping materials suggested by Stepanov was the basis of development of sets and continuous or semicontinuous processes for obtaining shaped metallic articles from their melts. Different shaping devices facilitate the pulling from the melt various shapes of cross section of profiles. Shaped articles can be obtained from various metals, alloys, and composite materials. Structure and properties of such articles were investigated. Rational application of shaped articles was revealed and realized for special aims.

Stepanov's method can find wide industrial application especially for obtaining Al-based alloy profiles. Currently, transition to industrial production from experimental processes is possible. For more success of large-scale competitive production, it is necessary to increase the productivity of the processing, which is possible because of more effective methods of cooling for pulling articles as well as using pulling group of articles from one crucible simultaneously.

There are facilities for further improving the mechanical characteristics of shaped profiles by means of heat treatment, covers, and reinforcement. Automatic regulation and stabilization of processing and using new alloys can perform improvement of surface quality.

It shows the outlook of production and application of shaped single crystals with shape memory effect and of various shaped composite materials both eutectic and reinforced by fiber composites as well as of dispersion reinforced articles.

Commercially pure effect of shaped articles can be realized owing to a number of ways [7]: decreasing net cost, using an article of most rational shapes by consumers, and using an article of the shape which can be obtained only by SM (e.g., complex shape) and which enables to create a new effective device or a new effective design.

The Stepanov method in the future will be used most of all for obtaining new polycrystalline shaped articles (or blanks) of various metals and alloys. Vanes of gas turbine or elements of magnetic of Ni or Co-based refractory alloys can be a good case in point.

Experience with growing and using of Cu–Al–Ni single crystals with shape memory effect shows facilities of SM for obtaining shaped single crystals of other metals and alloys. Shaped metal single crystals are a simpler subject of fundamental investigation than polycrystals and have preferences as compared with single-crystal samples cut out from ingots because they can be used without mechanical treatment. Thus they were used successfully for investigation of inelastic properties of Al single crystal with low-angle boundaries [86]. It is more important that they can be applied as profiled construction materials owing to higher plasticity and smaller brittleness as well as higher corrosion resistance than the same material in polycrystalline state. The possibility of obtaining novel shaped metal single crystals is based also on wide production by SM of shaped single crystals of semiconductors and dielectrics for electronics and optics.

REFERENCES

1. Balandin, G.F. *Casting by Freezing*; Mashgiz: Moscow, USSR, 1962; 262 pp.
2. German E. *Continuous Casting*; Metallurgizdat: Moscow, USSR, 1961, 814 pp. *in Russia*.
3. Czochralski, I. Ein neus Verfahren zur Messung der Krystallisationsgeschwindigkeit des Metalle. Zeitschr. Phys. Chem. 1917, *92*, 219.
4. Gomperz, E.V. Untersuchungen an Einkristalldrachten. Zeitschr. Phys. 1922, *8*, 184.
5. Mark, H.; Polanyi, M.; Schmid, E. Vorgange bei der Dehnung von Zinkkristallen. Zeitschr. Phys., Bd. 1923, *12*, 58–77.
6. Kapitza, P. The study of the specific resistance of bismuth crystals and its change in strong magnetic fields and some allied problems. Proc. R. Soc. London, Ser. A 1928, *119*, 358.
7. Antonov, P.I.; Zatulovskii, L.M.; Kostygov, A.S.; Levinson, D.I.; Nikanorov, S.P.; Peller, V.V.; Tatarchenko, V.A.; Yuferev, V.S. *Obtaining Shaped Single Crystals and Articles by Stepanov's Technique*; Nauka: Leningrad, USSR, 1981; 8–22 *in Russian*.
8. Shah-Budagov, A.L.; Stepanov, A.V. New technique obtaining articles (sheets, tubes, rods, different profiles etc.) from the melt directly. Zhurn. Tekhn. Fiz. 1959, *29*, 381. *in Russia*.
9. Stepanov, A.V. *Future of Metal Treatment*; Lenizdat: Leningrad, USSR, 1963; 129 pp. *in Russia*.
10. Stepanov, A.V. New technique obtaining articles (sheets, tubes, rods, different profiles etc.) from the melt directly. Zhurn. Tekhn. Fiz. 1959, *29*, 381–383. *in Russian*.
11. Gol'zman, B.M.; Stepanov, A.V. The technique of

obtaining sheet and tubes from the melt of aluminum and its alloys directly. Izv. Akad. Nauk SSSR. Metall. Fuel 1959, *N 5*, 49–53. *in Russian.*

12. Gol'dfarb, V.M.; Kostygov, A.S.; Yukhno, M.M.; Stepanov, A.V. Obtaining copper, brass and bronze rods directly from the melt. *Problems of Crystallization and Solid State Physics*; Science notes of the A.I. Gerzen Leningrad State Education Institute: Leningrad, USSR, 1965; 144–150. *in Russia.*

13. Donskoi, A.V.; Stepanov, A.V. Obtaining "tubes in the sheet" directly from the melt. *Problems of Crystallization and Solid State Physics*; Science notes of the A.I. Gerzen Leningrad State Pedagogic Institute: Leningrad, USSR, 1965; 33–41. *in Russia.*

14. Zivinskii, S.V.; Antonova, N.N.; Aleksandrov, B.N.; Prytkin, V.V.all. *Growing Mercury Crystals of Given Shape and Crystallographic Orientation. The Materials of the First Conference on the Production of Semiconductor Single Crystals by the Stepanov's Technique and Perspective of their Application for Device Production*; Ioffe Phys. Tech. Inst. Acad. Sci. USSR: Leningrad, USSR, 1968; 79–82. *in Russia.*

15. Tatarchenko, V.A.; Vakhrushev, V.V.; Kostygov, A.S. Obtaining profiles of high carbon alloys of iron. Izv. Acad. Nauk SSSR, Ser. Fiz. 1971, *35*, 511–513. *in Russian.*

16. Khohlov, G.G.; Kardashev, B.K.; Kostygov, A.S.; Nikanorov, S.P.; Peller, V.V.; Regel, V.R. Preparation and properties of model aluminum–steel composites. Bull. Acad. Sci. USSR, Ser. Phys. 1976, *40*, 54–56.

17. Vyatkin, P.I.; Peller, V.V.; Stolbova, A.D.; Kostygov, A.S.; Korchunov, B.N. Some features of crystallization of component consisting of magnesium alloys and aluminum by the Stepanov's method. Bull Acad. Sci. USSR, Phys. Ser. 1980, *44*, 105–108.

18. Vyatkin, I.P.; Korchunov, B.N.; Nikitina, N.I.; Peller, V.V.; Rokhlin, L.L.; Stolbov, A.D. Structure and mechanical properties of certain Mg alloys of component grown by the Stepanov's method. Bull. Acad. Sci. USSR, Phys. Ser. 1980, *44*, 109–112.

19. Vorobyev, M.A.; Nikanorov, S.P.; Peller, V.V. Microstructure of matrix aluminum reinforced by variation of fillers and factors influencing on it. Abstracts of IY All-Union Conference on composition materials. Moscow USSR, 1978; 103–104. *in Russia.*

20. Gol'zman, B.M. Pulling from the melt crystalline plates and tubes. Opt. Mech. Ind. 1958, *11*, 45–46. *in Russia.*

21. Koptev, Yu. I.; Stepanov, A.V. Obtaining tubes of germanium. Fiz. Tverd. Tela 1967, *9*, 3007–3008. *in Russian.*

22. MLa Belle, H.E.; Mlavsky, A.I. Growth of controlled profile crystals from the melt. Part 2. Mat. Res. Bull. 1971, *6*, 581.

23. Cullen, C.W., Surek, T., Antonov, P.I., Eds.; *Shaped Crystal Growth*; North-Holland Publ. Comp: Amsterdam, 1980; 399 pp.

24. Tatarchenko, V.A. Ustoichivii rost kristallov; p. 240 Nauka: Moscow, 1988 *In Russia*n, translation in: Shaped Crystal Growth, Kluwer, Dorbrecht, 1993.

25. Tatarchenko, V.A. Shaped crystal growth. Hurle, D.T., Ed.; *Handbook of Crystal Growth. Part 2.* North-Holland: Amsterdam, 1994; Vol. 2, 1011–1111.

26. Zatulovskii, L.M.; Hazanov, E.E.; Polischuk, Ya. A. General Problems of Equipment Design; Energiya: Moscow. USSR, 1973; 112–119. *in Russian.*

27. Chalmers, B. Principles of Solidifications; John Wiley & Sons, Inc.: New York-London-Sydney, 1964; 289 p.

28. Tiller, V.A. Multiphase crystallization. Pozdnyakova, G.L., Ed.;Liquid Metals and Their Solidifying; GNTI: Moscow, USSR, 1962; 307–354 *in Russian.*

29. Peller, V.V. Production of shaped items from aluminum and magnesium alloys by Stepanov's method as compared with alternative techniques. Bull. Russ. Acad Sci. Phys. 1994, *58*, 1559–1563.

30. Kostygov, A.S.; Korchunov, B.N.; Peller, V.V.; Khohlov, G.G.; Yakovlev, O.V. Construction of shaping devices: physico-technical aspects and unification. Bull. Russ. Acad. Sci. Phys. 1994, *58*, 1547–1551.

31. Korchunov, B.N.; Peller, V.V.; Kostygov, A.S.; Kholodkov, A.S.; Yakovlev, O.V.; Kazakevich, A.G. Production of large shaped aluminum alloy items in a semicontinuous-operation plant. Bull. Russ. Acad. Sci. Phys. 1994, *58*, 1564–1571.

32. Korchunov, B.N.; Kostygov, A.S.; Peller, V.V.; Khohlov, G.G.; Osipov, V.N. Effect on geometry and structure from heat and mass transfer in profiled aluminum alloy growth. Bull. Acad. Sci. USSR, Phys. Ser. 1988, *52*, 148–154.

33. Shashkov, Yu. M. Growing Single Crystals by Pulling Method; Metallurgiya: Moscow, USSR, 1982; 26–61. *in Russian.*

34. Kostygov, A.S.; Korchunov, B.N.; Khohlov, G.G. Investigation of heat regime of crystallization of aluminum alloy D16 tubes. Bull. Acad. Sci. USSR, Phys. Ser 1980, *44*, 100–104.

35. Kostygov, A.S.; Tatarchenko, V.A.; Stepanov, A.V. Convective cooling samples obtained from the melt by Stepanov's technique. Izv. Akad. Nauk SSSR, Ser. Fiz. 1972, *36*, 481–485. *in Russian.*

36. Kostygov, A.S.; Korchunov, B.N.; Fedorov, V. Yu.; Khohlov, G.G. The heat emission of a surface of the article pulled from the melt by Stepanov's method. Proceedings of the 9th Conference on the Preparing of Shaped Crystals and Parts by the Stepanov's Method and Their Application in the Economy, 10–12 March 1982; Antonov, P.I., Ed.; Ioffe Phys. Techn. Inst. Acad. Sci. USSR: Leningrad, 1982; pp. 300–303. *in Russian.*

37. Michael, A.B.; Bever, M.B. Solidification of Al-rich Al–Cu alloys. J. Met. 1954, *6*, 47–56. *in Russian.*

38. Kostygov, A.S.; Tatarchenko, V.A. Thermal regime of crystallization and its influence on the structure of aluminum alloys samples. Izv. Akad. Nauk SSSR, Ser. Fiz. 1973, *37*, 2315–2318. *in Russian.*

39. Tatarchenko, V.A.; Vladimirova, G.V.; Stepanov. A.V. Structure and properties of aluminum articles crystal-

lized from the melt by Stepanov's method. Izv. Akad. Nauk SSSR, Ser. Fiz. 1971, *35*, 499–503 *in Russian.*

40. Ioffe, O.B.; Peller, V.V.; Khohlov, G.G. Thin-wall shaped products: particulars of production and prospect of usage. Bull. Russ. Acad. Sci. Phys. 1994, *58*, 1552–1554.

41. Korchunov, B.N.; Kostygov, A.S.; Peller, V.V.; Fedorov, V. Yu.; Khohlov, G.G. Particulars of production of aluminum alloys tube articles by Stepanov technique. Proceedings of the Conference on the Preparing of Shaped Crystals and Parts by the Stepanov's Method and Their Application in the Economy, 4–6 March 1985; Krymov, V.M., Ed.; Ioffe Phys. Techn. Inst. Acad. Sci.: USSR, Leningrad, 1986; 312–317. *in Russian.*

42. Kostygov, A.S.; Korchunov, B.N.; Khohlov, G.G. Classification of shapers used in producing profiled aluminum alloys according to the rate of removal of heat from the working volume. Bull. Acad. Sci. USSR, Phys. Ser. 1980, *44*, 89–92.

43. Fedorov, V. Yu.; Antonov, P.I. Calculating and empirical determination of shaper thermal characteristics and choice of its construction. Bull. Russ. Acad. Sci. Phys. 1994, *58*, 1424–1429.

44. Kostygov, A.S.; Korchunov, B.N.; Fedorov, V. Yu.; Khohlov, G.G. Application of two-zone forced cooling in obtaining aluminum alloy tubes by Stepanov's technique. Proceedings of the 9th Conference on the Preparing of Shaped Crystals and Parts by the Stepanov's Method and Their Application in the Economy, 10–12 March 1982; Antonov, P.I., Ed.; Ioffe Phys. Techn. Inst. Acad. Sci.: USSR, Leningrad, 1982; 304–308. *in Russian.*

45. Kostygov, A.S.; Korchunov, B.N.; Fedorovof, V. Yu. Application of water–air cooling at preparing aluminum alloy tubes by Stepanov's technique. Proceedings of the Conference on the Preparing of Shaped Crystals and Parts by the Stepanov's Method and their Application in the Economy, 4–6 March 1985; Krymov, V.M., Ed.; Ioffe Phys. Techn. Inst. Acad. Sci. USSR: Leningrad, 1986; pp. 298–302. *in Russian.*

46. Goldfarb, V.M.; Golzman, B.M.; Stepanov, A.V. Homogeneous cooling of thin-wall article pulled from the melt. *Problems of Crystallization and Solid State Physics,* Ivanov, G.A., Ed.; LGPI: Leningrad, 1965; Vol. 265, 90–104. *in Russian.*

47. Goldfarb, V.M.; Donskoy, A.V.; Stepanov, A.V. Some problems of shaping in preparing products from the melt. In *Problems of Crystallization and Solid State Physics*; Ivanov, G.A., Ed.; LGPI: Leningrad, 1965; Vol. 265, 61–74. *in Russian.*

48. Peller, V.V.; Korchunov, B.N.; Fedorov, V. Yu.; Khohlov, G.G. Development of Stepanov technique: perspective classes of shaped aluminum articles, the control of their geometry, structure and properties. Izv. Akad. Nauk SSSR, Ser. Fiz. 1999, *63*, 1831–1838; there is a translation in Bull. Russ. Acad. Sci. Phys. 1999, 63.

49. Korchunov, B.N.; Peller, V.V.; Khohlov, G.G. Produc-

ing thick-wall aluminum alloys articles by Stepanov technique. Izv. Akad. Nauk SSSR, Ser. Fiz. 1999, *63*, 1854–1860; there is a translation in Bull. Russ. Acad. Sci. Phys. 1999, 63.

50. Artem'eva, I.N.; Plishkin, Yu. S.; Sedov, V.A.; Kostygov, A.S.; Korchunov, B.N. Physico-mechanical properties of Al–Mg alloy shaped articles. Bull. Acad. Sci. USSR, Phys. Ser. 1976, *40*, 1369–1375.

51. Goldfarb, V.M.; Donskoi, A.V.; Stepanov, A.V. Investigation of process preparing Al–Mn ribbon from the melt. *Problems of Crystallization and Solid State Physics,* Ivanov, G.A., Ed.; LGPI: Leningrad, 1965; Vol. 265, 50–60. *in Russian.*

52. Korchunov, B.N.; Kostygov, A.S. Quality of surface of aluminum alloy samples prepared from the melt by Stepanov's technique. Izv. Akad. Nauk SSSR, Ser. Fiz. 1973, *37*, pp. 2319–2321. *in Russian.*

53. Esin, V.O.; Brodova, I.G.; Tatarchenko, V.A.; Stepanov, A.V. Peculiarities of defect structure of aluminum products grown from the melt by Stepanov's technique. Izv. Akad. Nauk SSSR, Ser. Fiz. 1971, *35*, 504–510. *in Russian.*

54. Brodova, I.G.; Esin, V.O.; Korchunov, B.N.; Kostygov, A.S.; Osipov, V.N.; Peller, V.V.; Polenz, I.V.; Fedorov, V. Yu.; Khohlov, G.G. Structure of aluminum–alloy tubes maid by various forms of Stepanov's method. Bull. Acad. Sci. USSR, Phys. Ser. 1988, *52*, 142–144.

55. Brodova, I.G.; Borisova, I.A.; Kostygov, A.S.; Korchunov, B.N.; Peller, V.V.; Khohlov, G.G. Structure of aluminum alloy D-16 articles prepared by Stepanov's technique. Izv. Akad. Nauk SSSR, Ser. Fiz. 1985, *49*, 2435–2438. *in Russian.*

56. Fedorov, V. Yu. Dependence of structure dispersion and mechanical properties of Al–Mn alloy product grown by Stepanov's technique on productivity of the process. Bull. Russ. Acad. Sci. Phys. 1994, *58*, 1536–1542.

57. Fedorov, V. Yu. Structure dispersion of Al–Si alloy articles in relation to parameters of growing. Izv. Akad. Nauk SSSR, Ser. Fiz. 1999, *63*, 1847–1853; there is a translation in Bull. Russ. Acad. Sci. Phys. 1999, 63.

58. Esin, V.O.; Brodova, I.G.; Tatarchenko, V.A.; Stepanov, A.V. Peculiarities of structure of Al–Mn alloy flat ribbon grown from the melt by Stepanov's technique. Izv. Akad. Nauk SSSR, Ser. Fiz. 1972, *36*, 588–594. *in Russian.*

59. Goldfarb, V.M.; Donskoi, A.V.; Dyagilev, F.M.; Kostygov, A.S.; Stepanov, A.V. Structure and properties of Al and Cu based alloys articles grown by direct crystallization. *Casting Properties of Metals and Alloys*; Nauka: Moscow, USSR, 1967; 298–338. *in Russian.*

60. Zolotarevskii, V.S. Chapter 9. Fatigue and wearing. *Mechanical Properties of Metals*; Utkina, E.N., Ed.; Metallurgiya: Moscow, 1983; 298–338. *in Russian.*

61. Bairamov, A.H.; Peller, V.V.; Kostygov, A.S.; Korchunov, B.N.; Tagieva, S.M.; Mamedova, S.M. Corrosion and chemical behavior of aluminum alloys parts produced by Stepanov's method. Proceedings of the Conference on the Preparing of Shaped Crystals and Parts by

the Stepanov's Method and Their Application in the Economy, 4–6 March 1985; Krymov, V.M., Ed.; Ioffe Phys. Techn. Inst. Acad. Sci. USSR: Leningrad, 1986; pp. 335–340. *in Russian.*

62. Bairamov, A.H.; Peller, V.V.; Tagieva, S.M.; Kostygov, A.S.; Korchunov, B.N.; Sultanova, Z.B.; Bektashi, S.G Corrosion and chemical behavior of aluminum alloys parts produced by Stepanov's method. Proceedings of the Conference on the Preparing of Shaped Crystals and Parts by the Stepanov's Method and their Application in the Economy, 16–18 March 1988; Krymov, V.M., Ed. Ioffe Phys. Techn. Inst. Acad. Sci. USSR: Leningrad 1986; pp. 335–340. *in Russian.*

63. Achunov, P.M.; Baskin, B.L.; Vorob'ev, M.A.; Vyatkin I.P.; Ivanov, V.I.; Kardashev, B.K.; Lebedev, A.B.; Peller, V.V.; Fadin, Yu. A. Structure and physico-mechanical properties magnesium alloys for various conditions of their crystallization by Stepanov's method. Izv. Acad Nauk SSSR, Ser. Fiz. 1983, *47*, 1438–1451 there is a translation in Bull. Russ. Acad. Sci. Phys. 1983, 47.

64. Korchunov, B.N.; Peller, V.V.; Bairamov, A.H. Study of crystallization process and properties of shaped Mg alloys articles. Izv. Russ. Acad. Nauk, Ser. Fiz. 1999, *63*. 1861–1865 there is a translation in Bull. Russ. Acad. Sci Phys. 1999, 63.

65. Peller, V.V.; Korchunov, B.N. Features of crystallization and structure of magnesium alloys produced by Stepanov's method. Proceedings of the Conference on the Preparing of Shaped Crystals and Parts by the Stepanov's Method and their Application in the Economy, 4–6 March 1985; Krymov, V.M., Ed.; Ioffe Phys Techn. Inst. Acad. Sci. USSR: Leningrad, 1986; pp. 341–343. *in Russian.*

66. Nikulin, L.V.; Lipchin, T.N.; Zaslavskii, M.L. *Pressure-Die Casting of Magnesium Alloys*; Mashinostroenie Moscow, USSR, 1978; 182 pp. *in Russian.*

67. Rochlin, L.L.; Peller, V.V.; Vyatkin, I.P.; Drenov, N.N. Korchunov, B.N.; Nikitina, N.I.; Stolbova, A.D. Mg alloy shaped articles of advanced damping. Proceedings of the 9th Conference on the Preparing of Shaped Crystals and Parts by the Stepanov's Method and Their Application in the Economy 10–12 March 1982; Antonov. P.I., Ed.; Ioffe Phys. Techn. Inst. Acad. Sci. USSR Leningrad, 1982; pp. 342–347. *in Russian.*

68. Bairamov, A.H.; Peller, V.V.; Kostygov, A.S.; Korchunov, B.N.; Sultanova, S.A.; Vyatkin, I.P.; Tagieva. S.M.; Mamedova, S.M. Study electrochemical properties of Mg alloys samples produced by Stepanov's method. Proceedings of the 9th Conference on the Preparing of Shaped Crystals and Parts by the Stepanov's Method and Their Application in the Economy 10–12 March 1982; Antonov, P.I., Ed.; Ioffe Phys. Techn. Inst. Acad. Sci. USSR: Leningrad, 1982; pp. 336–341. *in Russian.*

69. Bairamov, A.H.; Peller, V.V.; Korchunov, B.N.; Stolbova, A.D.; Vyatkin, I.P.; Mamedova, S.M.; Tagieva. S.M. Electrochemical characteristics of Mg–Al alloys

samples produced by Stepanov's method. Proceedings of the Conference on the Preparing of Shaped Crystals and Parts by the Stepanov's Method and Their Application in the Economy, 4–6 March 1985; Krymov, V.M., Ed.; Ioffe Phys. Techn. Inst. Acad. Sci. USSR: Leningrad, 1986; 347–351. *in Russian.*

70. Bairamov, A.H.; Peller, V.V.; Korchunov, B.N.; Mamedova, S.M.; Sultanova, S.A. Electrochemical characteristics of a number of Mg alloys at samples produced by Stepanov's method. Proceedings of the Conference on the Preparing of Shaped Crystals and Parts by the Stepanov's Method and Their Application in the Economy, 16–18 March 1988; Krymov, V.M., Ed ; Ioffe Phys. Techn. Inst. Acad. Sci. USSR: Leningrad, 1989; 271–275. *in Russian.*

71. Lyublinskii, E.A. *Cathodic Protection of Sea Ships and Buildings from Corrosion*; Sudostroenie: Leningrad, 1979, 186 pp. *in Russian.*

72. Goldfarb, V.M.; Donskoi, A.V.; Dyagilev, F.M. Kostygov, A.S.; Stepanov, A.V. Structure and properties of Al and Cu based alloys articles grown by direct crystallization. Proceedings of 11th Conference on Theory of Casting. *Casting Properties of Metals and Alloys*; Nauka: Moscow, USSR, 1967; 143–147. *in Russia.*

73. Lihachev, V.A.; Kuzmin, S.L.; Kamenzeva, Z.P. *Shape Memory Effect*; Publishing house of Leningrad University: Leningrad, 1987; 216 pp.

74. Oozuka, K.; Sigidzu, K.; Sudzuki, Yu.; Sekiguti, Yu.; Tadaki, Z.; Homma, T. *Alloys with Shape Memory Effect*; Metallurgiya: Moscow, 1980; 221 pp. *in Russian.*

75. Tihonov, A.S.; Gerasimov, A.P.; Prohorova, I.I. *Application of Shape Memory Effect in Modern Machine Building*; Mashinostroenie: Moscow, USSR, 1981; 81 pp. *in Russian.*

76. Golyandin, S.N.; Kustov, S.B.; Pulnev, S.A.; Vetrov, V.V. Study of mechanical properties of shape memory effect CuAlNi alloys with automated equipment. Proceedings of the Conference on the Preparing of Shaped Crystals and Parts by the Stepanov's Method and Their Application in the Economy, 16–18 March 1988; Krymov, V.M., Ed.; Ioffe Phys. Techn. Inst. Acad. Sci. USSR: Leningrad, 1989; 245–248. *in Russian.*

77. Vetrov, V.V.; Korolev, M.N.; Lihachev, V.A.; Pulnev, S.A. Shape memory effect for twisting and bending of single crystal and polycrystalline CuAlNi alloys. Fiz. Met. Metalloved. 1989, *68*, 953–957. *in Russian.*

78. Belyaev, S.P.; Ermolaev, V.A.; Kuzmin, S.L. Leskina, M.L.; Lihachev, V.A.; Pulnev, S.A. Oriented transformation deformation and shape memory effect of materials with thermoelastic and explosion transformation. Fiz. Met. Metalloved. 1991, *8*, 171–175. *in Russian.*

79. Betehtin, K.B.; Kuzmin, S.L.; Pulnev, S.F. Anomalous behavior of shape memory effect Cu–Al–Ni single crystals. J. Tambov State Univ. 2000, *5*, 192–193. *in Russian.*

80. Pulnev, S.A.; Gulihandanov, E.L.; Vetrov, V.V.; Peller,

V.V. Structural formation in diffusion zones of Cu and Ni based single crystal and polycrystalline alloys during internal oxidation and nitration. Proceedings of the Conference on the Preparing of Shaped Crystals and Parts by the Stepanov's Method and Their Application in the Economy, 16–18 March 1988; Krymov, V.M., Ed.; Ioffe Phys. Techn. Inst. Acad. Sci. USSR: Leningrad, 1989; pp. 239–244. *in Russian.*

81. Viahhi, J.E.; Priadko, A.J.; Pulnev, S.A.; Yudin, V.J. Robototechnic constructions based on Cu–Al–Ni single crystal actuators. Proceedings of the Second International Conference on Shape Memory and Superelastic Technologies; Ashomar Conference Center: Pacific Grove, California, USA, 1997; 263–268.

82. Nikanorov, S.P.; Peller, V.V. Nikanorov, S.P., Leksovskii, A.M., Eds.; Features and Perspective of Shaped Composites Produced on the Based Stepanov's Method. *Kinetics of Deformation and Destruction of Composition Materials*; Ioffe Phys. Techn. Inst. Acad. Sci. USSR: Leningrad, 1983; pp. 133–148. *in Russian.*

83. Dobroumov, S.M.; Nikanorov, S.P. Leksovskii, A.M., Fadin, Yu. A., Eds.; Growing shaped single crystal articles of Al–Al$_2$Cu, AL–Al$_3$Ni eutectic by Stepanov's method. Physics of strength of composition materials; Ioffe Phys. Techn. Inst. Acad. Sci. USSR: Leningrad, 1980; pp. 45–48. *in Russian.*

84. Dobroumov, S.M.; Nikanorov, S.P. Growth and microstructure of thin strips of regular eutectic compositions based on aluminum. Bull. Acad. Sci. USSR. Phys. Ser. 1983, *47*, 185–187.

85. Dobroumov, S.M.; Nikanorov, S.P. The effect of the seed structure on eutectic composition structure at growing by Stepanov or Czochralski method. Proceedings of the Conference on the Preparing of Shaped Crystals and Parts by the Stepanov's Method and Their Application in the Economy 4–6 March 1985; Krymov, V.M., Ed.; Ioffe Phys. Techn. Inst. Acad. Sci. USSR: Leningrad, 1986; 273–278 pp. *in Russian.*

86. Khokhlov, G.G.; Kardashev, B.K.; Kostygov, A.S.; Peller, V.V. Models of aluminum–metal fiber compositions and their some properties. *Physics of Strength of Composition Materials*; Regel, V.R., Leksovskii, A.M., Kirienko, O.F., Eds.; Ioffe Phys. Techn. Inst. Acad. Sci. USSR: Leningrad, 1978; pp. 202–206. *in Russian.*

87. Vorob'ev, M.A.; Peller, V.V.; Khokhlov, G.G. Study of structure of reinforced aluminum compositions crystal-lized from the melt. Izv. Acad. Nauk SSSR, Ser. fiz. 1983, *47*, 1238–1242 there is a translation in Bull. Acad. Sci. USSR. Phys. Ser., 1983, 47.

88. Peller, V.V.; Vorob'ev, M.A.; Khokhlov, G.G. Features of the crystallization of metal matrix of reinforced compositions by Stepanov's method. Proceedings of the 9th Conference on the Preparing of Shaped Crystals and Parts by the Stepanov's Method and Their Application in the Economy, 10–12 March 1982; Antonov, P.I., Ed.; Ioffe Phys. Techn. Inst. Acad. Sci. USSR: Leningrad, 1982; pp. 313–316. *in Russian.*

89. Khokhlov, G.G.; Peller, V.V.; Nekrasov, A.A. Producing shaped B–Al compositions. Proceedings of the 9th Conference on the Preparing of Shaped Crystals and Parts by the Stepanov's Method and Their Application in the Economy, 10–12 March 1982; Antonov, P.I., Ed.; Ioffe Phys. Techn. Inst. Acad. Sci. USSR: Leningrad, 1982; pp. 332–335. *in Russian.*

90. Achunov, P.M.; Baskin, B.L.; Peller, V.V.; Fadin, Yu. A.; Khokhlov, G.G. Strength characteristics of boron–aluminum compositions. Izv. Acad. Nauk SSSR, Ser. fiz. 1985, *49*, 2447–2456; there is a translation in Bull. Acad. Sci. USSR, 1985, 49.

91. Peller, V.V.; Khokhlov, G.G. The matrix structure and strength characteristics of the reinforcement in shaped boron–aluminum. *Physical Problems of Prediction of Fracture and Deformation of Heterogeneous Materials*; Ioffe Phys. Techn. Inst. Acad. Sci. USSR: Leningrad, 1987; pp. 172–175. *in Russian.*

92. Khokhlov, G.G.; Peller, V.V.; Fadin, Yu. A. Structure of matrix and conserved strength of boron fibers in shaped composition boron–aluminum articles produced by Stepanov's method. Bull. Acad. Sci. USSR. Phys. Ser. 1988, *52*, 161–168.

93. Khokhlov, G.G.; Peller, V.V.; Korchunov, B.N.; Khokhlova, E.V. Macrostructure of matrix of locally reinforced boron–aluminum composition articles produced by Stepanov's method. Izv. Acad. Sci. Ser. Fiz. 1999, *63*, 1843–1846, there is a translation in Bull. Russ. Acad. Sci. Phys. 1999, 63.

94. Zabolozkii, A.A. Production and application of composition materials. Itogi nauki i texniki. Kompozitionnye materialy; VINITI: Moscow, 1979; 108 pp. *in Russian.*

95. Sokolovskaya, E.M.; Gusei, L.S. Physicochemistry of Composition Materials; MSU Publishing House: Moscow, 1978; 41–135. *in Russian.*

10

Production and Inspection of Quality Aluminum and Iron Sand Castings

William D. Scott
AAA Alchemy, Birmingham, Alabama, U.S.A.

Hanjun Li
City University of Hong Kong, Kowloon, Hong Kong

John Griffin and Charles E. Bates
University of Alabama at Birmingham, Birmingham, Alabama, U.S.A.

I. SCOPE

This chapter describes the technology for producing and inspecting cast aluminum and iron made by sand casting processes. This material is presented in three main sections: (1) sand casting processes, (2) microstructure and porosity development, and (3) nondestructive inspection.

This material is intended for design and process engineers and scientists who desire information about casting production techniques, process details, and inspection technologies. The casting techniques discussed include green sand, no-bake, and lost foam processes.

Metal microstructure begins to develop when the metal starts to solidify. The mode of solidification, phases formed, and especially the formation of macro- and microporosity has a significant effect on the properties of the metal. Microporosity is considered a part of the metallurgical structure. Attention is given to microporosity development and detection because microporosity can significantly degrade the properties.

There is a growing interest in nondestructive inspection of castings, and the third section describes nondestructive inspection technology with attention to its use on cast aluminum and iron. The techniques described include liquid penetrants, penetrating radiation, density measurements, and ultrasonic velocity, reflection, and attenuation. References are provided for additional reading.

II. SAND CASTING PROCESSES

The sand casting process involves making a mold, usually with silica sand, pouring molten metal into the cavity produced by a pattern, and allowing the metal to solidify. The result is a raw casting that is finished by abrasive blasting to remove adhering sand and ground to remove fins and blend surface contours [1]. In most cases, castings are machined to produce the surface quality and dimensions needed to mate with other parts of an assembly. Millions of tons of metal castings are made each year from aluminum, copper, nickel, cast iron, carbon and low-alloy steel, and stainless steel [2].

Other alloys such as the precious metals, titanium, and zinc are also cast but not usually in sand molds. These metals are commonly cast using die casting, permanent mold, and ceramic shell (investment casting) processes [3]. Wrought metals are also cast and then rolled, forged, or stamped to achieve the desired shape. These processes are beyond the scope of this discussion.

III. INTRODUCTION

The sand casting process begins with the production of a pattern that may be made from wood, plastic, or metal. The molds often have multiple cavities to produce several castings in each mold. If the casting is to contain internal cavities, a separate sand shape called a core is made and placed in the mold. In both mold and core making, sand is compacted against a pattern and the pattern removed and used to make another mold. After metal has been poured into the mold and allowed to solidify, the mold is broken open, the casting removed, and the sand processed for reuse.

A basic molding process is illustrated in Fig. 1. The process starts with the preparation of a pattern. The pattern has core prints that serve to locate one or more cores that may be placed in the mold. A flask, or metal box, is used to contain the sand during molding, but may be stripped off when molding is complete. The flask is often stripped off when resin bonded sands are used to produce "flaskless" molds. After mold making, the cores are set and the mold closed for pouring.

The advantage of the casting process is that intricate shapes can be produced having near-net dimensions. This process may simplify production parts historically made by forging, welding, and riveting. In addition, the mechanical properties of castings, such as strength and toughness, are generally isotropic [4].

The basic steps in producing a sand casting involve:

Making a pattern
Preparing molds and cores
Melting, refining, and pouring the metal
Cleaning
Inspection

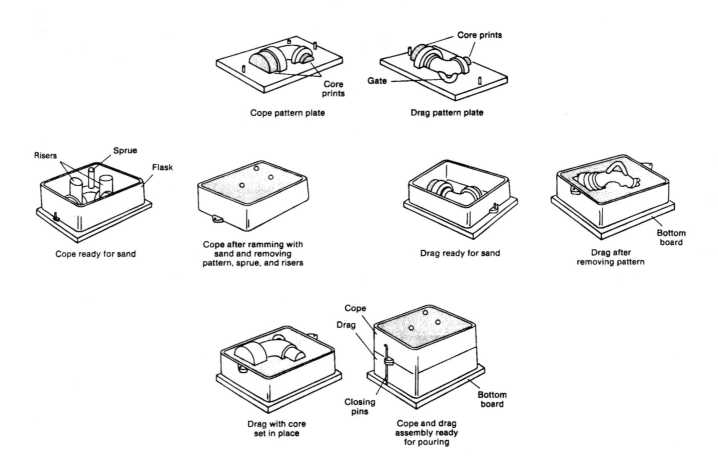

Figure 1 Schematic of molding process for green and no-bake sand. (From Ref. 3.)

Table 1 Common Nonferrous Alloy Melting Temperatures

Alloy	Nominal composition	Nominal liquidus (°C)	Ref.
Zinc	100% Zn	419	5
Magnesium	100% Mg	649	5
Magnesium alloys			
AZ series		595	6
HZ32A		650	6
Aluminum		660	5
Aluminum alloys			
380	8.5% Si, 0.5% Mn, 0.5% Ni, 3.5% Cu	520–590	
319	6% Si, 0.5% Mn, 1% Zn, 3.5% Cu	520–605	
356	7% Si, 0.35% Mg	560–610	
336	12% Si, 2.5% Ni, 1% Mg, 1% Cu	565	6
Copper	99.99% Cu	1083	5
Copper alloys			
Naval brass	62% Cu, 37% Zn, 1% Sn	900	6
Yellow brass	66% Cu, 34% Zn	930	6
Cartridge brass	70% Cu, 30% Zn	955	6
Red brass	85% Cu, 5% Zn, 5% Sn	1025	6
G bronze	88% Cu, 10% Sn, 2.5% Zn	980	6
Phosphorous bronze	80% Cu, 10% Sn, 10% Pb, 0.25% P	1030	
Aluminum bronze	86% Cu, 8% Al	1045	
Nickel–aluminum bronze		1045–1060	
Nickel silver	72% Cu, 18% Ni	1150	6
Cupronickel	90% Cu, 10% Ni	1150	6

Pattern making is a specialized form of tool and die making where the part is replicated as a three-dimensional (3-D) positive copy and slightly enlarged to accommodate linear contraction that occurs after solidification while the metal is cooling to room temperature. Risers or feeders are added at strategic locations to feed the liquid volumetric shrinkage that occurs as the molten metal transforms to a solid. Finally, channels or runners are added to the pattern to direct the molten metal into the risers and casting cavity. Collectively, these risers and runners are referred to as rigging.

Coreboxes are constructed as 3-D negative copies of casting cavities. Sand cores are made in the coreboxes, and the cores form the internal cavities of castings.

Metal is melted, refined, alloyed as necessary, the temperature adjusted, and the liquid metal poured into the mold cavity. Most metals are poured at temperatures of 100° to 200°C (200° to 400°F) above the liquidus temperature. The liquidus temperature is the temperature above which the alloy is completely liquid; values for several nonferrous alloys are presented in Table 1, and values for several ferrous alloys are presented in Table 2 [5,6]. Metal is usually transferred from the melting furnace to the mold in refractory-lined ladles.

After pouring, solidification, and some cooling in the mold, castings are "knocked out," which involves removing most of the sand from the interior and exterior surfaces. Many castings are heat-treated to enhance certain properties.

Cleaning involves blasting with an abrasive medium to remove adhering sand, grinding to remove gate and riser contacts, and perhaps grinding to remove surface imperfections [1]. Castings are then ready for heat treatment, inspection, packaging, and shipment. Inspection always involves visual examinations and may also involve dimensional analyses, hardness measurements, nondestructive evaluations, and destructive tests on selected parts.

IV. MOLDING PROCESSES

There are many types of casting processes and many different alloys considered castable. Casting processes are usually described in terms of the technique used to make the mold. The major processes include:

1. Green sand casting
2. No-bake casting (also known as airset and chemically bonded sand molding)

Table 2 Common Ferrous Alloy Melting Temperatures

Alloy	Nominal composition	Nominal liquidus (°C)	Ref.
Iron	100% Fe	1536	5
Gray iron	(eutectic, CE = 4.3%)	1150	
Ductile iron	(eutectic, CE = 4.3%)	1145	
White iron	(hypoeutectic)	1210–1320	
Steel			
Plain carbon	(.20 C)	1515	6
High-strength, low-alloy steel	(Ni/Cr/Mo)	1505	6
Austenitic stainless			
CF3M		1400	6
CF8		1425	6
17-4PH		1400–1440	6
Martensitic stainless			
CA6NM		1520	
Heat-resistant stainless			
HX		1290	6
HC		1495	6
Nickel		1453	6
Nickel-base alloys			
Monel S		1290	6
Monel K		1350	6
IN-625		1350	6
Hastelloy W		1315	6
Hastelloy C		1340	
Hastelloy C-4		1365	
Hastelloy B		1380	
CZ-10195% Ni, 4-1/2% C		1385	

3. Lost foam or expendable pattern casting (repli-cast and full mold processes are variations of the lost foam process) [7]
4. Die casting
5. Permanent mold casting
6. Investment casting (lost-wax process)

This material summarizes the technology for producing green sand, no-bake, and lost foam castings.

A. Green Sand Molding

Green sand casting refers to the production of castings in a mold where the sand is bonded with clay and water. The term "green" implies that the mixture has not been fired or cured. Additives to the basic sand–clay–water mixture are usually made to achieve a reducing mold atmosphere and control the sand expansion characteristics [8–11]. The basic constituents of green sand include refractory sand, usually quartz, 3% to 8% clay, 3% to 5% water to activate the clay, and organic and inorganic additives. Mixing green sand is done in a controlled sequence using a high-shear muller to distribute the additives throughout the sand and to plasticize the clay.

Green sands are mechanically compacted around patterns to form the mold cavities. Pneumatic jolting, jolting and squeezing, high-pressure hydraulic squeezing, or impacting may be used to compact the sand. The production of a mold on a ram-jolt machine is illustrated in Fig. 2. Green sand molds should be poured within 4–6 hr after preparation before moisture evaporation causes surface drying and friable mold surfaces.

B. Chemically Bonded Sand Molding

Thermosetting organic resins and the inorganic sodium silicate (water glass) binders have been used for many years to make molds [12,13]. Thermosetting resins were the first resins developed for producing castings and were particularly useful for making cores where strength was required for handling when the core was placed in a mold. Resin-bonded cores can be stored for extended periods, and the freedom from surface drying improves

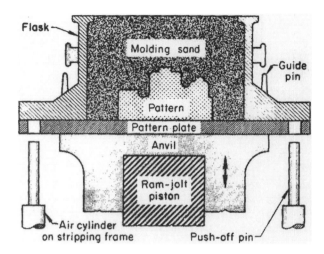

Figure 2 Jolt-squeeze machine for the production of green sand molds. (From Ref. 3).

casting quality. Sodium silicate is the most common inorganic self-setting binder in use.

C. Shell Molding

The shell process is commercially important for making relatively small castings in large quantities. The shell process uses thermoplastic novalac resins that are usually melted to coat the sand [12,14–16]. Hexamethylamine, often called hexa, is added to the sand to supply formaldehyde that will complete the polymerization, making the resin thermosetting [17].

The name "shell mold" is derived from the fact that molds are relatively thin, usually with a thickness of 10 to 15 mm, and hollow. Shell molds and cores are made on iron patterns preheated to about 225°C (475°F). Coated shell sand may be gravity fed or blown over the hot pattern, and the resin cures as formaldehyde from the hexa causes cross-linking. The resin content of the sand ranges from about 3% to 5% based on sand (BOS) weight, and the hexa addition ranges from about 10% to 17% based on resin (BOR) weight. Calcium stearate or other organic wax is commonly added in amounts of from 4% to 6% of the resin weight to serve as a lubricant and release agent so the core or mold can be drawn from the hot pattern box.

D. No-Bake Molding

The use of organic chemical binders that do not require heat or baking to cure has grown rapidly in the past 25

years because of their ease of use, ease of shakeout (removal of sand from the cast product), and reclaimability (the ability to reuse the sand). The use of no-bake binders has greatly simplified the production of dimensionally accurate castings.

The sand binder systems commonly used in foundries are listed in Table 3 [12,13]. The organic binders that are the subject of this section are formulations based on oils and resins including the alkyd, phenolic, and furan compounds.

Chemically bonded sand molds are produced by coating the sand with a resin system that will polymerize to develop a rigid bond. Where sand grains touch, capillary force causes the liquid resin to accumulate to form "binder necks" that, when cured, strengthen the physical bond.

The resin addition is normally about 1% to 1.2% of the sand weight, and higher resin contents produce higher aggregate strengths. Higher resin contents also produce proportionately more gas when the resin decomposes during pouring, and excess gas volumes contribute to casting defects. The resin and sand is mixed in high-intensity paddle or screw mixers, and the sand is compacted around patterns by vibration.

The same types of binders used to make molds can also be used to make pneumatically blown cores. Cores can be cured by passing a gas catalyst through the sand while it is still in the corebox. The size and shape of the sand grains used is important in assuring consistent densification of both molds and cores [18,19].

Table 3 Organic Sand Binder Systems

I. Core oils
II. Hot box
   A. Urea–formaldehyde
   B. Phenol–formaldehyde
     1. Resole
   C. Furan
   D. Modified furan
     1. UF/FA (urea–formaldehyde/furfuryl alcohol)
     2. PF/FA (phenol–formaldehyde/furfuryl alcohol)
     3. PF/UF (phenol–formaldehyde/urea–formaldehyde)
III. Acid no-bakes
   A. Furan
   B. Phenolic resole
IV. Urethane—no-bake
   A. Alkyd isocyanate
   B. Phenolic isocyanate
V. Alkaline phenolic

Note: Gaseous catalysts are available for many resin binders fast curing in a core-box. These "Cold-box" cures are favored for high volume core production.

The permeability of a cured sand–binder mixture and the geometry of the core or mold have important effects on casting quality because the sand must allow the gaseous decomposition products from the binder to leave the mold. The resins decompose to produce hydrogen, water vapor, and carbon oxides as well as a variety of hydrocarbon compounds. Low-permeability sands and a failure to properly vent molds and cores lead to gas porosity [20–24].

1. Acid-Cured Furans

Strong acids are required, and the acids of choice are phosphoric, toluene sulfonic (TSA), and benzosulfonic (BSA) acids [25,26]. The rate of cure is proportional to the amount of acid used. Acid additions of 25–50% of the binder weight are typical.

Furan resins readily coat sand grains because of their low viscosity, and this minimizes the amount of binder needed. The amount used varies from 0.9% to 1.3% based on sand weight. Furan polymerization is exothermic, and once the cross-linking begins, the temperature of the sand mass increases. The heat released speeds the curing process by helping evaporate water produced during polymerization. Moisture removal is important to both polymerization and casting quality. Residual water causes thick mold sections to cure more slowly, and stripping thick sections before enough strength has developed will cause mold distortion.

Cured furan sands readily decompose from the heat associated with iron and steel casting [27]. The residual binder after a casting is poured is measured by a loss-on-ignition (LOI) test conducted by weighing a sample of sand, heating to an elevated temperature, usually 975°C (1800°F), holding for sufficient time to burn the resin off, cooling to room temperature and reweighing. The weight loss is referred to as the loss on ignition or LOI.

The residual acid anhydrides, as measured by a pH or acid demand value (ADV), provide a measure of the residual acid catalyst [28]. The residual acid is usually the limiting factor determining how much reclaimed sand can be reused. Less acid is needed in subsequent mixtures because of the acidic residue in the sand.

2. Acid-Cured Phenolics

The second important class of acid-cured no-bake resins is based on phenolic resins. There are two types of phenolic resins depending on the conditions used to react with phenol. If the resin is produced with an excess of formaldehyde and a basic catalyst, the product is called a resole, and these resins are soluble in water [12]. If the phenol is reacted with a deficiency of formaldehyde, the product is called a novalac. Hybrid resins are formed if the starting phenol is a mixture containing other aromatic hydrocarbons [29].

Phenolic resin cross-linking is activated with the same strong organic acids used with furan resins. Binder additions are slightly above 1% of the sand weight, and the acid addition is usually about 0.4% of the sand weight. Sands bonded with acid-cured phenolic resins turn pink as the phenol oxidizes and generally behave much like acid-cured furan resins.

Phenolic resins naturally "age," which makes them somewhat more viscous and difficult to spread over sand grain surfaces compared to furan resins. However, they are less expensive, and phenolic resins are often added to furan resins to serve as less expensive "extenders."

3. Alkaline Phenolics

An important phenolic system was developed in Europe and is used extensively outside the United States. The pH of the resin is modified by the addition of an alkali such as potassium hydroxide to catalyze an organic ester, such as diacetin, and convert it to an intermediate ester.

The speed of cross-linking depends on the amount and type of ester coreactant present and increases in the order given below:

1. triacetin   slowest
2. ethylene glycol diacetate
3. butylpolactene
4. methyl formate   fastest

Sand temperature represents the greatest control problem. Temperature affects the reaction rate, bench life, work time, and strip time of each coreactant combination. If a different reaction speed is needed, the coreactant must be changed.

Because of the alkalinity required, these binders are compatible with impure silica sand and other minerals such as olivine that cannot be used with acid-cured binders. However, the ability to reuse this sand in subsequent casting cycles is limited by the accumulation of alkaline salts. Sand reuse is normally limited to not more than 50%.

4. Phenolic Urethanes

The phenolic urethane no-bake (PUNB) systems are the most commonly used chemical binders in the foundry today. Adjustments can be made to produce predictable and controllable behavior with cure times ranging from a minute to an hour [29].

Phenolic urethane resins are unique because they have a high molecular weight but a low viscosity and a high chemical stability. The low viscosity allows more efficient coating of the sand so binder concentrations as low as 0.8% based on sand weight can be used. These systems contain no water or sulfur.

The phenolic urethane resins are usually supplied in three parts. Part I is a resin in a solvent usually containing about 45% solvents and 55% solids by weight. Part II is a polymeric isocyanate. This material is sensitive to moisture and as little as 0.2% water in the sand can prevent a bond from developing. Part III is a catalyst. In most core applications, the catalyst used is phenyl propyl pyridine. Part III may be premixed with Part I or added separately in a solvent.

Phenolic urethane resins are used in concentrations (combined Parts I and II) of about 0.8% to 1.5% by weight of the sand. A 1% total addition is common. Clean, well-rounded sand grains require less resin than sands containing clay, silt, mica, and small sand grains. Fine material in the sand consumes all resins without contributing to strength. Foundries like to use less expensive materials such as the Lake Michigan sands, but these sands exhibit wide variations in grain size and contain iron oxides, silicates, and carbonates. These non-silicate materials require the use of more resin to achieve the desired strength.

Sands containing mica or calcium carbonates do not perform well with any resin. Acidic sands react with the resin and reduce the bond strength. Alkaline sands or sands containing alkaline impurities may cause premature resin cross-linking and a loss of binder strength.

The best results are obtained with washed and dried sand having at least 85% of the sand on four adjacent screens and an American Foundry Society grain fineness number (AFS GFN) of about 55. This sand distribution compacts readily and develops good strength with a minimum amount of resin.

Polymerization takes place simultaneously throughout the mold, which provides good control over strip time. The binder develops a high strength that minimizes the need for support wires and rods and lends itself to flaskless molding [30]. This chemically bonded sand has a sufficiently high strength to provide good mold dimensional stability during pouring.

Time must be given to allow solvents to evaporate before castings are poured. Solvents will evaporate at 45° to 60°C (110° to 140°F) and quickly evaporate at 70°C (160°F). Refractory mold coatings should not be immediately applied to a mold because they inhibit solvent evaporation, which may produce gas defects during pouring [31]. A good practice is to heat the mold surface to evaporate the solvents and then apply the mold coating.

The phenolic urethane no-bake sands have excellent collapsibility. Mechanical reclamation usually leaves some resin on the sand but is recommended for PUNB resins. The residual resin decreases the amount of resin needed for rebonding, but higher resin contents produce more gas during pouring. The best reclamation practice involves heating in an oxidizing environment. With thermal reclamation, 100% of this sand can be reused.

E. Procedural Effects

Chemical binders must be mixed with sand as rapidly as possible, molded, and allowed to cure. The highest PUNB binder strengths are obtained with set times of from 3 to 5 min. The temperature of the sand has a large effect on resin reaction rates. If the sand temperature is too high, binder solvents evaporate prematurely, the resin starts to cross-link prematurely, and the result is low binder strength. In high-production applications, it is important to maintain consistent sand, pattern, and corebox temperatures, and in general the temperature should not exceed 32°C (90°F) [32].

Heat can be applied to the sand surface after the mold or core is extracted from the mold box. The heat helps the curing process and aids in solvent removal. Heat is usually applied with a torch or by blowing hot, dry air over the surface.

A most important event occurs when metal hits the sand surface. Solvents flash to a vapor and incompletely dried mold washes produce gas [33]. These gases may form bubbles that enter the molten metal if the head pressure is not quickly developed [19,34].

If the metal is too hot when the metal hits the mold, gas bubbles form and may pass through the metal to produce oxidized bubble trails. If the metal is too cold, the bubbles may be trapped in the casting. Such bubbles may be exposed during grinding or machining.

Resin pyrolysis also causes gas-bubble damage. Generally, short-chain molecules decompose and are evolved faster than long-chain molecules. Hence, if the resin curing has not been consistent and the resin contains short-chain polymers, resin decomposition occurs rapidly and the gas bubbles through the liquid. These gas-forming events probably occur within a half second after the metal hits the mold surface [19,34]. Casting porosity is best controlled by minimizing the amount of binder used, thoroughly removing solvents including water, and properly venting the sand to allow the gasses to pass through the mold.

Heating silica sand above 585°C (1085°F) causes a phase change and an associated volumetric expansion, as illustrated in Fig. 3 [35]. Thermal expansion can cause several casting surface anomalies irrespective of the binder used, and there is a trend toward low- expansion sands such as mullite and zircon for the production of "precision molds." These sands do not expand as much as silica during the pouring and cooling process, and result in more dimensionally accurate castings.

If resin binders are pyrolized before the mold is filled, there may be a loss of mold integrity. For this reason, especially with the PUNB binders, an additive such as iron oxide may be required to provide a ceramic bond for hot strength. Additions may also be made to eliminate veining associated with sand thermal expansion.

Close attention must be paid to sand compaction. Compaction must be accomplished before the start of polymerization. Compaction can be accomplished by vibration, light tamping, or by blowing the sand into a mold, as when cores are blown. Inconsistencies in com-paction are a common problem in the foundry, and too much compaction can lead to sand expansion defects on the casting surface.

F. Lost Foam Casting

The lost foam casting (LFC) process is similar to the lost wax investment casting process in that a disposable pattern is needed for every casting made. It is different, however, in that the molds are made with unbonded sand, and the pattern is left in the mold to be decomposed by the hot metal entering the mold. The process was developed by H. F. Shroyer in 1958 and was referred to as the full mold process [36]. The patterns were initially machined from blocks of expanded polystyrene and molded in furan resin-bonded sand. The concept of casting into unbonded sand was patented by Smith in 1964 [37], but the users were required to be licensed by The Full Mold Process, Inc. until the patents expired in 1980.

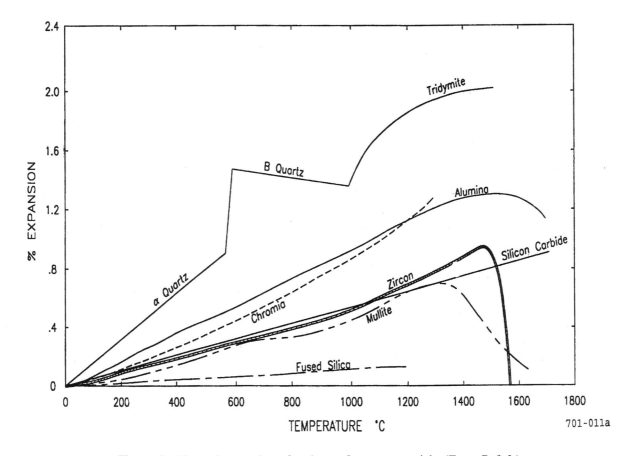

701-011a

Figure 3 Thermal expansion of various refractory materials. (From Ref. 3.)

The most popular use of the process places patterns in unbonded sand. Patterns are made with preexpanded beads, coated with refractory, placed in a flask, backed up with sand, and liquid metal poured to replace the foam.

The lost foam casting process has been referred to by several generic and proprietary names including expendable pattern casting [38], lost foam and full mold casting [39,40], castral process [41], policast process [42], and the replicast process [43,44]. While there are some variations in the proprietary processes, the generic term "lost foam casting" will be used throughout this chapter.

As presently practiced, the LFC process uses patterns produced from expanded polystyrene (EPS), expanded polymethyl methacrylate (EPMMA), blends of the two materials, or copolymerized materials. Experimental polycarbonate materials have been developed, but these have not yet reached the production stage [43,45,46].

1. The Lost Foam Casting Process Sequence

The steps involved in the LFC process are outlined in Table 4. The first step is to design a part, and this design must take advantage of the process capabilities. A tool is then designed to make the sections of foam required to form the pattern. Tool design is important because the tool fixes the initial foam dimensions and influences the foam pattern quality, surface and internal bead fusion, and production cycle time. Tool designs can be simple or quite complex using pullbacks to make undercuts and holes in the foam. Collapsible cores are sometimes used to make large internal cavities in foam sections.

Complex designs usually require that several foam sections be made and glued together to form the pattern. Each foam section requires a tool, and a fixture may be required to maintain section alignment during gluing

Table 4 Steps in the Lost Foam Process

   I. Part design
  II. Die design and production
 III. Prepare foam sections
       A. Preexpand Beads
       B. Mold beads
       C. Assemble parts
  IV. Assemble sections
   V. Coat patterns
  VI. Fill and compact sand
 VII. Pour
VIII. Shakeout and finish castings

operations. Tool design and construction may be costly and involve long delivery times.

Prototype parts are sometimes made from foam pieces cut from board stock and assembled by hand gluing. This technique is useful, but the resulting patterns have lower surface quality compared to blown and fused pattern sections and may have poor dimensional accuracy.

The production of molded foam sections is a two-step process. The foam materials most commonly used are polystyrene and polymethyl methacrylate containing about 5–8% pentane as an expansion agent [46]. These materials are used individually, as blends of the two materials and as copolymers. The first step in pattern making is to preexpand the beads to the desired use density. Aluminum casting producers prefer an EPS density of 22.4–25.6 g/L (1.4–1.6 lb/ft^3), and iron casting producers prefer a density of 20.8 g/L (1.3 lb/ft^3). Most PMMA is expanded to a density of about 24–29 g/L (1.5–1.8 lb/ft^3). Bead expansion is accomplished by rapid heating with steam, and after preexpansion, EPS beads are stabilized for a few hours before use.

Stabilized beads are transported to molding machines. The tool must be preheated to the operating temperature, dried, and closed to start the cycle. If pulls are used, they are normally pushed into their closed position when the tool is closed. The preexpanded beads are then blown into the empty cavity. The tool venting must be adequate to allow the air in the tool cavity to escape when the beads are blown in, and the fill guns must also be properly placed and regulated to completely fill the cavity. The process of blowing preexpanded beads into a tool is illustrated in Fig. 4A. When the tool cavity has been filled with beads, the steam cycle is begun. Steam is passed through the tool and beads to cause the beads to soften, the residual pentane to expand, and the beads to bond together. The pattern section is then cooled, as schematically illustrated in Fig. 4B and ejected from the tool as illustrated in Fig. 4C.

After the initial foam shrinkage caused by tool cooling, the foam briefly expands and then shrinks as pentane and water vapor diffuse out. Most of the shrinkage occurs within about 20 days and amounts to about 0.8% linear contraction in EPS patterns and 0.25% in PMMA.

The shrinkage of an EPS pattern as a function of time is illustrated in Fig. 5. The initial contraction and expansion as air diffuses into the pattern occurs in about 24 hr. Afterward, there is a continuous shrinkage that may last for 30 days or more (1500 hr). Shrinkage in the first 10 days usually amounts to about 0.5%. The final shrinkage of most parts is between 0.7% and 0.9% and

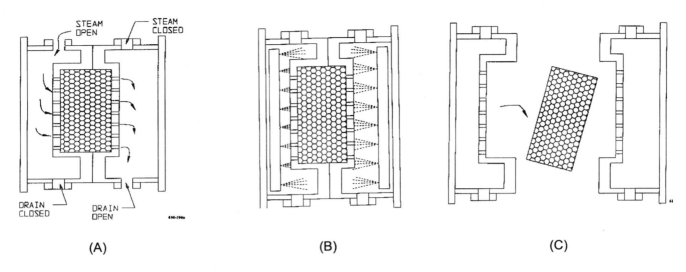

Figure 4 Steam cycle in the production of lost foam patterns: (A) bead expansion and fusion, (B) cooling the mold, (C) ejection of the part.

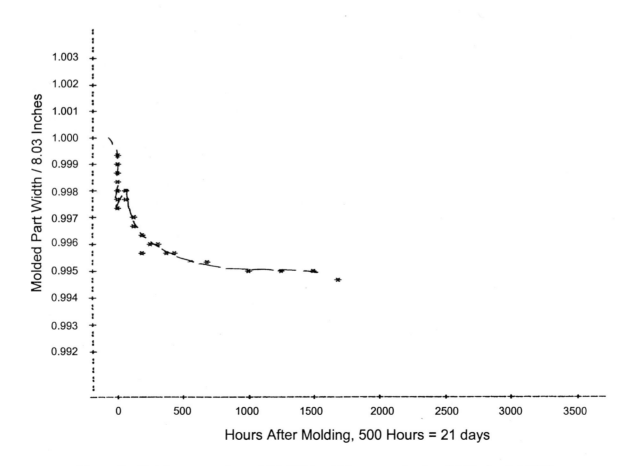

Figure 5 Shrinkage curve for an EPS "T" bead blown to a density of 0.02 g/cc (1.3 PCF).

is normally complete in 30 days [47,48]. Pentane content, bead density, section size, storage environment, and storage time all have measurable effects on pattern shrinkage and should be controlled in order to yield reproducible and dimensionally stable foam sections [49]. In general, more dense foams shrink less.

Understanding and controlling shrinkage is necessary to achieve dimensional consistency and gain the full benefit of the process. Patterns can also be artificially aged within a few hours in a circulating air furnace operated at 30–60°C (130–140°F).

After aging, the foam sections are ready for assembly. Sections are normally glued together on automatic machines using hot melt glue. Three parts to be glued to form a motor block are illustrated in Fig. 6. The glue joint must be sufficiently strong to hold the part together, be fast setting to maximize the production rate, seal joints against penetration of the ceramic coating, and present a minimum mass of glue that must be pyrolized by the molten metal [50]. Gating system sections may be molded or cut from board stock, but molded sections are preferred to minimize mold wash erosion. Attaching the gating system completes the preparation of the foam assembly or cluster for casting.

The foam cluster is coated with a ceramic to minimize sand erosion and metal penetration before a casting can be made. The coating also helps prevent sand from falling into the gap between the advancing liquid metal and the retreating foam. Finally, the coating stiffens the pattern to reduce distortion during compaction.

The coating is a slurry of refractory in a water carrier and is applied to the foam cluster by dipping. Careful control is necessary to obtain a uniform coating layer on each cluster because coating thickness affects coating permeability and performance. Coatings are usually dried by passing clusters through a circulating air oven. Drying takes from 2 to 10 hr at 40–50°C (100–140°F). The pattern weight, dipped (wet) weight, and the weight after drying provides useful information about coating consistency.

Coatings contain several ingredients such as refractory fillers, dispersants, binders, thixotropic agents, and a carrier [49]. The refractory and the water carrier comprise the majority of the coating. The permeability of the coating is usually controlled by the size and shape of the refractory. The refractories commonly used include silica, alumina, zircon, and aluminosilicates.

The refractory particles are normally held together with one binder to provide adhesion and cohesion before drying and a second binder to provide strength after drying and during pouring. In addition to the binder, the system may also include suspension agents for the refractor and surfactants to insure that the coating wets and coats the pattern. Coating formulation, application, and control are important to success with the LFC process [50]. Water is by far the most common carrier for the wash [51], but occasionally isopropyl-alcohol-based washes are used on PMMA patterns.

The ability of a coating to allow gaseous pyrolysis products to escape from the mold cavity is referred to as "permeability." Iron castings are poured at about 1370°C (2500°F) and permeability to gases is quite important. Aluminum castings are poured at about 760°C (1400°F) and the coating may need to allow more liquid products to be absorbed. The role of the coating in the removal of pyrolysis products is not well understood.

When the coated cluster has been dried, it is ready to be placed in a flask and backed up with sand. The process of compacting sand around a pattern is illustrated in Fig. 7. A "base" of sand is placed in the flask on which to set the pattern cluster, as illustrated in Fig. 7A. The cluster is then placed in the flask as illustrated in Fig. 7B and sand rained into the flask while the flask is vibrated, as illustrated in Fig. 7C.

Cluster placement, flask filling, and sand compaction must be accomplished such that the sand fills all pattern cavities and packs tightly around the cluster to provide support. The rate of filling and the distribution of sand in the flask have a significant effect on cavity filling and pattern deformation.

Compaction is achieved using a table vibrated with motors having eccentric counterrotating weights. The flask may be placed on locator pads or clamped to the

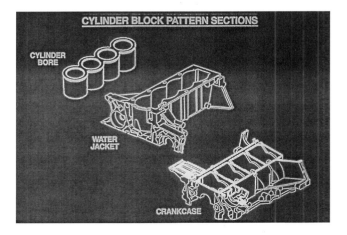

Figure 6 Sections to be glued to form a cylinder block.

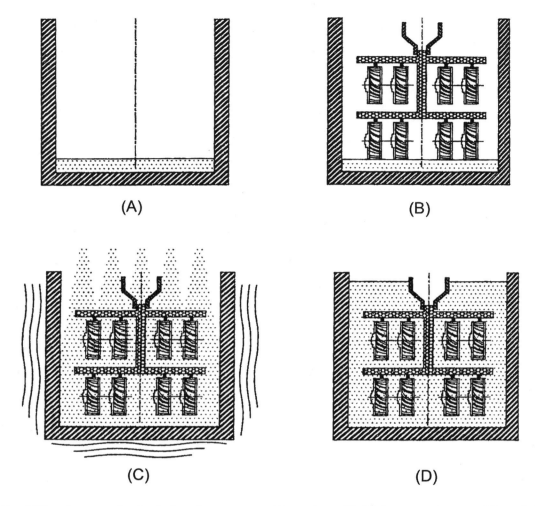

Figure 7 Sand fill and compaction in the lost foam process: (A) sand pre-fill, (B) insert pattern tree, (C) sand fill and flask vibration, (D) pattern and flask ready to be poured.

table. There is some controversy about the utility of vibrating clamped vs. unclamped flasks. System control and repeatability favors rigid clamping because this is more likely to produce a reproducible compaction cycle [52]. Compaction is usually done during filling to help sand flow into internal pattern cavities.

After a cluster is packed in a flask, as illustrated in Fig. 7D, the mold is ready to be poured. Pouring consistency is more critical with the LFC process than in conventional foundry practices, and automatic pouring is usually used to minimize pour-to-pour variations. If pouring is interrupted, the sand may collapse into the casting cavity. The pouring rate must be as rapid as possible to avoid mold collapse, but sufficiently slow to allow the foam pyrolysis products to escape. The gating system, coating properties, sand properties, and pour-

ing operation must operate as a system to allow both liquid and gaseous pyrolysis products to escape [53].

A vacuum is sometimes used during pouring to pull gases from the flask and prevent sand fluidization by pyrolysis gases. The vacuum system can also collect emissions from the pattern. However, the emissions are combustible and vacuum systems must be designed to withstand explosion pressure peaks of at least 10 atmospheres of pressure (150 psi) [54].

The process of replacing the foam with molten metal is illustrated in Fig. 8. Molten metal enters the cavity and replaces the foam by a process of both melting and vaporization. A kinetic zone is just ahead of the molten metal front that contains air initially contained in the foam pores, liquid styrene, and gaseous products from foam pyrolization. The rate of metal front movement is

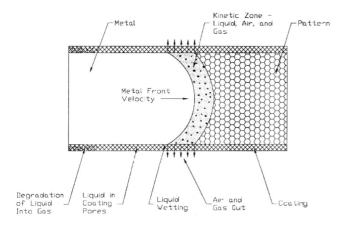

Figure 8 Current model of foam pattern replacement by molten metal in lost foam casting process.

dominated by the heat transfer from the molten metal to the foam.

After pouring, the casting cluster is allowed to solidify and cool in the flask before being dumped out. Shakeout involves dumping the loose sand and removing the castings. The sand can be repeatedly reused if the coating residue and other fine particles are removed and the organic condensates are periodically burned off [40] Cleaning LFC castings is easier and requires fewer operations because there are no fins or parting lines to grind or blend into the casting contour. The loose sand in the internal cavities is more easily removed than the lumps of bonded sand that remain in cores made with chemical binders.

Eliminating the foam pyrolysis products from the casting cavity during pouring is essential to success. These pyrolysis products produce carbon fold defects in iron, laps or folds in aluminum, and porosity in both metals. Eliminating fold defects is the subject of much of the current research on lost foam casting [55–63].

2. LFC Process Advantages

The use of foam patterns allows some design and process flexibility not possible with conventional casting techniques. The primary advantages of the LFC process are that bonded sand molds, bonded cores, and parting lines are eliminated, and casting accuracy is improved. The LFC process eliminates the need for mixing and molding of sand, ramming and jolting of the mold, core making, core setting, pattern drawing, mold closing, and sprue cutting [7,40,46]. Conventional mold closure may also lead to casting inaccuracies. A typical allowance for a shift between the cope and drag in a conven-

tional mold is 0.75 mm (0.030 in.). This allowance is also typical of the inaccuracy expected between molds and cores. The use of foam patterns eliminates shifting and this is one reason the process is more accurate.

A conventional mold cavity is made with a pattern that must be withdrawn from the mold after the sand has been compacted. Pattern removal requires a pattern with a sufficient draft to allow its removal. The presence of parting lines on conventional patterns often dictates the pattern orientation and restricts the number of castings in a flask. Multiple levels of castings around the sprue are not usually possible. The parting line also sometimes limits the placement of gates and risers and may cause gates or risers to be placed on surfaces that must be dimensionally precise. The removal of these gates and risers results in dimensional inaccuracies that must be corrected by machining.

The ability of foam patterns to eliminate cores allows complex internal passages and sometimes even multiple passages to be accurately cast. A cylinder block made without cores is illustrated in Fig. 9.

A foam pattern "stack-up" for a cylinder head is illustrated in Fig. 10A, and the head made from the stack-up is illustrated in Fig. 10B. No cores were used in this casting. A great deal of machining was eliminated by gluing up foam sections instead of having to face, drill, tap, add gaskets, and then join cast segments with bolts.

Casting walls can be curved or of variable thickness [64]. Pumps have been cast with sufficiently smooth internal surfaces that allowed the amount of water pumped to increase by 12% with a simultaneous reduction of about 25% in pump weight and with reduced product cost [65]. Eliminating cores also eliminates

Figure 9 Cast aluminum cylinder block made using the LFC process.

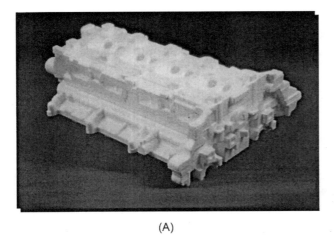

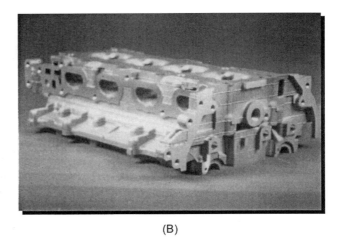

(A)   (B)

Figure 10 (A) Foam "stackup" for a cylinder head and (B) companion aluminum cylinder head.

separate tooling usually required for core production. The process of mixing core sand, ramming, curing, setting, and removing bonded-sand cores is eliminated. Core crush, core damage, core-related porosity, and other problems that result from setting cores is also eliminated.

Eliminating core prints, chaplets, fins, etc., also reduces the amount of cleaning required on the final casting. Cleaning room cost reductions of up to 80% have been reported, but cost reductions of 30% are more common [50,66–68].

The improved accuracy of the lost foam process allows machining operations to be reduced. The dimensional precision is reported to exceed that attainable from either green sand or chemically bonded sand molds and close to that possible from investment casting [69]. For example, holes that are greater than 1/4 in. (6 mm) in diameter that are normally drilled may be cast. The repeatability of the hole dimensions can be quite good, typically within 0.2% [59,60].

Design tolerances have been summarized by Troxler as follows [70]:

1. Linear dimensions between features located on the same piece of foam are normally controllable to ±0.3% in the first inch and ±0.2% for each subsequent inch.
2. Linear dimensions across glue joints can be expected to have less accuracy. An additional allowance of about 0.02-in. variation in thickness should be made for each glue joint.
3. Positional accuracy across glue joints is affected by shifting or mismatching foam pattern pieces.

An allowance of ±0.03 in. should be made for each glue joint.
4. Glue seams typically have an extruded bead of glue around the joint. The glue bead can normally be expected to be +0.04/−0.0 in [62].

The surface finish RMS values on LFC iron castings are typically in the range of 2.5–6.4 μm (100–250 μin.) compared to 6.4–25.4 μm (250–1000 μin.) for green sand and 3.2–12.7 μm (125–500 μin.) for shell sand castings. Aluminum castings have a measured RMS value as low as 3 μm (120 μin.) when special precautions are used to optimize the surface finish [71]. Process details need to control casting precision and data on casting precision and accuracy are provided in References 72–79.

3. LFC Process Disadvantages

There are also disadvantages to the LFC process. The lead times and costs of tooling are substantial. Simple tools can be cast to shape, but once the tooling is made, changes are neither easy nor cheap. Gating systems are larger than conventional processes and require more effort to remove. This reduces the casting yield in some cases. Foam patterns can distort during compaction and this results in distorted castings [50,67]. The most troublesome problems with lost foam castings seem to be distortion during compaction, laps and folds in aluminum castings, and carbon defects and penetration or burned-on sand on gray and ductile iron castings. These problems are either caused by or substantially aggravated by foam pyrolysis residue in the metal. A more thorough description of the LFC process, its advantages and disadvantages, has been prepared by Monroe [7], and Division 11 of the AFS is preparing an updated summary of the process.

V. CASTING SOLIDIFICATION AND MICROSTRUCTURE DEVELOPMENT

Casting solidification plays an important role in determining the microstructure and mechanical properties of cast metals. In this material, the microstructure refers to both metallurgical structural features such as grain size, phases present, and shrinkage pores. The solidification path influences the size and distribution of grains, eutectic cells, alloy segregation, and shrinkage. Subsequent processing of the casting cannot wholly remove the effects of solidification [80].

Producing a sound casting requires that metal begin to solidify at the edges and proceed toward the risers, a process known as directional solidification [81,82]. Pattern designers vary the rigging and molding parameters to aid directional freezing. Gate size and location, riser size and location, pouring rate, pouring temperature, tapered sections, chills, and special sands can be used to promote directional solidification [83].

A. Solidification and Grain Development

Metal freezes principally by heat loss to the mold with only small amounts of heat lost by radiation from risers [84]. There is a density increase of from about 4% to 6% during freezing of all commercial alloys, except gray and ductile cast iron. Precipitation of low-density graphite simultaneous with formation of austenite during eutectic solidification compensates for most of the density increase associated with austenite.

The density increase usually associated with solidification requires that risers be used to "feed" metal into the area where solidification is occurring and the density is increasing. If the shrinkage is not fed, castings may contain up to 6% void space. An insufficient volume of feed metal, or an inability of the feed metal to get to locations where freezing is occurring results in macro or microshrinkage. Macroshrinkage cavities are defined as shrinkage voids with a dimension of at least 5 mm (0.2 in.).

Microshrinkage voids are those with a linear dimension on a polished plane of up to 1 mm (0.04 in.), but the true length of the voids may be greater. Microshrinkage is usually interdendritic with a high length-to-thickness ratio. Macroshrinkage cavities are larger, usually more rounded, and often exhibit dendrites on the cavity

surface. Shrinkage can be tolerated in some low-stress designs, but areas that require pressure tightness and most areas that are to be machined must be free of shrinkage.

Most aluminum, nickel, carbon steel, low-alloy steel, and stainless steel castings freeze to form a skin with an equiaxed grain structure. Further from the surface, the grains take on a columnar pattern, but in the center, there may be another equiaxed grain zone. These changes in grain size and orientation are a function of the heat transfer rate during solidification and the composition of the alloy.

The macrostructure of a casting cross section showing equiaxed and columnar grains is illustrated in Fig. 11. The high cooling rate and numerous nucleation sites on the mold surface result in small and equiaxed grains along the mold wall. The cooling rate slows as the solidification front moves into the mold cavity. In most alloys, dissolved elements are rejected at the solid/liquid interface, which reduces the freezing temperature.

If the metal being solidified is pure or if the thermal gradient is steep, the liquid–solid interface is nearly planar. If, however, the interface velocity or the thermal gradient is low, grain growth becomes oriented. Grains grow parallel to, but in the opposite direction to heat flow from the casting. These grains can be cellular with smooth surfaces but usually grow as dendrites.

The region between a dendrite tip and dendrite base where solidification is complete is referred to as the "mushy" zone. The metal in this zone is a mixture of liquid and solid metal. The time required for solidification front to pass from the dendrite tip to the base is referred to as the local solidification time (LST).

Equiaxed grains often appear again in the center of castings. These grains have no preferential growth direction, and they grow much larger than surface grains. Dendrite arms broken from columnar grains may serve

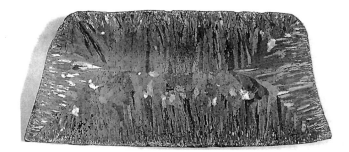

Figure 11 Casting cross section showing the equiaxed and columnar grains with some equiaxed center grains.

as nuclei for the equiaxed grains in the center of the casting.

Most alloy element concentrations increase from a casting surface toward the center and from dendrite cores toward grain surfaces. This change in composition with location is referred to as "coring" or "solidification segregation." Two important functions of heat treatment are to (1) reduce the composition gradients and (2) refine the as-cast grain size. Heat treatment allows elements to migrate from areas of high to low concentration and produce a more uniform alloy distribution.

B. Macro- and Microshrinkage Formation

In pure metals or alloys with a short solidification range, the solid–liquid interface is planar or has short dendrites extending from the solid into the liquid. A thin "mushy" zone allows feed metal from the body of the casting or riser to reach the solidification occurring between dendrites and "feed" the shrinkage. Thick "mushy" zones impede metal flow, especially when dendrites have secondary and tertiary arms that produce a tortuous flow path. Secondary arms grow transverse to the heat transfer direction and metal feed direction and significantly impede metal flow.

Progressive freezing of an alloy from the edge of a part is compared to a pure alloy without a dendritic structure in Fig. 12 [82]. The events illustrated in Fig. 12A represents freezing of a pure or nearly pure metal. Series (B) represents freezing of a metal that develops long dendrites, and Series (C) represents metals such as eutectic cast irons that freeze with approximately equiaxed grains or eutectic cells.

Interdendritic voids form in regions that are not completely fed, and these voids are referred to as "microshrinkage." Microshrinkage can be found in most castings, and the pores range in length from a few microns to 1 mm (0.04 in.) in length when viewed on a polished plane. Interdendritic voids are often interconnected, however, and are longer in three dimensions than they appear on a polished plane.

The head pressure produced by risers helps push liquid between the dendrites to fill pores, but the pressure is usually insufficient to cause complete feeding. This is especially true as the dendrites grow together and the feed paths become small. Machining can expose these microshrinkage cavities, and the surface may "bleed" when examined with liquid penetrants (LPs) or produce "indications" when examined using magnetic particle inspection (MPI) methods. Microshrink-

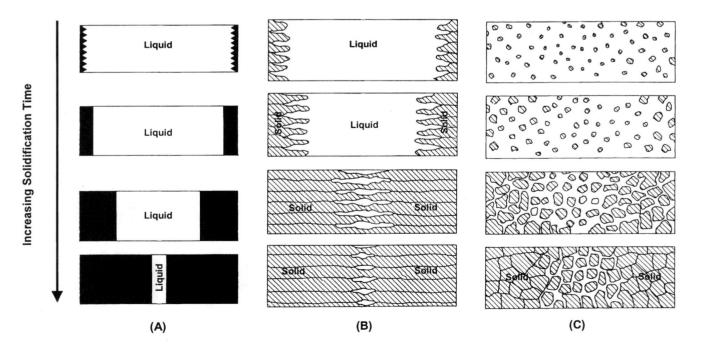

Figure 12 Schematic of the effect of cooling rate on grain morphology in a casting. (From Ref. 3.)

age has a significant effect on metal ductility, fatigue strength, ultimate strength, and toughness, and is more damaging if located on or near a part surface.

The equiaxed surface layer is almost completely free of porosity, but machining or heavy grinding can expose subsurface microshrinkage and produce leak paths for fluids. Leaving extra machining stock in an area prone to microshrinkage usually aggravates the problem. A pattern redesign or placing a chill in the mold to increase the local solidification rate is more effective in minimizing surface interdendritic shrinkage.

C. Aluminum Alloy Solidification and Structure Formation

About 90% of sand cast aluminum parts is based on aluminum-silicon alloys, with the majority containing about 7–8% silicon [85]. These alloys are melted in reverberatory and electric furnaces, heated to 750–800°C (1375–1475°F), and processed for pouring. Preparation for pouring involves making composition adjustments, degassing, adding grain refiners and eutectic modifiers, and adjusting the temperature.

These alloys solidify to form primary aluminum beginning at the liquidus temperature of about 615°C (1140°F). The addition of grain refiners reduces the undercooling needed to nucleate aluminum crystals by providing nuclei. After nucleation, the primary aluminum dendrites grow until the eutectic temperature of about 570°C (1060°F) is reached, at which point eutectic aluminum–silicon growth begins.

The size of the eutectic silicon particles may be reduced with the addition of eutectic modifiers such as strontium, sodium, or antimony. Effective modification depresses the silicon eutectic temperature a few degrees, usually from about 571°C (1060°F) to 568°C (1054°F). Other eutectics, including those formed by copper and magnesium, may form toward the end of solidification if the elemental concentrations are sufficiently high.

Cooling curves obtained by placing a thermocouple in A356 aluminum samples having a volume of about 20 cm^3 (1.2 in.^{3}) are illustrated in Fig. 13. One sample (curve A) contained about 0.015% titanium and the second (curve B) was treated with 0.02% TiB. The TiB addition nucleated primary aluminum and raised the liquidus by about 1.6°C (3°F). The effectiveness of TiB

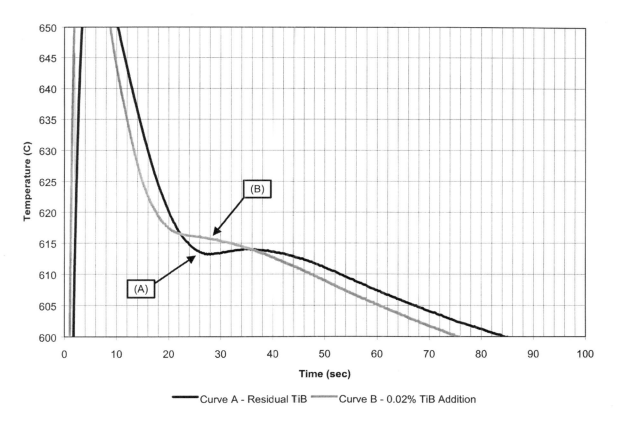

Figure 13 Cooling curves from A356 aluminum with (A) residual TiB and with (B) 0.02% fresh TiB grain refiner addition.

decreases with time and becomes ineffective after about 2 hr. However, titanium alone in concentrations above about 0.2% will refine grains, even if it has been in solution for an extended time.

The curves in Fig. 14 illustrate the effect of a strontium addition. Curve A represents an untreated metal, and curve B was obtained by making a strontium addition of 0.015% just before pouring. The strontium depressed the silicon eutectic from 571°C (1061°F) to 569°C (1056°F), as illustrated.

Small equiaxed grains obtained with grain refiners are beneficial in improving feeding, reducing porosity, and improving resistance to hot tearing [86]. Small grains improve the ability of liquid metal to flow between dendrites and feed solidifying metal. Reduced porosity improves ductility, ultimate tensile strength, and fatigue limit.

The silicon morphology developed during eutectic solidification has a substantial effect on tensile properties. Unmodified alloys contain coarse acicular silicon that results in low ductility. Proper modification with sodium, strontium, or rapid solidification converts the acicular silicon to a fibrous structure and improves the ductility.

Silicon particle morphology can be controlled with the addition of eutectic modifiers, using solution treating temperatures near the solidus, and using long solution treating times. Porosity and dendrite arm spacing are fixed during solidification and cannot be altered except by hipping (applying a high hydrostatic pressure at a temperature just below the solidus).

D. Shrinkage Porosity Formation

Solidification shrinkage in aluminum alloys ranges from 3.5% to 8.5% and is a major contributor to porosity [87]. Macroshrinkage is controlled by the size and location of gates and risers, casting section size, cooling rate, grain refiners, and pouring temperature. Minimum microporosity is obtained with high cooling rates to produce short dendrites, and this is usually accomplished with chills in the mold. In general, alloys that solidify over a narrow range are easier to feed and contain less microporosity [86].

Grain refiner additions usually reduce porosity, but additions of eutectic modifiers have been linked to increased porosity, often attributed to hydrogen. Modifying with sodium agitates the melt and may cause

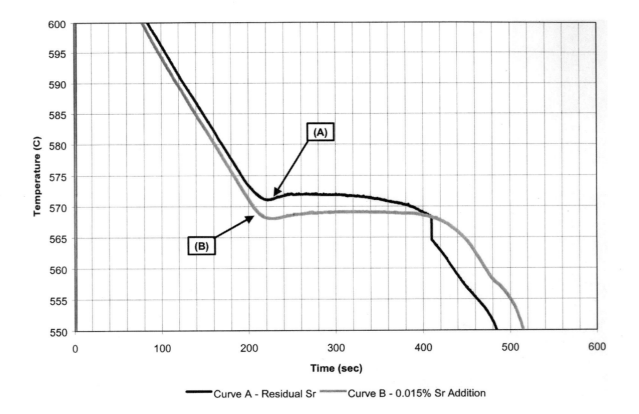

Figure 14 Cooling curves from A356 aluminum with (A) residual Sr and with (B) 0.015% fresh Sr eutectic modifier addition.

hydrogen absorption. Sodium fades rapidly, and it is impractical to degas after making the addition. Modifying with strontium may cause hydrogen absorption, but the melt can be degassed after a strontium addition without significant fading.

Some investigators report that eutectic modifiers redistribute and minimize the size of the pores formed during solidification rather than cause hydrogen absorption [88–91]. Strontium may reduce the gas content but increase porosity by facilitating pore nucleation or by changing the metal surface tension [92]. The effect of strontium on porosity formation and distribution is still the subject of some controversy.

Pouring temperature affects porosity, with more porosity usually associated with higher pouring temperatures. Volumetric metal shrinkage occurs in three regimes. First, there is liquid shrinkage as the metal cools from the pouring temperature to the liquidus. Then, solidification shrinkage occurs as the metal transforms from a liquid to a solid. Finally, thermal contraction occurs as the metal cools to room temperature.

Shrinkage associated with porosity formation occurs (1) as the liquid is cooled and (2) as the liquid transforms to a solid. Liquid aluminum volumetrically contracts about 1% for each 55°C (100°F) superheat above the liquidus [84]. If the metal pouring temperature could be lowered to near the liquidus without causing misruns or folds, volumetric shrinkage could be reduced. Sometimes, large runners are used to produce "hot" gates, and in such cases, pouring at moderately higher temperatures will preheat the gate and allow some feeding from the runner.

E. Hydrogen Effects on Porosity

Hydrogen is the only gas soluble in aluminum in appreciable quantities. Aluminum will absorb hydrogen from moisture in the air, combustion by-products, damp refractories, moisture on charge materials, and moisture in molds and mold binders [86,93]. Because aluminum readily reduces water to form aluminum oxide and free hydrogen, it is essential to minimize the contact of molten metal with sources of moisture or water.

Hydrogen solubility in aluminum is about 0.7 mL/100 cm^3 at the liquidus and 0.05 mL/100 cm^3 at the solidus. During cooling, dissolved hydrogen is rejected from solidifying metal into the liquid, and when the solubility limit is exceeded hydrogen pores nucleate and grow [94]. Nucleation is thought to occur primarily on oxides in the metal. If the oxide content could be substantially reduced, higher hydrogen concentrations might be tolerated [95,96].

If the hydrogen content is sufficiently high, hydrogen bubbles form throughout castings. At lower concentrations, hydrogen pores occur only in the last areas to solidify where segregation has occurred. However, dissolved hydrogen aggravates porosity at all concentrations above its solid solubility limit by segregating into and increasing shrinkage pore size [97]. Pore size is minimized by maintaining a low hydrogen concentration. Pressure can also be applied during solidification to minimize pore size [98].

The most common test for dissolved hydrogen is the reduced pressure test (RPT) [93,99–101]. A sample of molten aluminum is extracted from the molten metal, placed in a chamber, and a vacuum applied. The low pressure increases the pore size and reduces sample density. The RPT sample can be rated by the following: (1) when the first bubble appears, (2) the number of bubbles that appear during solidification, (3) apparent density of the sample, or (4) the size and distribution of the pores observed after the sample has been sectioned [102–107]. The amount of hydrogen present can be estimated by determining the density and comparing that to the density of a pore-free sample of similar composition [100]. However, pore nucleation is a function of the oxides present in the metal and absolute concentration values may not be obtained [95].

Dissolved hydrogen can be removed from molten aluminum by bubbling an inert gas through the metal. Hydrogen is absorbed into bubbles and carried from the bath [108]. The rate of gas removal depends on the bubble surface area, gas purity, and the presence of reactive gases. Small bubbles are the most effective in removing hydrogen and are most easily obtained by passing the gas through a rotary head immersed in the metal [109]. These heads shear and disperse the bubbles and can be used to disperse fluxing agents throughout the melt.

The rate of hydrogen removal can be increased by adding a small amount of reactive gas such as fluorine or chlorine to the gas stream [110,111]. Fluorine has been added as Freon, and chlorine is usually added in a dilute mixture of nitrogen or argon. Fluxes and reactive gases assist in removing aluminum oxides as well as hydrogen. Hydrogen control is essential to the production of high-quality castings [112,109]. More often than not, microshrinkage porosity and hydrogen interact to aggravate porosity formation.

Multiple regression analyses have been used to build models for predicting microporosity. Hydrogen was found to be the strongest single factor, but interactions with other variables including eutectic modifiers, solidification time, and solidus velocity made substantial contributions [113]. The rate of solidification has an important effect on the pore volume fraction, with the pore volume fraction decreasing at a constant hydrogen concentration with higher solidification rates [114].

F. Microstructural Effects on Properties

In general, the mechanical properties of sound aluminum castings are controlled by dendrite arm spacing, heat treatment, and silicon particle morphology [115–117]. All of these features must be considered when producing high-quality castings.

1. Dendrite Arm Spacing Effects

Flemings et al. established a strong relationship between properties and the dendrite arm spacing (DAS) in 195 and A356 aluminum alloys [115–117]. The results of his and other related publications were summarized in an AFS silver anniversary paper in 1991 [118]. The ultimate strength and elongation progressively increased as the dendrite cell size decreased from 115 to 25 μm (45×10^{-4} to 10×10^{-4} in.), but there was no observable effect on ductility over this range. The dendrite arm spacing followed the relation:

$$D = 7.5 \, \text{ST}^{0.39} \tag{1}$$

where D is the dendrite arm spacing in microns and ST is the solidification time in seconds.

Higher solidification rates were obtained with large chills in the mold to produce sounder metal and finer dendrites. The chills were not found to have a significant effect on grain size [118].

A similar relationship was found between the dendrite arm spacing and the ultimate strength of Al–4.5% Cu [119]. The dendrite arm spacing increased and the strength decreased with the logarithm of the solidification time. Porosity increased with solidification time and decreased with the logarithm of the dendrite arm spacing.

Boileau et al. examined the effect of solidification time on the properties of well-risered A356 wedge casting [120]. The pore size, secondary dendrite arm spacing, and pore distribution were affected by the solidification time and had significant effects on strength and ductility. Ultimate tensile strength and ductility were found to decrease linearly with increasing dendrite arm spacing while the yield strength remained unchanged.

Some castings were hipped but no differences were found in ultimate tensile strength between as-cast and hipped material. This suggested that the ultimate tensile strength was not significantly influenced by porosity in

the range of about 0.1% to 0.5% [120]. The pores at this level may have been sufficiently small so that their effects were lost in the effects of grain size and other variables.

Porosity did have a significant effect on fatigue limit. The fatigue limit of cast and heat-treated material was lower than hipped and heat-treated material by a factor of ten. The investigators concluded that pores have a significant effect, and the maximum pore size obtained with conventional metallographic techniques is understated [120].

The secondary dendrite arm spacing has been reported to be primarily controlled by composition, cooling rate, local solidification time, and temperature gradient [121]. The relationship between dendrite arm spacing and cooling rate was:

$$\log(dT/dt) = -[\log(\text{DAS}) - 2.37]/0.4 \tag{2}$$

where dT/dt is the cooling rate in °C/min and DAS is dendrite arm spacing in millimeters.

The secondary arm spacing was found to control the size and the distribution of pores and hence the mechanical properties.

The effects of microstructure on dendrite arm spacing and mechanical properties of a low hydrogen material resulted in the following relationship:

$$\text{DAS} = aV^n \tag{3}$$

where V is the average cooling rate in K/sec from the liquidus to the eutectic and a and n are constants determined for each alloy [122].

The form of the equation was similar to that developed by Flemings. Relationships between mechanical properties and dendrite arm spacing were also developed using an expression of the type:

$$Y = a = \text{DAS}^{n=} \tag{4}$$

where Y is the property of interest and $a=$ and $n=$ are constants determined for each alloy and property [122].

The published literature consistently reports a strong relationship between cooling rate and the dendrite arm spacing. Similarly, good relationships are found between dendrite arm spacing and ultimate tensile strength, yield strength, and ductility.

2. Porosity Effects on Properties

Pores form during solidification, whether from solidification shrinkage or hydrogen, and they have a substantial effect on mechanical properties. The tensile strength and elongation of 201 aluminum decreases with porosity associated with density decreases of up to 4% [123]. Test plates were cast in various lengths and thicknesses with a sprue, ingate, and riser on one end, and a steel chill on the opposite end. The effect of distance from the riser on the tensile strength and elongation in a 20-cm (7.9-in.)-long plate is illustrated in Fig. 15A and B, and the effect of porosity on strength and elongation in the alloy is illustrated in Fig. 16A and B [123]. Low porosity and the best properties were found in specimens solidified against the chill or near the riser.

In sand- and chill-cast A356-T6 aluminum, it was found that the yield strengths were similar, but the ultimate strengths were higher in the chilled castings [124]. The ultimate strengths were about 289 MPa (42 ksi) in chilled castings and 275 MPa (40 ksi) in sand castings. It was concluded that the dendrite spacing was important and that microcracks propagated through the Mg$_2$Si particles during fracture.

The spatial arrangement of pores is an important characteristic affecting properties [125]. Pore clusters of a given size and volume fraction have been found to be more deleterious to mechanical properties than uniformly dispersed pores having the same average pore size and volume fraction. Factors were developed to express the pore clustering tendency using a ratio of the average nearest neighbor distance [125].

The effect of hydrogen concentration and cooling rate on the strength and elongation of alloys containing 7% silicon, 10% silicon, and 3% copper was examined

over the hydrogen concentration range of from 0.16 to 0.37 mL H$_2$ per 100 g of aluminum at NTP. Both strength and elongation increased with higher cooling rates, at least up to 4°C/sec. At a given cooling rate, strength and elongation values decreased as the hydrogen content increased [103].

High-quality aluminum castings usually have less than 1% porosity in any section, and the pore size on a polished plane usually ranges from about 0.025 to 0.250 mm (0.001–0.01 in.). Since porosity cannot be completely eradicated, an acceptable level of porosity must be established in relation to the properties needed in a particular area. Pores act as stress concentrators, initiation sites for fatigue cracks and reduce tensile elongation. Low elongation values reduce the work hardening that occurs during deformation and reduce the ultimate strength. But, pore shape has been reported to have a minimum effect on tensile strength so long as the porosity is less than about 2% [126].

The effects of hydrogen, degree of modification, and cooling rate on the morphology and volume porosity in A356 and 319.2 aluminum as cast end-chilled plates has been examined [87,127]. In thinner castings, the maximum porosity occurred between the end of the casting and the riser, which is consistent with the results obtained by other investigators [125,123]. In thicker sections, the region having the highest porosity was

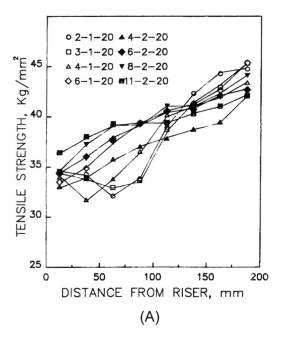
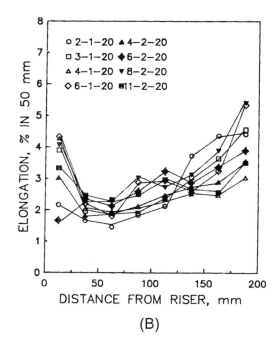

(A)   (B)

Figure 15 Effect of distance from the riser on (A) tensile strength and (B) elongation of specimens removed from 200-cm- long plates of A201 aluminum.

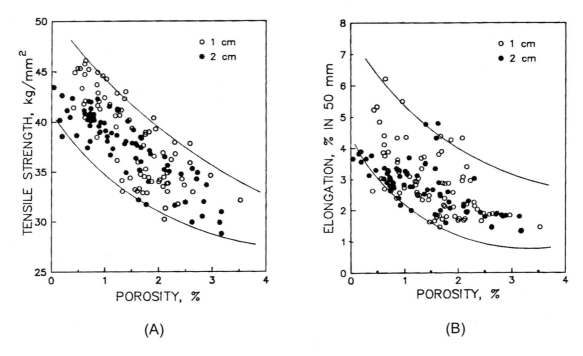

Figure 16 Effect of porosity in A201 aluminum on (A) tensile strength and (B) elongation.

shifted toward the riser, and there was less porosity in castings with larger risers. Ultimate strength and elongation values were sensitive to porosity and the relationship between porosity and properties was nonlinear [87,128] Porosity values below 0.5% were found to reduce tensile elongation by negligible values.

Tensile property measurements and metallographic analysis of A356 and A357 have been conducted to find factors that might be used in specifications or as a basis for design stress values [129]. The physical parameters considered were the cell count, aspect ratio of silicon particles and porosity. The properties of interest included 0.2% offset yield strength, ultimate strength, and total elongation. All three parameters had a significant effect on the tensile properties of the alloys. However, pores having diameters of less than 19 μm were not found to degrade ductility but rather to arrest the microcrack propagation between adjacent silicon particles.

An empirical expression describing the relationship between ultimate strength and porosity is expressed by Eq. (5) [130].

$$\sigma_p = \sigma_n (1 - \eta)^\theta, \qquad (5)$$

where

σ_p = the ultimate strength of the porous material
σ_n = the ultimate strength of the nonporous material,

η = the porosity
θ = an empirical constant varying between 3 and 6

Recent research on A356-T6 aluminum has given values of θ in the range of 2 to 3 [131].

G. Microstructure Development in Cast Iron

Cast iron is an iron-based alloy containing at least 2.1% carbon. Most commercial cast irons contain from 3.2% to 3.7% carbon, 1.8% to 2.4% silicon, 0.5% to 0.8% manganese, and smaller amounts of sulfur, chromium, copper, nickel, tin, phosphorous, and other minor elements.

The microstructure of most cast iron consists primarily of graphite in a matrix of either ferrite or pearlite or a combination of the two. The volume percent of graphite is usually in the range of 8% to 13% with the matrix phases (ferrite and pearlite) occupying from 87% to 92% of the volume. Minor amounts of carbides and nitrides may be present with volume fractions usually less than 0.5%.

When cast irons solidify, the phases formed can include austenite, graphite, iron carbide, and a ternary iron–phosphorus–carbon compound (steadite). Massive (eutectic) carbides must be minimized because their presence degrades the ductility and produces machining difficulties. Minimizing the massive carbide formation demands that graphite form during eutectic solidification.

Graphite is carbon precipitated as free carbon with a hexagonal crystalline structure. Graphite formation begins with nucleation near the austenite–graphite eutectic temperature of about 1150°C (2100°F). Compounds formed by reactions between magnesium, sulfur, and oxygen form nuclei on which graphite can grow, and initial graphite growth occurs by the addition of carbon atoms to the nucleus. Growth continues as carbon atoms diffuse through liquid metal in the case of gray iron, and through an austenite shell in the case of ductile iron, to become attached to the flake or nodule. The growth process controls the final graphite shape. Graphite acts like voids or cracks in iron, and higher strengths are favored by low carbon equivalent values, inoculating additions, and section sizes that minimize flake length [132].

When graphite precipitates as elongated flakes, the resulting product is gray iron. When graphite precipitates as nodules or spheres, the iron is called nodular or ductile iron. Graphite flakes in gray iron consist of planes that grow in the a-axis direction with basal planes in contact with austenite. The weak bonds between the basal planes permit relatively easy bending and twisting as flakes grow. Spheroidal graphite growth proceeds radially from the nucleation point along the c-axis direction. Basal planes are again in contact with solid austenite. Spheroidal graphite growth is the result of a high interfacial energy between iron and graphite that changes the graphite growth direction. Low oxygen and sulfur concentrations increase the interfacial tension and cause graphite to precipitate as nodules rather than as flakes.

If, during solidification, insufficient nuclei are present to provide graphite growth centers, the liquid metal cools to the metastable eutectic carbide formation temperature, and massive carbides form from the liquid. As iron cools after solidification, the carbon solubility in austenite decreases from about 2.1% at 1150°C (2100°F) to about 0.7% at 715°C (1320°F). The rejected carbon can precipitate onto existing graphite flakes or nodules, precipitate as microscopic iron carbides in the austenite, or become supersaturated in the austenite and transform to microcarbides during the eutectoid reaction.

At or near 715°C (1320°F), the eutectoid reaction occurs and austenite usually decomposes to produce ferrite and pearlite, although at low cooling rates, graphite can also be formed. Graphite growth requires sufficient time for carbon to diffuse from the austenite over distances of 5–25 μm (200–1000 μin.) to attach to existing graphite.

Pearlite is composed of thin plates of $(Fe,Mn,Cr)_3C$ and ferrite. Because the plates are quite small, the carbides in pearlite are sometimes referred to as microcarbides to differentiate them from eutectic carbides that are large enough to be seen by the unaided eye. Most gray irons have a matrix of pearlite, and ductile irons range from 100% ferrite to 100% pearlite depending on the desired tensile strength. Eutectic carbides may be present, but are almost always undesirable.

The mechanical properties of most interest to foundrymen have historically been the tensile strength and hardness of gray iron, and modulus, yield and ultimate strength, and hardness of ductile iron. The modulus and yield strength of gray iron are rarely measured.

High pearlite contents produce higher yield and ultimate strengths, higher hardness values, and lower elongation values in ductile iron. The ferrite content of ductile iron has a significant effect on ductility, with progressive increases in ductility occurring as the ferrite content increases. The strength of gray iron is dominated by the alloy content that controls the matrix strength and free carbon content [133–137]. The tensile elongation of gray iron is never high, typically between 0.2% and 0.3%, because of the presence of graphite flakes.

The influence of alloying elements on the tensile yield and ultimate strengths of ductile iron are primarily a result of the ferrite and pearlite contents. Most alloy elements including copper, manganese, nickel, chromium, and molybdenum function by increasing the microcarbide content or decreasing the pearlite spacing. Silicon is unusual because it tends to increase the ferrite content, but it also hardens and strengthens the ferrite. Silicon can also decrease the strength of irons intended to have pearlite structures by increasing the ferrite content, but at sufficiently high concentrations, silicon increases the strength by solid solution hardening [136].

VI. NONDESTRUCTIVE INSPECTION AND CORRELATIONS WITH PROPERTIES

Nondestructive evaluation (NDE) techniques have been used since the 1920s for part inspection, quality control, and process control. Properly implemented, NDE not only reduces parts rejected by the end user but provides feedback to the foundry for improving the casting process. The economic benefits of NDE lie in finding anomalies early in the manufacturing process, reducing customer rejects, reducing life-cycle costs, and improving part consistency and quality. When properly applied, process cost reductions and quality improvements outweigh the costs of inspection.

To effectively inspect a part, designers must consider the inspection process during the design stage. Adding NDE as an afterthought may increase inspection costs and inspecting critical areas might be impossible. For effective inspection, consideration must be given to the following [138]:

Fracture mechanics/NDT relationships
Safe life/fail-safe criteria
Critical area accessibility
Testing and verification procedures
Process anomalies

Designers should also understand the capabilities of various nondestructive techniques. No single technique provides complete inspection, and designers need to know whether particular techniques will reliably detect important anomalies in their parts.

The goal of NDE is to accurately identify parts that do not meet requirements. Requirements might focus on anomaly sizes in stressed areas, radiographic ratings, thickness at a particular location, or specific phases in the microstructure. Based on experience, code requirements, and experimental data, designers must anticipate and specify unacceptable anomalies. Anomalies may affect the modulus, yield strength, fatigue strength, corrosion resistance, pressure tightness, etc.

The ability to truncate populations of parts and remove parts with low properties has a significant effect on the design process. A histogram of tensile strength data from a population of A356 aluminum is illustrated in Fig. 17. These specimens contained microshrinkage that produced a skewed normal distribution. If it is

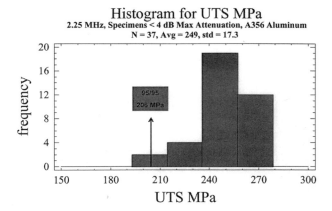

Figure 18 A histogram of ultimate strengths from 37 A356-T6 aluminum tensile specimens with ultrasonic attenuation less than 4 dB.

assumed that population is normal, there would be a 95% confidence that 95% of the population would meet a minimum strength of 23.7 ksi.

Ultrasonic attenuation responds to microporosity in cast aluminum [139]. If 2.25-MHz attenuation is used to truncate the population and remove specimens with attenuations above 4 dB, the strength distribution illustrated in Fig. 18 is obtained. The truncated distribution is narrower and normal. After truncation, there is 95% confidence that 95% of the specimens would have strengths above 30 ksi.

Truncation can also improve the reliability of parts. Distributions of calculated stresses under different duty cycles and fracture stress data from one material population are presented on the X axis of Fig. 19, and the frequency of occurrence is shown on the Y axis. The strength of a material being considered for an application must be known as from this data, estimated from experience, or available from published literature.

Designers sometimes use a bump-down factor on material strength to account for unexpected anomalies or a bump-up factor on the calculated stresses to account for unanticipated loads in service. The difference between the bump-up stress value and the bump-down strength value is sometimes referred to as the safety factor. If, through truncation, the strength can be raised and, more importantly if the distribution can be narrowed, the safety factor or the working stress can be increased without compromising part reliability.

The following material reviews several nondestructive techniques that are useful in evaluating castings. Data obtained during inspection of some aluminum and iron castings and relationships between NDE signals and properties are described in the following material.

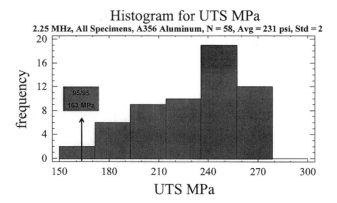

Figure 17 A histogram of ultimate strengths from 58 A356-T6 aluminum tensile specimens.

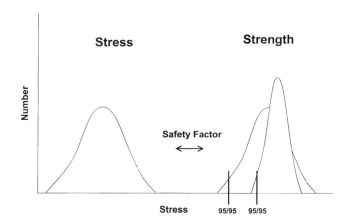

Figure 19 Interaction of part design, working stress, and strength.

Density is a material property and low densities reflect the presence of pores or holes but not the location of the hole. The function of liquid penetrant inspection is to locate cracks or holes extending to the surface. The function of penetrating radiation and ultrasonic beams provide information about the size and location of anomalies and some structural features in parts.

A. Density Measurements

Density is, by definition, a measure of mass per unit volume of material. While not frequently used, bulk density measurements can provide useful information about structure and soundness, especially in smaller parts. Low density is often responsible for low tensile and fatigue performance.

A generic part with a uniform shape is illustrated in Fig. 20. The part has one pore open to the surface (open pore) and one not open to the surface (closed pore). Three different values of density, referred to as bulk density, apparent density, and true density can be determined for this specimen based on dry weight, wet weight, and immersed weight.

Bulk density is defined as the part weight divided by the bulk volume. Bulk volume includes the total volume of the closed pore, open pore, and solid material. Bulk density is expressed as:

$$\rho_b = \frac{W_x}{V_b} = \frac{W_x}{V_s + V_{cp} + V_{op}} \qquad (6)$$

where

ρ_b = bulk density
W_x = specimen weight

V_b = bulk volume of the specimen
V_s = solid material volume that does not contain a pore
V_{op} = total volume of open pores
V_{cp} = total volume of closed pores.

Apparent density is defined as the part weight divided by its apparent volume, which is the total volume of the solid material and the closed pore. The apparent volume is equal to the bulk volume minus the open pore volume.

The apparent density can be calculated using Eq. (7).

$$\rho_a = \frac{W_x}{V_a} = \frac{W_x}{V_s + V_{cp}} = \frac{W_x}{V_b - V_{op}} \qquad (7)$$

where ρ_a is the apparent density and V_a is the apparent volume.

The true material density is the weight of the part divided by the volume of the solid material. True density is calculated using Eq. (8)

$$\rho_t = \frac{W_x}{V_s} \qquad (8)$$

where ρ_t is the true material density.

If there are no open pores, the bulk density and apparent density are the same. If, however, pores extend to the part surface, the two density values may be quite different, with the bulk density having the lower value.

There are several methods for measuring density, but the immersion method has the advantage of being useful when parts or specimens have an irregular shape. Immersion measurements require a reference liquid with known density as a function of temperature and atmospheric conditions. Accurate determinations require

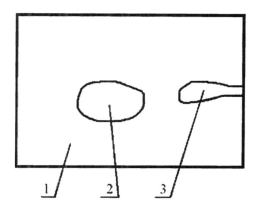

Figure 20 Schematic of a specimen used to define density measurement terminology. 1—material without pores, 2—closed pore, 3—open pore.

that buoyancy effects of air and the reference liquid be considered, as in Eq. (9).

$$\rho_b = \frac{W_x}{V_b}$$

$$= \frac{(W_d\rho_{ws} - W_s\rho_{ad})(\rho_{ww} - \rho_{aw})}{(W_w - W_s)(\rho_{ws} - \rho_{ad}) - (W_d - W_s)(\rho_{ws} - \rho_{ww})} \tag{9}$$

where

W_d = part weight in air or "dry weight"
W_s = part weight suspended in the reference liquid
W_w = "wet weight" or the part weight in air with the reference liquid filling open pores
ρ_{ad} = air density during dry weight measurements
ρ_{aw} = air density during wet weight measurements
ρ_{ww} = density of the reference liquid in wet weight measurements

Values of ρ_{ad}, ρ_{aw}, and ρ_{ww} can be found in published literature or measured [140].

The steps in making density measurements using a wetting reference liquid are summarized as follows:

1. Clean the part to remove cutting fluids or oils.
2. Bake to remove adsorbed moisture.
3. Cool to room temperature in a desiccator.
4. Weigh in air to determine the dry weight.
5. Immerse in the reference liquid and allow the liquid to penetrate the pores.
6. Suspend in the reference liquid and measure the suspended weight.
7. Lightly dry without removing liquid that has penetrated the surface pores.
8. Determine the wet weight.
9. Calculate density using Eq. (9).

B. Liquid Penetrant Inspection

Liquid penetrant inspection is used to locate discontinuities as thin as 1 μm so long as they are open to the surface [141]. This technique is simple and less expensive than most other nondestructive techniques and can be used on most castings. The effectiveness of LP inspection depends on the liquid wetting the surface and penetrating cavities. Penetration is a function of surface cleanliness, cavity shape and size, and wetting characteristics of the liquid. However, the surface being inspected must be relatively smooth and nonporous.

Both visible and fluorescent penetrants are available in a range of sensitivities. Visible penetrants can be viewed under white light, and fluorescent penetrants require a black light for inspection. The sensitivity range to be used depends on the size of the imperfection being sought and the part surface quality. Rough surfaces inspected with sensitive penetrants can produce false indications.

The first step in LP testing is surface preparation. Contaminants such as oil, water, or dirt must be removed or the penetrant will not wick into anomalies. The second step is to apply the penetrant and allow its absorption into surface anomalies. Penetrants are then washed from the surface. Some penetrants are water soluble and are removed by spraying with water. Oil-based penetrants require an emulsifier or solvent for removal. In all cases, care must be used to remove only the penetrant from the surface and not from the anomaly. A developer is then put on the part surface to aid penetrant seepage from anomalies to the part surface. The developer magnifies the cavity and allows it to be seen. Finally, the casting is ready to be examined and the anomalies rated [141].

C. Penetrating Radiation Inspection

Penetrating radiation inspection uses differences in absorption between sound metal and either inclusions or pores to produce images that reflect local density differences. The minimum density or thickness difference that can reliably be detected is about 2% of the part thickness transverse to the beam path. Three components are needed to produce a radiographic image—a radiation source, the test piece, and a method to record the penetrating radiation.

1. Radiation Sources

X-rays and gamma rays are the two most common types of penetrating radiation. X-rays are produced by the impact of high-energy electrons against an anode. When an electron strikes the nucleus of an anode atom, the electron energy is converted into a photon having the highest energy for the voltage applied. It is more likely, however, that an electron will strike an electron orbiting the nucleus, and when this happens, the electron loses some energy before reaching a nucleus. When it does strike the target nucleus, lower-energy photons are produced. The range of electron interactions produces a continuous spectrum of wavelengths, and the low-energy "soft" radiation tends to darken film without providing any useful information.

Gamma rays are produced by radioactive decay of isotopes such as thulium-170, iridium-192, cesium-137, cobalt-60, and radium. There is little difference between x-rays and gamma rays except that gamma rays may have distinct wavelengths and usually have higher energies. For example, iridium-192 emits gamma rays with 12 different wavelengths having energies ranging from 0.21 to 0.61 MeV [138].

2. Test Pieces

Each material absorbs and attenuates x- and gamma rays at rates that depend on the radiation source, material characteristics, and test piece density. Differences in attenuation provide the radiographic density variations observed on the recording medium. Absorption and attenuation depends not only on the density and atomic structure of the test piece but also on the energy, intensity, and type of radiation.

A successful exposure minimizes recording low-energy scattered radiation. Where possible, the tube voltage is increased to produce higher-energy radiation or the beam is "hardened" by placing a copper or lead filter between the tube and test piece. The filter blocks some of the soft radiation but allows higher-energy photons to pass through [138].

3. Recording Media

The recording media produces a permanent visible image of the variations in radiation intensity on the recording plane. The recording media can be film, radiographic paper, electrostatically sensitive paper, or a digital detector. Film is the most common recording media and consists of a thin plastic sheet coated on one or both sides with silver halide grains, usually silver bromide. When an x-ray reacts with silver halide, a latent image is formed that can be made visible by reacting the exposed grains with a developer that converts the exposed grains to black metallic silver.

When the film is immersed in a fixer, the unexposed silver halide is converted into a water-soluble compound that can be washed off to produce a transparent area [141]. The variation in metallic silver density produces the "gray levels" observed in the film. Film densities are usually measured using a densitometer.

Geometric unsharpness is the result of shadows produced by interactions of x-rays with edges of anomalies and test pieces [138]. Eliminating unsharpness requires an infinitely small anode spot size.

4. Radiographic Sensitivity

Radiographic sensitivity is a measure of the useable information contained on the x-ray detector and reflects both radiographic contrast and definition. Contrast is the ability to see differences in gray level from one area of the film to another, and radiographic definition is the sharpness of the transition from one contrast level to another [138].

An image-quality indicator (IQI) or penetrameter is always placed on a detector during an exposure to provide a record of the image quality. The IQI usually contains small geometric features, such as holes or wires, corresponding to numerical fractions of the test piece thickness.

ASTM E-142 describes an IQI with holes. The penetrameter should be made from the same or a radiographically similar material as the test piece and have a thickness corresponding to about 2% of the test piece thickness in the area of interest. For example, a 2% penetrameter for a 25.4-mm (1-in.)-thick section would be 0.51 mm (0.020 in.) thick.

The IQI usually contains three holes, referred to as 1T, 2T, and 4T holes. The thickness and hole diameter in the IQI are used to establish radiographic sensitivity. A common radiographic sensitivity level of 2-2T indicates that the 2T hole in a 2% IQI is visible on the finished radiograph. Penetrameters are basically go/no-go standards. Either the sensitivity meets the requirements or they do not [138].

5. Image Interpretation

Radiographic interpretation requires the interpreter to understand not only the process and limitations by which the image was produced but also the method of producing the test piece. This knowledge comes from experience and training, without which interpretation is difficult. Even with training, results vary between interpreters [138].

D. Ultrasonic Inspection

Mechanical vibrations at frequencies above about 20,000 Hz exceed the range of human hearing and are called ultrasound. Piezoelectric materials are used to produce ultrasonic transducers, and such transducers can generate and receive ultrasound. Intermittent voltage spikes are passed through transducers to bring them to resonance, and as a first approximation, the transducer operates as a piston to produce a high-frequency signal. The signal passes through the part being examined, is picked up by the same or a different transducer, and converted into an electrical signal that can be displayed and analyzed [139].

The passage of ultrasonic signals across an interface is a function of the angle of incidence and the properties of the medium. For normal incidence, the important parameter is the acoustic impedance (Z) of the transducer and the material being inspected. If there is no impedance mismatch across an interface, all of the energy is transmitted. If there is a mismatch, the energy transfer will be less than 100% [139].

The two most common ultrasonic techniques are referred to as "pulse–echo" and "through transmission" modes. In the pulse–echo mode, a transducer sends pulses into the test piece and then collects the returning signals (echoes). The pulse–echo method allows anomalies to be spatially located based on the signal velocity and the time of flight. The quantities of interest are the amplitude of the echo and the transit time of the pulse from the transmitter to the anomaly and back to the receiver.

The through transmission technique uses two transducers, one to generate the signal (pitch transducer), and the second to receive the signal (catch transducer). Discontinuities cause signal reflections that indicate high-impedance interfaces. Transducers can also be used to measure signal attenuation but these systems are less common than pulse–echo systems.

Ultrasound waves travel through most solids and liquids reasonably easily, but they are poorly transmitted by air. For this reason, a coupling agent such as water, oil, or special gels are used to eliminate the air gap between the transducer and the test piece. When a transducer is placed near a specimen surface with a couplant, the technique is referred to as the "contact method." When a bath of liquid, usually water, serves as the energy transmission medium, the technique is referred to as the immersion method [139].

Ultrasonic inspection methods examine a much larger volume of material than metallographic methods. Ultrasonic measurements are accurate, reliable, and relatively inexpensive so long as surface finish and coupling issues are considered [142–147].

Ultrasonic wave lengths are calculated using Eq. (10):

$$\lambda = c/f \tag{10}$$

where

c = sound velocity, m/sec
f = frequency, Hz
λ = wavelength, m

Some typical values of wavelengths in cast aluminum and iron as a function of frequency are presented in Table 5.

Ultrasonic inspection is usually conducted at frequencies of from 1 to 5 MHz, and anomalies having dimensions larger than one quarter of the wave length are usually detectable. Signal scattering (attenuation) caused by dispersed microporosity can be used but is not common in the casting industry.

Micropores usually range between 25 and 250 μm in length on a polished section, but cluster dimensions can be much larger. The degree of signal attenuation is proportional to the microporosity.

Ultrasonic velocity is related to the modulus and density of the material being inspected. Ultrasonic velocity has been used for many years to assess graphite nodularity in ductile iron, and to some extent, the tensile strength of gray iron [142–145,148–151]. The velocity of a longitudinal wave through a material can be expressed as [139]:

$$v_l = \sqrt{\frac{(1 - \mu)}{(1 + \mu)(1 - 2\mu)}} \sqrt{\frac{E}{\rho}} \tag{11}$$

where

v_l = longitudinal ultrasonic velocity
E = modulus
ρ = density
μ = Poisson's ratio

Table 5 Frequency Effects on Ultrasonic Wavelengths

Material	Velocity (m/sec)	Wavelength (mm) at frequency of			
		1 MHz	2.25 MHz	5 MHz	10 MHz
Gray iron	4830	4.3	1.9	0.87	0.43
Compacted graphite iron	5330	4.7	2.1	0.94	0.47
Ductile iron	5590	4.9	2.2	0.99	0.50
Aluminum	6350	5.6	2.5	1.10	0.55

Ultrasonic velocity numbers are approximate. The actual velocity in materials within each category varies with composition and density.

The engineering modulus of a material is usually determined from stress–strain curves, but can be determined from ultrasonic velocity and density measurements by rearranging Eq. (11).

$$E = K\rho v_l^2 \frac{(1+\mu)(1-2\mu)}{(1-\mu)} \qquad (12)$$

where K is a units conversion factor. K is 10^{-6} if E, ρ, and v_l are expressed as MPa, kg/m^3, and m/sec, respectively.

Pore-free A356 and A357 aluminum has a Poisson's ratio of about 0.35, and Eq. (12) reduces to:

$$E = 0.623 \times 10^{-6}\rho v_l^2 \qquad (13)$$

where, E, ρ, and v_l are in MPa, kg/m^3, and m/sec, respectively.

Poisson's ratio is about 0.23 to 0.24 in gray iron, and the equation for calculating modulus in iron reduces to Eq. (14):

$$E = 0.856 \times 10^{-6}\rho v_l^2 \qquad (14)$$

Again, E, ρ, and v_l are in MPa, kg/m^3, and m/sec, respectively.

The stress associated with an ultrasonic wave is quite low, and modulus values calculated from ultrasonic velocities are often referred to as "low-stress" modulus. These values are almost always higher than values obtained from tension or compression tests.

The modulus of most aluminum alloys is about 70 GPa (10 Mpsi) but may be lower if the material contains porosity. The modulus of cast iron is a function of the microstructure and primarily a function of the graphite volume and shape [146].

Ultrasonic attenuation measurements reflect energy dissipation as waves pass through a material. Ultrasonic attenuation or amplitude reduction (AR) is calculated using Briggs logarithmic equation as [139]:

$$AR = 20H \log (A_1/A_2) \qquad (15)$$

where A_1 is the amplitude of the input or reference signal and A_2 is the amplitude of the received signal measured in decibels (db).

The amplitude reduction is a property of a particular specimen, length, and surface finish. The attenuation coefficient has units of decibels per unit length and is a material property.

The two principal causes of attenuation are absorption and scattering. Absorption losses are caused by dislocation damping, anelastic hysteresis, relaxation, and thermoelastic effects. Absorption losses involve the conversion of acoustic energy to heat and are essentially independent of crystal grain size, shape, and volume [150,152,153].

The attenuation due to absorption is a function of frequency and can be expressed as:

$$\alpha_a = C_a f \qquad (16)$$

where

α_a = attenuation coefficient due to absorption
C_a = a constant unaffected by material grain size and anisotropy
f = frequency of ultrasound

Ultrasonic scattering occurs at grain and phase boundaries by reflection and refraction [153]. Signal attenuation is generally higher if there is a microstructural feature with a size ranging from about 0.25 to about 1.75 times the wavelength. Thus attenuation measurements can be used to estimate grain size, large inclusions, and eutectic cell sizes [146].

Generazio [154] summarized the expressions for ultrasonic attenuation caused by scattering as a function of frequency into three regimes referred to as Rayleigh, stochastic, and diffusion based on the grain size relative to wavelength:

1. Rayleigh scattering occurs when the wavelength, λ, greatly exceeds mean grain size $D(\lambda >> D)$

$$\alpha_s = C_r D^3 f^4 \qquad (17)$$

2. Stochastic scattering occurs where $\lambda \approx D$

$$\alpha_s = C_s D f^2 \qquad (18)$$

3. Diffusion scattering occurs where $(\lambda << D)$:

$$\alpha_s = C_d/D \qquad (19)$$

where α_s is the attenuation coefficient due to scattering and C_r, C_s, and C_d are constants that depend on material parameters including elastic constants. These equations are generally valid for homogeneous, regular, and equiaxed systems.

In general, all three types of scattering may occur simultaneously [154], and the total ultrasonic attenuation coefficient can be expressed as:

$$\alpha = \alpha_a + \alpha_s \qquad (20)$$

where α is the total attenuation coefficient.

Ultrasonic attenuation in cast aluminum is largely caused by porosity, but in some cases can be caused by grain boundaries. In cast iron, attenuation is caused by the interfaces between graphite and the iron matrix, pearlite structure, and by porosity. Ultrasonic attenuation is sensitive to small changes in graphite shape that may have a significant effect on tensile properties [150].

Several variables affect ultrasonic energy loss occurring as waves leave one transducer, enter a material, pass through the material, and exit to be picked up by either the same or a different transducer. Important factors include surface roughness, specimen soundness and shape, inspection frequency, and the sample microstructure. Considerable care is necessary because of possible variations in coupling, surface finish, and surface cleanliness. Ultrasonic attenuation caused by the material must be separated from energy losses that occur between transducers and the work piece before accurate correlations can be developed with microstructure.

E. Ultrasonic Measurements in Cast Aluminum and Iron

A typical through-transmission waveform is illustrated in Fig. 21. Three points are labeled including (1) first break, (2) first compressive peak, and (3) second compressive peak. The first break is the point where the transmitted signal causes the receiver crystal to begin to respond and create enough voltage to move off the base line. The positive peaks correspond to compressive waves that pass through the material and reach the receiver crystal [155].

1. Ultrasonic Velocity

The ultrasonic velocity through a material is determined by dividing the specimen thickness by the transit time of the signal. Small differences in calculated velocity are obtained depending on whether velocity is calculated from the first break, first peak, or second peak transit time.

The transit time is measured by comparing a wave passed through the specimen to a reference wave. A reference wave can be obtained by putting the pitch and catch transducers face to face or by putting a known material between pitch and catch transducers to create a time delay. By comparing the time difference between specific curve features obtained with a reference signal and a signal obtained after passage through the test piece, three velocities can be calculated. These values are referred to as the first-break velocity (FBV), first compressive peak velocity (FCPV), and the second compressive peak velocity (SCPV). The SCPV generally ranges from 5800 to 6200 m/sec (0.23 to 0.245 in./μsec) sec) in aluminum, from 4300 to 4800 m/sec (0.17 to 0.19 in./μsec) in gray iron, and from 4800 to 5300 m/sec (0.19 to 0.21 in./μsec) through ductile iron depending on the carbon equivalent and graphite nodularity [155].

The FBV is used in many commercial ultrasonic instruments such as thickness gauges. First-break values are based on the minimum pulse transit time through the sample and are relatively easy to measure if the signal has a minimum amount of noise and the material attenuation is not high. Porosity and other anomalies may not be seen using the first-break technique if there is a path for the wave through the specimen. Large pores may completely block wave transmission and be found, but small pores produce only small changes in FBV.

Some differences are found when velocity is measured using the different techniques because of energy losses within the specimen. Energy losses reduce the wave center frequency so the FBV is usually higher than FCPV, which is higher than SCPV. Specimens with higher attenuations create greater differences in the velocities, and SCPV values are affected more than the first-break values.

2. Ultrasonic Attenuation

Ultrasonic amplitude reduction (attenuation) can be calculated from the same signals used to calculate velocities by determining the peak height reduction after a signal has passed through the test piece. Attenuation is calculated by comparing the peak amplitudes from the reference signal and from the test piece, using Eq. (15).

F. Nondestructive Evaluation–Property Correlations in A356-T6 Aluminum

Ivan produced castings where systematic variations were made in the casting process and these castings

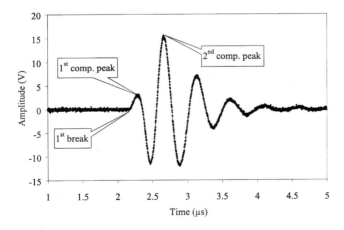

Figure 21 Typical ultrasonic wave indicating the first break, first compressive peak, and second compressive peak.

were inspected using density, x-ray, and ultrasonic techniques to correlate properties with NDE signals [156,157]. Experimental variables included casting thickness, the use of grain refiners and eutectic modifiers, and pouring temperature. The metal was well degassed, checked using an RPT procedure to ensure low hydrogen concentrations, and carefully poured and filtered in the mold to minimize dross in the castings.

The castings were made as plates with plate thicknesses of 12.6 or 16 mm (1/2 or 5/8 in.), and all castings were heat-treated to the T6 condition. Castings were rated radiographically according to ASTM Standard E-1472 using ASTM E-155. The majority of the plates contained only sponge porosity with grades ranging from 1 to 5. The bulk density of each plate was measured and found to correlate with the x-ray rating.

Tensile blanks were removed from various locations to obtain specimens with a range of porosity values. The density of each blank was determined, and each blank was radiographed using both film and a digital x-ray detector and ultrasonically inspected using both through transmission and pulse–echo methods. Tensile specimens were machined and tested and the ultimate strength, yield strength, and elongation values compared to the NDE signals. Metallography and fractographic examinations were made on fractured specimens to identify anomalies affecting strength and elongation.

1. Tensile Property Correlations with Density

Correlations between density and both yield and ultimate strengths are illustrated in Fig. 22. Specimen density ranged from 2.635 to 2.675 g/cm^3, and both the yield and ultimate strengths progressively increased with density. The correlation with ultimate strength was best with a correlation coefficient (R^2) of 0.67.

The relation between density and elongation is illustrated in Fig. 23. A slight decrease in density had a significant effect, and the elongation increased from 2%

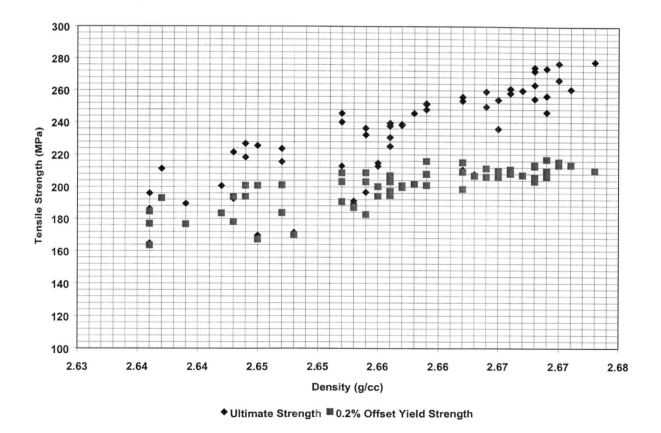

Figure 22 Effect of density on yield and ultimate tensile strength in A356-T6 cast aluminum.

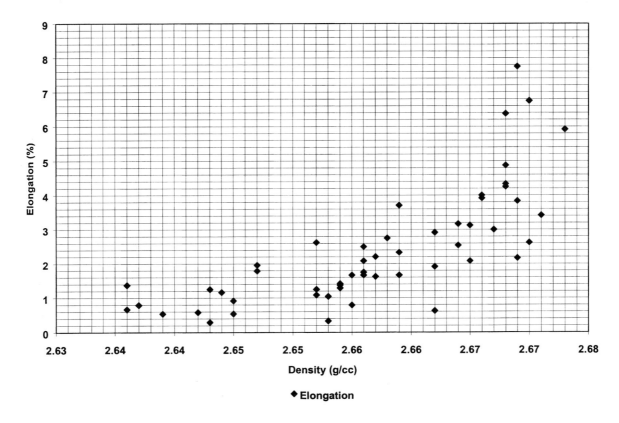

Figure 23 Effect of density on tensile elongation strength in A356-T6 cast aluminum.

to 7% as the density increased from 2.65 to 2.67 g/cm³. The porosity serves to nucleate cracks that propagate before much tensile elongation can occur.

The porosity of each specimen was calculated using Eq. (21). A density of 2.673 g/cm³ (0.097 lb/in.³) obtained from hipped material was used for the reference density:

$$P = \frac{100(\rho_{max} - \rho)}{\rho_{max}} \qquad (21)$$

where

P = percent porosity
ρ_{max} = reference density
ρ = bulk density of the specimen under consideration

The porosity in as-cast specimens generally ranged from about 0.04% to 1.38%.

The correlation between porosity and both ultimate and yield strength is illustrated in Fig. 24. Ultimate strength decreased linearly from 277 to 187 MPa (40 to 27 ksi), and the yield strength decreased from 215 to 180 MPa (31 to 26 ksi) over the porosity range. The correlation between porosity and specimen elongation is illustrated in Fig. 25. Specimen elongation dropped

from about 9% to 2.5% when porosity increased from 0% to 0.5%. Further increases in porosity above 0.5% decreased the elongation to values usually under 1%.

2. Property Correlations with X-Ray Data

Ivan also digitally radiographed castings and specimen blanks using a 12-bit gray-scale detector, and developed radiographic standards relating density changes to gray level [156]. Gray-level measurements were then made on the specimen blanks. One measurement made from the radiographs was the sum of the lengths of the major and minor axes of ellipses drawn around the anomalies. This measurement was made on indications with a density loss of 5% or greater.

The relationship between the sum of the major axis lengths with 5% or greater density loss and ultimate strength is illustrated in Fig. 26. The ultimate strength decreased linearly with increasing size and/or number of anomalies. Figure 27 presents similar data illustrating the effect of pores on elongation. The radiographic system resolution was not able to clearly define the transition from higher to lower elongation values.

Tensile elongation was more sensitive to porosity than tensile strength. In general, tensile properties de-

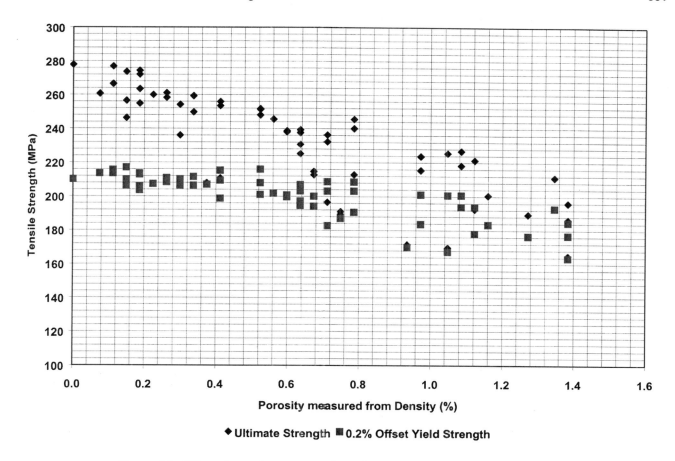

Figure 24 Effect of porosity measured from density on yield and ultimate strength.

creased as the sum of the anomaly axis lengths increased. The tensile response undoubtedly depends on both amount and distribution of porosity. Because x-ray radiography provides a 2-D representation of a 3-D structure, there may be situations where a particular porosity distribution is less harmful to the mechanical properties even when there is the same reduction in thickness.

3. Property Correlations with Ultrasonic Signals

Pulse–echo ultrasonic measurements were made in the gage area of each specimen and the echo voltage produced by the largest reflector compared to the tensile properties. The relationship between the reflected signal peak voltage at 2.25 MHz and tensile strength illustrated in Fig. 28. The yield and ultimate strength values decreased linearly with increasing echo peak voltage.

If porosity affects ultrasonic attenuation and echo peak height, there should be a relationship between ultrasonic measurements and density. Such relationships are illustrated in Fig. 29 where attenuation at 5

MHz is shown as a function of density, and in Fig. 30 where the maximum peak reflected voltage is shown as a function of porosity. An approximately linear relationship was found in both cases that porosity is a significant source of signal attenuation and reflection.

Attenuation measurements were made along the axis of specimen blanks, and the correlation between attenuation at 5 MHz and tensile strength is illustrated in Fig. 31. The ultimate strength decreased linearly with attenuation from 260 MPa (38 ksi) at 2 dB to 195 MPa (28 ksi) at 20 dB. Elongation decreased exponentially as illustrated in Fig. 32.

4. Metallography and Fractographic Correlations

Tensile fractures were examined and features on the fracture surface were measured. A fracture surface containing shrink porosity is illustrated in Fig. 33. The fraction of the surface containing shrinkage was compared to the tensile strength, and the results are illustrated in Fig. 34. Both ultimate and yield strengths decreased approximately linearly with fracture surface

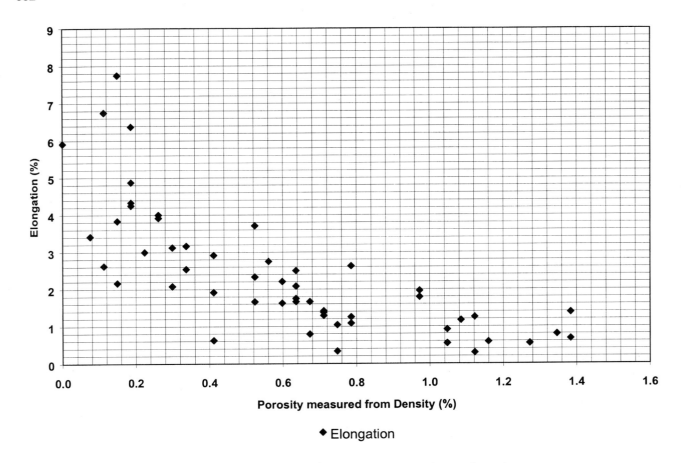

Figure 25 Effect of porosity measured from density on elongation.

porosity. The effect of fracture surface porosity on elongation is illustrated in Fig. 35, and elongation was found to decrease exponentially with porosity.

Metallographic examinations were conducted on tensile specimens along the tensile axis, and a section is illustrated in Fig. 36. From observations made on the fracture surface and from fracture mechanics principles, the pores in a cluster were considered to be interconnected, and the "flaw" size was the diameter of the cluster. Typical pore clusters are circled in Fig. 36.

The average diameter of each cluster on the polished cross sections was measured and relationships between cluster diameter and tensile properties were developed as illustrated in Figs. 37 and 38. Tensile strength decreased linearly and elongation decreased exponentially with cluster diameter.

A comparison of the metallographically measured flaw area and the tensile fracture flaw area is illustrated in Fig. 39. The area of fracture surface porosity was found to be about three times the area of the pore cluster area determined metallographically.

G. Nondestructive Evaluation Applications to Gray and Ductile Iron

Ultrasonic techniques are often used to locate relatively large internal discontinuities in castings, which may be a shrinkage, slag, dross, gas, voids or nonmetallic phases, and some investigators have reported that the yield and ultimate tensile strengths can be estimated using ultrasonic techniques. Ultrasonic NDE techniques are reported to be more sensitive to graphite parameters than eddy current and magnetic coercive force [142–146,148–151,158].

Kovacs [145] has shown that the energy-absorption capacity of cast iron is a function of the strain state in the crystal lattice and the presence of grain boundaries. A strained crystal lattice attenuates wave energy and

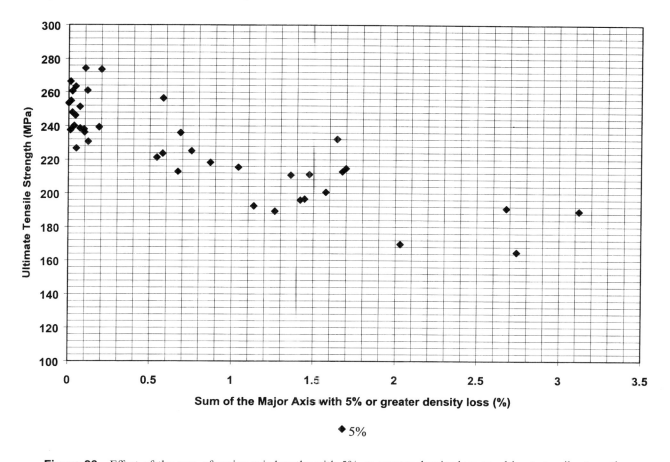

Figure 26 Effect of the sum of major axis lengths with 5% or greater density loss on ultimate tensile strength.

increases the energy absorption capacity. Kovacs found significant increases in damping capacity in high-strength, martensitic irons. The strains introduced during quenching substantially increased ultrasonic attenuation. Grain boundaries also absorb wave energy, so small grain sizes are expected to increase energy absorption.

1. Graphite Effect on Ultrasonic Nondestructive Evaluation

The use of ultrasonic velocity to assess the microstructure of cast iron is based on the fact that modulus and hence ultrasonic velocity is largely determined by the graphite shape and volume. Lerners and Vorobiev reported the modulus of cast iron to be quite sensitive to graphite volume and shape [144]. Graphite has a lower modulus than iron, and increasing the graphite volume fraction reduces the modulus. The modulus and corresponding ultrasonic velocity usually decreases

when iron is normalized or annealed because these treatments decompose the carbides and coarsen the graphite flakes and nodules [144,149,159].

Graphite decreases matrix continuity, adds compliance, and thus lowers the modulus compared to pure iron. Matrix continuity is highest in steel because the only phases present are ferrite, pearlite, and a small volume fraction of inclusions. Steels have a density of about 7.85 g/cm^3 (0.284 lb/in.^{3}), a modulus of about 208 GPa (30 Mpsi), and an ultrasonic velocity of about 5850 m/sec (0.23 in./μsec). Cast irons have carbon concentrations generally in the range of about 3.3% to 3.6% and silicon concentrations in the range of 1.8% to about 2.6% by weight. The free carbon content generally ranges from about 2.6% to 3.0% by weight and from 8% to 12% by volume.

Free carbon having a density of about 2.3 g/cm^3 (0.083 lb/in.^{3}), mixed with the matrix having a density of about 7.7 g/cm^3 (0.278 lb/in.^{3}), produces a density from 7.0 to 7.3 g/cm^3 (0.253 to 0.264 lb/in.^{3}) depending on the

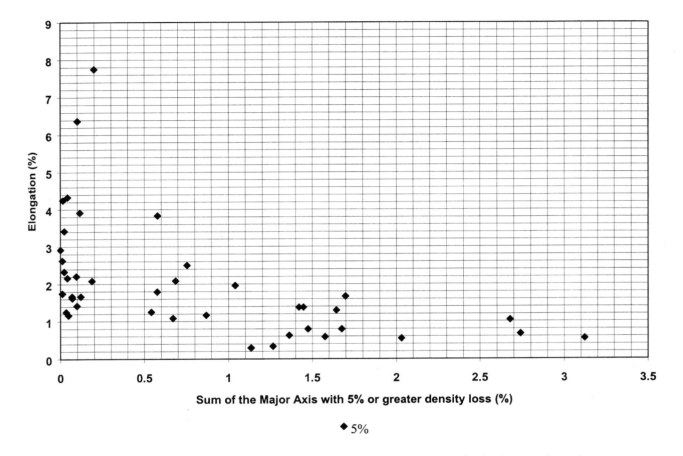

Figure 27 Effect of the sum of major axis lengths with 5% or greater density loss on elongation.

free carbon content. The ultrasonic velocity through cast iron ranges from about 4000 to 5500 m/sec (0.16 to 0.22 in./μsec) depending on the free carbon content and shape. Matrix continuity is highest in nodular iron, intermediate in compacted graphite (CG) iron and lowest in gray iron.

All graphite particles within a eutectic cell are interconnected and may serve as planes of weakness, so eutectic cells can be viewed as internal anomalies during the propagation of ultrasound [143,146]. Ultrasonic velocity increases in proportion to the fraction of ASTM Types D and E graphite present in gray iron, and decreases with the graphite surface area per unit volume [146,150].

Kovacs pointed out that modulus decreases when the crystal lattice is either strained or has more imperfections [145]. Irregular graphite shapes in ductile iron have sharper tips that serve as stress concentrators and affect the tensile strength and elongation. Irregular graphite

shapes promote localized plastic deformation and may lead to tensile fracture at relatively low strains, especially in pearlitic irons [142,144].

The degree of the graphite nodularity is an important factor affecting the properties of ductile iron. The nodularity can be estimated from measurements of the ultrasonic velocity when the matrix structure is relatively constant. Ultrasonic velocity is sensitive to the shape of all graphite present and not just the fraction that is nodular. Nodularity has been reported to have a smaller effect on ultrasonic attenuation than the matrix structure in ductile iron. A 1% decrease in nodularity was reported to produce a 0.5% increase in attenuation [146,150].

2. Matrix Effects on Ultrasonic Measurements

There is some confusion in the literature about the sensitivity of modulus to the pearlite and ferrite contents

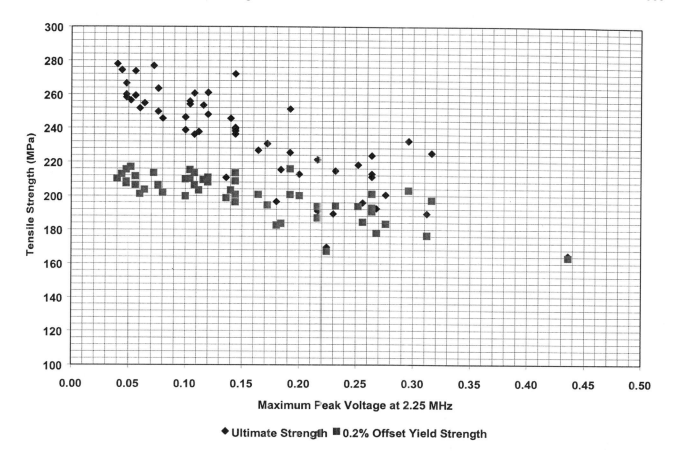

Figure 28 Effect of maximum peak voltage at 2.25 MHz on yield and ultimate strength.

in iron. Some investigators report very little difference in modulus between ferritic and pearlitic matrices, with ferrite producing slightly higher values of ultrasonic velocity and modulus. While ferrite itself may have a higher modulus than pearlite, heat treatments to convert pearlite to ferrite increase the graphite volume and this has the effect of decreasing the modulus. Even though ferrite has a higher modulus, the ultrasonic velocity may actually decrease because of the associated increase in graphite volume. Ultrasonic measurements have been reported to be incapable of assessing the matrix microstructure [143,144,146–149,151].

3. Microshrinkage Effects on Ultrasonic Signals

The results obtained when inspecting iron depends to a considerable extent on the technique for making the measurements. A ductile iron plate known to have a good nodularity from first-break ultrasonic velocity measurements was suspected to have microshrinkages due to the solidification characteristics of ductile iron and the section thickness. The surfaces of the plate were ground plane and parallel with a thickness of 15.62 mm (0.615 in.) and a surface finish of about 1.8 µm (70 µin.) RMS. Ultrasonic velocity and attenuation measurements were made using 5-MHz transducers to "map" the plate response.

First-break velocity is commonly used for commercial velocity or thickness gauges. There are only minor variations in FBV, with the values ranging from about 5560 to 5690 m/sec (0.219 to 0.224 in./µsec). No significant difference in FBV between the plate edge and center was found. No conclusions on the microshrinkage distribution could be drawn based on FBV, the technique normally used to estimate nodularity.

The 5-MHz SCPAR map is illustrated in Fig. 40, with the values ranging from about 10 to 80 db. High attenuation values are found at some locations that

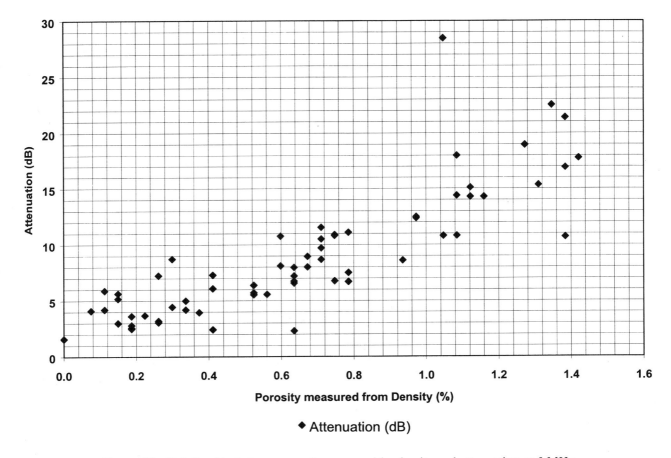

Figure 29 Relationship between porosity measured by density and attenuation at 5 MHz.

could not be explained by the anticipated microstructures. Specimens were removed from the plate for microstructural examination and bulk density measurements. Four specimens were removed from locations where the attenuation was 78, 62, 37, and 14 db. Microstructures showing microshrinkage are illustrated in Fig. 41A through D. More microshrinkage was found in locations where the attenuation was 78 and 62 db, as shown in Fig. 41A and B, respectively. Very little microshrinkage was found in where the attenuation was 37 db as shown in Fig. 41C. The materials were sound in locations close to the plate edge, as shown in Fig. 41D.

The bulk density is also an indication of microshrinkage. The bulk densities were 6.666 (0.241), 6.823 (0.246), 6.956 (0.251), and 7.097 g/cm³ (0.256 lb/in.³) for specimens whose ultrasonic attenuations are 78, 62, 37, and 14 db, respectively. Higher amounts of microshrinkage correspond to higher ultrasonic attenuation and lower bulk density.

The ultrasonic wavelength at 5 MHz was about 1.1 mm (0.044 in.) using an average ultrasonic velocity of about 5590 m/sec (0.22 in./μsec). The larger microshrink pores shown in Fig. 41A are about 1 mm in diameter, which is close to the ultrasonic wavelength. An appreciable ultrasonic scattering occurred at the microshrinkage sites that resulted in a significant amount of ultrasonic energy loss. Larger volumes of microshrinkage result in more energy loss and higher attenuation.

The SCPV velocities varied, with the values ranging from about 5160 to 5590 m/sec (0.203 to 0.220 in./μsec). SCPV velocities from locations A, B, C, and D designated in Fig. 40 on the ductile iron plate were much lower than the average, and these four locations all had higher attenuation values. SCPV is affected by ultrasonic attenuation. The higher the attenuation is, the lower the SCPV will be. The SCPV velocity is more sensitive to the microshrinkage than FBV. Conventional ultrasonic measurements made with a first-break technique will not

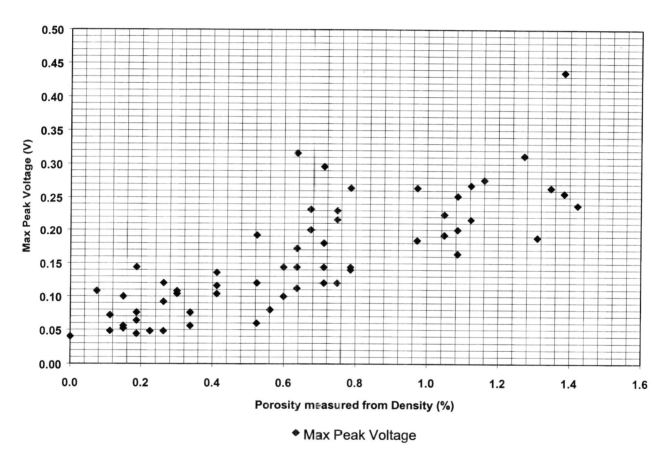

Figure 30 Relationship between porosity measured by density and maximum peak voltage at 2.25 MHz.

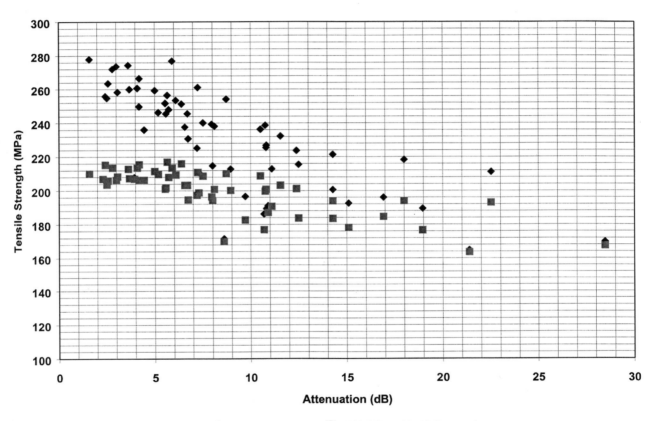

Figure 31 Effect of attenuation at 5 MHz on yield and ultimate strength.

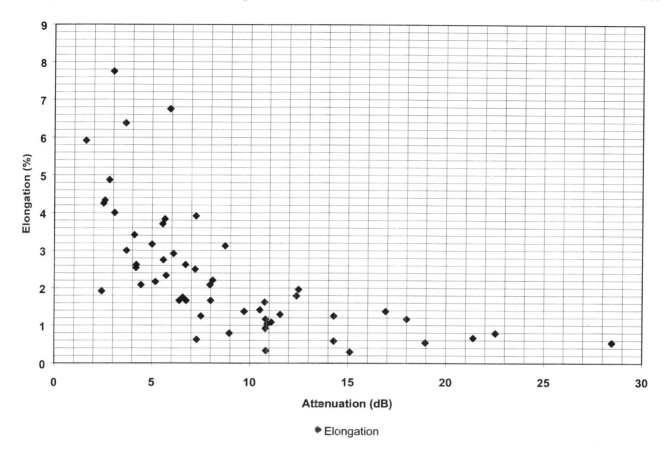

Figure 32 Effect of attenuation at 5 MHz on elongation.

Figure 33 Tensile fracture surface showing microshrinkage cluster (bright).

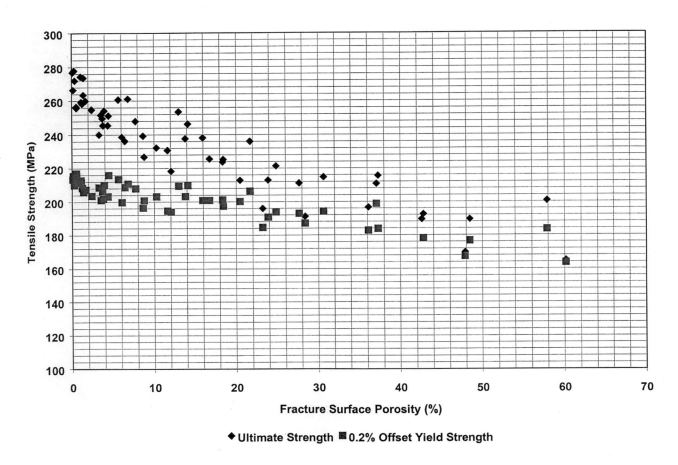

Figure 34 Effect of fracture surface porosity on yield and ultimate tensile strength in A356-T6 cast aluminum.

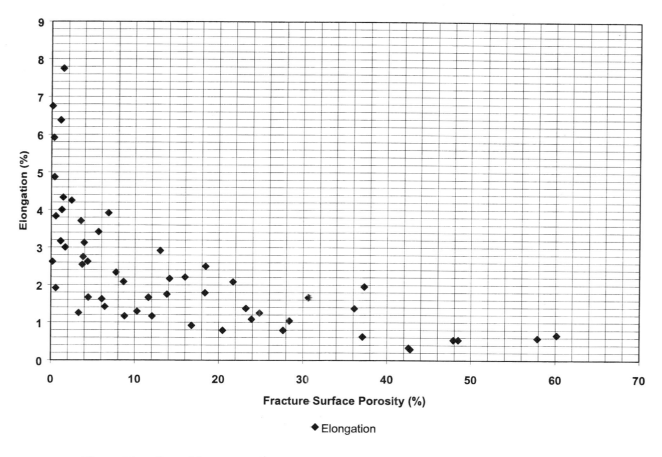

Figure 35 Effect of fracture surface porosity on tensile elongation in A356-T6 cast aluminum.

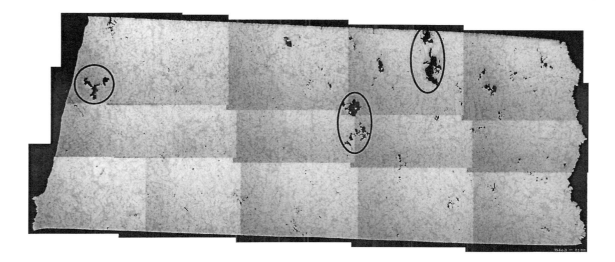

Figure 36 A typical metallographic section cut longitudinally through a tensile sample.

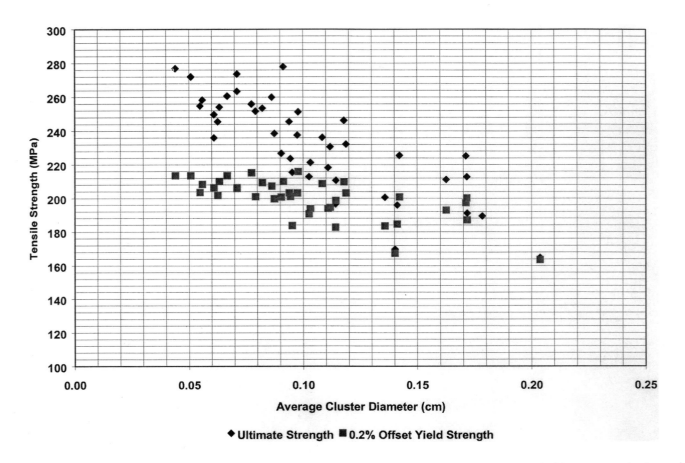

Figure 37 Effect of the average cluster diameter measured on a polished cross-section on the yield and ultimate strength of cast A356-T6 aluminum.

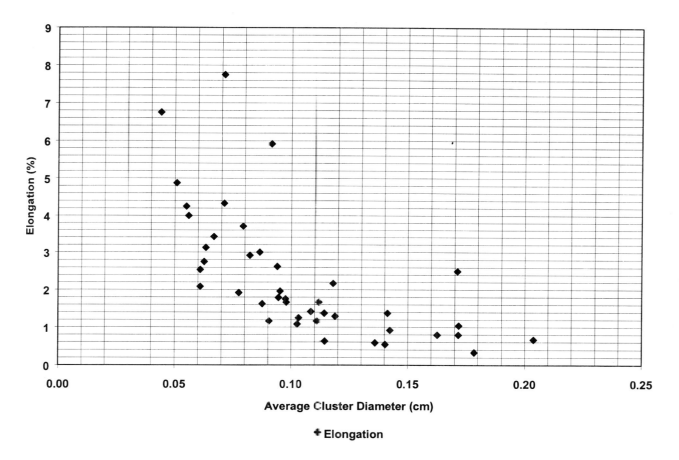

Figure 38 Effect of the average cluster diameter measured on a polished cross-section on the tensile elongation of cast A356-T6 aluminum.

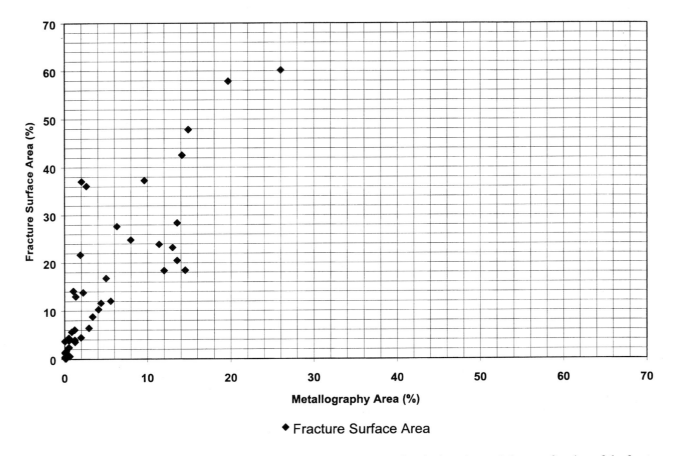

◆ Fracture Surface Area

Figure 39 Relationship between the area fraction of cluster from the metallurgical section and the area fraction of the fracture surface porosity.

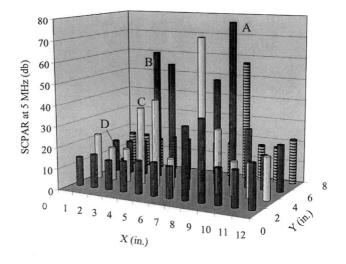

Figure 40 Ultrasonic attenuation map at 5 MHz from ductile iron plate. X and Y designate the location on the plate. The plate had an average of 70 µin. RMS surface finish on both sides. Specimens were removed from where designated as A, B, C, and D.

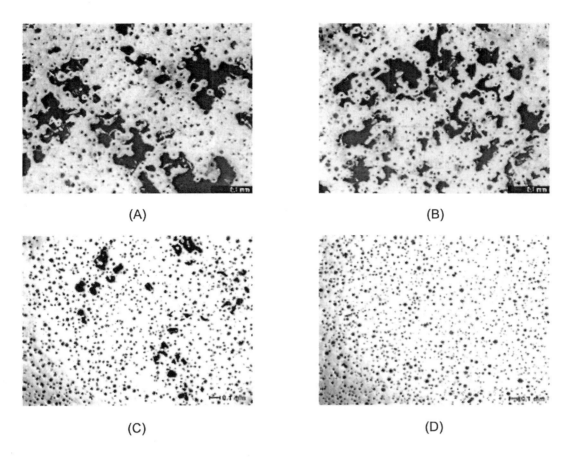

Figure 41 Shrinkage in ductile iron plate. (A) Specimen from location A in Fig. 3.28 having an SCPAR of 78 db. (B) Specimen from location B having an SCPAR of 62 db. (C) Specimen from location C having an SCPAR of 37 db. (D) Specimen from location D having an SCPAR of 14 db.

detect microshrink. The SCPV velocity and attenuation will, with attenuation being the best.

REFERENCES

1. Cleaning Castings. American Foundry Society: Des Plaines, IL, 1977 reprinted 1992.
2. Kirgin, K.H. AFS 2002 metalcasting forecast and trends. Mod. Cast. January 2002, *92* (1), 24–28.
3. *ASM Metals Handbook: Casting,* 9th Ed.; Stefanescu, D.M., Eds.; ASM International: Metals Park, OH, 1988; Vol. 15.
4. *SFSA Steel Casting Handbook*, 6th Ed.; Blair, M., Stevens, T.L., Eds.; Steel Founders' Society of America and ASM International: Metals Park, Ohio, 1995.
5. *Smithells Metal Reference Book*, 7th Ed.; Brandes, E.A., Brook, B.B., Eds.; Butterworth Heinemann: Oxford, England, 1998.
6. *Metals Handbook: Desk Edition*, 2nd Ed.; Davis, J.R., Ed.; ASM International: Metals Park, OH, 1998.
7. Monroe, R.W. *Expendable Pattern Casting*; American Foundry Society: Des Plaines, IL, 1992.
8. Sanders, C.A. *Foundry Sand Practice*, 6th Ed.; American Colloid Co.: Skokie, IL, 1973.
9. Guide to Sand Additives. American Foundrymen's Society: Des Plaines, IL, 1976.
10. The Application of Additives to Clay-Bonded Sand Systems. American Foundry Society: Des Plaines IL, 1980.
11. Green Sand Additives: Properties, Applications, and the Effects of Misuse. American Foundry Society: Des Plaines, IL, 1987.
12. Scott, W.D.; Feazel, C.E. A Review of Organic Sand Binder Chemistry. *Modern Casting*; American Foundry Society: Des Plaines, IL, June 1976; 25 pp. (Technical Report No. 765).
13. Carey, P., Ed.; Sand Binder Systems, Foundry M&T series (13 parts), Ashland Bulletin 2292, 1996.

14. Dietert, H.W. *Foundry Core Practice*; American Foundry Society: Des Plaines, IL, 1966.

15. Wile, L.E. Progress in coremaking: shell. Foundry March 1970, *98* (3), 50–57.

16. *Shell Process Foundry Practice*, 2nd Ed.; American Foundry Society: Des Plaines, IL, 1973.

17. Briggs, C.W. *Fundamentals of Core Sands and Binders*; Steel Founders Society of America: Cleveland, OH, 1961.

18. Vingas, G.J. Mold and core coating: Past, Present, and Future. AFS Trans. 1986, *96*, 463–475.

19. Otte, J.A. Jr.; Scott, W.D. No-bake molding—the practical side. *Transactions of the Australian Foundry Society*; Australian Foundry Society: Adelaide, Australia, October 19, 1998; 42–51.

20. Bates, C.E.; James, R.; Scott, W.D. *The Atmosphere at the Mold–Metal Interface*; AFS Sand Division Silver Anniversary Paper, Pittsburgh, May 2000. Available from the AFS as a Special Report.

21. Bates, C.E.; Scott, W.D. Decomposition of resin binders and the relationship between the gases formed and the casting surface quality. AFS Trans. 1975, *83*, 519–524.

22. Bates, C.E.; Scott, W.D. Decomposition of resin binders and the relationship between the gases formed and the casting surface quality: Part II. Gray iron. AFS Trans. 1976, *84*, 793–804.

23. Bates, C.E.; Scott, W.D. Decomposition of resin binders and the relationship between the gases formed and the casting surface quality: Part III. AFS Trans. 1977, *85*, 209–226.

24. Bates, C.E.; Monroe, R.W. Mold decomposition and its relation to gas defects in castings. AFS Trans. 1981, *89*, 671–686.

25. Wilkes, G.F.; Wright, R.L.; Hart, D.R.; Boswell, S.M. TSA–another catalyst for furan no-bakes. Foundry Mar. 1972, *100* (3), 81–84.

26. Nelson, B. An evaluation of toluenesulfonic acid as catalysts for furan no-bake foundry binders. AFS Trans. 1973, *81*, 153–157.

27. Delaney, J. Design and performance of a no-bake foundry. Steel Foundry Facts, Feb. 1975, *312*, 3–9.

28. *Mold & Core Test Handbook*, 2nd Ed.; The American Foundry Society: Des Plaines, IL, 1989.

29. Otte, J.A. Jr., Ed. Phenolic urethane chemistry. *Cold-Box Training Manual*; HA International: Westmont, IL, 2001.

30. Robins, J.; Toriello, L.J.; Hamilton, J. A new phenolic–urethane no-bake binder. Transactions of the American Foundry Society 1971, *79*, 105–114.

31. Scott, W.D.; Vingas, G.J. Mold wash quality control. AFS 1996, *104*, 551–556. Paper 96-48.

32. Scott, W.D. Sand reclamation: matching the systems to the processes—case studies in steel foundries. Presented by invitation at the AFS International Sand Reclamation Conference, Novi, MI, March 16, 1991. Published in the Conference Proceedings, AFS, Des Plaines, IL.

33. Penko, T. Measurement of emissions associated with application of flammable solvent-based core and mold coating. AFS Trans. 1998, *106*, 173–179. Paper 98-98.

34. Scott, W.D.; Goodman, P.A.; Monroe, R.W. Binder related casting defects. Address to annual meeting of the American Chemical Society—Symposium on New Developments in Foundry Binders, Miami, Fl., Sept 13, 1978. Published in the Conference Proceedings.

35. Scott, W.D.; Goodman, P.A.; Monroe, R.W. Gas generation at the mold-metal interface. AFS Trans. 1978, *86*, 99–609.

36. Shroyer, H.F. Inventor. Cavityless Casting Mold and Method of Making Same. US Patent 2,830,343, April 15, 1958; 10 pp.

37. Smith, T.R. Inventor. The Maytag Company, assignee. Method of Casting. US Patent 3,157,924, 1964; 4 pp.

38. Bailey, Robin. Understanding the evaporative pattern casting process. Mod. Cast. 1982; 58–61.

39. Wittmoser, Adalbert. Full mold casting. AFS Trans. 1964; 292–303.

40. Dieter, H.B.; Paoli, A.J. Sand without binder for making full mold castings. AFS Trans. 1967; 147–160.

41. Thomas, Jacques. Improvement to the lost foam process: casting–a new patented process developed by aluminum pechiney. 4th Evaporative Pattern Casting Conf., Rosemont, IL, June 6–7, 1989; 239–259. American Foundry Society, Des Plaines, IL.

42. Goria, C.A.; Remondino, M.; Seli, M. Developing equipment for policast lost foam moulding. Foundry Trade J. September 12, 1985; 188–196.

43. Ashton, M.C.; Sharman, S.G.; Brookes, A.J. The replicast FM (full mold) and CS (ceramic shell) process. AFS Trans. 1984; 271–280.

44. Grote, R.E. Replicast ceramic shell molding: the process and its capabilities. AFS Trans. 1986; 181–186.

45. Cannarasa, M.J.; Kesling, H.S., Jr.; Sun, H.N. Inventors. ARCO Chemical Co., assignee. Process for Preparing Polycarbonate Terpolymer Foam Suitable for Lost Foam Casting. US Patent 4,773,466, September 27, 1988; 5 pp.

46. ARCO Chemical Company. *The Evaporative Casting Process Using Expandable Polystyrene Patterns and Unbonded Sand Casting Techniques*. Newtown Square, PA. September 24, 1982; 8 pp.

47. Harsley, R. Tooling Requirements for the Evaporative Pattern Casting Process. AFS Trans. 1988, 787–792.

48. Weiner, S.A.; Piercecchi, C.D. Dimensional behavior of polystyrene foam shapes. AFS Trans. 1985, *93*, 155–162.

49. Clegg, A.J. Expanded-polystyrene moulding—a status report. Foundry Trade J. September 2, 1985; 177–187.

50. Goria, C.A.; Del Gaudio, G.; Caironi, Castek G. Molding of iron castings with evaporative polystyrene foam patterns. Casting World, (Winter 1986); 41–51.

51. Moll, N. EPC demands on pattern material. Private communication with Norman Moll, Dow Chemical Co., June, 1989.

52. Burchell, V.H. Production systems for lost foam castings. AFS Trans. 1984; 629–636.

53. McMellon, B.A. Design and operation of production lost foam casting systems. AFS Trans. 1988; 27–36.

54. Moll, N.; Johnson, D. Recent developments in the use of PMMA for evaporative foam casting. *Evaporative Foam Casting Technology III*; American Foundry Society: Rosemont, IL, June 7–8, 1988; 121–132.

55. Molibog, T.; Littleton, Harry. Experimental simulation of pattern degradation in lost foam. AFS Trans. 2001; 1–32.

56. Molibog, T.; Dinwiddie, R.B.; Porter, R.; Wang, W.D.; Littleton, H.E. Thermal properties of lost foam casting coatings. AFS Trans. 2000; 471–478.

57. Sun, W.L.; Littleton, H.E.; Bates, C.E. Real-time x-ray investigations on lost foam mold filling. AFS Trans. 2002; 1347–1356.

58. Warner, M.H.; Miller, B.A.; Littleton, H.E. Pattern pyrolysis defect reduction in lost foam castings. AFS Trans. 1998; 777–785.

59. Bennett, S.; Tschopp, M.; Vrieze, A.; Zelkovich, A.; Ramsay, C.; Askeland, D. Observations on the effect of gating design on metal flow and defect formation in aluminum lost foam castings: Part 1. AFS Trans. 2001; 1–15.

60. Tschopp, M.A., Jr.; Ramsay, C.; Askeland, D. Mechanisms of formation of pyrolysis defects in aluminum lost foam castings. AFS Trans. 2000; 609–614.

61. Buesch, A.; Carney, C.; Moody, T.; Wang, C.; Ramsay, C.W.; Askeland, D. Influence of sand temperature on formation of pyrolysis defects in al lost foam castings. AFS Trans. 2000; 7615–7621.

62. Hess, D.R.; Durham, B.; Ramsay, B.; Askeland, D. Observations on the effect of gating design on metal flow and defect formation in aluminum lost foam castings: Part II. AFS Trans. 2001; 1–16.

63. Liu, J.; Ramsay, C.W.; Askeland, D.R. Effects of foam density and density gradients on metal fill in the LFC process. AFS Trans. 1998; 435–442.

64. Getner, E.M. The applicability of polystyrene to the making of patterns. AFS Trans. 1965; 394–396.

65. Patz; Murray. Unique casting applications with lost foam. 4th Evaporative Pattern Casting Conference, June 6–7. American Foundry Society: Rosemont IL, 1989; 305–318.

66. Rodgers, Robert Automating lost foam casting. Foundry Manage. Technol., April 1988; 1–5.

67. Hubler, D. Evaporative pattern casting. The Crucible, March/April 1987; 2–5.

68. Hubler, D. Evaporative pattern casting. The Crucible, May/June 1987; 8–11.

69. Bates, Charles. State of the art and five year goals. *Evaporative Foam Casting Technology III*; American Foundry Society: Des Plaines, IL, June 7–8, 1988; 97–118 pp.

70. Troxler, John. The prospects for lost foam in 1995. AFS Trans. 1990, 371–378.

71. Toxler, J. Design advantages of the evaporative pattern casting process. Private communication with John Troxler, June 1989.

72. Littleton, H.E.; Vatankhah, B. Dimensional control parameters in lost foam casting. AFS Trans. 2001, 1–14.

73. Vatankhah, B.; Sheldon, D.; Littleton, H. Optimization of vibratory sand compaction. AFS Trans. 1993; 787–796.

74. Littleton, H.; Miller, B.; Sheldon, D.; Bates, C. Process control for precision lost foam casting. Part II. Foundry Manage. Technol. Feb 1997, *125* (2); 41, 43–45.

75. Littleton, H.; Miller, B.; Sheldon, D.; Bates, C. Process control for precision lost foam casting. Part I. Foundry Manage. Technol. Dec. 1996, *124* (12), 37–40.

76. Littleton, H.; Miller, B.; Sheldon, D.; Bates, C. Process control for precision lost foam casting. Part III. Foundry Manage. Technol. Mar. 1997, *125* (3), 34–37.

77. Littleton, H.; Miller, B.; Sheldon, D.; Bates, C. Lost foam casting—process control for precision. AFS Trans. 1996; 335–346.

78. Littleton, H.E.; Bates, C. Fill and Compaction Control. *Proceedings of Expendable Pattern Casting: Managing the Technology, Birmingham, AL, 7-1 to 7-29*; American Foundry Society: Des Plaines, IL, Sept 1993.

79. Bates, C.; Griffin, J.; Littleton, H. Accuracy and precision of iron and aluminum castings made by EPC nobake and green sand methods. AFS Trans. 1992; 323–334.

80. Trivedi, R., Kurz, W. Solidification of Single-Phase Alloys. *ASM Metals Handbook*, 9th Ed.; American Society for Metals: Metals Park, Ohio, 1988; 114–119.

81. Batty, George. Controlled directional solidification. Transactions of the American Foundrymen's Association 1934, *42*, 237–258.

82. Batty, G. The influence of temperature gradients in the production of steel castings. Transactions of the American Foundrymen's Association 1935, *43*, 75–102.

83. Ruddle, R.W. The solidification of castings, a review of the literature. Inst. Met. 1950; 10 pp.

84. Plutshack, L.A.; Suschil, A.L. Riser Design. *ASM Metals Handbook: Casting*, 9th Ed.; American Society for Metals: Metals Park, Ohio, 1988; *15*, 577–588.

85. Granger, D.A.; Elliott, R. Solidification of Eutectic Alloys. *ASM Metals Handbook: Casting*, 9th Ed.; American Society for Metals: Metals Park, Ohio, 1988; Vol. 15, 159–168.

86. Blackmun, E.V. Aluminum casting alloys-effect of composition and foundry practices on properties and quality. AFS Trans. 1971, *79*, 63–68.

87. Pan, E.N.; Lin, C.S.; Loper, C.R. Effects of solidification parameters on the feeding efficiency of A356 aluminum alloy. AFS Trans. 1990, *98*, 735–745.

88. Argo, D.; Gruzleski, J.E. Porosity in modified aluminum alloy castings. AFS Trans. 1988, *96*, 65–74.

89. Fang, Q.T.; Granger, D.A. Porosity formation in modified and unmodified A356 alloy castings. AFS Trans. 1989, 97, 989–1000.

90. Closset, B.; Gruzleski, J.E. Study on the Use of Pure Metallic Strontium in the Modification of Al-Si Alloys. AFS Trans. 1982, 801–808.

91. Gruzleski, J.E.; Handiak, N.; Campbell, H.; Closset, B. Hydrogen Measurement By Telegas in Strontium Treated A356 Melts. AFS Trans. 1986, 94, 147–154.

92. Shahani, H. Effect of hydrogen on the shrinkage porosity of aluminum copper and aluminum silicon alloys. Scand. J. Metal. 1985, 93, 14–18.

93. Neff, D.V. Nonferrous Molten Metal Processes. *ASM Metals Handbook: Casting*, 9th Ed.; American Society for Metals: Metals Park, Ohio, 1988; Vol. 15, 445–496.

94. Shivkumar, S.; Apelian, D.; Zou, J. Modeling of microstructure evolution and microporosity formation in cast aluminum alloys. AFS Trans. 1990, 98, 897–904.

95. Mohanty, P.S.; Samuels, F.H.; Gruzleski, J.E. Experimental study on pore nucleation by inclusions in aluminum castings. AFS Trans. 1995, 103, 555–564.

96. Laslaz, G.; Laty, P. Gas porosity and metal cleanliness in aluminum casting alloys. AFS Trans. 1991, 99, 83–90.

97. Rooy, E.L. Mechanisms of porosity formation in aluminum. Mod. Cast. Sept. 1992, 100, 26–34.

98. Fang, Q.T.; Anyalebechi, P.N.; O'Malley, R.J.; Granger, D.A. Effect of solidification conditions on hydrogen porosity formation in unidirectionally solidified aluminum alloys. Inst. Met. 1987, 95, 33–36.

99. Semersky, L.P. Detecting hydrogen gas in aluminum. Mod. Cast. August 1993, 91, 38–39.

100. Rosenthal, H.; Lipson, S. Measurement of gas in molten aluminum. AFS Trans. 1955, 63, 301–305.

101. Parmenter, L.; Apelian, D.; Jensen, F. Development of a statistically optimized test for the reduced pressure test. AFS Trans. 1998, 105, 439–452.

102. Neil, D.J.; Burr, A.C. Initial bubble test for determination of hydrogen content in molten aluminum. AFS Trans. 1961, 69, 272–275.

103. Chamberlain, B.; Sulzer, J. Gas content and solidification rate effect on tensile properties and soundness of aluminum casting alloys. AFS Trans. 1964, 7, 600–607.

104. Rooy, E.L.; Fischer, E.F. Control of aluminum casting quality by vacuum solidification tests. AFS Trans. 1968, 76, 237–240.

105. Church, J.C.; Herrick, K.L. Quantitative gas testing for production control of aluminum casting soundness. AFS Trans. 1970, 78, 277–280.

106. Rasmussen, W.M.; Eckert, C.C. RPT gauges aluminum porosity. Mod. Cast. March 1992, 100, 204–206.

107. Stahl, G.W. Twenty-five years tilt pouring aluminum. AFS Trans. 1986, 94, 793–796.

108. Sigworth, G.K. A scientific basis for degassing aluminum. AFS Trans. 1987, 95, 73–78.

109. King, Stephen; Reynolds, John. Flux injection/rotary degassing process provides cleaner aluminum. Mod. Cast., April 1995; 37–40.

110. Yamada, H.; Kitumura, T.; Iwao, O. Degassing media for molten aluminum: degassing capacity of Freon 12 (CC122F2) and a nitrogen–Freon mixture. AFS Cast Metals Res. J. March 1970, 78, 11–1404.

111. Rooy, E.L. Aluminum and Aluminum Alloys. *ASM Metals Handbook: Casting*, 9th Ed.; American Society for Metals, 1988; Vol. 15, 743–770.

112. Samuels, A.M.; Samuels, F.H. Various aspects involved in the production of low-hydrogen aluminum castings. J. Mater. Sci. 1992, 27, 6533–6563.

113. Tynelius, K.; Major, J.F.; Apelian, D. A parametric study of micro-porosity in the A356 casting alloy system. AFS Trans. 1993, 101, 401–413.

114. Shivkumar, S.; Wang, L.; Apelian, D. Lost foam casting of aluminum alloy components. J. Met. 1990, 98, 38–44.

115. Flemings, M.C.; Norton, P.J.; Taylor, H.F. Performance of chills on high strength-high ductility sand-mold castings of various section thicknesses. AFS Trans. 1957, 65, 259–266.

116. Flemings, Merton; Norton, Patrick J.; Taylor Howard, F. Rigging design of a typical high strength, high ductility aluminum casting. AFS Trans. 1957, 65, 550–555.

117. Flemings, M.C.; Uram, S.Z.; Taylor, H.F. Solidification of aluminum castings. AFS Trans. 1960, 68, 670–684.

118. Flemings, M.C.; Kattamis, T.Z.; Bardes, B.P. Dendrite arm spacing in aluminum alloys. AFS Trans. 1991, 99, 501–506.

119. Qingchun, L.; Yuyong, C.; Zhuling, J. Relationship between solidification thermal parameters, dendrite arm spacing, and ultimate tensile strength in Al–4.5% Cu. J. Less-Common Met. August, 1985, 110, 1–2.

120. Boileau, J.M.; Zindel, J.W.; Allison, J.E. The effect of solidification time on the mechanical properties in a cast A356-T6 aluminum alloy. Applications for Aluminum in Vehicle Design, Proceedings of the 1997 International Congress and Exposition, February 1997. SAE Special Publications: Warrendale, PA, January, 1991; 1251, 61–72.

121. Shivkumar, S.; Wang, L.; Apelian, D. Molten metal processing of advanced cast aluminum alloys. J. Metals, Metals Mater. Soc. January, 1991, 43 (1), 26–32.

122. Honma, U.; Kitaoka, S. Fatigue strength and mechanical properties of aluminum alloy castings of different structural fineness. Aluminum. English Ed. Dec 1984, 60 (12), 780–783.

123. Kuo, Y.-S.; Chang, E.; Lin, Y.-L. The feeding effect of risers on the mechanical properties of A201 Al alloy plate casting. AFS Trans. 1989, 97, 777–782.

124. Doglione, R.; Douziech, J.L.; Berdin, C.; Francois, D. Micro-structure and damage mechanisms in A356-T6 alloy. Mater. Sci. Forum 1996, 217–222, 3.

125. Tewari, A.; Dighe, M.; Gokhale, M.A. Quantitative

characterization of spatial arrangement of microporès in cast microstructures. Mater. Charact. February 1998, *40* (2), 119–132.

126. Eady, J.A.; Smith, D.M. The effect of porosity on tensile properties of aluminum alloy castings. Mater. Forum 1986, *9* (4), 217–223.

127. Samuel, A.M.; Samuel, F.H. Porosity factor in quality aluminum castings. AFS Trans. 1992, *100*, 657–665.

128. Samuel, F.H.; Samuel, A.M. Effect of melt treatment, solidification conditions, and porosity level on the tensile properties of 319.2 endchill aluminum castings. J. Mater. Sci. October 1995, *30* (19), 4823–4833.

129. McLellan, D.L.; Tuttle, M.M. Aluminum castings—a technical approach. AFS Trans. 1983, *91*, 243–252.

130. Pisarenko, G.S. *High Temperature Strength of Materials*; Published for the National Aeronautics and Space Administration. Jerusalem, Israel: Israel Program for Scientific Translations, 1966.

131. Griffin, J. Unpublished research, University of Alabama at Birmingham, 2002.

132. Griffin, R.D.; Scarber, P.; Janowski, G.M.; Bates, C.E. Quantitative characterization of graphite in gray iron. AFS Trans. 1996, *104*, 977–983.

133. Adewara, J.O.T.; Loper, C.R. Effect of carbides on crack initiation and propagation in ductile iron. AFS Trans. 1976, *84*, 507–512.

134. Adewara, J.O.T.; Loper, C.R. Effect of pearlite on crack initiation and propagation in ductile iron. AFS Trans. 1976, *84*, 513–526.

135. Adewara, J.O.T.; Loper, C.R. Crack initiation and propagation in fully ferritic ductile iron. AFS Trans. 1976, *84*, 527–534.

136. Venugopalan, D.; Alagarsamy, A. Effects of alloy additions on the microstructure and mechanical properties of commercial ductile iron. AFS Trans. 1990, *98*, 395–400.

137. Bates, C.E.; Tucker, J.R.; Starrett, H.S. *Composition, Section Size and Microstructural Effects on the Tensile Properties of Pearlitic Gray Cast Iron*; AFS Research Report, September 1991.

138. Bryant, L.E.; McIntire, P. *Non-Destructive Testing Handbook*, 2nd Ed. Radiography and Radiation Testing. American Society for Non-Destructive Testing: Columbus, Ohio, 1985; Vol. 3.

139. Krautkrämer, J.; Krautkrämer, H. *Ultrasonic Testing of Materials*, 3d Ed.; Springer-Verlag: Berlin. 1983; 135 pp.

140. Bowman, H.A.; Schoonover, R.M.; Jones, M.W. Procedure for high precision density determinations by hydrostatic weighing. J. Res. Natl. Bur. Stand. C Eng. Instrum. Jul–Aug 1967, *71* (3), 179–198.

141. Boyer, H.E.; Carnes, W.J. *Non-Destructive Inspection and Control*; 8th Ed. American Society for Metals: Materials Park, Ohio, 1976; Vol. 11.

142. Emerson, P.J.; Simmons, W. Final report on the evaluation of the graphite form in ferritic ductile iron by ultrasonic and sonic testing, and of the effect of graphite form on mechanical properties. AFS Trans. 1976, *84* (26), 109–128.

143. Fuller, A.G. Nondestructive assessment of the properties of ductile iron castings. AFS Trans. 1980, *88* (162), 751–768.

144. Lerner, Y.S.; Vorobiev, A.P. Nondestructive evaluation of structure and properties of ductile iron. AFS Trans. 1998, *106* (12), 47–51.

145. Kovacs, B.V. Prediction of strength properties in ADI through acoustical measurements. AFS Trans. 1993, *101* (80), 37–42.

146. Patterson, B.R.; Bates, C.E. Nondestructive property prediction in gray cast iron using ultrasonic techniques. AFS Trans. 1981, *89* (65), 369–378.

147. Papakakis, E.P.; Bartosiewicz, L.; Altstetter, J.D.; Chapman, G.B., II. Morphological severity factor for graphite shape in cast iron and its relation to ultrasonic velocity and tensile properties. AFS Trans. 1983, *91* (102), 721–728.

148. Lerner, Y.S. Evaluation of structure and properties of gray iron PM castings using NDT technique. AFS Trans. 1995, *103* (50), 151–155.

149. Kovacs, B.V.; Cole, G.S. On the interaction of acoustic waves with SG iron castings. AFS Trans. 1975, *83* (76), 497–502.

150. Kovacs, B.V. Quality control and assurance by sonic resonance in ductile iron castings. AFS Trans. 1977, *85* (76), 499–508.

151. Fuller, A.G. Evaluation of the graphite form in pearlitic ductile iron by ultrasonic and sonic testing and the effect of graphite form on mechanical properties. AFS Trans. 1977, *85* (102), 509–526.

152. Vary, A.; Kautz, H.E. Transfer function concept for ultrasonic characterization of material microstructures. In *Analytical Ultrasonics in Materials Research and Testing*, Proceedings of a conference held at the NASA Lewis Research Center, Cleveland, Ohio, Nov. 13–14, 1984; NASA: Cleveland, Ohio, 257–297.

153. Rosen, M. Analytical ultrasonics for characterization of metallurgical microstructures and transformations. In *Analytical Ultrasonics in Materials Research and Testing*, Proceedings of a conference held at the NASA Lewis Research Center, Cleveland, Ohio, Nov. 13–14, 1984, NASA: Cleveland, Ohio, 83–102.

154. Generazio, E.R. Ultrasonic verification of microstructural changes due to heat treatment. In *Analytical Ultrasonics in Materials Research and Testing*, Proceedings of a conference held at the NASA Lewis Research Center, Cleveland, Ohio, Nov. 13–14, 1984; NASA: Cleveland, Ohio, 207–217.

155. Li, H. *A Study on Characterization of Cast Iron Microstructure and Mechanical Properties Through Ultrasonic Nondestructive Evaluation*; Ph.D. dissertation, The University of Alabama at Birmingham, 2002.

156. Ivan, S. *Non-destructive Measures of Porosity in Cast*

A356 Aluminum and Correlations between Porosity and Tensile Properties. Master Thesis, The University of Alabama at Birmingham, 2001.

157. Kasumzade, F.; Ivan, S.; Griffin, J.; Bates, C. Effect of grain refinement, eutectic modification, and pouring temperature on soundness of A356 cast aluminum. Proceedings, 6th International AFS Conference on Molten Aluminum Processing. Orlando, FL. Nov. 11–13, 2001; American Foundry Society: Des Plaines IL, 293–302.

158. Schmerr, L.W., Jr. *Fundamentals of Ultrasonic Nondestructive Evaluation, A Modeling Approach*; Plenum Press: New York, 1998; 283–284.

159. Morooka, T.; Sugiyama, Y.; Ito, S. Effects of microstructure and residual stress on the sonic properties of gray cast iron. AFS Trans. 1969, *77*, 323–328.

11

Die Casting Process Design

Frank E. Goodwin
International Lead Zinc Research Organization, Inc., Research Triangle Park, North Carolina, U.S.A.

I. THE DIE CASTING PROCESS

Pressure die casting is a process in which molten metal is forced under high pressure into a cavity in a metal die in a few milliseconds, after which it rapidly solidifies. The die is then opened and the casting ejected from the die. This high-speed process allows complex shapes to be cast to net shape, or nearly net shape, requiring minimal finishing and machining. Thousands, and in some cases millions, of castings can be economically produced from a single die set.

A. Basic Die Casting Processes: Hot Chamber and Cold Chamber

There are two basic die casting processes: hot chamber and cold chamber. The hot chamber process is used for lower melting point metals such as zinc and lead. The cold chamber process is used for higher melting point metals such as aluminum, magnesium, and brass. However, some magnesium alloys can be hot chamber die cast, while zinc alloys with higher aluminum content are cold chamber die cast.

1. Hot Chamber Process

The basic components of a hot chamber die casting machine and die are illustrated in Fig. 1. In this process, the plunger and the cylinder are submerged in the molten metal in the holding furnace. The power to pump zinc into the die cavity is provided by a hydraulic or air powered accumulator, with hydraulic being the

most common. The hydraulic oil is supplied to the accumulator by a pump at a rate that will bring the accumulator pressure up to the desired operating level each time a casting (shot) is made.

The casting sequence in the hot chamber process is illustrated in Fig. 2. When a shot is made, the control valve opens, causing the shot cylinder to force the plunger down and molten metal through the nozzle, past the sprue pin, through the runners and gates, and into the die cavity. This molten metal displaces the gas contained in the runner system and the die cavity and this gas, together with some of the molten metal, flows through the die cavity and out through the vents, or into overflows. After the cavity is filled, the metal is allowed to solidify, the casting ejected, and the cycle is repeated. Because the gooseneck and plunger are submerged in the motel metal, the system automatically refills when the plunger is withdrawn. The length of the casting sequence (cycle time) is dictated by the ability of the die set to extract heat from the solidifying metal, and ranges from 1 sec for miniature castings to several minutes for larger, thick castings.

2. Cold Chamber Process

The casting sequence for the cold chamber process is illustrated in Fig. 3. In this process, molten metal is ladled into the cold chamber, either manually or automatically, after which the plunger advances to force the metal into the die cavity. Except for the manner in which molten metal is fed into the shot system, the casting sequences for the two processes are similar. The cold

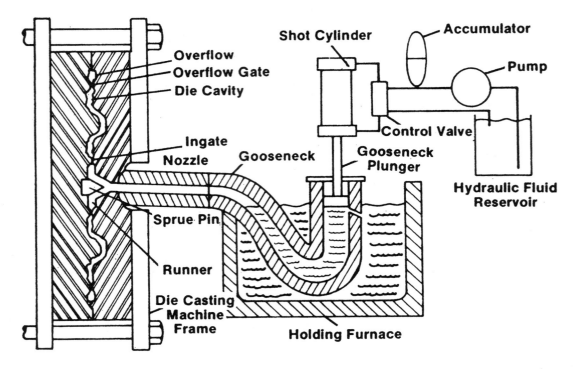

Figure 1   Basic components of a typical hot-chamber, die-casting system.

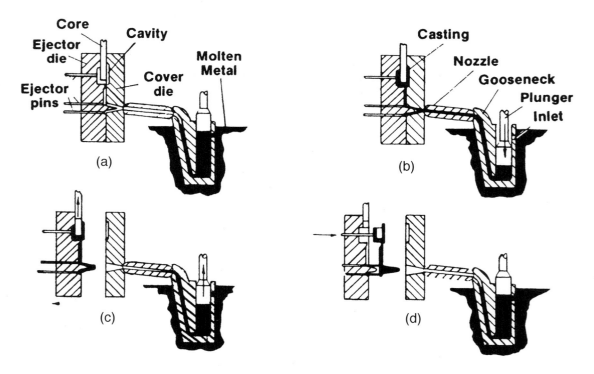

Figure 2   Operating sequence for the hot-chamber, die-casting process. (a) Die is closed and hot chamber (i.e., gooseneck) is filled with molten metal. (b) Plunger pushes molten metal through gooseneck and nozzle and into the die cavity. Metal is held under pressure until it solidifies. (c) Die opens and cores, if any, retract. Casting stays in ejector die. Plunger returns, pulling molten metal back through nozzle and gooseneck. (d) Ejector pins push casting out of ejector die. As plunger uncovers filling hole, molten metal flows through inlet to refill gooseneck.

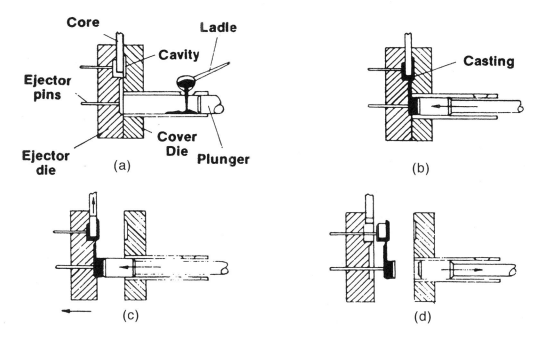

Figure 3 Operating sequence for the cold-chamber, die-casting process. (a) Die is closed and molten metal is ladled into the cold chamber. (b) Plunger pushes molten metal into die cavity. The metal is held under pressure until it solidifies. (c) Die opens and plunger advances to ensure casting stays in ejector die. Cores, if any, retract. (d) Ejector pins push casting out of ejector die and plunger returns to original position.

chamber process is used for casting metals that would attack injection system hardware, necessitating minimal contact between them. Cycle times are similar to the hot chamber process.

3. Die Casting Dies

Die casting dies consist of two parts, the cover (stationary) half and the ejector (moving) half, which meet at the parting line. The basic components of the hot chamber die casting die are shown in Figs. 1 and 2. The cover half of the die is secured to the front or stationary platen of the machine. The sprue for filling the die cavity is in this half and it is aligned with the nozzle of the hot chamber machine. The ejector half of the die is attached to the movable platen of the machine. It contains the ejection mechanism and in most cases the runners, as shown in Fig. 2.

The die cavity, which forms the part being cast, is machined into both halves of the die block or into inserts that are installed in the die blocks. The die is designed so that the casting remains in the ejector half when the die opens. The casting is pushed out of the die cavity with ejector pins that come through holes in the moving die half and are actuated by the ejector plate, which is powered by the machine. Guide pins extending from one die half enter holes in the other die half as

the die closes to ensure alignment between the two halves.

Dies that produce castings with complex shapes may contain stationary and movable cores. The movable cores are moved by cam pins or hydraulic cylinders and are locked in place when the die is closed.

Because die casting machines operate at high rates of speed, heat must be removed from the die at a sufficiently high rate. Removal of heat is accomplished by circulating water or other coolant, typically oil, through channels drilled in the die blocks. Heat is sometimes added to the die by electric or gas heaters, or by heating the oil, for warm up or for making thin sections that transmit insufficient heat to maintain the die at the proper operating temperature.

4. Die Casting System Design Considerations

To produce high-quality castings, the die casting system must be designed to have the following characteristics.

- The shot system must be large enough to deliver sufficient molten metal to fill the entire die cavity when a shot is made.
- The complete system must be designed so that the cavity is uniformly and rapidly filled so that the cavity is filled with casting metal that has not

started to solidify when flow stops. Otherwise, defects termed "cold laps" will result.

- The system must be designed to eliminate or at least minimize the entrapment of air in the die cavity when flow stops; otherwise, defects termed "blisters" will result.

II. DESIGN REQUIREMENTS FOR GOOD SHOT SYSTEM PERFORMANCE

To effectively utilize systematic design procedures for the runners, gates, and overflows required in die casting dies, knowledge of the shot system performance of the specific machine on which a die will be used is required. The shot system performance is characterized on a graph plotting the metal pressure in the system vs. the square of the flow rate. This information can be used with design guidelines described in this chapter to design the feed system for dies that are likely to produce first-shot success and more effectively utilize the metal delivery characteristics of the casting machine. This design information is based on research carried out at the Australian laboratory CSIRO [1] and the Canadian zinc producer Cominco Ltd. [2,3], together with early fundamental work carried out at Battelle Columbus Laboratories under sponsorship of International Lead Zinc Research Organization, Inc. [4].

A. The P–Q^2 Diagram

The application of fundamental principles of fluid mechanics, together with suitable assumptions, allows development of a relationship between the available metal pressure in the injection system (P) and the resultant metal flow rate into the die cavity (Q). The following assumptions are made for development of the P–Q^2 relationship:

- Steady state or constant velocity flow of injected casting metal
- Turbulent flow of casting metal.
- One-dimensional flow of cavity metal.
- The casting metal is incompressible and has constant density.
- The casting metal is fully molten and does not solidify during its injection into the die.
- There is no friction loss in the cylinder seals of the injection system.
- No variations in metal pressure occur because of metal inertia, inertia of mechanical components, or the dynamic characteristics of the hydraulic system.

- The accumulated pressure in the hydraulic system remains constant during a shot.

The invocation of these assumptions allows for a simple P–Q^2 analysis to be derived although they may only be partially valid during certain stages of the shot. The straight line P–Q^2 relationship usually applies over most of the shot stroke when the data have been averaged near specific points of interest to process designers. Discrepancies arise during times of high acceleration or deceleration, i.e., at the beginning or end of the shot.

By a plotting available casting metal pressure P on a linear scale vs. metal flow rate Q on a squared scale axis, an approximate straight-line relationship is established as is shown for line P_sQ in Fig. 4. That line represents the pressure that the machine can develop as a function of flow rate and is called the machine-pressure characteristic (pressure available). The machine-pressure characteristic is a function of the hydraulic system of the machine, including the shot cylinder and plunger.

Line OA on the Fig. 4 graph is called the gooseneck-nozzle characteristic. This line shows the pressure required to produce a specific flow out of the machine and into the die (resistive pressure). Its position is a function of the hydraulic characteristics of the gooseneck and nozzle. If the data are taken to include the sprue, runners, and gates, then the complete feed system characteristic line can be obtained as shown by line OB in Fig. 4. These parts of the injection system, in addition to the gooseneck and nozzle, cause flow resistances that reduce the ideal flow rate below rates that would be calculated based on the measured "dry shot" speed of the machine.

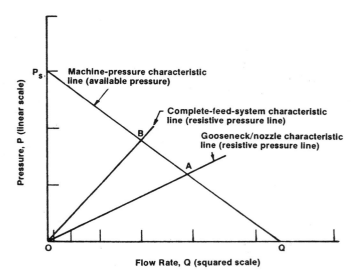

Figure 4 General P–Q^2 diagram configuration.

Metal pressure and flow rate for a given set of shot system parameters is defined by the intersection of the available and resistive pressure lines (points A and B in Fig. 4).

Depending on how the shot system is set up (hydraulic pressure, opening of shot valve, plunger size, nozzle opening diameter, and the sprue, runners, and gating configurations), the position of these lines can vary, as will be described in subsequent sections.

B. Shot System Instrumentation

1. Instrumentation Requirements

The data used to generate a P–Q^2 diagram can be obtained from shot traces, providing that proper instrumentation and transducers are used. Typical instrumentation required to obtain the necessary data is as follows:

(1) High speed digital acquisition system, or multichannel oscillograph UV (ultraviolet) recorder. Because the die casting shot is of only a few milliseconds, the information gathered will be high-frequency and transient in nature. This requires a data recorder with high-frequency response for shot system characteristic measurement, trouble shooting, and tuning. The following performance requirements are recommended for data recorders:

- Frequency range: 0–1500 Hz, flat within 5% variation, full scale.
- Maximum 10% overshoot to step input. For extensive research and troubleshooting of extremely high speed die casting machines, higher frequency response to about 4000 Hz or higher may be required. If only shot system characteristic measurements on larger machines are desired, a frequency range of 0–800 Hz is probably acceptable. However, with lower frequency response there may not be adequate information gathered for extensive troubleshooting of machine performance problems. High-speed digital transient or computer acquisition systems or FM tape recorders can also be used in conjunction with slower speed plotters or recorders to obtain the data. Data recording can also be performed on a two or more channel oscilloscope with a camera for permanent records. This method requires extreme care in setting up and can provide good results.

(2) Shot-Cylinder-Position (Displacement) Transducer. Several transducer types are available that meet the specifications below. Two that have been used are the cable-driven potentiometer and the linear-variable-differential transformer (LVDT). Other possible choices that might be used for this application are ultrasonic transducers, capacitive transducers, linear potentiometers, optical encoders, and magnetic encoders. Any position or displacement transducer should have:

- Frequency range 0–800 Hz flat within 5% variation.
- Maximum 10% overshoot to step input.
- 50 G acceleration—following capability.
- Transducer stroke slightly longer than shot cylinder stroke.
- Very rigid mounting to prevent transducer errors from relative motion.

(3) Velocity Transducer (optional). The requirement for velocity transducers are the same as those listed above for position (displacement) transducers. If a velocity transducer is not used, velocity information can be obtained by electronic differentiation of the displacement signal, or by graphically differentiating the recorded displacement signal.

(4) Pressure Transducer (two required): Pressure transducers must be rugged to withstand the high peak overpressures of the end of the shot, yet have good resolution and frequency response to measure the pressure drops and fluctuations during fill. Both strain gauge and piezoelectric types of transducers have been used and each has its own advantages. Strain gauge types work well with DC and are easily calibrated with a static pressure source and are reasonably tolerant of dirt and moisture. However, although the strain gauge types are not as rugged and tolerant of overpressure transients as the piezoelectric type, they have been successfully used. The piezoelectric type is rugged but has the disadvantages of not being tolerant of dirt and moisture and must be dynamically calibrated by rapidly applying or removing pressure [5]. Special low-drift electronics must be used with piezoelectric transducers and grounding switches must be used for measurement "standby." When the measurement is to be enabled, the switch is opened. This procedure minimizes drift and is used for both instrument calibration and shot system measurement. Either type can be used if they are selected and used properly. Pressure transducers should have:

- 0–5000 psi (34.5 MPa) working range minimum; 0–10,000 psi (69 MPa) working range preferred.
- 20,000 psi (138 MPa) minimum overpressure capacity preferred.
- Frequency range 0–1500 Hz within 5% variation for strain gauge or DC types, 1–1500 Hz within 5% variation for piezoelectric types.

- Drift rate less than 100 psi/min (0.7 MPa/min). This is particularly important for piezoelectric transducers and signal conditioners.

Other types of pressure transducers that might be used are capacitors, tuned-variable-frequency, and variable reluctance transducers.

Instrumentation packages can be purchased as total systems from manufacturers, or individual components may be purchased and assembled to form assistance.

2. Transducer Installation

The transducers used to obtain the necessary data must be properly installed on the die casting machine and connected to the high-speed recorder. Figure 5 schematically illustrates where the transducer should be installed. To install the pressure transducers, caps must be made available in the shot cylinder inlet and exhaust lines and a shut-off valve should be included for ease of installation and removal of the transducers. Bleed valves should also be incorporated to remove air from the hydraulic lines after installation. These transducers should be located on or very close to the shot cylinder. Drilling and tapping a port through the side of the pipe

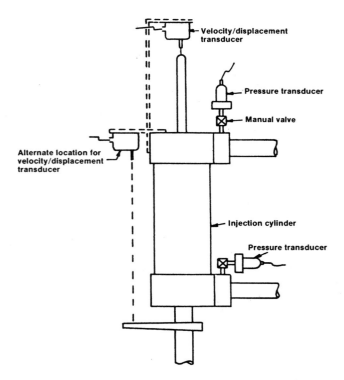

Figure 5 Typical transducer locations. (From Ref. 2.)

flange where it attaches to the cylinder is usually acceptable. The velocity and displacement of the shot cylinder rod can be measured with a single unit that provides both signals through two separate electrical outputs or with a displacement transducer only, with the velocity being calculated from the displacement data. For the cable-driven potentiometer, the cable can be attached to the shot cylinder tail rod, the side auxiliary rod, or to the plunger through the use of suitable brackets. The body of the unit is attached to any convenient nonmoving part, often through the use of a bracket directly attached to the shot cylinder housing. The mounting bracket and transducer must be very rigidly attached; otherwise, any bracket motion or vibration probably will be misinterpreted as plunger motion in the data. Care should be taken to locate the transducers where they will not be overheated or damaged by other moving parts. Similarly, the cables that connect transducers to the recorder should be protected.

3. Instrumentation, Calibration, and Check Out

After the instrumentation is assembled, it must be calibrated before use according to the manufacturer's instructions. Check out of the instrumentation is usually performed by comparing indicated singles with other measured parameters to know with certainty that the indicated measure corresponds with the measured parameter. Calibration and check out will assure that the measurement system is providing the necessary data to assess the performance of the shot system of the die casting machine.

4. Interpretation of Shot System Data

The information used to assess the shot system performance is obtained by recording the response of the transducers when castings are made. An example of the traces so obtained is shown in Fig. 6. Certain events during the shot are noted on this figure. However, determination of the start or the end of the shot are usually obvious from the response of the transducers to determine when the metal reaches the sprue or when the gate requires a less direct approach. First, the full casting shot is weighed, and then the sprue and runners are removed at the gate and the casting is reweighed. Those weights are converted to volumes and then to corresponding plunger displacements. Then, those displacements are measured back from the end of the shot along with the displacement trace to locate the two points. These points will identify the various portions of the shot that are used when obtaining data on pressure and velocity at impor-

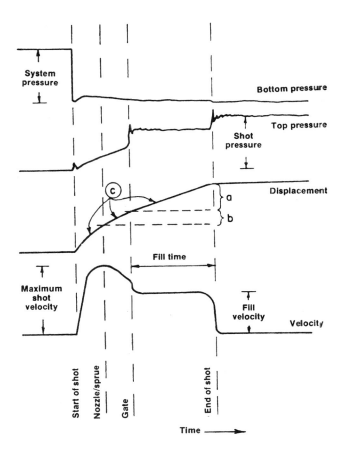

Figure 6 Simplified instrumentation traces of an injection cycle (shot valve on inlet, no slow shot). (a) Plunger displacement corresponding to filling the die cavity. (b) Plunger displacement to bring metal front from nozzle to gate. (c) Velocity can be obtained from the slope of the displacement curve ($\Delta X/\Delta t$) if a velocity transducer is not used.

tant stages of the shot for calibration purposes. They are also useful to obtain the cavity fill time, which is simply proportional to the distance between the gate point and the end of shot point if there are no overflows. If overflows are present, the cavity fill time can be determined by measuring the cavity volume and overflow volume and back calculating from the end of shot point. The plunger velocity during the cavity fill portion of the shot can be used for calculating the nominal gate velocity. That calibration is performed by multiplying the plunger velocity by the plunger area to obtain the volume flow rate of the metal and dividing the flow rate by the gate area.

Careful examination of the shot traces will often reveal malfunctions or anomalies in the machine oper-

ation. Thus the instrumentation provides a very useful troubleshooting tool for production problems and machine maintenance.

A specific method for interpreting shot system data has been developed that is able to be programmed into a computerized data collection system [6]. This system is used to calculate the dry shot speed and discharge coefficient of the shot system at either the nozzle or the gate. The discharge coefficient is a measure of the difference between actual and ideal flow. Factors that reduce the discharge coefficient of the feed system are surface roughness, sharp turns in the flow path, and sudden changes in the cross-sectional area of the flow path. This systematic method is similar to the method used by consultants to interpret shot traces.

C. Determining $P-Q^2$ Diagram for a Specific Machine

After the instrumentation described above has been attached to the machine and shot traces have been obtained, the $P-Q^2$ diagram for the machine can be determined.

1. Machine Characteristic Line

The line (P_sQ) in Fig. 4 is specified by two points: the static metal pressure P_s and dry shot speed flow rate Q. The metal pressure at any time during the shot is calculated from the following general relationship:

$$P_3 = (P_1A_1 - P_2A_2)/A_3 \qquad (1)$$

where P_1 = pressure at the top of the shot cylinder; P_2 = pressure at the bottom of the shot cylinder; A_1 = area of the top of the shot cylinder piston; A_2 = area of the bottom of the shot cylinder piston; A_3 = area of the plunger; P_3 = metal pressure.

Ideally, the pressure P_1 is the average accumulated pressure that is available during the shot. In most modern machines, the accumulator pressure should vary less than 10%; therefore the accumulator pressure could be measured with a calibrated gauge at any time and that value can be used for P_1 without introducing significant error. However, for best results, P_1 should be determined by using the pressure transducers to immediately measure the hydraulic system pressure before and after the shot and averaging the two values. The accumulator pressure before the shot is usually measured on the rod end of the cylinder (actually P_2 before the shot is made) and after the shot on the head of the cylinder (P_1). Usually, for purposes of plotting the $P-$

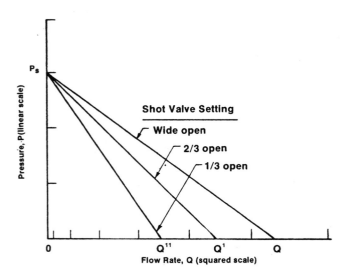

Figure 7 Effect of machine parameters on the P–Q^2 diagram. Effect of restricting hydraulic flow by the fast shot valve (reduces the metal flow rate).

Q^2 diagram, P_1 and P_2 are measured at the end of the shot at which time $P_2 = 0$. Then, Eq. (1) is simplified to:

$$P_s = P_1 A_1 / A_3 \qquad (2)$$

where P_s is the static metal pressure.

The dry shot speed can be determined by disconnecting the plunger and measuring the speed of the shot cylinder ram or rod and calculating a theoretical maximum flow rate for the plunger size from the following relationship:

$$Q = V A_3 \qquad (3)$$

where Q = maximum flow rate; V = shot cylinder ram velocity (dry shot speed); A_3 = area of plunger.

In most cases, dry shot speed is not directly measured, especially in larger machines, because of potential damage to the shot system. Instead, intermediate points are determined from information from the shot traces and the dry shot speed or flow rate is determined by extrapolation. For example, points A and B shown in Fig. 4 can be calculated from the measurements of the pressures and velocity when the metal leaves the nozzle or the metal reaches the gates, respectively. The data for those points are determined by measuring back appropriate distances on the shot trace from the end of the shot based on the volume of the casting and the volume of the runners and sprue. Corresponding pressure and flow rate values could be calculated at other points in the shot; however, for most machines, all of these points

should fall on the machine characteristic line. If the shot system has high pressure transients or significant internal effects (both of which violate the assumptions used to develop the P–Q^2 analysis), a die with a large cavity volume may have to be used during calibration to assure steady-state flow conditions can be achieved. Extrapolation of the line beyond points A and B then defines the dry shot flow rate (Q).

The slope of the machine pressure characteristic line ($P_s Q$) is a function of the flow area (or size) and the flow coefficient (or efficiency) of the hydraulic circuit. The partial closing of the shot valve acts to restrict the flow of hydraulic fluid, as illustrated in Fig. 7. Therefore when assessing the performance of the shot system, the shot valve should be "wide open." Increasing the hydraulic pressure causes the machine pressure characteristic line to shift upward, as illustrated in Fig. 8, but does not change the slope of the line. Changing the plunger size causes the position and slope of the machine pressure characteristic line to shift as shown in Fig. 9. Thus many parameters can affect the position and slope of the machine characteristic line, and changes in those parameters will influence the shot system performance.

2. Gooseneck/Nozzle Characteristic Line

Like any hydraulic system, the metal injection system of the die casting machine contains resistance to metal flow. The pressure developed on the metal in the gooseneck during injection is that necessary to overcome the

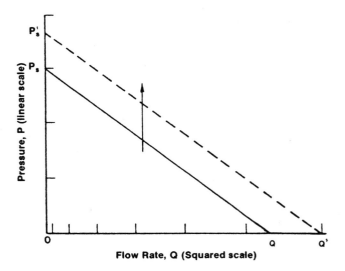

Figure 8 Effect of machine parameters on the P–Q^2 diagram. Effect of changing hydraulic pressure (increasing pressure).

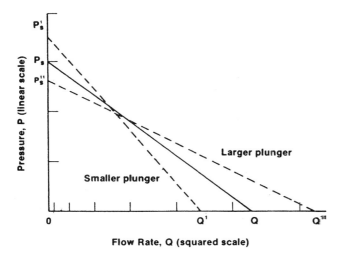

Figure 9 Effect of machine parameters on the P–Q^2 diagram. Effect of changing plunger size causes position and slope of machine-pressure characteristic line to shift.

flow losses in the system. The pressure losses due to the flow of metal mainly result from friction and changes in flow configuration, such as changes in cross-sectional area and direction (bends) of the passages. Rough surfaces in the gooseneck and nozzle passages contribute to pressure losses due to friction and increased turbulence and thus may contribute to porosity in the castings.

The data that were used to calculate point A on the machine characteristic line of Figs. 4 and 10 were taken from instrumentation traces at the point where the metal was first exiting the nozzle (see Fig. 6). The calculated metal pressure at that point is that required to force the metal through the gooseneck and nozzle at that flow rate. Because zero pressure is required at zero flow rate, a line from the origin through point A defines the metal pressure requirements of the gooseneck/nozzle system. Because pressure is proportional to the square of the flow, the relationship can be shown as a straight line on the same graph as for the machine pumping rate. Line OA in Fig. 10 then defines the gooseneck/nozzle characteristic line (simply referred to as the nozzle characteristic line) and basically defines the square relationship between pressure and flow rate of molten metal flowing through the gooseneck and nozzle. Thus Point A defines a maximum flow rate available through the gooseneck and nozzle on that machine and the pressure define by point A also establishes the power consumed in the gooseneck and nozzle to achieve that flow. If the gooseneck and nozzle losses could be reduced (through redesign or by increasing diameters),

the effective metal pumping capacity of the machine would be increased.

The pressure losses in a gooseneck can theoretically be calculated knowing the configuration and surface roughnesses of the internal passages, and methods for these calculations are described in the literature. It is much simpler to empirically describe these "efficiencies" from instrumentation data. While only one point (other than the origin) is required, normally a number of points are calculated from the data taken at various shot valve settings. Therefore, nozzle characteristic line is usually a line of best fit through a number of points such as A, A′, and A′ in Fig. 10.

As was the case with the machine characteristic line, the slope of line OA is also a function of the flow area (or size) and flow coefficient (or hydraulic efficiency). However, in this case, the efficiency relates to the metal flow system instead of the oil flow system of the machine. A modified Bernoulli equation has been shown to be useful in calculating the hydraulic efficiency of different gooseneck/nozzle systems. The relationship between pressure and flow is:

$$P = \gamma Q^2 / 2G C_d A \qquad (4)$$

where P = molten metal pressure in the gooseneck, psi (N/m² or Pa); γ = specific weight of the liquid metal alloy. For zinc, this is 0.221 lb/in.³ or 59,715 N/m³; G = gravity constant = 386 in./sec² or 9.8 m/sec²; Q = molten metal flow rate, cu in./sec (cu m/sec or 1000

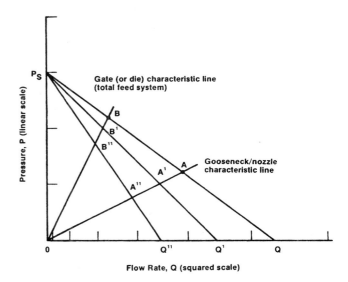

Figure 10 Nozzle and die characteristic lines defined by various experimental P and Q points. Line OB shows the effects of flow losses in the total feed system.

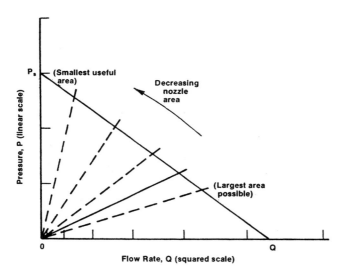

Figure 11 Characteristic lines calculated for varying nozzle sizes (with the same discharge coefficient).

L/sec); C_d = discharge coefficient, approximately 0.7–0.8 for well-designed system; A = area of the smallest part of the feed system, in.^{2} (m^2).

One of the key variables that affect the pressure/flow relationship in the gooseneck/nozzle system is the nozzle size. If the nozzle is enlarged, a greater flow results and conversely if the nozzle is reduced, then less flow occurs. Assuming that the discharge coefficient or value C_d of a gooseneck will remain constant with varying nozzle size, then Eq. (4) can be used to calculate nozzle characteristic lines for various nozzle sizes. This is accomplished by using the new nozzle area A and existing C_d value in the equation, assuming a Q value and solving for P. A line through the new pressure/flow rate point and the origin O determines the new nozzle characteristic line. This way, a series of characteristic lines can be calculated and drawn for a series of nozzle sizes as is illustrated in Fig. 11. The intersections of those lines with the machine characteristic line determine the flow rates that will be achieved for each nozzle.

3. Gate or Die Characteristic Line

A "gate characteristic line" is often included in the P–Q^2 diagram. It is similar to the nozzle characteristic line in derivation and meaning. It is also referred to as the "die characteristic line." Just as the gooseneck and nozzle offer resistance to metal flow, so also does this sprue–runner–gate system. The data that were used to plot point B on the machine characteristic line (Figs. 4 and 10) were taken from data traces at the time repre-

senting metal flowing through the gate or during the filling of the cavity (see Fig. 6). The calculated metal pressure at that stage represents the total pressure required to force the metal through the gooseneck, nozzle, sprue, runner, and gate at the flow rate that occurred. A line from the origin through point B (line OB in Fig. 10) defines the pressure/flow relationship requirements of the total metal system up to and including the gate. As was shown in Fig. 10, this gate characteristic line can also be a best fit for a number of points such as B, B′, and B′.

The full losses of a metal flow system up to and including the gate are always higher than for the same system up to and including the nozzle. Thus the gate (or die) characteristic line always has a greater slope than does the nozzle characteristic line, meaning that greater pressure is required when metal flows through the gate than is required before that event. At the same time, the flow rate through the gate is lower than the flow rate through the nozzle before metal reaches the gate.

Increasing the feed system size (gate area, runner, sprue, nozzle) achieves a higher flow rate. Also, increasing plunger size would change the machine characteristic line, as was shown in Fig. 9. Figure 12 illustrates the relationships between the feed system size and plunger size. As is shown, the use of a small feed system with a small plunger may increase flow rates in certain regimes of the diagram (point B′) in Fig. 12. The figure also shows that the use of a larger feed system (larger gate area) with a larger plunger (consistent with the design rules to be discussed in a subsequent section) will result in increased flow rates in other regimes (point A). For large, difficult to produce castings, it is desirable to fill

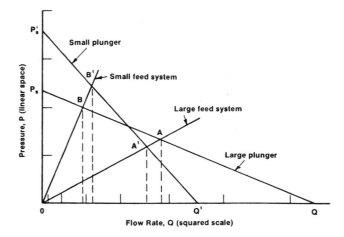

Figure 12 Shot-system performance as a function of machine characteristics and feed-system characteristics.

the cavity in the shortest possible time. Thus the shot system configuration that results in higher possible flow rates should be used. This might be achieved by using a large feed system and large gooseneck and plunger as shown in point A in Fig. 12.

The equation for calculating the nozzle discharge coefficient can also be used for calculating the gate discharge coefficient; however, area A used in the equation should be the smallest cross-sectional area of the complete metal flow system. The gate area must be the smallest of the system; therefore it should be used in the equation for calculating the gate discharge coefficient.

4. Metal-Velocity Lines

After constructing a P–Q^2 diagram, it is often useful for die design considerations to add a family of lines corresponding to metal velocities as illustrated in Fig. 13. Velocity values are arbitrarily selected but they should be within a reasonable and typical range such as 80–150 ft/sec (24–46 m/sec). A number of nozzle characteristic lines, for nozzle areas appropriate for the machine in question, are generated on the P–Q^2 diagram as was described earlier. By multiplying a selected velocity by each nozzle area, corresponding flow rates are determined. The points on the nozzle characteristic lines corresponding to the calculated flow rates are points on the line for the selected velocity.

These velocity lines indicate metal velocities in the nozzle over the range of nozzle sizes considered. If

certain assumptions are made, then the calculated metal velocities can be assumed to be gate velocities for various gate areas (substituting gate area for the nozzle area based on the use of designed systems where no cross-sectional area in the feed system beyond the nozzle will be larger then the nozzle area). Obviously, there will be some error in gate velocities determined in this manner, but if the die is built to minimize flow losses, the errors will be small and this becomes a useful technique for die design (sizing of the sprue, runners, and gates).

The shaded area on the P–Q^2 diagram of Fig. 13 shows how velocity lines together with the machine characteristic line and nozzle characteristic lines could serve as boundaries, defining a region within which the designer could work.

For example, on the diagram shown in Fig. 13, the machine characteristic line for the "normal" hydraulic system setup is represented by line P_sQ. $P_s'Q'$ shows the machine characteristic line obtained if the hydraulic system pressure were increased, if that is possible. Other machine characteristic lines could be plotted on the diagram. For example, lines obtained for different-sized goosenecks or lines obtained with different shot valve openings could be shown. Lines OA through OE or OA' through OE' represent the effect of increasing the nozzle size on the metal pressure and flow rate at the nozzle exit. The shaded area then represents an attempt to define practical ranges of operating conditions for the machine. That area is defined by (1) the upper limit of nozzle/gate size OE or OE', or (2) the lower limit of nozzle/gate size OA or OA'. The largest and smallest allowable gate areas will be determined by allowable gate thicknesses and practical gate lengths. A further restriction on the maximum gate area is that the gate area cannot exceed the nozzle or sprue inlet area.

Figure 13 shows that for the selected machine setup to achieve the machine characteristic line P_sQ, it would be impractical to achieve a gate velocity above 140 ft/sec (43 m/sec) with the smallest allowable gate area. If for the system it was determined that a 150 ft/sec (45 m/sec) gate velocity should be used, this could be achieved only by increasing, if possible, the hydraulic system pressure shown by line $P_s'Q'$ and by using a feed system with a small cross-sectional area.

With that higher system pressure, only low flow rates could be achieved with the small feed system. If it were determined that a very high flow rate was required to produce a large part with a short fill time, Fig. 13 shows the feed system would have to be large and only moderate gate velocities would be achieved with the large feed system.

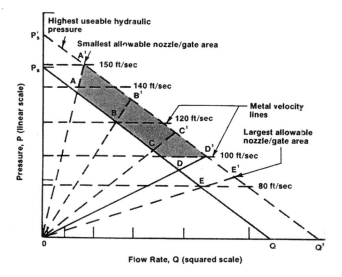

Figure 13 Calculated metal velocity lines and typical design boundary of P–Q^2 diagram shown shaded.

If in designing the system it were determined that both a high flow rate and high gate velocity were desired, a machine with higher capacity shot system would be required than the one hypothesized in Fig. 13. The results of the research conducted at Battelle Columbus Laboratories [4] indicated that to produce large castings with high surface quality, cavity fill time should be minimized (flow rate should be maximized). On the other hand, the Australian feed system design approach [1] attempts to achieve specific velocities and will accept longer fill times (lower flow rates) to achieve that velocity. Unfortunately, limits on acceptable and unacceptable cavity fill times at specific gate velocities have not been reported. However, the normal limits of gate velocity suggested are 100–150 ft/sec (30–45 m/sec) [7].

After the performance of the shot system of the die casting machine has been assessed, the information is used to design the runner and gating system (feed system) and overflow system for the casting. Two systematic approaches to design of the feed system will be described in the next section.

III. FEED SYSTEM AND OVERFLOW DESIGN

The objective of any method for designing the feed system of a die casting die is to produce high-quality castings with first-shot success. In any such system, great emphasis is placed on the control of molten metal flow and velocity through the feed system.

The first fundamental rule in designing the feed system is that the cross-sectional area of molten metal flow from the plunger to the gate should be constant or converge (become smaller). In the hot chamber die casting machine, traditional cast gooseneck design provides converging areas from the plunger to the gooseneck outlet. From that point, the cross-sectional area should remain constant or continue to converge; that is, the nozzle flow passage should be no larger than the gooseneck outlet, the sprue area should be no larger than the nozzle area, the total runner area should be no larger than the sprue area, and the total gate area should be no larger than the runner area. For cold chamber machines, the shot sleeve design is substituted for the gooseneck design consideration in this rule.

All systematic approaches attempt to provide uniform flow through the feed system by employing large and smooth radius turns rather than sharp corners. Sudden changes in cross-sectional area or sharp turns in the runner will cause flow separations with resulting turbulence and possibly soldering of casting metal to the die surface.

Such turbulence increases flow resistance and decreases the flow rate. It may also lead to die erosion. In addition, turbulence, especially when caused by sudden increases in cross-sectional area of the feed system, contributes to air entrapment in the metal and consequently less than desirable casting quality because of the formation of blisters and porosity in the casting. Turbulence also may result in cavitation, which reduces die life. The following design guidelines are aimed at efficiently controlling flow of the molten metal alloy through the feed system to the die cavity with minimal mixing of molten metal and air:

1. Determine machine shot system performance; this one of the bases for both machine selection and die design calculations.
2. Make the cross-sectional area of the feed system constant or convergent from plunger to gate.
3. Fill the die cavity in the shortest practicable time, using acceptable gate velocities.
4. Proportion the gate (distribute the metal flow) according to the metal requirements of various segments or sections of the cavity.
5. Use runner-type sprues or cross-sectional area sprues, tapered tangential runners, or fan gates (of constant or reducing areas). The dimensions of all components that are defined and discussed in subsequent sections are calculated to obtain desired cavity fill times and metal velocities.
6. Determine size and location of overflows according to metal flow in the cavity and the anticipated die temperature. Attempts to eliminate overflows or minimize their volume.

The following sections outline the procedures used to design the feed systems for die castings in accordance with these general design guidelines.

A. Selecting the Die Casting Machine

Selecting the die casting machine on which to run the die being designed is primarily based on the clamping force required to produce the casting and the molten metal delivery characteristics of the machine's shot system. To determine the required clamping force, it is first necessary to determine the injection force. The injection force F_i is obtained by multiplying the projected area (A_p) of the die cavity, including runners and overflows by the final injection pressure P_f:

$$F_i = A_p P_f \tag{5}$$

The clamping force (F_c) of the machine should exceed the injection force by the following ratios:

Low-inertia injection system: F_c/F_i = 1.25:1.
High-inertia injection system: F_c/F_i = 2.5:1.

or

$$F_c = 1.25 \text{ or } 2.5(F_i) \tag{6}$$

A low-inertia injection system is one in which the accumulator is located adjacent to the shot cylinder, which is typical of compact systems. A more traditional die casting machine (typically a high-inertia system) will have the accumulator located in the far end of the machine. In that case, high inertia is caused by the greater quantity of fluid in the longer hydraulic lines. That fluid must be suddenly stopped at the end of a shot stroke.

The machine should be able to deliver sufficient molten metal to achieve short cavity fill times and required gate velocities. Injection capability or metal flow rate is determined from the P–Q^2 diagram, as explained previously. The P–Q^2 diagram will show both the metal flow rate and metal velocity for the machine with several sizes of nozzles, as shown in Fig. 13.

The cavity fill time, which is the time required for molten metal to fill the cavity, assuming there are no overflows, is determined by the following relationship:

$$tf = \frac{V_c}{Q} \tag{7}$$

where t_f = cavity fill time, sec; V_c = casting volume, cm^3 or in.^{3}; Q = metal flow rate (cm^3/sec or in.^{3}/sec).

Recommended values for cavity fill time, assuming no overflows, range from 10 to 40 msec. The use of overflows increases the allowable fill time as discussed in a subsequent section.

The gate velocity that will be achieved with the shot system of the machine is calculated from:

$$V_g = \frac{KQ}{A_g} \tag{8}$$

where V_g = gate velocity (m/sec or ft/sec); Q = metal flow rate (L/sec or in.^{3}/sec); A_g = gate area (mm^2 or in.^{2}); K = constant, 100 for metric units, 0.083 for English units.

The recommended range for gate velocity (normal to the gate) is 30–40 m/sec. (about 100–150 ft/sec.). At this time, the die designer can determine the ranges of cavity fill times and gate velocities that can be achieved from the various nozzle sizes for the machine being considered. The cavity fill time is determined from Eq. (7) and by using the flow rates for the different size nozzles. The gate velocities can be determined from Eq. (8) and by using the nozzle bore cross-sectional areas or sprue areas for the gate areas if a constant area system is being designed, or the anticipated gate area, if a convergent feed system is being designed. Thus by considering the clamping force requirements and the shot system characteristics of the machines available relative to the

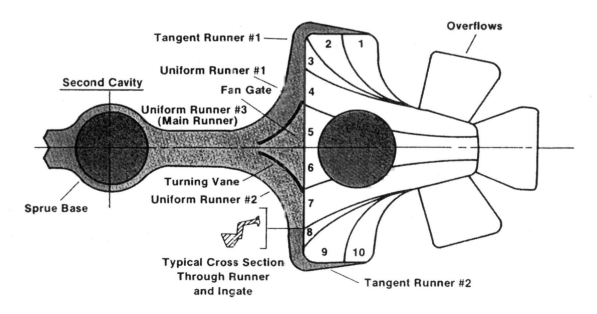

Figure 14 Sketch of general runner, gate, cavity, and overflow configuration of a faucet-fixture casting.

desired cavity fill time, and/or gate velocity, the designer can select the machine and nozzle bore diameter that is best suited to produce castings in the die being designed.

B. Selecting a Runner and Gate System

The first step in designing a die using the design guidelines is to select and sketch the general configuration of the runner, gating, and overflow system as shown in Fig. 14. There are usually several possible runner and gating configurations that might be used. Figure 15a,b

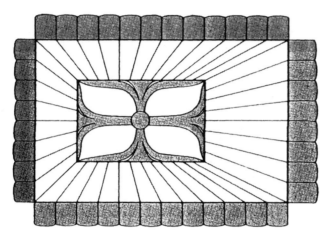

Center gated casting

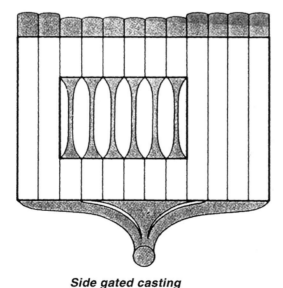

Side gated casting

Figure 15 Alternative methods of gating a large casting with a large center opening.

illustrates two alternatives for one casting and Fig. 16 illustrates alternatives for other casting shapes.

When considered as part of the overall feed system, the position, shape, and dimensions of the gate determine the direction and velocity of the metal flow into the cavity and the flow pattern through the cavity. Correct design and dimensioning of all gates is essential to achieving high casting quality and overall casting efficiency.

Decisions on the positioning of gates relative to die cavity are subjective and are based on factors such as the shape of the casting, production requirements, and tooling costs among others.

General rules to follow in selecting the feed system configuration are:

1. Minimize runner length.
2. Provide uniform flow across the cavity such that air is forced out of the cavity ahead of the molten metal into overflows and/or vents.
3. Feed from the side with the most critical surface requirements.
4. Flow across the short dimension of the cavity, as shown in Figs. 15 and 16.
5. Avoid flow across large openings (consider center gating) or provide runners across large openings.
6. Select runner and gate configurations that control flow direction through the cavity so that metal flow tends to follow desired paths. This will be described in the section on segmenting and Fig. 17.

Although several alternative configurations can be used, as shown in Figs. 15 and 16, none will satisfy all the guidelines listed. Therefore compromises must be made based on experience or other factors in the die design. To further illustrate how the metal feed systems might be selected, consider the casting shown in Fig. 15. For this casting, the die designer would have the following options for selecting the feed system:

1. Locating the sprue and runner system inside the opening (center gating) and locating the overflows and/or vents around the outside parameter as shown in Fig. 15a.
2. Feeding from one side of the casting, placing runners across the opening, and placing overflows and/or vents on the opposite side of the casting, as shown in Fig. 15b.

Either of these alternative configurations probably could be used to successfully produce this casting. The choice will depend to some extent on the casting size and

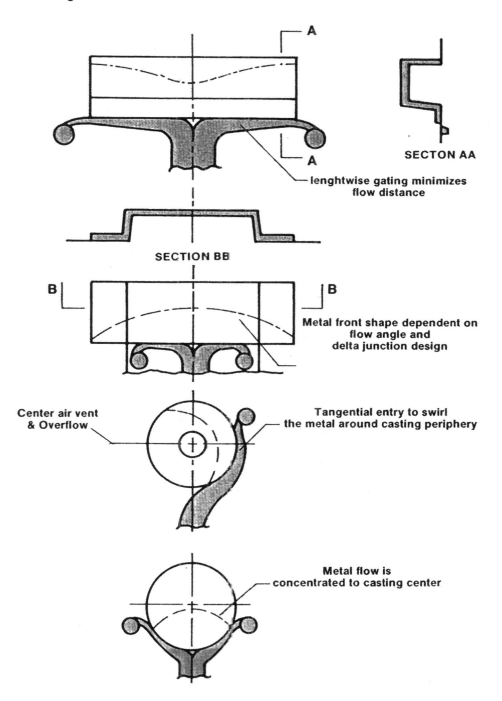

SECTON AA

lenghtwise gating minimizes
flow distance

SECTION BB

Metal front shape dependent on
flow angle and
delta junction design

Center air vent
& Overflow

Tangential entry to swirl
the metal around casting periphery

Metal flow is
concentrated to casting center

Figure 16   A general guide of how to select a gating configuration for various typical casting shapes.

1) Keep cavity flow distance to a minimum, i.e. gate on longest side if possible.

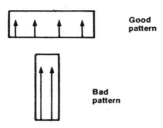

2) Match filling pattern to casting shape so as to reduce turbulence to minimum.

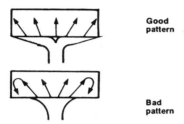

3) Plan filling pattern to feed the most important features.

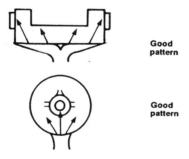

Figure 17 General rules for selecting the gate and runner system.

configuration, but usually the preferred method of gating a casting of this type is to use the center gating technique shown in Fig. 15a. This configuration has the advantages that both runner lengths and flow distances across the mold cavity are minimized and the flow direction out of the gates will tend to conform to the anticipated flow pattern.

The only disadvantages of center gating are that the velocity of the metal flow tends to decrease as it progresses across the cavity, and a multicavity die cannot be used. However, these are usually not significant problems. In some cases, the reduction in the volume of the metal that must be delivered to the die and cooled make it possible to achieve higher production rates with a

center-gated, single-cavity die than can be achieved with an outside or side-gated, two-cavity die.

It should be noted that the feed systems illustrated in Figs. 15 and 16 are either constant or convergent in cross-sectional area to control metal flow to the die cavity.

If there are two or more large openings in a casting, the sprue should usually be placed in one of the openings and runners should be placed across the other large openings to carry the metal across them with minimal heat loss as shown in Fig. 18. If the second opening is not too large, it may be possible to flash across it and fill the portion of the casting beyond the opening. The side gate configuration illustrated in Fig. 15b could also be used for such a casting.

The feed system illustrated in Fig. 19 shows a converging flow system for the nozzle to the gate. Sharp corners are avoided and the surfaces are, or should be, smooth and well finished. In addition, the bore of the gooseneck should be sufficiently large to ensure low velocity through the major bends, to minimize pressure losses. The bore of the nozzle should be only slightly smaller than the bore of the gooseneck or shot sleeve

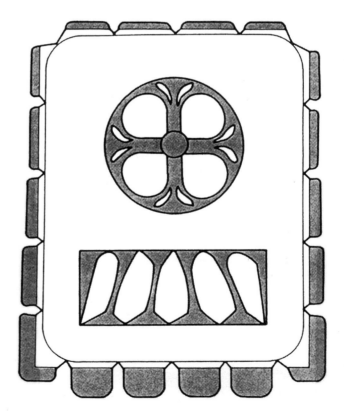

Figure 18 Preferred method of gating a large casting with large multiple openings.

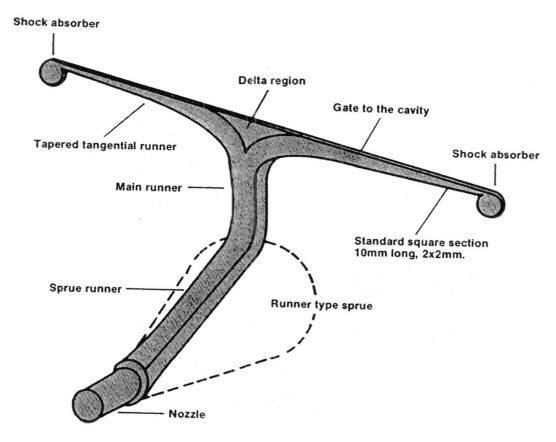

Figure 19 Converging flow system, showing single runner type sprue, main runner, tapered tangential runners, shock absorbers, and the gate to the die cavity.

outlet so that the velocity of the molten metal through the nozzle is maintained at low value. Again, low velocities, large radius turns, and smooth surfaces are used to keep pressure losses low.

C. Locating the Gate

Once the general configuration of the feed system is determined, the exact location of the gate should be established. Locating a gate involves the following variables:

1. Flow pattern in the cavity.
2. Flow distance in the cavity.
3. Casting and die configuration.
4. Surface finish requirements.
5. Air venting and/or overflows.
6. Trimming requirements.
7. Gate velocity.
8. Cavity fill time.
9. Die temperature.

The gate location and the gate area have a direct effect on the first eight factors. The machine injection system also affects the gate velocity and cavity fill time, and the die temperature has a significant affect on the casting surface finish. The main difficulty in selecting gate location is that these factors are so closely interdependent.

For example, the gate thickness, part thickness, and gate velocity can affect the metal flow behavior downstream from the gate. Depending on their values, the metal flow can be continuous or atomized, thus affecting the flow pattern in the cavity. In some cases, the complexity of the casting configuration and/or trimming requirements may be the only factors governing the location, length, and thickness of the gate.

D. Segmenting the Casting

As an aid to visualizing how metal will flow across the cavity, depending on the gate location, and to determine the metal requirements for various portions of the casting, a casting should be divided into segments. This segmenting should be performed for each alternative gate location considered. Typical examples of how casting might be segmented are shown in Fig. 20. General guidelines for segmenting are:

1. Use flow patterns that would tend to normally exist in the cavity, so that molten metal will tend to flow into a segment at an ingate, through that segment, and then flow out of that segment into an overflow. Because segment lines are only guidelines, flow totally within segment batteries will not exist during actual production except for the simplest casting shapes. In addition, the flow pattern in the cavity can be influenced by gate location and the angled metal flow out of the gate.
2. Minimize flow length in the segment.

In many cases, it will be impractical to follow all of the rules for selecting both runner and gate segment con-

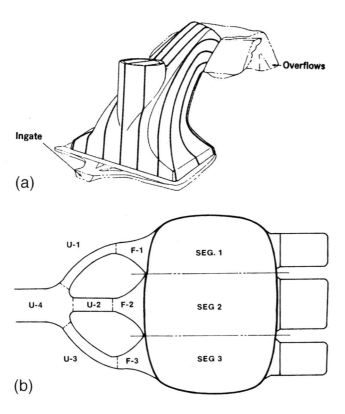

(a)

(b)

Figure 20 (a) Segmenting die castings for gate design calculations: Method of dividing water faucet casting into segments. (b) Segmenting die castings for gate design calculations: method of dividing mirror shell into segments.

figurations and so it will be necessary to make compromises. Although it may be necessary to sacrifice some of the desirable features in all but the simplest designs, following the design guidelines for segmenting above will, in general, provide flow conditions that will promote rapid and uniform filling of the die. Rapid and uniform filling of the cavity will minimize the formation of cold lap (cold shut) and blister defects. To control blister formation, flow across the cavity should be uniform to force the air out of the cavity ahead of the molten metal stream into overflows and/or vents. Of course, in practice, this condition is difficult to achieve because of changes in flow direction and section thickness that results from the configuration of the cavity.

E. Calculation of Gate Dimensions

To maintain the desired velocity of metal flow throughout the feed system and to fill a cavity in the shortest practical time, one should design the ingates so their total cross-sectional area is equal to or smaller than the smallest cross-sectional area in the die feed system from the gooseneck or shot sleeve to the die. This minimum area often occurs at the sprue. However, in some cases, it may occur at the nozzle in the hot chamber machine. When the latter condition exists, the nozzle size should be increased or the sprue size should be decreased to reduce turbulence and pressure loss between the nozzle and sprue. Based on the evaluation of the shot system performance and the desired fill time and/or gate velocity for the casting being considered, the nozzle diameter, hence the sprue or the die inlet area, will be defined, as was previously described in Sec. III.A. That area should be used as the basis for the sizing of the gates. In addition, the area of the ingate for each segment should be sized in proportion to the volume or weight of metal in that segment.

If the gate area calculations are made on a weight basis, the weight of the casting and weight of casting metal in each segment must be calculated first. Alternatively, if the part to be produced already exists as a casting, it can be weighed and the segment weight determined by drawing lines on the casting, selling the casting into the segments marked, and weighing the segments. The maximum allowable gate area for each segment is then given by:

Segment gate area

$$= \frac{\text{die inlet area* } \times \text{ segment weight**}}{\text{total casting weight**}} \quad (9)$$

Equation (9) defines the largest gates that can be used for the casting while at the same time keeping the gate small enough to control the pattern of metal flow into the cavity. Once gate location and length have been decided, the gate thickness should be made such that the desired gate velocity and cavity fill time are achieved. The limiting factors for gate thickness are that it can neither be thicker then the casting nor can it be too thick for efficient trimming. The minimum practical gate thickness depends on the accuracy of control of the die temperature and other process variables. For example, higher die temperatures allow thinner gates to be specified. In practice, the minimum gate thickness is often determined by manufacturing and tolerance considerations, but gates as thin as 0.007 in. (0.18 mm) have been found to satisfactorily work.

To determine desired metal flow velocities, gate thicknesses in the range of 0.008–0.024 in. (0.2–0.6 mm) are common and the length is in proportion to the volume or weight of the casting, or segment of casting being fed by that portion of the gate.

After the cross-sectional areas of the gates for each segment are computed, the gate thickness is calculated by dividing that area by the gate segment length. If making the total gate area equal to the die inlet area results in gates that are too thick for normal trimming, the die designer should reevaluate the runner-gate configuration to increase the gate length. This procedure will generally reduce the gate thickness.

If the gate thickness cannot be reduced by a change in the runner-gate configuration, then the thickest gate sections should be reduced to the maximum thickness considered practical for trimming, and all other gate and runner areas should be proportionately reduced. For example, if the maximum calculated gate thickness were a segment or several segments was 0.050 in. (1.27 mm) and the maximum gate thickness considered practical for trimming was 0.025 in. (0.64 mm), then the thicknesses of the gates for each segment should be reduced to one-half of their calculated thicknesses. At this point, it may be desirable to consider reducing the nozzle and sprue areas to match more closely the gate area and thus minimize air volume in the feed system. Alternatively, the designer could gradually decrease the cross-sectional areas of the various components of the feed system from the nozzle to gate using the guidelines for a converging area system. These guidelines will be described in a subsequent section. As a general guideline, the total gate area should not be less than 40% of the die inlet area, or in a hot chamber machine, the nozzle exit, in a converging system.

Making the total ingate areas equal to the die inlet area may produce gates that are too thin to be practical. The die designer then has a choice of redesigning the system to increase the component sizes in the die inlet area, selecting another gating configuration to use a shorter gate (usually not a desirable approach), or using what some die casters have called a comb gate. A comb gate is a series of short, moderately deep gates instead of a long continuous thin gate. Figure 21 shows how a gate of this type would be constructed. If comb gates are used, the spacing between the gates should be equal to or less than the length of the gate opening to minimize the effect of the flow separation through the individual comb gates. However, continuous gates as long as 0.007 in. (0.18 mm) have been successfully used, as previously noted.

Often, fan gates are used to distribute flow at the end of a runner. Figure 22 provides guidelines for designing

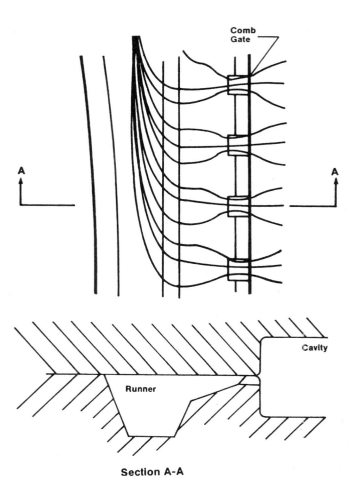

Figure 21 Details of comb gates.

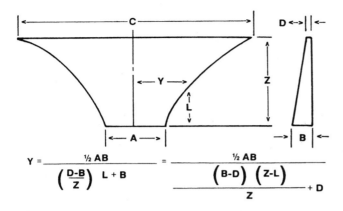

$$Y = \frac{\frac{1}{2}AB}{\left(\frac{D-B}{Z}\right)L+B} = \frac{\frac{1}{2}AB}{\frac{(B-D)(Z-L)}{Z}+D}$$

Figure 22 Design of fan gate with constant cross-sectional area.

a fan gate with the constant cross-sectional area. The design of a fan gate with decreasing cross-sectional areas is illustrated in Fig. 23. One of the advantages of the fan gate design shown in Fig. 23 is that it is relatively simple to machine. Also, the flow velocity is highest in the center of the gate, which is useful to achieving certain types of fill patterns. On the other hand, that design can lead to turbulence if carelessly performed and it is not suitable for wide fan gates.

F. Determining Gate Velocity

After the ingate areas have been determined, the nominal velocity normal to the gate that will be achieved can be calculated from Eq. (8) presented earlier.

Gate velocity will usually fall within the ranges recommended by the North American Die Casting Association which are 90–150 ft/sec (27–46 m/sec). If the gate velocity exceeds the maximum value of that range, die erosion or soldering problems may be encountered. Hence the gate design or metal flow rate should be modified to reduce the gate velocity to be within the recommended range.

Figure 24 illustrates a typical converging feed system with tapered tangential runners feeding the ingates and the recommended metal velocities in each component of that system. Those velocities are determined by selecting the shot system configuration based on the P–Q^2 diagram that yields the desired velocity at the nozzle exit and then by sizing the other components of the system in accordance with the previously given guidelines.

The true gate velocity through the gates fed by tapered tangential runners is higher than the value calculated from Eq. (8). This is because of the variation in flow angle (that is, the angle between the normal to the gate and the metal flow direction) resulting from the

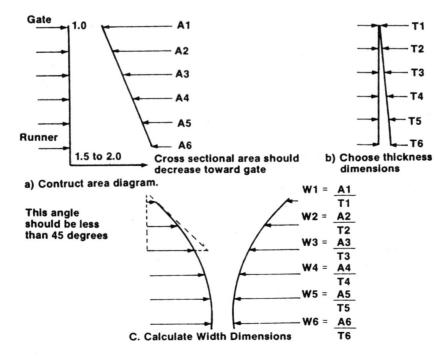

Figure 23 Design of a fan gate with a decreasing cross-sectional area. W6 is normally 2 to 3 times T6 to provide better fan feed shape at the expense of heat loss.

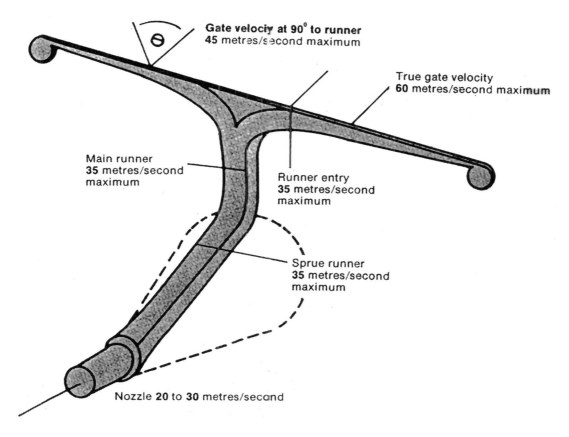

Figure 24 Converging flow system, showing molten alloy velocity upper limits.

dimensions of the tapered tangential runners as shown in Fig. 25. The true gate velocity can be approximated by the following relationships:

True gate velocity

$$= \sqrt{\text{runner velocity}^2 + \text{gate velocity}^2} \qquad (10)$$

where the runner velocity and gate velocity are calculated from Eq. (8), or:

$$\text{True gate velocity} = \frac{\text{gate velocity}}{\cos \theta} \qquad (11)$$

where the gate velocity is that calculated from Eq. (8) and θ is the flow angle from the gate, shown in Fig. 24.

Should the feed system be designed to operate at reduced injection pressures, the true gate velocities for all reduced- and full-injection pressure should be calculated and recorded. The true gate velocity that would result if the machine were running at full injection pressure should be noted on the instructions to machine operators. If the value of the true gate velocity exceeds the recommended range, under any injection pressures, the operator should be warned that excessive gate velocities might cause die erosion if design parameters are exceeded.

G. Calculating the Runner Dimensions

In fluid flow systems such as runners and die casting dies, the flow efficiency, described by the discharge coefficient C_d applies. The discharge coefficient, as it applies to flow in the shot system of the die casting machine, was discussed earlier.

The coefficient of discharge of the runner system is influenced by the configuration of the system including changes in cross-sectional area and changes in flow direction and friction between the flowing metal and the surfaces of the runner system. Sharp changes in flow direction, abrupt (stepped) changes in cross-sectional area, surface roughness, and flow distance all contribute to reducing the flow efficiency or discharge coefficient.

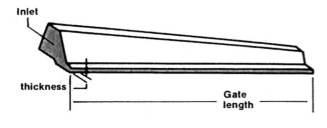

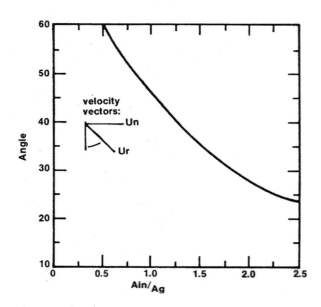

Figure 25 Approximate flow angle variation with the ratio of the runner inlet area to the gate area.

Therefore the runners should be designed with constant or gradually reducing cross-sectional areas, large radius turns, short flow distances, and smooth (polished) surface finishes. However, even when all of these guidelines are followed, the flow in the runner system will not be 100% efficient. Nevertheless, by following those general guidelines, the flow efficiency will be considerably better than that obtained in runner systems designed without following those guidelines. Unfortunately, many runner systems are still being cut in the dies with sharp turns, abrupt changes in flow direction, and no consideration for the cross-sectional area relative to those of the nozzle and the sprue or the gates.

For a constant area metal feed system, the drawing in Fig. 26 illustrates how the runner area calculations should be made for a tangent runner that is continuously feeding metal through the gate. As is shown, the runner cross-sectional area at any point is merely the sum of the cross-sectional areas of the segment gates being fed by that runner. The figure also shows the

location of the areas used in the calculations. This summing procedure is used for all runners back to the minimum die inlet area. Thus the total gate area cannot exceed the minimum die inlet area and the total cross-sectional area cannot exceed the minimum inlet area. Furthermore, as seen in Fig. 26, the use of this technique results in a natural taper in the tangent runners, and those runners taper to zero cross-sectional area at their ends. If correctly machined, the taper virtually eliminates high transient metal velocity and resulting shock during filling of the runner. Shock absorbers can also be used to eliminate shock during runner fill. Design guidelines for shock absorbers will be presented in a later section.

If the constant area runner system has long flow distances and several changes in flow directions, flow losses may occur such that the coefficient of discharge of the runner is in the range of 0.6–0.8. In those cases, the metal flow rate through the gate will be less than that calculated based on the metal flow rate exiting the nozzle. Thus the cavity fill time will be somewhat longer than the value calculated and the gate velocity will be lower than that calculated based on the flow rate exiting the nozzle. These factors could result in lower casting quality than anticipated.

Because of those possibilities, it may be preferable to design the runner system with a converging (decreasing) cross-sectional area. Figure 27 illustrates that type of runner system and provides guidelines for the reduction in cross-sectional area through the various sections of the runner system. General guidelines for the reductions in cross-sectional area for a converging system are as follows:

1. Decrease cross-sectional area 10% in straight sections such as along the sprue or along the main runner.
2. Decrease cross-sectional area 20–30% through turns, depending on the radius; the smaller the radius, the larger the reduction in area.
3. Short runners in the sprue or short main runners may be constant area.
4. The inlet area of the taper tangential runner should be 1.2 times the area of the gate fed by that runner. Depending on the desired flow angle into the cavity, this area ratio can be altered. See Fig. 25 for flow angle relationships.

Regarding the flow angle into the cavity, ratios of tangent runner areas to gate area of 1.1–1.2 will provide a flow angle into the cavity of 40–45° from the normal to the gate. If larger runner inlet areas are used, the runner velocity is decreased and the flow angle into the cavity

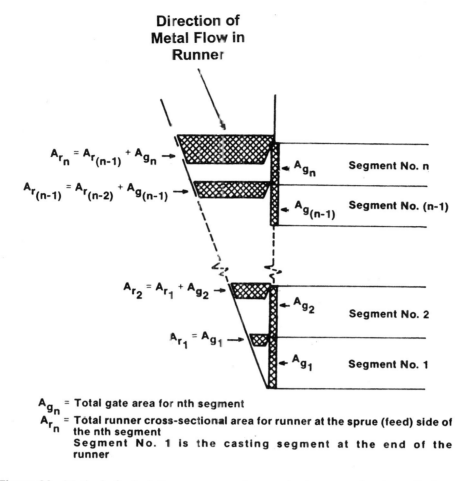

Figure 26 Method of calculating runner area in a constant cross-sectional area feed system.

will be closer to normal to the gate. For example, with a runner inlet area equal to twice the gate area, the flow angle will be about 30°. However, using a heavier runner will result in more metal to remelt after trimming.

Following the design guidelines for a converging cross-sectional area runner system will result in gate areas that typically are in the range of 40–60% of the nozzle area. The runner system with a decreasing cross-sectional area results in a greater latitude in achieving adequate die performance with the die casting machine whose flow efficiency discharge coefficient may be low or unknown.

There are two additional features that are typically used with converging runner systems as shown in Fig. 27. Those features are the shock absorbers and the delta region. The tangential runners typically terminate in 0.4 in. (10 mm) long sections with a 0.079 in. (2 mm) square cross section that enters disk-shaped shock absorbers. The use of shock absorbers is recommended to control

very high transient metal velocities that may be reached at the small square ends of the runners, with consequent high energy release (shock) through the extremities of the gate. That high transient velocity may cause severe erosion damage to the die. The shock absorbers also protect the die when the gate does not completely fill, as in a cold die during startup. The shock absorber should be cut 0.079 in. (2 mm) deep and its top (circular) area should be the same as the area at the entry to the tapered runner. Figure 28 shows the design guidelines for shock absorbers.

The delta region between the tangential runners shown in Fig. 27 feeds the central portion of the castings. The delta is often initially cut to the preselected gate depth. The flow in that region is then assessed during the initial casting trial. The delta region may then require modification to achieve proper flow in that region. The design guidelines for the delta region area are illustrated in Fig. 29.

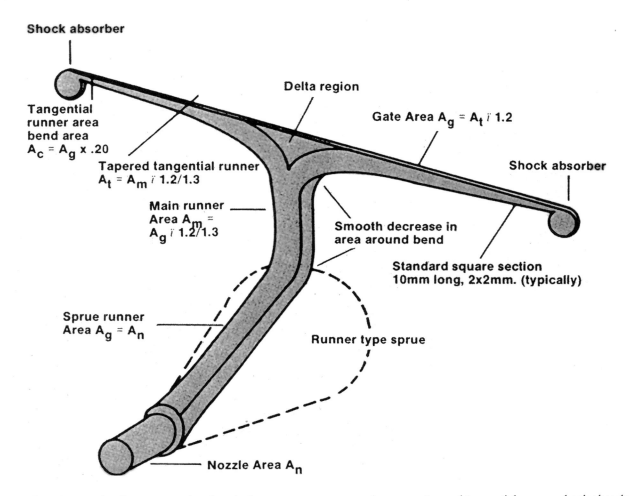

Figure 27 Converging flow system, showing single runner type sprue, main runner, tapered tangential runner, shock absorbers, the gate to the die cavity, and the typical areas of the various components of the runner system. General rules: (1) Decrease area 10% in straights (such as along sprue runner or main runner). (2) Decrease run 20–30% around bends, depending on severity. (3) Short runners in sprue or main runner may be constant area. A_n = Nozzle area; A_c = sprue area; A_m = main runner area; A_t = tangent runner area; A_g = gate area; A_c = tangent runner end area.

H. Determining Shape of Runners

After the runner areas have been calculated, runner cross-sectional configurations are selected that:

1. will minimize heat loss from the runners;
2. will cool rapidly enough so that high production rates can be obtained;
3. can be easily machined;
4. are easily designed;
5. have sufficient draft so that the casting can be easily ejected.

When long runners (at least 200 mm or 8 in., or longer) must be used, it is usually desirable to make the cross section of the long portion of the runners round to minimize heat losses. However, for practical reasons (economy, machinability), a rectangular (actually trapezoidal to allow draft for good ejection) configuration with depth equal to width is usually recommended. When short main runners are used, wide, shallow, rectangular (trapezoidal) runner sections are usually desirable to maximize the heat transfer from the runner and to maximize the production rate. Sec. IV, on die cooling systems and water line placement, gives information on determining the production rate that can be obtained with different size runners. At locations where runners split, it is usually desirable to make a rectangular (trapezoidal) runner section so as to minimize machining difficulties. Figure 30 shows how these features were employed in the design of one die and it shows one

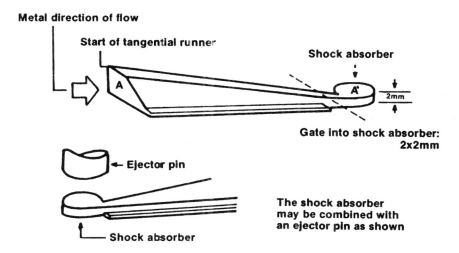

Figure 28 Design guidelines for runner shock absorber. The runner shock absorber prevents the "spurt" of metal that would occur at the end of the runner just as it fills. Shock absorber diameter should be sized to give area of shock absorber face (A') = area of start of tangent runner (A)—see diagram. The area of entry into the shock absorber should be 2×2 mm.

method of designing a runner with variable cross-section shapes.

Experience has shown that in most cases, the runners should be defined and described (at least the details shown in Fig. 31). In traditional practice, the runner system was first modeled in wood or other suitable medium first and then checked by the designer before the die was cut. This minimized costly machining errors. Use of computer-aided design techniques obviates the need for this modeling; however, careful checking of compatibility should be made before beginning die machining.

I. Selecting the Sprue Configuration

The constant or decreasing cross-sectional area rules should be followed through these sprue base and sprue. Standard sprue pin and bushing sets that have constant cross-section area are commercially available. Alternatively, a runner sprue can be designed and used to maintain constant area through the sprue. The use of runner sprues gives the designer added flexibility in that the runner channels can be made to various configurations, such as a round, half round, or rectangular (trapezoidal). Figure 32 illustrates a rectangular runner sprue and the cooling system normally applied. The shape of the flow path is good, metal flow characteristics are excellent, and the transitional curve into the runner has a relatively large radius. It is a simple design, but more expensive to make then a conventional fully turned sprue.

Runner sprues can be easily cut with a modified keyway (Woodruff) cutter, as shown in Fig. 33. To produce such a rectangular shape in a runner sprue, one should turn the cone shape of the sprue pin 0.010–0.020 in. (0.25–0.55 mm) less than the main cone of the sprue bushing. Channels are then cut into the sprue pin by use of the Woodruff cutter ground with appropriate draft angles on its sides. The included angle of the cone in the sprue bushing should be a minimum of 20°. Blank sprue pins in which the runners can be cut are also commercially available. Cooling of the sprue is carried out as with conventional sprues; however, cooling of the bushing may not be required. If cooling is required, it can usually be accomplished by a fountain under the runner region in the sprue pin.

J. Overflow System Design and Vents

Should overflows be necessary, as is usually the situation for parts where surface finish is critical, their locations should be determined by examination of the flow patterns that exist in the casting. Such overflows can serve four important functions in the casting:

1. They are reservoirs that receive and hold gasses flushed out of the cavity.
2. They are reservoirs for cold and oxidized metal that initially flows through the cavity.
3. They provide additional heat input to the die, therefore helping to maintain a satisfactory and stable die temperature.

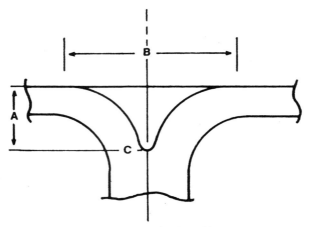

a. Length to width ration
Approximate A:B ratio 1:2.5.

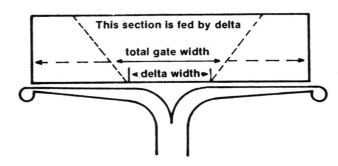

***If possible, use a larger eliptical or circular radius to produce a
delta width of 25% to 30% of the total gate width***

b. Width

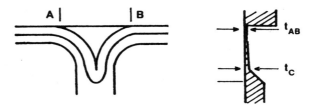

***Keep the gate thickness (at AB) constant. Increase thickness of
delta linearly towards point C by 30% to 40% to increase flow
through delta ($t_C = t_{AB}$ x 1.3 to 1.5).***

c. Thickness Variation

Figure 29   Design guidelines for the delta region.

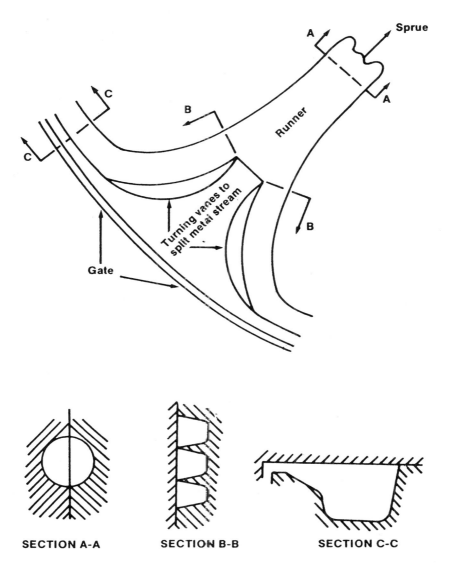

SECTION A-A   **SECTION B-B**   **SECTION C-C**

Figure 30 Cross sections through different segments of the runner-and-gating system used with the TV-bezel dies. Four runners were used to bring liquid metal into the diagonal corners of the die cavity.

4. They serve as locations for contact of ejector pins when the ejector pin marks are not permitted on any surface of the casting.

The first three functions help to eliminate blisters and cold laps. Overflow gates and overflows have been found to be very effective vents because large quantities of gasses flowing ahead of and mixed with the initial metal flowing into the cavity can be made to pass through the overflow gates, which are generally thicker and shorter than vents.

Traditionally designed vents are usually thin clearances with a long flow length cut between the die halves to allow the gasses to escape from the die cavity. However, as soon as a small amount of molten casting metal reaches the small vent openings, it often solidifies and prevents further flow through the vents. On the other hand, if the vents do operate, they can be very effective in allowing gasses to escape in front of the molten metal.

The use of constant area or converging area metal feed systems, with correctly designed runners, promotes the expulsion of air into the feed system ahead of the molten metal and through the cavity to the air vents. To promote expulsion of that air from the cavity, vents with total areas of the 20% of the gate

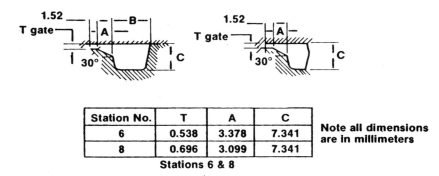

Station No.	T	A	C
6	0.538	3.378	7.341
8	0.696	3.099	7.341

Note all dimensions are in millimeters

Stations 6 & 8

Station No.	T	A	B	C
0&1	0.234	1.118	3.533	2.631
2	0.262	1.676	4.412	3.574
3	0.284	2.134	5.034	4.516
4	0.330	2.540	5.657	5.458
5	0.389	2.946	6.276	6.401
9	0.292	2.692	6.934	6.525
10	0.226	2.388	5.103	5.710
11	0.201	2.108	4.765	4.895
12	0.183	1.803	4.392	4.079
13	0.175	1.473	3.985	3.264
14&15	0.168	1.016	3.485	2.448

Stations 0-5 & 9-15

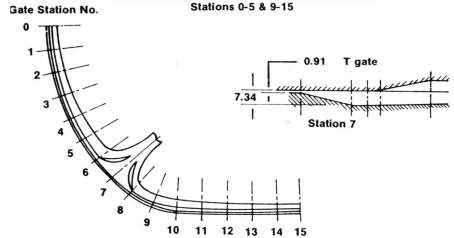

Figure 31 Dimensions of gates used with the modified TV-bezel die.

areas and with depths up to 0.039 in. (1 mm), as shown in Fig. 34, may be considered. Unfortunately, there has been very little research directed at optimizing vent design.

By proper design, the overflow gates and overflows can be used effectively as vents, and most of the molten metal that initially flows through the cavity and contains most of the entrapped gasses can be forced into the overflows. Experience has demonstrated that when the proper runner, gating, and overflow designs are used, blisters can be eliminated from die castings.

If the castings produced during the first trial run show poor surface finish or blisters, the following should be checked first:

1.   Machine operation and performance.
2.   Feed system design and dimensions.
3.   Die temperature.

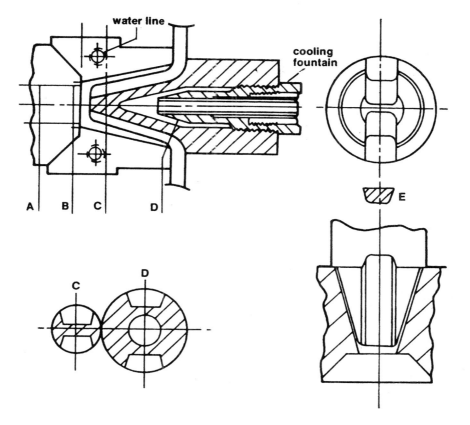

Figure 32 True runner type sprue, showing smooth flow path and large radius for transition into main runner. Area chart shows converging flow path.

This checking should indicate that both the machine and die are operating according to the values of parameters used in the design. If they are not, appropriate modifications should be made to ensure that the design values are being achieved.

The size of overflows will depend on the types of defects in the casting and on the degree of die heating required from the increased mass of molten metal. A further consideration is the distance of the molten metal travel through the feed system and cavity. Where long flow distances are unavoidable, large overflows may be necessary. Overflows may be necessary in the open center of ring-shaped castings feed around portions of their outer parameter from tapered tangential runners because air vents may be difficult to provide in such areas.

Some analytical studies have been conducted to aid the die caster in determining the size of overflows required to produce cold shut free castings. These studies show that if sufficient molten metal flows across the surface of the die and into an overflow, the surface of

the die is heated and only molten metal remains in the cavity when flow stops. Consequently, all the partly solidified metal will be flushed from the cavity and cold lap (cold shut) defects will be eliminated. As the casting thickness is reduced, the tendency for cold lap defects increases because less metal, and therefore less heat, is put into the die. Also, for a given surface area of casting in the die at a given temperature, the temperature of the metal in a thinner casting will be reduced more rapidly because of the shorter heat transfer path (reduced section thickness). These problems can be overcome by use of higher die temperatures, shorter cavity fill times, or the use of larger overflows.

Guidelines for determining the size of overflows required to produce cold lap free castings were an outgrowth of those studies that were followed by further laboratory and field studies. The first guideline that can be used when redesigning an existing casting to reduce its wall thickness is to take the volume or weight of the metal that was eliminated by reducing the section thickness and place it in the overflows. For example, if

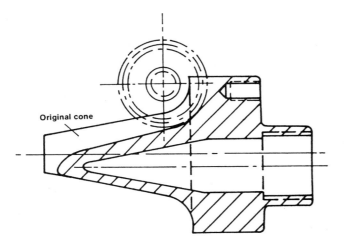

Figure 33 Use of a woodruff cutter to cut the runner into the sprue pin.

a casting with a wall thickness of 0.079 in. (2 mm) in weight of 2.2 lb (1 kg) has been successfully produced using 1.1 lb (0.5 kg) overflows, that casting can probably be successfully produced with a wall thickness of 0.040 in. (1 mm) and a weight of 1.1 lb (0.05 kg) by increasing the overflows to 2.2 lb (1 kg).

The general guideline was augmented by computer studies to determine the size of overflows required with a fixed set of die casting parameters. The casting parameters and the assumed values were:

1. The capacity of the die casting machine, assumed to be sufficient to pump 600 in.3 (9700 cm3) of zinc per second.
2. The die material, assumed to be P-20 hot work tool steel.
3. The runners, assumed to be four in number, round, and 0.50 in. (12.7 mm) in diameter.
4. No coating of any type on the die surface.
5. A holding temperature of 800°F (427°C).

With these parameters fixed on the indicated values, the cavity size, die temperature, runner length, and cavity fill time, which were varied in the corresponding overflow sizes, were calculated. Plots of the results show the required overflow size as a function of cavity fill time for a given die temperature. Two such plots are shown in Figs. 35 and 36. One of the interesting observations from these plots is that the runner length can be an important parameter. Heat loss from long runners is appreciable and is detrimental to casting quality. So the die caster should make every effort to keep runner lengths short.

These studies also uncovered many instances of inadequate and/or contradictory descriptions of the

thermal properties in the die casting system. Where this was the only kind of information available, worst-case assumptions were used. Therefore it is expected that by using the presented design curves, the die caster will be able to produce cold lap free castings with the overflow sizes calculated from these design curves. Because of the conservative assumptions that have been made, it may be practical to produce satisfactory castings with smaller overflows that would be calculated by using Figs. 35 and 36.

For example, a study conducted by investigators at the Gulf and Western Company's Natural Resources Group showed that virtually cold lap free castings 0.030 in. (0.7 mm) thick could be produced with overflows in zinc that were only 40% of the volume predicted when the curves given in Fig. 35 were used.

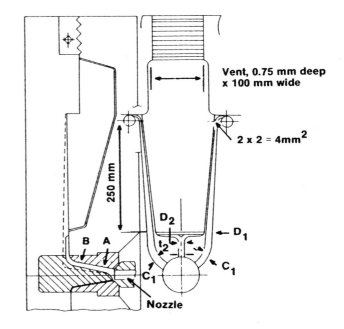

		Dimensions, mm	Total area, mm²
Nozzle		22 (Diameter)	380
Sprue runner	A		360
	B		330
Main runners	C_1	10.5x10.5 = 110	
	C_1	10.5x10.5 = 110	
	C_2	7x7 = 49	269
Tapered runners	D_1	10x10 = 100	
	D_1	10x10 = 100	
	D_2	7x7 = 49	249
Gate		600x0.335	201

Figure 34 Flow system details with dimensions and areas.

However, it is recommended that the die caster initially either construct dies with the size of overflows calculated according to these design curves, or at least allow room for larger overflows although initially smaller overflows (about one-half the size calculated) might be cut into the die. Then, during the die development period, adequate space will be available to enlarge the overflows to eliminate cold lap defects, if this approach is found to be necessary.

K. Discussion

Considerable attention has been devoted in this section to provide general and specific guidelines for designing feed systems for pressure die castings and a result was specifically given for zinc. As noted, they require a number of calculations to determine the dimension of the various components in the feed system. Adherence to the guidelines, and proper machining of the feed system in the die, should result in first-shot success and high-quality castings. Computer programs are now available from several sources such as the North American Die Casting Association that greatly simplify die design by eliminating many of the manual calculations and use of the graphs described in this chapter. However, they are based on the same principles.

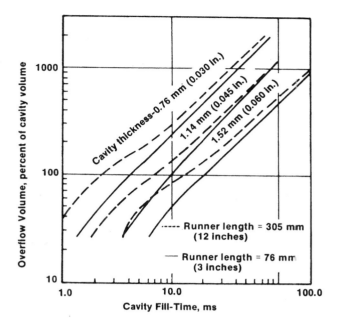

Figure 36 Calculated overflow size required to prevent cold-shut defects with a die temperature of 277°C (530°F). [Cavity fill time calculated from Eq. (7).]

IV. DESIGN OF THE COOLING SYSTEM FOR DIE CASTING DIES

A. Introduction

In addition to being a mold that produces a casting with a desired shape, a die casting die is a cooling device that extracts heat from the molten metal—the heat is then liberated while the metal cools to the solidification temperature; the latent heat of fusion is released during solidification and the heat from the solid metal is then cooled to the ejection temperature is released. Heat is removed from the casting by conduction through the steel die, either to coolant passages where the coolant transports heat away, or to the external surfaces of the die where heat is lost by radiation or convection, or through connections to the die holder and die casting machine that remove heat by conduction. The die performs the function of a heat exchanger in this way, and design considerations must be included in overall die design if good quality castings are to be made.

The die must rapidly remove the heat introduced during each shot so as to maintain an economical production rate; thus the heat-removal capacity of the die must balance the heat input to maintain the die temperature at the level required to produce satisfactory castings. The necessary cooling capacity, which may be different for various sections of the die, primarily

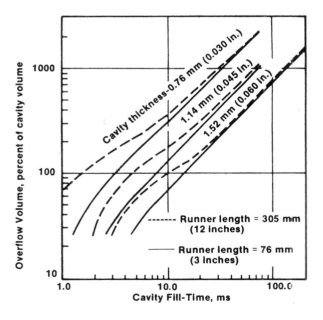

Figure 35 Calculated overflow size required to prevent cold-shut defects with a die temperature of 260°C (500°F). [Cavity fill time calculated from Eq. (7).]

depends on the section thickness of the casting, the production rate, and the type of surface finish required. When the optimum die temperature or die temperature balance has been achieved, it usually must be maintained within a relatively narrow range. In the production of thinner section castings with a high-quality surface finish, the control of die temperature may even be more critical than it is for castings with heavier sections.

To obtain the benefits inherent in automatic process controls, adequate cooling capacity must be designed into the die. To achieve that result, water line sizes and locations should be selected to provide more cooling capacity than will be required to cool the die when the die is at its maximum production. The die temperature then is maintained at the desired level by adjusting the flow of the coolant through the die by means of a temperature controller and solenoid operated valves.

In the die casting industry, the selection of the size and locations of cooling water lines has traditionally been left to the discretion of the die designer. In the majority of new dies constructed, that selection is based on the experience and judgment of the designer to provide proper cooling. The dies are then built and trial runs are made. If satisfactory castings cannot be made, water lines may then be added or deleted to lower or raise the temperature in selected portions of the die, or, in many cases, the machine production rate is decreased to compensate for inadequate cooling capacity of the dies. It is costly to conduct trial runs with the die, to add or eliminate water lines to achieve proper cooling, or to reduce the production rate because of inadequate cooling capacity. Also, the addition or elimination of water lines does not assure that optimum cooling conditions are obtained.

The determination of water line locations and sizes is a problem that can be solved through the use of computer analysis to simulate steady state heat transfer conditions. However, in many cases, the use of a sophisticated computer analysis is not necessary for determination of water line locations. High-quality die castings have been successfully produced with guidelines given below that ensure that adequate cooling capacity is available.

These guidelines depend on a steady state heat transfer analysis that is used to develop design criteria for the sizing and placement of cooling passages and dies. The configuration of die casting can be complex and vary from one type of casting to the next. Therefore rather than develop a program that would handle complex shapes, two simple casting sections were considered in development of these recommendations. It

was assumed that the casting was made of section that can be approximated by flat plates or strips. Even with those simplifications, the heat transfer analyses were formidable and required specialized computer programs. These programs used steady state heat transfer models whose description is beyond the scope of this chapter. The results of those computer calculations were analyzed to determine relationships among the imposed heating and cooling additions and resulting die surface temperatures. Depending on the particular casting and cooling configurations under consideration, the results were used to provide an equation for calculating the die surface temperature as a function of other parameters and graphs that relate die surface temperature and the other parameters [8].

Subsequently, to further simplify the effort required to determine the location of waterlines and dies, computer programs were written and are now widely available to the die casting industry through the North American Die Casting Association [9,10]. However, because all users of this chapter may not have access to these programs, the equations and graphical solutions initially developed from that work are presented in subsequent sections of this chapter. Although the use of the equations and graphs is more time consuming than the use of the computer programs, it is more cost-effective to spend the time to determine cooling passage locations before the die is constructed than to modify the die after initial production trials.

B. Heat Transfer Analysis

1. Casting and Cooling Passage Configurations Studied

Computer programs were established to model steady state heat transfer conditions in dies that produced simple casting shapes. The casting shapes selected for analysis were a plate with finite thickness (t) but infinite width and length, and a strip with finite width (w) and finite thickness (t) but infinite length. Four cases using these simple shapes were analyzed:

- Infinite plate casting cooled by water lines.
- Infinite strip casting cooled by a single water line.
- Infinite plate casting cooled by fountains.
- Infinite strip casting cooled by fountains.

Schematic diagrams of the casting and the cooling line configurations of these four cases are shown Figs. 37–40, respectively. These casting sections and cooling passage configurations can be used in various combinations and arrangements to represent many casting and cooling passage configurations employed in commercial dies. For example, many sections and

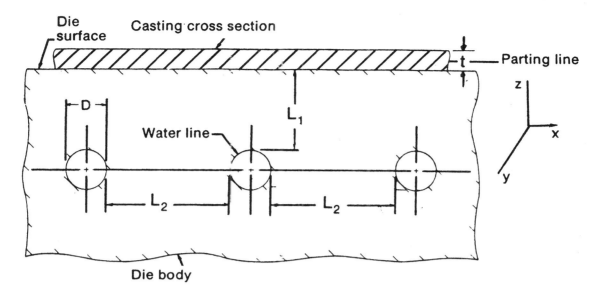

Figure 37 Water-line configuration used in a die section that produces a casting section in the shape of an infinite plate (case 1). Only one-half of the die assembly is shown.

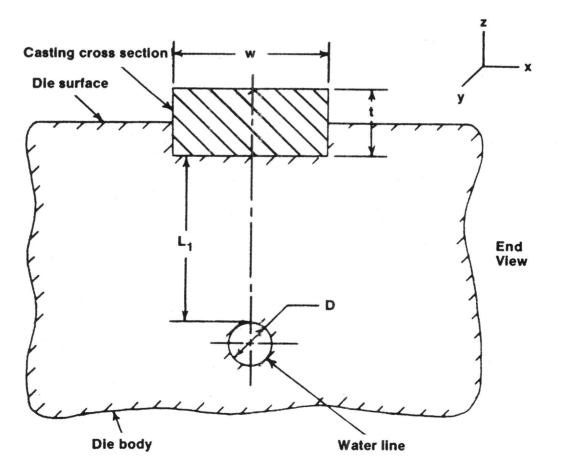

Figure 38 Water-line configuration used in a die section that produces a casting in the shape of a strip of infinite length (case 2). Only one-half of the die assembly is shown.

Top view of casting section

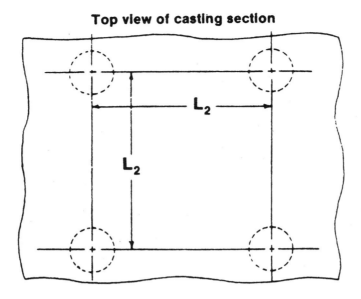

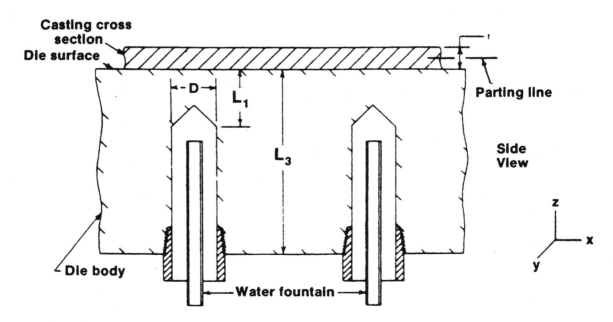

Figure 39 Water fountain configuration used in a die section that produces a casting in the shape of an infinite plate (case 3). Only one-half of the die assembly is shown.

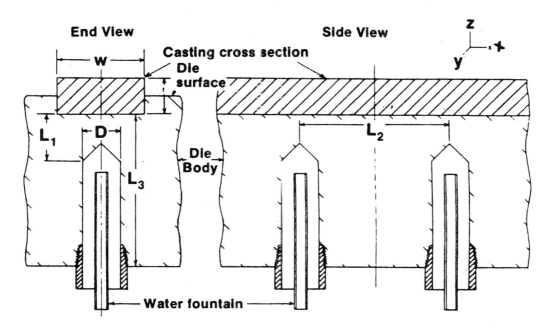

Figure 40 Water fountain configuration used in a die section that produces a casting section in the shape of a strip of infinite length (case 4). Only one-half of the die assembly is shown.

runners of frame and bezel castings are similar to the cases illustrated.

In the analysis, the effects of the following variables on the temperature distribution in the die were evaluated to determine the cooling capacities required for the selected casting sections:

1. The average die surface temperature (T_s) required to produce castings with a good surface finish.
2. Heat input $\dot{q}$, which is affected by the production rate, metal temperature, section size, latent heat of fusion, etc.
3. Thermal conductivity of the die steel (k).
4. Inlet temperature of the cooling water (T_c).
5. Flow rate of cooling water (V).
6. Distance of the water line or fountain from the die surface (L_1).
7. Distance between adjacent water lines or fountains (L_2).
8. Diameter of the water lines or fountains (D).

The results of the computer calculations were analyzed to determine relationships among the imposed heating and cooling conditions and the resulting die surface temperatures. Depending on the particular casting and cooling configurations under consideration, the results were correlated to provide equations (cases 1 and 2) for

determining the die surface temperature as a function of the other parameters, or as a series of graphs or charts (cases 3 and 4) relating die surface temperature and the other parameters.

C. Results

1. Equations for Cases 1 and 2, Infinite–Plate and Infinite–Strip Castings Cooled by Water Lines

For case 1, the infinite state casting cooled by water lines, illustrated in Fig. 37, the average die surface temperature is expressed as:

$$T_s = \dot{q}(D \times \phi K + B(1 + L_2/D)/4h_c) + T_c \qquad (12)$$

where T_s = die cavity surface temperature, °F (°C); $\dot{q}$ = heat input at the die surface, Btu/in.²/hr (cal/cm²/hr) [see Eqs. (15) and (16)]; D = diameter of the water line, in. (cm); K = thermal conductivity of the die steel, Btu/hr in. °F (cal/hr cm °C) (see Table 3); ϕ = dimensionless heat conduction parameter determined from a graph (Fig. 16); β = dimensionless heat convection parameter determined from a graph (Fig. 16); h_c = heat transfer coefficient between the surface of the waterline and the flowing water (coolant). This value is determined from an equation that expresses h as a function of the water properties of viscosity, density, thermal conductivity, and specific heat at the selected water temperature and pressure, and, also, of the water velocity [see Eq. (18) and Figs. 56 and 57]; L_2 = space in between water lines, in./cm; T_c = temperature of the water flowing through the water line, °F (°C).

For case 2, the infinite strip casting cooled by a single water line, illustrated in Fig. 38, the average die surface temperature can be calculated from the following equation:

$$T_s = \dot{q}(0.95)(DL_1/K^2)^{1/2}(1 + \ln(2w + t/3D)) \\ + (w + t)/h_c\pi D) + T_c \qquad (13)$$

where T_s, q, D, k, h_c, and T_c are defined as in Eq. (12); L_1 = distance between the die cavity surface and the top of the water line, in. (cm); w = width of the casting, inches (cm); and t = thickness of the casting, in. (cm).

Equation (13) can be rearranged to solve for the distance between the die surface and the top of the water line, L_1:

$$L_1 = k^2/D(((T_s - T_c)/\dot{q}_s \\ - (w + t)h_c\pi D)/(0.95)1 + \ln(2w + t)))^2 \qquad (14)$$

The mathematics involved in solving the equations for locating water lines is somewhat complex and lengthy. Thus the designer must spend a considerable amount of time to solve them. Therefore to facilitate the use of this information, computer programs have been written for cases 1 and 2 [9,10]. These programs are available from the North American Die Casting Association to die casters and die designers. With the use of these programs, the water line placement calculations can be performed in a matter of minutes.

Because all users of this manual may not have these computer programs, Eqs. (12) and (13) were solved for a number of conditions and the results were plotted to provide die designers with guidelines in locating water lines and dies. These graphs are presented in the next section.

2. Graphical Solutions of Eqs. (12) and (13)[*]

To aid the designer in using the results of the heat transfer analysis for locating water lines in die casting dies, the equations for cases 1 and 2 [Eqs. (12) and (13),

respectively] have been solved for various casting sizes and various locations of water lines, and the results are presented in graphical form.

3. Case 1: Infinite Plate Castings Cooled by Water Lines

The results of the computer analyses for die that employ 7/16 and 37/64 in. (11.1 and 14.7 mm) diameter cooling lines (the most commonly used sizes in the United States) to produce castings or casting sections can be approximated by an infinite plate or plotted in Figs. 41 and 42, respectively. In those figures, the distance between adjacent water lines (L_2) is plotted as a function of the heat input to the die cavity surface (q) for specific distances (L_1) of the water line below the cavity surface. The ends of each line on the graph represent the range of distances between the water lines that will result in a temperature difference across the die cavity surface of less than 20°F (11°C). The curves were determined for dies manufactured from P-20 steel. If a steel with a different thermal conductivity is used, the cooling capacity of the die would be increased or decreased by less than the percentage that the thermal conductivity of that steel exceeds or is less than, respectively, the thermal conductivity of the P-20 steel used in calculating the curves. For example, the thermal conductivity of H-13 steel is about 14% lower than that of P-20 steel; however, calculations using the value of k for H-13 steel showed that the die surface temperature for a given heat input and water line spacing is increased by 10% as compared with the die surface temperature obtained with the P-20 steel. Thus to use the graphs when a steel other than P-20 is used, a conservative approach would be to assume that the die surface temperature for each graph would be increased or decreased by the percentage by which the k value of the steel used in less than or greater than that of the P-20 steel, respectively. Thus if H-13 steel were used, the die surface temperature that would be maintained for Fig. 41 would be 570°F (296°C) rather than 500°F (260°C) because its k value is 14% lower then that of P-20 steel. Because the objective of the analysis is to insure that the die has adequate cooling capacity, using that approach will tend to assure that more than enough cooling capacity is installed in the die. Then, the desired surface temperature can be obtained and maintained by the use of commercially available die temperature control systems.

The curves presented in Figs. 41 and 42 were determined for cooling water temperature of 100°F (37.8°C). If, in practice, the cooling water temperature is different from that value, the average die surface temperature

---

[*] In the graphs, the water line diameter D, casting thickness and width (t and w), and water line spacings (L_1 and L_2) are expressed in mm, whereas those parameters were expressed in cm in Eqs. (12) and (13).

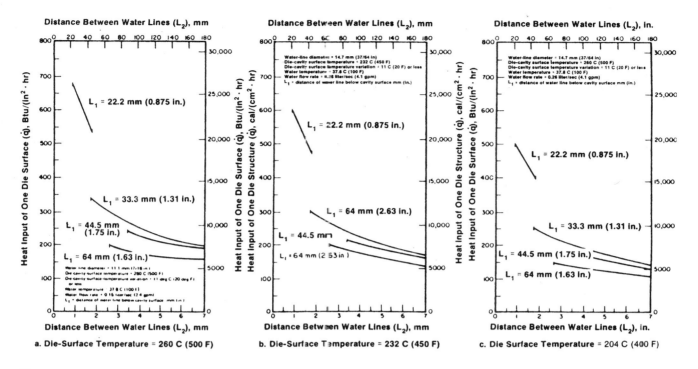

Figure 41 Graphs for determining the locations of 11.1 mm (7/16 in.) diameter water lines to maintain the die-cavity surface at the indicated temperatures for castings that can be approximated by a flat plate.

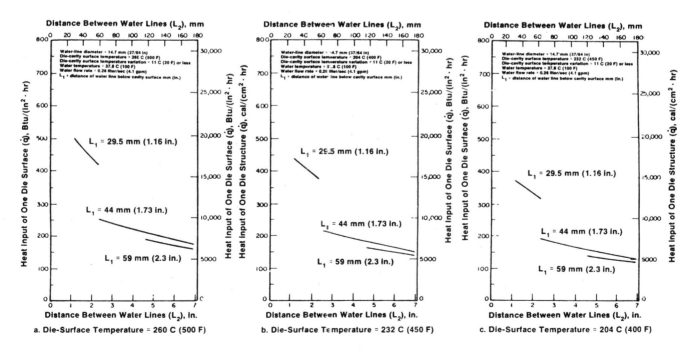

Figure 42 Graphs for determining the locations of 14.7 mm (37/64 in.) diameter water lines to maintain the die-cavity surface at the indicated temperatures for castings that can be approximated by a flat plate.

that can be maintained will be higher or lower than the value listed by approximately the amount that the actual water temperature used exceeds or falls below 100°F (37.8°C), except at very low coolant flow rates. For example, if the temperature of the coolant water flowing through the die is only 80°F (26.7°C), then the water line configurations based on Fig. 41 would be capable of maintaining the die surface temperature at 480°F (249°C) rather than 500°F (260°C). The water flow rates represented by these graphs is 2.4 gal/min (0.15 L/sec) for the 7/16 in. (11.1 mm) diameter water lines and 4.1 gal/min (0.26 L/sec) for the 37/64 in. (14.7 mm) diameter water line. Higher flow rates will not significantly increase the cooling capacity of the die. If the flow rate were to become infinitely large, the increase in the cooling capacity would be about 15%. However,

lower water flow rates will decreases the cooling capacity; for example, a flow rate of 1 gal/min (0.06 L/sec) will reduce for most cases the cooling capacity of the die with 7/16 in. (11.1 mm) lines by about 15% or less than that indicated by the graphs. For a die with 37/64 in. (14.7 mm) diameter water lines, the same flow rate will reduce the cooling capacity by about 20% or less than that indicated by the graphs.

4. Case 2: Infinite Strip Casting Cooling by a Single Water Line

The graphs for case 2, the infinite strip casting cooled by a single water line, are plotted in Figs. 43–45. In these graphs, the distance of the water line below the die cavity surface is plotted as a function of the heat input to

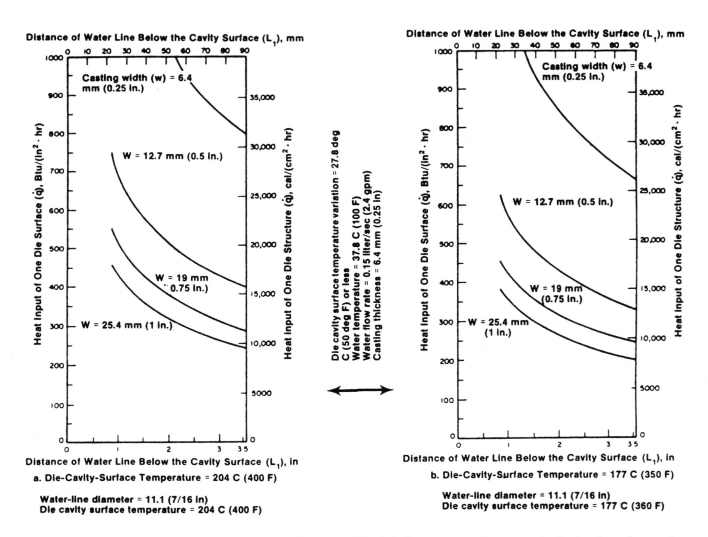

a. Die-Cavity-Surface Temperature = 204 C (400 F)

Water-line diameter = 11.1 (7/16 in)
Die cavity surface temperature = 204 C (400 F)

b. Die-Cavity-Surface Temperature = 177 C (350 F)

Water-line diameter = 11.1 (7/16 in)
Die cavity surface temperature = 177 C (360 F)

Figure 43 Graphs for determining the location of 11.1 mm (7/16 in.) diameter water lines to maintain the die surface at the indicated temperatures for a casting that can be approximated by a flat strip with a thickness of 6.4 mm (1/4 in.).

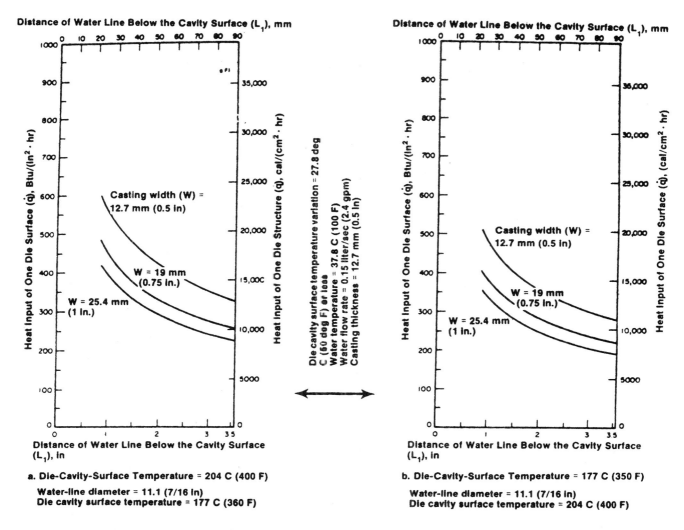

Figure 44 Graphs for determining the locations of 11.1 mm (7/16 in.) diameter water line to maintain the die surface at the indicated temperatures for castings that can be approximated by a flat strip with a thickness of 12.7 mm (0.5 in.).

one die half, q. Figure 43 is for a 0.24 in. (6.4 mm) thick strip. Figure 44 is for 0.5 in. (12.7 mm) thick strip and Fig. 45 is for 0.75 in. (19 mm) thick strip. Each figure includes curves for two specific die temperatures, 400°F (204°C) and 350°F (177°C). The effects of variations in die material, water temperature, and flow rate would be essentially the same as for case 1 discussed earlier.

5. Infinite Plate and Infinite Strip Castings Cooled by Fountains

The casting in the fountain configurations for cases 3 and 4 were illustrated in [Figs. 39 and 40], respectively. The results of the heat transfer analysis of the infinite

plate (case 3) are given in Figs. 46 and 47 and those for the infinite strip (Case 4) are given in Figs. 48–51. The graphs were prepared for a single fountain diameter, 7/16 in. (11.1 mm), and for a fixed value of the temperature of the water flowing through the fountains, 100°F (37.7°C). The data are plotted to show heat input, q, vs. the spacing between fountain centerlines, L_2. In Figs. 46 and 47, for the infinite plate, each cross-hatched area on the graph represents a particular fountain depth below the die/cavity surface. The graphs for the infinite strip have separate areas for different die thicknesses, t, and widths, w.

The cooling water flow rates were selected based on calculations that employed a wide ranges of values for the heat transfer coefficient between the water and

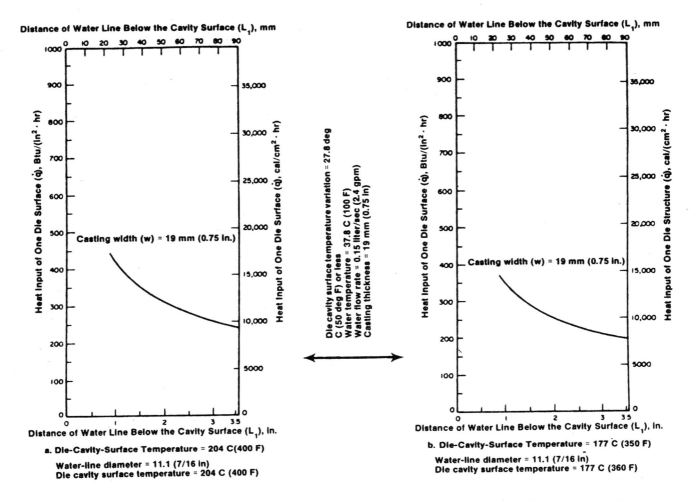

Figure 45 Graphs for determining the locations of 11.1 mm (7/16 in.) diameter water lines to maintain the die surface at the indicated temperatures for castings that can be approximated by a flat strip with a thickness of 19 mm (0.75 in.).

fountain surfaces, including the use of values for the fountain interior top surface different from those used for the fountain side walls. The results for the various flow combinations are presented in the graphs in terms of flow rates that reliably provide the cooling necessary to maintain die surface temperature, T_s, as either 500°F (260°C) or 400°F (204°C). The band of cooling conditions is bounded on the upper side by a maximum or medium cooling water flow rate and on the lower side by a minimum flow rate. The flow rate corresponding to maximum cooling is 10 gal/min (0.63 L/sec), the medium rate is 4 gal/min (0.25 L/sec), and the low rate is 1 gal/min (0.063 L/sec). In practice, most cooling water flow rates are expected to be near the low end of the range, 1–2 gal/min (0.063–0.126 L/sec). On the graphs for the strip casting with the fountains located 1 in. (25.4

mm) below the cavity surface (Figs. 49 and 50), the case for the casting 0.25 in. (6.4 mm) thick by 1 in. (25.4 mm) wide is represented by a single line for the low coolant flow rate. For higher water flow rates with that strip cross section, the temperature gradients on the die cavity surface exceeded 50°F (27.8°C).

For a fixed fountain spacing, the 500°F (260°C) or 400°F (204°C) die surface temperature, T_s, can be maintained by varying the coolant flow rate so that the rate of heat extraction equals the heat input, q, corresponding to a desired casting production rate and casting thickness. The value of q must be between the values corresponding to the two bounding coolant flow rates. The basic procedure is one minor adjustment in balancing the heat input (production rate for a particular casting) and the cooling water flow rate within the

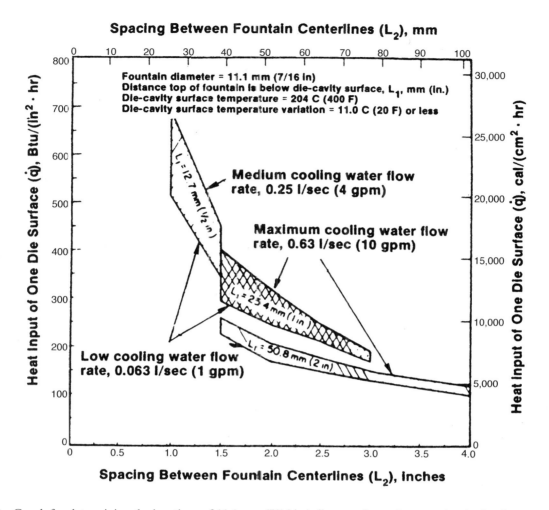

Figure 46 Graph for determining the locations of 11.1 mm (7/16 in.) diameter fountains to maintain the die-cavity surface at 204°C (400°F), for castings that can be approximated by a flat plate.

baths for a particular casting. The die designer is aided by the use of these graphs in that he can select a fountain spacing and know that he can maintain the required die surface temperature for a given production rate—he need only vary the cooling water flow rate, for example, by a temperature controller, or vary the production rate over a very small range.

6. General Procedures for Designing the Die Cooling System

To use the equations, computer programs, and graphs that had been developed, the designer must divide the casting into segments with shapes that are equivalent to flat plates or strips. For example, many frame and bezel castings consist of combinations of flat plates and strips,

and the runners in most castings can be considered as a strip. If the portion of the casting has a reinforcing lip or rib that is somewhat thicker than the remainder of the casting, then it is suggested that the average thickness of the section be used in the calculations. If there is a large boss present, a fountain may have to be placed immediately below the boss to ensure adequate cooling of that region. Possible cooling passage arrangements for a large shallow boss and a large deep boss on thin plate castings are illustrated in Figs. 52 and 53, respectively. The important point is that adequate cooling capacity must be designed into the die. Thus, for a given case, one should assume the worse conditions when considering the die cooling capacity required. With sufficient cooling capacity, an automatic die temperature controller will be capable of maintaining the proper die temperature in

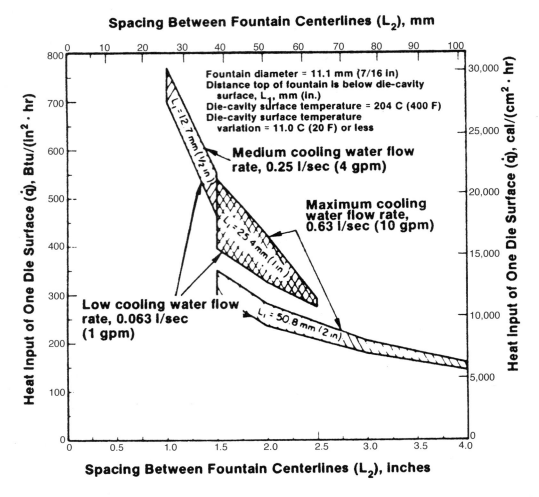

Figure 47 Graph for determining the locations of 11.1 mm (7/16 in.) diameter fountains to maintain the die-cavity surface at 260°C (500°F), for castings that can be approximated by a flat plate.

the various regions of the die by regulating the flow of coolant.

In addition, the die designer should obtain values or reasonable estimates of the values for the various input parameters that will be fixed or restricted to a narrow range by other design or die construction considerations, or by production conditions. For example, typical water line diameters (tap drill sizes) used in the United States are 7/16 in. (11.1 mm) and 37/64 in. (14.7 mm). Also, the temperature and velocity of the cooling water may be restricted by the temperature of the supply and the water system pressure. However, the values of these parameters should be measured at the point where the water enters the die. The flow rate can be determined by measuring the volume of flow in 1 min. Usually, these flow rates are less than 2.5 gal/min (0.16 L/sec).

Once the casting is divided into appropriate sections that can be approximated by plates and strips, the die designer must select values for the various parameters and solve the equations for T_s or locate the proper positions on the graphs. These parameters include the production rate, P, the thickness, t, or the thickness and width (t and w) of the casting section, the water line or fountain diameter, D, the distance between the water lines or fountains in the die cavity surface, L_1, the spacing between water lines or fountains, L_2, the temperature of the water in the cooling passage, T_c, and the flow rate of the water through these passages.

The value of the heat input, q, can then be determined from the casting geometry and the selected production rate as is illustrated in Sec. V.A using Eqs. (16) and (17) or Fig. 54.

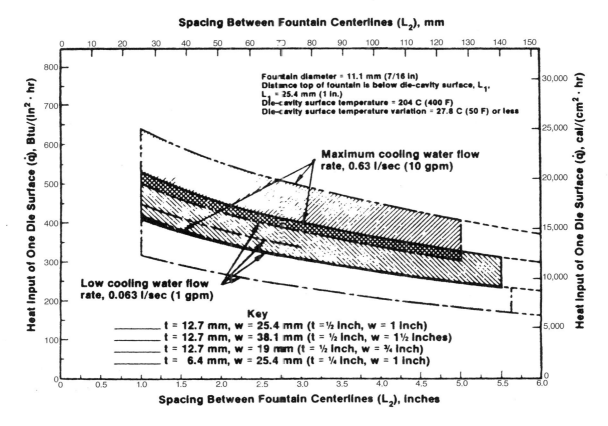

Figure 48 Graph for determining the spacing between 1_.1 mm (7/16 in.) diameter fountains located 25.4 mm (1 in.) below casting surface to maintain die-cavity surface temperature a: 204°C (400°F), for castings that can be approximated by a strip.

With the value of q and the cooling passage configuration, the designer can determine from the appropriate graph the cooling passage spacing that will mainta:n the die at the desired temperature. If the desired valu2 of T_s cannot be achieved with the initial values selected for the parameters, the die designer can repeat th2 procedure using different values for some of the parameters until the desired value for T_s is achieved. This procedure will allow the die designer to determine appropriate combinations of cooling passage spacings and diameters that will result in desired die surface temperature for specific values of production rat⊏, casting section configuration, cooling water flow rat⊏, and water temperature.

7. Example Calculations Using the Information Developed

To illustrate how the information developed can be utilized, sample calculations have been prepared. A⊏sume that the die caster is producing a flat plate-type zinc die casting 0.06 in. (1.52 mm) thick, and the casting is fed by a runner 0.5 in. (12.7 mm) thick and 0.75 in. (19 mm) wide. Assume that the water line or fountain arrangements illustrated in Figs. 37–40 will be used to cool the die. The production conditions for this hypothetical casting are listed in Table 1.

8. Plate Casting Cooled by Water Lines

The assumed cooling system configuration shown in Fig. 37 and the assumed casting conditions (Figs. 41 and 42) are used to determine the water line diameters and spacings. Assume that the die designer elects to use 7.16 in. (11.1 mm) diameter water lines that are distance L_1 of 1.75 in. (45 mm) beneath the die cavity surface and a distance L_2 of 3.5 in. (89 mm) apart. For $t = 0.06$ in. (1.52 mm) and $P = 300$ shots per hour, Fig. 16 shows that $q = 5460$ cal/cm² hr. Using Fig. 41, the values of L_1

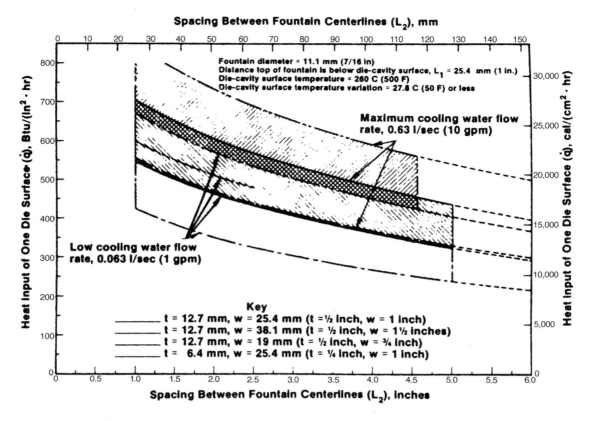

Figure 49 Graph for determining the spacing between 11.1 mm (7/16 in.) diameter fountains located 25.4 mm (1 in.) below the die casting surface to maintain die-cavity surface temperature at 260°C (500°F), for castings that can be approximated by a strip.

and L_2 for this case, and the heat input, it can be seen that the selected water line configuration can maintain the die surface at a temperature below 400°F (204°C). Thus the water line size and location selected will provide more than enough cooling capacity to hold the die surface temperature T_s at the desired temperature, 500°F (260°C). In addition, because scale may build up in the water lines and thus reduce heat transfer, the water line configurations determined in this example provide an extra margin of safety. The die temperature control system will regulate the flow of coolant to maintain the required die temperature.

Figure 38 illustrates the basic casting water line configuration for the runner feeding this casting, assuming that a single water line is used for cooling the runner. For the geometry $w/wt/(w + t) = 0.3$ in. (7.62 mm) and the desired production rate $P = 300$ shots per hour, Fig. 16 shows that $q = 700$ Btu/in²/hr (27,300 cal/cm²/hr). Should the die designer elect to place the water line 7/8 in. (22 mm) below the runner cavity surface, a check of Fig. 44 would reveal that a portion of the die

around the runners does not have sufficient cooling capacity to bring the die surface temperature in that region to the desired value of 400°F (204°C). Thus the die designer could elect to place the water line closer to the cavity or increase the water line diameter. Alternatively, the production rate would have to be lower to reduce the heat input at the die cavity surface.

9. Plate Casting Cooled by Fountains

Under the assumed casting conditions, $q = 140$ Btu/in²/hr (5460 cal/cm²/hr). Figure 47, the appropriate graph for this condition, shows that fountains located 2 in. (51 mm) below the cavity surface (L_1) and 4 in. (1.02 mm) apart (L_2) will maintain the die surface at 500°F (260°C). To provide added cooling margin, because the cooling efficiency will decrease as scale forms in the water line, Fig. 46 can be used to determine the cooling line spacing that will maintain the die surface at 400°F (204°C). From Fig. 47, $L_1 = 2$ in. (51 mm) and $L_2 = 2.85$ in. (72 mm).

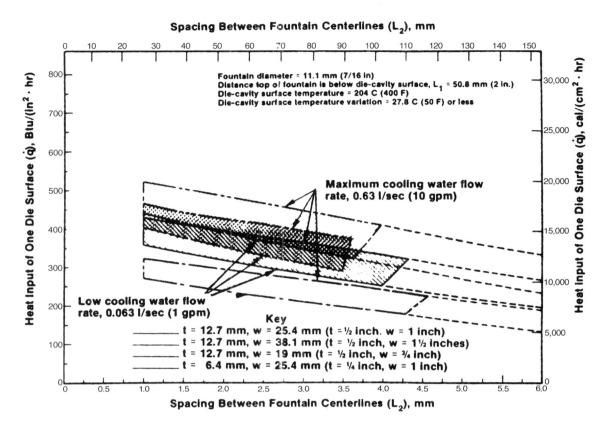

Figure 50 Graph for determining the spacing between 11.1 mm (7/16 in.) diameter fountains located 50.8 mm (2 in.) below the die-cavity surface to maintain die-cavity surface temperature at 204°C (400°F), for castings that can be approximated by a strip.

10. Strip Casting Cooled by Fountains

For the assumed geometry and casting conditions. $q = 700$ Btu/in.^{2}/hr (27,300 cal/cm^2/hr). As was previously indicated, the desired casting surface temperature for the runner is 400°F (204°C). Inspection of Figs. 48 and 50, the appropriate graphs for maintaining the die surface temperature at 400°F (204°C), shows that for the cooling water flow rate available with the 7/16 in. (11.1 mm) diameter fountains and within the limits of the graph, fountains cannot be used to maintain the desired die surface temperature and still maintain a surface temperature variation of less than 50°F (27.8°C). To maintain the die surface temperature at 400°F (204°C) in the runner area, the production rate would have to be decreased so that the heat input q is around 500 Btu/in^2/hr (19,500 cal/cm^2/hr); that value is obtained from Fig. 48 using a 1 in. (25.4 mm) fountain spacing beneath the die surface (L_1), the minimum spacing between fountains of 1 in. (25.4 mm), and taking the lower boundary of the cooling water flow rate for the strip size under consideration. Also, $t = 0.5$ in. (12.7 mm) and $w = 3/4$ in. (19 mm). For the assumed configuration of the runner, $wt/(w + t) = 0.3$ in (7.62 mm), the heat input value of 500 Btu/in.^{2}/hr (19,500 cal/cm^2/hr) corresponds to a production rate of $P = 200$ shots per hour (see Fig. 41).

11. Accuracy of the Procedure for Determining the Cooling Capacity of Dies

The procedures developed were designed to provide the die caster with a relatively quick and easy way to determine the approximate location of water lines and fountains in the dies that produce castings with configurations that could be approximated by flat plates or strips. In most cases, the procedures will provide estimates that are conservative; that is, more cooling capacity than is required will be installed in the dies. Dies designed with excess cooling capacity can make optimum use of automatic die temperature control systems.

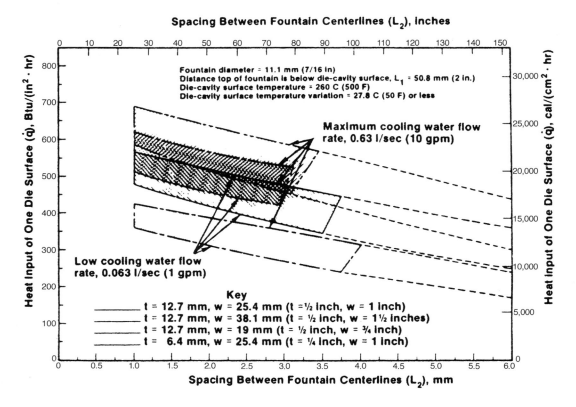

Figure 51 Graph for determining the spacing between 11.1 mm (7/16 in.) diameter fountains located 50.8 mm (2 in.) below the die-cavity surface temperature at 260°C (500°F), for castings that can be approximated by a strip.

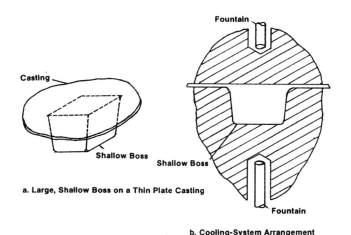

Figure 52 Possible method for using fountains to provide cooling for a large, shallow boss on a thin plate casting.

V. DETERMINING INPUT VALUES FOR WATER LINE PLACEMENT CALCULATIONS

A. Heat Input $\dot{q}$

The value of the heat input, $\dot{q}$, can be determined from the casting configuration and the selected production rate using the following equations.

For an infinite plate casting

$$\dot{q} = (C_p[T_i - T_e] + H_f)(t/2)XP \qquad (15)$$

And for an infinite strip casting

$$\dot{q} = (C_p[T_i - T_e] + H_f)(wtP)/(2[w + t]) \qquad (16)$$

where C_p = specific heat of the alloy being cast, cal/cm³ °C (Btu/in.³ °F); T_i = temperature of the metal at

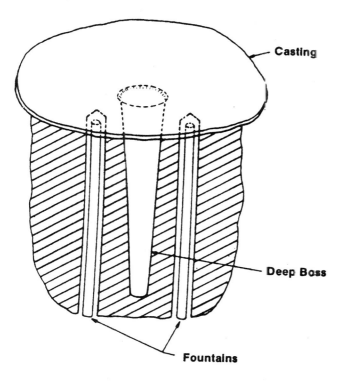

Figure 53 Possible fountain configuration for cooling a large deep boss on a thin-plate casting. For this case, the fountains should be treated as water lines cooling a strip casting.

injection °C (°F); T_e = temperature of the casting at ejection °C (°F); H_f = latent heat effusion for the alloy being cast, cal/cm³ (Btu/in.³); t = casting thickness cm (in.); w = casting width, cm (in.); P = production rate, shots per hour.

Values of the specific heat C_p and latent heat effusion H_f for zinc Alloy 3 are listed in Table 2.

In Eqs. (15) and (16), the casting thickness or the width and thickness parameters are divided by 2 because it was assumed in the calculations that half of the total heat flux would be removed by each die half, i.e., the cover and ejector die halves. The graph presented in Fig. 54 was prepared to simplify the calculation of $\dot{q}$. In this graph, the value of $\dot{q}$ is given for different combinations of casting thicknesses (or casting thickness and width) and production rate, assuming that the temperature difference ($T_i - T_e$) is 111°C (200°F). This graph can be used for both of the casting shapes considered, the infinite plate and the infinite strip, by using the proper geometry factor, t or $(w + t)/wt$, respectively.

B. Thermal Conductivity of Die Steel

The thermal conductivities, k, of P-20 and H-13 die steels in the temperature range of 300–500°F (149–260°C) are listed in Table 3. Single values of k are listed for the temperature range because the data indicate relatively little change in k over that temperature range.

C. Values for φ and β [For Use in Eq. (12)]

The values for φ and β are determined from the graph presented in Fig. 55. To determine φ, the designer locates the value of L_2/D on the ordinate. Using the left side of the graph, a line, parallel to the abscissa, is drawn from the L_2/D value until it intersects a line that represents the value of L_1/D calculated by the designer. A line is drawn from the intersection point on the L_1/D line perpendicular to the abscissa, and then the value of φ is read. The value of β is determined in exactly the same manner, except the right side of the graph presented in Fig. 55 is used. In Fig. 55, the short, horizontal lines drawn at the ends of the L_1/D curves indicate the upper and lower bounds of L_2/D for which the curves should be used and still maintain a thermal gradient of 11°C (20°F) or less on the die surface between the waterlines.

The values of h_c can be determined from the following general equation:

$$h_c = CXk_{T_c}/D_H(VD_H\rho_{T_c}/\mu_{T_c})^{0.08} \times (u_{T_c}XC_p/K_{T_c})^{0.4} \tag{17}$$

where h_c = heat transfer coefficient between the surface of the waterline and the flowing water (coolant) expressed here as cal/hr/cm² °C (Btu/hr/ft² °F). C = a constant, 9.6×10^{-4} for metric units and 2.3×10^{-2} for English units; K_{T_c} = thermal conductivity of the water flowing through the dies at an average temperature T_c, cal/hr/cm °C (Btu/hr/ft °F); D_H = the hydraulic diameter of the cooling channel for a channel of circular cross section (tube or pipe), D_H is equal to the inside diameter of the channel, m (ft); V = velocity of water through the cooling channel, m/sec (ft/hr); ρ_{T_c} = density of the flowing water taken at some average temperature T_c, kg/m² (lb/ft³); μT_c = absolute viscosity of water taken at some average temperature, kg/m/sec. (lbs./ft./hr); C_p = specific heat of water at some average temperature, T_c, cal/kg °C (Btu/lb °F).

The English unit of h_c as calculated by Eq. (17) is Btu/hr/ft² °F. This value must be divided by 144 to convert it to the proper units of h_c for use in Eqs. (12) and (13), i.e.,

Casting Thickness, t, or tw , millimeters
(t +w)

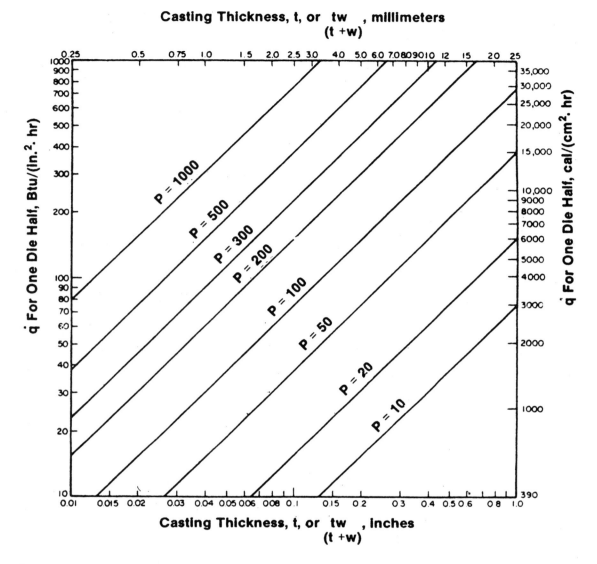

Figure 54 Graph for determining the heat input, $\dot{q}$, for each die half assuming $T_1 - T_\theta = 111\,°C\ (200\,°F)$. Graph is for Zinc Alloy No. 3. P = shots per hour; T_1 = temperature of the zinc at injection = $426\,°C\ (800\,°F)$; T_e = temperature of the casting at ejection = $316\,°C\ (600\,°F)$. The metric scale for casting thickness and the casting thickness and width parameter is expressed in millimeters whereas those parameters were expressed in centimeters in Eqs. (15) and (16).

Btu/hr/in.^{2} °F. The average water temperature T_c at which the various physical properties of water are determined is taken as the arithmetic mean of the inlet and exit water temperatures of the dies. To simplify the determination of h_c, the graphs presented in Figs. 56 and 57 were prepared for typical values of cooling line diameter, D_H, water velocity, V, and average temperature, T_c. In these curves, all necessary conversion factors

have been used to provide the value of h_c with the proper units for use directly in Eqs. (12) and (13).

D. Value of Water Temperature T_c

As was previously indicated, T_c is the average of the inlet and exit temperatures of the water flowing through the water lines in the die. Based on information obtained

Table 1 Production Conditions for Hypothetical Casting Considered in the Example Calculations

Parameter	Value
Casting thickness	1.52 mm (0.06 in.)
Runner thickness	12.7 mm (0.5 in.)
Runner width	19 mm (0.75 in.)
Production rate	300 shots/hr
Desired die-surface temperature, T_{s1}, for plate-casting area	260°C (500°F)
Desired die-surface temperature, T_{s1}, for runner area	204°C (400°F)
Water-line diameter, D	11.1 mm (7/16 in)
Fountain diameters, D	11.1 mm (7/16 in)
Water flow rate in water lines	0.15 L/sec (2.4 gpm)
Water flow rate in fountain	0.13 L/sec (2 gpm)
Water temperatures, T_c	37.7°C (100°F)
Die steel	P-20

Table 2 Thermal Properties of the No. 3 Zinc Alloy (SAE) 903 or ASTM Alloy AG40A, UNS Z33520

Property	Units	Value[a]
1. Specific heat, C_p	Btu/(in.^{3}, °F)	0.024
	Cal/(cm^3, °C)	0.661
2. Latent heat of fusion, H_f	Btu/in.^{3}	10.6
	cal/cm^3	163.1

Table 3 Thermal Conductivity (k) of Die Steels at Temperatures Between 149°C and 260°C (300°F and 500°F)

Steel	Units	Value of k
H-13	Btu/(hr in. °F)	1 38
	cal/(hr cm °C)	247 0
P-20	Btu/(hr in. °F)	1 60
	cal/(hr cm °C)	286 0

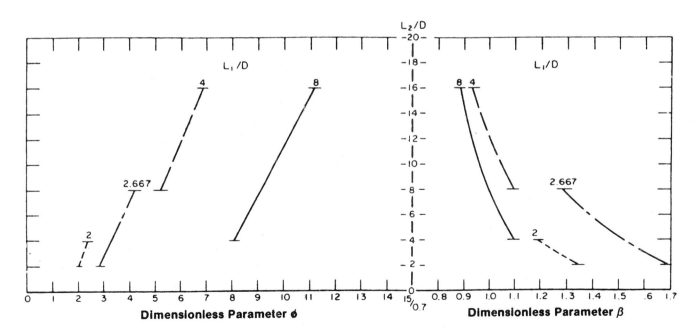

Figure 55 Graph for determining the value of ϕ and β in Eq. (12). The horizontal lines at the ends of each L_1/D curve indicate the upper and lower bounds of L_2/D for which the curves should be used and still maintain a thermal gradient on the die surface between water lines of about 11°C (20°F) or less.

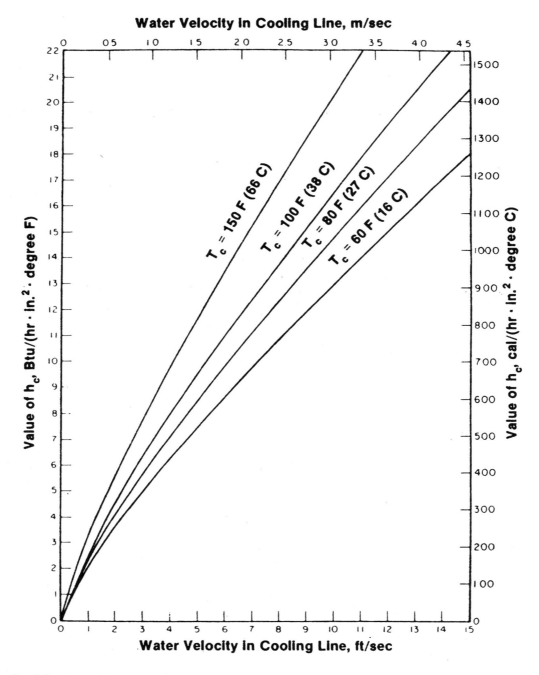

Figure 56 Graph for determining h_c at low water velocities and for a 11.1 mm (7/16 in.) diameter cooling line. $D_H = 0.0365$ ft of 0.0111 m. For this water-line size, a velocity of 1 m/sec (3.28 ft/sec) equals a flow rate of 0.097 L/sec (1.54 gpm).

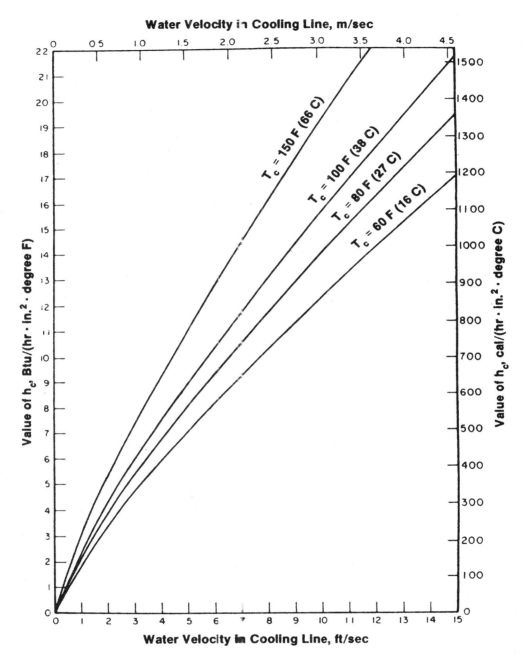

Figure 57 Graph for determining h_c at low water velocities and for a 14.7 mm (37/64 in.) diameter cooling line $D_H = 0.0482$ ft of 0.0147 m. For this water-line size, a velocity of 1 m/sec (3.23 ft/sec) equals a flow rate of 0.169 L/sec (2.69 gpm).

from field investigations, die casters typically run water through their dies so that sufficiently high velocity that usually there is a difference of less than 5°C (10°F) between the inlet and outlet water temperatures. Thus for practical purposes, T_c can be set equal to the inlet water temperature.

REFERENCES

1. Davis, A.J. Graphic method of analyzing and controlling the hot chamber die casting process. Die Casting Engineer 1978, 22 (2), 44–47.
2. Die Casting Machine Shot System Instrumentation: Description of Instrumentation, Transducer Connections and Interpretation. Cominco, Ltd., Product Technology Centre: Mississauga, Ontario, Canada, August 1980.
3. Die Casting Machine Calibration: Development and Interpretation of the Hydraulic Power Diagram. Cominco, Ltd., Product Technology Centre: Mississauga, Ontario, Canada, November 1980.
4. ILZRO Project ZM-132A. Die Casting Process Improvement; Project Reports, International Lead Zinc Research Organization, Inc.: Research Triangle Park, NC, 1979–1988.
5. Kaiser, W.D.; Groeneveld, T.P. *A Microcomputer Based Machine Calibration System, First South Pacific* Die Casting Congress, Paper 80–16, Australian Society of Die Casting Engineers; Melbourne, Australia, 1980.
6. Brawley, G.; Kaiser, W.D.; Groeneveld, T.P. A Method for Computer Analysis of Die Casting Machine Shot Calibration Data, 11th International Society of Die Casting Engineers Congress and Exhibition, Paper No. G-T81-115, Cleveland, Ohio, June 1981.
7. Cope, M.A. *Zinc Pressure Die Casting—The Metal Flow System*; Australian Zinc Development Association: Melbourne, 1979.
8. Groeneveld, T.P.; Robinson, R.; Kaiser, W.D. Design Data for Size, Number and Locations of Water Lines in Zinc Die Casting Dies, 8th International Society of Die Casting Engineers International Congress and Exhibition, Paper G-T75-085, Detroit, Michigan, 1975.
9. Doyle, G.R.; Kaiser, W.D.; Groeneveld, T.P. Computer Aided Water Line Placements for Zinc Die Casting Dies, International Society of Die Casting Engineers Exposition and Congress, Paper G-T81-054, Des Plaines, Illinois, 1991.
10. Herrschaft, D.C.; Nevison, D. In *Computer Aided Water Line Placement of Zinc Die Casting Dies*; International Lead Zinc Research Organization, Inc.: Research Triangle Park, North Carolina, 1980.

12

Heat-Treating Process Design

Lauralice Campos Franeschini Canale
Universidade de São Paulo, São Carlos, Brazil

George E. Totten
G. E. Totten & Associates, LLC, Seattle, Washington, U.S.A.

David Pye
Pye Metallurgical Consulting, Inc., Meadville, Pennsylvania, U.S.A.

Heat treating is defined by the International Federation for Heat Treating and Surface Engineering (IFHTSE) as: "a process in which the entire object, or a portion thereof, is intentionally submitted to thermal cycles and, if required, to chemical and additional physical actions, in order to achieve desired (change in the) structures and properties" [1]. Krauss [2] has added the additional caveat that "heat treatment for the sole purpose of hot-working is excluded from the meaning of this definition." The thermal cycles referred to in this definition are the various heat-treatment steps, which include stress relieving, austenitizing, normalizing, annealing, quenching, and tempering. Steel is heat-treated to control microstructure formation, to increase the strength and toughness, to release residual stresses and prevent cracking, to control hardness (and softness), to improve machinability, and to improve mechanical properties including yield and tensile strength, corrosion resistance, and creep structure. Each step of the heat-treatment process is performed for a particular purpose. Taken together, these heat-treatment steps are like "links in a chain" [3]. The acceptability of the final properties is limited by the weakest link.

In this chapter, an overview of the metallurgy involved in heat treatment will be provided which includes common microstructures and phase diagram interpretation. The use of time–temperature transformation (TTT) and cooling time transformation (CCT) curves relative to microstructure formation during heat treatment will be discussed. This will be followed by an overview of the different heat-treatment steps including austenitizing, stress relief, normalizing, annealing, quenching, and tempering. When possible, quantitative equations for estimating appropriate heat-treatment temperatures and times will be given.

I. GENERAL ASPECTS

The most part of steel applications are only possible after heat treatment. So in a project of component, it is necessary to consider the set steel + heat treatment.

There are many kinds of steels as summarized in Fig. 1.

Steels are usually standardized by the American Iron and Steel Institute (AISI), Deutsches Institut für Normung eV (DIN), the Society of Automotive Engineers (SAE), and others (different countries). It is useful to consult them to know specific properties helping to choose proper steel to a special application. There is a

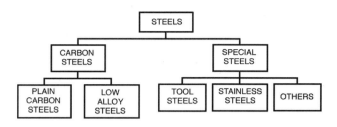

Figure 1 Summary of steel's types.

commercial database of standard steels that made this task easier where it is possible to find equivalence among the standards as well. Equivalent Steels (EQUIST) is one of which contains data of the standard steels and cast irons of 20 countries, as well as ISO and EN specifications [4].

Steel chemical composition has an important effect on the final properties obtained after heat-treatment process. Some effects of alloying elements are described in the following [5,6].

Manganese (Mn)—Increases strength and hardness; forms carbide; increases hardenability (especially in amounts greater than 0.8%); lowers the transformation temperature range; when in sufficient quantity, it produces austenitic steel; always present in a steel to some extent because it is used as deoxidizer.

Silicon (Si)—Strengthens ferrite and raises the transformation temperatures; has a strong graphitizing tendency; always present to some extent because it is used with manganese as a deoxidizer. It is relatively ineffective in low-carbon steels, but is very effective in high-carbon steels.

Chromium (Cr)—Increases strength and hardness; forms hard and stable carbides. It raises the transformation temperature significantly when its content exceeds 12%; increases hardenability; amounts in excess of 12% render steel stainless; good creep strength at high temperature.

Nickel (Ni)—Strengthens steel; lowers its transformation temperature range; increases hardenability and improves resistance to fatigue; strong graphite-forming tendency; stabilizes austenite when in sufficient quantity; creates fine grains and gives good toughness.

Nickel (Ni) and chromium (Cr)—Used together for austenitic stainless steels; each element counteracts disadvantages of the other.

Tungsten (W)—Forms hard and stable carbides; raises the transformation temperature range and

tempering temperatures. Hardened tungsten steels resist tempering up to 600°C.

Boron (B)—Improves considerably hardenability, the effect varying notably with the carbon content of steel. The full effect of boron on hardenability is obtained only in fully deoxidized (aluminum-killed) steels.

Molybdenum (Mo)—Strong carbide-forming element and also improves high-temperature creep resistance; reduces temper brittleness in Ni–Cr steels; improves corrosion resistance and temper brittleness; is effective in improving hardenability.

Vanadium (V)—Strong carbide-forming element; has a scavenging action and produces clean, inclusion-free steels; can cause reheat cracking when added to chrome molly steels.

Titanium (Ti)—Strong carbide-forming element; not used on its own, but added as a carbide stabilizer to some austenitic stainless steels.

Phosphorus (P)—Increases strength and hardenability; reduces ductility and toughness; increases machinability and corrosion resistance.

Sulfur (S)—Reduces toughness and strength and also weldability. Sulfur inclusions, which are normally present, are taken into solution near fusion temperature of the weld. On cooling, sulfides and remaining sulfur precipitate out and tend to segregate to the grain boundaries as liquid films, thus weakening them considerably. Such steel is referred to as burned. Manganese breaks up these films into globules of manganese sulfide; manganese to sulfur ratio >20:1, higher carbon and/or high heat input during welding >30:1, to reduce extent of burning.

II. STEEL TRANSFORMATION

In the heat treatment of steels, time and temperature are of critical importance on the formation of the micro-constituents on the steels, which determine properties such as hardness, strength, ductility, and toughness. Carbon and alloy content besides grain size have an important effect on the properties as well.

The first step in the heat-treatment process is to understand the decomposition of austenite for a given thermal history. Iron–Carbon (Fe–C) Phase Diagram and Time–Temperature Transformation (TTT) Diagram allow to verify, for each type of steel, which heating temperature is necessary to get austenite as well as which cooling rate is appropriate to obtain the desirable properties through the austenite transformation.

The most common transformation products that may be formed in quench-hardenable steels from austenite are in order of formation with decreasing cooling rate: martensite, bainite, pearlite (which is a mixture of ferrite and cementite), and ferrite (in same cooling rates, a mixture of these constituents can be formed). Each of these microstructures provides a unique combination of properties; especially, the relationship between ferrite and cementite in pearlite—depending on the carbon content and the cooling velocity—very strongly influences the mechanical properties. The definitions of the phases and microconstituents that can be found in steels are listed below [7,8].

Austenite—Designated by γ (gamma) is the solid nonmagnetic phase. Face-centered cubic iron with elements such as carbon and manganese dissolved to form a solid solution. In most steels, not stable at low temperatures (e.g., 25°C). Can dissolve up to about 2.0 wt.% C. Relatively soft. It is the desired solid solution microstructure produced prior to hardening. An austenite microstructure is illustrated in Fig. 2a (austenitic stainless steel).

Ferrite—Designated by α (alpha). Body-centered cubic iron with elements such as manganese dissolved to form a solid solution. The solubility of carbon in α is very low, 0.025 wt.% maximum. Relatively soft. Fully ferritic steels are only obtained when the carbon content is very low. Ferritic microstructure is illustrated in Fig. 2b.

Cementite (iron carbide)—A very hard and brittle compound of iron and carbon corresponding to the empirical formula of Fe_3C and possesses an orthorhombic lattice. Unstable phase, will decompose to iron and graphite, but takes a relatively long time and thus is a common phase in steels. In "plain carbon steels," some of the iron atoms in the cementite lattice are replaced by manganese and in "alloy steels" by other elements such as chromium and tungsten. Cementite will often appear as distinct lamellae together with α or as spheroids or globules of varying size in a ferritic matrix. The highest cementite contents are observed in white cast irons [9].

Pearlite—A metastable microstructure formed from austenite. The structure is an aggregate consisting of alternate lamellae of ferrite and cementite formed on slow cooling during the eutectoid reaction. In one alloy of given composition, pearlite may be formed isothermally at temperatures below the eutectoid temperature by quenching austenite to a desired temperature (generally above 550°C) and holding for a period of time necessary for transformation to occur. The interlamellar spacing is directly proportional to the transformation temperature; that is, the higher the temperature, the greater the spacing. An illustration of pearlitic microstructure is provided in Fig. 2c.

Bainite—Two-phase mixture of ferrite and cementite, which consists of fine roads of an iron carbide in acicular ferrite. Exact morphology depends upon temperature range of formation. This microstructure results from the transformation of austenite at temperatures between those that produce pearlite and martensite. Ordinarily, these structures may be formed isothermally at temperatures within the above range by quenching austenite to the desired temperature and holding for a specific period of time necessary for transformation to occur. If the transformation temperature is just below that at which the finest pearlite is formed, typically 350°C (660°F), the bainite (upper bainite) has a feathery appearance (Fig. 2d). If the temperature is just above that at which martensite is produced, the bainite (lower bainite; Fig. 2e) is acicular, slightly resembling tempered martensite and is formed below approximately 350°C (660°F).

Martensite—A supersaturated solid solution of carbon in alpha iron (ferrite) having a body-centered tetragonal lattice. It is a magnetic platelike construction formed by a diffusionless shear type of transformation of austenite below a certain temperature known as the M_s temperature (martensite start temperature). It is produced during quenching when the cooling rate of a steel, in the austenitic condition, is such that the pearlite or bainite, or both, transformation is suppressed. The amount of transformation depends on martensitic temperature range ($M_s–M_f$) attained since there is a distinct temperature where martensitic transformation begins (M_s) and ends (M_f) [9]. Three microstructural forms of martensite are lath (Fig. 2f), plate (Fig. 2g), and tempered (Fig. 2h) martensite. Table 1 provides a summary of the structural components of the iron–carbon system (adapted from Ref. 10). Details of the terminology of the phases and microconstituents of steels have been summarized by Davis [9].

It is important to observe that the morphology of the α-Fe_3C mixture is determined by heat treatment. Besides pearlite, there is spheroidite that consists of roughly spherical particles of Fe_3C dispersed in α (Fig. 2i) which exhibits improved machinability and formability.

III. IRON–CARBON (Fe–C) PHASE DIAGRAM

The fundamental elements of heat-treatment design are derived from the equilibrium phase diagram for the

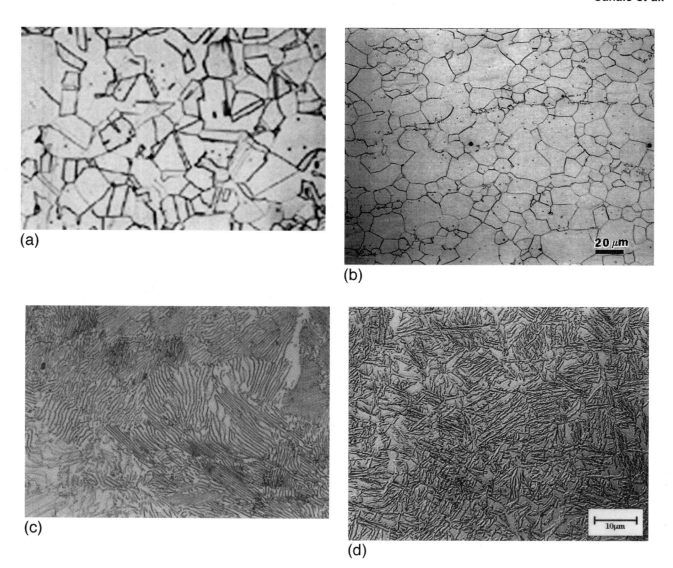

(a)

(b)

20 μm

(c)

(d)

10μm

Figure 2 Illustrations of microstructural transformation products. (a) Equiaxed austenite grains and annealing twins in an austenitic stainless steel. (b) Electrical iron (< 0.02% C) etched with 2% nital revealing a ferrite grain structure. (c) Coarse pearlitic structure isothermally annealed (780°C, 1 hr). SAE 1080 steel etched with 4% picral. Original at 1000×. (d) Upper bainite. (e) Lower bainite (dark plates). (f) Lath (low-carbon) martensite in SAE 8620 alloy steel (Fe–0.2%C–0.8%Mn–0.55%Ni–0.50%Cr–0.2%Mo) after heat treatment (954°C, 1 hr, water-quenched). Etched with 2% nital. (g) Microstructure of plate martensite obtained after improperly carburized surface. The white regions are retained austenite, but there are also white massive cementite particles present, as indicated by the arrows. Original at 1000×. (h) Microstructure of tempered martensite in alloy steel (Fe–0.33%C–0.48%Mn–2.46%Ni–0.64%Cr) after heat treatment (850°C, 1 hr, oil-quenched, tempered at 400°C). Etched with 2% nital. (i) Spheroidize-annealed SAE 52100 alloy steel for bearing. Etched with 4% picral revealing coarse spheroids of cementite in a ferritic matrix. Original at 1000×.

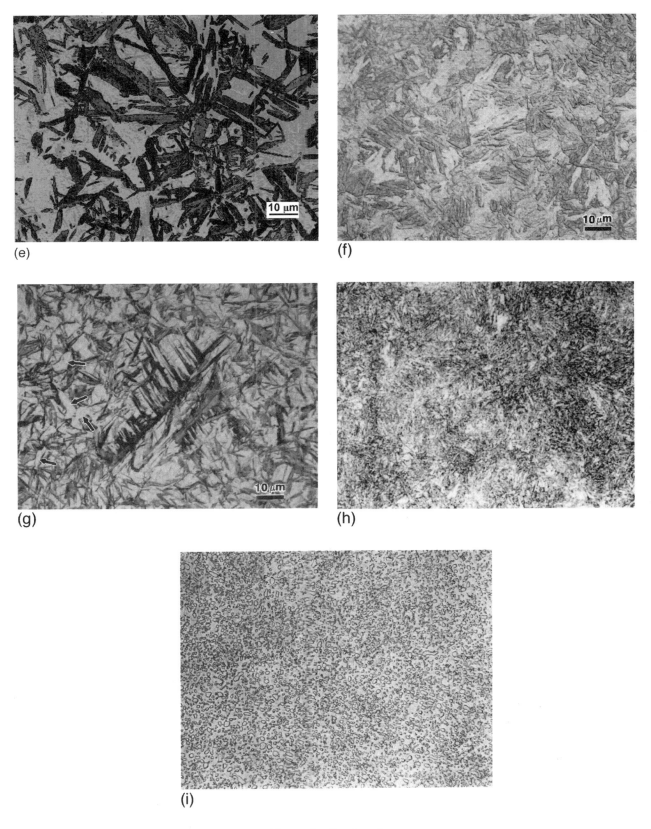

(e)

(f)

(g)

(h)

(i)

Figure 2 Continued.

Table 1 Components of an Iron–Carbon System

Phase or mixture of phases	Name
Solid solution of carbon in α-iron	Ferrite
Solid solution of carbon in γ-iron	Austenite
Iron carbide (Fe₃C)	Cementite
Mixture of carbon solid solution in α-iron with iron carbide	Bainite
Eutectoid mixture of carbon solid solution in α-iron with iron carbide	Pearlite

Source: Ref. 10.

steel of interest since the science of heat treatment is dependent on the formation by transformation of the desired phases and microstructures from austenite. For this discussion, the iron–carbon (Fe–C) phase diagram shown in Fig. 3 will be considered [10]. Although commonly used, the term iron–carbon phase diagram is fundamentally incorrect since the phase shown at the extreme right of the diagram is cementite. Therefore this diagram should be more properly designated as the iron–cementite (Fe–Cm) equilibrium diagram. The term iron–carbon diagram arises from the fact that cementite is not stable and degrades to iron and carbon over very long periods of time, typically much longer times than those encountered in heat treat processing. The solid lines in Fig. 3 show the equilibrium between Fe₃C and different phases of iron. Dashed lines show the equilibrium between iron and graphite, but graphitization rarely occurs in steel [2].

The phase diagram, such as the Fe–C diagram, shows the compositional limits of the different transformational phases formed by a steel alloy as a function of temperature that may exist during heating or cooling. If the compositional limits shown are in thermodynamic equilibrium, then this is an equilibrium phase diagram. Alternatively, this diagram may illustrate metastable conditions or phases. (The term "metastable" means that the material or phase is not actually stable with respect to transition but is stabilized by rapid cooling [9].)

In pure iron, there are four critical temperatures associated with phase changes, and the addition of car-

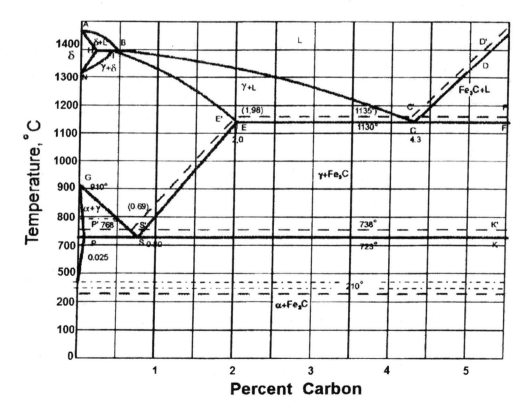

Figure 3 Iron–carbon equilibrium diagram up to 6.67 wt.% carbon. Solid lines indicate Fe–Fe₃C diagram; dashed lines indicate Fe–graphite diagram.

bon will affect these temperatures. From Fig. 3, it can be seen that carbon additions lower the melting point to a minimum of about 1133°C at about 4.3% C increasing for greater percentages. Line ABCD is the liquidus temperature (crystallization begins during cooling). It is related with the solubility limit of iron in the iron–carbon liquid. Line IECF is related with the final of solidification. GSE boundary represents the solubility limit of carbon in the γ phase with the maximum at about 2 wt.% of carbon. For greater values of carbon, formation of iron carbide in the austenite occurs and the alloy is classified as cast iron. So steels are alloys containing approximately 2 wt.% of carbon; most often, the total carbon content is less than 1 wt.%. Figure 4 (adapted from Ref. 11) shows part of the Iron–C Phase Diagram simplified corresponding to steel transformations.

The Fe–C phase diagram simplified depicted in Fig. 4 illustrates an important characteristic in the steel composition range which is called the "eutectic point" which refers to the composition of a solid phase which, upon cooling, undergoes a univariant transformation into two or more other solid phases [8]. For the carbon steel, the eutectic point occurs at about 0.77 wt.% carbon (0.8% C in Fig. 4). This is the basis of steel classification into hypoeutectoid, eutectoid, and hypereutectoid steels.

Hypoeutectoid steels are those steels with <0.80 wt.% carbon. Hypoeutectoid steels can, upon cooling, exist in two different phases, ferrite and austenite, each with different carbon contents. Upon further cooling, the microstructure of these steels typically exhibits ferrite grains in pearlite islands as shown in Fig. 4.

Eutectoid steels contain 0.77 wt.% carbon. (Practically, steels with 0.75–0.85 wt.% carbon are classified as eutectoid steels.) These steels exist as a solid solution at any temperature in the austenite range and all carbon is dissolved in the austenite. At the critical temperature (A_1) of the iron–cementite system (1340°F, 723°C), there is a transformation from austenite to cementite platelets in ferrite (pearlite, Fig. 4).

Hypereutectoid steels contain 0.8–2.0 wt.% of carbon. Upon cooling to A_{cm} (E′–S line) (Fig. 4), cementite separates from austenite. Below (A_1) 1340°F (723°C), austenite transforms to pearlite. At room temperature, the microstructure is pearlite areas in cementite.

The presence of carbon stabilizes austenite and expands the temperature ranged within which it is stable. Carbon is much more stable in austenite (2.11 wt.%) where it is in equilibrium with cementite at 1148°C than in ferrite (0.0218 wt.%) where it is in equilibrium with cementite at 723°C [12].

The austenite phase field shown in the phase diagram, such as Fig. 4, is the basis for selecting hot working and heat-treating temperature limits for carbon steels. Annealing, normalization, and austenitization processes are conducted in this region to facilitate the dissolution of carbon in iron. For example, Fig. 4 shows that if the steel is cooled slowly, the structure will change from austenite to ferrite and cementite. With faster cooling, martensite is formed. Austenite is only stable at elevated temperatures.

This temperature range is designated as the "critical temperature range" or "transformation range" which is defined as "those ranges of temperature within which austenite forms during heating and transforms during cooling" [9]. These two ranges may overlap but never coincide and are dependent on the alloy composition and the heating rate. Table 2 provides the recommended maximum hot working temperatures for various steels [13]. Table 3 shows the correlation of surface colors and approximate temperatures.

The transformation temperature indicates the limiting temperature of a transformation range. For irons and steels, the following standard terms are applied [9]:

A_{cm}—The temperature at which the transformation from austenite to cementite is complete in a hypereutectoid steel.

A_{c1} (A_1)—The temperature at which austenite begins to form during heating.

A_{c3} (A_3)—The temperature at which the transformation of ferrite to austenite is completed during heating.

A_{c4}—The temperature at which austenite transforms to delta ferrite during heating.

Ar_{cm}—The temperature at which the precipitation of cementite starts during cooling.

A_{r1}—The temperature at which transformation of austenite to ferrite or to ferrite and cementite is completed during cooling for hypereutectoid steels.

A_{r3}—The temperature at which austenite begins to transform to austenite during cooling.

A_{r4}—The temperature at which delta ferrite transforms to austenite during cooling.

A_r'—The temperature at which transformation from austenite to pearlite begins during cooling.

A''—The temperature at which transformation from austenite to martensite begins during cooling.

A summary of A_{c1}, A_{c3}, A_{r1}, and A_{r3} values for different steels is provided in Table 4 [13].

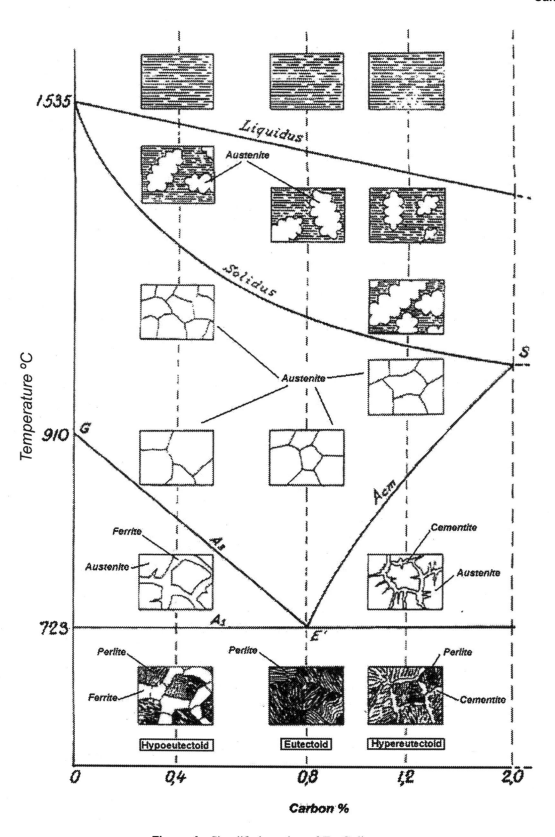

Figure 4 Simplified portion of Fe–C diagram.

Table 2 Recommended Maximum Hot Working Temperatures for Various Steels

SAE No.	Temperature (°F)	SAE No.	Temperature (°F)
1008	2250	4820	2250
1010	2250		
1015	2250	5060	2150
1040	2200		
		5120	2250
1118	2250	5140	2200
1141	2200	5160	2150
1350	2200	51100	2050
		52100	2050
2317	2250		
2340	2200	6120	2250
		6135	2250
2512	2250	6150	2200
3115	2250	8617	2250
3135	2200	8620	2250
3140	2200	8630	2200
		8640	2200
3240	2200	8650	2200
3310	2250	8720	2250
3316	2250	8735	2200
3335	2250	8740	2200
		9310	2250
4017	2300		
4032	2200	302	2200
4047	2200	303	2200
4063	2150	304	2200
		309	2150
4130	2200	310	2050
4132	2200	316	2150
4135	2200	317	2150
4140	2200	321	2150
4142	2200	347	2150
4320	2200	410	2200
4337	2200	416	2200
4340	2200	420	2200
		430	2100
4422	2250	440A	2100
4427	2250	440C	2050
		443	2100
4520	2250	446	1900
4615	2300	C–Mo	2300
4620	2300	DM	2300
4640	2200	DM-2	2300
4718	2250		

Source: Ref. 13.

Table 3 Correlation of Hot Steel Temperature with Color

Temperature		Hot Steel Color
°F	°C	
752	400	Red: visible in the dark
885	474	Red: visible in twilight
975	525	Red: visible in the daylight
1077	581	Red: visible in the sunlight
1292	700	Dull red
1472	800	Turning to cherry red
1652	900	Cherry red
1832	1000	Bright cherry red
2012	1100	Orange red
2192	1200	Orange yellow
2372	1300	White
2552	1400	Brilliant white
2732	1500	Dazzling white
2912	1600	Bluish white

Source: Ref. 13.

The M_s temperatures of many steels have been determined experimentally and have been approximated using several empirical formulas as shown in the following:

$$M_s \ (°\text{F}) = 1000 - 650x\%\text{C} - 70x\%\text{Mn} - 35x\%\text{Ni} - 70x\%\text{Cr} - 50x\%\text{Mo} \ [14].$$
$$M_s \ (°\text{C}) = 539 - 432x\%\text{C} - 30.4x\%\text{Mn} - 17.7x\%\text{Ni} - 12.1x\%\text{Cr} - 7.5x\%\text{Mo} \ [15].$$

All elemental concentrations are expressed in weight percent and considering all the carbides dissolved in the austenite.

IV. STEEL TIME AND TEMPERATURE TRANSFORMATION DIAGRAMS

Microstructures that are formed upon cooling and the proportions of each are dependent on the austenitization time (because of increasing solution of elements in austenite with increasing time), the time and temperature, cooling history of the particular alloy, and the composition of the alloy. The transformation products formed are typically illustrated with the use of transformation diagrams, which show the temperature–time dependence of the microstructure formation process for the alloy being studied.

Two of the most commonly used transformation diagrams are time–temperature transformation (TTT), which is also referred to as an isothermal transformation

Table 4 Approximate Critical Temperatures and M_s/M_f Points for Carbon and Alloy Steels

SAE no.	Heating (°F) A_{c1}	A_{c3}	Cooling (°F) A_{r3}	A_{r1}[a]	Quench temp. (°F)	M_s (°F)	M_f (°F)
1015	1370	1565	1545	1270	—	—	—
1020	1350	1555	1515	1270	—	—	—
1030	1350	1485	1465	1270	—	—	—
1035	1350	1475	1440	1270	—	—	—
1040	1350	1460	1420	1270	—	—	—
1045	1350	1440	1405	1270	—	—	—
1050	1340	1420	1390	1270	—	—	—
1065	—	—	—	—	1500	525	300
1090	—	—	—	—	1625	420	175
1330	1325	1470	1340	1160	—	—	—
1335	1315	1460	1340	1165	1550	640	450
1340	1340	1420	1310	1160	—	—	—
1345	1325	1420	1300	1160	—	—	—
2317	1285	1435	1265	1065	—	—	—
2330	1280	1360	1205	910/1050	—	—	—
2340	1285	1350	1185	1060	1450	580	400
2345	1265	1335	1125	1040	—	—	—
2512	1290	1400	1150	1060	—	—	—
2515	1260	1400	1160	1090	—	—	—
3115	1355	1500	1480	1240	—	—	—
3120	1350	1480	1445	1230	—	—	—
3130	1345	1460	1360	1220	—	—	—
3140	1355	1410	1275	1225	1550	630	440
3141	1355	1410	1300	1215	—	—	—
3150	1355	1380	1275	1215	—	—	—
3310	1335	1440	1235	1160	—	—	—
3316	1335	1445	1235	1160	—	—	—
4027	1360	1500	1400	1230	—	—	—
4032	1340	1500	1350	1250	—	—	—
4042	1340	1460	1340	1210	1500	610	—
4053	1310	1400	1320	1200	—	—	—
4063	1360	1390	1220	1190	1500	445	—
4068	1365	1395	1215	1195	—	—	—
4118	1385	1500	1410	1275	—	—	—
4130	1380	1475	1350	1250	1600	710	550
4140	1380	1460	1370	1280	1500	640	—
4147	—	—	—	—	1500	590	—
4150	1390	1450	1290	1245	—	—	—
4160	—	—	—	—	1575	500	—
4320	1355	1485	1330	840/1170	—	—	—
4340	1350	1425	1220	725/1210	1550	550	330
4342	—	—	—	—	1550	530	—
4615	1340	1485	1400	1200	—	—	—
4620	1300	1490	1335	1220	—	—	—
4640	1325	1400	1220	875/1130	1550	640	490
4695[b]	—	—	—	—	1550	255	—
4718	1285	1510	1410	1200	—	—	—
4815	1285	1450	1310	860/1110	—	—	—
4820	1290	1440	1260	825/1110	—	—	—
5045	1360	1430	1305	1255	—	—	—

Table 4 Continued

SAE no.	Heating (°F) A_{c1}	A_{c3}	Cooling (°F) A_{r3}	A_{r1}[a]	Quench temp. (°F)	M_s (°F)	M_f (°F)
5060	1370	1410	1305	1285	—	—	—
5120	1380	1525	1460	1305	—	—	—
5140	1360	1450	1345	1230	1550	630	460
51100	1385	1415	1320	1300	—	—	—
52100	1340	1515	1320	1270	1560	345	—
52100	—	—	—	—	1650	305	—
52100	—	—	—	—	1740	260	—
6117	1400	1560	1430	1270	1650	305	—
6120	1410	1530	1440	1300	1740	260	—
6140	—	—	—	—	1550	620	460
6150	1380	1450	1375	1275	—	—	—
8615	1360	1550	1455	1265	—	—	—
8620	1350	1525	1400	1200	—	—	—
8630	1350	1480	1340	1210	1600	690	540
8640	1350	1435	1275	1170	—	—	—
8650	1325	1390	1240	1195	—	—	—
8695[c]	—	—	—	—	1500	275	—
8720	1380	1520	1400	1200	—	—	—
8740	1350	1450	1300	1180	—	—	—
8750	1350	1410	1265	1190	—	—	—
9310	1315	1490	1305	830/1080	—	—	—
9317	1300	1455	1290	800	—	—	—
9395[c]	—	—	—	—	1700	170	—
9442	1350	1435	1280	1190	1575	620	410

[a] When two temperatures are given for A_{r1}, the higher temperature represents the pearlitic reaction and the lower temperature represents the bainitic reaction.
[b] Represents the case of 4600 grades of carburizing steels.
[c] Represents the case of 8600 and 9300 grades of carburizing steels, respectively.

(IT), and continuous cooling transformation (CCT) diagrams. When properly selected, either of these types of diagrams can be used to predict a steel microstructure and hardness after heat treatment, or they may be used to design a heat-treatment process when the desired microstructure and hardness are known.

A. Time–Temperature Transformation Diagrams

TTT diagrams are generated by heating small samples of steel to the desired austenitizing temperature and then rapidly cooling to a temperature intermediate between the austenitizing and the M_s temperature, and then holding for a fixed period of time until the transformation is complete at which point the transforma-

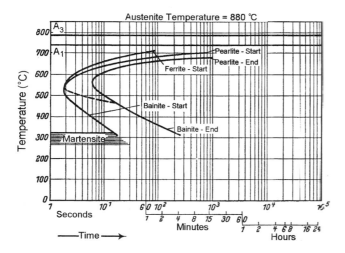

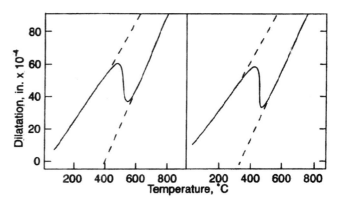

Figure 7 Dilatometer typical data showing the volume changes during continuous cooling upon the formation of decomposition products from austenite.

Figure 5 Time–Temperature Transformation (TTT) diagram of an unalloyed steel (AISI 1035 steel).

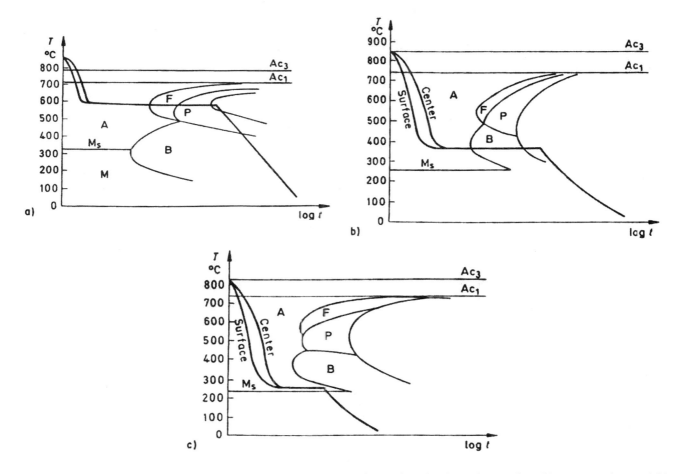

Figure 6 Isothermal processes for which only TTT diagrams may be used: (a) isothermal annealing, (b) austempering, and (c) martempering.

tion products are determined. This is done repeatedly until a TTT diagram is constructed such as that shown for an unalloyed steel (AISI 1035) in Fig. 5 (adapted from Ref. 15). TTT diagrams can only be read along the isotherms.

The fraction transformed to ferrite, pearlite, and bainite in isothermal processes can be calculated from [16]:

$$M = 1 - \exp\left(-bt^n\right)$$

where M is the fraction of the phase transformed, t is the time in sec, $b = 2 \times 10^{-9}$, and $n = 3$. By convention, the beginning of the transformation is defined as 1% of phase transformed and the ending is defined as 99% of phase transformed.

Only martensite formation occurs without diffusion. The Hougardy equation may be used to predict the amount of martensite formation for structural steels [16,17]:

$$M = 1 - 0.929 \exp\left[-0.976 \times 10^{-2}(M_s - T)^{1.07}\right]$$

where M = amount of martensite, M_s is the martensite start temperature, and T is the temperature below the M_s temperature.

The accuracy of TTT diagrams with respect to the isothermal positions on the diagram is typically accepted to be $\pm 10°C$ ($\pm 20°F$) or $\pm 10\%$ with respect to time.

Examples of heat treat processes where it is only appropriate to use a TTT diagram are isothermal annealing, austempering, and martempering. These processes are illustrated schematically in Fig. 6a–c, respectively [17].

B. Continuous Cooling Transformation Diagrams

CCT curves correlate the temperatures for each phase transformation, the amount of transformation product obtained for a given cooling rate with time, and the cooling rate necessary to obtain martensite. These correlations are obtained from CCT diagrams by using the different cooling rate curves.

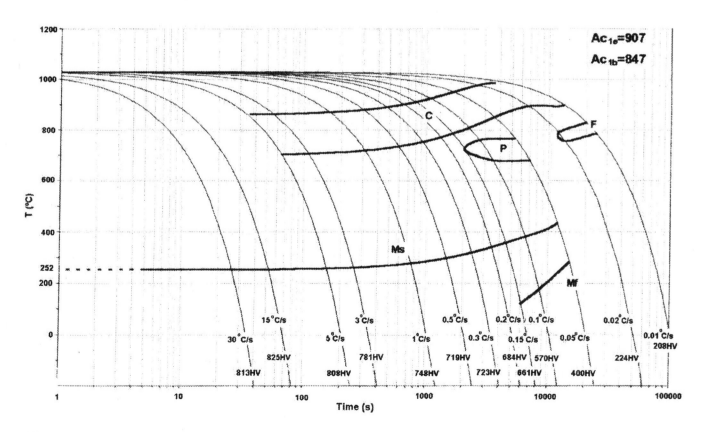

Figure 8 CCT diagram obtained from dilatometer data. High alloy steel (0.85%C–0.40%Mn–0.90%Si–8.5%Cr–2.0%Mo–0.60%V–0.15%Nb). (Courtesy of AcosVillares, Brazil.)

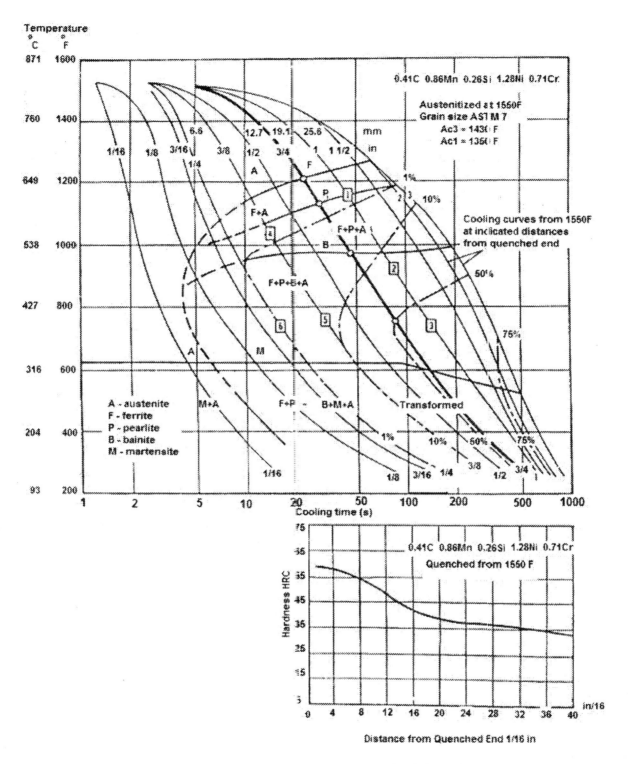

Figure 9 CCT diagram and Jominy curve for AISI 3140.

The "critical cooling rate" is the time required to avoid the formation of pearlite for the particular steel being quenched. As a general rule, a quenchant must produce a cooling rate equivalent to, or faster than, that rate indicated by the "nose" of the pearlite transformation curve to maximize the formation of martensite.

If the temperature–time cooling curves for the quenchant and the CCT curves of the steel are plotted on the same scale, then they may be superimposed to select the steel grade which will provide the desired microstructure and hardness for a given cooling condition [17]. This assumption is limited to bars up to 100 mm quenched in oil and bars up to 150 mm quenched into water.

CCT diagrams may be constructed in various forms. Steel may also be continuously cooled at different specified rates using a dilatometer and the proportion of transformation products formed after cooling to various temperatures intermediate between the austenitizing temperature and the M_s temperature and these data are used to construct a CCT diagram. In these processes, the length of the sample is recorded as a function of time at the transformation temperature. Typical data are shown in Fig. 7 (adapted from Ref. 7). This method of determining the CCT curves relies on detecting the volume expansion when the close-packed face-centered cubic austenite decomposes to the less dense products. These data are put together producing the CCT curves, as illustrated in Fig. 8. ASTM E228 standardizes procedures for these tests.

CCT diagrams can also be constructed as shown in Fig. 9 [17]. Figure 10 is a CCT diagram for an unalloyed carbon steel (AISI 1040) which provides curves for the beginning and the ending of the different phase transformations [18,19]. Figure 11, which was generated for a DIN 50CrV4 (AISI 6145) steel,

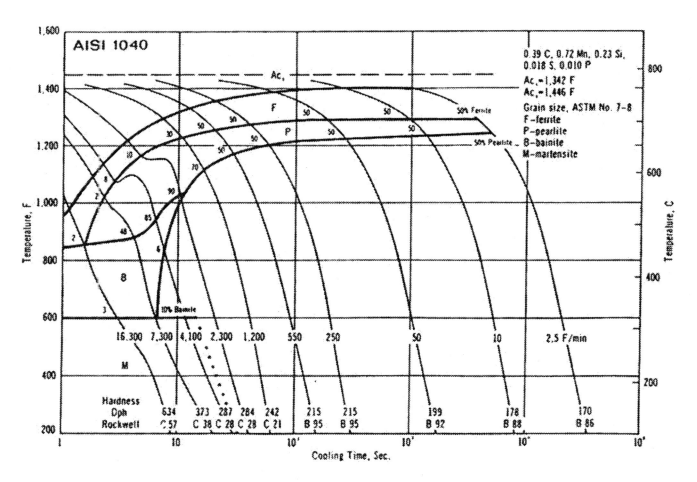

Figure 10   CCT diagram for an unalloyed steel (AISI 1040).

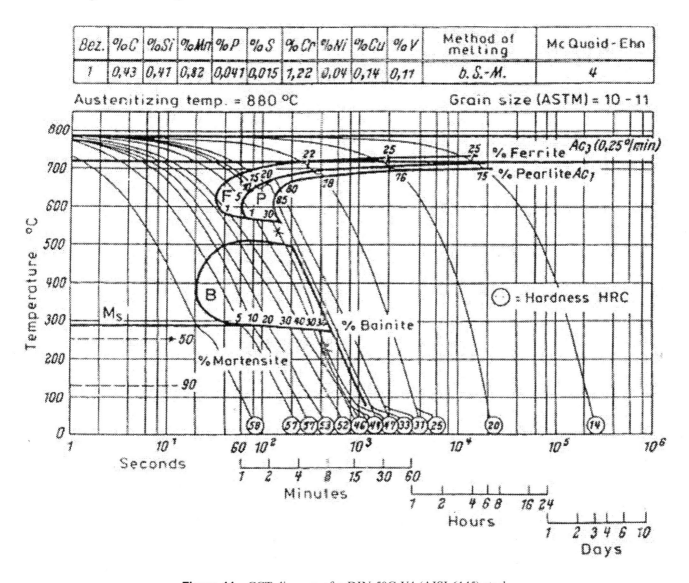

Bez.	%C	%Si	%Mn	%P	%S	%Cr	%Ni	%Cu	%V	Method of melting	Mc Quaid-Ehn
1	0,43	0,41	0,82	0,041	0,015	1,22	0,04	0,14	0,11	b.S.-M.	4

Austenitizing temp. = 880 °C   Grain size (ASTM) = 10 - 11

Figure 11   CCT diagram of a DIN 50CrV4 (AISI 6145) steel.

provides considerably more information [17,20]. In this figure, the fraction of the transformation product formed by a cooling curve is shown in the diagram and the resulting hardness is shown on the isotherm at the bottom of the diagram.

An alternative form of a CCT diagram is shown by Fig. 12 [17,20]. This curve was not generated using a dilatometer, but instead, cooling curves were measured at different distances from the end of a Jominy test bar. The corresponding Jominy curve is shown along with a diagram for a particular quenchant and agitation con-

dition which permits the prediction of cross-sectional hardness for a round bar [17,21].

Another form of CCT diagram, originally developed by Atkins [22], is illustrated in Fig. 13. This CCT diagram was generated by determining the cooling curves of round bars of the alloy represented in different quenchant media and then determining the corresponding transformation temperatures, microstructures, and hardnesses [17]. The data represented by these curves refer only to the center of the bar being quenched. A scale of cooling rates is provided at the

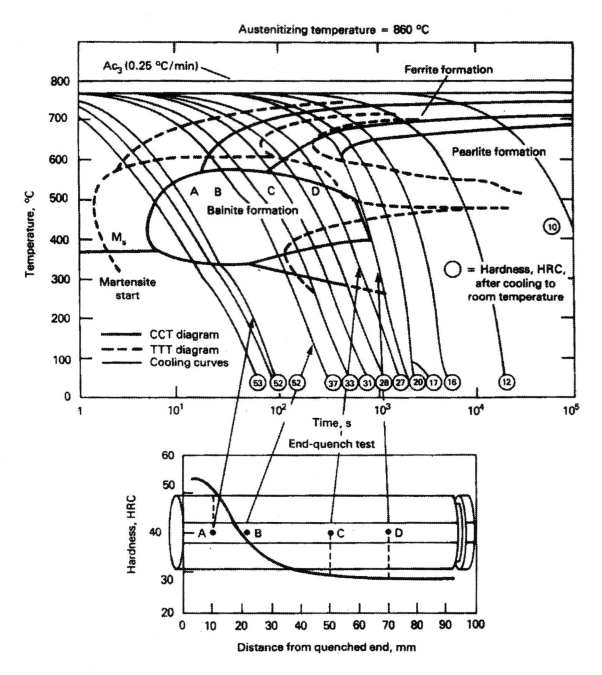

Figure 12 Experimentally determined CCT diagram (solid lines) for a DIN 42CrMo₄ steel. TTT diagram is also shown.

bottom of the diagram. These diagrams are read along vertical lines with respect to different cooling rates. This diagram is especially useful to quickly see the relative hardenability of different steels.

There are a number of heat-treatment processes where only the use of a CCT diagram is appropriate. These include continuous slow-cooling processes such as normalizing annealing by cooling in air, direct quenching to obtain a fully martensitic structure, and continuous cooling processes resulting in mixed microstructures as illustrated in Fig. 14 [17].

The Rose–Strassburg cooling law can be used to predict cooling times and temperatures for steels whose cross-section sizes are not excessively large

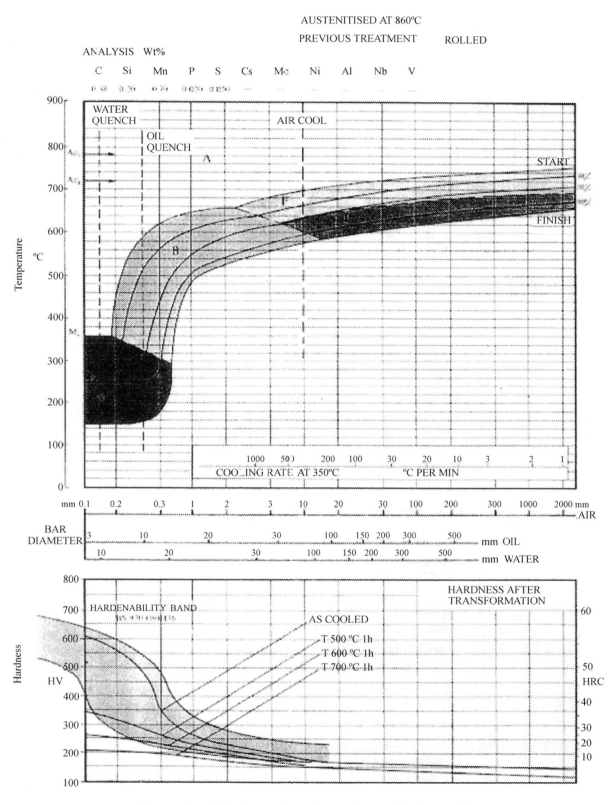

Figure 13 CCT diagram for rolled steel austenitized at 860°C.

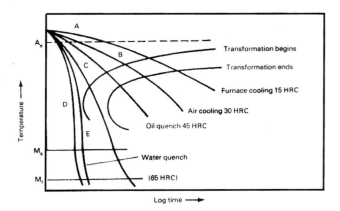

Figure 14 Heat-treatment processes where only CCT diagrams may be used. (A) Furnace cooling to obtain coarse ferrite-perlite microstructure. (B) Slow cooling to obtain ferrite-pearlite microstructure. (C) Continuous cooling for a mixed microstructure. (D), (E) Direct quenching to obtain fully martensitic microstructure.

and therefore their cooling is modeled by this relationship [20]:

$$T = T_O \exp\left[-\alpha t\right]$$

where T_O = austenitizing temperature, α = heat transfer coefficient, and t = time.

A number of points should be noted:

- The CCT diagram is only valid for the steel composition for which it was determined.
- It is not correct to assume that the area of intersection of a cooling curve with the transformation product is equivalent to the amount of that product that is formed.
- Scheil [23] has shown that transformation begins later in time for a continuous cooling process than for an isothermal process. This is consistent with TTT and CCT curve comparison.
- Since increasing the austenitizing temperature will shift the curves to longer transformation times, it is necessary to use CCT diagrams generated at the desired austenitizing temperature.

Caution: Although it is becoming increasingly common to see cooling curves (temperature–time profiles) for different cooling media (quenchants) such as oil, water, air, and others, superimposed on either TTT or CCT diagrams, this is not a rigorously correct practice and various errors are introduced into such analysis due to the inherently different kinetics of cooling used to obtain

the TTT or CCT diagrams vs. the quenchants being represented. A continuous cooling curve can be superimposed on a CCT, but not on a TTT diagram.

V. HARDENABILITY

Hardenability has been defined as the ability of a ferrous material to develop hardness to a given depth after being austenitized and quenched. This general definition comprises two subdefinitions; the first of which is the ability to achieve a certain hardness [24]. The ability to achieve a certain hardness level is associated with the highest attainable hardness which depends on the carbon content of the steel and, more specifically, on the amount of carbon dissolved in the austenite after austenitizing.

This is illustrated by considering the problem of hardening of high-strength, high-carbon steels. The higher the concentration of dissolved carbon in the austenitic phase, the greater the increase in mechanical strength after rapid cooling, and transformation of the austenite in the metastable martensite phase steels typically exhibits increasing hardness and strength with increasing carbon content, as shown in Fig. 15 [25], but they also exhibit relatively low ductility. However, with increasing carbon concentration, martensitic transfor-

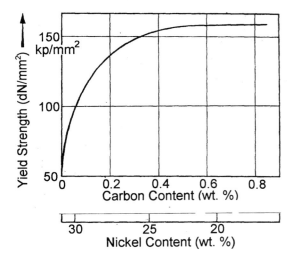

Figure 15 Influence of the carbon content in steel on the yield strength ($\sigma_{0.6}$) after quench hardening. The yield strength values were obtained from compression tests; the additional variation of nickel content causes a negligible solid solution hardening and was selected to obtain a constant M_s temperature for the start of martensitic transformation.

mation from austenite becomes more difficult resulting in a greater tendency for retained austenite and correspondingly lower strength.

The second subdefinition of hardenability refers to the hardness distribution within a cross section from the surface to the core under specified quenching conditions. It depends on the carbon content which is interstitially dissolved in austenite and the amount of alloying elements substitutionally dissolved in the austenite during austenitization. Therefore, as Fig. 15 shows, carbon concentrations in excess of 0.6% do not yield correspondingly greater strength [25]. Also, increasing the carbon content influences the M_f temperature relative to M_s during rapid cooling as shown in Fig. 16 [26]. In this figure, it is evident that for steels with carbon content above 0.6%, the transformation of austenite to martensite will be incomplete if the cooling process is stopped at 0°C or higher.

The depth of hardening depends on the following factors:

- Size and shape of the cross section
- Hardenability of the material
- Quenching conditions

The cross-section shape exhibits a significant influence on heat extraction during quenching and therefore on the hardening depth. Heat extraction is dependent on the surface area exposed to the quenchant. Bars of rectangular shape achieve less depth of hardening than round bars of the same cross-section size. Figure 17 can be used to convert square and rectangular cross sections to equivalent circular cross-section sizes [27].

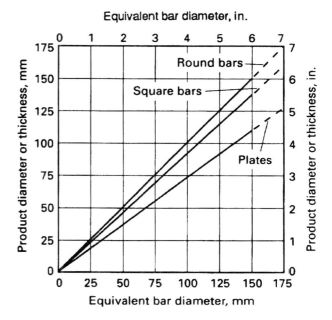

Figure 17 Correlation between rectangular cross sections and their equivalent round bar and plate sections.

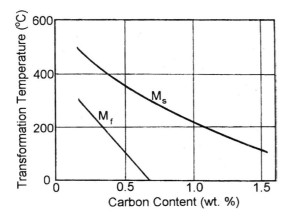

Figure 16 Influence of the carbon content in steels on the temperature of the start of martensite formation (M_s) and the end of martensite formation (M_f).

Hardenability of a steel is also dependent of the steel composition (carbon and alloy content) at the austenitizing temperature and the austenite grain size at the moment of quenching. In some instances, mainly in high-carbon and alloy carburizing steels, the chemical composition of the austenite may not be the same as that determined by chemical analysis because some carbide may be undissolved at the austenitizing temperature and so neither their carbon nor alloy content can contribute to hardenability. In fact, by nucleating transformation products, undissolved carbides can actively decrease hardenability. Consequently, such factors as austenitizing temperature, time at temperature, and prior microstructure are sometimes very important variables when determining the basic hardenability of a specific steel composition.

The effect of steel composition on hardenability may be calculated in terms of the "ideal critical diameter" or D_I which is defined as the largest bar diameter that can be quenched to produce 50% martensite at the center after quenching in an "ideal" quench, i.e., under "infinite" quenching severity. The ideal quench is one that reduces surface temperature of an austenitized steel to the bath temperature instantaneously. Under these conditions, the cooling rate at the center of the bar depends only on the thermal diffusivity of the steel.

The ideal critical diameter may be calculated from:

$$D_I = D_{I\ \text{Base}}(\text{carbon concentration and grain size})$$
$$\times f_{Mn}\, f_{Si}\, f_{Cr}\, f_{Mo}\, f_V\, f_{Cu}\, f_{Ni}\, f_X$$

where f_X is a multiplicative factor for the particular substitutionally dissolved alloying element. The base $D_{I\ \text{Base}}$ value and one set of alloying factors are provided in Table 5 [28] (note: this is not an exhaustive listing of alloying factors, but these are commonly encountered and they permit calculations to illustrate the effect of steel chemistry variation on hardenability). D_I values for a range of steels with differing hardenabilities are provided in Table 6 [27].

Grain size refers to the dimensions of grains or crystals in a polycrystalline metal exclusive of twinned regions and subgrains when present. Grain size is usually estimated or measured on the cross section of an aggregate of grains. Common units are (1) average diameter, (2) average area, (3) number of grains per linear unit, (4) number of grains per unit area, and (5) number of grains per unit volume. Grain size has an important influence on hardenability as shown in Fig. 18 [29].

Grain size may be determined according to ASTM Test Method E 112 [30]. The procedures in Test Method E 112 describe the measurement of average grain size and include the comparison procedure, the planimetric (or Jeffries) procedure, and the intercept procedures. Standard comparison charts are provided. These test methods apply chiefly to single phase grain structures, but they can be applied to determine the average size of a particular type of grain structure in a multiphase or multiconstituent specimen.

In addition, the test methods provided in ASTM E 112 are used to determine the average grain size of specimens with a unimodal distribution of grain areas, diameters, or intercept lengths. These distributions are approximately lognormal. These test methods do not cover methods to characterize the nature of these distributions. Characterization of grain size in specimens with duplex grain size distributions is described in Test Methods E 1181. Measurement of individual, very coarse grains in a fine-grained matrix is described in Test Methods E 930. These test methods deal only with the determination of planar grain size, that is, characterization of the two-dimensional grain sections revealed by the sectioning plane. Determination of spatial grain size, that is, measurement of the size of the three-dimensional grains in the specimen volume, is beyond the scope of these test methods.

These test methods described in E 112 are techniques performed manually using either a standard series of

Table 5 Hardenability Factors for Carbon Content, Grain Size, and Selected Alloying Elements in Steel

Carbon content (%)	Carbon grain size no.			Alloying element				
	6	7	8	Mn	Si	Ni	Cr	Mn
0.05	0.0814	0.0750	0.0697	1.167	1.035	1.018	1.1080	1.15
0.10	0.1153	0.1065	0.0995	1.333	1.070	1.036	1.2160	1.30
0.15	0.1413	0.1315	0.1212	1.500	1.105	1.055	1.3240	1.45
0.20	0.1623	0.1509	0.1400	1.667	1.140	1.073	1.4320	1.60
0.25	0.1820	0.1678	0.1560	1.833	1.175	1.091	1.54	1.75
0.30	0.1991	0.1849	0.1700	2.000	1.210	1.109	1.6480	1.90
0.35	0.2154	0.2000	0.1842	2.167	1.245	1.128	1.7560	2.05
0.40	0.2300	0.2130	0.1976	2.333	1.280	1.146	1.8640	2.20
0.45	0.2440	0.2259	0.2090	2.500	1.315	1.164	1.9720	2.35
0.50	0.2580	0.2380	0.2200	2.667	1.350	1.182	2.0800	2.50
0.55	0.273	0.251	0.231	2.833	1.385	1.201	2.1880	2.65
0.60	0.284	0.262	0.241	3.000	1.420	1.219	2.2960	2.80
0.65	0.295	0.273	0.251	3.167	1.455	1.237	2.4040	2.95
0.70	0.306	0.283	0.260	3.333	1.490	1.255	2.5120	3.10
0.75	0.316	0.293	0.270	3.500	1.525	1.273	2.62	3.25
0.80	0.326	0.303	0.278	3.667	1.560	1.291	2.7280	3.40
0.85	0.336	0.312	0.287	3.833	1.595	1.309	2.8360	3.55
0.90	0.346	0.321	0.296	4.000	1.630	1.321	2.9440	3.70
0.95	—	—	—	4.167	1.665	1.345	3.0520	—
1.00	—	—	—	4.333	1.700	1.364	3.1600	—

Table 6 Ideal Diameter (D_I) Values for Various Steels

Steel	D_I	Steel	D_I	Steel	D_I
1045	0.9–1.3	4135 H	2.5–3.3	8625 H	1.6–2.4
1090	1.2–1.6	4140 H	3.1–4.7	8627 H	1.7–2.7
1320 H	1.4–2.5	4317 H	1.7–2.4	8630 H	2.1–2.8
1330 H	1.9–2.7	4320 H	1.8–2.6	8632 H	2.2–2.9
1335 H	2.0–2.8	4340 H	4.6–6.0	8635 H	2.4–3.4
1340 H	2.3–3.2	X4620 H	1.4–2.2	8637 H	2.6–3.6
2330 H	2.3–3.2	4620 H	1.5–2.2	8640 H	2.7–3.7
2345	2.5–3.2	4621 H	1.9–2.6	8641 H	2.7–3.7
2512 H	1.5–2.5	4640 H	2.6–3.4	8642 H	2.8–3.9
1515 H	1.8–2.9	4812 H	1.7–2.7	8645 H	3.1–4.1
2517 H	2.0–3.0	4815 H	1.8–2.8	8647 H	3.0–4.1
3120 H	1.5–2.3	4817 H	2.2–2.9	8650 H	3.3–4.5
3130 H	2.0–2.8	4820 H	2.2–3.2	8720 H	1.8–2.4
3135 H	2.2–3.1	5120 H	1.2–1.9	8735 H	2.7–3.6
3140 H	2.6–3.4	5130 H	2.1–2.9	8740 H	2.7–3.7
3340	8.0–10.0	5132 H	2.2–2.9	8742 H	3.0–4.0
4032 H	1.6–2.2	5135 H	2.2–2.9	8745 H	3.2–4.3
4037 H	1.7–2.4	5140 H	2.2–3.1	8747 H	3.5–4.6
4042 H	1.7–2.4	5145 H	2.3–3.5	8750 H	3.8–4.9
4047 H	1.8–2.7	5150 H	2.5–3.7	9260 H	2.0–3.3
4047 H	1.7–2.4	5152 H	3.3–4.7	9261 H	2.6–3.7
4053 H	2.1–2.9	5160 H	2.8–4.0	9262 H	2.8–4.2
4063 H	2.2–3.5	6150 H	2.8–3.9	9437 H	2.4–3.7
4068 H	2.3–3.6	8617 H	1.3–2.3	9440 H	2.4–3.8
4130 H	1.8–2.6	8620 H	1.6–2.3	9442 H	2.8–4.2
3132 H	1.8–2.5	8622 H	1.6–2.3	9445 H	2.8–4.4

graded chart images for the comparison method or simple templates for the manual counting methods. Utilization of semiautomatic digitizing tablets or automatic image analyzers to measure grain size is described in Test Methods E 1382 [31].

The ASTM grain size number (G), referred to in Table 5, is a grain size designation bearing a relationship to average intercept distance at 100 diameters magnification according to the equation:

$$G = 10.00 - 2\log_2 L$$

where L = the average intercept distance at 100 diameters magnification. The smaller the ASTM grain size, the larger the diameter of the grains.

The effect of quenching conditions on the depth of hardening is not only dependent on the quenchant being used and its physical and chemical properties, but also on the process parameters such as bath temperature and agitation.

A. Hardenability Measurement

There are numerous methods to estimate steel hardenability. However, two of the most common are Jominy curve determination and Grossmann hardenability which will be discussed here.

1. Jominy Bar End-Quench Test

The most familiar and commonly used procedure for measuring steel hardenability is the Jominy bar end-quench test. This test has been standardized and is described in ASTM A 255, SAE J406, DIN 50191, and ISO 642. For this test, a 100-mm-long (4 in.) by 25-mm diameter (1 in.) round bar is austenitized to the proper temperature, dropped into a fixture, and one end rapidly quenched with 24°C (75°F) water from a 13-mm (0.5 in.) orifice under specified conditions as illustrated in Fig. 19 [24]. The austenitizing temperature is selected according to the specific steel alloy being studied; however, most steels are heated in the range of 870–900°C (1600–1650°F).

In the Jominy end-quench test, a cylindrical specimen is heated to the desired austenitizing temperature and well-defined time, which is material composition-dependent, and then it is quenched on one end with water. The cooling velocity decreases with increasing distance from the quenched end. After quenching, parallel flats are

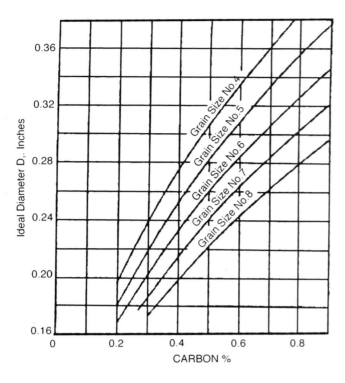

Figure 18 Influence of grain size (ASTM number) in the hardenability (D_I).

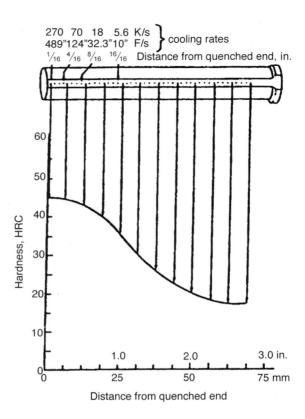

Figure 20 Measuring hardness on the Jominy test specimen and plotting hardenability curves.

ground on opposite sides of the bar and hardness measurements made at 1/16 in. (1.6 mm) intervals along the bar as illustrated in Fig. 20 [24]. The hardness, as a function of distance from the quenched end, is measured and plotted, and, together with the measurement of the relative areas of the martensite, bainite, and pearlite that is formed, it is possible to compare the hardenability of

different steels using Jominy curves. As the slope of the Jominy curve increases, the ability to harden the steel (hardenability) decreases. Conversely, decreasing slopes (or increasing flatness) of the Jominy curve indicates increasing hardenability (ease of hardening).

The Jominy end-quench is used to define the hardenability of carbon steels with different alloying elements like chromium (Cr), manganese (Mn), and molybdenum (Mo) and having different critical cooling velocities. Jominy curves for different alloy steels are provided in Fig. 21 [13]. These curves illustrate that the unalloyed, 0.4% carbon steel exhibits a relatively small distance for martensite (high hardness) formation. The 1% Cr and 0.2% Mn steel, however, can be hardened up to a distance of 40 mm. Figure 21 illustrates that steel hardenability is dependent on the steel chemistry, unalloyed steels exhibit poor hardenability, and that Jominy curves provide an excellent indicator of relative steel hardenability.

The Jominy test provides valid data for steels having an ideal diameter from about 25 to 150 mm (1 to 6 in.). This test can be used for D_I values less than 25 mm

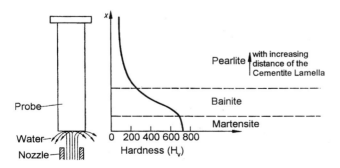

Figure 19 Schematic illustration of the Jominy end-quench test and microstructural variation with increasing distance from the quenched end.

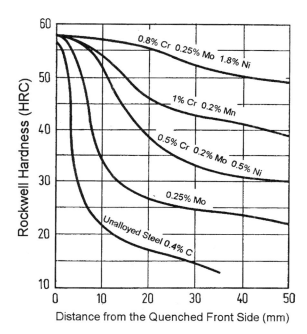

Figure 21 Jominy curve comparison of the hardenability of one unalloyed and a number of other different alloy steels. All alloy concentrations are in wt.%.

(1 in.), but Vickers or microhardness tests must be used to obtain readings that are closer to the quenched end of the bar and closer together than generally possible using the standard Rockwell "C" hardness test method [24].

The austenitizing time (shown in Fig. 22; Ref. 32) and temperature, extent of special carbide solution in the austenite and extent of oxidation or surface decarburization during austenitizing, care and consistency of surface flat preparation, and bar positioning prior to making hardness measurements are important factors that influence test results. Therefore all tests should be conducted in compliance with the standard being followed [24].

Because of the differences in chemical composition between different heats of the same grade of steel, the so-called *hardenability bands* have been developed using the Jominy end-quench test. According to American designation, the hardenability band for each steel grade is marked by the letter H following the composition code. Hardenability band for 1340 H steel can be seen in Fig. 23 [7,24]. The upper curve of the band represents the maximum hardness values, corresponding to the upper composition limits of the main elements, while the lower curve represents the minimum hardness values, corresponding to the lower limit of the composition ranges.

Using the composition of the steel, it is possible to calculate the Jominy end-quench curve for a wide range of steels with excellent correlation to experimental results. In many cases, calculation is preferred over experimental determination.

2. Grossmann Hardenability

The Grossmann method of measuring hardenability utilizes a number of cylindrical steel bars with different diameters, each hardened in a given quenching medium [17]. After sectioning each bar at mid-length and examining it metallographically, the bar that has 50% martensite at its center is selected, and the diameter of this bar is designated as the critical diameter D_{crit}. Other bars with diameters smaller than D_{crit} will have more martensite and correspondingly higher hardness values, and bars with diameters larger than D_{crit} will attain 50% martensite only up to a certain depth as shown in Fig. 24 [24]. The D_{crit} value is valid only for the quenching medium and conditions used to determine this value.

To determine the hardenability of a steel independently of the quenching medium, Grossmann introduced the term *ideal critical diameter* (D_I). D_I, defined above, is the diameter of a given steel bar that would produce 50% martensite at the center when quenched in a bath of quenching intensity $H = \alpha$. Here $H = \alpha$ indicates a hypothetical quenching intensity that reduces the temperature of heated steel to the bath temperature in zero time. Alternatively, excellent correlations with reported H-values are potentially achievable using cooling rates obtained by cooling curve analysis with 0.5, 1.0, 1.5, and 2.0 inches type 304 stainless steel probes [33]. Ideal diameters for various steels are provided in Table 6 [27].

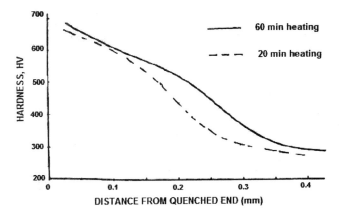

Figure 22 Influence of the austenitizing time in the hardenability of a lean steel.

AISI-SAE 1340H UNS H13400

Specified hardness limits

Distance from quenched surface		Hardness, HRC		Distance from quenched surface		Hardness, HRC	
1/16 in.	mm	Max	Min	1/16 in.	mm	Max	Min
1	1.58	60	53	13	20.54	46	26
2	3.16	60	52	14	22.12	44	25
3	4.74	59	51	15	23.70	42	25
4	6.32	58	49	16	25.28	41	24
5	7.90	57	46	18	28.44	39	23
6	9.48	56	40	20	31.60	38	23
7	11.06	55	35	22	34.76	37	22
8	12.64	54	33	24	37.92	36	22
9	14.22	52	31	26	41.08	35	21
10	15.80	51	29	28	44.24	35	21
11	17.38	50	28	30	47.40	34	20
12	18.96	48	27	32	50.56	34	20

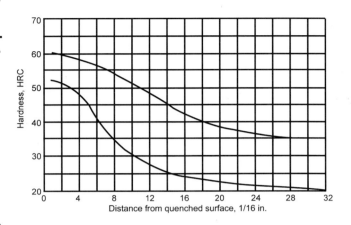

Figure 23 Hardenability band for SAE 1340 H.

To identify a quenching medium and its condition, Grossmann introduced the quenching intensity (severity) factor H. Table 7 provides a summary of Grossmann H factors for different quench media and different quenching conditions [34]. Although these data have been published in numerous reference texts for many years, it is of relatively limited quantitative value. One of the most obvious reasons is that quenchant agitation is not adequately defined with respect to mass flow rate, directionality, and turbulence and is often unknown, yet it exhibits enormous effects on quench severity during quenching.

The Grossmann value H is based on the Biot (Bi) number which interrelates the interfacial heat transfer coefficient (α), thermal conductivity (λ), and the radius (R) of the round bar being hardened:

$$\text{Bi} = \alpha/\lambda \times R = HD$$

$$H = \alpha/(2 \times \lambda)$$

Since the Biot number is dimensionless, this expression means that the Grossmann value, H, is inversely proportional to the bar diameter. This method of numerically analyzing the quenching process presumes that heat transfer is a steady-state, linear (Newtonian) cooling process. However, this is seldom the case and almost never the case in vaporizable quenchants such as oil, water, and aqueous polymers. Therefore a significant error exists in the basic assumption of the method.

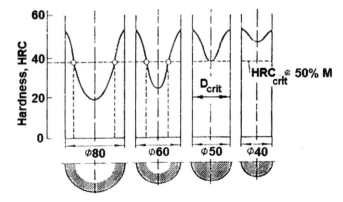

Figure 24 Determination of critical diameter D_{crit} according to Grossmann.

Table 7 Effect of Agitation on Quench Severity as Indicated by Grossmann Quench Severity Factors (H-Factors)

Agitation	Grossmann H-Factor			
	Air	Oil and salt	Water	Caustic soda or brine
None	0.02	0.25–0.3	0.9–1.0	2
Mild	—	0.30–0.35	1.0–1.1	2–2.2
Moderate	—	0.35–0.4	1.2–1.3	—
Good	—	0.4–0.5	1.4–1.5	—
Strong	—	0.5–0.8	1.6–2.0	—
Violent	—	0.8–1.1	4	5

Another difficulty is the determination of the H-value for a cross-section size other than one experimentally measured. In fact, H-values depend on cross-section size. Values of H do not account for specific quenching characteristics such as composition, oil viscosity, or temperature of the quenching bath. Tables of H-values do not specify the agitation rate of the quenchant either uniformly or precisely (see Table 7). Therefore although H-values are commonly used, more current and improved procedures ought to be used when possible. For example, cooling curve analyses and the various methods of cooling curve interpretation that have been reported [33,34] are all significant improvements over the use of Grossmann hardenability factors.

VI. AUSTENITIZATION

As indicated in the discussion thus far, the austenitization process refers to the formation of austenite by heating the steel above the critical temperature for austenite formation. It is important to note that the term austenitization means to completely transform the steel to austenite [2]. However, there are a number of critically important variables in the austenitization process, two of which are heat rate and holding (soaking) time.

The heating rate is critical. There is a specific heating rate that cannot be exceeded without causing warpage or cracking since steel typically possesses insufficient plasticity to accommodate increased thermal stresses in the temperature range of 250–600°C. Therefore this is a particularly critical temperature range especially when the component has both thick and thin cross sections. The heating rate is dependent on [17]:

- Size and shape of the component
- Initial microstructure
- Steel composition

For this reason, steel is often heated to the final austenitizing temperature in steps of 95–200°C/hr.

Shapes with corners and sharp edges are also susceptible to cracking, also known as the "corner effect." If heating rates are excessive or nonuniform, the re-

Table 8 SAE AMS 2759/1C Recommended Annealing, Normalizing, Austenitizing Temperatures, and Quenchants for Various Steels

Material designation	Annealing temperature[a] (°C)	Normalizing temperature (°C)	Austenitizing temperature (°C)	Quenching medium[b]
1025	885	899	871	w, p
1035	871	899	843	o, w, p
1045	857	899	829	o, w, p
1095[c]	816	843	802	o, p
1137	788	899	843	o, w, p
3140	816	899	816	o, p
4037	843	899	843	o, w, p
4130	843	899	857	o, w, p
4135	843	899	857	o, p
4140	843	899	843	o, p
4150	829	871	829	o, p
4330V	857	899	871	o, p
4335V	843	899	871	o, p
4340	843	899	816	o, p
4640	843	899	829	o, p
6150	843	899	871	o, p
8630	843	899	857	o, w, p
8735	843	899	843	o, p
8740	843	899	843	o, p

SAE AMS 2759/1C should be consulted for detailed description of the overall heat-treating requirements for these alloys.
[a] The cooling rate is not to exceed 111°C/hr to below 538°C except for 4330V, 4335V, and 4340 to below 427°C and 4640 to below 399°C.
[b] o = oil, w = water, and p = an aqueous polymer quenchant.
[c] 1095 parts should be spheroidize-annealed before hardening.

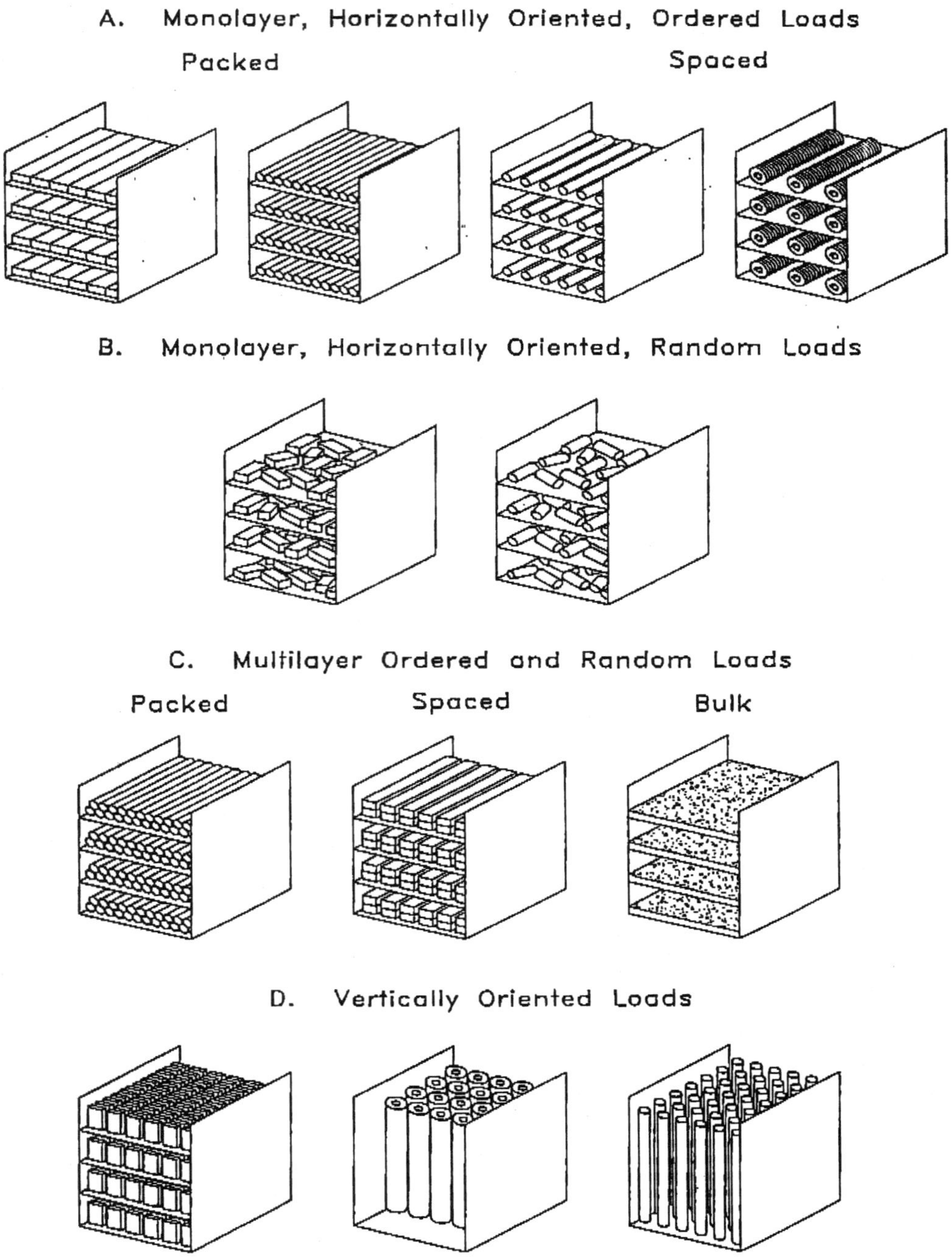

A. Monolayer, Horizontally Oriented, Ordered Loads

Packed   Spaced

B. Monolayer, Horizontally Oriented, Random Loads

C. Multilayer Ordered and Random Loads

Packed   Spaced   Bulk

D. Vertically Oriented Loads

Figure 25 Aronov load characterization diagram for soaking time calculation.

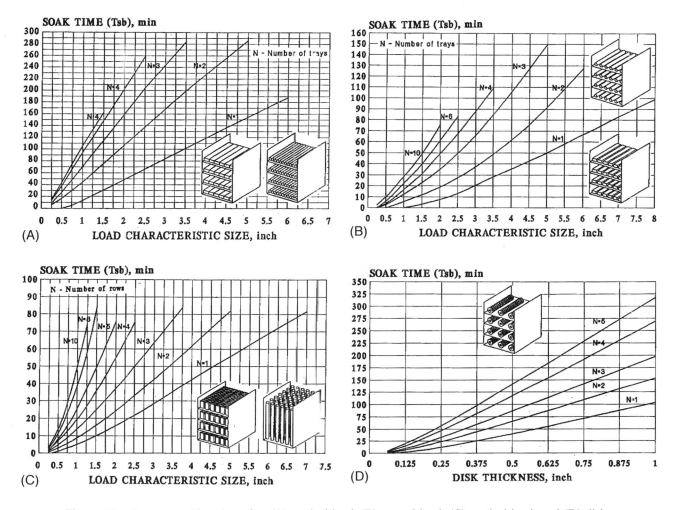

Figure 26 Aronov soaking times for: (A) packed load, (B) spaced load, (C) vertical load, and (D) disks.

sulting thermal stresses may be sufficient to cause cracking.

The propensity for steel to crack is dependent on composition. For example, increasing carbon content increases the potential for cracking. The effect of composition on the potential for cracking can be modeled by calculating the carbon equivalent (C_{eq}):

$$C_{eq} = C + \frac{Mn}{5} + \frac{Cr}{4} + \frac{Mo}{3} + \frac{Ni}{10} + \frac{V}{5} + \frac{Si - 0.5}{5} + \frac{Ti}{5} + \frac{W}{10} + \frac{Al}{10}$$

where the elements shown represent wt.% concentrations in the steel. The limits of this equation are $C \leq 0.9\%$, $Mn \leq 1.1\%$, $Cr \leq 1.8\%$, $Mo \leq 0.5\%$, $Ni \leq 5.0\%$, $V \leq 0.5\%$, $Si \leq 1.8\%$, $Ti \leq 0.5\%$, $W \leq 2.0\%$,

and $Al \leq 2.0\%$. Crack sensitivity increases with the C_{eq} value. The following general rules were reported by Liscic [17]:

$C_{eq} \leq 0.4$	Steel not sensitive to cracking, may be heated quickly
$C_{eq} = 0.4$–0.7	Moderate sensitivity to cracking
$C_{eq} \geq 0.7$	Steel is very sensitive to cracking and should be preheated to a temperature close to A_{c1} and held until the temperature was uniform throughout to minimize thermal stresses when austenitizing.

In addition to these effects, steel with high hardness and a nonuniform microstructure should be heated more slowly due to its crack sensitivity than a steel with low hardness and uniform microstructure.

The austenitization temperature for a given steel alloy is typically specified such as those values shown in Table 8. These values were selected to provide optimum hardness and grain size. As the austenitization temperature increases, the grain size increases. This is important because the grain size affects heat treatment and subsequent performance under various working conditions. For example, increasing grain size increases the impact transition temperature and increases propensity for brittle fracture. Fine grain steels have greater fatigue strength and coarse grain steels. However, coarse grain steels have better machinability than fine grain steels. The Hall–Petch equation predicts the effect of grain size on yield stress (σ_y) [35]:

$$\sigma_y = \sigma_i + K_y D^{-0.5}$$

Increasing the austenitizing temperature also [17]:

- Increases hardenability due to increased carbide solubilization and increased grain size
- Decreases the Ms temperature
- Increases the (incubation) time for isothermal transformation to pearlite or bainite to begin.
- Increases the amount of retained austenite

For unalloyed steels, the optimum austenitization temperature are 30–50°C above the A_{c3} temperature for hypoeutectoid steels and 30–50°C above the A_{c1} for hypereutectoid steels. The alloying elements in alloy steels may shift the A_1 temperature either higher or lower, and therefore appropriate references such as national standards must be consulted.

The well-known rule of thumb is often used to estimate the appropriate soaking time during austenitization. Another rule of thumb that is used is:

$$T = 60 + D$$

where T is the soaking time in min and D is the maximum diameter of the component in mm. However, these rules of thumb are imprecise. Soaking times are dependent on geometrical factors related to the furnace and the load, type of load, type of steel, thermal properties of the load, load and furnace emissivities, initial furnace and load temperatures, characteristic fan curves, and composition of the atmosphere.

Aronov et al. [36] developed a method for predicting furnace soaking times for batch loads based on "load characterization." Load characterization diagrams are provided in Fig. 25 and these models are

based on the generalized characterization equation for soaking time (T_S):

$$T_S = T_{Sb} K$$

where T_{Sb} is the baseline soak temperature condition taken from Fig. 26 and K is the correction factor for each type of steel ($K = 1$ for low-alloy steel and 0.85 for high alloy steel).

VII. ANNEALING

The primary purpose of the annealing is to soften the steel to enhance its workability and machinability. However, annealing may also be performed for [35]:

- Relief of internal stresses arising from prior processing including casting, forging, rolling, machining, and welding
- Improvement or restoration of ductility and toughness
- Improvement in machinability
- Grain refinement
- Improvement of the uniformity of the dispersion of alloying elements
- Achieving a specific microstructure
- Reduction of gaseous content within the steel

Annealing may be an intermediary step in an overall process or it may be the final process in the heat treatment of a component. Table 8 provides a summary of annealing temperatures recommended for some common steel alloys.

Annealing processes may be classified as full annealing, process (subcritical) annealing, isothermal annealing, recrystallization annealing, spheroidizing, and normalizing. Partial (intercritical) annealing is a subclass of full annealing. Figure 27 provides an illustrative summary of these annealing processes [35].

A. Full Annealing

Full annealing of steel involves heating the steel 30–50°C above the upper critical temperature (A_{c3}) for hypoeutectoid steels then furnace cooled through the critical temperature range at a specified cooling rate which is selected based on the final microstructure and hardness required [37]. The full annealing process breaks up the continuous carbide network microstructure of high-carbon steels into separated, spherical carbide particles (ferrite and cementite) with a larger grain size. This process is typically used for steels with

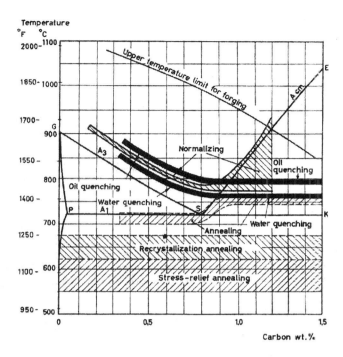

Figure 27 Temperature/carbon content correlations for different annealing processes.

carbon contents of 0.30–0.60% carbon to improve machinability. Unless otherwise noted, the term "annealing" usually refers to full annealing [38].

B. Partial (Intercritical) Annealing

Partial annealing is conducted by heating the steel to a point within the critical temperature range (A_{c1}–A_{c3}) followed by slow furnace cooling as illustrated in Fig. 28 [35]. Partial annealing, also known as "intercritical annealing," may be performed on hypereutectoid steels to obtain a microstructure of fine pearlite and cementite instead of coarse pearlite and a network of cementite at the grain boundaries as observed in the case of full annealing. For hypereutectoid steels, this results in grain refinement which usually occurs at 10–30°C above A_{c1}. Partial annealing is performed to improve machinability. However, steels with a Widmanstäten or coarse ferrite/pearlite structure are unsuitable for this process. Krauss [2] has noted that the term "partial annealing" is an imprecise term, and, to be meaningful, the type of material, time–temperature of the process, and the degree of cold working must be specified.

C. Process (Subcritical) Annealing

Process annealing is performed to improve the cold-working properties of low-carbon steels (up to 0.25% carbon) or to soften high-carbon and alloy steels to facilitate shearing, turning, or straightening processes [38,39]. Process annealing involves heating the steel to a temperature below (typically 10–20°C below) the lower critical temperature (A_{c1}) and is often called "subcritical" annealing. After heating, the steel is cooled to room temperature in still air. The process-annealing temperatures for plain carbon and low-alloy steels are typically limited to about 700°C to prevent partial re-austenitization and, in some cases, to about 680°C for steel compositions, such as high nickel-containing steels where the nickel further reduces the A_{c1} temperature [39].

This process can be used to temper martensitic and bainitic microstructures to produce a softened microstructure containing spheroidal carbides in ferrite [39]. Fine pearlite is also relatively easily softened by process annealing, while coarse pearlite is too stable to be softened by this process.

D. Recrystallization Annealing

Prior to cold working, steel microstructure is spheroidized or ferritic and is highly ductile. When steel is cold-worked, it becomes work-hardened and the prior microstructure becomes deformed due to imperfections

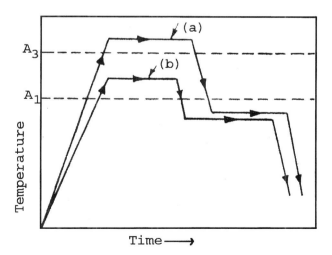

Figure 28 Schematic of a heat-treatment cycle for isothermal annealing of (a) hypoeutectoid steel and (b) a eutectoid steel.

within the grains which introduce high strain energy. The high strain energy in the deformed ferritic microstructure resulting from the cold-working process is reduced due to a recrystallization process which produces strain-free grains nucleation resulting in a ductile, spheroidized microstructure [2]. Nearly all steels, which are heavily cold-worked, undergo recrystallization annealing which reduces hardness and increases ductility [35].

Recrystallization annealing is performed by heating the steel for 30 min–1 hr at a temperature above the recrystallization temperature shown in Fig. 29 [40]. When heating is complete, the steel is cooled. As opposed to other annealing processes where the processing temperature is fixed, the recrystallization annealing temperature is composition-dependent, depends on prior deformation, grain size, and holding time, and therefore is not fixed [35]. Liscic [17] reported a correlation between recrystallization temperature (T_R) and the melting temperature (T_m) of the steel:

$$T_R = 0.4T_m$$

E. Isothermal Annealing

Isothermal annealing is conducted by heating the steel within the austenite transformation region (above A_{c3}) for a time sufficient to complete the solution process yielding a completely austenitic microstructure. At this time, the steel is cooled rapidly at a specified rate with-

in the pearlite transformation range indicated by the TTT diagram for the steel (less that A_{c1}, typically between 600°C and 700°C) until that the complete transformation into ferrite plus pearlite (lamellar pearlite) occurs at which time the steel is cooled rapidly to room temperature [1,35]. Caution: it should be noted that it is not rigorously correct to use TTT diagrams which are developed for hardening temperatures which are substantially greater than those used for annealing [40].

Isothermal annealing is used to achieve a more homogeneous microstructure within the steel and is faster and less expensive than full annealing. It is typically performed on hypoeutectoid steels, and, usually, it is not performed on hypereutectoid steels [35,41]. When isothermal annealing is used in continuous production lines for small parts or for parts with thin cross sections, it is called "cycle annealing."

F. Spheroidizing (Soft Annealing)

Spheroidizing involves the prolonged heating of steel at a temperature near the lower critical temperature (A_{c1}) as illustrated in Fig. 29 [1]. Eutectoid steels are heated 20–30°C above A_{c1}, and hypereutectoid steels are heated 30–50°C above A_{c1}. The A_{c1} temperature can be determined from Table 4, obtained from the appropriate TTT diagram or calculated from [17]:

$$A_{c1}(°C) = 739 - 22(\%C) + 2(\%Si) - 7(\%Mn)$$
$$+ 14(\%Cr) + 13(\%Mo) + 13(\%Ni) + 20(\%V)$$

(Hypereutectoid alloy steels may need to be heated at a higher temperature.)

Medium carbon steels may be spheroidized by heating just above or just below the A_{c1} temperature. Heating is followed by either furnace cooling to a temperature just above A_{r1} to produce a microstructure containing globular-shaped carbides which are uniformly distributed in a ferrite matrix (cementite of lamellar pearlite) and free cementite for hypoeutectoid and eutectoid steels, and in the case of hypereutectoid steels, the free cementite coalesces into small spheroids. The cooling rate is calculated from [17]:

1. For plain carbon and low-alloy steels up to 650°C (1200°F)—cooling rate = 20–25 K/hr (furnace cooling)
2. For medium alloy steels up to 630°C (1166°F)—cooling rate = 15–20 K/hr (furnace cooling)
3. For high alloy steels up to 600°C (1112°F)—cooling rate = 10–15 K/hr (furnace cooling)

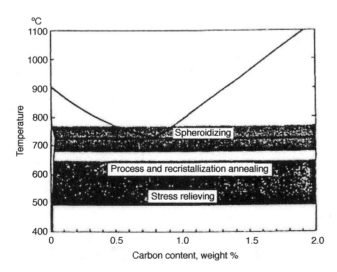

Figure 29 Correlation of typical temperatures ranges used for spheroidizing, process annealing, recrystallization, and stress relief with the iron–iron carbide phase diagram.

For alloy steels, the spheroidizing temperature (T) may be calculated from [42]:

$$T(°C) = 705 + 20(\%Si - \%Mn + \%Cr - \%Mo - \%Ni + \%W) + 100(\%V)$$

High-carbon and alloy steels are spheroidized to enhance their machinability and ductility. This process is desirable for cold-formed low-carbon and medium-carbon steels and for high-carbon steels that are machined prior to final machining.

G. Diusion (Homogenizing) Annealing

Diffusion annealing (homogenizing) is performed on steel ingots and castings to minimize chemical segregation. Chemical segregation defects occur as dendrites, columnar grains, and chemical occlusions. The presence of these defects produces increased brittleness and reduced ductility and toughness. The homogenization process, illustrated in Fig. 30 [35], is conducted by heating the steel rapidly to 1100–1200°C for 8–16 hr. The steel is then furnace-cooled to 800–850°C, and then cooled to room temperature in still air [38]. The defects are eliminated by chemical diffusion.

VIII. NORMALIZING

Some alloy steels produce martensitic microstructure even with air cooling. Therefore slower cooling rates are required to provide a uniform microstructure of ferrite plus pearlite (the same microstructure produced by annealing but in a smaller grains and finer lamellae). Hypoeutectoid steels are heated to a higher temperature (40–50°C above the A_{c3}) than that used for annealing. Hypereutectoid steels are heated above the A_{cm} temperature as illustrated in Fig. 30. The holding time depends on the size of the part, and when the heating step is completed and the part is completely austenitized, the part is cooled in still air [1,27,39]. However, the minimum time is 15 min at temperature with longer times for larger parts. Plain carbon and low-alloy steels should always be normalized [27]. Table 8 provides normalization temperatures for various commonly encountered steels.

Normalizing is conducted to [39]:

- Provide the desired microstructure, mechanical properties. For a given steel composition, normalized structures will be harder and stronger with lower ductility than if fully annealed.
- Improve hardening response by grain refinement and improved homogenization.

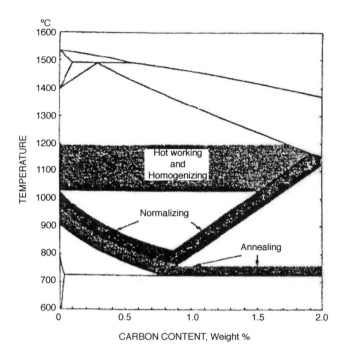

Figure 30 Illustration of typical temperature ranges for full annealing, normalizing, hot working, and homogenization on the relevant portion of the iron–iron carbide phase diagram.

- Improve machining characteristics, particularly for 0.15–0.40% carbon steels.
- Eliminate carbide networks in hypereutectoid steels.

There are various equations that may be used to calculate the hardness of normalized steel. One method is by using the Bofors equation [40]. The first step in this calculation is to determine the sum of the "carbon potentials"—C_p:

$$C_p = C[1 + 0.5(C - 0.20)] + Si \times 0.15$$
$$+ Mn[0.125 + 0.25(C - 0.20)]$$
$$+ P[1.25 - 0.5(C - 0.20)]Cr \times 0.2 + Ni \times 0.1$$

where C is the carbon concentration in %. The ultimate tensile strength (kp/mm^2) after normalization is:

$27 + 56 \times C_p$	for hot − rolled steel
$27 + 50 \times C_p$	for forged steel
$27 + 48 \times C_p$	for cast steel

For steels which may be used in subzero conditions, a double normalizing treatment may be specified [40]. In these cases, the steel is first heated to 50–100°C above the normalizing temperature. This will produce greater dissolution of the alloying elements. The second normalization step is conducted near the lower limit of the normalization temperature range for the purpose of producing a finer grain structure.

IX. STRESS RELIEVING

Stress relieving is typically used to remove residual stresses which have accumulated from prior manufacturing processes. Stress relief is performed by heating to a temperature below A_{c1} (for ferritic steels) and holding at that temperature for the required time to achieve the desired reduction in residual stresses, and then the steel is cooled at a rate sufficiently slow to avoid the formation of excessive thermal stresses. No microstructural changes occur during stress-relief processing. Nayar [38] recommends heating to:

- 550–650°C for unalloyed and low-alloy steels
- 600–700°C for hot-work and high-speed tool steels

These temperatures are above the recrystallization temperatures of these types of steels. Little or no stress relief occurs at temperatures < 260°C, and approximately 90% of the stress is relieved at 540°C. The maximum temperature for stress relief is limited to 30°C

below the tempering temperature used after quenching [43].

The results of the stress-relieving process are dependent on the temperature and the time which are correlated through Holloman's parameter (P) [10]:

$$P = T(C + \log t)$$

where T is the temperature in K, t is the time (hr), and C is the Holloman–Jaffe constant which is calculated from:

$$C = 21.53 - (5.8 \times \%C)$$

P is a measure of the "thermal effect" of the process, and those processes with the same Holloman's parameter exhibit the same effect. Another similar, commonly used expression used in evaluating the stress relief of spring steels is the Larson–Miller equation [17]:

$$P = T(\log t + 20)/1000$$

Stress relieving results in a significant reduction of yield strength in addition to reducing the residual stresses to some "safe" value. Typically, heating and cooling during stress relieving are performed in the furnace, particularly with distortion and crack-sensitive materials. Below 300°C, faster cooling rates can be used.

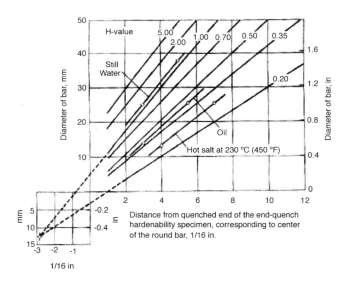

Figure 31 Correlation of bar diameter, steel hardenability (Jominy distance), and quench severity (*H*-value) where: 5.0 = brine quench with violent agitation, 2.0 = brine quench with no agitation, 1.0 = water quench at room temperature and no agitation, 0.7 = oil quench with strong agitation, 0.5 = oil quench with good agitation, 0.35 = oil quench with moderate agitation, and 0.20 = oil quench with no agitation.

X. QUENCHING

A. Quenchant Selection and Severity

Quench severity, as expressed by the Grossmann H-value, is the ability of a quenching medium to extract heat from a hot steel workpiece, expressed in terms of the Grossmann number (H). A typical range of Grossmann H-values (numbers) for commonly used quench media is provided in Table 7, and Fig. 31 provides a correlation between the H-value and the ability to harden steel as indicated by the Jominy distance (J distance) [28]. Although Table 7 is useful to obtain a relative measure of the quench severity offered by different quench media, it is difficult to apply in practice because the actual flow rates for "moderate," "good," "strong," and "violent" agitation are unknown.

Alternatively, the measurement of actual cooling rates or heat fluxes provided by a specific quenching medium does provide a quantitative meaning to the quench severity provided. Some illustrative values are provided in Table 9 [17,44]. Typically, the greater the quench severity, the greater the propensity of a given quenching medium to cause increased distortion or cracking. This usually is the result of increased thermal stress and not transformational stresses. Specific recommendations for quench media selection for use with various steel alloys are provided by standards such as AMS 2779. Some additional general comments regarding quenchant selection include the following [45,46].

Table 9 Comparison of Typical Heat Transfer Rates

Quench medium	Heat transfer rate $(W \cdot m^{-2}\ K^{-1})$	Reference
Furnace	15	10
Still Air	30	10
Compressed Air	70	10
Nitrogen (1 bar)	100–150	35
Salt bath or fluidized bed	350–500	35
Nitrogen (10 bar)	400–500	35
Air–water mixture	7	10
Helium (10 bar)	550–600	35
Helium (20 bar)	900–1000	35
Still oil	1000–1500	35
Liquid lead	1200	10
Hydrogen (20 bar)	1250–1350	35
Circulated oil	1800–2200	35
Hydrogen (40 bar)	2100–2300	35
Circulated water	3000–3500	35

Table 10 Suggested Carbon Content Limits for Water, Brine, and Caustic Quenching

Hardening method/shapes	Max. % carbon
Furnace hardening	
general usage	0.30
simple shapes	0.35
very simple shapes, e.g., bars	0.40
Induction hardening	
simple shapes	0.50
complex shapes	0.33

- Most machined parts made from alloy steels are oil-quenched to minimize distortion.
- Most small parts or finish-ground larger parts are free-quenched.
- Larger gears, typically those over 8 in., are fixture (die)-quenched to control distortion.
- Smaller gears and parts such as bushings are typically plug-quenched on a splined plug typically constructed from carburized 8620 steel.
- Although a reduction of quench severity leads to reduced distortion, it may also be accompanied by undesirable microstructures such as the formation of upper bainite (quenched pearlite) with carburized parts.
- Quench speed may be reduced by quenching in hot (300–400°F) oil. When hot oil quenching is used for carburized steels, lower bainite, which exhibits properties similar to martensite, is formed.
- Excellent distortion control is typically obtained with austempering, quenching into a medium just above the Ms temperature.
- Aqueous polymer quenchants may often be used to replace quench oils, but quench severity is still of primary importance.
- Gas or air quenching will provide the least distortion and may be used if the steel has sufficient hardenability to provide the desired properties.
- Low-hardenability steels are quenched in brine or vigorously agitated oil. However, even with a severe quench, undesirable microstructures such as ferrite, pearlite, or bainite can form.

It is well known that cracking propensity increases with carbon content. Therefore the carbon content of the steel is one of the determining factors for quenchant selection. Table 10 summarizes some steel mean carbon content concentration limits for water, brine, or caustic quenching [41,44].

B. Component Support and Loading

Many parts, such as ring gears, may sag and creep under their own weight when heat-treated which is an important cause of distortion. Proper support when heating is required to minimize out-of-flatness and ovality problems may result in long grinding times, excessive stock removal high scrap losses, and loss of case depth [47]. To achieve adequate distortion control, custom supports or press quenching may be required. Pinion shafts are also susceptible to bending along their length if they are improperly loaded into the furnace. When this occurs, they must then be straightened which will add to the production cost.

C. Surface Condition

Quench cracking may be due to various steel-related problems that are only observable after the quench, but the root cause is not the quenching process itself. Examples include prior steel structure, stress risers from prior machining, laps and seams, alloy inclusion defects, grinding cracks, chemical segregation (bonding), and alloy depletion [48]. In this section, three surface condition-related problems that may contribute to poor distortion control and cracking will be discussed: "tight" scale formation, decarburization, and the formation of surface seams or "nonmetallic stringers."

Tight scale problems are encountered with forgings hardened from direct-fired gas furnaces with high-pressure burners [41,46]. The effect of tight scale on the quenching properties of two steels, 1095 carbon steel and 18-8 stainless steel, is illustrated in Fig. 32 [34]. These cooling curves were obtained by still quenching into fast oil. A scale of not more than 0.08 mm (0.003 in.) increases the rate of cooling of 1095 steel as compared to the rate obtained on a specimen without scale. However, a deep heavy scale (0.13 mm, 0.005 in.) retards the cooling rate. A very light scale, 0.013 mm (0.0005 in.) deep, also increased the cooling rate of the 18-8 steel over that obtained with the specimen without scale.

In practice, the formation of tight scale will vary in depth over the surface of the part resulting in thermal gradients due to differences in cooling rates. This problem may yield soft spots and uncontrolled distortion and is particularly a problem with nickel-containing steels. Surface oxide formation can be minimized by the use of an appropriate protective atmosphere.

Another surface-related condition is decarburization which may lead to increased distortion or cracking [49]. At a given depth within the decarburized layer, the part

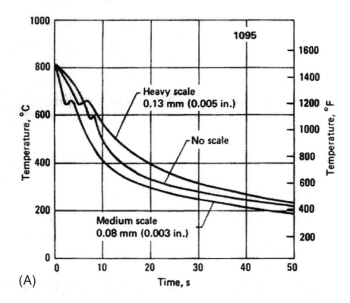

(A)

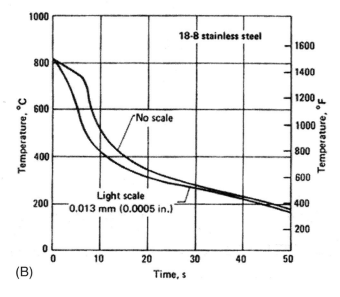

(B)

Figure 32 Cooling curves illustrating the effect of scale on cylindrical steel bars (13 mm diameter × 64 mm, 0.5 in. diameter × 2.5 in.) when quenched into an accelerated oil without agitation: (A) AISI 1095 steel—oil temperature = 50°C (125°F); (B) 18-8 stainless steel—oil temperature = 25°C (75°F).

does not harden as completely as it would at the same point below the surface if there was no decarburization. This leads to nonuniform hardness which may contribute to increased distortion and cracking because the decarburized surface transforms at a higher temperature than the core (the M_s temperature decreases with

Table 11 Minimum Recommended Material Removal from Hot-Rolled Steel Products to Prevent Surface Seam and Nonmetallic Stringer Problems During Heat Treatment

| | Minimum material removal per side[a] | |
Condition	Non-resulfurized	Resulfurized
Turned on centers	3% of diameter	3.8% of diameter
Centerless turned or ground	2.6%	3.4%

[a] Based on bars purchased to special straightness, i.e., 0.13 in. in 5-ft maximum.

carbon content) [46]. This will lead to high residual tensile stresses at the decarburized surface or a condition of unbalanced stresses and distortion. Since the surface is decarburized, it will exhibit lower hardenability than the core. This will cause the upper transformation products to form early nucleating additional undesirable products in the core. The decarburized side will be softer than the side that did not undergo decarburizing which is harder. The greater amount of martensite leads to distortion. The solution to this problem is to restore carbon into the furnace atmosphere or machine off the decarburized layer.

Surface seams or nonmetallic inclusions, which may occur in hot-rolled or cold-finished material, are defects that prevent the hot steel from welding to itself during the forging process. These defects act as stress risers. To prevent this problem with hot-rolled bars, stock should be removed before heat treatment. Recommendations made earlier by Kern [41] are provided in Table 11.

D. Heating and Atmosphere Control

An important source of steel distortion and cracking during heat treating is nonuniform heating without the appropriate protective atmosphere. For example, if steel is heated in a direct gas-fired furnace with high moisture content, the load being heated may adsorb hydrogen leading to hydrogen embrittlement and subsequent cracking which would not normally occur with a dry atmosphere [9,51].

Localized overheating is a problem for inductively heated parts [45,51]. Subsequent quenching of the part leads to quench cracks at sharp corners and areas with sudden changes in cross-sectional area (stress risers). Cracking is due to increases of residual stresses at the stress risers during the quenching process. The solution

to the problem is to increase the heating speed by increasing the power density of the inductor. The temperature difference across the heated zone is decreased by continuous heating or scanning of several pistons together on a single bar [51].

For heat-treating problems related to furnace design and operation, it is usually suggested that [50]:

1. The vestibules of atmosphere-hardening furnaces should be loaded and unloaded with purging. Load transfer for belt and shaker hearth furnaces should only occur with thorough purging to minimize atmosphere contamination.
2. Hardening furnaces typically contain excessive loads prior to quenching. If the steel at quenching temperature is greater than 20% of the distance from discharge to charge door, it is too much. Either the production rate can be increased or some of the burners can be turned off.

E. Retained Austenite

Dimensional changes may occur slowly or fast which are due to the volume composition of the transformation products formed upon quenching. One of the most important, with respect to residual stress variation, distortion and cracking is the formation and transformation of retained austenite. For example, the data in Table 12 [44] illustrates the slow conversion of retained austenite to martensite which was still occurring days after the original quenching process for the two steels shown [52,53]. This is particularly a problem since dimensional control and stability are the primary goals of heat treatment. Therefore microstructural determination is an essential component of any distortion control process.

F. Quenchant Uniformity

Quench nonuniformity is perhaps the greatest contributor to quench cracking. Quench nonuniformity can arise from nonuniform flow fields around the part surface during the quench or nonuniform wetting of the surface [44]. Both lead to nonuniform heat transfer during quenching. Nonuniform quenching creates large thermal gradients between the core and the surface of the part. These two contributing factors, agitation and surface wetting, will be discussed here.

Poor agitation design is a major source of quench nonuniformity. The purpose of the agitation system is not only to take hot fluid away from the surface and to the heat exchanger, but also to provide uniform heat

Table 12 Dimensional Variation in Hardened High-Carbon Steel with Time at Ambient Temperature

Steel type	Tempering temperature (°C)	Hardness (HRC)	Change in length (% × 10^3) after time (days)			
			7	30	90	365
1.1% C tool steel,	None	66	−9.0	−18.0	−27.0	−40.0
790°C quench	120	65	−0.2	−0.6	−1.1	−1.9
	205	63	0.0	−0.2	−0.3	−0.7
	260	61.5	0.0	−0.2	−0.3	−0.3
1% C/Cr,	None	64	−1.0	−4.2	−8.2	−11.0
840°C quench	120	65	0.3	0.5	0.7	0.6
	205	62	0.0	−0.1	−0.1	−0.1
	260	60	0.0	−0.1	−0.1	−0.1

removal over the entire cooling surface of all of the parts throughout the load being quenched. Although agitation is a critically important contributor to the performance of industrial quenching practice, relatively little is known about the quality and the quantity of fluid flow encountered by the parts being quenched. Recently, agitation in various commercial quenching tanks has been studied by computational fluid dynamics (CFD) and in no case was optimal and uniform flow present without subsequent modification of the tank [54]. Thus identifying sources of nonuniform fluid flow during quenching continues to be an important tool for optimizing distortion control and minimizing quench cracking.

The second source of nonuniform thermal gradients during quenching is related to interfacial wetting kinematics which is of particular interest with vaporizable liquid quenchants including, water, oil, and aqueous polymer solutions. Most liquid vaporizable quenchants exhibit boiling temperatures between 100°C and 300°C at atmospheric pressure. When parts are quenched in

these fluids, surface wetting is usually time-dependent which influences the uniformity of the cooling process and the achievable hardness and potential formation of soft spots. This is a problem with oil-contaminated aqueous polymer quenchants, sludge-contaminated oils, and foaming.

XI. TEMPERING

When steel is hardened, the as-quenched martensite is not only very hard, but is also brittle. Tempering, also known as "drawing," is the thermal treatment of hardened and normalized steels to obtain the desired mechanical properties which include improved toughness and ductility, lower hardness, and improved dimensional stability. During tempering, as-quenched martensite is transformed into tempered martensite which is composed of highly dispersed spheroids of cementite (carbides) dispersed in a soft matrix of ferrite resulting in reduced hardness and increased toughness. The objec-

Table 13 Metallurgical Reactions Occurring at Various Temperature Ranges and Related Physical Changes of Steel During Tempering

Stage	Temperature range	Metallurgical reaction	Expansion/contraction
1	0–200°C (32–392°F)	Precipitation of ε-carbide; loss of tetragonality	Contraction
2	200–300°C (392–572°F)	Decomposition of retained austenite	Expansion
3	230–350°C (446–662°F)	ε-carbides decompose to cementite	Contraction
4	350–700°C (662–1292°F)	Precipitation of alloy carbides. Grain coarsening	Expansion

tive is to allow hardness to decrease to the desired level and then to stop the carbide decomposition by cooling. The extent of the tempering effect is determined by the temperature and the time of the process.

The tempering process involves heating hardened steel to some temperature below the eutectoid temperature for the purposes of decreasing hardness and increasing toughness. (Tempering is performed as soon as possible after it has cooled to between 50–75°C and room temperature to reduce the potential of cracking. If a tool steel cannot be tempered immediately after quenching, it is recommended that it should be held at 50–100°C in an oven until it can be tempered [40]). In general, the tempering process is divided into four stages, which are summarized in Table 13 [55]. These include:

1. Tempering of martensite structure
2. Transformation of retained austenite to martensite
3. Tempering of the decomposition products of martensite and at temperatures >900°F
4. Decomposition of retained austenite to martensite [56]

Figure 33 illustrates the effect of microstructural variation during tempering on the volume changes occurring during the tempering of hardened steel [55].

The tempering process may be conducted at any temperature up to the lower critical temperature (A_{c1}). Figure 34 illustrates the effect of carbon content and tempering temperature on the hardness of carbon steels [55]. The specific tempering conditions that are selected are dependent on the desired strength and toughness.

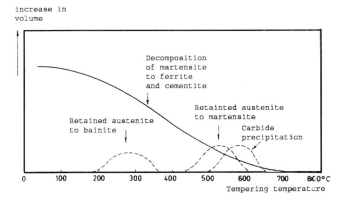

Figure 33 Effect of microstructural variation during tempering on the volume changes occurring during the tempering of hardened steel.

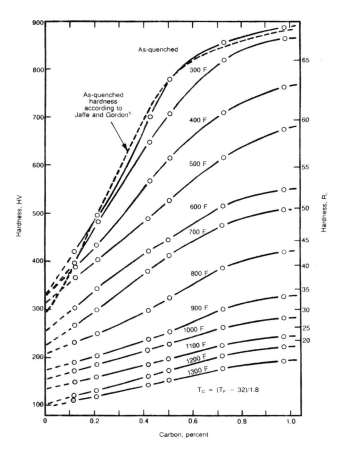

Figure 34 Correlation of carbon content of martensite and hardness of different Fe–C alloys at different tempering temperatures.

Nayar [38] has recommended the following tempering conditions:

- Heat the parts to 150–200°C to reduce internal stresses and increase toughness without a significant loss in hardness. This is often done with surface-hardened parts.
- To obtain the highest attainable elastic limit and sufficient toughness, temper at 350–500°C.
- For optimum strength and toughness, heat to 500–700°C. (Note: When tempering at high temperatures, 675–705°C, precautions must be taken not to exceed the Ac1 temperature, above which undesirable austenite may be formed which upon cooling would transform to pearlite. This is a particular concern for nickel-containing steels since nickel depresses the Ac1 temperature [56].)

When steel is tempered in air, the heated oxide film on the surface of the steel exhibits a color, known as "tempering colors," which is a characteristic of the surface temperature. Table 14 provides a summary of characteristic surface temperatures for tempering and their colors [35].

Retained austenite is usually present after quenching steel, and the amount is dependent on the carbon content of the austenite, as illustrated in Fig. 35 [57]. Then in addition to the four steps shown, there is another step referred to as "refrigeration" or "subzero" treatment. Subzero treatment is performed on steels to transform retained austenite to as-quenched martensite. Conversion of retained austenite in this way results in improved hardness, wear resistance, and dimensional stability. Subzero treatment is performed using dry-ice or liquid nitrogen and involves cooling the steel to a temperature less than the M_f temperature of the steels which is typically between $-30°C$ and $-70°C$. (Tool steels will grow between about 0.0005 and 0.002 in. per inch of original length during heat treatment [3].) An immediate tempering step is required to remove residual stresses imparted to the steel by this process. Subzero treatments are not effective on steels that have been held at room temperature for several hours, and therefore it is typically performed immediately after hardening [35]. Although the general rule is to allow 1 hr/in. of the thickest cross section, tool steels should be held at temperature for a minimum of 2 hr for each temper [3].

Martensitic stainless steels and alloy steels that contain >0.4% carbon and which exhibit M_s temperatures of about 300°C are particularly susceptible to cracking, especially if they are through-hardened [40]. In such cases, cooling may be interrupted at about 80°C followed by an immediate temper at about 170°C to stop the formation of martensite. However, significant amounts of untempered martensite remain after the steel is cooled to room temperature. Therefore the steel must be tempered a second time at the same temperature

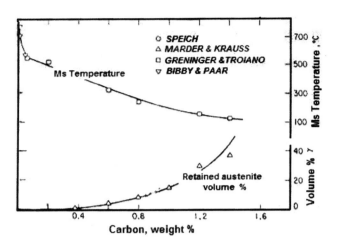

Figure 35 The amount of retained austenite upon quenching to 25°C as a function of carbon content of the austenite. M_s temperature is also shown.

to transform the hard and brittle as-quenched martensite to a softer and more ductile-tempered martensite. This is one form of "double-tempering" process.

Another form of double tempering occurs for high-alloy chrome steels and high-speed tool steels where significant amounts of retained austenite are transformed to martensite after tempering at about 500°C. Such steels should be retempered to obtain a tougher martensite structure. Generally, this second tempering step is performed at about 10–30°C below the original tempering temperature [40].

Steels that exhibit a high M_s temperature, ≥400°C, typically those that contain <0.3% carbon, form martensite which may be tempered during the remaining cooling (quenching) process. This is called "self-tempering" or "autotempering," and such steels are typically not crack-sensitive, particularly if the M_f temperature is ≥100°C [40].

Typically, tempering times are a minimum of approximately 1 hr. Thelning [40] has reported a "rule of thumb" of 1–2 hr/in. of section thickness after the load has reached a preset temperature. After heating, the steel is cooled to room temperature in still air.

The recommended tempering conditions, in addition to recommended heat-treating cycles, for a wide range of carbon and alloy steels are provided in SAE AMS 2759.

Tempering times and temperatures may also be calculated by various methods. One of the more common methods is to use the Larsen–Miller equation discussed above. The Larsen–Miller equation, although

Table 14 Temper Colors of Steel

Temperature (°C)	Temper color
220	Straw yellow
240	Light yellow
270	Brown
285	Purple
295	Dark blue
310	Light blue
325	Grey
350	Grey-purple
375	Grey-blue
400	Dull grey

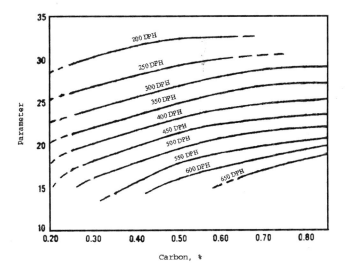

Figure 36 Variation of the Holloman parameter (P) at different hardness levels.

originally developed for the prediction of creep data has been used successfully for predicting the tempering effect of medium/high-alloy steels [39]. Bofors [3] reported that a Holloman–Jaffe constant ($C = 20$) was appropriate for all steels, but Grange and Baughman reported that $C = 18$ should be used [1].

Sinha [1] showed in Fig. 36 that the Holloman–Jaffe constant varies with carbon content and desired hardness. The incremental contribution to hardness of each alloying element of a steel may be determined from Table 15 [1]. The hardness (DPH) is calculated by multiplying the concentration of each of the alloying elements (within the range shown) times the factor for that element at a given constant (C), and

then all of these values are added together to provide the hardness (DPH). The interrelationship between tempering time and the Holloman–Jaffe parameter (P) at different tempering temperatures is shown in Fig. 37 [1].

The interrelationship between tempering temperature, time, and steel chemistry has been reported by Spies [17,58]:

$$HB = 2.84H_h + 75(\%C) - 0.78(\%Si) + 14.24(\%Mn)$$
$$+ 14.77(\%Cr) + 128.22(\%Mo) - 54.0(\%V)$$
$$- 0.55T_t + 435.66$$

where HB is the Brinell hardness after hardening and tempering, H_h is the Rockwell (HRc) hardness after hardening, and T_t is the tempering temperature in °C. This equation was developed for the following conditions: $H_h = 20-65$HRc, $C = 0.20-0.54\%$, $Si = 0.17-1.40\%$, $Mn = 0.50-1.90\%$, $Cr = 0.03-1.20\%$, and $T_t = 500-650$°C ($932-1202$°F).

The German DIN 17021 standard provides a relationship between as-quenched hardness (H_h) and as-tempered hardness (H_t) [17]:

$$H_h(HRc) = (T_t/167 - 1.2)H_t - 17$$

where T_t = tempering temperature in °C.

Tempering embrittlement may occur with some steels if they are tempered below 595°C which is observable by a reduction in notch toughness as illustrated in Fig. 38 for AISI 43200 (Ni–Cr–Mo steel) which was tempered at 260–370°C [43]. However, temper embrittlement may be avoided by tempering at higher temperatures with subsequent quenching to

Table 15 Factors for Predicting the Hardness of Tempered Martensite

Element	Range (%)	Factors at indicated parameter (C) value[a]					
		20	22	24	26	28	30
Mn	0.85–2.1	35	25	30	30	30	25
Si	0.3–2.2	65	60	30	30	30	30
Ni	≤4	5	3	6	8	8	6
Cr	≤1.2	50	55	55	55	55	55
Mo	≤0.35	40	90	160	220	240	210
		(20)[b]	(45)[b]	(80)[b]	(110)[b]	(120)[b]	(105)[b]
V[c]	≤0.2	0	30	85	150	210	150

[a] The Boron factor is 0.
[b] If 0.5–1.2% Cr is present, use this factor.
[c] May not apply if vanadium is the only carbide formerly present.

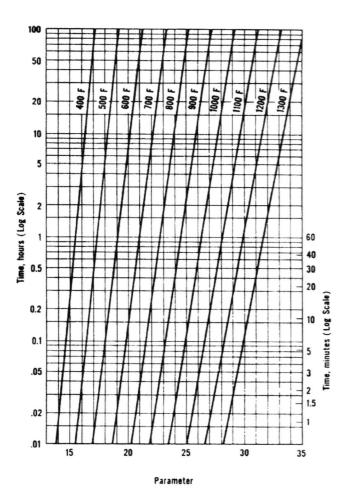

Figure 37 Time–temperature vs. Holloman parameter chart for $C = 18$.

minimize the time the steel will spend in the intermediate temperature range [43,56].

Tempering may be performed in convection furnaces, salt baths, or even by immersion in molten metal. Induction tempering and flame heating are also used but will not be discussed further here. Table 16 provides a comparative summary of the different heating media [59]. Of the different tempering systems shown, convection furnaces are the most common, and it is important that they should be equipped with fans and/or blowers to provide for uniform heat transfer when heating the load. Typically, convection-tempering furnaces are designed for use with a 150–750°C range.

Salt baths may also be used for various heating processes over the temperature range 150–1320°C [1], and they provide relatively rapid heat transfer compared to convection furnaces, although the actual use tempera-

ture is dependent on the composition of the salt bath. A comparison of heating rates for different steel cross sections is provided in Table 17 [40].

Sinha [1] has classified salt baths into three groups:

- Low-temperature salt baths may be used from 150°C to 620°C. These baths are of two types: a binary mixture of equal parts of potassium nitrate and sodium nitrite, which may be used for heating to 150–500°C, and a binary mixture of potassium nitrate and sodium nitrate, which may be used for heating in the range of 260–620°C. In addition to tempering, these baths may also be used for cooling. It is essential, however, that the baths are not contaminated with cyanides, organic compounds, or water.

- Medium-temperature neutral baths are suitable for use over the range of 650–1000°C. These baths are binary or ternary mixtures of the following salts: potassium chloride, sodium chloride, barium chloride, or calcium chloride. Two examples of typical binary compositions and working temperatures include NaCl (45%)/KCl (55%), which is suitable for use at 675–900°C, and NaCl (20%)/BaCl$_2$ (80%), which is suitable for use at 675–1060°C. BaCl$_2$, if used at 100%, has a relatively narrow use range of 1025–1325°C [40]. An advantage of these baths is that when they are freshly

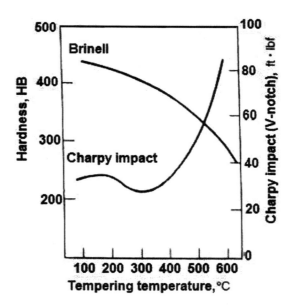

Figure 38 Loss in room-temperature toughness due to temper embrittlement of oil-quenched wrought Ni–Cr–Mo (AISI 4130) steel; section size = 13 mm (0.5 in.).

Table 16 Tempering Temperature Ranges Achievable with Different Tempering Equipment

Equipment type	Temperature range		Use conditions
	°C	°F	
Convection furnace	50–750	120–1380	For large volumes of similar parts; for variable loads, temperature control more difficult
Salt bath	160–750	320–1380	Rapid uniform heating; low to medium volume; should not be used for complex, hard-to-clean parts.
Fluidized beds	100–750	212–1380	Broad range of heat transfer rates is possible by varying choice of fluidizing gas, gas velocity, bed temperature, and the bed particle size. More energy-efficient than convection furnaces and they provide safe and ecological alternative to salts and lead with similar heat transfer rates.
Oil bath	≤250	≤480	Good if long exposure times are desired; special ventilation and fire control are required.
Molten metal bath	>390	>735	Very rapid heating; special fixturing required; molten metals may be toxic (Pb baths)

prepared, the steel surface will be clean without surface carburization or decarburization.

- High-temperature salt baths are used in the range of 1000–1300°C, and they typically contain mixtures of barium chloride, sodium tetraborate (borax), sodium fluoride, and silicates [1]. Since these baths may decarburize, steels as oxides build up after use.

Molten metal, most typically lead, has been used in the past, but due to its toxicity, its use is now limited to various heating operations where its outstanding heat transfer properties are demanded. Lead baths are used from 327°C, which is the melting point of lead, up to 900°C [1].

Fluidized beds are formed by passing a gas through solid particles such as aluminum oxide and silica sand which causes the particles to behave like a bubbling liquid. The particles are generally inert and do not react with metal parts but act to facilitate heat transfer between the fluidizing gas and the part being processed.

A broad range of heat transfer rates is possible over operating temperatures which may range from 100°C to 1050°C (212–1920°F) with fluidized bed furnaces for tempering operations typically ranging from 100°C to 750°C (212–1380°F) (see Table 15). Fluidized bed furnaces are not only more energy-efficient than convection furnaces, but also they exhibit heat transfer efficiency similar to salt baths and lead pots without the health and environmental safety hazards commonly associated with these systems.

Figure 39 provides a comparison of the relative heating rates that can be achieved with these different heating media [60].

XII. AUSTEMPERING

Austempering requires that the cooling process be fast enough to avoid the formation of pearlite as illustrated by the cooling curve for an austempering process which is superimposed onto a TTT diagram of a steel in Fig. 40 [40]. The steel is cooled below the nose of the pearlite transformation curve for the steel being quenched and just above the M_s temperature in a molten salt bath. The steel is held at this temperature until the transformation from austenite to bainite is complete. Upper or lower bainite may be formed depending on the molten salt temperature. Austempering eliminates the volumetric expansion due to martensite which helps to eliminate cracking and provides greater toughness and ductility at a particular hardness.

Table 17 Heat-Up Times for Different Cross Sections to 950–1000°C

Bar diameter (mm)	Salt bath (min)	Muffle furnace (min)
25	1	15
50	4	30
100	8	60

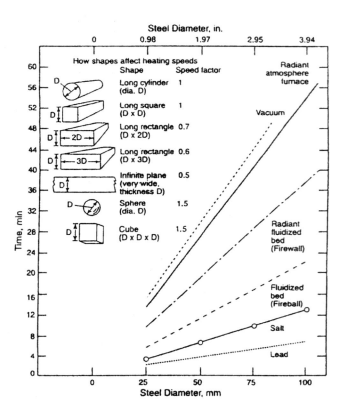

Figure 39 Recommended heating times for various heating media and shapes.

In practice, austempering is performed by heating the steel to 790–870°C and then quenched into a molten salt bath at a temperature just above the M_s temperature (260–400°C). The steel is kept at the molten salt bath until the transformation is complete at which time it is cooled to room temperature in air [38].

Steels that are suitable for austempering depend on [61]:

- The location of the nose of the TTT curve and the ability to cool the steel sufficiently fast to avoid pearlite formation
- The time required for complete transformation from austenite to bainite at the austempering temperature
- The Ms temperature of the steel

Figure 41 shows same examples of TTT curves with different characteristics.

The formation of retained austenite is a significant problem with austempering processes. Retained austenite is most pronounced where Mn and Ni are the major components. The best steels for austempering are plain carbon, Cr, and Mb alloy steels [45].

XIII. MARTEMPERING

Martempering (also known as marquenching) is a cooling process used to minimize distortion and cracking and to reduce the amount of residual stress formation of a part upon cooling. Typically, martempering involves austenitizing steel at a temperature above A_{c1} (815–870°C) and then quenching it into hot oil (marquenching oil) or molten salt at a temperature just above the M_s temperature (usually at a temperature ranging from 260°C to 400°C), and then holding the steel at this temperature until the entire steel piece is at the same temperature, at which time the steel is cooled to room temperature at a rate sufficiently fast to achieve a fully martensitic structure. Note that if the part is held at the martempering temperature too long, undesirable formation of bainite may occur [43]. Usually, the quenching step is performed in air which minimizes thermal gradients and related residual stresses throughout the part while cooling. This process is illustrated in Fig. 42 [40]. After quenching, the part is tempered as specified.

A variation of this process is to cool the part to a temperature intermediate between the M_s and M_f temperaure until the temperature at the surface and core have completely equilibrated at which time the steel is removed from the bath and air-cooled or placed directly in the tempering furnace. Table 18 provides a summary of M_s and M_f temperatures for a variety of common steel alloys [43].

Either martempering (hot quenching) oils or molten salt baths may be used for martempering. Molten salts were reviewed briefly above. Martempering oils will be discussed briefly here. Martempering oils are

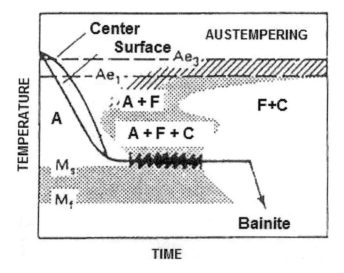

Figure 40 Illustration of austempering process.

used at temperatures between 95°C and 230°C. Molten salts are used for martempering operations performed at 204–400°C [61]. They are usually formulated with solvent-refined petroleum oils with a very high paraffinic fraction to maximize oxidative and thermal stability which is also enhanced by the addition of antioxidants. Accelerated and nonaccelerated martempering oils are available. Typical temperature ranges for commercially available martempering oils are summarized in Table 19 [62]. Because martempering

oils are used at relatively high temperatures, a protective, nonoxidizing atmosphere is often used which allows the use of temperatures to be much closer to the flash point than is generally recommended for open-air conditions.

More-hardenable alloy steels are martempered more often than less-hardenable plain carbon steels. Steels that are most often martempered to full hardness include AISI 1090, 4130, 4140, 4150, 4340, 4640, 5140, 6150, 8630, 8640, 8740, and 8745. Carburized steels

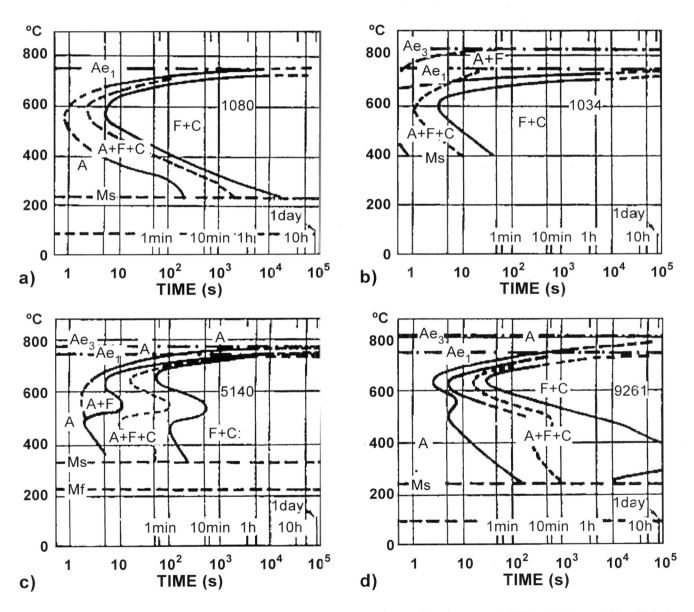

Figure 41 Examples of TTT diagrams where austempering process can be performed or not. a) Difficult, b) Impossible, c) Ideal, d) Not recommended (long time to get the bainitic transformation).

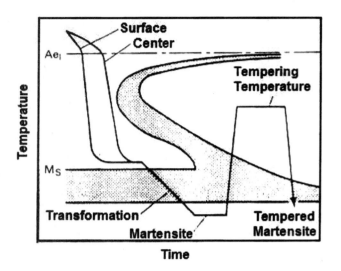

Figure 42 Martempering process.

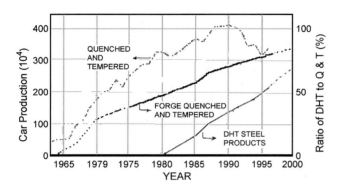

Figure 43 Trends of increasing FQ and DHT use.

from AISI 3312, 4620, 5120, 8620, and 9310 may also be martempered. Plain carbon steels AISI 1008 through 1040 are insufficiently hardenable to be martempered. Thin sections of some steels of borderline hardenability such as AISI 1541 can be martempered [61].

XIV. DIRECT HEAT TREATMENT TECHNOLOGY

Heat-treatment technology is currently facing various challenges such as environmental impact, energy conservation, and the more stringent market.

Processes as direct heat treatment that eliminate the hardening and tempering processes traditionally performed on forge or role-formed products contribute to the production efficiency, energy conservation, environmental issue, and improvement of components' performance [62].

There are a few types of direct treatment processes. They are the following:

Forge quench method (FQ or DQ): Forge components are directly quenched into oil or water after the forging operation [63,64].
Forge and/or heat treatment (DC): Forged or mild-rolled products are directly air-cooled or control-cooled to attain the strength same to the ordinary quench and temper process [63].

Table 18 Temperature Range of Martensite Formation of Several Carbon and Low-Alloy Steels

Steel grade AISI	M_s temperature		50% martensite formed		M_f—99% martensite formed	
	°C	°F	°C	°F	°C	°F
1030	343	650	293	560	232	450
1065	277	530	219	425	149	300
1090	219	425	157	560	232	450
2340	304	580	293	560	207	405
3140	335	635	288	550	221	430
4130	380	715	343	650	288	550
4140	343	650	299	570	227	440
4340	288	550	249	480	188	370
4640	343	650	299	570	249	480
5140	343	650	299	570	232	450
6140	327	620	293	560	232	450
8630	366	690	332	630	277	530
9440	330	625	282	540	207	405

Table 19   Typical Use Temperatures for Martempering Oils

Viscosity at 40°C, SUS	Minimum flash point		Use temperatures			
			Open air		Protective atmosphere	
	°C	°F	°C	°F	°C	°F
250–550	220	430	95–150	200–300	95–175	200–350
700–1500	250	480	120–175	250–350	120–205	250–400
2000–2800	290	550	150–205	300–400	150–230	300–450

Forge and roll and direct annealing (FA): Forged or mild-rolled products are directly annealed under the designed condition to give nearly the same formability same to ordinary normalizing or annealing processes [63].

These processes are enabling to give satisfactory strength and microstructure without reheat, quench, and temper processes by the use of thermal energy contained by forged products. However, to achieve the goals, it is necessary to have a good steel alloy design and to do an optimization of deformation rate in forging and roll condition for grain size refinement as well as the optimization of cooling condition. Usually, grain size refinement is performed by control of forging and rolling condition and the effective use of carbonitrides of Al, Ti, Nb, and also by the control of finely dispersed inclusions such as MnS and oxides [63].

The Forge Quench Method (FQ) has been applied at the automobile industry in Japan [63,64]. Several components in the automotive industry are undergone to direct heat treatment (DHT) process like FQ. They are crankshafts, front axle beams, knuckle arms, front lower arms, rack gear shafts, rear axle shafts, and others. DHT crankshafts have less distortion and machinability problems than traditional quenched and tempered one due to the elimination of quenching operation. Usually, properties of steel directly quenched after forging and then tempered are superiorly improved compared to steel air cooled after forging and then quenched and tempered.

DHT processes expanded widely since early 1970s and nearly 40% of the ordinary quenched and tempered components were converted to the FQ process, and more than 130 components are directly quenched after forging in the early 1980s. Figure 43 shows the tendency to use DHT steel products [62,63].

The fundamental concept of direct cooling methods as oil or water quenching just after the end of forging operation is expandable to other cooling methods to control microstructures of the treated parts, and various cooling methods are selected to get the desired strength and toughness. Quench oils have traditionally been the most commonly used quenchant media for the direct forge quenching of hot-forged parts. However, to obtain the desired mechanical properties and to obtain uniform product quality, recent works have suggested the use of certain aqueous polymer quenchants. These replacements have been made in conjunction with immersion time quenching system (ITQS) [62,65,66].

ITQS permit the continuous variation of agitation throughout the quenching processes using maximum agitation rate during the initial stages of the quench followed by minimizing the agitation rate as the M_s temperature is approached [67].

Track links made by AISI 15B37 were produced from the direct forge condition using ITQS process. In this case, an aqueous poly(alkylene glycol) quenchant at 10–12% was used. Schematic illustration of FQ process can be seen in Fig. 44. Excellent hardness uniformity as shown in Fig. 45 was achieved as well as reduced

Figure 44   Scheme of a FQ process.

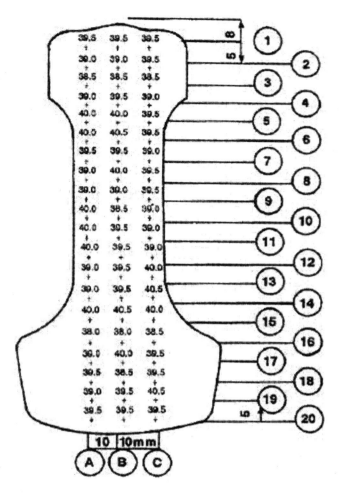

Figure 45 Hardness variation of track link after FQ and tempering at 450°C.

cracking, more uniform microstructure, and substantial cost reduction [65,67].

Similar results were obtained in engine valves and front wheel hubs, connecting rods produced by a direct forging process [65–67].

The use of DHT processes enables the improvement of manufacturing flexibility by reduced production steps and time which results in expressive cost reduction contributing to energy conservation as well.

XV. HEAT-TREATING PROCESS MODELING AND SIMULATION

Metal manufacturing companies continually waste time and resources dealing with the anticipated effects of heat treatment, particularly distortion. This waste would be reduced significantly if component response to heat

treatment would be predicted, in particular, the resulting material hardness and the distortion of the piece. Comparisons of different quenchants in heat processes of steels are greatly useful in order to control residual stresses, cracking, and distortion. The mathematical modeling of these processes is, nowadays, an indispensable tool for that purpose [68].

Researches have developed several computational models in order to predict component response of heat treatment. There are commercial computer software such as HEARTS [69,70], calc MAG [71], TRAST [72], MetalCore [73], INC. PHATRAN [74], INDUCTOR-B [75], OPTIBANC [76], ABAQUS/Standard [77], and others.

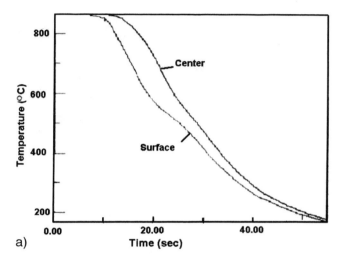

a)

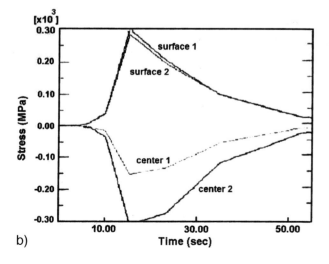

b)

Figure 46 (a) Temperature variation on the surface and on the center. Quenchant: aqueous polymer solution (25%). Diameter of the probe: 20.5 mm. (b) Residual stress variations in the same previous sample: 1—σ_θ and 2—σ_z.

Inverse conduction coupled with phase transforma-tion (INC-PHATRAN) is a program that can be employed to simulate a wide variety of heat-treatment processes, having plane geometry as well as axisymmetrical ones. The model is based on a numerical optimization algorithm, which includes a module responsible for calculating on time and space the temperature distribution and it is coupled to microstructural evolution. Whereas the temperature distribution in a two-dimensional domain with axial symmetry is calculated using finite element approximation, the time variation is approached using a Crank–Nicholson finite difference scheme. The temperature evolution, as measured by thermocouples at different positions in the component, is used as input for the program to calculate the time variation of the heat transfer coefficients together with the temperature and distribution of phases [78].

ABAQUS/Standard is a general program of finite element used to simulate the distortion and the residual stresses produced in samples as a consequence of the heat-treatment process [78,79].

A comparative study of residual stresses and distortion in steel cylindrical samples quenched at different bath temperatures using oil and aqueous polymer solutions can be found in Ref. 80. In this work, 12 AISI cylindrical samples of 20.5 and 13.5 mm in diameter were thermically treated in two different quenchants. Cooling curves kept in numerical files (obtained by thermocouples placed at the center of each sample) were the input to get the heat transfer coefficients as dependent of the temperature by means of the INC-PHATRAN. The simulations performed with this code used a finite element mesh containing nodes along the radial and longitudinal directions. The election of both the initial values for heat transfer coefficient and the quantity and length of the time intervals depended on each sample. In this step, the heat transfer coefficients were obtained for each sample in a range of temperature ($200°C$ to $800°C$) and used as input for ABAQUS/Standard in order to calculate the thermal field again as well as the corresponding distribution of thermal stresses depending on time during the heat-treating process. After the thermal field is known for each sample, ABAQUS/Standard solves the corresponding thermoelastic–plastic problem as well. In this case, calculations showed that such stresses were not high enough to generate residual stresses. Figure 46a [80] shows an example (for one of the samples) of the curve Temperature×Time in the surface and in the center of the sample. Figure 46b [80] shows the evolution of the longitudinal and hoop stress at the center and at the surface of the same previous sample. In this simulation, it was possible to conclude that there is no effect of the quenchants used in promoting residual stresses after heat treatment of the cylindrical probes of AISI 5160H spring steel with the diameters studied.

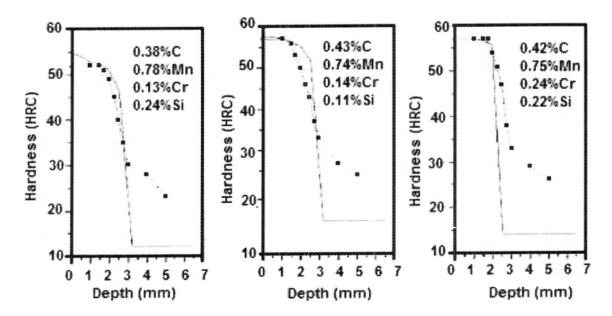

Figure 47 Comparison between measured and calculated hardness profiles for SAE 1040 steel bar with different chemical compositions.

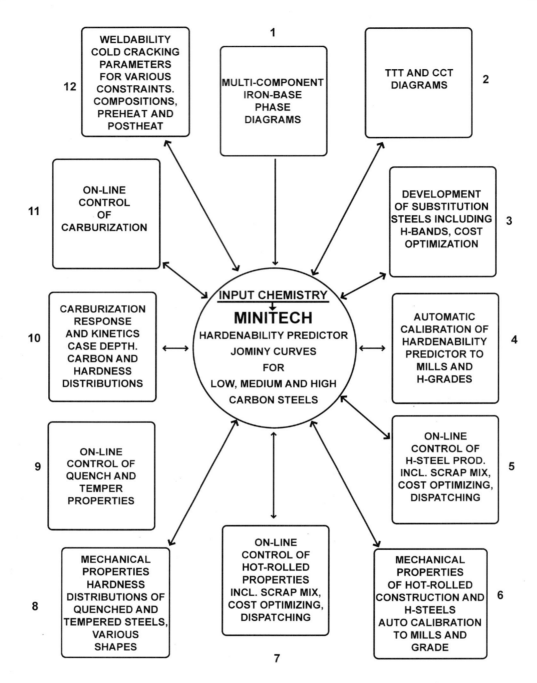

Figure 48   Block diagram of the Minitech Computerized Alloy Steel Information System.

Another example of residual stress simulation can be found in Ref. 81. In this study, the residual stress profile in a hardened layer of steel was measured and then compared with a finite element prediction. The residual hoop stress variation was measured using the crack compliance method and also using x-ray diffraction.

Simulation was made using DANTE (a heat-treatment simulation software package) interfacing with the finite element code ABAQUS.

Commercial software like DEFORM™-HT has the ability to simulate a wide range of heat-treatment cycles, including hardening, carburizing, tempering, aging

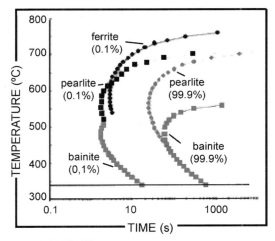

a) TTT CURVE

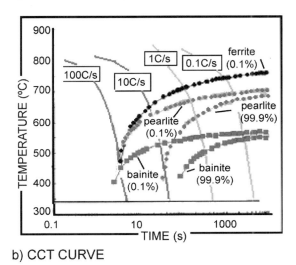

b) CCT CURVE

Figure 49 Calculated diagrams for Fe–0.37C–0.25Si–1.45Mn steel using JMatPro software. (a) TTT Diagram. (b) CCT Diagram.

spatial variation on the temperature and distribution of phases, as well as the hardness distribution throughout the steel bar. Workpieces with axisymmetrical geometry, coil moving at constant speed, and cooling with a water spray are hypotheses adopted in the model [84]. The program calculates, for a given set of process parameters, the hardness distribution in the workpiece, taking into account the influence of the chemical composition on the material hardenability. Examples of applications can be found in Sánches Sarmiento et al. [84,85].

Simulation of the heating of a bar by an induction coil moving upwards in the axial direction was made in order to verify the accuracy of the computer model. A set of numerical hardness profiles corresponding to samples of the same steel grade, but with different compositions, was compared to measured profiles. Figure 47 shows this comparison for 1040 steel bar [86].

MINITECH is another computational model that presents a set of related programs and subroutines, as illustrated in Fig. 48, which generate key phase diagram parameters, isothermal transformation diagrams (IT), and continuous cooling transformation diagrams (CCT) for multicomponent steels. The computers can be done for multicomponent steels with C, Mn, Ni, Cr, Mo, Cu, and V as alloying elements. Boron grades can also be accommodated. Eighteen different modules can be accessed, including Jominy Curve, Predict SAE H-Band, Normalized Properties, Quench and Temper Properties, Carburized Properties, TTT Diagram, and CCT Diagrams. For each module, different inputs are requested [87].

Using CCT Diagram module, the inputs are the steel grade, the chemical composition, the austenite grain size, and the austenitizing temperature. User must inform if the steel is a nonboron or boron steel and selects the cooling regime [87].

annealing, and solution treating. It is possible to predict the microstructure, grain size, residual stresses, distortion, and fracture resulting during these complex processes. Case depth can be predicted during carburizing and induction heating [82].

OPTBANC is a computer model developed to predict the number of cooling stands and their water flux rates, which are necessary to obtain specified properties in the final product during hot rolling processes [83].

INDUCTER-B is used to simulate induction heat-treating processes. It applies the same methodology as INC-PHATRAN and allows calculating the time and

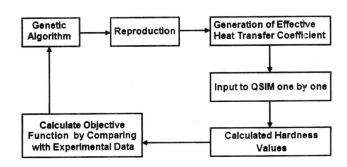

Figure 50 Heat transfer calculation procedure.

Module of normalized mechanical properties uses section size units and shape, chemical composition, and normalizing temperature as inputs. The outputs are Hardness (HRC, HB, and HV), Ultimate Tensile Strength (ksi and MPa), Yield Strength (ksi and MPa), % Elongation, and % Reduction [87].

Models to predict TTT and CCT diagrams are available in software package on the Internet as well [88–90]. Typical TTT and CCT diagrams calculated with JMat-Pro software are shown in Fig. 49a and b, respectively.

The improvement of computational fluid dynamics (CFD) computer programs has made possible to simulate the flow of fluid in a quench tank and to use these as tools to design better tanks and agitation system. This is important because achieving minimal distortion of hot metal parts immersed in a quench tank requires understanding the flow patterns that exist in the tank. Agitation of the working fluid that creates large uniformities in the quenchant flow can lead to thermal gradients across the immersed pieces, which results in a distortion of the parts. Usually, codes like NASTD and FLUENT are used to simulate the tank conditions. These programs are based on the solution of the compressible Navier–Stokes equations with a variety of turbulence models. To get confident results, the load to be quenched may be included in the calculations with an added complexity to the analysis. Simulations of this type can be used as a tool in production quench tank design in order to achieve regions of desired flow speed and direction [91,92].

Caterpillar has developed a reliable procedure which computes the heat extraction rate of production quenches by combining a numerical simulation program with experimental data obtained from quench characterization bars [93]. QSIM is a numerical heat treat simulation code that has been successfully applied to verify the interaction of the variables that affect the outcome of heat treating a part. In this program, a finite element model for thermal, microstructural, and structural analyses interacts with an extended database to characterize the steel behavior during heat treatment. Chuzhoy et al. [94] have done simulation studies using QSIM to verify the influence of quench characterization on simulation results. This paper shows a good application example of numerical methods helping to understand the heat treat process [94]. First, a numerical model is combined with a genetic algorithm to back calculate the heat transfer coefficient between the workpiece and quenchant. An interactive procedure is used to solve this inverse heat transfer problem, which requires obtaining boundary conditions based on a known response. In this study, a response is micro-

structure and hardness profile after quenching from a bar with a known chemistry and hardenability. The genetic algorithm uses the heat transfer coefficient (that leads to the best match between the calculated and measured hardness) as a starting point for the next generation of the heat transfer coefficient set, as shown in Fig. 50. After the heat transfer coefficient is determined, QSIM is used to compute microstructure, distortion, and residual stresses in a particular component and quench [94].

Procedures like that provide a quality-control tool in the heat-treating process, predicting component response and optimizing costs to the manufacturer and then to the customer.

REFERENCES

1. Sinha, A.K. Ferrous physical metallurgy. *Chapter 10— Basic Heat Treatment*; Butterworths: Boston, MA, 1989; 403–440.
2. Krauss, G. Glossary of selected terms. In: *Steels: Heat Treatment and Processing Principles*; ASM International: Materials Park, OH, 1990; 453–468.
3. Tarney, E. Heat treatment of tool steels. Tool. Prod. 2000, 102–104.
4. EQUIST 2001, User's Manual. SACIT Steel Advisory Centre Ltd. H-1221 Budapest, Leanyka u.7. X.63.(www.sacit.hu).
5. www.gowelding.btinternet.co.uk/met/alloys.html.
6. Metals Handbook, 10th Edition; *Properties and Selections: Irons, Steels and High-Performance Alloys*. ASM International: Materials Park, OH, 1990; Vol. 1.
7. Brooks, C.R. *Principles of the Heat Treatment of Plain Carbon and Low Alloy*; ASM International: Materials Park, OH, 1996; 490 pp.
8. ASTM E7-01 "Standard Terminology Relating to Metallography", ASTM International, West Conshohocken, PA.
9. Davis, J.R. *ASM Materials Engineering Dictionary*; ASM International: Materials Park, OH, 1992.
10. Sverdlin, A.V.; Ness, A.R. Fundamental concepts in steel heat treatment: Chapter 1. *Steel Heat Treatment Handbook*; Totten, G.E., Howes, M.A.H., Eds.; Marcel Dekker, Inc.: New York, NY, 1997; 1–44.
11. Colpaert, H. *Metalografia dos produtos siderurgicos comuns*. 3ed. Universidade de São Paulo, 1974; 412 pp. (in Portuguese).
12. Sinha, A.K. Ferrous physical metallurgy. *Chapter 1— Iron–Carbon Alloys*; Butterworths: Boston, MA, 1989; 1–42.
13. *Practical Data for Metallurgists*; The Timken Company: Canton, OH, 1996.
14. Grange, R.A.; Stewart, N.M. The temperature range of martensite formation. Trans. AIME 1946, *167*, 467.

15. Liscic, B.; Tensi, H.M.; Totten G.E.; Webster, G.M. Chapter 22-non-lubricating process fluids: steel quenching technology. In *Fuels and Lubricants Handbook: Technology, Performance and Testing*; Totten, G.E., Westbrook, S.R., Shah, R.J., Eds.; ASTM International: W. Conshohocken, PA, 2003; p. 589.

16. Hougardy, H.P. Die Darstellung des Umwandlungsverhaltens von Stählen in den ZTU-Schaubildern. Härterei-Tech. Mitt. 1978, *33* (2), 63–70.

17. Liscic, B. Chapter 8—Steel heat treatment. In *Steel Heat Treatment Handbook*. Totten, G.E., Howes, M.A.H., Eds.; Marcel Dekker Inc: New York, NY, 1997; 527–662.

18. Tensi, H.M.; Stich, A.; Totten, G.E. Chapter 4—Quenching and quenching technology. In *Steel Heat Treatment Handbook*. Totten, G.E., Howes, M.A.H., Eds.; Marcel Dekker Inc: New York, NY, 1997; 157–249.

19. Wever, F.; Rose, A.; Peter, W.; Strassburg, W.; Rademacher, L. Atlas zur Wärmebehandlung der Stähle *Verlag Stahleisen. Dusseldorf, Germany, 1956; Vols* I/II, 56–58.

20. Rose, A.; Strassburg, W. Anwendung des Zeit-Temperatur-Umwandlungs-Schaubildes für kontinuierliche Abkülung auf Fragen der Wärmebehandlung. Arch. Eisenhüttenwes. 1953, *24* (11/12), 505–514.

21. Thelning, K-E. Chapter 4—Hardenability. *Steel and Its Heat Treatment, Second Edition*; Butterworths: London, 1984; 144–206.

22. Atkins, M. *Atlas of Continuous Transformation Diagrams for Engineering Steels*; British Steel Corporation, BSC Billet, Bar and Rod Products: Sheffield, U.K., 1977.

23. Scheil, E. Arch. Eisenhüttenwes. 1934/1935, *8*, 565–567.

24. Liscic, B. Chapter 3—Hardenability. In *Steel Heat Treatment Handbook*. Totten, G.E., Howes, M.A.H., Eds.; Marcel Dekker Inc.: New York, NY, 1997; 93–156.

25. Winchell, P.G.; Cohen, M. Strength of Martensite. ASM-Trans. June 1962, *55*(2), 347–361.

26. Houdremont, E. Strength of Martensite. ASM-Trans. June 1962, *55* (2), 347–361.

27. Sinha, A.K. Ferrous physical metallurgy. *Chapter 11—Hardening and Hardenability*; Butterworths: Boston, MA, 1989; 441–522.

28. Totten, G.E.; Bates, C.E.; Clinton, N.A. Chapter 2—Measuring hardenability and quench severity. *Quenchants and Quenching Technology*; ASM International: Materials Park, OH, 1993; 35–68.

29. Grosmann, M.A.; Bain, E.C. *Principles of Heat Treatment*, 5th Ed.; American Society for Metals: Metals Park, OH, 1964.

30. ASTM Standard Test Method E 112-96e1 "Standard Test Methods for Determining Average Grain Size", ASTM International, West Conshohocken, PA.

31. ASTM Standard Test Method E1382-97, "Standard Test Methods for Determining Average Grain Size Using Semiautomatic and Automatic Image Analysis", ASTM International, West Conshohocken, PA.

32. Brown, G.T. Re-appraisal of the Jominy test and its applications. In *Hardenability Concepts with Applications to Steel*. Doanne, D.V., Kirkadly, J.S., Eds.; AIME, 1993; 69–128.

33. Totten, G.E.; Bates, C.E.; Clinton, N.A. Chapter 3—Cooling curve analysis. In *Quenchants and Quenching Technology*; ASM International: Materials Park, OH, 1993; 69–128.

34. Bates, C.E.; Totten, G.E.; Brennan, R.L. Quenching of steel. *ASM Handbook*; Heat Treating; ASM International: Materials Park, OH, 1991; Vol. 4, 67–120.

35. Rajan, T.V.; Sharma, C.P.; Sharma, A. Chapter 5—Heat treatment processes for steels. In *Heat Treatment: Principles and Techniques—Revised Edition*; Prentice Hall of India, Pvt. Ltd.: New Delhi, India, 1994; 97–123.

36. Aronov, M.A.; Wallace, J.F.; Ordillas, M.A. System for prediction of heat-up and soak times for bulk heat treatment processes. In *Heat Treating: Equipment and Processes, Proc. of 1994 Conference, Schaumburg, IL.*; Totten, G.E., Wallis, R.A., Eds.; ASM International: Materials Park, OH, 1994; 55–61.

37. Kasten, K. A primer of terminology for heat treat customers. Heat Treating, 32–39.

38. Nayar, A. The metal databook. *Chapter 3.4—Heat Treatment of Steel*; McGraw-Hill: New York, NY, 1997.

39. Naylor, D.J.; Cook, W.T. Heat treated engineering steels. Mater. Sci. Technol. 1992, *7*, 435–488.

40. Thelning, K.-E. Chapter 5—Heat treatment—General. *Steel and Its Heat Treatment—Second Edition*; Butterworths: London, 1984; 207–318.

41. Kern, R. Distortion and cracking III: How to control cracking. Heat Treating, April 1985; 38–42.

42. Ericsson, T. Principles of Heat Treating Steels. In *ASM Handbook*; Heat Treating. ASM International: Materials Park, OH, 1991; Vol. 8, 3–19.

43. Totten, G.E.; Bates, C.E.; Clinton, N.A. Chapter 1—Introduction to the heat treating of steel. *Handbook of Quenchants and Quenching Technology*; ASM International: Materials Park, OH, 1993; 1–33.

44. Narazaki, M.; Totten, G.E.; Webster, G.M. Hardening by reheating and quenching. In *Handbook of Residual Stress and Deformation of Steel*; Totten, G.E., Howes, M.A.H., Inoue, T., Eds.; ASM International: Materials Park, OH, 2002; 248–295.

45. Legat, F. Why does steel crack during quenching. Kovine Zlit. Technol. 1998, *32* (3–4), 273–276.

46. Kern, R. Distortion and cracking II: Distortion from quenching. Heat Treat., March 1985; 41–45.

47. Clarke, P.C. Close tolerance heat treatment of gears. Heat Treat. Met. 1998, *25* (3), 61–64.

48. Blackwood, R.R.; Jarvis, L.M.; Hoffman, D.G.; Totten, G.E. Conditions leading to quench cracking other than severity of quench. In *Heat Treating Including the*

Liu Dai Symposium, Proc. 18th Conference; Wallis, R.A., Walton, H.W., Eds.; ASM International: Materials Park, OH, 1998; 575–585.

49. Shao, H-H. Analysis of the causes of cracking of a 12% Cr steel cold die during heat treatment. Jinshu rechuli, 1995, (11), 43.

50. Kern, R.F. Thinking through to successful heat treatment. Met. Eng. Q. 1971, *11* (1), 1–4.

51. Sheng, X.; He, S. Analysis of quenching cracks in machine tool pistons under supersonic frequency induction hardening. Heat. Treat. Met. (China), 1991, (4), 51–52.

52. Cook, W.T. Review of selected steel-related factors controlling distortion in heat treatable steels. Heat Treat. Met. 1999, *26* (2), 27–36.

53. Toshioka, Y. Heat treatment deformation of steel products. Mater. Sci. Technol. Oct. 1985, *1*, 883–892.

54. Tensi, H.M.; Totten, G.E.; Webster, G.M. Proposal to monitor agitation of production quench tanks. In *Heat Treating Including the 1997 Induction Heat Treating Symposium Proc. 17th Conf.;* Milam, D.L., Poteet, D.A., Pfaffmann, G.D., Rudnev, V., Muehlbauer, A., Albert, W.B., Eds.; ASM International: Materials Park, OH, 1997; 441–443.

55. Krauss, G. Tempering of steel. *Steels: Heat Treatment and Processing Principles*; ASM International: Materials Park, OH, 1990; 206–261.

56. Grossmann, M.A.; Bain, E.C. Chapter 5—Tempering after quench hardening. *Principles of Heat Treatment*; American Society for Metals: Metals Park, OH, 1964; 129–175.

57. Speich, G.R.; Leslie, W.C. Metall. Trans. 1972, *3*, 1043.

58. Spies, H.J.; Münch, G.; Prewetz, A. Möglichkeiten der Optimierung der Auswahl vergütbarer Baustähle durch Berechnung der Härt-und-vergütbarkeit. Neue Hütte 1977, *8* (22), 443–445.

59. Heat treating of steel. In *Metals Handbook—Desk Edition,* Second Edition; Davis, J.R., Ed.; ASM International: Materials Park, OH, 1998; 970–982.

60. Totten, G.E.; Garsombke, G.R.; Pye, D.; Reynoldson, R.W. Chapter 6—Heat treating equipment. In *Steel Heat Treatment Handbook*, Totten, G.E., Howes, M.A.H., Eds.; Marcel Dekker Inc.: New York, NY, 1997; 293–481.

61. Boyer, H.E.; Cary, P.R. Chapter 4—Molten quenching methods: Martempering and austempering. *Quenching and Control of Distortion*; ASM International: Materials Park, OH, 1988; 71–88.

62. Funatani, K. Materials, heat treatment and surface modifications applied for automotive components. In Proceedings of the 1st International Automotive Heat Treating Conference, 13–15 July; Colás, R., Funatani, K., Stickels, C.A., Eds.; ASM International: Materials Park, OH, Puerto Vallarta, Mexico, 1998; 283–290.

63. Funatani, K. Forge quenching and direct heat treatment technology: Today and for the future. Proceedings from Materials Solutions '97 on Accelerated Cooling/ Direct Quenching Steels, 15–18 September, Indianapolis, Indiana, USA; ASM International: Materials Park, OH, 1997; 193–198.

64. Yamada, S.; Funatani, K. Strength and toughness on the forge quenched steels. Proceedings from Materials Solutions '97 on Accelerated Cooling/Direct Quenching Steels, 15–18 September, Indianapolis, Indiana, USA; ASM International: Materials Park, OH, 1997; 241–246.

65. Kang, S.H.; Han, S.W.; Totten, G.E.; Webster, G.M. Direct forge quenching with poly (alkylene glycol) polymer quenchants. Proceedings from Materials Solutions '97 on Accelerated Cooling/Direct Quenching Steels, 15–18 September, Indianapolis, Indiana, USA; ASM International: Materials Park, OH, 1997; 207–215.

66. Totten, G.E.; Webster, G.M.; Han, S.W.; Kang, S.H. Immersion time quenching technology to facilitate replacement of quench oils with polymer quenchants for production of automotive parts. In Proceedings of the 1st International Automotive Heat Treating Conference, 13–15 July, Puerto Vallarta, Mexico. Colás, R., Funatani, K., Stickels, C.A., Eds.; ASM International: Materials Park, OH, 1998; 449–455.

67. Han, S.W.; Kang, S.H.; Totten, G.E.; Webster, G.M. Principles and applications of the immersion time quenching system in batch and continuous processes. Proceedings of International Heat Treating Conference: Equipment and Processes, 18–20 April, Schaumburg, Illinois; ASM International: Materials Park, OH, 1994; 337–345.

68. Jiansheng, P.; Jiansheng, G.; Dong, T.; Totten, G.E.; Chen, X. Computer Aided Design of Complicated Quenching Process by Means of Numerical Simulation Method. In Proceed. 3rd International Conference on Quenching & Control of Distortion. Totten, G.E., Liscic, B., Tensi, H.M., Eds.; ASM International: Materials Park, OH, 1999; 251–259.

69. Inoue, T.; Uehara, T.; Ikuta, F.; Arimoto, K.; Igari, T. Simulation and experimental verification of Induction hardening process for some kinds of steels. In Proceed. 2nd Conference on Quenching & Control of Distortion; Totten, G.E., et al., Eds.; Cleveland, OH, USA. 4–7 Nov. 1996; 275–281.

70. Ikuta, F.; Arimoto, K.; Inoue, T. Computer simulation of residual stresses/distortion and structural change in the course of scanning induction hardening. In Proceed. 2nd Conference on Quenching & Control of Distortion; Totten, G.E., et al., Eds.; Cleveland, OH, USA, 4–7 Nov. 1996; 259–266.

71. Jacot, A.; Swierkosz, M.; Rappaz, J.; Rappaz, M.; Mari, D. Modeling of electro-magnetic heating, cooling and phase transformations during surface hardening of steels. J. Phys. IV Colloq. 1996, *C1*, 6.

72. Jarvstrat, N.; Sjostrom, S. Current status of TRAST: a material module subroutine system for the calculation of quench stresses in steel. In *ABAQUS User's Conference Proceedings*; 1993; 273–287.

73. Marchant, N.J.; Malenfant, E. Modeling of micro-structural transformation using MetalCore. In Heat Treating, Proceed. 16th ASM Heat Treating Conference and Exposition, Dosset, J.L., Luetje, R.E., Eds Cincinnati, OH, USA, 12–16 March 1996; 197–203.

74. Sanchez Sarmiento, G.; Barragan, C. INC-PHATRAN A computer model for the simulation of heat treating processes, *User manual, SOFT-ING Private Consulting* April 1997.

75. Sanchez Sarmiento, G.; Vega, J.; Gaston, A. INDUC-TOT-B: A computer model for the simulation of heat treating processes, *User manual, SOFT-ING Private Consulting*, December 1997.

76. Sanchez Sarmiento, G. OPTIBANC: Modelo Computacional para la optimización de los bancos de esfriamento de un laminador de chapas de acero en caliente, *Manual del Usuario, SOFT-ING Private Consulting*, April 1997.

77. Hibbitt, Karlsson and Sorensen, Inc; "*ABAQUS/Standard. User's Manual*". Versión 5.8-1, 1998.

78. Sanchez Sarmiento, G.; Coscia, D.M.; Jouglard, C.; Totten, G.E.; Webster, G.M. Distortion y tensiones residuales em probetas de alumínio sometidas a tratamento termico. Proceedings of the Second Argentinean Conference of Abaqus Users, Buenos Aires, September 10–11, 2001; 1–11.

79. Sanchez Sarmiento, G.; Coscia, D.M.; Jouglard, C.; Totten, G.E.; Webster, G.M.; Vega, J. Residual stresses, distortion and heat transfer coefficients of 7075 aluminum alloy probes quenched in water and polyalkylene glycol solutions. Funatani, K., Totten, G.E., Eds.; Heat Treating—Proceedings of the 20th Conference, 9–12 October 2000; Vol. 2, 1118–1224. ASM International Materials Park, OH, 2000.

80. Sanchez-Sarmiento, G.; Castro, M.; Totten, G.E.; Webster, G.M.; Jarvis, L.; Cabré, M.F. Modeling residual stresses in spring steel quenching. In Heat Treating—Proceedings of the 21st Conference. 5–8 November; Shrivastava, S., Specht, F., Eds.; ASM International: Materials Park, OH, 2001; 191–200.

81. Prime, M.B.; Prantil, V.C.; Rangaswamy, P.; Garcia, F.P. Residual stress measurement and prediction in a hardened steel ring. Mat. Sci. Forum 2000, 347–349, 6–228.

82. http://www.deform.com/ht_brochure.pdf.

83. Sanchez Sarmiento, G.; Tormo, J.; Schwarz y S.; Moriconi, J. "Design: An integrated system of computational models of hot rolling mills. *User's Manual*", *SOFT-ING Private Consulting*, December 1997.

84. Sánches Sarmiento, G.; Gastón, A.; Vega, J. IN-DUCTER-b: A finite element phase transformation model of induction heat treating of steels sensitive to its chemical composition. Proceed. Of the 1st International Conference on Induction heat treating, Indianapolis, USA, 5–18, September 1997.

85. Sánches Sarmiento, G.; Vega, J.; Barragán, C. "Predicción de la distribucíon de dureza en una pieza de acero posterior a un tratamiento térmico mediante simulación numérica y aplicación de la Norma SAE J406". *Anales de JORNADAS'97 DE LA Asociación Argentina de Materiales*, Tandil, Buenos Aires, 14–16 May 1997.

86. Sanchez Sarmiento, G.; Gaston, A.; Veja, J. Inverse heat conduction coupled with phase transformation problems in heat treating processes. In *Computational Mechanics*. Oñate, E., Idelsohn, S.R., Eds.; CIMNE: Barcelona, Spain, 1998; p. 56.

87. Buchmayer, B.; Kirkaldy, J.S. *Minitech User's Manual*: MINITECH Limited, 71 Paisley Ave N., Hamilton, Ontario, Canada, L8S 4H1.

88. http://www.msm.cam.ac.uk/map/ (Java Materials Property Software).

89. (http://www.thermotech.co.uk/).

90. http://engm01.ms.ornl.gov, "Modeling Microstructure Development in Welds", Internet online computational tool, Oak Ridge National Laboratory, Oak Ridge, Tenn.

91. Wallis, R.A.; Garwood, D.R.; Ward, J. The use of modeling techniques to improve the quenching of components. In Proceedings of the Heat Treating Conference: Equipment and Process-1994, Schaumburg, Illinois, 18–20 April; Totten, G.E. Wallis, R.A. Eds.; ASM International: Materials Park, OH, 1994; 105 pp.

92. Bower, W.W.; Cain, A.B.; Smith, T.D. Computational simulations of quench tank flow patterns. Proceedings of the 17th Heat Treating Society Conference and Exposition and the 1st International Induction Heat Treating Symposium, 15–18 September, Indianapolis, Indiana. Milam, D.L., Poteet, D.A., Jr., Pfaffmann, G.D. Rudnev, V., Muehlbauer, A., Albert, W.B. ASM International: Materials Park, OH, 1997; 389–393.

93. Ready, A.V.; Akers, D.A.; Chuzoy, L.; Pershing, M.A.; Woldow, R.A. Development of a method to evaluate commercial quenches. In 20th ASM Heat Treating Society Conference Proceedings, October 9–12, 2000; Funatani, K., Totten, G.E., Eds.; ASM International: Materials Park, OH, p. 854–857.

94. Chuzhoy, L.; Cai, J.; Sharma, R.; Rivera, E.; Li, M.; Burris, K.; Johnson, M. Quantitative characterization of production quenches using numerical methods. Proceedings of 13th Congress IFHTSE, Columbus, OH. USA; ASM International: Materials Park, OH, October 2002 scheduled April 2003.

13

Design of Carburizing and Carbonitriding Processes

Małgorzata Przyłęcka and Wojciech Gęstwa
Poznan University of Technology, Poznań, Poland

Kiyoshi Funatani
IMST Institute, Nagoya, Japan

George E. Totten
G. E. Totten and Associates, LLC, Seattle, Washington, U.S.A.

David Pye
Pye Metallurgical Consulting, Inc., Meadville, Pennsyvania, U.S.A.

I. INTRODUCTION

Various surface modification technologies, also known as surface engineering and case hardening [1], are used to improve the properties of mechanical components. These include thermochemical, electrochemical, and mechanical processes. Thermochemical processes are characterized by diffusion of carbon and/or nitrogen, oxygen, or boron into the material surface, after which the parts are thermally treated to form a hardened case. Thermochemical processes include carburizing, carbonitriding, nitriding, ferritic nitrocarburizing, and boronizing. The typical process conditions for these methodologies are compared in Fig. 1 [1]. In this chapter, the design principles of two thermochemical processes, carburizing and carbonitriding, will be discussed. The primary focus will be on carburizing because it is more widely used in the heat treating industry.

II. CARBURIZING

A. Thermodynamics of Carburizing Processes

The gaseous atmosphere capable of carburizing steel objects when maintained in a carbon-rich-rich-rich atmosphere, at an appropriate process temperature, is known as a controlled carburizing atmosphere. For the creation of a carbon-rich-rich diffusion zone diffusion, the indispensable carbon will react in the following manner [2]:

$$2CO \leftrightarrow C^{\gamma} + CO_2 \tag{1}$$

using the following equilibrium constant of:

$$K_B = p_{CO}^2 / (p_{CO_2} a_{cg}) \tag{2}$$

When using methane as the process enrichment gas, the following reaction/dissociation will occur:

$$CH_4 \leftrightarrow C^{\gamma} + 2H_2 \tag{3}$$

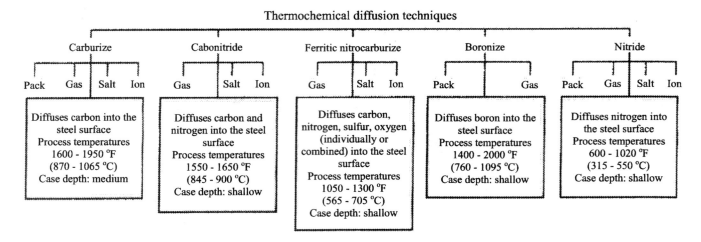

Figure 1 Comparison of common thermochemical diffusion process methods.

Using the equilibrium constant:

$$K_M = p_{H_2}^2 / (p_{CH_4} a_{cg}) \qquad (4)$$

As well as a reaction of exchange gas:

$$CO + H_2 \Leftrightarrow C + H_2O \qquad (5)$$

of equilibrium constant:

$$K_W = p_{CO} p_{H_2} / (p_{H_2} a_{cg}) \qquad (6)$$

where K_B, K_M, and K_W are the equilibrium constant with graphite; a_{cg} is the carbon activity in gas medium; and p_{CO}, p_{H_2}, and p_{H_2O} are the partial pressures of component atmosphere.

The furnace atmosphere composition during carburizing depends on the chemical composition of the gas. Table 1 shows the composition of various common

Table 1 Composition of Various Atmospheres Used for Carburizing

Gas Type	CO_2	CO	H_2	CH_4	N_2
RX/NG	0.0	20.2	38.3	0.2	41.3
RX/TG	0.0	24.5	35.5	1.5	38.5
RX/C_3H_8	0.0	23.8	31.7	0.3	44.2
RX/C_4H_{10}	0.0	24.2	30.3	0.4	45.1
CO_2/C_4H_{10}	1.0	30.7	53.7	11.7	0.3
N_2/CO/C_3H_8	0.8	28.2	47.8	18.5	3.2
N_2/C_2H_5OH	0.3	20.5	41.0	4.5	37.5
FC35/NTG[a]	0.2	23.5	56.6	4.3	15.4
FC35/C_4H_{10}[b]	0.2	27.3	47.1	2.8	22.6
C_2H_5OH	1.0	30.7	53.7	11.7	0.3
$(CH_3)_2CHOH$	0.8	28.2	47.8	18.5	3.2

[a] FC35/NTG (Natural Town Gas) = $CO_2 + CH_4$.
[b] FC35/C_4H_{10} = $CO_2 + C_4H_{10}$.

atmospheres employed for carburizing [3]. The furnace atmosphere is a result of the decomposition of the gas during carburizing, which is similar to thermal decomposition, except that the reactivity of the gas during carburization is dependent on the adsorption character of the gas and subsequent diffusion into the steel, which governs the partial pressure of the gaseous components. The thermal decomposition of a gas is related to the adsorption and diffusion of the gas itself. The difference between the two processes was studied by Okumura and Iwase, who found that the dissociation processes of ethane and propane were similar. However, there was a significant difference between the thermal decomposition and carburizing processes for acetylene. This is illustrated in Fig. 2 [4].

The level of chemical absorption rate of gases exhibits the following order on reactivity:

$$O_2 > C_2H_2 > C_2H_4 > CO > H_2 > CO_2 > N_2 > AR.$$

The affinity of the gaseous component with steel or adsorption affects, dissociation rate, and subsequent diffusion govern the efficiency of the carburizing process [5].

Figure 3 shows the equilibrium curves for different furnace gas ratios (% CO_2/% CO and % H_2O/% H_2) as a function of furnace atmosphere temperature for iron, carbon, and steel. One method of measuring furnace gas concentration is to measure the water concentration by "dew point." The relationship between dew point and % CO_2 and % carbon at different furnace atmosphere temperatures for "endo gas" is shown in Figs. 4 and 5, respectively. Alternatively, an "oxygen probe" may be used to determine the furnace atmosphere composition [6]. Figure 6 illustrates the relationship between the

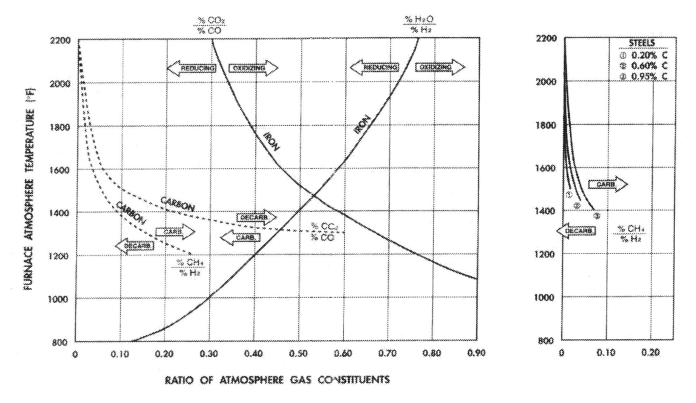

Figure 2 Equilibrium curves for furnace gases on iron, carbon, and steel. (Source: Surface Combustion Inc., Maumee, OH.)

composition of the gas, % CO_2, and the carbon potential as measured by a properly calibrated oxygen probe. The "carbon potential" is defined as: "the measure of the ability of an environment containing active carbon to alter or maintain, under prescribed conditions, the carbon level of a steel. In any particular environment, the carbon level attained will depend on such factors as temperature, time, and steel composition" [7].

Recently, work has been conducted to establish in situ control measures of the dissociation process for vacuum carburizing processes [6,8–11]. Quadra-pole mass-spectro-analyzer methods were shown to be effective in following the gas reaction connected with sooting and loss of carbon potential [8].

The most advanced carburizing processes consume only one-tenth of the gas volume compared with more conventional carburizing methods, and the carburizing reaction is very different from the old carrier gas method. However, it is more difficult to achieve a strong control of the atmosphere composition. Figure 7 illustrates the relationship between the carbon potential and the gas composition of such an advanced atmosphere (F35) [3,12].

The proportion of specific components within the selected atmosphere is selected in relation to the gas phase and the carbon activity, a_{cg}. Conversely, carbon transfer to austenite may occur in the atmosphere within the furnace. Depending on the carbon concentration within the atmosphere (if it is at a high potential), the carbon can be present in the form of soot (graphite). The soot/graphite will precipitate out of the atmosphere and begin to concentrate on the work piece, walls of the furnace, and the mechanical parts within the furnace, such as rails, hearth, rollers, burner tubes, and elements.

Individual equilibrium reactions of the carburizing process occurring in the solid state are expressed under isobaric conditions:

$$\log K_B = -8750/T + 9.022 \tag{7}$$

$$\log K_M = -4768/T - 5.767 \tag{8}$$

$$\log K_W = -6908/T + 7.457 \tag{9}$$

The mechanism of transfer of carbon from a gaseous atmosphere into a solid solution within the steel being processed will occur during the activity of carbon in the gaseous stage when the steel is in the austenite phase at

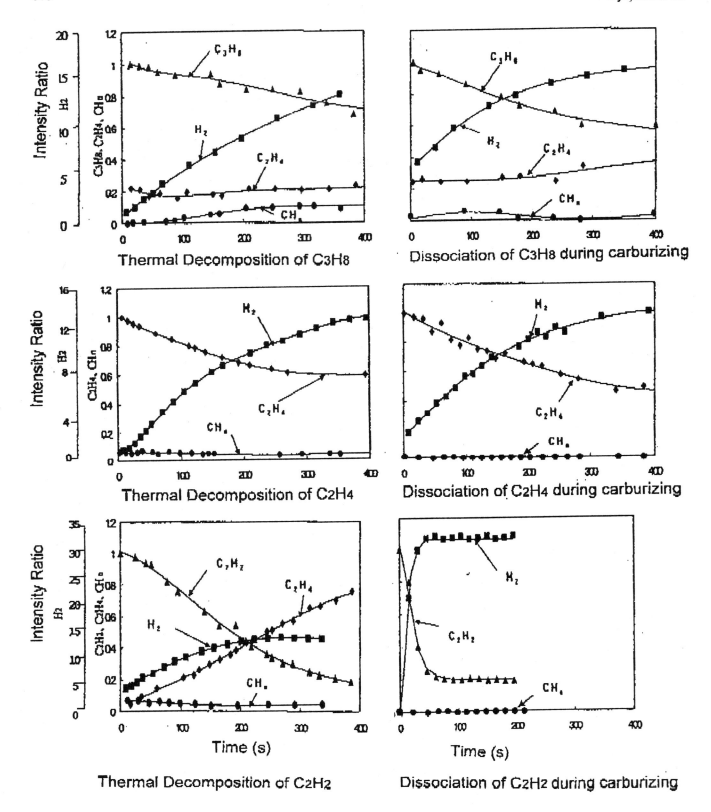

Figure 3 Dissociation character of carburizing gases: ethylene (C_2H_4), propylene (C_3H_8), and acetylene (C_2H_2).

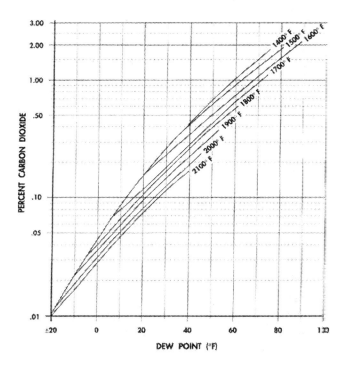

Figure 4 Equilibrium chart for % CO vs. dew point for RX (endothermic gas) 20% CO and 40% H₂. (Source: Surface Combustion Inc., Maumee, OH.)

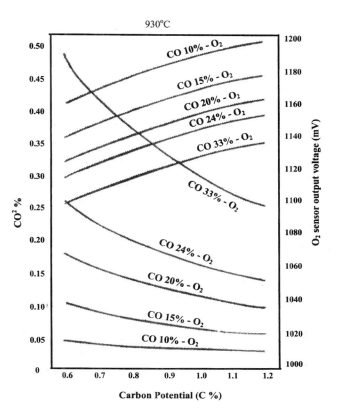

Figure 6 Relationship between % CO, CO₂, and O₂ with oxygen probe sensor mv reading and carbon potential.

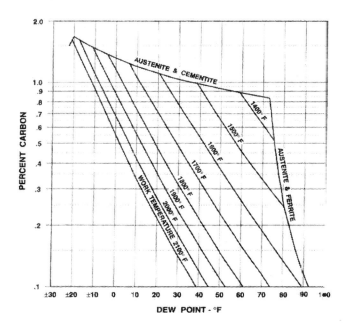

Figure 5 Equilibrium chart for dew point vs. % carbon in austenite for RX (endothermic gas). (Source: Surface Combustion Inc., Maumee, OH.)

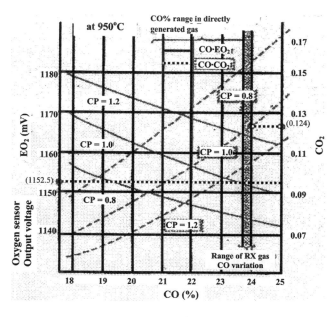

Figure 7 Relationship between O₂ and CO gas concentration and carbon potential of FC35 atmosphere.

the surface of the steel. The activity of carbon in austenite is a function of carbon potential within the atmosphere as well as the steel temperature, and the value is always smaller than unity.

We now come to the question of the carburizing phenomena within the gas phase, or at the interface between the steel surface and the gas phase, as well as within the steel. At the steel/gas phase interface, the carburization reaction will be dependent on the difference between the carbon potential (carbon concentration) of the atmosphere and the steel surface.

Carbon will diffuse from the atmosphere to the steel surface when the carbon activity in the atmosphere is higher than the carbon activity in austenite, which is depndent on furnace temperatures and carbon concentration.

The diffusion rate (number of carbon atoms that penetrate surface area (S) per unit time) is dependent on the difference in carbon activity in gas medium, α_{cg}, and carbon activity on surface steel, α_{cp}:

$$\frac{M}{Sdt} = -\beta(\alpha_{ag} - \alpha_{cp}) \tag{10}$$

If carbon concentration is substituted for activity, then:

$$\frac{M}{Fdt} = -\beta(C_A - C_P) \tag{11}$$

where M is the carbon mass in grams; F is the surface area penetrated by carbon mass (M), dt is the time of carbon penetration; C_A is the carbon potential of the atmosphere; C_P is the surface carbon content; and β is the surface reaction rate constant (carbon transport coefficient).

The carbon transfer coefficient (carbon potential) is defined in European norm EN10052:1993 [13] as: "the transfer of a carbon mass from the carburizing medium to the steel surface, through individual carbon molecules to the steel surface by the difference of potentials between carbon potential of medium, and the steel carbon potential at the surface at any given moment. The mass transfer stream of carbon from atmosphere to steel is quantifiable in the following formula":

$$J_C = \beta_C(C_P - C_S)F \tag{12}$$

where β_C is the carbon transfer coefficient [cm/sec]; C_P is the carbon potential of atmosphere [g/cm^3]; C_S is the carbon content by surface steel [g/cm^3] (carbon potential); and F is the surface diffusion [cm^2].

Fick's first law of diffusion defines carbon diffusion into steel:

$$J_C = D_C \frac{dC}{dx}F \tag{13}$$

where D_C is the carbon diffusion coefficient [cm^2/sec], and dC/dx is the carbon concentration gradient [g/cm].

To determine the continuity of carbon mass flow from atmosphere to steel, and then into the steel surface, it is possible to use the following mass transfer formula [2]:

$$\frac{D_C}{\beta_C} = \frac{C_P - C_S}{dC/dx} \tag{14}$$

The value of carbon transfer coefficient β depends on the source and the carbon potential of the medium used for carburizing, and is therefore dependent on the chemical reactions occurring at the steel atmosphere interface. The smallest carbon transfer coefficient β is 0.54×10^{-7} for the mixture CO–CO$_2$. A somewhat larger value of 1.79×10^{-7} has been reported for the mixture CH$_4$–H$_2$–H$_2$O. Mixtures of CO–CO$_2$–CH$_4$–H$_2$–H$_2$O exhibit higher β transfer coefficients: 3.1×10^{-5} 50% CO–CO$_2$ and 50% CH$_4$–H$_2$–H$_2$O. An industrial endothermic atmosphere composed of fuel gases exhibits a carbon transfer coefficient β of 1.2×10^{-5}. Typically, β values vary between [14,15]

- $\beta = 1.25 \times 10^{-3}$cm/sec for atmospheres generated from endothermic gases; and
- $\beta = 2.5 \times 10^{-5}$ cm/sec for atmospheres generated from thermally cracked alcohols.

β-Values may be easily calculated for various atmospheres using the equation [16]:

$$\beta = \frac{D}{C_g - C_S}\left[\frac{dc}{dx}\right]$$

where D is the diffusion coefficient of carbon in austenite (cm^2/sec); C_g is the carbon concentration of the gas corresponding to the carbon activity of the gas (g/cm^3); C_S is the carbon concentration at the steel surface (g/cm^3); and $\partial c/\partial x$ is the carbon concentration gradient tangent of the concentration gradient of the surface.

During the carburizing process, the dissociated/decomposed carbon to material transports from the atmosphere to the steel surface. The proportional content of carbon in the atmosphere is referred to as "carbon potential." Increasing β transfer coefficients indicates increasing carburizing potential.

B. Kinetics of Carburizing Processes

Fick's first law of diffusion defines the driving force for diffusion, change in concentration/change in time of a dilute solution of atoms in a stream $J(\partial c/\partial t)$, which, in a function of gradient concentration in one dimension (parallel to the x axis) $\partial c/\partial x$, has the form [2]

$$J = -D\frac{\partial c}{\partial x} \tag{15}$$

where J is the rate of diffusion in atoms/area/unit time (flux), D is the coefficient of diffusion (diffusivity) expressed in cm^2/sec or m^2/sec, c is the concentration, and x is the distance (case depth). In Fick's second law, the value J relates the *time rate of change* of the concentration c at any point to the *gradient* of the concentration gradient. It should be noted that D and J are a function of temperature.

The use of Fick's first law for describing the diffusion coefficients of J is limited by the difficulty of quantifying J relative to a constant gradient concentration, because it is assumed that there is no change in the concentration gradient. However, in practice, there is a variation in the concentration gradient with respect to time and position. Therefore, Fick's second law of diffusion is used to determine the value of an "effective diffusion coefficient" D from a concentration change in the diffusion rate $\partial c/\partial \tau$ as a function of the concentration gradient if D is independent of concentration.

$$\frac{\partial c}{\partial \tau} = D\frac{\partial^2 c}{\partial x^2} \tag{16}$$

Calibration of the carbon potential value is possible by correlating the carbon concentration (%) and electrical resistance, or chemical analysis of foils, or coils in the furnace atmosphere. The effect of carbon potential

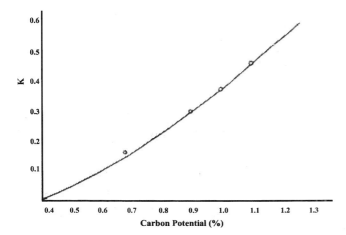

Figure 8 Correlation between carbon potential and the effective diffusion constant (K). Diffusion rate calculated relative to effective case depth having 0.4% carbon.

Table 2 Effect of Temperature on the Diffusion Constant (D)

Temperature (°F)	Temperature (°C)	Diffusion constant (D)
1400	760	0.008
1450	788	0.010
1500	815	0.012
1550	843	0.015
1600	871	0.018
1650	899	0.021
1700	927	0.025
1750	954	0.029
1800	982	0.034
1850	1010	0.040
1900	1038	0.046
1950	1065	0.052
2000	1093	0.060
2050	1121	0.068

on the "effective" diffusion constant is illustrated in Fig. 8 [2].

The effect of temperature on the effective diffusion coefficient (D) is calculated using the Arrhenius equation:

$$D = D_0\exp\left(-\frac{Q}{RT}\right) \tag{17}$$

where D_0 is the element-dependent diffusion constant, Q is the energy of diffusion activation (kJ/mol), R is the ideal gas constant, and T is the absolute temperature.

A common expression that illustrates the dependence of the diffusion constant (D) on temperature is [14]:

$$D = 0.2\ \exp\left[\frac{-16,608}{T}\right]$$

where T is the carburizing temperature in K, and D is the carburizing diffusion coefficient in γ iron in cm^2/sec.

Diffusion rates increase as the temperature rises, as illustrated by Table 2 [17]. Using these values and Harris's equation [18]:

Case Depth $(x) = Dt^{1/2}$

where t is the time (in hours) at temperature, and D is the coefficient of diffusion. It is evident that increasing the carburizing temperature will greatly reduce the carburizing time (t), but because of the higher temperature, the operating cost of the furnace is increased (Fig 9). Under actual operating conditions, a balance must be established between the rate at which parts are needed, maintenance costs of the furnace, etc. The carburized

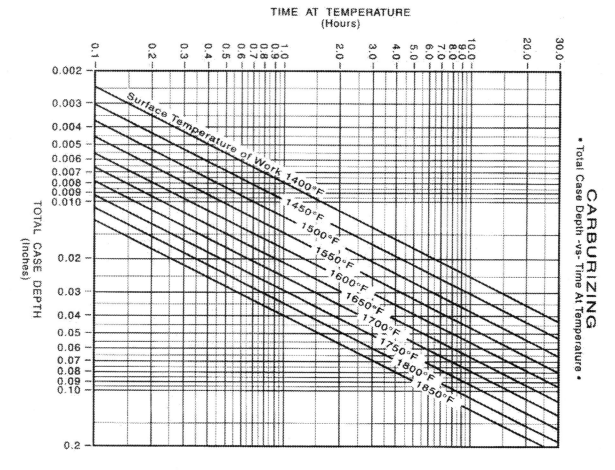

Figure 9 General carburizing curves for various times at temperatures. (Source: Surface Combustion Inc., Maumee, OH.)

case depth, as a function of furnace temperature, and time at temperature is shown in Fig. 9 [1].

The case depth (x) can also be calculated from:

$$x = \frac{0.79(Dt)^{1/2}}{0.24 + \left[\frac{C_{x,t}-C_s}{C_s-C_0}\right] - 0.7\frac{D}{\beta}}$$

where x is the case depth (cm), D is the carbon diffusion coefficient in austenite (cm²/sec), t is the carburizing time (sec), $C_{x,t}$ = carbon content at case depth x after carburizing for time t, C_0 is the initial carbon content in the steel, C_s is the carbon content on the surface of the steel during carburization, β is the carbon transport coefficient from the atmosphere to the steel surface (cm/sec).

Accurate carbon potential measurements are important in establishing effective processing conditions to yield optimum surface carbon concentration and to minimize carburizing times, as shown in Fig. 10 [3,19,20]. The interrelationship between diffusion time, carbon potential, and carburizing time is shown in Fig. 11 [3,19].

The amount of carbon that will dissolve in austenite is dependent on the composition of the steel. The influence of alloying elements on carbon solubility may be estimated from [9,10]:

$$\log \frac{C_C}{C} = 0.013[4\text{Si} + 16\text{N} + \text{Ni} \\ -(3\text{Cr} + 4\text{V} + \text{Mn} + \text{M}_0 + \text{Al})]$$

where C_C is the carbon content equivalent to the binary Fe–C system, and C is the carbon content in the complex Fe–C–Mn–Si–etc. system.

Values of C_C/C for common carburizing steels are provided in Table 3 [9]. Typically, if the ratio of C_C/C is less than 1, the steel is easily carburized.

Carbon solubility in austenite may be calculated from [14]:

$$\log a_c = \log\left[\frac{14C_C}{300 - 59C_C}\right] + \frac{2084}{T} - 0.6387$$

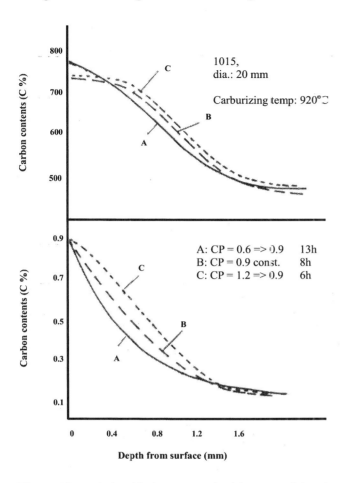

Figure 10 Relationship between carburizing potential and carburizing time.

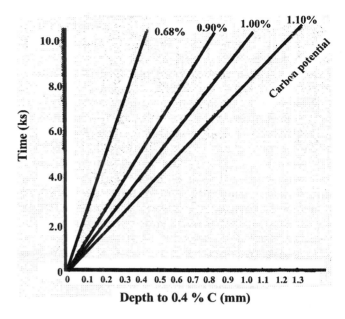

Figure 11 Correlation between carburizing potential and diffusion depth to 0.4% carbon.

where a_C is the carbon activity in austenite for the binary Fe–C system, C_C is the carbon content in the binary Fe–C system, and T is carburizing temperature in K.

Typically, a value of $a_C = 0.9$ is used because a reducing factor of 10% is often used in practice, where it is assumed that carbon activity refers to graphite and that carbon is in the form of cementite in steel. Thus the equation to calculate the maximum solubility of carbon in steel is [14]:

$$\log 0.9 = \log\left[\frac{14C_C}{300 - 59C_C}\right] + \frac{2084}{1173} - 0.6387$$

Carburizing temperature and strength of carburizing gas influence carbon diffusion. Generally, it is believed that the vacuum or partial pressure carburizing processes is superior to traditional gas carburizing [3]. However, their capability should be equally compared in the same condition. Figure 12 illustrates the relationship between carburizing temperature and the effective dif-

fusion constant [3]. The base line, which was investigated by Harris [21], was determined up to up to 970°C, and the results reported by Sugiyama et al. [8] are on the same line showing the equivalency of the diffusion rate of the "old" pack carburizing process and more current vacuum carburizing methods. The relationship of the effective case depth, shown for 0.3% or 0.4% C, indicates that partial pressure carburizing is comparable with that of gas carburizing at selected carbon potential atmosphere.

When the carbon potential of gas carburizing atmosphere at 920°C is 1.1–1.2% C, which is almost the A_{cm} carbon concentration, where time is the same for partial pressure carburizing by acetylene that forms cementite at the steel surface. A high carbon potential in the early

Table 3 C_c/C Ratios for Common Carburizing Steels

Steel Grade	C_c/C Ratio
AISI 1010	1.035
AISI 3310	0.981
AISI 4320	0.979
AISI 5115	0.939
AISI 8620	0.975
AISI 9315	1.002
DIN 16MnCr5	0.931
DIN 17CrNiMo6	0.942

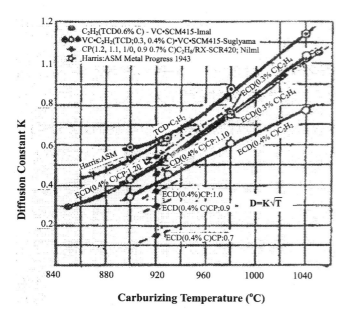

Figure 12 Correlation of effective diffusion constant with different carburizing methods.

stage of the process can reduce carburizing treating time by increasing the carbon diffusion mass [19,20]. Toyota Motor has been using the boost and diffuse carburizing method since 1960 and has been successfully performing good carburizing work.

These results show that, to develop more effective carburizing methods and case hardening, steels cannot depend on traditional approaches without additional metallurgical mechanisms because the carburizing time cannot be reduced further without increasing the effective diffusion rate in iron and steels. One future possibility is the use of rare-earth metal alloying, and another is the development of new alloyed steels [22–25].

C. Classification of Carburizing Processes

The most common processes that are encountered industrially are as follows.

(1) Pack Carburizing. The pack carburizing process is typically conducted by surrounding the steel in a pit furnace or steel box furnace with granules of charcoal or charcoal plus coke. An "activator" for the charcoal, such as barium borate ($BaBO_3$), is added to facilitate the release of CO_2, which then reacts with excess CO_2 to form CO, which in turn reacts with the low-carbon steel surface to form carbon, which then diffuses

into the steel. Pack carburizing is typically conducted at 920–940°C for 2–36 hr.

(2) Liquid Carburizing. Liquid carburizing is typically conducted in internally or externally heated molten salt pots containing a cyanide salt, such as sodium cyanide (NaCN). There are generally two types of liquid carburizing processes. One type is a low-temperature process (840–900°C), which is conducted when low case depths (0.003–0.03 in.) are required. The second liquid carburizing process is conducted at a high temperature (900–950°C) when case depths of 0.03–0.12 in. are desired. In either case, process times may be 1–4 hr.

(3) Gas Carburizing. Currently, the most common carburizing process is gas carburizing, which may be potentially performed with any carbonaceous gas such as methane, ethane, propane, and natural gas. Carburizing times of 4–10 hr are typical. The carburizing temperature is higher than the upper critical temperature (in the austenite transformation region, >954°C). Case depths are typically less than 0.05 in. Conventional gas-carburizing process allows operators to measure the gaseous carbon activity within the furnace process chamber by using the shim test method, dew point test method, CO/CO_2 test method, and oxygen probe.

(4) Vacuum Carburizing. Vacuum carburizing is a clean method used to introduce carbon into the surface of the steel, and also prevents grain boundary oxidation. Vacuum or low-pressure carburizing is carried out in a vacuum furnace at pressures below that of normal atmospheric pressure. The principle of carburizing is exactly the same as that of the gas carburizing process, the main difference being the use of subatmospheric pressure.

(5) Ion Carburizing. As discussed above, carbonaceous gases such as methane have been used for vacuum carburizing because of their widespread availability and current use in gas carburizing. Although methane is reactive and controllable when used with endothermic atmospheres, methane alone is extremely stable, even at elevated temperatures. Therefore, when methane is used in vacuum carburizing, relatively high pressures of 250–400 Torr are required. Furthermore, carburizing processes conducted at these relatively high pressures produce significant amounts of carbon sooting. However, these deficiencies may be overcome by using an ion process. Methane

ionization is produced with a high voltage (approximately 1000 V) at a relatively low pressure of 10 Torr. When methane ionizes through this "ion carburizing" process, a reactive gas blanket is formed in close proximity to the workpiece without concurrent soot formation. Other hydrocarbons can be similarly used.

(6) Fluidized Bed Carburizing. Steel carburizing processes may also be conducted in fluidized bed furnaces. Various atmospheres may be used including conventional endo gas/hydrocarbon mixtures or nitrogen/methanol/hydrocarbon mixtures. Depending on the carburizing atmosphere, fluidized bed temperatures of 850–975°C for 30 min–3 hr may be used. Case depths of up to 0.7 mm are typical.

Table 4 provides a comparative summary of the relative advantages and disadvantages of these processes.

D. Steel Grades Used for Carburizing

Construction alloy steels that can successfully be carburized are many and varied. When selecting a steel for carburizing, it is important to consider such factors as machinability, resistance to overheating, susceptibility to deformations during thermal processing, hardenability relative to the cross-section size, and geometrical features in addition to mechanical strength; not only in the case but also in the core.

Typically, steels selected for carburizing contain <0.25% carbon. The alloy composition is selected to provide case and core hardenability. Plain carbon steels may be carburized, but the carburizing response is limited due to the lack of alloying elements. This is illustrated by the selected listing of carburizing steels provided in Table 5 [26].

Case depths of 0.003–0.250 in. with surface hardness of $R_C = 58$–62 are usually specified. An "effective case depth" is typically specified which may be determined in various ways. One method is to measure the case depth on a metallographic sample of the part or of a test bar by determining the microhardness at various depths from the surface. The desired case depth is determined by the application, such as those shown in Table 6 [27]. The Common International Carburizing Grades of Steel are presented in Table 7.

E. Heat Treatment After Carburizing

After carburizing, austenitization is necessary. If the workpiece is quenched directly after carburizing tem-

perature, retained austenite will form, along with a course grain size (depending on the carburizing time at temperature). Postheat treatment is necessary to provide high surface hardness that will resist abrasion and wear. Depending on the requirements of the carburized part, the postheat treatment temperatures are selected to provide not only the high surface hardness, but the appropriate core strength. The method of heat treatment after carburizing is shown in Fig. 13.

The hardening methods summarized in Fig. 13 are the most commonly used. However, the following procedures are also used, although less commonly:

• Hardening after pearlitic transformation will provide greater energy efficiency and will assure grain reduction.
• Hardening after initially normalizing the structure of the outer layer and core; the method is similar to double hardening, except that the steel is cooled slowly after carburizing. The core posses a normalized structure with less strength but more ductility.

Depending on the steel alloy and component shape, the carburized steel may be quenched in water, oil, or water polymer solution. Specific recommendations are provided in the Engineering Society for Advancing Mobility on Land, Sea and Space (SAE) Standard Aerospace Materials Specification (AMS) 2759. Because of the various limitations of oils or water, there is continuing research into alternative quench media in the heat treating industry. For example, there has been a long search for quenchants that would exhibit faster cooling rates than those exhibited by many quench oils to avoid the pearlitic microstructure formation of many steels. This has led to the development and use of aqueous polymer quenchants as an alternative to many quench oils

Water quenching of gas carburized parts will provide cooling rates in excess of the critical velocity of martensitic structure formation. However, water quenching is likely to cause hardening cracks. A new quenching procedure has been developed, using water as the medium and it is known as Intensive Quenching.

Quenching in oil, on the other hand, reduces the risk of cracking, but it is not often possible to achieve the desired martensitic microstructure throughout the entire carburized case. Only water-soluble polymer quenchant solutions provide cooling rates ranging between those attainable with water and those attainable with oil. The Grossman H-factor obtained for polymer solutions may vary from 0.2 to 1.2. (Water typically exhibits an H-factor of 0.9–2, and the H-factors for oil may vary from 0.25 to 0.8.) Therefore, the use of an aqueous polymer quenchant will provide a relatively "mild"

Table 4　Comparison of Carburizing Methods

4A.　*Carburizing with Solids (Pack Carburizing)*

Description of method	Disadvantage	Advantage
1. The basic component of carburizing mediums is ground wood charcoal, granulation of about 3-5 mm and mixed with carbonates of barium, soda, calcium, lithium, or potassium. 2. Temperature of carburized about 900°C. 3. It is necessary to place the components into a steel box with a spacing of 25 mm between components. The box design is very simple and requires a lid that can be sealed with clay to contain the liberated gas. 4. Starting the reactions during heating can be accomplished by 　(a) burning of small quantity of the wood charcoal by the introduction of oxygen; 　(b) reactions of activator with wood charcoal. 5. After carburizing, the carburizing boxes can be cooled down with the furnace. Or, the boxes can be removed from the furnace, while hot-air-cooled.	• Long heat-up time necessary to reach the process temperature and to achieve temperature uniformity throughout the box. • Decarburization of surface will occur if the components are allowed to air-cool without protection, or removed from the process box. • It is difficult (but possible) to harden directly from the carburizing box. It is the usual practice to allow the box to cool down and reheat to the required ausenitize temperature. • The method is not reliable in terms of repeatability and cannot be accurately controlled. It is a slow production method. • Grinding is necessary after the procedure, because of a slight surface porosity potential.	• Low capital equipment cost. • Simple procedure. • Inexpensive operating costs

4B.　*Liquid Carburizing*

Description of method	Disadvantage	Advantage
1. Mixture of molten salt is the carburizing medium. Usually, mixtures of carbonates of chlorides and cyanides of alkaline metals, sometimes with an addition of SiC (silicon carbide). A typical mixture would be: 75% Na_2CO_3, 15% NaCl, 10% SiC. 2. Temperature of carburizing is usually between 900 and 950°C. 3. The steel components are directly placed into the molten salts after preheating. A cocoon of salt will immediately adhere to the steel surface, thus offering some thermal protection.	• There is a high amount of sludge that collects in the bottom of the bath. It is mandatory to clean out the sludge on a frequent basis. • High costs, due to postcleaning, effluent disposal, labor intensive, long prewiring times of the components. • Carburizing stop off is difficult.	• Possibility of direct hardening. • Components needs a long preheat time. • Uniform carburizing of clean surface, • The probability of distortion remains.

Table 4 Continued

4B. Continued

Description of method	Disadvantage	Advantage
4. In presence of iron, the cyanide salt decomposes at high temperature of the bath and cyanate (CN) is liberated, which further decomposes to provide carbon to diffuse into the steel.		

4C. *Gas Carburizing*

Description of method	Disadvantage	Advantage
1. Atmospheres for carburizing are produced in special generators, which produce a process gas from natural gas blended with air. The generators are known as endothermic generators. The natural gas will contain a large proportion of methane, plus lower concentrations of other hydrocarbon gases. The air that is mixed with the natural gas will contain moisture, which will assist in controlling the carbon potential of the endothermic gas. 2. Temperature of carburizing approximately 870–950°C. 3. The furnace heating system will aid the gaseous atmosphere to ensure good temperature uniformity within the furnace process chamber. In addition to this, the carbon potential is usually very uniform throughout the process chamber.	• Limited speed of diffusive satiating, resulting from limitations of the furnace construction, as well as carbon potential of previous carburizing atmospheres. • Carburizing atmospheres will contain oxygen in the form of moisture, which will cause integranular oxidation and also create the potential for grain boundary corrosion. This will cause deterioration of fatigue strength of the carburized case. • Considerable emission to atmosphere of harmful substance (oxides of carbon and heated quench oil effluent resulting from quenching). It is advisable to install fume extraction systems to ensure adequate shop ventilation. • Oil quenching will lead to distortion and possibly the risk of cracking potential	• Ease in changing the carbon potential of the furnace atmosphere, simply by adjusting the enrichment gas in relation to either the moisture present in the atmosphere or by the presence of free oxygen. • Small waste of energy and economy of time. • Parts are relatively clean, except from oil quenching if the furnace has an oil quench system, like the ones seen on an integral quench furnace. • Possibility of hardening directly after carburizing. However, great care must be taken when considering this factor.

4D. *Vacuum Carburizing*

Description of method	Disadvantage	Advantage

1. The process is conducted at subatmospheric (partial) pressures. Process gases of methane, propane, or acetylene are introduced into the process chamber.
2. Atomic carbon is accomplished as a result of the break-up of gases. The process of vacuum carburizing comes from:
  (a) phases of saturated carbon at the process working pressure of the atmosphere;
  (b) phases of diffusive transportation of excess carbon into the steel surface in high vacuum.
	• High capital cost of equipment.
	• High operating costs.
	• Difficult process control in terms of determining and controlling the atmosphere carbon potential
		• Fast processing time at conventional carburizing temperatures.
		• Advantage can be taken of high-temperature carburizing, up to 1075°C.
		• The carburized layer mark bests mechanical proprieties.
		• Clean, finished work surfaces that do not require postcleaning.

Table 4 Continued

4D. Continued

Description of method	Disadvantage	Advantage
3. Single-chamber vacuum furnaces can be used in any configuration. Also, front or rear cooling chamber can be fitted that can be fitted that can facilitate controlled cooling of the processed batch after austenitizing. High-pressure gas quenching can be accomplished when using blended gaseous mixtures of nitrogen and helium, of blended mixture using nitrogen and hydrogen. Blended gas quenching can (depending on the gas blend and delivery pressure), equal the quench speed of oil		• Environmentally friendly, with no toxic gas emissions. Mechanical handling equipment can be easily installed into the equipment for part transportation within the furnace and outside of the furnace. • Lower volumes of effluent gas. • Effective and energy-saving process • Higher carburizing process temperature's can be used with temperatures of approximately 150°C higher than conventional gas carburizing temperatures. This will result in a faster gas-phase transportation time, leading to much faster carburizing times. The superficial concentration of carbon as a result of unequal process of break-up of hydrocarbons is, as a rule, very high. Both of these factors will considerably accelerate the diffusive saturation. • Hydrocarbons are carriers of carbon exclusively. Therefore, there is no risk of grain boundary oxidation. • Cooling of the workload can be accomplished, if necessary, under nitrogen. This will eliminate the need for postwashing. • The quench gas direction can be manipulated to suite the part geometry, thus reducing the risk of distortion. In addition to the gas flow, speed can be adjusted when using a two-speed gas circulation drive motor. The risk of distortion is less than when quenching into oil.

4E. *Ion Carburizing*

Description of method	Disadvantage	Advantage
1. It depends on the steel in the vacuum stove, in atmosphere of hydrocarbons about low pressure, with simultaneous application of high solid tension on heating.	• High capital investment. • Difficulty in analyzing the carbon potential within the process chamber. • Carbon potential is usually accomplished by prior data acquisition of previously carburized loads	• Possible control of thickness and structures of diffusive layer. • It has been possible to carburize and successfully treat on a repeatable basis, constantly, stainless steels as well as heat-proof and acid-resisting steels.

Table 4 Continued

4E. Continued

Description of method	Disadvantage	Advantage
2. The furnace wall is the anode while the workload sits on the furnace hearth at cathode potential. A voltage is applied (dependent on the chamber pressure and the workload surface area) in the region of 450–800 V, which will cause a glow discharge in the process chamber, very much like a fluorescent light. With the workload at cathode potential, the process gas is immediately ionized and diffusion will begin to take place. The steel surface is not acting as a catalyst, as what occurs in the more conventional carburizing procedure. The process gases used are still methane, propane, or acetylene, along with nitrogen and hydrogen.		• Case uniformity is excellent, irrespective of the part geometry.

4F. *Carburizing in Fluidized Beds*

Description of method	Disadvantage	Advantage
1. A fluidized bed is usually created as a result of the activation by a gas passing through a bed of particles, such as sand or aluminum oxide. The particles of the bed are kept in suspension by hot satiating gas passing upwards and through the particles. 2. Because of the particle movement, the coefficient of heat transfer is high. 3. The fluid-bed furnace can be heated directly or indirectly by electrical or by gas, with the enrichment gases being added with the heating gas. 4. The method of operation is exactly the same as with the salt bath method of heat treatment, with the parts simply being immersed into the fluidized particles.	• The parts are not wet as in a salt bath. There is no slag, or desludging of the bed to do. • Large coefficient of heat transfer makes gives good heat transfer up to process temperature. • It has been possible to apply direct hardening of components by reducing the carburizing temperature to the appropriate austenitizing temperature.	• Simple, but efficient furnace design. • Low operating cost • Ease of operation • Not labor-intensive

Table 5 Selected List of Steel Commonly Used for Carburizing and Their Features

Steel Grade	Features and Benefits
4620	Lower cost, chrome/nickel/ molybdenum steel. Used only where nominal hardenability and core response are required.
8620	Most commonly specified steel for carburizing. Excellent carburizing response, with good hardenability for most section sizes.
4320	Higher hardenability for improved core response in thicker cross sections.
4820	Increased nickel content for improved core toughness, slower response results in longer processing times.
9310	Maximum nickel content for maximum core toughness, slower response results in longer processing times.

quench severity sufficient to provide the desired martensitic transformation for both case and core [28–33].

After hardening the carburized case by quenching, it is necessary to temper the steel by selecting a low tempering temperature in the region of 180–275°C to reduce the residual stress caused by the phase transformation from austenite to martensite. Reducing quantity of

Table 6 Required Case Depths of Selected Applications for Carburized Parts

Application	Case Depth (in.)
High wear resistance, low to moderate loading. Small and delicate machine parts subject to wear.	≤0.020
High wear resistance, moderate to heavy loading. Light industrial gearing.	0.020–0.040
High wear resistance, heavy loading, crushing loads or high magnitude alternate bending stresses. Heavy-duty industrial gearing.	0.040–0.060
Bearing surfaces, mill gearing, and rollers.	0.060–0.250

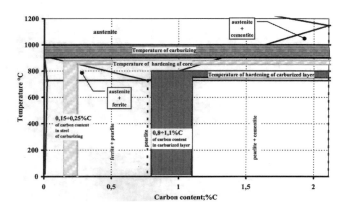

Figure 13 Diagram of heat treatment after carburizing.

retained austenite may be achieved by cryogenic treatment before tempering, especially after hardening variants "2" and "4" with Fig. 13 in the case of alloy steels.

Steels that are hardened after vacuum or iron carburizing are often quenched in a high-pressure gas. The high-pressure gas may also be a blended (helium/nitrogen) to achieve the desired cooling rates for the alloy being quenched and geometry of the part. The heat transfer coefficient of gas quenching medium depends on the gas type, delivery pressure, velocity, turbulence, and directionality. In addition to producing clean parts, a major advantage of gas quenching is that is does not harm the environment.

F. Relationships Among Structure and Properties

Despite the progress and application range of carburizing technology, the influence of case structure on properties continues to be controversial. One of the most significant issues is the influence of retained austenite and carbides in the structure of case-hardened layers and carbon concentration in the surface zone of those layers on properties. For example, an improvement in bending resistance has been observed for 10–30% retained austenite. However, others have reported an unfavorable influence without specifying the percentage values [34–41]. Peyre and Tournier [34] state that fatigue strength increases as the amount of retained austenite increases, and recommends that the minimum retained austenite content should be about 25%. Other works [36–41] reported a detrimental influence without stating the bending resistance limit.

The differences of opinion on the influence of retained austenite are attributable to a lack of understanding of conditions and mechanisms involved in microstructure formation as a function of the properties of case-hardened layers. Variable process methodologies that have been utilized also contribute to the different results obtained. The problem is related to the ways retained austenite forms in the surface zone of carburized layers, which does not account for the influence of changes of other layer parameters on properties of case-hardened steels including carbon content in hardened structures, chemical composition of the matrix, size of former austenite grain, martensite morphology, internal stresses state, etc.

There have been reports that carbide participation does not unfavorably influence properties [34–36,37–39]. However, others have reported detrimental effects. These reports should be regarded with great caution because lower properties than expected may also be due to an overabundance of carbon or tensile stress in the surface zone of the layer related to it, or the manner in which the carburized part is loaded. The occurrence of carbides in the structure of carburized cases, with appropriate hardness, machine parts, and tools, will have high abrasive wear and bending resistance. The carbide form and fraction in the case structure will depend on the carburizing process conducted, especially on the carbon potentials of the atmosphere, as well as the temperature and time of the process (see Table 8).

One of the factors that exhibit a great influence on the properties of machined parts and tools is grain size. Austenite grain formation in the diffusion layer and in the core is related to the morphology of martensite being formed [34,38,40,42]. Austenite grain formation is also related to the plastic properties of the carburized steel. Grain size variation during carburization influences the increase in bending fatigue [43], or the intensity factor, K_{LC} [44]. The small effect of austenite grain size in the case or in the core can be examined by controlling the formation of carbides or chemical composition of the steel, or through process control.

According to Refs. 45 and 46, austenitizing from a two-phase austenite–cementite region lowers the phenomenon of plate martensite microcracking. Austenitizing in the temperature range where the diphase structure austenite–carbide is formed, leads to the appearance of cracking of the martensite. Austenitizing at this temperature is favored for small austenite grain material. The favorable influence of austenite temperatures less than A_{cm} have been reported [47,48]. The influence of austenization temperature on properties exhibited by the carburized case has also been addressed [35,49–51] (see Table 8). In addition, the effect of mechanical loading on the properties and

structure of the diffusion layer has been addressed as well [38] (see Table 8).

A low martensite (Ms) temperature results in a reduction of the temperature range for which self-tempering of martensite may occur, which favorably affects many properties. In other words, a low Ms temperature leads to significant inhibition of self-tempering processes occurring during hardening, and, therefore, to high carbon content in solid solution (martensite), which will increase strength and hardness. A minimum Ms value corresponds with the highest compressive stresses (Fig. 14).

Carbides do not reduce hardness. Carbides are themselves hard. Carbide microhardness in carburized carbon steel has been measured at 1000 HV, and in 2% Ni–Cr steel ~880 HV [52]. Therefore, carburized and hardened surfaces containing 25–50% of dispersed carbides should exhibit higher hardness than carburized surface without carbides: 65–67 HRC (830–900 HV) [53].

The occurrence of carbides in the case also exerts an indirect influence over occurring internal stresses. The extent of carbide's influence on the matrix depends on the quenching method. Slow cooling results in microinfluence of carbides on matrix, while fast cooling results in macroinfluence. (Microinfluence is the influence on chemical composition, and macroinfluence is the influence on all structure elements. The micro- and macroinfluence are important, and yes, self-important.) If the matrix of a case containing carbides transforms into martensite, macrostresses will probably be compressive. Their magnitude depends on the amount of retained austenite and on martensite type (plate or lath) (see Table 8). Slow formation of large carbides will produce eutectoidal carbon content in the matrix, which, for high-alloy steels, is a relatively low content. The formation of a higher amounts of plate martensite in the matrix during quenching is favored, which results in higher matrix toughness and compressive macrostresses.

It has also been shown [54] that surfaces containing large amounts of carbides result in the formation of lower compressive stresses (−60 MPa) than a surface without carbides (−500 MPa). Another work [55] has shown how various carbides influence surface internal stresses (see Table 8). In that case, carbides led to lower compressive stresses, what is good and bad in depending on the operating conditions of hardening carburizing element.

Contact fatigue life for four different carburized surfaces is presented in Ref. 26, based on the results of sliding and rolling tests on carburized and tempered 2%

Table 7 Summary of Common International Carburizing Grades of Steel

Steel group	Steel sign	Polish norms	Steel sign	ASTM Norms	Average concentration of elements [%]						Temperature [°C]		Minimum mechanical property				Application
					C	Mn	Cr	Na	Mo	Inne	Hardening	Tempering	R_m [MPa]	R_e [MPa]	A_s [%]	KCU [J/cm]	
Chromium	15H	PN-89/H-84030/02	~5117	ASTM A 322-91	0.15	0.7	0.9	—	—	—	880/800	180	690	490	12	70	On small-dimension parts which do not carry heavy loads such as camshafts, reel, bolts, sleeves, spindles.
	20 H	PN-89/H-84030/02	~5120	ASTM A 322-91	0.20	0.7	0.9	—	—	—							On small-dimension parts which do not carry heavy loads such as camshafts, reel, bolts, sleeves, spindles.
Chromium–manganese	16HG	PN-89/H-84030/02	~5120 H	ASTM A 534-94	0.16	1.2	1.0	—	—	—	860	180	830	590	12	—	On smaller-dimension parts such as cogs, camshafts, perpetual screws, bolts
	20 HG	PN-89/H-84030/02			0.20	1.3	1.2	—	—	—	880	180	1080	740	7	—	On medium-dimension cogs, shafts, and other parts which take bigger stress and variable load
	18 HGT	PN-89/H-84030/02			0.18	1.0	1.2	—	—	—	870/820	200	980	830	9	80	On high-load parts with high durability of core like cogs, shafts, etc.
Chromium–manganese–molybdenum	15 HGM	PN-89/H-84030/02			0.15	1.0	1.0	—	0.2	—	840	180	930	780	15	80	On medium-dimension parts such as cogs, shafts, and other parts which take significant stress and variable load
	15 HGMA	PN-89/H-84030/02			0.15	1.0	1.0	—	0.2	—	840	180	930	780	15	80	On bigger-dimension parts such as cogs, shafts, and other parts which take big stress and variables load
	18 HGM	PN-89/H-84030/02			0.18	1.1	1.1	—	0.2	—	860	190	1080	880	10	90	
Chromium–nickel–manganese	15 HGN	PN-89/H-84030/02			0.15	0.9	1.0	1.5	—	—	850	175	880	640	10	59	High-load, smaller-dimension cogs, gears which work at full capacity, shafts
	17 HGN	PN-89/H-84030/02			0.18	1.2	1.0	0.8	—	—	860	160	1030	830	11	70	On significant-load, very small cogs, shafts, perpetual screw, bolts

Material	Grade	Standard	Equivalent standard												Application
Chromium–nickel	15 HN	PN-89/H-84030/02		0.15	0.6	1.6	1.6	—	860	190	980	830	12	80	High-load, smaller-dimension cogs, gears which work at full capacity, shafts
	15 HNA	PN-89/H-84030/02		0.15	0.6	1.6	1.6	—	860	190	980	830	12	80	On high-load cogs and rolls, gears work at full capacity and others which take variable load
	18 H2N2	PN-89/H-84030/02		0.18	0.6	2.0	2.0	—	860	190	1230	890	7	—	
	12 HN3A	PN-74/H-84032		0.12	0.5	0.8	3.0	—	860/790	180	930	690	11	90	To make equipment which are especially loaded such as air equipment and parts of internal combustion engine
	12 H2N4A	PN-74/H-84032	~E 3310 ASTM A 837-91	0.12	0.5	1.5	3.5	—	860/790	180	1130	930	10	90	To make equipment which are especially loaded such as air equipment and parts of internal combustion engine
	20 H2N4A	PN-74/H-84032		0.20	0.5	1.5	3.5	—	860/780	180	1270	1080	9	80	To make equipment which are especially loaded such as air equipment and parts of internal combustion engine
Chromium–nickel–molybdenum	17 HNM	PN-89/H-84030/02		0.17	0.6	1.7	1.6	0.3	860	170	1180	830	7	—	High-load cogs, camshafts work at full capacity, shafts, and other parts which take big stress
	20 HNM	PN-89/H-84030/02	8620 H ASTM A 534-94	0.20	0.8	0.5	0.6	0.2	880/820		—[a]	—[a]	—[a]	—[a]	On medium-load cogs, shafts, and other parts which take big stress
	22 HNM	PN-89/H-84030/02	8622 H ASTM A 304-95	0.22	0.8	0.5	0.6	0.2	880/820		—[a]	—[a]	—[a]	—[a]	On medium-load cogs, shafts, and other parts which take big stress
Chromium–nickel–tungsten	18 H2N4WA	PN-74/h-84032		0.18	0.4	1.5	4.2	—	950/850	180	1130	830	12	100	To make equipment which are especially loaded such as air equipment and parts of internal combustion engine

[a] Based on purchase specification data.

Table 8 Influence of Different Parameters of Structure on Characteristic Mechanical Property of Carburized Layers

Structural parameters	Mechanical property				
	Hardness, H	Fatigue strength, σ_D	Contact of fatigue strength	Intensity of cracking	Wear abrasive
Thickness of layer (t_c)	Hardness is suitably highest for the whole thickness of layer. It has been possible to reach the aimed value of 700–900 HV.	Reported definition for samples onto bending (it was bent-rotatory) are in agreement with the results of Dawes and Cooksey (0.08) [56], Tauscher (0.014–0.21) [57], Weigand and Tolash (0.07—0.075) [58]. Large dispersion of the result is attributable to own stress, resistance of core, and internal oxygenations in different samples [59]. It appears that the optimum thickness of layer to the thickness of part correlates with the σ_D maximum value, and so the value of 0.06–0.07 accompanies the endurance of core 1080 N/mm² [59].	Generally, the double depth of covering of maximum stress yields effective thickness (taking into consideration possible geometrical defects, bad carrying capacity, overload…)	It falls with the level of thickness. Cracking depends on numerous parameters: species of steel, thickness of carburized layer, preliminary processing of detail. We measured cracking at burdens ranging from 38,000 to about 58,000 [N] on carburized samples [60,61]. Parameters of leading processing to elevation of intensity of cracking are as follows [60,61]: content of nickel in steel (increase in profitable nickel content); small thickness of carburized layers; hardening in warm oil (80–160 °C); tempering after carburizing and hardening.	The thickness of layer is dependent on the value of wear abrasive
Resistance of core	Unequivocal relation between the resistance of core and hardness was not observed	Optimum for: R_m content 1080–1240 N/mm² [58]	Optimum for: R_m content 850–1080 N/mm² [58]	It drops when the resistance of core increases	Unequivocal relation between the resistance of core and wear abrasive was not observed.

Residual austenite, γ_R	When γ_R increases, hardness diminishes	γ_R not tolerated; tolerance = 25%. (Influence of residual austenite in carburized layers onto fatigue strength is controversial. Some authors affirm improvement of resistance onto bending with proportional content of residual austenite between values 10% and 30% [62]. Some authors will pass unfavorable influence γ_R on σ_D. Content of $\gamma_R = 80$–85% induces lower value σ_D of about 25% [63] and about 30% [64]. Agreement with this result accepts content of residual austenite government 15–25% [65] or 20–30%[66].)	γ_R tolerated for 50% content. (Different works show how residual austenite influences profitably onto fatigue border at rolling or rolling with slide [67,72–74].)	Content of residual austenite has little influence [64]	It falls when γ_R climbs
Carbides	Sensitivity of the value of hardness depends on the form and the quantity of carbides	σ_D falls [67]; globural in net: small influence. (Causes itself meshes from produced carbides of lower fatigue strength [68]. Presence of dispersal carbides or net has little profitable influence [69]	Globular, admissible small influence. (Fatigue strength appears to be tolerant to the superficial presence of globural carbides [74].)	Unequivocal relation between carbide part and intensity of cracking was not observed.	Carbides admissible part in structure, but not in form of net.
Internal oxygenation	Hardness drops when thickness of internal oxygenation increases	σ_D drops when the thickness of internal oxygenation increases. (Change σ_D is proportional to the thickness of internal oxidation. It has been possible to accept that they fell for thickness 6–10 μm [70]. For thickness of 13 μm, reduction of border of fatigue can reach 20–25%, and 45% for depth 30 μm [71].	Thickness of internal oxygenation exerts little influence	Unequivocal relation between internal oxygenation and intensity of cracking was not observed.	Internal oxygenation has no favorable influence on wear abrasive.

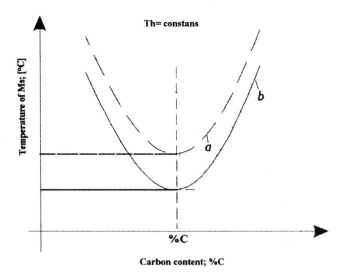

Figure 14 Influence of carbon content on the Ms temperature of steel.

Cr–Mn steel (see Table 8). The tests showed that coarse carbides can also (such as coarse-grain martensite) influence contact fatigue (see Table 8).

Carbides are hard. Therefore, it is expected that sufficient amounts will guarantee that surfaces containing them will have abrasive wear resistance (because of their hardness) and adhesive wear resistance. It can be generally concluded that increasing the surface carbide content will increase the abrasive wear resistance. This relationship is true when spheroidized carbides appear in a nonmartensitic matrix. When the structure of the matrix consists of martensite and retained austenite in various amounts, the amount of carbides present does not significantly influence wear resistance [75]. A microstructure that produces best contact fatigue resistance also increases abrasive wear resistance [76]. It should be further noted that contact fatigue and abrasive wear resistance are also influenced by the quality and condition of the system lubrication.

Network continuity, carbide coarseness, and penetration depth reduce fatigue resistance. High manganese fraction in low-alloy steel may be the cause of carbide networks that occur in carburized cases. The form and amount of carbides are influenced not only by the diffusion process temperature and time, but also by temperature and time of subsequent processes (e.g., annealing, hardening) (see Table 8).

Table 8 presents some relation between factors of structural and working properties for carburized structures [34].

G. Influence of Residual Stress on Fatigue Strength

Carburizing is the most favored process to improve fatigue strength and wear resistance of automobile gears, as well many other machine components. Surface hardness, case depth, and residual stress are the primary factors that influence the strength and durability of carburized and hardened steels. Case depth and core hardness vary extensively because of the variability of chemical composition, steel hardenability, carburizing conditions, and quenching methods even if the parts are of the same quality. Moreover, those values vary widely, depending on the hardenability of selected steels, even when the same type of steel is treated at the same temperature [77–80].

1. Charpy Impact Strength

Charpy impact test were conducted with electrically assisted analysis of energy absorbing curve to analyze impact fracture mechanism. The Charpy impact value decreases remarkably as the case depth increases. This trend roughly shows negligibly small influence of core hardness. However, results of analysis in energy absorbing curve and observation of microfractographs indicated minor influence of core hardness with the total deformation of the test piece during fracturing [78,82]. The tendency of total absorbed energy and the peak load to initiate fracture are different when total energy is directly related with the amount of deformation with respect to case depth and core hardness. The peak load value seems to be influenced by the strength of core, and higher core hardness tend to result in higher fracture load even though the total energy decreases, as observed in Fig. 15 [82].

2. Impact Bending Fatigue Life

The life of impact bending test measured at 1.5×10^4 and 5×10^4 are influenced by carburizing time and case depth. If the test pieces have no hardened case, core hardness exhibits a profound influence on their impact fatigue life, but when test pieces have a hardened case, strengths are affected by case depth. In this impact fatigue test samples, which have 5 R notch, strength exhibits a peak at very thin case zone, such as those exhibited by the samples with less than 1-hr carburizing time, as observed in Fig. 16 [82]. However, when the impact load is decreased and as life time increases, the influence of case depth becomes different and reveals two peaks.

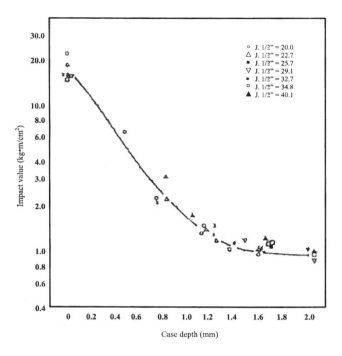

Figure 15 Influence of case depth on core hardness and Charpy impact strength.

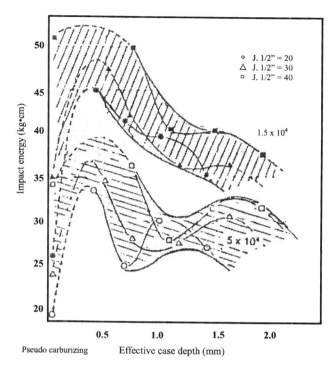

Figure 16 Effect of case depth on impact fatigue strength.

3. Rotating Beam Fatigue Strength

Fatigue test results using the rotating beam shows the important influence of case depth. At first, fatigue strength increases as case depth increases but after the peak value, it decreases as case depth increases, as shown in Fig. 16 [82]. The influence of core hardness is somewhat different and not simple when carburizing time is between 5 and 8 hr. Fatigue strength remains high without much influence of the core hardness. This trend seems to relate with residual stress and influence of intergranular oxidation, as discussed in a later section. The influence of case depth and core hardness on fatigue strength is different when the test piece possesses stress concentrators such as grooves or notches.

The endurance limit of rotating beam bending fatigue test is influenced by case depth which exhibits a peak at about 1 mm case depth, as shown in Fig. 17 [82]. The endurance limit of the rotating beam test changes with core hardness, as observed in Fig. 18 [82]. In particular, when the carburizing time is short (2.5 hr) and very long (20 hr), the influence of core hardness is evident and the peak is around core strength of 140–150 kg/mm.

4. Intergranular Oxidation

When the carburizing process uses endothermic gas containing CO in its atmosphere, it will influence intergranular oxidation at the carburized surface. The depth of intergranular oxidation increases as the carburizing time increases. The secondary influence of inter granular oxidation is that it results in the decrease of case

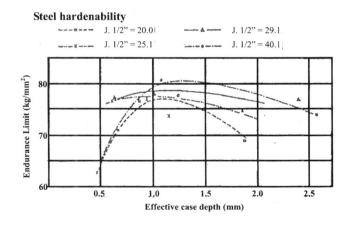

Figure 17 Effect of case depth on fatigue limit of rotating beam test.

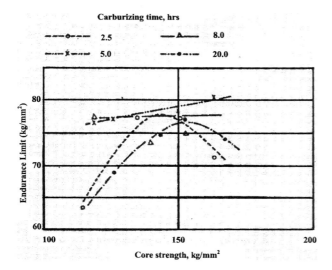

Figure 18 Influence of core hardness on fatigue strength of a carburized test specimen.

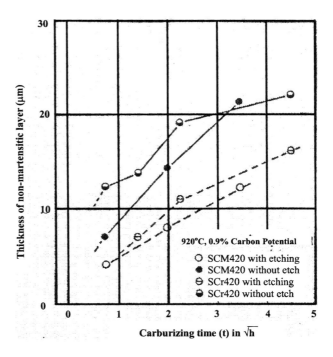

Figure 19 Effect of carburizing time on intergranular oxidation.

hardenability and causes the formation of nonmartensitic transformation phases, which decrease hardness and residual compressive stress. In Fig. 19, the influence of oxidation is observed in two ways—as a nonetched state and underetched by 5% natal solution [82]. As plotted in Fig. 19 [82], the influence depth increases as carburizing time increases. The depth is always thinner under nonetched condition. This difference is caused by the amount of nonmartensitic phase, such as upper or lower bainite microstructure [78,83].

However, contrary to the general opinion, this influence on fatigue strength or life is easily eliminated by postshot blasting or peening, as observed in Fig. 20 [82]. The postquenching and tempering process, such as steel shot blasting, is used to remove colored contamination or soot. The shot-blasting process is more effective than postelectrochemical etching, which removes surface layers of nonmartensitic phase. Recent peening processes with small harder shots and with higher speed can easily eliminate the influence of oxidation and resultant nonmartensitic surface layer. However, care should be taken to minimize the formation of nonmartensitic surface layer by optimum atmosphere carbon potential control, and to prevent the delay of quench operation by the quench operation system and selection of the optimum quench media.

5. Tooth Bending Fatigue Testing of Gears

Spur gears of $m = m$, illustrated in Fig. 21 [82], are used for a gear tooth bending fatigue test for evaluating and

correlating those results to understand and develop a design recommendation for automotive gears [83]. Tooth bending fatigue test is conducted by hydraulic pulsating load on the weakest contact position of gear teeth. Case depth influences the endurance limit and lifetime; when the endurance limit exhibits double peaks, the lifetime will exhibit only one peak at very short carburizing times, as observed in Fig. 22 [82]. Similar to impact fatigue test results, the fatigue life of

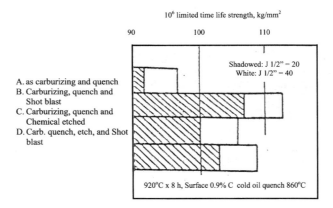

Figure 20 Influence of postfinishing methods on rotating beam bending fatigue life.

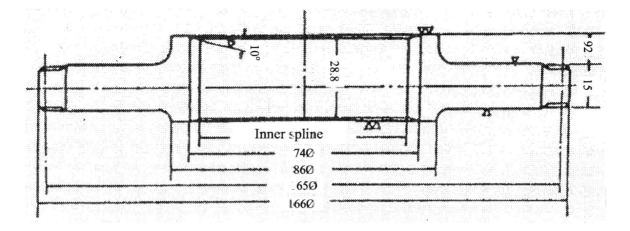

Gear Module: 3, Center distance 162.5,
Pressure angle: 20o, Pitch diameter: 165 mm

Figure 21 Design of test gears used for the tooth bending fatigue test.

gear tooth show difference tendency. The limited time fatigue life decreases as the carburizing time increases, but the endurance limit exhibits a difference trend having double peaks. The highest peak is in the thin case zone and second peak in thicker case, which is similar to the peak of rotating beam fatigue test. Compared with the influence of case depth, that of core hardness is smaller but shows similar tendencies.

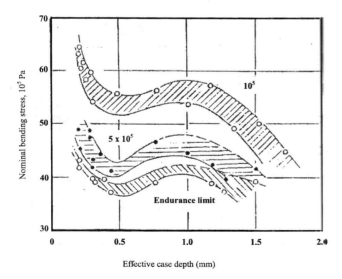

Figure 22 Relationship between effective case depth and tooth bending strength as determined using the hydraulic pulsating testing machine described in Ref. 83.

Fatigue test of gears are also tested by four gears tester with a normal tooth contact condition. The fatigue test results shown in Fig. 23 exhibits a similar tendency as single gear tooth bending fatigue tests [82]. The fatigue life decreases as case depth increases. However, fatigue life and endurance limit show no clear double peaks, which is clearer with a single teeth bending tests. The difference between Figs. 21 and 22 seems to be caused by the lack in the number of test gears and case depth levels. Another difference of test condition is in the loading condition. Single-teeth bending test is under highest loading stress condition at the foot of a teeth, while the load of the four gear tests are normal contact condition and all of the gear tooth are loaded in rotating condition, where the accuracy of gear tooth pitch error and mounting shaft stiffness have considerable influence on fatigue behavior.

6. Residual Stress of Carburized Steels

The carburizing process enables the formation of a hard case, which is necessary for wear resistance and high fatigue strength. Because of the martensite transformation of high carbon case during the end of quenching, the case area results in high compressive residual stress if carburized surface have enough case hardenability and the quenching operation is adequate. If the quench operation is not adequately controlled, the case will not achieve a sufficiently high compressive stress, as observed in Fig. 24, and sometimes, it be-

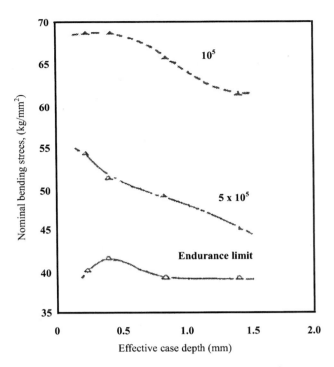

Figure 23 Relationship between effective case depth and gear tooth fatigue life determined using the four-gear testing machine described in Ref. 84.

comes a tensile residual stress and reduces the fatigue strength [23].

The influence of carburizing and hardening methods on residual stress is not simple. Carburizing elements such as carbon potential of atmosphere, carburizing time, resulting surface carbon, and their distribution have a profound influence on the level and distribution of residual stress of carburized and hardened steels. Intergranular oxidation has significant influences on residual stress and sometimes the excessive formation of nonmartensitic phase could result in residual tension stress at the surface. Figure 24 shows the variation of surface residual stress in relation to the effective case depth, which are heat-treated by different methods. Influence of core hardness and case depth on surface residual stress depends both on transformation timing, and balance or core strength and case depth. Generally, the peak position of residual stress against case depth is nearly at the same area as observed in Fig. 24 even with different core hardness.

Residual stresses formed in case-hardened steels exhibit a profound influence and contribute to the improvement of fatigue strength [84–86]. Although the

distribution of residual stress varies during the fatigue process, it does not disappear even after the fatigue fracture. Figure 25 shows some examples of their variation during the fatigue process [82].

7. Influence of Shot Blasting and Peening

Shot blasting or peening after carburizing is an effective means to eliminate the influence of intergranular oxidation by increasing compressive residual stress. However, the influence on the surface mechanical properties is not only caused by intergranular oxidation, but also is largely affected by surface carbon concentration and quench condition. When the thickness of the nonmartensitic layer is large, this leads to low surface hardness, and the residual stress seriously damages surface mechanical strength [87–91]. Recently, new peening technologies made a great advancement to fulfill the need of improving surface properties to meet increasing engine power and demands for weight reduction of the power train.

Residual compressive stress at near surface and its distribution varies by shot and peening conditions, namely, shot materials and their hardness, shape, and mass, and the speed to hit the surface during peening. As the hardness, weight or mass of shot, and shot speed increase, the resulting residual stress increases. Small

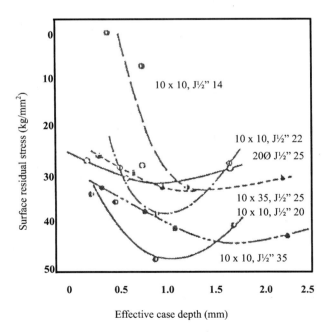

Figure 24 Correlation of case depth with surface residual stress.

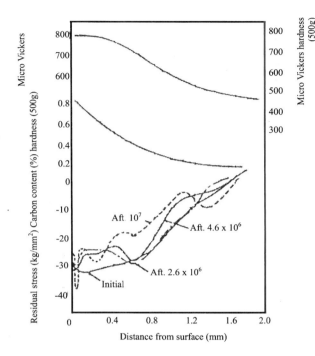

Figure 25 Variation of residual stress during the fatigue test.

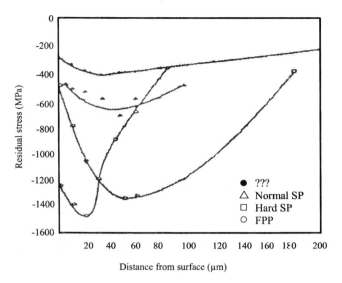

Figure 26 Influence of shot peening methods on residual stress distribution.

shot sizes cause the residual stress peak to move to near surface [88–90]. Combing the influence of those factors, a double peening process is favored to improve fatigue strength of transmission gears. Figure 26 shows the influence of the shot peening media and methods and difference of residual stress peak at near the surface of carburized hardened steels [82].

8. Influence of Residual Stress and Intergranular Oxidation on Fatigue Strength

The depth and their influence of intergranular oxidation increases as the elongation of carburizing time. The existence of carburized hard case deteriorates the static, impact, and limited time fatigue strength [83]. However, the residual stress possesses an independent tendency that attains its highest value when a certain case depth is reached at the surface layer. These two factors combine together and exert complex influence on fatigue process. Apparently, although oxidation has deleterious effects, the importance of residual stress is more profound and should be taken into account to design for gear durability. The effect of intergranular oxidation on residual stress is shown in Fig. 27 [82].

In addition to the basic influences of residual compressive stress results from optimum case depth, the advancement of postpeening technologies can further improve the gear fatigue strength through near tooth root surface bending strength and also, the surface pitting fatigue life by very high compressive residual stress [56–58,91].

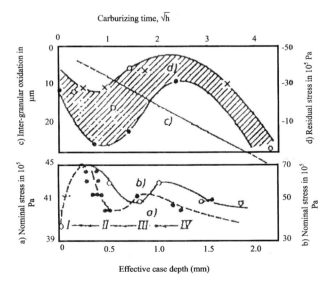

Figure 27 Influence of intergranular oxidation on residual stress.

III. CARBONITRIDING

A. Physical–Chemical Basis of Carbonitriding Processes

1. Carbon and Nitrogen Diffusion into the Steel Surface

Carbonitriding is dependent on the simultaneous diffusion of carbon and nitrogen into the immediate surface layers of the steel. The process is usually conducted in the temperature range of 850–880°C. The process gas for the treatment is based on the use of either a nitrogen/methanol blended gas or an endothermic atmosphere with the addition of a hydrocarbon enrichment gas for the carbon plus ammonia for supply of the nitrogen.

The carbonitriding process is a process of surface-treating a steel using the method that combines both nitriding and carburizing simultaneously, and while the two processes are both separate, they are—to some degree—dependent on each other.

The process of diffusion of carbon from the process atmosphere into the steel surface is controlled by Fick's law of diffusion and is expressed in Eqs. (18) and (19). The equations describe the penetration of carbon into the steel.

The diffusion of both carbon and nitrogen into a solid solution of the steel, each with their own mutual influences on the process and each being derived from the process atmosphere for adsorption into the steel surface raises the question, "What influence (if any) does nitrogen have to facilitate the process, and does nitrogen assist carbon absorption into the steel surface?"

At a given temperature, the chemical reactions progress until equilibrium is achieved. This means that each equilibrium reaction will level itself, and the reaction will move from left to right, and right to left. This means that when in equilibrium, the thermodynamic potentials of both atmospheric products (prod) and the substrate (sub) material (steel) are equal. That it is:

$$\Delta G^\circ = \sum G^\circ_{prod} - \sum G^\circ_{sub} = 0 \qquad (18)$$

The equilibrium constant of any reaction of the type:

$$A + B \Leftrightarrow C + D \qquad (19)$$

may be expressed by partial pressures of the reactants and products:

$$K_p = p_c^* / p_a^* p_b \qquad (20)$$

where p_a, p_b, p_c, and p_d represent the partial pressure of reactants A and B, and products C and D, respectively.

In isobaric equilibrium, the equilibrium constant, K_p, is related to thermodynamic equilibrium by:

$$\Delta G^\circ = -RT \ln K_p \qquad (21)$$

where ΔG° is the thermodynamic potential of reaction [cal/mol], R is the constant gas levels 1.987 [cal/(mol K)], T is the temperature [K], K_p is the constant equilibrium.

At the carbonitriding process temperature, the primary equilibrium thermal disassociation processes producing carbon and nitrogen are:

$$C + CO_2 \Leftrightarrow 2CO \qquad (22)$$
$$CH_4 \Leftrightarrow C + 2H_2 \qquad (23)$$
$$C + H_2O \Leftrightarrow CO + H_2 \qquad (24)$$
$$2NH_3 \Leftrightarrow N_2 + 3H_2 \qquad (25)$$
$$NH_3 \Leftrightarrow N + 3/2H_2 \qquad (26)$$
$$1/2N_2 \Leftrightarrow N \qquad (27)$$
$$HCN \Leftrightarrow C + N + 1/2H_2 \qquad (28)$$
$$CO + 2NH_3 \Leftrightarrow CH_4 + H_2O + N_2 \qquad (29)$$
$$CO + NH_3 \Leftrightarrow HCN + H_2O \qquad (30)$$
$$CO_2H_2 \Leftrightarrow CO + H_2O \qquad (31)$$
$$CH_4 + H_2O \Leftrightarrow CO + 3H_2 \qquad (32)$$
$$CH_4 + CO_2 \Leftrightarrow 2CO + 2H_2 \qquad (33)$$

Reactions (5)–(10) play a direct and active part in the production of both carbon and nitrogen. In creating atomic carbon for the diffusion reactions (5)–(7) have the greatest significance. The atomic nitrogen is created in reactions (8)–(10) and is available to the steel for diffusion, combined with carbon. In accelerating the penetration of both carbon and nitrogen, part of the process will create very small quantities of prussic acid onto the steel surface. The remaining reactions or indirect reactions are inconsequential. At the process temperature of the carbonitriding procedure, molecular nitrogen will create active nitrogen (N) in accordance with reaction (27).

Atomic hydrogen is emitted and is directed to the surface of the steel, and will intensify both the reduction and dissociation of surface oxides, as well as convert the oxygen emitted as a result of the carburizing reaction. When using furnace atmospheres, oxygen will always be present as a result of moisture from the endothermic gas, or from natural furnace leakages.

The source of nitrogen for diffusion into the steel is derived from the introduction of ammonia together with the hydrocarbon enrichment gas. At the process temperature in ammonia, dissociation will rapidly occur, which begins in reaction (25) and in the final decomposition, as described in reaction (26). Atomic nitrogen is produced almost immediately as result of the decomposition of ammonia, followed by decomposition into atomic nitrogen. However, the amount of molecular nitrogen produced is too great a volume for diffusion into the iron of the steel, and it will not dissolve into the steel as molecular nitrogen (dissolubility of up to 0.025%). It should be noted that the volume of ammonia should be between 4% and 8% of the total gas flow into the process furnace.

It is only the active or atomic nitrogen resulting from the dissociated ammonia that will penetrate and diffuse into the steel surface. The quantity of dissociated ammonia, which is available for diffusion, is proportional to the undissociated ammonia present in the furnace atmosphere. It is not possible to quantify the undissociated ammonia on the basis of thermodynamic dependence, resulting from the conditions of equilibrium between ammonia and other products present in the furnace atmosphere. This is because such equilibrium will only occur in the range of temperatures between 400 and 600°C. Between this range of temperatures, dissociation is almost total, because the quantity of undissociated ammonia does not exceed 0.1% ammonia.

The influencing factors on the degree of dissociation of ammonia, which occurs during the carbonitriding process, that will in turn influence of the process of dissociation, are:

- Process temperature,
- Atmosphere changes per hour within the furnace,
- Circulation and distribution of the process gas within the furnace,
- Furnace size (process chamber volume),
- Furnace heating elements or a gas heating system.

Figure 28 provides a correlation of the case depth as a function of time at temperature for various processing temperatures.

The surface properties of the carbonitrided layers are essential in exerting the concentration of nitrogen in the superficial zone layer of the surface of the steel. It will depend directly on the concentration of the early reaction stages of the ammonia (gas–metal) and indirectly from other influencing factors such as the speed of the reaction. Therefore, for the control of the process, the superficial concentration of nitrogen is usually a function of the concentration of ammonia in the atmosphere for different temperatures.

At even higher temperatures of 850–930°C, the volume of ammonia will be increased due to the temperature. This means that the addition of ammonia will increase, to a small degree, the concentration of nitrogen in the surface layer of the steel. If this is allowed to occur, then the potential for nitride networking is extremely high. Conversely, if the process of temperature is reduced, the opposite will occur. Thus even small additions of ammonia will cause a dramatic increase in the surface layer nitrogen concentration. Great care should be taken with the addition of ammonia as a source of atomic nitrogen, as this can cause nitride networks to occur within the formed case. The limit of solubility of nitrogen in iron at a temperature of 600°C maximum is approximately 6%.

Steels that include the stable nitride-forming elements will readily react with nitrogen. The alloying elements that will react favorably with nitrogen to form the stable nitrides are:

- Chromium
- Vanadium
- Tungsten
- Aluminum
- Molybdenum
- Boron
- Silicon

It should be noted that chromium, when combined with manganese, will display higher concentrations of nitrogen in the surface layers by comparison to steels that do not include these elements. Furnace atmospheres with higher carbon potentials can saturate the steel surface excessively. Consideration must be given not only to the gaseous reactions, but also to suppression of the potential for retained austenite formation by the nitrogen activity at the steel surface. Excess carbon will also encourage carbide formation in the surface of the steel, and nitrogen will assist in lessening the retention of retained austenite by allowing a lower austenitizing temperature to be selected. An added benefit of the nitrogen reaction (particularly with the nitride forming alloying elements) is that there is an increase in the resulting surface hardness, as well as the reduction of the potential for distortion.

2. Role of Nitrogen in the Carbonitriding Process

The influence of carbon on the diffusion of nitrogen into a solid solution is comparatively insignificant, although so many other very essential influences of carbon on the process of diffusion of nitrogen in the steel will occur. In the process of carbonitriding, the role of nitrogen restrains (as a rule) the changes of conditions of carbon diffusion into the structure of the formed layer. Nitrogen facilitates itself in dissolving carbon in iron, and will also contribute to a significant increase in the speed of the diffusion process.

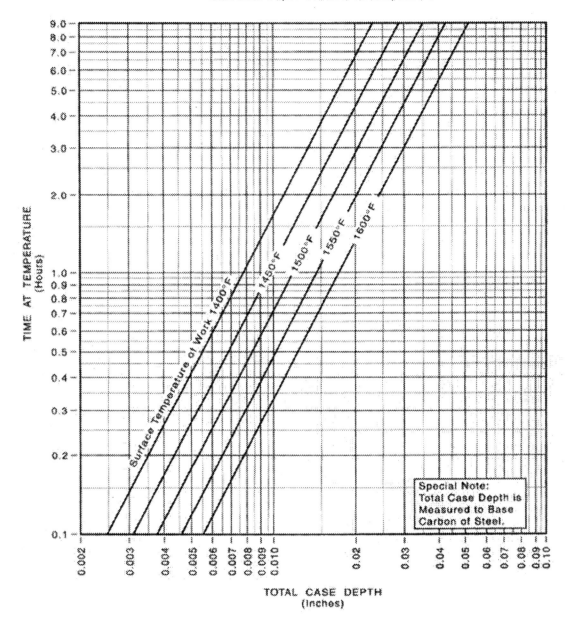

Figure 28 General carbonitriding curves for various times at temperatures shown. (Source: Surface Combustion Inc., Maumee, OH.)

Changes of Carbon Potential in the Furnace Atmosphere

Ammonia and the products of ammonia dissociation will change the conditions of equilibrium of the main carburizing reaction, thus continually changing the carbon potential of the furnace process atmosphere. All products for the dissociation of ammonia (nitrogen and hydrogen) are in accordance with reaction (25), in relation to the partial pressures of carburizing components to noncarburizing changes in the direction of raising the carbon potential of the furnace atmosphere.

Ammonia will also react with the oxides of carbon in accordance with reaction (29), and will produce water vapor, which will lower the carbon potential of the furnace atmosphere. The complete reaction, along with its function to change the atmosphere carbon potential, is very definite.

Ammonia will influence the dependence between the carbon potential of the endothermic atmosphere (and enrichment gas) and the temperature of the dew point. Insignificant additions of ammonia to a carburizing atmosphere will cause a decrease in the carbon potential of the furnace atmosphere, which also means a reduction of the atmosphere dew point.

Increasing Carbon Diffusion Rate in Austenite

The diffusion of nitrogen into steel raises the coefficient of diffusion of carbon into austenite, and thus increases the speed of complete diffusion. This creates the possibility of utilizing low process temperatures, because carbonitriding will diffuse into the surface layers through the approximate thickness of the total carburized layer in temperatures of approximately $50\,^\circ C$ or higher.

The diffusion and creation of the carbonitrided layer will begin below $850\,^\circ C$, and that the case will begin to form even faster than at carburizing temperatures of $900\,^\circ C$.

Activity of Carbon in Austenite

Nitrogen, within the diffused layers, will amplify the activity of carbon in austenite, which is as a result of the superficial concentration of carbon in the primary layer from the furnace atmosphere. In comparison with carburizing, for example, the process of carbonitriding utilizes lower process temperatures and thus reduces the risk of excessive concentrations of carbon within the formed surface layers.

Changes of Equilibrium Conditions in the Arrangement of Fe–Fe₃C

The changing conditions of equilibrium in the iron cementite arrangement is caused by the presence of nitrogen, and is dependent on the direction of the lower concentrations of carbon to expand the area of occurrence of austenite through the lower temperatures of transformation of the A1 and A3 points (iron–carbon equilibrium diagram) as well as the movement of the eutectoid line. This promotes the diffusion of carbon into the steel surface and yields the possibility of apply-

ing lower process temperatures for carbonitriding as well as utilizing lower temperatures of hardening. This is created by the nitrogen that will also assist in the suppression of retained eustenite conditions, thus the speed at which carbon diffuses is increased.

Enlargement of Durability of Over-Cooling from Austenite

The introduction of nitrogen into the surface of the steel improves the resistance of the steel when exposed to a rapid rate of cooling, by moving the critical cooling curve of the Isothermal Cooling diagram to the right. This enables the steel to respond better to the hardening process by intensive cooling.

By the utilization of lower austenitizing temperatures, the use of oil quenching can be successfully utilized as the quench medium of choice, thus considerably reducing the potential for distortion. Because of the nitrogen diffusion into the steel surface, the dimension with stability of the treated part is greatly improved.

Quenching conditions will determine the surface hardness, the type and properties of induced internal stresses, and mechanical properties such as torque and tensile strength. Because of potential defects related to water quenching, it was generally recommended that carbonitrided steels be quenched in oil, particularly blended quench oils that would facilitate martensitic transformation of the surface and a pearlitic core structure within a specific range of oil cooling speeds. In addition, it is typically necessary to provide an exhaust ventilation for gaseous exhaust effluent, and there is often a concern for environmental impact related to the use of quench oils.

It has been demonstrated that aqueous solutions of polyalkylene glycols (PAG)-based quench media will reduce the potential environmental impact and, by varying the concentration of the polymer quenchant in water, a wide range of cooling speeds is possible. General impact of quenchants and cooling rates on carbonitrided steel properties will be discussed subsequently in addition.

Reduction of the Austenite Transformation Temperature to Martensite

The presence of nitrogen in the carbonitrided layer will reduce the temperature of transformation at the start of the formation of martensite (Ms) as well as a raise the martensite finish point (Mf) in a very positive manner. The transformation temperature of the carbonitrided steel at the point of hardening will be lower than if the same steel was carburized. This is due to the presence of nitrogen within the formed case and will also cause a reduction in the compressive stress through the formed case due to the reduction of the martensite start point (Ms).

Reduction of Retained Austenite

The lowering of the martensite start temperature will reduce the amount of residual austenite in the carbo-nitrided layer. The residual austenite will normally be found when the surface carbon potential is in excess of 0.7–0.9% with concentrations of nitrogen in solution. If the concentration of both carbon and nitrogen are controlled in the structure of the formed layer, then the retention of any untransformed austenite will be minimal, thus producing a more dimensionally stable carbonitrided case.

Control of Carbon and Nitrogen Within the Diffusion Layer

The production of available carbon for the process is principally derived from the enrichment gas, which is added to the endothermic gas, or the nitrogen/methanol plus enrichment gas. The objective is to ensure a surface carbon potential around the eutectoid line of the steel (0.7–0.9% carbon). The carbon potential can be controlled via the accepted methods of control, such as dew point, shim analysis, oxygen probe, or CO/CO_2 analysis. The nitrogen is normally derived from the ammonia enrichment gas to the furnace. This is extremely difficult to control. There is simply no commercial analysis system to accurately control the decomposition of ammonia to release nascent nitrogen. Therefore, one must rely on the relationship of the volumetric flow of the process gas (carrier gas and enrichment gas) in relation to ammonia. Generally, the ammonia flow should be between 4% and 8% (by a volume) for all the total gas flow into the furnace. In reality, as well as in ideal conditions, the flow rate should be between 4% and 8% by volume to produce an effective formed carboni-trided case formation.

3. Carbonitriding Procedure

The carbonitriding process very closely resembles the process of gas carburizing. The procedure follows the following technological actions:

- Conditioning of the furnace,
- Loading the work into the furnace,
- The carbonitriding process,
- Quenching the carbonitrided work.

The carbonitriding process times are much shorter than the soaking times experienced with the carburizing process.

During the heating cycle of the batch, the introduction of the carbonitriding atmosphere into the furnace takes place. It is not necessary to introduce the process gas into the furnace during the heat up cycle. This is because during the heat up of the batch, and particularly in the temperature range of 550–700°C, ammonia can cause the formation of nitrides in the steel surface, so that on completion of the carbonitriding process, the formed nitrides may not be stable or be able to undergo rearrangement in the formed case of the carbonitriding process and at the carbonitriding process temperature.

The selection of the carbonitriding process requires careful consideration for process at temperature selection, for this atmosphere composition as well as the heating method of the furnace. This is to ensure a well-conducted process that is able to produce the required case depth formation, as well as the appropriate carbon and nitrogen concentrations within the formed case. When selecting low carbonitriding process temperatures, or long cycle times to produce deep case of formations, the probability of high surface nitrogen concentrations is the extremely high.

It is not usual to select deep carbonitrided formed case depths. Generally, the formed case for carbonitriding is usually at a maximum of 0.5 mm (0.020 in.). Deep case formations will undoubtedly give rise to the high probability for the retention of untransformed austenite (retained austenite). If ammonia is introduced into the furnace during, say the last one-third of the process cycle, then the risk of high surface nitrogen concentrations is reduced, almost to the point of elimination. On completion of the workload in the furnace atmosphere, after spending sufficient time at the process temperature to form the necessary case, then the load is quenched at the appropriate austenitizing temperature. Care must be taken when selecting the austenitizing temperature because of the potential risk of retained austenite and, of course, the potential for the occurrence of distortion.

4. Heat Treatment After Carbonitriding

Once the process of the carbonitriding is completed, it is usually followed by a cooling procedure down to the hardening temperature (Fig. 29). The light tempering

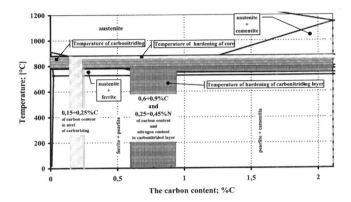

Figure 29 Diagram of heat treatment processes after carbonitriding.

process in the region of 180–200°C usually follows this step. This is to temper the formed martensitic case. and reduce the possibility of case/surface cracking to occur. Tempering is often omitted without consideration that untempered martensite is now formed in the newly formed case.

The cool-down procedure is to accomplish the required properties and structures that is required by the engineering design for core and surface properties of the steel. The steel can be:

- Cooled within the furnace down to the case austenitizing temperature in a controlled manner, followed by quenching.
- Cooled down to, say 500°C, removed from the furnace and cooled externally, followed by a reheating to the austenitizing temperature, then followed by the quenching procedure, which will harden the formed carbonitrided case.

The diffusion of nitrogen into steel raises the coefficient of diffusion of carbon into austenite and thus increases the speed of complete diffusion. This creates the possibility of utilizing low process temperatures, because carbonitriding will diffuse into the surface layers through the approximate thickness of the total carburized layer in temperatures of approximately 50°C or higher. The diffusion and creation of the carbonitrided layer will begin below 850°C, and the case will begin to form even faster than at carburizing temperatures of 900°C.

Table 9 Influence of Structural Factors on Use Property of Carbonitrided Layer

	Use property				
Structural factors	Hardness	Fatigue strength	Contact fatigue	Cracking resistance	Wear resistance
Layer thickness, t_c	Hardness of surface 700–900 HV	Optimum $t_c/t_p \approx 0.7$	$t_c \approx 2t_z$	Drops when thickness of layer grows	Thickness limits admissible waste
Core strength, R_m	—	Optimum when $R_m\varepsilon$ 1300–1500 MPa at con. C in core 0.2–0.25%	Optimum when $R_m\varepsilon$ 850–1080 MPa	Drops when resistance of core grows	—
Retained austenite, γ_R	If γ_R enlarges, this hardness diminishes	It favors in borders 25–60% γ_R	Favors to $\gamma_R = 70\%$	—	—
Carbonitrided	Little sensitivity		Small influence	—	Enlarge resistance
Internal oxidation	Hardness reduced	—	Small or harmful influence	—	Harmful influence
Pearlite or bainite	Hardness reduced	It falls when the thickness of layer of pearlite or bainite grows	—	—	Harmful influence
Porosity	Hardness reduced	Harmful	—	—	—

Source: Refs. 69 and 90–92.

Table 10 The Chemical Composition of the Used Steels in Carbonitriding Processes

| Steel sign | Average concentration of elements [%] | | | | | Country |
	C	Mn	Ni	Cr	Mo	
20MoCr4	0.20	0.70		0.40	0.45	Germany
25MoCr4	0.25	0.70		0.40	0.45	
20CrMo2	0.20	0.70		0.60	0.35	
4028	0.24	0.82			0.25	
20NiMoCr6	0.20	0.70	1.60	0.50	0.45	
23CrMoB3	0.23	0.80		0.70	0.35 + B	
8620	0.20	0.80	0.55	0.50	0.20	Great Britain
4024/28	0.23/0.28	0.80			0.25	
Mod8822	0.22	0.85	0.55	0.50	0.55/0.65	
18CD4	0.18	0.75		1.05	0.22	France
8620	0.20	0.80	0.55	0.50	0.20	
8620	0.20	0.80	0.55	0.50	0.20	Italy
18NiCrMo5	0.18	0.70	1.35	0.80	0.20	

An overview of the influence of structural factors on the properties of carbonitrided layers is summarized by the data provided in Table 9 [34]. Some typical steels used for carbonitriding are shown in Table 10.

REFERENCES

1. Anon. Surface hardening. Industrial Heating December 2002, No. LXIX, Vol. 12, 50–56.
2. Krauss, G. *Surface hardening*. Steels: Heat Treatment and Processing Principles. ASM International: Materials Park, OH, 1990; 281–318.
3. Funatani, K. In *Key features in carburizing and carbonitriding*. Proceedings of the Second International Conference on Carburizing and Nitriding with Atmospheres; Grosch, J., Morral, J., Schneider, M., Eds.; ASM International: Materials Park, 1995; 255–260.
4. Funatani, K. In *Diffusion constant in carburizing processes*, Proceedings of the 8th Seminar of the International Federation for Heat Treatment and Surface Engineering; Liscic, B., Ed.; IFHTSE: London, UK, 2001; 193–200.
5. Okumura, N.; Iwase, A. *Vacuum Carburizing Using Acetylene Gas*. Journal of the Japanese Society for Heat Treatment 1998, *38* (4), 194–197.
6. Imai, N. (NACH) Recent trends in vacuum carburizing. Chubu Chapter Meeting, Reasume, Japanese Society for Heat Treatment, Nagoya, Japan, December 12, 2000; 1–7.
7. Davis, J.R. *ASM Materials Engineering Dictionary*; ASM International: Materials Park, OH, 1992; 59 pp.
8. Sugiyama, M. Journal of the Japanese Society for Heat Treatment 1997, *37* (3), 154–160.
9. Sugiyama, M.; Ishikawa, K.; Iwata, H. Vacuum Carburizing with Acetylene. Advanced Materials and Processing 1999, *155* (4), H-29–33.
10. Okumura, N.; Iwase, A. *Vacuum Carburizing Using Acetylene Gas*. Journal of the Japanese Society for Heat Treatment 1998, *38* (4), 194–197.
11. Kawata, K.; Asai, S.; Sato, H. *Vacuum Carburizing Method with Atmosphere Control*. Proceedings of the Japanese Society for Heat Treatment, May 23, 2001, *22*, 49–50.
12. Inuzuka, M.; Sasaki, N.; Funatani, K. In *Energy conservation of continuous gas carburizing furnaces*. Conference Proceedings of Heat Treating: Equipment and Processes. Totten, G.E. Wallis, R.A. Eds.; ASM International: Materials Park, OH, 1994; 29–41.
13. PN-EN 10052:1999 (EN10052: 1993), "Dictionary of term of heat treatment of iron alloy IDT EN 10052: 1993," Polish Committee for Standardization, Warsaw, Poland.
14. Hirschheimer, L.R. In *The mathematical basis for carburizing*; Proceedings of the Second International Conference on Carburizing and Nitriding with Atmospheres; Grosch, J. Morral, J., Schneider, M., Eds.; ASM International: Materials Park, 1995; 129–131.
15. Hirschheimer, L.R. Aspectos matematicos da cementacao gasosa. Metallurgia (Brazil) 1995(September), *31* (214), 557.
16. Naito, T.; Ogihara, K., In *Examined the direct carburizing Method*. Proceedings of the Second International Conference on Carburizing and Nitriding with Atmospheres; Grosch, J., Morral, J., Schneider, M., Eds.; ASM International: Materials Park, 1995; 43–51.
17. Lutz, J.A. Carburizing at High Temperatures, Heat Treating Progress; Advance Materials and Processes, June 1997, *151* (6), 68AA–68CC.

18. Lutz, J.A. Carburizing at High Temperatures. Heat Treating Progress; Advance Materials and Processes, June 1993, *151* (6), 68AA–68CC.

19. Niimi, L. Speed up carburizing by controlling gas atmosphere in gas carburizing, Toyota Technical Review 1962, *13* (4), 339–343.

20. Niimi, L. Control of Gas Carburizing Atmosphere by Measurement of Carbon Potential, Toyota Technical Review 1959, *11* (2), 82–85.

21. (a) Harris, F.E. Metal Progress, 1111; (b) *Case Depth – an attempt at a practical definition.* Metal Progress August 1943; 265–272; (c) *Carburizing and Diffusion Data.* Metal Progress May 1944, 910.

22. Juyi, W.; Lin, P.; Hui, Z. *Effect of Rare Earth on Ionic Nitriding Process,* Proceedings of the 1st Asian Conference on Materials (Beijing) May 1998; 57–61.

23. Hu, L.; Wang, X.; Wei, W.; Wang, Y.; Huang, Y. *Influence of Rare Earth Contents on Carburizing Rate of Steels and Morphology of their Carbides,* Proceedings of the 1st Asian Conference on Materials (Beijing) May 1998; 68–72.

24. Yang, M.; Liu, Z. *Advantage of Rare Earth Chemical Heat Treatment,* Proceedings of Carburizing and Nitriding, ASSM, 1995; 120 pp.

25. Mayan, J.; Yan, J.; Liu, Z. *Numerical Simulation of Carbon Concentration Profiles in Case Layer of the Steel ZORE-Carburized in Multipurpose Furnace with Drip Feed,* Proceedings of the II CTPM and CS, Shanghai, International Federation for Heat Treatment and Surface Engineering, March 2000; 310 pp.

26. Geller, A.L.; Lozhushnik, L.G. Contact Fatigue Limit of Carburised 25Kh2GNTA Steel. Met. Sci. Heat Treat. (USSR), No. 6, June 1968; 474 pp.

27. Source: Treat All Metals Inc. Website: http://www.treatallmetals.com/gas.htm, November 4, 2002.

28. Totten, G.E.; Dakins, M.E.; Jarvis, L.M. How H-factors can be used to characterize polymers. Journal of Heat Treating December 1989, *21* (12), 28–29.

29. Grum, J.; Božiè, S.; Lavriè, R. Influence of mass of steel and a quenching agent on mechanical properties of steels. In 18th Conference, Heat Treating, Including the Liu Dai Memorial Symposium; Wallis, R.A., Walton, H.W., Eds; ASM Int. Materials Park Ohio, USA, October 1998. First Printing, April 1999; 645–654

30. Kobasko, N.I.; Totten, G.E.; Webster, G.M.; Bates, C.E. Compression of cooling capacity of aqueous poly (alkylene glycol) quenchants with water and oil. In 18th Heat Treating Society Conference Proceedings; Walton, H., Wallis, R., Eds.; ASM International: Materials Park, OH, 1988; 559–567.

31. Dakins, M.E.; Bates, C.E. Estimating quench severity with cooling curves. Journal of Heat Treating, April 1992, *24* (4), 24–26.

32. Totten, G.E.; Dakins, M.E.; Heins, R.W. Cooling curve analysis of synthetic quenchants—a historical perspective. Journal of Heat Treating 1988, *6* (2), 87–95.

33. Totten, G.E.; Sun, Y.; Webster, G.M.; Jarvis, L.M.; Bates, C.E. Quenchants selection. In *18th Heat Treating Society Conference Proceedings*; Walton, H., Wallis, R., Eds.; ASM International: Materials Park, OH. 1998; 183–191.

34. Peyre, J.P.; Tournier, C. *Choix des Traitements Thermiques Superficiels*; Center Technique des Industries Mécaniques: Paris, 1985.

35. Przyłęcka, M. Materiałowo-technologiczne aspkty trwałościżysk tocznych., Politechnika Poznańska, Seria Rozprawy Nr202, Poznań, 300 str, 1988.

36. Przyłęcka, M. The modeling of structure and properties of carburized low-chromium hypereutectoid steels. *Journal of Materials Engineering and Performance*; ASM International, Materials Park: OH, Vol. 5 (2), 165–191.

37. Lee, H-Y. Stress measurement for retained austenite phase in iron and steel by X-ray diffraction. Journal of the Korean Institute of Metals and Materials (South Korea) Feb. 1996, *34* (2), 150–157.

38. Parrish, G. "Chapter 5: Influential microstructural features" and "Chapter 6: Core properties and case depth." *Carburizing: Microstructure and Properties*; ASM International: Metals Park, OH, 1999; 99–170.

39. Burakowski, T.; Wierzchoń, T. Surface Engineering of Metals. *CRC Series in Materials Science and Technology*; CRC Press: New York, 1999.

40. Murai, N.; Takayama, T., et al. Effect of phosphorous and carbon segregation and grain size on bending strength of the carburised and induction hardened steel. Journal of the Iron and Steel Institute of Japan Mar 1997, *83* (3), 215–220.

41. Moon, W.J.; Kano, C.Y.; Suno, J.M. A study on the formation of retained austenite and tensile properties in Fe-Mn-Si-P steel. Journal of the Korean Institute of Metals and Materials Mar 1997, *35* (3), 297–304.

42. Tikhonov, A.K.; Palagin, Y.U.M. Method of testing gear wheels in impact testing. Metal Science and Heat Treatment May 1995, *36* (11–12), 655–657.

43. Balter, M.A.; Dukarevics, J.S. Vliyaniye kachyestva matyeriala na nadyeznost I dolgoviechnost zubchatih kolyes. Metallovedenie i Termicheskaya Obrabotka Metalloy 1985, *7*, 50–53.

44. Gulalev, A.P.; Serebriusikov, L.N. *Method of testing gear wheels in impact testing.* Vliyanie raznozyernistosti na myehanichyeskiye svoystva stali 18H2N4MA. Metallovedenie i Termicheskaya Obrabotka Metalloy 1977, *4*, 2–5.

45. Mendiratta, M.G.; Sasser, J.; Krauss, G. Effect of dissolved carbon on microcracking in martensite of an Fe–1.39%C alloy. Metal Transactions 1969, *62*, 351–353.

46. Grange, R.A. On the nature of microcracks in high-carbon martensite. Transactions of American Society for Metals 1969, *62*, 1024–1027.

47. Apple, F.A.; Krauss, G. Microcracking and fatigue in carburized steel. Metallurgical and Materials Transactions 1973, *4*, 1195–1200.

48. Krauss, G. The microstructure and fatigue of carburized steel. Metallurgical and Materials Transactions 1978, *9A*, 1527–1535.

49. Przyłęcka, M.; Gęstwa, W. The modeling of residual stresses after direct hardening of carburized and carbonitrided low-chromium hypereutectoid steels (£H15), "MAT-TEC 97—*Analysis of Residual Stresses from Materials to Bio-Materials,*" *IITT International*; 1997; 117–124.

50. Lesage, J.; Chicot, D.; Przylecka, M.; Kulka, M.; Gestwa, W. Role du chrome sur la cementation Hyper-Austenitique d'un acier a roulement. Traitement Thermique 1994, *276*, 42–46.

51. Przyłęcka, M.; Gęstwa, W. In *The modelling of structure and properties of carburizing or carbonitriding layers, as well as hardening in different quenching mediums.* 20th ASM Heat Treating Society Conference Proceedings, 9–12 October 2000; ASM International: St. Louis, MO, 2000; 624-634.

52. Vinokur, B. The composition of the solid solution, structure and contact fatigue of case hardened layer. Metallurgical and Materials Transactions A May 1993, *24*, 1163–1168.

53. Kern, R.F. Supercarburising. Journal of Heat Treatment Oct 1986; 36–38.

54. Gyulikhandanov, E.L.; Khoroshailov, V.G. Carburising of heat resistant steels in a controlled endothermic atmosphere. Metal Science and Heat Treatment (USSR) Aug 1971, *13* (8), 650–654.

55. Wang, J.; Qia, L.; Zhau, J. In *Formation and properties of carburised case with spheroidal carbides,* International Congress on 5th Heat Treatment of Materials Proceedings, Budapest, Hungary, International Federation for the Heat Treatment of Materials (Scientific Society of Mechanical Engineers), 20-24 1986; 1212 pp.

56. Miyasaka, Y.; Pat. J. H7-188738.

57. Miyasaka, Y. U.S. Patent 592840, January 14, 1997.

58. Miyasaka, Y. EP 0687 739 B1, June 14, 1995.

59. Rengstorff, G.W.P.; Bever, M.B.; Floe, C.F. The carbonitriding process of case hardening steel. Transaction of American Society for Metals 1951, *43*, 342–371.

60. Luty, W. The parts of LH15(52100) steel carbonitrided in atmosphere with liquid of organic compounds, The work of bearing industry no. 1/23/, Poland, 1972.

61. Vasilewa, E.V.; Sawiceve, C.H.; Krjukowa, J.V. Porysenie iznosojkosti stali SCH15 ionnoj implantacjej. Metalowedienie i Termiceskaja Obrabotka Metallov 1987,(1), 59–62.

62. Dawes, C.; Cooksat, R.S. Surface treatment of engineering components. *Heat Treatment of Metals. Special Report 95, 77, 92*; The Iron and Steel Institute, 1966.

63. Tauscher, H. In *Relationship between carburised case depth stock thickness and fatigue strength in carburised steel,* Symposium on fatigue Damage in Machine Parts, Prague, 1960. translation BISI 11, 340.

64. Weigand, H.; Tolasch, G. Fatigue behaviour of case hardened samples. Translation BISI 6329 from Härterei Technische Mitteilungen déc. 1962, *22* (4), 330–338.

65. Parrish, G. The influence of microstructure on the properties of case carburised components: Part 6. Core properties and case depth. Heat Treatment of Metals 1977, *2*, 45–54.

66. Lacoude, M. Propriétés d'emploi des aciers de cémentation pour pignonerie. Aciers Spéciaux 1970, (12), 21–29.

67. Beumelburg, W. Comportement d'éprouvettes cémentées présentant divers états superficiels, des teneurs en carbone variables en surface, lors d'essais de flexion rotative, flexion statique et resilience, Exposé présenté á la réunion ATTT 26/3/75.

68. Parrish, G. The influence of microstructure on the properties of case hardened components: Part 4. Retained austenite. Heat Treatment of Metals 1976, *4*, 101–109.

69. Champin, B. Commentaires sur la conférence de M. BEUMELBURG. Réunion ATTT 26/3/75.

70. Sheehan, J.P.; Howes, M.A. *The effect of case carbon content and heat treatment on the pitting fatigue of 8620 steel SAE 720 268*; Automotive Engineering Congress: Detroit, MI, January 10–14, 1972.

71. Razim, C. Einfluss des Randgefüge einsatzgehärteter Zahnräder auf die Neigung zur Grübehenbildung. Härterei Technische Mitteilungen 1974, *22 (Heft 4)*, 317–325.

72. Champin, B.; Seraphin, L.; Tricot, R. Effets comparés des traitements de cémentation et de carbonitru-ration sur les propriétés d'emploi des aciers pour engrenages. Mémoires Scientifiques. Revue de Métallurgie 1977; 77–90.

73. Wyszykowski; Preignitz, H.; Gozdzik, E.; Ratliewcs, A. Influence de l'austénite résiduelle sur quelques propriétés de l'acier cémenté. Revue de Métallurgie juin 1971, *68* (6), 411–422.

74. Razim, C. Influence de l'austénite résiduelle sur la résistance mécanique d'éprouvettes cémentées soumises á des efforts alternés. Revue de Métallurgie 69,147–157.

75. Sagaradze, V.S. Effect of heat treatment on the properties of high carbon alloyed steels. Metal Science and Heat Treatment (USSR) December 1964, (12) 720–724.

76. Przyłęcka, M. In *The effect carbon on utility properties of cemented bearing steel.* International Congress on 5th Heat Treatment of Materials Proceedings, Budapest, Hungary, 20–24 October 1986; International Federation for the Heat Treatment of Materials (Scientific Society of Mechanical Engineers), 1986; 1268–1275.

77. Funatani, K.; Nakamura, N. The impact Strength of carburized and Hardened parts. Toyota Technical Review 1965, *17* (2), 146–153.

78. Funatani, K.; Nakamura, N. The effects of case depth and core hardness on the fatigue strength of carburized and hardened steels. Toyota Technical Review 1966, *17* (4).

79. Funatani, K. Relation between hardenability of steels and distortion of differential gears. Toyota Technical Review 1966, *18* (1).

80. Funatani, K. Fatigue and impact strength of carburized chrome molybdenum steels. Transactions of Japan Institute of Metals 1968, *9*, 1025–1031.

81. Funatani, K.; Noda, H.; Tsuzuki, Y. Fracture surface appearance of various alloys with electron microscope Report Number 1, Toyota Technical Review 1970; Vol. 21, No. 4

82. Funatani, K. In *The influence of residual stress on fatigue strength of carburized gears*; Heat Treating—Proceedings of the 20th Conference, Funatani, K. Totten, G.E. Eds.; ASM International: Materials Park, OH, 9–12 October, 2000; Vol. 2, 418–425.

83. Funatani, K. Einfluss von Einsatzhaertungstiefe und Kernhaerte auf de Biegedauerfestigkeit von Aufgekohlten Zahnraedern. Härterei Technische Mitteilungen 1970, *25* (2), 92–97.

84. Funatani, K.; Noda, F. The change of residual stress of case hardened steels during fatigue tests. *Proceedings Symposium on X-ray Study of Materials Strength, SMSJ*; 1967; 87–91.

85. Funatani, K.; Noda, F. On the residual stress of carburized steel and their fatigue strength. *Proceedings of the 7th Symposium on X-ray Study of Materials Properties, SMSJ, Kyoto*; July 1968; 76–79.

86. Funatani, K.; Noda, F. The influence of residual stress on fatigue strength of carburized hardened steels. Journal of SMSJ 1968, *17* (183), 1124–1128.

87. Hisamatsu, S.; Kanazawa, T. Improvement of carburized gear strength by shot peening. Journal of JSAE 1987, *41* (7), 722–728.

88. Aihara, H. Shot peening methods for gears. JSME, Symposium on New Manufacturing Technologies for Gears, April 1992; 125 pp.

89. Namiki, K. Recent development of case hardening technology and materials. Journal of JISI 1994, *80* (5), 233–239.

90. Tanaka, H.; Kobayashi, T., et al. Effects of alloying elements and shot peening on impact fatigue strength of carburized steels. Journal of JISI 1993, *79* (1), 90–97.

91. Ueda, N. The effect of fatigue property on fine particle peening. Nihon Parkurzing Technical Report, 2000, (12), 73–81.

92. Totten, G.E.; Garsombke, G.R.; Pye, D.; Reynoldson, R.W. Chapter 6—Heat treating equipment. In *Steel Heat Treatment Handbook*; Totten, G.E., Howes, M.A.H., Eds.; Marcel Dekker, Inc.: New York, NY, 1997; 293–481.

93. Turbalter, M.A.; Turovskh, M.L. Résistance of case hardened steel to contact fatigue. Metal Science and Heat Treatment March 1966; (3) 177–180.

94. Diament, A.; El Haik, R.; Lafont, R.; Wyss, R. Tenue en fatigue superficielle des couches carbonitrurées et cémentées en relation avec la répartition des contraintes résiduelles et les modification du réseau cristallin apparaissant en cours de fatigue. Traitement Thermicque, 1974; (87), 87–97.

95. Shepelyakovskii, K.Z.; Kal'ner, V.D.; Mikonov, V.F. Technology of heat treating steel with induction heating. Metal Science and Heat Treatment, November 1970, (11), 902–908.

96. Robinson, G.H. The Effect of Surface Condition on the Fatigue Resistance of Hardened Steel. *Fatigue Durability of Carburised Steel 11.46*; American Society for Metals, 1957.

97. Beumelburg, W. The effect of surface oxidation on the rotating bending strength and static bending strength of case hardened specimens. Härterei Technische Mitteilungen oct. 1970, *25* (3), 191–194.

98. Parrish, G. The influence of microstructure on the properties of case carburised components. Heat Treatment of metals 1976, *2*, 49–53.

14

Design of Nitrided and Nitrocarburized Materials

Michel J. Korwin, Christopher D. Morawski, and George J. Tymowski
Nitrex Metal, Inc., St. Laurent, Quebec, Canada

Witold K. Liliental
Nitrex Metal Technologies, Inc., Burlington, Ontario, Canada

I. INTRODUCTION

Nitriding is one of diffusion treatments applied to the surface of machine parts, tools, and other metallic objects to enhance their surface hardness and to improve a number of other useful properties. Although the reactions between nitrogen and iron at elevated temperatures were already observed in the 19th century, a surge of interest in the possible hardening of ferrous surfaces, as well as in resulting industrial applications, occurred only in the early 1900s [1–7]. In the beginning of industrial implementation, the nitrided layers were obtained in ammonia atmospheres, in the course of a long process, on parts manufactured from special nitriding steel [4]. Because of the long duration and high aggressiveness of the process, a thick, porous, and brittle zone of iron and alloy nitrides formed at the surface, which had to be removed by mechanical means (grinding). Over the years, the process was adapted to other materials and substantially shortened.

Salt bath nitriding in molten cyanides was first described in 1929 [8]. Development in this area allowed one to obtain, in a short process (up to 3 hr), hardened layers that did not require grinding away any brittle zones. This process was applied to carbon, alloyed, and high-speed tool steels.

In the 1930s, the possibilities of nitriding in the plasma of a glow discharge were already studied. In the 1940s, two- and three-stage gas nitriding processes were developed [9], with varying ammonia dissociation rates, reducing the thickness and brittleness of the compound layer, thus improving the useful properties of treated parts. In the 1960s, the practice of diluting ammonia with nitrogen or with dissociated ammonia was begun. As another development, the first patent for the nitrocarburizing in gas was granted in 1961 [10].

The introduction of the concept of nitriding potential K_N [11,12], representing the nitriding capability of an atmosphere, began the period of development of controlled gas nitriding in NH_3–N_2 and NH_3–dissociated NH_3 mixes. Between 1970 and 2000, a number of further modifications of gas, salt bath, and plasma nitriding processes have appeared, aiming at improving both the properties of treated materials, as also the technological advantages of the process itself. An overview of principal variations of nitriding can be found in Sec. II.C of this chapter (see Table 3).

II. NITRIDING AS A COMPETITIVE OPTION OF SURFACE HARDENING

In its applications, nitriding has to compete with other surface-hardening methods. For comparison, three main groups of these methods are shown in Table 1.

Table 1 Main Groups of Surface-Hardening Processes

Group	Description	Examples
1	Methods that do not introduce additional elements into the surface	Flame hardening Induction hardening Laser hardening
2	Methods depositing a coating on top of the original surface of a part	Plating, e.g., chrome CVD PVD
3	Methods in which additional elements are diffused into the surface and modifications take place at and under the original surface	Carburizing Carbonitriding Nitriding Nitrocarburizing

The economic benefits of nitriding are not always obvious. In particular, process duration is often perceived as a negative factor. However, in many cases the benefits are revealed when an evaluation based on a complete cost analysis is carried out, comparing a manufacturing process including nitriding against any of the competitive options.

A. Groups of Surface-Hardening Methods

1. Group 1—Surface Heat Treatment

These methods can be used only on steels with sufficient carbon to harden by quenching. In one method (*flame hardening*), the steel surface is heated with a flame and subsequently quenched in a suitable medium. Another method, *induction hardening*, utilizes eddy currents, induced in the steel surface by a high-frequency coil. The depth of their penetration is inversely proportional to the frequency of the current in the coil. In this method, the temperature of the affected zone rises rapidly, reaching austenitization levels within seconds. The part is quenched by dropping into a tank or by spraying the surface with a quenching medium. A third method, *laser heating*, provides very fast localized heating that can be applied to larger surfaces by appropriate scanning movement of the laser gun. Quenching occurs through rapid heat transfer to the cold substrate.

2. Group 2—Hard Coatings

The oldest known representative method belonging to this group is *electroplating*, mainly chrome. The disadvantage of wet plating methods is waste disposal and neutralization in order to meet today's high ecological standards. Modern, high-tech methods include chemical vapor deposition (CVD) and physical vapor deposition (PVD). The result of such deposition is referred to as a *plating* or *coating*.

In chemical vapor deposition, hard coatings are produced by chemical reactions at elevated temperatures, 800–1100°C (1472–2012°F) from gaseous media. The disadvantage is high temperature, which may cause transformations in the core and distortion. Low-temperature CVD coatings for wear protection are successfully carried out by some providers but, generally, this is not yet a mass-scale technology.

In physical vapor deposition, metal vapor is produced and directly deposited on the steel substrate, in the presence of a reactive gas, to form a hard coating. The process is characterized by low substrate temperatures, down to 200°C (392°F), preventing loss of core hardness. A typical PVD application is the titanium nitride (TiN) coating, used for diverse cutting tools and components. It provides an attractive golden-yellow appearance. Limiting factors of this technique are high capital equipment and operational costs. Gaining popularity are multilayered, so-called *duplex coatings*, which comprise hard PVD or CVD layers over a nitrided substrate. The nitrided layer provides a high hardness support for the thin (1–3 µm) coating, resulting in exceptional wear resistance. Utilization of duplex coatings has opened a new field of application for nitriding.

3. Diffusion Layers

In these processes, various elements may be released by an active medium, absorbed by the surface and diffused into the material of the part. The resultant surface modifications are termed hardened *cases* or *layers*. The most popular processes are carburizing, nitriding, and their combinations, i.e., carbonitriding or nitrocarburizing, depending on processing conditions.

The most important features characterizing the main methods of surface hardening are summarized in Table 2.

Table 2 Advantages and Limitations of Four Case-Hardening Methods

Method ⟹	Flame hardening	Induction hardening	Carburizing	Nitriding
Material compatibility	Only steels above 0.25% C	Only steels above 0.25% C	Low- to medium-carbon steels. No high alloy grades	Practically all steel grades
Temperature range	730–850°C (1346–1562°F)	730–950° (1346–1742°F)	850–950°C (1562–1742°F)	450–600°C (788–1112°F)
Technique	Primitive	Sophisticated	Sophisticated	From primitive to highly sophisticated (several methods and many generations of equipment in use)
Advantage	Simple and cheap	Very rapid	Well controlled, capable of yielding deep cases	Low temperature minimizes distortion. No requirements to harden and temper or finish grind after nitriding
Disadvantage	No control	Expensive, limited to simple geometrical shapes, incapable of hardening internal bores	High temperature may cause distortion. Need to quench and temper after carburizing	Process times longer than in carburizing. Traditional nitriding encumbered with many control problems (e.g., brittle white layer)

B. Carburizing and Carbonitriding

Only steels containing minimum 0.25% C can be practically hardened by austenitization and quenching. Enrichment of the surface with carbon makes the subsurface zone of the steel eligible to harden upon subsequent quenching. Carburizing is carried out at 850–980°C (1562–1796°F), with higher temperatures applicable depending on the steel characteristics and available equipment. The case depth is a function of time and temperature. Carbon diffuses in the direction of the core and hardness shows a gradual decline from the surface inward.

Presently, it is accomplished in a gaseous medium with control of surface carbon concentration, determining case properties. Oversaturation causes precipitation of brittle carbides along grain boundaries under the surface. Carbon concentration profile is critical and modern carburizing systems feature sophisticated and precise carbon concentration control methods. The disadvantage is high temperature and the need to quench, causing distortion and the necessity of correcting it by costly mechanical finishing. Adding ammonia to a carburizing atmosphere promotes a faster simultaneous diffusion of carbon and nitrogen into the austenitic structure of the treated surface. This allows one to lower the temperature to 820–900°C in a process called

carbonitriding. This process is applied to medium-carbon steel for case depths thinner than those obtained in carburizing. Quenching is still necessary, to harden both the diffusion layer and the core.

C. Nitriding and Nitrocarburizing

1. General

Nitriding is typically carried out at 420–630°C (788–1166°F), this range most often narrowed down to 500–590°C (932–1094°F). Temperatures significantly lower than in carburizing, and absence of volume transformations in the core, substantially minimize distortion. See also Sec. III.F, "Austenitic Nitriding and Nitrocarburizing."

Nitriding is often carried out on hardened and tempered steel and core hardness depends on the tempering temperature. In order not to further temper the core, nitriding temperature is selected lower than that of preceding tempering. Figure 1 compares typical hardened case profiles obtained by induction hardening, carburizing, and nitriding.

Nitrided cases are sometimes compared to those obtained by carburizing without understanding the differences between the two processes. This has often led to demands for nitrided cases of depths similar to those used in carburizing. Obviously, extremely long

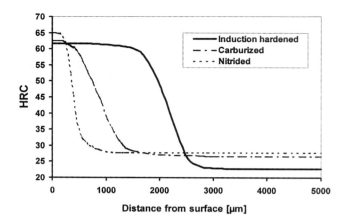

Figure 1 Typical hardness profiles obtained by different surface treatments.

times required to produce such layers would render the process impractical. Experience shows that a nitrided layer of 30–40% of the effective carburized case depth can successfully replace carburizing in many applications requiring good wear resistance.

There are three main nitriding methods, i.e., gas, salt bath, and plasma. Gas nitriding, being the earliest method applied in industry, has been used for decades in what is now called "traditional" or "conventional" form. Because of poor or even nonexistent process control of the early days, it has often been associated with the formation of an oversaturated, thick, and brittle white layer, which had to be removed. Since the 1960s, steady improvement in the understanding of the nitriding process has led to the development of gas nitriding controlled by the nitriding potential (see Sec. III).

In the quest for improved properties of treated parts, and looking for a more economical and reliable process, several modifications of nitriding have been developed over the years. A brief review of these methods is given in Table 3.

2. Conventional Gas Nitriding

Traditionally, control of the gas nitriding process has been exercised by monitoring the ammonia dissociation rate, which, in turn, depends both on temperature and on atmosphere flow. Although there are still many units in service in which control is effected manually, more modern systems rely on computers to maintain either a constant flow of the atmosphere or a preset dissociation rate. Neither of these systems offers full, comprehensive

Table 3 Types of Nitriding

| | | Nitriding medium | |
Process	Definition	State	Example
Nitriding	Thermochemical treatment to enrich the surface zone of parts with nitrogen (depending on nitriding conditions, a diffusion layer, or a nitride plus diffusion layer may be formed).	Gas Plasma	NH_3 $NH_3 + N_2$ (or H_2) $N_2 + H_2$
Nitrocarburizing	Thermochemical treatment, to enrich the treated surface with nitrogen and carbon. Both elements are present in the outer nitride layer. The underlying diffusion layer is enriched mostly with nitrogen. The process may take place below 590°C (ferritic nitrocarburizing) or above (austenitic nitrocarburizing).	Molten salt bath Gas Plasma	Cyanide/cyanate salt mixes $NH_3 + CO/CO_2$ $N_2 + H_2 + CH_4$
Oxynitriding	Thermochemical treatment to enrich the treated surface with nitrogen and oxygen, with the formation, depending on conditions, of a diffusion layer, or of a nitride plus diffusion layer.	Gas	$NH_3 + H_2O$ (or air)
Sulfonitriding	Thermochemical treatment to enrich the treated surface with nitrogen and sulfur, with the formation of a nitride layer containing these elements. The underlying diffusion layer is enriched mostly with nitrogen.	Gas (Plasma)	$NH_3 + H_2S$ (or sulfur vapor)
Sulfonitrocarburizing	Thermochemical treatment to enrich the treated surface with carbon nitrogen, and sulfur, with the formation of a nitride layer containing these elements. The underlying diffusion layer is enriched mostly with nitrogen.	Molten salt bath (Plasma)	Cyanide/cyanate salt mixes + sulfur

Based on Table 1 in Chatterjee-Fischer, R. Härterei-Tech. Mitt. 1983, 38, 35.

control of the nitriding process. Dissociation rates can only be selected empirically to suit a particular application. Most important, same dissociation rate, but obtained from two different compositions of the nitriding atmosphere (because if diluting by a neutral gas) corresponds to two different nitriding capabilities.

3. Gas Nitriding Controlled by the Nitriding Potential

In the early 1970s, an advanced form of gas nitriding began to establish itself [11,12], resulting a decade later in fully developed industrial form [13]. This method harnessed the nitriding potential as the principal controlling process parameter. The nitriding potential determines the equilibrium concentration of nitrogen in the steel surface, which cannot be exceeded. It is an analogy to the carbon potential in carburizing and its proper control serves to prevent oversaturation and brittleness. Thus, processing conditions are directly linked to the concentration of nitrogen in the surface of the treated object, hence to layer properties and behavior.

4. Nitrocarburizing

Nitrocarburizing, sometimes specifically termed "ferritic nitrocarburizing" to distinguish it from the higher-temperature process of carbonitriding, is a fast expanding spin-off of nitriding. In the 1960s, gas nitrocarburizing was developed as a replacement for salt bath nitriding, which, in essence, as explained in Sec. III, introduces both nitrogen and carbon into the surface layer. The main advantage of simultaneous absorption of nitrogen and carbon lies in the accelerated growth of the nitride layer with a dominant epsilon (ε) structure, required in most applications where wear is a major requirement. Good performance of treated parts, combined with the possibility of running the process in conventional furnaces at a relatively low cost, have contributed to the rapid acceptance of the method.

Properties obtained by nitrocarburizing are, for a comparable layer thickness, similar to those obtained through nitriding. The results of some studies indicate that nitrocarburizing brings advantages in wear and/or corrosion resistance, superior to those of plain nitriding. Introducing carbon, and quenching, both increasing the percentage of the ε phase in the surface layer, appear to have a marked effect on the friction coefficient and on the wear rate. However, the available evidence is incomplete and a more systematic investigation is necessary. This concerns, in the first place, the benefits of a high ε-phase content in the compound layer. From experimental observations [13] it appears that modification of the ratio of C to N in the ε phase, i.e., reducing

the nitriding potential, can be used to reduce the amount of porosity in the superficial zone, with positive results in some applications (e.g., resistance to corrosion).

5. Oxynitriding

The observation that industrial-grade ammonia produces faster-growing nitrided layers on iron than the pure product dates back to the early years of the 20th century. This information remained irrelevant until gas nitriding began to develop qualitatively in the 1960s. Then came the recognition that preliminary oxidation of the surface, or addition of oxygen-bearing compounds (air, CO_2, H_2O, etc.) to the process atmosphere can accelerate the nitrogen intake and improve the uniformity of the nitrided layer [14]. Industrial heat-treating practice has taken advantage of preoxidation and CO_2 additions, without inquiring into the reasons for the observed phenomena. Only in recent years have studies begun to appear [15], looking for a better understanding of the mechanisms involved in various methods of surface activation.

Oxynitriding is carried out in atmospheres of ammonia with continuous additions of water vapor or of air. (Adding CO_2 brings the process into the category of nitrocarburizing.) It was claimed [16] that in carbon steel, oxynitriding produces comparatively thicker nitride layers and higher ε-phase contents, as determined by x-ray phase analysis. A beneficial effect of oxynitriding has also been found in the treatment of austenitic and ferritic stainless steels [17]. However, this process, widely used in East Germany before 1989, has found only very reluctant acceptance in other countries.

6. Other Modifications of Nitriding

In the course of development of improved wear properties of nitrided surfaces the additions of sulfur and phosphorus to the hardened layers have been investigated. This resulted in currently used processes such as *sulfonitriding* and *sulfonitrocarburizing*. These are modifications of gas nitriding, during which the atmosphere is supplemented with sulfur vapor or H_2S gas. As a result, finely dispersed iron sulfide become incorporated into the growing nitride layer, reducing adhesive wear on surfaces exposed to friction and allowing higher contact loads on all nitrided materials [18].

A similar improvement of wear properties is obtained through enrichment of surface layers with phosphorus in phosphonitriding. Addition of gaseous phosphorus to an ammonia atmosphere leads to the formation of Fe_3P and Fe_2P phosphides at the surface, resulting in an increase of hardness and of wear resistance [19].

III. GROWTH AND STRUCTURE OF THE NITRIDED LAYER

A. The Nitrogen Equilibrium

Every physical system, solid, liquid, or gas, has a natural tendency to reach *equilibrium*, representing its lowest state of energy. Imbalance between any two bodies is the driving force of physical movement, chemical reactions, exchange of energy, and transformation of state. In the case of nitriding, nitrogen will pass into the metal from a surrounding medium if its activity (a thermodynamical term) is higher than that of nitrogen in the metal itself. This will take place until the activity on both sides of the interface is equal. The state of equilibrium determines, for a given set of conditions (temperature, pressure, etc.), a certain surface concentration of nitrogen in the metal. Phases existing in conditions of equilibrium in the Fe–N system are shown in Fig. 2.

From the fact that enrichment of the metal by nitrogen from the surrounding medium is thermodynamically possible, it does not yet follow that nitriding will take place. In practice, nitrogen absorbable by ferrous alloys has to exist in atomic form, obtained in gas nitriding through catalytic cracking of ammonia at the steel surface. In plasma nitriding, positively charged ions strike the surface and through a series of collisions with sputtered Fe atoms, form nitrides that precipitate on the steel surface.

Kinetics of nitriding, i.e., the buildup of surface concentration, the formation and growth of a nitride layer, depends on several factors. These are initial microgeometry and structure of the superficial zone, presence or absence of surface impurities and films that can act as barriers, and characteristics of the nitriding medium, process parameters, e.g., temperature and

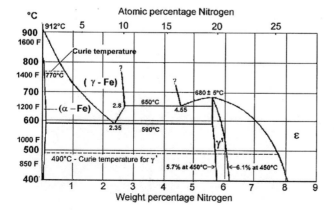

Figure 2 Fe-N phase diagram.

pressure. In no case can nitrogen concentration at the surface exceed the thermodynamically justified value.

B. Control of the Nitriding Process

Controlling the nitrogen activity in the medium enables the control of the concentration of nitrogen at the steel surface. The purpose of such control is to prevent exceeding acceptable levels. The efficiency of the salt bath is controlled by its chemical composition and corrected at intervals (usually once per work shift or once per day), by the addition of fresh salt and/or additive chemicals, to compensate for nitrogen depletion during the process. Many years' experience has led to practical knowledge of how to repeatedly obtain viable results, but the method does not lend itself to automatic control based on physical laws governing the process. In plasma nitriding, the activity of the medium is controlled by the chemical composition of the gas mix and by current density at the surface of the nitrided steel. The latter is a function of load surface area, and although an experienced operator can exercise effective control by making in-process adjustments, based on observation through a viewing port, no automated closed-loop, self-correcting control system has been developed for this method to date.

The method lending itself to automated closed-loop control is gas nitriding, because basic reactions taking place at the gas/steel interface are known and controllable.

Historically, gas nitriding has been controlled by the flow rate of the nitriding atmosphere, typically 100% raw ammonia (NH_3). It was observed that the faster the flow rate, the more severe the nitriding effect and vice-versa. This method was not accurate.

The first control method based on thermodynamics and used to this day is based on the measurement of ammonia *dissociation rate*.

Atomic nitrogen [N] is obtained through the partial catalytic cracking of ammonia, according to the reaction:

$$NH_3 = N + 3/2H_2$$

Dissociation rate represents the percentage of ammonia dissociated into hydrogen and nitrogen. The higher it is, the lower the surface nitrogen concentration, or, in other words, the "gentler" the nitriding activity of the atmosphere and hence, the weaker the nitriding response, in terms of white layer thickness and nitrogen concentration. Because dissociation rate decreases inversely to the atmosphere flow rate through the retort, the gas nitriding process is usually controlled by flow to yield a desired rate of dissociation, measured, e.g., by a *dissociometer* (Fig. 3). Placed in the furnace exhaust gas line, it can indicate the volume fraction of undissociated ammonia due to rapid absorption of the latter in water.

C. Nitriding Potential in Gas Nitriding

Control of dissociation rate does not always ensure obtaining fully predictable results, especially when mixed-gas atmospheres, e.g., diluted by nitrogen, are used. Two atmospheres of different initial composition may be controlled to achieve the same dissociation rate, but surface concentrations of nitrogen in the steel may differ. Differences in surface nitrogen concentration result in differences in layer thickness and properties. Insufficient control is the primary reason why, despite many advantages, nitriding to this day is viewed by some with reservations.

The modern, controlled, gas nitriding process employs pure ammonia or ammonia with one or two additive gases and is controlled not by the dissociation rate but by the nitriding potential, K_N. This thermodynamic quantity has been harnessed to automatically control the gas nitriding process [13], and since the late 1980s has been implemented in industrial gas nitriding

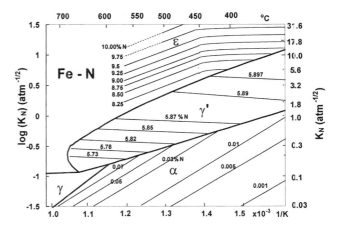

Figure 4 The K_N-T equilibrium (Lehrer) diagram, updated by L. Maldzinski. (From Ref. 20.)

systems. This modern method involves fully integrated and automated control of all processing parameters and functions needed to run the process, including entirely automated safety procedures.

The nitriding potential is determined by the formula:

$$K_N = \frac{p_{NH_3}}{(p_{H_2})^{3/2}}$$

where p_{NH_3} is the partial pressure of ammonia and p_{H_2} the partial pressure of hydrogen in the gas atmosphere exiting the furnace retort. There is a direct correlation between the nitriding potential and the concentration of nitrogen in the steel surface, assuming that a thermodynamical nitrogen equilibrium exists at the surface.

Progress in understanding the thermodynamics of the nitriding process led to an updated Lehrer's equilibrium diagram, linking the nitriding potential with temperature and nitrogen concentration for pure iron [20].

This diagram in Fig. 4 shows the areas of existence of nitride phases and lines of equal nitrogen concentrations. To each combination of nitriding potential and temperature, a concentration of nitrogen in iron, in equilibrium with the atmosphere, can be ascribed. According to the same source, shifts of phase boundaries in the diagram caused by most alloying elements (with the exception of high carbon and nickel contents) are relatively small and the diagram can be practically used for most low alloyed steels.

D. Kinetics of the Nitriding Process

The nitrided layer is formed by the diffusion of nitrogen atoms into the steel, driven by the difference in concentrations at the surface and below it. As nitrogen diffuses

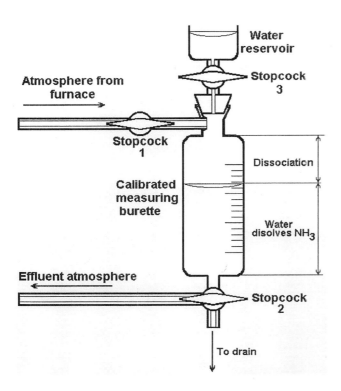

Figure 3 Burette dissociometer for measuring the ammonia dissociation.

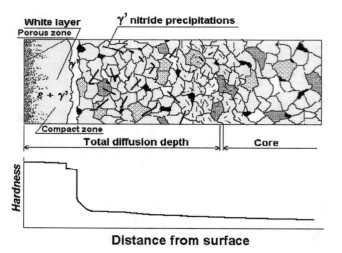

Figure 5 Typical features of the nitrided layer on iron and/ or non-alloyed steels (schematic).

inward, it forms a concentration profile, dropping in the direction of the core. At low concentrations, it forms a solid solution in α-Fe. When the concentration exceeds maximum solubility limits, the first nitride to form in pure iron is γ' (Fe$_4$N). Isolated nitrides grow and coagulate, finally forming a continuous compound or "white" layer, so called because it resists etching by common reagents like Nital. This indicates that it has a higher chemical resistance. Higher nitrogen concentrations lead to the formation of the ε nitride—Fe$_2$N$_{(1-x)}$ where $0 < x < 0.5$. On some alloyed steels, the sequence of nitride phase formation may differ, e.g., the ε nitride may be the first to form.

Seen as a function of time, the kinetics of growth of the nitrided layer depends on parameters affecting diffusion rates: nitriding potential (determining the nitrogen concentration at the surface) and temperature (determining diffusion coefficients in the different phases).

Growth models exist [21] that allow simulating the development of a nitrided layer in given conditions. Simulations can predict the phase composition of the layer, the thickness and hardness of its particular zones, and nitrogen concentration profiles, depending on processing conditions. They are easier to develop for unalloyed low-carbon steel. Growth kinetics of the nitrided layer on alloyed steels is affected in a complex manner by alloying elements and by higher carbon content. Nitride-forming elements strongly reduce the depth of nitrogen penetration into the matrix. Other elements (such as Ni or Si), which are non-nitride forming, by remaining free in the solid solution affect the activity and diffusion of nitrogen through it.

E. Formation and Microstructures of Nitrided and Nitrocarburized Layers

1. Microstructure and Features of Nitrided Layers on Nonalloyed Steel and on Cast Iron

The presence of the white layer is the main factor determining the properties obtained through nitriding of iron and unalloyed steels. The depth of nitrogen diffusion below the white layer is manifested, upon slow cooling, by the precipitation of nitride needles visible under the microscope, but with a very limited hardening effect. Typical features of the nitrided layer on pure iron and/or nonalloyed steels are shown schematically in Fig. 5. Examples of applications of nitrided nonalloyed steel include automotive door sectors, seat slide rails, clutch hubs, and plates.

The mechanism of layer formation and growth on nonalloyed cast iron is similar to that on nonalloyed steel. With absence of nitride-forming elements, the hardness response is practically limited to the white layer. The microhardness profile below the white layer is irregular and depends on the type of grain (ferrite or pearlite) where the indentation is made. Features of the nitrided layer on cast iron are shown in Fig. 6. Examples of nitrided cast iron applications are differential housings, compressor journals, and shafts.

2. Microstructure and Features of Nitrided Layers on Alloyed Steel

Nitride forming elements in steel, e.g., Cr, Mn, Ti, V, and Al, harden the white layer and in particular the

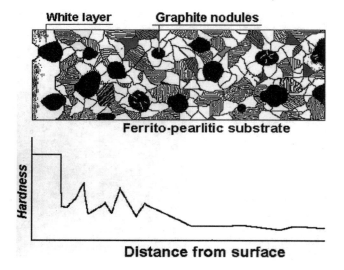

Figure 6 Typical features of the nitrided layer on cast iron (schematic).

diffusion zone below, creating a profile in which hardness drops with decreasing nitrogen concentration.

Figure 7 shows typical features of the nitrided layer on alloyed steel. On most grades, when etched, the diffusion layer appears as a darker zone, eventually blending into the substrate. On alloy steels, the white layer is primarily, although not solely, responsible for wear resistance. Its other role is that of corrosion protection. The diffusion layer provides support for the white layer, and a transition between the white layer and the core. Its hardness and depth also have a significant effect on fatigue properties. Fatigue strength also depends on core hardness, which is determined by prior heat treatment. Surface hardness and case depths of nitrided layers may vary for different alloy steels. Applications of nitrided alloyed steels include gears, pinions, crankshafts, rocker arms, as well as tooling, e.g., forging and extrusion dies.

3. Microstructure and Features of Nitrided Layers on Stainless Steels

Nitrided layers on stainless steels exhibit sharp, well-defined borders between the diffusion case and core. Core hardness on austenitic stainless steels, which do not harden by heat treatment, is independent of the temperature of the nitriding process.

The nitrided layer on austenitic stainless steels is shown in Fig. 8. Hardness ranges from 800 HV to ca. 1200 HV. On martensitic stainless steels (Fig. 9), the nitrided layer exhibits similar features but core hardness may vary, depending on preceding heat treatment. A

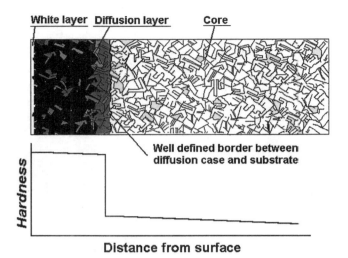

Figure 8 Typical features of the nitrided layer on austenitic stainless steels (schematic).

transition zone exists between the diffusion case and the substrate. Examples of application:

Austenitic stainless steels: engine valves (exhaust), expanders
Martensitic stainless steels: piston rings, injection nozzles, aerospace gearing
Precipitation-hardening stainless steels (components of aircraft landing gear assemblies)

Nitriding at temperatures above 450°C (842°F) causes chromium nitrides to precipitate in the nitro-

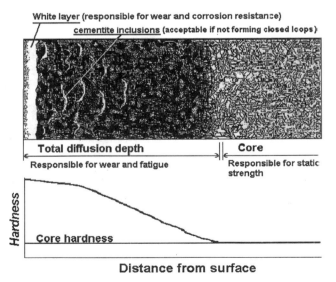

Figure 7 Typical features of the nitrided layer on alloyed steels (schematic).

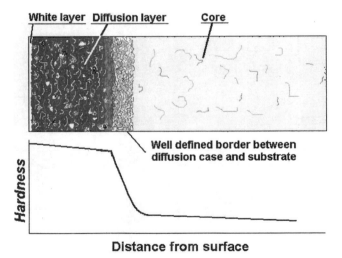

Figure 9 Typical features of the nitrided layer on martensitic stainless steels (schematic).

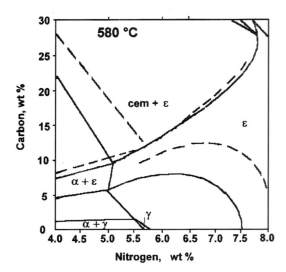

Figure 10 Part of the ternary Fe-N-C phase diagram at 580°C (1076°F). (From Ref. 23.)

gen-enriched surface zone. This reduces the chromium content in the matrix, which as a result loses its corrosion resistance (its capability to create a protective passive surface film in oxidizing environments). Recent research has shown that hardening without loss of corrosion resistance is possible when gas or plasma nitriding is carried out at temperatures below 450°C and down even to 300°C (572°F—in plasma) [22]. In these conditions the outer zone of the nitrided layer consists of so-called S-phase, also called 'expanded austenite'. This is supersaturated austenite, free from nitride precipitation, appearing as a bright zone under the microscope. The surface hardness in austenitic steel can reach HV01 = 1200 – 1700. The corrosion resistance is the same as in non-nitrided steel. The downside of this approach is that the process has to be very long.

4. Microstructure and Features of Nitrocarburized Layers

Carbon can dissolve in the ferrite phase as well as in the ε nitride, converting it to carbonitride. As seen from a portion of the ternary Fe–N–C phase diagram in Fig. 10 [23], the presence of carbon expands the range of the ε domain to lower N contents. In consequence, it is easier to obtain compound layers with close to 100% of the ε phase, especially if quenching is applied at the end of the diffusion process, to avoid partial transformation of ε to γ'.

Gas nitrocarburizing is performed in an atmosphere containing ammonia, optionally diluted with nitrogen, with additions of carbon-bearing gases, such as exo-

thermic and endothermic atmospheres, CO_2, CO, or hydrocarbons, such as propane. Ammonia mixed with endogas (40 vol.% N_2, 40% H_2, 20% CO) represents a popular type of nitrocarburizing atmosphere. The interaction of CO_2 and of CO with ammonia leads to a multiplicity of chemical reactions among the atmosphere constituents and with the metallic substrate. The most significant of these are [24]:

- $2NH_3 \Rightarrow 2N + 3H_2$ nitriding reaction
- $H_2 + CO_2 \Leftrightarrow CO + H_2O$ the water–gas shift reaction, raising K_N with reduction of the H_2 content in the atmosphere
- $NH_3 + CO \Leftrightarrow HCN + H_2O$ where HCN is also a source of N and C
- $2CO \Rightarrow CO_2 + C$ carburizing reaction, which establishes the carburizing potential K_C, $K_C = (p_{CO_2})/(p_{CO_2})$
- $2NH_3 + CO_2 + H_2O \Leftrightarrow (NH_4)_2CO_3$ forming ammonium carbonate or
- $NH_3 + CO_2 + H_2O \Leftrightarrow NH_4HCO_3$ ammonium carbamate, both of which condense to solid precipitates on cooling

Other reactions can produce methane (CH_4). Controlling the nitrocarburizing process to maintain stable values of K_N and K_C is not easy, because both are interrelated but not in a manner that could be described by a mathematical formula [25]. With varying ammonia dissociation rate, nitriding and carburizing potentials change in opposite directions (see Fig. 11).

In unalloyed low-carbon steel, nitrocarburizing produces accelerated growth of the white layer (see Fig. 12a).

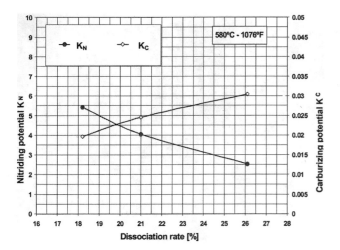

Figure 11 Variation of potentials K_N and K_C vs. ammonia dissociation rate in NH_3-N_2 atmosphere with 5% CO_2, at 580°C (1076°F).

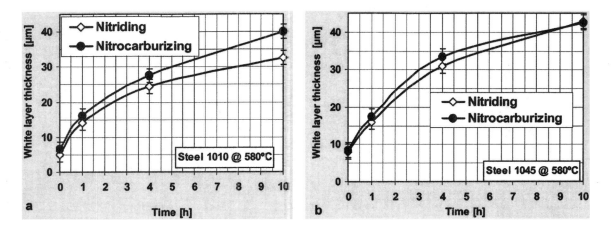

Figure 12 Comparison of growth kinetics of the compound layer between gas nitriding and gas nitrocarburizing (NH$_3$-N$_2$-CO$_2$ atmosphere), for steel grades 1010 (a) and 1045 (b), at 580°C (1076°F).

This effect of nitrocarburizing on growth kinetics is due to several factors: in the presence of oxygen-bearing compounds, hydrogen from the atmosphere is absorbed, increasing K_N in an uncontrolled atmosphere. Moreover, there is an activating effect of the same compounds on the metallic surface [26,27] (see also Sec. II.C.5, "Oxynitriding"). This accelerating effect of nitrocarburizing on compound layer growth is less apparent on medium-carbon (e.g., 1045) unalloyed (or alloy) steels (see Fig. 12b).

Nitrocarburizing can be carried out in the plasma (ion) process. The hydrogen–nitrogen atmosphere is, in this case, enriched with methane (natural gas), acetylene, or other hydrocarbons. In this case, similarly as in the case of salt bath nitriding, there is no direct control over the reaction mechanisms.

F. Austenitic Nitriding and Nitrocarburizing

The term "austenitic nitriding" or nitrocarburizing describes processes carried out in the temperature interval between the eutectoid (A$_1$) point of the Fe–N phase diagram (590°C, 1094°F) and the corresponding point in the Fe–C system (721°C, 1330°F). The increase in the process temperature raises the diffusion rate, allowing one to shorten process times or to obtain thicker layers. In the course of the process, under the outer compound layer, an intermediate austenitic zone is formed. Rapid cooling results in this austenitic layer being retained. It can be transformed into a hard martensitic structure by reheating to a temperature of about 300°C (572°F). Slower cooling produces in the intermediate zone a pearlite-like structure called braunite.

There is no doubt that compound layers on carbon steel can be grown faster in the austenitic range. It has

been demonstrated [28] that producing a 20-μm-thick compound layer on a 1035-type steel by salt bath nitriding requires 120 min at 580°C (1076°F), and, in the same bath, 50 min at 610°C (1130°F) and 40 min at 630°C (1166°F). At the same time, the increase in process temperature shifts the hardness profile to lower values and thus reduces the effective diffusion layer thickness. This negative effect on hardness is one reason why process temperatures above 630°C (1166°F) are rarely applied in practice. The other is the tendency for pores in the compound layer to form deep channels [28] facilitating the ingress of corrosive electrolytes.

There is a lack of data on the advantages of the presence of the intermediate layer. The reduction of the effective case depth (due to lowering of hardness) has an adverse effect on resistance to fatigue, therefore austenitic nitriding and nitrocarburizing are applied more often in circumstances where wear and corrosion resistance are the main requirements.

G. Effect of Pressure: High- and Low-Pressure Nitriding

The effect of altered pressure on the nitriding potential is the result of two opposing factors. First, because of the exponent on H$_2$ partial pressure in the formula for K_N, higher pressures reduce the K_N value, lower pressures raise it for the same ammonia/hydrogen ratio. On the other hand, according to Le Chatelier's law, higher pressure reduces the ammonia dissociation rate, yielding, in otherwise constant conditions, higher K_N values. In practice the second effect dominates; thus higher pressures produce a higher nitriding potential, and vice versa. Also, the availability of ammonia at the metal surface rises with higher pressure, allowing faster saturation of the initial solid solution. Nitriding above 0.2 MPa (above 2 bar of absolute pressure) was patented in the 1990s [29]. It has been reported [30] that especially in unalloyed steels, higher pressure (up to 12 bar) yields a substantially increased compound layer depth and an increase in porous zone thickness, as may be expected with a higher K_N and, consequently, higher nitrogen concentration. This is less pronounced in alloy steels such as H11.

Other reports [31] also indicate that the diffusion layer depth is increased by higher pressure. In Nitralloy 135M-type steel, for the same conditions of 6 hr at 540°C (1004°F) in a given atmosphere, increasing the pressure from 0.2 KPa (0.002 bar) to 30–50 KPa (0.3–0.5 bar) resulted in:

Reduced dissociation rate, from 35% to 25%
Increased thickness of the compound layer, from 0 to 15 μm (0 to 0.0006 in.)
Increased depth of the diffusion layer, from 0.19 to 0.31 mm (0.0076 to 0.0124 in.)

It has been claimed [32] that high-pressure nitriding facilitates depassivation of the surface of alloy steels. This, if proved, would enhance the growth kinetics and uniformity of nitrided layers on materials like stainless and heat-resisting steel.

At pressures lower than atmospheric, conversely, the effect of passive layers is more of a hindrance to the diffusion process and requires suitable methods of surface activation. One possibility is the introduction of an oxidant, e.g., nitrous oxide (N_2O), to promote activation of the metallic surface, improving the growth kinetics and uniformity of the nitrided layer. As an advantage, a lower partial pressure of hydrogen in the atmosphere weakens the interaction of this gas with the treated material, producing higher ductility of both the compound and diffusion layers [18].

H. Nitriding of Nonferrous Alloys

Besides iron, aluminum and titanium form the base of alloys that can benefit from surface hardening through nitriding.

Nitrogen is practically insoluble in aluminum. However, at elevated temperatures aluminum can react with nitrogen, forming a superficial layer of aluminum nitride (AlN), characterized by high hardness (HV1 about 1600). Because of the affinity to oxygen, the aluminum surface is normally covered with an oxide (Al_2O_3) film, which prevents diffusion of nitrogen. Therefore, the only method suitable for nitriding of aluminum and its alloys is the plasma process, in which a sputtering stage can be applied to remove preexisting oxides. Even then, the formation of oxides cannot be fully avoided, as the oxi-

dation limit lies at an oxygen partial pressure of 1.6×10^{-70} bar [33]. Also, a problem results from the large difference between thermal expansion coefficients of Al and AlN (Al: 23×10^{-6}, AlN: 5×10^{-6} 1/K. This produces very high compressive stresses in the nitride layer, easily leading to cracking and spalling in service. Another fact, complicating the application of nitriding is that to reduce the electrical resistance of the nitride layer (important in the plasma process), the process is usually run at temperatures close to 90% of the lower melting point of alloys. As a result, risk of local melting appears and the core strength is considerably decreased, unless a solution anneal and aging can be performed after nitriding.

Unlike the case of aluminum, the nitriding of titanium and its alloys belongs today to widely applied industrial practice. Compared with aluminum, titanium involves fewer difficulties. The thermal expansion and electrical resistivity of titanium and of the TiN are practically identical. The oxidation limit, at 2.3×10^{-33} bar, although still very low, is much easier to handle in industrial conditions.

The structures of nitrided layers depend on the type of the alloy and the process temperature. Pure Ti exists at lower temperatures in a hexagonal α structure, which is also stabilized by nitrogen. Depending on the composition, alloys can exhibit the α structure, a body-centered cubic β structure or an α + β mix. Typical high-strength Ti alloys belong to the latter category.

In the popular Ti–6Al–4V alloy, contact of the surface with atomic nitrogen (whether in a gas or plasma process) leads to the formation of a compound layer, composed of δ-TiN_{1-x} and ε-Ti_2N nitrides. Because they do not dissolve Al, the latter is displaced toward the diffusion zone, which contains an increased proportion of the α phase. With decreasing temperature, dispersed nitrides can be precipitated in this zone, contributing to a rise in hardness. The growth of the diffusion layer is dependent primarily on the temperature and duration of the process, regardless of whether this is gas or plasma nitriding. There is, however, an effect of the composition of the process atmosphere on the growth of the compound layer. An increase of the H_2 content in the plasma process and of NH_3 in the gas process accelerates the growth of the nitride zone [33].

The affinity of titanium to hydrogen causes an excess of this element in the surface layers of alloys treated in H_2-containing atmospheres, with detrimental effect on ductility. However, hydrogen can be removed through heating in vacuum or an inert-gas atmosphere, after nitriding.

Because of the relatively low diffusion coefficient of nitrogen in titanium, nitriding is carried out at relatively high temperatures—800° to 950°C (1472° to 1742°F), which can lead to a considerably reduced core strength, unless a more complex process is adopted, such as, e.g., a combination of nitriding and solution treatment, with a quench from the nitriding temperature, followed by a low-temperature aging anneal [34].

I. Duplex Treatments

As mentioned above, nitriding can be very useful in providing a strong substrate for the very thin, wear-resistant hard coatings, e.g., of titanium nitride, carbonitride, or aluminide. The duplex layer improves the performance of tools, both cutting tools as well as of press- and die-casting dies. An important requirement is very good adherence of the hard coating to the nitrided surface. The nitrided case, suitable for the subsequent hardfacing, should not contain any, or only thin nonporous, compound layers. This is easier obtained through plasma nitriding, with the additional advantage of the possibility of carrying out the hard-facing operation in the same installation, and the same production cycle, as a PVD or CVD process.

IV. CATEGORIES OF NITRIDING APPLICATIONS

A. Nitriding Applications Where Only Wear Resistance Is Required

In nonalloyed steels and cast irons, nitriding produces practically only a compound layer composed of γ' or

Figure 13 Nitrided clutch hub, low carbon steel.

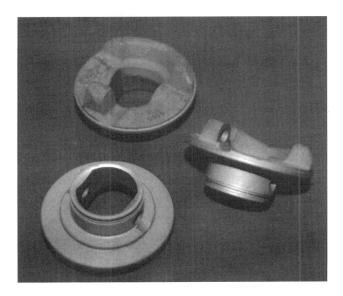

Figure 14 Nitrided automotive journal, cast iron.

$\varepsilon + \gamma'$ nitrides. Surface hardness is not very high (HV1 = 320–400 for mild steel, 400–580 for cast iron), but sufficient to considerably improve resistance to wear and galling in conditions of insufficient lubrication. An example is a low-carbon-steel clutch hub (Fig. 13). A 6-μm (0.00024-in.) minimum compound layer and passing of a specified wear test are required.

Moving (sliding, rotating) automotive components made of cast iron belong to the same category. The usually specified 8–12 μm (0.0003–0.00047 in.) compound layer seen in Fig. 14 is a compromise between distortion behavior and wear properties of the finished part. Other applications in this category include cams, camshafts, crankshafts, seat screws, components of solenoid actuators, rubber or plastic molds, and forming dies.

B. Combined Wear and Corrosion Resistance in Unalloyed and Low-Alloyed Steel

For improved corrosion resistance, a thicker compound layer is recommended. A typical application is the car window sector (Fig. 15). In this example, a 100,000-cycle wear validation test and the required minimum 96-hr corrosion test in a salt-spray chamber were both passed with a nitrided layer thickness of 15 μm (0.0006 in.) minimum.

C. Combined Wear and Corrosion Resistance in Martensitic Stainless Steel

Piston rings require high surface hardness at elevated temperatures, good sliding wear resistance, a low coeffi-

Figure 15   Nitrided door sector, mild steel.

cient of friction and good corrosion resistance, on account of the medium with which parts come in contact. Also required are good ductility and uniformity of the layer, a prerequisite for prevention of distortion. Figure 16 shows nitrided rings, used by both the compressor and the automotive industries.

D. Wear and Fatigue Resistance in Alloy Steels

Properly controlled nitriding satisfies simultaneous need for improved wear and fatigue resistance. Fatigue strength is improved by nitriding, due to favorable compressive stresses in the surface zone, making it less sensitive to stress peaks (the notch effect). However, fatigue damage is most often initiated below the nitrided layer. Therefore, the fatigue limit of nitrided parts depends primarily on core strength (or hardness). As typical examples, automotive and marine rocker arms, made of AISI 4140 steel, with 28–32 HRC core hardness, are nitrided to produce a compound layer of 8–12 μm (0.0003–0.00047 in.), and an effective diffusion case (>44 HRC) of 0.2 mm (0.0079 in.) minimum (Fig. 17). Other applications include crankshafts and gears.

E. Wear and Fatigue Resistance (with Fatigue of Primary Importance)

Spiral springs (Fig. 18) used in clutch and valve assemblies of automotive engines must retain high flexibility after enhancement of wear resistance. Because of pulsating tension during service, fatigue is of primary importance. Retention of high flexibility depends on high core hardness. Nitriding yields a rise of wear resistance and an increase of up to 35% in fatigue life, expressed by the number of cycles endured without deforming.

F. Wear Resistance, Fatigue Resistance, and Impact Strength

All three of the above properties, with an additional requirement of resistance to contact fatigue are required for gears. In this case, the thickness of the compound layer is usually kept to a low value of 5 to 10 μm (0.0002 to 0.0004 in.), enough to improve the resistance to sliding or rolling wear. Resistance to contact loads is ensured by a relatively thick diffusion layer.

Figure 19 shows a view of a typical nitrided gear. Such gears, made of 4140 steel, are given an effective case of 300–350 μm, (0.013–0.0138 in.); the white layer is limited by the specification to 15 μm (0.0006 in.) but comes out under 10 μm (0.0004 in.). Surface hardness of 85 HR15N is easily reached starting with a 28–32 HRC core.

G. Reduction of Overall Manufacturing Costs

A major gear company manufactured gears from carburized 8620 steel. Because of the high process temperature and the need to austenitize and quench, gears had to be manufactured with extra allowance for removal, and finish grinding was a necessity. To reduce costs, the manufacturer decided to try out nitriding as a replacement

Figure 16   Nitrided piston rings, 440-B martensitic stainless steel.

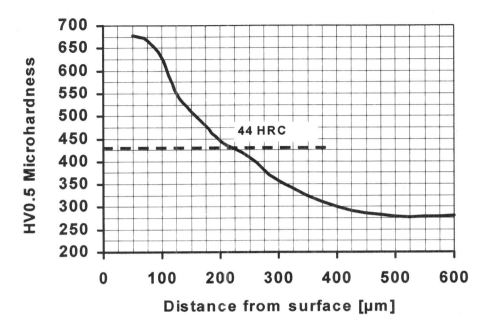

Figure 17 Microhardness profile on a nitrided rocker arm. 4140 steel. The effective case depth for the selected evel of HRC = 44 is 225 μm.

for carburizing. To replace the 8620 carburizing grade, 4340-grade steel was selected, and tests proved full compliance of nitrided gears with requirements regarding performance. Today, a major proportion of all gears manufactured by this company is nitrided. The added advantage is elimination of finish grinding or shaving, a

major item in overall manufacturing costs. Figure 20 shows gears made of 4340 steel, 41 in. in diameter, nitrided in a gas process controlled by the nitriding potential to a case depth of 0.625 mm (0.0246 in.) and a surface hardness of 58 HRC. Such gears were previously carburized.

Figure 18 Nitrided automotive spiral springs, high Si steel.

Figure 19 Nitrided gear, 4140 steel.

Figure 20 Nitrided gear, 4340 steel, previously manufactured from 8620 and carburized.

V. NITRIDING MATERIALS AND THEIR TYPICAL PROPERTIES

A. Nitride-Forming Elements and Characteristics of Nitrides

The interaction between the metallic material and atomic nitrogen, absorbed by its surface leads first (as described in Sec. III) to the formation of a solid solution of nitrogen in the metallic matrix—ferrite in the case of steel—next to the formation of nitrides or carbonitrides. These can exist in two structural forms:

1. A continuous nitride layer at the surface (compound layer; see Sec. III.D), and/or

2. Dispersed nitrides or carbonitrides of alloying elements precipitated in the diffusion zone, under the progressively advancing compound layer/matrix interface

In alloy steels and some nonferrous materials (e.g., Al- or Ti-based), the formation of alloy nitrides produces surface layers with improved useful properties.

Nitride-forming elements are those with a high trend to interact with nitrogen. This is reflected by the value of the heat of formation, i.e., the amount of heat absorbed in the formation reaction. The more heat absorbed, the greater the tendency to form nitrides (see Table 4) [35].

In steel, chromium is the principal alloying element, increasing the hardness of the diffusion layer and its thickness. Other elements such as V and Ti, with a high heat of formation, can form very stable and hard nitrides. However, having also a high affinity to carbon, in annealed or tempered structures, they exist bound in stable carbides. Therefore, even if carbides are converted to nitrides through nitriding, there is little hardening in the diffusion layer because new precipitates, instead of being dispersed, crystallize to a large extent on the already existing grains, only increasing their size. A similar segregation of chromium to carbides may adversely affect hardness in Cr-containing steels [36] by leaving less chromium in solution to form nitrides.

The magnitude of the hardening effect, because of the presence of alloying elements in the nitrided layer, depends on the affinity of the element to nitrogen (heat of formation of nitrides) and on the facility of nucleation of nitride phases. Higher heat of formation prefers nucleation of alloy nitrides over iron nitrides. However, nucleation of more complex, crystallographic lattices can cause very high hardness increases—as in the case of aluminum.

Table 4 Structure and Heat of Formation of Some Nitrides Occurring in Industrial Materials

Element	Nitride	Lattice	Unit cell dimensions (nm)	Heat of formation (kJ/mol)
Fe	Fe_2N_{1-x} (ε)	Hexagonal	$a = 0.2764^a$ $c/a = 1.599$	−3.8
	Fe_4N (γ')	fcc (Fe atoms)	$a = 0.3795$	−10.5
Mo	Mo_2N	fcc	$a = 0.4169$	−81.6
Cr	CrN	fcc	$a = 0.4149$	−124.7
Mn	Mn_3N_2	Tetragonal	$a = 0.2974$ $c = 1.2126$	−204.2
V	VN	fcc	$a = 0.4139$	−217.1
Al	AlN	Hexagonal	$a = 0.3110$ $c = 0.4975$	−318.0
Ti	TiN	fcc	$a = 0.4244$	−338.0

With the exception of Al, nitride-forming elements also form carbides.
Of common alloying components of steel only Si and Ni do not form nitrides.
^{a} Unit cell for the hcp Fe atom arrangement at the Fe_2N limit.

Table 5 Nitridable Steels (Selection)

Grade AISI	European equivalent grade	Main alloy components	Core hardness HV/HRC	Surface hardness HV1	Surface hardness HV10
A. Nitriding steels					
N135M	38CrAlMo7	C = 0.38, Cr = 1.7, Mo = 0.3, Al = 1	350/36	1190–1290	950–1150
Nitralloy N		C = 0.35, Cr = 1.15, Ni = 3.5, Mo = 0.25	350/36	1100–1200	930–1120
	34CrAlNi7	C = 0.34, Cr = 1.7, Ni = 1 Al = 1	350/36	1190–1290	
	42CrMoV12	C = 0.42, Cr = 3, Mo = 1.2, V = 0.20	490/48	1050–1150	
	31CrMo12	C = 0.32, Cr = 3, Mo = 0.4	320/32	870–920	
	30CrMoV9	C = 0.30, Cr = 2.5, Mo = 0.2, V = 0.15	320/32	900–930	
	32CDV13	C = 0.32, Cr = 3.00, Mo = 1.00, V = 0.20			
	25CrMo20	C = 0.25, Cr = 6, Mo = 0.20	280/27	1000–1100	
B. Other hardenable alloy steels					
5140		C = 0.40, Cr = 0.8, Mn = 0.8	300/30	580–650	460–600
4130	30CrMo4	C = 0.30, Cr = 1, Mo = 0.2	300/30	640–680	480–620
4140	42CrMo4	C = 0.40, Cr = 1, Mn = 0.9, Mo = 0.2	300/30	650–700	500–650
4340		C = 0.40, Cr = 0.8, Ni = 1.8, Mo = 0.25	310/31	650–700	550–670
	30NiCrMo12	C = 0.30, Cr = 0.8, Ni = 2.8, Mo = 0.12	310/31	600–650	
	35NiCrMo15	C = 0.35, Cr = 1.7, Ni = 3.8, Mo = 0.15	330/33	800–850	
C. Carbon steels					
1010	C10	C = 0.10, Mn = 0.50	160/<20	320–380	320–370
1020	C20	C = 0.20, Mn = 0.50	180/<20	320–380	
1030	C30	C = 0.30, Mn = 0.7	180/<20	380–420	
1045	C45	C = 0.45, Mn = 0.7	200/<20	420–470	390–440
1060	C60	C = 0.60, Mn = 0.7	250/22	525	
D. Free machining steels					
1144		C = 0.45, Mn = 1.5, S = 0.28	<20	420–470	350–370
12L14		C = 0.15, Mn = 1.0, P = 0.07, S = 0.3, Pb = 0.25	<20	420–460	340–360
12B15		C = 0.09, Mn = 0.9, P = 0.07, S = 0.3, B = 0.001	<20	420–460	340–360
E. Carburizing alloy steels					
5115	16MnCr5	C = 015, Cr = 0.8, Mn = 0.9	180/<20	660–720	
	20 MnCr5	C = 0.20, Cr = 1.15, Mn = 1.3	240/21	750–800	620–680
8620		C = 0.20, Cr = 0.5, Ni = 0.5, Mo = 0.2	190/<20	500–520	540–630
	18NiCrMo5	C = 0.18, Cr = 0.9, Mn = 0.8, Ni = 1.3, Mo = 0.25	210/<20	700–750	580–690
F. Tool steels					
D2		C = 1.5, Cr = 12, Mo = 1, V = 1.1, Co = 1	580/54	1270–1370	990–1050
	X150CrMo12	C = 1.5, Cr = 12, Mo = 0.8	580/54	1150–1370	
H11	AFNOR Z38	C = 0.35, Cr = 5.0, Mo = 1.5, V = 0.4	580/54	1150–1370	980–1120
H13	AFNOR Z40	C = 0.40, Cr = 5.2, Mo = 1.5, V = 1	480/48	1150–1280	980–1120
	X38CrMoV5.1	C = 0.40, Cr = 5.2, Mo = 1.1, V = 0.4	480/48	1180–1280	

Table 5 Continued

Grade AISI	European equivalent grade	Main alloy components	Core hardness HV/HRC	Surface hardness HV1	Surface hardness HV10
	40CrMnMo7	C = 0.40, Cr = 2, Mn = 1.5, Mo = 0.2	340/34	870–930	
M2	DMo5	C = 0.9, Cr = 4, V = 2, Mo = 5, W = 6	800/65	1280–1390	1120–1280
M7	DIN 1.3348	C = 1, Cr = 3.8, Mo = 9, W = 1.8, V = 2	787/63	1100	
M42	DIN 1.3247	C = 1.1, Cr = 3.9, Mo = 9.5, W = 1.5, V = 1.1, Co = 8	865/66	1100	
G. Stainless steels					
	X5CrNiMo18.10	C = 0.7, Cr = 18, Ni = 10, Mo = 2	230/20	1150–1250	
316L	X2CrNiMo18.10	C = 0.3, Cr = 18, Ni = 12, Mo = 2	230/20	1150–1250	410–480
440B	X90CrMoV18	C = 0.0, Cr = 17, Mn = 1, Si = 1, Mo = 0.75	420/43	1200–1350	650–850
17-4PH		C = .07, Cr = 16, Mn = 1, Si = 1, Ni = 4, Cu = 4, Nb = .3	300/30	950–1100	850–1000
15-5PH		C = .07, Cr = 15, Mn = 1, Si = 1, Ni = 5, Cu = 3		950–1100	850–1000
13-8Mo		C = .05, Cr = 13, Mn = 0.1, Si = 0.1, Ni = 8, Mo = 2.25		950–1100	850–1000

B. Typical Nitriding Materials

Chemical compositions and hardness of selected nitridable steels are given in Table 5.

1. Special Nitriding Steels

These steels are formulated specially for nitriding and always contain nitride-forming elements, optimized to enhance the hardening effect. Historically, the first such alloy was Nitralloy, used to this day as Nitralloy 135M. Its aluminum content produced a high surface hardness of over HV = 1200, but also, in an uncontrolled process, a very brittle white layer that had to be removed. Modern process control is capable of nitriding such alloys with a hard but ductile white layer or without it, as required. A typical microhardness profile obtained on nitrided Nitralloy 135M is shown in Fig. 21.

2. Low- and Medium-Alloyed Hardenable Steels

This group, although not formulated specifically for nitriding, lends itself well to the process, achieving surface hardness up to approx. 850HV. Nitriding is finding broadening acceptance in application to machine parts, where wear and fatigue resistance are needed. Low distortion, reducing finishing costs, represents the most visible advantage. But with modern technology allowing precise control of depth, structure, and phase composition of the surface zone, the performance of nitrided parts can in many instances come out equal, if not superior, to competing technologies of carburizing, carbonitriding, or induction hardening. Because of the link between surface hardness after nitriding and core hardness, nitriding is applied to hardenable steels with a medium (0.3–0.5%) carbon content. The hardness increase produced by nitriding, as well as the level of compressive residual stresses, are more pronounced for a lower carbon content. Hence the tendency to nitride steels with less than 0.3% C, where reduced hardenability and core strength can be tolerated. Core hardness before nitriding is generally HRC = 28–35 and does not drop if the temperature of the process

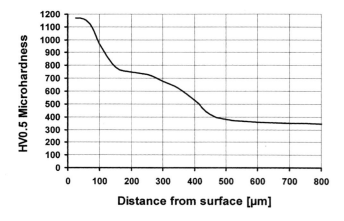

Figure 21 Typical microhardness profile obtained on nitrided Nitralloy 135M.

stays within the usual range of 520–570°C (968–1058°F).

Typical hardenable steels used for nitriding contain chromium, manganese, and molybdenum. This corresponds to the 5140, 4140, and 4340 AISI-SAE grades. A material used both for carburizing and nitriding is 8620. Nickel does not contribute to the hardness increase due to nitriding and slows the growth rate of the diffusion layer. However, it improves the hardenability and so will have a beneficial effect on core hardness.

In most applications where wear resistance of the surface must be supplemented by high fatigue strength, the compound layer thickness is limited to a maximum of 5–10 μm (0.0002–0.0004 in.). The resistance to contact loads is ensured by a sufficiently deep diffusion layer, usually in the range of 0.2–0.5 mm (0.008–0.020 in.). A typical microhardness profile is shown in Fig. 22.

In cases where nitriding is implemented to replace carburizing, a tendency is sometimes observed to specify a case depth comparable to that obtained in carburizing, as, e.g., 0.8 mm (0.032 in.). However, the advantages of such deep cases have not been demonstrated. It is important to control the process so that the joint nitrogen and carbon concentration in the superficial compound zone does not exceed 8.5 wt.%. Otherwise, a brittle nitride layer forms that may produce spalling, unless removed by mechanical or chemical means.

3. Carbon Steels

In unalloyed carbon steel, surface hardness is relatively low (see Table 4) The hardening effect is practically limited to the white layer, which is typically 10–25 μm (0.0004–0.001 in.) thick, depending on the application Hardening by iron nitrides precipitated from the super-saturated solid solution in the diffusion case during cooling can be neglected. More on the properties of the nitrided layer on carbon steels is to be found in Sec. IV of this chapter.

4. Free-Machining Steels

These steels, with additions of sulfur, lead, boron, or bismuth or with calcium modification for improved machinability, behave in nitriding similarly to unalloyed steels.

A typical specification requirement of 8–20 μm (0.0003–0.0008 in.) white layer thickness is relatively easy to meet. In industrial applications, large volumes of small parts are nitrided, usually by the gas or salt-bath method, mainly for wear and corrosion protection.

5. High-Alloy Tool Steels

Depending on the application, case depths may vary from 25–40 μm (0.001–0.0016 in.) for cutting tools, to 85–100 μm (0.0034–0.004 in.) for extrusion dies, to 200–300 μm (0.008–0.012 in.) for forging dies. Nitriding is carried out mainly for wear resistance but load-bearing capability and sometimes resistance to fatigue require a relatively high core hardness. For this reason, nitriding of tools is usually carried out in the lower end of the temperature range. Because tools are usually renitrided after some wear has occurred in service, white layer thickness must be well controlled. A problem encountered in some cases in high alloy steels is segregation of carbides in the underlying microstructure. This nonuniformity may lead to cracking in service that will sometimes mistakenly be blamed on nitriding.

6. Stainless Steels

Nitriding, enhancing corrosion resistance on most steels, impairs it on stainless steels. This is because chromium, being a strong nitride-forming element, readily combines with diffusing nitrogen, thus depleting the steel matrix of its chromium content in solution, necessary for the formation of a protective passive film. Alleviation of this effect can be obtained by nitriding at low temperatures, involving, however, long processing times [37]. Nitrided layers on precipitation-hardening steels exhibit similar features to those on martensitic stainless grades.

Gas nitriding of stainless steels requires surface activation to break down the compact oxide film, which creates a barrier for nitrogen atoms. Stainless steels are nitrided mainly to enhance wear resistance in cases where non-ferromagnetic properties are required but where corrosion resistance is not a major requirement. Gas nitriding at 350°C (662°F) produces in 304 steel a total nitriding

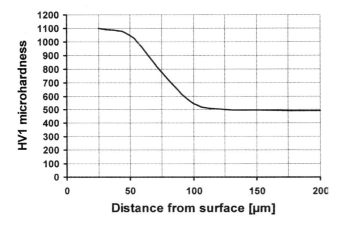

Figure 22 Typical microhardness profile on H13 steel (aluminum extrusion die).

depth of 5 μm after 24 hours, at the same temperature plasma nitriding produces 7.5 μm after 16 hours [38].

Faster nitriding rates, and deeper cases, can be obtained through a high temperature process (1050–1150°C, 1922–2102°F) in atmospheres containing a high proportion of molecular nitrogen [39]. The proposed name for the process is *solution nitriding*. At this temperature level, thermal dissociation is sufficient to supply atomic nitrogen that can diffuse into the surface. The nitrogen concentration at the surface is governed by the partial pressure of nitrogen in the atmosphere, which for some alloys will be <1 bar, for others up to 3 bar (in an appropriate furnace). A quench is necessary to avoid chromium nitride precipitation. For best results some optimization of material composition and process conditions is advisable. Stable ferritic steels cannot be treated in this manner—too fast grain growth in the core. A typical result for a 0.2%C, 13%Cr martensitic steel (X20Cr13), after nitriding (15h?) at 1150°C, quenching, sub-zero treatment at −80°C, and tempering at 170°C, is a surface hardness of HV2 = 675 (HRC = 59) for a core of HRC 50-51, and a total 1.5 mm diffusion depth. The nitride (Fe,Cr)$_2$N precipitated during tempering and increasing the hardness as a result of secondary hardening does not deplete the matrix of Cr, and therefore does not impair the corrosion resistance.

7. Cast Iron

The emphasis here is on obtaining a sufficiently thick white layer, which is the only usable part of the nitrided case on this material, unless the cast iron is alloyed with nitride-forming elements. It must be sufficiently thick to seal graphite inclusions exposed to the surface of the casting. Because castings have a relatively rough surface, causing a high rate of ammonia cracking, this tendency must be compensated for by higher atmosphere flows to sustain a potential required to obtain a

Figure 23 Typical nitrided layer on unalloyed ferrito-pearlitic cast iron. Etched by 3% Nital. Magn. 400×.

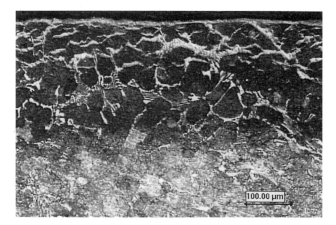

Figure 24 Microstructure of a nitrided layer on alloyed cast iron. Etched by 3% Nital.

white layer. Typically, unalloyed cast irons are nitrided 4–8 hr at 540–580°C (1004–1076°F). Depending on nitriding conditions, the obtained white layer thickness ranges from 8 to 15 μm, (0.0003–0.00059 in.), whereas typical surface hardness is 420–500 HV1. Figure 23 shows a typical nitrided layer on unalloyed ferrito-pearlitic cast iron. As can be seen, on this type of cast iron only a white layer is observed.

Nitriding of alloyed cast iron yields a diffusion case with a hardness profile. Figure 24 shows the microstructure of a nitrided layer on alloyed cast iron containing 12–15% Cr and up to 2.5% Mo. The visually measured diffusion case depth obtained is approx. 220 μm (0.0087 in.), the white layer thickness is 2–4 μm (0.00008–0.00016 in.), and the surface hardness is above 1000 HV1.

Because castings usually contain surface contaminants, e.g., core mix binders, which may inhibit nitriding, baking out is recommended prior to nitriding as a surface activation measure.

8. Powdered Metal Components

Gas nitriding of powdered metal (P/M) components does not present problems when their density is 7 g/cm^3 or higher. Lower densities mean higher porosity, facilitating rapid diffusion of nitrogen from the surface inward, along canals from pore to pore, lining grain boundaries and, in some cases, the inner surfaces of pores at a certain distance from the surface. As a result, no sufficient buildup of nitrogen concentration takes place at the surface for the compound layer to form. No appreciable surface-hardening effect is obtained. Figure 25 shows the microstructure of a gas-nitrided, low-density P/M part. Experience has shown that low density P/M parts respond better to plasma nitriding.

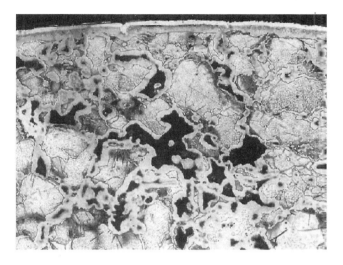

Figure 25 Microstructure of a gas nitrided layer on a P/M part of low density. Etched by 3% Nital.

This is due to the different mechanism of nitrogen deposition on the surface. Plasma nitriding of P/M materials gives best results at high temperatures, e.g., 570–590°C (1058–1094°F), with a high nitrogen/hydrogen proportion (70/30–90/10) and an addition of methane. Figure 26 shows the microstructure of a plasma-nitrided P/M part with a density of 6.8–6.9 g/cm³.

9. Nonferrous Materials

In the case of wrought aluminum alloys, successful nitriding experiments have been conducted in all kinds of plasma, from ordinary direct current (DC) to high-frequency pulse-plasma nitriding [33]. However, the nitriding results are very much dependent on the chemical composition of the alloy. The process temperature is maintained usually in the 400–500°C (752–932°F) range, while the lower melting point for alloys like 5083 (Al–Mg–Mn) lies close, at about 574°C (1065°F). This shows the need for a very reliable and accurate control of temperature.

In the absence of a supporting diffusion layer, nitrided Al alloys featuring a very hard, thin layer (up to 10 μm) on a soft substrate may be subjected to only low contact loads in service, unless a duplex process is employed to provide a more resistant composite structure.

Among titanium alloys, the alpha–beta age hardenable grades, of which Ti–6Al–4V is a typical representative, are the most often nitrided. Solution treatment at 955–970°C (1750–1775°F) is followed by a water quench and aging is carried out at 480–595°C (900–1100°F). Nitriding at 800–950°C (1472–1742°F) results, as previously mentioned, in loss of strength in

the core. This can be prevented by running the process at lower temperatures, in this case preferably as intensified plasma nitriding [40].

VI. DESIGN OF THE NITRIDING PROCESS

A. Significance of the Nitriding Process

Nitriding represents in most cases the final operation of the manufacturing process. It is preceded by all the preliminary (volume) heat treatment and shaping operations. Successful results depend on correct preparation of the surfaces to be treated, correct setup of the process cycle, and its faultless execution. In some cases, as with plasma nitriding, the arrangement of the load in the furnace is also an important factor. The following stages in the process design have to be considered:

B. Surface Preparation

Intensive and uniform build-up of the nitrided layer requires an unimpeded flux of atomic nitrogen from the surface into the metallic matrix. Because the process takes place at relatively low temperatures, surface barriers created by residues of previous forming or machining operations can effectively obstruct nitrogen absorption. These barriers can be organic (oil, grease) or inorganic (residues of drawing compounds, oxides). The goal of surface preparation is to remove these barriers. This may be combined with operations enhancing the activity of the surface (see further: "Activation").

Figure 26 Microstructure of a plasma nitrided layer obtained on a P/M part with a density of 6.8–6.9 g/cm³. Etched by 3% Nital. Magn. 400×.

1. Cleaning

Organic residues and mechanical particulate soils are removed by vapor degreasing, by spraying or by immersion in liquid solutions belonging to one of three categories:

Organic solvents
Aqueous emulsions of hydrocarbons
Alkaline anionic or nonionic aqueous solutions

Vapor degreasing with chlorinated hydrocarbons, although most effective in removing of all surface soils, has lately receded into insignificance due to pollution-related legislation. The use of organic solvents containing isoparaffins and naphtenes remains limited because of fire hazards. They are more acceptable when used as emulsions dispersed in water. Most frequently used are alkaline or neutral solutions with surface-active compounds (surfactants).

The choice of a cleaner, preparation of the solution, and maintenance of the cleaning bath can have a profound effect on the kinetics and results of subsequent nitriding. It has been experimentally demonstrated [41] that residues of cleaning solutions, especially those of inorganic components such as phosphates, silicates, borates, and carbonates, as well as chlorides and silicones, themselves form barriers to nitriding, substantially reducing the growth rate of the compound and diffusion layers and lowering surface hardness. When present in the form of dried-up adsorption layers, such residues cannot be fully removed even by intensive rinsing with demineralized water. The use of silicates in cleaning baths should be particularly avoided. The retarding effect of surface residues on nitriding increases with alloy content, from carbon steels, to Cr- and Al-bearing grades to Cr-rich tool steels. For example, when nitriding at $580°C$ ($1076°F$), 4 hr, $K_N = 0.7$, a reduction of surface hardness by 10% is caused by 2 g/m^2 of sodium phosphate residue in pure iron, 0.2 g/m^2 in 1015 steel, and 0.1 g/m^2 in a 4140 equivalent [42]. Light organic residues are most often removed from the steel surface by evaporation during the heating-up stage.

2. Activation

In gas nitriding, initiation of formation of the nitride layer can be accelerated by a number of methods covered by the term "activation." These include:

Mechanical means: roughening of the surface, e.g., by sandblasting, glass beading or brushing
Etching in acid solutions

Conversion coatings: preliminary phosphating or oxidation
Chemical reactions with chlorinated or fluorinated compounds within the furnace retort

Roughening of the surface by mechanical or chemical means, besides removing a large proportion of oxides and contamination films, increases the active surface area, thus facilitating transport of nitrogen into the metal. The mechanisms of other activation methods are not well understood [42]. More systematic investigations are necessary.

Salt-bath nitriding does not require a separate activation step. The chemical action of the bath is sufficiently aggressive. However, the quality of the salt-bath is subject to rapid deterioration, proportional to temperature, time and surface of parts being nitrided, and will require frequent stops for regeneration. In plasma nitriding, an initial sputtering step in higher vacuum is used to remove any remaining oxide films and to thus activate the surface. Also in this case, even though cathode sputtering is able to remove surface contamination, it is very important to clean parts thoroughly because all nonmetallic matter removed from the surface causes deterioration of the partial vacuum and has a profoundly negative effect on the nitriding results.

C. Heating Rate and Initial Atmosphere

One of the main advantages of correctly executed nitriding is the practical elimination of distortion of treated parts. However, to maintain this advantage, it is necessary to provide a sufficiently controlled heating-up to avoid temperature nonuniformity in the load or within large treated objects. Uniform, steady heating is relatively easy to accomplish when gas nitriding in a modern furnace with controlled temperature ramping and adequate atmosphere circulation. In plasma nitriding, heating is less controllable because a major portion of the heat-transferred energy comes from glow discharge. The energy coming from heating elements in hot-wall units is transferred mainly by radiation with convection playing a minor role. In salt-bath nitriding, only stepwise preheating to a fixed temperature setting, involving separate baths maintained at different temperatures, can be used.

For most loads, a heating rate of $180°C$ ($324°F$) per hour is adequate. For very large objects this heating rate must be substantially reduced—even to $50°C$/hr ($90°F$/hr).

In gas nitriding, air is removed from the furnace before heating begins, through purging or by evacuating with a vacuum pump. Active nitriding atmosphere is usually admitted before process temperature is attained, because the change in atmosphere composition, even with substantial gas flows, has to take a certain time. It is necessary to ensure a relatively high initial nitriding potential to quickly saturate the superficial ferritic zone and to initiate the formation of the nitride layer. This begins to take place from approx. 400°C (752°F), before reaching the usual temperature set point of 500–590°C (932–1094°F). An atmosphere with a high nitriding potential is usually applied at this stage. Hence, by the time the formal process stage begins, a well-developed continuous compound layer is often already formed.

D. Configuration of Nitriding Stages

In cases where the objective of nitriding is only to develop a sufficiently thick compound layer, a *single-stage* process is normally employed. The load is heated to the process temperature and soaked for a time necessary to obtain the specified layer depth. The process is, as a rule, short, with the constant-temperature diffusion stage not exceeding 4 hr. The temperature is selected to maintain the required core hardness, but at the same time the highest possible to enhance the diffusion rate and shorten the process cycle. For the same reason, intensive cooling is implemented at the end of the process. Means of increasing the cooling rate are described in Sec. IX.B. Recent advances in gas nitriding and nitrocarburizing tend to aim at increasing the ε-phase content in the surface zone through fast cooling of the parts, instead of cooling them down with the furnace. Figure 27 is a schematic representation of a single-stage process.

A *two* (or more)-*stage process* is used when a substantial diffusion layer is required while the thickness of the compound layer must be limited. In this case the first

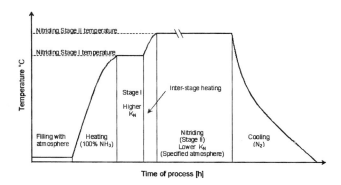

Figure 28 Schematic representation of a double-stage nitriding process.

stage is run with a more aggressive atmosphere (high K_N value) to create an initial, relatively thin compound layer. The temperature of the first stage is usually lower, to increase the density of nucleation and improve the uniformity of the nitride layer. In the second stage, a low K_N atmosphere is introduced, with the potential adjusted to maintain a constant surface concentration of nitrogen. With a correctly selected and properly controlled nitriding potential the thickness of the compound layer can be maintained at a desired thickness or even reduced to zero. At the same time, diffusion progresses to build up the required case depth. Two-stage processes are, as a rule, longer, with the total duration of both stages in excess of 12 hr. In extreme cases, when very deep diffusion layers are required, the process cycle may exceed 100 hr. A schematic representation of a two-stage process is given in Fig. 28.

Cooling at the end of a multistage process, not affecting the established microhardness profile, is less important. Some process time reduction by faster cooling is always desirable.

E. Process Atmosphere

Ammonia is the active component of atmospheres used in gas nitriding. In earlier decades, it was the only component. Since then, mixed gas atmospheres have been developed, comprising ammonia and nitrogen (to reduce process costs and to improve process control) or ammonia and dissociated ammonia (to facilitate obtaining low potentials). In the late 1980s, automatic flow rate and potential control industrial nitriding systems were developed and implemented by Nitrex Metal Inc.

The phase diagram shown in Fig. 2 (see Sec. III) shows the areas of stability of the different nitride phases

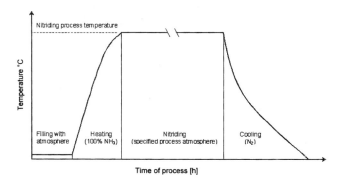

Figure 27 Schematic representation of a single-stage nitriding process.

as a function of nitriding potential and temperature. It should, however, be observed that the diagram has been developed for iron, and for conditions of equilibrium between the atmosphere and the metallic surface. In practice, this equilibrium is reached only after a certain time, if ever, during a short process, and the presence of carbon as well as some alloy elements may shift the phase boundaries. As a result, selection of the process atmosphere and of other parameters depends still to a large extent on the experience of the process designer.

For modified atmospheres with additions of oxygen (oxynitriding), compounds see Sec. VIII.

F. Process Duration—Growth of the Nitrided Layer

The growth of both the compound as well as the diffusion layer depends predominantly on the diffusion rate of nitrogen through the matrix. Given a constant nitrogen concentration at the surface, the evolution of thickness of the compound layer δ will follow the parabolic law, i.e., $\delta^2 = kt$, where t is time. The same can usually be observed with regard to the effective diffusion case, measured as the depth corresponding to an arbitrarily defined hardness level in the microhardness profile (e.g., equivalent to 50HRC or core hardness + 50 HV, etc.).

Specifications for nitrided parts include surface hardness at a determined test load (e.g., HV1 for 1-kg load, HV10 for 10 kg, or HR15N for 15 kg) and the thickness of the compound and diffusion layers. More recently, the phase composition, i.e., a given percentage of the ε-phase in the compound layer, begins to be specified as a requirement. There is still, however, insufficient evidence

as to the relationship between phase composition and usable properties of a part. Extending the time to increase total case depth can, and frequently does, lower surface hardness, as recombination of nitrogen into the molecular form occurs, with formation of a porous zone at the surface. In addition, in low-potential atmospheres, the compound layer may regress until its total disappearance. In alloy steel, an evolution of the microhardness profile occurs as the diffusion depth increases. Because of coagulation of the dispersed alloy nitrides, in some areas hardness may actually decrease even as the total layer grows (see Fig. 29). This evolution should be considered when a specific microhardness profile is required.

VII. DESIGN OF NITRIDED PARTS

A. Factors Critical to Final Quality of Nitrided Parts

When designing parts that are to be nitrided, the following factors should all be taken into consideration:

> Selection of material
> Geometry of parts and configuration of the nitrided layer
> Preliminary heat treatment and core hardness after nitriding
> Dimensional changes occurring during nitriding
> Effect of nitriding on surface roughness
> Design considerations ensuring uniformity of the nitrided layer
> Selective nitriding and masking of nonhardened surfaces, where applicable

B. Selection of Material

Material selection for a nitrided part requires consideration of several factors:

1. *Required bulk properties of the part:* mechanical strength, hardness, fatigue resistance, toughness. These properties characterize the core of the nitrided part and depend on the material's chemical composition and the heat treatment preceding nitriding. For improved core strength (hardness), steel carbon content is critical. Typical values for machine parts are 0.25–0.4%C. Lower content is acceptable where higher strength is not required, and, as in the case of alloy steels, lower carbon can actually improve surface properties (e.g., level of compressive stresses) after nitriding. In

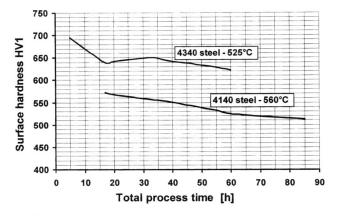

Figure 29 Decrease in surface hardness with time of process for two steel grades and process temperatures.

some steels, the alloy content may compensate for lower carbon. Alloying elements, e.g., Cr, Mn, Ni, and Mo, are added to improve hardenability and, in specific cases, the steel's resistance to tempering (hot strength).

2. *Nitridability:* the ability of the material to develop, in the nitriding process, a high surface hardness to enhance wear resistance, a sufficiently deep case to support contact loads, and a sufficiently high level of residual compressive stresses to improve fatigue resistance. Nitricability depends on the presence of nitride-forming elements, predominantly chromium, then manganese, molybdenum, aluminum, titanium, and vanadium. Their effect is related to the carbon content, as nitride-forming elements also, for the most part, combine with carbon to form carbides. The characteristics of nitrides and their role in nitrided steels are reviewed in Sec. V of this chapter. In the selection of material composition for increased surface hardness and case depth, the chromium content should be the primary guiding factor. With its increase to about 1.5–2%, both surface hardness and effective case depth increase linearly, except for compositions with strong nitrogen binders (e.g., Al and Ti), whose effect is more dominating. For a more detailed description of nitriding materials and their properties, see also Sec. V.

3. *Machinability:* In many cases machinability is an important economical factor in determining the choice of the material. Free-machining steels (e.g., leaded or resulfurized/rephosphorized, of the AISI-SAE 1100 and 1200 series) are selected where machinability is of primary importance, overshadowing mechanical strength. On the other hand, modern, less environmentally hazardous steelmaking practices, like calcium modification, allow the improvement of machinability of low- and medium-alloy steels without loss of mechanical properties.

C. Geometry of Nitrided Parts and Configuration of the Nitrided Layer

There are no practical limitations to the shape of objects treated with gas nitriding. Complex shapes, stacked parts, or loose bulk loads can be treated with a high degree of uniformity. The same applies to salt-bath nitriding, except that in the latter case the limiting factor is the dimension of the pot employed. In plasma nitrid-

ing, problems may arise when the parts contain holes, cavities, or recesses. Depending on the pressure in the treatment chamber, glow-discharge seams may overlap, creating areas of intense heat concentration and catastrophic damage (see Sec. IX).

The nitriding process is usually the final manufacturing operation, carried out on the fully finished part. Light polishing or honing may be used where surface roughness is of critical importance. The nitrided layer is usually allowed to cover the whole surface of the treated part. This allows for equal distribution of residual stresses in the superficial zone, thus avoiding distortion. In cases where design considerations or subsequent assembly operations demand it, some surfaces may be left non-nitrided. In gas nitriding this can be obtained through masking (see Sec. VII.H), in the plasma process through shielding, and in salt-bath nitriding, through coating or, where possible, partial immersion.

Wherever possible, sharp corners should be avoided in the design of the part, particularly when a deeper case is required, to prevent oversaturation. Where they cannot be avoided, the optimum nitriding method should be selected for the given application and a shallower case should be specified. Gas process should be run at lower nitriding potential values to limit nitrogen concentration (see Sec. XI).

Internal surfaces of small holes less than 1 mm (0.040 in.) in diameter, narrow gaps of the same size, can develop a uniform nitrided layer in gas nitriding; however this is not possible to achieve with plasma nitriding.

D. Preliminary Heat Treatment and Core Hardness After Nitriding

Nonalloyed carbon steels, in which nitriding produces essentially only the wear- and corrosion-resistant compound layer, can be nitrided in the normalized state. This is usual, especially with nonhardenable, low-carbon steels. Hardening through austenitizing, quenching, and tempering, in the case of steels containing sufficient carbon, yields a uniform microstructure, enhancing the build-up of the compound layer (see Table 6). Its main function, however, is improvement of bulk mechanical properties, such as higher strength and toughness of the core. Higher core hardness yields higher hardness of the nitrided layer, and better fatigue strength (Fig. 30). In alloy steels, preliminary heat treatment is carried out, as a rule, to make the best use of properties of the nitrided layer.

The temperature of the diffusion stage of the nitriding treatment is usually 30°C (or 50°F) below the temperature of the last tempering. This ensures that

Table 6 Effect of Preliminary Heat Treatment on Results of Nitriding Chromium Steel 40Cr4 (Equivalent of AISI5140)

Heat treatment	Compound layer, μm (in.)	Effective diffusion layer, mm (in.)	Hardness at surface (HV0.1)	Core hardness (HV0.1)	Hardness increase (HV0.1)
Normalized	25 (0.001)	0.54 (0.0021)	630	205	425
Quenched and tempered 400°C (752°F)	19 (0.00075)	0.50 (0.00197)	715	280	435
Quenched and tempered 600°C (1112°F)	21 (0.00083)	0.50 (0.00197)	600	255	345

nitriding is performed on a stable structure, thus avoiding bulk transformations in the core. In consequence, dimensional and shape changes of treated parts are substantially reduced.

On the other hand, the requirement that the tempering temperature exceed that of subsequent nitriding narrows down the range of core strength values that can be maintained in nitrided parts. Because nitrogen diffusion rate depends exponentially on temperature, lower process temperatures involve substantially longer times necessary to obtain the same case depth. In consequence, lowest practical temperatures of gas nitriding rarely drop below 500°C (932°F). As a result, lowest preliminary tempering should not be carried out below 530°C (986°F). There are cases where the nitriding temperature must be substantially lower than 500°C (932°F). This may be required in materials with insufficient tempering resistance or those susceptible to stress relaxation (e.g., springs). The process may still be economical in the case of very thin nitrided layers. Otherwise, it would carry the penalty of high cost. Summarizing, the selection of material for a given part should involve not only considerations of hardenability, but also those of the tempering curve and the resulting limitations on hardness and strength.

If the above precautions with regard to the tempering temperature, short nitriding processes (up to 12 hr) have no effect on the final core hardness.

E. Dimensional Changes Occurring During Nitriding

Changes of dimensions and shape of nitrided parts can occur for the following reasons:

1. Increase of volume of the superficial zone as a nitride (compound) layer is being formed. This is due to a greater specific volume of iron nitrides as compared with the ferrite matrix. This increase will always accompany the formation of the compound layer.
2. Decrease of bulk volume, occurring during nitriding if the nitriding temperature exceeds that of preceding tempering. With correctly adjusted temperatures this will be avoided.
3. Distortion of shape due to mechanical creep may occur, especially at higher process temperatures and during longer nitriding times. This risk can be eliminated by a properly designed support system (racking) of the furnace load.
4. Distortion may also occur to some extent if the nitriding process is followed by quenching, which is typical for salt-bath and fluidized bed processes, but may sometimes take place in some gas nitriding and nitrocarburizing processes.

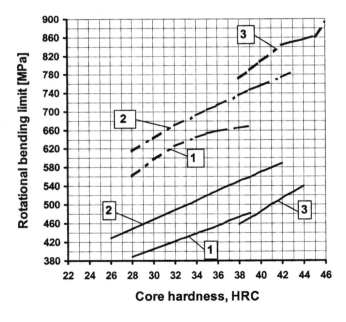

Figure 30 Variation of rotational fatigue limit with core hardness on heat treated and nitrided samples. 1—Nitralloy 135M; 2—hot work tool steel; 3—0.4%C Cr-V nitriding steel. Solid lines: heat treated only; dashed lines: heat treated and nitrided.

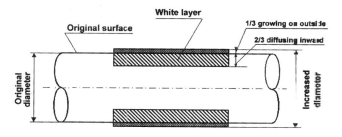

Figure 31 Dimensional growth of a nitrided part, due to white layer formation.

In general, nitriding, especially the short-time process, produces minute dimensional change and distortion. Dimensional increase due to compound layer formation equals, as a rule of thumb, one third of the compound layer thickness for each surface, thus two thirds across a diameter (see Fig. 31). Still, for precision parts this has to be taken into account. In reproducible process conditions, as in correctly controlled nitriding, the increase will be repetitive and allowance for it can be made during prior machining operations. Unpredicted growth may occur if uncontrolled nitriding potential is excessive, resulting in gross precipitation of nitrides along grain boundaries in the diffusion zone.

F. Effect of Nitriding on Surface Roughness

Surface roughness always increases to some extent after nitriding. One factor causing an increase in roughness is the nonuniform nucleation and growth of nitride grains in the white layer during the unstable period of heating. Roughness is also increased by porosity formed in the outer zone of the compound layer by trapped molecules of nitrogen that recombine to N_2. The tendency to form porosity by this mechanism grows with the temperature of the nitriding process, as well as with the nitriding potential. It should be borne in mind that in the case of surfaces subjected to friction wear or corrosion, some porosity may actually be desired because it promotes absorption of a lubricant or a corrosion inhibitor, introduced by immersion. Surface roughness prior to the nitriding process has a significant influence on final roughness, as can be seen in Fig. 32.

G. Design Considerations Ensuring Uniformity of the Nitrided Layer

To ensure uniformity of the nitrided layer on the part surface it is essential to ensure uniform surface condi-

tions prior to nitriding. This means that the given part surface must be machined in a uniform manner and have the same roughness; otherwise, the intake rate of nitrogen will vary producing differences in layer thickness and structure. The entire part should be thoroughly washed or cleaned and properly degreased. The effect of some deficiencies in operations prior to nitriding and in the execution of the process are indicated in Sec. XI.

Small holes, narrow slits, and other constricted spaces in treated objects are no obstacle to maintaining uniform layer thickness in gas and to some extent in salt bath nitriding. In plasma nitriding there are limitations to the penetration of the nitriding effect into such narrow spaces and each case should be considered individually.

H. Selective Nitriding and Masking of Nonhardened Surfaces

Surfaces of parts that are to be left unnitrided must be protected. In salt-bath treatment, selective partial nitriding can be obtained by partial immersion, possible only on objects where the entire lower portion is to be nitrided and the entire upper portion left unnitrided. An example of this may be a tool (drill, reamer, etc.) where the shank is to be protected.

In gas nitriding, and, where necessary, in salt-bath nitriding, there are essentially two methods of masking employed, i.e., electroplating and special coatings (stop-off paste).

Electroplating, typically alkaline copper or bronze. The coating must be fine-grained and not porous and have a thickness of not less than 0.025 mm

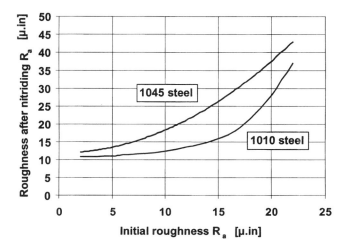

Figure 32 Effect of prior roughness on final roughness after nitriding, for 1010 and 1045 steel.

(0.001 in.) in the case of copper and 0.0125 mm (0.0005 in.) in the case of bronze.

Stop-off paste, comprising a tin or lead base and a filler, applied thinly with a brush on the surfaces to be protected from nitriding, prevents atomic nitrogen from being absorbed by the steel surface. Usually, stop-off pastes are formulated in such a way that they peel off from the part surface after the completed nitriding process. Caution: large amounts of stop-off paste on a nitrided batch have been known to lower the surface hardness, even of unprotected surfaces.

Nitriding specifications usually require that maskant strippers be nonembrittling and do not cause pitting.

In plasma (ion) nitriding, protection against nitriding is accomplished by simple mechanical masking (shielding). Surfaces that are not exposed to the glow are not nitrided. Covering of surfaces to be protected with steel foil is effective in preventing nitriding. Surfaces that touch (remain in close contact) are also effectively protected from undesired nitriding. One example of this is stacking cylindrical gears in which case only the crowns are nitrided while the touching side surfaces remain unnitrided. Another example is nitriding of only the inner bore of a cylinder, which is accomplished by placing the cylinder inside a bigger cylinder, covering, and thus protecting the outside diameter. Conversely, an inner bore can be easily protected by filling it with a cylinder of appropriate diameter or by placing a flat piece of sheet (a "cap") over the opening.

VIII. MODIFICATIONS AND EXTENSIONS OF NITRIDING

A. Oxynitriding

In technological experiments, additions of up to 25 vol. % of air to ammonia, or water injection, have been tried [17], taking care to maintain a nonoxidizing character of the atmosphere with regard to iron. It has been confirmed that oxynitriding, even in conditions of a constantly controlled nitriding potential, leads to a higher surface concentration of nitrogen and a faster build-up of the compound layer. The effect of continuous addition of oxygen to the process atmosphere appears to be stronger than that of preoxidation. This was found in experiments with 1010-type [17], 1015-type [26], and alloy [17] steels (see Table 7).

Experiments with still higher oxidation potentials produced, on the surface of the steel, oxide layers that hinder the nitriding process. In Table 7, the strong effect of oxynitriding on the alloy steel should be noted. This is interpreted as the result of depassivating action because of formation of Fe_3O_4.

Oxygen additions during the process have also been applied to nitrocarburizing and to a plasma process [43].

Table 7 Effect of Oxygen Additions (Measured as the Level of Oxygen Potential K_O) on the Thickness and Structure of the Compound Layer

K_N	K_O	% N surface (μm)	Compound layer (μm)	Phase composition (%) of compound layer (information from depth 2.5–3.5 μm)			
				ε	γ'	α-Fe	Fe_3O_4
1010-type steel							
6	0	10.4	1.5–2.0	34	47	19	
	0[a] pre-oxid.	10.3	2.9–3.5	67	25	8	
	0.28	11.3	2.5–3.5	67	25	5	3
1015-type steel							
6	0	10.2	1.8–2.3	45	38	17	
	0[a] pre-oxid.	10.3	1.9–2.7	48	37	15	
	0.28–0.30	11.2	2.9–3.6	62	20	9	9
31CrMoV9 (European nitriding steel: 0.3% C, 3% Cr, 0.7% Mo, 0.25% V)							
6	0	3.2–10.7	0–2.5	18	15	67	
	0.28–0.30	11.3	3.9–5.5	85	2	67	13

[a] One hour at 350C (662°F).
Adapted from Refs. 17 and 25.
Process temperature 550°C (1022°F), time 30 min.

No particular benefits in useful properties have been claimed in comparison with standard nitriding or nitro-carburizing.

B. Postnitriding Oxidation

In 1985 and again in 1986, patents were granted to Lucas Industries, UK [44,45], for a nitrocarburizing process involving a subsequent oxidation stage. This stage, designated here as *postnitriding oxidation*, was introduced for two reasons: (1) to improve the corrosion resistance of the surface while maintaining its antiwear characteristics and (2) to produce surfaces with a pleasing black finish, enhancing the attractiveness of many treated parts. The concept constituted a reproduction of an earlier practice of oxidizing in a molten hydroxide bath, where the surfaces of parts were salt-bath nitrided in a prior stage.

In the original patented process, oxidation took place by exposing the load to air for 2 to 120 sec during transfer from the nitrocarburizing furnace to a quench bath. In further developments, the oxidation stage assumed a more controlled form, as a separate operation following gas nitriding (or nitrocarburizing), or integrated with the latter into a continuous process. A number of proprietary treatments have been proposed. Typically, postnitriding oxidation is carried out in a water-vapor-based medium or in untreated exothermic gas. The nitrided parts are cooled down to room temperature and, before oxidation, can optionally be lapped or polished where a surface finish of R_a about 0.2 μm (8 μin) is required. Typically, oxidation is carried out in the temperature range of 450–550°C (842–1022°F) and is often followed by quenching if quality requirements demand it. In the case of oxidation integrated with nitriding, gaseous (exogas, CO_2, H_2O vapor base), or liquid oxidizing media are admitted for a limited time at the end of the diffusion period. Above 570°C (1058°F), FeO may be formed, with detrimental effects because of increased specific volume (cracking and flaking).

The oxygen-bearing components of the atmosphere react with the nitrided surface to displace nitrogen and to produce iron oxides. Depending on the oxygen potential (or partial pressure) in the atmosphere, and on temperature, the oxides formed are Fe_3O_4 (magnetite) or α-Fe_2O_3 (hematite) (see Fig. 33) [46].

At low oxygen potential values, as typically in the case of water vapor base treatment, Fe_3O_4 predominates. It forms a dense and adherent coating on top of the iron nitride and fills pores in the surface zone, improving corrosion resistance. It has been noted [47] that the adherence of the oxide to a nitrided surface is better, compared with an analogous layer grown directly on ferrite. With increasing time of oxidation, the out-

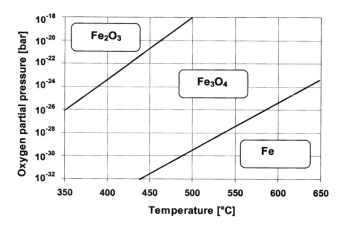

Figure 33 Stability ranges of iron oxides in the Fe-O system, as a function of temperature and oxygen partial pressure. (From Ref. 46.)

ward diffusion of iron atoms produces vacancies coalescing into pores at the metal–oxide interface. This may lead to loss of adherence of thicker oxide films. The recommended oxide thickness on carbon steel, producing the best corrosion resistance, was originally given as 0.2 to 0.7 μm, with an optimum at 0.5 μm (0.00002 in.). With the application of stricter atmosphere control, it is possible to obtain oxide layer thickness of 1–2 μm without spalling (the ONC® process). In a Mn–Cr alloy steel (European 20MnCr5) with 0.2%C, 1.7% Mn, and 1.2% Cr, it has been found [46] that the optimum passivation effect and, therefore, best resistance to pitting corrosion, is obtained at each temperature within the range of 450–550°C (842–1022°F) in conditions of a sufficiently high oxygen potential, if the oxide layer is about 0.5 μm at 450°C, 1 μm at 500°C, and 2.1 μm at 550°C. Solid-electrolyte sensors (oxygen probes) can be applied to control the oxygen potential during the process, allowing one to optimize the treatment results in a reproducible manner.

Improvement of wear characteristics, in addition to corrosion resistance, by postnitriding oxidation has also been reported [47]. Certain further enhancement in this respect can be obtained by introducing an addition of up to 1% SO_2 to the oxidizing atmosphere (see Refs. 44 and 45). This results in the formation of friction-reducing sulfides in the oxide layer. The effect on fatigue resistance appears to depend less on the degree of surface oxidation than on the cooling rate after the treatment [47].

C. Duplex Processes

As described in Sec. III.I, nitriding followed by an application of hard coatings is used to improve the performance of tools and other parts in which the wear resistance at elevated temperatures is required.

The optimum case and process characteristics are strongly dependent on application. For tools where a relatively thin (50–80 μm) nitrided layer is required (and no compound layer) the best solution is represented by integrated, continuous duplex processes taking place consecutively in the same retort: plasma nitriding + PVD, or plasma nitriding + plasma-assisted CVD (PACVD). Gas nitriding, if used, has to be carried out separately in conditions that prevent the formation of continuous compound layer; otherwise the adhesion of subsequently applied hard coatings is strongly impaired—especially in the presence of porous outer zones. Polishing after nitriding improves adhesion and eliminates the differences between results of plasma and gas processes. PVD methods are superior to PACVD because of a wider variety of obtainable coatings and because of lower temperatures, affecting less the properties of the combined case system.

The requirement of strong support by the underlying material, typical in applications of hardened and tempered low-alloy steels, can be satisfied only by relatively thick (>0.25 mm) nitrided cases [48]. In practice it appears useful to separate the nitriding from the hard-coating process, running each in a suitably optimized installation. The nitrided case, suitable for subsequent hardfacing, should not contain any, or only thin non-porous, compound layers, and also low surface roughness. These requirements can be satisfied both by plasma as well as by gas nitriding.

Application of hard coatings on nitrided substrates with a compound layer necessitates taking steps to avoid denitriding. When highly uniform distribution of residual stresses and precise hardness profiles are required, as for instance in parts subjected to contact fatigue, it is recommended to use a nitriding process producing a compound layer, with a subsequent mechanical removal of the latter [48].

IX. INDUSTRIAL IMPLEMENTATIONS OF THE NITRIDING PROCESS

A. Classification

In present day nitriding technology, the following methods can be distinguished:

Conventional gas nitriding and nitrocarburizing (controlled by dissociation rate)
Controlled potential nitriding and nitrocarburizing
Salt-bath nitriding and nitrocarburizing

Plasma (also known as "ion" or "glow discharge") nitriding and nitrocarburizing
Fluidized bed—a modification of the gas process

These methods differ by the mechanism transporting nitrogen from the medium to the steel surface. The diffusion of nitrogen atoms from the surface into the core obeys the same laws, formulated by A. Fick, regardless of the nitriding method adopted.

B. Gas Nitriding

The principle of gas nitriding is described in detail in Sec. III. In industrial applications, the surface of nitrided parts and the inner surface of the retort act catalytically, promoting the cracking of ammonia. Dissociation rate is directly related to the size of the retort, load surface area, and temperature and inversely proportional to the atmosphere flow rate. Figure 34 shows the dependence of dissociation rate on temperature for several flow rates in a typical medium-size, pit-type nitriding furnace.

The atmosphere entering the furnace retort may be ammonia only or may be diluted by, e.g., nitrogen or cracked ammonia. Dilution changes the relationship between atmosphere flow rate and ammonia dissociation rate.

Addition of carbon-bearing gases to the nitriding atmosphere, as in gas nitrocarburizing, increases the complexity of gas reactions and, for a controlled process, requires a system to maintain both the nitriding as well as the carburizing potentials. It has been indicated in Sec. III that addition of carbon from the atmosphere enhances growth kinetics of the white layer and stabilizes the ε phase. Its positive effect on fatigue strength of the layer has been reported, but because of insufficient

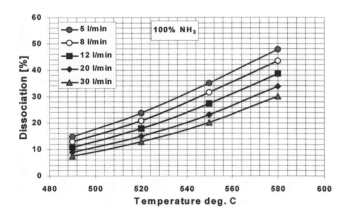

Figure 34 Effect of temperature on ammonia dissociation rate for flow rates. Medium-size pit nitriding furnace.

structural data, the advantages of nitrocarburizing over nitriding are not in clear evidence.

An advantage of the gas nitriding or nitrocarburizing process over other methods, e.g., plasma or salt bath, is flexibility in load arrangement inside the retort. Thus, it is possible to treat loosely heaped and stacked parts whose surfaces touch, with same results as those obtained on parts that are spaced from one another.

Listed below are several well-known proprietary versions of the gas nitriding process.

- Floe process—a double-stage, gas nitriding process in which the first stage is characterized by a dissociation rate of 15–30% and the second stage by dissociation rate increased to 65–80%, regardless of temperature.
- Malcomizing®—gas nitriding process involving a proprietary method of activation of stainless steel surfaces by PVC pellets (trademark of Crane Corp., U.S.A.).
- Nitrotec®—process involving postnitriding oxidation and quench, resulting in a protective black surface, composed of Fe_3O_4 (trademark of Lucas Industries, UK).
- Nitemper®—nitrocarburizing process, employing 50% NH_3 and 50% endogas (trademark of Ipsen Industries, U.S.A.)
- Nitreg®—utilizing total integrated automatic control of nitriding potential and all other process parameters, with in-process surface activation for stainless steels. (trademark of Nitrex Metal Inc., Canada)
- ONC®—the Nitreg® process combined with integral postnitriding oxidizing
- Nitreg®—C—a gas nitrocarburizing process with K_N and K_C control
- High-pressure and low-pressure processes [5]

C. High- and Low-Pressure Processes

All versions of nitriding have their specific limitations that encourage further attempts at diversifying technological solutions. Related to this quest for improvement are gas nitriding processes employing pressures lower or higher than atmospheric.

1. High-Pressure Nitriding

High-pressure nitriding requires expensive equipment, the cost rising with retort dimensions and process temperature. Existing units, therefore, feature relatively small work zones. The potential-control system, involving constant, or pulsed pressure is also more complex and expensive than that for nitriding at atmospheric pressure.

2. Low-Pressure Nitriding

Several proprietary low-pressure nitriding processes [18,49] have been implemented. The atmosphere, of the NH_3–N_2 type, usually contains a minor amount of an oxidant, e.g., N_2O, to promote activation of the metallic surface, improving growth kinetics and uniformity of the nitrided layer. The process is run in the temperature range typical for gas nitriding, and at pressures of 200–400 mbar. For this technology a vacuum furnace is required. Among the benefits of low-pressure nitriding, lower gas consumption in comparison with atmospheric-pressure gas nitriding has been claimed but this has not been adequately demonstrated in industrial practice. A denser, less porous white layer has also been claimed.

D. Salt-Bath Nitriding

Because salt-bath nitriding causes simultaneous diffusion of nitrogen and carbon into the steel, it is in reality a version of ferritic nitrocarburizing. The salt-bath process is carried out in a pot, in which powdered salt is melted by gas or electrical heating. Parts to be nitrided are suspended in the molten salt. The first salt baths contained predominantly sodium cyanide (NaCN), potassium cyanide (KCN), and potassium carbonate Na_2CO_3. This process is known as Tufftride in the United States and Tenifer in Europe. At the treatment temperature, typically 580°C (1076°F), the cyanide is oxidized to cyanate by the action of metered oxygen, introduced to the bath through a tube. The typical reaction is:

$$4NaCN + 2O_2 \rightarrow 4NaCNO$$

At the steel surface, the cyanate breaks down, releasing nitrogen and carbon, which are adsorbed and diffuse inward. This is described by the reaction:

$$8NaCNO = 2Na_2CO_3 + 4NaCN + CO_2 + C + 4N$$
$$\searrow \qquad \searrow$$
Diffuse into steel

The salt-bath process offers rapid heating and the possibility of rapid quenching because parts may be removed from the salt bath and immersed into any cooling medium of choice. The advent of the salt-bath process demonstrated for the first time that nitriding could be successful on carbon and low-alloy steel, provided the process was short (maximum 12 hr) to ensure limitation of white layer growth. On plain carbon and low-alloy steels, this white layer exhibited good hardness and wear resistance when not exceeding approx. 20 μm (0.0008 in.). Keeping process time short is also a technological necessity. During the process, the salt bath becomes progressively depleted of its active constituents. Besides, the level of the salt in the pot drops due to dragout on parts. Therefore, the salt pot must be periodically stopped for regeneration and refilling to its former level.

Short cycle times, due in part to rapid heating of the load immersed in the salt bath, can be viewed as an advantage of this process. The disadvantage is the aggressiveness that may cause oversaturation. Moreover, salt-bath processes are carried out at temperatures often higher than in gas, typically 570–590°C (1058–1094°F). Nitriding at these temperatures is successful on plain carbon and some low-alloy steels but for many grades with higher alloy contents these temperatures cause excessive tempering of the core and a resultant drop of hardness. Therefore, the salt-bath process can be used only on such steels and applications where core hardness is not an issue.

In the 1950s, problems related to health hazard and waste disposal posed by salt-bath nitriding were not adequately addressed. A new generation of nitriding salt baths, formulated in the late 1960s, was based mainly on sodium and potassium cyanates and carbonates. These baths are cyanide-free prior to use but the reaction at the surface of the nitrided steel itself produces cyanide, generally less than 3%. An example of this new technology is the Melonite process, also known in Europe as TF1 (trademark of Degussa, Germany). To convert the cyanide back to cyanate, an oxidizing quench bath operated at 400°C (750°F) is placed in the processing line. An off-shoot of Melonite is the quench–polish–quench (QPQ) process in which parts coming out of the oxidizing quench bath are polished and once again placed into the oxidizing bath for approx. 10 min, where they acquire a black finish.

Process control of salt baths is sometimes claimed in commercial literature; however, once a part is placed in the bath, processing conditions depend on the cyanate level and no in-process control is possible. The rate of its depletion depends on load size. Nevertheless, it can be said that many applications have been and are successfully nitrided by the salt-bath method, especially in those instances where deep cases are not required and process time must be short to meet productivity demands.

E. Plasma (Ion) Nitriding

Plasma nitriding, also known as "ion" or "glow discharge" nitriding, utilizes the effect of voltage applied between two electrodes in a sealed chamber to ionize gas under low pressure. Positive ions are accelerated toward the cathode where the potential drop emits visible radiation, popularly called "glow." The process gas contains nitrogen and its positive ions strike the treated surface connected to the cathode when a potential of several hundred volts is applied. The inner wall of the chamber is the anode. A nitrogen-bearing gas mixture (usually nitrogen and hydrogen) is introduced into the chamber while a pump continuously evacuates the gas to maintain a soft vacuum of 0.1 to 10 mbar. The principle of operation of the plasma nitriding furnace is shown in Fig. 35.

Besides time and temperature, plasma nitriding implements new control parameters: pressure, voltage, and current density. At higher pressures the current density is higher but the energy of individual ions is reduced. An advantage of this process is the possibility of nitriding stainless steels, which in gas require special activation of the surface. Cathode sputtering, i.e., removal of atoms from the workpiece surface by the force of striking ions, occurs throughout the process. Its intensity of material removal from the surface, depending on several factors, is of the order of 0.03–0.5 μm (1.1–19.6×10^{-6} in.) per hour. Greater intensity occurs in the initial stages of the process when the pressure is lowest and the mean free path of positive ions is longest. Accelerated ions develop higher velocities and strike the surface of the steel with greater kinetic energy, removing contamination, including oxide films, leaving the surface

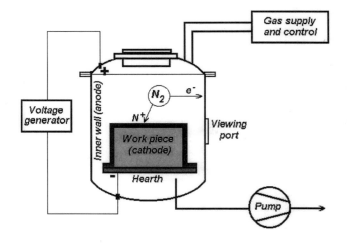

Figure 35 Principle of operation of a plasma nitriding furnace.

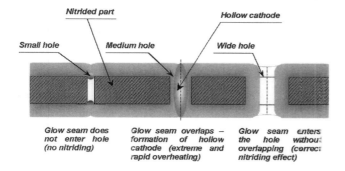

Nitrided part

Hollow cathode

Small hole   Medium hole   Wide hole

Glow seam does not enter hole (no nitriding)   Glow seam overlaps – formation of hollow cathode (extreme and rapid overheating)   Glow seam enters the hole without overlapping (correct nitriding effect)

Figure 36 Generation of the hollow cathode effect in a plasma nitriding process.

depassivated and ready to admit nitrogen atoms. However, at low pressures, the current density may be insufficient to attain the required surface temperature. Such cases are referred to as "weak nitriding response."

Plasma nitriding also presents certain specific difficulties. Problems exist with temperature uniformity and measurement because the process is carried out in partial vacuum where the only viable means of heat transfer is radiation. Thus, temperature readings by a thermocouple in spaces between parts are not representative. Thermocouples may not touch electrically charged nitrided surfaces, unless measures are taken to isolate the thermocouple circuit from the load. Temperatures may be read by an optical pyrometer but correction factors must be applied, and these vary with the time of process because of changing emissivity with progressing surface coverage by nitrides.

Results critically depend on part geometry and arrangement in the chamber. Certain minimum distances between parts must be maintained to prevent overheating.

The rate of nitrogen adsorption depends on current density that itself depends on the thickness of the "glow seam." At low pressures (below 3 mbar) this glow seam is thick and diffuse and carries a low current density, generating lower surface temperatures and a weaker nitriding response. A diffuse glow does not enter narrow and deep holes. Higher pressures (3–10 mbar) generate progressively thinner glow seams with greater energy. In order to enter holes or grooves it is often necessary to make the glow seam thinner but then some exterior exposed surfaces risk being overheated.

The pressure and corresponding glow seam thickness must be closely monitored through a viewing port throughout the process. If necessary, manual adjustments of pressure are made. Failure to monitor and to control the pressure may result in local overheating due to a phenomenon known as "hollow cathode."

The hollow cathode effect is demonstrated on a steel bar with three holes: small, medium, and big, placed inside the plasma nitriding furnace. The combination of parameters will determine a certain thickness of the glow seam. Figure 36 shows the principle of generation of a hollow cathode and concentration of heat energy, accompanied by rapid local overheating. This phenomenon may occur with different pressure–current combina-

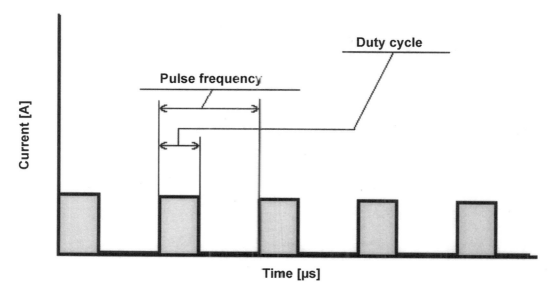

Duty cycle

Pulse frequency

Current [A]

Time [μs]

Figure 37 Principle of operation of the pulse-plasma system.

tions, when another size of hole (or gap between neighboring parts) favors the overlapping of two glow seams.

The problem is alleviated by the new-generation pulsed-plasma equipment, where the direct current is cut off for controlled periods of the order of microseconds. The system generates pulses and the heat energy dissipated at critical surfaces is reduced.

By controlling the ratio of the duty cycle to pulse frequency it is possible to avoid the hollow cathode effect in critical gaps and holes. Figure 37 shows the principle of operation of the pulse–plasma system.

Generally, plasma nitriding may yield excellent results but always requires a highly skilled, knowledgeable, and experienced operator. If it is required to obtain a white layer consisting of the ε phase, an addition of approx. 3–4% CH_4 is made to the nitriding gas mixture, thus accomplishing plasma nitrocarburizing. Moreover, because of the ease of removing passive films the plasma

Table 8 Comparison of Advantages and Disadvantages of Nitriding Methods

Method	Advantages	Disadvantages
Salt-bath nitriding	Rapid heating and processing Ease of obtaining good quality layers on low-carbon and low-alloy steels (if bath composition well-controlled) Possibility of quenching parts immediately after process	No in-process control Limited to those applications that can be heated to higher temperatures without losing core hardness Short processes only Requires thorough washing to remove salt residues Health hazard and expensive waste-disposal problems No possibility of masking
Plasma nitriding	Simple mechanical masking of surfaces to be protected from nitriding Ease of surface activation of stainless steels by cathodic sputtering Low-temperature nitriding possible	Difficult temperature control and measurement Poor temperature uniformity Process requires frequent human monitoring Results critically sensitive to part geometry and arrangement in furnace
Fluidized bed nitriding	Fast heating and cooling possible without the problems connected with salt-bath nitriding	High gas consumption due to necessity of running high flow rates to overcome sand bed resistance Difficult process control
Traditional gas nitriding	Low temperature in comparison with carburizing Simple control techniques	Controlling parameter (ammonia dissociation rate) inadequate for precise process control Problems with meeting zero white layer (where required)
Nitreg gas nitriding controlled by nitriding potential	Precise control of white layer thickness and phase composition Full automation Total integrated control of all process variables and functions Predictability and repeatability of nitriding results Meeting zero white layer (where required) possible Ease and simplicity of operation No finishing grinding required Low-temperature nitriding possible	Masking requires plating or application of protective pastes Stainless steels require special activation techniques

process is also suitable for treating Al and Ti alloys. Clean surfaces allow the duplex processes to be run, e.g., a PACVD process depositing TiN or TiC following a prior nitriding treatment in the same enclosure.

F. Comparison of Main Nitriding Methods

Advantages and disadvantages of nitriding methods are compared in Table 8.

G. Typical Industrial Furnaces

1. Gas Nitriding Furnaces

A pit-type gas nitriding furnace is shown schematically by Fig. 38. It comprises an outer metal shell (1) with a flanged, water-cooled lid (2). A fan (3) is powered by a motor (4). The furnace is fitted with a retort (5), inside which an insert (6) is placed. The retort is usually made of a heat-resistant alloy, e.g., Inconel 600. A deflector (7) directs the atmosphere into the space between the insert and the retort wall. Resistance elements (8) are situated between the retort and the outer shell. The atmosphere flows down, around the insert, and back up through the workspace, over the rack (9) with the load (10). The atmosphere inlet (11) and outlet (12) are situated in the furnace lid. Thermocouples (13) control the furnace temperature. The lid is usually sealed by high-temperature-resistant rubber seals (14).

Nitriding furnaces have been built over the years in a variety of shapes, in all of them, the objectives being:

Sealing of gases inside the vessel
Heating and cooling efficiency of the equipment

Stability of control parameters
Ease and low cost of installation
Convenience of use

Design features of other types of gas nitriding furnaces are given in Table 9.

2. Plasma (Ion) Nitriding Furnaces

Plasma nitriding furnaces are similar in shape to gas types—most are vertically oriented cylinders. In some cases the shell of the furnace is composed of segments that may be stacked, allowing loads of different heights to be accommodated easily. Plasma nitriding furnaces are electrically heated. Electric power is used to simultaneously ionize the gas and heat the load. In older units heating was accomplished by the kinetic energy of ionized particles striking the steel surface. Modern units have auxiliary resistance heating elements to accelerate heating to process temperature. The furnace must be extremely well sealed, as most of the processing is conducted in partial vacuum.

3. Molten Salt Nitriding Furnaces

A suitably designed vessel of any shape may serve as a salt-bath furnace. Salt pots are often covered but not sealed. A typical salt-bath furnace in cross-section is shown in Fig. 39.

H. Auxiliary Equipment

Depending on the installation, an appropriate collection of auxiliary equipment is necessary to efficiently and safely process the components. Such equipment and services may include the following: water-cooling systems, gas supply with pressure control, crane or hoist, electric power with suitable transformers or generators, air compressors, racking, washers, exhaust neutralizers, and ammonia dissociators.

I. Process Control Systems

1. Functioning/Hardware Controls

Designed to perform their specified functions, the furnaces are equipped with such features as loading devices, lid or door closing/locking mechanisms, load-moving transport systems in flow-through furnaces, gas supply valves, pressure switches, cooling water controls, power supply controls, overtemperature protection units, etc. Many of these are wired to provide a variety of alarms in the event of a malfunction.

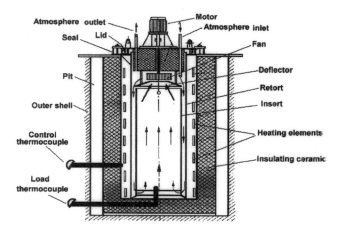

Figure 38 Pit-type furnace used for gas nitriding (schematic).

Table 9 Comparison of Main Features of Features of Gas Nitriding Furnaces of Different Design

Feature	Pit-type retort	Bell	Horizontal	Semicontinuous	Fluidized bed
Vessel	Vertical cylinder	Vertical cylinder	Horizontal cylinder with box-like work zone	Horizontal cylinder or box, multichamber	Generally a vertical cylinder
Loading	Lid raised and swung out; basket or fixture lowered into retort from top	Entire outer shell lifted off the load, which rests on base. Load placed on base	Front door on hinges	Front and back doors on hinges	Top, lid usually with some lifting mechanism (chain, levers), some installations may require overhead crane
Fan	Vertical shaft, mounted below lid	In the base	In any wall or in door	In any wall or in the doors	None, gas circulation achieved by feeding it through a sand bed, with parts immersed in the sand
Sealing	Gasket, compression seal (high temperature rubber type)	Gasket, compression seal (high temperature rubber type) or sand	Gasket, compression seal (high temperature rubber type)	Gasket, compression seal (high temperature rubber type)	Gasket, compression seal (high temperature rubber type), oil channel
Heating	Generally electric. Elements mounted around retort	Electric or gas	Generally electric, elements mounted in removable outer shell	Electric or gas	Electric or gas
Cooling	Forced cooling by nitrogen circulation. May be accelerated by turbo (forced extraction of atmosphere)	Accomplished by removing the outer shell and allowing the retort-covered parts to cool down in ambient air	None, blower (external), or forced (internal)	None, blower (external), forced (internal)	None, blower (external), forced internal possible
Installation	Sunk in pit with lid approx. at floor level or placed on floor with access by mezzanine	Base flush with shop floor	Body of furnace usually on its own legs, floor mounted	Body of the furnace usually on its own legs, floor mounted	Usually in a pit except for vary small units located on the shop floor

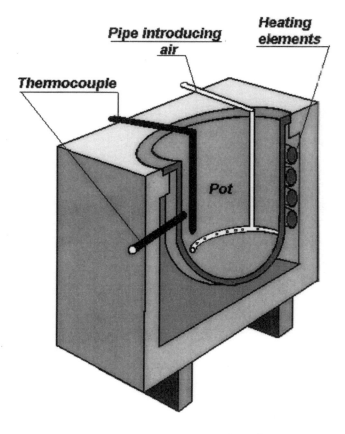

Figure 39 Cross-section of a salt-bath furnace.

2. Heating and Cooling Controls

The main objective of the temperature control system is to heat the load while preventing temperature from overshooting the set point by more than a few degrees, and to stabilize the temperature at the given setpoint. This is accomplished by algorithms programmed in dedicated controllers or as part of the control software for the entire system. PID controllers are frequently used, with fuzzy logic being recently employed by certain manufacturers. In some cases, limiting the heating rate to avoid distortion is desirable. This function is programmed into the more sophisticated systems. Temperature uniformity throughout the furnace, essential for uniform results, depends on design features and good insulation. Subdividing the furnace into several heating zones, with independent temperature control, improves temperature uniformity.

3. Flow Controls/Gas Supply

In gas and plasma installations control of gas flow into the vessel is critical. In primitive, older installations it may be a manually operated on/off valve that opens gas

flow at a steady rate. Modern installations are equipped with flowmeters, flow rates being controlled by solenoid valves, or electronically operated mass-flow controllers.

4. Pressure Controls

Changing the temperature and gas flow in a sealed vessel produces variations in pressure, requiring control. In most gas nitriding furnaces this is applied on the exhaust end. Motorized, computer-controlled exhaust valves utilizing pressure transducers are used in modern equipment. The main objective of pressure control is to prevent ammonia from escaping the vessel, and air from being sucked in. This may be accomplished by devices as simple as an oil trap, with computer-controlled valves at the other end of the spectrum. In certain proprietary nitriding methods partial vacuum or elevated pressures are used, requiring sophisticated control systems and vacuum pumps.

Because plasma is generated in a partial vacuum and pressure control is critical, vacuum pumps and relatively sophisticated pressure controls are required.

X. INSPECTION AND TESTING

A. General

A control system ensuring that all nitrided parts or tooling leaving the facility are of good quality and that all process results meet specification requirements, encompasses in-process control, which may vary, depending on furnace design and nitriding method used, as well as postnitriding inspection, which is basically the same, regardless of the method of nitriding employed. While the first is aimed at ensuring proper processing, the latter verifies compliance of obtained results with specification requirements.

Nitriding specification requirements may range from a simple surface hardness test to determination of white layer thickness, case depth, and microstructure. Specified sometimes are surface roughness, dimensional changes, and service properties, e.g., wear and/or corrosion resistance, fatigue, and impact strength. In critical cases, the specification may call out the phase composition of the nitrided layer. Such in-depth testing is, in most cases, outside of the scope of an industrial-quality inspection laboratory and delegated to specialized sources. This section describes equipment and methods used to test nitrided objects in industrial metallurgical laboratories. Because metallographical testing is destructive, quality inspection is, in most cases, carried out on samples or coupons placed in the nitriding furnace together with the load. Such samples or coupons must be made of the same material (preferably from the same heat) and have the same prior heat treatment history as the nitrided batch. In some cases, the customer allows one or more parts from the nitrided batch to be destroyed for quality inspection.

B. Quality Control Equipment and Methods

Most contemporary industrial metallurgical laboratories have facilities for verification of the following basic parameters, constituting specification requirements:

Surface hardness
White layer thickness
Nitrided case depth

Some laboratories are additionally equipped to measure surface roughness, dimensional changes, and corrosion resistance.

Industrial facilities required to carry out basic quality inspection must possess the necessary equipment for *preparation* of specimens and for the actual inspection.

The sequence of operations carried out from the moment of removal of the nitrided specimen from the furnace to the certification of results obtained is as follows:

1. Sectioning of specimen, in conditions avoiding overheating

2. Nickel-plating for edge protection (optional)
3. Mounting of specimen, to facilitate preparation of selected cross-section
4. Grinding and polishing of surface
5. Etching, to reveal the microstructure of the nitrided layer and/or of core
6. Drying
7. Metallurgical evaluation
8. Certification

The above operations are described in the ASTM Standard E3-95 "Standard Practice for Preparation of Metallographic Specimens."

Not requiring any special preparation, except for light cleaning with 360 or finer grit emery paper, surface hardness may be measured on a portion of the specimen at any time.

Table 10 gives a list of basic equipment required to prepare specimens and to carry out quality inspection of nitrided parts in an industrial metallurgical laboratory.

C. Inspection

1. Surface Hardness Testing

Surface hardness testing is carried out on surfaces cosmetically cleaned by fine emery paper to facilitate a good contrast view of the indentation. The measure-

Table 10 Basic Equipment and Supplies for a Metallurgical Laboratory

Designation	Operation	Name of equipment	Consumables
Preparation	Specimen sectioning for inspection	Abrasive saw	Abrasive wheels, coolant
	Edge protection	Nickel plating bath	Chemicals, electrodes
	Specimen mounting	Mounting press, Mold	High hardness thermosetting compound
	Mount grinding and polishing	Polishing stand or polishing machine	Emery papers, cloth, diamond paste, alumina suspension
	Specimen washing	Exhaust hood, spray bottles	Water, alcohol
	Specimen etching	Exhaust hood, spray bottles, Petri dish	Chemicals, water, alcohol
	Specimen drying	Exhaust hood, specimen dryer	
Inspection	Surface hardness measurement	Vickers or Knoop superficial hardness tester	
		Rockwell superficial hardness tester	320–600 grit emery paper
		Portable hardness tester	
	Surface roughness measurement	Roughness tester	
	White layer thickness measurement	Metallographic microscope with calibrated reticle	
	Microstructure evaluation	Metallographic microscope	
	Effective case depth	Microhardness tester (Vickers or Knoop)	

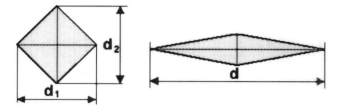

Figure 40 Appearance of the Vickers and Knoop micro-hardness indentations.

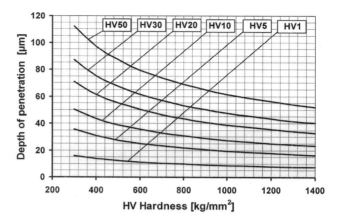

Figure 42 Correlation between hardness and penetration depth for various loads in the Vickers method.

ment can be made using a Vickers or a Knoop indentor under a load of 1 to 30 kg (9.81 to 294.3 N) and measuring the diagonals of the indentation.

The Knoop method is based on the same principle, but the indentor is of a different shape, penetrating shallower zones. The Vickers indentor is a square pyramid with an angle of 136°. The Knoop indentor is a flat pyramid with the cross-section of an elongated diamond. Figure 40 shows the appearance of the Vickers and Knoop indentations.

Vickers and Knoop hardness values deviate slightly from each other. Figure 41 compares Vickers and Knoop hardness relative to the HRC scale. Tables with conversion between HRC, HR15N, HR30N, HV, and HK hardness can be found in most handbooks [24].

The depth of penetration in the Vickers and Knoop methods depends on hardness of the tested material and on the applied load. Measurements taken on nitrided surfaces under low loads (especially microhardness readings below 1 kg) are usually higher because they reflect the hardness of the external portion of the layer, while measurements taken under increasing loads reflect

the effect of lower hardness of the deeper lying zones. Figure 42 shows the correlation between hardness and penetration for various loads in the Vickers method. In the Knoop method, the small size of the minor diagonal permits indentations to be made closer to one another. Its penetration depth is also shallower than that in the Vickers method (Fig. 43).

2. Portable Hardness Testing

In some cases the use of a regular superficial hardness tester may not be possible. This occurs in field conditions and on surfaces of big parts that do not fit on the hardness tester stage. For such needs, portable hardness testers are used. These are usually compact assemblies, comprising an indentor moving inside a cylinder, and

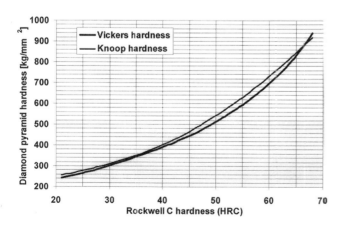

Figure 41 Comparison between Vickers and Knoop hardness, relative to the HRC scale.

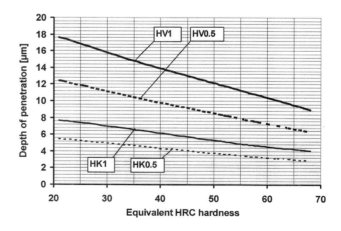

Figure 43 Comparison between depth of penetration for the Vickers and Knoop methods.

connected electrically to a circuit converting electronic signals to hardness values, displayed on an LCD panel. The measuring cylinder with the indentor is pressed by hand against the measured surface. This tester requires some practice with positioning of the indentor normal to the surface for proper results. On flat surfaces, this is ensured by a special attachment ("foot") resting on the measured surface and fitted with a hole through which the indentor is pressed. The advantage of this equipment is compactness and possibility of field use. The disadvantage is low accuracy.

3. Measurement of Effective Case Depth

The determination of effective case depth consists of taking microhardness measurements at known intervals from the surface in the direction of the core, thus creating a microhardness profile, and measuring the depth at which the hardness is at a specified value. Readings are taken with Vickers or Knoop indentors, but with lighter loads, from 1 kg down to 10 g. The resultant indentations are accordingly smaller and measured with the accurate eyepiece of a microhardness tester (mounted on or integrated with an incident-light microscope). Such small indentations can be made relatively close to one another, thus obtaining microhardness traverses with test points separated by intervals as little as 25 μm (0.01 in.). Different standards may be used for determining the effective diffusion (hardened) case depth. These are usually defined by the part designer, but all spell out a depth at which the hardness drops to an arbitrarily defined value, e.g., 50 HRC (equivalent) or 50 HV above core hardness. Microhardness measurements are usually made on unetched sections. Measurements of particular phases, usually taken under smallest loads in order to fit the indentation within a limited area, are made on etched sections to distinguish the microstructure. A typical microhardness tester is shown in Fig. 44.

4. Surface Roughness Testing

Surface roughness testing is carried out by very sensitive instruments called roughness testers or profilometers. Such instruments are equipped with a sharp-pointed stylus that precisely follows surface asperities over a standardized measuring length. Its movements are amplified and recorded. An electronic processor calculates the required roughness parameters, e.g., center line average (R_a), root mean square roughness (R_q), or maximum peak-to-valley height (R_t).

Modern roughness tester models feature several options, e.g., measurement in metric or imperial units. If required, the equipment is capable of presenting the

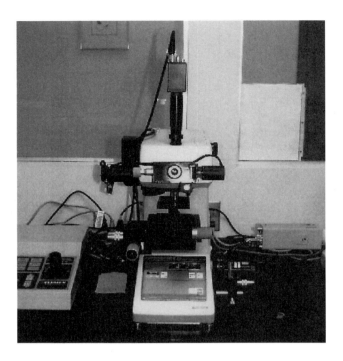

Figure 44 Microhardness tester.

roughness profile graphically, with asperities magnified electronically.

5. Metallographic Evaluation

Metallographic evaluation of a nitrided surface entails measurement of the thickness of the white layer, depth of the diffusion case, where revealed by etching, and an assessment of microstructure. This may involve a broader scope of possible issues, as needed, such as:

Uniformity and continuity of white layer thickness
Uniformity and continuity of diffusion case
Presence of nitrides along grain boundaries
Presence of structural flaws, e.g., microcracks
Surface decarburization
Grain size
Verification that the nitrided material is the one specified

All of the above assessments can be accomplished with the aid of a metallurgical microscope. Typical specifications require all evaluations to be carried out at magnifications of not less than 400X. Modern metallographic microscopes are equipped with filters used for image enhancement, as well as additional optical systems, like the Nomarski differential interference contrast (DIC), enabling a three-dimensional view of microstructural features. Metallurgical microscopes are also equipped with a camera for recording microstructures. Digital cameras transmit the viewed image to a TV monitor allowing quantitative image analysis, simultaneous viewing by more observers, and storing of images as digital files.

XI. TYPICAL NITRIDING FAULTS AND THEIR PREVENTION

A. General

Heat treatment is often indiscriminately blamed for many nonconformances. Nitriding, in particular, often receives blame for faults both caused by this process and those "inherited" by it. Many faults, evident only upon completion of the nitriding process, originate during earlier stages of manufacturing. These cannot be corrected by nitriding. Results of the nitriding process are also very sensitive to preconditions; hence some faults may occur during nitriding, although not caused by it. Some faults are attributable to poor process control. A new approach, based on control of the nitriding potential, is able to eliminate some of the most common problems related to the process itself.

B. Classification of Faults

Table 11 classifies faults most commonly encountered in nitrided parts and tooling, originating before, during and after the nitriding process.

C. Faults Originating During and Attributable to Nitriding

Faults that originate during nitriding may be associated with the adopted method, i.e., salt bath, plasma (ion), or gas.

In the salt bath process, a typical problem is the possibility of growing an excessively thick white layer because of the very aggressive and rapid nitriding reaction taking place in that medium. Such a layer usually contains a higher concentration of nitrogen, conducive to brittleness. Even when the salt-bath composition is well engineered, once a part is immersed in the bath, there is practically no in-process control. For that reason, salt-bath nitriding has developed mainly as a short-cycle process.

In plasma (ion) nitriding, among the most commonly encountered problems is nonuniformity of layer thickness along contours of nitrided parts. This is not usually

Table 11   Classification of Faults in Nitrided Steel Components

Faults manifest during service	Fault originating	Operation during which fault originates	Root causes
Nonuniform diffusion layer, microcracks at grain boundaries, surface cracks, lips	Before nitriding	Steel making process	Ferrite banding, carbide and/or sulfide segregation, laps, pipes, etc.
Cracks, low surface hardness, low core hardness, exfoliation of layer from surface, excessively thick white layer	Before nitriding	Heat treatment preceding nitriding	Decarburization, overheating, temper embrittlement
Surface cracks, portions of surface unnitrided, oversaturation in fillets and corners, nonuniform and noncontinuous layer	Before nitriding	Machining prior to nitriding, grinding and/or EDM	Grinding cracks and burns, burrs, sharp corners, machining stresses, recast layer on surface
Soft spots on surface, partial or complete surface left unnitrided	Before nitriding	Cleaning and activation	Solid precipitations on the surface, greasy film, oxidation
Nonuniform and low surface hardness, brittleness, generally poor service properties, poor lubricity at surface, excessive white layer	During nitriding	Nitriding process	Insufficient or ineffective process control
Flaking, cracking	After nitriding	Postnitriding preheating of dies	Excessive oxidation

encountered in gas nitriding, as seen in Fig. 45. Another problem is the difficulty and sometimes even inability to nitride inner surface of small and deep bore holes.

Plasma nitriding requires close control of pressure and current density at the surfaces of the nitrided parts that directly affect the thickness of the glow seam. More on the subject can be found in Sec. IX of this chapter.

In conventional gas nitriding, the often-encountered practice of using raw ammonia as the only gas is conducive to high surface nitrogen concentrations, resulting in thick and brittle white layers. Many nitriding shops have adopted atmospheres diluted by nitrogen or by dissociated ammonia. Such atmospheres have proven themselves to be advantageous, giving the process more flexibility and producing layers with greater ductility. The most commonly encountered faults originating during gas nitriding and their root causes are summarized in Table 12.

In potential-controlled gas nitriding, most of the typical faults connected with conventional gas nitriding are overcome by the control of the nitriding potential.

D. Examples of Typical Faults Encountered in Gas Nitriding

1. Nonuniformity of the Nitrided Layer

In particular, the following factors affect the uniformity of the nitrided layer and the general results of the nitriding process:

> *Local decarburization* from the forging or rolling process, not removed due to shallow machining ("skinning"). This may cause the nitrided layer to become thicker in affected areas, as well as induce undesired tensile residual stresses. Deep local decarburization causes the resultant white layer to be excessively thick and brittle. This fault, originating in an earlier stage of the manufacturing process, equally affects subsequent nitriding, regardless of method used.

> *Overheating during heat treatment prior to nitriding*, due to poor thermal control, usually affects corners and thin cross-sections, causing local flaking of the subsequently nitrided surface. It may also cause undesired tensile residual stresses. As in the case described above, overheating during heat treatment affects results of subsequent nitriding, regardless of method. Overheating may also occur in uncontrolled plasma nitriding due to cathodic sputtering.

> *Grinding burns* may cause severe local oxidation and inhibition or, at least, retardation of nitriding, resulting in uneven or discontinuous layers, regardless of method.

> *Electrical Discharge Machining (EDM)* usually leaves a defected, recast layer on the surface, which may totally or partially inhibit nitriding. For uniform nitriding results, this defected layer must be removed, usually by grinding, prior to nitriding.

For uniformity of nitriding from part to part within a batch, it is necessary to ensure:

> Temperature uniformity within $\pm 5°C$ ($\pm 10°F$) throughout furnace retort.

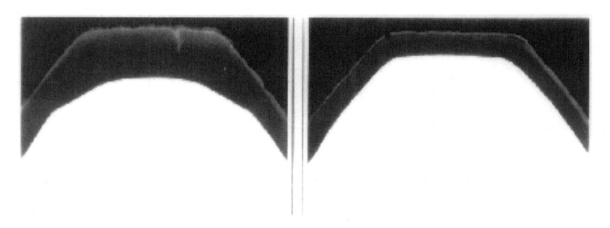

Figure 45 Nitrided case on a gear tooth: non-uniform in plasma nitriding (left) and uniform after potential-controlled Nitreg® process, (right).

Table 12 Typical Faults Attributable to the Gas Nitriding Process and Their Root Causes

Feature regulated by specification requirements	Type of fault	Root cause
White layer (WL) thickness	WL excessively thick	Nitriding conditions too aggressive
		Process time too long
	WL uneven/discontinuous	Incorrect prenitriding conditions
		Bad surface condition
		Insufficient surface cleaning
Porous zone of WL	Porous zone excessive	Prenitriding or nitriding conditions too aggressive, (nitrogen concentration too high)
		Process temperature chosen too high
Diffusion layer thickness	Diffusion layer uneven/discontinuous	Uneven/discontinuous white layer
Microstructure of diffusion case	Network of nitrides in diffusion layer	Nitriding conditions too aggressive, nitrogen concentration too high
Surface hardness	Surface hardness too low	Improper process temperature
		Porous zone too thick
		Incorrect process time (too short or too long)
		Nitriding conditions insufficiently active
		Incorrect atmosphere circulation
	Soft spots on nitrided surfaces	Diffusion layer uneven/discontinuous
		Bad surface condition
Case microhardness profile	Failure to meet specified hardness at given depth	Nitriding conditions insufficiently active
		Incorrect process temperature
		Incorrect process time (too short or too long)
Shape and size of part	Deformation	Incorrect fixturing or location of parts in retort
		Nonuniform layer thickness on different areas of surface
		Nitriding temperature too high
Surface roughness	Excessive roughness	Porous zone too thick
		White layer too thick

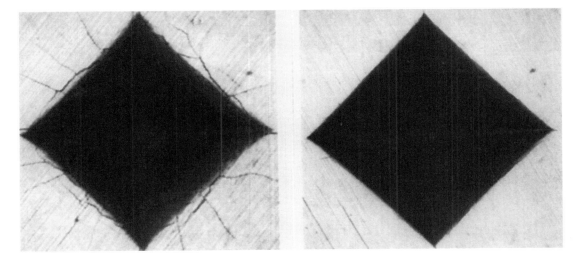

Figure 46 Vickers indentations. Left: conventional gas nitriding process (brittleness of surface revealed by cracks); right: potential-controlled Nitreg® process.

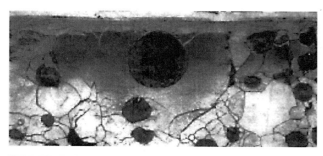

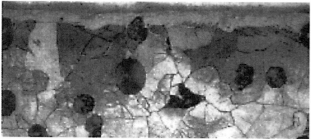

Figure 47 Comparison of microstructure of cast iron parts, obtained with two different nitriding potentials resulting in different degree of porosity. Etched by 3% Nital. Magn. 400×.

Uniformity of atmosphere chemistry within furnace retort. This requires adequate atmosphere recirculation within the retort.

Cleanliness of retort internals and load surface. Contamination of some parts or of the furnace retort may result in precipitation of deposits on

the surface to be nitrided, causing local barriers for the absorption of atomic nitrogen.

Prevention consists of avoiding sharp corners, or where they cannot be avoided, of running the nitriding process at lower nitriding potential values, to limit nitrogen concentration.

2. *Brittleness of the White Layer*

For a nitrided layer to have good load-bearing capacity and not exhibit brittleness, the joint nitrogen and carbon concentration at the surface should not exceed 8.5 wt.% [25]. In a conventional gas process, where the dissociation rate is the only measured parameter, white layer brittleness is common. Such layers often require expensive removal.

A good measure of layer brittleness is given by the density and length of cracks formed at the corners of a microhardness indentation. It is known from fracture mechanics that when a hard indentor is driven into relatively softer material, the material deforms to counteract stresses formed around the indentation, until its plastic deformability is exhausted. A more ductile material will thus deform while a brittle material will crack.

Figure 46 shows a comparison between two indentations, both on 4340 steel, nitrided to the same specification. In both cases the underlying diffusion case is of same depth and the white layer of same thickness. The size of the indentations, both made under a 30-kg load is the same, indicating same hardness. It is seen how

Figure 48 Corner effect on H13 steel. Left: conventional gas nitriding; right: potential-controlled Nitreg® process. Etched by 3% Nital. Magn. 100×.

brittleness, a commonly encountered white layer fault, is prevented through control of the nitriding potential.

3. Uncontrolled Porous Zone in White Layer

The porous zone in the white layer is associated with the formation of the ε nitride. Because of thermodynamic instability of the ε phase at elevated temperatures, partial recombination to molecular nitrogen occurs, forming clusters of trapped gas, which create porosity. If uncontrolled, this porosity will grow to the point where the white layer becomes spongy and fragile, losing its wear resistance characteristics. This is a serious fault of the nitriding process whose main objective is to enhance wear resistance. However, some porosity is considered benign, especially in cases requiring surface lubricity and antiseizure characteristics.

Figure 47 shows two examples of a white layer obtained on a cast iron differential housing. The one with little to no porosity and a surface hardness of 620 HV1 (top) failed in field tests. By contrast, the one with porosity covering approx. 45–60% of the white layer thickness and a hardness of only 520 HV1 (bottom) performed well. The extent of porosity depends on nitriding conditions, hence on nitriding potential control. In this particular case, better results were obtained with a higher nitriding potential value.

4. Corner Effect

The "corner effect" is caused by simultaneous nitrogen diffusion from two convergent directions, resulting in local oversaturation. This causes local embrittlement, hence spalling.

Although convergent diffusion at corner areas is unavoidable, appropriate control of the nitriding potential in gas nitriding makes it possible to avoid oversaturation in nitrided parts of tooling while maintaining sufficient diffusion of nitrogen in areas remote from corners.

Figure 48 shows two examples of corners of a nitrided extrusion die, and the difference caused by appropriate control of the nitriding potential.

REFERENCES

1. Chizhevski, N. The occurrence and influence of nitrogen on iron and steel. J. Iron Steel Inst. 1915, *92*, 47–105.
2. Comstock, G.F.; Ruder, W.F. The effect of nitrogen on steel. Chem. Metall. Eng. 1920, *22*, 399–405.
3. Fry, A. Nitrogen hardening of steel (in German). Stahl Eisen 1922, *42*, 1656.
4. Fry, A. Nitrogen in iron, steel and alloy steel; a new surface hardening process. Stahl Eisen 1923, *43*, 1371–1379. *in German.*
5. Vanick, J.S. Behavior of metals in hot ammonia gas. Proc. ASTM 1924, *24*, 348–372.
6. Guillet, L. The nitriding of steel and its technical applications. Comptes Rendus 1927, *185*, 818–821. *in French.*
7. McQuaid, H.W. Surface hardening by the nitriding process. Am. Mach., 28 Feb. 1929.
8. Kinzel, A.B.; Egan, J.J. Nitriding in molten cyanides. Trans. Am. Soc. Steel Treat. 1929, *16*, 175–182.
9. Floe, C.F. Method of nitriding. US Patent 2,437,249, 1948.
10. Mitchell, E.; Dawes, C. Surface treatment of mild steel. British Patent 1011580, 1961.
11. Lightfoot, B.J.; Jack, D.H. Kinetics of nitriding with, and without white layer formation. *Heat Treatment '73*; The Metals Society: London, 1973; 59–65.
12. Bell, T.; Birch, B.J.; Korotchenko, V.; Evans, S.P. Controlled nitriding in ammonia-hydrogen mixtures. *Heat Treatment '73*; The Metals Society: London, 1973; 51–57.
13. Czelusniak, A.; Morawski, C.D.; Liliental, W.K. Automatic nitriding potential control in gas nitriding. *Proceedings of the International Heat Treatment Conference, Equipment and Processes,* Schaumburg, Illinois, April 1994; ASM International, Materials Park, OH, pp. 449–454.
14. Eckstein, H.-I.; Lerche, W. Investigation of acceleration of nitriding in the gas phase. Neue Hütte 1968, *13*, 210–215. *in German.*
15. Stiles, M.; Dong, J.; Haase, B.; Haasner, T.; Bauckhage, K. Acceleration of the gas nitriding process by a pretreatment in a reactive gas phase. Härterei-Tech. Mitt. 1998, *53* (4), 211–219. *in German.*
16. Spies, H.-J. Progress in gas nitriding and nitrocarburizing of ferrous materials. Härterei-Tech. Mitt. 2000, *55* (3), 135–140.
17. Spies, H.-J.; Vogt, F. Gas oxynitriding of highly alloyed steels. Härterei-Tech. Mitt. 1997, *52*, 342–349. *in German.*
18. Has, Z.; Kula, P. The new Polish nitriding and nitriding-like processes in the modern technology. *Proceedings of the International Conference on Carburizing and Nitriding with Atmospheres*; Cleveland, Ohio, ASM International: Materials Park, OH, 1995; 227–231.
19. Nowacki, J. Morphology and properties of phosphonitrided layers. *Proceedings 11th Congress IFHT and Surface Engineering*; Associazione Italiana di, Metallurgia: Florence, Italy; 253–262.
20. Przyleck, Z.; Maldzinski, L. Carbides, nitrides, borides. 4th International Conference, Poznan, Poland, Politechnika Poznanska, 1987; 153–162.
21. Maldzinski, L.; Liliental, W.; Tymowski, G.; Tacikowski, J. New possibilities of controlling the gas nitriding process by utilizing simulation of growth kinetics of

nitride layers. Proceedings 12th International Conference on Surface Modification Technologies, Rosemont, IL, Oct. 12–14; ASM International: Materials Park, OH, 215–227.

22. Bell, T.; Li, Chen X. Stainless steel, low-temperature nitriding and carburizing, Adv. Mat & Proc. June 2002, 49–51.

23. Somers, M.A.J. Thermodynamics, kinetics and microstructural evolution of the compound layer; a comparison of the states of knowledge of nitriding and nitrocarburizing. Heat Treat. Met. 2000, 4, 92–102.

24. Sproge, L.; Slycke, J. Control of the compound layer structure in gaseous nitrocarburizing. J. Heat Treat. 1992, 9 (2), 105–112.

25. Edenhofer, B.; Lerche, W. Monitoring and control of nitriding and nitrocarburizing processes for predictable layer formation. Härterei-Tech. Mitt. 1997, 52 (1), 21–27. in German.

26. Spies, H.-J.; Schaaf, P.; Vogt, F. Effect of oxygen additions on gas nitriding. Mat-Wiss. Werkstofftech. 1998, 29, 588–594. in German.

27. Wünning, J. NITROC- a new process and installations for nitriding with an ε compound layer. Härterei-Tech. Mitt. 1974, 29 (1), 42–49.

28. Mädler, K.; Bergmann, W.; Dengel, D. "Austenitic nitrocarburizing with post-oxidation of non-alloyed steel." Härterei-Tech. Mitt. 1996, 51(6), 338–346. in German.

29. Preisser, F.; Seif, R. Method of nitriding work pieces of steel under pressure. US Patent 5,211,768, 1993.

30. Jung, M.; Hoffmann, F.; Mayr, P.; Minarski, P. High pressure nitriding. Proceedings of the 2nd International Conference on Carburizing and Nitriding with Atmospheres, Cleveland; ASM International, Materials Park, OH, 1995; 263–268 pp.

31. Chen, Tao; Chen, Binnan Jinshu Rechuli (Heat Treatment of Metals—in Chinese) 1998, No. 3, 5–8.

32. Jung, M.; Walter, A.; Hoffmann, F.; Mayr, P. High-pressure nitriding of austenitic stainless steel. Proceedings 11th Congress IFHT and Surface Engineering, Associazione Italiana di Metallurgia, Florence, Italy; 1998; 253–262 pp.

33. Spies, H.-J. The status and development of nitriding of aluminum- and titanium alloys. Härterei-Tech. Mitt. 2000, 55 (3), 141–150. in German.

34. Spies, H.-J.; Wilsdorf, K. Gas- and plasma nitriding of titanium and titanium alloys. Härterei-Tech. Mitt. 1998, 53, 294–305. in German.

35. Based on data from Handbook of Chemistry and Physics, CRC Press, 71st ed, 1990–1991, and from HJ Goldschmidt, Interstitial Alloys, Butterworths, London, 1967

36. Spies, H.; Goedicke, H. On the effect of the initial structure on the nitridability of steel. Neue Hütte 1984, 29 (3), 97–99. in German.

37. Larisch, B.; Spies, H.-J.; Brusky, U.; Rensch, U. Plasma nitriding of stainless steels at low temperatures. Proceedings 1st International Automotive Heat Treating Conference, Puerto Vallarta; July 1998; ASM International: Materials Park, OH, 221–228 pp.

38. Spies, H.-J.; Reinhold, B.; Wilsdorf, K.; Gas nitriding–process control and nitriding non-ferrous alloys. Surf. Eng. 2001, 17 (1), 41–53.

39. Berns, H.; Juse, R.L.; Bouwman, J.W.; Edenhofer, B. Solution nitriding of stainless steels–a new thermochemical heat treatment process. Heat Treatment of Metals, No.2. 2000, 39–45.

40. Muraleedharan, T.M.; Meletis, E.J. Surface modification of pure titanium and Ti–6Al–4V by intensified plasma ion nitriding. Thin Solid Films 1992, 221, 104–113.

41. Irretier, O.; Dong, J.; Haase, B.; Klümper-Westkamp, H.; Bauckhage, K. Effect of cleaner residues on gas nitriding. Härterei-Tech. Mitt. 1997, 52 (1), 32–38. in German.

42. Dong, J.; Haase, B.; Stiles, M.; Irretier, O.; Klümper-Westkamp, H.; Bauckhage, K. The effect of films of reaction products at steel surfaces on short-time gas nitriding. Härterei-Tech. Mitt. 1997, 52 (6), 356–364. in German.

43. Vermesan, G.; Lieurade, H.P.; Duchateau, D.; Ghiglione, D.; Peyre, J.P. Corrosion resistance of oxynitrided and oxynitrocarburized layers. Trait. Therm. April 1998, 307, 29–34. in French.

44. Dawes, C.; et al. Corrosion resistant steel component and method of manufacturing thereof. US Patent 4,496,401, 1985.

45. Dawes, C.; et al. Corrosion resistant steel component and method of manufacturing therefore. US Patent 4,596,611, 1986.

46. Ebersbach, U.; Vogt, F.; Naumann, J.; Zimdars, H. Effect of water vapor treatment on the corrosion behavior of nitrided and nitrocarburized 20MnCr5 steel, Part 2. Härterei-Tech. Mitt. 1998, 53 (1), 56–62. in German.

47. Khani, M.K. Post-nitriding oxidizing treatment of nitrocarburized layers—effect on fatigue resistance. Mat-Wiss. Werkstofftech. 1996, 27, 190–198.

48. Spies, H.J.; Höck, K.; Larisch, B. Duplex surface layer in the combined process: nitriding—hard coating. Härterei-Tech. Mitt. 1996, 51 (4), 222–231.

49. Foissey, S.; Atale, O.; Deramaix, C.; Jacquot, P. Low-pressure nitriding: Nitral Nitralox, Carbonitral. Proceedings 11th Congress IFHT and 4th ASM Heat Treatment and Surface Engineering, Associazione Italiana di Metallurgia, Florence, Italy 291–300.

15

Design Principles for Induction Heating and Hardening

Valentin S. Nemkov and Robert C. Goldstein
Centre for Induction Technology, Inc., Auburn Hills, Michigan, U.S.A.

I. INDUCTION TECHNIQUE IN METAL PROCESSING

A. Introduction

The induction technique has been around for almost 100 years, but only now have all of the pieces in place reached their full potential. Induction heating (IH) established itself as one of the key manufacturing technologies, especially in the automotive, metal production, forging, tube welding, and other metal processing industries. The current growth of induction heating includes the medical, food, chemical, electronics, and agricultural industries, to name a few.

The first industrial applications of induction heating were for melting and were developed in the United States near the dawn of the 20th century. Induction heating then became the method of choice for new metal melting installations. However, it was not until the mid-1930s that the first induction heat treating process was developed. Nearly simultaneously, TOCCO USA and Russia developed commercial processes for the induction hardening of crankshaft pins and journals. This very successful development helped to spur a new age of induction heating, and a great number of new heating and heat treating applications emerged throughout the world.

Despite the complicated mathematical description of induction heating, intensive efforts yielded the development of the theory and methods for the calculation of induction heating devices. The first profound technical book on induction heat treating, titled *Surface Harden-*ing by Induction Method* [1], was published in 1939 in Russia. Shortly after, several other books on the theory and application of induction heating were published in the United States and Russia [2–5]. These publications described the physical effects in induction systems and provided design guidelines for heating time, power, and frequency selection for typical applications, including mass heating, melting, and hardening (gears and other parts). They established a good basis for increased penetration of the induction technique into different industries such as automotive, military, railway transport, metallurgy, and some others.

Many new revolutionary processes appeared in the 1950s including crystal growth and induction tube welding. A long period of technology evolution followed the initial pioneering period [5–12]. At the end of the 20th century, a new period of rapid expansion emerged due to the following four factors:

Increased demand for advanced manufacturing techniques
Microprocessor control and intelligent control systems
Induction equipment improvements
Computer simulation of induction heating systems.

The increased demand from industry is based mainly on requirements for highly automated, controllable, and environmentally friendly manufacturing processes. Induction heating has special features, which help meet all of these requirements better than competitive technologies in many applications.

Microprocessor-based process control systems are replacing control systems of previous generations in all industries. Induction heating processes are ideal for intelligent control due to very fast responses to control signals, well-defined process characteristics, and individual handling of each part.

There have been major strides made in induction heating equipment including heat treating machines and power supplies. The most significant of these improvements has been in power supplies. Up until the early 1980s, there were three types of power supplies for induction heating processes: old motor generators, thyristor inverters, and tube oscillators. Motor generators generated frequencies up to 9600 Hz. Special motor generators were made for higher frequencies, but these were expensive and had low efficiency. Tube oscillators were used for frequencies 66 kHz and above. The drawbacks of tube oscillators are low efficiency, very high output voltages (typically around 10 kV), large size and weight, and limited tube lifetime.

Thyristor (SCR) solid state power supplies were introduced at the end of the 1960s. They were the main power sources for frequencies up to 30 kHz. New "transistorized" power supplies based on IGBT and MOSFET transistors expanded the range of frequency available with solid state generators to over 1 MHz. Modern units have smaller dimensions, greater flexibility, and higher efficiency. Some of them offer frequency variation in the ratio of 3:1, making the new power supplies much more versatile than their ancestors. Information about this equipment may be found easily via the Internet, periodicals (*Industrial Heating, Heat Treating Progress, Heat Treatment of Metals*, etc.), and other sources.

The last factor, computer simulation of induction heating systems, is also very significant. The complicated methods of calculation were very time-consuming and did not grant accurate results for most cases. Laborious and expensive experiments were necessary to verify calculations and to adjust the system performance for the design of induction heating devices. Only a few people at leading power supply manufacturers and in the academia were able to provide correct results, combining calculations, intuition, and their own experience.

That is no longer the case as the old ways are now replaced by computer simulation. The role of computer simulation is growing continuously throughout the world for system study, design, and optimization, and induction heating is no exception. The induction system design process is now faster, more accurate, and open to a much greater segment of the public. The results of computer simulation can be easily disseminated even to those who do not have a technical background. Instead of equations, generic numbers or graphs, two-dimensional (2-D) and three-dimensional (3-D) graphs, or color shades generated for each individual case provide a clear depiction of the process dynamics, which anyone can understand.

Due to the new situation, the goal of this chapter is to give the readers an overview of the techniques, basics, advantages, and specific features of induction heating that make this method unique. This method, often described by people in the industry as more art than science, is now accessible to a much wider population. For a more detailed understanding of induction heating theory, the following books are recommended for the basic [8] and more advanced [13,14] levels. Good sources for descriptions of induction heating applications are contained in the following books: *Electromagnetic Induction and Electric Conduction in Industry* [15], *Practical Induction Heat Treating* [16], and *Steel Heat Treatment Handbook* [17,18].

Many of the referenced books are relatively old and several of them are available only in Russian. These references were selected to give credit to those who have made significant, really pioneering contributions to the induction technique. Besides just recognition, there are also many interesting findings contained within these books, which did not receive wide-scale implementation at that time. Now, these ideas may be used successfully due to the new industrial situation and an improved level of technology.

B. Typical Applications of Induction Heating in Metal Processing

Induction heating is used in a wide variety of applications in the metal processing industry, from very large melting installations (tens of megawatts) to heat treating of small pins (hundreds of watts). Some of the processes induction heating is used for include the following:

Melting (metal production, refining, secondary melting, alloying, pouring, and material research)

Hot forming (rolling, pressing, forging, extrusion, drawing, and bending)

Heat treating (surface hardening, through hardening, tempering, annealing, and sintering)

Surface treatment (hard facing, coating, remelting, and curing)

Bonding/joining (tube and profile welding, contour welding, soldering, brazing, shrink fitting, and adhesive bonding)

Special applications (induction plasma, spectrometry, medicine, crystal growing, and cooking)

The processes of most interest for this handbook are found within the heat treating group.

The number of induction heating installations within the metal processing industry continues to grow and its use expands into other areas due to improvements in design tools and strategies, high frequency equipment, and general awareness of its advantages over competitive technologies.

C. Special Features of Induction Heating

The induction heating technique has several special inherent features, which are beneficial in many applications. Induction heating:

Has internal heat sources
Is a noncontact heating method
Can provide high power densities
Has high selectivity of heating in the depth and along the surface
Can work in any processing atmosphere (air, protective gas, and vacuum)
Has very low standby losses
Does not emit any physical pollution.

These physical and technical features give the following technological and economical advantages:

Short heating cycles and high production rates
Better metallurgical results (hardness, strength, and ductility)
Good control and high repeatability of the process
Small or negligible surface oxidation and decarburization
Possibility to treat parts "one by one" with individual cycle control
Low distortion due to localized heating in-depth and along the surface
Energy costs reduction due to local heating and low losses
Labor cost reduction in automated processes
Very short start-up times
Simpler and more economical steels and quenching media (water or polymer solutions instead of oil) may often be used with the same, or better, final part properties
A process that is very friendly to the industrial environment (no exhaust gases and other emissions, small size of equipment with high reliability, and excellent automation ability).

There are also some difficulties in the implementation of the induction heating technique compared to other heating methods. For effective heating, the part must be closely "coupled" with the inductor, which usually requires each part or family of similar parts to have its own induction coil. The laws of physics and technical limitations set some restrictions on achievable heat pattern. Induction process and coil design involve more complicated phenomena than other heating techniques, and are more knowledge-demanding than traditional methods. Computer simulation and modern manufacturing technology with computer aided design (CAD) and prototyping methods can provide an accurate design of induction processes, tooling, and equipment if the user has specific knowledge of the induction heating techniques [19].

II. BASICS OF INDUCTION HEATING

For an effective use of induction processes and equipment, knowledge of the basics is mandatory. At the present time, this knowledge must be in the understanding of phenomena, rather than in the quantitative description of dependencies and calculation methods, because computer simulation can provide very quickly the required solution. However, the user must know what changes to make and how to make changes to improve the process.

A. What Is Induction Heating?

Induction heating is electromagnetic heating based on energy absorption from an alternating magnetic field, generated by an inductor (induction coil or coil). Only conductive bodies can be directly heated by induction. Nonconductive bodies may only be heated indirectly. In this case, an intermediate conductive body (susceptor) is heated by induction, which in turn heats the nonconductive material by heat diffusion, convection, or radiation.

There are two mechanisms of electromagnetic energy absorption in the process of induction heating: eddy current losses and hysteresis losses. Hysteresis losses occur only in ferromagnetic materials. These losses are caused by the alternating magnetic field forcing the elementary magnetic areas (domains) of the material to turn back and forth following the direction of the magnetic lines. The "internal friction" of these domains generates heat. The contribution of hysteresis losses in the heating of compact materials is typically very small, less than 5% at low frequencies, and gets smaller as frequency rises. In practice, they are not usually taken into account. In some special cases, the contribution of hysteresis losses could be significant (e.g., in the heating of metal powders, ferromagnetic particulate materials, and ferrites).

Eddy currents are induced in a conductive body by an alternating magnetic field flowing through the body's cross section (i.e., by the field, "coupled" to the body). According to the laws of physics, eddy currents must always be closed, so there must be a conductive path or loop for the current flow within the body. If there is no conductive loop (e.g., in a thin, non-magnetic cut ring), there are no eddy currents and hence no heating. If we close the ring, there will be heating. A user of induction heating and a designer of induction heating devices must have good understanding of what are the possible paths of eddy currents, which could provide an effective heating of the required area [20].

Understanding induction heating requires knowledge of a group of mutually coupled processes: electrical processes in power supplying circuitry, electromagnetic processes in induction system (loaded coil), and electrothermal processes inside the workpiece.

B. Power Flow in Induction Heating Installation

An induction heating installation consists of alternating current (AC) power source (RF tube generator, transistor, or thyristor power supply), heat station, and induction heating coil or device (Fig. 1). Power supplies convert line frequency (50 or 60 Hz) electrical energy into a well-controlled power of "high frequency" current. This current transfers energy from the power supply to a heat station containing capacitors and a matching transformer. The role of the heat station is to match an induction coil to the power supply. A power supply can deliver rated power to the load when the load impedance is in a narrow range close to rated voltage divided by the rated current of the generator. When the induction coil impedance does not fit this value and we need high power from the power supply, precise matching must be made. The heat station usually contains

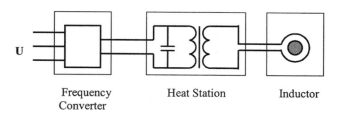

Figure 1 Conceptual layout of an induction heating installation.

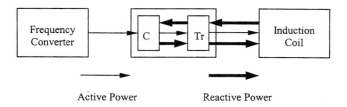

Figure 2 Power flow among power supply, heat station, and induction coil.

a matching transformer and capacitor battery. The matching transformer changes the load impedance proportionally to the square of its ratio. For multiturn coils or large single turn radio frequency coils, their impedance may be high enough and the coil may be connected to the power supply without a matching transformer. For surface heat treating coils, a transformer is almost always required.

The other component of heat station, capacitor battery, is required for compensation of the coil reactive power. The induction coil cannot transfer all the electrical power supplied by the heat station into thermal energy as resistance heaters do. A significant part of the total power (a.k.a. apparent power, or kilovolts amperes) is "reflected" from the induction coil and returns back to the supplying circuitry. This reflected power is called reactive power (inductive reactive power). There is no sense to allow the reactive power to overload an expensive power supply. Capacitors help to solve this problem. When alternating voltage is applied to capacitors, they generate capacitive reactive power, which compensates the inductive power. Thus, reactive power oscillates between the coil and capacitors (Fig. 2). This part of high frequency circuitry is called the resonant or tank circuit. Ideally, the power supply must deliver only active power to the tank circuit. However, some solid state power supplies require a small portion of reactive, usually capacitive, power for their operation, too. This power may be provided by additional capacitors inside the power supply, or by the heat station capacitors. Typically, the capacitors are installed on the primary side of the transformer in parallel to the primary winding. When the coil voltage and frequency are high enough, it is better to install capacitors on the secondary side. In the first case, the transformer must be designed for the whole apparent power of the coil (big transformer). In the second case, the transformer is small, but capacitors may be larger than in the first case because of the lower voltage applied to them.

The amount of apparent power absorbed by the part (workpiece or charge) depends on the coil current, frequency, coil design, size, and material of the workpiece and their mutual position. A ratio of absorbed (active) power to apparent power is called power factor or "cosine phi" (cos phi) because it equals the cosine of the phase angle shift between the coil voltage and the current waveforms. The ratio of reactive power to active power is called the quality factor Q; this term comes from the radio technique where low losses and resulting large Q values are beneficial. In induction heating, the lower the Q values are, the better. Quality factor is approximately equal to the inverse value of the power factor. Power factor usually drops when frequency or coupling gap between the coil and the workpiece grows. Power factor is typically 0.15–0.35 for medium frequency range (up to 20 kHz) and 0.05–0.2 for radio frequency. It means that at radio frequency, reactive power may be 20 times higher than active power. However, in some applications, the power factor may be as high as 0.5, even at 1 MHz. In many cases, especially in small and medium sized installations, the capacitors and matching transformer are inside the power supply cabinet and the coil is connected directly to the power supply. Recently developed, small, handheld or portable matching transformers with attached coils may be connected to the power supply with a flexible cable. This configuration, which is available now for power up to 60 kW, is becoming more and more popular in the industry.

Almost all components of induction heating installations are "standard" because they may be found in the market. Only the induction coil should be developed individually for a particular application. Induction coils may be designed by the original equipment manufacturer (OEM), by the final user, or by a "coil company" specializing in induction coil manufacturing. In many applications, the induction coil is the most critical component that affects the whole installation performance (efficiency, product quality, equipment lifetime, etc.).

C. Electromagnetic Processes in Induction System

The heat station must deliver to the coil electrical energy at voltages and current levels dictated by the coil. Currents flowing in the coil turns (ampere turns) generate alternating magnetic fields in surrounding space. The flux of magnetic fields (magnetic flux) flows around the coil conductors in a closed loop (magnetic field path). In simple cylindrical coils (Fig. 3), magnetic flux density B (magnetic induction), or corresponding magnetic field strength H is much higher inside the coil than in the external space, where magnetic fields return more widely distributed in space (magnetic field back path).

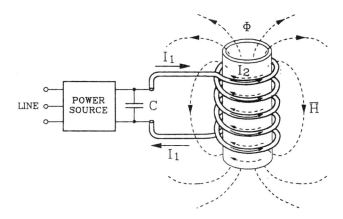

Figure 3 Magnetic field lines for a simple cylindrical induction coil. (Courtesy of UIE; from Ref. 20.) Other Figures, which are provided courtesy of UIE, will be referenced in the Acknowledgements.

If a conductive body is placed inside the coil, it gets heated by induced eddy currents. A tube is the best body shape to demonstrate how induction heating works. In the general case, magnetic flux flows in the gap between the tube and coil internal diameter (ID), in the tube wall, and inside the tube. According to the law of electromagnetic induction (Faraday's law), alternating magnetic flux flowing through any contour generates an electric field (electromotive force) of the same frequency in this contour. If the contour is formed by nonconductive materials (e.g., a plastic tube), this electric field is distributed along the contour and there is no manifestation of electromotive force. If the contour is made of conductive materials but is not closed, the electromotive force drives the electric charges of the conductor to the edges of the contour and voltage appears between the sides of the opening. This voltage, which may be measured by a voltmeter, is proportional to magnetic flux, flowing through the contour and frequency of the field variation:

$$U = k \, \Phi f = k \, BAf, \tag{1}$$

where Φ is magnetic flux, B is magnetic induction, A is the area encircled by the contour, f is frequency, and k is a coefficient depending on the unit system.

If we connect the ends of the contour, thus forming a ring or a tube, the electromotive force will drive the current around the contour. The value of this induced (eddy) current is the ratio of the electromotive force (or induced voltage) to the resistance of the ring or tube (Eq. (2)):

$$IR = U = kfBA, \tag{2}$$

where I is the eddy current per unit of tube length, and R is tube resistance per unit of length, calculated in the same manner as for a direct current (DC) flowing around the tube circumference.

Eddy currents generate heat according to Joule's law, similar to traditional resistance heating:

$$P_\text{w} = I^2 R = (kfBA)^2/R, \tag{3}$$

where P_w is the workpiece power per unit of length.

Sometimes induction heating is called "resistance heating by induced currents," which is true, but the resistance heating stage is only a small part of a long chain of processes taking place in an induction heating system. Equation (3) and the process description correspond to induction heating at "low frequency." At low frequency, the absorbed power is proportional to frequency squared and is inversely proportional to the workpiece resistance (material resistivity). This heating mode is not favorable for industrial application due to low efficiency. We can say that the workpiece is transparent to the magnetic field and power absorption is low. These conditions are typical for fixtures and other system or structural components, which must not be heated when located in the magnetic field of the coil.

If frequency increases, absorbed power should grow quickly according to Eq. (3). However, an opposite process takes place simultaneously when the frequency grows. Eddy currents generate their own magnetic field (field of reaction), which, inside the body, is in the opposite direction to the coil field. This field of reaction reduces magnetic flux inside the workpiece and hence diminishes the eddy currents and absorbed power. A simultaneous influence of these two phenomena provides a complicated dependence of absorbed power with frequency variation. However, if the magnetic field strength on the surface is constant, the absorbed power always grows with frequency, very fast at low frequency and much slower at high frequency.

It is important to note that a part of the magnetic field of reaction is coupled to the coil turns, where it induces an electromotive force of reaction. As a result, the coil impedance changes, usually drops, and power factor always grows. Only in the case of heating of magnetic materials at relatively low frequency can the impedance of the loaded coil be higher than for the empty coil. Change of the coil impedance during the heating process, or from one heating cycle to another can provide valuable information for process monitoring. Numerous coil monitors work on the principle of measuring coil impedance variation due to the load reaction field.

D. Skin Effect and Reference Depth

In the threshold case of "high" frequency, the field of reaction compensates for the field of the coil inside the workpiece. A magnetic field attenuates in the workpiece material before it reaches the tube ID, or the central part of a solid cylinder. As a result, eddy currents flow in a surface layer of the workpiece only. This phenomenon is usually called "skin effect." When skin effect is well pronounced, relatively simple relationships exist between the frequency and magnetic field intensity, on one hand, and the induced current and absorbed power, on the other hand. Because electromagnetic processes are concentrated in the surface area of the conductor, it is convenient to consider specific surface and volumetric values instead of total current and power values. A magnetic field with density B_0 on the workpiece surface induces current I_0 in the surface layer (per unit of the body length along the magnetic field lines):

$$I_0 = B_0/\mu_0 = H_0, \tag{4}$$

where μ_0 is the permeability of air (magnetic constant) and H_0 is magnetic field strength.

Power density per unit of surface equals:

$$P_0 = I_0^2 \, \rho/\delta = H_0^2 \, \rho/\delta, \tag{5}$$

where ρ is the electrical resistivity of the material, and δ is the reference (penetration) depth.

Reference depth depends on the frequency and material properties and does not depend on the body shape:

$$\delta = 5030\sqrt{(\rho/\mu f)} \ \text{[cm]} \tag{6}$$

where ρ is in ohm centimeters and f is in hertz. The permeability of material μ is a relative dimensionless value.

Table 1 and Fig. 4 show reference depth values for different materials and frequencies. They range from several centimeters for non-magnetic materials at low frequencies to fractions of a millimeter at high frequencies, especially for magnetic materials.

Reference depth plays a fundamental role in induction heating theory. It allows us to evaluate easily the skin effect in an induction system, to select the proper frequency for heat treating and other processes, to predict the power distribution inside the workpiece cross section and the power variation in the process of heating. If characteristic dimensions (cylinder diameter or plate thickness) are smaller than the reference depth, it is considered a low frequency case. If these dimensions are larger than 4δ, it is considered a high frequency case. For tubes, there may be a case where the tube diameter is much larger than δ, but the wall thickness is much

Table 1 Reference Depth Values for Different Materials and Frequencies (in Centimeters)

Material	T °C	$\rho \times 10^6$ Ohm.cm	Frequency, kHz				
			0.05	1.0	10.0	50.0	500.0
Steel	1000	130	8.1	1.80	0.64	0.256	0.081
Copper	20	1.84	0.96	0.22	0.070	0.030	0.010
Brass	20	7.0	1.90	0.42	0.13	0.06	0.019
Silver	20	1.63	0.91	0.20	0.065	0.029	0.010
Aluminum	20	2.95	1.20	0.27	0.085	0.038	0.012
Nickel	1000	47.0	4.90	1.10	0.35	0.154	0.050
Titanium	1200	175	9.40	2.10	0.67	0.30	0.094
Graphite	600	1000	22.5	5.0	1.60	0.71	0.225

less. In this case, a more complicated criterion must be used. If the product of wall thickness and diameter is less than δ^2, the frequency is low; if the product is greater than $4\delta^2$, the frequency is high.

Another example that is important for an understanding of the laws of induction heating is the heating of a thin disc. Consider a case where the disc thickness t is less than the reference depth δ but the disc diameter D is several times larger than δ. Is it possible to heat the disc at this frequency effectively? For the correct answer, the disc orientation in magnetic field must be considered (Fig. 5). If the disc surface is parallel to magnetic lines

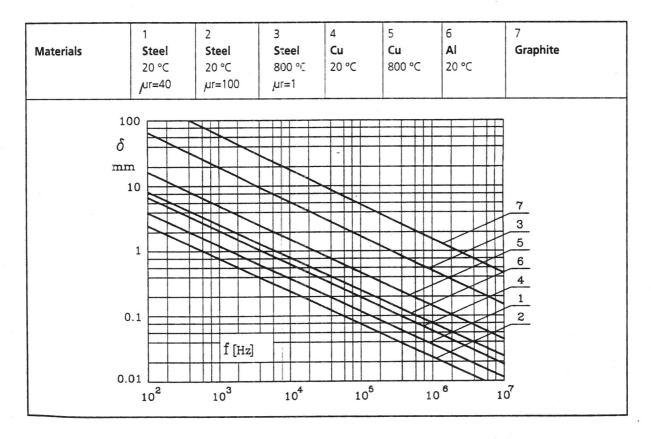

Figure 4 Reference depth values for different materials and frequencies. (From Ref. 20.)

Figure 5 Longitudinal (right) and transversal (left) heating of a disc.

(longitudinal heating), the characteristic dimension of the disc is its thickness t and the process efficiency will be very low because of too low a frequency ($t < \delta$). If the disc is located perpendicular to magnetic lines (transversal heating), both dimensions are important and a criterion Dt/δ^2 must be calculated. If this criterion is more than 2, heating is efficient; if it is less than 1, heating is low efficient.

It is clear from the above considerations that the definition of frequency as "low" or "high" is very relative. From the point of view of induction theory, a frequency of 50 or 60 Hz may be "high" (e.g., when heating aluminum ingots or slabs). At the same time, a frequency of 100,000 Hz may be "low" for effective heating of thin plates such as knife blades or saw teeth above the Curie temperature. However, in the induction heating industry terminology, there are three or four main frequency ranges: low, up to 3 kHz; medium, from 3 to 20 kHz; high, from 30 to 100 kHz; and RF, above 100 kHz. This classification or similar classifications are based on tradition and certain technical differences in equipment at different frequencies, and not on electromagnetic effects in the induction system.

E. Workpiece Power

Equation (5) for the absorbed power calculation is the same as the formula for the power generated by a current flowing uniformly in a layer δ. This simple formula is very convenient and widely used in practice. However, it is necessary to know that in reality, current and power are not distributed uniformly in the layer δ. Current density J and power density P_v are distributed exponentially with a distance x from the surface (Fig. 6):

$$J(x) = J_0 \, \exp(-x/\delta); \qquad P_v = P_{v_0} \, \exp(-2x/\delta), \tag{7}$$

where J_0 and P_{v_0} are current and power density on the surface, respectively.

Only 63% of current and 86% of power are located in the layer δ. These exponential distributions are correct only for a flat thick body with constant electromagnetic properties (ρ and μ). For multilayer ferromagnetic materials and nonflat bodies, the current and power distributions are complicated and may be correctly found by computer simulation, or, in some cases, by sophisticated analytical methods. Some examples of real power distribution are given further.

Total power absorbed by a workpiece at high frequency may be found as a product of specific power P_0 and the body surface S_w exposed to a magnetic field. This method of power calculation may be used at any frequency when the power absorption coefficient (power transfer factor) K is added (Eq. (8)):

$$P_w = H_0^2 \, S_w \, K \, \rho/\delta. \tag{8}$$

As it was shown earlier, at low frequency, the absorbed power is proportional to frequency squared. Because the reference depth is inversely proportional to the square root of frequency, coefficient K at low frequency is proportional to frequency to the power 1.5. For a well-pronounced skin effect in the body with constant properties, the coefficient K is always equal to one. It means that at high frequency, the absorbed power is always proportional to the root square of frequency for the constant magnetic field strength on the surface.

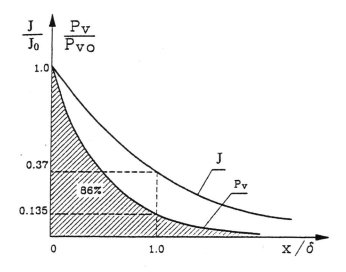

Figure 6 Ideal current and power density distribution in the depth of a workpiece. (From Ref. 20.)

In order to keep low electrical losses in induction coil conductors, their thickness must be higher than the reference depth. We do not consider here multilayer coils or coils made of special stranded cables (Litzendraht or Litz) [14,15], where the optimum thickness of individual strips or wires is less than the reference depth, whereas the overall copper cross section is much thicker. Electrical losses in induction coils are approximately proportional to the root square of frequency. It means that at high frequency, the power and coil losses vary with frequency in the same manner and the coil efficiency reaches its threshold value, which no longer depends on frequency. This is a general rule, which must be slightly corrected in particular cases by account of end effects and other phenomena.

The power transfer factor depends on the workpiece shape and the ratio of characteristic dimensions of the workpiece cross section to the reference depth. For a simple geometry, such as a solid or hollow cylinder, plate, square, or rectangular profile, coefficient K is tabulated. A transfer factor for a solid non-magnetic cylinder with diameter d is drawn in Fig. 7 vs. the ratio d/δ, often called "electrical diameter." Electrical diameter is proportional to the root square of frequency. The transfer factor for a solid cylinder grows continuously with frequency, asymptotically approaching the value $K = 1$. Practically, efficiency reaches its threshold value at $d = 8\delta$. The transfer factor for a plate with thickness a has a maximum $K = 1.12$ at $d/\delta = 3.14$ and then tends to $K = 1$ with increasing frequency. It means that for longitudinal heating of plates, there is a small maximum of efficiency, after which efficiency slightly drops and then remains almost constant as frequency grows.

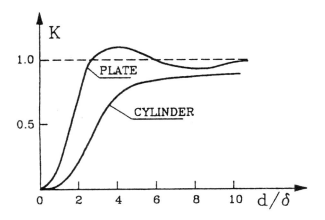

Figure 7 Transfer factor for plate and solid non-magnetic cylinder vs. their characteristic electrical dimensions. (From Ref. 20.)

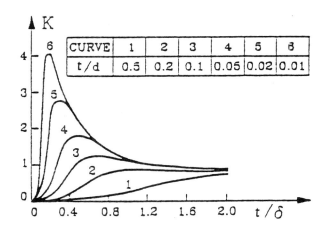

Figure 8 Transfer factor vs. electrical thickness for tubes. (From Ref. 20.)

For tubes, the transfer coefficient K depends on both the electrical diameter d/δ and the electrical wall thickness t/δ (Fig. 8.). If the wall thickness is less than half of the radius ($t/d < 0.25$), the transfer factor curve has a maximum, whose value and location are frequency-dependent. The location of the maximum corresponds approximately to the relation $td/\delta^2 = 4$. The thinner the tube is, the higher is the maximum value of K and the greater is the frequency at which the maximum occurs. It means that thin tubes may be heated with higher efficiency than thick tubes or solid cylinders with proper frequency selection.

F. Electromagnetic Effects in Induction Systems

The complicated real distribution of magnetic field, current, and power in the induction system may be described by an interaction of different so-called effects. Besides the already discussed skin effect, there are proximity effect, coil effect, effect of magnetic concentrator, and end and edge effects of the coil and the workpiece.

1. Proximity Effect

Proximity effect describes a fact that induced current tends to flow as close as possible to the inducing current. When a coil loop is positioned above the flat workpiece, the induced current tends to follow the coil geometry (Fig. 9). For higher frequency and smaller coupling gap, the proximity effect is more pronounced. The same effect takes place in the busswork. With high enough skin effect, the oppositely directed currents flow

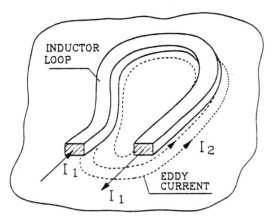

Figure 9 Induced current flow for a coil above a flat workpiece demonstrating the proximity effect. (From Ref. 20.)

close to each other (i.e., on internal sides of the buss-bars) (Fig. 10).

2. Coil Effect

High frequency current is distributed uniformly on the conductor circumference only in the case of a single straight round rod or tube. In a cylindrical coil, the current will flow mainly on the internal surface (ID) of the winding. This phenomenon is called the coil effect. Actually, the coil effect is a manifestation of proximity effect in the coil turns. The currents on the opposite sides of ID attract each other, whereas the currents of the same direction in neighbor conductors push one an-

other out from the sides of the profile. If the straight conductor is bent, current tends to flow on the side of the lower radius of bending. A widely used saying that "current takes the shortest way" is closely related to the coil effect. In external induction coils, the proximity and coil effects work in one direction (Fig. 11), whereas in internal (ID) coils, they work in opposite directions. Coil effect forces the current to flow on an ID surface and proximity effect moves current to the coil outer diameter (OD), closer to the workpiece surface. The resulting distribution depends on a combination of the system dimensions, material property, and frequency. The application of magnetic flux concentrator (magnetic core) is strongly recommended for all ID coils in order to assist the proximity effect in overcoming the coil effect.

3. Effect of Magnetic Flux Concentrator

The effect of the concentrator (Magnetic Slot Effect) is extremely important in induction technique [19]. A C-shaped concentrator made of nonconductive ferromagnetic materials (laminations, ferrite, and magnetodielectric composite), attached to the conductor, pushes a current to the open side of C (Fig. 12). The influence of concentrators on the coil parameters and power distribution in the workpiece will be discussed in Section IV.

4. End Effects in Cylindrical System

End effects are caused by magnetic field irregularity near the ends of the workpiece or the coil. Figure 13 shows both effects for a relatively long induction coil with a cylindrical load inside. On the right side (zone b), the magnetic field is weaker due to the coil end effect. It

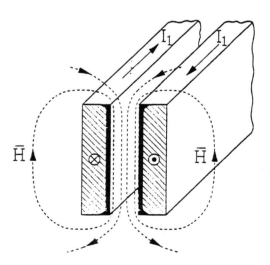

Figure 10 Induced current flow for parallel bussbars demonstrating the proximity effect. (From Ref. 20.)

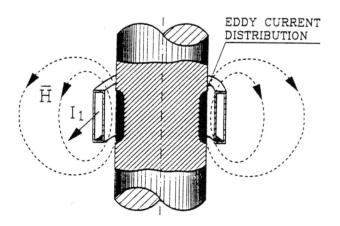

Figure 11 Current distribution for an external induction coil showing the proximity and coil effects. (From Ref. 20.)

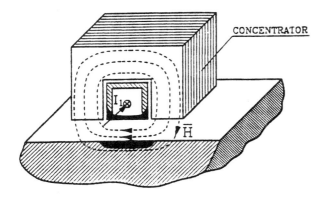

Figure 12 Magnetic slot effect. (From Ref. 20.)

results in a smooth decline of power along the workpiece. Near the workpiece end (zone a), magnetic field lines tend to "cut" the edge of non-magnetic cylinder. It leads to more flux coupled to the metal, higher current induced, and higher absorbed power in this area. This effect gets stronger when frequency grows. Simultaneously, the length of the end effect zone gets shorter. The zone of end effect for a non-magnetic cylinder is one diameter or five reference depths, whichever is smaller. For non-magnetic material, the workpiece end effect is always positive (power increases to the end). For magnetic material, the end effect may be positive or negative (power decreases), depending on frequency, cylinder diameter, and material properties [14,20,21].

At low frequency, it is always negative. It means that for low temperature heating such as for tempering operation on tubes or shaft ends, underheating of the end area may be a problem and increasing frequency is recommended to achieve uniform heating. When heating ferromagnetic material to high temperature, the end effect can change from negative to positive in the process of heating mainly at the Curie point, resulting in a quick temperature redistribution along the workpiece end. In a real induction system, the coil and workpiece end effects interfere. Balancing a proper selection of frequency and the coil position or length allows the designer to provide a uniform or required gradient heating of the workpiece end. A coupled electromagnetic and thermal 2-D simulation program such as Flux 2D provides an accurate description of end effect dynamics during the heating process.

5. Electromagnetic Effects in Slab

Electromagnetic effects in bodies with a rectangular cross sections are more complicated due to two types of effects: end effect and edge effect.

Edge Effect in Slab

When a long body with a rectangular cross section is heated in an oval induction coil, the magnetic field strength is constant along the slab perimeter. Induced currents flow around the cross section as shown in Fig. 14. In the central area of the cross section, the current distribution corresponds to the regular skin effect phenomenon. Approaching the slab edges, the currents tend to cut the edges and turn to the back path, taking "the shortest path." The skin effect opposes this tendency and, as a result, different current patterns take place at different frequencies. When frequency is low (slab thickness is less than two reference depths), the current turns back relatively far from the edges and the whole edge area is underheated. For high frequency (high skin effect), currents flow in a thin surface layer and the edge areas are overheated because of heating from three sides, instead of two sides, in the regular zone (Fig. 14a). It was shown that the most uniform power distribution in the slab width occurs when the reference depth is 3.14 times less than the slab thickness (Fig. 14b). For high temperature heating, a higher frequency is required for the most uniform heating because additional power is necessary to compensate for additional thermal losses in the edge area [14]. If the slab is not wide enough, the edge effects from both sides of the body interact and there is no regular zone corresponding to heating of a very wide plate. A body with a square cross section may be substituted for a cylinder with the same cross section for the calculation of absorbed power and temperature distribution. The single difference is that for a square cross section, the very corners are always underheated (local minima) because of higher thermal losses and reduced heat source densities (currents cut the corner,

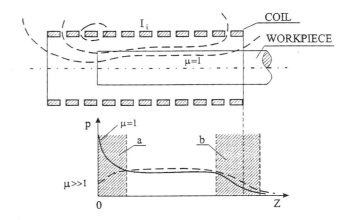

Figure 13 End effects for a long induction coil for heating a cylinder.

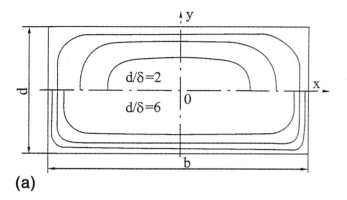

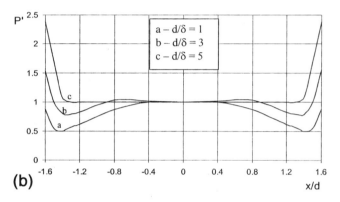

Figure 14 Edge effect for slab heating. (a) Induced current flow near the edge of the slab (edge effect) for low and high frequencies. (b) Power distribution in width at different frequencies.

taking the shortest path). Using a higher frequency results in better heating of the corners.

End Effect in Slab

End effects in slab are principally the same as for cylindrical workpieces, with power increase in the end area at any frequency for non-magnetic bodies and possible power decrease for magnetic bodies when frequency is low. The single difference is that for a wide slab (plate), the end effect is more pronounced than in a cylindrical system.

There are also special areas (the 3-D corners) where end and edge effects interfere. The electromagnetic field structure in this area is very complicated and difficult for analysis due to the 3-D nature. Some results (Fig. 15) have been obtained by means of computer simulation only recently. Figure 15 shows the power density in the 3-D corner area for a non-magnetic stainless steel slab with a ratio d/δ equal to 4.1. The areas in white correspond to the highest power density and those in

black correspond to the lowest. With this ratio, there is relatively strong end effect in the system and the edge effect is positive also. In the 3-D corner, these two effects combine to make this the area of highest power absorption [22].

In bodies with small cross section especially for materials with high thermal conductivity such as aluminum alloys, these 3-D areas are not important due to fast temperature equalization by heat diffusion. However, for heating of big slabs of materials with low thermal conductivity, such as titanium alloys, the 3-D end–edge effects must be taken into consideration.

G. Electromagnetic and Thermal Properties of Materials

In the process of heating, the electromagnetic and thermal properties of materials vary. The electrical resistivity of all metals grows with temperature, resulting in reference depth increase and higher power absorption (Fig. 16). For carbon steel, resistivity at 900°C is six to eight times higher than at room temperature. For austenitic stainless steel, the difference in resistivity levels is much lower, less than two times. The resistivity of graphite remains almost constant with temperature variation in very wide range. Many nonconductive-at-room-temperature materials such as salts, oxides, carbides, and glasses become conductive at high temperature and may be heated by induction. External start heating is required in order to start the process [15,20].

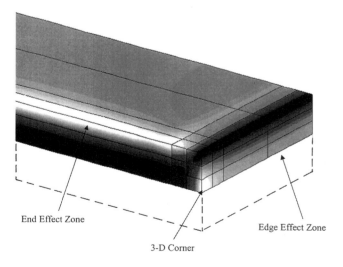

Figure 15 A 3-D computer simulation of slab heating showing power density distribution in a 3-D area.

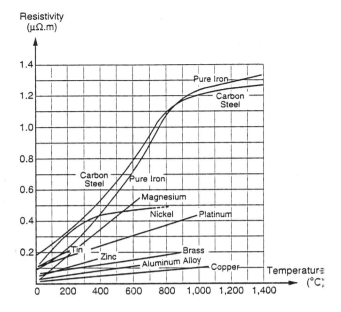

Figure 16 Resistivity of metals as a function of temperature.

The magnetic permeability of steels and other ferromagnetic materials drops significantly before the temperature reaches Curie point, after which the material becomes completely non-magnetic. Figure 17 shows the variation in the electromagnetic properties of steel as a function of temperature, with μ_1 and μ_2 representing surface permeabilities at two different levels of saturation. The corresponding values of reference depth δ_1 and δ_2 grow gradually with temperature until the Curie point, where they increase sharply with temperature. Above the Curie point, the reference depth continues to grow slightly due to resistivity increase with temperature rise. Resistivity and permeability variation with temperature influence the coil parameters and power source distribution during heating. The thermal properties of materials also depend on the temperature, especially near the points of structural or phase transformation (austenization point, melting point, etc.). Material property temperature dependence may be accurately accounted for by means of a computer simulation of the process.

For magnetic materials, the permeability depends on magnetic field strength. On the workpiece surface, the field strength is high and permeability is relatively low: typically 6–25 for surface hardening and 30–60 for through heating. The magnetic field strength drops with distance from the surface and the permeability grows correspondingly, reaching a maximum (600–1000 for carbon steels) and then declines to the initial permeability value.

H. Temperature Distribution in Induction Heating Processes

There are two major applications of induction heating: surface hardening and through or mass heating for forging, rolling, extrusion, and other operations. In surface hardening, the goal is to heat a surface layer above the austenization temperature before quenching. The core of the workpiece may remain relatively cold. In mass heating, the entire cross section should be heated uniformly, or with a specified gradient.

With flame heating, all the heat power should flow through the surface from outside to inside, resulting in large temperature gradients near the surface, which restrict power density and heating speed. The maximum temperature is always on the surface. Induction heating is heating by internal sources. Therefore, on the workpiece surface, the heat flow is always directed outside the body due to heat losses. Maximum temperature is always inside the body. In surface hardening processes, these heat losses are very small compared to the absorbed power density, and maximum temperature is very close to the surface (Fig. 18). In high temperature mass heating, these losses may be of the same order as the absorbed power and maximum temperature shifts inside the body (Fig. 18). Finally, in temperature holding processes, the surface losses must be equal to induced power. After a long holding process, the temperature distribution is uniform in the body core with

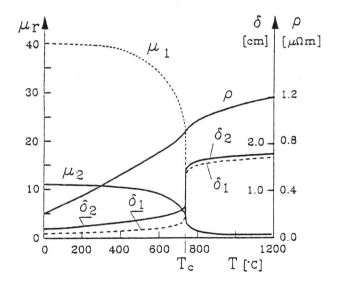

Figure 17 Variation of electromagnetic properties of steel with temperature and corresponding reference depth. (From Ref. 20.)

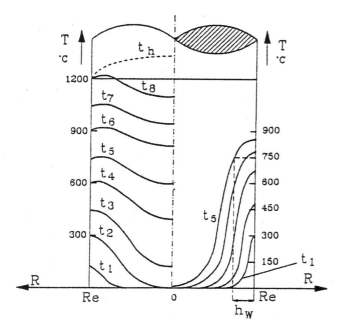

Figure 18 Temperature distribution in workpiece radius for surface hardening (right side) and mass heating and holding (left side). (From Ref. 20.)

reduction near the surface. The higher the losses are, the higher is this "inverse" temperature gradient between the workpiece surface and the center in the process of holding. At higher frequencies, overheating of the core is less than that at low frequency.

The goal of the designer is to provide the optimum combination of power, frequency, and time in order to obtain the required temperature distribution at the end of the process (e.g., before quenching or extrusion).

I. Heat Source Distribution Dynamics in the Process of Induction Heating

In the process of heating, the material properties vary with temperature, and heat source distribution varies, too. This phenomenon is very favorable for surface heat treating of steels. At the beginning of heating, the steel is magnetic and its resistivity is relatively low. The reference depth is small and the power density drops quickly with distance from the surface (Fig. 19). As temperature rises, the power penetrates deeper into the material due to increasing resistivity. When surface temperature reaches Curie point, the outer layer of steel becomes non-magnetic. It leads to a dramatical change in power distribution due to electromagnetic field "re-

flection" from the magnetic core. Power density drops slightly in the non-magnetic layer, grows (reaching a maximum in the magnetic transient zone), and then drops very quickly inside the magnetic core. The marked curves in Fig. 19 correspond to the end of the heating process showing this phenomenon. This power distribution is very favorable for fast and efficient heating for steel austenitization.

The fastest and most efficient heating takes place when a non-magnetic layer is 0.3–0.6 of the reference depth value calculated for hot steel. This ratio is usually used for the selection of optimal frequency for induction hardening (Table 2). With higher ratios of hardness depth and reference depth (i.e., for too-high frequency), power and temperature gradients inside the hardness depth are larger and it is necessary to reduce power and to increase heat time in order to avoid surface overheating. The temperature of the core and the total energy for the heating cycle also increase due to longer heating. With too-low frequency, power penetrates too deep, resulting again in a higher temperature of the core and increased energy demand. In addition, at lower frequency, power density is higher, resulting in larger power supply. Plus the coil lifetime may become problematical due to heavy loading of the copper. The Curie point for carbon steels, which are typically hardened by induction, is 710–780°C (i.e., just below the temperature of austenization). Additional temperature rise is required for the completion of the austenization process.

When heating completely non-magnetic materials, power distribution is less favorable. It is parabolic in the workpiece radius or thickness at low frequency, and drops exponentially from the surface at high frequency.

J. Induction Coil Parameters Variation During the Heating Process

In the process of heating, the coil parameters can significantly change due to changes in the workpiece material properties with temperature. This effect is especially strong when heating a magnetic part to a temperature above the Curie point. Variation of the coil parameters in the process of heating is very important for the coil matching to the power supply and for power delivery to the workpiece. Coil impedance and resistance variation are especially strong for surface hardening coils because they usually work with a small coupling gap. Multiturn inductors for high temperature through heating have a big coupling gap with thermal insulation. Their impedance does not change much, but the coil resistance and power factor variation may be large.

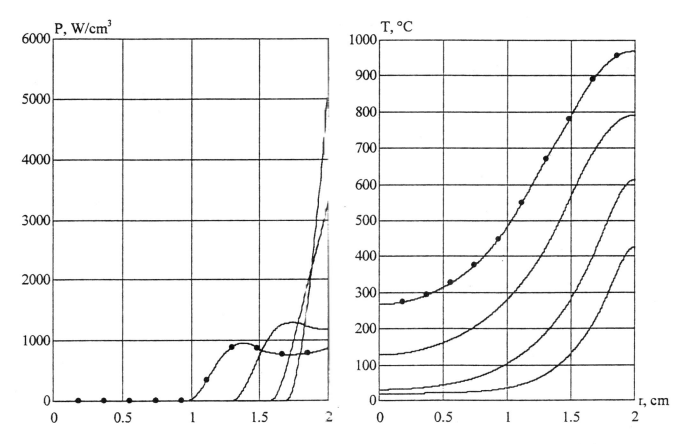

Figure 19 Power density (left) and temperature distribution (right) at different stages in a surface hardening process.

Figure 20 shows the variation in the coil impedance, reactance, and resistance for induction heating of carbon steel cylinder with diameter of 4 cm at 3 kHz. The coil inner diameter is 6 cm. At the beginning of the process, the coil impedance grows due to increasing resistivity, but then decreases as permeability starts to decline and the impedance drops to the level corresponding to a non-magnetic workpiece. The coil and workpiece power variation depend not only on the coil parameter variation, but also on the mode of power supply operation. Many modern power supplies can work in three operation modes: constant coil current, constant coil voltage, or constant power. For the case of constant voltage applied to the coil of the previous example, the coil power remains almost constant for the first 20 sec (Fig. 21). When the workpiece surface reaches Curie point, the coil power grows and then drops to a final level. The overall power variation is less than 1.7 times. For constant current operation mode, the power variation is more than three times. The induction coil parameter variation is very important for both the heating process and the power supply selection and performance.

Table 2 Optimal Frequency for Induction Surface Hardening Processes

Case Depth mm	Frequency, kHz					
	1.0	3.0	10	30	100	300
Min	5.0	2.7	1.5	1.0	0.6	0.5
Opt	8.0	4.5	2.5	1.6	1.0	0.8
Max	15	9.0	5.0	3.0	1.6	1.2

K. Electrodynamic Forces in Induction Heating Systems

Electrodynamic forces are caused by the interaction between the magnetic field and currents (Lorentz forces), or between magnetic field lines and magnetic masses. Both kinds of forces may be described as a side pressure between the magnetic lines and tension along

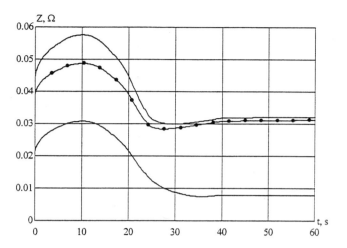

Figure 20 Variation in coil resistance (bottom line), reactance (marked line), and impedance (top line) in an induction hardening process.

the lines. This rule, proposed by the father of electromagnetism, J. Maxwell, is very simple and illustrative. In a generic induction system, there are three types of "active" bodies that influence the magnetic field distribution and in its turn are influenced by the magnetic field: induction coil turns (copper), nonconductive magnetic pieces (magnetic flux controllers), and conductive magnetic or non-magnetic workpiece. Magnetic field lines enter the concentrator perpendicular to its surface and the forces applied to the poles tend to move

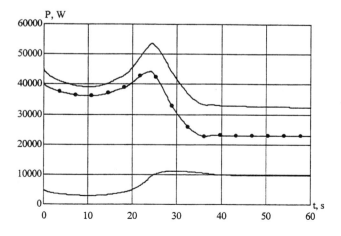

Figure 21 Variation in inductor power for a constant voltage output mode of operation: coil losses (lower curve), workpiece power (marked curve), and total coil power (upper curve).

them inside the area with high magnetic field strength (i.e., closer to the surface of the workpiece and inside the coupling gap of the coil). On the contrary, under the coil face, the field lines are almost parallel to the surface and there are forces of pressure applied to the copper surface current-carrying layer. Forces applied to magnetic workpieces (Fig. 22) are the forces of repulsion under the coil face and the forces of attraction under the concentrator poles. The sum of the forces applied to all the components of the system equals zero at any moment of time. The designers of induction systems often do not pay attention to electrodynamic forces and believe that they are unimportant for the process of heating and for induction system performance. This is true for small-power, high frequency induction heating installations, but there are many processes where electrodynamic forces must be accounted for.

Electrodynamic forces are proportional to the square of the magnetic field strength and therefore have a permanent (static) component F_c and alternating component with magnitude F_a. Alternating force varies in time at double the frequency of the coil current:

$$F = C \sin^2 \omega t = F_a \cos 2\omega t + F_c, \qquad (9)$$

where ω is an angular frequency $\omega = 2\pi f$, C is a coefficient, and F, F_a, and F_c are forces per unit of body surface.

If the skin effect is well pronounced, the static force F_c and the magnitude of alternating force F_a do not depend on frequency. It means that with higher frequency, the ratio of absorbed power to applied force gets smaller and, at radio frequency, forces are almost always small. Electrodynamic forces are especially high when heating materials with low electrical resistivity (aluminum and copper alloys) at low frequency.

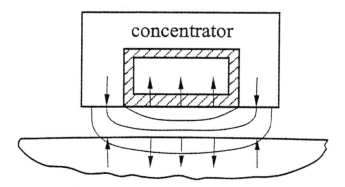

Figure 22 Electrodynamic forces in an induction heating system. (From Ref. 20.)

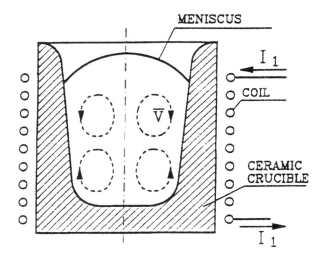

Figure 23 Electrodynamic forces causing metal stirring in an induction crucible furnace. (From Ref. 20.)

There are several effects caused by electrodynamic forces in induction systems:

(a) Induction coil deformation caused by the static component of electrodynamic force
(b) Additional distortion of heated workpiece
(c) Coil turn vibration, which can damage electrical insulation and cause failure of brazed seams.
(d) Liquid metal stirring and shape deformation
(e) Strong vibration of the system components and noise
(f) Attraction of magnetic chips or bits of scale from the workpiece to the coil turns, causing short circuiting.

These effects are mainly negative for the process and coil. At the same time, liquid metal stirring may be positive, providing better uniformity of the melt temperature and composition in an induction furnace (Fig. 23). Small quantities of molten metal may even be levitated by the magnetic field (levitation melting) without physical contact to the crucible [20].

III. INDUCTION COILS

A. Induction Heating Coils

The term "induction coil" or "work coil" denotes a current-carrying conductor in close proximity to a workpiece to be heated [23]. The induction coil is the most critical component of an induction heating installation, which defines process efficiency and product

quality. An ideal induction coil must meet the following requirements:

Provide a specified heat pattern
Have parameters favorable for matching to the power supply (high impedance and power factor)
Have a high efficiency
Provide easy loading and unloading of workpieces
Have a satisfactory lifetime
Have low sensitivity to changes in the part dimensions and positioning within the specified range
Meet special needs such as quenchant supply, atmosphere, material handling, incorporation into the machine structure, etc.

The importance of each of these requirements in a given application depends upon the process. For heat treating pins or small shafts, having a high efficiency is not so important. For larger heat treating applications, such as single-shot axle hardening, which require high power, the efficiency and matching to the power supply are the most important after heat pattern quality [24].

B. Induction Coil Styles

All induction heating coils can be classified into two families based upon the relationship between the magnetic flux direction and the part surface, longitudinal or transversal inductors. Longitudinal inductors create magnetic field with lines flowing along the main axis of the part. The field lines are perpendicular to the main axis or part surface for transversal ones. Within these two families, there is a large number of induction coil geometries used, but only a few types of them are common in the metal processing industry.

For through heating of long tubes, bars, and billets with round, square, and other cross sections, cylindrical-type inductors are typically used (Fig. 24). These induction coils have good electrical efficiency, high power factor, good reliability, and good uniformity of heating. In addition, this style of inductor often does not need a transformer because it is possible to change the number of turns to match the coil directly to the power supply. In high temperature applications, cylindrical inductors have thermal insulation to protect the coil turns from radiation and to obtain good thermal efficiency (Fig. 3).

Another type used for through heating is the oval inductor (Fig. 25). Oval inductors are often used for heating slabs as well as bars and billets with reasonable lengths. The difference between the performance of oval induction coils and traditional cylindrical inductors is that the axis of the part may or may not be the same as the axis of the workpiece(s). The bars may be located

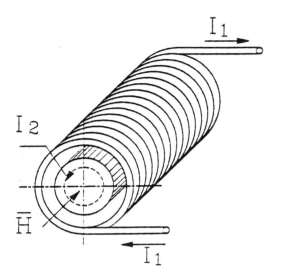

Figure 24 Solenoidal-type inductor heating a cylindrical rod. (From Ref. 20.)

either parallel or perpendicular to the magnetic field lines inside the coil. Therefore, oval inductors can be members either of the longitudinal or the transversal family.

For through heating of thin bodies such as thin slabs, plates, and strips, a flat transverse flux inductor (Fig. 26) is often used. As its name implies, this type of inductor belongs to the transversal family. Magnetic flux controllers (see Section IV) are almost always used for this type of induction coil. Transverse flux inductors have a much lower frequency demand and higher electrical efficiency than solenoidal-type inductors. However, generated power is distributed nonuniformly in the workpiece width, and special measures must be taken in order to obtain good temperature uniformity.

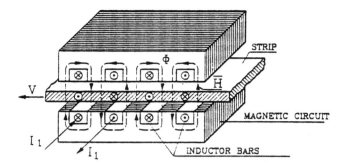

Figure 26 Flat transverse flux inductor for continuous heating of strip. (From Ref. 20.)

For through heating of rings, small plates, wires, or discs, a transformer-type inductor may be used. There are two types of transformer inductors: those with an open magnetic circuit (Fig. 27) and those with a closed magnetic circuit (Fig. 28). Inductors with open magnetic circuits are used for heating discs or other parts, which are placed in the gap in the magnetic circuit. Inductors with closed magnetic circuits are used for heating rings and short tubes.

Local heating applications (mainly represented by surface hardening) require different styles of induction heating coils. The goal is to heat uniformly only a specified area on the surface of the part with a defined temperature gradient in-depth. The heating of adjacent areas is not required and is actually undesirable in most cases.

For local heating of shafts, single-turn cylindrical inductors (Fig. 29) are often used. Single-turn inductors provide heating mainly underneath the face of the inductor. They almost always require a transformer for matching to the power supply.

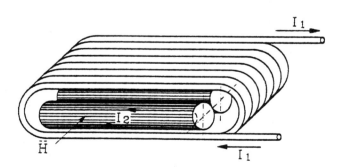

Figure 25 Transversal heating of cylindrical bars in an oval-type inductor. (From Ref. 20.)

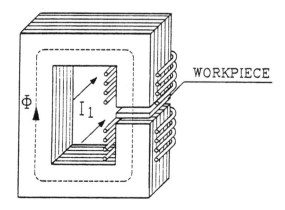

Figure 27 Transformer-type inductor with open magnetic circuit for heating of small plates. (From Ref. 20.)

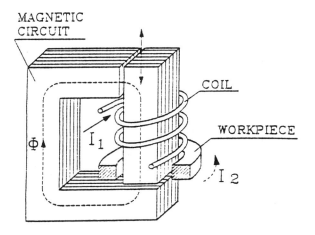

Figure 28 Transformer-type inductor with closed magnetic circuit for heating of rings. (From Ref. 20.)

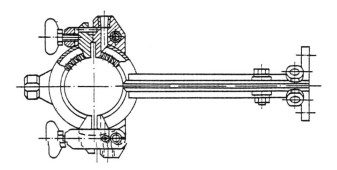

Figure 30 Clamshell-type inductor with manual operation for crankshaft hardening. (From Ref. 10.)

Another type of inductor used for local heating applications is called a clamshell or clam inductor (Fig. 30). When the current is flowing through them, clamshell inductors behave like single-turn inductors. The difference is that clamshell inductors can be opened for part loading and unloading. Clamshell inductors are for parts that cannot be loaded inside cylindrical inductors, or unloaded from them after a joining operation. A typical example is the hardening of crankshaft pins and journals. Special attention must be paid to the contact area design and maintenance for a reliable operation of these inductors.

Another type of local heating inductor is the single-shot inductor (Fig. 31). Single-shot inductors are used for heating a large area of a rotating cylindrical workpiece, such as a shaft. Magnetic flux controllers are used on single-shot inductors for local temperature control and improvement of the coil electrical efficiency and power factor.

For local heating of flat surfaces, a hairpin (Fig. 32) induction coil is often used. Like single-shot inductors, the electrical efficiency of hairpin inductors is dramatically improved through the application of magnetic flux controllers. For hairpin inductors, due to the laws of physics, the induced current and corresponding power density at the centerline must be zero, so temperature along the centerline is lower than under the conductor faces. Therefore, with hairpin-style inductors, temperature uniformity under the conductor face is not as good as for other choices. However, they are very effective when used for induction scan hardening operations. Single-shot coils are similar in design to hairpin coils for heating rotating cylindrical parts.

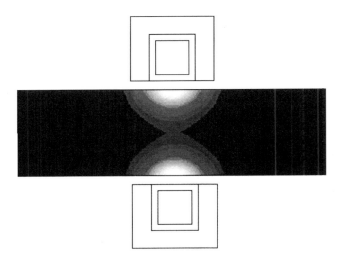

Figure 29 Single-turn inductor for the local hardening of shafts.

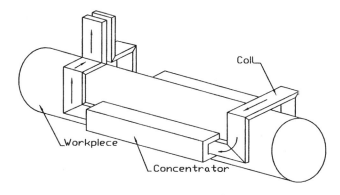

Figure 31 Single-shot style inductor for heating of rotating shafts.

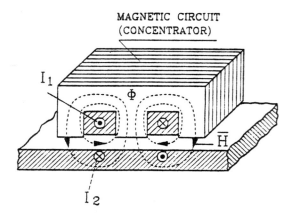

Figure 32 Hairpin-type inductor used for heating a flat part. (From Ref. 20.)

Pancake inductors (Fig. 33) produce a circular heat pattern. The temperature at the center is less than that under the face of the inductor because the power density at the center is zero. They are often used for surface hardening, or joining of discs or plates (e.g., for brazing of copper or aluminum heat dispensers to the bottom of cookware). Magnetic flux controllers improve significantly the performance of pancake inductors. They reduce the underheating of the central area, and increase efficiency and power factor.

If a single path needs to be heated, such as a bearing race or machine way, a vertical loop (Fig. 34) or split-n-return inductor (Fig. 35) is typically used. The heat pattern produced by both of these induction coils is concentrated under the face of the main (central) leg of the inductor. Magnetic flux controllers are almost always used for this style of induction coil because they increase coil impedance, improve efficiency, and lead to increased concentration and better utilization of power in the workpiece.

For heating of internal surfaces, such as sleeves, internal gears, or bearing races, internal diameter inductors are used. There are three common types of ID inductors: single-turn ID (Fig. 36), multiturn ID, and hairpin ID (Fig. 37). Magnetic flux controllers significantly improve the electrical efficiency, power factor, and coil impedance of ID inductors. With small ID inductors, there is limited space for magnetic flux to return inside the inductor. If the same coil current is supplied, the inductor with magnetic core generates a much higher magnetic flux and subsequent power transferred to the workpiece. For small multiturn ID coils, the return leg for the current passes through the coil ID, which reduces the space for magnetic flux to

flow even further. A magnetic flux controller must be used in such coils for the process to work at all.

Many real-world inductors for heat treating are combinations of different types of inductors described above. All coils have leads of various designs (busswork and flexible cable) for connection to the power supply or heat station. Real coil assembly may also include components for part fixturing, quenchant supply, sensors for process monitoring and control, shields, thermal insulation, and other devices [19,20,24].

For example, Fig. 38 shows a multiturn ID coil that has three water systems. One is for the copper winding cooling. The second is for permanent cooling of the part outer diameter. The third is for quenchant supply. The quenchant is delivered through channels machined in the magnetic core, which is made of Fluxtrol® B magnetodielectric material (MDM) [19].

C. Induction Heating Coil Manufacturing

The technology for induction coil manufacturing can have a significant influence upon the induction coil performance and lifetime. Work coils are typically made from copper tubing or solid machined blocks and are almost always water cooled [16,23]. Induction coil failures are caused by one of the following reasons:

Damage due to mechanical impact
Copper cracking due to thermal stresses from cyclical overheating
Braze joint failure due to overheating and/or mechanical stresses

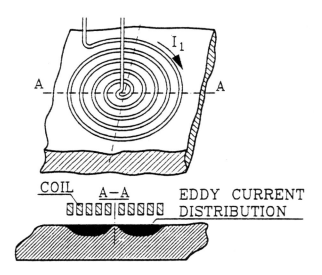

Figure 33 Pancake-style inductor for heating a circular area on a flat plate. (From Ref. 20.)

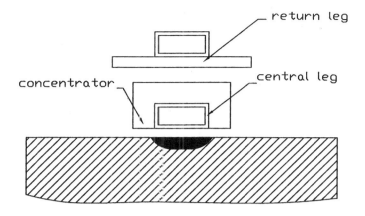

Figure 34 Vertical loop inductor for local heating of a flat plate.

Over heating due to insufficient cooling flow

Deformations caused by electrodynamic forces

Arcing due to contact with the part

Short circuiting between turns on multiturn inductors due to electrical insulation damage, or metallic chips from the workpiece

Magnetic flux controller deterioration, or failure due to overheating

Arcing in or overheating of connections due to poor contact.

The induction coil manufacturing technique must address all of these issues to provide reliable operation. Some of the guidelines for induction of coil manufacturing are given below.

The prevention of failure due to mechanical impact is the most difficult to address in the coil manufacturing stage. The best way to prevent "crashes" is to use good handling equipment and to ensure good dimensional tolerances of the incoming parts. If it must be addressed from the coil side, then the induction coil must have a robust design and/or incorporate protective

components. To minimize mechanical damage, protective components can include a sacrificial or protective material between the coil face and the workpiece to absorb the impact. Robustness can be provided by a structural nonconductive material, such as G-10, or use a heavy-walled copper tubing. Another danger in the event of impact is electrical damage caused by arcing between the part and the coil. To prevent this, there is a special grounding control device, which automatically shuts down the system in the case of contact, preventing significant damage due to arcing. Both mechanical and electrical damages can be limited in some cases through the use of a slightly larger coupling gap. Using a larger coupling gap will have a negative impact upon heat pattern controllability, power factor, and electrical efficiency, so this option should be used as a last resort.

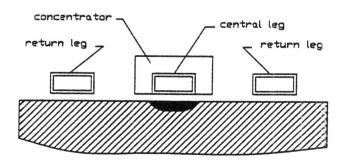

Figure 35 Split-n-return inductor for local heating of a flat plate.

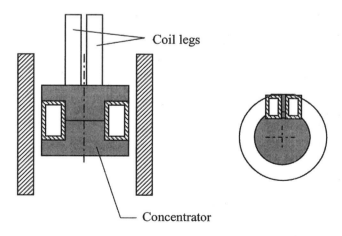

Figure 36 Layout of a single-turn ID induction coil (left) and coil view without the upper piece of a magnetic flux controller.

Copper overheating occurs due to insufficient cooling in heavily loaded inductors or high temperature applications. For heavily loaded inductors, oxygen-free high-conductivity (OFHC) copper should be used in coil manufacturing. It has better service properties than standard copper. Both OFHC and standard copper should be annealed prior to bending or forming and after for removal of the cold work induced in the manufacturing process. For applications with rectangular copper cross section, the corners should be round, like in standard round-cornered square copper tubing. For high temperature applications, a protective layer, such as a refractory spray coating or protective liner, should be applied to the surface of the inductor.

Braze joint overheating is one of the most common forms of induction coil failure. Power losses tend to concentrate in the copper conductor corners, where the braze joint is located. The best way to eliminate this failure is to not use braze joints wherever possible, using bending or forming instead. When brazing is necessary, attention must be paid to joint design and proper selection of the filler metal(s). Copper silver alloys are best for induction coil brazing. The use of electrical solders instead of brazing is not recommended.

In low to middle frequency applications, coil deformations or insulation damage due to electrodynamic forces can occur. Although not part of the coil manufacturing technology, one way to limit the potential for deformations is to increase the frequency. With regards to the coil manufacturing technology, there are two main choices for reducing deformation. The first is to use heavy-walled copper tubing, which gives the induction coil more rigidity. The other choice is to provide additional support through a nonconductive structure, made, for example, of G-10 or coil potting in cementlike or resin materials.

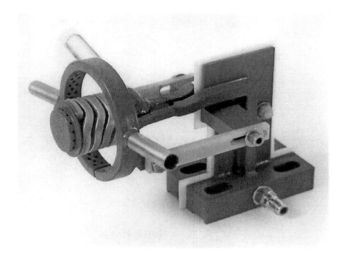

Figure 38 Multiturn ID coil with separate quenching and cooling systems.

Short circuiting between the turns is common for high frequency, high voltage inductors. For these induction coils, two levels of insulation may be used. The first level should be either an electrical varnish or a nylon coating applied on the entire copper coil head, typically in a hot dipping process. The second level should be a solid dielectric material between the inductor turns. Teflon is the most popular material for the second level, but other materials such as Mica, Kapton, or Mylar may be used.

IV. MAGNETIC FLUX CONCENTRATION AND CONTROL

A. Principles of Magnetic Flux Control

In induction heating, the heat pattern is dependent upon the distribution of the magnetic flux in the workpiece. The magnetic field distribution is a function of the induction coil copper geometry, frequency, workpiece (geometry and material(s)), and other materials in close proximity to the induction coil.

Magnetic flux controllers are a powerful tool in induction heating technology [25,26]. Recently, a better understanding of their role in different induction heating systems has been obtained through the use of computer simulation with experimental validation. The simulation results show that the proper use of a magnetic flux controller is always beneficial in an induction heating system. The degree and the type of system improvement depend upon the coil type and dimensions

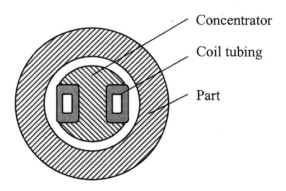

Concentrator

Coil tubing

Part

Figure 37 Layout of a hairpin ID induction coil.

[25]. They can play different roles in induction heating installations and, in various applications, are being named as concentrators, controllers, diverters, cores, impeders, or screens.

1. Magnetic Field Control

A highly conductive material, such as copper block, in close proximity to the induction coil will divert the magnetic flux away from it. These highly conductive materials are often used as diverters or screens to prevent or to reduce unintended heating of induction system components, such as parts of the material handling machine or workpiece zones adjacent to the area to be heated. For example, if a copper ring (often called a Faraday or "robber" ring) is located near the end of a cylindrical induction coil, it will prevent magnetic field penetration into the area beyond it. Faraday rings always reduce the induction coil efficiency because their action is based on generating eddy currents. These eddy currents "repel" the magnetic field. The eddy currents generate heat and, oftentimes, the Faraday ring requires water cooling. These highly conductive materials can be referred to as deconcentrators.

A magnetic material in close proximity to the induction coil will attract the magnetic field to it. Materials that reduce the reluctance of the path for the magnetic field in induction heating systems are called magnetic flux concentrators. When a magnetic flux controller alters the path of the magnetic field, it also alters the induction heating coil parameters. The concentrator "attracts" the magnetic field and directs it toward the desired heating zone of the workpiece. Concentrators usually have a positive effect on the induction coil efficiency. At the same time, the magnetic flux in the concentrator causes what are called magnetic or core losses. Core losses are heating due to hysteresis and eddy currents in the concentrator. Every magnetic material has core losses when exposed to an alternating magnetic field. Core losses increase with increasing frequency and flux density. The heating that occurs is usually small compared to the power in the workpiece and losses in the copper coil, and is removed through thermal contact with the water-cooled induction coil.

Of the two types of magnetic flux controllers, concentrators should be used instead of deconcentrators whenever possible in induction system design. Deconcentrators have a negative effect on many parameters of the induction installation, such as coil impedance and electrical efficiency reduction, whereas concentrators tend to improve the system performance. Concentrators are also the most commonly used in the industry for magnetic field and heat pattern control. In the rest of this chapter, deconcentrators will not be discussed further and only the role of concentrators as magnetic flux controllers will be covered.

2. Power Distribution

Heat pattern control, process repeatability, and coil reliability are more important than energy savings in many processes, such as local case hardening. On many parts with complicated geometries (tulips, bearing rings with grooves, etc.), the required hardness pattern may not be obtained without flux controllers.

When coils or parts of a coil are connected in series, the concentrator strongly increases the magnetic flux and specific power in the place where it is applied. The power density under the coil may increase by more than two times for external cylindrical coils [27], and by over three times for a flat one. In coils with parallel current circuits, the application of controllers can redistribute currents smoothly between sections and can effectively control temperature distribution.

Proper application of magnetic materials can also protect machine components against unintended heating. Magnetic flux shielding makes it possible to incorporate induction coils into small spaces inside the production lines without undesirable heating of adjacent machine components and without interference with the control systems and sensors. Flux shielding can also protect parts of the workpiece from undesired heating. For example, thin plates of magnetic materials can solve a problem of backtempering of previously hardened area adjacent to the zone to be heated [25].

Figure 39 illustrates the shielding effect for a single-turn coil in an ID bearing ring race hardening application. The coil has a complex machined profile in order to produce an approximately uniform heat pattern. The magnetic flux controller in this application fulfills three roles: improvement of the heat pattern under the coil, improvement of the coil electrical parameters (efficiency, power factor, and current demand), and shielding of the flange against heating. Without a concentrator (Fig. 40), strong overheating of the flange takes place with less uniform and shallower heating [25].

3. Induction Coil and System Parameter Improvement

Oftentimes, when a magnetic flux controller is applied, the induction coil parameters are improved dramatically. One example of this is a local heating application. In this application, the goal is to harden an area

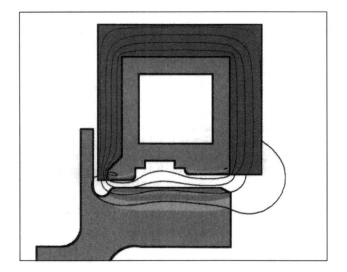

Figure 39 Magnetic field lines and power sources density map produced by coil with Fluxtrol A concentrator. Black corresponds to the highest power density; white corresponds to low power density; and grey corresponds to negligible power density.

on the face of a flat part. Typical local heating applications are seam annealing, bearing race hardening, machine ways hardening, etc. Usually, one of two induction coil styles is used for this process: split-n-return or vertical loop [27]. In both of these coil styles, the use of magnetic flux concentrators is very beneficial.

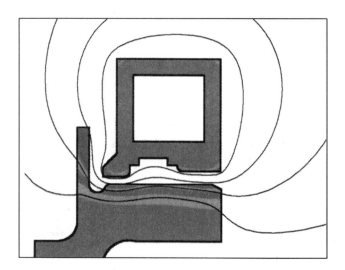

Figure 40 Magnetic field lines and power sources density map produced by bare coil. Same legend as in Fig. 39.

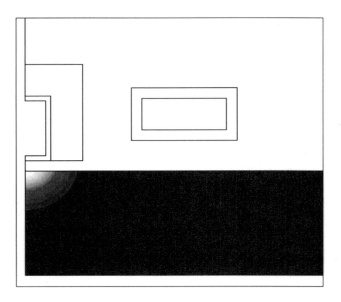

Figure 41 Final temperature distribution, split-n-return coil with concentrator.

Figure 41 shows the temperature profile produced by a split-n-return-type coil with concentrator and Fig. 42 shows the temperature profile for an inductor with the same dimensions, but no concentrator. The heat pattern for the coil with concentrator is more focused, whereas the bare inductor produces a wider temperature distri-

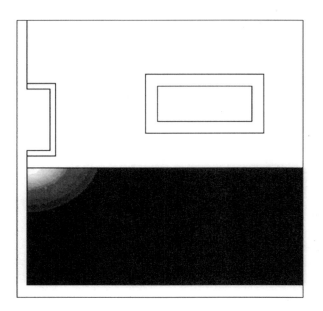

Figure 42 Final temperature distribution, split-n-return coil without concentrator.

Table 3 Electrical Parameters for Split-n-Return and Vertical Loop-Style Induction Coils with and Without Concentrator in Local Heating of Flat Bodies

Style	Concentrator	Current	Voltage	Efficiency	Power
SNR	Yes	2200	26.3 V	88.5%	27.1 kW
SNR	No	3400	28.6 V	74%	42.7 kW
VL	Yes	2200	30.2 V	87.7	26.0 kW
VL	No	8000	36.8 V	67.1	45.0 kW

bution. The heat penetration depth for both inductors is approximately the same. The more focused heating leads to better power utilization in the workpiece and reduced power demand. Table 3 shows a summary of the electrical parameters for the two cases. It is clear that the application of a magnetic flux controller on split-n-return inductors provides drastic improvements in terms of efficiency, required power, coil impedance, and power supplying circuitry performance.

Figure 43 shows the temperature profile produced with the vertical loop coil with concentrator and Fig. 44 shows the temperature profile for the inductor without concentrator. Similar to the split-n-return inductors, the coil with concentrator produces a more focused heat pattern in width than that without a concentrator, whereas the depth of heating is close to the same. The electrical parameters of the vertical loop coil show even greater improvement by the concentrator than for the split-n-return one (Table 3) [27].

B. Materials for Magnetic Flux Concentrators

Requirements for magnetic flux controlling materials in induction heating processes can be very severe in many cases. They must work in a very wide range of frequencies, and possess sufficient permeability and saturation flux densities. The concentrator permeability required is strongly dependent upon the application.

Unlike transformers or electrical motors, almost all induction coils have an open magnetic circuit. The magnetic field path includes not only the area with concen-

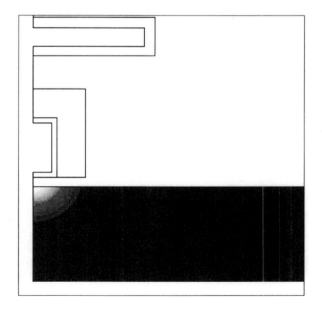

Figure 43 Final temperature distribution, vertical loop coil with concentrator.

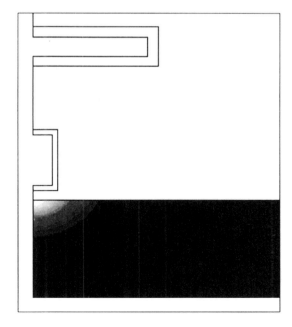

Figure 44 Final temperature distribution, vertical loop coil without concentrator.

trator, but the workpiece surface layer and the air between the surface and the concentrator, which cannot be changed. Therefore, the reluctance of the magnetic path is only partially dependent upon the concentrator permeability. For the same coil current, the workpiece power increases with increasing concentrator permeability very quickly at first, but then it approaches asymptotically the threshold value. At the same time, losses in the induction coil increase slowly with increasing concentrator permeability. Computer simulation shows that for most induction heating applications, the threshold value for workpiece power occurs when concentrator permeability is less than 100. For high frequency induction heating applications (greater than 50 kHz), the threshold value occurs at lower levels of permeability [20–40]. Therefore, increasing the permeability to higher values will not improve the coil parameters.

Besides electrical and magnetic properties, magnetic flux controllers must have stable mechanical properties and exhibit resistance to elevated temperatures from losses in the concentrator and radiation from the heated part surface. In heat treating and brazing, the material must withstand hot water attack, quenchants, and active technological fluxes. Machinability is also a very important property for successful application of flux controllers because it allows you to obtain the exact concentrator dimensions necessary to assure consistent results and precise magnetic field control.

1. Material Types

Three groups of magnetic materials may be used for magnetic flux control in induction heating systems: laminations, ferrites, and magnetodielectric materials [27].

Laminations

Laminations manufactured from silicon steel are a traditional material for low frequency electrical equipment including induction melting or forging furnaces, matching transformers, cores, etc. The main frequency range of their application is below 10 kHz. The technology of manufacturing and the application of traditional laminations are mature techniques and it is difficult to expect significant improvements of their properties. New amorphous laminations have much lower losses than laminations, but they are expensive and the technology of manufacturing is more difficult.

Laminations have high permeability and high saturation flux density $B_s = 1.8$–2 T (18,000–20,000 G) (Fig. 45). The Curie point is about 700°C. For that reason, laminations can theoretically work at elevated temper-

atures if they are protected against chemical degradation such as intensive oxidation in air, or rusting in water.

Laminations of different types including grain-oriented silicon steels are available in thickness of 0.05–0.2 mm. Each individual lamination is coated with an insulating material several microns thick to prevent shorting between adjacent laminations and to reduce chemical degradation. A thickness of 0.1–0.2 mm is used usually for audio frequencies of 1–10 kHz. Very thin laminations (0.05 mm) may be used up to 100 kHz under conditions of intensive cooling [27].

An inherent feature of laminations, very important for induction heating applications, is that they work well in plane-parallel fields only when a magnetic flux passes along the sheets. If a component of the magnetic field is perpendicular to the sheet plane, significant eddy currents are induced and intensive heating of the material occurs in laminations of any thickness, resulting in rapid concentrator destruction.

When comparing laminations to other magnetic materials, one must consider the space factor (ratio of the length of magnetic material to the total length of a stack of laminations). Usually copper separators are placed between 6 and 12 mm stacks of laminations to remove more easily the heat generated in laminations by eddy currents and hysteresis losses. The insulation and the stacking factor (ratio of the length of lamination to the total length) cause a reduction in the space factor. Real space factor accounting sheet insulation and separators are about 0.75–0.85 for simple linear coils and notably less for cylindrical coils, especially ID coils where it can be as low as 0.5. This leads to a corresponding reduction of equivalent saturation flux density to 1.3–1.5 T, instead of 1.8–2 T, and a proportional decrease in equivalent permeability.

Mechanical properties of laminations are not favorable for induction heating applications. Typically, they require special fixtures and cooling systems, and may not be used as a construction material. In addition, the application of laminations to induction heating coils is a labor-intensive operation as all laminations must be stacked individually. Furthermore, machining lamination is a relatively difficult operation.

Ferrites

Ferrites are ceramic-like materials with magnetic properties varying in a wide range depending on the composition and technology of manufacturing. They are widely used in electronics and electrical engineering due to their high permeability (in weak fields), high electrical resistivity, and low magnetic losses. Different kinds of ferrites may cover a very wide range of fre-

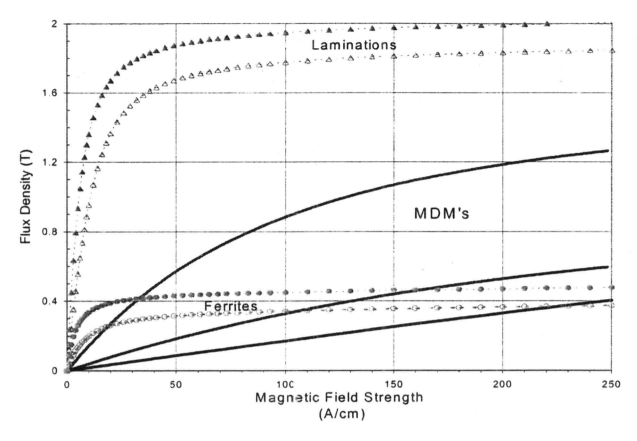

Figure 45 Magnetization curves for laminations, ferrites, and magnetodielectrics.

quencies up to and above 100 MHz. These materials are generally isotropic and can work in three-dimensional fields.

Different types of ferrites are widely used in communications and electronics due to their high permeability in weak magnetic fields and relatively low magnetic losses. However, ferrites have some properties that limit their application in the induction heating technique. Ferrites have relatively low saturation flux densities, $B_s < 0.5$ T (5000 G), and the Curie points are typically less than 250°C. Above the Curie temperature, the material becomes non-magnetic. At elevated temperatures (e.g., 160°C), the saturation flux density and permeability decrease, resulting in strong temperature dependence of the concentrator performance. Although chemically inert, ferrites are sensitive to thermal and mechanical shocks. Due to their high brittleness, they crack easily under mechanical impact, or when temperature shocks or gradients occur in their volume. They are also very hard and not easily machinable, requiring careful grinding or diamond cutting. This property is

very unfavorable for induction heating applications, where a small quantity of concentrators with different shapes is the typical situation.

Magnetodielectric Materials (MDM)

These materials are made from soft magnetic powders and dielectric materials, which serve as binders and electrical insulators of the particles [26]. The magnetic properties of these composites depend on the properties of their constituents and the technology of manufacturing. MDMs have a valuable combination of electromagnetic, thermal, and mechanical properties, adjustable in a wide range to match particular process requirements. Modern materials and technologies permit us to provide a complex of magnetic, mechanical, thermal, and chemical properties for effective use in the induction heating systems.

MDMs are well known in power electronics, communications, and radio electronics due to their thermal stability and quasi-linear properties. The components for these industries usually have simple shapes (rings, cups, and rods) and have very large production volumes. Similar to ferrites and laminations, these MDMs are not easily machined and do not fit well to induction heating requirements. Machinable MDMs, represented mostly by the Fluxtrol and Ferrotron families of materials, now cover almost all of the world's induction heating market demands. They are produced by pressing magnetic powder mixed with a binder, which then undergoes a thermal treatment according to special technology.

MDMs for induction heating applications have varying degrees of anisotropy. Permeability and saturation flux density are minimal in the direction of pressing and maximum in perpendicular directions. The degree of anisotropy depends on material composition and technology of manufacturing. Anisotropy is significant in low frequency materials (Fig. 46, curves A and A′) and is less significant or negligible in high frequency materials. A coil designer must take anisotropy into consideration to achieve the optimum results in MDM application.

The Fluxtrol and Ferrotron materials have been designed specifically for use in the induction heating industry. These materials all have excellent machinability, chemical stability, temperature resistance, and good magnetic properties. The family of materials contains materials specifically designed for the low frequency (below 30 kHz), medium frequency (30–200 kHz), and high frequency (above 200 kHz) ranges. The high frequency materials generally have lower permeability and saturation flux density. Sometimes, a combination of materials with different permeabilities may be used for smooth heat pattern control in challenging applications.

MDM Application Technique

In the case of single-turn coil, net shape or machined MDM pieces may be applied to the water-cooled coil tubing directly. Soft soldering of MDM to the coil copper provides the best heat transfer. This method is possible only for Fluxtrol A. A thin layer of thermally conductive epoxy filled with alumina, silica, or metal powder, when properly applied to the copper, provides excellent bonding and superior heat removal.

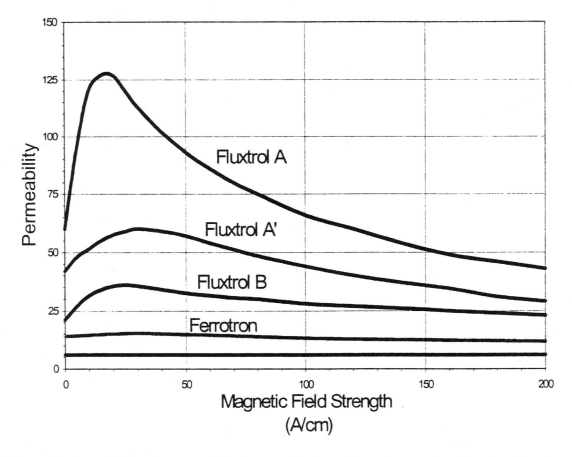

Figure 46 Permeabilities of magnetodielectric materials: Fluxtrol A′–Fluxtrol A in a direction of pressing.

In multiturn coils, the MDM pieces must be insulated from the coil copper in order to avoid electrical shorts and subsequent thermal breakdown. Low frequency MDMs have electrical resistivity sufficient to prevent the origin of induced eddy currents in the concentrator body but not enough to withstand high external electrical voltages. One millimeter of silica-filled epoxy compound or silicon resin is usually sufficient for attaching the concentrator to multiturn coils for heat treatment.

For high voltage coils (>500 V), better electrical insulation, such as mica sheets, must be provided and a separate means of MDM cooling may be required. Properly applied MDMs have been working on some induction heating coils for many years without degradation (Fig. 38).

V. INDUCTION PROCESS AND COIL DESIGN TECHNOLOGY

The design of induction processes and coils may be very different even when the parts to be treated look similar (e.g., axles or shafts). The same axle can be hardened by a static single-shot or scanning process with many variations in coil design and operating conditions (frequency, power, time, scanning speed, and quenching method).

The traditional practice of process and coil design is based mainly on an empirical approach and "rules of the thumb." Many induction coils are improperly designed. In some installations, this results in poor heat pattern control, low efficiency, low power factor, poor metallurgical results, and short induction coil lifetime. The situation with induction processes and coils may be explained by the complexity of the involved processes (electromagnetic, thermal, mechanical, and hydraulic). Solving some problems, such as coil copper or part cracks, requires a fundamental study with a multidisciplinary approach. In the meantime, the development of the methods and programs for the mathematical simulation of induction systems and new materials and technologies constitutes a good base for process and coil enhancement.

A. Designed-for-Induction-Heating Strategy

Sometimes, small changes in heat treating specifications and part manufacturing process flow lead to a drastic simplification or complication of the process and coil design. Hence, a reasonable compromise must be found to adapt the specifications to the nature of induction heating. A detailed examination of the process specifications must include hardness pattern, tolerance of specified parameters, part handling method, sequence of operations, etc. Sometimes, the part material and geometry can be changed for better matching to the induction heating technique.

To take full advantage of induction heating, some companies have begun using a Designed-for-Induction-Heating (DFIH) strategy [28,29]. It is important to consider that induction hardening can provide higher hardness and better mechanical properties than traditional hardening of the same steel. In addition, many materials respond differently to induction hardening than furnace hardening. DFIH is a new process design strategy where the part and heat treatment specifications are formulated with an account of the special features of the induction technique [28,29]. It is necessary that the part designer and the process engineer understand the specifics of induction heating in order to simplify the manufacturing process. Dimensional tolerances, basic surfaces, and the sequence of mechanical and heat treating operations all must be considered in the DFIH strategy.

The DFIH strategy includes the following:

1. Material selection: The selection of a material that fits well to the conditions of very fast heat treating cycles is the first consideration in the DFIH strategy. A fine-grained structure is required for fast diffusion process completion during austenitization. It is desirable to avoid materials that require oil quenching. Oil does not fit well to the induction heating treating lines. Water and polymer solution spray quenchants are best for induction processes. Because only a portion of the part volume is being heated, it is easier to quickly extract the heat during quenching. Some alloyed steels can even be quenched with forced air. In some applications with small case depth for alloyed steels, the possibility of self-quenching due to heat soaking inside the part should be considered.
2. Part geometry: In the part geometry design, it is necessary to avoid sharp edges and internal corners. External sharp corners will tend to overheat, causing stress concentration, which may lead to possible part cracking. In addition to edges and corners, orifices or holes can cause complications in the induction heating process and coil design. Too steep variations of shaft diameter can also lead to problems with heating processes.
3. Sequence of operations: The sequence of the mechanical and heat treating operations can have

an influence upon the induction heat treating process. One example could be a gearshaft with a relatively thin wall, which needs to have both internal and external splines machined during the manufacturing process and the internal spline must be surface hardened. It is more simple to heat treat the internal spline before the external spline has been machined because the equivalent wall thickness will be greater, which helps to avoid through hardening.

4. Hardness pattern specification: With induction heating, the induced current lines must be closed or continuous. For this reason, it is difficult to harden uniformly a rectangular spot on a flat surface. One solution for this problem is to move the heat source loop across the spot for uniform heating. Another possibility is to slowly heat the perimeter of the closed loop area and wait until the heat diffuses across the whole area. By specifying smooth hardness patterns (Figs. 47 and 48), the design process can be greatly simplified.

The examples used above were mainly focused on surface hardening. In other applications, such as brazing or soldering, the DFIH strategy may be even more important.

B. Modern Induction Coil and Process Design Technology

In the past, induction heat treating processes and coils were designed based upon "rules of thumb" and the "trial and error" method. Now, there are new tools and opportunities available to induction coil and process designers such as computer simulation, possibility to use power supplies with almost any frequency between

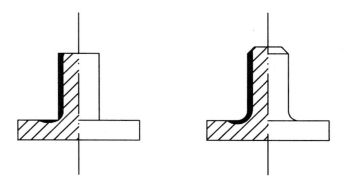

Figure 47   Favorable (right side) and unfavorable (left side) geometry for induction hardening of an axle fillet area.

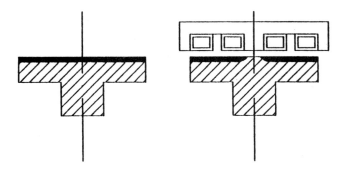

Figure 48   Favorable (right side) and unfavorable (left side) hardness pattern specifications for an axle flange.

50/60 Hz and 500 kHz, and a wider selection of materials for magnetic flux control. To fully utilize these improvements requires a different approach to induction process and coil design. The authors recommend the following five-step induction coil design procedure for complex applications:

1. A detailed analysis of engineering specifications and industrial conditions
2. Induction coil style and process-type selection
3. Computer-aided design of the coil
4. Coil manufacturing
5. Coil tests and validation.

1.   Detailed Analysis of the Engineering Specifications and Industrial Conditions

The first stage of the coil design is a detailed analysis of process specifications. There are many different situations regarding the part that must be heat treated, which can simplify or hinder the induction heat treating process design. Often, the initial specifications to the part hardening are created without an account for the particular features of induction technique and cannot be reached because of restrictions laid by the laws of physics or technical limitations on power density, active materials properties, etc. In other cases, the specifications are very favorable for induction heating because the part has been designed using the DFIH strategy. Another case could be when the process exists already, but has problems such as part cracking or undesired through heating, low production rate, etc. Oftentimes, better parts and an improved process can be achieved with only minor modifications to the specifications.

The laws of continuity of magnetic flux (div $\mathbf{B} = 0$, where $\mathbf{B}$ is the magnetic flux density vector) and of electrical current (div $\mathbf{J} = 0$, where $\mathbf{J}$ is the current density vector) are the main "restrictive" laws in the induction technique. Due to the law of magnetic flux continuity, the magnetic field lines must always be closed. After they pass through the "work area" under the face of the coil, they must return back around the coil turns that carry the current, the driving force of the magnetic field. The currents and power sources induced in the workpiece have bell-shaped distribution in a direction perpendicular to the coil current flow (Fig. 49). By varying the coil profile, frequency, and coupling gap, and especially by applying magnetic flux concentrator, a designer can control this distribution within certain limits, but cannot change its nature completely. Figure 49 shows the influence a concentrator has on power distribution in a two-layer (non-magnetic/magnetic) workpiece imitating the part at the end of a surface hardening process.

According to the law of current continuity, eddy currents induced in a certain area must return back somewhere on the surface layer of the workpiece, causing material heating on this "current back path" area. In cylindrical induction systems, this law does not impose any restrictions because the heat pattern forms a natural close loop. For local heating with hairpin, "split-n-return," and some other coil styles, this additional heating in unintended areas may be a serious problem. The

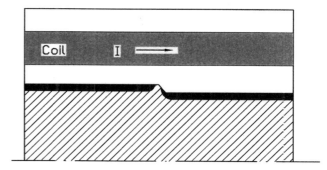

Figure 50 System layout and heat sources pattern. Heating with a single-shot style coil. Frequency, 20 kHz.

variation of the coil copper geometry and the application of magnetic flux controllers are sufficient for the elimination of this unintended heating in some cases. In others, local air, water, or quenchant shower cooling may be used during the heating cycle for cooling the back path area.

2. Induction Coil Style and Process Type Selection

The second stage of the process and coil design is choosing the heating method and the coil style, in combination with a preliminary choice of frequency and power. In many cases, the same specifications may be met using different coil styles. For example, a relatively short shaft with a variation in diameter may be surface hardened using a scanning, single-shot, or static heating technique with corresponding coil designs. Each method has its own advantages and disadvantages. The single-shot method permits us to heat relatively easily the zone of fillet due to the favorable natural current density distribution (Figs. 50 and 51).

Power density is highest at the root where heat soaking into the part core is high, and much less on the corner where heat soaking is small. Additional heat pattern control by application of magnetic flux controllers and by variation of coupling gap or coil bar width can provide the desired hardness pattern on both diameters and in the zone of diameter change [24].

It is much more difficult to do this with a scanning or static coil (Fig. 51). Here, the maximum power density is in the corner, where the energy demand is smallest and there is a lack of power at the root. A special coil profile and significant coupling gap variation are required to obtain acceptable results with the static heating method. Single-shot coils are also much more favorable for automatic part handling fit well to high production rates. At

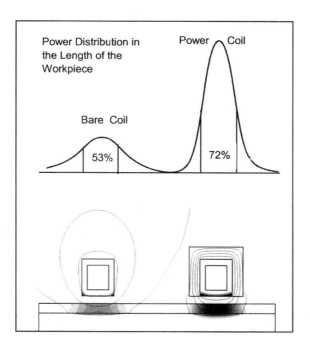

Figure 49 Influence of magnetic flux concentrator on power density distribution (top) and magnetic field lines (bottom) in a workpiece.

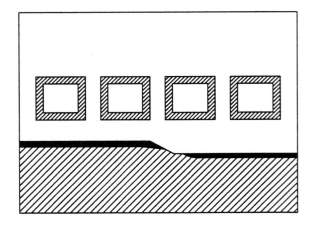

Figure 51 System layout and heat sources pattern. Heating with a multiturn static style coil. Frequency, 20 kHz.

the same time, these coils are more heavily loaded because the coil's copper covers only a part of the surface to be heated. Special attention must be, therefore, paid to the coil design in order to prevent overheating of the copper and coil failure. In addition, a designer must consider the electrodynamic forces, which can cause coil deformation. Single-shot coils need to be rigid enough to withstand these forces and coil reinforcement is usually required.

Both static and single-shot methods require high power and provide short heating times. Compared to these methods, the scanning method requires much less power and is widely used for "smooth" shaft hardening. When there is "smooth" shaft diameter variation in a scanning process, either the speed, the power level, or both can be changed to achieve the specified hardness depth. However, if there is a significant and steep diameter variation, these control methods are insufficient. A scanning process with a short linear coil, similar to an abbreviated single-shot coil, is a relatively new technique and can help in these cases. The short linear coil combines the benefits of the single-shot and scanning methods and should be considered for complex geometries if the production rate is relatively low.

The first and second stages of coil development require good experience from the designer in aspects of induction heating, metallurgy, and material processing techniques. Certain calculations are necessary for a correct choice of the coil style, frequency, power, and production rate. At these stages, it is very difficult to formalize the problem and solve it mathematically. The experience and the intuition of the designer are very important here. The application of magnetic flux con-

trollers for heat pattern control and/or coil performance enhancement must be considered at these stages as well [24].

3. Computer-Aided Design and Engineering of the Coil

The third stage is determining the coil dimensions and operating conditions. The goal of this stage is to find the optimal combination of coil design and operating conditions (frequency, power density, heat time, scanning speed, etc.) to ensure an efficient, high-quality heating of the specified area. The optimization of the coil geometry is a complex problem. In addition to high efficiency and correct heating pattern, a coil must be mechanically strong and reliable. For this reason, many designers make scanning machined integral quench (MIQ) coils with a big radial dimension. This leads to high current demand and low efficiency of the coil, mainly due to poor utilization of the power transferred into the workpiece and increased losses in supplying circuitry. A solution may be found in the application of magnetic flux controllers, which concentrate power in a desired area of the workpiece and reduce the current demand.

Computer simulation is a very effective tool at this stage. Due to the large variety of heat treating processes and part geometries, it is difficult to expect that one single universal program will cover all the needs of the coil design. Coupled (electromagnetic plus thermal) 3-D programs may be considered as the most universal and adequate codes for this job. However, there are no 3-D coupled packages on the market that could be able to handle general induction heat treating systems at this time. Some software packages hope to add this capability soon. ID coupled and 2-D coupled or electromagnetic programs may be successfully used for practical needs in coil design [22,30].

This stage of design requires knowledge of induction heating techniques, and practical experience in computer simulation and coil manufacturing technology. It provides input data for the coil engineering (theoretical copper profile, magnetic flux controller type and dimensions, etc.) and process parameters (cycle time, power, coil current, voltage, efficiency, and power factor) required for a choice of supplying circuitry components and power supply. The real induction coil with all the required constructive components (feet, concentrators, quick change adapter if required, quenchant supply for MIQ coils, etc.) must be engineered after the copper coil head cross section is finalized. The technology of coil manufacturing must also be developed at this stage. A final choice of materials, machining and joining operations, and electrical and thermal insulation must be made.

Usually, coil designers do not use computer simulation at this stage, but rely only on existing technology and experience. However, computer simulation may also be very useful at this stage. It can provide valuable information on maximum temperatures of copper and concentrator materials under different conditions of heating and cooling to assure coil reliability. Even if water cooling is intensive and the average temperature of copper during the treatment cycle is not high, local overheating can occur. This causes significant stresses and an accumulation of defects. The resulting surface microcracks lead to further increases in the local concentration of losses and stresses. The cracks grow, causing coil copper failure. Mechanical loads due to electrodynamic forces and machine operation accelerate the process of failure. The calculation of electrodynamic forces affecting the workpiece and coil components may also be important. The computer simulation of copper heating with local optimization of the coil profile can help to reduce high local temperatures without notable affects on the heat pattern and coil parameters.

Coil engineering drawings and manufacturing specifications are the outputs of this stage. Cost analysis and long-term experience clearly demonstrate that the best available materials and solutions, even if more expensive, must be used if they provide better coil performance and longer coil lifetime. The coil design is now completed, but not the coil and process development.

4. Coil Manufacturing

The fourth step is coil manufacturing. The level of manufacturing technology and discipline is as important for coil quality as its design. The parameter most sensitive to manufacturing technology is the coil life. Although the main technological solutions must be established at the engineering stage, some changes are often required to fulfill or simplify a manufacturing process. Special attention must be paid to the technology of magnetic flux controller application. In addition, tests should be made to ensure the hydraulic, electrical, and mechanical soundness of the inductor.

5. Coil Testing and Validation

The final stage of development is the coil tests. They include a test of coil performance and an analysis of the results (metallurgy, distortions, etc.). In the case of unsatisfactory results, a minor coil or operating condition adjustment may be done on site; otherwise, a return to one of the previous stages may be necessary. One of the most effective methods of coil adjustment and final tuning is the modification of magnetic flux controllers. Although changes in copper profile require significant coil rebuilding, the modification of some types of controllers may be easily done during testing. For example, controllers made of magnetodielectric materials may be removed, reshaped, and reinstalled onto the coil. When the required results are achieved, the controllers may be permanently fixed using soldering, gluing, or mechanical fastening. Sometimes, minor changes to the concentrator profile by means of cutting or shaving without removing it from the coil are sufficient for heat pattern adjustment. More and more induction companies use this method in the development of single-shot, hairpin, and other coils.

The described design methodology has proven to be effective for the development of challenging processes and induction coils. For simpler cases, some stages may be omitted. For example, stages 2 and 5 are not necessary if the design has already been validated and used in the industry.

VI. COMPUTER SIMULATION OF INDUCTION HEATING PROCESSES

The authors use computer simulation as an everyday tool for the study and design of induction heating systems. The purpose of this section is to help the reader to understand the general situation and existing opportunities for simulation in the induction heating technique. This chapter is not intended to compare different computer simulation packages available on the market. Although the description of programs and examples of their application are contained in this section for the software used by the authors, there is no intention to say that this software is the best, or the only software for induction heating computer simulation for a given application.

A. Current Status of Computer Simulation for Induction Heating

Computer simulation of induction heating systems, which was once a tool used only by the academia or at various technical centers, is now penetrating into the industrial world. Computer simulation is used for induction project evaluation, process development, setup, maintenance, marketing, and business presentations [22]. There is still some resistance mainly from the old generation, but even this sector is warming to the modern simulation tools with the ever-increasing evidence of their advantages. This increased acceptance has happened faster than most would have predicted and can be attributed to vast improvements in both the computer hardware and software available on the market today.

1. Computer Hardware and Software

The most obvious factor that has contributed to the growth of computer simulation is the increased speed and memory of personal computers. The computational power of the personal computer has risen exponentially over the past decade. What 3 years ago would have taken minutes, hours, days, or even weeks on the average personal computer now takes several seconds, minutes, hours, or possibly days for the most complicated tasks.

Besides the increased processing power of today's computers, great strides have been made in the complementary hardware such as data storage, communications (Internet), and peripherals. In 1998, a good personal computer had a 1-GB hard drive and a 1.44-MB floppy drive. By 2001, the average new computer came with a 20- to 40-GB hardware and a read/write CD ROM (550 MB) drive, making it possible to store and to transfer much larger amounts of information including pictures, computer animations, and even movies.

The amount of information exchanged between computers is also much greater. Through the Internet, it is possible to send papers, drawings, simulation results, and pictures via email. It is also possible now to remotely operate another computer through the Internet. One example is a big automotive parts supplier that has people using Flux 2D in the United States, Germany, and France. Through a secure Internet site, the users at one of the offices can let someone from one of the other sites take control of their computer and demonstrate how to perform certain operations in Flux 2D [30].

2. Types of Software

Programs for the simulation of induction heating and quenching processes may be classified as 1-D coupled (electromagnetic + thermal); 2-D electromagnetic, thermal, or coupled; and 3-D electromagnetic or thermal. Other programs also have the possibility to calculate structural transformations, stresses, and distortions.

The majority of 2-D and 3-D programs used for induction heating simulation originate from companies specialized in the development of general purpose software for the calculation of physical fields in electrical engineering. For induction heating simulation, only a part of the whole package is used, usually an "eddy current block," a "thermal block," or a combination of these blocks with an electrical circuit block (power supply circuitry). The programs usually do not have the proper databases necessary for induction heating, or the explicit means for simulation of the induction heating machine performance (scanning, pushing of billets, etc.). Many of them must be adapted for the simulation of real induction heating processes. Flux 2D is probably the exception to the rule because it has been developed by taking into account the special features of induction heating processes [31,32].

The 2-D and 3-D programs for induction heating simulation are mostly based upon the finite element analysis (FEA) method. However, some computer simulation programs use other techniques, such as the finite difference method (FDM) or integral equations method, for problem solution. The FEA method is the most universal technique, providing accurate results for any properly described case. There are other methods available, which are developed for specific types of systems, such as transverse heating of strips. They have reduced computation time compared to FEA and problem-oriented interfaces.

3. Rule of Pyramid Strategy

Induction heating coil designs and operating conditions are very diverse, and different programs are necessary in order to meet practical needs (Fig. 52). Theoretically, we

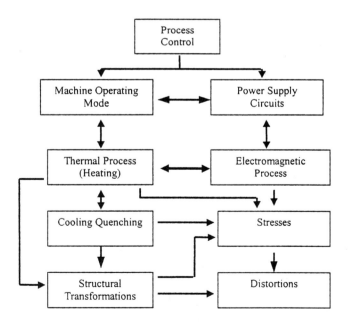

Figure 52 Different phenomena involved in induction heat treating process.

can assume that a single, universally coupled 3-D program would be able to meet all the major needs of the simulation of induction heating systems, including power supplying circuitry and machine operating mode. However, at the present time, no program like this exists despite tremendous progress in computer hardware and software tools. Any 3-D code requires powerful computers and the process of simulation is knowledge-consuming and time-consuming. Workstations or the most powerful personal computers must be used.

From 3 days to 1 week of work for a skilled operator may be used as an estimate for the average time for one case study using 3-D electromagnetic simulation. This estimate takes into account problem formulation, laborious preparation of input data, checking and correction of the almost inevitable errors or inaccuracies, geometry and mesh construction, physical property and boundary condition description, calculation itself, analysis of results, parameter variation (design iterations), and preparation of documentation (reports and graphs). The complexity of 3-D analysis and the required skill of the user make direct 3-D optimization of induction heating processes and systems rather difficult and inefficient. A strategy based on a hierarchical use of programs is most effective [31].

This strategy is called the "rule of pyramid" for induction heating computer simulation (Fig. 53). For most cases, the first stage in the simulation of the induction heating system should be done with a 1-D coupled program. This allows you to study the influence of frequency, power level, quench type, and temperature dynamics on the process. The ranges of interest can be studied quickly and effectively. A good estimate for the heating time, coil power, coil voltage, and coil current

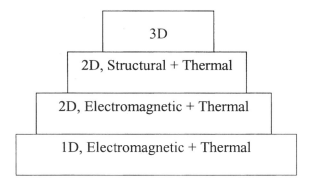

Figure 53 Rule of pyramid for induction heating computer simulation.

level can be made from the results of the 1-D simulation [32]. In most practical cases, 1-D simulation is sufficient; in other cases, a higher level of software must be used.

The second stage in the simulation may be done using 2-D electromagnetic or coupled code. The coil current, tube profile, coupling distance, and magnetic flux concentrator dimensions that furnish the required field and power intensity may be determined. For nearly all cases, the use of 1-D plus 2-D coupled simulation is sufficient for induction heating process and coil design. With this strategy, 3-D simulation and/or experiments are only required for the coil design correction in zones of irregularity where 3-D effects are significant, such as the corner area of a slab (Fig. 15).

B. Simulation vs. Experimental Method

For many processes, computer simulation provides very exact results and experimental tests are unnecessary. Through heating of billets before forging or extrusion is an example of this type of process. The system geometry is simple, material properties are relatively well defined, and the magnetic permeability is equal to one during the majority of the process. In this case, only the electromagnetic and thermal processes need to be simulated. For other processes, the accuracy of the results obtained from simulation may be lower. Surface hardening is a good example of this because of the more complex nature of the process. More phenomena are involved in the simulation to obtain the final results (hardness and mechanical properties). The initial part of the coil design problem and process simulation may be rather far from the final metallurgical result. The case depth and hardness are separated from the coil design and power supply setup by electrical circuit, electromagnetic and thermal process, quenching, and structural transformations. Inaccuracies during each stage of computer simulation can accumulate, resulting in more significant final discrepancies.

There are three sources of inaccuracy in computer simulation: errors of computation, mathematical description of the process, and material properties or process parameters. The errors of computation depend strongly on the algorithm of the process simulation and the discretization of the problem in space and time. Modern simulation programs can control these errors inherently and reduce them to a negligible value. Electromagnetic and thermal processes are quite accurately described by the Maxwell and diffusion (Fourier) equations, respectively. With correct physical properties of materials, the solution of these equations is accurate. The processes of structural transformations and quenching have less accurate mathematical descriptions. Existing descriptions of these processes are mainly approximations of experimental data, with parameters depending on the material composition and operating conditions [31]. These parameters and inaccuracies in the physical properties of the material to be treated are the main sources of error in simulation in most induction processes.

Additional phenomena and factors are involved in some other induction processes. For example, in induction brazing, besides electromagnetic and thermal processes, an important role is played by chemical, capillary, gravitational, and magnetohydrodynamic effects.

Specific heat and thermal conductivity as functions of temperature are necessary for the simulation of any heating process. For induction heating, we also need to know electrical resistivity vs. temperature and magnetic permeability vs. temperature and field strength. The errors in temperature field prediction resulting from material properties inaccuracy depend on the process type. For through heating or surface hardening, they may be rather small because the whole workpiece or a surface layer of steel is non-magnetic during the final stage of the cycle and the material properties are well defined. Correct temperature prediction in the simulation of the tempering process, where the steel is magnetic and the material properties are not as well defined, may be more difficult. Errors in the prediction of the results of structural transformations, residual strengths, and distortions may be more significant.

However, the results of computer simulation are very valuable even if they are not quite accurate. With a series of calculations, the user can find the system's response to the intended (process design) or unintended (input data tolerance) variation of parameters. One can then modify the process without experimental tests.

A simulation's accuracy may be improved by using the results of experiments, or by a study of material properties for the different stages and parameters of the process. The existing data on structural transformation in steel during heating and cooling refer mainly to furnace heating conditions. The local heat concentration, high heating speeds, and short processing cycles typical of the induction method influence material response to heat treatment. A special study of steels' properties for the purposes of induction heating computer simulation is strongly desired. It is especially important for the powder metal industry, where public information on electromagnetic and thermal properties, which is required for induction heating computer simulation, is very poor.

C. Simulation of Axle Scan Hardening Process Using a 1-D Approach

One very common application that was previously designed using rules of thumb supported by the empirical method is the hardening of an axle shaft. Figures 54 and 55 show the temperature distribution for an induction scan hardening of a shaft with an OD of 5 cm for the automotive industry. Figure 56 shows the cooling curves for the quenching part of the scan hardening process. The curves begin when the temperature at a given radius passes through the Ar level and phase transformations begin. For the process simulation, the ELectroThermal Analysis (ELTA), 1-D coupled electromagnetic plus thermal program was used. ELTA has a special scanning application, which is able to make fast and accurate simulation and optimization of this actually 2-D process. The results of simulation provide the required

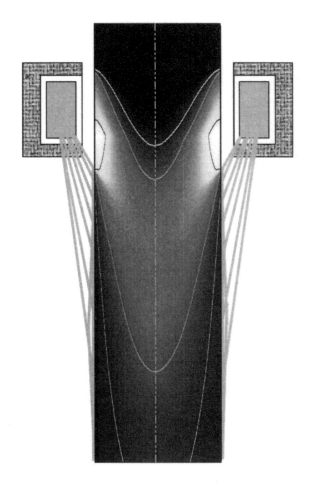

Figure 54 Temperature distribution map for an induction scan hardening process of an automotive axle.

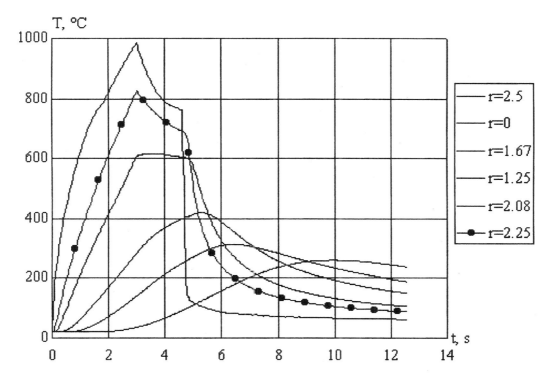

Figure 55 Temperature graphs at different radii vs. time for the axle hardening process. The marked curve corresponds to the hardness depth (2.5 mm).

power, scanning speed (5 mm/sec), coil face width, and frequency (10 kHz) for the necessary hardness depth (2.5 mm).

D. 2-D Simulation of Seam Annealing Process

Induction heating is used for many operations in the tube manufacturing industry: tube welding, seam annealing, heat treating, cutting, bending, etc. The following example shows how computer simulation can help to optimize and to study the process and coil design for seam annealing of a 0.4% carbon steel welded tube. The tube is 2 in. in diameter with a 1/8-in. wall thickness. In the welding seam area, the tube needs to be annealed (heated to above 800°C) to improve the microstructure. The production rate is 3.8 ft/sec.

Typically, one of two induction coil styles is used for this process: split-n-return or vertical loop [22]. These two coil styles provide the majority of their heating under a central or main leg with magnetic flux concentrator. The difference between the two coils is the return path for the induction coil current. Vertical loop inductors return the coil current above the main leg of the

induction coil (Fig. 57). In split-n-return coils, the coil current is split in two halves and returned along the sides of the part (Fig. 58). Before 2000, there were no published studies on which induction coil style performs better.

To study the performance of these two induction coil styles for seam annealing, the 2-D coupled software Flux 2D was used for computer simulation. Half of the geometry was simulated due to a plane of symmetry. The same current is applied to both coils such that the losses in the busswork, transformer, and power supplying circuitry will be the same. The frequency used for simulation is 10 kHz. In the seam annealing process, there is a continuous feed of tube underneath the inductor. For simulation, the time corresponds to the position of the cross section of the tube under the coil.

The maximum temperature exiting in the coil in the vertical loop (Fig. 57, Table 4) and the split-n-return (Fig. 58, Table 4) inductors with the same current applied is within 2% (959°C and 934°C, respectively) and their distributions are close to the same. The highest temperature is directly under the center of the main leg of the respective inductors for both cases. The heating outside of the area under the face of the main leg is

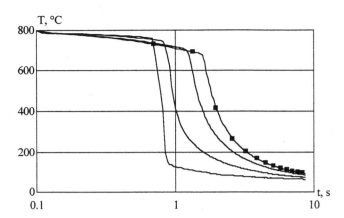

Figure 56 Cooling curves at different radii (0, 1, 2, and 2.5 mm from the surface, respectively) for the quenching process.

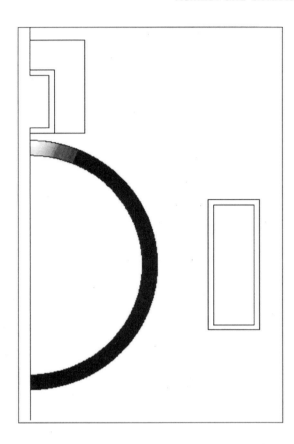

Figure 58 Final temperature distribution for split-n-return inductor.

negligible for both induction coil styles. The power required for the vertical loop inductor is slightly less than for the split-n-return one, but the electrical efficiency is slightly lower (Table 4). This paradox is due to better utilization of the power in the workpiece for the vertical loop inductor. When adjusted to the same maximum temperature, the vertical loop inductor will require 4% less power.

1. Vertical Loop Inductors

Figure 59 shows the magnetic field lines at the end of heating for the vertical loop inductor. The field lines from the main leg of the inductor penetrate into the non-magnetic area under the face of the main leg of the inductor and are concentrated in the magnetic material near the magnetic/non-magnetic boundary (temperature of 750°C). There are less field lines surrounding the return leg of the induction coil, but they do interact with the magnetic flux concentrator.

In seam annealing processes, both the coil copper and the magnetic flux controller are heavily loaded. The interaction of the top turn with the magnetic flux con-

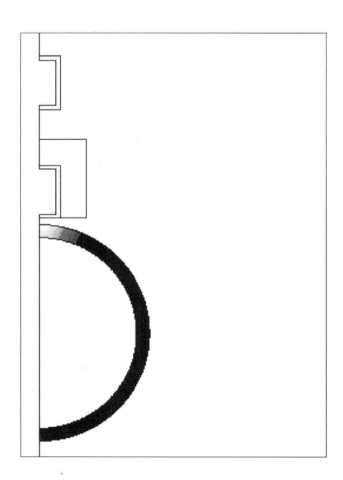

Figure 57 Final temperature distribution for vertical loop inductor.

Table 4 Summary of the Results of the Study

Style	Coil V	Coil I	Coil Loss	Tube P	Max T(°C)	Efficiency	Total Power
VL Same	378	2200	36.0	216	934	85.7%	252
VL Wide	321	2200	31.8	216	938	87.0%	248
SNR	272	2200	31.1	236	959	88.0%	267

troller causes a wide area of high flux density (greater than 0.5 T) in the top portion of the concentrator where it is farthest from the copper, which provides the cooling. This can lead to concentrator overheating and eventual failure.

If the return leg of the induction coil is made wider, the magnetic field lines around the return leg are much flatter. This leads to less interaction between the top leg and the concentrator and less field lines around the top leg (Fig. 60), which means that less coil voltage is required (321 V compared to 378 V). The wider return leg also leads to reduced flux density in the concentrator, which improves the induction coil performance. It also leads to lower losses in the return leg (6.9 kW compared to 7.6 kW), but this amount is insignificant when compared to the total power (248 kW). This inductor pro-

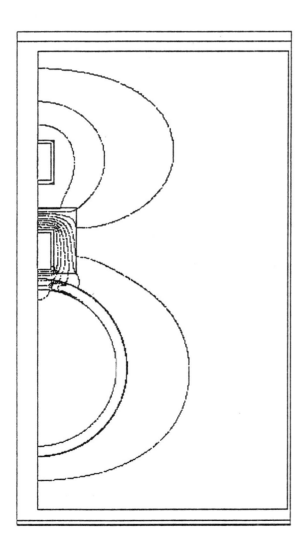

Figure 59 Magnetic field lines at the end of heating for the vertical loop inductor with the same return leg cross section as the main leg.

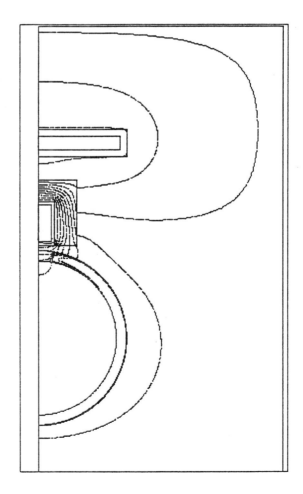

Figure 60 Magnetic field lines at the end of heating for vertical loop inductor with a wider return leg cross section.

duces an almost identical heat pattern as the one with the same size tube for the return leg as for the main leg.

For both versions of vertical loop inductor, the induced currents are focused under the main leg of the inductor and return relatively uniformly around the surface of the tube with the highest density in the magnetic material near the magnetic/non-magnetic boundary.

2. Split-n-Return Inductors

Figure 61 shows the magnetic field lines at the end of heating for the split-n-return inductor. The field lines from the main leg penetrate into the non-magnetic area of the tube and are concentrated in the magnetic material near the magnetic/non-magnetic boundary (temperature 750°C). The return leg has even less field lines than the vertical loop inductors, which leads to a lower coil voltage (272 V) and higher coil power factor. The flux density in the magnetic flux concentrator for the split-n-

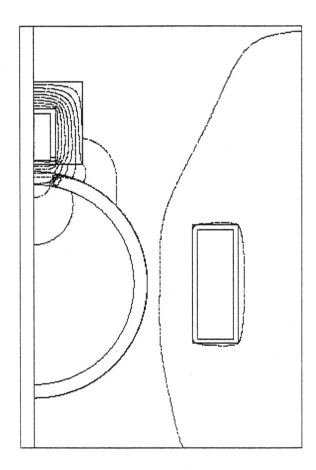

Figure 61 Magnetic field lines at the end of heating for the split-n-return inductor.

return coil style is much less than for the vertical loop (around 0.3 T). With this lower flux density, there are no questions about the induction coil reliability because the losses in the concentrator increase at a rate slightly higher than the square of flux density.

The induced currents are focused under the main leg of the inductor with the highest density at the magnetic/non-magnetic boundary. The currents return along the surface of the tube with the highest density under the face of the return leg. However, this peak is much less than under the main leg and produces a small amount of heating.

3. Results of the Study

The split-n-return and the vertical loop induction coil styles perform similarly for seam annealing for steel tubes. The vertical loop induction coil requires slightly less power than the split-n-return one. However, the split-n-return inductor has a higher power factor and reliability (very important factors in tube production), which is usually runs continuously, and any shutdown causes a great deal of lost production. The vertical loop induction coil's performance can be improved using a wider return leg. Both induction coil styles can be effectively used when properly designed and the final decision must be made based on mechanical and maintenance considerations.

All induction coils compared in this study used Fluxtrol A magnetic flux controlling material. It has already been established that without a magnetic flux controller, the induction coil performance would be extremely poor for this type of system, especially in the case of the vertical loop inductor [22]:

> VL same = vertical loop with the same return tube
> VL wide = vertical loop with a wide return tube
> SNR = split-n-return.

VII. INDUCTION HEATING APPLICATIONS

A. Induction Melting

Induction melting was the first industrial application of induction technique (see Section I.A). There are three main types of induction melting furnaces: channel, crucible, and special furnaces.

1. Channel Furnaces

Channel furnaces operate almost always at a line frequency of 50 or 60 Hz. They are used for the production of large quantities of molten metal with relatively low melting temperature such as cast iron, copper, aluminum, zinc, and their alloys for casting and other purposes. These furnaces have the capacity from several tons to hundred of tons, and a power up to 4 MW. An induction channel furnace (Fig. 62) consists of a refractory covered vessel containing molten metal, one or several melting units (inductors), and mechanical equipment. Installations also contain power supplying transformer, compensating capacitor battery, and control and auxiliary systems. Channel furnaces have better efficiency and require less capital investments than crucible ones. However, they do not work reliably when the metal temperature exceeds 1350°C and have difficulties with the start-up process [15,20].

The melting inductor is essentially a transformer with magnetic core, water-cooled primary winding(s), and a refractory channel forming a loop with the ends merged into the bath. The channel filled with molten metal acts as the secondary of the transformer. The current induced in the channel heats the molten metal. The rest of the metal heats due to mass exchange between the channel and the bath caused by thermal convection or stirring due to electrodynamic forces. These forces are created in some channel furnaces by a special design of the inductor. The furnace must work continuously and the channel has to always be filled with metal. For start-up, the channel must be filled with metal melted in another furnace [15,20].

2. Crucible Furnaces

There is a large variety of crucible furnaces for induction melting. These furnaces are called also coreless because unlike channel furnaces, they have no magnetic core.

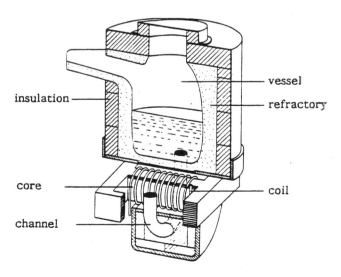

Figure 62 Induction channel furnace. (From Ref. 20.)

Small coreless furnaces are used for metal alloying and refining in research and industrial laboratories, and for metal casting in jewelry, dentistry, and other applications. They have capacity ranges from several grams to several kilograms, and operate at frequencies from 10 to 400 kHz. Smaller furnaces generally work at higher frequencies. Crucible furnaces are very versatile apparatuses. They can start from the cold condition, work at any temperature acceptable for the refractory, operate in a vacuum or protective atmosphere, and provide uniform metal composition due to intensive stirring caused by electrodynamic forces. Small furnaces usually have no magnetic circuit at all, but computer simulation and new practical experiments show that the application of magnetic concentrators to these furnaces is beneficial in many applications. Furnaces with magnetic circuit in the form of shunts or an external shell have higher power factor and efficiency and strongly reduced stray magnetic fields. The last factor is especially important for vacuum and protective atmosphere furnaces.

The furnace consists of water-cooled winding, a crucible containing the melt, and a mechanical system. The power supplying circuitry consists of a tube generator or solid state power supply with compensating capacitor battery and control system. Coreless furnaces with a conductive crucible, made of metal or graphite, may be used for the processing of nonconductive materials.

Large crucible furnaces are used in the industry for the production of high-quality steels such as stainless steels and special alloys, and for casting purposes. These furnaces have the capacity of up to hundreds of tons and operate at frequencies from 50/60 Hz up to several kilohertz. The rated power of these installations ranges from 100 kW to 20 MW. Large crucible furnaces are supplied either directly from the line transformer, or from a thyristor frequency converter. Crucible furnaces have a low power factor, so large compensating capacitor batteries are required. The size of the battery grows with the furnace capacity due to higher power and lower power factor (refractory is thicker), and lower frequency is usually required.

A typical furnace design (Fig. 23) consists of a multi-turn coil made of a rigid copper profile; one or several layers of refractory, magnetic shunts (yokes) made of laminations in the form of stacks distributed around the coil; mechanical structure; tilting mechanism; and other equipment. Large furnaces must have a rigid case or shell, which holds all of the furnace components together. Magnetic shunts work as structural elements and as concentrators, improving efficiency and power factor and providing a screen, protecting the furnace shell and mechanisms from unintended heating by stray magnetic field. Rigid construction is very important because the furnace must support the mass of the charge and withstand strong electrodynamic forces applied to the winding turns and magnetic shunts [15,20].

3. Special Furnaces

In advanced technological processes, an important role is played by special crucible furnaces. These special furnaces include cold crucible and levitation furnaces.

Levitation Melting

Levitation "furnaces" are mainly research tools used in laboratories for the production of very pure metals and special alloys. These devices melt metal samples levitated by electrodynamic forces. These forces are produced by the melting inductor, or by an additional inductor with a current of the same or different frequency. Levitation is necessary to avoid contact of the molten metal to the ceramic or metal crucible, and thus avoid any contamination. For levitation melting, three different criteria must be met simultaneously:

The power transferred to the charge must be sufficient for melting.

The total electrodynamic force affecting the charge must be equal to its weight.

The electromagnetic force distribution on the surface of the charge, together with the surface tension of molten metal, must keep the melt integrity.

Due to the complicity of these requirements, only small quantities of metals (from several grams to tens of grams) may be levitation-melted. This quantity depends on material properties, induction system design, and operating conditions (frequency and current).

Larger quantities of metals may be melted without contamination in furnaces for pedestal melting. In this device, the melt has contact with the support on the bottom only. The side surface of the melt is shaped and supported by the combined action of electrodynamic forces, surface tension, and gravitational forces [15].

Cold Crucible Furnaces

Cold crucible furnaces found various applications in the research area and the industry, ranging from the synthesis of materials to oxide crystal growth. A cold crucible furnace consists of a multiturn coil containing a "crucible" made of isolated water-cooled metal sections called often "fingers" (Fig. 63). The crucible is usually made of copper alloys with or without an insulating coating. The magnetic field generated by the coil pene-

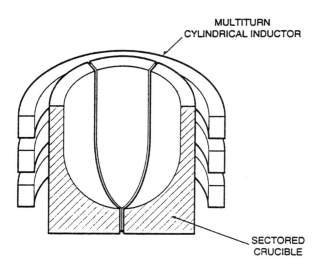

Figure 63 Cold crucible melting furnace with a special design for melt levitation. (From Ref. 15.)

trates inside the crucible through the gaps between the sections and melts the material. The advantage of these furnaces is that they can melt, with negligible contamination, large quantities of materials including high temperature and reactive materials. The cold surface of the crucible does not react with the melt and contaminate it. However, cold crucible furnaces have low thermal efficiency when melting metals due to very large heat losses from the melt to the crucible sections. With proper frequency selection and power level, the electrodynamic forces are strong enough to push the molten metal out of the gaps between the sections, reducing the area of contact to the cold crucible fingers. This factor and other techniques, such as ceramic coating, are used to reduce losses.

Another case where cold crucible furnaces are used is the melting of nonmetallic materials such as glasses, oxides, carbides, etc. These materials are nonconductive at room temperature. However, when preheated, they become conductive enough to be heated and melted by induction. These processes require high frequency (200 kHz–5.28 MHz). The electrical efficiency of glass and oxide melting may be very high due to the high resistivity of the material. Thermal efficiency is relatively high also because a nonconductive layer (skull) of frozen materials is formed on the surface of the crucible. This layer reduces heat losses. There are many other technologies where cold crucible furnaces are used [15,20]. They include the extraction of precious metals from slams and other wastes, the production of high-quality abrasive materials and monocrystals for technical purposes and jewelry (e.g., cubic zirconia), and the treatment of radioactive wastes, etc.

B. Induction Heat Treating

There is a huge variety of induction heat treating processes used in the industry. A short summary of some of the main areas of application is provided below.

1. Continuous Induction Heat Treating of Wires, Rods, Tubes, and Profiles

Wire, tube, and rods are usually heat treated in continuous processes. The advantages of induction heating in these applications are fast heating with no decarburization and very low oxidation, small length of heating zone, no environment pollution, fast start-up, excellent controllability, and better mechanical properties due to small grain size. All heat treating operations (austenization, quenching, tempering, and cooling) may be incorporated into the production line with other operations.

Two main coil types are used for this application. Multiturn solenoidal induction coils are very robust and reliable and may be easily matched to power supply by means of the turn number variation. Solenoidal coils are used for tubes and profiles with relatively large cross sections. The typical frequency range is 3–20 kHz. The required power may reach thousands of kilowatts (e.g., for tube heating). To boost production rate and to improve efficiency, sometimes oval coils are used for the simultaneous continuous heating of several wires.

When heating non-magnetic loads with small cross section (rods and wires), solenoidal coils have low efficiency and power factor. In addition, solenoidal coils cannot provide selective heating of the profile cross section that is required in some applications. Tunnel coils with magnetic flux concentrators (Fig. 64) have better

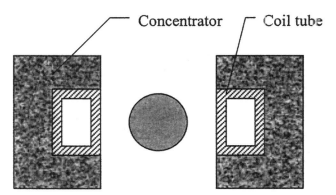

Figure 64 Tunnel coil with flux concentrator.

efficiency and power factor, allow convenient material loading, and provide selective heating of the load cross section. The typical frequency range for hardening is 10–300 kHz (for tempering, the range is 10–50 kHz), with power requirement from tens to thousands of kilowatts.

2. Surface Hardening of Machinery Parts

The surface hardening of machinery parts was the second, after melting, industrial application of the induction method [1,2]. The advantages of induction hardening compared to carburizing and other surface treating technologies are as follows:

Short heating times with the possibility of in-line treatment
Better treatment quality than with furnace operations
Selective heating in-depth and along the surface
Low distortions due to local heating and repeatable, individual handling of each part
Cheaper steels used to achieve the same final mechanical properties of the parts
No additional operations required, such as masking for carburizing technology
Immediate start-up and very low standby losses
Small space (footprints) required
No, or very low, environmental pollution
Protective atmosphere used for bright heat treating
Possibility to monitor treating process for each part and to provide continuous quality control.

Some of these advantages are evident, whereas others require some explanations. The most important issue for discussion is the improvement in metallurgical results. Typically after induction hardening, the material has higher hardness (2–3 HRc) compared to furnace hardening and better mechanical characteristics. especially ductility and impact strength. There are several factors that contribute to quality improvement:

1. No surface decarburization and low oxidation due to very short heating. The material is subjected to temperature above AC3 for several seconds, or even a fraction of a second.
2. Fine martensite structure after hardening with uniform carbon and alloying element distribution.
3. Favorable distribution of residual stresses with a properly designed process. In the process of martensitic transformation, the surface layer tends to expand due to the lower density of martensite

compared to other structures. The core resists this expansion and strong compressive stresses appear in the hardened layer. These forces increase the mechanical properties of the part including the equivalent surface hardness.

A large variety of parts are subjected to induction hardening with many types of induction coils and installations developed for this purpose. Several typical applications are described below as examples of induction hardening technology for machinery components.

Crankshaft Hardening

The first part hardened by induction was a crankshaft in the early 1930s [1,2]. There are three main types of processes for induction hardening of crankshaft bearing pins and journals.

The first one employs single-turn MIQ clamshell inductors, which encircle a pin or journal (Fig. 30). The induction coil face may be profiled in order to provide a more uniform hardness pattern. The crankshaft is stationary in the hardening machine with several clamshell inductors. A pneumatic, mechanical, or hydraulic system closes the inductors, providing sequential hardening of journals one after another. This method was the first industrial application of induction hardening and still remains competitive due to fast heating, local heat pattern control, and relatively low coil copper loading. If only one journal is being hardened, the crankshaft may rotate inside the inductor; however, a complicated mechanical machine is required in this case. Contact problems and some limits on heat pattern are the drawbacks of this type of hardening process.

The second type of crankshaft hardening process was developed in the 1960s for the mass production of automotive crankshafts [33]. In this process, the crankshaft rotates in centers. Several U-shaped induction coils with magnetic flux concentrators are applied to the journals, which must be hardened simultaneously (Fig. 65). Ceramic separators installed on the coil structure provide fixed coupling gaps. Local concentrators, made of laminations or magnetodielectric materials, are applied to these inductors to provide the required heat pattern and to improve coil parameters. Because the coil face surface is much less than that of the journal, the coil is very heavily loaded. Inductors are connected to special matching transformers with "flat" designs. These transformers are attached to the second crankshaft drive, which rotates synchronously with the shaft to be hardened. This rotation allows the coils to follow the surface of the journals. The hardening machine is completely automated.

Figure 65 A U-shaped inductor for crankshaft hardening. (Courtesy of Robotron Corporation.)

The third type of crankshaft hardening process employs two semicircular induction coils for the hardening of each journal. In this process, the crankshaft does not rotate. These coils may be energized from the same or two different power supplies. When coils are attached to the journal from two sides, they form a coil system similar to a single-turn cylindrical coil, but without electrical contact. The original version of this method was developed by Dr. A. Demichev, who utilized two semicircular inductors attached to two adjacent journals, which were connected in series to the secondary of the output matching transformer. The two opposite semicircular inductors were also connected in series, forming a short-circuited contour. The second contour was magneticly coupled to the first contour by means of a special magnetic circuit located on the connecting busses [12]. The advantages of the nonrotational method are the same as for clamshell inductors, but without the weakpoint of the electrical contact [34]. Instead of the physical contact, there is magnetic "contact" of the two halves of the magnetic coupler, which is completely reliable. This method has been successfully used for the hardening of crankshafts and camshafts.

Gear Hardening

Many types of gear hardening processes and machines have been developed, and their number continues to grow due to more rigid hardening specifications and new possibilities opened by advancements in the induction technique. Process type selection depends on the gear type, material, module and dimensions, hardening specifications, production rate, and other factors.

Small gears with modules less than 2.5 mm are usually hardened in single-turn cylindrical inductors. Frequency selection plays a very important role in this case [3]. When the reference depth for hot steel exceeds the tooth thickness, intensive tooth heating takes place only until the Curie temperature is reached. After that, the tooth becomes transparent to magnetic field and its temperature remains almost constant and insufficient for austenization. The root area initially heats slower than the teeth, but then the temperature continues to grow above Curie point. As a result, with fast heating, the teeth remain unhardened (Fig. 66). For longer processes, small teeth will be heated by heat diffusion and the whole gear crown will be hardened. If frequency is high enough (tooth thickness greater than three times the reference depth for the hot steel), the teeth will be austenitized, whereas the roots will be at low temperature and only the teeth will be hardened.

By balancing frequency, power, and heating time, it is possible to obtain the required hardness depth at the root with through hardening of the teeth without overheating in the process of austenitization [3].

CONTOUR HARDENING. Contour hardening (Fig. 67) provides the best service properties of gears; however, it is also the most difficult hardness pattern to achieve. Several methods have been proposed for contour hardening. It is clear from the above considerations that there must be a frequency, which could provide more or less uniform hardening of both gear teeth and roots. Theoretically, any gear may be contour-hardened by heating in cylindrical coil with the proper combination of frequency, time, and specific power. Optimal frequency, time, and power density depend mainly on the gear module M [3]:

$$F_{\mathrm{opt}} = 400/M^2; \qquad t = 0.04M^2$$

For small modules, high frequency and specific power are required and the heating time must be very short. The generation of high power at a certain fixed

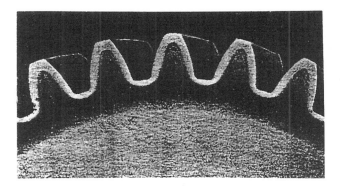

Figure 67 Gear with contour hardness pattern. (From Ref. 9.)

frequency, especially in the range from 10 to 60 kHz, was for many years a factor limiting this technology. Now, modern solid state power supplies can generate power at almost any frequency and are even able to change it in the same machine in the ratio of 3:1. However, this method has not found wide application because different combinations of high power and frequency are required for different gears.

Another method called dual frequency or "dual pulse" was developed for the contour hardening of gears [16,17,35]. The dual frequency method employs powers at two frequencies applied one after another to the same inductor, or to two different inductors. The first pulse heats the gear at a lower frequency (3–10 kHz, depending on gear dimensions). In this stage, the teeth are heated below Curie point, with the roots heated to a slightly higher temperature. In the second stage, a shorter pulse of power at high frequency (50–450 kHz) provides surface austenitization along the whole contour. Fast quenching by water or polymer solutions immediately after the pulse provides contour gear hardening.

This method is very flexible. By varying the power and the duration of pulse, it is possible to harden a wide range of gears. There are special installations for contour hardening with rated high frequency power up to 700 kW [35]. Typically, gears with modules from 2.5 to 6 mm are contour-hardened using the dual frequency method.

Another method for contour hardening is called simultaneous dual frequency induction heating. Currents of two frequencies may be supplied to the induction coil simultaneously, providing contour hardening. There are two potential means for simultaneous dual frequency heating, using two power supplies (high and low frequency) or one power supply. A version of this

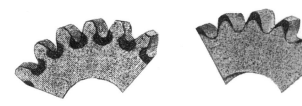

Figure 66 Gear hardness pattern for low frequency (left) and high frequency (right). (From Ref. 3.)

method based on a simultaneous application of power from a motor generator and a tube oscillator was known for a long time, but did not find practical implementation. Using two power supplies requires the installation of special filters in order to prevent interference between the low frequency and high frequency generators.

A new technology was developed recently [36] with a special solid state power supply capable of generating "lower" frequency voltage modulated by a high frequency signal. Thus, one power supply generates simultaneously the power of two frequencies. Power transferred to the gear by each frequency may be controlled independently, assuring the process flexibility.

DEEP HEAT CONTOUR HARDENING. There is one more method of contour hardening of gears and many other parts with complex geometries. If the parts are made of special steel with controlled (usually reduced) hardenability, contour hardening may be achieved through deep induction heating of the whole tooth area above the austenization temperature followed by quenching. After quenching with a severe water spray, only the surface layer will have a martensite structure. Between the hardened layer and the soft core, there is a layer with a bainite structure (Fig. 67). The bainite layer provides additional strength to the gear. High hardness, up to 65 HRc, may be achieved with steels containing 0.6% carbon. Several carbon and low-alloy steels with different hardenabilities have been developed and successfully used for gears, truck springs, big bearing rings, and others [9].

TOOTH-BY-TOOTH HARDENING. For gears with large modules and dimensions, high power is necessary for contour hardening. For very large gears with diameters reaching several meters, simultaneous hardening is not possible at all. High power installations require significant capital investments. When a large investment is not justified by high production rates, the tooth-by-tooth method is used. If the hardening of the tooth flanks is sufficient, a simple coil in the form of a loop around the tooth may be used. Magnetic flux concentrators made of laminations or MDMs significantly improve the coil parameters and prevent the overheating of the tooth ends where the induced current flows from one tooth to another. This method provides a good wear resistance of gear teeth, but not bending strength.

For better results, the hardness pattern must cover the whole tooth valley and sides. An unhardened strip along the tooth tip does not influence the gear service characteristics. The most common method of "con-

tour" tooth-by-tooth treatment is scan hardening with a coil moving along the valley (Fig. 68). The coil is a vertical loop-style inductor with profiled coil "leads" according to the valley shape. The coil must have a magnetic flux controller in order to be effective. This method is widely used for the hardening of gears with dimensions up to 3 m and tooth modules of 40 mm or more.

A quenching head usually follows the induction coil in this method. However, there is a version of this method where the gear and the induction coil are submerged into a water or oil bath. During the heating process, heat losses are relatively small due to the formation of a static vapor blanket in the small gap between the coil and the gear. After heating, the coil is removed and intensive cooling occurs, providing quenching.

3. Quenching for Induction Heat Treating

Induction surface hardening is an organized process, with individual part positioning and orientation. The area of high temperature is confined to the hardened zone (surface area), higher rates of heat extraction are possible due to the relatively cold part core, and more severe quenchants are often used. These factors result in better process repeatability, higher part quality, and lower distortion.

There are four main types of quenchants used for surface hardening applications:

Water
Polymer solutions
Oils
Air, mist, or high-pressure gas.

Water quenchants are the most severe followed by polymer solutions, oils, and air quenchants, respectively.

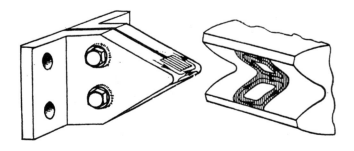

Figure 68 Induction coil and induced current heat pattern for tooth-by-tooth induction scan hardening. (From Ref. 12.)

Besides the four types of quenchants, there are three different quenching techniques: dipping, spray, or self-quenching. In dipping quench processes, the heated workpiece is submerged in a fluid, which may be still or agitated. Dipping quench processes have lower heat transfer coefficients than their spray counterparts. This technique is popular for furnace hardening applications to provide milder quenching, but is not often used in induction hardening processes.

Spray quenching is the most widely used method for induction hardening. In some applications, the spray is delivered from the induction coil itself. These inductors are called MIQ inductors and have holes drilled in the induction coil copper or magnetic flux controller (if an MDM is applied). In other applications, the spray is delivered from a separate nozzle or "quench head." In induction hardening applications with multiple heat treating stations, the spray quench may be delivered at separate stations. In static hardening installations, the part is often indexed from the coil directly to the spray quench located under the coil head. For induction scanning applications, the quench head, made of copper or plastic, is typically located just after the induction coil head. Figures 55, 56, and 57 show a computer simulation of a typical induction scan hardening process for an automotive shaft. The pictures and curves were generated by the ELTA computer simulation program. Water spray quench with a $1.2 \, \text{m}^3/\text{sec m}^2$ velocity was used as a quenchant. The results provide a good illustration of the temperature dynamics for a typical induction scan hardening process.

Sometimes, quenching in induction hardening processes can be done through the part wall from the side opposite of the heating. This is used for parts with relatively thin walls, or for alloyed steels that do not require fast cooling rates. Water is sprayed on the cold side of the part throughout the process to prevent through heating of the wall, and to provide final cooling for surface hardening after heating. For high-alloyed steels, it is sometimes possible to substitute an air quenchant for water in this process.

In some applications, the quenching could be made through heat conduction within the part itself. This process is called "self-quenching" and is widely used in laser heat treating processes. Self-quenching with induction heating is only possible for alloyed steels and shallow case depth. Very fast heating (pulse heating 50–500 msec) and high frequency are required for this technology [37]. The highest frequency used for hardening with self-quenching was 27 MHz.

A great deal of attention is given to quenching and distortions in steel heat treatment processes [38], but not much information is available on the quenching characteristics for induction at this time. Due to the stability of forced quenchant flow, the characteristics of induction quenching are more consistent and computer simulation may be made more accurately than for batch furnace heat treating [39].

Heat transfer coefficients K depend upon the quenchant composition, part surface temperature, quenchant flow rate or velocity, and method of quenchant supply (radial, axial, and rotating axial). Besides these factors, the heat transfer coefficient depends most strongly upon the part surface temperature. Figures 69 and 70 show the heat transfer coefficients for a water spray, oil spray, and agitated oil bath, respectively [11]. The water spray quench has much higher K values between the temperatures of 150°C and 250°C than in the rest of the temperature range due to intensive boiling. At temperatures below the boiling point, cooling is due to convection only and K is low. At higher temperatures, the value of K is low again due to the formation of a "vapor blanket," but remains high enough for fast heat extraction.

The zone of maximum cooling for agitated oils is in the temperature range of 350–600°C with very low values for K outside of this zone. The oil spray quench provides much higher K values, especially at higher temperatures. In addition, the maximum K value for the oil spray quench occurs at a higher temperature and declines more slowly with increasing temperature than for the agitated oil bath. This may be explained by the action of the spray disrupting the formation of the vapor blanket.

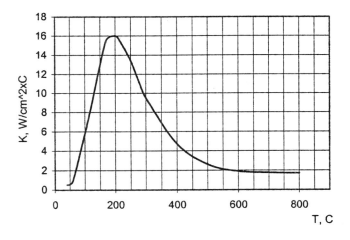

Figure 69 Heat transfer coefficient for water spray quench as a function of temperature. (From Ref. 11.)

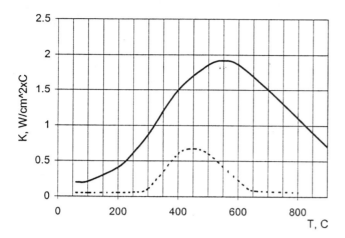

Figure 70 Heat transfer coefficient for oil spray and agitated oil quench as a function of temperature.

For accurate CAD of the quenching process, more experimental studies are required to build a database for heat transfer coefficients for induction heat treating conditions, especially for polymer solutions.

4. Steel Selection for Induction Hardening

The steels for induction hardening must have a favorable composition and microstructure for rapid heating conditions and low tendency for crack formation due to stresses caused by temperature gradients and structural transformations.

Heating times in induction hardening installations range from several seconds to 20–30 sec. As of this time, the material is above the austenitization temperature (AC3) for a very short time, from a fraction of a second at the depth to several seconds on the surface. Because the obtainable time for diffusion processes is very short, a fine-grain microstructure with a uniform distribution of carbon and alloying elements is required for best results. The short diffusion time may be partially compensated for by using a higher temperature, which for induction hardening is usually 50–100°C higher than for furnace hardening. The short heating times limit oxidation and grain growth despite the use of higher processing temperatures. Due to this, cheaper carbon or low-alloyed steels may be used for induction hardening processes compared to steels used for furnace heating.

The steel types traditionally used for induction hardening include the following:

Plain carbon steels with 0.3–0.5% carbon
Carbon steels with up to 0.8% carbon
Low-alloy steels with chromium, manganese, molybdenum, and vanadium

High-alloy steels, such as bearing steels, tool steels, high-speed steels, martensitic stainless steels, etc.

Besides steel, some types of cast iron may also be induction hardened, such as pearlitic or malleable cast iron. There have been recent developments for induction heat treating of nonferrous materials, such as aluminum and titanium alloys.

The applications discussed in this section represent only a few of the most common areas where induction heating is used and some of the key considerations such as steel selection and quenching methods. More information on induction heating processes may be found in the following books: *Electromagnetic Induction and Electric Conduction in Industry* [15], *Practical Induction Heat Treating* [16], or *Steel Heat Treatment Handbook* [17,18]. Another good source for applications of induction is periodicals (e.g., Refs. 40 and 41). Good guidelines for induction system maintenance may be found in periodicals also, especially the two article series published in the *Journal of Industrial Heating* [42,43].

VIII. CONCLUSION

The induction technique is currently in a new period of rapid expansion. One of the main driving forces for this expansion is that it is very well suited to the new industrial environment. The application of induction to high value-added processes is not sensitive to such drawbacks as higher cost of electrical energy, compared to gas and initial capital investments, compared to traditional heating methods. Improvements in power supplying equipment, computer simulation, materials for magnetic flux control, and thermal insulation open up many new possibilities for the induction technique. The vision for the future of the heat treatment industry described in the *Roadmap for Process Heating Technology* (Vision 2020) [44] identifies induction heating as one of the key technologies moving forward.

"Bienergy" processes are one of the areas where induction may be used especially efficiently. These processes include induction boosters before gas or resistance furnaces in metal product heat treatment or hot forming. The other example of bienergy heating is the combination of induction and laser heating, where laser assists induction, heating local areas, which it is difficult to reach with induction techniques. New emerging technologies include bright hardening with heating in a protective atmosphere, induction carburizing in liquid active media [45], hot gas forming [40], and other processes. The Designed-For-Induction-Heating strategy promises many improvements especially the development and application of steels with controlled properties favorable for induction heat treating.

Together with the development of more universal software packages, problem-oriented simulation codes are expected with more vast and reliable databases. The promotion of induction techniques and educational efforts is also necessary for a wider and more effective use of induction heating.

ACKNOWLEDGMENTS

The authors would like to thank the International Union for Electricity Application (UIE) for kind permission to use figures from the book, *Induction Heating Industrial Applications* [20]. The following figures were taken from this book: Figs. 3–4, 6–12, 17–18, 22–28, 32–33, and 62–63. The authors would also like to thank the Centre for Induction Technology, Inc. for support.

REFERENCES

1. Vologdin, V.P. *Surface Hardening by Induction Method*; Gosmetallurgizdat: Leningrad, Russia, 1939; 244 pp. *in Russian.*
2. Curtis, F.W. *High Frequency Induction Heating*; McGraw-Hill: New York, 1944; 235 pp.
3. Vologdin, V.P. *Surface Induction Hardening*; Oborongiz: Moscow, Russia, 1947; 291 pp. *in Russian.*
4. Stansel, N.R. *Induction Heating*; 1st Ed.; McGraw-Hill: New York, 1949; 212 pp.
5. Baker, R.M. Transverse flux induction heating. AIEE Trans. 1950, *69*, 711–719.
6. Lozinskii, M.G. *Industrial Applications of Induction Heating*; Pergamon: London, 1969.
7. Simpson, P.G. *Induction Heating Coil and System Design*; McGraw-Hill: New York, 1960; 295 pp.
8. Tudbury, C.A. *Basics of Induction Heating*; John F. Rider Publishers. Inc.: New York, 1960; Vol. 1–2, 132.
9. Shepelyakovsky, K.Z. *Surface Induction Hardening of Machinery Parts*; Mashinostroyeniye: Moscow, Russia, 1972; 287 pp. *in Russian.*
10. Slukhotskii, A.E.; Ryskin, S.E. *Inductors for Induction Heating*; Energia Publ.: Leningrad, Russia, 1974. *in Russian.*
11. Golovin, G.F.; Zimin, N.V. *Technology of Metal Heat Treatment with Induction Heating*; Mashinostroyenie: Leningrad, Russia, 1990; 87 pp. *in Russian.*
12. Demichev, A.D. *Induction Surface Hardening*; 2nd Ed.; Mashinostroyenie: Leningrad, Russia, 1979; 80 pp.
13. Davis, E.J. *Conduction and Induction Heating*; Peter Peregrinus: London, 1990.
14. Nemkov, V.S.; Demidovich, V.B. *Theory and Calcu-*

lation of Induction Heating Devices; Energoatomisdat: Leningrad, Russia, 1988; 300 pp. *in Russian.*
15. *Electromagnetic Induction and Electric Conduction in Industry*; Bialod, D., Ed.; Centré Francais de l'Electricité: France, 1997; 765 pp.
16. Haimbaugh, R.E. *Practical Induction Heat Treating*; ASM Publication: Materials Park, OH, 2001; 332 pp.
17. Rudnev, V.I.; Cook, R.L.; Loveless, D.L.; Black, M.R. Induction heat treatment. In *Steel Heat Treatment Handbook, Chapter 11A*; Marcel Dekker: New York, 1997; 765–871.
18. Loveless, D.L.; Cook, R.L.; Rudnev, V.I. Induction heat treatment. In *Steel Heat Treatment Handbook, Chapter 11B*; Marcel Dekker: New York, 1997; 873–911.
19. Ruffini, R.S.; Ruffini, R.T.; Nemkov, V.S. Power inductors for heat treating processes. Proceedings of the 3rd International Conference on Quenching and Control of Distortion, Prague, Czech Republic, March 1999, 564–569.
20. Lupi, S.; Mulbauer, A.; Bose, D.; Charette, A.; Geominne, P.; Gonsales, V.; Eranov, V.; Nacke, B.; Nemkov, V.; Paskins, A.; Reboux, I.; Tatero, M. *Induction Heating Industrial Applications*; UIE: France, 1992; 144 pp.
21. Nemkov, V.S.; Demidovich, V.B.; Rudnev, V.; Fishman, O. Electromagnetic End and Edge Effects in Induction Heating. Proceedings of UIE Congress, Montreal, 1991.
22. Nemkov, V.S.; Goldstein, R.C. Computer simulation for fundamental study and practical solutions to induction heating problems. Proceedings of the International Seminar on Heating by Internal Sources, Padua, Italy, September 2001; 435–442.
23. Sinha, A.H. *Physical Metallurgy Handbook*; Marcel Dekker: New York, 2002.
24. Ruffini, R.S.; Ruffini, R.T.; Nemkov, V.S. Advanced design of induction heat treating coils: Part I. Design principles. J. Ind. Heat. June 1998; 59–63.
25. Ruffini, R.S.; Ruffini, R.T.; Nemkov, V.S. Advanced design of induction heat treating coils: Part II. Magnetic flux concentration and control. J. Ind. Heat. November 1998; 69–72.
26. Ruffini, R.S.; Nemkov, V.S. Induction Heating Systems Improvement by Application of Magnetic Flux Controllers. Proceedings of the International Induction Heating Symposium, Padua, Italy, May 1998; 133–140.
27. Ruffini, R.T.; Nemkov, V.S.; Goldstein, R.C. Prospective for improved magnetic flux control in the induction heating technique. Proceedings of the 19th ASM Heat Treating Society Conference, St. Louis, November 2000.
28. Pearson, E. Designing powertrain components to optimize induction heat treatment performance. Proceedings of the 17th ASM International Heat Treating Society Conference, Indianapolis, September 1997; 801–810.
29. Nemkov, V.S.; Ruffini, R.T.; Goldstein, R.C.; Grant, C.N.; Wakade, S.G. Induction heating in the power-

train industry. Proceedings of the 3rd Annual Global Powertrain Congress, Detroit, October 2000.

30. Brunotte, X. *The Future of Flux: The Flux Project—2002*; 2001 Magsoft Users Meeting: Saratoga Springs, NY, May 2001.

31. Nemkov, V.S. Role of computer simulation in induction heating technique. Proceedings of the International Induction Heating Symposium, Padua, Italy, May 1998; 301–308.

32. Nemkov, V.S.; Goldstein, R.C. Computer simulation of induction heating processes. Proceedings of the 20th ASM Heat Treating Society Conference, St. Louis, November 2000.

33. Madeira, R.J.; Schwarz, H.R. Induction hardening technique ensures precise case-hardened zone. J. Mod. Appl. News January 2001, 38–39.

34. Loveless, D.L.; Rudnev, V.I.; Desmier, G.; Lankford, L.; Medhanie, H. Nonrotational induction crankshaft hardening capabilities extended. J. Ind. Heat. June 2001; 45–47.

35. Storm, J.M.; Chaplin, M.R. Dual frequency induction gear hardening. J. Gear Technol. 1993, *10* (2), 22–25.

36. Hammond, M. Simultaneous dual frequency gear hardening. J. Ind. Heat. June 2001; 41–42.

37. Di Pieri, C.; Lupi, S.; Cappello, A.; Crepaz, G. *Capacitors Discharge Induction Heating Installations for High Frequency Pulse Hardening*; Xth UIE Congress: Stockholm, Sweden, June 1984.

38. Tensi, H.M.; Tich, A.S.; Totten, G.E. Quenching and quenching technology. *Steel Heat Treatment Handbook*; Marcel Dekker: New York, 1997; 157–251.

39. Nemkov, V.S.; Goldstein, R.C.; Bukanin, V.A.; Zenkov, A.; Koutchmassov, D.V. Computer simulation of induction heating and quenching processes. Proceedings of the 3rd International Conference on Quenching and Control of Distortion, Prague, Czech Republic, March 1999; 370–377.

40. Pfaffman, G.D.; Wu, X.; Dykstra, W.K. Hot metal gas forming of auto parts. Heat Treat. Prog. February 2000; 35–38.

41. Spain, K. Induction heating offers many processing opportunities. J. Ind. Heat. June 2001; 49–52.

42. Specht, F.R.; Welch, G. Preventive maintenance keeps induction system in peak condition: Part I. J. Ind. Heat. June 2001; 37–39.

43. Specht, F.R.; Welch, G. Preventive maintenance keeps induction system in peak condition: Part II. J. Ind. Heat. August 2001; 35–38.

44. Roadmap for Process Heating Technology, March 2001.

45. Gugel, S. Applications increasing for induction carburizing technique. J. Ind. Heat. November 2001; 43–48.

16

Laser Surface Hardening

Janez Grum
University of Ljubljana, Ljubljana, Slovenia

I. INTRODUCTION

Lasers represent one of the most important inventions of the 20th century. With their development it was possible to get a highly intensive, monochromatic, coherent, highly polarized light wave [1,2]. The first laser was created in 1960 in Californian laboratories with the aid of a resonator from an artificial ruby crystal. Dating from this period is also the first industrial application of laser which was used to make holes into diamond materials extremely difficult to machine. The first applications of laser metal machining were not particularly successful mostly due to low capabilities and instability of laser sources in different machining conditions. These first applications, no matter how successful they were, have, however, led to a development of a number of new laser source types. Only some of them have met the severe requirements and conditions present in metal cutting. As the most successful among them, CO_2 laser should be mentioned. The high-intensity CO_2 lasers have proved to be extremely successful in various industrial applications from the point of view of technology as well as economy. A great number of successful applications of this technique have stimulated the development of research activities which, since 1970, have constantly been increasing.

Laser is becoming a very important engineering tool for cutting, welding, and to a certain extent also for heat treatment. Laser technology provides a light beam of extremely high-power density acting on the workpiece surface. The input of the energy necessary for heating up the surface layer is achieved by selecting from a range of traveling speeds of the workpiece and/or laser-beam source power.

The advantages of laser materials processing are as follows [3–7]:

- Savings in energy compared to conventional surface heat treatment welding or cutting procedures.
- Hardened surface is achieved due to self-quenching of the overheated surface layer through heat conduction into the cold material.
- As heat treatment is done without any agents for quenching, the procedure is a clean one with no need to clean and wash the workpieces after heat treatment.
- The energy input can be adapted over a wide range with changing laser source power, with having focusing lenses with different focuses, with different degrees of defocus (the position of the lens focus with respect to the workpiece surface), and with different traveling speeds of the workpiece and/or laser beam.
- The guiding of the beam over the workpiece surface is made with computer support.
- It is possible to heat-treat small parts with complex shapes as well as small holes.
- The optical system can be adapted to the shape or complexity of the product by means of different shapes of lenses and mirrors.
- Small deformations and/or dimensional changes of the workpiece after heat treatment.
- Repeatability of the hardening process or constant quality of the hardened surface layer.

- No need for/or minimal final machining of the parts by grinding.
- Laser heat treatment is convenient for either individual or mass production of parts.
- Suitable for automation of the procedure.

The use of laser for heat treatment can be accompanied by the following difficulties:

- Nonhomogeneous distribution of the energy in the laser beam;
- Narrow temperature field ensuring the wanted microstructure changes;
- Adjustment of kinematic conditions of the workpiece and/or laser beam to different product shapes;
- Poor absorption of the laser light in interacting with the metal material surface.

Engineering practice has developed several laser processes used for surface treatment:

- Annealing;
- Transformation hardening;
- Shock hardening;
- Surface hardening by surface layer remelting;
- Alloying;
- Cladding;
- Surface texturing;
- Plating by laser chemical vapor deposition (LCVD) or laser physical vapor deposition (LPVD).

For this purpose, besides CO_2 lasers, Nd:YAG and excimer-lasers with a relatively low power and a wavelength between 0.2 and 1.06 μm have been successfully used. A characteristic of these sources is that, besides a considerably lower wavelength, they have a smaller focal spot diameter and much higher absorption than CO_2 lasers.

In heat treatment using laser light interaction, it is necessary to achieve the desired heat input, which is normally determined by the hardened layer depth. Cooling and quenching of the overheated surface layer is in most cases achieved by self-cooling as after heating stops, heat conducted into the workpiece material is so intensive that the critical cooling speed is achieved and thus also the wanted hardened microstructure. Figure 1 illustrates the dependence of power density and specific energy on laser light interaction time and on the workpiece surface in order to carry out various metalworking processes. Diagonally there are two processes, i.e., scribing and hardening, for which quite the opposite relationship between power density and interaction time has to be ensured [5–6]. In scribing material vaporiza-

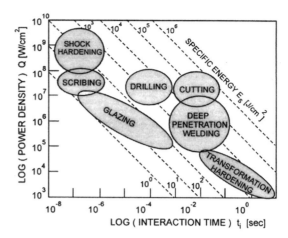

Figure 1 Dependence of power density, specific energy, and interaction time at laser metalworking processes.

tion at a depth of a few microns has to be achieved, ensuring the prescribed quality and character resolution. On the other hand, for hardening a considerably lower power density per workpiece surface unit is required, but the interaction times are the longest among all the mentioned metalworking processes.

Thus different power densities and relatively short interaction times, i.e., between 10^{-1} and 10^{-3} sec, are related to transformation hardening and remelting of the material. This group thus includes the processes in which either the parent metal alone or the parent metal and the filler material are melted. The highest power densities are applied in the case that both the parent metal and the filler material are to be melted. Such processes are laser welding and alloying. With cladding, however, the lowest power densities are required as the material is to be deposited on the surface of the parent metal and only the filler material is to be melted. The power density required in both welding and alloying of the parent metal, however, depends on the materials to be joined or alloyed. In hardening of the surface layers by remelting it is necessary, in the selection of the power density, to take into account also the depth of the remelted and modified layer. In laser cutting a somewhat higher power density is required than in deep welding. In the cutting area the laser-beam focus should be positioned at the workpiece surface or just below it. In this way a sufficient power density will be obtained to heat, melt, and evaporate the workpiece material. The formation of a laser cut is closely related to material evaporation, particularly flowing out of the molten pool and its blowing out due to the oxygen auxiliary gas, respectively.

Lasers produce a collimated and coherent beam of light. The almost parallel, single-wavelength, light rays that make up the collimated laser beam have a considerably higher power density profile across the diameter of the beam and can be focused to a spot size diameter on the workpiece surface [8–10].

Various types of lasers are classified with reference to a lasing medium [15–19]. In laser treatment processes, the most frequently used gaseous lasers are CO_2, lasers [20].

The transverse power density distribution emerging from the laser source is not uniform with reference to the optical laser axis. It depends on the active lasing medium, the maximum laser power, and the optical system for transmission and transformation of the laser beam [10–14].

Figure 2 shows different mode structures of the laser beam, i.e., TEM_{00} (Gaussian beam), and multimode beam structures TEM_{01} (b), TEM_{10} (c), TEM_{11} (d), and TEM_{20} (e), respectively [11].

The transverse power density distribution of the laser beam is very important in the interaction with the workpiece material. The irradiated workpiece area is a function of the focal distance of the convergent lens and the position of the workpiece with reference to the focal distance. The transverse power density distribution of the laser beam is also called the transverse electromagnetic mode (TEM). Several different transverse power density distributions of the laser beam or TEMs can be shaped. Each individual type may be attributed a different numeral index. A higher index of the TEM indicates that the latter is composed of several modes, which makes beam focusing on a fine spot at the workpiece surface very difficult. This means that the higher

the index of the TEM is the more difficult it is to ensure high power densities, i.e., high energy input. For example in welding, mode structures TEM_{00}, TEM_{01}, TEM_{10}, TEM_{11}, TEM_{20}, and frequently combinations thereof are used. Some of the laser sources generate numerous mode structures, i.e., multimode structures [8–12].

The most frequently used lasers are continuous lasers emitting light with the Gaussian transverse power density distribution, TEM_{00}. A laser source operating 100% in TEM_{00} is ideal for cutting and drilling. A TEM_{00} beam can be focused with a convergent lens to a very small area thus providing a very high power density. In practice TEM_{01}, having energy concentrated at the periphery of the laser beam with reference to the optical axis, is used as well. This mode is applied primarily to drilling and heat treatment of materials because it ensures a more uniform elimination of the material in drilling and a uniform through-thickness heating of the material in heat treatment. For welding and heat treatment, very often mixtures of multimode structures giving an approximately rectangular, i.e., top-hat-shaped, energy distribution in the beam are used [13–20].

Heat treatment requires an adequate laser energy distribution at the irradiated workpiece surface. This can be ensured only by a correct transverse power density distribution of the laser beam. The power density required for heat treatment can be achieved by the multimode structure or a built-in kaleidoscope or segment mirrors, i.e., special optical elements.

Heating of the material for subsequent heat treatment requires an ideal power density distribution in the laser beam providing a uniform temperature at the sur-

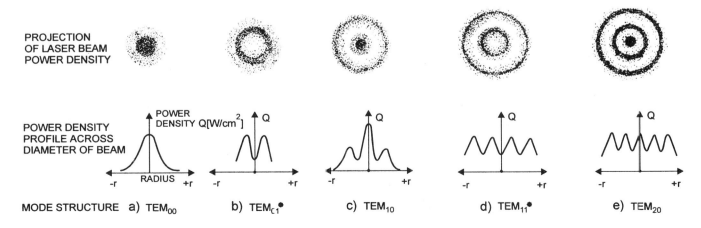

Figure 2 Basic laser beam mode structures. (From Ref. 11.)

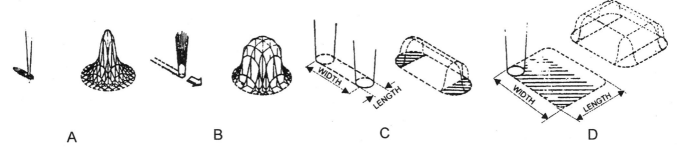

Figure 3 Basic shapes of transverse power density distribution. (A) Focused laser beam suitable for deep penetration welding and cutting, (B) defocused laser beam suitable for heat treatment, (C) unidirectional oscillating laser beam for wide paths and intensive heating, and (D) bidirectional oscillating laser beam for wide paths and medium heating. (From Ref. 21.)

face, and below the surface to the depth to which the material properties are to be changed.

Figure 3 shows different shapes of transverse power density distribution which can be accomplished by both the defocusing of the laser beam and the laser-beam transformation through the mirrors or differently shaped optical elements [21]. When the focused laser beam with the Gaussian power density distribution is used, the irradiated spot (A) is smaller than the one obtained with the defocused beam (B). To fulfill the requirements of a uniform power density distribution, the top-hat mode of the power density should be ensured. Figure 3 shows two solutions both giving a rather uniform power density distribution. Such a power density distribution can be accomplished by the unidirectional (C) or bidirectional (D) oscillating laser beam that is obtained by being guided through appropriately shaped optical elements showing adequate kinematics [21]. It is characteristic of the two basic shapes of the laser beam that they are suitable for heat treatment and can be applied to heating of larger areas. Thus with the unidirectional oscillating laser beam, heating of the material is relatively fast whereas with the bidirectional oscillating beam heating is less intense, but it provides a larger and rectangular heated surface. Both types of laser beam are very suitable for fast hardening of larger areas. In any case it should be taken into account that such a transformation of the laser beam is feasible only with a laser source of sufficiently high maximum power.

II. LASER SURFACE HARDENING

A. Laser Heating and Temperature Cycle

A prerequisite of efficient heat treatment of a material is that the material shows phase transformations and

is fit for hardening. Transformation hardening is the only heat-treatment method successfully introduced into practice. Because of intense energy input into the workpiece surface, only surface hardening is feasible. The depth of the surface-hardened layer depends on the laser-beam power density and the capacity of the irradiated material to absorb the radiated light defined by its wavelength. Laser heating of the material being characterized by local heating and fast cooling is, however, usually not suitable to be applied to precipitation hardening, steroidization, normalizing, and other heat-treatment methods [3–6,22].

Kawasumi [23] studied laser surface hardening using a CO_2 laser and discussed the thermal conductivity of a material. He took into account the temperature distribution in a three-dimensional body taken as a homogeneous and isotropic body. On the basis of derived heat conductivity equations, he accomplished numer-

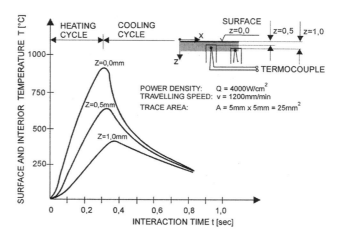

Figure 4 Temperature cycle on the workpiece surface and its interior versus interaction time. (From Ref. 23.)

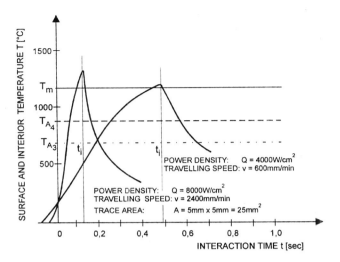

Figure 5 The effect of laser light interaction time on the temperature cycle during heating and cooling at various power densities and traveling speeds. (From Ref. 23.)

ous simulations of temperature cycles and determined the maximum temperatures achieved at the workpiece surface.

Figure 4 shows temperature cycles in laser heating and self-cooling. A temperature cycle was registered by thermocouples mounted at the surface and in the inside at certain workpiece depths [23]. In this case the laser beam with its optical axis was traveling directly across the centers of the thermocouples inserted in certain depths. A thermal cycle can be divided into the heating

and cooling cycles. The variations of the temperature cycles in the individual depths indicate that:

- The maximum temperatures have been obtained at the surface and in the individual depths.
- The maximum temperature obtained reduces through depth.
- The heating time is achieved at the maximum temperature obtained or just after.
- The greater is the depth in which the maximum temperature is obtained the longer is the heating time required.
- In the individual depths the temperature differences occurring are greater in heating than in cooling.
- Consequently, in the individual depths the cooling times are considerably longer than the heating times to obtain, for example, a maximum temperature.

Figure 5 shows two temperature cycles in laser surface heating [23]. In each case the maximum temperature obtained at the surface is higher than the melting point of the material; therefore remelting will occur. The remelting process includes heating and melting of the material, fast cooling, and material solidification. The maximum temperature at the surface being higher than the melting point, a molten pool will form in the material surrounding the laser beam. Because of a relative travel of the laser beam with reference to the workpiece, the molten pool travels across the workpiece as well whereas behind it the metal solidifies quickly. The

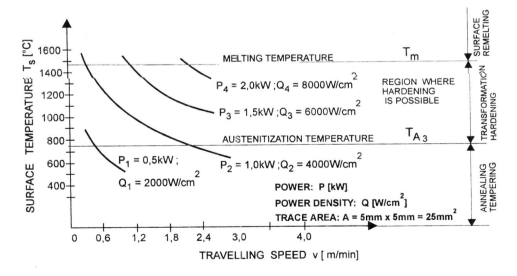

Figure 6 The effects of power density and traveling speed on hardening. (From Ref. 23.)

depth of the remolten material is defined by the depth in which the melting point and the solidification temperature of the material have been attained.

The depth of the remelted layer can be determined experimentally by means of optical microscopy or by measuring through-depth hardness in the transverse cross section.

Figure 6 shows the dependence of the maximum temperature attained at the workpiece surface in laser heating on the given power density and different traveling speeds v [23]. The four curves plotted indicate four different power densities Q_i, i.e., 2, 4, 6, and 8 kW/cm^2. With the lowest power density Q_1, transformation hardening can be performed if the traveling speed is varied between 0.3 and 0.5 m/min. With the highest power density Q_4 hardening by remelting of the thin surface layer can be efficiently performed as the traveling speed v of the workpiece should not exceed 2.0 m/min. With the power densities between the highest and the lowest, Q_2 and Q_3, it is transformation hardening which can be efficiently performed because the wide range of the traveling speeds, i.e., from 0.4 up to 2.0 m/min, makes it possible to obtain a hardened microstructure. The wide range of traveling speeds of the workpiece and/or the laser beam enables hardening of the surface layers of different thicknesses.

III. METALLURGICAL ASPECT OF LASER HARDENING

Prior to transformation hardening, an operator should calculate the processing parameters at the laser system. The procedure is as follows. Some of the processing parameters shall be chosen, some calculated. The choice is usually left to operators and their experience. They shall select an adequate converging lens with a focusing distance f and a defocus z_f taking into account the size of the workpiece and that of the surface to be hardened, respectively. Optimization is then based only on the selection of power and traveling speed of the laser beam. The correctly set parameters of transformation hardening ensure the right heating rate, then heating to the right austenitizing temperature T_{A_3}, and a sufficient austenitizing time t_A. Consequently, with regard to the specified depth of the hardened layer, in this depth a temperature a little higher than the transition temperature T_{A_3} should be ensured. Because of a very high heating rate the equilibrium diagram of, for example, steel does not suit; therefore it is necessary to correct the existing quenching temperature with reference to the heating rate. Thus with higher heating rates a higher austenite transformation temperature should

be ensured in accordance with a time-temperature-austenitizing (TTA) diagram. The left diagram in Fig. 7 is such a TTA diagram for 1053 steel in the quenched-and-tempered state (A) whereas the right diagram is for the same steel in the normalized state (B) [24]. As the steel concerned shows pearlitic–ferritic microstructure, a sufficiently long time should be ensured to permit austenitizing. In fast heating, austenitizing can, namely, be accomplished only by heating the surface and subsurface to an elevated temperature. For example with a heating time t of 1 sec, for total homogenizing a maximum surface temperature T_s of 880°C should be ensured in the first case and a much higher surface temperature T_s, i.e., 1050°C, in the second case. This indicates that around 170°C higher surface temperature ΔT_s should be ensured in the second case (normalized state) than in the first case (quenched and tempered state).

Figure 8 shows a space TTA diagram including numerous carbon steels with different carbon contents. The TTA diagram gives particular emphasis to the characteristic steels, i.e., 1015, 1035, 1045, and 1070 steels, and their variations of the transition temperature T_{A_3} with reference to the given heating rate and the corresponding heating time [24].

When the laser beam has stopped heating the surface and the surface layer, the austenitic microstructure

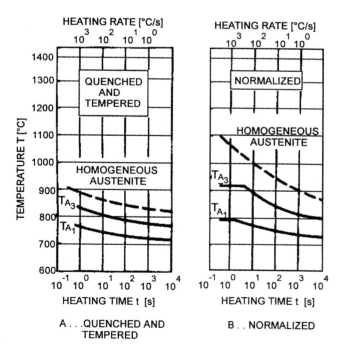

Figure 7 TTA diagram of steel 1053 for various states. (From Ref. 24.)

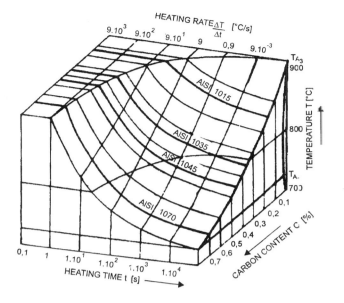

Figure 8 Influence of heating rate and carbon content on austenitic transformation temperature. (From Ref. 24.)

should be obtained. Then the cooling process for the austenitic layer begins. To accomplish martensite transformation, it is necessary to ensure a critical cooling rate that depends on the material composition. Figure 9 shows a continuous cooling transformation (CCT) diagram for EN 19B steel including the cooling curves. As carbon steels have different carbon contents, their microstructures as well show different contents of pearlite and ferrite. An increased carbon content in steel decreases the temperature of the beginning of martensite transformation T_{M_s} as well as of its finish T_{M_F}. Figure 10 shows the dependence between the carbon content and the two martensite transformations. Consequently, the increase in carbon content in steel results in the selection of a lower critical cooling rate required. In general, the microstructures formed in the surface layer after transformation hardening can be divided into three zones, i.e.:

- A zone with completely martensitic microstructure,

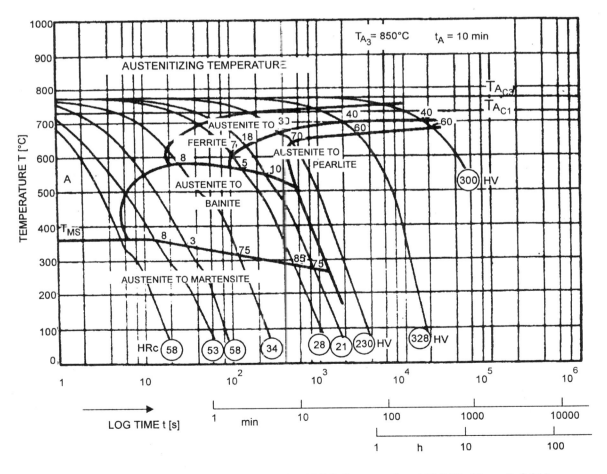

Figure 9 Continuous cooling transformation (CCT) diagram of steel EN19B. (From Ref. 24.)

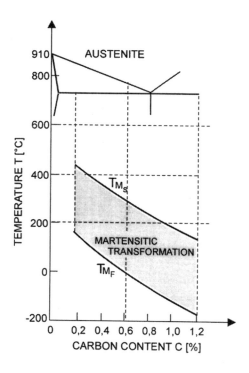

Figure 10 Influence of carbon content in steels according to start and finish temperature of martenzitic transformation.

- A semimartensitic zone or transition microstructure,
- A quenched-and-tempered or annealed zone with reference to the initial state of steel.

Sometimes, particularly in the martensitic zone, retained austenite occurs too due to extremely high cooling rates and the influence of the alloying elements present in steel.

Figure 11 shows three calculated depths of the hardened layer of C45 heat-treatment steel obtained with three different shapes of the laser beam, i.e., square, top-hat, and Gaussian distributions [25]. Different energy inputs selected permitted to achieve different maximum temperatures at the material surface. In the calculations two cases of laser heating, i.e., a steady beam, the traveling speed v being 0 mm/sec (Fig. 11A), and a beam traveling with a speed v of 20 mm/sec (Fig. 11B), were taken into account. The greatest depth of the hardened layer was obtained with the square power density distribution in the beam cross section. It was followed by the depths of the hardened layer with the top-hat and Gaussian distributions. With the stationary hardening, the depth of the hardened layer was by approximately 30% greater than with the top-hat distribution. In kinematic or scan hardening, i.e., when the laser beam is traveling at 20 mm/sec, and with the square energy density distribution, a 2.3-times smaller depth of the hardened layer was obtained at the maximum surface temperature T_{Smax}, i.e., 1420°C, a temperature just below the melting point of the given steel.

Figure 12 shows a through-depth variation of the maximum temperature of transformation-hardened SAE 1045 steel [26]. Two heating conditions were selected, they were defined by different power densities and different traveling speeds of the laser beam across the material surface. It is true in both cases that with the same transformation temperatures, T_{A_1} in T_{A_3}, different depths of the hardened layer and different widths of the transition zone consisting of the hardened microstructure and the microstructure of the base metal were obtained. The results of measurements indicated that the depth of the transition zone amounted to around 20% of the depth of the hardened layer. This relation-

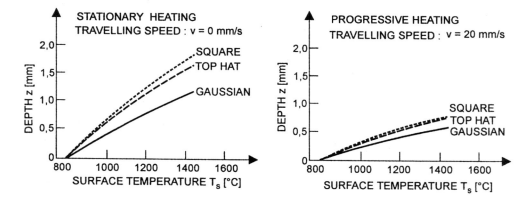

Figure 11 Calculated hardening depths for C45 steel at stationary and progressive laser beam acting for a square beam of 5×5 mm, a uniform distributed beam of 5 mm, and a Gaussian beam $w = 5$ mm. (From Ref. 25.)

ship between the depth, i.e., the width of the transition zone and the depth of the hardened layer depends primarily on thermal conductivity of the material. The materials with higher thermal conductivity require a higher energy input so that the same maximum temperature in the same depth may be obtained, which, however, results in greater depths of the hardened layers and a larger transition zone. Figure 13 shows a complete diagram including possible processing parameters and the depths of the hardened layers obtained in transformation hardening [26]. The data are valid only for the given mode structure of the laser beam (TEM), the given area of the laser spot (A), and the selected absorption deposit. In this case the processing parameters are selected from the laser-beam power P [W] and the traveling speed of the laser beam v [mm/sec]. The upper limit is a power P of 8 kW. This is the limiting energy input which permits steel melting. Although the data collected are valid only for a very limited range of the processing parameters, conditions of transformation hardening with the absorptivity changed due to the change in the deposit thickness of the same absorbent, or other type of absorbent, can efficiently be specified as well. Greater difficulties may occur in the selection of the laser trace that can be obtained in different ways

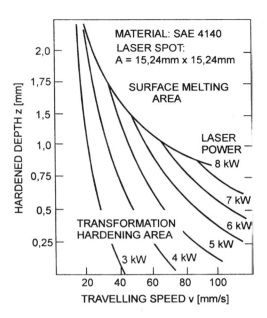

Figure 13 Influence of laser power and traveling speed on depth of hardened layer at a given laser spot. (From Ref. 26.)

and can be varied too. In case the size of the laser spot changes due to optical conditions, it is recommended to elaborate a new diagram of the processing conditions of transformation hardening.

IV. MICROSTRUCTURAL TRANSFORMATION

In transformation hardening of steel we start from its initial microstructure, which is ferritic–pearlitic pearlitic–ferritic, or pearlitic. In steel heating, transformation into a homogeneous austenitic microstructure should be ensured.

We have thus a controlled diffusion transformation of the microstructure in the surface layer of the workpiece material of a certain thickness. The thickness of the hardened layer required can be obtained if the heating process is well known. Heating is, namely, defined by the heating rate, the maximum temperature attained at the surface, heating to the depth required, and the time of heating of the material above the austenitizing temperature T_{A_3}. The time required for homogenizing of austenite depends on the type of the initial microstructure and the grain size. It is a basic prerequisite to ensure a homogeneous distribution of carbon in austenite, which results in homogeneous martensite showing a uniform hardness after quenching and self-

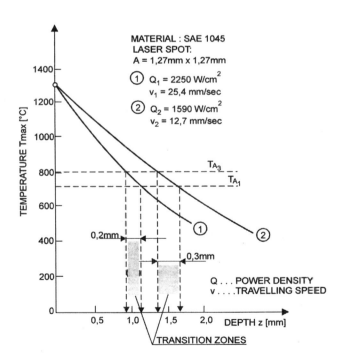

Figure 12 Influence of processing parameters on hardening depth in laser surface transformation hardening. (From Ref. 26.)

cooling, respectively. Account should be taken that the martensite transformation is a nondiffusion transformation occurring during the cooling process only at a sufficiently high cooling rate.

Figure 14 shows two thermal cycles, i.e., the one at the surface, $z = 0.0$ mm, and the one below the surface, $z = 0.5$ mm, in laser heating of C45 heat-treatment carbon steel [25]. Figure 14A shows two thermal cycles of a laser beam with a spot diameter D of 6.0 mm traveling at a speed v of 25 mm/sec. Figure 14B shows the operation of a static laser beam of the same diameter.

For efficient homogenizing, the temperature difference ΔT is important. It is defined as a difference between the maximum temperature T_{max} and the austenitizing temperature T_{A_3} occurring in a depth z of 0.5 mm. Such a temperature difference ensures, with regard to the heating and cooling conditions of the specimen, the time required for austenite homogenizing t_a in the given depth.

Figure 15 shows a shift of the transformation temperature, which ensures the formation of inhomogeneous and homogeneous austenite within the selected interaction times [25]. A shorter interaction time will result in a slightly higher transformation temperature T_{A_1} and also a higher transformation temperature T_{A_3}. To ensure the formation of homogeneous austenite with shorter interaction times, considerably higher temperatures are required. Figure 15A shows a temperature/ time diagram of austenitizing of C45 steel. The isohardnesses obtained at different interaction times in heating to the maximum temperature that ensures partial or complete homogenizing of austenite are plotted. Figure

15B shows the same temperature/time diagram of austenitizing of 100 Cr6 hypereutectoid alloyed steel. The diagram indicates that with short interaction times, which in laser hardening vary between 0.1 and 1.0 sec, homogeneous austenite cannot be obtained; therefore the microstructure consists of austenite and undissolved carbides of alloying elements that produce a relatively high hardness, i.e., even up to 920 $HV_{0.2}$. After common quenching of this alloyed steel at a temperature of homogeneous austenite, a considerably lower hardness, i.e., only 750 $HV_{0.2}$, but a relatively high content of retained austenite was obtained. Retained austenite is unwanted as it will produce unfavorable residual stresses and reduce wear resistance of such a material.

V. MATHEMATICAL MODELING

A. Mathematical Prediction of Hardened Depth

The classical approach to modeling the heat flow induced by a distributed heat source moving over the surface of a semi-infinite solid starts with the solution for a point source, with integrations over the beam area [27]. This method requires numerical procedures for its evaluation. These solutions are rigorous, so their computations are complex and the results difficult to be applied. Bass [28] gives an alternative approach by presenting temperature field equations for various structural beam modes. Analytical results show good response to various materials. Ashby and Easterling [29] and Li et al. [30] developed further analytical approach

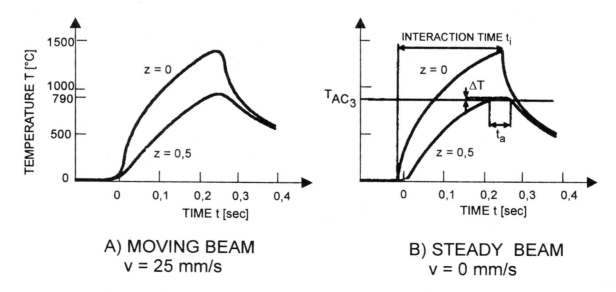

Figure 14 Temperature cycles during laser heating and cooling of steel C45 at the same interaction time. (From Ref. 25.)

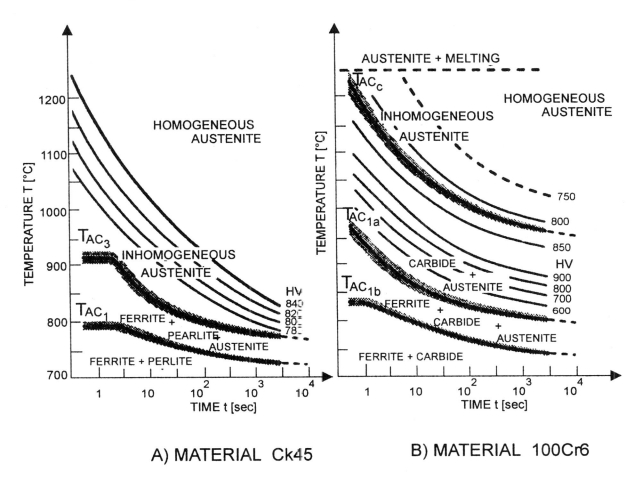

Figure 15 Temperature–time–austenite diagrams with lines of resulting hardness for various steels. (From Ref. 25.)

and an approximate solution for the entire temperature field. Comparison of the analytical results with numerical calculations shows adequate description of laser transformation hardening. Variability of laser transformation hardening parameters and changing material properties with temperature result in some scatter. With the use of dimensionless parameters to simplify computation, results general to all materials are obtained. There exist many such examples in the analysis of welding [31] and in laser surface treatment [32–34].

In mathematical modeling the following assumptions were considered:

- The surface absorptivity A is constant.
- The latent heat of the α- to γ transformation is negligible.
- The thermal conductivity λ and thermal diffusivity of steel are a constant.
- The eutectoid temperature T_{A_1} is as given by the phase diagram.

- The radius of Gaussian beam rB is the distance from the beam center to the position where the intensity is 1/e times the peak value.

The origin of the coordinate system is the beam center. The laser of total power P moves in the x direction with traveling speed v, with the y axis across the track, and z axis is the distance below the surface.

The temperature field equation from Ashby and Easterling [29] is valid for the Gaussian line source:

$$T - T_0 = \frac{Aq}{2\pi\lambda v[t(t + t_0)]^{1/2}} \cdot \exp - \frac{1}{4a}$$

$$\times \left[\frac{(z + z_0)^2}{t} + \frac{y^2}{(t + t_0)} \right]$$

The equation contains two reference parameters, defined by $t_0 = r_B^2/4a$ and z_0 is a characteristic length, as a function to limit the surface temperature.

Shercliff and Ashby [33] define the following dimensionless parameters:

$T^* = (T - T_0)/(T_{A_1} - T_0)$ is the dimensionless temperature rise,

$q^* = Aq/r_B\lambda (T_{A_1} - T_0)$ is the dimensionless beam power,

$v^* = vr_B/a$ is the dimensionless traveling speed,

$t^* = t/t_0$ is the dimensionless time, and

$(x^*,y^*,z^*) = (x/r_B, y/r_B, z/r_B)$ is the dimensionless x, y, z coordinates.

The distance z_o is normalized as follows

$$z_o^* = z_o/r_B$$

and dimensionless temperature parameter is then

$$T^* = \frac{(2/\pi)(q^*/v^*)}{[t^*(t^*+1)]^{1/2}} \exp - \left[\frac{\left(z^* + z_0^*\right)^2}{t^*} + \frac{y^{*2}}{(t^*+1)} \right].$$

The time-to-peak temperature t_p^* at an (x^*,y^*,z^*) position is found by differentiating with respect to the time.

$$t_p^* = \frac{1}{4} \left[2\left(z^* + z_0^*\right)^2 - 1 + \left[4\left(z^* + z_0^*\right)^4 + 12\left(z^* + z_0^*\right)^2 + 1 \right]^{1/2} \right]$$

At stationary laser beam of uniform intensity Q produces a peak surface temperature given by Bass [28]:

$$T_p - T_0 = \frac{2Aq}{\pi^{1/2}\lambda} (a\tau)^{1/2}.$$

The average intensity of a Gaussian beam is $Q = q/\pi r_B^2$ so the previous equation can be rewritten:

$$T_p - T_0 = \frac{2Aq}{\pi^{3/2}r_B^2\lambda} (a\tau)^{1/2}$$

or in dimensionless form τ being $2r_B/v$ and

$$\left(T_p^*\right)_{z^* = 0} = (2/\pi)^{3/2} q^*/(v^*)^{1/2}.$$

The conditions for the first hardening $T_p = T_A^*$ or the onset of melt $T_p = T_m$ are thus defined by a constant value of the process variables as follows:

$$q^*/T_p^*(v^*)^{1/2} = (\pi/2)^{3/2} = \text{constant}$$

with T_p^* taking the value appropriate to the peak temperature of interest T_{A_1} or T_m.

Bass [28] gives a more general solution for the peak surface temperature at stationary Gaussian beam acting in dimensionless form

$$\left(T_p^*\right)_{z^* = 0} = (1/\pi)^{3/2} q^* \tan^{-1}(8/v^*)^{1/2}$$

A constant value of a single dimensionless parameter defines the first hardening and the onset of melt as follows for all v^*:

$$(q^*/T_p^*)\tan^{-1}(8/v^*)^{1/2} = \pi^{3/2} = \text{constant}.$$

Four dimensionless parameters define laser hardening with a Gaussian beam. The aim is to produce a diagram from which process variables may be readily selected. A convenient plot is the dependent variables z_c^*, v^*, q^*, and $T_p^* = 1$ which give $T_p = T_{A_1}$ as shown in Fig. 16A. As surface melting is generally undesirable, the contours are dashed if the surface melts $T_p^* = (T_m - T_0)/(T_{A_1} - T_0)$ at $z^* = 0$. Figure 16A shows the depth at which surface melt commences for a 0.4-pct carbon steel at $T_p^* \approx 2.12$.

A non-Gaussian source may be simulated by superposing a number of Gaussian sources. The total power is shared between the Gaussian sources, to give the best fit to true energy profiles. For particular location and given time the temperature rises due to each source contribution to heating. This is valid if the thermal properties of the material are independent of the temperature, meaning that the differential heat flow according to the equation is linear. This method will be presented by the same authors for laser-beam heating with rectangular sources defined as laser beam spot ratio. A laser-beam ratio R is defined by length l in travel direction x, and the width w, of track in cross direction y ($R = 1/w$). In practice fixed width is normally used and the length is varied, so it is sensible to normalize the process variables using the beam width as follows:

$$q_R^* = Aq/w\lambda q (T_A - T_0)$$

$$v_R^* = vw/a$$

$$z_R^* = z/w$$

where the subscript R refers to rectangular laser beam.

Figure 16B shows a dimensionless diagram for medium carbon steel according to various ratio power to track width q/w [W/mm] and spot ratio $R = 1/w$ [-].

The diagram shows the nondimensional ratio of the hardened-layer depth to the laser-spot width for a

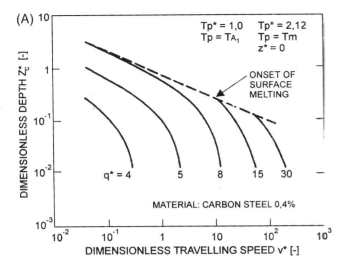

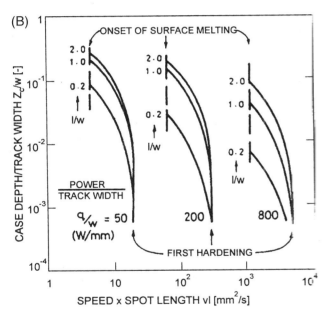

Figure 16 (A) Dimensionless hardened depth Z_a^* against laser beam traveling speed v* with curves of constant laser beam power q* for Gaussian power density cross section. (B) Dimensionless hardened depth Zc/w against laser beam traveling speed parameter $v\,l$ with curves of q/w = const. for rectangular power density and spot ratio l/w = const. (From Ref. 29.)

rectangular laser beam as a function of a product of the traveling speed and the laser-trace length. Three characteristic power densities per unit of laser-spot width, i.e., q/w = 50, 200, and 800 W/mm, were chosen, which equals a ratio of 1 to 4 to 16. In the diagram there are curves plotted for individual power densities q/w valid

with certain ratios of the laser-spot dimensions of the rectangular beam, i.e., l/w = 0.2, 1.0, and 2.0. In the individual cases laser-hardening conditions and boundary conditions of laser hardening, i.e., laser remelting, were known too. The individual curves in the diagram indicate that by increasing the laser-beam power density per unit of laser-spot width the boundary conditions will be achieved with smaller depths of the hardened layer. The diagram shown is of general validity and permits the determination, i.e., prediction, of the hardened-layer depth based on the variation of the power density per unit of laser-spot width and the traveling speed.

It can be summarized that the approximate heat flow model of Ashby and Easterling [29] for laser transformation hardening has been presented, describing Gaussian and non-Gaussian sources over a wide range of process variables.

The following advantages can be noted:

- Simplification by choosing dimensionless parameters.
- Surface temperature calibration, extending the approximate Gaussian solution to all laser-beam traveling speeds.
- High traveling speed solution was found to be acceptable for rectangular, uniform sources.
- General Gaussian solution and enabling extension of the model to non-Gaussian sources.
- Identification of constant dimensionless parameters containing all of the process variables for both sources, which determine the position of the first hardening or the onset of melting.
- Process diagrams for rectangular sources allow the choice of the process variables such as the beam power, track width, traveling speed, and spot dimensions.

B. Mathematical Modeling of Microstructural Changes

Ashby and Easterling [29] presented their results as laser-processing graphs, which show the microstructures and hardnesses regarding process variables.

In their experiments they studied two steels, i.e., a Nb microalloyed and a medium-carbon steel. They varied the laser power P, the beam spot radius r_B, and its traveling speed v.

They carried out laser surface heat treatments using a 0.5- and a 2.5-kW continuous wave CO_2 laser using Gaussian and "top hat" energy profiles.

The microhardness of individual martensitic and ferritic regions in the low-carbon steel could be estimated fairly well from a simple rule-of-mixtures

$$H_p \text{ (mean)} = f_m H_m + (1 - f_m) H_f$$

where H_m is the mean microhardness of the martensite and H_f that of the ferrite at that depth.

Figure 17 shows the hardness profiles for the two steels with different energy densities $q/v r_B$ and interaction time r_B/v. It can be noted that worse hardening is obtained when the carbon content of the steel is low. For the high-carbon steel, complete carbon redistribution occurs within the austenitization process, which gives a uniform high hardness.

They developed simple models for pearlite dissolution, austenite homogenization, and martensite formation.

The heat cycle $T(t)$ at the depth causes microstructural changes if high enough temperature is achieved. Some of the microstructure changes are diffusion controlled such as the transformation of pearlite to austenite and the homogenization of carbon in austenite. The microstructure changes depend on the total number of diffusive jumps that occur during the temperature cycle. It is measured by the kinetic strength, I, of the temperature cycle, defined by

$$I = \int_0^\infty \exp - \frac{Q_A}{RT(t)} \, dt$$

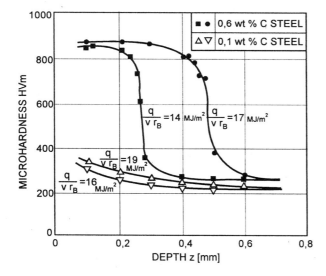

Figure 17 Hardness profiles for the two steels with different energy densities, $E = q/v r_B$ and interaction time, $t_i = r_B/v$. (From Ref. 29.)

where Q_A is the activation energy for the typical microstructural transformation and R is the gas constant. It is more convenient to write it as

$$I = \alpha \tau \, \exp - \frac{Q_A}{RT_p}$$

where T_p is the maximum temperature and τ is the thermal constant. The constant α is well approximated by

$$\alpha = \sqrt[3]{\frac{RT_p}{Q_A}}.$$

At rapid heating, the pearlite first transforms to austenite, which is followed by carbon diffusion outward and increases the volume fraction of high-carbon austenite.

If the cementite and ferrite plate spacing within a colony distance is λ, it might be thought that lateral diffusion of carbon to austenite occurs. In an isothermal heat treatment, this would require a time t given by $\lambda^2 = 2Dt$, which is

$$\lambda^2 = 2D_0 \, \alpha \tau \, \exp - \frac{Q_A}{RT_p}$$

where D is the diffusion coefficient for carbon. In a temperature cycle $T(t)$ the quantity Dt is where the maximum temperature (T_p) is found to cause the transformation.

The most important to the understanding of laser transformation hardening is the modeling of the carbon redistribution in austenite. When a hypoeutectoid, plain-carbon steel with carbon content c is heated above the T_{A_1} temperature, the pearlite transforms instantaneously to austenite. The pearlite transforms to austenite containing $c_e = 0.8\%$ carbon and the ferrite becomes austenite with negligible carbon content c_f. Thereafter, the carbon diffuses from the high to the low concentration regions, to an extent which depends on temperature and time. On subsequent cooling of steel from austenitic temperature, the austenite with carbon content greater than the critical value of 0.05 wt % C transforms to martensite and the rest of austenite with carbon content less than 0.05 wt % C transforms into pearlite.

The volume fraction occupied by the pearlite colonies is as follows:

$$f_i = \frac{c - c_f}{0.8 - c_f} \approx \frac{c}{0.8}$$

where c_f is the carbon content of the ferrite.

The volume fraction of the martensite is as follows [29]:

$$f = f_m - (f_m - f_i) \exp - \left[\frac{12 f_i^{2/3}}{\sqrt{\pi g}} \ln \left(\frac{c_e}{2 c_c} \right) \sqrt{Dt} \right]$$

where g [m] is mean grain size.

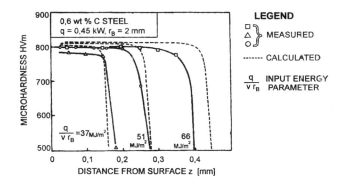

Figure 18 Measured and calculated hardness profiles for the 0.6 wt.% carbon steel compared with those predicted. (From Ref. 29.)

The hardness of the transformed surface layer depends on the volume fraction of martensite and its carbon content. The authors [29] calculated the hardness of the martensite and ferrite mixture by using a rule of mixtures.

$$H = fH_m + (1 - f)H_f$$

and suggested the following formula for calculating hardness

$$H = 1667c - 926c^2/f + 150.$$

Figure 18 shows three measured hardness profiles for different energy parameters (full lines) and the calculated profiles (dashed lines). Figure 19 shows a laser-processing diagram for a 0.6 wt.% plain carbon steel. The horizontal axes present energy density q/vr_B and beam spot radius r_B. These variables determine the temperature cycle in the transformed layer. The vertical axis is the depth below the surface. Within the shaded region, melting occurs and outside the transformation hardening process. The diagram also shows the contours of martensite volume fraction. The volume fraction and carbon content of the martensite are used to calculate the hardness HV after laser surface transformation hardening.

The author [29] showed that:

- Steels with a carbon content below about 0.1 wt.% do not respond to transformation hardening.
- Optimum combination of process variables gives maximum surface hardness without surface melting.

The method could be used for laser glazing and laser surface alloying.

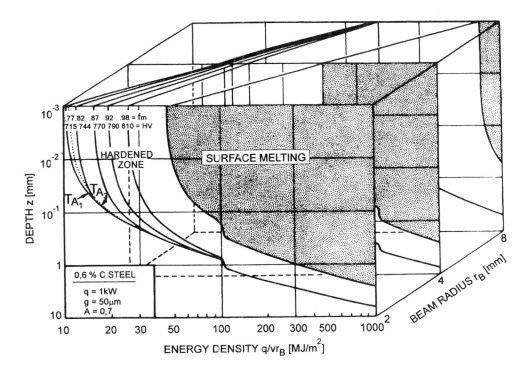

Figure 19 Laser processing diagram for 0.6 carbon steel. (From Ref. 29.)

VI. COMPUTING METHOD FOR CALCULATING TEMPERATURE CYCLE

Several methods exist to solve the heat conduction equations for various conditions; interesting descriptions are given by Carslaw and Jaeger [27]. Most of the computing method to calculate temperature cycles are based on one of the many cases which are modified to suit the particular case [35].

Gregson [36] discussed a one-dimensional model using a semi-infinite flat-plate solution for idealized uniform heat source which is constant in time. Expressions used for temperature profile separated for heating and cooling are as follows.

Heating temperature–time profile:

$$T(z,t) = \frac{\varepsilon z Q_{\mathrm{AV}}}{\lambda} \sqrt{at} \cdot \mathrm{ierfc}\left\{\frac{z}{2\sqrt{xt}}\right\},$$

$$Q(t) = \left\{\begin{array}{l} Q \text{ for } t > 0 \\ 0 \text{ for } t < 0 \end{array}\right\}.$$

Cooling temperature–time profile:

$$T(z,t) = \frac{2 Q_{\mathrm{AV}} \sqrt{a}}{\lambda} \left\{ \sqrt{t} \cdot \mathrm{ierfc}\, \frac{z}{2\sqrt{at}} \right.$$

$$\left. - \sqrt{t - t_{\mathrm{L}}} \cdot \mathrm{ierfc}\left(\frac{z}{2\sqrt{a(t - t_{\mathrm{L}})}}\right) \right\}$$

$$f(t) = \left\{\begin{array}{l} Q \text{ for } 0 < t < t_{\mathrm{L}} \\ 0 \text{ for } 0 > t > t_{\mathrm{L}} \end{array}\right\}$$

where T [°C] is the temperature, z [cm] is the depth below the surface, t [sec] is the time, $\varepsilon \cong 1$ is the emissivity, Q_{AV} [W/cm^2] is the average power density, λ [W/cm °C] is the thermal conductivity, a [cm^2/sec] is the thermal diffusivity, t_0 [sec] is the time start for power on, t_{L} [sec] is the time for power off, and ierfc is the integral of the complementary error function.

These equations for description of laser heating and cooling process are valid if the thickness of the base material is greater than $t \geq \sqrt{4at}$ and they could be approximately described for the hardened layer.

These one-dimensional analyses may be applied to laser transformation hardening process with idealized uniform heat sources, which are produced by using optical systems such as laser-beam integrator or high-power multimode laser beam with top-hat power density profile. These equations present one-dimensional

solutions and provide only approximate temperature time profile. For better description of thermal conditions, a two- or three-dimensional analysis considering actual input power density distribution and variable thermophysical properties treated material is required.

Sandven [37] presented the model that predicts the temperature time profile near a moving ring-shaped laser spot around the periphery of the outer or inner surface of a cylinder. This solution can be applied to the transformation hardening processes using toric mirrors. Sandven [37] developed his model based on a flat-plate solution and assumed that the temperature time profile $T(t)$ for cylindrical bodies can be approximated by

$$T = \theta I$$

where θ depends on workpiece geometry and I is the analytical solution for a flat plate. The final expression for a cylindrical workpiece, which is derived from this analysis, is:

$$T \approx (1 \pm 0.43\sqrt{\phi})\frac{2 Q_0 a}{\pi \lambda v} \int_{x-B}^{x+B} \mathrm{e}^{u} \cdot K_{\mathrm{o}}(z^2 + u^2)^{1/2} \mathrm{d}u$$

where the $+$ sign means the heat flow into a cylinder, the $-$ sign means the heat flow out of a hollow cylinder, Q_{o} is the power density, v is the laser-beam traveling speed in the x direction, K_{o} is the modified Bessel function of the second kind and 0 order, u is the integration variable, $2b$ is the width of the heat source in the direction of motion, and z is the depth in radial direction,

$$B = \frac{v_{\mathrm{b}}}{2a}, \quad Z = \frac{v_{\mathrm{z}}}{2a}, \quad X = \frac{v_{\mathrm{x}}}{2a}$$

where $\phi = at/R^2$, and R is the radius of the cylinder.

Sandven [37] provided graphical solutions for $Z = 0$ for various values of B. To estimate an approximate depth of hardness, maximum temperature profile across the surface layer is the only item to be interested.

Cline and Anthony [38] presented most realistic thermal analysis for laser heating. They used a Gaussian heat distribution and determined the three-dimensional temperature distribution by solving the equation:

$$\partial T / \partial t - a\nabla^2 T = Q_{\mathrm{AV}}/c_{\mathrm{p}}$$

where Q_{AV} is the power absorbed per unit volume and c_{p} is the specific heat per unit volume.

They used a coordinate system fixed at the workpiece surface and superimposed the known Green function solution for the heat distribution; the following temperature distribution is:

$$T(x,y,z) = P(c_{\mathrm{p}} a r_{\mathrm{B}})^{-1} f(x,y,z,v)$$

where f is the distribution function

$$f = \int_0^\infty \frac{\exp(-H)}{(2\pi^3)^{1/2}(1+\mu^2)}\,d\mu \quad \text{and}$$

$$H = \frac{\left(X + \frac{\tau\mu^2}{2}\right)^2 + Y^2}{2(1+\mu^2)} + \frac{Z^2}{2\mu^2}$$

where $\mu^2 = 2at'/r_B$, $\tau = vr_B/a$, $X = x/r_B$, $Y = y/r_B$, $Z = z/r_B$, P is total power, r_B is the laser-beam radius, t' is the earlier time when laser was at (x', y'), and v is the traveling speed.

The cooling rate can be calculated as follows:

$$\partial T/\partial t = -v[x/\gamma^2 + v/2a\ (1+x/\gamma)]T$$

where $\gamma = \sqrt{x^2 + y^2 + z^2}$.

The given cooling rate is calculated only when point heat source is used.

This three-dimensional model is a great improvement over one-dimensional models because it includes temperature-dependent thermophysical properties of the material used for numerical solutions.

Grum and Šturm [39] obtained a relatively simple mathematical model describing the temperature evolution $T(z,t)$ in the material depending on time and position, where we distinguished between the heating cycle and the cooling cycle.

For reasons of simplifying the numerical calculations, it is necessary to make certain assumptions:

- The latent heat of material melting is neglected.
- The material is homogeneous with constant physical properties in the solid and liquid phase.

So we assume that material density, thermal conductivity, and specific heat are independent of temperature.

- Thermal energy is transferred only through transfer into the material, thermal radiation and transfer into the environment are disregarded.
- The laser light absorption coefficient to workpiece material is constant.
- Limiting temperatures or transformation temperatures are assumed from phase diagrams.
- The remelted surface remains flat and ensures a uniform heat input.

Thus we obtained a relatively simple mathematical model describing the temperature evolution $T(z,t)$ in the material depending on time and position, where we distinguished between the heating cycle and the cooling cycle.

(1) The heating cycle conditions in the material can be described by the equation:

$$T(z,t) = T_0 + \frac{A \cdot P}{2 \cdot \pi \cdot A \cdot v_B \cdot \sqrt{t \cdot (t_i + t_0)}}$$

$$\times \left[e^{-\left(\frac{(z+z_0)^2}{4 \cdot a \cdot t}\right)} + e^{-\left(\frac{(z-z_0)^2}{4 \cdot a \cdot t}\right)} \right]$$

$$\times \operatorname{erfc}\left(\frac{z+z_0}{\sqrt{4 \cdot a \cdot t}} \right)$$

for $0 < t < t_i$.

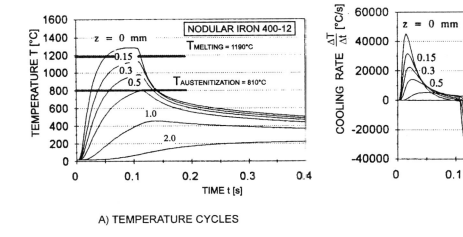

Figure 20 Temperature cycles and cooling rate versus time at various depths. (From Ref. 39.)

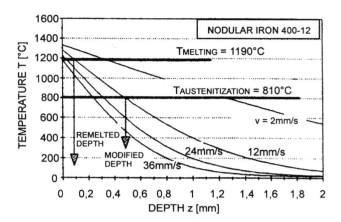

Figure 21 Maximum temperature drop as a function of depth in nodular iron 400-12. (From Ref. 39.)

(2)   The cooling cycle conditions in the material can be expressed by the equation:

$$T(z, t) = T_0 + \frac{A \cdot P}{2 \cdot \pi \cdot A \cdot v_B \cdot \sqrt{t \cdot (t_i + t_0)}}$$

$$\times \left[e^{-\left(\frac{(z + z_0)^2}{4 \cdot a \cdot t}\right)} + e^{-\left(\frac{(z - z_0)^2}{4 \cdot a \cdot t}\right)} - e^{-\left(\frac{(z - z_0)^2}{4 \cdot a(t - t_1)}\right)} \right]$$

$$\times \mathrm{erfc}\left(\frac{z + z_0}{\sqrt{4 \cdot a \cdot t}} \right)$$

for $t > t_i$.

Where variable t_0 represents the time necessary for heat to diffuse over a distance equal to the laser-beam radius on the workpiece surface and the variable z_0 measures the distance over which heat can diffuse during the laser-beam interaction time [29]. C is a constant, in our case defined as $C = 0.5$. Figure 20A presents the time evolution of temperatures calculated according to equations at specific depth of the material in the nodular iron 400-12 at a laser-beam traveling speed $v_B = 12$ mm/sec. Figure 20B illustrates the variation of heating and cooling rates during the process of laser remelting in the remelted layer and in deeper layers of the material. The temperature gradient is at the beginning of the laser-beam interaction with the workpiece material, i.e., on heating up, very high, on the surface achieving values as high as 48,000°C/sec. The results show that the highest cooling rate is achieved after the beam has passed by half the value of its radius r_B across the measured point. Knowing the melting and austenitization temperatures, they can successfully predict the depth of the remelted and modified layer (see Fig. 21). Considering the fact that on the basis of limiting temperatures it is possible to define the depth of particular layers and that these can be confirmed by microstructure analysis, we can verify the success of the proposed mathematical model for the prediction of remelting conditions. Thus a comparison is made in Fig. 22A and B between the experimentally obtained results for the depth of particular zones of the modified layer and the results calculated according to the mathematical model. We can see that the calculated depths of

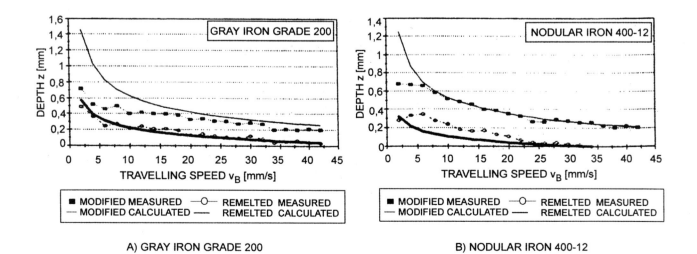

A) GRAY IRON GRADE 200   B) NODULAR IRON 400-12

Figure 22 Comparison of experimentally measured, remelted, and hardened zone depths with those calculated with the mathematical model. (From Ref. 39.)

the remelted and hardened zones correlate well with the experimentally measured values. Large deviations in the depth of the modified layers are found only in gray iron at insufficient workpiece traveling speeds, and they are probably due to the occurrence of furrows on the workpiece surface.

VII. LASER LIGHT ABSORPTIVITY

With the interaction of the laser light and its movement across the surface, very rapid heating up of metal workpieces can be achieved, and subsequent to that also very rapid cooling down or quenching. The cooling speed, which in conventional hardening defines quenching, has to ensure martensitic phase transformation. In laser hardening the martensitic transformation is achieved by self-cooling, which means that after the laser light interaction the heat has to be very quickly abstracted into the workpiece interior. While it is quite easy to ensure the martensitic transformation by self-cooling, it is much more difficult to deal with the conditions in heating up. The amount of the disposable energy of the interacting laser beam is strongly dependent on the

Table 1 Typical Values of Surface Reflectivity for Various Materials and Surface States

Material	Surface state	Reflectivity R [%], $\lambda = 10.6$ μm
Titanium 6A1-4V	Polished 300C	85
Aluminum	Polished	98
Stainless steel 304	Polished	85
Steel, mild	Polished	94
	Roughened with sand paper	
	to 1 μm	92
	to 19 μm	32

Source: From Ref. 43.

absorptivity of the metal. The absorptivity of the laser light with a wave length of 10.6 μm ranges in the order of magnitude from 2% to 5%, whereas the remainder of the energy is reflected and represents the energy loss. By heating metal materials up to the melting point, a much higher absorptivity is achieved with an increase of up to 55%, whereas at vaporization temperature the absorptivity is increased even up to 90% with respect to the power density of the interacting laser light.

Figure 23 illustrates the relationship between laser light absorptivity on the metal material surface and temperature or power density [40,41]. It is found that, from the point of view of absorptivity, laser-beam cutting does not pose any problems, as the metal takes the liquid or evaporated state, and the absorptivity of the created plasma can be considerably increased. Therefore it is necessary to heat up the surface, which is to be hardened, onto a certain temperature at which the absorptivity is considerably higher and enables rapid heating up onto the hardening temperature or the temperature which, for safety reasons, is lower than the solidus line. This was successfully used in heat treatment of camshafts [42].

The heating up of the workpiece surface material by the laser beam is done very rapidly. The conditions of heating-up can be changed by changing the energy density and relative motion of the workpiece and the laser beam.

In surface hardening this can be achieved without any additional cooling and is called self-quenching. The procedure of laser surface hardening is thus simpler than the conventional flame or induction surface hardening as no additional quenching and washing is required. Absorptivity always depends on the wavelength of light and the surface preparation of the material that interacts. In Table 1 typical values of reflectivity of polished

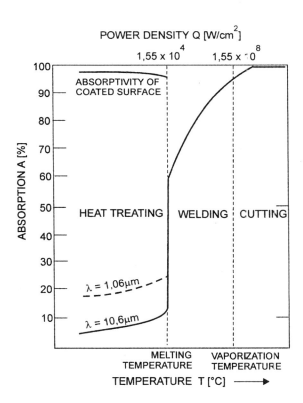

Figure 23 The effect of temperature on laser light absorptivity. (From Refs. 40 and 41.)

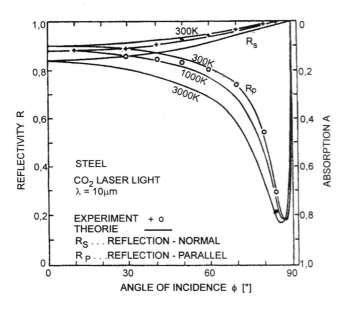

Figure 24 Variation of reflectivity with angle and plane of polarization. (From Ref. 53.)

surfaces of a titanium alloy, aluminum, 304 austenitic stainless steel, and soft steel in interaction with CO_2 laser light are stated [43]. The reflectivity of the polished surface of the given materials depends on the temperature and wavelength of the incident light. Bramson [44] defined the dependence between electric resistance and emissivity $\varepsilon_\lambda (T)$ for the light radiation striking the material surface at a right angle.

$$\varepsilon_\lambda(T) = 0.365\left(\frac{\rho_r(T)}{\lambda}\right)^{1/2} + 0.0667\left(\frac{\rho_r(T)}{\lambda}\right)$$
$$+ 0.006\left(\frac{\rho_r(T)}{\lambda}\right)^{3/2}$$

where ρ_r [Ω cm] is the electrical resistivity at a temperature T [°C], $\varepsilon_\lambda (T)$ is the emissivity at T [°C], and λ [cm] is the wavelength of incident radiation. Reflectivity depends on the incident angle of the laser beam with reference to the polarization plane and the specimen surface [45–53]. Figure 24 shows the reflectivity of CO_2 laser light from a steel surface at different incident angles

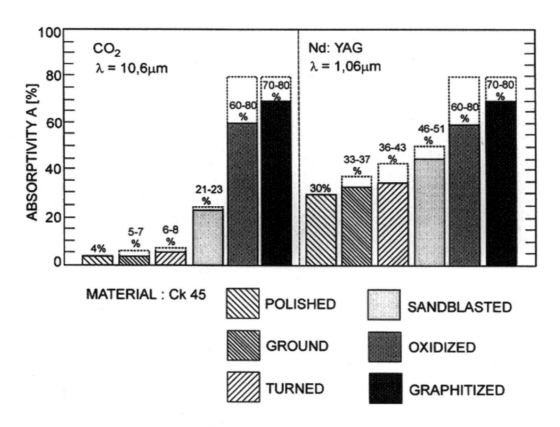

Figure 25 Influence of various steel Ck 45 treatments on absorption with CO_2 or Nd:YAG laser light. (From Ref. 53.)

and different temperatures [43,53,54]. The diagram combines experimental data (plotted dots) on reflectivity and absorptivity, and theoretically calculated values of reflectivity (uninterrupted lines). The variations of absorptivity indicate that absorptivity strongly increases at elevated temperatures due to surface oxidation and at very high temperatures due to surface plasma absorption.

Figure 25 shows the influences exerted on absorptivity of CO_2 and Nd:YAG laser light in the interaction with specimens made of Ck45 steel [53]. The steel specimens were polished, ground, turned, and sandblasted. To Ck45 heat-treatment carbon steel, various methods of surface hardening, and particularly laser hardening, are often applied.

From the column chart it can be inferred that absorptivity of steel specimens subjected to different machining methods is considerably lower with CO_2 laser light than with Nd:YAG laser light. The lowest absorptivity was obtained with the polished specimens. It varied between 3% and 4% with reference to the laser-light wavelength. Absorptivity was slightly stronger with the ground and then turned surfaces. It turned out in all cases that the absorptivity of Nd:YAG laser light is seven times that of CO_2. If absorptivity of the two wavelengths is considered, smaller differences may be noticed with the sandblasted surfaces. The oxidized and graphitized surfaces showed the same absorptivity of laser light regardless of its wavelength. The latter varied between 60% and 80%.

VIII. INFRARED ENERGY COATINGS

To increase laser-beam absorptance at metal surfaces various methods were used:

- Metal surface painted with absorbing coatings followed by laser processing,
- Chemical conversion coatings, and
- Uncoated metal surfaces processed by a linearly polarized laser beam [55].

Infrared energy coatings having high absorptance must have the following features for increased efficiency during laser heating at heat treatment:

- High thermal stability,
- Good adhesion to metal surface,
- Chemically passive to material heat conduction from coating to material,
- Proper coating thickness referring to type of material and the way of heat treatment,
- High heat transfer coefficient for better heat conduction from coating to material,
- Easily applied and removed, and
- As lower expenses for coatings as possible.

A. Paint and Spray Coatings

Several commercially available paints exhibit low normal spectral reflectance for CO_2 laser with wavelength $\lambda = 10.6\ \mu m$. These paints in general contain carbon black and sodium or potassium silicates and are applied to metal surfaces by painting or spraying. Thicknesses of paint coatings range from 10 to 20 μm.

B. Chemical Conversion Coatings

Chemical conversion coatings, such as manganese, zinc, or iron phosphate, absorb infrared radiation. Phosphate coatings are obtained by treating iron-base alloys with a solution of phosphoric acid mixed with other chemicals. Through this treatment, the surface of the metal surface is converted to an integrally bonded layer of crystalline phosphate. Phosphate coatings may range

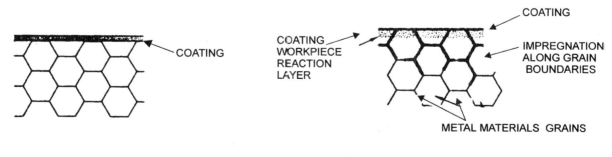

Figure 26 Potential reaction of infrared energy-absorbing coating after laser-treated metal material. (From Ref. 55)

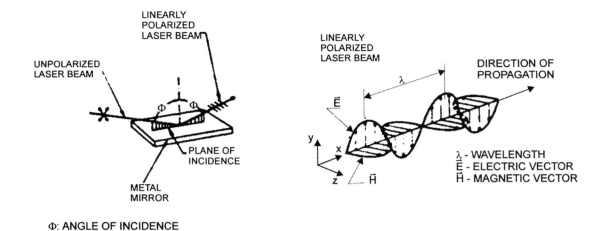

Figure 27 Conversion of an unpolarized laser beam to a linearly polarized beam by reflection at a specific angle. (From Ref. 55.)

in thickness from 2 to 100 μm of coating surface. Depending on the workpiece geometry, phosphating time can range from 5 to 30 min regarding temperature and concentration of solution. Phosphate coatings on the metal surface can be prepared with a fine or coarse microstructure.

In terms of chemical passiveness and ease of coating application on metal surfaces silicates containing carbon black are more effective than the phosphate coatings. Figure 26 schematically illustrates the reaction of manganese phosphate with metal surface and subsequent formation of low melting compounds, which can penetrate along the grain boundaries over several grains below the surface of metal material. This reaction can be prevented by using chemically inert coatings.

C. Linearly Polarized Laser Beam

Metals have lower reflectance for linearly polarized electromagnetic radiation. The basis of this optical phenomenon has been applied to the CO_2 laser heat treatment of uncoated iron-base alloys. An unpolarized laser beam can be linearly polarized by using proper reflecting optical elements. Figure 27 shows unpolarized laser beam with specific incident angle referring to metal mirror and reflected beam is linearly polarized. This angle of incidence is called the polarizing angle. When the laser beam is linearly polarized, the dominant vibration direction is perpendicular to the plane of incidence. The plane of incidence is defined as the plane that contains both the incident laser beam and the normal to the reflecting surface.

The electric vector **E** of the linearly polarized beam has components parallel E_p and perpendicular E_s to the plane of incidence. Figure 28 illustrates absorptance as a

function of the angle of incidence for iron [55]. At an angle of incidence between 70° and 80°, the absorptance is between 50% and 60% for E_p and 5% to 10% for E_s. Thus by directing a linearly polarized laser beam at an angle of incidence greater than 45°, substantial absorptance by iron-base alloys is possible. Possible weakness of this method is the important laser-beam power loss during conversion of an unpolarized to a linearly polarized laser beam.

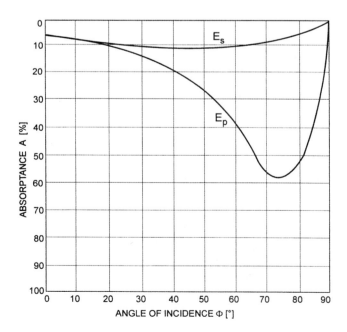

Figure 28 Effect of the angle of incidence of a linearly polarized laser beam on absorptance by iron base alloys. (From Ref. 55.)

IX. INFLUENCE OF DIFFERENT ABSORBERS ON ABSORPTIVITY

Rothe et al. [56] determined absorptivity of metal surfaces on the basis of calorimetric measurements. Kechemair et al. [57] studied the influence of the graphite absorption deposit on the surface of the specimens of 35 NCD16 steel.

Woo and Choo [58] analyzed the general effects of coating thickness of absorber on the depth and width of hardened layer. Hardening experiments were carried out at various laser energy input. Trafford et al. [59] studied absorptivity in the case when the traveling speed of the laser beam across the workpiece surface varied. They studied various absorption deposits, e.g., carbon black and colloidal graphite, zinc and manganese phosphates, which were deposited on carbon steel with a carbon content of 0.4%. The first two absorbents behaved very favorably at lower surface temperatures, but at higher temperatures their effect weakened. Another deficiency of depositing carbon black or colloidal graphite is that in the manipulation of articles the deposit will get damaged and its effect will therefore be somewhat poorer. Consequently, it can show in a nonuniform depth of the hardened layer.

Phosphate deposits are wear and damage resistant and quite easy to apply to the surface. Consequently, they are considered to be better and more useful. They absorb both the light of the IR spectrum and visible light; therefore these deposits look black as well. When hardening was performed only at a limited area of the workpiece surface, it could be assessed from the rest of the absorption deposit at the boundary between the hardened and unhardened surfaces how the heating process proceeded.

The diagram in Fig. 29 shows the results obtained in calorimetric measurements of absorptivity shown by steel with the carbon content of 0.4%, when coated with the respective four deposits, as a function of the traveling speed of the laser beam. The other laser-hardening parameters were a square power density distribution of the laser beam Q of 4.5×10^3 W/cm^2, which allows both transformation hardening and hardening by remelting. A difficulty encountered in hardening by remelting is that temperatures occurring are very high and the absorption deposit will get damaged, which will deteriorate the absorption effect. In the figure there is a hatched area separating low traveling speeds of the laser beam across the workpiece surface that in connection with the other parameters ensures surface hardening by remelting. The boundary zone of the traveling speeds is comparatively wide, i.e., $v = 20 \pm 1.5$ mm/sec, which confirms how difficult it is to ensure the repeat-

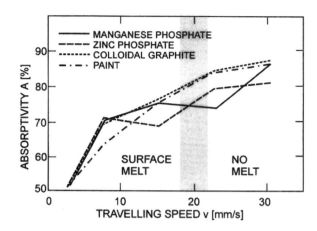

Figure 29 Influence of different absorbers and traveling speeds on absorptivity. (From Ref. 59.)

ability of the depth of hardening in both transformation hardening and hardening by remelting. The zone at the extreme left end of the hatched area corresponds to low traveling speeds that produced temperatures at the surface exceeding the melting point suitable for remelting. The traveling speed can thus be used to control the depth of the remelted and hardened layer. The zone at the extreme right corresponds to the temperatures exceeding the limiting speed with which no remelting but only transformation in the solid state could be obtained. The traveling speed defines the through-depth heating conditions and thus determines the depth of the hardened layer and the size of the transition zone. The size of the transition zone and the variation of hardness in it are extremely important as they permit to fulfil requirements set for dynamically loaded parts.

Inagaki et al. [60] studied various types of deposits in order to increase the depth of the hardened layer and reduce laser-beam energy losses by means of improved absorption in the interaction. Relevant tests were performed with a CO$_2$ laser with a maximum power of 2.0 kW and the square power density distribution ensuring a uniform depth of the hardened layer after the interaction. A quantitative analysis of absorptivity and the conditions of transformation hardening showed that the most fit-for-purpose absorption deposit was a blend of mica and graphite powder, which provided a 1.8-mm-deep and 10.0-mm-wide single hardened spot. The following substances were used: graphite, mica, and various oxides, e.g., SiO$_2$, TiO$_2$, Al$_2$O$_3$, which were added to graphite. The binder was acrylic resin. The different ratios between the absorbent and the binder were tested too. A ratio between the contents of "powder" and "binder" was selected among the ratios of 2 to

1, 1 to 1, 1 to 2, 1 to 3, and 1 to 4. The authors focused their studies on a relationship between the size of the graphite particles and its content and the content of binder. The graphite particles were defined by their average size, which amounted to around 0.5 µm, around 4.0 µm, and around 10 µm.

With the graphite-oxide type of deposit, the particles were very fine and showed an average size of graphite particles of 0.5 µm and an average size of oxides of 1.0 µm. In this case, the grains being very tiny, the ratio between the content of the two substances and the binder was very limited. Thus only the ratio of 2 to 1 was selected.

Figure 30 shows the testing of graphite deposited on SK3 tool steel with a carbon content of 1.1%, taking into account all the above variables in the deposit composition [60]. It turned out that the absorbent having smaller graphite grains had a better effect, while a higher ratio between the graphite particles and the binder had no particular influence and the largest particles even diminished the effect. At the extreme right end the effect of zinc phosphate as absorbent on the depth of the hardened layer is shown. The tests showed that with lower traveling speeds of the laser beam, i.e., with a higher surface temperature, the zinc-phosphate-based deposit was more efficient than the graphite one.

Figure 31 shows the effect of the second type of absorbents consisting of graphite particles with an average grain size of 0.5 µm and an oxide with an average grain size of 1.0 µm [60]. The ratio of the two constituents was 1 to 1 and the ratio between the solid substances and the binder 2 to 1. Testing of the absorptivity of laser light was performed with different graphite deposits with oxides or mica deposited on the same steel surfaces. The depth and the area of the obtained hardened layer, respectively, were somewhat greater

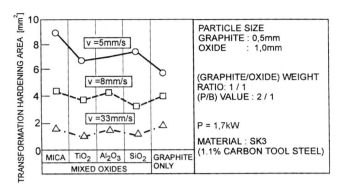

Figure 31 Effect of adding various oxides to graphite on the transformation hardening area. (From Ref. 60.)

with graphite than with the mica-based deposit. The absorptivity was somewhat lower with the deposits consisting of different oxides. Testing was performed using SK3 tool steel with a carbon content of 1%, a laser-beam power of 1.7 kW, and different traveling speeds v of the laser beam, i.e., 33.8 and 5 mm/sec.

Figure 32 shows a comparison of the depth of a single hardened trace obtained with different surface preparations, i.e., with mica and graphite, the binder, and a zinc-phosphate coating [60]. As this comparison of absorptivity was made using the same tool steel, it can be stated that in laser transformation hardening the best results were obtained with the graphite deposit with the addition of mica.

Grum et al. [61] studied microstructures of various aluminum–silicon alloys after casting and the influence of laser hardening by remelting on the changes in microstructure and hardness of a modified surface layer. Their aim was to monitor changes in the thin, remelted surface layer of a specimen material in the form of thin plates with regard to remelting conditions. The authors gave special attention to the preparation of specimen surface by an absorbent.

Aluminum and its alloys absorb laser light poorly; therefore it is needed that the surface of the aluminum alloys be suitably prepared prior to laser remelting. This can be done by surface treatment or by deposition of a suitable absorbent. To this end a high-temperature coating medium based on silicon resin with the addition of metal pigment was applied. It was a sluggish resin. The silicon resin containing pigment was applied to the surface as an about 10-µm-thick layer. The absorbing coating permitted a considerable increase in laser light absorption and energy input, which almost agrees with the calculated values [62–65].

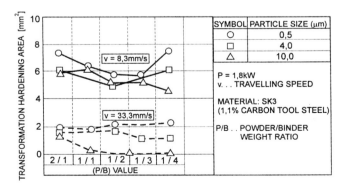

Figure 30 Effect of graphite particle size and powder/binder weight ratio on the transformation hardening area in cross section. (From Ref. 60.)

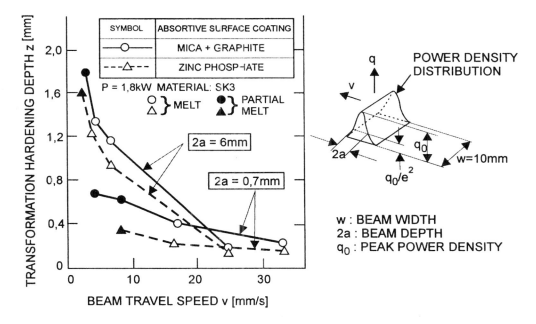

Figure 32 Effect of beam irradiating conditions on the transformation hardening depth for various absorbers. (From Ref. 60.)

Hardening of the coating medium was carried out at different temperatures, i.e., at the ambient temperature (T_{AM}) and temperatures of 100°C, 150°C, 200°C, and 250°C. In accordance with recommendations of the manufacturer, a single hardening time was selected, i.e., 1 hr at the temperature selected. The aim of the different hardening conditions for the thermostable resin coating was to study its absorptivity regarding laser light and thus establish the influence on the size of the remelted layer after remelting. The surface appearance seemed not to change after hardening. In accordance with the manufacturer's assurance, the absorbent evaporates during the remelting process and does not affect formation of the remelted layer.

Figure 33 shows the influence of hardening temperature of absorbent on the width and depth of the remelted layer, which is different with different alloys. The largest remelted trace is obtained after absorbent hardening at a temperature between 100°C and 150°C. With alloys AlSi12 and AlSi8Cu the largest trace is obtained, provided the absorbent has been hardened at a temperature of around 100°C and with alloys AlSi5 and AlSi12CuNiMg at a hardening temperature of the absorbent of around 150°C.

Gay [66] conducted an investigation on the absorption coefficients obtained with various surface treatment processes, including different absorbents and different fine machining processes.

The investigations were conducted on SAE 1045 heat-treatment steel. An absorption coefficient was determined by the calorimetric method [67,68].

Guangjun et al. [69] studied exacting laser transformation hardening of precision V-slideways. Experiments with conventional hardening and laser hardening were performed. For this purpose four steels were selected, i.e., 20, 45, 6Cr15, and 18Cr2Ni4WA. Table 2 refers to all the steels concerned and both heat-treatment methods and gives detailed data on the preliminary heat treatment and execution of surface hardening. Data on the hardness achieved with the two heat-treatment methods are given as well. In order to improve the absorptivity of laser light by the individual steels, carbon ink and manganese phosphate $Mn_3 (PO_4)_2$ were used as coatings.

Laser irradiation of metal surfaces is a very complex phenomenon, which is described by the three-dimensional thermal conductivity in the material. There are several equations available for simple calculations of the effects of heat conduction in terms of the temperature obtained inside the hardened layer or the heat-affected zone. It is known from the theory of thermal conductivity that the heating temperature of the irradiated material is, in approximation, dependent on the incident laser-beam power and the irradiation time. Thus it can be presumed that the depth of the hardened layer depends on parameter $P/\sqrt{D_B \cdot v}$, which is valid only

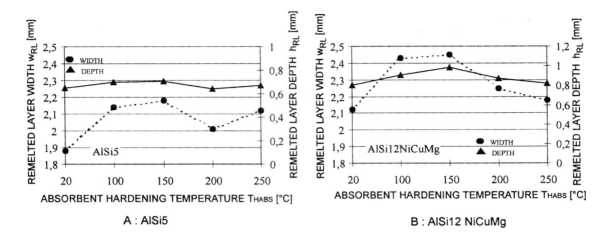

Figure 33 Size of the remelted layer obtained after laser remelting AlSi$_5$ (A) and AlSi12NiCuMg (B) versus hardening temperature of the absorber in the given remelting conditions: $P = 1500$ W, $v_b = 350$ mm/min and $z_s = 9$ mm. (From Ref. 61.)

in the case that no melting of the surface occurs. In the equation, P is the power, D_B is the laser-spot diameter at the specimen surface, and v is the laser-beam traveling speed. The laser-beam diameter is defined as a diameter with which the power is exponentially reduced by a factor of $1/e$. It turned out that the hardness achieved was in direct proportion to $H \alpha P \sqrt{vv}$.

Figure 34 indicates that the strongest influence on surface heating and the temperature variation was exerted by the incident laser-beam power, and a reverse influence by the laser-beam traveling speed [69]. Both parameters of laser transformation hardening affect the changes of microstructure and, consequently, changes in hardness. The dependence shown in the diagram can generally be valid only with the same defocus and the same surface preparation preceding laser hardening. Meijer et al. [70] studied the possibility of hardening smaller workpieces or smaller workpiece surfaces by a CO$_2$ laser delivering a continuous energy (CW) of low power in the TEM$_{00}$ mode structure. Two ways of line

hardening, i.e., the one with parallel hardened traces and the one with crossing hardened traces including an unhardened space, were selected. Hardening was applied to C45 heat-treatment carbon steel by (1) transformation hardening and (2) hardening by remelting. The authors [70] discussed the behavior of the colloidal graphite as an absorbent from the viewpoint of:

- Deposits of different thickness, i.e., ranging between 2 and 14 μm,
- Different traveling speeds of the laser beam,
- The influence of the absorbent condition, i.e., wet (immediately after deposition), partially dried, and dry.

Based on calorimetric measurement a relation between absorptivity and the thickness of the absorbent deposited as well as different traveling speeds of the beam defining surface temperature was determined. Figure 35 shows the absorptivity achieved with reference to the deposit thickness and the temperature

Table 2 Comparison Between Laser-Hardened and Various Conventional Heat-Treated Structural Steel Hardness

Material	Laser hardened	Conventional heat-treated hardness structure	Heat treatment prior to laser transformation hardening
Steel C20	547-529 HK, 51-54 HRC	< 40–45 HRC	Annealing
Steel C45	712-889 HV, 60-66 HRC	HRC 45-50 (oil-quenched) HRC 52-60 (water quenched)	
Steel 6Cr15	880-939 HK, 66-68 HRC	HRC 64-66 (oil-quenched)	Quenching and tempering
Steel 18Cr2Ni4WA	524-620 HV, 51-56 ~ HRC	HRC 37-39 (air-quenched) HRC 41-42 (oil-quenched)	

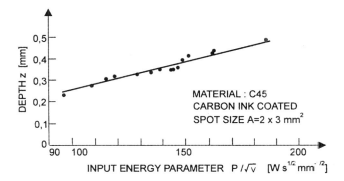

Figure 34 Dependence of laser beam power and traveling speed on hardened depth of steel C45. (From Ref. 69.)

obtained at the specimen surface [70]. The results obtained in measurements made after line transformation hardening confirmed that the differences in absorptivity were generally amounting up to 20%, that the optimum deposit thickness of colloidal graphite was 8 μm with the higher traveling speeds v, i.e., 4 or 8 mm/sec, and as much as 14 μm with the lower speeds. The optimum absorbent deposit with lasers of greater powers amounted even up to 50 μm, which meant that the deposit had to be made in several layers, depending on the absorbent used, to obtain the thickness required.

For the determination of the deposit thickness they used the optical method, i.e., by measurement of the light transmitted through the absorbent and glass to a photocell. This was accomplished simultaneously with the deposition of absorbent on the specimen and glass under the assumption that a uniform deposit thickness was obtained. Figure 36 shows the results of the transmission of light of a certain intensity (U_0) through the absorbent and glass (U) showing a dependence

$U = U_0 \cdot e^{-\alpha t}$ [70]. For colloidal graphite, it follows that $1/\alpha = 1.06$ μm, which is valid during or after drying of the deposited absorbent.

Arata et al. [71] state that phosphate coating from zinc phosphate—$Zn_3 (PO_4)_2$—and manganese phosphate—$Mn_3(PO_4)_2$—have outstanding absorptive properties, extreme heat resistance, and good adhesion with a constant thickness of the deposit on the base material. Therefore phosphate coatings are widely used in industry. In testing phosphates in terms of absorption, the optic conditions were chosen so that, using a laser source $P = 1.5$ kW, the spot ran in the direction of the y axis and had a size of $Dy = 3$ mm.

The absorptivity was analyzed in the air and argon atmosphere in different kinematic conditions defined by the traveling speed of the workpiece or the laser beam. The results have confirmed that the absorptivity is almost independent of the atmosphere, which immediately brought about a simplification of the conditions to be maintained during hardening.

The absorptivity dependence on the workpiece traveling speed was more significant. The tests were carried out at traveling speeds of 1 and 8 m/min The results have confirmed that at lower traveling speeds the absorptivity

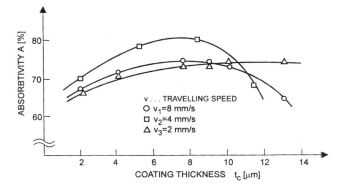

Figure 35 Absorptivity of a sprayed colloidal graphite coating, measured during line hardening. (From Ref. 70.)

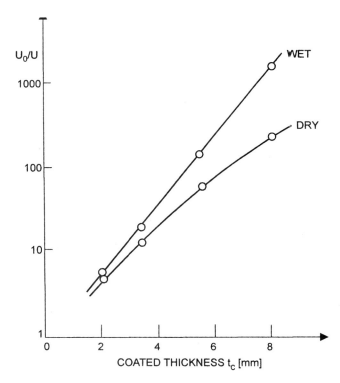

Figure 36 Light transmittance (inverse) as a measure of coating thickness. (From Ref. 70.)

is less due to heat transfer to the cold workpiece material and also into the surrounding area. When the traveling speed is increased from 1 to 8 m/min, an absorptivity increase from 40% to almost 70% is achieved (Fig. 37).

Arata et al. [71] also studied the effects of optical conditions on phosphate absorptivity. The starting point for this study was the spot size in the direction of the y axis, which was denoted by Dy. The spot size was changed from 1 to 6 mm for the purpose of studying the absorptivity dependence on different workpiece traveling speeds. Thus at the spot size of $Dy = 6$ mm and traveling speed of 1 m/min, the absorptivity was $A = 65\%$; at the traveling speed of 8 m/min, it was $A = 80\%$ (Fig. 38). The absorptivity of metal surfaces coated with phosphates is essentially bigger than that of metal surfaces that are only polished.

Research by Grum and Žerovnik [72] has confirmed that laser-beam hardening can be successfully carried out despite the laser source's extremely low power.

The effects they had chosen to study include the following:

- Influences of alloying elements on heat treatment results,
- Influences of different kinds of absorbers on heat treatment results,
- Influences of energy input on the microstructure and on the magnitude of residual stresses and their nature,
- Influences of various laser-beam paths on heat treatment results and the magnitude of residual stresses.

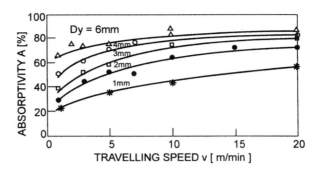

Figure 38 The spot size was changed from 1 to 6 mm in order to study the absorptivity dependence on different workpiece traveling speeds. (From Ref. 71.)

Therefore four different types of surfaces prepared in various ways were chosen:

- Surface machined by grinding
- Surface treated with absorber A
- Surface treated with absorber B
- Surface treated with Zn3(PO4)2

Absorber A—Miox, a PVK medium is made from polymer base with additions. It was applied by submersion to a thickness of not more than 20 μm, which was achieved by hardening at room temperature.

Absorber B—Melit Email was deposited by submersion, and hardening was carried out at a temperature between 120°C and 150°C for 30 to 40 min. This product is made from an alkaline base and alumina resins that have to be air-dried for 10 to 15 min before hardening for the solvents to evaporate and the paint deposit to set. The instructions given by the manufacturer must be strictly observed, and special attention should be paid to the cleanliness of the surface, paint deposition technique, and drying method [72].

The zinc phosphate absorber—$Zn_3(PO_4)_2$—was deposited on the workpiece surface simply and efficiently. The specimens for laser hardening were first degreased in alcohol or trichloroethylene and then submerged in a zinc phosphate bath at a temperature of 50°C/122°F for 4 to 5 min. This was followed by flushing in hot water and drying. We obtained a uniform and high-quality zinc phosphate coating with good absorptivity.

Grum and Žerovnik [72] show in Figs. 39 and 40 differences in microhardness at a given measurement position for all kinds of steel, using absorber A or B after hardening with a laser beam traveling along a square spiral line.

Figure 39 shows the microhardness profiles subsequent to heating with an energy input of the laser source

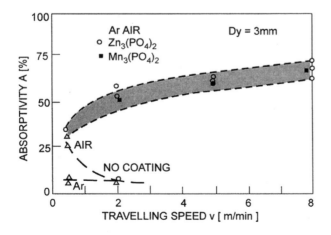

Figure 37 The absorptivity of specimens coated with zinc and manganese phosphates was measured in air and argon atmospheres. (From Ref. 71.)

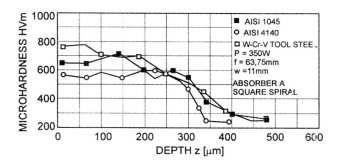

Figure 39 The microhardness profiles for given steels after using absorber A and a square-spiral laser path. (From Ref. 72.)

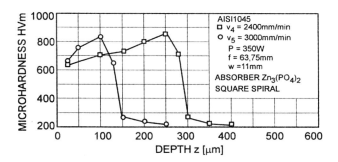

Figure 41 The microhardness for AISI-1045 after using a zinc phosphate absorber and a square-spiral laser path.

with a power of $P = 350$ W and workpiece traveling speed $v = 1000$ mm/min. The optical conditions are defined by laser-beam diameter ($d = 8$ mm) before the focusing lens, distance to the focus ($f = 63.5$ mm) and by the defocusing degree ($w = 11$ mm). The defocusing degree describes the distance of the lens focus from the workpiece surface. The microhardness was measured in the middle of the hardened track and into the depth until the hardness of the base material is reached. As expected the highest hardness profile is displayed by W–Cr–V tool steel with the highest proportion of carbon and alloying elements having a positive effect on hardenability or hardness. The lowest hardness on the surface, ranging between 500 and 600 HVm, was achieved on AISI 4142 steel with the shallowest hardened track. This surprising finding points to the disturbances in heating, which may have been caused by the laser source or the poor quality of the absorber deposit. The steels were chosen carefully to permit the study of the effects of alloying elements in terms of their kind and amount. According to heat treatment instructions, the chosen steels have equal hardening temperatures; therefore the effects due to

these differences might be excluded. Heat conductivity might have exerted a certain influence on the surface as well as on the differences in the depth and width of the hardened track.

Figure 41 shows the microhardness profile for different hardened track depths on AISI 1045 steel using the zinc phosphate absorber, with the beam traveling via a square spiraling path. The workpiece traveling speeds were considerably higher than when absorber A or B was used. These tests have shown that the workpiece traveling speed when using zinc phosphate was up to three times higher than those with absorber A or B.

Gutu et al. [73] studied transformation hardening of large gears made of alloyed 34MoCrNi15 steel and two heat-treatment steels, OLC45 and OLC60, respectively. The absorbents used were carbon black and colloidal graphite. The effects of the absorbents used were monitored by measuring the depth and the through-depth hardness of the relevant hardened layers.

Ursu et al. [74] studied the influence of the irradiation of a copper surface by CO_2 laser light. During irradiation, the temperature of the specimen was measured with a thermocouple positioned at the back side of the specimen. The irradiation test was performed twice on the same specimen. Prior to and after the irradiation of the same specimen, the absorptivity, A_0 and A_1, was determined and analyses of the surface with transmission- (TEM) and scanning electron microscopes (SEM) were made. Some zones at the specimen were chosen for crystallographic analysis by x-ray diffraction as well.

X. LASER-BEAM HANDLING TECHNIQUES

At laser transformation hardening the laser beam passes over the surface so fast that melting cannot occur [75]. Traveling speed or scan speed is one of the important

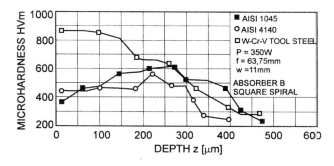

Figure 40 The microhardness profiles for steels after using absorber B and a square-spiral laser path. (From Ref. 72.)

laser transformation hardening parameters. There are two scanning methods:

- Linear traverse using a defocused beam and
- Transversely oscillating a focused beam and moving to the direction of traverse.

When the laser beam is defocused to a spot dimension according to the surface of the workpiece, a compromise between hardened depth and traveling speed at surface hardening has to be achieved [75]. Figure 42A shows the general shapes of the cross section of the hardened tracks produced in the metal by this method. The center of the hardened track represents the deepest area at the cross section for laser beam with Gaussian energy distribution. To increase the width of the heat-treated single track, multiple overlapping tracks may be used. However, as shown in Fig. 42B this method still does not give uniform hardened depth and interface soft streaks named heat-affected zone (HAZ) may develop at the overlap region [75]. Figure 42C shows the desirable heat-treated shape [75]. The hardened depth is uniform where the surface width

represents the width of the hardened track and the HAZ is comparatively small.

The method of oscillating a laser beam requires more hardware, the added expense, and the increase in the number of process parameters. Figure 43 illustrates the general principle of the laser-beam oscillating method [75]. In the simplest case, the beam is oscillated transverse to the direction of part or beam traverse, as shown in the right side of the figure. The temperature time diagram is the combination of the heating and cooling cycle as superimposed on transformation curve. The laser beam is focused to produce power densities of at least 10^{+5} W/in.^{2}. This power density assures high heating rate or fast temperature rise times. The oscillating frequency and traveling speed are selected so that the temperature of a given laser-beam spot on the surface varies between the melting temperature (T_m) and the austenitic transformation temperature (T_{A_3}). This condition is allowed to exist until the required volume of metal below the surface reaches the transformation temperature. At transformation temperature the bulk of the material provides the necessary heat sinking to

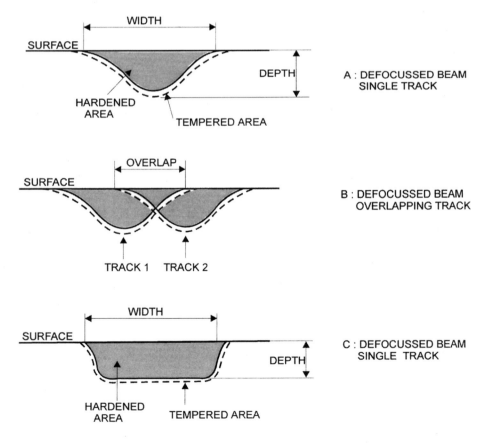

Figure 42 Laser heat treating patterns. (From Ref. 75.)

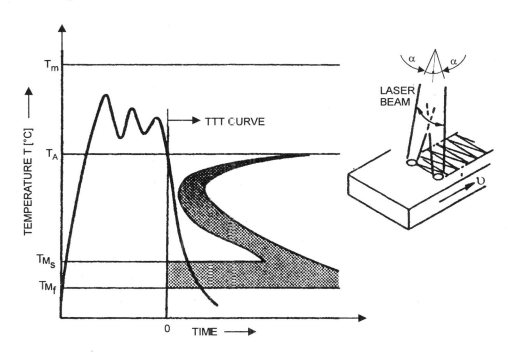

Figure 43 General concept of laser heat treating by the use of oscillating laser beam. (From Ref. 75.)

transform austenite to a hardened matrix of martensite. Uniform hardened depth is achieved when the beam oscillates so fast that the metal, because of its relatively slow conductivity, sees the oscillating tracks as a solid line of energy. Appropriate overlapping of the single tracks will produce a minimum of overtempered zone in the hardened depth, as shown in Figure 42C [75].

If the oscillating frequency is too slow at transformation surface hardening for a given laser power density, the surface temperature will be driven into the melting region [75]. If the power density is too low for a given frequency, the material will not be heated up to its transformation temperature and no hardening will be produced. If the beam is allowed to oscillate too long in a given area in an attempt to increase hardened depth, the temperature in the surface layer will be raised to the point where efficient self-quenching no longer takes place. The cooling curve will then cross the "nose" of the transformation curve resulting in the transformation of austenite into pearlite (annealed, soft microstructure) rather than martensite.

XI. PRESENTATION OF CO_2 LASER MACHINING SYSTEMS

Considering numerous practical applications of laser in electrical, metal working, and chemical industry, we can see that laser technology offers several advantages over the previously known technologies. Let us just mention one of the special plus factors which is contactless machining where there is no tool in the conventional sense but just the laser beam that never wears off. Thus frequent problems caused by tool wear such as workpiece deformation, surface layer damage, and damage of vital machine tool elements have all been avoided. However, there are other requirements that have to be considered in laser machining, as for example the quality of the laser source, modes of guidance of the workpiece or laser beam. All this requires interdisciplinary research and development work if we want to meet the needs of end users. The high investments required for machining systems have to be justified by high-quality products, increased productivity and possibilities of integration into manufacturing cells. The need to integrate laser machining systems into a manufacturing cell requires further development of mechatronic and electronic systems which would, together with computer and systems engineering, enable a more intensive introduction of this technology into the factories of the future. Considering the difficulties encountered in microdrilling, cutting, welding, and heat treatment, laser technology is the only technology that can, in the near future, as part of an integrated system, provide all sorts of manufacturing possibilities: from the blank to the end product. Laser technology will no doubt play an important role in the development of the factory of the future. This has already been recognized by a number of leading producers, e.g., in car industry, where it is impossible to imagine the assembly of car body parts without laser cutting and welding [76,77].

Up to now the development of laser technology has dealt with the following technological innovations:

- Guidance of the laser beam to different working locations in the space;
- Guidance of the laser beam over larger distances;
- Possibility of dividing the laser beam into several working places;
- Timing of the laser beam;
- Kinematic conditions within the workpiece/laser-beam relation.

These technological innovations offer extreme possibilities of adaptation and flexibility and allow machining of geometrical elements that previously could not be machined by conventional technologies and processes. Among the machining advantages we can mention:

- Possibility of welding thin parts;
- Possibility of drilling small holes;
- Adaptation of heat conditions to parts geometry in heat treatment.

In addition to that, the high flexibility of laser technologies is supported also by high machining capabilities of the laser beam on a variety of materials with very different properties, i.e.:

- Possibilities of machining plastics;
- Possibilities of machining ceramics;
- Possibilities of machining metal parts;

- Possibilities of machining different composite materials.

Figure 44 shows the main components of a CO_2 laser machining system manufactured by Toshiba, type LAC 554 with a power of 3 kW [76]. By guiding the laser beam and by guiding the workpiece in x–y plane we can achieve a great variety of machining by cutting and produce very different geometric elements. The maximum length of the laser beam is 18 m and is led through an optic system of prisms and mirrors along the longitudinal and traverse direction. A given configuration of optic and kinematic systems allows machining to go on only on one place. Previous laser machining systems only allowed the possibility of moving the operating table or the workpiece, which also required very large working areas, higher investment, and resulted in lower product quality. Technologies based on "time sharing method" were also known allowing the possibility of guiding the laser beam to different working places where activities are going on (Fig. 45) [77].

Figure 46 shows the production line equipped with transport belts supplying parts with a device for prepositioning and positioning within the reach of the laser beam [78]. This is followed by removal, deposition, and further transport on the belt until the next station. The whole equipment enables continuous running of the machining process. In Fig. 46, we can see a laser machining system illustrating a special aspect of guidance of the laser beam through a prism to the working place. The working place rotates, in this way enabling that the work is done in sequences, e.g., inserting, laser machining, cooling, taking off, and deposition on the conveyer belt. To achieve a higher utilization rate of the

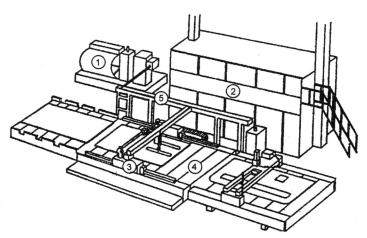

LASER TYPE LAC554

1 LASER OSCILLATOR

2 SYSTEM MONITORING ROOM
   LASER CONTROL BOARD
   NC EQUIPMENT FOR LASER
   SYSTEM CONSOLE

3 BEAM SCANNER

4 WORK TRANSPORTING CONVEYOR

5 BEAM TRANSMISSION ROUTE (18m)

Figure 44 A Toshiba laser machining system with a laser source power of 3 kW. (From Ref. 76.)

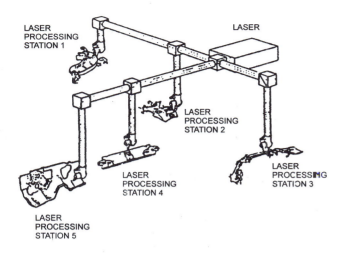

Figure 45 Guiding the laser light to five working positions. (From Ref. 77.)

laser system and increase of productivity, it is possible to introduce another working place by redirecting the laser beam.

Another technological change is introducing robots for guiding the laser beam. New types of devices for dividing/guiding the laser beam have been developed known as "Robolasers." Example 1 in Fig. 47 illustrates a robot for feeding the workpiece/product under a fixed laser beam [78]. In order to achieve a constant focal distance from the workpiece surface, the laser system has to have a suitable degree of freedom with a corresponding software provided by mechatronic and electronic systems. Example 2 (Fig. 47) shows a robotized system which separately guides the laser beam and the nozzle for guiding the assistant gas. The third example, by using a special design, ensures a coordinate motion of the laser beam and auxiliary gas to the working place.

In Fig. 48 we can see a system for guiding the laser beam, manufactured by Spectra-Physics, having the base plate fixed at the exit of the system/laser beam [78]. The laser beam travels through the optic shoulder to the optic joint and further to the focusing lens and goes out through the symmetric nozzle. On the upper side of the robot there is a gas tube through which the assistant gas is led from the gas station. The whole system is highly upgraded and flexible, enabling a very successful machining on very complex products as well as on large products. Because it is so highly flexible the system got the name Laserflex.

Figure 49 shows a laser machining system enabling a line-surface hardening of camshafts for the car industry [79]. Because of problems with absorptivity, the laser

system was added a place for induction heating or preheating of parts on to the temperature of 400°C. This is followed by heating up with the laser beam to the hardening temperature. After hardening follows the cooling down of parts in the cooling chamber and their deposition on the conveyer belt leading to the next working place.

(1)

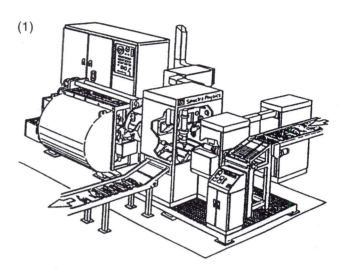

(2)

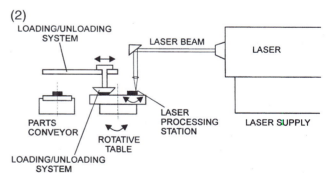

(3)

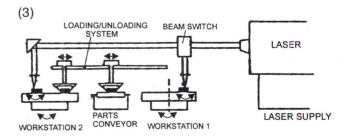

Figure 46 (1) Laser machining system with components for guiding the laser beam and the workpiece. (2) Workstation with loading/unloading system and laser processing station. (3) Laser machining system and parts handling/feeding double workstation. (Spectra Physics; From Ref. 78.)

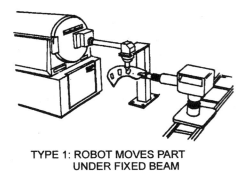

TYPE 1: ROBOT MOVES PART
UNDER FIXED BEAM

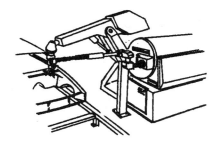

TYPE 2: ROBOT MOVES BEAM
DELIVERY SYSTEM

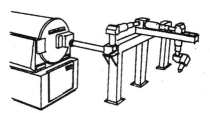

TYPE 3: ROBOT HAS BUILT-IN
BEAM DELIVERY SYSTEM

Figure 47 Three types of robolasers (laser + robot). (From Ref. 78.)

Figure 50 shows a line-beam of laser light, obtained from a system of optic elements such as a prism and a spherical, concave mirror [79]. A convex mirror ensures a constant distance of the optic system focus from the object surface/camshaft. The entire operation of surface heat treatment takes 2 to 5 sec for one camshaft assembly and at the maximum 29 sec for one camshaft depending on the size and kind of the latter.

A most important place in building laser machining systems with computer support is now given to mechatronic and electronic systems as the quality of the whole system depends on them [80]. Modern highly flexible laser technology represents a universal tool suitable for small and large products of simple or complex shape. Another fact which strongly favors this new technology is the possibility of guiding the laser

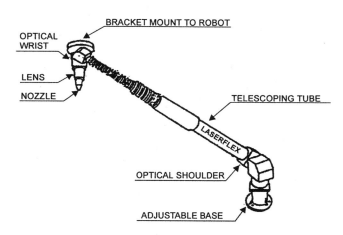

Figure 48 Laserflex for guiding the laser beam to the working place. (Spectra Physics; From Ref. 78.)

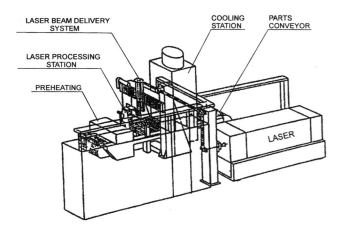

Figure 49 Laser system for line-surface hardening of camshafts. (From Ref. 79.)

beam to different, even very distant working places. Finally, using the same tool but in different working conditions it is possible to carry out different heat processes from heat treatment to welding. Thus we can establish that the advent of laser technology will finally solve the problems of conventional heat treatment and welding and will ease the transition to the manufacturing systems structure necessary for the factory of the future [81–83].

A. Possibilities of Kaleidoscope Use for Low-Power Lasers

Laser surface hardening is desirable to utilize the available energy distribution for heating and microstructural changes; however, the Gaussian distribution with large differences in energy distribution across the laser beam does not provide this possibility. In practice, therefore, we tend to use other laser-beam energy distributions, namely, the square top hat and rectangular. A very simple device, kaleidoscope, is therefore suitable for practical use. A kaleidoscope is a device in the form of a chimney that is a hollow body with a square or rectangular cross section [84,85]. It is necessary for the beam to reflect at least two or three times along the length of the kaleidoscope. To achieve this high reflectivity a material of high heat conductance and mirror-smoothed material should be used. Reflectivity depends on the wave-

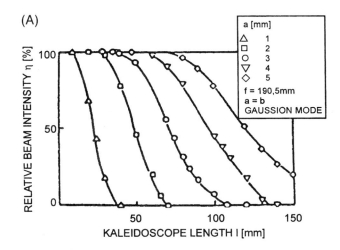

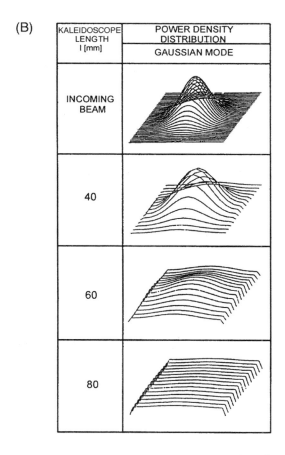

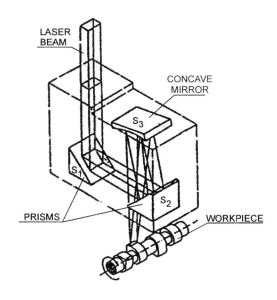

Figure 50 Optic system for redirecting the laser beam to the working place. (From Ref. 79.)

Figure 51 (A) Influence of kaleidoscope length and cross section on the relative change in beam intensity. (B) Power density distribution of laser beam versus kaleidoscope length. (From Ref. 84.)

length of the laser light and the material from which the kaleidoscope is made [86].

To assess the uniformity of beam intensity, the relative change in beam power density is considered irrespective of the effect of interference. Figure 51A shows the influence of kaleidoscope length and cross section on the relative change in beam intensity [84]. Figure 51B [84] shows the uniformity of beam power density distribution for different kaleidoscope lengths.

It is evident from the figure that with the increase in kaleidoscope length l and decreasing cross-dimensions "axa" or "axb," the uniformity of the beam increases. In the case of the rectangular cross section, it then follows that the kaleidoscope length must be chosen according to the magnitude of the longer side, which means larger length of the kaleidoscope than in the case of the square cross section. The size of the kaleidoscope depends also on the focal length of the lens at its entrance side. Figure 52 shows the dependence of focal length of the entrance lens and the kaleidoscope dimensions l/a on the uniformity of the laser-beam power density at the exit side. From the focal length of the entrance lens, the kaleidoscope dimensions can therefore be determined and this also ensures the highest uniformity of the beam power density at the exit side.

In the experiments, a longer than the minimum length necessary for the square cross section was chosen in order to enable the use of the same kaleidoscope for both types of cross section [87]. A kaleidoscope with a length of 75 mm was applied, which is more than the recommended length of the rectangular cross section

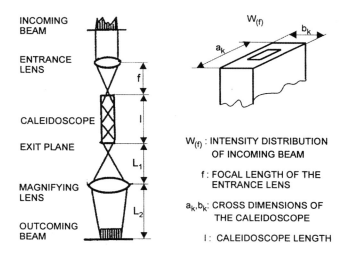

Figure 53 Kaleidoscope structure. (From Ref. 84.)

(Fig. 53). This proved disadvantageous later as the interference raster was not distinct. The too long kaleidoscope for the square cross section of 3×3 mm^2 enabled an interference of most of the elementary laser light beams despite the fact that the optical axes of the laser system and the kaleidoscope may not have fitted due to the assembly error.

On the upper side of the kaleidoscope with a cross section of 4×2 mm^2, an opening was made for the laser light to enable the best possible interference of elementary laser beams. The kaleidoscope consists of perpendicular copper walls, polished to a high gloss, thus enabling the creation of mirroring surfaces reflecting laser light. The walls are fastened to one another by bolts, which enables dismounting of the kaleidoscope and repeated polishing.

In the experimental part of the research, Grum [87] tried to choose such laser surface hardening conditions that may have shown the advantages of using a kaleidoscope in laser heat treatment with a low-power source. The main feature of the kaleidoscope is that it considerably lowers the traveling of the workpiece, still granting the desired uniformity of heating and also uniformity of hardening depth.

Figure 54 presents the hardened tracks and the measured microhardness on heat-treatable carbon steel AISI 1042. The hardening was carried out with the aid of a kaleidoscope with a defocusing degree defined by the size of the bright spot on the workpiece surface, which was 1.65 mm. The workpiece traveling speeds changed in the same way as those in the treatment of tool steel, i.e., 0.10, 0.12, and 0.14 m/min [87].

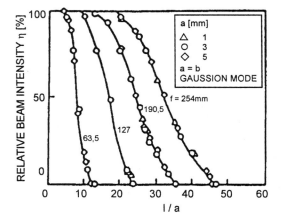

Figure 52 Influence of local length of entrance lens on the relative change in beam power density. (From Ref. 84.)

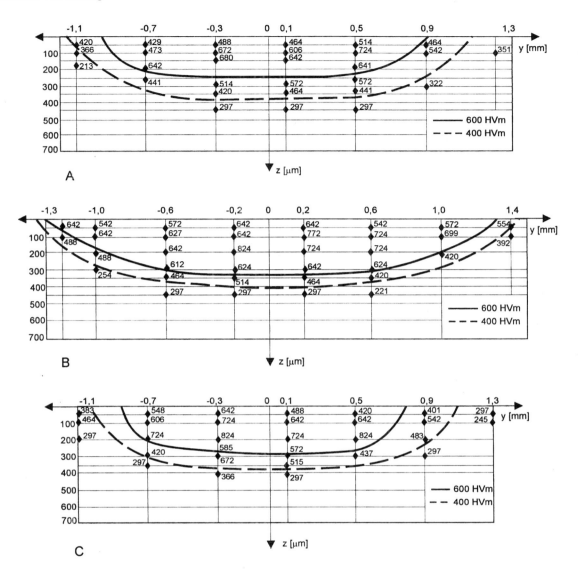

Figure 54 Measured hardnesses in laser surface transformation hardening 1042 AISI (0.46% C, 0.65% Mn) steel, using a kaleidoscope, traveling speed $v = 0.10 \, \text{m/min}$ (A), $v = 0.12 \, \text{m/min}$ (B), $v = 0.14 \, \text{m/min}$ (C), and beam dimension $a = b = 1.65 \, \text{mm}$. (From Ref. 87.)

XII. RESIDUAL STRESS AFTER LASER SURFACE HARDENING

From the technology viewpoint, various surface hardening processes are very much alike as they all have to ensure adequate energy input and the case depth required. In the same manner, regardless of the hardening process applied, in the same steel, the same microstructure changes, very similar microhardness variations,

and similar variations of residual stresses within the hardened surface layer may be achieved.

Normally, the surface transformation hardening process also introduces compressive stresses into the surface layers, leading to an improvement in fatigue properties. Hardened parts—always ground because of their high hardness—require a minimum level of final grinding. This can only be achieved by a minimum oversize of a machine part after hardening, thus shortening the final grinding time and reducing cost of the final grinding to a minimum. With an automated manufacturing cell, one should be very careful when selecting individual machining processes as well as machining conditions related to them. Ensurance of required internal stresses in a workpiece during individual machining processes should be a basic criterion of such a selection. In those cases when the internal stresses in the workpiece during the machining process exceed the yield stress, the operation results in workpiece distortion and residual stresses. The workpiece distortion, in turn, results in more aggressive removal of the material by grinding as well as a longer grinding time, higher machining costs, and a less-controlled residual-stress condition. The workpiece distortion may be reduced by subsequent straightening, i.e., by material plasticizing, which, however, requires an additional technological operation, including appropriate machines. This solution is thus suited only to exceptional cases when a particular machining process produces the workpiece distortion regardless of the machining conditions. In such cases the sole solution seems to be a change of shape and product dimensions so that material plasticizing in the machining process may be prevented.

It is characteristic of laser surface hardening that machine parts show comparatively high compressive surface residual stresses due to a lower density of the martensitic surface layer. The compressive stresses in the surface layer act as a prestress which increases the load capacity of the machine part and prevents crack formation or propagation at the surface. The machine parts treated in this way are suited for the most exacting thermomechanical loads as their susceptibility to material fatigue is considerably lower. Consequently, much longer operation life of the parts can be expected.

Residual stresses are the stresses present in a material or machine part when there is no external force and/or external moment acting upon it. The residual stresses in metallic machine parts have attracted the attention of technicians and engineers only after manufacturing processes improved to the level at which the accuracy of manufacture exceeded the size of deformation, i.e., distortion, of a machine part.

The surface and subsurface layer conditions of the most exacting machine parts, however, are monitored increasingly by means of the so-called surface integrity. This is a scientific discipline providing an integral assessment of the surface and surface layers and defined at the beginning of the 1960s. More detailed information on the levels of surface integrity description may be obtained [88,89].

XIII. THERMAL AND TRANSFORMATION RESIDUAL STRESSES

Yang and Na [90] proposed the model for determination of thermal and transformation residual stresses after laser surface transformation hardening. Thermal stresses were induced due to thermal gradient and martensitic phase transformation causing prevailing residual stresses. The dimensions of the specimen for simulation were 6 mm in length, 2 mm in thickness, and the width of the sliced domain was 0.2 mm. To carry out heat treatment laser power $P = 1.0$ kW, traveling speed $v = 25$ mm/sec, absorptivity $A = 0.15$, and characteristic radius of heat flux $r = 1.0$ mm were chosen. At rapid heating austenitic transformation temperature started at $830\,°C$ and ended at $950\,°C$. Cooling rate was also high enough in the heated material and austenite transformed to martensite in the temperature range from $360\,°C$ to $140\,°C$. In the simulation the volume changes at phase transformation base microstructure to austenite at heating ($\alpha_{F+P \to A} = 2.8 \times 10^3$) and then also volume changes at phase transformation of austenite to martensite at cooling ($\alpha_{A \to M} = 8.5 \times 10^{-3}$) were considered. Figure 55 shows longitudinal stress profiles in the direction of the laser-beam travel at various time moment at heating, namely, at $t_1 = 1.35 \times 10^{-2}$ sec, $t_2 = 2.85 \times 10^{-2}$ sec, and $t_3 = 4.65 \times 10^2$ sec [90]. From the graph it can be seen that the compressive stresses are in immediate vicinity of the laser beam. Nevertheless, the thermal stresses at heating change to tensile stresses according to the distance from the laser beam.

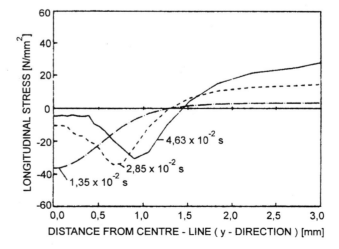

Figure 55 Longitudinal stress distributions at the surface during laser heating. (From Ref. 90.)

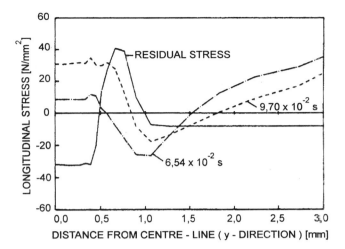

Figure 56 Longitudinal stress distributions and residual stress distributions at the surface during cooling in laser surface hardening. (From Ref. 90.)

As the laser provides more energy according to the heating time, the temperature increases, so that the compressive thermal stresses become progressively lower in the vicinity of the laser-beam center.

Figure 56 shows longitudinal stress profiles in the direction of the laser-beam travel at various cooling times during laser surface hardening [90]. The residual stress in laser surface hardening is induced from temperature gradient and phase transformations. During cooling process longitudinal stresses are always tensile and change to compressive during phase transformation. This means that longitudinal residual stress is always of compressive nature. As shown in Fig. 56 the compressive residual stress was generated in the hardened depth within 0.5 mm from the center line of the exposed laser beam.

XIV. MATHEMATICAL MODEL FOR CALCULATING RESIDUAL STRESSES

Li and Easterling [91] developed a simple analytical model for calculating the residual stresses at laser transformation hardening. Following the temperature cycle of the laser beam travel, a certain volume of the heat-affected material expands as a result of the martensitic transformation. The created martensite microstructure has higher specific volume than matrix microstructure, which causes residual stresses in the surface layer.

The magnitude of this residual stress is calculated as a function of the laser input energy and the carbon content of the steel.

To simplify the calculations, the following assumptions can be made:

- Stresses induced by thermal expansion near the surface can be neglected.
- Plastic strains caused by the martensitic transformation can be considered negligible.
- Dilatation in the HAZ is completely restrained along the x and y directions during laser transformation hardening.

From the author's equations, different factors and parameters are influential on residual stress calculations, despite the simplifying assumptions made. These factors include:

- The dimensions of the specimen,
- The properties of the material and composition,
- The laser processing variables.
- The residual stress σ_{xx} should equal

$$\sigma_{xx} = -\frac{\beta f E}{1-v} + \frac{1}{bc}\int_0^c dz \int_{-b/2}^{+b/2} \frac{\beta f E}{1-v} dy$$
$$+ \frac{12(0.5c-z)}{bc^3}\int_0^c dz \int_{-b/2}^{+b/2} \frac{\beta f E}{1-v}(0.5c-z)dy.$$

This equation is a complete description of the residual stress in the x direction, where the first term is negative and decreases with depth below the surface, and the second and third terms are both positive. It shows that these terms resulted in a compressive residual stress at the surface and changed to a tensile residual stress at some depth below the surface.

The y component of the residual stress σ_{yy} can be expressed as

$$\sigma_{yy} = -\frac{\beta f E}{1-v}.$$

At several laser runs overlap in the y direction there is some stress relief in the overlapping zone. The maximum σ_{yy} given by the equation is still valid.

Assuming that expansion in the z direction proceeds freely, the z component of residual stress σ_{zz} is zero

$$\sigma_{zz} = 0.$$

It is impossible to change the volume because of phase transformation, therefore shearing strains do not occur, and the shear components of residual stresses are zero

$$\sigma_{xy} = \sigma_{yz} = \sigma_{zx} = 0.$$

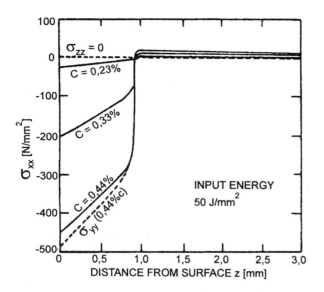

Figure 57 Predicted residual stress distribution σ_{xx} as a function of depth below surface and steel carbon content C. (From Ref. 91.)

The given equations present a simplified way of calculating residual stresses at laser transformation hardening.

Figure 57 shows the effect of carbon content on σ_{xx} through depth residual stress profile [91].

It can be seen that the value of the compressive residual stress is very dependent upon carbon content, increasing by about 200 N/mm² for each 0.1% C. The

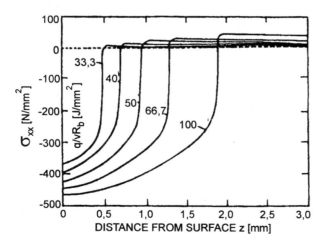

Figure 58 Predicted residual stress σ_{xx} as a function of depth below surface and laser input energy density ($q/v(R_b)$) for 0.44% C steel. (From Ref. 91.)

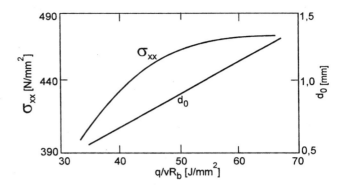

Figure 59 Predictions of surface compressive residual stress σ_{xx} and depth d_o as a function of laser input energy density for 0.44% C steel. (From Ref. 91.)

crossover from compressive to tensile residual stress in the x direction occurs approximately at the depth where the steel has been heated to the T_{A_1} temperature. The figure shows also the profile for σ_{yy} in the steel with 0.44% C. As can be seen σ_{yy} lies at slightly higher compressive stresses than the corresponding profile for the x direction. The effect of changing the input energy density of the laser process on residual stress profile σ_{xx} is shown in Fig. 58. It can be seen that, for the 0.44% C steel, changes in laser input energy do not substantially affect the value of the compressive residual stress but do affect the depth d_o of the compressive layer. Figure 59 shows changes in compressive residual stress σ_{xx} and depth of compression layer d_0 as a function of laser input energy density. In practical laser transformation hardening, a surface may be hardened not by a single laser track but by a number of overlapping tracks. This obviously results in local stress relaxation in the y directions in the overlapping zones, although the residual stresses calculated at the center of each laser track will not be affected and remain as presented above. If the overlapping tracks cover the entire surface, end effects, as discussed in conjunction with σ_{xx}, should be included in the calculation of σ_{yy} for completeness.

XV. INFLUENCE LASER SURFACE TRANSFORMATION HARDENING PARAMETERS ON RESIDUAL STRESSES

Solina [92] presented residual stress distribution below the surface after laser surface transformation hardening at various heating and cooling conditions of different steels.

For experimental laser surface transformation hardening they used a CO_2 laser. The absorptivity of the material for the laser light with a wavelength of $\lambda = 10.6$ μm was increased by an absorption coating of Zn-phosphate. In the tests two types of steel were used, i.e., carbon steel AISI 1045 and a low-alloy steel AISI 4140, both in normalized state. The specimen was self-cooled (mode 1) and quenched with liquid jet (mode 2).

The ranges for laser hardening parameters were 10 to 38 W/mm^2 sec for the energy density and from 0.024 to 1.12 sec for the beam–metal interaction time. From residual stress profiles in the surface layer, it is found that the residual stresses in the more superficial layers may vary, with changing laser hardening conditions, from compressive residual stress values $+200$ MPa to strong tensile residual stress values up to $+800$ MPa.

Figure 60 shows the residual stress profiles as a function of the depth below the surface in two specimens of carbon steel AISI 1045 (C43) and AISI 4140

(40CoMo 4) alloyed steel [92]. The specimens were heated with the same power density $Q = 27.5$ W/mm^2 and at interaction time $t_i = 0.56$ sec at beam spot area 14×14 mm^2 and self-quenched (mode 1). The residual stress profiles for both steels show the same laser surface hardening results in very similar changing residual stresses from the hardened layer. Maximum values of the compressive residual stress are about -200 N/mm^2, and very similarly residual stress droops to neutral plane with small shift for alloyed steel.

Figure 61 shows residual stress profiles through the surface layer after laser surface hardening of alloyed steel AISI 4140 at very high power density $Q_1 = 121.9$ W/mm^2, at interaction time $t_{i1} = 0.16$ sec and $Q_2 = 418.7$ W/mm^2, at interaction time $t_{i2} = 0.024$ sec, with cooling in liquid jet (cooling mode 2) [92]. In both laser surface conditions very drastic tensile residual stress profiles are obtained. Residual stress profiles differ only in absolute values. At laser surface hardening conditions for both steels tensile residual stresses were found on the surface layer from $+200$ to $+450$ N/mm^2.

Com-Nogue and Kerrand [93] analyzed the through-depth variations of residual stresses and microhardness of a laser-hardened layer on S2 chromium steel. Heating was carried out by a laser beam with a power P of 2 kW and a traveling speed v of 7 mm/sec. A semielliptical power density distribution in the laser beam with defocusing was chosen so that a constant depth of the hardened layer d, i.e., 0.57 mm, was obtained with a width W of 10.8 mm. Figure 62 shows the through-depth variations of the measured hardness and experimentally determined residual stress in the hardened layer. The variations of both hardness and residual stresses in the surface hardened layer depended on the microstructure of the latter. Microstructural changes can be affected by the heating conditions which, in laser heating, can be determined by the selection of the laser power and the area of the laser-beam spot diameter d_B or A_B with a given mode structure. The traveling speed of the laser beam was then suitably adapted. The measured average surface hardness was 725 HV. It slowly decreased to around 300 HV, i.e., hardness of steel in a soft state, in a depth d of 0.57 mm.

The variation of microhardness permitted an estimation that heating was correctly chosen to achieve the maximum surface hardness and a very distinct width of area with the transition microstructure (a mixture of hardened and base microstructure).

The through-depth variation of residual stresses of the hardened layer depended on the residual stresses occurring in the material prior to heating. Then followed relaxation of the existing stresses due to laser

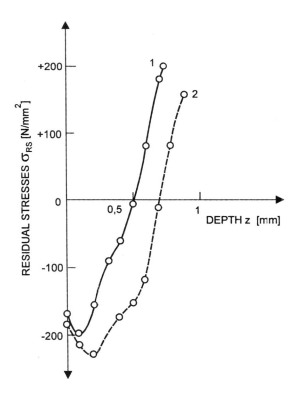

Figure 60 Residual stress distribution below the surface for given specimens: 1: C43, $P = 27.5$ W/mm^2, $t_i = 0.56$ sec, cooling mode 1. 2: 40CoMo4, $P = 27.5$ W/mm^2, $t_i = 0.56$ sec, cooling mode 1. (From Ref. 92.)

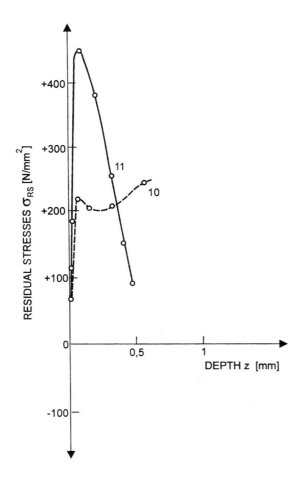

Figure 61 Residual stress distribution below the surface for given specimens: 10: 40CoMo4, $P = 121.9$ W/mm^2, $t_i = 0.16$ sec, cooling mode 2. 11: 40CoMo4, $P = 418.7$ W/mm^2, $t_i = 0.024$ sec, cooling mode 2. (From Ref. 92.)

sign in a depth of 0.57 mm and became the maximum tensile stress in a depth of 0.9 mm. In the range between the maximum compressive and compressive tensile residual stresses, i.e., between the depths of 0.57 and 0.9 mm, a linear variation of the residual stresses occurred.

Mor [94] studied measurement of residual stresses in laser surface hardened specimens made of AISI 4140 steel using the x-ray diffraction method. Figure 63 shows the results of the measurements performed with two power densities Q, i.e., 10 and 14 W/cm^2, and three respective times of interaction t_i, i.e., 2.0, 1.8, and 1.0 sec. The residual stresses were measured in the direction perpendicular to the hardened traces. Microhardness of the base material, i.e., 200 HVm, increased up to around 700 HVm in the hardened trace and reduced to around 400 HVm in the zone of overlapping of two adjacent traces. The residual stresses in the hardened traces showed the compressive character due to an increased specific volume of the martensitic microstructure formed, but at the edge of the hardened traces their character gradually turned into the tensile one. Tensile residual stresses occurred also in the case of two parallel laser-hardened traces. In the hardened traces the maximum compressive residual stresses amounted to -300 N/mm^2, whereas in the zone of overlapping of the hardened traces they varied between $+220$ and $+300$ N/mm^2. At the edge of the hardened traces the tensile residual stresses amounted on the average up to 1–0 N/mm^2. The variations of residual stresses obtained in the hardened traces were as expected. Surprising, however, is the variation of residual stresses having the tensile character in the zone of overlapping of two adjacent traces, which may be a critical zone with dynamically loaded parts.

heating for further hardening. Finally is the variation of the residual stresses after laser heating affected also by the initial microstructure and the cooling rate of steel. A higher austenitizing temperature and an ensured homogeneity of carbon in austenite and as long as possible heating of the material for the given depth of the hardened layer ensured efficient tempering of the preceding residual stresses and introduced new low-tensile stresses. Thus after laser surface hardening, comparatively high compressive residual stresses could be ensured. They could be confirmed by computations as well. The through-depth variation of residual stresses of the hardened layer was, as expected, coordinated with the variation of hardness. It was very important that the maximum residual stress occurred just below the surface and amounted to -530 N/mm^2. Then it changed the

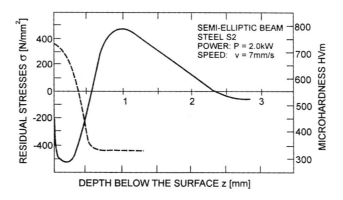

Figure 62 Residual stress and microhardness profiles through the laser surface hardened layer. (From Ref. 93.)

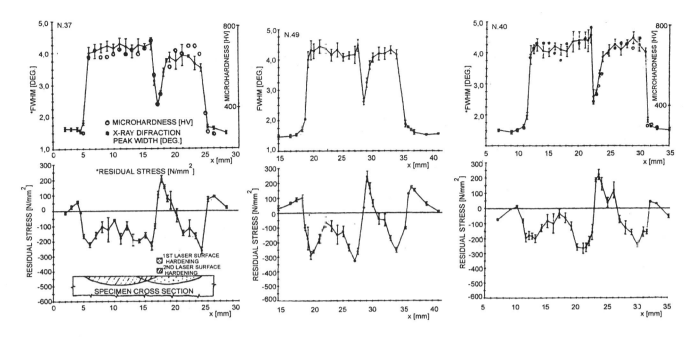

Figure 63 Microhardness, residual stress, and x-ray diffraction along the laser hardened steel AISI 4140 at the surface. (From Ref. 94.)

XVI. RESIDUAL STRESSES, MICROSTRUCTURES, AND MICROHARDNESSES

Ericsson et al. [95] studied residual stresses, the content of retained austenite, and the microstructure supported by microhardness measurements. The experiment consisted in making a laser-hardened trace on a cylinder with a diameter of 40 mm and a length of 100 mm. In the experiment AISI 4142 and AISI 52100 steels in quenched and tempered (320 HV) and fully annealed conditions (190 HV) were used. Laser surface hardening was carried out with a CO_2 laser with a power of 3 kW in continuous-wave mode. Several laser surface hardening parameters were chosen. The laser power, the laser-beam spot diameter, and the traveling speed of the workpiece, however, were varied. Figure 64 shows the results of calculations in laser surface hardening with reference to through-depth distribution of austenite in the heated layer after 24 sec and then the through-depth variations of martensite and residual stresses in the hardened layer after quenching. In the calculations of the austenite and martensite contents and the variation of residual stresses in surface hardened AISI 4142 steel with 55.2 mm in diameter, a power density Q of 6.5 MW/m^2, a traveling speed v of 0.152 m/min, and width of hardened trace W of 8.175 mm were taken into

account. After hardening, up to 35% of martensite was found in the surface layer to a depth of 0.5 mm. Then the martensite content was decreasing in a linear manner to a depth of 1.0 mm. Very similar was the through-depth variation of residual stresses of the compressive character with a maximum value of around $-150 \ N/mm^2$ in the axial and tangential directions. This is followed by a transition to tensile residual stresses in the unhardened layer amounting to $+350 \ N/mm^2$. This was followed by a transition to the central part of the specimen with a constant stress of around $-200 \ N/mm^2$. The radial residual stresses were almost all the time constant, i.e., $\sigma_r = 0$, from the surface to a depth z of 20 mm, then they increased to around $-200 \ N/mm^2$ in the center of the cylinder. An efficient indicator of the variation of residual stresses was the martensite content. The martensite content showed the variation of the residual stresses occurring in the axial and tangential directions, which means that the martensite transformation had a decisive role in the determination of the variation of residual stresses.

Cassino et al. [96] studied the variation of residual stresses after laser surface hardening of 55 CI heat-treatment carbon steel with a carbon content of 0.52%.

The investigations were conducted using flat specimens of $100 \times 80 \times 20$ mm in size and a square spot area of 5×15 mm. Heating was carried out with a laser power varying between 3.0 and 3.4 kW, and a traveling speed of the workpiece v of 5 mm/sec. The specimen surfaces were coated with absorbents such as graphite and manganese phosphate that ensured absorptivity in the range from 65% to 85%. The specimens were subjected to forced cooling with argon (A) and water (W). Residual stresses were measured in the longitudinal and transverse directions using the x-ray diffraction method.

X-ray of 2A wavelength from synchrotron radiation was employed to obtain diffraction peaks from (211)-lattice planes of martensite at nine different ψ angles (the angle between the normal to the diffracting planes and the normal to the surface of the specimen) in the range of −40.0° to 35.0°. The strains were measured in the regions corresponding to the center of the laser traces and along directions parallel and perpendicular to them.

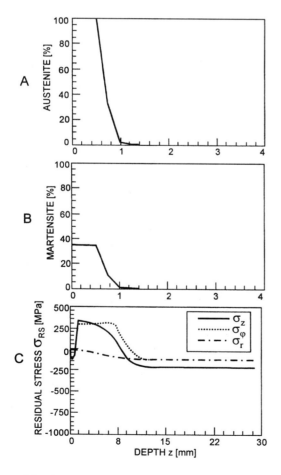

Figure 64 Austenite distribution after 24 sec (A), martensite distribution at the end of cooling (B), and σ_z, σ_φ, and σ_r residual stress profiles (C). (From Ref. 95.)

Table 3 Residual Stress Values Along Longitudinal and Traverse Directions

Specimen	$\sigma\,[\phi=0]$ MPa	$\sigma\,[\phi=\pi/2]$ MPa	Peak width [°]
A3	−610 ± 230	−230 ± 160	−2.1 ± 0.6
A4	−540 ± 370	−350 ± 100	−2.3 ± 0.3
W2	−200 ± 130	−100 ± 150	−2.0 ± 0.2
W4	−300 ± 160	−60 ± 20	−2.2 ± 0.2
W5	−440 ± 1200	−100 ± 120	−2.2 ± 0.3

Source: From Ref. 96.

The elastic constants $E = 0.21$ TPa and $v = 0.28$ were used for stress calculations.

Table 3 shows residual stress values along longitudinal and transverse direction. For quenching of the workpieces with argon (A_3, A_4), the longitudinal compressive residual stress ranged between 540 and 610 N/mm^2, whereas the transverse one ranged between 230 and 350 N/m^2. With forced quenching of the specimens with water (W_2, W_4, and W_5), the longitudinal compressive residual stress ranged between 200 and 400 N/mm^2 and the transverse one between 200 and 100 N/mm^2.

The dissipation of the measured longitudinal and transverse stresses was exceptionally strong and represented up to 50% of their measured value. The results of measurement of the residual stresses confirmed that in the given case a sufficient cooling rate was achieved by cooling with argon. It was necessary to use any more strong cooling medium.

XVII. SIMPLE METHOD FOR ASSESSMENT OF RESIDUAL STRESSES

Grevey et al. [97] proposed a simple method of assessment of the degree of residual stresses after laser hardening of the surface layer. The method proposed was based on the knowledge of the parameters of interaction between the laser beam and the workpiece material taking into account laser power and the traveling speed of the laser beam across the workpiece and the thermal conductivity of the material concerned. The authors of the method maintained that in the estimation of residual stresses with low-alloy and medium-alloy steels, the expected deviation of the actual variation from the calculated variation of residual stresses in the thin surface layer did not exceed 20%.

On the basis of the known and expected variation, the authors divided the residual-stress profiles through the workpiece depth into three areas as shown in Fig. 65 [97].

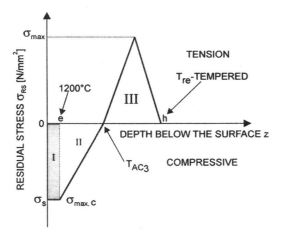

Figure 65 Residual stress profile, classified in typical areas. (From Ref. 97.)

- Areas I and II make up a zone from the surface to the limiting depth with compressive stresses defined in accordance with a TTA diagram (Orlisch diagram) for the given steel.
- Area III is the zone extending from the limiting depth, thus being an adjoining zone where the specific volume of the microstructure is smaller than that at the surface and, consequently, acts in the sense of relative contraction, which produces the occurrence of tensile stresses in this zone. The zone showing tensile residual stresses is strongly expressed in the specimens that were quenched and tempered at 400°C or 600°C and then laser surface hardened. Because of a thermomechanical effect occurring during surface heating and fast cooling, certain residual stresses persist in the material after cooling as well. This thermomechanical effect was related to the state of the material prior to laser surface hardening.

Area III is known as the retempered zone of the material at the transition between the laser-hardened surface layer and the base metal, which is in a quenched-and-tempered state in the given case. The thermal effects in laser heating occurring in the retempered zone were calculated by the authors from the variation of the temperature in the individual depths of the specimen on a half infinite solid plate:

$$T(z) = T_s \left[1 - \mathrm{erf}\left(\frac{z}{2\sqrt{t_i}} \right) \right]$$

where T_s [°C] is the surface temperature, $T(z)$ is the temperature at depth z, $T_s = \rho P_o a^{1/2} t_i^{1/2}/S\lambda$, ρ [%] is

total energy efficiency, P_o [W] is the average power required, S [cm^2] is the spot area of the beam, a [cm^2/sec] is the thermal diffusivity, t_i [sec] is the interaction time, and v [cm/sec] is the traveling speed.

An error function was approximated with (1-exp $(\sqrt{\pi u})$, which, with $0.2 < u < 2.0$, provided a favorable agreement between the theoretical and experimental results. The depth in which the transition from compressive residual stresses to tensile residual stresses occurred was calculated using the following equation:

$$z = -\frac{4}{\pi} \sqrt{at_i} \ln \frac{T(z)v^{1/2}\pi \cdot r_o^{3/2} \cdot \lambda}{\rho a^{1/2} \cdot P_o}$$

The mathematical description of the variation of thermal cycle permitted them to predict the depth in which the austenite transformation T_{A_3} occurred and the temperature range in which retempering occurred.

The difficulty of the theoretical method consists in the requirement for knowledge of three parameters: thermal conductivity λ [W cm^1 °C^{-1}], thermal diffusivity a [cm^2 sec^{-1}], and total energy efficiency ρ [%].

Concerning the first two parameters, they depend on the type of material used, on its initial microstructural state, and also on its temperature. They are linked by the relationship:

$$a = \lambda/d \cdot c_p$$

where $d = 7.8$ g cm^{-3} is the material density and c_p [J g^{-1} °C^{-1}] is the specific heat which depends on the temperature.

In the calculations only the effective energy was taken into account; therefore the efficiency of material heating and the global coefficient ρ [%], including optical losses along the hardened trace and the reflection of the laser beam from the specimen, were taken into account. Thus they treated the display power P_0 and calculated, by means of the global coefficient of efficiency ρ, the so-called effective power, $P_e = \rho \cdot P_0$. The authors experimentally verified the effective power and confirmed the linear relationship as shown in Figure 66A and B.

Figure 67 shows the results of calculations and measurements of longitudinal residual stresses. The deviations could be defined with reference to the depth of the transition of the compressive zone into the tensile zone ($\Delta h = h_{EXP} - h_{EST}$) and the deviation of the size of the maximum tensile stress in the subsurface ($\Delta\sigma_{maxT} = \sigma_{maxEXP} - \sigma_{maxEST} < 4\%$).

In the calculations the deviation between the experimental and estimated powers ($\Delta P = P_{EXP} - P_{EST} < 3\%$) was taken into account.

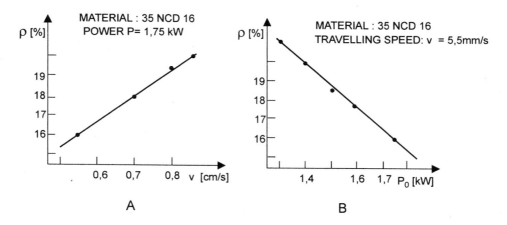

Figure 66 Variation of efficiency recording traveling speed (A) and laser beam power (B). (From Ref. 97.)

A very simple and practical method of determination of the through-depth variation of longitudinal residual stresses in the laser-hardened trace was proposed. The procedure was based on measurement of longitudinal residual stress at the surface σ_s and measurement of different characteristic depths by means of metallo-graphic photos or from the through-depth hardness profile, which are defined by e, p, and h, i.e.:

$$\sigma_m = -\frac{p + e}{h - p}\sigma_s.$$

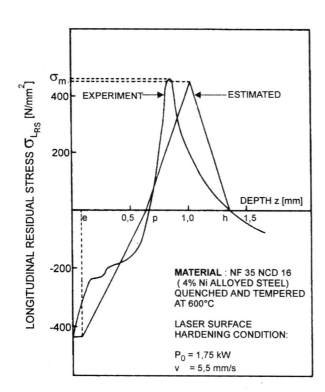

Figure 67 Comparison of the experimental and estimated longitudinal residual stress profiles. (From Ref. 97.)

XVIII. INFLUENCE OF PRIOR MATERIAL HEAT TREATMENT ON LASER SURFACE HARDENING

Chabrol and Vannes [98] analyzed the variations of hardness and residual stresses in the thin surface layer of alloyed heat-treated steels, i.e., steel A (Cr – Ni), steel B (Cr – Mo), steel C (Cr – Ni – Mo), and micro-alloyed steel D (V). All steels were quenched and tempered at temperatures of 400°C and 500°C. First measurements of hardness and residual stresses were made at a single trace. The specimens showed a variation of Vickers hardness $HV_{0.5}$ of steel B that was annealed and then laser transformation hardened in the first case, and quenched and tempered at 400°C and 500°C and then laser surface hardened in the second case. They represent dissipation of hardness, which varied with different specimens between 610 and 750 $HV_{0.5}$, after laser surface hardening. The initial state of steel was soft.

In the second case steel was quenched and tempered and then laser surface treated, which produced a considerable change of variation of hardness in the surface layer. Dissipation of hardness in the thin surface layer was reduced too to vary between 640 and 695 $HV_{0.5}$. The depth of the hardened layer thus increased by approximately 30%. The variation of hardness at the transition between the hardened layer and the quenched-and-tempered base metal where softening, i.e., quenching and tempering, of the material occurred was interesting as well.

This study based the identification of residual stresses on the relaxation method, which consists of measuring specimen strain. The relaxation was induced by electrochemical removal of the stresses surface layer, causing a breakdown in the existing equilibrium state. The restoration of the equilibrium is accompanied by specimen strains. The strains were measured by means of resistance strain gauges and calculated into residual stresses using a mathematical model.

Figure 68 shows the variations of the longitudinal and tangential residual stresses in the hardened trace on steel A that was preliminarily annealed. The two variations are very similar, i.e.:

- Compressive residual stresses of 330 N/mm2 to a depth of 0.15 mm.
- Then follows an almost linear variation of the residual stresses up to +200 N/mm2 in a depth of 1.5 mm including a transition from the compressive zone to the tensile one occurring in a depth of 1.10 to 1.20 mm.
- In the depths exceeding 1.5 mm the two stresses are of the tensile character and almost constant.

The variation of residual stresses in multipass laser surface hardening was very similar to the one obtained with a single trace. The variations of the longitudinal and tangential residual stresses were very similar as well.

The measurement of residual stresses in the x-ray diffraction method showed surface stresses in the hardened traces as well as at the transition between the hardened traces.

Based on the experimental results, it is possible to make some conclusions on laser surface transformation hardening without melting of hypoeutectoid steels:

1. Microstructures observed in the hardened layer are martensitic, characteristic of conventional hardening but with some peculiarities:

   - Martensite is heterogeneous as a result of the nonuniform austenite. This heterogeneity is related to the initial microstructure and the transformation curves which are displaced to higher temperatures.
   - It is impossible to dissolve all the carbides with a heat treatment not involving melting process. Incomplete dissolution of carbides changes the hardenability. Nevertheless, undissolved carbides can be considered hard inclusions which are favorable in wear applications.

2. The analysis of residual stress profiles through the hardened layer confirms microstructural observations. Their profiles change with the type of steel and laser surface hardening conditions, but they correlate with microstructural transformations.

Compressive stresses, always observed in the laser surface hardened layer, improve the mechanical behavior. Tensile residual stresses that appear below the hard-

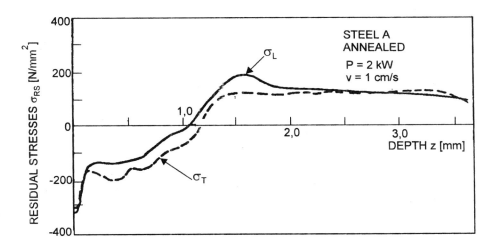

Figure 68 Longitudinal and traversal residual stress distribution in annealed steel A at given laser hardening parameters. (From Ref. 98.)

ened layer are heterogeneous. Quenched and tempered steels, however, are more homogeneous and the tensile residual stresses much more pronounced and potentially dangerous because they facilitate crack initiation. Based on the microstructural and residual stress analysis, laser surface transformation hardening leads to better results in a plain medium carbon steel than in alloyed steel. Multipass laser surface hardening shows softening and tensile residual stresses at the surface. Both the softening effect and tensile residual stresses should be avoided in practice with a chosen higher overlapping degree.

XIX. INFLUENCE OF LASER SURFACE HARDENING CONDITIONS ON RESIDUAL STRESS PROFILES AND FATIGUE PROPERTIES

Ericsson and Lin [99] presented the influence of laser surface hardening on fatigue properties and residual stress profiles of notched and smooth specimens.

The authors studied the effect of laser surface hardening on two Swedish steels, SS 2225 (0.26% C, 1.13% Cr, 0.1% Mo) and SS 2244 (0.45% C, 1.02% Cr, 0.16% Mo). Both steels were delivered as hot rolled bars in quenched and tempered condition with a hardness of 300 HV. Smooth and notched fatigue specimens in quenched and tempered condition were laser surface hardened. By using x-ray diffraction technique residual stress profiles in the hardened layer were measured.

Figure 69 shows axial residual stress profiles in the center of the laser-hardened track. For laser-hardened smooth specimen of SS 2225 steel, compression residual stress profiles were found in both axial and tangential

directions. Compressive residual stresses were also found in the hardened layer of smooth specimen of SS 2244 steel. The values of axial and tangential residual stresses are very similar. Figure 69 shows that residual stress profiles in the track center were more compressive in comparison to those at the overlap tracks. The difference became much larger at depth between 0.4 and 0.6 mm, which could be explained by the difference in microstructures.

Figure 70 shows that the compressive residual stress profiles were found in the hardened notches specimen of SS 2225 steel and the some shape specimen of SS 2244 steel. The value of axial residual stress was higher than that of tangential residual stress especially for SS 2225 steel. This might be due to the different constraint to deformation at the notch. Although the input heat from hardening the second track had little tempering effect, relaxed the first track, and changed residual stress profiles. As shown in Fig. 70, the value of residual stress was lower in the first track.

As can be seen, the fatigue strength of quenched and tempered specimens was greatly improved after laser surface hardening. Both low and high cycle fatigue lives were increased. The evaluated bending fatigue limits are summarized in Table 4, together with the values of notch sensitivity index q. Notch sensitivity index q is a measure of the effect of a notch on fatigue and is defined as $(K_f-1)/(K_t-1)$, where K_f is the ratio of the fatigue limit of smooth specimen to that of notched specimen.

Figure 71 shows the S–N curves for smooth and notched specimens. The curves 1 and 2 represent fatigue strength for smooth specimens after quenched and tempered (Q&T) conditions, curves 3 and 4 represent fatigue strength for smooth and notched specimens after laser surface hardening. It can be concluded that:

- The bending fatigue limits of smooth specimens were increased by 31% for SS 2225 steel and 29% for SS 2244 steel, while those of notched specimens were increased by 64% for SS 2225 steel and 69% for SS 2244 steel.
- The notch sensitivity index q decreased from 0.66 to 0.16 for SS 2225 steel and from 0.71 to 0.13 for SS 2244 steel.
- Important increase in fatigue properties was caused by compressive residual stresses in the surface layer after laser surface hardening.

Bohne et al. [100] analyzed X 39 CrMo17 1 (1.4122) steel which was quenched and tempered at temperatures of 520°C and 200°C. The hardened specimens were rapidly reheated with a Nd:YAG laser under such conditions that a maximum surface temperature rang-

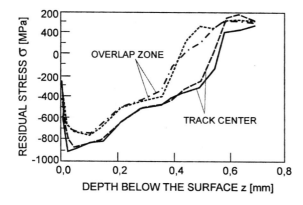

Figure 69 Axial residual stress profiles at the center of the track on smooth specimen of SS 2244 steel. (From Ref. 99.)

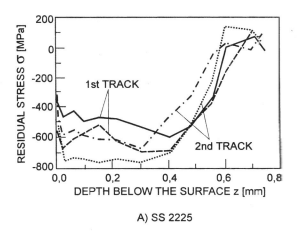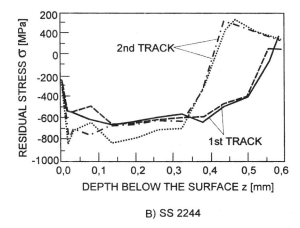

A) SS 2225

B) SS 2244

Figure 70 Residual stress profiles after laser surface hardening notched specimens. Full and open symbols refer to axial and tangential residual stress, respectively. (From Ref. 99.)

ing between 860°C and 1390°C was obtained, and then quenched with a cooling rate ranging between 480 and 1160°C/sec. Depending on the conditions of the preceding heat treatment and various parameters of laser processing, differences in the microstructure of a laser trace, in the hardness, and in the profile of residual stresses in the laser-hardened trace were obtained. Depending on the heating conditions, different contents of retained austenite, i.e., between 2% and 62%, were found in the laser-hardened traces. These great differences in the measured contents can be attributed to the height of the maximum surface temperature, the cooling rate, and the preceding heat treatment, i.e., tempering. It was found that the major influence on the formation of retained austenite was exerted by the tempering temperature in the preceding hardening so that a lower tempering temperature resulted in a higher content of retained austenite. The content of retained austenite, however, increased with an increase in the maximum

temperature in laser surface heating as well as by an increase in the rate of laser heating.

In Table 5 the results of microhardness measurements obtained in a depth of 30 μm below the surface of the laser-hardened trace are stated in dependence of the maximum surface temperature T_{Smax} [°C] and the rate of laser heating, $v = \Delta T/\Delta t$ [°C/sec]. It is worth mentioning the difference in the microhardness attained after steel tempering at the temperatures of 520°C and 200°C, respectively. In the case of steel tempering from 520°C and then laser hardening, the microhardness decreased from 700 to 636 $HV_{0.05}$ with the increase in the content of retained austenite. In case of steel tempering from 200°C the microhardness increased with the increase in the content of retained austenite, which was to be attributed to precipitates that additionally hardened austenite and martensite.

Residual stresses were measured in the longitudinal and transverse directions in the central part of the

Table 4 Bending Fatigue Limits and Notch Sensitivity

Material	Specimen type	Condition	Fatigue limits [MPa]	Notch sensitivity index
SS 2225	Smooth	Q&T	480 ± 41	0.66
	Notched	Q&T	350 ± 62	
SS 2225	Smooth	Q&T + Laser hard	627 ± 10	0.16
	Notched	Q&T + Laser hard	575 ± 21	
SS 2244	Smooth	Q&T	481 ± 7	0.71
	Notched	Q&T	343 ± 10	
SS 2244	Smooth	Q&T + Laser hard	621 ± 33	0.13
	Notched	Q&T + Laser hard	581 ± 29	

Source: From Ref. 99.

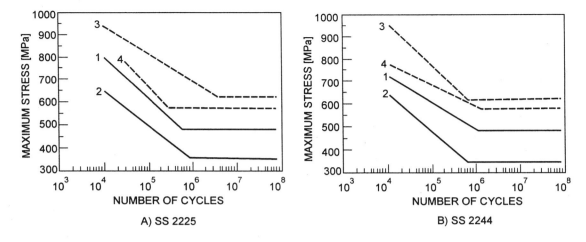

Figure 71 S–N curves for smooth (1,2) and notched (3,4) specimens. (From Ref. 99.)

hardened trace. The variation of the residual stresses is shown for the previously hardened steel that was tempered at 520°C and then rapidly laser heated to T_{Smax}, which provided lower contents of retained austenite, i.e., between 2% and 7%.

Figure 72 shows the variation of the residual stresses in the longitudinal direction with reference to the orientation of the hardened traces and Fig. 73 the variation in the transverse direction. The specimens were, prior to surface hardening, quenched and tempered at 520°C. Both the longitudinal and transverse residual stresses show the compressive character at the surface and then get balanced through depth where they finally obtained the tensile character. The through-depth variations of

the longitudinal and transverse residual stresses are similar. The compressive residual stresses can be attributed to austenite/martensite transformation whose magnitude and variation depend on the content of retained austenite and the precipitates that could additionally harden the surface layer. The magnitude and variation of the residual stresses were affected also by internal stresses occurring in the cooling process. If the latter were higher than the yield stress at the given temperature, plastic deformation and certain internal stresses accompanied by transformation stresses occurring in further cooling was obtained. The stresses occurring during the cooling process are called residual stresses after hardening. They are accompanied by

Table 5 Microhardness and Retained Austenite (RA) Content After Laser Surface Hardening in Tempered Steel at 520°C

Specimen	$T_{\max}$ [°C]	T/t [°C/sec]	RA [%]	Microhardness [HV$_{0.05}$]
Tempered at 520°C				
520-2r	860	620	<2	540 ± 14
520-3r	900	620	3	451 ± 7
520-3l	980	775	7	700 ± 18
520-4r	1020	670	9	706 ± 13
520-2l	1100	700	6	700 ± 17
520-1l	1125	920	24	653 ± 12
520-5r	1399	1160	62	636 ± 19
Tempered at 200°C				
200-6l	800	480	<2	397 ± 10
200-7l	840	5202	20	487 ± 14
200-7r	980	665	24	480 ± 9
200-6r	1399	1050	60	646 ± 14

Source: From Ref. 100.

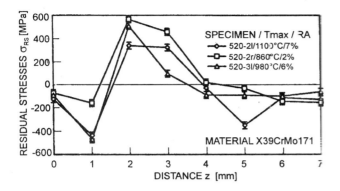

Figure 72 Longitudinal residual stress profiles after laser surface heat treating of quenched and tempered steel at 520°C. (From Ref. 100.)

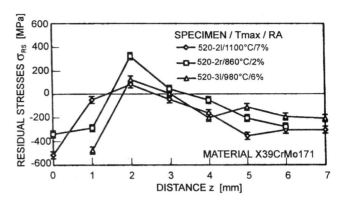

Figure 73 Transverse residual stress profiles after laser surface heat treating of quenched and tempered steel at 520°C. (From Ref. 100.)

specimen deformation. Account should be taken also of the residual stresses due to the preceding conventional hardening of the specimen.

Mordike [101] reported on several applications of laser transformation hardening, laser remelting, laser nitriding, and carbonitriding as well laser cladding and alloying. The results show the variation of residual stresses in the longitudinal and transverse directions in pearlitic gray cast iron, for two sets of conditions after laser remelting of the surface layer. Both through-depth profiles of the residual stresses confirmed great differences in the energy input obtained by varying the power density and the traveling speed of the laser beam. Although the two profiles of the residual stresses were very similar, it was found that the depths of the remelted layers differed, i.e., $d_1 = 0.2$ mm and $d_2 = 1.0$ mm. In both cases there were compressive residual stresses in the surface layer. They differed, however, in the depth in which the transition from compressive to tensile stresses occurred. In the first case the transition was outside the remelted depth whereas in the second case it was within the remelted zone. The compressive stresses at the surface amounted to around 400 N/mm^2 under the conditions of shallow remelting and to around 700 N/mm^2 in the subsurface, i.e., in a depth z of 0.3 mm, under the conditions of deep remelting.

Grum and Žerovnik [102] presented the experimental results of residual stress analysis after previously mentioned laser transformation hardening of various steels by using different absorbers. Their study based the identification of residual stresses on the relaxation method, which consists of measuring specimen strain. The relaxation was induced by electrochemical removal of the stressed surface layer, causing a breakdown in the

existing equilibrium state. The restoration of the equilibrium is accompanied by specimen strains. The strains were measured by means of resistance strain gauges and calculated into residual stresses using a mathematical model.

Figures 74–76 show the variation of residual stresses as a function of depth, the distance of the observed location from the surface. In laser transformation hardening, tracks have very small widths and depths that are dependent on the optical and kinematic conditions and laser source power. In the investigation the specimen traveling speed was changed with respect to the absorber. Figure 74 shows three residual stress profiles of the investigated W–Cr–V tool steel, and 4140, 1045 structural steels with the laser beam traveling along a zigzag

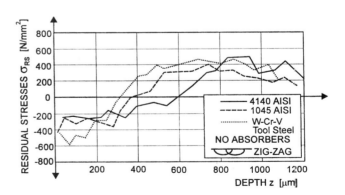

Figure 74 Residual stresses profiles in steels after laser transformation hardening at zigzag traveling path; no absorber. (From Ref. 102.)

path and the ground surfaces were free from any of the absorbers. The residual stress profiles are very similar to one another, displaying the following characteristics:

- In all three cases, the surface is subjected to compressive residual stresses.
- At a depth between 300 and 600 μm, the compressive stresses transform themselves into tensile residual stresses.
- Tensile residual stresses are present in deeper layers and have an almost constant value around 500 N/mm2.

From the residual stress results, it can be seen that both the heat treatable steels have low compressive stresses on the surface and a less explicit transition into tensile stresses. The alloyed W–Cr–V tool steel has a considerably greater content of carbon than the heat treatable steels; therefore we can maintain that carbon plays a decisive role in residual stresses size and transformation.

Residual stresses W–Cr–V in tool steel, to which absorbers A and B were applied and the laser traveled a square-spiral path, are further illustrated in Fig. 75. Thanks to light absorption, shorter interaction times are necessary, and as a result, the compressive residual stresses on the surface are lower. In addition, the depth of the transition from compressive into tensile stresses is lower using absorber A. The latter yields lower compressive residual stresses and a steeper transition into tensile stresses. Both features increase the sensitivity of parts to dynamic loads. Therefore the better choice is absorber B.

Figure 76 illustrates the variation of residual stresses with an intermittent laser trace [103]. The residual stresses in the compressive and tensile zone are low, a

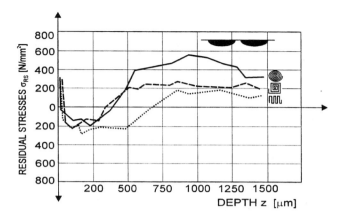

Figure 76 Residual stresses in W–Cr–V tool steel for interspaced hardening traces, concentric circle, square spiral, and zigzag hardening methods. (From Ref. 103.)

special feature being a thin surface layer with tensile stresses. Because after laser heat treatment grinding should be applied, in each of the discussed cases compressive stresses in the surface are a usual result.

An interesting study was made by Yang and Na [104] in his paper titled "A study on residual stresses in laser surface hardening of a medium carbon steel" by using two-dimensional finite element model. By using the proposed model, the thermal and residual stresses at laser surface hardening were successively calculated. The phase transformation had a greater influence on the residual stress than the temperature gradient. The simulation results showed that a compressive residual stress region occurred near the hardened surface of the specimen and a tensile residual stress region occurred in the interior of the specimen. The maximum tensile residual stress occurred along the center of the laser track in the interior region.

The compressive residual stress at the surface of the laser-hardened specimen has a significant effect on the mechanical properties such as wear resistance, fatigue strength, etc.

The size of the compressive and tensile regions of the longitudinal residual stress for various spot ratios of the square beam mode is shown in Fig. 77. It should be observed that with increasing beam width the compressive region becomes wide but shallow. From the comparison of the results, it is recommended that wide laser-beam spots are used for obtaining the desirable heat-treated region.

Figure 78 shows the sizes of the compressive and tensile regions of the longitudinal residual stress for various

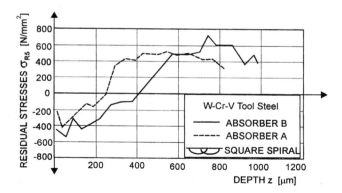

Figure 75 Residual stress profiles in steel S1 after laser transformation hardening at square spiral traveling path; absorbers A and B. (From Ref. 102.)

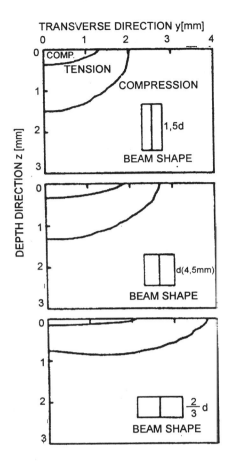

Figure 77 Compressive and tensile regions of the longitudinal residual stress for various laser beams of square beam structural mode at constant laser power. (From Ref. 104.)

laser-beam power and traveling speed at a given input energy. Although the input energy is constant, the compressive residual stress region increases according to increased laser power and traveling speed. That means that it is desirable to use the high power beam and high traveling speed at laser surface hardening.

Estimation and optimization of processing parameters in laser surface hardening was explained by Lepski and Reitzenstein [105]. Optimum results were obtained if the processing was based on temperature cycle calculations, taking into account the material properties and input energy distribution. A user-friendly software is required in industry application of laser surface hardening. The software should fulfill the following criteria:

- Ability to check any given hardening problem;
- Ability to estimate without experiments which laser power beam shaping or beam scanning system is selected;

- Ability to predict the laser hardening results at a given application;
- Ability to calculate the processing parameters with minimum cost at desired hardening and annealing zone size;
- Graphically present the relationship between the processing parameters and the hardening zone characteristics.

The integration of laser hardening in complex manufacturing systems requires hardening with high traveling speeds. In order to get a sufficient hardening depth neither the surface maximum temperature nor the laser interaction time must fall below certain limits even for high traveling speeds. This may be achieved to a certain degree by laser spot stretching along the traverse direction. In Fig. 79 the track depth as well as the track width is represented for the steel C45 as functions of the spot

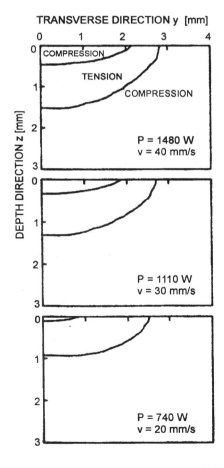

Figure 78 Compressive and tensile regions of the longitudinal residual stress for various laser powers and traveling speeds at a given input energy. (From Ref. 104.)

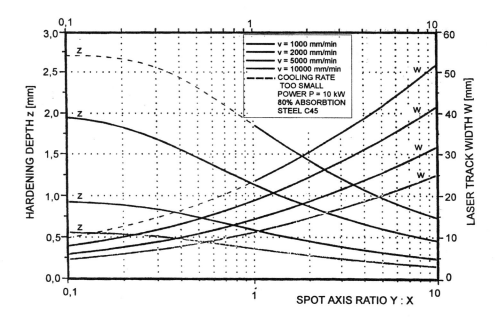

Figure 79 Influence of laser spot axis ratio and various traveling speeds on obtained depth and width of single hardened track. (From Ref. 105.)

axis ratio (SAR) (SAR = y/x = 1...10) for various values of the traveling speeds (v = 1–10 m/min and a laser power of 10 kW). Values less than unity of the ratio SAR correspond to a spot stretched along the traverse direction.

Meijer et al. [106] presented a very interesting and practical subject of transferability of the same testing parameters in laser hardening using different laser systems used in several laboratories.

The chosen experimental conditions show that the same material was tested, the same absorption deposit and the same working method in a shielding atmosphere were applied, but different degrees of defocus, small power densities, different interaction times, and different mode structures of the laser beam were used.

The deviations occurring are relatively strong; therefore it is dubious whether with different laser systems the same and reproducible results with the hardened surfaces can be obtained. As the machining conditions were uniformly selected, the same depths of the hardened traces could be assumed, yet this was not the case. The differences were exceptionally strong. They can be attributed to the chosen mode structure of the laser beam and other characteristics selected such as the degree of defocus and the travel speed of the laser beam. Consequently, the results of the investigation should serve only as a proof that the influence of energy input is strong. As far as the energy input is concerned, not only

the mode structure of the laser beam but also the selected and interdependent parameters such as the travel speed and the degree of defocus should be taken into account.

Results of the second set of joint measurements are given in dependence of the width of a single hardened trace on the material surface and the depth of the hardened trace. The deviations occurring can be attributed only to the mode structures selected in the individual laboratories. The differences in the measured widths and depths of the hardened single traces are, as expected, smaller with shorter interaction times. They gradually increase with increasing interaction times.

The deviation occurring in hardening of carbon steel can be attributed to the available mode structures of the laser beam and the selection of different interaction times.

Kugler et al. [107] reported on the latest developments of powerful diode lasers permitting the manufacture of new laser hardening systems. Advantages of the diode laser are as follows:

- Optics adjustable and adaptable to the workpiece shape.
- Adjustable optics permits shallow or deep heating of the surface layers.
- Absorptivity of laser light with a double wavelength, $\lambda 1 = 0.808$ and $\lambda 1 = 0.94$, ranges, as

assessed, between 50% and 70%, which means that no particular preparation of steel surface is required for heat treatment.

- It has a pyrometer mounted to measure the temperatures at the workpiece surface surrounding the laser beam.
- A computer-aided system for temperature setting makes a comparison with the reference temperature and automatically sets a higher or lower laser-source power in the very course of heat treatment.
- A robotic hand and optical transmission of laser light to the workpiece make it possible to perform heat treatment of large parts simply and efficiently.
- It is suitable for both small-scale and large-scale productions.

Such a way of temperature monitoring, i.e., by changing the power input, is extremely important at the workpiece edge, along the notches in the workpiece, or at the transition from a larger diameter to a smaller one.

Figure 80 shows the dependence between the power density distribution across the beam diameter, the specific energy, and the depth of the modified layer. In the diagram above, different distributions, i.e., Gaussian (A), cylindrical (B), prismatic (C), and sharp-edged (D) distributions, are shown. The individual characteristic

power density distributions across the beam diameter ensure, with the same heating conditions, small differences in the energy input, and great differences of the depth of the hardened layer attained. The lowest specific energy is obtained with the cylindrical (B), prismatic (C), and sharp-edged (D) power density distributions. It ranges between 13.0 and 15.0 J/mm³. The highest specific energy is attained with the Gaussian power density distribution. On the average, it is 100% higher than the lowest one. A more important piece of information is the one on the depth of the hardened layer attained. The latter equals 0.75 mm with the Gaussian power density distribution, 1.17 mm with the cylindrical distribution, and over 2.10 mm with the prismatic and sharp-edged distributions. This proves that the distributions most suitable for heat treatment are the prismatic and sharp-edged power density distributions.

XX. PREDICTION OF HARDENED TRACK AND OPTIMIZATION PROCESS

Marya and Marya [108] reported about prediction hardened depth and width and optimization process of the laser transformation hardening. They used a dimensionless approach for Gaussian and rectangular

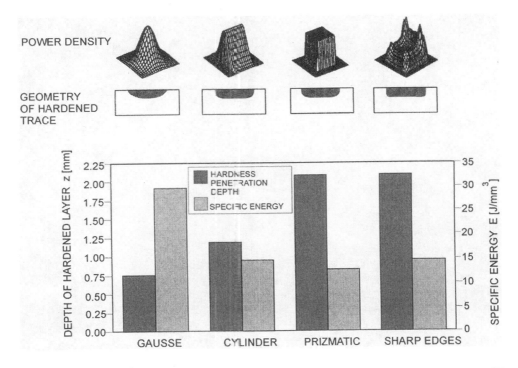

Figure 80 Influence of various laser beam shapes and specific energy on the depth of the hardened layer. (From Ref. 107.)

sources to find laser heating parameters at given dimensions of the hardened layer.

The laser transformation hardening has been performed on a 0.45% carbon steel, coated with a carbon to maintain the surface absorptivity to about 70%. Figure 81 shows the author's results according to their model of predicted dimensions of the hardened layer and process optimization. The diagram shows that decreasing power and laser-beam traveling speed at Gaussian distribution of laser power density in cross section which defines interaction times are required to obtain specific data. Thus any useful combination of laser processing parameters must maximize heat diffusion to required depth of hardened layer. It is necessary that surface melting temperature is reached. Figures 81 and 82 show a low dimensionless traveling speed v^* that is necessary to allow the heat conduction in depth and achieve high dimensionless depth $Z_h^* = Z_h/R$ and dimensionless width $W_h^* = W_h/R$. The dimensionless power parameter (q^*) is determined with respect to (v^*) to reach the melting on set.

If the laser beam moves faster, greater values (q^*) must be selected to reach surface melting. Similar calculations were realized for square laser-beam power density. Hardened widths should correspond rather well to beam spot diameter because the step energy gradient of the beam edge should produce an evenly steep temperature gradient.

Results in stationary laser-beam hardening show that melting cannot be achieved at dimensionless power below 7.6. Similar conclusions have already been drawn for Gaussian beams. Indeed, as the beam speed approaches

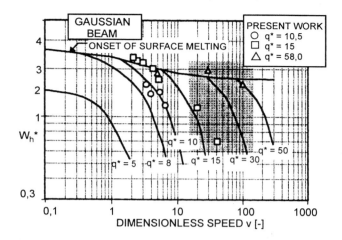

Figure 82 Influence of dimensionless power (q^*) and traveling speed (v^*) on hardened width (w^*) for Gaussian beam. (From Ref. 108.)

zero, the beam profile contribution decreases as heat tends to dissipate more uniformly.

Figure 83 shows optimized beam spot dimensions at the surface which are very well predicted by the theoretical analysis. Moreover, the results show that hardened depth increases with heat input energy. They experimentally verified that melting conditions are proportional to the spot dimension. Although increasing powers and spot beam diameter produced wider hardened layer the heating rates decreased significantly. Figure 83 shows the variation in surface hardness according to

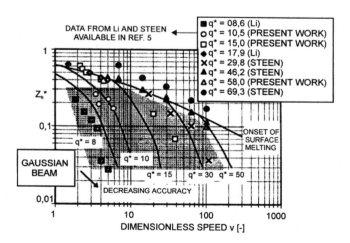

Figure 81 Influence of dimensionless power (q^*) and traveling speed (v^*) on hardened depth after (z_n^*) for Gaussian beam. (From Ref. 108.)

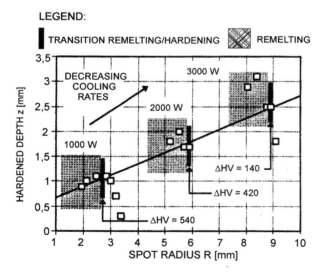

Figure 83 Influence of spot radius and laser-beam power on hardened depth. (From Ref. 108.)

the hardness of base material (HV = 205). In an optimization process, a compromise between a high quenched depth and a significant hardness increase must therefore be found.

XXI. MICROSTRUCTURE AND RESIDUAL STRESS ANALYSIS AFTER LASER SURFACE REMELTING PROCESS

Ductile iron is commonly used in a wide range of industrial applications because if its good castability, good mechanical properties, and low price. By varying the chemical and microstructure composition of cast irons, it is possible to change their mechanical properties as well as their suitability for machining. Ductile irons are also distinguished by good wear resistance, which can be raised even higher by additional surface heat treatment. With the use of induction or flame surface hardening, it is possible to ensure a homogeneous microstructure in the thin surface layer; however, this is possible only if cast irons have a pearlite matrix. If they have a ferrite–pearlite or pearlite–ferrite matrix, a homogeneous microstructure in the surface hardened layer can be achieved only by laser surface remelting.

After the laser beam had crossed flat specimen, a microstructurally modified track was obtained, which was shaped like a part of a sphere (Fig. 84). To achieve a uniform thickness of the remelted layer over the entire area of the flat specimen (Figure 85), the kinematics of the laser beam were adapted by 30% overlapping of the neighboring remelted traces [109].

The microstructure changes in the remelting layer of the ductile iron are dependent on temperature conditions during heating and cooling processes. In all of the

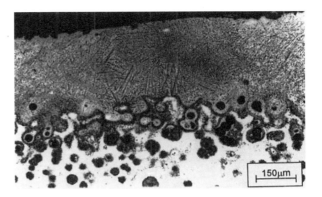

Figure 85 Laser surface modified layer at 30% overlap of the width of the remelted traces; remelting conditions: $P = 1.0$ kW, $z_s = 22$ mm, and $v_b = 21$ mm/sec. (From Ref. 109.)

cases of the laser surface remelting process two characteristic microstructure layers were obtained, i.e., the remelted layer and hardened layer. Figure 86 shows that the microstructure in the remelted surface layer is fine grained and consists of austenite dendrites, with very fine dispersed cementite, together with a small portion of coarse martensite. X-ray phase analysis of the remelted layer showed the average volume percentages of the particular phases as follows: 24.0% austenite, 32.0% cementite, 39.0% martensite, and 5.0% graphite. Figure 87 shows the microstructure of the hardened layer consisting of martensite with a presence of residual austenite, ferrite, and graphite nodules. Graphite nodules are surrounded by ledeburite and/or martensite shells [109].

Grum and Šturm [109] showed the results (Fig. 88) of the calculations of principal residual stresses in the flat

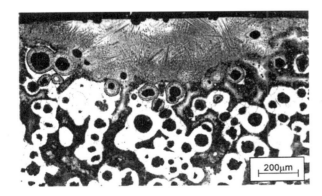

Figure 84 Cross section of a single laser-modified trace; remelting condition: $P = 1.0$ kW, $z_s = 22$ mm, and $v_b = 21$ mm/sec. (From Ref. 109.)

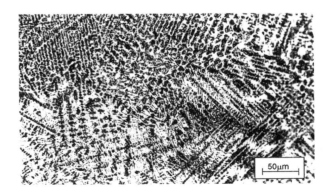

Figure 86 Microstructure of the remelted layer. (From Ref. 109.)

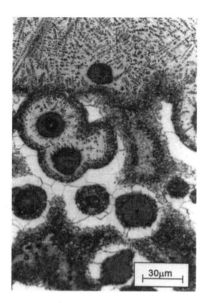

Figure 87 Microstructure of the hardened layer. (From Ref. 109.)

the laser beam turning round outside the specimen to achieve more uniform thermal conditions in the material. In this way, it was possible to achieve different thermal conditions in the thin flat specimen during the remelting process as well as during cooling, which influence the preheating of the specimen prior to remelting and the tempering of the created modified microstructure. They stated the following:

- Residual stresses $\sigma 1$ are directed in the longitudinal direction and $\sigma 2$ directed in the transverse direction of the flat specimen.
- During the process of laser surface remelting, the specimen bends more in the longitudinal direction causing additional lowering of tensile residual stresses.
- At 0% overlapping of the remelted layer, the compressive residual stresses were achieved between $+100.0$ and -5.0 MPa. A 30% higher overlapping degree induces the occurrence of tensile residual stresses of $+90$ MPa in the remelted layer and compressive stress of 50 MPa in the hardened layer.

The same authors in Figure 89 showed the results of principal residual stresses in the thin surface layer with a laser beam guiding in the shape of a square spiral at a given laser-remelting conditions, without overlapping traces.

specimen after laser remelting with the zigzag laser beam guiding at the 0% and 30% overlapping degree of the remelted layer. Different modes of guiding the laser beam over the specimen surface were selected, i.e., zigzag (A), square-shaped spiral toward the center (B), and square-shaped spiral away from the center (C), with

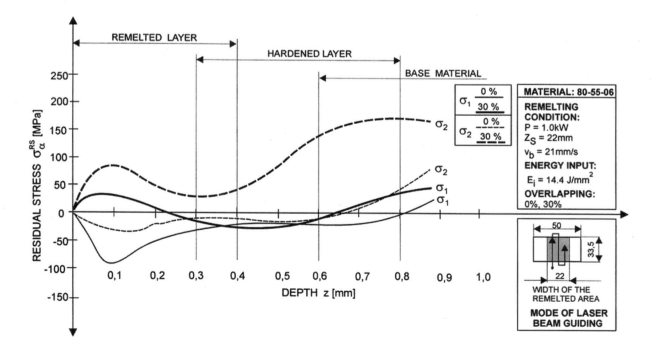

Figure 88 Residual stresses in ductile iron 80-55-06; zigzag laser beam guiding laser-remelting conditions. (From Ref. 109.)

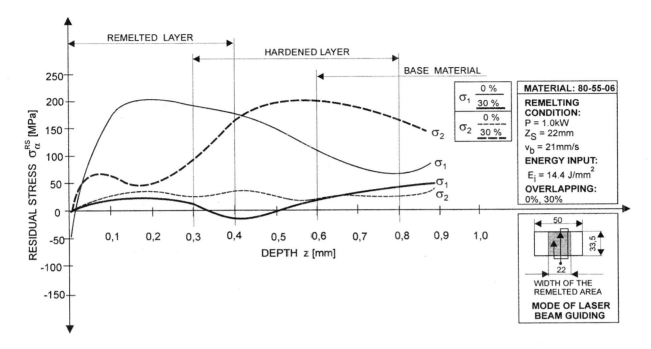

Figure 89 Residual stresses in ductile iron 80-55-06 with a circular laser beam guiding in the shape of a square spiral beginning on the edge of the remelted area and ending in the middle of the specimen at the above indicated laser-remelting condition. (From Ref. 109.)

They showed the calculated principal residual stresses profiles through the depth and the direction angles of the principal residual stresses. The residual stresses were measured to the depth of 0.9 mm, which means that they were measured to the transition into the matrix.

From the results of calculated residual stress profiles the following conclusions can be drawn:

- The principal residual stresses $\sigma 1$ are directed perpendicularly to the direction of the laser path.
- The principal residual stresses $\sigma 2$ acting in the direction of laser remelting and being of tensile nature are much higher than the residual stresses $\sigma 1$.
- The principal residual stresses $\sigma 1$ are reduced because of the temperature field, and lowering the yield strength of the material.
- Residual stresses in the modified surface layer in the longitudinal direction of the specimen are compressive and range between 80.0 and 5.0 MPa when, in the transition area from the hardened layer to the matrix, they change into tensile residual stresses.
- In the transverse direction of the specimen, residual stresses are always tensile and range from + 50.0 to + 200.0 MPa in the modified layer.

All of these laser-remelting conditions, each in its own way, change the amount of the energy input and can have an important effect on the size and quality of the modified layer and residual stresses. A greater amount of input energy into the specimen results in a higher increase of temperature in it and higher overheating of the specimen, which, on the other hand, lowers the cooling rate in the modified layer and gives rise to the occurrence of small microstructure residual stresses in this layer.

A. Dimensions of the Remelted Track

Hawkes et al. [110] studied laser-remelting process on ferrite and pearlite gray iron at various melting conditions. The results are presented by microstructure, microhardness analysis, and size data of the remelted tracks. Figure 90 shows remelted depth according to traveling speed and beam diameter. Depth of remelting layer increased significantly by approaching the focussed beams at lower traveling speed of 100 mm/sec due to the keyholing mechanism operating in the molten pool. This occurred at sufficient laser-beam power density to vaporize the metal under the beam center. The pressure of the expanding vapor held the cavity

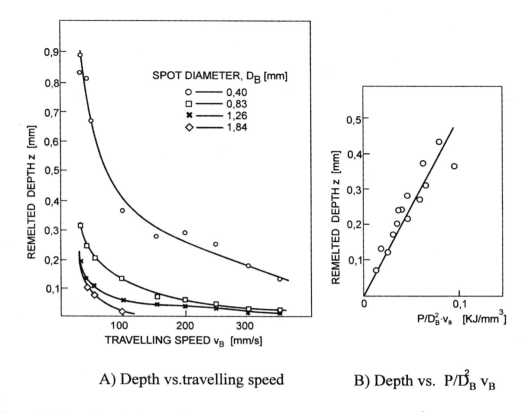

A) Depth vs.travelling speed   B) Depth vs. $P/D_B^2 v_B$

Figure 90   Remelting depth according to traveling speed v and input energy $P/D_B^2 \cdot v_b$. (From Ref. 110.)

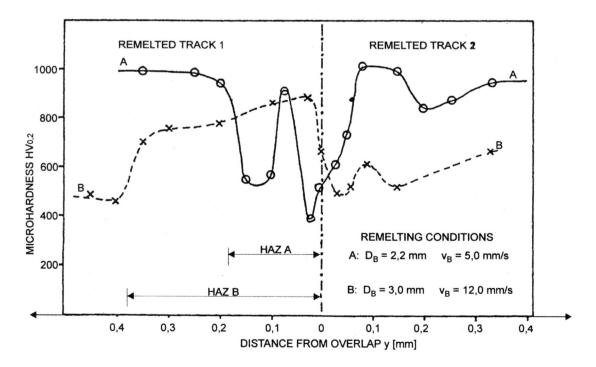

Figure 91   Microhardness profiles of tempered microstructures in overlap multitracks. (From Ref. 110.)

open forming a black body, thus transferring more energy deeper into the material.

The remelting width was greater at smaller areas of beams at low traveling speeds due to the intense vapor and liquid convection. Figure 90A shows the dependence of melting depth according to traveling speed at various beam spot diameters. Figure 90B shows the depth of remelting, assuming depth/width is constant, proportional to $P/D_B^2 \cdot v_B$.

Greater depth of remelting was achieved in the ferritic gray iron as compared to the pearlitic gray iron with higher beam spot diameters at lower traveling speeds. Figure 91 shows microhardness profiles across overlapping tracks at two different laser-remelting conditions. Laser track B shows a much wider HAZ at increased microhardness, and laser track A shows a greater HAZ at increased microhardness. The differences in HAZ width and microhardness profiles depend on the austenite decomposition temperature in the center of overlapping tracks.

B. Mathematical Modeling of Localized Melting Around Graphite Nodule

Roy and Manna [111] described in their paper mathematical modeling of localized melting around graphite nodules during laser surface hardening of austempered ductile iron. Similar findings were presented by Grum and Šturm [112] at laser surface remelting in the transition zone, while heating at low power beam $P = 700$ W and traveling speed v of 60 mm/sec, where dissolution of the graphite nodules at depth $z = 100$ μm below the surface occurred. At heating process in the region with

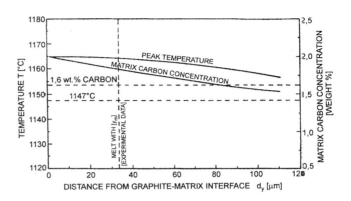

Figure 93 Variation of the peak temperature and matrix carbon concentration according to the distance from graphite matrix interface. (From Ref. 111.)

dissoluted graphite in austenite, lowering melting temperature was reached which resulted in local remelting. Figure 92 gives the variation of carbon concentration as a function of the graphite nodule distance from the surface. From the diagram it can be concluded that carbon concentration in the matrix is higher on the surface and lower with decreasing depth below the surface. The dashed line shows 1.6 wt.% carbon, which determines the minimum level of carbon enrichment and maximum width y_m of the localized remelted zone around the graphite nodule.

Figure 93 gives the matrix carbon concentration and maximum temperature as a function of graphite interface distance d_y, the dashed horizontal line denotes the effective temperature and maximum carbon solubility 1.6 wt.% for austenite, and the vertical one represents the given remelt width (y_m) from the graphite surface.

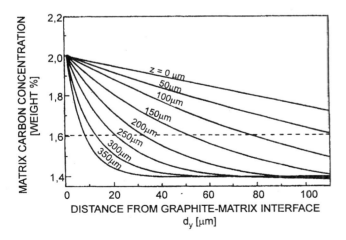

Figure 92 Variation of matrix carbon concentration as a function of distance from graphite matrix interface. (From Ref. 111.)

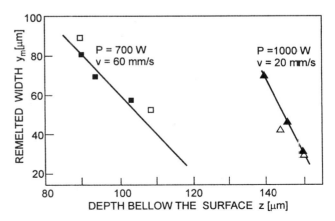

Figure 94 Variation of remelted width y_m according to remelting depth z. (From Ref. 111.)

Figure 94 shows the changing of remelt width (y_m) as a function of the depth below the surface (z) at given laser surface hardening conditions. Solid symbols in the diagram show theoretically predicted maximum melt width y_m around the graphite nodule as a function of the depth below the surface. The open symbols note the experimental data at given laser hardening conditions and depth below the surface as well.

C. Transition Between the Remelted and Hardened Layers

Grum and Šturm [112] studied rapid solidification microstructure in the remelting layer and microstructure changes in the hardened layer.

The application of laser surface remelting to nodular iron 400-12 causes the material to undergo microstructural changes. A newly created austenite–ledeburite microstructure with the presence of graphite nodules in the remelted layer and a martensite–ferrite micro-

structure with graphite nodules in the hardened layer have been observed. Microscopy of the hardened layer was used to analyze the occurrence of ledeburite shells and martensite shells around the graphite nodules in the ferrite matrix. The thickness of the ledeburite and martensite shells was supported by diffusion calculations. The qualitative effects of the changed microstructures were additionally verified by microhardness profiles in the modified layer and microhardness measurements around the graphite nodules in the hardened layer.

The tests involved the use of an industrial CO_2 laser with a power of 500 W and Gaussian distribution of energy in the laser beam. The optical and kinematic conditions were chosen so that the laser remelted the surface layer of the specimen material. The specimens were made from nodular iron 400-12 with a ferrite–pearlite matrix that contained graphite nodules.

A characteristic of the transition area between the remelted and the hardened layer is that local melting

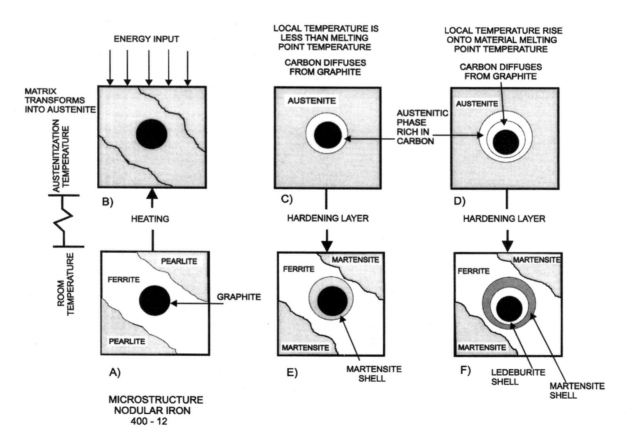

Figure 95 Schematic presentation of the microstructural changes in the transition area between the melted and the hardened zones. (From Ref. 112.)

occurs around the graphite nodules. Grum and Šturm [112] showed in Fig. 95 a schematic representation of the sequence of processes occurring in the phases of heating and cooling:

- The matrix transforms into a nonhomogeneous austenite.
- Then diffusion of carbon from graphite nodules into austenite.
- Increased concentration of carbon in the austenite around the graphite nodule lowered the melting-point temperature in local melting of part of the austenitic shell around the graphite nodule.
- After rapid cooling, a ledeburite microstructure is formed locally. This is then further surrounded by a martensite shell.

The transition area is very narrow and very interesting from the microstructural point of view. However, it is not significant in determining the final properties of the laser-modified surface layer.

Figure 96 shows the measured and calculated thickness of martensite shells around graphite nodules with respect to their distance from the remelted layer. Our estimation is that the differences between the measured and the predicted martensite shell thickness are within the expected limits as the data on heat conductivity and diffusivity were chosen from the literature. The calculations confirmed the validity of the mathematical model for the determination of temperature T and time t remaining in a given temperature range, which enabled us to define the diffusion path of the carbon or, in other words, define the thickness of the martensite shell.

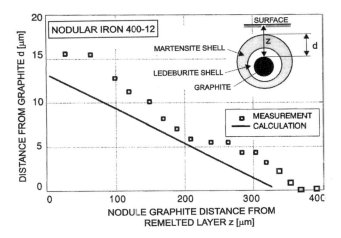

Figure 96 Comparison of measured and calculated thicknesses of martensite shells around graphite nodules in the hardened zone. (From Ref. 112.)

D. Circumstances for Rapid Solidification Process of Cast Iron

Three features are significant for the rapid solidification process after laser remelting [113]:

- The cooling rate $\dot{\varepsilon} = dT/dt$,
- The solidification rate $R = dx/dt$, which characterizes crystal grain growth per time unit in the liquid/solid interface,
- The temperature gradient $G = dt/dx$ across the liquid/solid interface to a given location.

These parameters are connected by the equation

$$\dot{\varepsilon} = RG$$

Depending on the values of the above variables, different microstructures can be produced after solidification process. It is important that different microstructures can be obtained in the same material at various solidification conditions. In Fig. 97, one set of parallel lines represent lines equal to G/R ratios and the other rectangular to the former one, which represents RG products (equal to $\dot{\varepsilon}$) [113].

The G/R parallel lines give similar solidification conditions before solidification increases with ε. Therefore an increasing nucleation frequency is obtained resulting in a finer microstructure of the same morphology. For the set of lines where ε is constant, the same grain size occurs, with changed solidification. At laser remelting it is difficult to verify experimentally the two limiting cases of solidification. Therefore special conditions must occur at rapid solidification:

- Substantial superheating of the melt which influences the heterogeneous nucleation,
- Extreme temperature gradients which assure rapid, directional solidification, and
- Epitaxial growth on substrate crystals.

In Fig. 98, cooling rate, remelted depth, and dendrite arm spacing are correlated [113].

E. Evaluation of Residual Stresses After Laser Remelting of Cast Iron

Domes et al. [114] gave estimation of microstructure, microhardness, and residual stress distribution after laser surface remelting of gray cast iron and nodular iron.

The authors carried out experiments with an 18-kW CO_2 laser. The laser beam was shaped by a line inte-

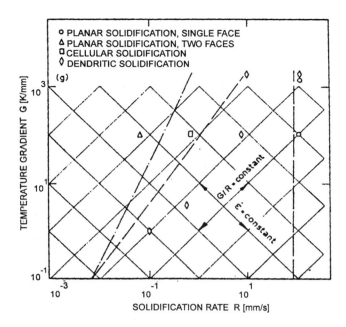

Figure 97 Variation of microstructure of a cast iron with solidification conditions. (From Ref. 113.)

grating mirror to the laser spot area of 2 × 15 mm. The residual stresses were measured by x-ray diffractometer.

Figure 99 shows the microstructure of laser-remelted nodular iron consisting of primary austenite dendrites, which can solve more carbon at increasing solidification rate and eutectic ledeburite. Therefore with increasing solidification rate the ledeburitic portion decreases because of the formation of supersaturated austenite solid solution. A portion of austenite transforms, depending on the cooling rate, into martensite with residual austenite (RA), bainite, or pearlite. Figure 99A shows the microstructure of laser-remelted nodular iron after remelting with different preheating temperatures ($T = 500°C$ or $600°C$). The non-preheated specimens are characterized by a high content of residual austenite at preheating temperature, while after preheating to 400°C the microstructure consists mainly of martensite and bainite. Further raising of T_p generates higher portion of bainitic and/or pearlitic microstructure. The formation of residual austenite can be observed at preheating temperature up to a T_p of 500°C and at traveling speed of 1 m/min (Fig. 99E). The residual austenite can be avoided either by further increasing T_p or by decreasing the traveling speed or by a post heat treatment.

The surface hardness of laser-remelted nodular iron is about 700 to 800 HV. As shown in Fig. 99B and C,

higher traveling speed and a lower preheating temperature T_p increase the microhardness. The microhardness of ledeburitic surface layers results from a law of mixture, where the hardness of the transformed austenite dendrites and eutectic ledeburite contributes at a given solidification rate. The microhardness in the remelted layer increases with raising temperature gradients and quenching solidification rate, respectively.

The formation of cracks in the remelted layers depends strongly on the process parameters as well as on melt depth and preheating treatments, respectively. To achieve crack-free surfaces a martensitic transformation in the remelted or heat-affected zone has to be avoided. Figure 99E shows that preheating to more than $600°C$ at traveling speed $v = 1$ m/min is necessary to avoid retained austenite and therefore martensitic transformation. A smaller traveling speed allows a lower preheating temperature T_p, because the specimen is heated during surface remelting, so that the critical cooling rate is not achieved. A similar effect can be reached by preheating on temperature $T > T_{MS}$ if post heat treatment is added.

Figure 100A and B shows the influence of a preheat treatment on the longitudinal and transversal residual stress profiles at the surface after remelting of nodular iron. Because of high contents of retained austenite for low preheating temperatures only the residual stress after preheating to $500°C$ and $600°C$ is analyzed. In both cases, compressive residual stresses in the modified layer can be observed, which reach higher values for

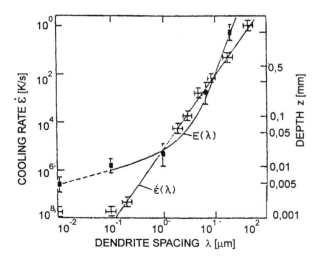

Figure 98 Variation of dendrite arm spacing with solidification parameters. (From Ref. 113.)

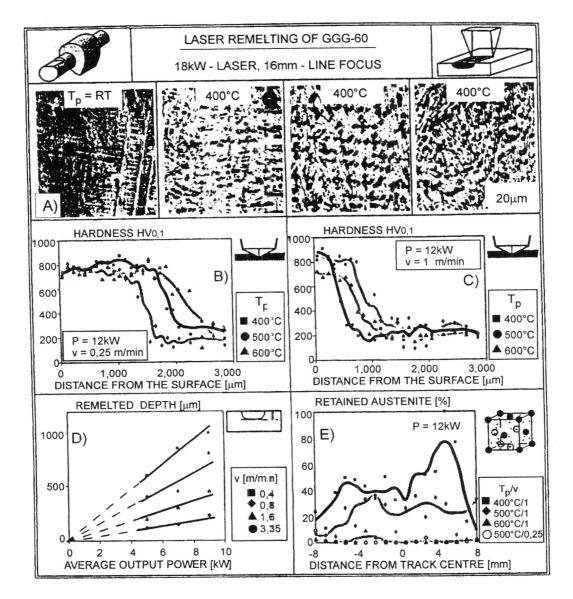

Figure 99 Microstructure and microhardness profiles after laser remelting of nodular iron at given laser-remelting conditions. (From Ref. 114.)

higher preheating temperature. The reasons for this are as follows:

- The martensitic transformation in the HAZ can be avoided, and
- The difference in heat expansion coefficient is rising because of the higher content of cementite in the remelted layer with increasing preheating temperature Tp.

The influence of the specimen geometry on the resulting longitudinal residual stress profiles in the modified layer is shown in Fig. 100B and C. After laser remelting of the flat specimens tensile residual in longitudinal direction can be found, while remelting of the web specimens leads to compressive residual stresses. A possible explanation for the difference in residual stress profiles is the stress relaxation during the laser-remelting process. The compressive residual stresses reach into the depth of the remelted layer and into the HAZ, which leads to tensile residual stresses (Figure 100D).

Figure 100E and F shows the deviations of the residual stresses in the center of the modified track.

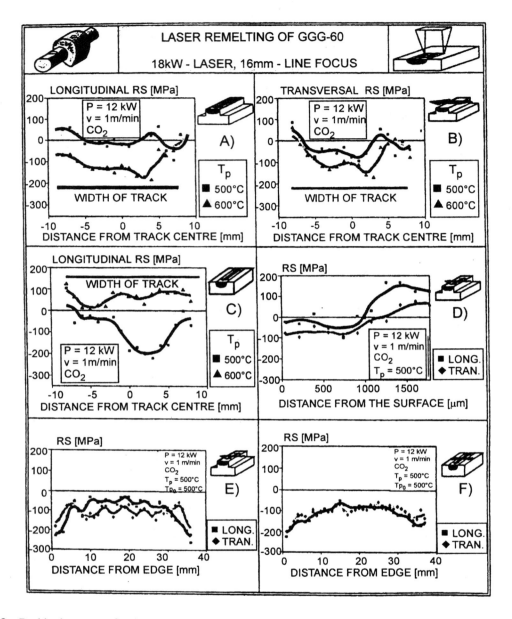

Figure 100 Residual stresses after laser remelting of nodular iron at given laser-remelting conditions. (From Ref. 114.)

These specimens were preheated to $T_p = 500\,°C$ and post heat treated at the same temperature. For both types of specimens compressive residual stresses of about -100 MPa are measured at the surface. Differences between longitudinal and transversal residual stresses can only be observed for both types of specimens.

Pre- and post heat treatment of the laser-treated materials has an influence on the portion of residual austenite in the layers as well as on the portion of martensite in the HAZ, which causes tensile residual stresses and the possible formation of cracks in the modified layer. To create a complete compressive residual stress profiles in the modified layer a suitable heat treatment has to be applied considering the specimen geometry.

Grum and Šturm [115] gave estimation of residual stress distribution after laser-remelting gray cast iron Grade 250 at different laser-beam guiding. Heating and cooling conditions in a relatively thin workpiece are very much dependent on laser-beam power, laser-beam

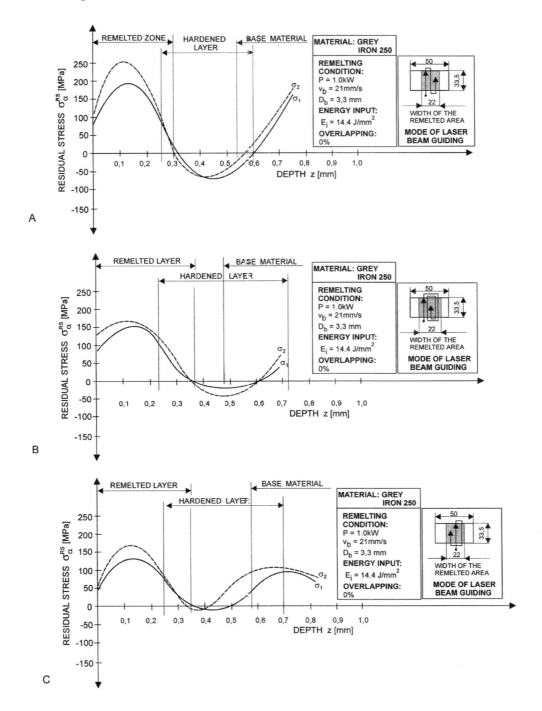

Figure 101 Residual stresses in gray iron grade 250 after different paths of laser beam travel: $P = 1$ kW; $v_B = 21$ mm/sec; DB = 3.3 mm; overlapping 0%; σ1, solid line and open symbols; σ2, broken line and closed symbols. (From Ref. 115.)

diameter on the workpiece surface, interaction time, and the degree of overlapping of the remelted layer.

Heating and cooling conditions are also very much dependent on the paths along which the laser beam travels over the flat specimen surface. Figure 101 shows the variation of residual stresses as a function of the depth of the modified layer for gray iron Grade 250 after different paths of laser-beam travel [115,116]. The graphs present two measured curves, i.e., main stresses σ_1 and σ_2.

From the graphs we can conclude the following:

- Residual stresses have, in all cases of different laser-beam travel paths, a very similar profile differing only in absolute values (Fig. 101A–C). In the surface remelted layer, tensile residual stresses were found in the range 50–300 MPa.
- The change from tensile to compressive residual stress takes place in the transition area between the remelted and hardened zones. Maximum compressive residual stress values were found in the middle of the hardened zone in the range 25–150 MPa. But there is one exception. In the case when the laser-beam travel path starts in the center of the workpiece (Fig. 101C) no compressive stresses were found.
- By x-ray diffraction the amount of constituents present in the remelted layer was analyzed. From the above results it can be concluded that in the case when it is possible to produce the remelted surface layer without any residual austenite in the microstructure, tensile residual stresses are minimized. The higher the amount of martensite in the remelted layer the lower the tensile residual stresses.
- The remelted surface layer cracked in the direction of graphite flakes in the gray iron. It is believed that during the cooling process, owing to high temperature differences, extremely high tensile stresses were generated which exceeded the yield point of the material at increased temperature in the remelted surface layer.

With laser-remelting procedure a sufficient depth of the modified layer is achieved, in addition to desirable microstructure changes and good microhardness profiles of the modified layer. Surface remelting a cast microstructure with under-cooled graphite results in an extended HAZ in which the carbon diffuses into the ferrite. Success of the carbon penetration depends on the temperature or the distance from the surface. In the remelted layer crystals are supersaturated and grow initially epitaxially and with a planar front. This is then followed by dendritic growth of crystals. Depending on the cooling rate, carbide particles or carbide eutectic can remain between the dendrites. The graphite eutectic is completely dissolved in the short heating cycle.

After rehardening, the fine dendrites in the remelted layer at overlapping regions coalesce and the striped appearance of cementite is replaced by a random arrangement of martensite between the cementite regions. In the transition zone, the microstructure consists of finely dispersed carbides and martensite. Great attention has to be given to the selection of optimal laser remelting as different structure of the microstructure matrix and the size of flaky graphite can substantially affect the thermal condition of the material.

After laser remelting of gray and nodular irons, the measured residual internal stresses show a similar variation as well as absolute values at different kinds of guiding of the laser beam across the surface. Residual stresses are in all the cases of tensile type on the surface and then decrease to the depth of 0.3 mm where they transform into compressive stresses. The relatively great depths of transformation of tensile into compressive internal stresses confirm that even after finish grinding it is not possible to achieve the desired stress state in the material with more or less high compressive residual stresses on the surface.

XXII. ASSESSMENT OF DISTORTION AND SURFACE CHANGES

Careful assessment is being made of component distortion in view of the attractions of treating finish-machined components and of the stringent dimensional specifications. The present indications are that dimensional changes resulting from laser treatment are minimal, provided that a high-temperature stress relief is employed at the appropriate stage in machining and that the volume of transformed material is low.

Grum and Šturm [117] presented a new method of simultaneous measuring of the strain on the bottom side of the flat specimen during laser surface remelting process.

In the experiment, the chosen thickness of the specimen was 5.5 mm and was such that the maximum achieved temperature on the bottom side of the flat specimen was always less than 350°C. This temperature was dictated by strain measuring rosette used to measure the specimen strain.

Figure 102 shows the experimental system for continuous measurements of the strain and temperature on the bottom side of specimen during the process of

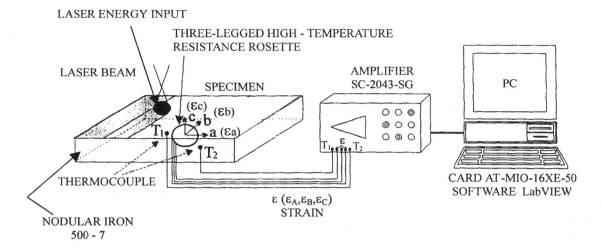

Figure 102 Experimental system for measuring the temperatures and strain of the flat specimen during the laser-remelting process. (From Ref. 117.)

laser remelting, going on the top side of the flat specimen. On the bottom side of the specimen were placed a three-legged, 45° high-temperature self-compensating resistance-measuring rosette and two thermocouples.

The thermocouples, placed in the longitudinal direction of the specimen on the left and right sides next to the high-temperature resistance-measuring rosette, continuously measure the temperature and thus define the temperature cycles on the bottom side of the flat specimen induced on the specimen's top side. This kind of placement enables the continuous monitoring of the strain and specimen temperature during the remelting process.

The measurements of specimen strains during the remelting and cooling process and subsequent calculations of the main residual stresses were made on the nodular iron 500-7 (ISO) with a pearlite–ferrite matrix.

Figure 103 presents the results of the measured temperature cycles in the middle of the specimen bottom side during the remelting process for three different laser traveling ways. The temperature gives information on the temperature changes in heating and cooling of the material on the bottom side of the specimen. To measure the effects of different laser-beam traveling ways and the different number of laser-beam passes across the surface, thermocouples registered partial temperature during the heating process. In the phase of heating the specimen, partial temperature occurs with a period of laser-beam passage across the specimen surface. Partial temperature occurs with a period of laser-beam passage across the specimen surface. In the phase of heating, the

highest peaks of partial temperature can be noted with the zigzag laser-beam traveling ways. The process of cooling can be described in general from the moment the maximum temperature in the specimen material on the bottom side was reached. It is possible to conclude that the time at maximum temperature depends on the laser-beam traveling way. For the chosen three laser-beam traveling ways at 0% overlap, the same amount of energy $E = 14.4 \ J/mm^2$ was provided. Each laser-beam passage across the specimen surface induces gradual heating of the material, the result of which is preheating of the material before the next laser-beam passage. The increased temperature of the specimen material makes the yield point of the material slightly lower, which may, with the given internal stresses, result in strain of the specimen. How much the specimen will preheat depends on laser-beam traveling way. Considering the three different laser-beam traveling ways across the specimen surface, the following can be stated:

- The lowest maximum temperature is reached at circular beam traveling way, starting in the middle, of the remelted area. In this case, a considerable high cooling rate in the remelted layer is ensured, which has influence on the amount of the residual austenite.
- Higher temperatures are achieved at laser-beam traveling way. From known volume changes in phase transformations, a smaller amount of residual austenite has an influence on the size of tensile residual stress profiles in the modified layer.

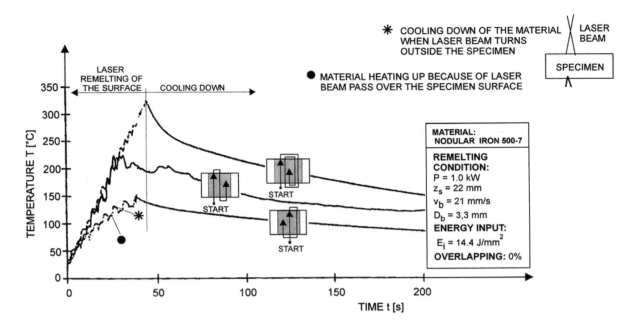

Figure 103 Temperature measured on bottom side of the specimen during laser surface remelting at various laser beam travels. (From Ref. 117.)

- Maximum temperature in the material on the bottom side of the specimen is achieved at circular mode of laser-beam traveling way beginning on the edges of the remelted area of the specimen. On completion of cooling, the remelted layer contains a smaller proportion of residual austenite, which strongly lowers the residual stress profiles.

In cooling, the just-remelted surface layers solidifies, and cooling is continued in solid state involving phase transformations, which are reflected in a characteristic microstructure in the lower area of the modified surface layer. As a result of microstructural changes in solid state, i.e., due to the austenite–martensite transformation, an increase in the volume of

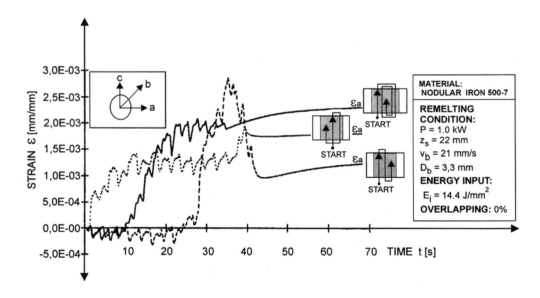

Figure 104 Time dependence of specimen strain. (From Ref. 117.)

the layer takes place. The volume of the remelted layer decreases slightly, which causes the occurrence of tensile stresses in it during the process of cooling or at ambient temperature. The decrease in the volume of the remelted surface layer is greater than the increase in the volume of the hardened layer. Therefore the result is the strain of the specimen.

In Fig. 104, results of strain measurements with different travelling directions are presented. The following can be seen:

- A very different progress of strain changes in the particular directions.
- The largest strain ε_a is found in the direction of the longer side of the specimen.
- In the direction of the shorter side of the specimen, i.e., in the transverse direction, the strain ε_c is at first of tensile nature but after six passes it changes into a compressive one.

A very important conclusion is that the size of tensile strain on the bottom side of the specimen is increasing with the increase in temperature of the specimen material.

Continuous measurement of specimen strain during laser surface remelting with a resistance-measuring rosette is a new method, not yet reported in the literature, of describing the strain events in the specimen. It has been found that the information about the time-dependent changes in specimen strain contributes to better knowledge of the conditions during laser remelting and thus better process of optimization.

XXIII. OPTIMIZATION OF THE LASER SURFACE REMELTING PROCESS

For making decisions in determining the optimal conditions Grum and Šturm [118] worked out a set of descriptive evaluation criteria regarding the condition of the remelted layer, hardened layer, and the surface. This enabled them to define the highest and lowest possible specimen traveling speed which will still ensure a sufficient thickness of the modified layer of acceptable quality.

Optimization of the laser surface remelting process is related to the selection of the remelting conditions such as laser power, the spot diameter at the surface of the workpiece, and the relative speed of movement between the laser beam and the workpiece. If the power of laser radiation is directed to the surface of the workpiece, then the radiated area is defined with the spot diameter and power density. The calculated power density can represent a satisfactory relative comparison of the

remelting processes, which occur at different remelting conditions. Actually, it is the energy input that is important for the evaluation of remelting, which is dependent on the power density as well as on the traveling speed of the laser beam. The effect of the input energy is related to material absorptivity.

Optimization of the laser surface-remelting process for a thin-surface layer can be accomplished in four ways, i.e.:

- Selection of heating conditions for a given remelting depth;
- Selection of heating conditions and the mode of guiding the laser beam by minimizing the strain of the machine parts;
- Selection of heating conditions for the formation of compressive or minimum tensile residual stresses at the surface and in the thin-surface layer;
- Selection of heating conditions for the formation of compressive or minimum tensile residual stresses at the surface and in the thin-surface layer and a requirement for a minimum strain of the machine part.

The optimization of the laser-remelting process has to be performed for the lamellar and nodular iron separately [119].

XXIV. FATIGUE PROPERTIES OF LASER SURFACE HARDENED MATERIAL

The fatigue properties of ferritic S.G. iron were studied with the pull–pull test [113,120]. The mean tensile stress was 50% of the yield point, i.e., ≈ 150 N mm^2. A 10-Hz frequency of various amplitudes was then applied to the specimens and the fatigue limit determined. Untreated specimens were ground after machining. As there is no relevant data in the literature, optimized processing parameters for laser surface hardening had to be defined. This was done by keeping the depth of the hardened layer as a constant at about 10% of the thickness. Compared to the untreated S.G. iron, a decrease in fatigue limit was found at laser surface hardened specimens. In addition, a spiral laser track gives better results than a longitudinal multitrack. Grinding the surface, i.e., smoothing small amounts of roughness produced at laser surface hardening specimens, leads to a small decrease in fatigue limit as compared with untreated specimens.

The most favorable values (Fig. 105) were found when the specimens were annealed at 240°C for 2 hr after laser surface hardening with helium and subse-

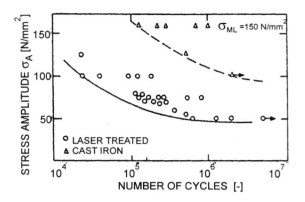

Figure 105 Fatigue behavior of laser-remelted S.G. iron with ferrite matrix. (From Ref. 120.)

quently ground [113,120]. It can be seen that the untreated specimens for a medium mean load can carry a higher amplitude. However, with increasing load the advantage of strengthening due to laser remelting allows operation at mean loads over the yield point of the untreated material.

In Fig. 106 the corresponding Smith diagram gives the fatigue life as a function of the mean load [120]. Pure fatigue behavior was found in laser surface hard-

ened specimens, which may partly depend on the surface finish and partly on residual stresses. Grinding at the surface improves the fatigue behavior but is not able to give properties equivalent to the untreated specimens.

XXV. WEAR PROPERTIES OF LASER SURFACE HARDENED MATERIAL

The advantage of laser-remelted S.G. iron was demonstrated for rolls which run dry against each other with a fixed relative slip, one wheel being driven and the other partially braked (Fig. 107). Various stresses were applied (e.g., 500 and 1000 N mm^{-2}) and the humidity was controlled. For comparison, ~ 1% slip was used. It is obvious that the wear properties of laser-remelted cast iron are superior to those of conventional hardened steels. Combinations of lasered irons and steels must be avoided, as significant deformation of the steel occurs. When laser treatments were carried out under He, better results were found than with other gases. Excellent results were obtained when laser-remelted S.G. irons were used in combination with TiN- or TiC-coated,

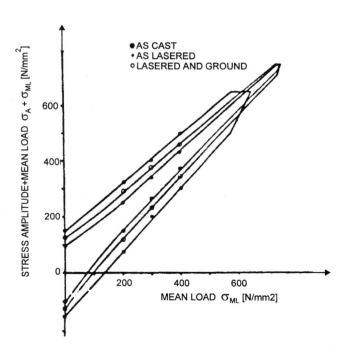

Figure 106 Smith diagram. (From Ref. 120.)

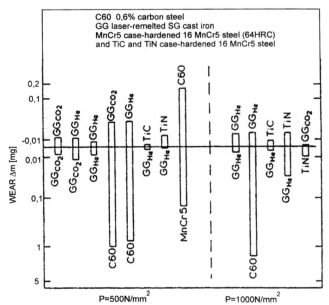

Figure 107 Comparison of wear properties of laser and TIG surface remelted S.G. iron with those of various steels, with and without surface treatments: P = load on rolls during test. (From Ref. 120.)

hardened steels, the TiN/TiC surface exposed to the wear. After the tests, no deformation is visible on the ledeburitic surface. Good wear properties are obtained in the combination of laser-treated S.G. iron with nitrided or borided steels. A comparison between laser-treated and TIG-remelted ledeburitic surfaces favors the laser hardening process.

Grum and Šturm [121] presented research results of laser surface remelting on different qualities of gray and nodular irons. The effects of laser hardening are expressed in the qualitative changes in the microstructure resulting in higher hardness and better wear resistance of the modified machine part surface [122,123].

In the gray iron, an extremely high hardness around 1000 $HV_{0.1}$ on the surface of the remelted area can be noted, which is falling uniformly to the value 620 $HV_{0.1}$ in the transition layer. The drop in microhardness then reoccurs at a depth of 250 μm (depth of the remelted area), which is attributed to changes in size and concentration of the residual austenite. Then follows hardened layer with a predominantly martensitic microstructure ensuring an increased hardness ranging from 700 to 830 $HV_{0.1}$. On the other hand, in the nodular iron, the remelted area displays a rather homogeneous distribution of the austenite and ledeburite giving very uniform profile of the measured microhardnesses ranging from 730 to 880 $HV_{0.1}$. In the heat-affected or hardened layer there is a fine-grained martensitic microstructure with a ferrite surrounding, the microhardness ranging from 540 to 680 $HV_{0.1}$.

Figure 108 presents a comparison of microhardness values subsequent to surface remelting by a single trace and by several overlapping traces. The results of microstructure and the measured microhardness in the heat-affected layer confirm that in surface remelting by overlapping, the remelted traces of tempering martensite into the fine bainite microstructure in the nodular iron come into effect. This effect is seen in Fig. 108B in the lowering of microhardness values. In the gray iron, on the other hand, the overlapping of the remelted traces leads to the effect of material preheating, and for this reason we can achieve greater depth of the heat-affected area. A comparison of the graphs in Fig. 108A and B shows that the effect of preheating is significant only in case of unfavorable flaky graphite.

In laser surface remelting, traces of heat treatment remain on the specimen. Their dimensions depend on laser-beam parameters, conditions of rotational and translatory motion of the specimen, and kind of material. Table 6 shows the average values of the measurements of the depth of the remelted and heat-affected layer measured from the workpiece surface, as shown on the right side of the table.

The wear resistance test was performed on an Amsler machine, the specimens being made of gray and nodular irons with the remelted surface in sliding lubricated contact with a hardened heat-treatable steel C55 (ISO). The normal loading force was 700 N at a sliding pressure of 700 N/cm^2. The wear resistance of the discussed casts with a laser-remelted surface was determined by the measurements of mass losses and

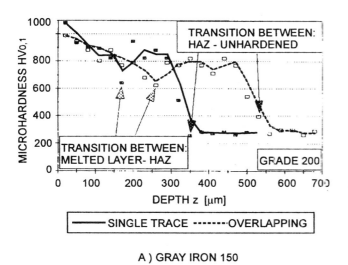

A) GRAY IRON 150

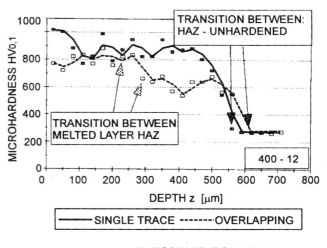

B) NODULAR IRON 400-12

Figure 108 Microhardness profiles in modified layer for single and overlapping traces. (From Ref. 121.)

Table 6 Measured Depths of Remelted and Modified Layer

	Depth z [mm]			
	Remelted layer		Modified layer	
Material	$Z_{R_{min}}$	$Z_{R_{max}}$	$Z_{M_{min}}$	$Z_{M_{max}}$
Grade 200	0.12	0.29	0.37	0.57
400-12	0.19	0.3	0.47	0.62

Source: From Ref. 121.

by defining the wear coefficient k, which is expressed as [124,125]:

$$k = \frac{W}{Fn \cdot L} \, [mm^3/Nm]$$

where W [mm³] is the volume of wear, Fn [N] is the applied load, and L [m] is the operating path.

Figure 109 is a graphical presentation of the results of cumulative loss of mass and calculated values of wear coefficient versus operating path. From the bar chart, we can see that the nodular iron wears down more than the gray iron and that the wear coefficient after the initial running-in period drops significantly in all materials (Fig. 109A). Considering the recommendation in the literature, we can, however, say that the wear coefficient is, in all the cases, substantially lower than the one allowable for machine parts ($k < 10^{-6}$ mm³/Nm)

[125]. This shows that the conditions of the discussed sliding pair are very favorable.

The cumulative loss of mass (Fig. 109B) presents the growing loss of mass after different operating paths chosen as measuring spots. The remelted surface of the flake graphite gray iron has, by almost 100 HVm, higher hardness on the surface than the nodular iron, which is reflected, as the test results have shown, in the wear resistance. Therefore the loss of mass on nodular iron specimens is much bigger, even by two or three times. On the other hand, from the experimental results of microhardness measurements and structure analysis of the different qualities of nodular iron, we can note no differences in the quality of the remelted layer. However, it was found that a bigger loss of mass on nodular iron specimens may be due to the following reasons:

• A fall-out of graphite nodules of the surface of the remelted layer during wear test.
• A smaller amount of chemically bound carbon in the cementite or ledeburite, which results in a lower hardness of the remelted layer.
• A higher ductility under the remelted layer, which results in higher deformation of the surface layer in loaded condition during the wear test.

The experimental results have confirmed that with a low-power laser source it is possible to achieve a sufficient thickness of the modified layer if surface remelting is applied. Because of the morphology of its graphite, gray iron is a very demanding material for heat treat-

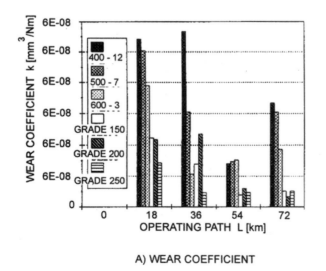

A) WEAR COEFFICIENT

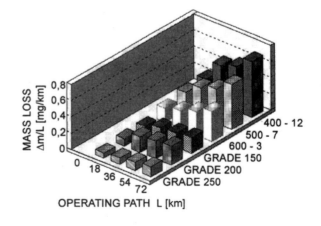

B) CUMULATIVE LOSSES OF MASS

Figure 109 Wear of gray iron 200 and nodular iron 400-12 with laser-remelted surfaces in sliding lubricated contact with a hardened heat-treatable C55 steel. (From Ref. 121.)

ment even at suitably chosen machining conditions, the obtained surface was uneven, and had small craters, cracks, and gaseous porosity in the remelted layer. On the other hand, in nodular iron no such irregularities could be observed. Yet it should be mentioned that, because of incomplete dissolution of graphite nodules, a ledeburite microstructure is obtained with a slightly lower hardness and thus a lower wear resistance compared to gray iron.

Fukuda et al. [126] carried out the wear test on the pin-and-disc machine (Fig. 110). It was found that laser-treated iron shows remarkable wear resistance by comparison with non-heat-treated steel, and also good wear resistance compared to induction surface hardened cast steel. There exists a quantitative relationship between the surface hardness of laser-treated nodular iron and its martensite portion. The surface hardness of laser-treated nodular iron is lower than that of induction surface hardened cast steel.

Transformation hardening of steels or cast irons having the pearlitic microstructure is very efficient despite the comparatively short interaction time in laser heating. Cementite lamellae in the ferritic matrix disintegrate very fast. The change into the face-centered cubic lattice with the austenitic microstructure takes a very short time, too. Moreover, some time is available for the migration of carbon atoms. The migration of carbon

atoms increases in alloyed steels with carbides of alloying elements and in gray or nodular cast irons having a larger content of carbon that will dissolve in austenite when heated and contribute to an increased hardness of the surface layer after cooling. Thus it often occurs that the hardness at the surface and in the surface layer obtained in transformation hardening is higher than that obtained after induction hardening of the same steel. Many authors monitored the variations of hardness in the hardened surface layer after laser hardening and induction surface hardening. It is extremely important that in laser hardening a higher hardness of the hardened surface layer is achieved. Then follows a very fast variation of hardness in the transition zone from the hardened to the unhardened layer. In induction hardening this transition from the hardened to the unhardened zone is very gentle, which allows a better behavior of the material of dynamically loaded parts. This indicates that induction hardening provides more favorable hardness profiles in the hardened layer and thus also a better in-operation behavior of machine parts. It seems that the martensite surrounding the graphite nodules and in the matrix, which has been transformed by the laser hardening, prevents the ferrite and the graphite nodule from deforming under the wear.

Magnusson et al. [127] in his paper titled "creating tailor made surfaces with high power CO_2-lasers"

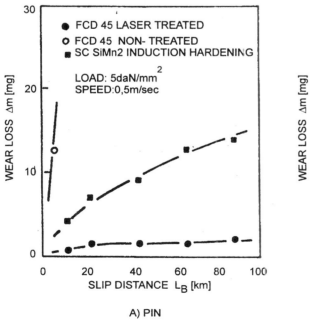

A) PIN

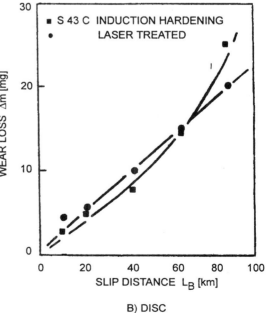

B) DISC

Figure 110 Wear loss on pin and disc machine as a function of slip distance. (From Ref. 126.)

Table 7   Conditions Used in the Dry Wear Tests

Type of specimen	Pin type	Load F [daN]	Surface preparation	
As cast	Steel*	2.5	As received (ground)	*Hardness 927 HV grid 1000
Hardened	SI_3N4	5.0	Ground on SiC paper	
Remelted	SI_3N4	5.0	As received (ground)	

Source: From Ref. 127.

connected various types of surface modification on wear resistance. They investigated laser hardening, remelting, and alloying of cast irons and the influence on microstructure, wear properties, and resistance to tempering.

Untreated specimens and specimens with laser-hardened single circular tracks were exposed to dry sliding wear in a conventional pin-on disc machine. Additional specimens were treated with overlapping hardened as well as remelted tracks. Test conditions are given in Table 7.

For laser-treated specimens, the wear rates are in all cases small but lower for surface remelted than for surface hardened specimens. The results of the wear tests are presented in Fig. 111. For the laser-treated specimens, the wear process leads to tempering of the surface layer, which gives a decrease in hardness of 100 HV.

The wear rate as well as the wear mechanism depends on the type of cast iron. The lowest wear rate of the surface hardened and surface remelted condition was found out. On surface hardened gray cast iron and also in the case of ferritic–pearlitic nodular iron, the wear loss appears to take place by brittle fracture in the material at the graphite flakes or nodules.

The surface hardness of cast irons can be increased from about 200 to 500–800 HV by surface hardening, remelting, and alloying by a CO_2 laser. The wear rate of different cast irons will decrease by a factor of approximately 10 after surface hardening or remelting. Finally, results have shown that it is possible to combine wear resistance and tempering resistance by an appropriate selection of cast iron and laser hardening process.

XXVI.   REMELTING OF VARIOUS ALUMINUM ALLOYS

Coquerelle and Fachinetti [128] presented friction and wear of laser-treated surface, i.e., laser remelting at various aluminum alloys. Hypoeutectic, eutectic, and hypereutectic aluminum silicon alloys have been tested for identification of tribological properties at assuring the lowest possible thermal expansion coefficient. All mechanical properties were desired for such application as cylinder blocks and piston for automotive industry.

Before laser treatment the face of the flat specimens was phosphated or blacked with a graphite, so that the laser-beam absorption could be increased. The single track is larger at blacked surface conditions than after phosphatation. The remelted surface layer of the alloys when the power density was reached 10^5 W/cm^2 at only 1-mm laser-beam spot. The remelted depth increases with the portion of added silicon in the alloy at the same power density but decreases with higher travelling speed. The microstructure was determined with interdendritic space, which depends on the interaction time and chosen traveling speed. Figure 112 is the influence of the traveling speed on the interdendritic spaces in eutectic alloy (12% Si) presented. A substantial reduction of the silicon particle size has been achieved on hypereutectic

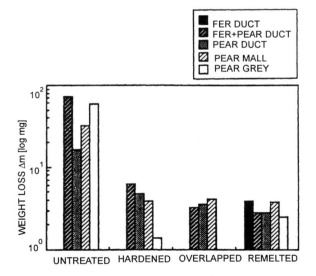

Figure 111   Wear test results for different gray cast and nodular irons. The untreated specimens were tested with a 2.5-daN load on a steel pin, the other with a 5-daN load on a Si3N4 pin. (From Ref. 127.)

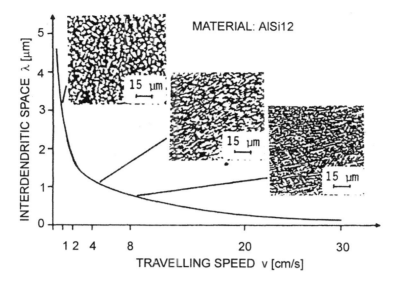

Figure 112 Influence of the traveling speed on interdendritic spaces (Si12 wt.%). (From Ref. 128.)

alloy at traveling speed $v = 1$ cm/sec below 5 μm, and at traveling speed $v = 100$ cm/sec it was smaller than 1 μm.

Figure 113A and B shows the influence of laser-beam traveling speed on surface hardness after remelting alloy with 12% Si or 17% Si. For the tribological test the remelted surface at different degrees of overlapping was made. Table 8 shows the Vickers microhardness value at loading 3 N for eutectic aluminum silicon alloy after remelting with given conditions. The microhardness of

AlSi12 alloy is 83 $HV_{0.3}$ before treatment and after single-track remelting is 160 $HV_{0.3}$. The hardness of the overlapping tracks is always lower than that of the single track.

Coquerelle and Fachinetti [128] conducted sliding and wear experiments by a pin-on-disk machine at ambient temperature. Figure 114 shows the wear rate for three aluminum silicon alloys before and after laser remelting. Wear rate depends on the silicon portion and

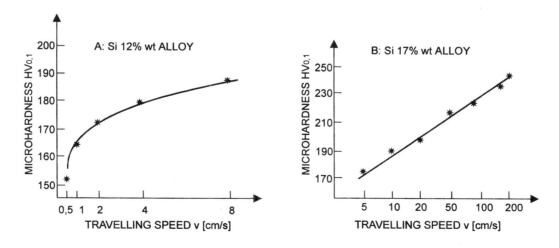

Figure 113 Influence of the traveling speed on the microhardness Al–Si alloys. (From Ref. 128.)

Table 8 Vickers Hardness Values

			Surface generation		
Eutectic alloy	Before treatment	Single scan	MODE a/2	MODE a/2	MODE β
Power: 3.0 kW	83 $HV_{0.3}$	160 $HV_{0.3}$	140 $HV_{0.3}$	100 $HV_{0.3}$	150 $HV_{0.3}$
Speed: $v = 1.0$ cm/sec		+93%	+70%	+20%	+80%

Source: From Ref. 128.

silicon particle size. The wear rate is determined by volume mass loss of the particle [mm³] on sliding path [km]. The wear rate in mm³/km is about two times greater at soft material than at fine-grained microstructures after remelting in hypoeutectic alloy. With the increase of the silicon portion in the alloy, wear rate decreased. Figure 115 shows the influence of silicon particle size on wear rate at various remelting conditions for hypereutectic alloy. If the silicon particle size decreases from 100 to 2 μm, the wear rate decreases four times. The authors presented also significant wear mechanisms for various types and states of alloys, such as the following:

- A work-hardened layer can be observed on hypoeutectic alloys (5% Si) at soft state. The depth of this layer is about 25 μm and transverse cracks are often present.
- The wear mechanism in eutectic alloy is not the same as in hypoeutectic alloy. Before laser surface

remelting, three phases—Si, $Al_8Si_6Mg_3Fe$, and Al_3Ni—are very large and harder than a matrix. Under Hertzian pressure, the harder phases were fractured and they additionally hardened the soft matrix in the surface layer. Subsequently, the wear rate is reduced. Meanwhile, the harder phases in contact with other surface react abrasively and the wear rate is still high.

- After laser remelting the size of the dendrites is lower than in the alloy or the soft state. For the eutectic alloy, almost all the phases of silicon have been melted again and after solidification the size of the particles is lower than 2 to 4 μm. The wear rate decreases by about 100%.

For the hypereutectic alloy the mechanism is similar to the one observed in the eutectic alloy. The wear rate in hypereutectic alloy is lower than that in eutectic alloy. The wear rate increases with the size of the silicon

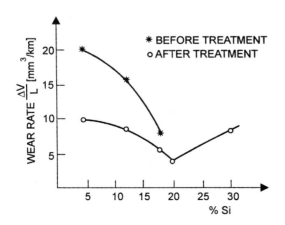

Figure 114 Influence of silicon percentage on wear rate. (From Ref. 128.)

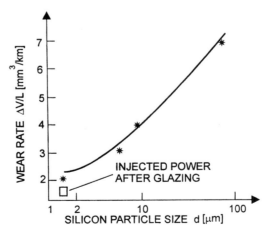

Figure 115 Influence of silicon particle size on wear rate (hypereutectic alloy). (From Ref. 128.)

particles because fracture is more often found on specimens at higher particle size.

Antona et al. [129] tested AlSi7Cu3 aluminum cast alloy at various remelting conditions. The surface layer of the individual flat specimens was assessed on the basis of metallographic inspection to point out the maximum depth, hardness, and microstructure. Figure 116 shows the depth of the remelted single laser track with three rectangular laser spot sizes and three amounts of traveling speeds. The remelted depths were measured at the center of the section taken at 20 and 40 mm from the starting point of the laser remelting. From microstructure analysis it can be concluded that the most remarkable effect of rapid solidification is the refinement of the dendritic microstructure. Figure 117 shows a microstructure at a typical area of the remelted layer.

In general the authors distinguished three areas with different microstructure in the remelted surface layer as follows:

- An interface between remelted area and base material with partially coarse microstructure and microporosities;
- An area with a very fine microstructure near the surface with typical equiaxial growth of the dendrites;
- An intermediate area between the two previously mentioned areas with column type of dendrites.

Luft et al. [130] reported on laser surface remelting of eutectic aluminum alloys.

In the past, surface remelting was researched only using eutectic or near eutectic Al–Si alloys for automotive pistons. It was observed that rapid remelting resulted in a very fine grained microstructure. Much research on rapid solidification has shown that many other alloys can have mechanical properties improved. In general nonferrous alloys were investigated in which the solubility in solid state was limited or where at high melting temperature point intermetallic compounds could be formed. Typical promising solute elements are Fe, Cr, Mn, and Ni, which have influence on the refining mechanical properties of alloys. Rapid solidification at remelting surface layer shows that the range of solid solubility can be extended. The second method for improving mechanical properties is precipitation hardening of remelted surface layer. The electron beam heat treatment must be carried out in a vacuum, which limits the size of the workpiece.

In Table 9 the composition of the investigated aluminum alloys after laser surface remelting process and the melting point of the intermetallic compounds are presented.

The laser conditions for the microstructural investigations were the same for all alloy: focused beam with 2-kW power and laser-beam traveling speed rate of 4 mm/min. The influence of the traveling speed on the remelted area for laser and electron beam is given in Fig. 118. As would be expected, laser-beam heating is less efficient than electron beam heating.

The microstructures showed fractured intermetallic compounds aligned in the rolling direction in an aluminum matrix. At laser surface remelting process these depend on the size of the melting point of the intermetallic compound and the interaction time, which have influence on the degree of dissolving, and their redistribution depends on the size of the hardened layer.

In general, segregation lines can be observed, reflecting solidification conditions and the stepwise control of

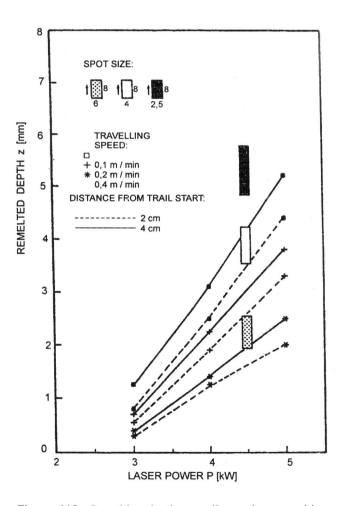

Figure 116 Remelting depth according to laser-remelting conditions. (From Ref. 129.)

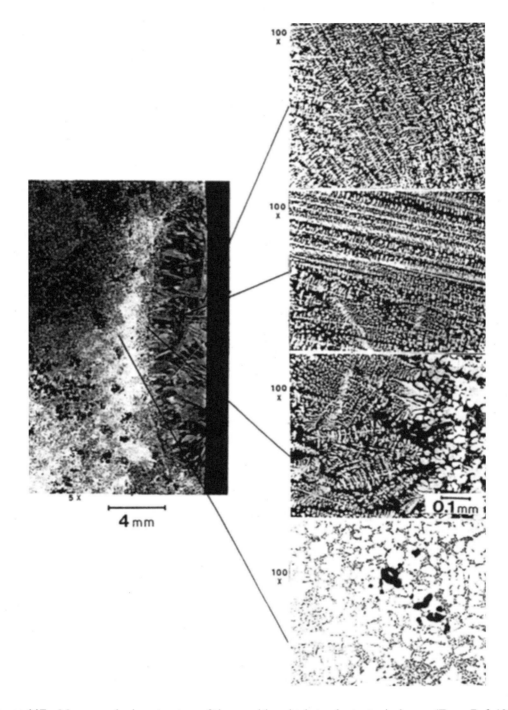

Figure 117   Macro- and microstructure of the remelting single track at a typical area. (From Ref. 129.)

Table 9 Various Compositions of Aluminum Alloys and Specific Data About Solubility, Type of Intermetallic Compounds, and Melting Temperature

Solute element	Content [wt.%]	Solubility [wt.%]	Intermetallic compounds	T_m [°C]
Cr	6	0.72	$CrAl_7$	725
			$CrAl_{11}$	940
			$CrAl_4$	1010
			$CrAl_3$	1170
Mn	6	1.40	$MnAl_6$	710
			$MnAl_4$	822
			$MnAl_3$	880
Fe	6 and 8	0.00	$FeAl_3$	1160
Ni	6	0.00	$NiAl_3$	854
Ti	1.5	0.20	$TiAl_3$	1340
Zr	1.5	0.28	$ZrAl_3$	1580
Si	11.8	1.65	—	—

Source: From Ref. 130.

the specimen movement. The degree of solubility extends according to traveling speed at given remelting conditions. In Fig. 119 the microhardness profiles across the remelted layer of various types of aluminum alloys are presented.

Based on given data for laser surface hardening by remelting various nonferrous alloys at given laser-remelting conditions the following can be concluded:

• The alloys of aluminum and transition metals such as Ti and Zr form intermetallic compounds

with the highest melting point. The high melting point of $ZnAl_3$ or $TiAl_3$ prevents complete solution in the matrix. The small amount of precipitated intermetallic compounds on point boundary or segregation line does not offer an improvement of the hardness after laser remelting.

• The aluminum–cromium alloys after remelting and subsequent cooling include precipitates of $CrAl_7$ that took place in the matrix, which improve the hardness from 45 to 100 HV. The size of the dendrites was about 1–2 μm, but it is finest at the substrate.

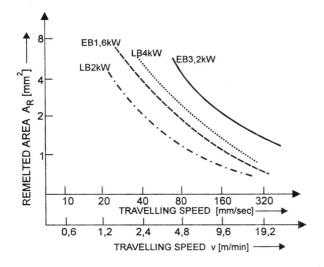

Figure 118 Remelted area as a function of traveling speed at constant power obtained by laser or electron beam. (From Ref. 130.)

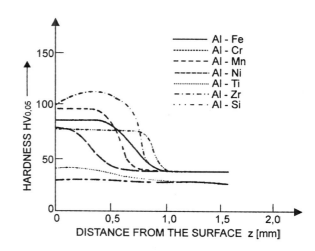

Figure 119 Microhardness profiles of laser-remelted layer of various Al alloys. (From Ref. 130.)

- The aluminum manganese alloys appear to consist of a relatively coarse lamellar eutectic with a spacing in the range from 0.3 to 0.5 μm. Close to the base material intermetallic compounds $MnAl_7$ can be found as interdendritic eutectic phase. After laser surface remelting of the thin surface layer, a hardness of 120 HV was obtained;
- Dissolution of Al_3Fe in Al–Fe alloys at laser surface remelting was complete in the remelted layer; only near the base material, i.e., in the transition layer, could it be incompletely dissolved; fine intermetallic dendrites and interdendritic eutectic are found in the transition layer. The hardness of the remelted material was 90–100 HV and the hardness of the base material is about 40 HV. After laser remelting by overlapping single laser tracks a microstructure is formed in eutectic crystals of the size between 8 and 12 μm, which contains Al_3Fe particles of the size between 1 and 2 μm. Fine intermetallic dendrites and interdendritic eutectic are found. The hardness in the remelted layer is 90–100 HV and in the base material about 40 HV.
- The aluminum and nickel alloys could only be remelted using graphite absorber coatings. The microstructure consists of fine feathery eutectic with grain size from 4 to 10 μm and lamellar spacing of 0.3 μm. Measured hardness is approximately 80 HV of the remelted layer. After annealing of the remelted specimen at 500°C for 1 hr there is no effect on changing the hardness profile through the remelted layer. After annealing of the remelted specimen at 580°C for 1 hr in Al–Fe and Al–Ni alloys a reduction in hardness was obtained.

Vollmer and Hornbogen [131] investigated various aluminum silicon alloys as follows:

- Hypoeutectic alloy with 8.0% Si
- Eutectic alloy with 12.5% Si and
- Hypereutectic alloy with 17.0% Si.

Remelting process was realized with two remelting conditions. At various interaction times, i.e., in the first case $t_i = 7.4 \times 10^{-3}$ sec and in the second case $t_i = 0.4 \times 10^3$ sec. Rapid solidification occurred during subsequent cooling by a eutectic reaction, giving a much finer eutectic spacing as from the base material. This investigation has clearly shown that the laser-remelting process depends on energy density, on the duration of laser treatment, and, finally, on chemical composition and type of base material microstructure.

The various base microstructures may be exposed to different laser heating conditions as follows:

- The temperature at the liquid–solid interface may be higher than the melting temperature of eutectic $T_E = 577°C$.
- The interface temperature may be higher than the melting temperature of aluminum $T_{Al} = 660°C$.
- The interface temperature may be higher than the melting temperature of silicon $T_{Si} = 1410°C$.

Surface remelting is of special interest for possibilities of improving surface wear resistance. The main purpose of this research was to describe melting and rapid solidification with special attention to liquid–solid interface at various heating conditions. For various heating conditions, various temperature cycles were achieved, i.e., maximum temperature solidification rate.

However, if an alloy with primary crystallization of silicon or aluminum is exposed to a temperature above T_E but below the melting temperature of pure components, mixing of aluminum and silicon is required to acquire the low melting temperature of the eutectic. Consequently, long-range diffusion is necessary until the front of melting can propagate.

As a result of this the authors expected a high velocity vm of the melting zone according to

$$v_m \sim \frac{D_f}{b}$$

for the first mentioned case, the mobility is determined principally by one atomic hop of about the atomic spacing b. The last mentioned case requires high velocity of the melting zone vm ~ Df/S long-range diffusion, which implies a diffusion path S of the order of magnitude of the size of primary crystallized particles.

The second process occurred during the melting of the hypo- and hypereutectic alloy with subsequent laser treatment. As a consequence, an inhomogeneous liquid containing undissolved crystals has formed. Evidently, there is no formation of glasses and only small amounts of supersaturated homogeneous phase form surface are present by given laser treatment conditions and alloy compositions.

Grum et al. [132] studied microstructures of various aluminum–silicon alloys after casting and the influence of laser hardening by remelting on the changes in microstructure and hardness of a modified surface layer. Their aim was to monitor changes in the thin, remelted surface layer of a specimen material in the form of thin plates with regard to remelting conditions. With the selected remelting conditions for the thin surface layer, a sufficiently high energy input into the surface of individual specimen was ensured. It varied between 165 and 477 J/mm². A comprehensive study provides a good insight into the circumstances of laser remelting of aluminum alloys and permits an efficient prediction of the microstructure, hardness level, and residual stresses in the remelted surface layer of the specimen in different remelting conditions.

As it is known, the magnitude and variation of residual stresses exert a decisive influence on the operating performance of machine parts; therefore constructors very often set requirements regarding the magnitude of residual stresses in the most stressed surface layer. To this end it is necessary to study the influences producing residual stresses, particularly in the surface layer. In order to be able to ensure the wanted properties of the surface layer, the influences on the generation of residual stresses are to be known. In the hardening of the thin surface layers by laser remelting, the following should be additionally taken into account [132]:

- Control mode for the laser-beam travel across the workpiece surface.
- Separate influences of the laser-beam power density and travel speed across the workpiece surface are to be known. By changing each of the parameters the same energy input can be ensured. Energy input may influence the size of the remelted trace and, which is even more important, overheating of both the remelted and the non-remelted specimen parts.
- Because of a relatively small width of the remelted trace on the surface in comparison to the work-

piece size, laser beam has to travel across the workpiece surface several times. With regard to the Gaussian distribution of energy in the beam, overlapping of the remelted traces has to be ensured so as to ensure also a uniform depth of the remelted layer across the entire workpiece. Consequently, in a relatively narrow range also the degree of overlapping between two neighboring remelted traces may be varied. In our case a 30% degree of overlapping of the two neighboring remelted traces was selected.

The graph in Fig. 120 presents the main stresses and the depth of the remelted layer of AlSi12CuNiMg alloy, defined by measurements of strain in a given direction and by calculating the main stresses and defining the directions of the main axes.

In the conditions of unbalanced state, i.e., at higher cooling rates, copper remains dissolved in the aluminum with a concentration higher than the balanced concentration. After laser remelting, the aluminum crystals are oversaturated with atoms of copper, which results in the segregation of copper in the form of fine inclusions and thus higher hardness of the matrix.

Residual stresses are a result of temperature and microstructural stresses occurring in the workpiece material directly after the process of remelting a thin surface layer. During the process of rapid cooling, when the process of solidification is going on, the volume of machine parts contracts, resulting in temperature stresses. However, the variation and size of residual stresses in the remelted layer depend also on the composition and homogeneity of the melt and conditions of

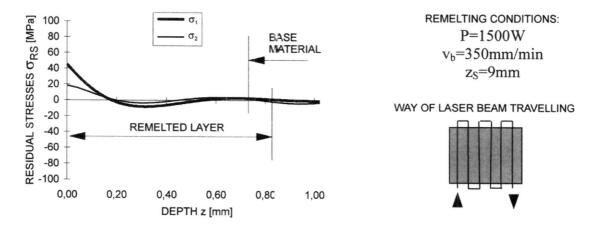

Figure 120 Residual stresses versus depth of the modified layer after laser remelting of AlSi12CuNiMg alloy. (From Ref. 132.)

cooling. The cooling conditions are very important because, at higher cooling rates, it is possible to achieve an ever finer distribution of silicon in the matrix, i.e., in the solid solution of aluminum and silicon.

Hardening of the thin surface layer may be influenced by:

- As fine as possible distribution of the silicon particles in the oversaturated solid solution of aluminum and silicon with other alloying elements;
- As fine as possible and uniform distribution of intermetallic compounds such as Al_2Cu and Ni_3Al. Regarding the nature of laser-remelting process, in which there is a melt pool, which mixes due to hydrodynamic and electromagnetic forces, around the laser beam, a rather homogeneous melt and, after rather rapid cooling, quite uniform hardness in the remelted specimen layer may be established.

Measurements of residual stresses were made on the specimen on which laser beam was led at a 15% overlapping of the remelted layer. After laser surface remelting, residual stresses of a magnitude of 20–60 MPa were identified into a depth of 0.2 mm. In the depth greater than 0.2 mm from the specimen surface, compressive and tensile residual stresses were intermittently present ranging up to 10 MPa. The variation and size of residual stresses depend on the cooling rates or time necessary for the remelted layer to cool down to ambient temperature.

Microhardness of aluminum alloys prior to laser remelting ranged, depending on the content of silicon and other intermetallic compounds, between 52 $HV_{0.1}$ with AlSi5 alloy and 10 $HV_{0.1}$ with AlSi12NiCuMg alloy. The diagram in Fig. 121 shows the variations of microhardness as a function of the depth of the laser-remelted layer of the four aluminum alloys under the same remelting conditions [132].

The diagram indicates the following findings:

- The lowest microhardness in the remelted layer was obtained in AlSi5 alloy and ranged between 60 and 65 $HV_{0.1}$. The microhardness of the hypoeutectic alloy increased from 52 $HV_{0.1}$ in the soft state to 62 $HV_{0.1}$ in the hardened condition. The microhardness increased only by around 20% with reference to the initial soft state of the alloy.
- After remelting of AlSi12 and AlSi8Cu3 eutectic alloys, very similar variations of microhardness were obtained although the microhardnesses of the two alloys differed in the soft state, i.e., it was 63 HV0.1 with AlSi12 alloy and 76 $HV_{0.1}$ with

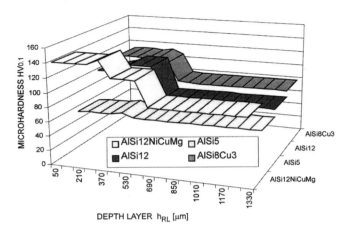

Figure 121 Through-depth variations of microhardness of four aluminum alloys; remelting conditions. $P = 1500$ W; $v_b = 400$ mm/min; $z_s = 8$ mm. (From Ref. 132.)

AlSi8Cu3 alloy. In the remelted layer a microhardness of around 100 $HV_{0.1}$ was achieved, i.e., the microhardness of the alloys increased by 30% to 40% with reference to their initial soft state. The important increase in microhardness is attributed to both fine distribution of silicon in the solid solution and Al_2Cu intermetallic compound.

- In AlSi12NiCuMg compound a maximum microhardness of up to 160 HV0.1 was obtained after remelting of the thin surface layer. Other intermetallic compounds (Al_3Ni, Mg_2Si), which were present in the matrix in a relatively fine form, contributed to a major increase in microhardness as well.

The microhardness measurements performed at specimens of different materials confirmed that the increase in microhardness depended on the type of aluminum alloy. Laser surface hardening by remelting of the thin surface layer provides a homogeneous fine-grained microstructure consisting of finely distributed silicon and intermetallic compounds in the solid solution of aluminum and silicon.

XXVII. CORROSION PROPERTIES OF LASER SURFACE REMELTED IRON

During metallographical preparation it is a common feature that laser-remelted iron in the as-quenched condition etches less than the substrate. This indicates that

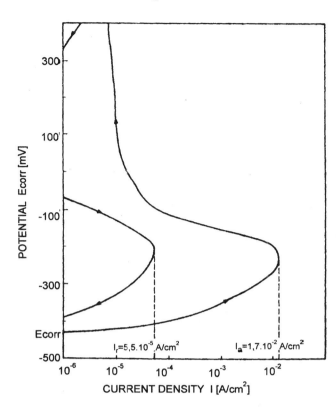

Figure 122 Polarization curve for 304 stainless steel as received. (From Ref. 133.)

the remelted material exhibits better corrosion properties [120].

From the current–density–potential curves one can derive that the remelted material is more noble than the unremelted one. The fact that this is true even for white cast substrates can only be interpreted by the fact that the higher supersaturation and small grain size prevent the $\gamma \rightarrow \alpha$ transformation almost completely so that the difference in the potential corresponds to the difference in γ- and α-iron [120]. This would explain the etching behavior in the as-quenched condition and is consistent with the fact that corrosion behavior does change when the laser-treated material is annealed. The second thing that is obvious is that for all potentials the current density is about a factor 3–6 times smaller than for the untreated material [120].

De Damborena et al. [133] studied the elimination of intergranular corrosion susceptibility of austenitic stainless steel 304 after laser surface remelting.

The typical composition of a given austenite steel is 18% Cr, 8% Ni, 0.03–0.20% C. The chromium content and crystal lattice are responsible for its good anticorrosion characteristics.

The use of this type of steel is limited by sensitization, which results in intergranular corrosion. The sensitization of stainless steel is due to the precipitation of chromium carbides at grain boundaries at temperatures of 450–900°C. The corrosion process is accelerated due to the formation of galvanic couples in affected areas, which has influence on the corrosion rate.

The solutions to this problem are as follows:

- Use of a steel with low carbon content;
- Stabilization of the steel material with titanium and niobium;
- Solution of the carbides at ≈ 1050°C, followed by a cooling rate which assures a typical microstructure.

Before the laser-remelting application of the beam the specimens were shot peened to remove any impurities and to minimize laser-beam reflection from the metal surface.

The microstructures after laser remelting are a typical rapid solidification process, mainly composed of dendritic cellular growth of austenite.

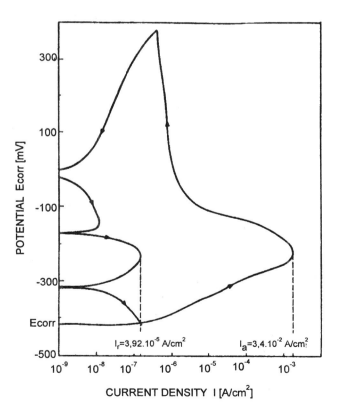

Figure 123 PPR test for material after laser surface melting. (From Ref. 133.)

Table 10 Results of Corrosion Tests

Condition	$E_{corr.}$ [mV]	I_r/I_a
As received	−429	0.0031
Sensitized	−432	0.60
Sensitized and laser treated	−400	6.6×10^{-5}
Single track	−420	0.0032
Double track: overlap 100%	−410	0.0011
Triple track: overlap 50%	−443	0.0011
Double track: overlap 75%	−405	0.0032

Source: From Ref. 133.

Three typical types of solidified microstructure are stated:

- The dendritic cellular microstructure of remelted austenite.
- The precipitates of chromium carbides at austenitic grain boundaries in HAZ have disappeared in it.
- The last region contains a fully sensitized microstructure.

Figure 122 shows the polarization curve for the austenitic stainless steel 304 before laser remelting. The I_r/I_a quotient is 0.0031, which means that the given austenitic steel is immune to intergranular corrosion.

Figure 123 shows curves obtained following laser surface remelting, showing that the remelted surface has better resistance to corrosion than that of the sensitized state of steel, and also better than the one in the received state.

Table 10 includes the data about corrosion potentials, which are similar in each case, and the I_r/I_a quotient for each case investigated.

From the research it can be concluded that conditioning the surface of a given previously sensitized austenitic stainless steel after laser remelting has a beneficial effect on intergranular corrosion resistance properties.

REFERENCES

1. Daley, W.W. Laser processing and analysis of materials. *Chapter 1: Lasers and Laser Radiation*; Plenum Press: New York, 1983; 158–162.
2. Koebner, H. Overview. *Chapter 1: Industrial Applications of Lasers*; Koebner, H., Ed.; John Wiley & Sons Ltd.: Chichester, 1984; 1–68.
3. Gregson, V.G. Chapter 4: Laser Heat Treatment. In *Laser Materials Processing*. Bass, M., Ed.; Materials Processing Theory and Practices; North-Holland Publishing Company: Amsterdam, 1983, Vol. 3, 201–234.
4. Sridhar, K.; Khanna, A.S. Laser surface heat treatment. In *Lasers in Surface Engineering*; Dahotre, N.B., Ed.; ASM International: Materials Park, Ohio, 1998; 69–179.
5. Rykalin, N.; Uglov, A.; Kokora, A. Laser melting and welding. *Chapter 3: Heat Treatment and Welding by Laser Radiation*; Mir Publisher: Moscow, 1978; 57–125.
6. Steen, W.M. Laser material processing. *Chapter 6: Laser Surface Treatment*; Springer-Verlag: London, 1996; 172–219.
7. Migliore, L. Heat treatment. In *Laser Materials Processing*. Migliore, L., Ed.; Marcel Dekker: New York, 1996; 209–238.
8. Migliore, L. Considerations for real-world laser beams. In *Laser Materials Processing*. Migliore, L., Ed.; Marcel Dekker: New York, 1996; 49–64.
9. Luxon, J.T. Propagation of laser light. In *Laser Materials Processing*. Migliore, L., Ed.; Marcel Dekker: New York, 1996; 31–48.
10. Bass, M. Chapter 1: Lasers for Laser Materials Processing. In *Laser Materials Processing*; Bass, M., Ed.; Materials Processing Theory and Practices; North-Holland Publishing Company: Amsterdam, 1983; Vol. 3, 1–14.
11. Dawes, C. *Laser Welding*; Ablington Publishing and Woodhead Publishing in Association with the Welding Institute: Cambridge, 1992; 1–95.
12. Luxon, J.T. Laser optics/beam characteristic. In *Guide to Laser Materials Processing*. Charschan, S.S., Ed.; CRC Press: Boca Raton, 1993; 57–71.
13. Bolin, S.R. Chapter 8: nd-YAG laser application survey. In *Laser Materials Processing*; Bass, M., Ed.; Materials Processing Theory and Practices; North-Holland Publishing Company: Amsterdam, 1983; Vol. 3, 407–438.
14. Nonhof, C.J. Materials processing with nd-lasers. *Chapter 1: Introduction*; Electrochemical Publications Limited: Ayr, Scotland, 1988; 1–40.
15. Steen, W.M. Laser material processing. *Chapter 1: Background and General Applications*; Springer-Verlag: London, 1996; 7–68.
16. Steen, W.M. Laser materials processing. *Chapter 2: Basic Laser Optics*; Springer-Verlag: London, 1996; 40–68.
17. Schuöcker, D. High power lasers in production engineering. *Chapter 3: Beam and Resonators*; Imperial College Press: London and Singapore: World Scientific Publishing Co. Pte. Ltd, 1999; 39–72.
18. Schuöcker, D. High power lasers in production engineering. *Chapter 4: Laser Sources*; Imperial College Press: London and Singapore: World Scientific Publishing Co. Pte. Ltd, 1999; 73–150.
19. Migliore, L. Theory of laser operation. In *Laser Mate-

rials Processing. Migliore, L., Ed.; Marcel Dekker: New York, 1996; 1–30.

20. Steffen, J. Lasers for micromechanical, electronic, and electrical tasks. Chapter 10. In *Industrial Applications of Lasers.* Koebner, H., Ed.; John Wiley & Sons: Chichester, 1984; 209–221.

21. Seaman, F.D.; Gnanamuthu, D.S. Using the industrial laser to surface harden and alloy. In *Source Book on Applications of the Laser in Metalworking*; Metzbower, E.A., Ed.; American Society for Metals: Metals Park, Ohio, 1981; 179–184.

22. Charschan, S.S.; Webb, R. Chapter 9: considerations for lasers in manufacturing. In *Laser Materials Processing*; Bass, M., Ed.; Materials Processing Theory and Practices; North-Holland Publishing Company: Amsterdam, 1983; Vol. 3, 439–473.

23. Kawasumi, H. Metal surface hardening CO_2 laser. In *Source Book on Applications of the Laser in Metalworking*; Metzbower, E.A., Ed.; American Society for Metals: Metals Park, Ohio, 1981; 185–194.

24. Amende, W. Transformation Hardening of Steel and Cast Iron with High-Power Lasers. In *Industrial Applications of Lasers.* Koebner, H., Ed.; John Wiley & Sons: Chichester, 1984; 79–99; Chapter 3.

25. Meijer, J.; Kuilboer, R.B.; Kirner, P.K.; Rund, M. Laser beam hardening: transferability of machining parameters. Proceedings of the 26th International CIRP Seminar on Manufacturing Systems—LANE'94. In *Laser Assisted Net Shape Engineering*; Geiger, M. Vollertsen, F., Eds.; Meisenbach-Verlag: Erlangen, Bamberg, 1994; 234–252.

26. Belforte, D., Levitt, M., Eds.; *The Industrial Laser Handbook,* Section 1, 1992–1993 Ed. Springer-Verlag: New York, 1992; 13–32.

27. Carslaw, H.C.; Jaeger, J.C. Conduction of heat in solids. Chapter II: Linear Flow of Heat: The Infinite and Semi-infinite Solid, 2nd Ed. 50–91 Oxford University Press:1986; 50–91.

28. Bass, M. Laser Heating of Solids. In *Physical Processes in Laser–Materials Interactions*; Bertolotti, M., Ed.; Plenum Press: New York and London, 1983; 77–116. Published in Cooperation with NATO Scientific Affairs Division.

29. Ashby, M.F.; Easterling, K.E. The Transformation hardening of steel surface by laser beam: I. hypoeutectoid steels. Acta. Metall. 1984, *32*, 1935–1948.

30. Li, W.B.; Easterling, K.E.; Ashby, M.F. Laser transformation hardening of steel: II. hypereutectoid steels. Acta. Metall. 1986, *34*, 1533–1543.

31. Kou, S. Welding glazing, and heat treating—a dimensional analysis of heat flow. Metall. Trans., A. 1982, *13A*, 363–371.

32. Kou, S. Sun, D.K. A fundamental study of laser transformation hardening. Metall. Trans., A. 1983, *14A*, 643–653.

33. Shercliff, H.R.; Ashby, M.F. The prediction of case depth in laser transformation hardening. Metall. Trans., A. 1991, *22A*, 2459–2466.

34. Festa, R.; Manza, O.; Naso, V. Simplified thermal models in laser and electron beam surface hardening. Int. J. Heat Mass Transfer. 1990, *33*, 2511–2513.

35. Mazumder, J. Laser heat treatment: the state of the art. J. Met. 1983; 18–26.

36. Gregson, V. Laser heat treatment. Paper no. 12. Proc. 1st USA/Japan Laser Processing Conf., LIA Toledo, Ohio, 1981.

37. Sandven, O.A. Heat Flow in Cylindrical Bodies during Laser Surface Transformation Hardening. In *Laser Application in Materials Processing*; Proc. of the Society of Photo-Optical Instrumentation Engineers, SPIE, Washington; Ready, J.F., Ed.; San Diego, California, 1980; Vol. 198, 138–143.

38. Cline, H.E.; Anthony, T.R. Heat treating and melting material with a scanning laser or electron beam. J. Appl. Phys. 1977, *48*, 3895–3900.

39. Grum, J.; Šturm, R. Calculation of temperature cycles heating and quenching rates during laser melt-hardening of cast iron. In *Surface Engineering and Functional Materials,* Proc. of the 5th European Conf. on Advanced Materials and Processes and Applications, Materials, Functionality & Design; Maastricht, NL, Sarton, L.A.J.L., Zeedijk, H.B., Eds.; Published by the Netherlands Society for Materials Science Aj Zwijndrecht; 3/155–3/159.

40. Engel, S.L. Section IV. Surface hardening—basics of laser heat treating. In *Source Book on Applications of the Laser in Metalworking*; Metzbower, E.A., Ed.; American Society for Metals: Metals Park, Ohio, 1981; 149–171.

41. Tizian, A.; Giordano, L.; Ramous, E. Laser surface treatment by rapid solidification. In *Laser in Materials Processing.* Metzbower, E.A., Ed.; Conference Proceedings; American Society for Metals: Metals Park, Ohio, 1983; 108–115.

42. Mordike, S.; Puel, D.R.; Szengel, H. Laser Oberflächenbehandlung - ein Productionsreifes Verfahren für Vielfältige Anwendungen. In *New Technology for Heat Treating of the Metals, Conference Proceedings,* Liščić, B., Ed.; Croatian Society for Heat Treatment: Zagreb, Croatia, 1990; 1–12.

43. Steen, W.M. Laser cladding, alloying and melting. In *The Industrial Laser Annual Handbook 1986*; Belforte, D. Levitt, M., Eds.; Penn Well Books, Laser Focus: Tulsa Oklahoma, 1986; 158–174.

44. Bramson, M.A. Infrared radiation. *A Handbook for Applications*; Plenum Press: New York, 1968.

45. Nonhof, C.J. Materials Processing with Nd-Lasers. *Chapter 5: Absorption and Reflection of Materials;* Electrochemical Publications Limited: Ayr, Scotland, 1998; 147–163.

46. Rykalin, N.; Uglov, A.; Zuer, I.; Kokora, A. Laser and electron beam material processing handbook. *Chapter*

1: Lasers and Laser Radiation; Mir Publisher: Moscow, 1988; 9–73.

47. Rykalin, N.; Uglov, A.; Kokora, A. Laser melting and welding. *Chapter 1: Basic Physical Effects of Laser Radiation on Opaque Mediums*; Mir Publisher: Moscow, 1978; 9–40.

48. Rykalin, N.; Uglov, A.; Kokora, A. Laser melting and welding. *Chapter 2: Techniques for Studying Laser Radiation Effects on Opaque Materials*; Mir Publisher: Moscow, 1978; 4156.

49. Ready, J.F. Absorption of laser energy. In *Guide to Laser Materials Processing*. Charschan, S.S., Ed.; CRC Press: Boca Raton, 1993; 73–95.

50. Migliore, L. Laser-Material Interactions. In *Laser Materials Processing*, Migliore, L., Ed.; Marcel Dekker: New York, 1996; 65–88.

51. Schuöcker, D. *High Power Lasers in Production Engineering*; Imperial College Press and World Scientific Publishing: London, 1987; 1–448.

52. von Allmen, M.; Blatter, A. *Laser-Beam Interactions with Materials: Physical Principles and Applications*; Springer-Verlag: Berlin, 1999; 6–48.

53. Wissenbach, K.; Gillner, A.; Dausinger, F. Transformation Hardening by CO_2 Laser Radiation, Laser und Optoelektronic; AT-Fachferlach, Stuttgart, 1985; Vol. 3, 291–296.

54. Beyer, E.; Wissenbach, K. *Oberflächenbehandlung mit Laserstrahlung*; Springer-Verlag: Allgemaine Grundlagen, Berlin, 1998; 19–83.

55. Guanamuthu, D.S.; Shankar, V. Laser heat treatment of iron-base alloys. In *Laser Surface Treatment of Metals*. Draper, C.V. Mazzoldi, P., Eds.; NATO ASI, Series-No. 115, Martinus Nijhoff Publishers: Dordrecht, 1986; 413–433.

56. Rothe, R.; Chatterjee-Fischer, R.; Sepold, G. Hardening with Laser Beams. Proceedings of the 3rd International Colloquium on Welding and Melting by Electrons and Laser Beams, Organized by Le Commisariat a l'Energie Atomique l'Institut de Soudure; Contre, M., Kuncevic, M., Eds.; Lyon, France, 1983; Vol. 2, 211–218.

57. Kechemair, D.; Gerbet, D. Laser metal hardening: models and measures. In Proc. of the 3rd Int. Conf on Lasers in Manufacturing (LIM-3). Quenzer, A., Ed.; IFS Publications: Bedford; Springer-Verlag, Berlin, 1986; 261–270.

58. Woo, H.G.; Cho, H.S. Estimation of hardened laser dimensions in laser surface hardening processes with variations of coating thickness. Surf. Coat. Technol. 1998, *102*, 205–217.

59. Trafford, D.N.H.; Bell, T.; Megaw, J.H.P.C.; Bransden, A.S. heat treatment using a high power Laser. Heat Treatment'79; The Metal Society: London, 1979; 32–44.

60. Inagaki, M.; Jimbou, R.; Shiono, S. Absorptive Surface Coatings for CO_2 Laser Transformation Hardening, Proceedings of the 3rd International Colloquium on Welding and Melting by Electrons and Laser Beam, Organized by Le Commisariat a l'Energie Atomique l'Institut de Soudure; Contre, M., Kuncevic, M., Eds.; Lyon, France, 1983; Vol. 1, 183–190.

61. Grum, J.; Božič, S.; Šturm, R. Measuring and analysis of residual stresses after laser remelting of various aluminium alloys. In Proc. of the 7th Int. Seminar of IFHT, Heat Treatment and Surface Engineering of Light Alloys, Budapest, Hungary; Lendvai, J., Reti, T., Eds.; Hungarian Scientific Society of Mechanical Engineering (GTE); 507–516.

62. von Allmen, M. Laser-beam interactions with materials: Physical Principles and Applications. *Chapter 2: Absorption of Laser Light*; Springer-Verlag: Berlin, 1987; 6–48.

63. von Allmen, M. Laser-beam interactions with materials: physical principles and applications. *Chapter 3: Heating by Laser Light*; Springer-Verlag: Berlin, 1987; 49–82.

64. von Allmen, M. Laser-beam interactions with materials: physical principles and applications. *Chapter 4: Melting and Solidification*; Springer-Verlag: Berlin, 1987; 83–145.

65. Daley, W.W. Laser processing and analysis of materials. *Chapter 1: Lasers and Laser Radiation*; Plenum Press: New York, 1983; 1–110.

66. Gay, P. Application of mathematical heat transfer analysis to high-power CO_2 laser material processing: treatment parameter prediction, absorption coefficient measurements. In *Laser Surface Treatment of Metals*. Draper, C.W. Mazzoldi, P., Eds.; Martinus Nijhoff Publishers in cooperation with NATO Scientific Affairs Division: Boston, 1986; 201–212.

67. Rykalin, N.; Uglov, A.; Zuer, I.; Kokora, A. Laser and electron beam material processing handbook. *Chapter 3: Thermal Processes in Interaction Zones*; Mir Publisher: Moscow, 1988; 98–167.

68. Steen, W.M. Laser Material Processing. *Chapter 5: Heat Flow Theory*; Springer-Verlag: London, 1996; 145–171.

69. Guangjun, Z.; Qidun, Y.; Yungkong, W.; Baorong, S. Laser transformation hardening of precision v-slideway. In *Proceedings of the 3rd Int. Congress on Heat Treatment of Materials*, Shanghai, 1983. Bell, T., Ed.; The Metals Society London: 1984; 2.9–2.18.

70. Meijer, J.; Seegers, M.; Vroegop, P.H.; Wes, G.J.W. Line hardening by low-power CO_2 lasers. In *Laser Welding, Machining and Materials Processing*. Proceedings of the International Conference on Applications of Lasers and Electro-Optics "ICALEO'85", San Francisco, 1985; Albright, C., Eds.; Springer-Verlag, Laser Institute of America: Berlin, 1986; 229–238.

71. Arata, Y.; Inoue, K.; Maruo, H.; Miyamoto, I. Application of laser for material processing—heat flow in laser hardening. In *Plasma, Electron & Laser Beam Technology, Development and Use in Materials*

Processing: Arata, Y., Eds.; American Society for Metals: Metals Park, Ohio, 1986; 550–557.

72. Grum, J.; Žerovnik, P. Laser hardening steels, Part 1. Heat treating. Vol. 25-7, July 16–20, 1993.

73. Gutu, I.; Mihâilescu, I.N.; Comaniciu, N.; Drâgânescu, V.; Denghel, N.; Mehlmann, A. Heat treatment of gears in oil pumping units reductor. In *Proceedings of SPIE— The International Society for Optical Engineering*; Fagan, W.F., Ed.; Washington: Industrial Applications of Laser Technology, Geneva, 1983; Vol. 398, 393–397.

74. Ursu, I.; Nistor, L.C.; Teodorescu, V.S.; Mihâilescu, I.N.; Apostol, I.; Nanu, L.; Prokhorov, A.M.; Chapliev, N.I.; Konov, V.I.; Tokarev, V.N.; Ralchenko, V.G. Continuous Wave Laser oxidation of Copper. In *Industrial Applications of Laser Technology*. Fagan, W.F., Ed.; Proc. of SPIE, The Int. Society for Optical Engineering: Washington, 1983; Vol. 398, 398–402.

75. Engel, S.L. Basics of Laser Heat Treating. In *Source Book on Applications of the Laser in Metalworking. A Comprehensive Collection of Outstanding Articles from the Periodical and Reference Literature*; Metzbower, E.A. American Society for Metals: Metals Park, Ohio, 1979; 49–171.

76. Toshiba CO_2 Laser Machining System; Toshiba Corp. Shiyodo-ku: Tokyo, 18 pp.

77. Schachrei, A.; Casbellani, M. Application of high power lasers in manufacturing. Keynote papers. Ann. CIRP 1979, 28, 457–471.

78. Carroz, J. Laser in high rate industrial production automated systems and laser robotics. In *Laser in Manufacturing*, Proc. of the 3rd Int. Conf. Paris; IFS: France, Bedford, 1986 Quenzer, A. Springer-Verlag, Berlin, 1986; 345–354.

79. Mordike, S.; Puel, D.R.; Szengel, H. Laser overflächenbehandlung - ein Productionsreifes Verfahren für Vielfältige Anwendungen. In *New Technologies in Heat Treating of Metals*; Croatian Society for Heat Treatment: Zagreb, Croatia, Liščić, B., Ed.; 1–12.

80. Marinoin, G.; Maccogno, A.; Robino, E. Technical and economic comparison of laser technology with the conventional technologies for welding. In *Proc. 5th Int. Conf. Lasers in Manufacturing*, Birmingham. Steen, W.M., Eds.; IFS Publication: Bedford Springer-Verlag, Berlin, 1989; 105–120.

81. Pantelis, D.I. Excimer laser surface modification of engineering metallic materials: case studies. In *Lasers in Surface Engineering*. Dahotre, N.B. Ed.; ASM International: Materials Park, Ohio, 1998; 179–204.

82. Steen, W.M. Laser material processing. *Chapter 7: Laser Automation and In-Process Sensing*; Springer-Verlag: London, 1996; 220–243.

83. Sona, A. Lasers for surface engineering: fundamentals and types. In *Lasers in Surface Engineering*. Dahotre, N.B., Ed.; ASM International: Materials Park, Ohio, 1998; 1–33.

84. Shono, S.; Ishide, T.; Mega, M. Uniforming of Laser Beam Distribution and Its Application to Surface Treatment, Takasago Research & Development Center, Mitsubishi Heavy Industries, Ltd, Japan; Institute of Welding; IIW-DOC-IV-450-88, 1988; 1–17.

85. Kreutz, E.V.; Schloms, R.; Wissenbach, K. Absorbtion von Laserstrahlung. In *Werkstoffbearbeitung mit Laserstrahlung: Grundlagen - Systeme - Verfahren*; Herziger, G. Loosen, P., Eds.; Carl Hanser Verlag: München, 1993; 78–87.

86. von Allmaen, M. Laser-beam interactions with materials. *Physical Principles and Applications. Chapter 2: Absorbtion of Laser Light*; Springer-Verlag: Berlin, 1987; 6–48.

87. Grum, J. Possibilities of kaleidoscope use for low power lasers. In *Conf. Proc. Heat Treating: Equipment and Processes*; Totten, G.E. Wallis, R.A., Eds.; ASM International: Materials Park, Ohio, 1994; 265–274.

88. Field, M.; Kahles, J.F. Review of surface integrity of machined components. Ann. CIRP 1970, 20, 107–108.

89. Field, M.; Kahles, J.F.; Cammet, J.T. Review of measuring method for surface integrity. Annals. CIRP 1971, 21, 219–237.

90. Yang, Y.S.; Na, S.J. A study on the thermal and residual stress by welding and laser surface hardening using a new two-dimensional finite element model. Proc. Inst. Mech. Eng. 1990, 204, 167–173.

91. Li, W.B.; Easterling, K.E. Residual stresses in laser transformation hardened steel. Surf. Eng. 1986, 2, 43–48.

92. Solina, A. Origin and development of residual stresses induced by laser surface hardening treatment. J. Heat Treat. 1984, 3, 193–203.

93. Com-Nougue, J.; Kerrand, E. Laser surface treatment for electromechanical applications. In *Laser Surface Treatment of Metals*. Draper, C.W. Mazzoldi, P., Eds.; Martinus Nijhoff Publishers in Cooperation with NATO Scientific Affairs Division: Boston, 1986; 497–571.

94. Mor, G.P. Residual Stresses Measurements by means of x-ray diffraction on electron beam welded joints and laser hardened surfaces. In *Proceedings of the 2nd International Conference on Residual Stresses "ICRS2"*. Beck, G. Denis, S., Simon, A., Eds.; 696–702 Elsevier Applied Science: Nancy, London, 1988; 696–702.

95. Ericsson, T.; Chang, Y.S.; Melander, M. Residual Stresses and Microstructures in Laser Hardened Medium and High Carbon Steels. In Proceedings of the 4th International Congress on Heat Treatment of Materials, Berlin, Int. Federation for the Heat Treatment of Materials; Vol. 2, 702–733.

96. Cassino, F.S.L.; Moulin, G.; Ji, V. Residual stresses in water-assisted laser transformation hardening of 55C1 steel. In *Proceedings of the 4th European Conference on Residual Stresses "ECRS4"*; Denis, S. Lebrun, J.L., Bourniquel, B., Barral, M., Flavenot, J.F., Eds.; Vol. 2, 839–849.

97. Grevey, D.; Maiffredy, L.; Vannes, A.B. A simple way

to estimate the level of the residual stresses after laser heating. J. Mech. Work. Technol. 1988, *16*, 65–78.

98. Chabrol, C.; Vannes, A.B. Residual stresses induced by laser surface treatment. In *Laser Surface Treatment of Metals*; Draper, C.W. Mazzoldi, P., Martinus Nijhoff Publishers in Cooperation with NATO Scientific Affairs Division: Boston, 1986; 435–450.

99. Ericsson, T.; Lin, R. Influence of laser surface hardening on fatigue properties and residual stress profiles of notched and smooth specimens. In *Proceeding of the Conference "MAT-TEC 91"*, Paris; Vincent, L., Niku-Lari, A., Eds.; Technology Transfer Series, Published by Institute for Industrial Technology Transfer (IITT) Int.; Gowruay-Sur-Marne, France, 1991; 255–260.

100. Bohne, C.; Pyzalla, A.; Reimers, W.; Heitkemper, M.; Fischer, A. *Influence of rapid heat treatment on microstructure and residual stresses of tool steels*. Eclat—European Conf. on Laser Treatment of Materials, Hanover, 1998; Werkstoff - Informationsgesellschaft GmbH: Frankfurt, 1998; 183–188.

101. Mordike, B.L. Surface treatment of materials using high power lasers, advances in surface treatments, technology–applications–effects. In *Proceedings of the AST World Conf. on Advances in Surface Treatments and Surface Finishing, Paris 1986*. Niku Lari, A. Ed.; Pergamon Press: Oxford, 1996; Vol. 5, 381–408.

102. Grum, J.; Žerovnik, P. Laser Hardening Steels, Part 2. Heat Treating; Chilton Publication Company, August, 1993; Vol. 25, No. 8, 32–36.

103. Grum, J; Žerovnik, P. Residual stresses in laser heat treatment of plane surfaces. In *Proc. of the First Int. Conf. on Quenching & Control of Distortion, Chicago, Illinois*; Totten, G., Ed.; ASM International: Materials Park, Ohio, 1992; 333–341.

104. Yang, Y.S.; Na, S.J. A study on residual stresses in laser surface hardening of a medium carbon steel. Surf. Coat. Technol. 1989, *38*, 311–324.

105. Lepski, D.; Reitzenstein, W. Estimation and Optimization of Processing Parameters in Laser Surface Hardening. In Proceedings of the 10th Meeting on Modeling of Laser Material Processing, Igls/Innsbruck; Kaplan A., Schnöcker D., Eds.; Forschungsinstitut für Hochcleistungsstrahltechnik der TüW Wien, 1995; 18 pp.

106. Meijer, J.; Kuilboer, R.B.; Kirner, P.K.; Rund, M. Laser beam hardening: transferability of machining parameters. Manuf. Syst. 1995, *24*, 135–140.

107. Kugler, P.; Gropp, S.; Dierken, R.; Gottschling, S. Temperature controlled surface hardening of industrial tools—experiences with 4kW-diode-laser. In *Proceedings of the 3rd Conference "LANE 2001": Laser Assisted Net Shape Engineering 3, Erlangen*; Geiger, M., Otto, A., Eds.; Meisenbach-Verlag GmbH: Bamberg, 2001; 191–198.

108. Marya, M.; Marya, S.K. Prediction & optimization of laser transformation hardening. In *Proceedings of the 2nd Conference "LANE'97": Laser Assisted Net Shape Engineering 2, Erlangen*; Vollersten, F., Ed.; Meisenbach-Verlag GmbH: Bamberg, 1997; 693–698.

109. Grum, J.; Šturm, R. Residual stress state after the laser surface remelting process. J. Mater. Eng. Perform. 2001, *10*, 270–281.

110. Hawkes, I.C.; Steen, W.M.; West, D.R.F. Laser Surface Melt Hardening of S.G. Irons. In Proceedings of the 1st International Conference on Laser in Manufacturing, Brighton, UK; Kimmit, M.F., Ed.; Co-published by: IFS (Publications), Bedford, UK, Ltd. and North-Holland Publishing Company: Amsterdam, 1983; 97–108.

111. Roy, A.; Manna, I. Mathematical modeling of localized melting around graphite nodules during laser surface hardening of austempered ductile iron. Opt. Lasers Eng. 2000, *34*, 369–383.

112. Grum, J.; Šturm, R. Microstructure analysis of nodular iron 400–12 after laser surface melt hardening. Mater. Charact. 1996, *37*, 81–88.

113. Bergmann, H.W. Current status of laser surface melting of cast iron. Surf. Eng. 1985, *1*, 137–155.

114. Domes, J.; Müller, D.; Bergmann, H.W.Evaluation of Residual Stresses after Laser Remelting of Cast Iron. In Deutscher Verlag fuer Schweisstechnik (DVS), 272–278.

115. Grum, J.; Šturm, R. Residual stresses on flat specimens of different kinds of grey and nodular irons after laser surface remelting. Mater. Sci. Technol. 2001, *17*, 419–424.

116. Grum, J.; Šturm, R. Residual stresses in gray and nodular irons after laser surface melt-hardening. In *Proceedings of the 5th International Conference on Residual Stresses "ICRS-5"*; Ericsson, T., Odén, M., Andersson, A., Eds.; Institute of Technology, Linköpings University: Linköping, 1997; Vol. 1, 256–261.

117. Grum, J.; Šturm, R. Deformation of specimen during laser surface remelting. J. Mater. Eng. Perform. 2000, *9*, 138–146.

118. Grum, J.; Šturm, R. Optimization of laser surface remelting process on strain and residual stress criteria. Mater. Sci. Forum 2002, *404–407*, 405–412.

119. Grum, J.; Šturm, R.; Žerovnik, P. Optimization of Laser Surface Melt-Hardening on Gray and Nodular Iron. In *Surface Treatment: Computer Methods and Experimental Measurements*; Aliabadi, M.H., Brebbia, C.A., Eds.; Computational Mechanics Publications: Boston, 1997; 259–266.

120. Bergmann, H.W. Laser surface melting of iron-base alloys. In *Laser Surface Treatment of Metals*; Draper, C.W., Mazzoldi, P., Eds.; Series E: Applied Science—No. 115, NATO ASI Series; Martinus Nijhoff Publishers: Dordracht, 1986; 351–368.

121. Grum, J.; Šturm, R. Laser surface melt-hardening of gray and nodular iron. In Proceedings of the Interna-

tional Conference on Laser Material Processing, Opatija; Croatian Society for Heat Treatment, 1995; 165–172.

122. Hawkes, I.C.; Steen, K.M.; West, D.R.F. Laser surface melt hardening of S.G. Irons, In Proceedings of the 1st International Conference on Laser in Manufacturing, Brighton, UK; Kimmit, M.F., Ed.; Co-published by: IFS (Publications), Bedford, UK, Ltd. and North-Holland Publishing Company: Amsterdam, 1983; 97–108.

123. Ricciardi, G.; Pasquini, P.; Rudilosso, S. Remelting surface hardening of cast iron by CO_2 laser. In Proceedings of the 1st International Conference on Laser in Manufacturing, Brighton, UK; Kimmit, M.F., Ed.; Co-published by: IFS (Publications), Bedford, UK, Ltd. and North-Holland Publishing Company: Amsterdam, 1983; 87–95.

124. Czichos, H. Basic tribological parameters. ASM Handbook, Volume 18; Friction, Lubrication, and Wear Technology; Volume Chairman PJ Blau; ASM International: Materials Park, Ohio, 1992; 473–479. Printed in the United States of America.

125. Czichos, H. Presentation of friction and wear data. ASM Handbook, Volume 18, Friction, Lubrication, and Wear Technology; Volume Chairman PJ Blau; ASM International: Materials Park, Ohio, 1992; 489–492. Printed in the United States of America.

126. Fukuda, T.; Kikuchi, M.; Yamanishi, A.; Kiguchi, S. Laser Hardening of Spheroidal Graphite Cast Iron. In *Proc. of the Third Int. Congress on Heat Treatment of Materials, Shanghai 1983*; Bell, T., Ed.; The Metals Society: London, 1984; 2.34–2.44.

127. Magnusson, C.F.; Wiklund, G.; Vuorinen, E.; Engström, H.; Pedersen, T.F. Creating Tailor-Made Surfaces with High Power CO_2-Lasers. Proceedings of the 1st ASM Heat Treatment and Surface Engineering ConferenceMater. Sci. Forum; 1992, Vol. 102–104, 443–458.

128. Coquerelle, G.; Fachinetti, J.L. Friction and wear of laser treated aluminium–silicon alloys. Paper Presented at the European Conf. on Laser Treatment of Materials, Bad Nauheim, 1986. In *Laser Treatment of Materials*; Mordike, B.L., Ed.; DGM Informationsgesellschaft Verlag: Oberursel, 1987; 171–178.

129. Antona, P.L.; Appiano, S.; Moschini, R. Laser furface remelting and alloying of aluminum alloys. Paper Presented at the European Conf. on Laser Treatment of Materials, Bad Nauheim, 1986. In *Laser Treatment of Materials*; Mordike, B.L., Ed.; DGM Informationsgesellschaft Verlag: Oberursel, 1987; 133–145.

130. Luft, U.; Bergmann, H.W.; Mordike, B.L. Laser Surface Melting of Aluminium Alloys. In *Paper Presented at the European Conf. on Laser Treatment of Materials, Bad Nauheim, 1986*. Mordike, B.L., Ed.; Laser Treatment of Materials; DGM Informationsgesellschaft Verlag: Oberursel, 1987; 147–161.

131. Vollmer, H.; Hornbogen, E. Microstructure of Laser Treated Al–Si-Alloys. In *Paper Presented at the European Conf. on Laser Treatment of Materials, Bad Nauheim, 1986*. Mordike, B.L., Ed.; Laser treatment of Materials; DGM Informationsgesellschaft Verlag: Oberursel, 1987; 163–170.

132. Grum, J.; Božič, S.; Šturm, R. Measuring and Analysis of Residual Stresses after Laser Remelting of Various Aluminium Alloys. In *Proc. of the 7th Int. seminar of IFHT, Heat Treatment and Surface Engineering of Light Alloys, Budapest, Hungary*; Lendvai, J., Réti, T., Eds.; 1999; 507–516.

133. de Damborena, J.; Vazquez, A.J.; Gonzalez, J.A.; West, D.R.F. Elimination of intergranular corrosion susceptibility of a sensitized 304 steel by subsequent laser surface melting. Surf. Eng. 1989, 5, 235–238.

17

Design of Steel-Intensive Quench Processes

Nikolai I. Kobasko and Wytal S. Morhuniuk
Intensive Technologies Ltd., Kiev, Ukraine

Boris K. Ushakov
Moscow State Evening Metallurgical Institute, Moscow, Russia

I. INTRODUCTION

The development of new technology of heat treatment based on the intensification of heat transfer processes during phase transformations and on the existence of the optimal depth of hard layer is of great interest. As is known, in the case of the intensification of the cooling processes for machine parts to be quenched, great thermal and structural stresses appear, which often results in the destruction of the material. For this reason, it is important to investigate current and residual stresses with regard to cooling conditions and character of phase transformations that occur in accordance with CCT or TTT diagrams for the transformation of overcooled austenite.

Here there are results of numerical investigations of current and residual stresses made on computers based on well-known methods, and the main attention is paid to the optimal layer. On the basis of investigations made, common recommendations have been given, which can be used for the improvement of the technology of heat treatment of machine parts and equipment and can be applied in machine construction and other fields of engineering. Besides, the main attention is paid to the maximum compressive stresses which are formed at the surface. These data are based on the work done in 1979–1983, results of which published in 1983 [1]. This chapter also discusses the advantages of using controlled-hardenability steels, so-called shell hardening. The practical use of controlled-hardenability steels and their great advantages have been provided by Professor Ushakov, which are described in detail in Refs. 2–4. Additional calculations have been made, which prove the existence of optimal hard layer corresponding to the maximum compressive stresses at the surface.

It should be noted that different authors dealing with intensive quenching at which the surface hard layer is formed introduced different names: shell hardening [5] and through-surface hardening (TSH) [3].

Regularities of intensive quenching using controlled-hardenability steels are also discussed, which are described in detail in Ref. 6. The optimal hard layer corresponding to maximum compressive stresses at the surface is also related to shell hardening or TSH.

The chapter has the following plan:

1. Study of the effect of the intensification of the cooling process upon the value of residual stresses during quenching. Determination of the notion of optimal depth of hard layer.
2. Investigation of the effect of size upon thermal and stress–strain state of parts to be quenched. Similarity in the distribution of residual stresses.
3. Some recommendations concerning the heat treatment of machine parts.

II. MATHEMATICAL MODELS AND METHODS OF CALCULATION OF THERMAL AND STRESS–STRAIN STATE

The coupled equation of nonstationary thermal conductivity, as known, is given in the form [1,7,8]:

$$c\rho \frac{\partial T}{\partial \tau} - \text{div}(\lambda \text{ grad } T) - \sigma_{ij}\dot{\varepsilon}_{ij}^p + \sum \rho_I l_I \dot{\varepsilon}_I = 0. \quad (1)$$

with corresponding boundary conditions for film boiling

$$\frac{\partial T}{\partial r} + \frac{\alpha_f}{\lambda}(T - T_s)|_{r=R} = 0 \quad (2)$$

and initial conditions:

$$T(r, 0) = T_0. \quad (3)$$

The transition from film boiling to nucleate boiling is made when the following is fulfilled:

$$q_{cr2} = \alpha_f(T_{per} - T_s) \quad (4)$$

where $q_{cr2} = 0.2q_{cr1}$.

At the stage of the nucleate boiling, the boundary conditions have the following form:

$$\frac{\partial T}{\partial r} + \frac{\beta^m}{\lambda_T}(T - T_s)^m|_{r=R} = 0, \quad (5)$$

$$T(r, \tau_f) = \varphi(r). \quad (6)$$

Finally, at the area of convection heat transfer, the boundary conditions are analogous to those for film boiling.

$$\frac{\partial T}{\partial r} + \frac{\alpha_{cn}}{\lambda_T}(T - T_c) = 0, \quad (7)$$

$$T(r, \tau_{nb}) = \psi(r).$$

At intensive quenching, the film boiling is absent and the main process is nucleate boiling and convection; that is, boundary conditions (5) and (7) are used.

For the determination of current and residual stresses, plasticity theory equations are used, which are presented in detail in Refs. 9–11 and have the following form:

$$\dot{\varepsilon}_{ij} = \dot{\varepsilon}_{ij}^p + \dot{\varepsilon}_{ij}^e + \dot{\varepsilon}_{ij}^T + \dot{\varepsilon}_{ij}^m + \dot{\varepsilon}_{ij}^c, \quad (8)$$

where

$$\varepsilon_{ij}^e = \frac{1-\nu}{E}\sigma_{ij} - \frac{\nu}{E}\sigma_{kk}\delta_{ij}; \quad \varepsilon_{ij}^T = \alpha_T(T - T_0)\delta_{ij},$$

$$\varepsilon_{ij}^m = \sum_{I=1}^{N} \beta_I \xi_I \delta_{ij},$$

$$\dot{\varepsilon}_{ij}^p = \Lambda\frac{\partial F}{\partial \sigma_{ij}} = \hat{G}\left(\frac{\partial F}{\partial \sigma_{kl}}\dot{\sigma}_{kl} + \frac{\partial F}{\partial T}T + \sum_{I=1}^{N}\frac{\partial F}{\partial \xi_I}\dot{\xi}_I\right)\frac{\partial F}{\partial \sigma_{ij}},$$

Here $F = F(\sigma_{ij}, \varepsilon^p, k, T, \xi_I)$,

$$\frac{1}{\hat{G}} = -\left(\frac{\partial F}{\partial \varepsilon_{mn}^p} + \frac{\partial F}{\partial k}\sigma_{mn}\right)\frac{\partial F}{\partial \sigma_{mn}}.$$

The creep strain rate is as follows:

$$\dot{\varepsilon}_{ij}^c = \frac{3}{2}A_c^{1/m}\sigma^{(n-m)/m}\varepsilon^{c(m-1)/m}s_{ij}. \quad (9)$$

A. Kinetics of Phase Transformations and Mechanical Properties of Material

Kinetics of phase transformations is described by CCT or TTT diagrams; therefore in our program, when temperature fields were computed at each step by the space and time, results of calculations were compared with thermal-kinetic diagrams with regard to the location, and the corresponding law of phase transformations was chosen at the plane of diagram.

In other words, the program is constructed in such a way that the data entered into computer include the coordinates of all areas of main phase transformations, and for each separate area, the corresponding formulas are provided for the description of the process of the phase transformations. As for this approach, there is a good agreement between data computed and results of experiments.

The regularities of phase transformations are considered in detail in works of many authors [12–16].

It should be noted that the determination of mechanical steel properties is a quite difficult and serious problem since phases are not stable and the process is fast at high temperatures. Let us note that the coefficient of linear expansion α_T is negative within the martensite range since the expansion of the martensite is observed, which is presented in Fig. 1b. Within the martensite range, superplasticity is also observed, which is reflected in Fig. 1a in the form of decline of plasticity. In calculations, these declines result in oscillations and this causes difficulties such as lower precision of numerical calculations.

The phase distribution was calculated on the basis of CCT diagram, presented in Fig. 2.

The process of the calculation is divided into time steps. At each time step, a heat conductivity problem is solved. The resulting temperature field determines mechanical characteristics and temperature load for the thermal plasticity problem. The initial conditions for

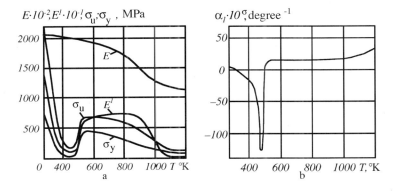

Figure 1 Variation in strength properties, elastic modulus E, hardening modulus E' (a), and coefficient of linear expansion α_T, ultimate stress σ_u, yield stress σ_y of steel AISI 52100 (ShKh15) with allowance for increase in specific volume as austenite transforms into martensite (b), as functions of temperature.

thermal conductivity and thermal plasticity problems are taken from results of solving the corresponding problems at previous steps. Besides, at each time step, the calculation results are compared with the diagram of the supercooled austenite transformations, and new thermal, physical, and mechanical characteristics for the next step are chosen depending on the structural components. In every separate region of the diagram, a specific transformation law is established, if necessary. The calculations are made until the part is cooled.

The time of reaching the maximum compressive stresses at the surface is sought by solving the same problem, in intensive quenching conditions, while the temperature of the core reaches a certain low value.

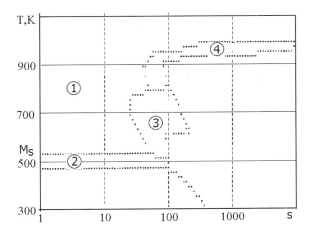

Figure 2 Diagram of the martensite–austenite transformations for steel 52100: 1—austenite, 2—martensite, 3—intermediary phase, 4—perlite.

Then, stress distribution fields are analyzed by computer or visually and the time is found when the stress distribution is optimal. After reaching maximum compressive stresses, self-tempering is made by air-cooling and it is conducted until the temperature fields are equalized at cross sections.

III. BASIC REGULARITIES OF THE FORMATION OF RESIDUAL STRESSES

The calculation was fulfilled of the thermal and stress–strain states of steel parts on the basis of finite element methods using software "TANDEM-HART" [8].

At each time and space step, the calculation results were compared with the CCT or TTT diagrams of the supercooled austenite transformation, and new thermophysical and mechanical characteristics for the next step were chosen depending on the structural components. The principal block diagram of the computing complex "TANDEM-HART" is shown in Fig. 3 [8]. The calculation results are temperature field, material phase composition, migration of points in the volume which is calculated, components of stress and strain tensors, intensities of stress and strain, and the field of the safety factor, which means the relation of stresses for which the material will be destroyed. These values are presented in the form of tables and isometric lines that allow to observe the kinetics of phase changes in the process of heating and cooling.

Using the potentialities of the computing process mentioned above, the current and residual stresses were determined depending on the cooling intensity for cylindrical specimens made of different steel grades. Sim-

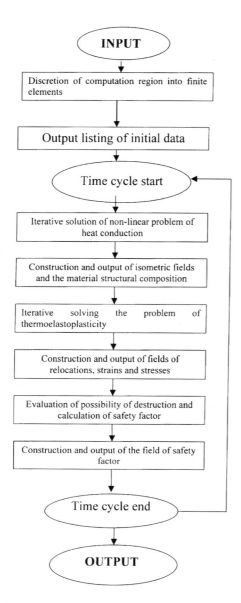

Figure 3 Chart describing the algorithm of the software package.

ilar calculations were fulfilled for quenching the parts of complex configuration.

The most important results used at developing the intensive methods of quenching were as follows.

It has been stated [1] that with the quench intensity increase, the residual stresses grow at first, then lower, with the further increase of Biot number and transfer to compression stresses (Fig. 4).

The dependence of residual stresses on the cooling rate of the specimen core at temperature of 300°C is readily represented.

It appeared that the maximum probability of quench cracks formation and maximum tensile stresses, which depend on the specimen's cooling rate, coincided [17–20].

The absence of quench cracks at quenching alloyed steels under intensive heat transfer can be explained by high compression stresses arising at the surface of the parts being quenched.

The mechanism of arising high compression stresses in the process of intensive heat transfer is described in Refs. 1, 20, and 21.

The results obtained were confirmed experimentally by measuring residual stresses at the surface of quenched parts (specimens) with the help of x-ray crystallography technique [22].

The intricate character of residual stresses dependence on the cooling rate can be explained by "super-plasticity" and variation of the phase specific volume at phase changes [23]. Under the conditions of high-forced heat transfer ($Bi \rightarrow \infty$), the part surface layer is cooled down initially to the ambient temperature, while the core temperature remains practically constant.

In the process of cooling, the surface layers must compress. However, this process is hampered by a heated and expanded core. That is why compression is balanced out by the surface layer expansion at the moment of superplasticity. The higher the temperature gradient and the part initial temperature are, the greater the surface layer expansion is. At the further cooling, the core is compressed due to which the surface layer begins to shrink towards the center and compression stresses arise in it. At the core cooling, transformation of austenite into martensite takes place. The martensite specific volume is higher than that of austenite. For this

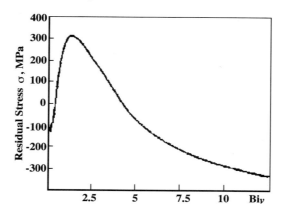

Figure 4 Residual hoop stresses at the surface of a cylindrical specimen vs. generalized Biot number.

reason, the core swelling is taking place that causes the surface layer extension at moderate cooling.

At Bi→∞, the surface layer is stretched to a maximum and therefore despite swelling, it cannot completely occupy an additional volume formed due to the external layer extension. It is just under the conditions of high-forced heat transfer that compression stresses occur in the surface layer. More detailed information about the calculation results can be found in Refs. 23–25.

A. Modeling of Residual Stress Formation

The mechanism of the formation of compressive stresses on the surface of steel parts is very important to the development of new techniques for thermal strengthening of metals, such as intensive quenching.

The reason why intensive quenching results in high compressive stresses can be explained using a simple mechanical model (Fig. 5) consisting of a set of segments (1) joined together by springs (2) to form an elastic ring. The segments are placed on a plane surface and connected with rigid threads (3), which pass through a hole (4) in the center of the ring and are attached to the opposite side of the plane surface.

Now consider the processes that take place while quenching a cylindrical steel specimen and how they would affect the behavior of the model. Assume that the specimen is being quenched under conditions of intensive cooling. In this case, the cylinder's surface layer is cooled to a certain depth, while the core remains at almost the austenizing temperature and considerably expanded in volume. Let the cooled surface layer correspond to the model's segmented ring.

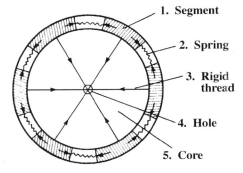

1. Segment

2. Spring

3. Rigid thread

4. Hole

5. Core

Figure 5 Mechanical model that can be used to explain the formation of hoop stresses at the surface of a cylindrical specimen during intensive quenching.

Because metals contract when cooled, the ring's segments (1) also will contract. The springs (2) will then extend by an amount that corresponds to the increase in tangential tensile stresses. However, when the surface layer is further cooled, austenite transforms to martensite, which has a high specific volume. That is why the cooled layer enlarges or swells.

Now imagine that the segments expand. In this case, resulting compression of the springs corresponds to the appearance of tangential compressive stresses on the surface of the part. With additional time, the temperature of the specimen's core drops, and its diameter decreases. In the model, the core is represented by the smaller blank circle, which is held in tension by the rigid threads.

When the threads are taut, the springs also will compress. The level of hoop compressive stresses will increase until the austenite in the core of the part transforms to martensite. The core will then start to swell because the specific volume of martensite is greater than that of austenite, which causes the compressive stresses to decrease. In the model, this would be reflected by an enlargement of the blank circle and a resulting decrease in the springs' compressive power.

B. Why Compressive Stresses Remain in the Case of Through-Hardening

At intensive quenching, the temperature of the core almost does not change at the first period of time, and the temperature at the surface instantly drops to the martensite start temperature M_s. At the surface, tensile stresses are formed. At the beginning of martensite transformations, the phenomenon of superplasticity takes place. Due to tensile stresses and superplasticity, the surface stress layer obtains the shape of the part, for example, a cylinder; that is, the surface layer becomes essentially extended. Then, martensite transformations take place, as a result of which the surface layer has increased volume. The core cools to the martensite start temperature, and because of great specific volume of the martensite, the core starts to expand. However, this volume is not enough to fill that initial volume formed by the shell. It looks like the formation of the empty space between the core and shell, and the core pulls this surface layer to itself. In Fig. 38, showing segments, this process can be illustrated by threads pulling the shell to the core. Due to it, compressive stresses are formed at the surface.

In the case of conventional slow cooling, the difference of temperatures between the surface and core at the time of reaching martensite start temperature is not large. Therefore the initial volume of the shell is not big either. In this case, when the core expands, the volume of the core becomes greater than the initial volume of the shell, and the core expands the surface layer and it causes fracture. It is similar to ice cooling in a bottle, which leads to cracking due to the expansion of the ice. The calculations of the linear elongation factor and changes in the surface layer and volume of the core support this fact. It is accounted that the specific volume of the martensite is greater than those of austenite by 4%.

C. Similarity in the Distribution of Residual Stresses

1. Opportunities of Natural and Numerical Modeling of Steel Part Quenching Process

The numerical calculation of current and residual stresses in accordance with the method described above was made for cylindrical bodies of different sizes, i.e., for cylinders having diameter of 6, 40, 50, 60, 80, 150, 200, and 300 mm. Besides, the calculations were made for 45 steel and for cases when CCT diagram is shifted to the right by 20, 100, and 1000 sec. This allowed to simulate the quenching process for alloyed steels, where the martensite formation is observed on all cross sections of parts to be quenched.

After making investigations, it has been established that in the case of fulfillment of certain conditions, the distribution of current and residual stresses is similar for cylinders of different sizes. This condition is meeting the following correlation:

$$\theta = F(\overline{\mathrm{Bi}}, \overline{\mathrm{Fo}}, r/R), \qquad (10)$$

where

$$\theta = \frac{T - T_\mathrm{m}}{T_0 - T_\mathrm{m}}; \quad \mathrm{Bi} = \frac{\alpha}{\lambda} R = \mathrm{idem};$$

$$\mathrm{Fo} = \frac{a\tau}{R^2} = \mathrm{idem}.$$

For this case, average values of heat conductivity and thermal diffusivity of the material within the range from T_m to T_0 are used. Despite of this, there is a good coincidence of the character of the distribution of current stresses in cylinders of different sizes.

Thus Fig. 6 represents the results of computations made for a cylinder of 6- and 60-mm diameter. In both, cases the martensite was formed through all cross section of the cylinder, which was fulfilled through the shift of the CCT diagram by 100 sec. For the comparison of current stresses, the first time moment was

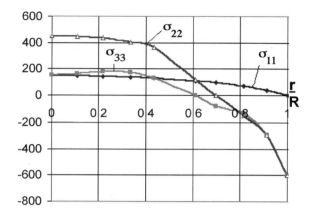

Figure 6 The distribution of stresses on the cross section of cylindrical sample of diameter of 6 and 60 mm at the time of reaching maximum compressive stresses on the surface ($\varepsilon_1 = 7$; Fo $= 0.7$); 1—sample of 6-mm diameter; 2—sample of 60-mm diameter.

chosen when compressive stresses on the surface of the cylinder to be quenched come to their maximum values. For cylinder of 6-mm diameter, this time was 0.4 sec, and for the cylinder of 60-mm diameter, the maximum compressive stresses on the surface are reached after 40 sec provided that Bi $= (\alpha/\lambda)R = \mathrm{idem}$. The latter was reached due to calculating current and residual stresses for 6-mm-diameter cylinder with $\alpha = 300\,000$ W/m^2 K, and for 60-mm-diameter cylinder, with $\alpha = 30\,000$ W/m^2 K. For both cases, Bi $= 45$. Correspondingly, for both cases, maximum compressive stresses were reached at Fo $= 0.24$, i.e., for 6-mm-diameter cylinder at $\tau = 0.4$ sec and for 60-mm-diameter at $\tau = 40$ sec.

The same values of hoop and tangential values are reached at the same correlation of r/R (Fig. 6). Thus hoop stresses for both cases are zero at $r/R = 0.65$; that is, for 6-mm-diameter cylinder, hoop stresses are zero at $r = 1.95$ mm, and for 60-mm-diameter cylinder, hoop stresses are reached at $r = 19.5$ mm correspondingly.

More accurate modeling of the hardening process can be fulfilled with the use of water–air cooling, which allows to change the heat transfer coefficient by a law set in advance. Knowing the cooling conditions, for example, for a turbine rotor, one can make them so that for the rotor and its model, the function Bi $= f(T)$ have the same value. In this case, there will be similarity in the distribution of residual stresses. In practice, it is advisable to investigate the distribution of residual stresses in large-size power machine parts by models made in accordance with theorems of similarity with regard to necessary conditions of cooling and appropriate CCT diagrams.

IV. REGULARITIES OF CURRENT STRESS DISTRIBUTION

It is of high practical interest to study the character of stress distribution in the case when the end phase through all sections is martensite. Calculations showed that at the initial moment of time on the surface, tensile stresses appear, which, while moving to the core, are changed to compressive ones (Fig. 7). While martensite appears in the surface layer, tensile stresses start to reduce, and while martensite moves to the core of the part, they become compressive. In some time, compressive stresses on the surface reach the maximum (Fig. 8), at that time inside the part tensile stresses are in force. When the part cools completely, depending on the Biot number, tensile or compressive stresses appear.

The characteristic feature of the residual stress distribution in parts quenched thoroughly is that on surface layers, there are always tensile stresses, although on the surface itself, tangential and hoop stresses are

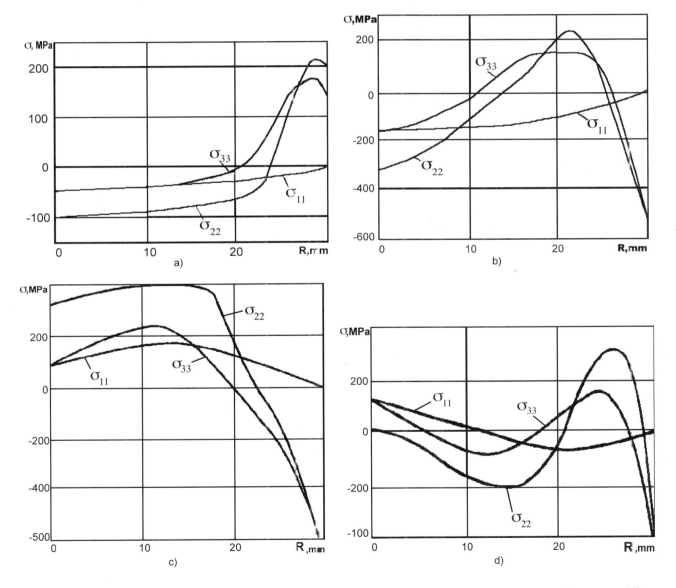

Figure 7 Stress distribution on the cross section of a cylindrical sample quenched thoroughly of 60-mm diameter at different moments of time ($\alpha = 25\,000$ W/m^2 K): (a) 1 sec; (b) 10 sec, (c) 40 sec; (d) 100 sec.

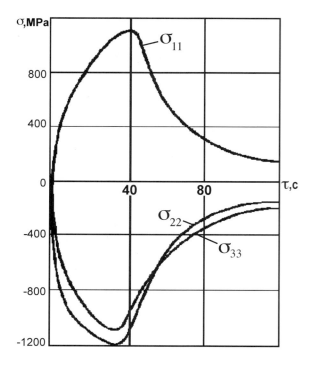

Figure 8 Stresses on the surface of cylindrical sample (diameter of 60 mm) at through-hardening (Bi = 40) vs. time.

high compressive stresses on the surface of bodies quenched. For this reason, it is of great practical interest to investigate the regularities of the distribution of current and residual stresses at parts made out of the mentioned steel grades. The depth of the hard layer can be regulated in two ways. One can either enlarge the part or shift the CCT diagram by time. In both first and second cases, a different depth of hard layer is observed, which is regulated, in practice, for the account of changes in the chemical composition of the steel. At modeling, one must consider the relative depth of hard layer. With regard to these facts, the calculation of current and residual stresses was made for cylinders of different sizes with appropriate time shifts in CCT diagrams. As a result, it has been established that at the initial moment of cooling on the surface of parts, tensile stresses appear, and in inner areas, compressive stresses appear. As the part cools and martensite begins to form in the surface layer, stresses begin to reduce and become negative. Continued martenite formation moves to the core, compressive stresses gradually increase, and if the Biot number is sufficiently large, they may be as large as −1400 MPa. It has been noted that while the Biot number increases, the compressive stresses in-

compressive. Figure 7 shows results of the stress distribution calculation in a cylinder at different moments of time which prove the above-said statement (Fig. 9).

In the case of thoroughly heating of samples or parts in which the martensite layer is formed after full cooling at a not big depth, a completely different situation is observed. On the surface of part or sample, high compressive tangential and hoop stresses appear, which change gradually to tensile stresses only while moving to inner layers.

It becomes obvious when high-alloy steel parts are quenched into water leading to quench-cracking. However, no cracking is observed for carbon steel parts quenched in the same manner.

In this connection, it is of great practical interest to study the characteristics of stress distribution of parts with different configurations as a function of the conditions of heat transfer and depth of hardness. Analyses of these results permits the development of advanced steel quenching technology.

It is well-known that one of the factors having effect on residual stress distribution is the depth of hard layer. In practice, steel grades of controlled hardenability have become widely used. Parts made out of these steel grades have longer service life for the account of the creation of

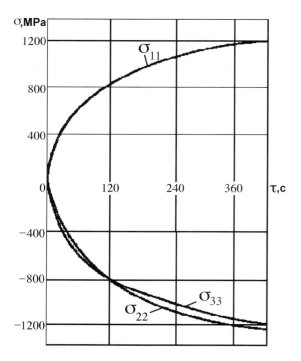

Figure 9 Stresses on the surface of a cylindrical sample (diameter of 300 mm) vs. time. The sample is made out of 45 steel; the quenching is not thorough (Bi > 100).

crease. However, there is an optimal depth of hard layer for which compressive stresses on the surface of part reach maximum values. It is well known that compressive stresses on the surface of parts to be quenched have greater strength and resistance to wearing, and as a result, they have longer service life in the case of cyclical loads. For this reason, machine parts, for example, car axles, should be made out of such steel grade that provides optimal depth of hard layer. The calculations given above show that compressive stresses in parts can be reached while using simple carbon steels without using steels of controlled hardenability like 47GT, melting of which is a very complicated process. In particular, at plant "GAZ" (Gorkogo, Russia), half-axles for GAZ cars are made out of 40G steel. The quenching process of those axles is made in intensive water flows under not big superfluous pressure.

The results obtained can be also used for making machine parts, in which it is necessary to create high compressive stresses in surface layers of the steel, which essentially prolongs the service life.

In addition to creating high compressive stresses on the surface of parts to be quenched, it is important to know the character of stress distribution on cross sections of parts (Fig. 10). As one can see from this figure, in inner layers, quite high compressive stresses are observed, which change to compressive stresses while moving to the surface. This is a distinctive feature for axial σ_{22} and hoop σ_{33} stresses. Radial stresses on the surface of quenched samples are zero; on the axis of cylinder, they are equal to hoop stresses.

The investigations made show that when quenching parts such that the Biot number is large (this is usually done for large-size machine parts), quench cracks can form inside parts.

To avoid quench cracks, for example, while quenching large-size turbine rotors, a hollow is formed at the axis of rotor, through which the quenchant penetrates in some time. In this case, in inner layers, compressive stresses appear, which are the factor preventing the quench cracking. The shortcoming of this method is essential costs for drilling hollows in large-size parts. For this reason, it is of great practical interest to develop the technology of hardening large-size parts preventing quench cracking in solid parts without axial hollows.

Joint investigations have been started between various companies for the introduction of intensive quenching methods.

A. Some Advice on Heat Treatment of Machine Parts

It is of practical interest to investigate the effect of the heat transfer intensification process while austenite is transformed into martensite. Such investigations can be used in two-step quenching, when at the first step, the transformation of austenite into martensite is delayed because parts are quenched in hot oils or liquid medium under pressure, or in high-concentration aqueous salt solutions having high temperature of boiling.

The second step of cooling is made in the area of convective heat transfer, where the transformation of austenite into martensite mainly takes place. The process of the transformation during the convective heat transfer can occur in cases of high (large Biot numbers) and low cooling rates. During the transformation of

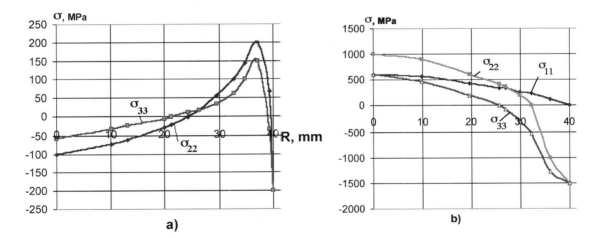

Figure 10 Stress distribution on the cross section of cylindrical sample (diameter of 80 mm) made out of 45 steel in the case of nonthorough hardening when Bi = 50: (a) after 2 sec; (b) after quenching.

austenite into martensite, the cooling process is usually slowed down in practice because of the fear that in the case of high cooling rates in the martensite range, there will be the fracture of material and big distortions. The investigations made earlier have shown that the heat transfer intensification in the martensite transformation range has a good effect upon the distribution of residual stresses and even results in the reduction of distortions. In particular, the example of two-step cooling of bearing rings made out of ShKh15 steel has shown that during the intensive cooling in the martensite range, the cone distortion of rings is significantly reduced [26]. The result obtained is practically important not only because the cooling intensification results in increase in cost and time savings, but also because in the case of intensive cooling in the martensite range, one can get high mechanical properties of material, the strength becomes higher, and plastic properties of material are improved [13].

For further proving the truth of the established fact, the simulation of carburized pin hardening has been made for a pin made out of 14KhGSN2MA steel. The carburized pin was firstly cooled at slow cooling rate in hot oil MS-20 to temperature of 220°C, and then after allowing temperature to become even on the cross section, it was quenched in quenchants having different heat transfer coefficients.

In accordance with the described technology, the simulation has been made for hardening process with setting different heat transfer coefficient at the second stage of cooling. By results of investigations, it has been established that while cooling is intensified, compressive stresses in the carburized layer grow and conical distortion of the pin is reduced. In this case, tensile stresses are changed insignificantly. Thus it has been established that cooling intensification in the martensite range has a good effect upon the distribution of residual stresses and reduces the distortion of parts quenched. This can be used for the development of advanced technology of bearing ring hardening allowing to reduce the oval distortion of rings and conical distortions while the heat transfer process is intensified; it can be also used for the development of advanced technology of heat treatment for tools and machine parts, in particular, cooling treatment. Thus, for example, carburized pins of gas-turbine aviation engine after their oil quenching were treated by cooling at temperature of 70°C. This treatment is usually made by air cooling in special refrigerators. It is advisable in the case of cooling treatment to cool carburized parts as intensively as possible, which will have a positive effect for the reasons mentioned above.

V. SHELL HARDENING OF BEARING RINGS (THROUGH-SURFACE HARDENING)

The material given above proves the existence of the optimal depth of hard layer for bodies of simple shape. It has been emphasized that there is a similarity in the distribution of hardness on the cross section for bodies of simple shape with regard to the size, i.e., $\Delta r/R =$ const, where Δr is the depth of the martensite layer (shell) and R is the radius of cylinder or ball (or half-width of the plate). It should be noted that the optimal hard layer corresponds to the best stress distribution on the surface. In this case, compressive stresses are much higher than at hardening for martensite on all the cross section. In this paragraph, we will show that the same regularities are true for the bodies of complicated shape. In particular, we will consider the process of quenching for three different bearing rings presented in Figs. 11 and 12a–c.

Let us investigate the stress distribution of these rings made of steel 52100, using mechanical properties steel as initial data. At the same time, the hardenability controlling we will make for the account of the shift on the diagram. In our case, we shifted the CCT diagram by 22 sec to the left, which is shown in Fig. 13.

The results of calculations of the stress state in the case of through-surface hardening and shell quenching for the bearing ring shown in Fig. 11 are presented below.

So Fig. 14a shows the character of the phase distribution in the case of through-surface hardening and shell quenching. Besides, the through-surface hardening was made in oil where heat transfer coefficient $\alpha = 1800$ W/m^2 K. In the case of shell quenching [Fig. 14b], the quenching was made by intensive cooling with $\alpha = 60\,000$ W/m^2 K.

Now consider the character of hoop stress distribution for these two cases, namely, though-surface hard-

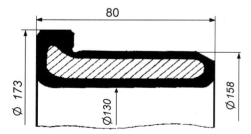

Figure 11 Dimensions and the configuration of hardened layer after TSH in inner ring of bearing for boxes of railway cars.

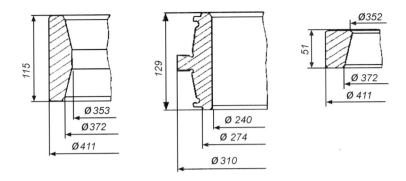

Figure 12   Dimensions of large bearing rings hardened by TSH at furnace or induction heating.

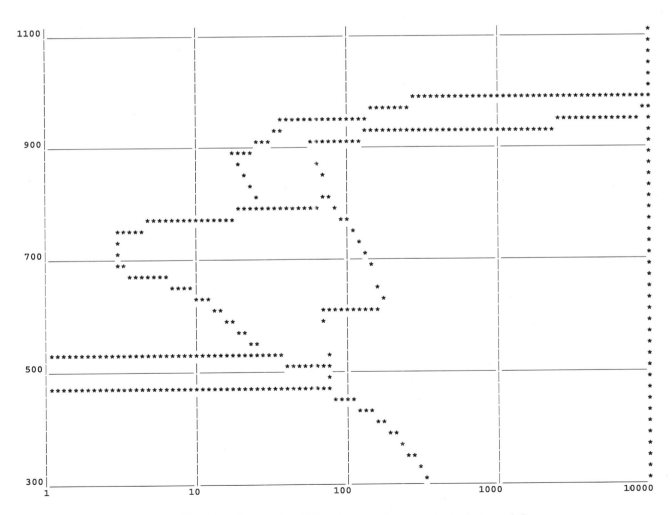

Figure 13   CCT diagram for 52100 steel with lesser content of Mn and Cr.

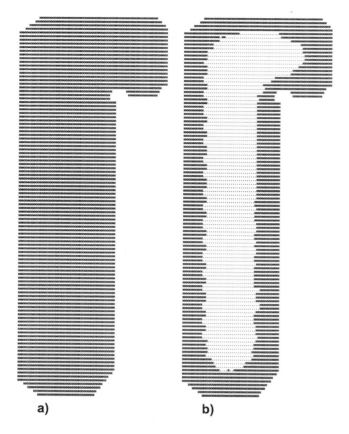

a)   **b)**

Figure 14 Phase distribution fields: (a) through-hardened, $\alpha = 1800$ W/m^2 K; (b) shell-hardened, $\alpha = 60,000$ W/m^2 K.

ening in oil [Fig. 14a] and shell hardening [Fig. 14b] In the case of through-surface hardening in oil at the surface of bearing rings, residual hoop stresses are formed reaching the following values: 180 (on the left) and 260 (on the right) (see the middle of the cross section in Fig. 15).

In the case of shell quenching with $\alpha = 60\,000$ W/m^{2}K, the stress distribution at the corresponding time is as follows: -800 (left) and -900 (right), so they are compressive, and in the core, $+500$ (tensile) (Fig. 16).

Now consider the residual hoop stresses for bodies of complicated shapes for two cases: intensive through-surface hardening and intensive shell quenching. In both cases, $\alpha = 40\,000$ W/m^2 K. Let us study the difference between the stresses at the surface.

Phase distribution fields for through-hardening and shell quenching for part shown in Fig. 12b are presented in Fig. 17.

In the case of through-surface hardening of bearing ring [Fig. 17a] at $\alpha = 40\,000$ W/m^2 K at the surface (in

the middle of the cross section at the left and at the right), not big compressive stresses are formed, which are about -120 at the left and -160 at the right (Fig. 18).

At the same time, in the case of shell quenching, huge compressive stresses are formed at both sides in the middle of the cross section (Fig. 19). These compressive stresses reach the value -1050 at the left and -1450 at the right.

Thus through-surface hardening results in not big compressive stresses at the surface, while shell quenching results in huge compressive stresses at the surface. The calculations show that when the hard layer is closer to the optimum, higher compressive stresses are at the surface.

Consider one more example illustrating this very important regularity. Let us show this fact with the rings shown in Fig. 20.

The cooling is in the same conditions: $\alpha = 40\,000$ W/ m^2 K, but in the first case, it is through-surface harden-

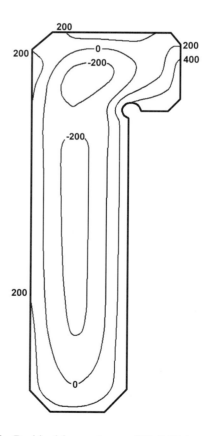

Figure 15 Residual hoop stresses S33 (MPa), $\alpha = 1800$ W/ m^2 K.

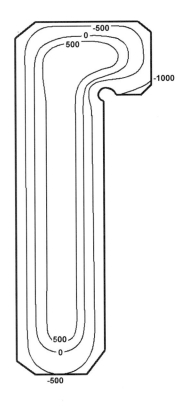

Figure 16 Residual hoop stresses S33 (MPa), $\alpha = 60,000$ W/m^2 K.

ing, and in the second case, it is shell quenching. In the case of through-surface hardening, hoop stresses in the middle are not big tensile: $+80$ (at the left) and $+480$ (at the right) (Fig. 21). At the same time, in the case of shell quenching, these stresses are big compressive: from -1200 to -1500 MPa (Fig. 22).

It should be noted that in the case of intensive through-hardening for bodies of big size and not complicated shape, it is possible to reach uniform compressive stresses at all the surface (Fig. 23).

As one can see from calculations presented above, in the case of intensive cooling, if the steel is hardened through to lesser degree, higher compressive stresses are formed at the surface. This factor must have effect upon the increase in the service life of the part. To prove the above-mentioned, inner rings of railway bearing were made of different steel grades, namely, ShKh15SG, 18KhGT, and ShKh4.

(a) Steel ShKh4 (GOST 801-78) contains 0.95–1.05%C, 0.15–0.30%Si, 0.15–0.30%Mn, 0.35–0.50%Cr, 0.015–0.050%Al, <0.020%S, <0.027%P, <0.25 %Cu, <0.30%Ni, and

(Ni + Cu) < 0.50%. The steel is intended for bearing rings with wall thickness of 18 mm and more and rollers with diameter of more than 30 mm.

(b) Steel ShKh4 (TU 14-19-33-87) is distinguished from the previous version by the regulation of alloying chromium addition depending upon the real content of nickel which is not added especially but comes to the steel from scrap at smelting. The regulation proceeds by the use of the following ratios of nickel and chromium contents:

Ni (%)	<0.10	0.08–0.20	0.18–0.30
Cr (%)	0.35–0.50	0.27–0.42	0.20–0.35

These ratios were obtained on the basis of the following equation inferred by statistical treatment of data on hardenability for 80 industrial heats of ShKh4 steel:

$$D = 13.213[1 + (\text{Ni})]^{0.764}[1 + (\text{Cr})]^{0.895}$$

where $D = $ ideal critical diameter (mm) and Ni and Cr = percentage of nickel and chromium, respectively.

The steel has a little lower and more stable hardenability level and is used for rings of railway bearings with wall thickness of 14.5 mm; steel ShKh2 has been elaborated and tested in laboratory conditions. The steel contains 1.15–1.25%C, 0.15–0.30%Si, 0.15–0.30%Mn, 0.015–0.030%Al, <0.020%S, <0.027%P, <0.15%Cr, <0.10%Ni, <0.12%Cu, and <0.03%Mo.

A. Heating During Through Surface Hardening

For TSH, comparatively slow induction heating with isothermal holding at the hardening temperature was used. The required specific high-frequency generator power is usually about 0.05–0.2 kW per 1 cm^2 of surface area of heated parts. For example, for rings of railway car bearings, total heating time is about 3 min including isothermal holding at hardening temperature (820–850°C) for 50–60 sec.

The heating procedure provides nearly uniform heating of rings of complex shape, required degree of carbides dissolution, and saturation of austenite by carbon (0.55–0.65%), the fine austenitic grain being the same (not worse than No.10, GOST 5639-82—the diameter of grains is an average of 0.01 mm).

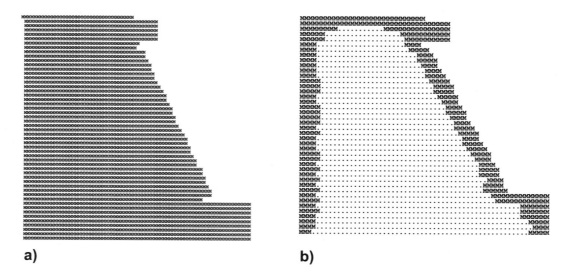

a)   **b)**

Figure 17 Phase distribution fields: (a) through-hardened, $\alpha = 40,000$ W/m^2 K; (b) shell-hardened, $\alpha = 40,000$ W/m^2 K.

B. Quenching During Through Surface Hardening

The quenching is carried out in special devices by intense water stream or shower from pumps with pressure 1–4 atm. Total water consumption is not large because the closed-circuit water cycle is used (water tank–water pump–quenching device–water tank). A small amount of cold water is added into the tank to prevent water from heating above 50°C.

The design of quenching devices should guarantee velocity of water with respect to the surface of quenched parts in the order of 10–15 m/sec. Water inside the devices should be under pressure (1.5–3 atm.). Time of intense water quenching should be limited to allow self-tempering of parts at 150–200°C. One of typical designs

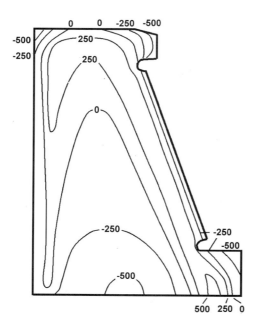

Figure 18 Residual hoop stresses S33 (MPa) through-hardened, $\alpha = 40,000$ W/m^2 K.

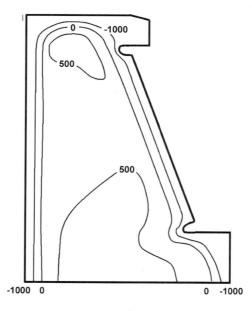

Figure 19 Residual hoop stresses S33 (MPa) shell-hardened, $\alpha = 40,000$ W/m^2 K.

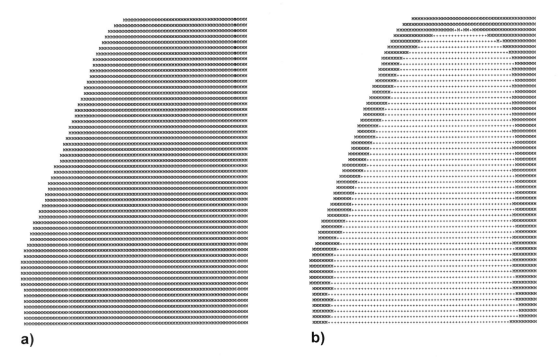

Figure 20 Phase distribution fields: (a) through-hardened, (40,000 W/m² K); (b) shell-hardened (40,000 W/m² K).

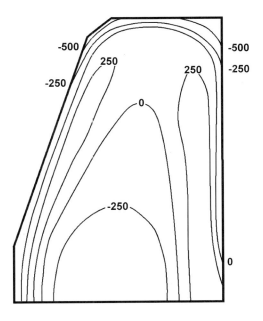

Figure 21 Residual hoop stresses S33 (MPa), through-hardened (40,000 W/m² K).

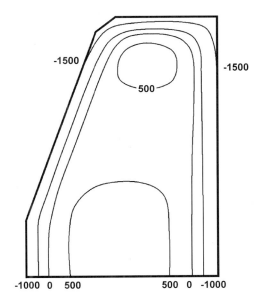

Figure 22 Residual hoop stresses S33 (MPa) at shell hardening (40,000 W/m² K).

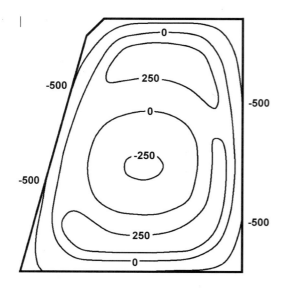

Figure 23 Residual hoop stresses S33 (MPa), through-hardened (40,000 W/m² K).

of the device for quenching of bearing rings is shown in Fig. 24.

1. Through-Surface Hardening of Inner Rings of Bearings for the Boxes of Railway Cars

In the past, the bearing rings were made of traditional bearing steel ShKh15SG (1.0%C, 1.5%Cr, 0.6%Si, and 1.0%Mn) and hardened by a standard method to obtain HRC 58–62. After a short period in service, some rings failed because of brittle cracking and fatigue.

Since 1976, TSH has been used for bearings. The inner rings are made of steel ShKh4. Dimensions of the rings and the typical configuration of hardened layer after TSH are given in Fig. 11.

At present, several millions of such bearings are successfully used in operation on car boxes in the rail-roads of Russia and other states of CIS. Automatic machine for TSH of the inner rings is shown in Fig. 26.

In order to show advantages of TSH for bearings, the results of mechanical tests are given below for the inner rings of railway bearings made of different steels and strengthened by various heat treatments including:

1. Standard steel ShKh15SG (52100 Grade 4) with electroslag refining, traditional through-hardening with furnace heating and quenching in oil; low tempering.
2. Standard carburized steel 18KhGT (GOST 4543-71) containing on the average 0.2%C,

1.1%Cr, 1.0%Mn, and 0.05%Ti, carburizing, quenching, low tempering.
3, 4. Steel ShKh4, TSH, low tempering.

These numbers of steels and heat treatment variants are referred in Table 1 which contains some structural characteristics and hardness of the rings. Characteristic curves of residual stresses in the cross section of hardened rings and their strength under static, cyclic, and impact loads are given in Figs. 25 and 26.

The technology developed has been implemented at GPZ-1 plant (Moscow).

2. Through-Surface Hardening of Rollers

Through-surface hardening is very suitable and profitable for bearing rollers. It allows to obtain roller properties similar to those of carburized rollers made of expensive chromium–nickel steels, but at lower steel cost and production expenses.

Through-surface hardening was successfully used at GPZ-9 for tapered rollers of 40–50 mm diameter made of ShKh4 steel.

Recently, new technology of TSH was developed for hollow rollers of railway car box bearings. Application of a design of hollow rollers with a central hole is a promising direction in bearing production. In this case, contact area between the ring and the roller increases because of a small elastic deformation of the rollers, contact stresses decrease, and durability of the bearing increases by several times.

According to service tests, durability of railway bearings with hollow rollers increases by 2.8 times. The rollers had the following dimensions: external diameter of 32 mm, length of 52 mm, and diameter of

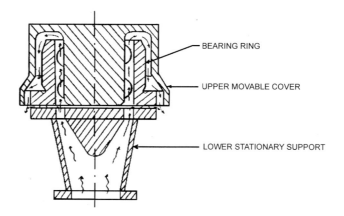

Figure 24 Schematic design of a quenching device for inner rings of railway bearings (water stream is shown by arrows).

Table 1 Structure and Hardness of Inner Rings of Railway Bearings Made of Various Steels

Parameters	Values for rings made of the following steels			
	ShKh15SG	18KhGT	ShKh4	ShKh4
1. Number of steel and heat treatment variant	1	2	3	4
2. Surface hardness (HRC)	60–61	59–61	62–64	62–64
3. Depth of hardened layer, mm:	Through hardening			
HRC > 58		0.7–0.9	1.5–2	2.5–4
HRC > 55		1.8–2.1	2.4–2.7	3–4.5
4. Core hardness (HRC)	60–61	32–35	36–40	40–45
5. Microstructure in the core	Martensite, carbides	Low-carbon martensite	Sorbite, troostite	Troostite
6. Percentage of retained austenite in surface	14–16	8–10	6–8	6–8
7. Number of austenitic grain size[a]	9–10	9	10–11	10–11

[a] GOST 5639-82.

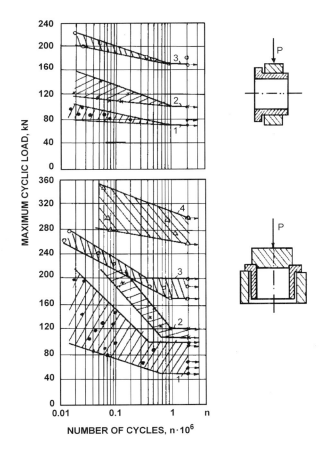

Figure 25 Fatigue testing results for inner rings of railway car bearings made of various steels (the numbers of the curves correspond to numbers of steel and heat treatment variants in Table 1).

the central hole of 12 mm. Macrostructure and hardness of the rollers hardened by TSH are shown in Fig. 27. While zones on the surface correspond to the martensitic structure, the dark core has a troostite and sorbite structure.

The primary condition for heat treatment of hollow rollers consists in providing surface hardening on the external and internal surfaces. This leads to high compressive internal stresses in the hardened layers which prevents formation of fatigue cracks in service. It should be noted that through-hardening is dangerous for hollow rollers. This is demonstrated by fatigue tests of the rollers under maximum load of 200 kN. Through-hardened rollers made of 52100 steel were destroyed after 15 000–30 000 cycles, while TSH-hardened rollers made of ShKh4 steel survived after 10 million cycles (Fig. 28).

More detailed information about results of experimental studies can be found in Refs. 2 and 27–33.

VI. DESIGNING OF STEEL-INTENSIVE QUENCH PROCESSES IN MACHINE CONSTRUCTION

A. Through-Surface Hardening of Small-Size Driving Wheels Made Out of 58 (55PP) Steel

The original technology of through-surface quenching has been developed for driving wheels with module of 4–6 mm, which are typical for construction and road machines, metal-cutting machines, transport devices,

Figure 26 Automatic machines for TSH of inner rings for railway car box bearings designed and put into operation at GPZ-1 (Moscow) in 1995.

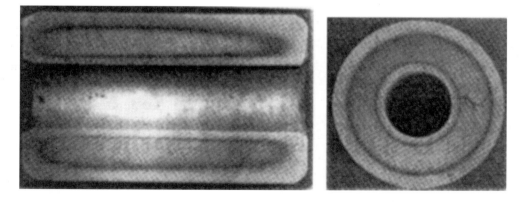

Figure 27 Macrostructure of through-the-wall sections of hollow rollers for railway car box bearings.

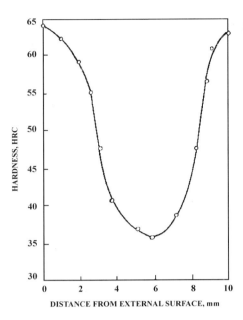

Figure 28 Hardness of through-the-wall sections of hollow rollers for railway car box bearings.

and other machines. They are produced in big varieties but in small lots. The peculiarities of the technology are the use of carbon steel with the low content of admixtures, and thus, low hardenability; untraditional choice of current frequency and conditions of induction heating; performance of high-quality quenching of driving wheels of various kinds and sizes with the use of a single inductor and cooling device; minimum quenching distortion; and ecological purity of the quenching process. The design of universal easily readjustable quenching installment has been developed.

In modern engineering, a great number of driving wheels are used that have diameter of 80–120 mm and module of 4–6 mm. A part of them, satellite driving wheels, require hardening for only teeth crown, while the other body and the hole cannot be hardened. The greatest number of driving wheels having a hole, according to operation conditions, must be hardened to high hardness on all surfaces, including teeth crown, flange surface, and surface of hole.

At present, such driving wheels are hardened by either surface induction under a layer of water, when they are produced out of medium-carbon construction steels, or carburizing with the further quenching and low tempering with the use of carburized steels.

Quenching under a layer of water very often results in the appearance of quench cracks on the surface, with which one can do nothing but allow; however, it is very

difficult to determine and control the limits of such allowance. The use of carburizing and nitrocarburizing on many plants is prevented by the absence and high cost of modern thermal equipment and automatic regulation devices, especially as for the regulation of components of gaseous saturating atmosphere. Making carburizing and nitrocarburizing at out-of-date universal equipment does not guarantee the required level and stability of the quality of heat treatment.

Due to the stated above, one of the processes having high prospects for hardening driving wheels is through-surface quenching with induction heating. The advantages of this method are as follows:

1. Opportunity of high-quality and reliable hardening of steel parts.
2. Opportunity to produce equipment for through-surface quenching directly at machine-construction plants.
3. Electric power saving, absence of quench oil in the production process, and gaseous-controlled atmospheres, which cause ecological problems.
4. Lesser production area in comparison with chemical heat treatment, better labor conditions, and opportunity to place induction installments in shops of mechanical treatment.

In a laboratory of Moscow State Evening Metallurgic University, engineers developed the technology of through-surface quenching for driving wheels with module of 4–6 mm. The wheels have been made out of 58 (55PP) steel having the following chemical content: 0.56%C, 0.30%Si, 0.09%Mn, 0.10%Cr, 0.08%Cu, 0.04%Ni, 0.004%S, 0.006%P, and 0.008%Al. The low content of admixtures in the steel is due to production at Oskolskyi Steel Works, which is the main producer of this steel.

For production of driving wheels, rolling 130-mm diameter circle was used. The specimens were obtained through forging, further cutting, and preparation for induction. Five kinds of driving wheels presented in Fig. 29 were used in tests, which are used at excavator plants and having a module of 4.6 mm.

Heating of all wheels has been made at one-spiral cylindrical inductor with the inner diameter of 150 mm and height of 110 mm; cooling has been performed by quick water flows provided at outer, inner, and flange surfaces of driving wheels.

In the experiment, heating induction was at current frequencies of 2.5 and 8 kHz from machine generators of 100-kW power.

On the basis of traditional principles of current frequency selection for teeth crowns of driving wheels

Figure 29 Macrostructure of driving wheels with module of 4–6 mm after through-surface quenching: light surfaces are the surface hard layer having martensite structure and hardness greater than HRC 60; dark surfaces are areas with the structures of troostite or sorbite of quenching and hardness of HRC 45–30.

with module of 4–6 mm, with regard to uniform heating of teeth and holes while heating in ring inductors, the optimal current frequency must be as follows:

$$f_{opt} = 600/M^2 = 600/(4-6)^2 = 40\ldots20$$

where f_{opt} is optimal current frequency (kHz) and M is a module of a driving wheel (mm).

For this reason, making a choice out of available industrial induction devices for such driving wheels, more preferable are devices for 8 kHz than 2.5 kHz.

This was proved to be true for satellite driving wheels for which surface quenching of hole is not necessary, in connection with which the through-surface quenching is fulfilled with not thorough heating but quite deep induction heating of teeth crown only.

The induction heating of such cylindrical driving wheels of 115-mm diameter on the circle of teeth tips and 48-mm height has been performed at the current frequency of 8 kHz for 30 sec, and the teeth have been heated thoroughly, and cavities at the depth of 3.4 mm. The cooling time, cooling being performed by intensive water flows, was 4 sec. After quenching on the surface of teeth and cavities, the uniform hardness within the range of HRC 63–66 has been received. The macrostructure of driving wheels after through-surface quenching with detected hard layer by itching with "triple" agent is given in Fig. 29 (lower line of wheels). Tempering of hard driving wheels was made in electric furnace equipped with ventilator, at temperature of 165°C, for 3.5 hr.

The results of measuring dimensions of driving wheels on the mean line of the common normal and diameter of the hole in the original state before quenching, after quenching, and after quenching with tempering are given in Fig. 30. The reduction of mean length of common normal is in average as follows: 0.014 mm in quenching, 0.021 mm in tempering, and the total change in quenching and tempering is 0.035 mm in comparison with the original state after mechanical treatment. The diameter of the hole reduced in average by 0.019 mm in quenching, 0.026 mm additionally in tempering, and the total reduction in quenching and tempering was 0.045 mm. The values of distortion obtained are quite small

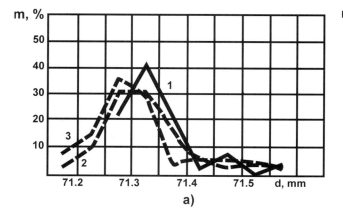

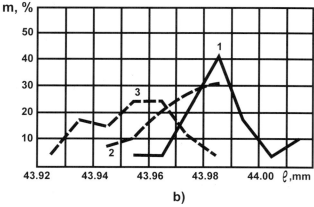

Figure 30 Changing in the diameter (d) of hole for placement (a) and length (l) for common normal (b) of satellite wheels in the case of through-surface quenching (m is a group frequency); 1—sizes of driving wheels in the original state after mechanical treatment; 2—after through-surface quenching; 3—after through-surface quenching and tempering.

and completely meet the requirements for the heat treatment of satellite diving wheels.

The majority of standard driving wheels are driving wheels with inner-gear hole, where it is necessary to harden at high hardness not only teeth crown, but also surface of inner-gear hole, and in this case, the allowance of distortion on the diameter of inner-gear hole is quite severe and usually it is not more than 0.12 mm.

It has been noticed earlier that proceeding from conditions of uniformity of heating for cavities and teeth of outer teeth crown for driving wheels with module of 4–6 mm, it is more preferable to use induction heating with the current frequency of 8 kHz. However, experiments fulfilled showed that for such driving wheels with regard to their geometry and peculiarities of the work conditions, the use of 2.5-kHz frequency gives a number of advantages and allows to expand the range of effective application of through-surface quenching and 58 (55PP) steel for the production towards the reduction of module of driving wheels to 4 mm while providing the clearly expressed effect of outline quenching of teeth crown.

This is explained by the following circumstances.

1. Using the current frequency of 8 kHz, even in the case of the computer step-by-step control of the voltage generator and time of heating of 3 min, since it is necessary to reach the minimum quenching temperature of 790–800°C on the surface of inner-gear hole, the result is the considerable overheating of teeth crown to 850–

Figure 32 Macrostructure of cross section of a driving wheel with module of 4 mm after through-surface quenching with induction heating at the current frequency of 8 kHz.

880°C (Fig. 31a), due to which the teeth are hardened to high depth or even thoroughly (Fig. 32). Although such quenching can be done in many cases, outline quenching is more preferable.

2. Using current frequency of 2.5 kHz, the heating is more uniform with respect to thickness of driving wheels, and by time of finishing heating, the temperature in cavities of teeth crown is 850°C (Fig. 31b), which provide the 1.7-mm thickness of hard layer in cavity with up to HV600 hardness. At the same time, the temperature on the tip of teeth is essentially lower, 790–800°C, which is allowed for quenching with regard to constructive hardness of driving wheels since the tips of teeth are areas with the lowest load. The specified difference of temperatures on the outline of teeth crown gives advantages for receiving the optimal configuration of hard layer as for the height of the teeth (see Fig. 29, upper line of driving wheels).

With heating at current frequency of 8 kHz, through-surface quenching has been performed for batches of three types of driving wheels, macrostructure of which after quenching is presented in Fig. 29 (upper line of driving wheels). The time of heating for different driving wheels was within the range of 2–2.5 min; cooling was performed directly in inductor by quick water flows within 4–8 sec, providing self-tempering at temperature of 150–200°C. The driving wheels after quenching were tempered at 165°C for 3.5 hr.

Figure 33 presents results of hardness measurement in various zones on the cross section of a driving wheel with module of 4 mm after through-surface quenching and tempering in comparison with driving wheels made

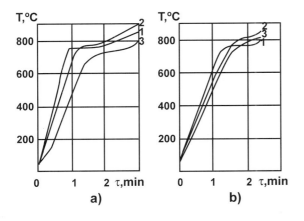

Figure 31 Thermal curves of induction heating for driving wheels with module of 4 mm at current frequency of 8 kHz (a) and 2.5 kHz (b). Specifications of the driving wheel: outer diameter is 91.1 mm, inner diameter is 45 mm: 1 is the tip of a tooth; 2 is the cavity of teeth crown; 3 is the surface of inner-gear hole.

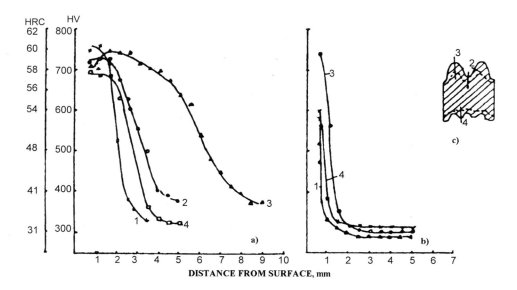

Figure 33 Hardness in different zones on the cross section of a teeth crown and inner-gear hole: (a) is for a driving wheel made out of 58 (55PP) steel after through-surface quenching and tempering; (b) is for a driving wheel made out of 18KhGT steel after carburizing and heat treatment; (c) is for cross section of a teeth crown and inner-gear hole; 1–4 are zones of the cross section.

out of 18KhGT steel after carburizing and heat treatment by serial technology of TEQ. As one can see, through-surface quenching through one session of thorough heating and cooling has provided efficient hardening of both teeth crown and zone of inner-gear hole with the hardness of HRC 58–60 and thickness of hard layer within 2–5 mm in different cross sections and, at the same time, hardening of the wheel core to HRC 30–40. This combination of properties is advantageous for high service life and operational reliability.

Table 2 gives results of measuring the distortion of wheels in the case of through-surface quenching with low tempering. As one can see, the technology of through-surface quenching of driving wheels provided quite low values of distortion not exceeding 0.05 mm on the mean length of common normal and 0.075 mm on the diameter of inner-gear hole, which meets the highest requirements for the distortion.

Batches of driving wheels hardened by this through-surface quenching technology were used in the serial production of excavators. The design of industrial induction heating installment for through-surface quenching of driving wheels is outlined in Figs. 34 and 35. The installment is universal for a wide range of

Table 2 Distortion Characteristics for Driving Wheels with Inner Inner-Gear Hole after Through-Surface Quenching

Dimensions of driving wheels (mm)	Mean length of common normal (mm)		Diameter of inner hole (size of inner gear for rolls) (mm)	
OD = 91.1, ID = 45,	31.77–31.84	31.805	31.27–31.30	31.285
H = 37, M = 4	31.78–31.84	31.81	31.14–31.28	31.28
OD = 91.1, ID = 51,	31.80–31.84	31.82	40.13–40.17	40.15
H = 52, M = 4	31.80–31.85	31.825	40.07–40.11	40.09
OD = 89.6, ID = 45,	39.78–39.80	39.79	40.12–40.17	40.145
H = 50, M = 5	39.80–39.81	39.805	40.06–40.14	40.10

ID is the inner diameter; OD is the outer diameter; H is height of the wheel; M is the module of teeth crown. The data above line are sizes of wheels in the original state after mechanical treatment; and data below line are sizes of those wheels after through-surface quenching with low tempering.

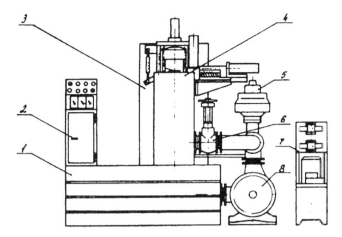

Figure 34 Installation for through-surface quenching of driving wheels through induction heating: 1—water tank; 2—control panel; 3—quenching block; 4—case of quenching device; 5—pneumatic-hydraulic valve; 6—valve; 7—pump; 8—hydraulic station.

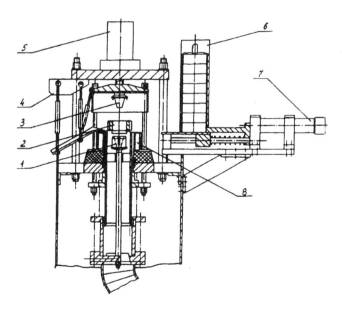

Figure 35 Quenching device: 1—changeable mandrel; 2—driving wheel to be quenched; 3—cone; 4—removing device; 5—hydraulic cylinder for vertical movement and wheel fixing; 6—receiver of parts, changeable; 7—hydraulic cylinder for wheel feeding; 8—inductor.

driving wheels of different sizes and easily readjusted for different kinds of driving wheels through changing upper and lower mandrels (Table 3).

Thus the typical technology of through-surface quenching of driving wheels with diameter of 80–120 mm and module of 4–6 mm includes as follows:

1. Production of driving wheels out of 58 (55PP) steel at Oskolskyi Steel Works.
2. Deep (for satellite wheels) or thorough (for wheels with inner gear hole) induction heating for 0.5–3 min in cylindrical one-spiral inductor. The optimal current frequency is 8 kHz for satellite wheels and 2.5 kHz for wheels with an inner-gear hole, which are to be hardened.
3. All-side intensive cooling at a quenching device linked directly to the inductor by quick water flows fed to outer, inner, and flange surfaces of driving wheels.
4. Low tempering at 160–170°C for 3–4 hr.

The technology developed provided the optimal configuration of hard layers on all work surfaces (Fig. 29), providing high constructive durability of driving wheels with the minimum quench distortion of no more than 0.1 mm along the common normal and diameter of inner-gear hole.

The cooling rate of the core of the specimen also is high under these conditions, but will drop when the compressive stress reaches its maximum value. The core's microstructure consists of bainite or tempered

Table 3 Installation Specifications

Dimensions of driving wheels to be quenched:	
Maximum outer diameter	145
Maximum height	90
Parameters of frequency transformation	
Power (kW)	100
Current frequency (kHz)	2.5–10
Specific consumption of electric power (kW h/Mt)	300–500
Quenching water-supply pump:	
Output (m³/hr)	300
Pressure (MPa)	0.1...0.4
Water consumption (m³/hr)	
For cooling electrical elements	3–4
For cooling wheels during quenching	1–1.5
Type of work drives	Hydraulic
Dimensions (mm):	
Length	3000
Width	2800
Height	2600

Table 4 Mechanical Properties at the Core of Conventional (Oil) and Intensive (Chloride Medium) Quenched Steels (Cylindrical 50-mm Diameter Specimen Used)

Steel	Quenchant	Tensile strength, R_m (MPa)	Yield strength, $R_{p0.2}$ (MPa)	Elongation, A (%)	Reduction in area, Z (%)	U-notch impact energy, KSU (J/cm^2)	Hardness (HRB)
40×	Oil	780	575	21	64	113	217
	Chloride	860	695	17	65	168	269
35 × M	Oil	960	775	14	53	54	285
	Chloride	970	820	17	65	150	285
25 × 1M	Oil	755	630	18	74	70	229
	Chloride	920	820	15	68	170	285

Source: Ref. 34.

martensite, and its mechanical properties are higher than those of oil-quenched steel (Table 4).

The work of Shepelyakovskii and Ushakov [35] supports these observations. They developed steels: 58 (55PP), 45C, 47GT, and ShX4—having limited hardenability (LH) and controlled hardenability (CH). These steels contain small amounts of aluminum, titanium, and vanadium. They are less expensive than conventional alloy grades because they use two to three times less total alloying elements. The researchers emphasize that the depth of the hard layer is determined by steel hardenability.

One of the new steels is first induction through-surface hardened, and then slowly heated (2–10°C/sec) either through the section or to a depth beyond that required for a fully martensitic layer. The slow-heating time is usually 30 to 300 sec. Parts are then intensively quenched in a water shower or rapidly flowing stream of water. Production parts heat-treated using this method are listed in Table 5. Note that the durability of intensive-quenched, noncarburized low-alloy steel parts is at least twice that of carburized and conventionally oil-quenched alloy steel parts.

Other production applications for these new low-alloy steels include [35]:

- Gear, 380-mm diameter and 6-mm modulus, steel 58 (GOST 1050-74). Note: modulus = pitch diameter/number of teeth
- Cross-head of truck cardan shaft, steel 58 (GOST 1050-74)
- Bearing for railway carriage, 14-mm ring thickness, steel ShKh4 (GOST 801-78)
- Truck axle, 48-mm bar diameter, steel 47GT

The optimum depth of hard layer is a function of part dimensions. Therefore it is recommended that the steel (and its hardenability) be tailored to the part to ensure that it will be capable of hardening to this depth [36], making reaching the optimum depth a certainty.

With the right steel, this method can provide an optimum combination of high surface compressive stress, a high-strength, wear-resistant quenched layer of optimum depth, and a relatively soft but properly strengthened core. The combination is ideal for applications requiring high strength and resistance to static, dynamic, or cyclic loads.

Table 5 Production Applications of Intensive-Quenched Limited-Hardenability Steels

Applications	Former steel and process	New steel and process	Advantages
Gears, modulus, $m = 5–8$ mm	18KhGT	58 (55PP)	No carburizing; steel and part costs decrease; durability increases.
Large-modulus gears, $m = 10–14$ mm	12KhN3A	ShKh4	No carburizing; durability increases 2 times; steel cost decreases 1.5 times.
Truck leaf springs	60C2KhG	45S	Weight decreases 15–20%; durability increases 3 times.
Rings and races of bearings thicker than 12 mm	ShKh15SG and 20Kh2N4A	ShKh4	No sudden brittle fracture in service; durability increases 2 times; high production rate.

Source: Ref. 35.

VII. NEW METHODS OF QUENCHING

There are two methods of reaching the optimal depth of hard layer at the surface.

1. For each specific steel part, a special steel grade is selected, which provides the optimal hard layer and maximum compressive stresses at the surface. When the sizes of the part are changed, different steel grades are selected providing meeting the condition $\Delta r/R = \text{const}$.
2. Steel part quenching is made so that $0.8 \leq \text{Kn} \leq$ and the process of intensive cooling is interrupted at the time of reaching maximum compressive stresses at the surface. In this case, the optimal depth of hard layer is reached automatically. This method was protected by inventor's certificate in 1983, which became a patent of Ukraine in 1994 [36].

In order to confirm the above statements, one can refer to Ref. 37 where a method of IQ-quenching steel parts by using intensive jet cooling is described. The author points out that the method of intensive cooling is applied to superficial hardening of small parts (shafts, axes, pinions, etc.) made of alloy steels. Here a very high intensity of cooling is achieved that allows to obtain a 100% martensite structure in the outer layer and high residual compression stresses. It should be noted that under the conditions of very intensive cooling, the strain decrease is observed. While treating the parts

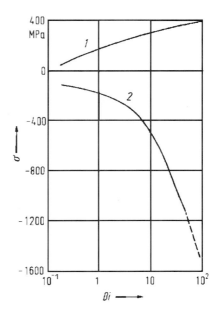

Figure 37 Maximum compressive hoop stresses at the surface (2) and tensile stresses in the center (1) of the cylindrical specimen of 6 mm in diameter vs. Biot number.

of complex configuration, it is necessary to use several combined jets in order to prevent the steam jacket formation. A disadvantage of this method is a high cost of the equipment.

Let us consider in more detail the character of changes in the current stress at the surface of the parts being quenched depending on various intensities of cooling. It has been stated that in the course of time, small tensile stresses occur at first at the surface of the specimen to be quenched, and then during the martensite layer formation, these stresses transform into the compressive ones that reach their maximum at a certain moment of time and then decrease (Fig. 36). The current stresses become the residual ones that can be either tensile or compressive depending on the cooling intensity. The maximum tensile stresses correspond to the maximum compressive stresses in the control layer of the part being quenched (Fig. 37).

The mechanism of the current stress formation is as follows. When the part is completely in the austenite state, there arise tensile stresses that transfer into the compressive ones in the process of the martensite phase formation and due to increase of its specific volume. The larger part of austenite is transformed to martensite and the larger the martensite layer is, the higher the compression stress is. The situation goes on until a sufficiently thick martensite crust is formed resembling a rigid vessel that still contains the supercooled austenite

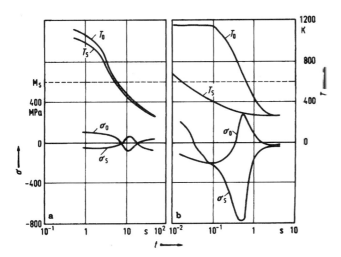

Figure 36 (a–b) Change of current hoop stresses at the surface (σ_s) and in the center (σ_c) of the cylindrical specimen being cooled under various heat transfer conditions (a) Slow cooling; (b) intensive cooling. (From Ref. 24.)

in the supercooled phase. The further advance of martensite inside the part causes the effect of water freezing in a glass vessel [23]. Due to the core volume increase at martensite transformations, either decrease of compressive stresses in the surface layers or destruction of the external layer will take place if the phase specific change is large enough and the external layer is insufficiently stretched and strong. Under such conditions, compression stresses in the surface layer change over to stress state that cause destruction in the surface layer.

Reduction of compressive stresses at the further advance of martensite into the part to be quenched is caused by the parting action that is attributed to variation in the phase specific volume in the core.

If the process of intensive cooling is stopped at the moment of achieving the maximum compression stresses, and isothermal holding is realized at the temperature of the martensite start (M_s), then the martensite phase advance will cease and sufficiently high compression stresses can be fixed. They will slightly decrease due to the isothermal holding at which stress relaxation takes place.

An optimal depth of the quenched layer that depends on part dimensions corresponds to maximum compression stresses.

Using the calculation methods developed and the potentialities of the software package "TANDEM-HART" [1,8], it is easy to find with the help of computer the time of achieving the maximum for bodies of arbitrary axisymmetric form being quenched under various heat transfer conditions.

The degree of intensive cooling can be characterized by Bi_V number or by Kondratjev number Kn. There is a universal interconnection between these numbers

$$Kn = \Psi Bi_V = \frac{Bi_V}{\sqrt{Bi_V^2 + 1.437 Bi_V + 1}} \qquad (11)$$

which is valid for bodies of various configurations.

The author of the well known handbook "Theory of Heat Conduction" Lykov [38] has called Eq. (1) an important relation of the theory of regular conditions. Criterion $Kn = \Psi Bi_V$ is the main value determining the heat transfer mechanism of the body. It was named Kondratjev number (criterion) in honor of the outstanding thermal scientist G.M. Kondratjev.

It appeared that the curves $Kn = f(Bi_V)$ for geometrically different bodies (sphere, parallelepiped, cylinder, etc.) were located so close to each other that practically all the family could be replaced by a single averaged curve [38].

The parameter criterion Ψ characterizing the temperature field nonuniformity is equal to the ratio of the body surface excess temperature to the mean excess temperature over the body volume. If the temperature distribution across the body is uniform ($Bi_V \rightarrow 0$), then $\Psi = 1$. The higher is the temperature nonuniformity, the less Ψ is. At $\Psi = 0$, the temperature distribution nonuniformity is the highest ($Bi \rightarrow \infty$, while $T \rightarrow T_\infty$).

Thus Kondratjev number characterizes not only the temperature field nonuniformity, but also the intensity of interaction between the body surface and the environment.

Kondratjev number is the most generalizing and the most universal value, which may serve to describe the cooling conditions under which compression stresses occur at the surface of various bodies.

For rather high compression stresses to occur at the surface of the part being quenched, it is sufficient to meet the following condition:

$$0.8 \leq Kn \leq 1.$$

On the basis of regularities mentioned above, a new method of quenching was elaborated. The essence of the new method is that alloy and high-alloy steel parts are cooled under the conditions of high-intensive heat transfer ($Kn \geq 0.8$) up to the moment of reaching maximum compression stresses at the surface with the following isothermal holding under temperature M_s [39]. A year later, similar quenching method was proposed in Japan [40].

In accordance with the method mentioned, alloyed steel parts are quenched in such a way that a very hard surface layer of the given depth and an arbitrarily hard matrix are obtained. An example of such method realization is given below. An alloy steel specimen containing (in %) 0.65–0.85 C, 0.23–0.32 Si, 0.4–0.9 Mn, 2Ni, 0.5–1.5Cr, and 0.1–0.2Mo is heated up to 800–850°C and spray-quenched with water fed under pressure of 0.4–0.6 MPa during 0.2–0.8 sec. The specimen is subject further to isothermal heating at 150–250°C for 10–50 min [40].

It is obvious that the spray quenching under high pressures provides intensive cooling ($Kn > 0.8$) that is completed when a certain depth of the composition-quenched layer is achieved.

For the steel composition cited, the temperature of the martensite start is within the range of 150–200°C.

The isothermal holding time at this temperature (about 10–15 min) is chosen from CCT diagrams of supercooled austenite dissociation in such a way to provide this dissociation into intermediate components in the part central layers.

The analysis of the methods described shows that various authors have come independently to an identical conclusion that is a rather pleasant coincidence because it testifies to urgency and authenticity of the technology being studied.

Structural steel transformations during quenching are accounted through dependencies of thermal–physical and mechanical properties of the material on the temperature and time of cooling in accordance with CCT diagram for the transformation of overcooled austenite. The method has been proved by a number of test problems [41]. The error of calculations was $\leq 3\%$ for temperature and $\leq 12\%$ for stresses, which provides grounds for using this method for the study of regularities of changes in thermal and stress–strain state of parts to be quenched with regard to cooling conditions and character of structural steel transformations.

The calculation of current and residual stresses for cylindrical sample of 6-mm diameter made out of 45 steel was made for different heat transfer coefficients, so that Bi changed from 0.2 to 100. The temperature of sample heating is 1300 K.

The investigations have shown that as far as the process of cooling is intensified, the residual stresses on the surface of cylindrical sample firstly increase reaching the maximum value at Bi = 4, and then when Bi = 18, they become negative, and as far as Bi grows, they become compressive. At Bi = 100, the hoop stresses σ_{33} reach the value of 600 MPa.

The facts observed can be grounded as follows: when Bi is small, there is insignificant temperature gradient in the body. As far as austenite is transformed into martensite, due to the great specific volume of martensite, the stresses appearing first on the surface are not big and compressive. However, when the martensite forms at the center of the sample, big forces moving aside appear which result in the tensile stresses on the surface. In the case of intensive cooling (Bi > 20), martensite transformations start in thin surface layer of the sample while the temperature at its other points is high. The greater the Bi number is, the greater the gradient in the surface layer is, and the further from the axis the layer of freshly formed martensite is. As far as inner layers become cooler, two processes fight against each other: process of shrinking for the account of the temperature reduction and process of expansion of the material for the account of the formation of martensite having big specific volume in comparison with austenite. In the case Bi > 20, the process of shrinking prevails in inner points of the sample. Thus in the cooled sample, the surface layer appears to be shrunk because shrunk inner layers of the sample try to move initially formed layer of

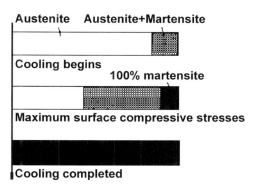

Figure 38 Relative amounts of microstructural phases present at the optimal hard depth in a steel specimen at the beginning and end of intensive quenching, and at the time when the surface compressive stress reaches its maximum value.

martensite closer to the axis. In the case of not big temperature gradient (Bi ≪ 18), the outer layer of freshly formed martensite, in comparison with cold state, is lesser shifted from the axis; for this reason, in this case, tensile residual stresses will appear for the account of increase in the specific volume of the material during martensite transformations in inner layers. It is obvious that there exist such value of Bi that forces connected with material shrinking compensate each other. In this case, on the surface of quenched sample, residual stresses are zero (Bi = 18–20).

Let us note that the quenching method [36,42] provides the optimal depth of the hard layer for any alloy steel.

The optimum depth of hard layer is that which corresponds to maximum surface compressive stresses.

To obtain very high surface compressive stresses, it is sufficient to meet the condition $0.8 \leq Kn \leq 1$, where Kn is the Kondratjev number. This condition can be satisfied by intensive quenching using water jets or rapidly flowing water. Additional strengthening (superstrengthening) of the surface layer will also result. The high compressive stresses and superstrengthening both help enhance the durability and prolong the service life of machine parts (Fig. 38).

VIII. DISCUSSION

The results of investigations made evidence that shell hardening is one of advantageous direction since it brings big economical savings for the account of the following.

Reaching the optimal depth of hard surface layer can be done with lesser amount of alloying elements, and, in this case, the cost of the steel is reduced. For shell hardening, simple water is used. Since in the case of optimal hard layer, optimal compressive stresses are reached at the surface, the cyclical durability of ready steel products increases, which is proved in Ref. 2.

Despite all of these facts and that idea of the existence of the optimal hard layer has been published in 1983 [1], by the present time, this issue has not been studied carefully yet. Probably this is because of the lack of software and databases for these calculations. Experimental way of studying this is almost impossible since it is a huge load of work and requires a lot of finances. Only on the basis of our software TANDEM-HART and generalization of big experimental data of many authors it became possible to ground the existence of the optimal hard layer, which can be widely used in practice for designing of steel-intensive quench processes.

In our opinion, designing of steel-intensive quench processes should be done as follows:

1. Depending on the shape and size of the part, conditions of intensive quenching are selected by calculating the water flow speed or intensity of the shower cooling at the surface of the part to be quenched.
2. The proper steel grade is selected which provides the optimal depth of hard layer and maximum compressive stresses at the surface or the intensive cooling process is interrupted at the time of reaching this optimal depth.
3. On the basis of software TANDEM-HART, calculations of parameters of intensive cooling and thermal and stress–strain state are made.
4. Special equipment and devices are designed for the implementation of intensive cooling.

The final steps after designing are as follows:

1. The equipment and devices designed are produced.
2. The quenching conditions and equipment are adjusted in industrial conditions on site to prevent possible failure with the full automation of the whole industrial hardening process. The automation is very important because the work of the line must be stable. Cracks can be formed in the case of failure to meet all the conditions designed.

A steel quenching method where the depth of the quenched surface layer is controlled, which increases the service life, is described in publication [43]. The above-mentioned steels have the following characteristics: low depth of hardened layer and fine grain with arrested growth of austenite grains at high temperatures. Due to limited hardenability, on the surface of parts, residual compressive stresses appear and the fine grain has the effect of high strength of material. In addition to the increase in the service life, there is an opportunity to replace expensive materials with cheaper materials and fire-dangerous quench oils with simple water. Steel quenching where the depth of the hardened surface layer is controlled is made in intensive water jets. The service life of parts made of steels where the depth of the hardened surface layer is regulated increases by some times [43] in comparison with oil quenching. The weak aspect of this quenching method where the depth of the hardened surface layer is regulated is that steel having the effect of optimal depth of hardened surface layer must be chosen each time for definite shape and dimensions. In the case of the change in the shape or dimensions, it is necessary to change the steel grade to have the effect of high compressive stresses on the surface. As known, the changes in the dimensions of the part and the optimum depth have the following correlation [13]

$$\frac{\Delta\delta}{D} = \text{idem},$$

where idem means constant. When the depth of the hardened layer is greater or less, the compressive stresses are lower. If the layer is thin, in the transition zone, high extensive stresses can appear, which results in cracking. In this quenching method, there are no criteria allowing to calculate the rate of water flow for each concrete part. High water flow rate is chosen for all kinds of parts which is not always justified and results in unnecessary energy spending and makes the industrial process more complicated. The high service life of parts where the depth of hardened surface layer is regulated is considered as an advantage of these steel grades; however, for various steel grades, the effect of superstrengthening and high residual compressive surface stresses can be reached. In this method, the induction heating is mainly used and there are no data regarding oven heating, including such data for carburized parts. The industrial regimes are not optimized. In the method mentioned above, it is advised to use only those steels where the depth of the hardened surface layer is regulated; however, there are some engineering problems related to melting of such steels where the depth of the hardened surface layer is regulated.

There is a steel quenching method [5] dealing with shell hardening, which means uniform quenching of all the surface to the insignificant depth until reaching high hardness on the basis of using intensive jet cooling. In this method, the examples of the application of mean-alloy steel 45 are given. The main advantage of this method is the opportunity to increase the service life of steel parts while using usual carbon steels, not using steels where the depth of the hardened surface layer is regulated. This method also has weaknesses the same as in the method above, namely, no consideration is given to conditions optimizing the depth of the hardened surface layer, and the following correlation is ignored

$$\frac{\Delta \delta}{D} = \text{idem},$$

(that is, in the case of the change in dimensions of parts, the depth of the hardened surface layer should be changed correspondingly). This method does not have criteria allowing to calculate the rate of cooling quench flow which would prevent the development of self-regulated thermal process. The technological process is not optimized.

An intensive steel quenching method has been also developed in Japan [40]. In accordance with this method, alloyed steel parts are quenched in such a way that a very hard surface layer of the given depth and an arbitrarily hard matrix are obtained. For given steel grades, according to this method, ranges for hardening regimes are found by experiments to increase the service life of such parts. An example of such method realization is given below. An alloy steel specimen containing (in %) 0.65–0.85 C, 0.23–0.32 Si, 0.4–0.9 Mn, 2 Ni, 0.5–1.5 Cr, and 0.1–0.2 Mo is heated up to 800–850°C and spray-quenched with water fed under pressure of 0.4–0.6 MPa during 0.2–0.8 sec. The specimen is subject further to isothermal heating at 150–250°C during 10–50 min [40]. The weakness of the method above is that it considers just high-carbon alloyed steels. The depth of the hard surface layer is not optimal for various dimensions of parts. Because of this steel, superstrengthening is not reached. It does not consider conditions of the optimization of the quenchant circulation rate.

The analysis of the existing methods of steel quenching used in various countries (in Russia, United States, Japan, and Ukraine) shows that the intensive steel quenching with the formation of hard surface layer of the given depth has greater advantages than for thorough quenching. The common weakness of all these methods is that there is no change in the optimum depth of the hard surface layer in cases of the changes in dimensions of parts, and that the quenchant circulation rate is not optimized for the prevention of the development of self-regulated thermal process and reaching material superstrengthening.

The intensive steel quenching method [36] has been chosen as a prototype of this invention, including heating, intensive cooling until the appearance of maximum compressive surface stresses, isothermal heating, and tempering. The method is based on the following: intensive cooling is formed in the range of $0.8 \leq \text{Kn} \leq 1$, where Kn is the Kondratjev number, until reaching maximum compressive surface stresses, and then it is isothermally heated at martensite start temperature M_s until the complete transformation of the overcooled austenite of the matrix, then it is tempered.

The basic weakness of this method is that it deals just with alloyed steels. To reach the maximum compressive stresses on the surface, the intensive cooling is stopped and due to it, the effect of superstrengthening does not show itself in full. There is no concrete method of the calculation of optimal rate of quenchant flow to ensure reaching material superstrengthening.

The proposed quenching method is new since in the part quenched for any steel grades, alloyed and not alloyed, high, mean, and low carbon, the depth of the hard surface layer reached is optimal that the maximum compressive stresses are formed. If the depth of the hard surface layer is greater or less, the compressive stresses are lower.

The creation of conditions to reach maximum compressive stresses is reached due to greater effect of additional strengthening (superstrengthening) of material. In the compressed surface layer during the quench process, martensite transformations take place. Due to greater specific volume of martensite plates (than for the resulting phases), the plastic deformation of austenite occurs which is located between the martensite plates. The higher the compressive surface stresses in the layer being hardened and the higher the cooling rate is in the martensite area, the greater the deformations in the austenite which is between martensite plates. In this case, martensite plates function like "microhammers" due to which high density dislocations are reached under high pressure. While cooling is fast, these dislocations are "frozen" in the material.

In such conditions, the effect of low-temperature mechanical heat treatment is present. After this treatment, the material hardened can have higher mechanical and plastic properties in comparison with usual hardening. Thus the optimal depth of the quenched layer is necessary for not only reaching maximum compressive surface stresses, but also for the formation of optimal conditions under which the effect of additional material strengthening (superstrengthening) is present in full. The additional strengthening (superstrengthening) of material and high compressive stresses in the surface layer quenched result in the increase in their service life.

In conclusion, it should be noted that the main idea of this chapter is the existence of the optimal depth of hard layer for which the maximum compressive stresses are reached at the surface. In our opinion, it is a principally new approach which can be used for designing of steel-intensive quench processes. By the present time, the progress has been made in two ways. One of them is through-surface hardening, at which special steel grades are used that provide shell hardening [2–4,46]. In this case, it does not deal with the consideration of the optimal depth of hard layer, and when changing to other sizes with the same steel grade, the optimal depth of hard layer is never reached.

The other way lies in the interruption of intensive cooling when maximum compressive stresses are formed at the surface, as mentioned above [36,39]. In this case, the optimal depth of the hard layer is reached automatically.

The first method was used with induction heating and steel grades of controlled hardenability. It found wide application in Russia [2–4,6,46]. The second method was used with furnace heating and interrupted cooling at the time of reaching maximum compressive stresses at the surface. It is widely used in Ukraine, other countries of the former Soviet Union, and the United States [36,39,49–51]. The purpose of this chapter is to unite efforts of these two methods for the optimization of intensive quenching processes. For the optimization of these processes, TANDEM-HART software has been developed, which belongs to Intensive Technologies Ltd. and distributed by it. The software package consists of two parts: TANDEM-HART package and TANDEM-HART ANALYSIS. TANDEM-HART package makes numerical calculations of fields of temperature, stress, deformation, and safety factor. TANDEM-HART ANALYSIS finds the time of reaching optimal stress distribution and makes quick calculations of quenching conditions. For this software to produce good results, it is necessary to acquire data of mechanical properties, CCT and TTT diagrams for new steel grade of controlled hardenability, and boundary conditions. We believe that uniting efforts of many specialists will help in solving these problems.

Thus the application of the proposed steel quenching method allows as follows:

1. To reach the effect of material superstrengthening and high compressive surface stresses when using arbitrary steel grade.

2. Alloyed and high-alloyed steels can be replaced with simple carbon steels that has the effect that the depth of the hard layer is optimal. In this case, the effect of material superstrengthening is greater. Due to it, the service life of such parts increases.

3. Expensive and fire-dangerous oils can be replaced by water and water solutions.

4. The labor efficiency increases.

5. The ecological state of the environment is improved.

IX. CONCLUSIONS

As a result of investigating the kinetics of phase transformations in bodies of complicated configuration, the following regularities have been found:

1. When the Biot number increases, the axial and hoop residual stresses on the surface of a part to be quenched firstly grow reaching the maximum at $Bi = 4$ and then reduced and become negative at $Bi \geq 20$.

2. The distribution of residual stresses in parts with full and partially controlled hardenability while the heat transfer is highly forced ($Bi > 20$) has different character. In parts of controlled hardenability on the surface, high compressive stresses appear, which gradually change to tensile stresses at the center of the part. In the case of thorough hardening, while $Bi > 20$ in the surface layer, there are high tensile stresses changing to compressive stresses on the surface. With the elapse of the time, parts made out of steel of controlled hardenability have compressive residual stresses on the surface growing all the time, while parts hardened thoroughly have stresses that are compressive and grow until a certain moment of time at which they reach the maximum, then they are reduced.

3. Methods of numerical investigation of the kinetics of phase transformation in bodies of arbitrary shape has been developed. Having CCT diagrams with physical and mechanical properties of structural components for these diagrams, one can determine the structure, hardness, and hardenability of parts having a complicated configuration and forecast the mechanical properties of the material [26,44–48]. For this purpose, TANDEM-HART software has been developed.

4. It has been established that there is optimal depth of hard layer for a part at which compressive stresses on the surface reach the maximum.

5. There is similarity of the distribution of current and residual stresses in bodies having different sizes and the conditions when this similarity is observed are given above.

6. The practical application of shell hardening for car box rollers and wheels is described.

7. For the wide application of these methods, it is necessary to improve further the software and to develop databases of initial data for solving the problem of calculating the optimal depth of hard layer for various steel grades, which would provide the optimal distribution of compressive stresses at the surface and in the core.

8. The optimal depth of the hard layer can be reached for the account of either proper selection of steel grade or interrupting the cooling at the time of reaching the optimal maximum compressive stresses at the surface.

ACKNOWLEDGMENTS

The authors would like to present their thanks to V.V. Dobrivecher for the help in the preparation of the material and translation of this chapter into English.

REFERENCES

1. Kobasko, N.I.; Morhuniuk, W.S. *Investigation of thermal and stress–strain state at heat treatment of power machine parts (Issledovanie teplovogo i napryagenno-deformirovannogo sostoyaniya pri termicheskoy obrabotke izdeliy energomashinostroyeniya)*; Znanie: Kiev, 1983; 16 pp.

2. Ouchakov, B.K.; Shepelyakovskii, K.Z. *New Steels and Methods for Induction Hardening of Bearing Rings and Rollers, Bearing Steels: Into the 21st Century, ASTM STP 1327*; Hoo, J.J.C., Ed.; American Society for Testing and Materials: 1998.

3. Ushakov, B.K.; Lyubovtsov, D.V.; Putimtsev, N.B. Volume-surface hardening of small-module wheels made of 58 (55PP) steel produced at OEMK. Mater. Sc.. Trans. Mashinostroenie: Moscow, 1998, (4), 33–35.

4. Beskrovny, G.G.; Ushakov, B.K.; Devyatkin, N.E. Raising of longevity of car box bearings with use of hollow rollers hardened by volume-surface hardening. *Vestnik*; VNIIZhT: Moscow, 1998, (1), 40–44.

5. Kern, R.F. Intense quenching. Heat Treat. 1986. (1), 19–23.

6. Shepelyakovskii, K.Z. *Hardening Machine Parts by Surface Quenching Through Induction Heating*; Mashinostroenie: Moscow, 1972; 288 pp.

7. Kobasko, N.I.; Morhuniuk, W.S.; Gnuchiy, Yu.B. *Investigation of technological machine part treatment (Issledovanie tekhnologicheskikh protsesov obrabotki izdeliy mashinostroeniya)*; Znanie: Kiev, 1979; 24 pp.

8. Kobasko, N.I.; Morhuniuk, W.S. *Investigation of thermal stress state in the case of heat treatment of power machine parts (Issledovanie teplovogo i napryagennogo sostoyaniya izdeliy energomashinostroyeniya pri termicheskoy obrabotke)*; Znanie: Kiev, 1981; 16 pp.

9. Inoue, T.; Arimoto, K.; Ju, D.Y. *Proc. First Int. Conf. Quenching and Control of Distortion*; ASM International: 1992; 205–212.

10. Inoue, T.; Arimoto, K. Development and implementation of CAE system "HEARTS" for heat treatment simulation based on metallo- thermo- mechanics. JMEP 1997, 6 (1), 51–60.

11. Narazaki, M.; Ju, D.Y. Simulation of distortion during quenching of steel—Effect of heat transfer in quenching. Proc. of the 18th ASM Heat Treating Society Conference & Exposition, Rosemont, Illinois, USA, October 12–15, 1998.

12. Reti, T.; Horvath, L.; Felde, I. A comparative study of methods used for the prediction of nonisothermal austenite decomposition. JMEP 1997, 6 (4), 433–442.

13. Kobasko, N.I. *Steel quenching in liquid media under pressure (Zakalka stali v zhidkikh sredakh pod davleniem)*; Naukova Dumka: Kiev, 1980; 206 pp.

14. Kobasko, N.I.; Morhuniuk, W.S.; Lushchik. L.V. Investigation of thermal stress state of steel parts in the case of intensive cooling at quenching. *Thermal and Thermomechanical Steel Treatment (Termicheskaya i termo-mekhanicheskaya obrabotka stali)*; Metallurgy: Moscow, 1984; 26–31.

15. Kobasko, N.I.; Morhuniuk, W.S.; Dobrivecher, V.V. Calculations of cooling conditions of steel parts during quenching. Proc. of the 18th ASM Heat Treating Society Conference & Exposition, Rosemont, Illinois, USA, October 12–15, 1998.

16. Totten, G.E., Howes, A.H., Eds.; *Steel Heat Treatment Handbook*; Marcel Dekker, Inc.: New York. 1997; 1192 pp.

17. Kobasko, N.I.; Prokhorenko, N.I. Cooling rate effect of quenching on crack formation in 45 steel. Metalloved. Term. Obrab. Metall. 1964, (2), 53–54.

18. Kobasko, N.I. Crack formation at steel quenching. MiTOM 1970, (11), 5–6.

19. Bogatyrev, JuM.; Shepelyakovskii, K.Z.; Shklyarov, I.N. Cooling rate effect on crack formation at steel quenching. MiTOM 1967, (4), 15–22.

20. Ganiev, R.F.; Kobasko, N.I.; Frolov, K.V. On principally new ways of increasing metal part service life. Dokl. Akad. Nauk USSR 1987, 194 (6), 1364–1473.

21. Kobasko, N.I. Increase of service life of machine parts

and tools by means of cooling intensification at quenching. MiTOM 1986, (10), 47–52.

22. Kobasko, N.I.; Nikolin, B.I.; Drachinskaya, A.G. Increase of service life of machine parts and tools by creating high compression stresses in them. Izvestija VUZ (Machinostrojenie), 1987, (10), 157.

23. Kobasko, N.I. Increase of steel part service life and reliability by using new methods of quenching. Metalloved. Term. Obrab. Metall. 1989, (9), 7–14.

24. Kobasko, N.I.; Morhuniuk, W.S. Numerical study of phase changes, current and residual stresses at quenching parts of complex configuration. Proc. of 4th Int. Congr. Heat Treatment Mater, Berlin, 1985; 466–486.

25. Kobasko, N.I. On the possibility of controlling residual stresses by changing the cooling properties of quench media. *Metody povyshenija konstruktivnoi prochnosti metallicheskikh materialov*; Znanije RSFSR: Moscow, 1988; 79–85.

26. Kobasko, N.I.; Morhuniuk, W.S. *Investigation of Thermal and Stress State for Steel Parts of Machines at Heat Treatment*; Znanie: Kyiv, 1981; 24 pp.

27. Shepelyakovskii, K.Z.; Devjatkin, V.P.; Ouchakov, B.K. Induction surface hardening of rolling bearing parts. Metalloved. Term. Obrab. Metall. January 1974, (1), 17–21. *in Russian*.

28. Shepelyakovskii, K.Z. Surface and deep and surface hardening of steel as a means of strengthening of critical machine parts and economy in material resources. *Metal Science and Heat Treatment (A translation of Metallovedenie i Termicheskaya Obrabotka Metallov)*; Consultants' Bureau: New York, November–December, 1993, (11,12), 614–622.

29. Ouchakov, B.K.; Efremov, V.N.; Kolodjagny, V.V. New compositions of bearing steels of controlled hardenability. Steel October 1991, (10), 62–65. *in Russian*.

30. Shepelyakovskii, K.Z.; Devyatkin, V.P.; Ushakov, B.K.; Devyatkin, V.F.; Shakhov, V.I.; Bernshtein, B.O. Induction surface hardening of swinging bearing parts. Metalloved. Term. Obrab. Metall. 1974, (1), 17–21.

31. Devyatkin, V.P.; Shakhov, V.I.; Devin, R.M.; Mirza, A.N. Application of hollow rollers for the prolongation of service life of cylindrical rolling bearings. Vestn. VNIIZhT 1974, (3), 20–22.

32. Polyakova, A.I. Comparative tests of car bearings 42726 and 232726 with solid and hollow rollers. *Bearing Industry, Issue 8*; NIINAvtoprom: Moscow, 1974; 1–10.

33. Rauzin, Ya.P. *Heat Treatment of Chrome-Containing Steel*; Mashinostroenie: Moscow, 1978; 277 pp.

34. Mukhina, M.P.; Kobasko, N.I.; Gordejeva, L.V. Hardening of structural steels in chloride quenching media. Metalloved. Term. Obrab. Metall. 1989, (9), 32–36.

35. Shepelyakovskii, K.Z.; Ushakov, B.K. Induction surface hardening—Progressive technology of XX and XXI centuries. Proceedings of the 7th International Congress on Heat Treatment and Technology of Surface Coatings, Moscow, Russia 11–14 Dec. 1990, 2 (11–14), 33–40.

36. NI Kobasko. Patent of Ukraine: UA 4448, Bulletin No. 6-1, 1994.

37. Sigeo, O. Intensive cooling. Kinzoku Metals Technol. 1987, 57 (3), 48–49.

38. Lykov, A.V. *Theory of Heat Conduction*; Vysshaya Shkola: Moscow, 1967; 560 pp.

39. Kobasko, N.I. Method of part quenching made of high-alloyed steels, Inventor's certificate 1215361 (USSR), Bulletin of Inventions No. 12., Applied 13.04.1983., No. 3579858 (02-22), 1988.

40. Naito, Takeshi. Method of steel quenching. Application 61-48514 (Japan), 16.08.1984, No. 59-170039.

41. Loshkarev, V.E. Thermal and stress state of large-size pokovok at cooling in heat treatment. Dissertation abstract. Sverdlovsk, 1981; 24 pp.

42. Kobasko, N.I. *Intensive Steel Quenching Methods, Theory and Technology of Quenching*; Liscis, B. Tensi, H.M., Luty, W., Eds.; Springer-Verlag: New York, NY, 1992; 367–389.

43. Shepelyakovskii, K.Z.; Bezmenov, F.V. New induction hardening technology. Adv. Mater. Process. October 1998; 225–227.

44. Morhuniuk, W.S. Thermal and stress–strain state of steel parts with complicated configuration at quenching. Dissertation abstract, Kyiv, 1982; 24 pp.

45. Morhuniuk, W.S.; Kobasko, N.I.; Kharchenko, V.K. On possibility to forecast quench cracks. Probl. Procn. 1982, (9), 63–68.

46. Shepelyakovskii, K.Z. Through-surface quenching as a method of improving durability, reliability and service life of machine parts. MiTOM 1995, (11), 2–9.

47. Bashnin, Yu.A.; Ushakov, B.K.; Sekey, A.G. *Technology of Steel Heat Treatment*; Metallurgiya: Moscow, 1986; 424 pp.

48. Kobasko, N.I. Self-regulated thermal process at steel quenching. Prom. Teploteh. 1998, 20 (5), 10–14.

49. Kobasko, N.I. Generalization of results of computations and natural experiments at steel parts quenching. J. Shanghai Jiaotong Univ. June 2000, E-5 (1), 128–134.

50. Kobasko, N.I. Thermal and physical basics of the creation of high-strength materials. Prom. Teploteh. 2000, 22 (4), 20–26.

51. Aronov, M.A.; Kobasko, N.I.; Powell, J.A. Practical application of intensive quenching process for steel parts. *Proc. of the 12th Int. Federation of Heat Treatment and Surface Engineering Congress*, (Melbourne, Australia), 29 Oct.–2 Nov. 2000; 51 pp.

18

Design of Quench Systems for Aluminum Heat Treating

D. Scott MacKenzie

Houghton International, Inc., Valley Forge, Pennsylvania, U.S.A.

I. INTRODUCTION TO ALUMINUM PHYSICAL METALLURGY

The principal alloying additions to aluminum are copper, manganese, silicon, magnesium, and zinc. Other elements are also added in smaller amounts for grain refinement and to develop special properties. Because there is a variety of aluminum alloys, special designation systems were developed by the Aluminum Association to distinguish the alloys in a meaningful manner and, further, to indicate what metallurgical condition, or temper, has been imparted to the alloy.

Aluminum and its alloys are divided into two classes according to how they are formed: wrought and cast. The wrought category is indeed a broad one, because virtually every known process can form aluminum. Wrought forms include sheet and plate, foil, extrusions, bar and rod, wire, forgings and impacts, drawn and extruded tubing, and others. Cast alloys are those specially formulated to flow into sand or permanent mold, to be die cast, or to be cast by any other process where the casting is the final form.

Each wrought or cast aluminum alloy is designated by a number to distinguish it as a wrought or cast alloy and to broadly describe the alloy. A wrought alloy is given a four-digit number. The first digit classifies the alloy series or principal alloying modification in the basic element. The second digit, if different than 0 (zero), denotes a modification in the basic alloy. The third and fourth digits form an arbitrary number that identifies the specific alloy in the series. A cast alloy is assigned a three-digit number followed by a decimal.

Here again the first digit signifies the alloy series or principal addition; the second and third digits identify the specific alloy; the decimal indicates whether the alloy composition is for final casting (0.0) or for ingot (0.1 or 0.2). A capital letter prefix (A, B, C, etc.) indicates a modification of the basic alloy. The designation systems for aluminum wrought and cast alloys are shown in Tables 1 and 2, respectively.

A. Temper Designation System

Specification of an aluminum alloy is not complete without designating the metallurgical condition, or temper, of the alloy. A temper designation system, unique for aluminum alloys, was developed by the Aluminum Association and is used for all wrought and cast alloys. The temper designation follows the alloy designation, the two being separated by a hyphen. Basic temper designations consist of letters. Subdivisions, where required, are indicated by one or more digits following the letter. The basic tempers are shown in Table 3 (see also Tables 4 and 5).

Wrought alloys are divided into two categories. Non-heat-treatable alloys are those that derive strength from solid solution or dispersion hardening and are further strengthened by strain hardening. They include the 1xxx, 3xxx, 4xxx, and 5xxx series alloys. Heat-treatable alloys are strengthened by solution heat treatment and controlled aging, and include the 2xxx, some 4xxx, 6xxx, and 7xxx series alloys.

Casting alloys cannot be worked-hardened and are either used in the as-cast or heat-treated conditions.

Table 1　Wrought Alloy Designation System

Alloy series	Description or major alloying element
1xxx	99.00 Minimum aluminum
2xxx	Copper
3xxx	Manganese
4xxx	Silicon
5xxx	Magnesium
6xxx	Magnesium and silicon
7xxx	Zinc
8xxx	Other element
9xxx	Unused series

Typical mechanical properties for commonly used casting alloys range from 20 to 50 ksi for ultimate tensile strength, from 15 to 40 ksi tensile yield and up to 20% elongation.

B.　Heat-Treatable Alloys

Heat-treatable aluminum alloys will naturally age at room temperature following quenching and will be strengthened by precipitation hardening. Natural aging following quenching from a high-temperature forming process, for example casting or extruding, is designated T1. More commonly, natural aging follows solution heat treatment (T4). Artificial aging is accomplished by heating the product to a temperature of roughly 400°F for several hours (time and temperature depend on the alloy) to accelerate the precipitation process and to further increase the strengthening effect. Here again, artificial aging may follow quenching from a high-temperature forming process (T5) or more commonly following solution heat treatment (T6). The T7 temper indicates overaging from a T6 temper of maximum strength to improve characteristics such as resistance to corrosion. The other T tempers indicate that strain

Table 2　Cast Alloy Designation System

Alloy series	Description or major alloying element
1xx.x	99.00 Minimum aluminum
2xx.x	Copper
3xx.x	Silicon plus copper and/or magnesium
4xx.x	Silicon
5xx.x	Magnesium
6xx.x	Unused series
7xx.x	Zinc
8xx.x	Tin
9xx.x	Other element

Table 3　Basic Temper Designations

F	As fabricated: Applies to products of forming processes in which no special control over thermal or work hardening conditions is employed. Mechanical property limits are not assigned to wrought alloys in this temper, but are assigned to cast alloys in "as cast," F temper.
O	Annealed: Applies to wrought products that have been heated to effect recrystallization, produce the lowest strength condition, and cast products that are annealed to improve ductility and dimensional stability.
H	Strain-hardened: Applies to wrought products that are strengthened by strain hardening through cold working. The strain hardening may be followed by supplementary thermal treatment, which produces some reduction in strength. H is always followed by two or more digits (see Table 4).
W	Solution heat-treated: Applies to an unstable temper applicable only to alloys that spontaneously age at room temperature after solution heat treatment. This designation is specific only when the period of natural aging is specified. For example, W 1/2-hr solution heat treatment involves heating the alloy to approximately 1000°F to bring the alloying elements into solid solution, followed by rapid quenching to maintain a supersaturated solution to room temperature.
T	Thermally treated: Applies to products that are heat treated, sometimes with supplementary strain hardening, to produce a stable temper other than F or O. T is always followed by one or more digits (see Table 5).

hardening has been employed either to supplement the strengths achieved by precipitation or to increase the response to precipitation hardening.

1.　Mechanism of Precipitation Hardening

Some heat-treatable alloys, especially 2xxx alloys, appreciably harden at room temperature to produce the useful tempers T3 and T4. These alloys that have been naturally aged to the T3 or T4 tempers exhibit high ratios of ultimate tensile strength/yield strength. These alloys also have excellent fatigue and fracture toughness properties.

Natural Aging

Natural aging and the increase in properties occur by the rapid formation of GP (Guinier–Preston) Zones from the supersaturated solid solution and from

Table 4 Subdivisions of the H Temper

First digit indicates basic operations:

H1	Strain hardened only
H2	Strain hardened and partially annealed
H3	Strain hardened and stabilized

Second digit indicates degree of strain hardening:

HX2	Quarter hard
HX4	Half hard
HX8	Full hard
HX9	Extra hard

Third digit indicates variation of two-digit temper.

quenched-in vacancies. The strength rapidly increases, with properties becoming stable after approximately 4–5 days. The T3 and T4 tempers are based on natural aging for 4 days. For 2xxx alloys, improvements in properties after 4–5 days are relatively minor, and become stable after 1 week.

The Al–Zn–Mg–Cu and Al–Mg–Cu alloys (7xxx and 6xxx) harden by the same mechanism of GP Zone formation. However, the properties from natural aging are less stable. These alloys still exhibit significant changes in properties even after many years.

The natural aging characteristics change from alloy to alloy. The most notable differences are the initial incubation time for changes in properties to be observed, and the rate of change in properties. Aging effects are suppressed with lower than ambient temperatures. In many alloys, such as 7xxx alloys, natural aging can be nearly completely suppressed by holding at $-40°C$.

Because of the very ductile and formable nature of as-quenched alloys, retarding natural aging increases scheduling flexibility for forming and straightening operations. It also allows for uniformity of properties during the forming process. This contributes to a quality part. However, refrigeration at normal temperatures does not completely suppress natural aging. Some precipitation still occurs. Table 6 shows typical temperature and time limits for refrigeration.

Often, refrigeration systems are inadequate to cool thick gage parts quick enough to rapidly cool parts. In this case, several heat treaters immerse the parts in Stoddard's Solvent at $-40°C$ immediately after quenching. Alternatively, the use of dry ice and methanol has also been used. However, either solution is very flammable and requires special precautions for operating and disposal of organic wastes. Immersion of the parts in very cold liquid ensures that the parts will rapidly cool to the desired temperature. The parts are then transferred to the normal refrigeration system.

Interestingly, the electrical conductivity decreases with the progression of natural aging. Generally, the reduction of solid solution content would indicate an increase in the conductivity. This decrease in conductivity indicates that GP Zones are forming, instead of "true" precipitates. This decrease in conductivity is related to the consumption of vacancies by the GP Zones.

Besides conductivity changes, dimensional changes also occur during natural aging. The dimensional change observed is not consistent with a reduction in the amount of solute in solid solution. However, it also suggests the formation of GP Zones, or the formation of a precipitate during natural aging.

Precipitation Heat Treatment (Artificial Aging)

Precipitation hardening (aging) involves heating the alloyed aluminum to a temperature in the 200–450°F range. At this temperature, the supersaturated solid solution, created by quenching from the solution heat-treating temperature, begins to decompose. Initially, there is a clustering of solute atoms near vacancies. Once sufficient atoms have diffused to these initial

Table 5 Subdivisions of the T Temper

First digit indicates sequence of treatments:

T1	Naturally aged after cooling from an elevated temperature shaping process
T2	Cold worked after cooling from an elevated temperature shaping process and then naturally aged
T3	Solution heat treated, cold worked, and naturally aged
T4	Solution heat treated and naturally aged
T5	Artificially aged after cooling from an elevated temperature shaping process
T6	Solution heat treated and artificially aged
T7	Solution heat treated and stabilized (overaged)
T8	Solution heat treated, cold worked, and artificially aged
T9	Solution heat treated, artificially aged, and cold worked
T10	Cold worked after cooling from an elevated temperature shaping process and then artificially aged

Second digit indicates variation in basic treatment:
Examples: T42 or T62—Heat treated to temper by user
Additional digits indicate stress relief:
Examples: TX51—Stress relieved by stretching
TX52—Stress relieved by compressing
TX54—Stress relieved by stretching and compressing

Table 6 Typical Time and Temperature Limits for Refrigerated Parts Stored in the As-Quenched Condition

Alloy	Maximum delay time after quenching	Maximum storage time for retention of the AQ condition		
		−12°C (10°F) max.	−18°C (0°F) max.	−23°C (−10°F) max.
2014	15 min	1 day	30 days	90 days
2024				
2219				
6061	30 min	7 days	30 days	90 days
7075				

vacancy clusters, coherent precipitates form. Because the clusters of solute atoms have a mismatch to the aluminum matrix, a strain field surrounds the solute clusters. As more solute diffuses to the clusters, the matrix eventually can no longer accommodate the matrix mismatch. A semicoherent precipitate forms. Finally, after the semicoherent precipitate grows to a large enough size, the matrix can no longer support the crystallographic mismatch, and the equilibrium precipitate forms. What follows is a brief description of the precipitates and precipitation sequence of the most common precipitation-hardenable aluminum alloys.

Precipitation hardening is the mechanism where the hardness, yield strength, and ultimate strength dramatically increase with time at a constant temperature (the aging temperature) after rapidly cooling from a much higher temperature (solution heat treat temperature). This rapid cooling or quenching results in a supersaturated solid solution, and provides the driving force for precipitation. This phenomenon was first discovered by Wilm [1], who found that the hardness of aluminum alloys with minute quantities of copper, magnesium, silicon, and iron increased with time, after quenching from a temperature just below the melting temperature.

The first rational explanation for this effect was by Merica et al. [2], who explained that the hardening that occurred over time was because the solid solubility increased at higher temperatures, the lower aging temperature enabled a new phase to occur by precipitation from the initially solid solution.

The concept of precipitation hardening opened up a new field of physical metallurgy, and was the primary focus of research in the 1920s and 1930s. This concept was very difficult to validate because the precipitates during the initial and intermediate stages of aging were too small to be observed with the instruments of that era.

Mehl and Jelten [3] describe the history and progression of thought by the 1930s on the mechanisms of precipitation hardening in the review article. It is interesting to note that in the review article, the concept of dislocations was not mentioned, although dislocations were discovered and discussed in the early 1930s.

The earliest attempt at explaining precipitation hardening by dislocations [4] thought that the strength increases derived from precipitation hardening was from the interaction of dislocations and the internal stresses developed by coherent particles and the resulting misfit.

Orowan [5] developed his famous equation relating the strength of an alloy containing hard particles to the particle shear modulus and the interparticle spacing. The Orowan equation is a remarkable achievement and is the basis for the theory of dispersion hardening.

Precipitation hardening strengthens alloys by coherent particles sheared by dislocations, with a drastic effect on properties. The mechanisms of precipitation hardening all have in common the method in which dislocations are impeded through the particle and matrix and the description of that motion.

Six primary mechanisms of precipitation hardening have been described in the literature. They are chemical strengthening, stacking fault strengthening, modulus hardening, coherency strengthening, order strengthening, and spinoidal decomposition. Briefly, chemical strengthening provides hardening from the formation of an additional matrix–precipitate interface from the dislocation shearing the particle. Stacking fault strengthening hardens from the different stacking fault energies of the matrix and the precipitate. In modulus hardening, the increased strength is because the shear modulus of the matrix and precipitate differ. In coherency strengthening, there is an elastic interaction between the strain fields of the dislocation and the coherent particle. Order strengthening is when the precipitate is a superlattice, and the matrix is a relatively disordered solid solution. Spinoidal decomposition is a special case where the lattice changes in solute concentration, yielding a periodic variation in the elastic strength from changes in composition. Nearly all the above mechanisms are based on dislocation/particle interactions, with the exception of spinoidal decomposition.

2. Effect of Alloying Elements

Table 7 shows the maximum solid solubility of the principal alloying additions in aluminum and the temperature of maximum solubility. These values are for binary systems, and the presence of other elements in the alloy will usually affect the solubility. Additions greater than maximum solubility are often made, especially in the case of silicon, and this results in the presence of element particles in the solid alloy.

Copper is one of the most important additions to aluminum. It has appreciable solubility and a substantial strengthening effect through the age-hardening characteristics it imparts to aluminum. Many alloys contain copper either as the major addition (2xxx or 2xx.x series), or as an additional alloying element, in concentrations of 1–10%.

Manganese has limited solid solubility in aluminum, but in concentrations of about 1% forms an important series of non-heat-treatable wrought aluminum alloys (3xxx series). It is widely employed as a supplementary addition in both heat-treatable and non-heat-treatable alloys and provides substantial strengthening.

Silicon lowers the melting point and increases the fluidity of aluminum. A moderate increase in strength is also provided by silicon additions.

Magnesium provides substantial strengthening and improvement of the work-hardening characteristics of aluminum. It has a relatively high solubility in solid aluminum, but Al–Mg alloys containing less than 7%

Mg do not show appreciable heat-treatment characteristics. Magnesium is also added in combination with other elements, notably copper and zinc, for even greater improvements in strength.

Zinc is employed in casting alloys and in conjunction with magnesium in wrought alloys to produce heat-treatable alloys having the highest strength among aluminum alloys.

Magnesium provides substantial strengthening and improvement of the work-hardening characteristics of aluminum. It has a relatively high solubility in solid aluminum, but Al–Mg alloys containing less than 7% Mg do not show appreciable heat-treatment characteristics. Magnesium is also added in combination with other elements, notably copper and zinc, for even greater improvements in strength.

Copper and silicon are used together in the commonly used 3xx.x series casting alloys. Desirable ranges of characteristics and properties are obtained in both heat-treatable and non-heat-treatable alloys.

Magnesium and silicon are added in appropriate proportions to form Mg_2Si, which is a basis for age hardening in both wrought and casting alloys.

Tin improves the antifriction characteristic of aluminum, and cast Al–Sn alloys are used for bearings.

Lithium is added to some alloys in concentrations approaching 3 wt.% to decrease density and increase the elastic modulus.

Copper and silicon are used together in the commonly used 3xx.x series casting alloys. Desirable ranges of characteristics and properties are obtained in both heat-treatable and non-heat-treatable alloys.

Magnesium and silicon are added in appropriate proportions to form Mg_2Si, which is a basis for age hardening in both wrought and casting alloys.

There are also miscellaneous additions to aluminum wrought and cast alloys. Tin improves the antifriction characteristic of aluminum, and cast Al–Sn alloys are used for bearings. Lithium is added to some alloys in concentrations approaching 3 wt.% to decrease density and increase the elastic modulus.

3. Effect of Dispersoids

Dispersoids are small intermetallic particles whose size depends on the homogenizing temperature before hot rolling. Their main purpose is to pin subgrains formed during recovery and to control recrystallization during hot rolling. They are unshearable particles and suppress localized shear.

Chromium as a dispersoid former increases the rate of aging at elevated temperatures, but retards aging at

Table 7 Maximum Solid Solubility of Principal Alloying Additions in Aluminum

| | Maximum solubility | | |
Addition	Weight percent	Atomic percent	Temperature (°F)
Cu	5.6	2.5	1013
Mg	14.9	16.3	842
Mn	1.8	0.9	1217
Si	1.6	1.6	1071
Zn	82.8	66.4	720
Mg_2Si	1.8	1.9	1103
$MgZn_2$	16.9	9.6	887

30°C and below [6]. Chromium inhibited recrystallization while favoring the development of a fine subgrain structure [7]. This subgrain structure inhibited hardening behavior. When compared to other dispersoid formers, Cr shows the greatest quench sensitivity [8,9].

Zirconium is also used as a dispersoid former. The precipitation of Al_3Zr phase from the solid solution occurs during soaking of the ingot prior to rolling [10–12]. It was observed [13] that the Al_3Zr dispersoids raised the recrystallization temperatures considerably in an Al–0.5% Zr alloy. The dispersoids are heterogeneously distributed, with the sphere and the rod shape most prevalent. They were found with higher density at subgrain boundaries and dislocation tangles. Zirconium increases the precipitation of S during aging at temperatures greater than at 210°C, and increases the precipitation of the intermediate η' phase and eventually the equilibrium phase η ($MgZn_2$), in 7xxx alloys during aging at 120°C [14].

There is some confusion regarding the structure of the Al_3Zr precipitates. A tetragonal structure, conforming to the crystal structure DO_{23} [10], a coherent ordered FCC structure with a space group Pm3m approximately 300–700 Å in diameter [11], and a simple cubic structure [12] have been reported. In addition, a metastable ordered LI_2 structure has been observed [15]. This metastable phase can form at low temperatures from supersaturated solid solution, and will transform to the equilibrium tetragonal DO_{23} at elevated temperatures. The cubic LI_2 structure has a lattice mismatch with the aluminum matrix of 0.58%, while the tetragonal DO_{23} structure has a lattice mismatch of 2.91% [16].

Holl [9] found that Cr, V, Mn, and Zr increase quench sensitivity of these alloy systems, with Cr showing the greatest effect. Silicon and iron also increases the quench sensitivity and should be minimized. In addition, Fe and Si also form large intermetallic particles that are detrimental to fatigue and fracture properties. The quench sensitivity of the alloys containing Zr was found [8] to be governed by heterogeneous precipitation of η or T phase on Zr-rich compounds. These heterogeneous precipitates did not have an L_{12} structure like Al_3Zr, and were incoherent with the matrix. Precipitation sites increased with the percent reduction from cold rolling after hot extrusion and prior to solution heat treatment. Cold rolling introduced a higher dislocation density. It is probable that metastable Al_3Zr was present. Normally, Al_3Zr is coherent in the matrix, but becomes incoherent because of recrystallization from plastic strain and later solution heat treatment [8]. On subsequent solution heat treatment, the Al_3Zr transformed to DO_{23}, creating additional lattice mismatch,

resulting in additional heterogeneous precipitation sites.

Other Group IVB elements forming dispersoids have been examined [8]. It was shown that, only in the alloys containing Zr or Hf, quench sensitivity increased with increasing reduction. This was most severe when the alloys were cold-worked. The coherency at the dispersoid interface, which remained between matrix and Al_3Zr (or Al_3Hf) particles after working, disappeared as a result of recrystallization during solution heat treatment, so that the particles acted as heterogeneous nucleation sites for the η phase.

4. Alloy System Al–Cu–Mg (2xxx)

The Al–Cu system has been reviewed in detail [17]. The equilibrium phase diagram [18] is a eutectic, in equilibrium with $CuAl_2$ (θ) at 548°C at approximately 32% copper. The extent of solution solubility at the aluminum-rich end is approximately 5.7% copper. Commercial alloys of this type are 2219, 2011, and 2025.

The precipitation sequence was originally established by Guiner [19,20] and Preston [21,22]. Hornbogen [23] further examined the precipitation in Al–Cu and confirmed the results of Guiner and Preston. The precipitation sequence after rapid quenching has been accepted as being Guiner–Preston Zones (GPZ) plates parallel $\{001\}_{Al}$, transforming to the coherent precipitate θ, followed by semicoherent θ' plates parallel $\{001\}_{Al}$. The final equilibrium precipitate is θ (Cu_2Al). Silcock et al. [24] examined this progression of precipitates and show multiple stages in precipitation, as evidenced by changes in hardness and Laue reflections.

The coarsening behavior of θ and θ' in Al–Cu alloys was examined by Boyd and Nicholson [25], and found to follow the theory of Lifshitz and Slyozov [26] and Wagner [27]. In this theory, originally applied to the dispersion of spherical particles in a fluid, the rate of coarsening of is controlled by the diffusion of solute through the matrix. The variation of the mean radius, r, with respect to time, t, of spherical particles in a matrix is given by:

$$\bar{r}^3 - \bar{r}_0^3 = \frac{8}{9}\frac{\gamma D c_0 V_m^2}{RT}(t - t_0) \qquad (1)$$

where r_0 is the mean particle radius when coarsening begins at $t - t_0$. D and V_m are the diffusivity and the molar volume of the precipitate, while c_0 is the equilibrium molar concentration and γ is the precipitate/matrix interfacial energy. Boyd and Nicholson [25] found that the measured coarsening kinetics of θ was in good agreement with the Lifshitz–Wagner theory. However, the coarsening of θ' occurred at a much higher rate than expected, and did not follow the Lifshitz–Wagner theory. They attributed this difference to short circuiting diffusion and particle coalescence.

Aluminum–copper–magnesium alloys were the first precipitation hardenable alloys discovered [28]. The first precipitation hardenable alloy was a precursor to alloy 2017 (4% Cu, 0.6% Mg, and 0.7% Mn). A very popular alloy in this group is 2024.

The addition of magnesium greatly accelerates precipitation reactions. In general, the precipitation sequence is:

$$SS \rightarrow GP \; Zones \rightarrow S'(Al_2CuMg) \rightarrow S(Al_2CuMg) \tag{2}$$

The GP Zones are generally considered to be collections of Cu and Mg atoms collected as disks on the $\{110\}_{Al}$ planes. S' is incoherent, and can be directly observed in the TEM. S' heterogeneously precipitates on dislocation. These precipitates appear as laths on the $\{210\}_{Al}$, oriented in the $<001>$ direction [29]. Because S' precipitates on dislocations, cold working after quenching increases the number density of S' and produces a fine distribution of precipitates in the matrix.

5. Alloy System Al–Mg–Si (6xxx)

This alloy system forms the basis for the 6xxx series aluminum alloys. In this heat-treatable alloy system, magnesium is generally in the range of 0.6–1.2% Mg, and silicon is in the range of 0.4–1.3% Si. The sequence of precipitation is the formation of GP Zones, followed by metastable β' (Mg_2Si), followed by the equilibrium β (Mg_2Si). The GP Zones are needles oriented in the $<001>$ direction, with β' and β showing similar orientations.

6. Alloy System Al–Zn–Mg–Cu (7xxx)

In 7xxx Al–Zn–Mg–Cu alloys, several phases have been identified that occur in Al–Zn–Mg–Cu alloys as a function of precipitation sequence. Four precipitation sequences have been identified. This is schematically shown below:

$$\alpha_{sss} \Rightarrow S$$
$$\alpha_{sss} \Rightarrow T' \Rightarrow T$$
$$\alpha_{sss} \Rightarrow VRC \Rightarrow GPZ \Rightarrow \eta' \Rightarrow \eta \tag{3}$$
$$\alpha_{sss} \Rightarrow \eta$$

In the first precipitation sequence, the S phase, Al_2CuMg, is directly precipitated from the supersaturated solid solution. This phase has been identified [30] as a coarse intermetallic that is insoluble in typical Al–

Zn–Mg–Cu alloys at 465°C and as a fine lath precipitate in Al–4.5% Zn–2.7% Cu–2.2% Mg–0.2% Zr alloys.

In the second precipitation sequence, an intermediate phase T' occurs in the decomposition of the supersaturated solid solution. Bernole and Graf [31] first identified this phase. Further in the second precipitation sequence, the equilibrium T phase forms. It has been proposed by Bergman et al. [32] that the chemical formula $Mg_{32}(Al,Zn)_{49}$ was appropriate. It was found incoherent with the aluminum matrix. Several orientation relationships have been reported [33] between the T phase and the aluminum matrix. This phase has been rarely reported in substantial quantities, although commercial heat treatments up to 150°C lay in the Al + $MgZn_2$ + $Mg_{32}(Al,Zn)_{49}$ phase field. In general, the T phase only precipitates above 200°C [33].

In the third sequence of precipitation, the supersaturated solid solution decomposes to form vacancy-rich clusters, Guinier–Preston Zones, η' and then η. Guinier–Preston Zones have been inferred in Al–Zn–Mg alloys, and is based on small increases in electrical conductivity and an increase in hardness during the initial stages of aging [31].

The η' phase is an intermediate step toward the precipitation of the equilibrium phase η ($MgZn_2$). Direct evidence of η' is rare, and difficult to obtain. Review of the literature indicates that there is a dispute over the occurrence of η' and its nucleation [30,32,34–36]. Investigations with similar compositions have found discrepancies regarding the presence of η'. This leads to the speculation that the formation of η' is path dependent, and subject to local chemical variations. Mondolfo et al. [34] indicated that nucleation of η' occurs by the segregation of alloying elements to stacking faults, gradually losing coherency until the ordered η phase develops. Others indicate that the formation of η' is the result of vacancy-rich clusters (VRC) [35]. GP Zones are also thought to nucleate η' [36].

The equilibrium precipitate in the third sequence is the hexagonal η ($MgZn_2$) phase [37]. This phase is the prototype of the hexagonal Laves phase, with 12 atoms to the unit cell and belonging to the space group $P6_3/mmc$. There are 12 orientation relationships between the precipitate and the aluminum lattice. It has been suggested that the orientation is related to the type of nucleation during or after quenching [38].

Electron and x-ray diffraction of an Al–Zn–Mg alloy revealed that η', η and T' were present in –T6 condition and only η and T' were present in the –T73 condition [39]. Only the precipitates η and η' were detected in the Al–Zn–Mg–Cu alloy in the –T73 condition. The presence of copper suppresses the formation of T' in favor of η.

Copper also stabilizes the η phase resulting in little strength loss during overaging compared to significant strength loss of the ternary alloy during overaging. It was observed that the η phase in the quaternary alloy was multilayered and interpreted in terms of the $MgZn_2$ Laves Phase. The size, interparticle spacing and volume fraction of the precipitated metastable phase (η') were evaluated [40] on the effect of artificial aging time. It was found that the amount of η' increases with aging time, but that the electron density remained constant. A strong correlation between yield strength and the structure of the fine precipitates was found. If precipitates were less than 2 nm in average radius, dislocations cut through the precipitates. When precipitates grew in size to approximately 50–60 Å, the yield stress was governed by the Orowan mechanism. Hirsch and Humpherys' [41] theory provided a quantitative explanation. The Langer–Schwartz model was accurate for predicting precipitation as long as a time-dependent nucleation rate term was added [42]. The elastic strain increases the work of formation of a critical radius, and lowers the nucleation rate.

It was found [40] that rates of precipitation of η and η' were limited by reaction kinetics. Dissolution of η' is dominated by diffusion, while the dissolution of η is dominated by thermodynamic equilibrium between precipitate and the matrix.

C. Quench Sensitivity and Effect of Quench Rate

An understanding of heterogeneous precipitation during quenching can be understood by nucleation theory applied to diffusion-controlled solid state reactions [47]. The kinetics of heterogeneous precipitation occurring during quenching is dependent on the degree of solute supersaturation and the diffusion rate, as a function of temperature. So, as an alloy is quenched, there is greater supersaturation (assuming no solute precipitates). But the diffusion rate increases as a function of temperature. The diffusion rate is greatest at elevated temperatures. When either the supersaturation or the diffusion rate is low, the precipitation rate is low. At intermediate temperatures, the amount of supersaturation is relatively high, as is the diffusion rate. Therefore the heterogeneous precipitation rate is the greatest at intermediate temperatures. This is schematically shown in Fig. 1. The amount of time spent in this critical temperature range is governed by the quench rate.

The quantification quenching and the cooling effect of quenchants have been extensively studied [43–46]. The first systematic attempt to correlate properties to

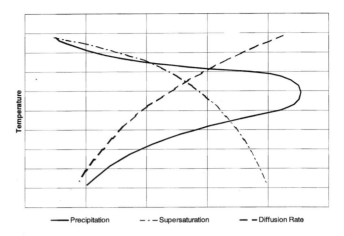

Figure 1 Schematic showing the effect of supersaturation and diffusion rate on precipitation during quenching.

the quench rate in Al–Zn–Mg–Cu alloys was performed by Fink and Wiley [47] for thin (0.064 in.) sheet. A time–temperature–tensile property curve was created and was probably the first instance of a TTT diagram for aluminum. It was determined that the critical temperature range for 75S is 400–290°C. This is similar to the critical temperature range found for Al–Zn–Mg–Cu alloys [48]. At quench rates exceeding 450°C/sec, it was determined that maximum strength and corrosion resistance were obtained. At intermediate quench rates of 450–100°C/sec, the strength obtained is lowered (using the same age treatment), but the corrosion resistance was unaffected. Between 100 and 20°C/sec, the strength rapidly decreased, and the corrosion resistance is at a minimum. At quench rates below 20°C/sec, the strength rapidly decreases, but the corrosion resistance improved. However, for a given quenching medium, the cooling rate through the critical temperature range was invariant no matter the solution heat treat temperature.

One method that quantifies the quench path and material kinetic properties is called the "quench factor" and was originally described by Evancho and Staley [49]. This method is based on the integration of the area between the time–temperature–property curve and the quench path. Wierszykkowski [50] provided an alternative explanation of the underlying principles of the quench factor. However, his discussion is more generally applied to the thermal path prior to isothermal transformation. The procedures for developing the quench factor have been well documented [51–56]. This procedure could be used to predict tensile properties [57], hardness [58], and conductivity [59]. It was found that the quench factor could not be used to predict elongation because of its strong dependence on grain size [59]. This method tends to overestimate the loss of toughness [60]. This method also can be used to determine the critical quench rate for property degradation [61].

1. Precipitate-Free Zone Formation During Quenching

Taylor [62] found that the width of precipitate-free zones (PFZ) in aluminum alloys vary as a function of the solution heat-treating temperature, and the aging temperature. It was found that the width of the PFZ decreased as the solution heat-treating temperature was increased from 410 to 490°C (at a constant age temperature). Hardness remained constant above 440°C. indicating that the solute atoms were completely in solution above 430°C. It was also found that the width of the precipitate-free zone increased as the aging temperature was raised from 120 to 200°C. The width of the PFZ is inversely proportional to the quench rate [63]. A typical precipitate-free zone (PFZ) formed during quenching is shown in Fig. 2.

The decrease in the PFZ width as the solution heat-treating temperature was increased was explained by the increase of vacancies, which in turn expedited diffusion, limiting the width of the PFZ. The decrease in the width

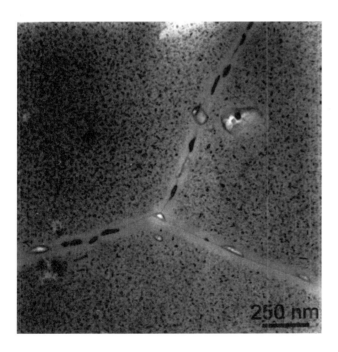

Figure 2 Precipitate-free zone formed in a 7050 alloy during slow quenching.

of the PFZ at lower aging temperatures was explained by a higher degree of solute supersaturation and a change in the volume free energy. This reduces the critical value of vacancies required for nucleation, and reduces the width of the PFZ.

Several authors have investigated the effect of quench rate on the width of the precipitate-free zone in Al–Zn–Mg alloys [64–68]. The results of these investigations are summarized in Table 8.

Embury and Nicholson [66] proposed that there was a critical concentration of vacancies necessary for precipitation to occur. This was extended by Chang and Morral [64] to show that the precipitate-free zone width was inversely proportional to the square root of the quench rate:

$$w = 4r_c T_s \left(\frac{RD_v}{\Delta H_m Q} \right)^{1/2} \tag{4}$$

where r_c is a constant related to the critical concentration of vacancies necessary for precipitation to occur; R is the gas constant; D_v is the diffusion coefficient of vacancies at the solution heat treating temperature, T_s; ΔH_m is the activation enthalpy for vacancy movement; and Q is the quench rate from the solution heat treating temperature.

Work by Embury and Nicholson [69] and Shastry and Judd [68] did not confirm this inverse square root relationship. However, the work presented in Ref. 65 confirms that the precipitate-free zone width follows an inverse square-root relationship with quench rate. However, neither Embury and Nicholson [69], Shastry and Judd [68], nor Chang and Morrall [64] established the critical vacancy concentration for precipitation to occur.

From the analysis by Nicholson and Embury [69], the concentration of vacancies at a distance x from a grain boundary is similar to the classic diffusion case of non-steady state diffusion of a semi-infinite medium. This is similar to the diffusion of carbon in steel, causing decarburization [70]:

$$C(x, t) = C_s \left[\mathrm{erf} \left(\frac{x}{2\sqrt{Dt}} \right) \right] \tag{5}$$

where C_s is the initial vacancy concentration, $C(x,t)$ is the concentration of vacancies at a distance x from the grain boundary at time t, and D is the diffusion coefficient of vacancies at temperature T.

During quenching, the temperature and the diffusion coefficient vary. At the beginning of quenching, the diffusion rate is greatest, and decreases as a function of temperature. Because of the difficulty of determining

Table 8 Comparison of Previous Work Relating the Precipitate-Free Zone (PFZ) Width to Quench Rate in Al–Zn–Mg Alloys

Alloy	Quenchant	Quench rate (°C/sec)	PFZ width (nm)	Reference
Al–5.9Zn–2.9Mg	Oil		1,000	66
	Water		500	
Al–5.9Zn–2.9Mg	Oil	2,000	1,280	67
	Water	30,000	500	
Al–6.9Zn–2.34Mg	Air		340	68
	Oil	2,000	310	
	Water	10,000 (est.)	290	
	Brine		210	
Al–5.9Zn–2.2Mg	Furnace cool	0.3	390[a]	64
	Air	42	8,150[a]	
	Liquid N_2	200	4,990[a]	
	Oil	3,340	1,000[a]	
	Brine	25,000	550[a]	
	Furnace cool	0.3	—	
	Air	42	190[b]	
	Liquid N_2	200	130[b]	
	Oil	3,340	230[b]	
	Brine	25,000	170[b]	

[a] Aged at 200°C (above GP Solvus) for 3 hr.
[b] Aged at 180°C (below GP Solvus) for 3 hr.

the diffusion coefficient, the expression $\sqrt{Nb}$ is substituted, where b is the mean jump distance and N is the number of jumps made by a vacancy. Using the expression for N, given by Cottrell [71]:

$$N = Azv \int_{T_s}^{T_0} \exp\left(\frac{-E_m}{kT}\right) dt \qquad (6)$$

where A is the entropy term, z is the coordination number, v is the atomic vibration frequency, E_m is the activation energy for vacancy migration, and k is the Boltzmann constant. T_s and T_0 are the solution heat-treating temperature and the quench temperature. The quench path and the temperature (T) of the specimen is known as a function of time from quenching.

Letting $\phi = E_m/kT$, the quench rate is assumed to follow a linear path of the form $T(t) = T_s - Qt$. This assumption is not quite accurate, except at the very fast quench rates. However, this assumption readily facilitates a numerical analysis. By substituting the above into the expression for the number of jumps made by a vacancy, the equation becomes:

$$N = \frac{AzvE_m}{kQ} \int_{\phi_1}^{\phi_2} \frac{e^{-\phi}}{\phi^2} d\phi \qquad (7)$$

If $\phi \gg 1$ and $\phi = \infty$ (meaning that $T_0 = 0$ K), Flinn [72] indicates that this equation can be solved. However,

this introduces an error because it is evaluated at 0K. Integration yields:

$$N = \frac{AzvE_m}{kQ} \left[\frac{\exp\left(\dfrac{-E_m}{kT_s}\right)}{\left(\dfrac{E_m}{kT_s}\right)^2} \right] \qquad (8)$$

Because $\sqrt{Nb}$ is the same order of magnitude and have the same units as $\sqrt{Dt}$, this is considered an equivalent expression. Rewriting the original expression for the

Table 9 Values of Constants Used to Calculate the Vacancy Concentration as a Function of Quench Rate

Constant	Value
A	1
γ	10^{13}/sec
z	12
k	8.6×10^{-5} eV/K
T_s	753 K
E_m	0.58 eV
E_F	0.68 eV [74]
E_B	0.31 eV (half of energy required to form divacancy, 0.62 eV [75])
c	2.5 at.%

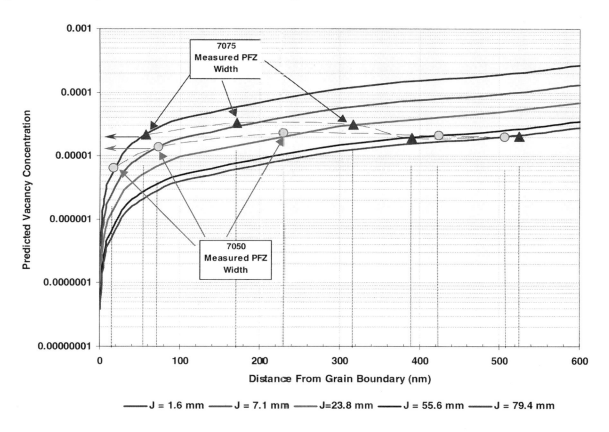

Figure 3 Vacancy concentration as a function of quench rate (distance from the quenched end of the Jominy End Quench) and the distance from a grain boundary.

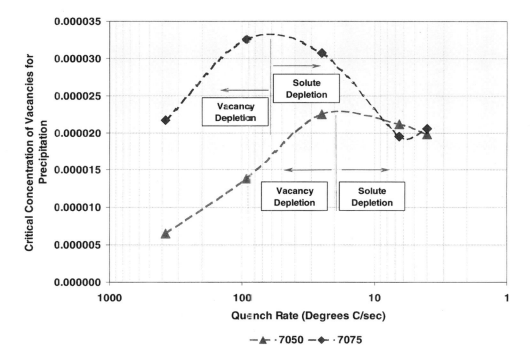

Figure 4 Critical vacancy concentration, C_{crit}, as a function of quench rate.

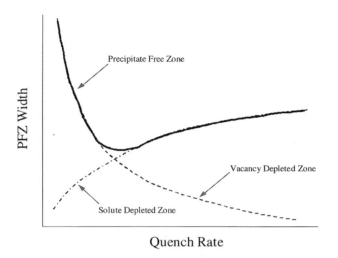

Figure 5 Relationship between vacancy and solute depleted precipitate-free zones (PFZ). (From Ref. 67.)

concentration of vacancies as a function of the distance and time from a grain boundary:

$$C(x, t) = C_s \operatorname{erf}\left(\frac{x}{2\sqrt{Nb}}\right) \quad (9)$$

It is difficult to obtain good values for C_s, the initial concentration of vacancies, because of the lack of measurements for Al–Zn–Mg–Cu alloys. However, the initial concentration of vacancies can be estimated using Lomer's expression [73]:

$$C_s = A \exp\left(\frac{-E_F}{kT}\right)\left(1 - 12c + 12c \exp\left(\frac{E_B}{kT}\right)\right) \quad (10)$$

where E_F is the formation energy of vacancies, c is the concentration of solute, and E_B is the binding energy between solute atoms and vacancies.

From these expressions, the vacancy profile as a function of the distance from a grain boundary can be

estimated as a function of the quench rate. Table 9 shows the values that were used in this calculation. The results of this calculation are shown in Fig. 3.

However, these calculations do not indicate the critical concentration of vacancies required for precipitation. Using the measured precipitate-free zone width shown in Figs. 4 and 5 (from Ref. 65), the critical concentration of vacancies, C_{crit}, necessary for precipitation to occur was determined. The results are summarized in Table 10.

The results of the critical vacancy concentration suggest that the critical concentration of vacancies reaches a peak at a quench rate of approximately 60°C/sec for 7075, and 20°C/sec for 7050. The critical vacancy concentration then decreases.

Precipitate-free zones form because of either vacancy depletion or solute depletion. It has been found [69] that precipitate-free zones that have formed because of vacancy depletion tend to shrink with aging time, while precipitate-free zones that form as a result of solute depletion tend to expand as the result of aging time [76]. Because of this, and the relative mobility of vacancies and solute atoms, it is likely that vacancy depletion would tend to dominate at fast quench rates, while solute depletion would dominate at slow cooling rates. Based on this, it is likely that at quench rates faster than 60°C/sec (20°C/sec for 7050), the PFZ width is governed by vacancy depletion. At quench rates slower than 60°C/sec (20°C/sec for 7050), the PFZ width is governed by solute depletion.

II. QUENCHANTS

For any quenchant, there are typically three phases that occur while the part cools: the vapor phase; nucleate boiling; and finally convection. Each of these stages has very specific characteristics and heat transfer mechanisms.

Table 10 Results of the Calculations to Determine the Critical Vacancy Concentration, C_{crit}, as a Function of Quench Rate

Distance from quenched end of JEQ (mm)	Quench rate (°C/sec)	7050 PFZ width (nm)	7050 Critical vacancy concentration C_{crit}	7075 PFZ width (nm)	7075 Critical vacancy concentration C_{crit}
1.6	380	17[a]	6.49×10^{-6}	57[a]	2.17×10^{-5}
7.1	93	73	1.38×10^{-5}	172	3.25×10^{-5}
23.8	25	230	2.25×10^{-5}	317	3.07×10^{-5}
55.6	6.5	424	2.12×10^{-5}	390	1.95×10^{-5}
79.4	4.0	507	1.98×10^{-5}	525	2.06×10^{-5}

[a] Extrapolated.

In the vapor phase, a stable gas film of superheated quenchant surrounds the part. The stability of this vapor film depends on several factors, including surface roughness, the boiling temperature of the quenchant, and viscosity of the quenchant. Heat transfer is very slow through this film, as heat transfer primarily occurs by radiation and conduction. Because of the temperatures involved at aluminum solution temperatures, radiation heat transfer through the film is negligible. Conduction is also negligible because of the conduction heat transfer characteristics of gases. As the part cools, the stability of the vapor film also decreases, until the collapse of vapor film occurs. At this point, nucleate boiling occurs.

The transition between stable film boiling and nucleate boiling is called the "Ledenfrost temperature."

Nucleate boiling is the fastest regime of cooling. This is where the vapor stage starts to collapse and all liquid in contact with the component surface erupts into boiling bubbles. This is the fastest stage of quenching. The high heat extraction rates are due to carrying away of heat from the hot surface and further transferring it into the liquid quenchant, which allows cooled liquid to replace it at the surface. In many quenchants, additives have been added to enhance the maximum cooling rates obtained by a given fluid. The boiling stage stops when the temperature of the component's surface reaches a temperature below the boiling point of the liquid. For many distortion-prone components, high boiling temperature oils or liquid salts are used if the media is fast enough to harden the steel, but both of these quenchants see relatively little use in induction hardening.

The final stage of quenching is the convection stage. This occurs when the component has reached a point below that of the quenchant's boiling temperature. Heat is removed by convection and is controlled by the quenchant's specific heat and thermal conductivity, and the temperature differential between the component's temperature and that of the quenchant. The convection stage is usually the slowest of the three stages. Typically, it is this stage where most distortion occurs. An example showing the three stages of quenching is shown in Fig. 6.

Obtaining properties and low distortion is usually a balancing act. Often, optimal properties are obtained at the expense of high residual stresses or high distortion. Low distortion or residual stresses are usually obtained at a sacrifice in properties. Therefore the optimum quench rate is one where properties are just met. This usually provides the minimum distortion.

A. Water

Water is the most common quenchant in all materials. It is easy and inexpensive to obtain, and it is readily disposed unless severely contaminated.

In general, as the temperature of water is raised, the stability of the vapor phase increases, and the onset of nucleate boiling in a stagnant fluid is suppressed. The maximum rate of cooling is decreased, and the overall rate of cooling is also decreased.

Cold water quenching is the most severe of commonly used quenchants. In an early study using cooling curves [77], it shown that quenching into still water caused ra-

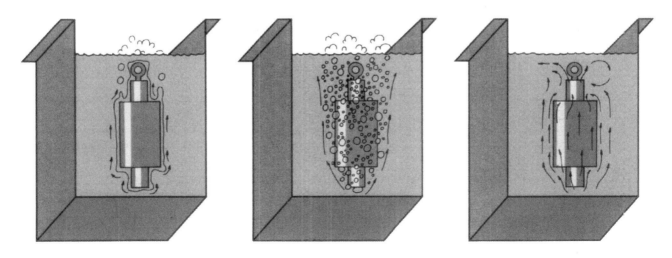

Figure 6 Schematic representation of the three stages of quenching.

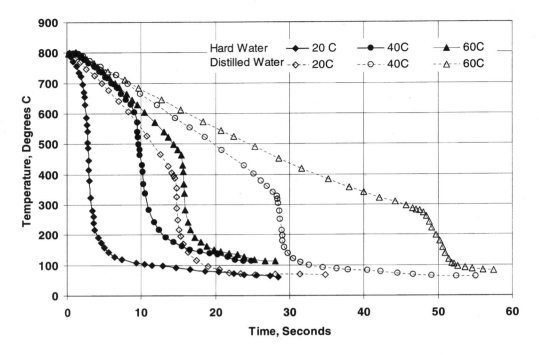

Figure 7 Comparison of the cooling rates of distilled water and normal "hard" water.

pid heat transfer. This study showed that heat transfer at the surface of the part was very turbulent at the metal/water interface. This study also showed that there was a marked difference between hard water and distilled water. Distilled water showed an extensive vapor blanket that extended to very low temperatures (Fig. 7).

The cooling rate of water quenching is independent of material properties such as thermal conductivity and specific heat. It is primarily dependent on water temperature and agitation [78]. Water temperature is the largest primary variable controlling the cooling rate. With increasing water temperature, the cooling rate de-

creases. The maximum cooling rate also decreases as the water temperature is increased. In addition, the temperature of maximum cooling decreases with increasing water quench temperature. The length of time and stability of the vapor barrier increases with increasing water temperature. This is shown in Table 11.

Quenching into water at <50–60°C often produces nonuniform quenching. This nonuniformity manifests itself as spotty hardness, distortion, and cracking. This nonuniformity is caused by relatively unstable vapor blanket formation. Because of this difficulty, it was necessary to develop an alternative to water quenching. Polyalkylene glycol quenchants (PAG) were devel-

Table 11 Effect of Water Temperature on Cooling Rates

Water temperature (°C)	Maximum cooling rate (°C/sec)	Maximum cooling rate temperature (°C)	Cooling rate (°C/sec) at T		
			704°C	343°C	232°C
40	153	535	60	97	51
50	137	542	32	94	51
60	115	482	20	87	46
70	99	448	17	84	47
80	79	369	15	77	47
90	48	270	12	26	42

oped to provide a quench rate in between that of water and oil. By control of agitation, temperature, and concentration, quench rates similar to water and thick oil can be achieved (Fig. 8).

B. Polymer Quenchants

There are two types of polymer quenchants on the market. The first quenchant is polyvinyl alcohol (PVA). This quenchant is resistant to bacterial attack. However, because of its chemical makeup, it is prone to degrade and change its heat transfer characteristics over time. It also produces a hard plastic-type lacquer finish on the parts, which is difficult to remove.

The other type of polymer quenchant is polyalkyene glycol (PAG). For the past 40 years, it has captured the largest market share. It is a copolymer of ethylene oxide and propylene oxide. It exhibits an inverse solubility with water. In other words, as the water temperature is increased, the solubility of PAG quenchants in water is decreased. A two-phase system results as the temperature of the water is raised. The lighter phase is water, which floats to the top. A second phase, denser than water, sinks to the bottom. Each region contains a bit of the other in solution. In other words, the glycol-rich region contains some water, while the water-rich region contains some PAG quenchant. However, as the temperature is increased, the partitioning of PAG and water increases.

The temperature at which separation occurs is called the cloud point. The cloud point is effected by pH, %PAG, and other contaminants in the system. As the pH is increased, the cloud point decreases. As the concentration of PAG increases, the cloud point also decreases.

Water is one of the most severe quench media. Because of the severity of the quench, this quench media presents problems with residual stresses and distortion. Residual stresses in thick sections are caused by differential thermal stresses that occur during quenching. The magnitudes of the stresses also increase as the thickness of the part increases. Because large sections are often machined, a redistribution of the residual stress occurs. This redistribution can cause warpage. Residual stresses can also impact the fatigue of the part.

Slow cooling minimizes temperature differences between the surface and center of a part. This reduces the residual stresses because residual stresses are the result of large differential thermal strains. However, slow cooling also results in heterogeneous precipitation during quenching. This decreases properties by decreasing the amount of supersaturated solute. So a balancing act

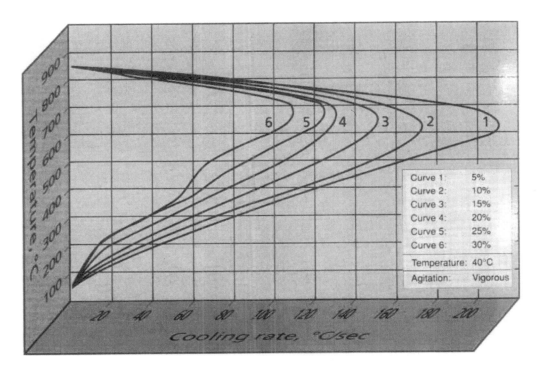

Figure 8 Effect of PAG concentration on cooling rate.

between residual stresses and acceptable properties occurs. To achieve the best balance, it is necessary to use the slowest possible quench that will still achieve properties, with an appropriate safety factor.

Increasing the PAG concentration decreases the quench rate Q. Because of a concept called "zero delta," concentrations of PAG quenchants are limited in the aerospace industry to those that only produce equivalent properties to a cold water quench. These typical concentrations are shown in Table 12. This results in products that have a reduced residual stress and warpage, but have properties in excess of those required.

This technique has great flexibility, but has not been adopted on a commercial basis. In general, this technique is only used in plate and sheet mills because of the simple quenching geometry, enabling sprays to be placed on either side of the plate. In addition, aerospace specifications require extensive documentation and testing to ensure that the spray quenching system is operating at peak performance at all times. All spray nozzles must be monitored during the quench for any intermittent or faulty operation. A daily log of quench temperatures and pressures are usually maintained. In addition, a detailed first article inspection is required, documenting the placement and location of all spray nozzles used. Aerospace forgings and machined parts, because of their complex shape, are generally good candidates for spray quenching. However, because of the equipment

and documentation cost, most heat treaters opt for more conventional immersion quenching.

III. QUENCH FACTOR ANALYSIS

Historically, the average quench rate has been used to predict properties and microstructure after quenching [79–81]. However, average quench rates are not sufficient to provide accurate property data, and serve as a predictive tool [82]. The first systematic attempt to correlate properties to the quench rate in Al–Zn–Mg–Cu alloys was performed by Fink and Wiley [83] for thin (0.064 in.) sheet. A time–temperature–tensile property curve was created and was probably the first instance of a TTT diagram for aluminum. It was determined that the critical temperature range for 75S is 400–290°C. This is similar to the critical temperature range found for Al–Zn–Mg–Cu alloys [84]. At quench rates exceeding 450°C/sec, it was determined that maximum strength and corrosion resistance were obtained. At intermediate quench rates of 450–100°C/sec, the strength obtained is lowered (using the same age treatment), but the corrosion resistance was unaffected. Between 100 and 20°C/sec, the strength rapidly decreased, and the corrosion resistance is at a minimum. At quench rates below 20°C/sec, the strength rapidly decreases, but the corrosion resistance improved. However, for a given

Table 12 Typical Concentration Limits for Quenching in PAG

| Alloy | Form | Maximum thickness | | Concentration volume percent |
		Inches	Millimeters	
2024	Sheet	0.040	1.02	34 Max.
		0.063	1.60	28 Max.
		0.080	2.03	16 Max.
2219	Sheet	0.073	1.85	22 Max.
6061	Sheet, Plate	0.040	1.02	34 Max.
7049		0.190	4.83	28 Max.
7050		0.250	6.35	22 Max.
7075				
6061	Forgings	1.0	25	20–22
7075		2.0	50	13–15
		2.5	64	10–12
7049	Forgings	3.0	76	10–22
7149				
7050	Forgings	3.0	76	20–22
6061	Extrusions	0.250	6.35	28 Max.
7049				
7050	Extrusions	0.375	9.52	22 Max.
7075				

quenching medium, the cooling rate through the critical temperature range was invariant no matter the solution heat-treating temperature. An illustration of the effect of average cooling rate from the solution heat treating temperature on tensile strength is shown in Fig. 9.

One method that quantifies the quench path and material kinetic properties is called the quench factor and was originally described by Evancho and Staley [82]. This method is based on the integration of the area between the time–temperature–property curve and the quench path. Wierszykkowski [85] provided an alternative explanation of the underlying principles of the quench factor. However, his discussion is more generally applied to the thermal path prior to isothermal transformation. The procedures for developing the quench factor have been well documented [59,60,86–89]. This procedure could be used to predict tensile properties [90], hardness [91], and conductivity [59]. It was found that the quench factor could not be used to predict elongation because of its strong dependence on grain size [59]. This method tends to overestimate the loss of toughness [89]. This method also can be used to determine the critical quench rate for property degradation [92].

The quench factor [82] was developed to quantitatively predict properties. This quench factor depends on the rate of precipitation during quenching. The rate of precipitation during quenching is based on two competing factors: supersaturation and diffusion. As temperature is decreased during quenching, the amount of supersaturation increases, providing increased driving force for precipitation. This was shown in Fig. 1. In addition, at the beginning of quenching, the temperature is high, increasing the rate of diffusion. The Avrami

precipitation kinetics for continuous cooling can be described by [82]:

$$\zeta = 1 - \exp(k\tau)^n \qquad (11)$$

where ζ is the fraction transformed, k is a constant, and τ is defined as:

$$\tau = \int \frac{dt}{C_t} \qquad (12)$$

where τ is the quench factor, t is the time (sec), and C_t is the critical time. The collection of the C_t points, also known as the C curve, is similar to the time–temperature–transformation curve for continuous cooling.

In general, the C_t function is described by [86]:

$$C_t = K_1 K_2 \left[\exp\left(\frac{K_3 K_4^2}{RT(K_4 - T)^2} \right) \right] \exp\left(\frac{K_5}{RT} \right) \qquad (13)$$

where C_t is the critical time required to precipitate a constant amount of solute, K_1 is a constant that equals the natural logarithm of the fraction that was not transformed during quenching, and $K_1 = \ln(0.995)$ or -0.00513. K_1 is chosen that for $\tau > 1$, a decrease in properties is observed. K_2 is a constant related to the reciprocal of the number of nucleation sites, and K_3 is a constant related to the energy required to form a nucleus. K_4 is a constant related to the solvus temperature, K_5 is a constant related to the activation energy for diffusion, R is the universal gas constant, and T is the temperature in Kelvin.

To determine the parameters K_1, K_2, K_3, K_4, and K_5, it is first necessary to have the C curve. C curve data is scarce and of limited availability. Table 13 shows some previously published data.

The quench factor is determined by graphically integrating the above equation. This is schematically shown in Fig. 10. Mathematically, this is shown by:

$$\tau = \frac{\Delta t_1}{C_1} + \frac{\Delta t_2}{C_2} + \frac{\Delta t_3}{C_3} + \ldots + \frac{\Delta t_{n1}}{C_n} \qquad (14)$$

where $C_1 \ldots C_n$ are the critical times of the C curve, and $\Delta t_1 \ldots \Delta t_n$ are the incremental times described by the quench path. But, to do this integration, the C curve must be known.

Typically, it is necessary to measure the quench path of several sheets of material and then measure the properties after processing. The quench factor is determined for each quench path and associated with the measured properties. Typically, hardness and tensile properties have been used [88]. Examples of calculated quench factors and their associated quench paths are

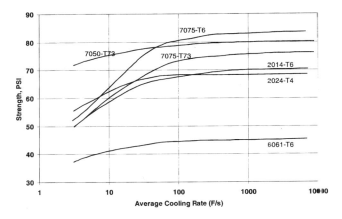

Figure 9 Tensile strengths of six alloys as a function of average cooling rate during quenching.

Table 13 Coefficients for Calculating Quench Factors at 99.5% of Attainable Strength

Alloy	K_2 (sec)	K_3 (cal/mol)	K_4 (K)	K_5 (cal/mol)	Reference
7050-T76	2.2×10^{-19}	5190	850	1.8×10^5	82
7075-T6	4.1×10^{-13}	1050	780	1.4×10^5	82
2024-T851	1.72×10^{-11}	45	750	3.2×10^4	93
7075-T73	1.37×10^{-13}	1069	737	1.37×10^5	88
2219-T87	0.28×10^{-7}	200	900	2.5×10^4	94

Data for 2024-T851 was evaluated using $R = 1.987$. All others were evaluated with $R = 8.3143$.

shown in Table 14. Properties are then related to the quench factor by the equation [86]:

$$p = p_{max} \cdot \exp(K_1 \tau) \qquad (15)$$

where p is the property of interest, p_{max} is the maximum property attainable with infinite quench rate, and K_1 is -0.005013 (natural log of 0.995).

There are two difficulties with this method. First, it is necessary to know the specific quench path that the part experienced. This is often difficult to measure and requires specialized equipment to achieve repeatable results [88]. Secondly, it is also necessary that the C curve is known with sufficient precision. As previously indicated, this data is often not available for the specific conditions of interest. The lack of having detailed information regarding the C curve has limited the applicability of the use of the quench factor.

A. Experimental Determination

There are two primary methods to determine the quench factor of a specific quenchant and alloy processing condition. The first method is somewhat labor intensive, and requires the measurement of multiple cooling curves, using small specimens in different quenchants. The second method utilizes the Jominy End Quench [65].

1. Multiple Quenchant

Figure 10 illustrates the superposition of a cooling curve on a C curve [9]. Experimentally, cooling curves are obtained by acquiring time–temperature data over finite time steps (Δt), which is taken to be the data acquisition rate used to obtain the cooling curve being analyzed. The average temperature (T) between each time step

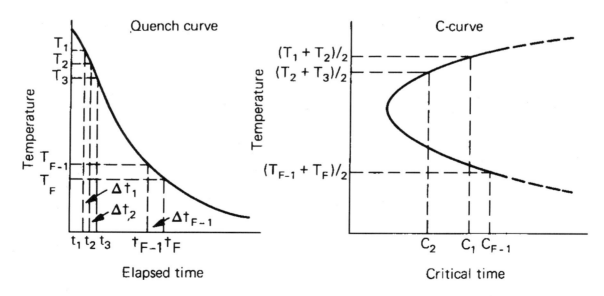

Figure 10 Determination of the quench factor (Q) by superimposing an experimentally determined cooling curve on the C curve for the alloy of interest.

Table 14 Quench Factors and Measured Yield Strength for 1.6-mm-Thick 7075-T6 Sheet

Quench path	Quench factor	Measured yield strength (MPa)
Cold water, strongly agitated	0.464	73.4
Denatured alcohol to 290°C, then cold water	8.539	69.1
Boiling water to 315°C, then cold water	15.327	66.4
Still air to 370°C, then cold water	21.334	67.9

Source: Ref. 95.

interval is then calculated. The C_t value is then calculated for each average temperature using the above equation. The ratio of the time step length used for data acquisition (Δt_i) is divided by the C_{ti} value at the temperature to provide an "incremental quench factor" (q) [9].

$$q = \frac{\Delta t}{C_t}$$

The C_t function for the alloy of interest is calculated from the following equation [22]

$$C_t = K_1 K_2 \left[\exp\left(\frac{K_3 K_4^2}{RT(K_4 - T)^2}\right) \exp\left(\frac{K_5}{RT}\right) \right] \quad (16)$$

where C_t is the critical time required to precipitate a constant amount (the locus of the critical line is the C curve), K_1 is the constant that equals the natural logarithm of the fraction untransformed (1 − fraction defined by the C curve). K_2, K_3, K_4, and K_5 are taken from Table 3, $R = 8.3143$ J K^{-1} mol^{-1} and $T =$ temperature in K.

To obtain the overall quench factor, Q, the incremental quench factor values are progressively summed as the probe or part is cooled through the precipitation range, normally about 800–300°F (425–150°C) [9].

$$Q = \sum_{300}^{800} q$$

The quench factors for 7075-T73 quenched in 100°F (38°C) water at 50 ft/min (0.25 m/sec) have been studied to determine the effect of the size of the time step on the quench factor calculation. The results of this study are shown in Table 4 [20]. These data show that time step changes in the range of 0.1–0.4 sec caused no appreciable change in the calculated quench factor. However, time step variations between 0.5 and 0.8 sec caused considerable scatter in the calculated quench factor (Q). Excessively long time steps may result in an inadequate number of data points to properly calculate transition in the critical portion (knee) of the C curve. It is suggested that the time step interval should be selected such that the average temperature drop is not greater than 75°F (25°C) over the critical cooling range for the alloy of interest (Table 15).

The quench factor obtained is only valid for this specific quench path. To be valid for a series of quenchants, or to predict properties as a function of different quench paths, then multiple quenchants must be used, and the quench factor must be determined for each quench path. Obviously, this method is tedious and requires extensive experimental work.

2. Jominy End Quench

The Jominy End Quench [97–100] provides a method of determining the quench factor for different processing conditions. There has been extensive work utilizing the Jominy End Quench for aluminum alloys [101–108].

Table 15 Effect of Time Step Magnitude on Quench Factor Calculation

Time Step	0.1	0.2	0.3	0.4	0.5	0.6	0.7	0.8
Quench Factor	1.19	1.19	1.17	1.14	1.30	1.52	1.53	1.33

The values were calculated for a cooling curve obtained by quenching an aluminum probe into 38°C water flowing at 0.25 m/sec.
Source: Ref. 96.

The quench rates achievable in the Jominy End Quench have been found to exceed 500°C/sec at the quenched end, to approximately 2°C/sec at 80 mm from the quenched end.

The relationship between properties and the quench factor is:

$$\frac{p}{p_{max}} = \exp(K_1 \tau) \qquad (17)$$

where p is the property of interest. In this case, it is Vicker's hardness (1.0-kg load). By rearranging the above equation, the quench factor, τ, can be determined for hardness on the Jominy End Quench for a specific processing condition:

$$\tau = \frac{1}{K_1} \ln\left(\frac{H_{VN}}{H_{max}}\right) \qquad (18)$$

where H_{VN} is the hardness at a specific location on the Jominy End Quench bar, and H_{max} is the maximum hardness. The maximum hardness, H_{max}, is generally an average of the first several hardness indentations.

Therefore for a specific set of processing conditions, the quench factor can be determined for multiple quench rates using the Jominy End Quench. The generated data can be used to predict properties occurring in similar material and similar operating conditions. This is illustrated in Fig. 11. In this figure, the percentage of the attainable property from hardness (7075-T6 and 7050-T6 from the Jominy End Quench) is compared with that obtained by Hatch [109] by quenching individual panels and measuring the yield strength. The data for quenching individual panels is shown in Table 14. From the above discussion, the quench factor, τ, can be determined for any position on the Jominy End Quench with hardness measurements. Note that there is scatter in the data from quenching individual panels and measuring the tensile properties, while the hardness data from the Jominy End Quench follows the predicted response without significant scatter. In addition, the Jominy End Quench is capable of producing quench rates much slower (or much larger quench factors) than quenching individual panels. This enables predictive property models to be developed for much larger spans

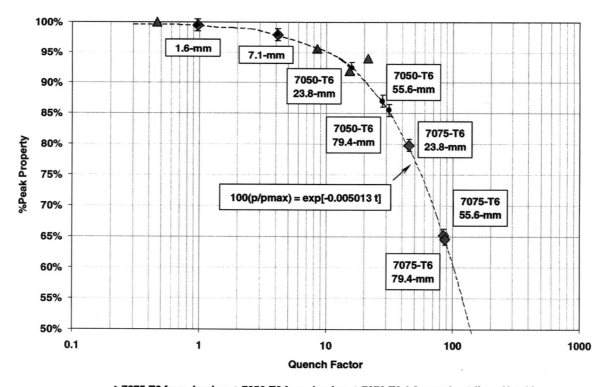

Figure 11 Comparison of historical methods of determining quench factors, and using the Jominy End Quench.

in quenching and heat treating conditions. This is accomplished without complicated equipment or sophisticated instrumentation. This application of the Jominy End Quench is new and has not been reported elsewhere.

3. C Curves

One of the problems with the use of the quench factor is that it is only valid for a specific set of processing conditions. The quench factor depends not only on the quench path, but also on subsequent processing. In addition, time–temperature–property curves for many alloys are scarce, and often proprietary. So the difficulty arises on how to obtain the T–T–P curve for the specific processing conditions of interest.

Historically [49], the C curve is constructed by quenching multiple specimens to specific temperatures, and isothermally holding the specimens at this temperature for a period of time. The specimens are then rapidly quenched and aged. The properties of interest are then measured. The methodology is similar to constructing an isothermal time–temperature curve for steel. This is a very time- and labor-intensive process.

As discussed earlier, the C curve is a collection of points described by the equation:

$$C_t = K_1 K_2 \left[\exp\left(\frac{K_3 K_4^2}{RT(K_4 - T)^2} \right) \exp\left(\frac{K_5}{RT} \right) \right] \quad (19)$$

From the above discussion, the quench factor, τ, can be determined for any position on the Jominy End Quench with hardness measurements. Because the quench factor is related to the C curve by the relationship:

$$\tau = \int \frac{dt}{C_t} \quad (20)$$

or

$$\tau = \frac{\Delta t_1}{C_1} + \frac{\Delta t_2}{C_2} + \frac{\Delta t_3}{C_3} + \cdots + \frac{\Delta t_n}{C_n} \quad (21)$$

it should be possible to calculate the time–temperature–property curve (C curve) from the Jominy End Quench. In the Jominy End Quench, an infinite number of quench rates are available, and the path is known from the relationship described by:

$$\theta_T = \text{erfc}\left[\frac{x}{2\sqrt{\alpha t}} \right] - e^\gamma \text{erfc}\left[\frac{x}{2\sqrt{\alpha t}} + \frac{h}{k}\sqrt{\alpha t} \right]$$

$$\gamma = \frac{h}{k}\sqrt{\alpha t}\left[\frac{x}{2\sqrt{\alpha t}} + \frac{h}{k}\sqrt{\alpha t} \right] \quad (22)$$

Based on this relationship, the time increments, Δt_1, Δt_2, Δt_3, through Δt_n, can be determined by examining the quench path, and dividing the critical temperature range (400–200°C) into intervals at specific temperatures. Because the C curve is independent of the quench path, a series of nonlinear equations can be established to solve for the critical time, C_t, for different quench factors:

$$\tau_1 = \frac{\Delta t_1}{C_1} + \frac{\Delta t_2}{C_2} + \frac{\Delta t_3}{C_3} + \cdots + \frac{\Delta t_n}{C_n}$$

$$\tau_2 = \frac{\Delta t_1}{C_1} + \frac{\Delta t_2}{C_2} + \frac{\Delta t_3}{C_3} + \cdots + \frac{\Delta t_n}{C_n}$$

$$\tau_3 = \frac{\Delta t_1}{C_1} + \frac{\Delta t_2}{C_2} + \frac{\Delta t_3}{C_3} + \cdots + \frac{\Delta t_n}{C_n} \quad (23)$$

$$\tau_4 = \frac{\Delta t_1}{C_1} + \frac{\Delta t_2}{C_2} + \frac{\Delta t_3}{C_3} + \cdots + \frac{\Delta t_n}{C_n}$$

where τ_1, τ_2, τ_3, ..., τ_n are the quench factors from locations on the Jominy End Quench specimen; Δt_1, Δt_2, Δt_3, through Δt_n are the temperature intervals from the quench path at specific locations on the Jominy End Quench and C_1, C_2, ..., C_n are the critical times on the C curve. These equations can be solved using MathCad™, Mathmatica™, or other software programs.

To minimize false and unrealistic answers, it is necessary to constrain the results of the solution to the series of nonlinear equations solved above. Examples of the constraints used are the stipulation that all solutions for C_t must be positive and not equal to zero. Further, the results are constrained to yield results C_t less than 5000. Examination of the data available [82,92,93] indicates that this is a reasonable assumption. Better solutions for the shape of the C_t curve would be obtained with more nonlinear equations.

Solution of this set of equations will provide the critical time as a function of temperature, i.e., the C curve. Once the C curve is obtained, calculation of the constants K_2, K_3, K_4, and K_5 is difficult because of the very nonlinear nature of the equations. FORTRAN programs are available [94,110] that will calculate the necessary constants. Typically, these programs use an adaptive nonlinear least squares routine [111]. The program CT [110] was used to calculate these constants.

To illustrate a specific use of the Jominy End Quench to determine the C curve, the hardness data for a Jominy End Quench heat-treated to the T6 condition was used (Fig. 12). The critical range of 400–200°C was divided into 10 intervals of 25°C each. The quench paths were determined for positions on the Jominy End Quench corresponding to 1.6, 5, 7.1, 10, 14, 21, 30, 40, 50, and 60 mm from the quenched end. The time

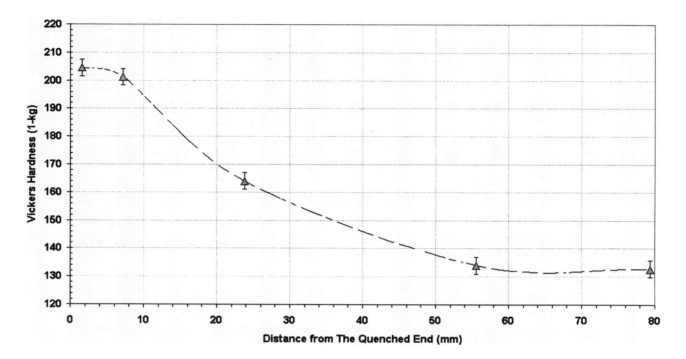

Figure 12   Hardness data from 7075 Jominy End Quench heat treated to the −T6 condition.

intervals through each of the 10 temperature intervals were determined. A series of 10 nonlinear equations of the form:

$$\tau = \frac{\Delta t_1}{C_1} + \frac{\Delta t_2}{C_2} + \frac{\Delta t_3}{C_3} + \ldots + \frac{\Delta t_n}{C_n} \qquad (24)$$

were established, with the time interval Δt corresponding to specific temperatures. These nonlinear equations were solved for least error using MathCad™. Once the values for the critical time, C_t, were obtained, the constants K_2, K_3, K_4, and K_5 were calculated using the software program CT [110]. The specific values of C_t, and the calculated C curve from the program CT (from the individual values calculated) were compared to previously published C curves for 7075-T6 [112]. The results are shown in Fig. 13.

The results shown in Fig. 13 show that the individual calculated C_t points correspond well to the published data of Totten et al. [112]. The calculated values of K_2, K_3, K_4, and K_5, when plotted as a C curve, also fit the published C curve well. However, the data is optimistic at the "knee" of the C curve. This can be from several sources. First, only 10 values of C_t were calculated. Bates [88] indicated that at least 40 data points are necessary to ensure the optimal accuracy in calculation of the constants. The few data points calculated could

contribute to inaccuracy of the C curve. This would also be reflected in the calculation of the constants, and the resultant calculated C curve. Further, the chemistry of the 7075-T6 C curve of Totten et al. [112] was not reported. This could also contribute to errors if the chemistry of the published curve was leaner than that used in the Jominy End Quench.

B.   Property Prediction

As described above, properties can be predicted from the quench factor by use of:

$$\frac{p}{p_{\max}} = \exp(K_1 \tau) \qquad (25)$$

where p is the predicted property, $p_{\max}$ is the maximum property after an infinite quench; K_1 is ln(0.995), and Q is the quench factor [49]. In terms of hardness, this equation can be expressed as:

$$H_{\mathrm{VN}} = H_{\max} \exp(K_1 \tau) \qquad (26)$$

For tensile properties, the relationship is:

$$\sigma = \sigma_{\max} \exp(K_1 \tau) \qquad (27)$$

These relationships can be exploited to predict properties, either as part of quenchant selection, or during the

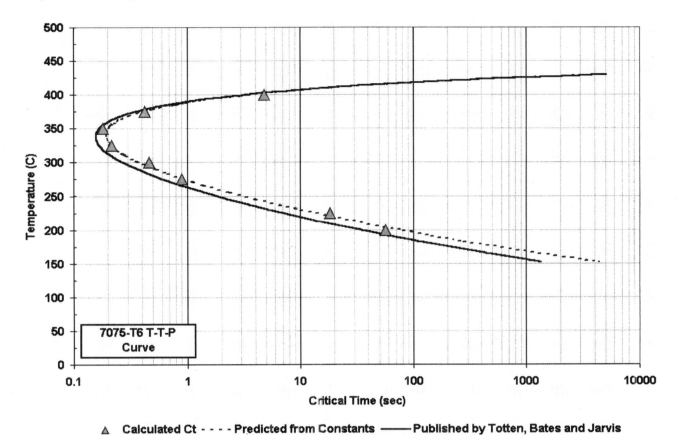

Figure 13 Calculated and predicted *T–T–P* curve from 7075-T6 Jominy End Quench hardness.

Table 16 Relationship Between Quench Factor and Yield Strength in Aluminum Alloy (σ_{max} = 475.1 MPa)

Quench factor (Q)	Percent attainable yield strength	Predicted yield strength (MPa)	Quench factor (Q)	Percent attainable yield strength	Predicted yield strength (MPa)
0.0	100.0	475.1	26.0	87.8	417.2
2.0	99.0	470.2	28.0	86.9	413.0
4.0	98.0	465.4	30.0	86.0	408.9
6.0	97.0	461.3	32.0	85.2	404.7
8.0	96.1	456.5	34.0	84.3	400.6
10.0	95.1	451.6	36.0	83.5	396.5
12.0	94.2	447.5	38.0	82.7	393.0
14.0	93.2	442.7	40.0	81.8	388.9
16.0	92.3	438.5	42.0	81.0	384.7
18.0	91.4	434.4	44.0	80.2	381.3
20.0	90.5	429.6	46.0	79.4	377.2
22.0	89.6	425.4	48.0	78.6	373.7
24.0	88.7	421.3	50.0	77.8	396.5

Source: Ref. 113.

design process, to ensure that properties are achieved after heat treatment (Table 16).

IV. SUMMARY AND COMMENTS

In this chapter, we have briefly examined the basics of aluminum metallurgy, showing the importance of the quench rate to achieving proper properties. We have also shown the effect of the different quenchants on the cooling rate. Finally, we showed several techniques for calculating quenchant performance and property prediction using the quench factor.

REFERENCES

1. Wilm, A. Metallurgie 1911, *8*, 225.
2. Merica, R.; Waltenburg, W.; Scott, T. Trans. AIME 1920, *64*, 41.
3. Mehl, R.; Jeten, T. *Age Hardening of Metals*; ASM: Cleveland, 1940; 342 pp.
4. Mott, N.; Nabarro, R. Proc. R. Soc. Lond. A 1940, *145*, 362.
5. Orowan, *Symposium on Internal Stresses in Metals and Alloys — Session II Discussion*; Inst. Metals: London, England, 1948; 51 pp.
6. Holl, H.A. J. Inst. Met. 1964, *93*, 364.
7. Fischer, G.; Lynker, F.W.; Markworth, M. Aluminum 1972, *48* (6), 413.
8. Suzuki, H.; Kanno, M.; Saitoh, H. Keikinzoku 1983, *33* (7), 399.
9. Holl, H.A. *Metallurgy Note #59*, Australian Defense Scientific Service, 1968; 10 pp.
10. Nes, E.; Billdal, H. Acta Met. 1977, *25*, 1039.
11. Thundal, B.; Sundberg, R. J. Inst. Met. 1969, *97*, 160.
12. Nes, E. Acta Met. 1972, *20* (4), 499.
13. Ryum, N. Acta Met. 1969, *17*, 269.
14. Peel, C.J.; Poole, P. Grain Bound 1976, 7.
15. Palmer, I.G.; Thomas, M.P.; Marshall, G.J. *Dispersion Strengthened Aluminum Alloys*; Kim, Y., Griffiths, W., Eds.; TMS: Warrendale, PA, 1988; 217 pp.
16. Das, S.K. *Intermetallic Compounds*; Westbrook, J. Fleischer, R. Eds.; John Wiley and Sons, 1994; Vol. 2, 175.
17. Mondolfo, L. *Aluminum Alloys, Structure and Properties*; Butterworths: London, 1976.
18. Metals Handbook; 8th Ed.; ASM: Metals Park, 1973; Vol. 8, 259.
19. Guinier, A. Compt. Rend. 1937, *204*, 1115.
20. Guinier, A. Nature 1938, *142*, 669.
21. Preston, G.D. Proc. R. Soc. A 1934, *166* (6), 572.
22. Preston, G.D. Nature 1938, *142*, 570.
23. Hornbogen, E. Aluminum 1967, *43*, 115.
24. Silcock, J.M.; Heal, T.J.; Hardy, H.K. J. Inst. Met. 1953, *82*, 239.
25. Boyd, J.D.; Nicholson, R.B. Acta Met. 1971, *19*, 1379.
26. Lifshitz, I.M.; Slyozov, V.V. Sov. Phys. JETP 1959, *35*, 331.
27. Wagner, Z. Electrochemistry 1961, *65*, 581.
28. Wilm, A. Metallurgie 1911, *8*, 225.
29. Wilson, R.; Partridge, P. Acta Met. 1965, *13*, 1321.
30. Hyatt, M.V. Proc. Int. Conf. Aluminum Alloys, Torino, Italy, October 1976.
31. Bernole, M.; Graf, R. Mem. Sci. Rev. Metall. 1972, *69*, 123.
32. Bergman, G.; Waugh, L.; Pauling, L. Nature 1952, *169*, 1057.
33. Ryum, N. Z. Met.kd. 1975, *66*, 377.
34. Mondolfo, L.F.; Gjostein, N.A.; Lewisson TAIMME 1956, *206*, 1378.
35. Ryum, N. Z. Met.kd. 1975, *65*, 338.
36. Pashley, D.W.; Jacobs, M.H.; Vietz, J.T. Philos. Mag. 1967, *16*, 51.
37. Laves, F. *Theory of alloy phases*; ASM Symp: Cleveland, 1956; 124 pp.
38. Auger, P.; Raynal, J.M.; Bernole, M.; Graf, R. Mem. Sci. Rev. Metall. 1974, *71*, 557.
39. Peel, C.J.; Clark, D.; Poole, P., et al. RAE Technical Report 78110; 1978.
40. Osamura, K.; Ochai, S.; Uehara, T.J. Inst. Met. 1984, *34* (9), 517.
41. Hirsch, P.B.; Humphereys, F.J. *Physics of Strength and Plasticity*; MIT Press, 1969; 189 pp.
42. Sundar, G.; Hoyt, J.J. Phys. Rev. B 1992, *46* (12), 266.
43. Grossman, M.A. Met. Prog. 1938, *4*, 373.
44. Scott, H. Quenching mediums. In *Metals Handbook*; ASM, 1948; 615 pp.
45. Wever, F. Arch. Eisenhüttenwes. 1936, *5*, 367.
46. Dakins, M. Central Scientific Laboratory, Union Carbide, Report CSL-226A.
47. Fink, W.L.; Wiley, L.A. Trans. AIME 1948, *175*, 414.
48. Suzuki, H.; Kanno, M.; Saitoh, H. Keikinzoku 1983, *33* (1), 29.
49. Evancho, J.W.; Staley, J.T. Metall. Trans. 1974, *5* (1), 43.
50. Wierszykkowski, I.A. Metall. Trans., A 1991, *22A*, 993.
51. Bates, C.E.; Totten, G.E. Heat Treat. Met. 1988, *4*, 89.
52. Swartzenruber, L.; Beottinger, W.; Ives, I., et al. National Bureau of Standards Report NBSIR 80-2069, 1980.
53. Bates, C.E.; Landig, T.; Seitanakis, G. Heat Treat. 1985, *12*, 13.
54. Bates, C.E. *Recommended Practice for Cooling Rate Measurement and Quench Factor Calculation*, ARP 4051 Aerospace Materials Engineering Committee (SAE) 1987, 1.
55. Staley, J.T.; Doherty, R.D.; Jaworski, A.P. Metall. Trans. A 1993, *24A* (11), 2417.

56. Staley, J.T. Mater. Sci. Technol. 1987, *3* (11), 923.

57. Hall, D.D.; Mudawar, I. J. Heat Transfer 1995, *117* (5), 479.

58. Kim, J.S.; Hoff, R.C.; Gaskell, D.R. *Materials Processing in the Computer Age*; Vasvey, V. R., 1991; 203 pp.

59. Swartzenruber, L.; Beottinger, W.; Ives, I., et al. National Bureau of Standards Report NBSIR 80-2069, 1980.

60. Staley, J.T.; Doherty, R.D.; Jaworski, A.P. Metall. Trans. A 1993, *24A* (11), 2417.

61. Bates, C.E. *Quench Factor-Strength Relationships in 7075-T73 Aluminum*; Southern Research Institute 1987; 1 pp.

62. Taylor, J.L. J. Inst. Met. 1963, *92*, 301.

63. Newkirk, J.W.; MacKenzie; D.S., Ganapathi, K. TMS: San Diego, 1999.

64. Chang, S.; Morral, J.E. Acta Met. 1975, *23*, 685.

65. MacKenzie, D.S. *Quench Rate and Aging Effects in Aluminum-Zinc-Magnesium-Copper Aluminum Alloys*; Ph.D. Dissertation, University of Missouri-Rolla, December 2000.

66. Embury, J.D.; Nicholson, R.B. Acta Met. 1965, *13*, 403.

67. Unwin, P.N.; Lorimer, G.W.; Nicholson, R.B. Acta Met. 1969, *17*, 1363.

68. Shastry, C.R.; Judd, G. Trans. AIME 1969, *62*, 724.

69. Embury, J.D.; Nicholson, R.B. Acta Met. 1965, *13*, 403.

70. Darken, L.S. *Atom Movements*; ASM, 1959.

71. Cottrell, A.H. *Vacancies and Other Point Defects in Metals and Alloys*; Institute of Metal, 1958. 1 pp.

72. Flinn, P.A. *Strengthening Mechanisms in Solids*; ASM 1962; 17 pp.

73. Lormer, W.M. *Vacancies and Other Point Defects in Metals and Alloys*; Institute of Metals, 1958; 79 pp.

74. Shewmon, P. *Diffusion in Solids*; TMS, 1989. 78 pp.

75. Prabhu, N.; Howe, J.M. Metall. Trans. A 1992, *23A* (1), 135.

76. Smith, W.F.; Grant, N.J. Trans. ASM 1969, *62*, 724.

77. Speith, K.; Lange, H. Mitt. Kaiser-Wilhelm-Inst. Eisenforssch. 1935, *17*, 175.

78. Rose, A. Arch. Eisenhullennes. 1940, *13*, 345.

79. Grossman, M.A. Met. Prog. 1938, *4*, 373.

80. Scott, H. Quenching mediums. In *Metals Handbook*; ASM, 1948; 615 pp.

81. Wever, F. Arch. Eisenhüttenwes. 1936, *5*, 367.

82. Evancho, J.W.; Staley, J.T. Metall. Trans. 1974, *5* (1), 43.

83. Fink, W.L.; Wiley, L.A. Trans. AIME 1948, *175*, 414.

84. Suzuki, H.; Kanno, M.; Saitoh, H. Keikinzoku 1983, *33* (1), 29.

85. Wierszykkowski. I.A. Metall. Trans., A 1991 *22A*, 993.

86. Bates, C.E.; Totten, G.E. Heat Treat. Met. 1988, *4*, 89.

87. Bates, C.E.; Landig, T.; Seitanakis, G. Heat Treat. 1985, *12*, 13.

88. Bates, C.E. Recommended Practice for Cooling Rate

89. Measurement and Quench Factor Calculation; ARP 4051 Aerospace Materials Engineering Committee (SAE) 1987, 1.

89. Staley, J.T. Mater. Sci. Technol. 1987, *3* (11), 923.

90. Hall, D.D.; Mudawar, I. J. Heat Transfer 1995, *117* (5), 479.

91. Kim, J.S.; Hoff, R.C.; Gaskell, D.R. *Materials Processing in the Computer Age*; Vasvev V. R., Ed.; 1991; 203 pp.

92. Bates, C.E. Quench Factor-Strength Relationships in 7075-T73 Aluminum; *Southern Research Institute*, 1987; 1 pp.

93. Processing/Microstructure/Property Relationships in 2024 Aluminum Alloy Plates, U.S. Department of Commerce, National Bureau of Standards Technical Report NBSIR 83-2669, January 1983.

94. Nondestructive Evaluation of Nonuniformities in 2219 Aluminum Alloy Plate—Relationship to Processing, U.S. Department of Commerce, National Bureau of Standards Technical Report NBSIR 80-2069, December 1980.

95. Hatch, J.E. *Aluminum: Properties and Physical Metallurgy*; ASM: Metals Park, 1984; 260 pp.

96. Polyalkylene glycol heat treat quenchant. *Aerospace Material Specification, AMS 3025B*; Society of Automotive Engineers Inc: Warrendale, PA, July 2000.

97. Jominy, W.E.; Boegehold, A.L A hardenability test for carburizing steel. ASM Trans. 1939, *27* (12), 574.

98. Jominy, W.E. A hardenability test for shallow hardening steels. ASM Trans. 1939, *27* (12), 1072.

99. ASTM A255. Jominy Test, standard method for end-quench test for hardenability of steel. In *Annual Book of ASTM Standards*; ASTM: Philadelphia, PA.

100. SAE J406c. Methods of determining hardenability of steels. In *Annual SAE Handbook*; SAE: Warrendale, PA.

101. Newkirk, J.W.; MacKenzie, D.S. Faster methods of studying the quenching of aluminum using the Jominy end quench. Proc. Heat Treating, Rosemont, IL, 1998.

102. Totten, G.E.; MacKenzie, D.S. Aluminum quenching technology: a review. Proc. 7th Intl. Conference ICAA7, Charlottesville, Virginia, April 9–14. Mat. Sci. Forum 2000, 331–337, 589. Part 1.

103. MacKenzie, D.S. Proc. Intl. Federation of Heat Treatment and Surface Engineering, Budapest, Hungary, 12 September, 1999.

104. Newkirk, J.; MacKenzie, D.S. J. Mater. Perform. Eval. ASM 2000, *9* (4), 408.

105. Newkirk, J.W.; MacKenzie, D.S.; Ganapathi, K. *Light Metals IV*; TMS: San Diego, 1999.

106. Loring, B.M.; Baer, W.H.; Carlton, G.M. The use of the Jominy test in studying commercial age-hardening aluminum alloys. Trans. Am. Inst. Min. Metall. Eng. 1948, *175*, 401.

107. 'tHart, W.G.J.; Kolkman, H.J.; Schra, L. The Jominy

End-Quench Test for the Investigation of Corrosion Properties and Microstructure of High Strength Aluminum, National Aerospace Laboratory, NLR, Netherlands, NLR TR 80102U, 1980.

108.   'tHart, W.G.J.; Kolkman, H.J.; Schra, L. National Aerospace Laboratory, NLR, Netherlands, NLR TR 82105 U, 1982.

109.   Hatch, J.E. *Aluminum: Properties and Physical Metallurgy*; ASM: Metals Park, 1984; 51 pp.

110.   Orszak, K.B.; Totten, G.E. Union Carbide Chemicals and Plastics, Central Scientific Laboratory Report CSL-293, February 1990.

111.   Wilson, W.; Swartzendruber Comput. Phys. Commun. 1974, *7*, 151.

112.   Totten, G.E.; Bates, C.E.; Jarvis, L.M. Heat Treat. December 1991, 16.

113.   Bates, C.E.; Totten, G.E. Procedure for quenching media selection to maximize tensile properties and minimize distortion on aluminum alloy parts. Heat Treat. Met. 1988, (4), 89–97.

19

Surface Engineering Methods

Paul K. Chu, Xiubo Tian, and Liuhe Li
City University of Hong Kong, Kowloon, Hong Kong

The optimization of the bulk properties of materials can be difficult and the development of such protocols can be time-consuming, but in many applications, modification of the surface properties of materials or industrial components will suffice. Surface modification/engineering is generally easier, more flexible, and more economical compared to bulk counterparts. Another advantage is that surface techniques can sometimes be extended beyond classical thermodynamics and equilibrium constraints, and so excellent surface modification schemes and modified materials can be produced. In this chapter, the three common surface engineering techniques—physical vapor deposition (PVD), chemical vapor deposition (CVD), and ion implantation—are reviewed and current applications are described. Hybrid processes in which the material properties can be further optimized by employing a combination of techniques are also discussed.

I. PHYSICAL VAPOR DEPOSITION

A. Introduction

Physical vapor deposition (PVD) is a generic term applied to coating processes in which the transport of materials to the substrate is facilitated by a physically driven mechanism such as sputtering, evaporation, ion plating, or ion-assisted sputtering [1]. As a versatile technique, many types of thin films, including elemental layers (e.g., Cu films) and compound films (e.g., TiN, Al_2O_3), can be deposited on different substrates [2]. During the synthesis of compound films, chemical reactions usually take place, but the material transport process does not always rely on chemical reactions. In some processes, the energetic ions upon impact cause ion mixing in the substrate, thereby improving layer adhesion and surface properties. Efficiency can usually be enhanced if low-temperature plasma is introduced to the deposition process. Compared to conventional plasma nitriding or chemical vapor deposition (CVD) (plasma or nonplasma), PVD processes are typically performed at temperatures below 500°C, making it feasible for aluminum and some steel substrates without significantly affecting the structural characteristics. Recent innovations in PVD techniques, especially the use of plasmas and unbalanced magnetron, have led to wider industrial acceptance [1].

B. Techniques and Equipment

The primary mechanism used to introduce the source materials is evaporation or sputtering. There are many variations of evaporation and sputtering and the choice depends on the applications, requirements, costs, and convenience.

1. Evaporation

In the evaporation process, vapors are produced from a source heated resistively or radiatively, eddy currents, electron beams (EBs), laser beams, or an arc discharge. The process is usually carried out in vacuum (typically 10^{-5}–10^{-6} Torr) so that the evaporated atoms travel in an essentially collisionless line-of-sight trajectory prior

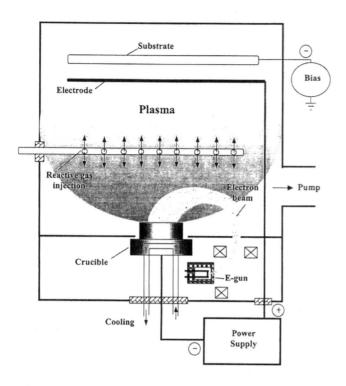

Figure 1 Reactive evaporation using an EB evaporation source. (From Ref. 5.)

Bunshan [4], the family of materials that are deposited by evaporation includes metals, semiconductors, alloys, intermetallic compounds, refractory compounds (oxides, carbides, nitrides, borides, etc.), and mixtures thereof. An important point is that the source material should be pure and free of gases and/or inclusions to forestall the problem of molten droplet ejection from the pool, commonly called spitting [4]. So far, when compound films are formed, plasma-assisted evaporation (also called activated reactive evaporation, ARE) is preferred due to high reactivity, as shown in Fig. 1 [5]. Compared to reactive sputtering or plasma-assisted chemical vapor deposition (PACVD) processes, ARE is the most efficient, as indicated in Table 1 [4].

2. *Sputtering*

Sputtering is an effective tool, although the deposition rate is typically smaller than that of evaporation and ion plating. An important parameter is the energy distribution of the sputtered species that usually has a maximum of about 4–7 eV and a long tail of over 50 eV. These energies are more than an order of magnitude higher than those of the evaporated species and allow the depositing species more penetration and surface mobility to form dense, well-bonded films [1]. The sputtering yield is related to the nature of ions and target, temperature, incident angle, bombardment energy, etc. [6–8]. The simplest technique is direct current (DC) glow discharge sputtering (planar diode sputtering), but owing to low efficiency, this type of sputtering is generally enhanced by external means such as magnetic fields and high-frequency electric fields. In contrast to processes in which the substrate is immersed in plasma, the film purity achieved by ion sputtering can be considerably enhanced due to the separation of the ion source and the sample chamber. The most frequently used technique is magnetron sputtering, in which a combination of the electric fields and magnetic fields (50–500 G) causes the secondary electrons to drift in a closed circuit or magnetron tunnel in front of the target surface. The magnetic field concentrates the plasma in the space immediately above the target-trapping electrons in this

to condensation onto the substrate, which is usually held at a ground potential. The lateral film thickness variation usually follows the cosine rule, and more uniform film deposition can be accomplished by rastering the sample or by operating at a higher pressure to enhance collisions [3]. A reasonable uniformity of ±10% can be attained in gas scattering evaporation or pressure plating. For best purity, an electron beam heating system should be used as the material can then be evaporated from the center of the crucible containing the charge and reactions with the crucible wall are mitigated. The energy of the species arriving at the substrate is on the order of 0.05–0.1 eV, implying that they have limited energy for surface migration to find good nucleation sites. As reviewed by Deshpandey and

Table 1 Deposition Rate of Some Compound Films Produced by ARE, Reactive Sputtering, and PACVD

Compound	ARE (nm min^{-1})	Reactive sputtering (nm min^{-1})	PACVD (nm min^{-1})
TiC, HfC, ZrC	200–300	40–50	15–40
TiN, HfN, ZrN	200–300	30–40	2–15
TiO$_2$, ZrO$_2$, Al$_2$O$_3$	100–200	20–80	20–30
TiS$_2$, MoS$_2$, MoS$_3$	100–200		

vicinity. Electron confinement significantly increases the efficiency and, consequently, magnetrons can be operated at low pressure (e.g., 1–3 mTorr) and low voltage (e.g., 350 V). There are many variations of magnetron sputtering systems, as shown in Fig. 2 [1].

3. Vacuum Arc

A vacuum arc source features both evaporation and higher ion energy, and the technique has attracted much attention. The average energy of ions is about 40 eV per particle compared to 0.1 eV in evaporation and 5–10 eV in typical sputtering processes [9]. The ionization efficiency in cathodic arc processes is high (approaching 100%), whereas the ionization efficiency in ion plating (~20%) and sputtering (<10%) is much smaller. The high ionization efficiency provides highly reactive metallic and gaseous species to form compounds such as nitrides and carbides with excellent properties [10]. Cathodic arc deposition can be conveniently optimized by adjusting substrate bias and pulse duration (when operated in the pulsed mode), and, to a certain extent, is preferred over magnetron sputtering or e-beam evaporation because the arc process has a much broader process window for stoichiometric nitrides [11]. The biggest drawback of the cathodic arc process is the emission of macroparticles. They vary in size depending on cathode composition and temperature, and are believed to arise from the spattering of molten materials from the edges of the arc crater. The plasma has to be filtered if a high-quality film is needed [12,13]. The curved magnetic duct is the most frequently used filtering device, but plasma transport efficiency is usually quite small. In order to enhance plasma transport, the duct or a plate inserted in the duct is biased with respect to the plasma to about several tens, and plasma transport efficiency can be elevated by about 25% in an optimized 90° duct [14,15]. The plasma behavior of a cathodic arc has been widely investigated and some useful data are summarized in Table 2.

Cathodic arcs are operated in either a pulsed mode or DC mode. As reviewed by Sanders and Anders [11], the decision on whether to use a continuous mode or pulsed mode largely depends on the objective. Large-area, high-throughput deposition of relatively thick films is better performed using DC as this mode is capable of very high deposition rates. This is particularly true if the process or coating can tolerate the presence of macroparticles and most commercial cathodic arc systems are operated in DC. In contrast, pulsed systems have a smaller deposition rate due to their lower duty cycle. The advantage of pulsed operation is mainly related to

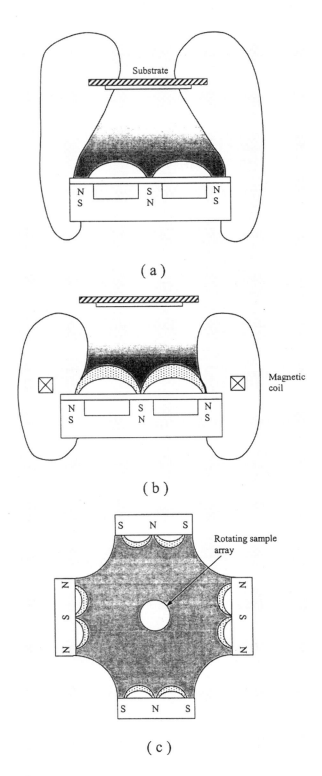

Figure 2 Magnetron sputtering systems: (a) unbalanced magnetron; (b) unbalanced magnetron with controlled field; and (c) closed field unbalanced magnetron. (From Ref. 1.)

Table 2 Mean Charge State, Ion Velocity, and Burning Voltage of Cathodic Arc Materials

Atomic symbol	T_{melt} (°C)	Mean charge state (Z) [16,17]	Drift velocity (10^4 m sec^{-1}) [18]	Burning voltage (V) [19]
Li	180.5	1.00	2.38	23.5
B	2300	1.11		
C	3550	1.00	2.97	29.6
Mg	648.8	1.54	3.06	18.8
Al	660.4	1.73	2.76	23.6
Si	1410	1.39	2.58	27.5
Ca	839	1.93	2.59	23.5
Sc	1514	1.79		
Ti	1660	2.03	2.22	21.3
V	1890	2.14	1.93	22.5
Cr	1857	2.09	1.94	22.9
Mn	1244	1.53	1.08	22.0
Fe	1535	1.82	1.18	22.7
Co	1495	1.73	1.18	22.8
Ni	1453	1.76	1.09	20.5
Cu	1083	2.06	1.28	23.4
Zn	419.6	1.20	1.04	15.5
Ge	937.4	1.40	1.10	17.5
Sr	769	1.98		18.0
Y	1522	2.28	1.43	18.1
Zr	1852	2.58	1.57	23.4
Nb	2468	3.00	1.55	27.0
Mo	2617	3.06	1.74	29.3
Pd	1552	1.88		21.3
Ag	961.9	2.14	1.04	23.0
Cd	320.9	1.32	0.68	16.0
In	156.6	1.34	0.55	17.5
Sn	232	1.53	0.75	17.5
Sb	630.7	1.00	0.52	15.8
Ba	725	2.00	0.67	18.3
La	921	2.22	0.70	17.2
Ce	799	2.11	0.70	17.9
Pr	931	2.25	0.87	20.0
Nd	1021	2.17		19.7
Sm	1077	2.13	0.74	14.6
Gd	1313	2.20	0.74	21.6
Dy	1412	2.30	0.74	19.8
Ho	1474	2.30	0.83	20.0
Er	1529	2.36	0.82	19.0
Tm	1545	1.96	0.83	21.7
Yb	819	2.03		14.4
Hf	2227	2.89	0.92	24.3
Ta	2996	2.93	1.14	28.7
W	3410	3.07	1.05	31.9
Ir	2410	2.66		24.5
Pt	1772	2.08	0.68	22.5
Au	1064	2.97	0.58	19.7
Pb	327	1.64	0.54	15.5
Bi	271.3	1.17	0.42	15.6
Th	1750	2.88	0.99	23.3
U	1132	3.18	1.14	23.5

the reduced requirements for cooling because the average power can be kept low (e.g., < 1 kW). Film properties can also be better and process control is easier in the pulsed mode. For instance, amorphous carbon (a-C) films deposited by pulsed carbon vacuum arcs tend to have a slightly higher sp^3 content than their DC-deposited counterparts. However, pulsed operation does not always lead to enhanced properties. Many metal films produced by pulsed deposition with a low duty factor possess higher amounts of unwanted oxygen and hydrogen, but pulsed systems can be of great value if ultrathin a-C and oxide films are to be deposited.

C. Coating Materials

Materials selection for coatings, albeit complicated, is of great importance in the industry. Many pragmatic considerations must be taken into account, including coating materials, substrates, interfaces, and diffusion. Hard coating materials can be divided into three groups depending on chemical characteristics: metallic, covalent, and ionic [4,20]. Holleck [20] reviewed the typical properties of representative materials from these three groups and the results are summarized in Table 3. Comparing these and other relevant properties of the hard materials groups (Table 4), the following conclusions can be drawn [20]:

1. Each group has advantages and disadvantages in hard coating applications.
2. Metallic hard materials seem to be the most suitable and versatile as layers and substrates.
3. Ionic (ceramic) hard materials are suitable, in particular, on the surface because of high stability and low interaction tendency.

These are true for single coatings, but optimal wear resistance may be achieved using multiphase or multilayer coatings instead. For example, Grimberg et al. [21] observed extremely high microhardness (47–51 GPa) from multicomponent (Ti,Nb)N coatings. Andrievsky et al. [22] reported that the hardness of coatings a few micrometers thick consisting of 20 alternating TiN and NbN layers was close to 50 GPa, much higher than that of pure TiN or NbN. Multilayered coatings consisting of alternating layers of two hard phases (e.g., TiN with VN or NbN) exhibit considerable hardness enhancement when the thicknesses of the individual layers are on the order of several nanometers. Differences in shear moduli of the phases and low-energy interfaces have been identified as prerequisites for enhanced hardness [23].

Table 3 Properties of Coating Materials with Metallic Bonds

Type of coating		Density (g cm^{-3})	Hardness (H_V)	Melting point (°C)	Thermal conductivity (J (cm sec K)$^{-1}$)	Coefficient of linear expansion (10^{-6} K^{-1})	Young's modulus (kN mm^{-2})	Resistivity (μW cm)
Nitrides	TiN	5.40	2100–2400	2950	0.289	9.35–10.1	256–590	18–25
	VN	6.11	1560	2050–2177	0.113	9.1–9.2	460	85
	ZrN	7.32	1600–1900	2980	0.109	7.9	510	7–21
	NbN	8.43	1400	2200–2300	0.0374	10.1	480	58
	TaN		1300	2090	0.096	5.0		128
	HfN		2000	2700	0.113	6.9		28
	CrN	6.12	1100	1050		2.3	400	640
Carbides	TiC	4.93	2800–3800	3070–3150	0.172–0.35	7.61–8.6	460–470	51
	VC	5.41	2800–2900	2650–2830	0.043	6.5–7.3	430	60
	HfC		2700	3890	0.063	6.73–7.2	359	37
	ZrC	6.63	2600	3445–3530	0.205	6.93–7.4	355–400	42
	NbC	7.78	1800–2400	3480–3610	0.142	6.84–7.2	345–580	19–35
	WC	15.72	2000–2400	2730–2776	0.293	3.8–6.2	600–720	17
	W$_2$C		2000–2500					
	TaC	14.48	1550–1800	3780–3985	0.22	6.61–7.1	291–560	15–20
	Cr$_3$C$_2$	6.68	1500–2150	1810–1850	0.188	10.3–11.7	400	75
	Mo$_2$C	9.18	1660	2517		7.8–9.3	540	57
Borides	TiB$_2$	4.50	3000	3225		7.8	560	7
	VB$_2$	5.05	2150	2747		7.6	510	13
	NbB$_2$	6.98	2600	3036		8.0	630	12
	TaB$_2$	12.58	2100	3037		8.2	680	14
	CrB$_2$	5.58	2250	2188		10.5	540	18
	Mo$_2$B$_5$	7.45	2350	2140		8.6	670	18
	W$_2$B$_5$	13.03	2700	2365		7.8	770	19
	LaB$_6$	4.73	2530	2770		6.4	400	15

Source: Ref. 20.

Table 4 Properties of Hard Coating Materials with Covalent Bonds

Type of coating		Density (g cm^{-3})	Hardness (H_V)	Melting point (°C)	Coefficient of linear expansion (10^{-6} K^{-1})	Young's modulus (kN mm^{-2})	Resistivity (μW cm)
Nitrides	BN (cubic)	2.52	3000–5000	2730		660	10^{13}
	BN (FCC)		4700	1200–1500			
	AiN	3.26	1230	2250	4.2–5.7	350	10^{15}
	Si$_3$N$_4$	3.19	1720	1900	2.5–3.9	210	10^{13}
Carbides	B$_4$C	2.52	300–4100	2450	4.5–6.1	441	0.5×10^6
	C (diamond)	3.52	~8000	3800	1.0	910	10^{20}
	SiC	3.22	2600	2760	5.3–6.1	480	10^5
Borides	B	2.34	2700	2100	8.3	490	10^{12}
	TiB$_6$	2.43	2300	1900	5.4	330	10^7
	AlB$_{12}$	2.58	2600	2150		430	2×10^{12}

Source: Ref. 20.

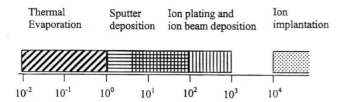

| Thermal Evaporation | Sputter deposition | Ion plating and ion beam deposition | Ion implantation |

Figure 3 Energy ranges of physical vapor deposition processes. (From Ref. 24.)

D. Structure and Property Optimization

In PVD, optimization of both the coating materials and processing parameters is important as they are interrelated. Variations of the parameters can lead to considerable differences in film properties such as film density, structure, mechanical properties, chemical properties, and adhesion. Generally speaking, bombardment energy, working pressure, and sample temperature are controlled to optimize the films.

1. Energy

Deposition energy is related to generation techniques. As summarized in Fig. 3, the atoms produced by evaporation possess energy in the range of 0.1–0.6 eV. The energy of sputtered atoms is from 4.0 to 10 eV, whereas the species generated in vacuum arc can reach several tens to hundreds of electron volts. The energy can also be controlled by an external field or sample bias. In most cases, deposition is assisted by external energy, which can more effectively optimize the film properties. As reviewed by many authors [1,25,26], bombardment energy determines the processing temperature, film density, stress, adhesion, structure, morphology, and phase.

Processing Temperature

A lower processing temperature can be used in concert with ion bombardment. The deposited energy provides the adatoms with the ability to move further across the surface, or to adjacent sites in the bulk. For example, vacuum evaporation at temperatures exceeding 1300°C can be replaced by ion plating at temperatures below 300°C [25].

Film Density

With ion bombardment, the film becomes more compact and has a higher density. For instance, the density of ZrO_2 can be increased by 14% using 600 eV

oxygen ion bombardment compared to that of films formed by pure Zr evaporation [27].

Film Stress

The internal stress of the film can be correlated to film density. Neglecting crystalline effects, a density that is lower than the nominal density results in tensile stress, whereas an excessive one results in compressive stress [26]. Stress may be increased or decreased by ion bombardment. In the case of refractory materials, ion bombardment tends to increase film stress.

Film Adhesion

There are many examples of improved adhesion resulting from energetic ion irradiation. However, there is no clear evidence as to the prime cause of this improvement, but it is believed to be a synergistic result of atomic mixing across the interface, surface cleaning, and enhanced surface mobility, eliminating voids in the interfaces and leading to better bonding [26,28].

Film Structure

The preferred orientation of TiN films, for example, depends on the energy of the bombarding ions. As shown in Fig. 4 [29], at energies below 0.3 keV, the preferred orientation is (111). A higher energy favors the (200) orientation and local saturation is observed above 1 keV. This saturation is, however, not maintained when the ion energy is further raised.

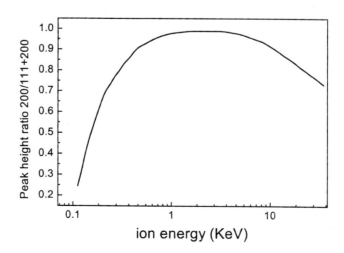

Figure 4 Orientation coefficient as a function of ion energy of TiN deposited by nitrogen ion bombardment with an arrival ratio I/A of unity. (From Ref. 29.)

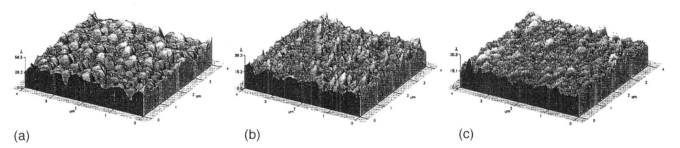

(a)   (b)   (c)

Figure 5 Surface morphology revealed by atomic force microscopy: (a) 8 kV ion bombardment; (b) 16 kV bombardment; and (c) 23 kV bombardment. (From Ref. 30.)

Film Morphology

Ion bombardment affects early film growth, leading to epitaxial, porous, or columnar film formation; surface atomic mobility; and surface sputtering. Hence, the surface morphology of the film depends on the ion energy. As shown in Fig. 5, with increasing energy, roughness decreases and morphology varies dramatically [30].

Film Phases

In addition to stoichiometric phases, new metastable phases can be formed. For example, Cu_2O_5 is found in addition to Cu_2O and CuO during 100 eV oxygen bombardment of an evaporated copper film [31].

2. Pressure

Gas pressure or flow rate is an important parameter to control process and film quality. It affects the arrival rate of the deposition species, thickness uniformity, film color, surface roughness, and internal stress. Pressure effects vary with different PVD methods. In evaporation, the film thickness variation usually follows the cosine rule, and more uniform film deposition can be accomplished by scanning the sample, or by operating at a higher pressure to enhance collisions. In hollow cathode discharge (HCD)-type ion plating [32], the deposition rate and film structure are related to the gas flow rate (or pressure), as shown in Fig. 6. A maximum AlN deposition rate is observed at a nitrogen flow rate of 5 sccm, and then it decreases precipitously. At nitrogen flow rates between 10 and 30 sccm, the fracture and surface features of the films are very fine and dense, but slight columnar structure growth can be observed. The growth rate decreases for nitrogen flow rates above 30 sccm because there are more collisions and scattering among vaporized Al, N_2, Ar, and other particles in the chamber. In reactive sputtering, there exists a typical hysteresis loop of the experimental

parameters upon the introduction of the reactive gas into the system [33]. The sputtering rate substantially decreases when the pressure (flow) of the reactive gas reaches a certain point. This is similar to cathodic arc plasma deposition in which a slower deposition rate may result from a higher gas pressure. The ion charge state also decreases, which surely affects the deposition energy and resultant surface structure and morphology. Different metals possess different sensitivities to gas variation. For example, the color of ZrN coatings is more sensitive to the nitrogen partial pressure compared to TiN coatings, but the surface morphology of TiN coatings is more sensitive [34]. The partial nitrogen pressure affects both the structure and internal stress [35]. For instance, amorphous TiAlCr (N) metallic films formed at low P_N (growth at high P_N leads to polycrystalline nitrides) are composed of a mixed cubic $Ti_{1-x}Al_xN$ and $Cr_{1-x}Al_xN$ structure. Stress behavior is significantly influenced by the addition of nitrogen. The amorphous films exhibit lower strength and larger

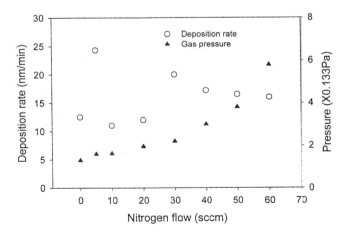

Figure 6 Deposition rate and gas pressure as a function of nitrogen flow rate. (From Ref. 32.)

stress relaxation, whereas the polycrystalline TiAlCrN films fabricated at 20 at.% P_N are much stronger.

3. Sample Temperature

The sample temperature has a critical influence on the growth, structure, and properties of deposited films. A higher processing temperature can lead to a larger sticking force between the substrate and the deposited films. A higher temperature also gives rise to crystallinity improvement of the compound films. The intensity of the ZrN (200) x-ray line increases with temperature (by approximately 30 times) when Zr–Y–N films are sputtered onto an unheated substrate ($T_s = RT$ and $T_s = 500°C$, respectively) [36]. An increase in the intensity of the of ZrN (200) reflection line with increasing T correlates well with: (a) a decrease in the full width half maximum (FWHM) of the Zr (200) reflection line; and (b) an increase in the microhardness of the Zr–Y–N films, as shown in Fig. 7. Cremer et al. [37] have also observed enhanced crystallinity and higher hardness of Al_2O_3 films with increasing sample temperature.

It should be mentioned that many parameters affect film formation, but the substrate temperature, gas pressure, and bombardment energy are most important. Figure 8 illustrates parameter-dependent film structures, indicating that a high discharge power and low gas pressure may lead to the formation of a layered structure [38].

E. Film Adhesion

A crucial property of a thin film or coating is its adherence strength to the substrate, and interface engineering is important to achieve a film with high quality in this respect. Adhesion is determined by two factors: the strength of the interface between the film and substrate, and the stress residing in the film.

1. Stress

Stress is directly related to the film type and processes. As an example, the residual stress in a TiN film produced by plasma-enhanced physical vapor deposition (PEPVD) is generally highly compressive and affected by the deposition conditions as well as the nature of the substrate [39]. It should be noted that such a high level of stress is undesirable because the potential buckling of the film has an adverse effect on the adhesive and cohesive properties. However, some degree of compressive stress can be advantageous in order to protect the substrate film from transverse rupture and fatigue inherited from the often defective, brittle coating. The amount of stress can be controlled by ion energy, interlayer, gradient layer, doped layer, and other parameters.

Ion Energy

It is well established that there is a rapid increase in lattice parameters and compressive residual stress in a PVD TiN film when the substrate bias exceeds a critical value, usually on the order of −100 V, which depends on general deposition conditions and inert sputtering gas [39].

Interlayer

It has been demonstrated that the presence of a Ti interlayer between TiN and steel can dramatically reduce the thermal stress in the TiN coating [40].

Gradient Layer

Films 400 μm thick have been formed by cathodic arc ion plating on a gradient Si–Ni interlayer on WC–Co substrates. The interlayer is believed to reduce internal stress and mismatch, thereby improving adhesion [41].

Doped Coatings

Compared to undoped diamond-like carbon (DLC) or ta-C (tetragonal amorphous carbon), doped films have smaller internal compressive stress (N, Si metal incorporation), lower surface energy, smaller friction coefficients (F, Si–O incorporation), and better electrical properties [42].

Ion Bombardment

Ion implantation after deposition may decrease [43] or increase internal stress, depending on the nature of

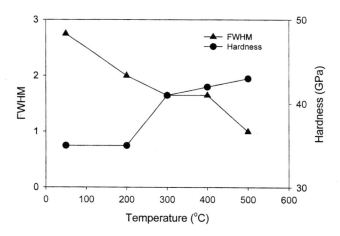

Figure 7 FWHM of ZrN (200) and microhardness of Zr–Y–N as a function of substrate temperature. (From Ref. 36.)

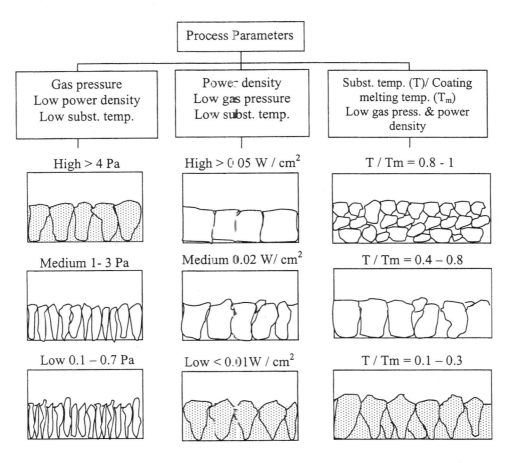

Figure 8 Effects of ion plating process parameters on coating structure. (From Ref. 38.)

the coating, the ions used, the implantation energy, as well as the coating thickness [44].

2. Interface Optimization

Adhesion is determined by interfacial conditions, and some of the important issues are discussed here.

1. Surface cleaning: Prebombardment has the effect of cleaning the surface of oxides and other contaminants normally detrimental to film adhesion, but care must be exercised. It is possible that the recontamination rate can exceed the rate of contaminant removal, and the prebombardment time must be carefully controlled, particularly in argon ion sputtering of aluminum alloys due to the different sputtering rates of Al_2O_3 and Al [45].

2. Chemical state of the interface: Adhesion has been reported to be stronger after argon bombardment in the presence of oxygen, rather than using argon bombardment alone. In both cases,

adhesion strength is higher than that measured from the unirradiated sample [46].

3. Degree of disorder in the interface and ion-to-atom ratio [47]: A certain degree of disorder can result in good adhesion, whereas films with highly ordered structure tend not to adhere well.

4. Interfacial mixing or blurring: Ion bombardment and the resulting implantation/sputtering effects provide atomic scale mixing and, consequently, better adhesion.

F. Summary

PVD coatings are used in many types of applications [1] such as optical (laser optics, mirrors, and selective solar absorber), chemical (corrosion-resistant coatings), mechanical (wear-resistant coatings), electrical (electrical contact and interconnectors), and decorative (to provide attractive finish on many products) components. PVD is undoubtedly a versatile surface metallurgical design technique. By introducing species that are different or similar to the substrate materials, the surface properties can be selectively enhanced to cater to specific applications, whereas the bulk attributes can in general be retained. As aforementioned, PVD can be a metastable process and synthesize materials not restricted by equilibrium constraints. Flexible control of processing parameters such as sample temperature, gas pressure, ion bombardment energy, ion-to-neutral ratio, etc. render PVD a more user-friendly, efficient, and versatile technique compared to techniques used in bulk metallurgy.

II. CHEMICAL VAPOR DEPOSITION

A. Introduction

Vapor deposition can be classified into two basic families: PVD, when films are prepared from pure condensation processes; and CVD, when deposits are formed by chemical reactions only [48]. Chemical vapor deposition is a process whereby solid films are synthesized from the gaseous phase via chemical reactions [49,50]. As a result of the rapid development of both techniques, they are often combined in practice and the separation is less distinct. Many physical processes such as laser beam and glow discharge have been incorporated into CVD processes.

A large number of coatings are now produced commercially by CVD in the electronics, optical, and metallurgical industries, and a big database as a result of intensive research has been accumulated. In this section, common processes such as thermal chemical vapor deposition (TCVD), PACVD, and photo-assisted CVD are described.

B. Fundamentals of Chemical Vapor Deposition

Owing to the diversity of CVD, thorough understanding of the processes must go beyond simple reaction chemistry, particularly for plasma CVD and laser CVD. A broader range of scientific and technological concepts is needed for a better understanding of the steps involved in CVD, thereby making the field highly interdisciplinary.

1. Reaction Types

The versatility of the CVD technique is demonstrated by the variety of films synthesized by various reaction

schemes such as pyrolysis, reduction, oxidation, disproportionation of the reactants, and compound formation. In principle, any controllable reaction with one or more vapors leading to a solid reaction product may be used to produce thin films. Even though the search and design of better reactions are constantly in progress, some of the more well-known processes are discussed here [51–71].

Pyrolysis

The pyrolysis process usually comprises thermal decomposition of metal carbonyls, hydrides, halides, and organometallic compounds. For example:

$$(C_8H_{10})_2Cr \rightarrow Cr + 2C_5H_{10} + 6C \quad (400-600\,°C)$$

In this reaction, HI is used as a catalyst to suppress the codeposition of C.

$$SiH_4(g) \rightarrow Si(s) + 2H_2(g) \quad (650\,°C)$$
$$TiI_4 \rightarrow Ti + 2I_2 \quad (1200\,°C)$$

This is known as the Van Arkel process.

$$Ni(CO)_4 \rightarrow Ni + 4CO \quad (150-220\,°C)$$

This is known as the Mond process and used in Ni smelting for about a century.

Reduction

Hydrogen is usually used as the reducing agent, but sometimes, active metals such as K, Na, Mg, Zn, Cd, etc. can also be used as reducing agents. Halides, carbonyl halides, oxide halides, and other oxidizing agents are often used. The speed of the reduction reaction in which metals are used as the electron donor is usually difficult to control. Important examples of reduction reactions are [52,54–58]:

$$SiCl_4(g) + 2H_2(g) \rightarrow Si(s) + 4HCl(g) \quad (1200\,°C)$$
$$SiHCl_3(g) + H_2(g) \rightarrow Si(s) + 3HCl(g) \quad (1125\,°C)$$

These reactions are used in the deposition of epitaxial Si, which is essential to silicon technology. Refractory metal films such as W, Mo, Re, and Ir usually used in microelectronics can also be deposited by reduction reactions.

$$CrCl_2(g) + H_2(g) \rightarrow Cr(s) + 2HCl(g) \quad (927\,°C)$$
$$WF_6 + 3H_2 \rightarrow W + 6HF \quad (350-1000\,°C)$$
$$2WF_6(g) + 3Si(s) \rightarrow 2W(s) + 3SiF_4(g) \quad (300\,°C)$$
$$TaCl_5 + 5/2H_2 \rightarrow Ta + 5HCl \quad (700-1100\,°C)$$

$$MoF_6(g) + 3Si(s) \rightarrow 2Mo(s) + 3SiF_4(g)$$
$$(200 - 500\,°C)$$
$$MoF_6(g) + 3H_2(g) \rightarrow Mo(s) + 6HF(g)$$
$$(750 - 840\,K)$$
$$ReF_6(g) + 3H_2(s) \rightarrow 2Re(s) + 6HF(g)$$
$$(200 - 500\,°C)$$
$$IrF_6(g) + 3H_2(s) \rightarrow 2Ir(s) + 6HF(g) \quad (775\,°C)$$

An example of the reduction reaction where a metal is used as an electron donor is:

$$TiCl_4 + 2Mg \rightarrow Ti + 2MgCl_2 \quad (900\,°C)$$

This method is known as the Kroll process.

Oxidation

Two important oxidation reaction examples are:

$$SiH_4(g) + O_2(g) \rightarrow SiO_2(s) + 2H_2(g) \quad (450\,°C)$$
$$PH_3(g) + O_2(g) \rightarrow 2P_2O_5(s) + 2H_2(g) \quad (450\,°C)$$

Amorphous SiO_2 (a-SiO_2) films are widely used in microelectronic devices as passivation layer, charge storage layer of metal–nitride oxide semiconductor (MNOS) memory devices, and gate dielectrics [59–61]. P_2O_5 is used to lower the reflow temperature of SiO_2 passivating films in silicon devices.

A new CVD technique, atmospheric pressure combustion chemical vapor deposition (CCVD), has recently been introduced [62,63]. The precursors necessary for CVD reaction are provided by means of a flame [64]. So far, CCVD has been used to deposit high-melting-point materials such as Al_2O_3, Cr_2O_3, SiO_2, CeO_2, $MgAl_2O_4$, $NiAl_2O_4$, and yttrium-stabilized zirconium (YSZ) [63].

Compound Formation

Silicon carbide (SiC) is a promising material in electronic and optical devices due to its high thermal conductivity, high melting point, high breakdown field, high drift velocity, small dielectric constant, and wide bandgap [65,66]. It can be deposited via compound formation reactions. Other kinds of films such as carbide, nitride, and boride can also be deposited similarly. Examples are:

$$SiCl_4(g) + CH_4(g) \rightarrow SiC(s) + 4HCl$$
$$(400 - 2400\,°C)$$
$$TiCl_4(g) + CH_4(g) \rightarrow TiC + 4HCl(g) \quad (1000\,°C)$$

$$2TiCl_4(g) + N_2(g) + 4H_2(g)$$
$$\rightarrow 2TiN(s) + 8HCl(g) \quad (900 - 1000\,°C)$$
$$BF_3 + CH_4(g) \rightarrow BN(s) + 3HF(g) \quad (1100\,°C)$$
$$3SiCl_2H_2(g) + 4NH_3(g) \rightarrow Si_4N_4(g) + 6H_2(g)$$
$$+ 6HCl(g) \quad (750\,°C)$$

Tantalum pentoxide thin films [68–71] can also be deposited by this method:

$$2TaCl_5 + 5H_2O \rightarrow Ta_2O_5 + 10HCl \quad (300 - 800\,°C)$$

Disproportionation

In some compounds, especially halides in which the metal species have mixed valence, disproportionation reactions take place at elevated temperature. For example:

$$GeI_2 \underset{600\,°C}{\overset{300\,°C}{\rightleftharpoons}} Ge(s) + GeI_4(g)$$

Other metal films such as Al, B, Ga, In, Si, Ti, Cu, Zr, Be, and Cr can be deposited using similar reactions.

2. Thermodynamic Criteria

Although CVD is a nonequilibrium process controlled by chemical kinetics and transport phenomena, an equilibrium analysis is still useful in understanding the CVD process [54]. Thermodynamic criteria can be used to predict whether the CVD reaction can take place and to calculate the gas partial pressure in the equilibrium state.

The free energy of a substance A is related to its temperature (T), activity (i.e., its effective concentration) (a_A), and concentration. For 1 mol of A, its free energy (G_A) is given by:

$$G_A = G_A^0 + RT \ln a_A \tag{1}$$

where R is the gas constant and G_A^0 represents the standard molar free energy of A under standard conditions, and:

$$nG_A = n(G_A^0 + RT \ln a_A) = nG_A^0 + RT \ln a_A^n \tag{2}$$

For a reaction of the type:

$$jB + kC = lE + mF$$

where B, C, E, and F represent chemical species and j, k, l, and m represent the numbers of moles, the free energy change (ΔG) of the reaction is a thermodynamic function given by:

$$\Delta G = G_{\text{products}} - G_{\text{reactants}} \tag{3}$$

where G_{products} and $G_{\text{reactants}}$ are the total free energies of the products and the reactants, respectively. From Eq. (2), we obtain:

$$\Delta G = \Delta G^0 + RT \ln \frac{a_E^l a_F^m}{a_B^j a_C^k} \tag{4}$$

where

$$\Delta G^0 = \left(l G_E^0 + m G_F^0 \right) - \left(j G_B^0 + k G_C^k \right) \tag{5}$$

ΔG^0 is the standard free energy change of the reaction when the initial and final activities are all equal to unity. At equilibrium $\Delta G = 0$, Eq. (4) becomes:

$$\Delta G^0 = -RT \ln \frac{a_{E(\text{equ})}^l a_{F(\text{equ})}^m}{a_{B(\text{equ})}^j a_{C(\text{equ})}^k} \tag{6}$$

where $a_{E(\text{equ})}$, $a_{F(\text{equ})}$, $a_{B(\text{equ})}$, and $a_{C(\text{equ})}$ are the equilibrium activities of substances B, C, E, and F, respectively. Therefore

$$\Delta G^0 = -RT \ln K$$

$$K = \frac{a_{E(\text{equ})}^l a_{F(\text{equ})}^m}{a_{B(\text{equ})}^j a_{C(\text{equ})}^k} \tag{7}$$

The thermodynamic equilibrium constant (K) is the equilibrium constant of the reaction. Based on Eq. (6), Eq. (4) becomes:

$$\Delta G = RT \ln \left\{ \frac{\left(\dfrac{a_E}{a_{E(\text{equ})}}\right)^l \left(\dfrac{a_F}{a_{F(\text{equ})}}\right)^m}{\left(\dfrac{a_B}{a_{B(\text{equ})}}\right)^j \left(\dfrac{a_C}{a_{C(\text{equ})}}\right)^k} \right\} \tag{8}$$

where $a_i / a_{i(\text{equ})}$ represents the supersaturation (>1) or subsaturation (<1) of reaction substance i.

Because the equilibrium mixture of the reaction corresponds to a state of minimum free energy, a reversible reaction will take place when $\Delta G < 0$, whereas if $\Delta G > 0$, a reaction does not take place spontaneously.

In a CVD reaction, at least one of the reaction substances is solid (films) and the others are gases. For practical purposes, it is usually assumed that the pure solid activity is unity and the partial pressure is used in lieu of the gas activity. The thermodynamic criteria of reactions discussed previously can be used to not only predict if a CVD reaction can take place, but also judge if the CVD reaction takes place spontaneously and

calculate the partial pressure in equilibrium. Even though the thermodynamic criteria provide useful information, it does not mean that the reactions that take place with an increase in free energy ($\Delta G > 0$) will not occur under any circumstances. Such a reaction can occur if energy is provided by an external source, and this is the reason why many industrial CVD processes are conducted at elevated temperature. In addition, many chemical reactions in which $\Delta G < 0$ require the initial application of heat for the necessary activation energy. In making practical predictions on the likelihood of a reaction, the activation energy as well as free energy change must be taken in account.

3. Gas Transport

Transport phenomena such as gas flow, heat radiation and transfer, and mass transfer are crucial to CVD processes. They govern the deposition process, film microstructure, composition uniformity, and impurity levels. Among these transport phenomena, gas delivery plays a particularly critical role because other transport phenomena usually depend on it. In the design of a CVD system, gas transport phenomena such as laminar flow, turbulent flow, and convection must be considered thoroughly.

C. CVD Reactor Systems

A CVD system usually consists of a gas handling and delivering mechanism, reactor, exhaust and by-product disposal unit, and sometimes a gas recycling system. The reactor is usually the most important component of a CVD system.

1. Reactors

There are many different types of CVD reactors and the common ones are schematically displayed in Figs. 9–11. Figure 9a–c shows the traditional horizontal reactors whereas vertical reactors are depicted in Fig. 10a–d [72–77]. Horizontal and vertical reactors are mainly used in atmospheric and reduced pressure CVD. Their symmetrical design enables heat, mass, and gas transport uniformity. The horizontal reactors have the advantages of higher productivity and convenience compared to the vertical ones, but the films deposited in horizontal reactors are usually less uniform. Figure 9a shows a conventional horizontal atmospheric pressure chemical vapor deposition (APCVD) reactor. The cross-sectional shape of horizontal reactors can be rectangular or circular [72]. The susceptors in these reactors are usually

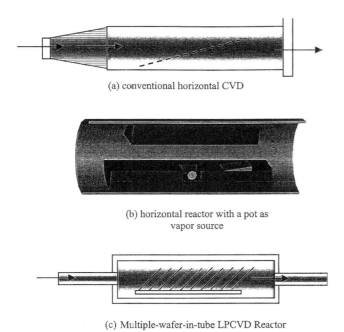

(a) conventional horizontal CVD

(b) horizontal reactor with a pot as vapor source

(c) Multiple-wafer-in-tube LPCVD Reactor

Figure 9 Traditional horizontal reactors: (a) conventional horizontal CVD reactor; (b) horizontal reactor with a pot as vapor source; and (c) multiple wafer in tube LPCVD reactor (From Refs. 72–77.)

slanted at an angle to achieve more uniform film deposition rates.

To accomplish uniform deposition and film composition at reasonable rates, or to avoid contamination the introduction of precursors is quite important. The horizontal reactor can be designed such that the precursors can be introduced in the form of a molten solid vapor [74,77]. Figure 9b displays an example of this kind of reactor for the growth of thick GaN by the direct reaction method. Before the reaction, the solid Ga in the susceptor is melted by radiofrequency (RF) at a frequency around 100 kHz generated by a solenoid coil [77]. Figure 9c exhibits a multiple-wafer, low-pressure chemical vapor deposition (LPCVD) reactor tube. At low pressure, the large mean free path reduces gas phase interactions that would be more severe under isothermal conditions, and so multiple samples can be deposited at the same time.

Figure 10a and b presents typical vertical CVD reactors operated at atmospheric pressure [78,80]. The precursor gases are usually fed either from the top or bottom of the reactors. However, at the relatively high temperature used in CVD, buoyancy effects driven by free convection can become important and disturb the laminar flow pattern required for deposition uniformi-

ty. This is especially true when large temperature gradients are present (cold wall systems), unless the forced flow component is large enough (i.e., high carrier flow and/or low pressure). In this situation, the reactors can be designed to have a parallel-flow chimney geometry (Fig. 10c). The chimney reactor possesses the advantage of reducing the boundary thickness above the susceptor from $x^{1/2}$ to $x^{1/4}$ (x is the distance from the leading edge of the susceptor) by utilizing the synergistic effect of the free convection and forced convection flows [73–80]. This allows the use of smaller carrier gas flows compared to other reactor configurations. In addition,

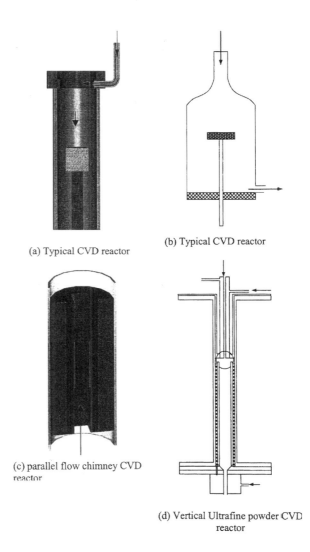

(a) Typical CVD reactor

(b) Typical CVD reactor

(c) parallel flow chimney CVD reactor

(d) Vertical Ultrafine powder CVD reactor

Figure 10 Typical vertical CVD reactors: (a, b) typical vertical CVD reactors; (c) a parallel flow chimney CVD reactor; and (d) CVD reactor used for the production of composites ceramic particles. (From Refs. 78–81.)

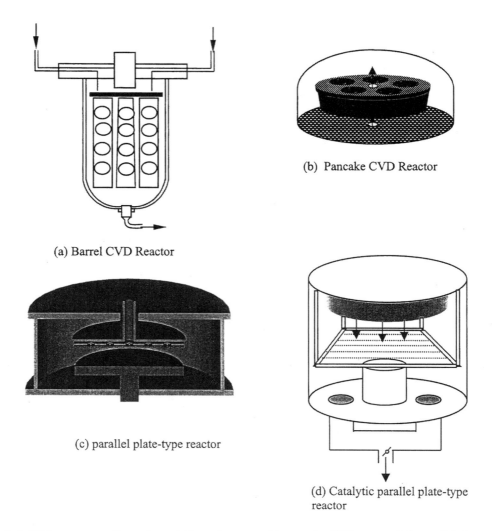

(a) Barrel CVD Reactor

(b) Pancake CVD Reactor

(c) parallel plate-type reactor

(d) Catalytic parallel plate-type reactor

Figure 11 Typical CVD reactor configurations: (a) barrel reactor; (b) pancake CVD reactor; (c) parallel plate-type reactor; and (d) catalytic parallel plate-type reactor. (From Refs. 50, 82, and 83.)

under identical processing conditions, this enhances diffusion and uniformity.

Figure 10d shows a more modern vertical, double-tubular CVD reactor used mostly for ceramics [81]. The component inside the inner tube works as the reactor, and the part between the outer and inner tubes serves as a heat exchanger for the introduction of heated inert gas. Deposition is usually conducted with two streams of reactant gas and solid, one comprising fine particles suspended in a reactant gas mixture diluted with Ar and the other composed of a heated inert gas. The cool fine particles/reactant gas mixture and heated inert gas are mixed at the top of the reactor and a homogeneous nucleation reaction occurs to produce ultrafine powders in the vicinity of the particles.

The horizontal and vertical reactors are typical especially in APCVD, and the heat, mass, and gas flow must be considered in the hardware design. There are many variations of basic configurations and they are also displayed. Figure 11a depicts a barrel reactor used primarily for silicon epitaxy [50]. Figure 11b is a pancake CVD reactor capable of multiwafer planetary motion [50]. It is also extensively used in silicon technology. Figure 11c and d shows the schematic diagrams of parallel plate-type reactors [82]. In Fig. 11c, the upper plate consists of a showerhead with small holes where gases enter the reactor and a baffle that contains larger holes is installed in the middle of the showerhead. This type is one of the most widely used reactors in CVD. The transport phenomena are governed by a variety of mechanisms including convection, diffusion, external forces, and inertia.

Figure 11d depicts a recently developed catalytic chemical vapor deposition (Cat-CVD) apparatus for Cat-CVD, often called hot wire CVD [75,83–85]. Catalytic chemical vapor deposition is a relatively new technology to obtain device-quality thin films at low substrate temperature without using plasma. In this method, gases are decomposed by catalytic cracking reactions with a heated catalyzer placed near a substrate. These species are transported to the substrate and film deposition is performed at low substrate temperature. This process can also accommodate large samples simply by enlarging the area of the catalyzer. The reactor used in Cat-CVD is usually a modified reactor of other traditional CVD reactors. As shown in Fig. 11d, a showerhead is attached to the top flange of the chamber to introduce the gaseous species and a ceramic frame with a tungsten wire used as the catalyzer is also installed on this top flange.

2. Reactor Design

The reactor is the heart of a CVD system and its malfunction or failure has disastrous consequences [50], especially for thermal CVD reactors operated at atmospheric or near-atmospheric pressure. The most important criterion of CVD reactor design is film uniformity. Hence, the following factors must be considered:

1. Avoiding nucleation in the gas phase: Substrate temperature, reactor pressure, and gaseous composition influence nucleation.
2. Efficient transport of reacting gases by either diffusion or flow to the substrate surface: Laminar flow, turbulent flow, and convection of reacting gases must be considered.
3. Heat transport and radiation: It must be efficient and uniform.
4. By-products: They must be vented properly without deleterious effects on the deposition and the environment.

There have been numerous modeling studies on CVD reactors [50,55,86–110]. Modeling is an effective way to combine experimental results with physics to optimize the design of reactors. Another means is to combine CVD models with computational fluid dynamic models and results of smaller-scale experiments to design industrial reactors. This is analogous to chemical engineers using results from individual experiments to determine the kinetics of catalytic reactions in order to scale up packed-bed reactors [111].

D. Metal–Organic Chemical Vapor Deposition

Nearly all kinds of metal and nonmetal elements and their compounds can be deposited by CVD. Metal–organic chemical deposition (MOCVD) is an important method used for the deposition of metals and their compounds. MOCVD has also been called organometallic chemical vapor deposition (OMCVD). The more widely accepted term MOCVD is used to emphasize that the metals are transported by organometallics in the deposition process [50].

With the exception of special precursor gases, the reaction thermodynamics and reaction mechanism are similar to those in thermal CVD process. The mostly used metal–organic precursor gases are alkyls and hydrides, and their physical properties are as important as their chemical properties.

E. Plasma-Enhanced Chemical Vapor Deposition

Plasma-enhanced chemical vapor deposition (PECVD), which is also called PACVD, has been used for many years. It has many advantages:

1. Lower substrate temperature: Lower substrate temperature is one of the most utilized merits. Conventional CVD commonly requires high temperature on the substrate and sometimes in the gas as well. It is thus normal that CVD processes are carried out at temperatures higher than 600°C. However, high temperature deposition may not be compatible with polymers, heat treatment bearings, aiguilles, and some cutting tools that cannot bear high temperature. Some low-melting-point metal films can also not be deposited at high temperatures. In this case, PEVCD, taking advantage of the high reactivity of plasma, can enhance the reaction without the need to resort to a high temperature.
2. Higher deposition rate and better deposition rate control: When the precursors are in the plasma, they become more reactive and some of them are ionized. A negative bias applied to the substrate will significantly enhance the arrival rate of the ionized precursors. Thus, sample voltage can be used to optimize the process and PECVD boasts of a higher deposition rate than LPCVD.
3. Fabrication of novel films and improvement of film microstructure: PECVD can improve the film microstructure and produce novel structures

due to special plasma properties and plasma–materials interactions. The plasma environment in PECVD performs two basic functions. Reactive chemical species are formed by electron impact or other mechanisms to overcome the kinetic limitations intrinsic to nonplasma CVD processes. The plasma supplies the energy to produce positive ions, metastable species, electrons, and photons, and all of them can take part in the CVD reactions. Ion bombardment also changes the surface chemistry, and the combined processes enable the synthesis of novel films and different microstructures.

The typical disadvantages of PECVD include lack of substrate selectivity, poorer conformability, and possibly plasma-induced substrate damage.

1. Classification of PECVD

PECVD is conducted in concert with PVD in many applications because of the same plasma creation methods. Classification of the PECVD methods is usually based on plasma production methods. The most widely used plasma creation method is by subjecting gases to an alternating electric field at high frequency. Frequency is an important factor influencing the characteristics of plasma.

By subjecting the gas to an RF electric field, the PECVD process is called RF-PECVD (Fig. 12a and b) and if microwave is used, it is called microwave plasma-enhanced chemical vapor deposition (MW-PECVD) (Fig. 12c). By applying a powerful magnetic field of 800–1200 G in conjunction with the MW discharge, the plasma generated by the resonance of MWs and electrons has a high dissociation rate even at lower pressure. This process is called electron cyclotron resonance chemical vapor deposition or ECR-PECVD (Fig. 12d) [50,54,112,113].

RF-PECVD is usually operated at frequencies between 50 kHz and 13.56 MHz and at pressures between 0.1 and 2.0 Torr. The plasma density is typically between 10^8 and 10^{12} cm^{-3}, and the fastest electrons may possess energies as high as 10–30 eV [54,112,114]. According to the power coupling methods, it can be subclassified into capacitively coupled RF-PECVD and

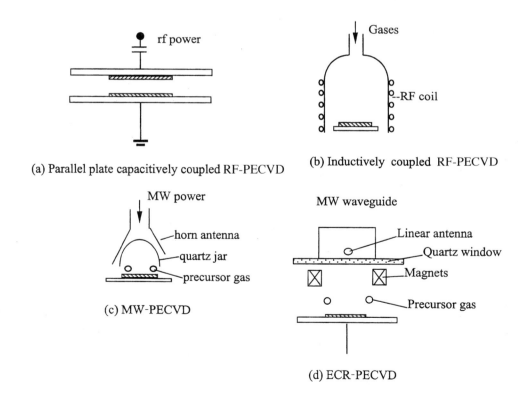

(a) Parallel plate capacitively coupled RF-PECVD

(b) Inductively coupled RF-PECVD

(c) MW-PECVD

(d) ECR-PECVD

Figure 12 Typical PECVD reactors: (a) parallel plate capacitively coupled RF-PECVD; (b) inductively coupled RF-PECVD; (c) MW-PECVD; and (d) ECR-PECVD. (From Refs. 50, 54, 112, and 113.)

inductively coupled RF-PECVD. Figure 12a shows an asymmetric capacitively coupled system whereas Fig. 12b is an inductively coupled RF-PECVD reactor.

MW-PECVD is typically operated at a microwave frequency of 2.54 GHz. Figure 12c shows a MW plasma reactor with horn antennae used as the MW-PECVD excitation component [113]. The excitation component can be an antenna, standing wave applicator, or traveling wave applicator, and it can be operated in a continuous wave or pulsed mode. The cost of power generation is usually not high [112] because microwave sources (similar to those used in microwave ovens) operated at 2.45 GHz are mass-produced. In general, it can be operated between 400 MHz ($2R = 100$ mm; R is the discharge radius) and 2.45 GHz ($2R = 10$ mm) [112,115,116]. The gas pressure ranges from a few tenths of a pascal to above atmospheric pressure. The plasma density in surface wave discharge can be as low as 10^3 cm^{-3} in the low-pressure and low-frequency range, and can be as large as 10^{15} cm^{-3} at atmospheric pressure [112,115,117].

The ECR conditions take place at the standard frequency $\omega/2\pi = 2.54$ GHz and resonance field strength of 87.5 mT. The plasma produced in the ECR region drifts along the magnetic field lines in the chamber where the substrate can be biased independently. Plasma densities between 10^{10} and 10^{12} cm^{-3} can be achieved in the source region at an argon pressure of 0.1 Pa [118].

PECVD reactors can be designed to operate under different modes. Owing to the difference in working pressure, the deposition apparatus can be designed to be a MW-PECVD and ECR-PECVD system [119] by moving the substrate heater holder (Fig. 13a) up or down (i.e., the apparatus can be employed as an MW-PACVD system in the medium-pressure to high-pressure region as well as an ECR-PACVD system in the low-pressure region). By applying an RF-biased voltage to the substrate holder, a dual-mode PECVD system can be developed. Figure 13a and b shows the schematic diagrams of the MW/RF and ECR/RF-PECVD systems, respectively [113,120].

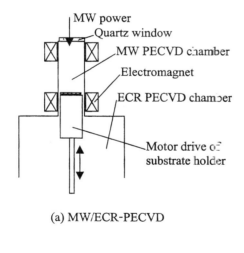

(a) MW/ECR-PECVD

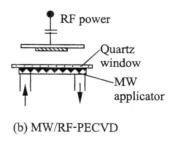

(b) MW/RF-PECVD

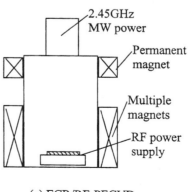

(c) ECR/RF-PECVD

Figure 13 Dual-mode PECVD systems: (a) MW/ECR-PECVD; (b) MW/RF-PECVD; and (c) ECR/RF-PECVD. (From Refs. 113, 119, and 120.)

Depending on the electric field exerted on the precursor gases, PECVD can be subclassified. For example, considering the substrate position, PECVD can be subclassified into remote PECVD and direct PECVD [54,121]. The schematic diagrams of remote PECVD and direct PECVD are shown in Fig. 14a and b, respectively. When the substrate is placed outside the plasma generation region (Fig. 14a), it is called remote PECVD or indirect PECVD [54]. In this mode, the precursor gases are fed into the plasma generation region, excited, ionized, and finally dissociated in the plasma. Other precursor gases can be fed directly to the surface of the substrate (i.e., not all of the processing gases are subject to direct plasma excitation) [113, 122,123]. This technique is expected to produce promising results because it can take advantage of the wide variation of plasma chemistry and does not induce any damage to the substrate surface due to ion bombardment.

Direct PECVD is illustrated in Fig. 14b. All the gases (precursor gases and carrier gases) are fed into the chamber, and the substrate is placed inside the plasma. In both remote PECVD and direct PECVD, RF, MW, or other kinds of plasma excitation can be

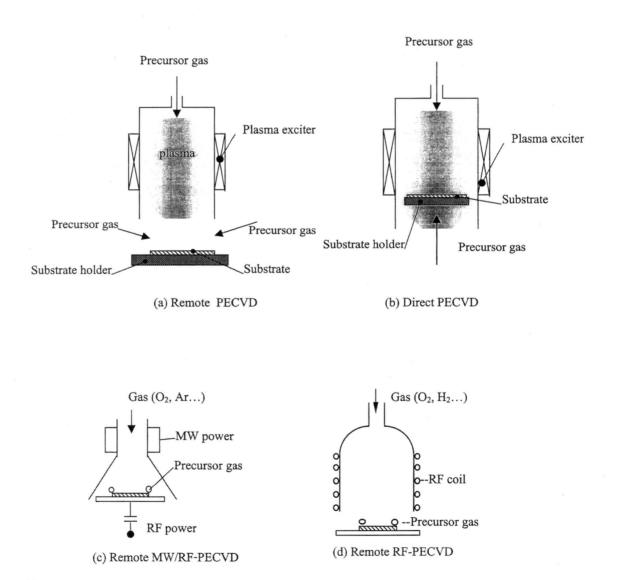

Figure 14 Schematic graphs of remote and direct PECVD: (a) remote PECVD; (b) direct PECVD; (c) remote MW/RF-PECVD; and (d) remote RF-PECVD. (From Ref. 113.)

Table 5 Properties of Hard Coating Materials with Ionic Bonds

Type of coating		Density $(g\ cm^{-3})$	Hardness (H_V)	Melting point (°C)	Coefficient of linear expansion $(10^{-6}\ K^{-1})$	Young's modulus $(kN\ mm^{-2})$	Resistivity $(\mu W\ cm)$
Oxides	Al_2O_3	3.98	1800–2500	2047	8.4–8.6	400	10^{20}
	Al_2TiO_3	3.68		1894	0.8	13	10^{16}
	TiO_2	4.25	1100	1867	9.0	205	
	ZrO_2	5.76	1200–1550	2677	7.6–11.1	190	10^{16}
	HfO_2	10.2	780	2900	6.5		
	ThO_2	10.0	950	3300	9.3	240	10^{16}
	BeO_2	3.03	1500	2550	9.0	390	10^{23}
	MgO	3.77	750	2327	13.0	320	10^{12}

Source: Ref. 20.

used. Figure 14c and d illustrates two examples of remote PECVD systems where the plasma are excited by MW and RF, respectively [113].

2. Thin Film Deposition by PECVD

PECVD utilizes chemical reactions under nonequilibrium conditions to fabricate coatings by means of direct current glow discharges or high-frequency plasmas. Many types of films including dielectric and refractory materials can be deposited, and they are used in many applications and industries such as solar cells, semiconductors, and superhard and lubricant coatings. A summary is presented in Tables 5–7.

F. Photo-Assisted Chemical Vapor Deposition

1. Mechanism of Photochemical Vapor Deposition

Thermal CVD and PECVD have been most widely investigated and used in the industry. Either thermal energy or plasma energy, or both are used in these CVD processes. Another source of energy, light energy, can

also been used in CVD. The technique is called photoassisted CVD. The photo reactions in PACVD can be classified as photothermal reactions in which pyrolysis is achieved by localized heating of the substrate or photochemical reactions in which photolysis of the vapor phase or absorbed constituents plays an important part.

The photothermal reaction mechanism is shown in Fig. 15. The substrate is immersed in a gas or mixture of gases. The substrate is heated optically and the precursor molecules are decomposed when the surface temperature exceeds a critical value that varies with the materials and gases. The decomposed gases form the thermal dissociation layer and the film is formed on the substrate surface. Because a laser is usually used to provide the thermal dissociation heat energy, this approach is also known as laser chemical vapor deposition (LCVD) [154]. The films can only be deposited in the local heated area and LCVD is similar to conventional CVD in this respect. LCVD is subject to the physical properties of the substrate and precursor gases. The substrate must have a melting point higher than the decomposing temperature of the gases, and the substrate should have a higher laser absorption rate and

Table 6 Physicochemical Properties of Hard Coating Materials

Value	Hardness	Brittleness	Melting point	Stability	Coefficient of linear expansion	Adhesion to metallic substrate	Reactivity	Suitability for multiplayer
High level	C	I	M	I	I	M	M	M
↓	M	C	C	M	M	I	C	I
Low level	I	M	I	C	C	C	I	C

M = metallic bond; C = covalent bond; I = ionic bond.
Source: Ref. 20.

Table 7 Examples of Films Fabricated by RF-PECVD

Compound	Method	Power density	Gases	Pressure	Substrate temperature	Reference
a-Si:H	RF-PECVD	35 mW cm^{-2}	SiH$_4$, Ar	0.5–1 Torr	250°C	124
a-Si	Magnetically confined RF-PECVD	7.1–21.4 mW cm^{-3}	SiH$_4$	0.2 Torr	250°C	125
a-C:H	RF-PECVD	—	CH$_4$	0.075 Torr	—	126
a-C(N):H	RF-PECVD	300 W	CH$_4$, NH$_4$OH	10 mTorr	—	127
Nitrogen-doped DLC	RF-PECVD	—	CH$_4$, N$_2$	10 mTorr	—	128
c-Si	RF-PECVD	—	SiH$_4$/SiH$_2$, Cl$_2$/H$_2$		200°C, 350°C	129
BN	PECVD		B$_2$H$_6$, N$_2$, H$_2$		(<400°C)	130
Hydrogenated carbon–nitrogen	RF-PECVD	200 W	CH$_4$, He, N$_2$	200 mTorr	100°C, 250°C, 400°C	131
a-SiC:H	Parallel RF-PECVD	20–30 mW cm^{-2}	SiH$_4$, CH$_4$, H$_2$	0.7 Torr	200°C	132
DLC	RF-PECVD	30–210 W	CH$_4$, He, Ar, N$_2$	100 mTorr	20°C	133
SiO$_2$	Inductively coupled RF-PECVD	50–125 W	TMOS	37.5–112.5 mTorr	—	134
Films containing —C$_3$N$_4$ microcrystallites	Capacitively PECVD (13.56 MHz)	0.53 W cm^{-2}	N$_2$, CH$_4$, H$_2$	2 mTorr	—	135
Water-repellent thin films	RF-PECVD	—	FAS	112.5 mTorr	<50°C	136
Low defect a-SiGe:H	RF-PECVD	30–60 mW cm^{-2}	Silane and germane diluted by helium or by hydrogen	0.5–0.8 Torr	250°C	137
a-B$_{1-x}$C$_x$:H	Capacitively PECVD (13.56 MHz)	0.65 W cm^{-2}	CH$_4$, B$_2$H$_6$, H$_2$	15 mTorr	330 K	138
TiN	Parallel plate RF-PECVD	50 W	TiCl$_4$, N$_2$, H$_2$	2 Torr	400–620°C	139

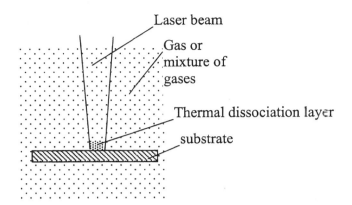

Figure 15 Schematic graph of photothermal CVD reaction.

lower thermal conductivity. The gases in the reactor should be transparent at the selected laser wavelength [54,154]. They should not be dissociated optically and eventually fragmented by thermal processes.

LCVD has the advantage of being a direct writing technique. By focusing the laser beam to microspots or nanospots, small patterns can be produced by LCVD directly [155,156]. By taking advantage of the pulsing nature of laser, the film can be deposited on thermally sensitive substrates on which deposition by other thermal CVD processes is difficult.

The photochemical vapor deposition mechanism is shown in Fig. 16. The energy of ultraviolet (UV) and visible photons ($\sim$ 2–6 eV) is often sufficiently large that chemical bonds in polyatomic molecules can be broken by absorption of one or two photons in a localized area. Although photochemical reactions can occur using UV lamps, UV lasers are more efficient and accurate. The precursor gas is selected according to its absorption cross-section at the wavelength of the laser. In order to obtain reasonable deposition rates, a one-photon absorption mechanism is recommended. The absorption of optical radiation by a gas of number density N (per cubic centimeters) and path length L is, at low intensities, governed by the Beer–Lambert law [154]:

$$I_t = I_i e^{-N\sigma L} \tag{9}$$

where I_i and I_t are optical intensities (W cm^{-2}) incident upon and transmitted by the gaseous medium, respectively, and σ is the absorption cross-section (generally expressed in square centimeters) of the gas. Equation (9) is valid only when the precursor molecule absorbs a single photon from the optical source beam. Multiphoton absorption varies nonlinearly with the source intensity I and is usually significant only for I 10^5–10^6 W

cm^{-2} [54,154]. Many gases have a large absorption cross-section and can be used in photochemical CVD.

The main merit of photochemical CVD is that the photochemical reaction can be carried out at a low temperature because it is excited by photons. Besides, the substrates, whether transparent or not, can be used in the process. The disadvantage is that the deposition rate of photochemical CVD is usually low. However, with the introduction of more powerful and reliable lasers such as excimer lasers, photo-assisted CVD is becoming a more attractive technique in thin film deposition.

In photo-assisted CVD, because photons are used to supply energy, liquids can be used as precursors. This method is called laser-induced chemical liquid-phase deposition (LCLD) when the laser is used as the photon source. LCLD offers many advantages such as laser-induced forward transfer (LIFT), pulsed laser deposition (PLD), and laser-assisted chemical vapor deposition (LCVD). LCLD is time-effective and cost-effective because vacuum tools and special pretreatments are not required. The liquid chemicals used in precursors are usually harmless and easy to handle [155].

2. Films Deposited by Photo-Assisted Chemical Vapor Deposition

Although photo-assisted CVD is still in its infancy, the quality of the metal, semiconductor, and dielectric films fabricated by this technique is being improved at a rapid pace [54]. This is especially true for semiconductor and optical films. This process offers deposition of thin films with good quality at a low temperature and with better control of film composition and deposition area. For example, silicon nitride can be deposited at 200°C by PACVD, whereas the deposition temperature is 850°C and 400°C by thermal CVD and PECVD, respectively. A partial list of PACVD studies in recent years is shown in Table 8.

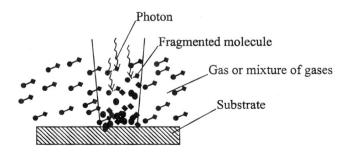

Figure 16 Schematic illustration of photochemical CVD.

Table 8 Examples of Films Fabricated by MW-PECVD

Compound	Method	Power density	Gases	Pressure	Substrate temperature	Reference
DLC and other ordered carbon nanostructures	MW-PECVD (2.45 GHz), RF bias (13.56 MHz)	25–250 W	C_2H_2, Ar	—	—	140
Si-containing crystalline carbon nitride	MW PECVD	2500–3500 W	CH_4, H_2, NH_3	80 Torr	1000–1200°C	141
Diamond	MW-PECVD (2.45 GHz), DC bias (300 V)	1000 W	CH_4, H_2	30 Torr	—	142
Diamond	MW-PECVD	1400 W	CH_4, H_2	37.5–50 Torr	800–950°C	119,143

G. Summary

CVD processes are not restricted to traditional chemical reactions. Different mechanisms such as photon, phonon, plasma, and heat can be used to dissociate the precursor molecules. Traditional CVD processes are important to modern industries such as electronics, mechanical, biomaterial, chemical, aeronautic, and astronautic industries. It should be emphasized that reactions in CVD are not limited to the gas phase, and reactions can also take place on the substrate surface. Many intermediate and transitional reactions may occur. Although most CVD reactions are in nonequilibrium, the thermodynamic criteria facilitate the understanding of the reactions and reaction directions. Because the CVD reactor is the heart of the CVD system, different types of reactors are summarized in this section.

PECVD and PACVD are important commercial CVD processes. PACVD is usually subclassified according to plasma generation methods. Various plasma reactors are discussed in this section. PACVD can also be subdivided based on different reaction mechanisms. The reactions in PACVD can be classified as photothermal, when pyrolysis is achieved by localized heating of the substrate, and photochemical, in when photolysis of the vapor phase or absorbed constituents plays an important part. The reaction mechanisms of these two reactions are briefly described.

III. ION IMPLANTATION

A. Fundamental

Compared to vapor deposition processes such as PVD and CVD, ion implantation has some unique advan-

tages as a metallurgical design tool. During vapor deposition, the processes of adsorption, nucleation, island formation, and film growth are more or less dictated by equilibrium dynamics. Nonstoichiometric metastable phases are sometimes difficult to produce. In contrast, ion implantation is independent of chemical solubility, and usually being a room-temperature or low-temperature process, dopant diffusion tends to be minimal. During ion implantation, atoms or molecules are ionized, accelerated by an electrostatic field, and implanted into a solid. In this way, a myriad of ion–substrate combinations is possible [170]. The acceleration energy is typically between a few hundred electron volts (for shallow junction formation in semiconductors) and several million electron volts. The penetration depth of the incident ions depends upon not only their energy, but also upon mass, charge state, and atomic mass of the solid. The in-depth distribution of implanted ions can be approximated by a Gaussian distribution with a projected range R_p and standard deviation ΔR_p described as follows:

$$N(x) = \frac{N_{dose}}{\sqrt{2\pi}\Delta R_p} \exp\left[-\frac{(x - R_p)^2}{2\Delta R_p^2}\right] \quad (10)$$

The peak or maximum dopant concentration is given by:

$$N_{max} = \frac{N_{dose}}{\sqrt{2\pi}\Delta R_p} \quad (11)$$

where N_{dose} is the implant dose.

B. Techniques and Equipment

Ion implantation introduces energetic ions to modify the structure of materials or to fabricate new materials.

The ions are generated using duoplasmatron-type ion guns or other forms of plasma [5]. Depending on the source of ions and the ways implantation is conducted, the technique can be categorized as beam line or plasma immersion [171–173]. Both techniques have pros and cons, and the former technique is more widely used, especially in the microelectronics industry.

Figure 17 depicts a schematic illustration of a typical beam line ion implanter used in the semiconductor industry [170]. It consists of several major components: ion source, extracting and ion analyzing mechanism, accelerating column, beam scanning system, and end station. The ion source contains the implant species in a solid, liquid, or gaseous form, as well as an excitation system to ionize the species. The ion source produces an ion beam with a small energy spread enabling high mass resolution. Ions are extracted from the ion source and mass-selected using either a quadrupole or magnetic mass analyzer. In the preanalysis mode, only the mass (m/e)-selected ions are injected into the accelerating column. In a system operating in the postanalysis mode, the ions are accelerated to full energy before being mass-separated. To ensure uniform implantation into the substrate, the ion beam is scanned by x,y electric fields, or the substrate is mechanically rastered

while the ion beam remains stationary. The ion current is measured by a Faraday cup consisting of a mechanism to compensate for the emitted secondary electrons from the detector.

Figure 18 depicts a prototype plasma immersion ion implanter [174]. The system is designed for general R&D applications, metallurgy, tribology, surface modification, and fabrication of novel materials. Using the RF plasma source in conjunction with the internal antenna system, the plasma density achieves excellent uniformity both radially and axially. Hot filament glow discharge can also be used to produce nonreactive gaseous plasma such as nitrogen. This system incorporates two types of metal sources, including four metal arc sources and a sputtering electrode, so that multiple metal deposition and implantation steps can be performed in succession in the same equipment without exposing the samples to air. The implantation dynamics can be described by a series of processes. When a negative voltage $-V^0$ relative to the grounded chamber wall is applied to the sample immersed in a plasma sustained by an external source, electrons near the surface are driven away on the time scale of the inverse electron plasma frequency ω_{pe}^{-1}, leaving ions behind to form an electron-depleted ion sheath. Subsequently, on

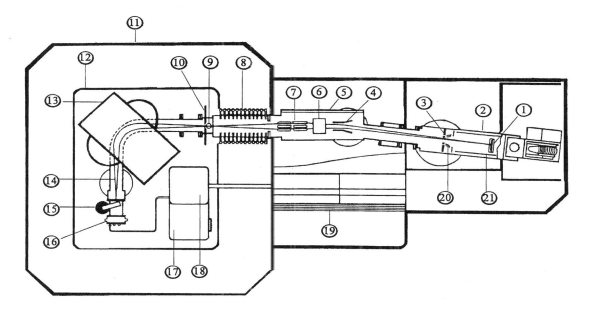

Figure 17 Schematic illustration of an Extrion medium current implanter: maximum energy is 200 keV and maximum current is 1.5 mA: (1) wafer (target position); (2) target chamber; (3) beam mask; (4) lens and scanner box; (5) X-scan plates; (6) Y-scan plates; (7) quadrupole lens; (8) acceleration tube; (9) variable slit; (10) resolving aperture; (11) terminal enclosure; (12) high-voltage terminal; (13) analyzer magnet; (14) ion beam; (15) source magnet; (16) ion source; (17) ion source power supplies; (18) gas box; (19) control console; (20) corner Faraday cups; and (21) main Faraday cup. (From Ref. 170.)

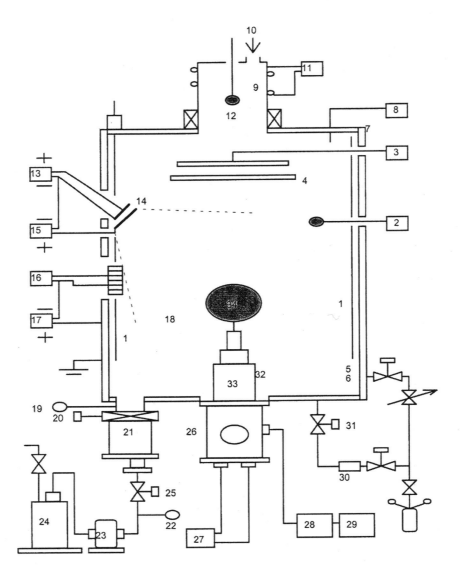

Figure 18 Schematic of the CityU PIII equipment: (1) RF antenna; (2) radial Langmuir probe; (3) sputtering electrode; (4) shutter; (5) liners; (6) chamber wall; (7) permanent magnets; (8) temperature measurement module; (9) RF plasma chamber; (10) gas inlet; (11) RF generator; (12) axial Langmuir probe; (13) trigger; (14) MEVVA plasma sources; (15) arc power supplies for MEVVA plasma sources; (16) filament power supply; (17) discharge bias power supply; (18) chamber; (19) ionization gauge; (20) gate valve; (21) turbo pump; (22) CT; (23) roots pump; (24) rotary pump; (25) valve; (26) oil barrel and HV feedthrough; (27) oil cooling system; (28) pulse modulator; (29) DC HV power supply; (30) mass flow meter; (31) gas feeding system; (32) Pyrex cylinder; (33) HV target stage; and (34) target. (From Ref. 174.)

the time scale of the inverse ion plasma frequency ω_{pi}^{-1}, ions within the sheath are accelerated into the biased samples. The consequent drop in the ion density in the sheath drives the sheath edge further away, exposing new ions on the way and causing them to be accelerated toward the sample and implanted. The time evolution of

the transient sheath determines the implantation current and ion energy distribution. On a longer time scale, the system evolves into a steady-state Child law sheath [175].

Conventional ion beam ion implantation (IBII) and plasma immersion ion implantation (PIII) are fundamentally different and have different applications. PIII has the following unique features compared to IBII:

1. Target shape: Due to the immersion characteristics, PIII is suitable for samples with irregular geometries without the need for complex beam scanning or sample manipulation. In contrast, IBII is more suitable for planar target due to line-of-sight nature, and implant uniformity across the samples is determined by sophisticated beam or sample rastering. Some geometrically complex samples, such as the inner surface of a long cylindrical bore, are very difficult to implant using IBII. On the other hand, experiments and simulations have demonstrated the feasibility of inner surface PIII [176,177].

2. Sample throughput: Being a parallel process (i.e., implantation time independent of sample size), the sample throughput of PIII is better than that of IBII, particularly at high plasma density. Multiple targets can be simultaneously implanted in a single PIII process.

3. Surface reaction and adsorption: The overlying plasma in PIII gives rise to surface reactions and adsorption that do not usually occur in IBII. This may be favorable for processes such as nitriding, but may cause deleterious effects if the plasma contains chlorine or other undesirable elements.

4. Ion species: With the absence of ion filtration, all ions in the plasma are coimplanted in PIII. Therefore, quality control and process optimization can be relatively difficult. In comparison, a single species is usually implanted in IBII.

5. Dose control and monitoring: It is easier to control and measure the implant dose in IBII, but it is more difficult in PIII in which the measurement current is composed of not only the incident ion current but also the secondary electron current, which is very difficult to determine accurately. To exacerbate the situation, the incident dose also depends on the pulse shape voltage, ion mass, plasma density, and other parameters [178].

6. Ion distribution: The ion depth distribution in IBII is approximately Gaussian because the ion beam is mass-to-charge and energy-selected. However, broadening of the ion energy spectrum inevitably occurs in PIII due to plasma sheath expansion as well as the finite voltage pulse rise and fall times. Consequently, a non-Gaussian distribution with a significant low-energy component is frequently observed.

Ion acceleration in IBII is through biased grids or lenses, whereas ion acceleration in PIII occurs via the plasma sheath formed by the negative voltage on the sample. Hence, implantation of insulating samples can be difficult. In addition, although high-energy IBII (MeV) is quite routine, hardware limitation makes PIII above 250 kV a big challenge [179]. Higher-voltage operation in PIII also requires more substantial x-ray shielding, larger sample chamber, and large power supplies that can cope with the high secondary electron yield at high voltage.

C. Ion Species and Dose

1. Ion Species

Appropriate changes of the surface properties of metallic materials are obtained by the proper selection of elements, implant dose, energy, as well as temperature of the implanted surface [5,180–182]. Ion implantation can modify the mechanical, chemical, and/or electrical properties of the materials.

When conducted properly, ion implantation has a positive effect on tribological properties of the implanted materials. Table 9 shows the different choices of implant species for different substrates with respect to the enhancement of tribological properties. Among the various elements, nitrogen is most widely investigated in metallurgical applications such as steels due to the simplicity of the process, effectiveness, and relatively well-understood mechanism. The improvement in tribological properties can be attributed to multiple effects described as follows:

1. Surface hardening effect: The surface hardness in the near-surface region of materials can be substantially increased after proper ion implantation. It is typically due to the formation of compressive stresses, impediment of dislocation movement, and formation of precipitates. For nitrogen implantation, this surface region generally consists of a large volume fraction of precipitates dispersed in a matrix of the substrate material. The precipitate volume fraction varies with depth below the surface in accordance with the atomic concentration profile of the implant species. The high hardness of these precipitates translates significantly to the strength of the surface [112].

2. Effect of ductility improvement [5]: Heavy metal ions (e.g., Sn and Mo) can smoothen the surface without spalling, form a solid lubricating zone (e.g., implanting Sn, Mo, S, Mo + S, N + Ca, and

Table 9 Examples of Films Fabricated by ECR-PECVD

Compound	Method	Power density	Gases	Pressure	Substrate temperature	Reference
CN_x, CN_xH_y	ECR-PACVD, a magnetic field of 1200 G	100–200 W	CH_4, N_2	7.5 mTorr	$< 80\,°C$	144
DLC films	ECR-PECVD, a magnetic field of 875 G	380 W	CH_4, Ar	0.6–0.75 mTorr	$80\,°C$	119
3C–SiC	ECR-PECVD	Microwave power of 720 W	H_2, SiH_4, CH_4	25 mTorr	$930–1100\,°C$	145
a-SiN	ECR-PECVD	1200 W	SiH_4, He, N_2	2 mTorr	$80\,°C$	120
SiOF	ECR-PECVD	—	SiF_4, O_2, Ar	1 mTorr	to $400\,°C$	146
SiOF	ECR-PECVD	900 W	SiH_4, SiF_4, O_2	5.5 mTorr	$25\,°C$	147
SiO_2	ECR-PECVD (2.45 GHz), RF bias (0–100 W)	50–500 W	SiH_4, O_2, $GeCl_4$	0.75 Torr	$< 120\,°C$	148
SiN_x:H	ECR-PECVD (2.45 GHz)	900 W	N_2, Ar, SiH_4	4 mTorr	$30–50\,°C$	149
(Pb, La) (Zr, Ti)O_3	ECR-PECVD	520 W	$Pb(DPM)_2$, $La(DPM)_3$, ZrTB, TiIP, O_2	1.4 mTorr	$490\,°C$	150
PZT	ECR-PECVD	200 W	$Pb(DPM)_2$, $Pb[C_{11}H_{19}O_2]_2$, ZrTB, $Zr(OC_4H_9)_4$ TiIP, $Ti[OC_3H_7]_4$, Ar	3 mTorr	$470\,°C$, $500\,°C$	151
TiN	ECR	1300 W	$TiCl_4$, N_2, H_2	3 mTorr	$350–500\,°C$	139
AlN	ECR-PECVD	250–450 W	TMA, N_2, H_2	0.3 mTorr	$300–500\,°C$	152
Ta_2O_5	ECR-PECVD	300 W	$Ta(OC_2H_5)_5$, O_2, Ar	0.5 mTorr	$300\,°C$	153

N + Mo), or actually produce a layer of high lubricity to combat wear (e.g., Cr).

The corrosion resistance of metallic materials can also be improved by ion implantation, and the main elements are N, Cr, Al, Ta, Y, and Sn, as shown in Table 10. Ar, He, Xe, Cu, Ni, and Mo are also beneficial [5]. Wolf [182] has reviewed the development of ion implantation with respect to corrosion resistance. It can be chronologically divided into three periods: (1) gold age or age of optimism (1973–1984); (2) iron age or age of pessimism (1985–1991); and (3) new era or realism (1991–present). The conclusions are as follows: (1) nitrogen implantation (1×10^{17} cm^{-2}) is not very effective for corrosion reduction; (2) elements with more negative corrosion potentials than iron such as Zn are ineffective; (3) chromium is only useful in concentrations of 20% or higher, corresponding to 1×10^{17} cm^{-2}; and (4) only Zr and Ti are candidates for long-term

corrosion protection with peak current densities below 1 mA cm^{-2}.

2. Ion Dose

In addition to the ion species, the ion dose plays a crucial role in ion implantation. The ion dose affects the properties of the materials and there is a maximum retained dose on account of surface sputtering. The optimal dose is determined by materials requirements as well as implantation physics. For metals, a nitrogen dose of 10^{17} cm^{-2} is required for tribological applications, but a higher dose is required for the improvement of corrosion resistance. As another example, the highest surface hardness of nitrogen implanted WC–Co may be achieved at a dose of 1.5×10^{17} cm^{-2}, and lower or higher doses lead to a decrease in hardness, as shown in Fig. 19 [187]. The surface properties are thus a function of incident ion dose. As reported, for doses

Table 10 Films Gained by PACVD

Materials deposited	Substrate	Precursor gas(es)	Pressure	Laser	Laser parameter	Reference
SiN	Si and glass	SiH_4/NH_3	8.1–8.4 Torr	Excimer laser	24 nsec, 1–80 Hz, 100–170 mJ pulse^{-1}	157
a-$Si_{1-x}N_x$:H	Si (100), corning glass, Al	SiH_4/NH_3	2.5 Torr	ArF	0.5 W cm^{-2}	158
nc-Si grains	—	Si_2H_6/NH_3, SiH_4/Ar	10 Torr, 14–250 mTorr	Excimer laser	24 nsec, 2–80 Hz, 193 nm	159,160
CN_x	Silicon wafer	CCl_4/NH_3	757 Torr	CuBr vapor laser	510 and 578 nm, 60 nsec, 20 Hz	161
SiN	Si and glass	SiH_4/NH_3	—	Excimer laser	193 nm 24 nsec, 1–80 Hz	162
SiO_2		SiH_4/N_2O				
TiN	Mild steel	$TiCl_4$, N_2, H_2	210 Torr	CW-TEM$_{00}$, CO_2	400–700 W, 1.88–2.26 W cm^{-2}	163
TiC	Fused silica plate	$TiCl_4$, CH_4, H_2	157.5 Torr	CW-TEM$_{00}$, CO_2	125–250 W, 80–160 W cm^{-2}	164
$B_xN_yC_z$	Quartz	B_2H_6,C_2H_4,NH_3, DMA, Ar	—	CW CO_2	10.6 μm	165
a-Ge:H	Si (100)	Germane/helium	40 Torr	ArF excimer laser	1 W	166
GaN	Sapphire (AlN)	NH_3, TMG	—	ArF excimer laser	80 mJ pulse^{-1}, 10 Hz, 193 nm	167
B_4C	Fused silica	BCl_3, CH_4, H_2	—	CW-TME$_{00}$, CO_2	10.6 μm, 125–250 W, 80–160 W cm^{-2}	168
W	—	WF_6, H_2	—	Argon ion laser	514 nm	156
Ta_2O_5	p-type (100) Si wafers	N_2O, Ta (DMAE), argon	—	Krypton and chlorine UV radiation	222 nm	169

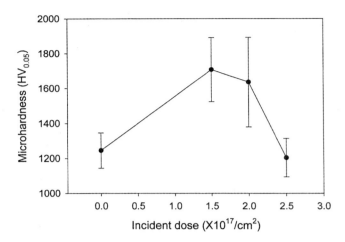

Figure 19 Surface microhardness of nitrogen-implanted WC–Co substrate. (From Ref. 187.)

Table 11 Implanted Ions Improving Tribological Properties

Substrates	Implanted ions	Reference
Be alloys	B	5
Cu alloys	B, N, P	5,183
Ti alloys	N, C, B	5
Al alloys	N	184
Zr alloys	C, N, Cr + C	5
High-alloy steels	Ti + C, Ta	5
Low-alloys steels	N	5
Stainless steels	N, O	5,185
Tool steels	N	5
Bearing steels	Ti, Ti + C	5
Superalloys	Y, C, N	5
Cobalt, coated with WC	N, Co, B, C	5
Cr alloys	N	181,186

up to approximately 10^{16} cm^{-2}, only marginal improvements in wear resistance are observed. In this dose regime, solid solution strengthening and dislocation entanglements are the primary hardening mechanisms. However, their effects are mild, at least for the gaseous species used in PIII. Significant wear resistance improvements are observed in the retained dose range of 10^{17}–10^{18} cm^{-2}, a regime in which second-phase strengthening becomes important. For doses approaching 10^{18} ions cm^{-2}, however, substrate sputtering limits the maximum implant concentration. Excessive sputtering may also lead to surface roughening, which may enhance wear debris formation. At these high doses, precipitates may also coalesce to form a film on the surface. Delamination or brittle fracture of the film can become dominant, reducing the wear resistance, especially if the substrate is soft compared to the film. In general, an implant dose in the range of 3–6×10^{17} cm^{-2} results in optimal wear resistance improvements [112].

The retained dose, the net dose after taking into account sputtering, is a critical parameter and is related to the ion species, substrate materials, incident angles, sputtering rates, and other factors. Zhang and Wu [188] have proposed a simple analytical model without considering chemical reaction effects:

$$\Phi_0 = (6.36 \Delta R_p + R_p) n_0 / Y \qquad (12)$$

where Φ_0 is the saturation dose, n_0 is the atomic density of substrate material, and Y is the sputtering coefficient. A higher implantation energy can increase the retained dose due to a larger R_p and ΔR_p and smaller Y [189], as shown in Table 11.

The retained dose of nitrogen implanted into metals is also affected strongly by the formation of compounds [190]. As shown in Fig. 20, the more negative the heat of nitride formation is, the higher is the retained dose.

D. Implantation Energy

Implantation energy is a very important parameter. It affects the depth profile of the implanted species, defect formation, and thermal dynamics. In metallurgical applications, high energy is usually preferred due to

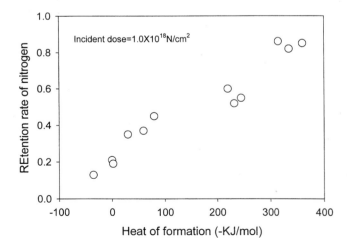

Figure 20 Relationship between the retention rate of nitrogen and the heat of formation of nitride for various nitrogen-implanted metals. (From Ref. 190.)

the inherently shallow implantation layer, but it is not universally true [e.g., in synthesizing compound films using ion beam-assisted deposition (IBAD)]. Hence, implantation energy must be chosen carefully by considering the interrelationship between material properties and ion beam interactions with the substrate as well as limitations in implantation hardware. The role of the ion energy in different implantation techniques is discussed here.

1. Low-Temperature Ion Implantation

The ion energy determines the location, distribution, and concentration of implanted ions. The ion range and distribution can be calculated using simulation programs such as TRIM [191]. In order to achieve the best results such as thicker modified zone and multiple ions, energies may be needed [192]. Although this process can be time-consuming for conventional ion beam ion implantation, it is a niche advantage of PIII in which multiple species usually exist in the plasma and variation of the implantation energy is relatively easy. The finite rise and fall times of the voltage pulse in PIII may be beneficial in some applications, and as shown in Fig. 21, a single PIII process can yield a broad distribution that would require multiple steps in IBII.

2. Elevated Temperature Ion Implantation

Elevated temperature is frequently used to enhance thermal diffusion in order to thicken the modified layer if the substrate can tolerate a higher temperature. In this case, the depth profile is more influenced by the thermal

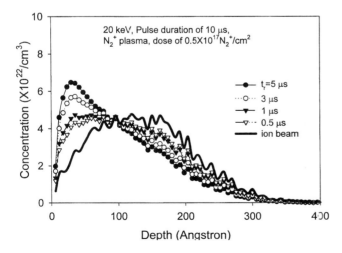

Figure 21 Nitrogen depth profile in iron by PIII for different voltage pulse rise times: 0.5, 1.0, 3.0, and 5.0 μsec.

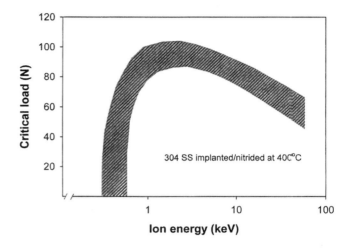

Figure 22 Effect of implantation energy on critical load of treated stainless steel. (From Ref. 194.)

budget than implantation energy, and a lower implantation voltage is sometimes preferred provided that the ions possess sufficient energy to penetrate the surface barrier to reach the bulk of the materials [193]. Experimental results have demonstrated that the optimal voltage is about 1 keV for ion beam nitriding, as shown in Fig. 22 [194].

3. Ion Mixing

Ion mixing plays a particularly important role in film adhesion. Thus, the ion energy must be adjusted with respect to the thickness so that the projected range is comparable to the depth of the interface region below the surface layer [195]. The thickness of the mixed layer increases with higher ion energy ions up to a threshold energy, after which saturation is observed. This threshold energy varies with ion species, but the rule of thumb is that the thickest ion mixed zone is formed when the projected range is about twice that of the film thickness. Heavier ions are also more effective in mixing the interface [196].

4. Plasma Immersion Ion Implantation

The depth profile in PIII does not usually resemble that of IBII because of multiple species in the plasma and the high voltage waveform (finite rise and fall times) in the pulsed mode. A depth profile skewed toward the surface and exhibiting a broader peak can be attributed to plasma-induced surface reactions such as adsorption and low-energy ion implantation [197]. Hence, optimization of the surface profile is more difficult, but as

Table 12 Implanted Ions Improving Corrosion Resistance

Implanted material	Ions
Al alloys	Mo
Cu alloys	Cr, Al
Zr alloys	Cr, Sn
High-alloy steels	Cr, Ta, Y
Low-alloy steels	Cr, Ta
Superalloys	N, C, Y, Ce
Copper	N
Surgical alloys	N

Source: Ref. 5.

aforementioned, there are some advantages with a broadened distribution.

E. Formation of Thicker Layers

Although ion implantation possesses many advantages in metallurgical design, one of its major drawbacks is the shallowness of the implanted ions, and consequently, the thinness of the modified layer as shown in Tables 12–14 using nitrogen implantation as an example. Hence, the acceptance of ion implantation by nonsemiconductor industry has been limited. A feasible solution is to combine ion implantation and other techniques, for instance, at elevated temperature and/or to combine ion implantation with ion beam-assisted deposition. Recently, a "long-range" effect favoring the formation of a thicker modified layer has attracted attention in high-energy-density ion implantation [198,199].

1. Ion Beam-Assisted Deposition

Deposition processes such as PVD and CVD are more effective in synthesizing a thicker layer, but the weaker film–substrate adhesion strength may hamper its performance. Ion stitching during the deposition processes has experimentally been proven to be an effective approach. Atomic mixing blurs the interface and consequently increases the sticking force. The ion flux, ion species, and bombardment energy must be optimized in order to achieve the desired surface morphology, internal stress, structure, and mixed zone.

2. Elevated Temperature

The modified layer thickness is dictated by the ion energy in low-temperature ion implantation, but in elevated-temperature ion implantation, it can be increased substantially. For example, the thickness of the

plasma-nitrided zone can be on the order of millimeters in spite of the low ion energy. As shown in Fig. 23, the high temperature spurs thermal diffusion of the implanted ions. Direct nitrogen incorporation under an electric field instead of surface reaction and adsorption rapidly increases the nitrogen surface concentration, which in turn accelerates nitrogen diffusion into the bulk of the substrate. This is very important for rapid strengthening, and high efficiency can be accomplished even at low temperature. This technique is particularly important for metals with a lower temper temperature and metals that are difficult to nitride using the conventional method such as aluminum and stainless steel.

3. Intense Pulsed Ion Beam (IPIB)

This technique combines the advantages of flash annealing and ion implantation [198,199]. The typical operating parameters are as follows: irradiation energy of 10^5–10^6 eV, pulse width of 10–1000 nsec, power intensity of 10^6–10^9 W cm^{-2}, pulse energy flux of 1–50 J cm^{-2}. The irradiation dose is generally less than 1.0×10^{14} cm^{-2}. Figure 24 shows the phenomena occurring on the near-surface of the irradiated substrate [198]. Using a high power intensity, the surface heating and cooling rates may be up to 10^{10} K sec^{-1}, which is sufficiently high to promote mixing, rapid diffusion, formation of nanocrystalline or amorphous surface layers [199], and generation of shock stress and defect structures. Consequently, the hardened zone can be as thick as several hundred microns [200] that is orders of magnitude larger than the ion range. It has been observed that a "long-range" implantation effect occurs in metallic materials with a low yield point or high plasticity and with a small dislocation density in their initial state prior to ion im-

Table 13 Maximum Retained Implant Dose of Nitrogen into Some Metals

Energy (kV)	Saturation fluence ($\times 10^{17}$ ions cm^{-2})				
	Al	Ti	Fe	Zr	Hf
1	0.4	0.6	0.3	0.6	0.6
2	0.6	0.8	0.4	1.0	0.9
5	1.3	1.5	0.75	1.5	1.6
10	2.0	2.6	1.0	2.0	2.0
20	4.0	4.5	2.0	3.5	3.0
50	8.0	8.5	3.2	6.0	5.5
100	14.0	14.0	6.0	10.0	9.5
200	24.0	22.0	9.0	16.0	15.0

Table 14 Calculated Ion Range of Nitrogen (N^+) Implantation into Different Materials (in Nanometers)

Substrate	Implantation energy (keV)				
	10	20	50	80	10⁻
Al	25.7	49.3	118.8	184.6	226.3
Ti	18.7	34.9	82.4	127.8	156.8
Fe	11.7	22.2	54.0	85.7	106.3
Mo	11.4	21.1	50.9	80.9	100.8
Ta	9.3	16.8	39.3	62.5	78.3

Source: Ref. 191.

plantation [200]. Due to a small ion dose and weak effect of conventional ion implantation, IPIB is more suitable for surface modification of materials vulnerable to structural defects.

F. Contamination

Ion implantation-induced contamination is a critical issue particularly in semiconductor applications. Reduction of contamination levels has been an area of active research in silicon technology [201,202]. As reviewed by Ryssel and Frey [202], there are different possible contamination sources in a traditional ion beam implanter. Ions impacting apertures or walls of the vacuum system sputter atoms from these surfaces and they can be redeposited or reimplanted into the wafers. These also include the sample holder and parts that hold the wafers. Another contamination source is the ion beam itself. A number of undesirable events can

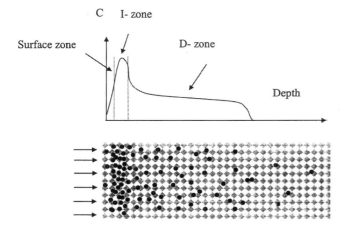

Figure 23 Elevated temperature ion implantation/nitriding (I = ion implantation; D = diffusion; C = concentration).

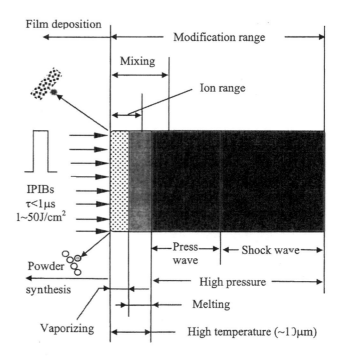

Figure 24 Schematic representation of the IPIB process. (From Ref. 198.)

happen before the ion beam impinges into the wafers. For instance, a portion of the beam can be neutralized, ions can change their charge states, and molecular ions can dissociate due to scattering. This results in energy contamination, giving rise to inhomogeneous doping both laterally and in-depth. Particles are another major contaminants in ion implanters.

Compared to conventional IBII, PIII possesses more serious contamination problems. The immersion nature of PIII increases the sputtered surfaces including all negatively biased areas of the sample chuck and, to a smaller extent, the inner wall of the vacuum chamber [203–205]. Another contamination source is the reaction of immersion plasma with the top surface of the substrate. For example, surface oxidation of silicon is inevitable in an overlying oxygen plasma.

The same can be said about metals although the surface properties are not as sensitive to traces of contamination elements as semiconductors. In metal ion implantation, the biggest contamination-related problem is surface oxidation or carburizing particularly in PIII due to the non-ultra-high vacuum (UHV) operation and plasma environment [205,206]. The impact of surface C and O contamination depends on the vacuum and substrate [207]. Depending on the conditions, a large amount of surface carbon can be incorporated into Fe, Ni, Co, Al, and Fe–4.5 at.% C substrates. High surface oxygen incorporation on Al, Ti, and Ta, and traces of carbon or/and oxygen introduction into Cu, Au, Pt, and Pd have also been observed.

G. Summary

Ion implantation is a versatile tool in metallurgical applications. Compared to PVD and CVD, forced incorporation of impurities circumvents thermodynamic and solubility constraints. Even though the thickness of the implanted zone is typically quite small, there are techniques to increase it, including the use of elevated temperature and hybrid techniques. Both conventional ion beam ion implantation and plasma immersion ion implantation have found applications in niche areas.

IV. HYBRID PROCESSES

A. Introduction

When performed properly, plasma surface modification drastically enhances the tribological properties of materials and industrial components. However, it is a complex science because the mechanism depends on many factors. One single coating or surface modification technique may not yield the best mechanical, chemical, and tribological properties. For example, a coating consisting of mechanical and/or chemical multiple layers may exhibit surface characteristics that bode well for certain applications. A well-designed hybrid process can, in principle, combine the advantages of several independent techniques while minimizing their negative impacts. However, a synergistic combination of different techniques can be quite complicated, and one must also consider the cost as well as compatibility of equipment and processes. For instance, a pressure range of

10–1000 Pa is usually used in plasma nitriding but plasma implantation demands a much better vacuum in the regime of 0.1–0.01 Pa. Nonetheless, successful implementation of hybrid processes can offer unmatched technical and economical benefits [208] and some hybrid techniques are described in Fig. 25 [209].

B. Hybrid PBPVD

An early application of the hybrid plasma-based PVD concept is the combination of arc evaporation and magnetron sputtering with deposit homogeneous or graded (Ti,Al)N coatings [210]. The process uses an aluminum cathodic arc source and titanium sputtering electrode in a nitrogen/argon atmosphere to prevent oxidation. Coatings with graded Al concentrations are deposited by changing the magnetron power in the desired direction.

C. PBPVD and PBCVD

It has been shown that metals at a concentration below 10 at.% can lower the stress of DLC and increase film adhesion [210]. The ideal Me:C coating should possess a layered structure with smooth transition from pure metal to Me:C at the interface. A PVD process is used to deposit the first metallic interface layer giving the high film adhesion. In a second step, a mixture of hydrocarbon gas, argon, and hydrogen is used. Under proper conditions, the process is dominated by plasma CVD because the sputtering rate of carbon from the poisoned target is too low due to target poisoning.

D. Plasma Nitriding and PBPVD

Plasma nitriding and physical vapor deposition techniques are sometimes used together to extend the life-

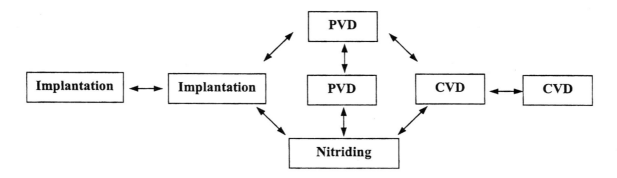

Figure 25 Examples of plasma-based hybrid processes.

time of industrial components. This hybrid process has been applied to tool steel, stainless steel, construction steel, aluminum alloy, and titanium alloy [211,212].

E. Nitriding and PBCVD

Like hybrid plasma nitriding/PVD, nitriding in conjunction with PECVD has received attention. For example, a DLC coating has been deposited on steel substrate without an interlayer [213]. A laboratory-size plasma nitriding system (300 mm in diameter, 300 and 400 mm tall) equipped with a 33-kHz unipolar power supply is used. After the nitriding treatment, Ar sputtering is conducted at 450°C for 30 min to clean and roughen the surface to enhance film adhesion. The DLC film is deposited by means of RF-PACVD in a $CH_4:N_2:H_2 = 66:24:10$ (by volume) ambience for 20 min at room temperature. The RF power is 80 W. The surface roughness after treatment is about the same as that of the surface that has undergone plasma nitriding alone. Film adhesion is noticeably improved on account of the formation of the thin transition layer.

F. Nitriding and Ion Implantation

Hybrid processes combining nitrogen ion implantation and plasma treatment such as ion nitriding and ECR plasma nitriding have been employed in concert to treat aluminum and Al–Mg alloys [214]. The treated samples exhibit better hardness, friction coefficient, and wear. In particular, ion implantation followed by ion nitriding is found to be most effective in decreasing the wear volume in aluminum.

G. PBPVD and Ion Implantation

Nitrogen PIII of TiN films deposited by magnetron sputtering increases the lifetime of the HSS (M series) cutting tools [215]. The sample having a 5-μm predeposited TiN film is implanted at 95 kV with a nitrogen dose of $2.5 \times 10^{17} \, cm^{-2}$. The wear rate of the TiN-coated cutters is reduced by four to eight times compared to that of a pure TiN coating. An apparatus incorporating both PVD and PIII has been produced to synthesize TiN films [216]. The plasma is generated by an ECR plasma confined in the upper part of the vacuum chamber by a magnetic field generated by a solenoid. In the lower part of the vacuum chamber, a magnetron sputtering cathode and resistively heated evaporator are mounted for film deposition. An instrument for in situ or sequential plasma immersion ion beam treatment, in combination with RF sputtering deposition or triode

DC sputter deposition, has also been built [217]. In this apparatus, several coating processes in conjunction with PIII can be conducted. The first step is usually precleaning of the substrate by low-voltage DC sputtering, and then it is followed by deposition using DC or RF/DC triode sputtering. PIII is carried out either during the coating process or afterward in one step, or sequentially in multiple steps. In this way, a transition layer between the substrate and the coating is first formed before the deposition of the final coating.

H. Plasma Immersion Ion Implantation–Deposition (PIII–D)

Because of the capability to deposit films containing both metallic and gaseous elements, PIII–D has received attention from materials scientists and engineers. In this mode, a vacuum arc plasma source in a pulsed mode or DC mode is operated in the presence of gas plasma typically made of nitrogen [218,219]. In the pulsed mode, the cathodic arc pulses are synchronized with the substrate bias pulses to achieve either deposition or implantation. Implantation occurs when the sample is biased, whereas deposition results when no bias is applied to the sample. The coexistence of gas plasma enables the synthesis of compound films. For instance, TiN films can be fabricated using a Ti cathodic arc and nitrogen plasma [220]. In the experiments, the DC cathodic arc current and voltage are 70 A and 20 V, respectively. The nitrogen pressure is 0.27 Pa. The power modulator applies a pulse voltage of −40 kV with a width of 20 μsec and repetition rate of 400 Hz to the sample surrounded by the nitrogen plasma.

REFERENCES

1. Colligon, J.S. Physical vapor deposition. In *Non-Equilibrium Processing of Materials*; Suryanarayana, C., Ed.; Pergamon: New York, 1999; 225–253.
2. Subramanian, C.; Strafford, K.N.; Wilks, T.P.; Ward, L.P. On the design of coating systems: metallurgical and other considerations. J. Mater. Process. Technol. 1996, 56 (1–4), 385–397.
3. Khandagle, M.J.; Gangal, S.A.; Karekar, R.N. Analysis of the properties of radio-frequency ion-plated Cu films based on the calculation of the ion energy available film atom. J. Appl. Phys. 1993, 74, 6150–6157.
4. Deshpandey, C.V.; Bunshan, R.F. Evaporation processes. In *Thin Solid Films II*; Vossen, J.L.; Kern, W., Eds.; Academic Press, Inc.: New York, 1991; 79–132.
5. Burakowski, T.; Wierzchon, T. Surface Engineering of Metals. Principles, Equipment, Technologies; CRC Press: New York, 1998; 506 pp.

6. Thornton, J.A.; Greene, J.E. Sputter deposition processes. In *Handbook of Deposition Technologies for Film and Coatings*; Bunshah, R.F. Ed.; Noyes Publications: New Jersey, 1994.

7. Yamamura, Y.; Itikawa, Y.; Itoh, N. Angular Dependence of Sputtering Yields of Monatomic Solids. Research Information Center, Institute of Plasma Physics, Nagoya University: Japan, 1983.

8. Mahan, J.E. *Physical Vapor Deposition of Thin Films*; John Wiley and Sons, Inc.: New York, 2000.

9. Martin, P.J. Cathodic arc deposition. In *Handbook of Thin Film Process Technology*; David A., Glocker., Ed.; Institute of Physics Publishing: London, 1995.

10. Bunshah, R.F. Critical issues in plasma-assisted vapor deposition processes. IEEE Trans. Plasma Sci. 1990, *18*, 846–854.

11. Sanders, D.M.; Anders, A. Review of cathodic arc deposition technology at the start of the new millennium. Surf. Coat. Technol. 2000, *133/134*, 78–90.

12. Anders, A. Approaches to rid cathodic arc plasmas of macro and nanoparticles: a review. Surf. Coat. Technol. 1999, *120–121*, 319–330.

13. Karpov, D.A. Cathodic arc sources and macroparticle filtering. Surf. Coat. Technol. 1997, *96*, 22–33.

14. Aksenov, I.I.; Belous, V.A.; Padalka, V.G.; Khoroshikh, V.M. Tranport of plasma streams in a curvilinear plasma–optics system. Sov. J. Plasma Phys. 1978, *4*, 425–428.

15. Malik, M.M.M.; Yin, Y.; Mckenzie, D.R. A study of filter transport mechanisms in filtered cathodic vacuum arcs. IEEE Trans. Plasma Sci. 1996, *24*, 1165–1173.

16. Brown, I.G.; Godechot, X. Vacuum arc ion charge-state distributions. IEEE Trans. Plasma Sci. 1991, *19*, 713–717.

17. Brown, I.G. Vacuum arc ion sources. Rev. Sci. Instrum. 1994, *65*, 3061–3082.

18. Yushkov, G.Y.; Anders, A.; Oks, E.M.; Brown, I.G. Ion velocity in vacuum arc plasmas. J. Appl. Phys. 2000, *88*, 5618–5622.

19. Anders, A.; Yotsombat, B.; Binder, R. Correlation between cathode properties, burning voltage, and plasma parameters of vacuum arcs. J. Appl. Phys. 2001, *89*, 7764–7771.

20. Holleck, H. Material selection for hard coatings. J. Vac. Sci. Technol. A. 1986, *4* (6), 2661–2669.

21. Grimberg, I.; Zhitomirsky, V.N.; Boxman, R.L.; Goldsmith, S.; Weiss, B.Z. Multicomponent Ti–Zr–N and Ti–Nb–N coatings deposited by vacuum arc. Surf. Coat. Technol. 1998, *109*, 154–159.

22. Andrievsky, R.A.; Anisimova, I.A.; Anisimov, V.P. Structure and microhardness of TiN compositional and alloyed films. Thin Solid Films 1991, *205*, 171–175.

23. Sproul, W.D. New routes in the preparation of mechanically hard films. Science 1996, *273*, 889–892.

24. Takagi, T. Role of ions in ion-based film formation. Thin Solid Films 1982, *92*, 1–17.

25. Johnson, P.C. The cathodic arc plasma deposition of thin films. In *Thin Solid Films II*; Vossen, J.L., Krern, W., Ed.; Academic Press, Inc.: New York, 1991; 209–282.

26. Nastasi, M.; Moller, W.; Ensinger, W. Ion implantation and thin-film deposition. In *Handbook of Plasma Immersion Ion Implantation and Deposition*; Anders, A., Ed.; John Wiley and Sons, Inc.: New York, 2000; 125–242.

27. Martin, P.J.; Netterfield, R.P.; Sainty, W.G. Modification of the optical and structural-properties of dielectric ZrO$_2$ films by ion-assisted deposition. J. Appl. Phys. 1984, *55*, 235–241.

28. Baglin, J.E.E. Ion beam effects on thin film adhesion. In *Ion Beam Modification of Insulators*; Mazzoldi, P., Arnold, G.W., Eds.; Elsevier: Amsterdam, 1987; 585–630.

29. Ensinger, W. On the mechanism of crystal growth orientation of ion beam assisted deposited thin films. Nucl. Instrum. Methods Phys. Res. B. 1995, *106*, 142–146.

30. Tian, X.B.; Wang, L.P.; Zhang, Q.Y.; Chu, P.K. Dynamic nitrogen and titanium plasma ion implantation/deposition at different bias voltages. Thin Solid Films 2001, *390*, 139–144.

31. Rossnagel, S.M.; Cuomo, J.J. Ion-beam bombardment effects during film deposition. Vacuum 1988, *38*, 73–81.

32. Kishi, M. Low-temperature synthesis of aluminium nitride film by HCD-type ion plating. Jpn. J. Appl. Phys. 1992, *31*, (Part 1), 1153–1159.

33. Rousselo, C.; Martin, N. Influence of two reactive gases on the instabilities of the reactive sputtering process. Surf. Coat. Technol. 2001, *142–144*, 206–210.

34. Nose, M.; Zhou, M.; Honbo, E.; Yokota, M.; Saji, S. Colorimetric properties of ZrN and TiN coatings prepared by DC reactive sputtering. Surf. Coat. Technol. 2001, *142–144*, 211–217.

35. Huang, F.; Wei, G.H.; Barnard, J.A.; Weaver, M.L. Microstructure and stress development in magnetron sputtered TiAlCr(N) films. Surf. Coat. Technol. 2001, *146–147*, 391–397.

36. Polakova, H.; Kubasek, M.; Cerstvy, R.; Musil, J. Control of structure in magnetron sputtered thin film. Surf. Coat. Technol. 2001, *142/144*, 201–205.

37. Cremer, R.; Witthaut, M.; Neuschutz, D.; Erkens, G.; Leyendecker, T.; Feldhege, M. Comparative characterization of alumina coatings deposited by RF, DC and pulsed reactive magnetron sputtering. Surf. Coat. Technol. 1999, *120/121*, 213–218.

38. Ahmed, N.A.G. Ion plating: optimum surface performance and material conservation. Thin Solid Films 1994, *241*, 179–187.

39. Perry, A.J. The state of residual stress in TiN films made by physical vapor deposition methods: the state of the art. J. Vac. Sci. Technol., A. 1990, *8*, 1351–1358.

40. Shieu, F.S.; Cheng, L.H.; Shiao, M.H.; Lin, S.H. Effects of Ti interlayer on the microstructure of ion-plated TiN coatings on AISI 304 stainless steel. Thin Solid Films 1997, *311* (1–2), 138–145.

41. Murakami, Y.; Kuratani, N.; Nishiyama, S.; Imai, O.; Ogata, K. Study on the effect of the interlayer on the adhesion of 400 μm thick film. Nucl. Instrum. Methods Phys. Res. B. 1997, *121*, 212–215.

42. Grill, A. Diamond-like carbon: state of the art. Diam. Relat. Mater. 1999, *9*, 428–434.

43. Nowak, R.; Yoshida, F.; Morgiel, J. Postdeposition relaxation of internal stress in sputter-grown thin films caused by ion bombardment. J. Appl. Phys. 1999, *85*, 841–852.

44. Perry, A.J.; Geist, D.E.; Rafaja, D. Residual stress in cemented carbide following a coating process and after an ion implantation post-treatment of the coating. Surf. Coat. Technol. 1998, *109*, 225–229.

45. Quast, M.; Mayr, P.; Stock, H.R.; Podlesak, H.; Wielage, B. In situ and ex situ examination of plasma-assisted nitriding of aluminium alloys. Surf. Coat. Technol. 2001, *135*, 238–249.

46. Martin, P.J.; Netterfield, R.P.; Sainty, W.G. Enhanced gold film bonding by ion-assisted deposition. Appl. Opt. 1984, *23*, 2668–2669.

47. Kellock, A.J.; Baglin, J.E.E.; Bardin, T.T.; Pronko, J.G. Adhesion improvement of Au on GaAs using ion beam assisted deposition. Nucl. Instrum. Methods Phys. Res., B. 1991, *59/60*, 249–253.

48. Teyssandier, F.; Dollet, A. Chemical vapor deposition. In *Non-Equilibrium Processing of Materials*; Suryanarayana, C., Ed.; Pergamon: New York, 1999; 257 pp.

49. Möhl, W. Chemical vapor deposition. In *Thin Films on Glass*; Bach, H., Krause, D., Eds.; Springer: Berlin, 1997; 59–60.

50. Hitchman, M.L.; Jensen, K.F. *Chemical Vapor Deposition Principles and Applications*; Academic Press, Inc.: San Diego, 1993.

51. Gordon, R.G. Recent advances in the CVD of metal nitrides and oxides. In *Metal–Organic Chemical Vapor Deposition of Electronic Ceramics*; Desu, S.B., Beach, D.B., Wessels, B.W., Gokoglu, S., Eds.; Materials Research Society Symposium Proceedings, 1994, Vol. 335, 9–20.

52. William Lee, W.; Reeves, R.R. Near-room temperature deposition of W and WO_3 thin films by hydrogen atom assisted chemical vapor deposition. In *Chemical Vapor Deposition of Refractory Metals and Ceramics II*; Besmann, T.M., Gallois, B.M., Warren, J.W., Eds.; Materials Research Society Symposium Proceedings, Materials Research Society: Pittsburgh, 1992; 137–142.

53. Hocking, M.G.; Vasantasree, V.; Sidky, P.S. *Metallic and Ceramic Coatings: Production, High Temperature Properties and Applications*; Longman Scientific and Technical: New York, 1989; 103–172.

54. Vossen, J.L., Kern, W., Eds.; *Thin Film Processes II*; Academic Press, Inc.: New York, 1991.

55. Habuka, H. Hot-wall and cold-wall environments for silicon epitaxial film growth. J. Cryst. Growth 2001, *223* (1–2), 145–155.

56. Björklund, K.L.; Heszler, P.; Boman, M. Laser-assisted growth of molybdenum rods. Appl. Surf. Sci. 2002, *186* (1–4), 179–183.

57. Creighton, J.R. Surface chemistry and kinetics of tungsten chemical vapor deposition and selectivity loss. Thin Solid Films 1994, *241*, 310.

58. Pérez, F.J.; Hierro, M.P.; Pedraza, F.; Gómez, C.; Carpintero, M.C.; Trilleros, J.A. Kinetic studies of Cr and Al deposition using CVD-FBR on different metallic substrates. Surf. Coat. Technol. 1999, *122*, 281–289.

59. Powell, M.J.; Easton, B.C.; Hill, O.F. Amorphous silicon–silicon nitride thin-film transistors. Appl. Phys. Lett. 1981, *38*, 794.

60. Street, R.A.; Tsai, C.C. Fast and slow states at the interface of amorphous silicon and silicon nitride. Appl. Phys. Lett. 1986, *48*, 1672–1674.

61. He, L.N.; Hasegawa, S. A study of plasma-deposited amorphous SiO_2 films using infrared absorption techniques. Thin Solid Films 2001, *384* (2), 195–199.

62. Hunt, A.T.; Carter, W.B.; Cochran, J.K., Jr. Combustion chemical vapor deposition: a novel thin-film deposition technique. Appl. Phys. Lett. 1993, *63* (2), 266–268.

63. Hampikian, J.M.; Carter, W.B. The combustion chemical vapor deposition of high temperature materials. Mater. Sci. Eng., A. 1999, *267* (1), 7–18.

64. Hunt, A.T.; Cochran, J.K.; Carter, W.B. Combustion Chemical Vapor Deposition of Films and Coatings. US Patent 5,652,021, July 29, 1997.

65. Lee, W.-H.; Lin, J.-C.; Lee, C.; Cheng, H.-C.; Yew, T.-R. Effects of CH_4/SiH_4 flow ratio and microwave power on the growth of –SiC on Si by ECR-CVD using $CH_4/SiH_4/Ar$ at 200°C. Thin Solid Films 2002, *405* (1–2), 17–22.

66. Foti, G. Silicon carbide: from amorphous to crystalline material. Appl. Surf. Sci. 2001, *184* (1–4), 20–26.

67. Leonhardt, A.; Liepack, H.; Bartsch, K. CVD of TiC_x/a-C-layers under d.c.-pulse discharge. Surf. Coat. Technol. 2000, *133–134*, 186–190.

68. McKinley, K.A.; Sandler, N.P. Tantalum pentoxide for advanced DRAM applications. Thin Solid Films 1996, *290–291*, 440–446.

69. Siodmiak, M.; Frenking, G.; Korkin, A. On the mechanism of chemical vapor deposition of Ta_2O_5 from $TaCl_5$ and H_2O. An ab initio study of gas phase reactions. Mater. Sci. Semicond. Process. 2000, *3* (1–2), 65–70.

70. Kukli, K.; Ritala, M.; Matero, R.; Leskelä, M. Influence of atomic layer deposition parameters on the phase content of Ta_2O_5 films. J. Cryst. Growth 2000, *212* (3–4), 459–468.

71. Forsgren, K.; Hårsta, A. Halide chemical vapor deposition of Ti$_2$O$_5$. Thin Solid Films 1999, *343/344*, 111–114.

72. Johnson, E.J.; Hyer, P.V.; Culotta, P.W.; Clark, I.O. Evaluation of infrared thermography as a diagnostic tool in CVD applications. J. Cryst. Growth 1998, *187* (3–4), 463–473.

73. Stockhause, S.; Neumann, P.; Schrader, S.; Kant, M.; Brehmer, L. Structural and optical properties of self-assembled multilayers based on organic zirconium bisphosphonates. Synth. Met. 2002, *127* (1–3), 295–298.

74. Ottosson, M.; Carlsson, J.-O. Chemical vapour deposition of Cu$_2$O and CuO from CuI and O$_2$ or N$_2$O. Surf. Coat. Technol. 1996, *78* (1–3), 263–273.

75. Matsumura, H. Summary of research in NEDO Cat-CVD project in Japan. Thin Solid Films 2001, *395* (1–2), 1–11.

76. Zhu, D.; Brown, P.H.P.; Sahai, Y. Characterization of silicon carbide coatings grown on graphite by chemical vapor deposition. J. Mater. Process. Technol. 1995, *48* (1–4), 517–523.

77. Yang, S.H.; Ahn, S.H.; Jeong, M.S.; Nahm, K.S.; Suh, E.-K.; Lim, K.Y. Structural and optical properties of GaN films grown by the direct reaction of Ga and NH$_3$ in a CVD reactor. Solid-State Electron. 2000, *44* (9), 1655–1661.

78. Leycuras, A. Optical monitoring of the growth of 3C SiC on Si in a CVD reactor. Diam. Relat. Mater. 1997, *6* (12), 1857–1861.

79. Stock, L.; Richter, W. Vertical versus horizontal reactor: An optical study of the gas phase in a MOCVD. J. Cryst. Growth 1986, *77*, 144–150.

80. Ellison, A.; Zhang, J.; Henry, A.; Janzén, E. Epitaxial growth of SiC in a chimney CVD reactor. J. Cryst. Growth 2002, *236* (1–3), 225–238.

81. Hanabusa, T.; Uemiya, S.; Kojima, T. Production of Si$_3$N$_4$/Si$_3$N$_4$ and Si$_3$N$_4$/Al$_2$O$_3$ composites by CVD coating of fine particles with ultrafine powder. Chem. Eng. Sci. 1999, *54* (15–16), 3335–3340.

82. Setyawan, H.; Shimada, M.; Ohtsuka, K.; Okuyama, K. Visualization and numerical simulation of fine particle transport in a low-pressure parallel plate chemical vapor deposition reactor. Chem. Eng. Sci. 2002, *57* (3), 497–506.

83. Nozaki, Y.; Kongo, K.; Miyazaki, T.; Kitazoe, M.; Horii, K.; Umemoto, H.; Masuda, A.; Matsumura, H. Identification of Si and SiH in catalytic chemical vapor deposition of SiH$_4$ by laser induced fluorescence spectroscopy. J. Appl. Phys. 2000, *88* (9), 5437–5443.

84. Nozaki, Y.; Kitazoe, M.; Horii, K.; Umemoto, H.; Masuda, A.; Matsumura, H. Identification and gas phase kinetics of radical species in Cat-CVD processes of SiH$_4$. Thin Solid Films 2001, *395* (1–2), 47–50.

85. Van Veen, M.K.; Schropp, R.E.I. Amorphous silicon deposited by hot-wire CVD for application in dual junction solar cells. Thin Solid Films 2002, *403–404*, 135–138.

86. Kelkar, A.S.; Mahajan, R.L.; Sani, R.L. Real-time physiconeural solutions for MOCVD. Trans. ASME 1996, *118*, 814–821.

87. Dollet, A.; Casaux, Y.; Chaix, G.; Dupuy, C. Chemical vapour deposition of polycrystalline AlN films from AlCl$_3$–NH$_3$ mixtures; analysis and modelling of transport phenomena. Thin Solid Films 2002, *406* (1–2), 1–16.

88. Hwang, G.-J.; Onuki, K. Simulation study on the catalytic decomposition of hydrogen iodide in a membrane reactor with a silica membrane for the thermochemical water-splitting IS process. J. Membr. Sci. 2001, *194* (2), 207–215.

89. Krumdieck, S. Kinetic model of low pressure film deposition from single precursor vapor in a well-mixed, cold-wall reactor. Acta Mater. 2001, *49* (4), 583–588.

90. Knutson, K.L.; Carr, R.W.; Liu, W.H.; Campbell, S.A. A kinetics and transport model of dichlorosilane chemical vapor deposition. J. Cryst. Growth 1994, *140*, 191–204.

91. McMaster, M.C.; Hsu, W.L.; Coltrin, M.E.; Dandy, D.S. Experimental measurements and numerical simulations of the gas composition in a hot-filament-assisted diamond chemical-vapor-deposition reactor. J. Appl. Phys. 1994, *76*, 7567–7577.

92. Zumbach, V.; Schäfer, J.; Tobai, J.; Ridder, M.; Dreier, T.; Schaich, T.; Wolfrum, J.; Ruf, B.; Behrendt, F.; Deutschman, O.; Warnatz, J. Experimental investigation and computational modeling of hot filament diamond chemical vapor deposition. J. Chem. Phys. 1997, *107*, 5918–5928.

93. Goodwin, D.G.; Gavillet, G.G. Numerical modeling of the filament-assisted diamond growth environment. J. Appl. Phys. 1990, *68*, 6393–6400.

94. Ruf, B.; Behrendt, F.; Deutschmann, O.; Warnatz, J. Simulation of reactive flow in filament-assisted diamond growth including hydrogen surface chemistry. J. Appl. Phys. 1996, *79*, 7256–7263.

95. Kondoh, E.A.; Tanaka, K.; Ohta, T. Reactive-flow simulation of the hot-filament chemical-vapor deposition of diamond. J. Appl. Phys. 1993, *74*, 4513–4520.

96. Mankelevich, Y.A.; Rakhimov, A.T.; Suetin, N.V. Three-dimensional simulation of a HFCVD reactor. Diam. Relat. Mater. 1998, *7*, 1133–1137.

97. Ji, W.; Lofgren, P.M.; Hallin, C.; Gu, C.-Y.; Zhou, G. Computational modeling of SiC epitaxial growth in a hot wall reactor. J. Cryst. Growth 2000, *220* (4), 560–571.

98. Carra, S.A.; Cavallotti, C.; Masi, M. Modeling of growth processes in epitaxial reactors. Mater. Sci. Forum 1998, *276/277*, 135–152.

99. Kleijn, C.R. Computational modeling of transport phenomena and detailed chemistry in chemical vapor deposition—a benchmark solution. Thin Solid Films 2000, *365* (2), 294–306.

100. Kersch, A.; Schafbauer, T. Thermal modeling of RTP

and RTCVD processes. Thin Solid Films 2000, *365* (2), 307–321.

101. Merchant, T.P.; Gobbert, M.K.; Cale, T.S.; Borucki, L.J. Multiple scale integrated modeling of deposition processes. Thin Solid Films 2000, *365* (2), 368–375.

102. Vorob'ev, A.N.; Karpov, S.Yu.; Bord, O.V.; Zhmakin, A.I.; Lovtsus, A.A.; Makarov, Yu.N. Modeling of gas phase nucleation during silicon carbide chemical vapor deposition. Diam. Relat. Mater. 2000, *9* (3–6), 472–475.

103. Lu, S.-Y.; Lin, H.-C.; Lin, C.-H. Modeling particle growth and deposition in a tubular CVD reactor. J. Cryst. Growth 1999, *200* (3–4), 527–542.

104. Tsang, R.S.; May, P.W.; Ashfold, M.N.R. Modelling of the gas phase chemistry during diamond CVD: the role of different hydrocarbon species. Diam. Relat. Mater. 1999, *8* (2–5), 242–245.

105. Lee, Y.L.; Sanchez, J.M. Simulation of chemical-vapor-deposited silicon carbide for a cold wall vertical reactor. J. Cryst. Growth 1997, *178* (4), 505–512.

106. Hofmann, D.; Eckstein, R.; Kölbl, M.; Müller, St.G.; Schmitt, E.; Winnacker, A.; Makarov, Y.; Rupp, R.; Stein, R.; Völkl, J. SiC-bulk growth by physical-vapor transport and its global modelling. J. Cryst. Growth 1997, *174* (1–4), 669–674.

107. Durst, F.; Kadinski, L.; Makarov, Yu.N.; Schäfer, M.; Vasil'ev, M.G.; Yuferev, V.S. Advanced mathematical models for simulation of radiative heat transfer in CVD reactors. J. Cryst. Growth 1997, *172* (3–4), 389–395.

108. Park, M.H.; Cho, D.H. Low dimensional modeling of flow reactors. Int. J. Heat Mass Transfer 1996, *39* (16), 3311–3323.

109. Yuferev, V.S.; Vasil'ev, M.G.; Kadinski, L.; Makarov, Y.N.; Schfer, M.; Makarov, Yu.N. Development of advanced mathematical models for numerical calculations of radiative heat transfer in metalorganic chemical vapour deposition reactors. J. Cryst. Growth 1995, *146* (1–4), 209–213.

110. Bismo, S.; Duverneuil, P.; Pibouleau, L.; Domenech, S.; Couderc, J.P. Modelling of a new parallel-flow CVD reactor for low pressure silicon deposition. Chem. Eng. Sci. 1992, *47* (9–11), 2921–2926.

111. Komiyama, H.; Shimogaki, Y.; Egashira, Y. Chemical reaction engineering in the design of CVD reactors. Chem. Eng. Sci. 1999, *54* (13–14), 1941–1957.

112. Anders, A. *Handbook of Plasma Immersion Ion Implantation and Deposition*; Wiley-Interscience: New York, 2000.

113. Martinu, L.; Poitras, D. Plasma deposition of optical films and coatings: a review. J. Vac. Sci. Technol. A 2000, *18* (6), 2619–2645.

114. Hollahan, J.R., Bell, A.T., Eds.; *Techniques and Applications of Plasma Chemistry*; John Wiley and Sons, Inc.: New York, 1974.

115. Margot-chaker, J.; Moisan, M.; Chaker, M.; Glaude, V.M.M.; Lauque, P.; Paraszczak, J.; Sauvé, G. Tube diameter and wave frequency limitations when using

the electromagnetic surface wave in the $m = 1$ (dipolar) mode to sustain a plasma column. J. Appl. Phys. 1989, *66*, 4134–4148.

116. Moisan, M.; Hubert, J.; Margot, J.; Zakrzewski, Z. The development and use of surface-wave sustained discharges for applications. In *Advanced Technologies Based on Wave Beam Generated Plasmas*; Schluter, H., Shivarova, A., Eds.; Kluwer: Dordrecht, 1999; 23–64.

117. Moisan, M.; Margot, J.; Zakrzewski, Z. Surface wave plasma sources. In *High Density Plasma Sources: Design, Physics and Performance*; Popov, O.A, Ed.; Noyes Publications: Park Ridge, 1995; 191–250.

118. Carl, D.A.; Williamson, M.C.; Lieberman, M.A.; Lichtenberg, A.J. Axial radio frequency electric field intensity and ion density during low to high mode transition in argon electron cyclotron resonance discharges. J. Vac. Sci. Technol. B 1991, *9*, 339–347.

119. Buchkremer-Hermanns, H.; Ren, H.; Weiss, H. A combined MW/ECR-PACVD apparatus for the deposition of diamond and other hard coatings. Surf. Coat. Technol. 1995, *74–75*, 215–220.

120. Flewitt, A.J.; Dyson, A.P.; Robertson, J.; Milne, W.I. Low temperature growth of silicon nitride by electron cyclotron resonance plasma enhanced chemical vapour deposition. Thin Solid Films 2001, *383* (1–2), 172–177.

121. Kulisch, W. Remote plasma-enhanced chemical vapour deposition with metal organic source gases: principles and applications. Surf. Coat. Technol. 1993, *59*, 193–201.

122. Nagel, H.; Metz, A.; Hezel, R. Porous SiO_2 films prepared by remote plasma-enhanced chemical vapour deposition—a novel antireflection coating technology for photovoltaic modules. Sol. Energy Mater. Sol. Cells 2001, *65* (1–4), 71–77.

123. Lauinger, T.; Moschner, J.; Aberle, A.G.; Hezel, R. Optimization and characterization of remote plasma-enhanced chemical vapor deposition silicon nitride for the passivation of p-type crystalline silicon surfaces. J. Vac. Sci. Technol., A 1998, *16*, 530–543.

124. Ray, P.P.; Chaudhuri, P.; Chatterjee, P. Hydrogenated amorphous silicon films with low defect density prepared by argon dilution: application to solar cells. Thin Solid Films 2002, *403–404*, 275–279.

125. Lavareda, G.; Nunes de Carvalho, C.; Amaral, A.; Conde, J.P.; Vieira, M.; Chu, V. Properties of high growth rate amorphous silicon deposited by MC-RF-PECVD. Vacuum 2002, *64* (3–4), 245–248.

126. Balachova, O.V.; Swart, J.W.; Braga, E.S.; Cescato, L. Permittivity of amorphous hydrogenated carbon (a-C:H) films as a function of thermal annealing. Microelectron. J. 2001, *32* (8), 673–678.

127. Chen, S.-Y.; Lue, J.-T. The characterization of amorphous carbon nitride films grown by RFCVD method. J. Non-Cryst. Solids 2001, *283* (1–3), 95–100.

128. Choi, W.; Kim, Y.D.; Iseri, Y.; Nomura, N.; Tomokage, H. Spatial variation of field emission current on nitrogen-

doped diamond-like carbon surfaces by scanning probe method. Diam. Relat. Mater. 2001, *10* (3–7), 863–867.

129. Guo, L.; Kondo, M.; Matsuda, A. Microcrystalline Si films deposited from dichlorosilane using RF-PECVD. Sol. Energy Mater. Sol. Cells 2001, *66* (1–4), 405–412.

130. Vilcarromero, J.; Carreño, M.N.P.; Pereyra, I. Mechanical properties of boron nitride thin films obtained by RF-PECVD at low temperatures. Thin Solid Films 2000, *373* (1–2), 273–276.

131. Anguita, J.V.; Silva, S.R.P. Semiconducting hydrogenated carbon–nitrogen alloys with low defect densities. Diam. Relat. Mater. April–May 2000, *9* (3–6), 777–780.

132. Iftiquar, S.M.; Barua, A.K. Control of the properties of wide bandgap a-SiC: H films prepared by RF PECVD method by varying methane flow rate. Sol. Energy Mater. Sol. Cells 1998, *56* (2), 117–123.

133. Clay, K.J.; Speakman, S.P.; Morrison, N.A.; Tomozeiu, N.; Milne, W.I.; Kapoor, A. Material properties and tribological performance of RF-PECVD deposited DLC coatings. Diam. Relat. Mater. 1998, *7* (8), 1100–1107.

134. Inoue, Y.; Takai, O. Mass spectroscopy in plasma-enhanced chemical vapor deposition of silicon-oxide films using tetramethoxysilane. Thin Solid Films 1998, *316* (1–2), 79–84.

135. He, J.L.; Chang, W.L. Preparation and characterization of RF-PECVD deposited films containing C_3N_4 microcrystallites. Surf. Coat. Technol. 1998, *99* (1–2), 184–190.

136. Takai, O.; Hozumi, A.; Sugimoto, N. Coating of transparent water-repellent thin films by plasma-enhanced CVD. J. Non-Cryst. Solids 1997, *218*, 280–285.

137. Hazra, S.; Middya, A.R.; Ray, S. Low defect density amorphous silicon germanium alloy (1.5 eV) deposited at high growth rate under helium dilution in RF-PECVD method. J. Non-Cryst. Solids 1997, *211* (1–2), 22–29.

138. Jacob, W.; Annen, A.; Von Keudell, A. Erosion of amorphous hydrogenated boron–carbon thin films. J. Nucl. Mater. 1996, *231* (1–2), 151–154.

139. Lee, W.-J.; Kim, J.-S.; Jun, B.-H.; Lee, E.-J.; Hwang, C.-Y. A comparative study on the properties of TiN films prepared by chemical vapor deposition enhanced by r.f. plasma and by electron cyclotron resonance plasma. Thin Solid Films 1997, *292* (1–2), 124–129.

140. Kumar, S.; Rauthan, C.M.S.; Dixit, P.N.; Srivatsa, K.M.K.; Khan, M.Y.; Bhattacharyya, R. Versatile microwave PECVD technique for deposition of DLC and other ordered carbon nanostructures. Vacuum 2001, *63* (3), 433–439.

141. Chen, L.C.; Chen, K.H.; Chen, C.K.; Bhusari, D.M.; Yang, C.Y.; Lin, M.C.; Huang, Y.F.; Chuang, T.J. Si-containing crystalline carbon nitride derived from microwave plasma-enhanced chemical vapor deposition. Thin Solid Films 1997, *303* (1–2), 66–75.

142. Fu, Y.; Yan, B.; Loh, N.L.; Sun, C.Q.; Hing, P. Characterization and tribological evaluation of MW-PACVD diamond coatings deposited on pure titanium. Mater. Sci. Eng. A. 2000, *282* (1–2), 38–48.

143. Avigal, Y.; Hoffman, A.; Glozman, O.; Etsion, I.; Halperin, G. [100]-Textured diamond films for tribological applications. Diam. Relat. Mater. 1997, *6* (2–4), 381–385.

144. Shaginyan, L.R.; Onoprienko, A.A.; Vereschaka, V.M.; Fendrych, F.; Vysotsky, V.G. Role of ion bombardment in forming CN_x and CN_xH_y films deposited by r.f.-magnetron reactive sputtering and ECR plasma-activated CVD methods. Surf. Coat. Technol. 1999, *113* (1–2), 134–139.

145. Mandracci, P.; Chiodoni, A.; Ciero, G.; Ferrero, S.; Giorgis, F.; Pirri, C.F.; Barucca, G.; Musumeci, P.; Reitano, R. Heteroepitaxy of 3C–SiC by electron cyclotron resonance-CVD technique. Appl. Surf. Sci. 2001, *184* (1–4), 43–49.

146. Kim, S.P.; Choi, S.K. The origin of intrinsic stress and its relaxation for SiOF thin films deposited by electron cyclotron resonance plasma-enhanced chemical vapor deposition. Thin Solid Films 2000, *379* (1–2), 259–264.

147. Byun, K.-M.; Lee, W.-J. Water absorption characteristics of fluorinated silicon oxide films deposited by electron cyclotron resonance plasma enhanced chemical vapor deposition using SiH_4 SiF_4 and O_2. Thin Solid Films 2000, *376* (1–2), 26–31.

148. Zhang, J.; Ren, Z.; Liang, R.; Sui, Y.; Liu, W. Planar optical waveguide thin films grown by microwave ECR PECVD. Surf. Coat. Technol. 2000, *131* (1–3), 116–120.

149. Bae, S.; Farber, D.G.; Fonash, S.J. Characteristics of low-temperature silicon nitride (SiN_x:H) using electron cyclotron resonance plasma. Solid-State Electron. 2000, *44* (8), 1355–1360.

150. Shin, J.-S.; Lee, W.-J. A comparative study on the nucleation and growth of ECR-PECVD PLZT ((Pb,La)(Zr,Ti)O_3) thin films on Pt/SiO$_2$/Si substrates and on Pt/Ti/SiO$_2$/Si substrates. Thin Solid Films 1998, *333* (1–2), 142–149.

151. Chung, S.O.; Kim, J.W.; Kim, S.T.; Lee, W.J.; Kim, G.H. Microstructure and electric properties of the PZT thin films fabricated by ECR PECVD: the effects of an interfacial layer and rapid thermal annealing. Mater. Chem. Phys. 1998, *53* (1), 60–66.

152. Lee, W.-J.; Soh, J.-W.; Jang, S.-S.; Jeong, I.-S. *C*-axis orientation of AlN films prepared by ECR PECVD. Thin Solid Films 1996, *279* (1–2), 17–22.

153. Lee, W.-J.; Kim, I.; Chun, J.-S. Effects of bottom electrodes on dielectric properties of ECR-PECVD Ta$_2$O$_5$ thin film. Mater. Chem. Phys. 1996, *44* (3), 288–292.

154. Eden, J.G. *Photochemical Vapor Deposition*; Wiley-Interscience: New York, 1992; 5–9.

155. Kordás, K.; Békési, J.; Vajtai, R.; Nánai, L.; Leppävuori, S.; Uusimáki, A.; Bali, K.; George, T.F.; Galbács, G.; Ignácz, F.; Moilanen, P. Laser-assisted

metal deposition from liquid-phase precursors on polymers. Appl. Surf. Sci. 2001, *172* (1–2), 178–189.

156. Tóth, Z.; Piglmayer, K. Laser-induced local CVD and simultaneous etching of tungsten. Appl. Surf. Sci. 2002, *186* (1–4), 184–189.

157. Tamir, S.; Berger, S.; Shakour, N.; Speiser, S. Correlation between photoluminescence in the gas phase and growth kinetics during laser induced chemical vapor deposition of silicon nitride thin films. Appl. Surf. Sci. 2002, *186* (1–4), 251–255.

158. Banerji, N.; Serra, J.; Chiussi, S.; León, B.; Pérez-Amor, M. Photo-induced deposition and characterization of variable bandgap a-SiN:H alloy films. Appl. Surf. Sci. 2000, *168* (1–4), 52–56.

159. Tamir, S.; Berger, S. Laser induced deposition of nanocrystalline Si with preferred crystallographic orientation. Appl. Surf. Sci. 1995, *86*, 514–520.

160. Tamir, S.; Berger, S. Electroluminescence and electrical properties of nano-crystalline silicon. Mater. Sci. Eng. B. 2000, *69–70*, 479–483.

161. Popov, C.; Bulir, J.; Ivanov, B.; Delplancke-Ogletree, M.-P.; Kulisch, W. Inductively coupled plasma and laser-induced chemical vapour deposition of thin carbon nitride films. Surf. Coat. Technol. 1999, *116–119*, 261–268.

162. Tamir, S.; Berger, S.; Rabinovitch, K.; Gilo, M.; Dahan, R. Laser induced chemical vapor deposition of optical thin films on curved surfaces. Thin Solid Films. 1998, *332* (1–2), 10–15.

163. Silvestre, A.J.; Conde, O. TiN films deposited by laser CVD: a growth kinetics study. Surf. Coat. Technol. 1998, *100–101* (1–3), 153–159.

164. Paramês, M.L.F.; Conde, O. Growth of TiC films by thermal laser-assisted chemical vapour deposition. Appl. Surf. Sci. 1997, *109–110*, 554–558.

165. Oliveira, M.N.; Conde, O.; Botelho do Rego, A.M. XPS investigation of $B_xN_yC_z$ coatings deposited by laser assisted chemical vapour deposition. Surf. Coat. Technol. 1998, *100–101* (1–3), 398–403.

166. León, B.; Chiussi, S.; González, P.; Serra, J.; Pérez-Amor, M. Amorphous germanium layers prepared by UV-photo-induced chemical vapour deposition. Appl. Surf. Sci. 1996, *106*, 75–79.

167. Tansley, T.L.; Zhou, B.; Li, X.; Butcher, K.S.A. Microwave plasma assisted LCVD growth and characterization of GaN. Appl. Surf. Sci. 1996, *100–101*, 643–646.

168. Oliveira, J.C.; Oliveira, M.N.; Conde, O. Structural characterisation of B_4C films deposited by laser-assisted CVD. Surf. Coat. Technol. 1996, *80* (1–2), 100–104.

169. Yu, J.J.; Zhang, J.Y.; Boyd, I.W. UV annealing of ultrathin tantalum oxide films. Appl. Surf. Sci. 2002, *186* (1–4), 57–63.

170. Ryssel, H.; Ruge, I. *Ion Implantation*; John Wiley and Sons, Inc.: New York, 1986.

171. Conrad, J.R.; Radtke, J.L.; Dodd, R.A.; Worzala, F.J.; Tran, N.C. Plasma source ion-implantation technique for surface modification of materials. J. Appl. Phys. 1987, *62*, 4591–4596.

172. Tendys, J.; Donnelly, I.J.; Kenny, M.J.; Pollock, J.T.A. Plasma immersion ion implantation using plasmas generated by radio frequency techniques. Appl. Phys. Lett. 1988, *53*, 2143–2145.

173. Mizuno, B.; Nakayama, I.; Aoi, N.; Kubota, M.; Komeda, T. New doping method for subhalf micro trench sidewalls by using an electron resonance plasma. Appl. Phys. Lett. 1988, *53*, 2059–2061.

174. Chu, P.K.; Tang, B.Y.; Wang, L.P.; Wang, X.F.; Wang, S.Y.; Huang, N. Third-generation plasma immersion ion implantation for biomedical materials and research. Rev. Sci. Instrum. 2001, *72*, 1660–1665.

175. Lieberman, M.A. Model of plasma immersion ion implantation. J. Appl. Phys. 1989, *66*, 2926–2930.

176. Sheridan, T.E. Sheath expansion into a large bore. J. Appl. Phys. 1996, *80*, 66–69.

177. Zeng, X.C.; Tang, B.Y.; Chu, P.K. Improving the plasma immersion ion implantation impact energy inside a cylindrical bore by using an auxiliary electrode. Appl. Phys. Lett. 1996, *69*, 3815–3817.

178. Tian, X.B.; Chu, P.K. Modeling of the relationship between implantation parameters and implantation dose during plasma immersion ion implantation. Phys. Lett. A. 2000, *277*, 42–46.

179. Matossian, J.N.; Wei, R.H. Challenges and progress toward a 250 kV, 100 kW plasma ion implantation facility. Surf. Coat. Technol. 1996, *85* (1–2), 111–119.

180. Goode, P.D.; Baumvol, I.J.R. The influence of implantation parameters on the surface modification of steels. Nucl. Instrum. Methods 1981, *189*, 161–168.

181. Onate, J.I.; Alonso, F.; Garcia, A. Improvement of tribological properties by ion implantation. Thin Solid Films 1998, *317* (1–2), 471–476.

182. Wolf, G.K. An historical perspective of ion bombardment research for corrosion studies. Surf. Coat. Technol. 1996, *83*, 1–9.

183. Wang, S.B.; Zhu, P.R.; Wang, W.J. The microstructure and tribological properties of copper surfaces implanted with carbon ions. Surf. Coat. Technol. 2000, *123* (2–3), 173–176.

184. Rodriguez, R.J.; Sanz, A.; Medrano, A.; Garcia-Lorente, J.A. Tribological properties of ion implanted aluminum alloys. Vacuum 1999, *52* (1–2), 187–192.

185. Evans, P.J.; Vilaithong, T.; Yu, L.D.; Monteiro, O.R.; Yu, K.M.; Brown, I.G. Tribological effects of oxygen ion implantation into stainless steel. Nucl. Instrum. Methods, B. 2000, *168* (1), 53–58.

186. Fischer, G.; Welsch, G.E.; Kim, M.C.; Schieman, R.D. Effects of nitrogen ion-implantation on tribological properties of metallic surfaces. Wear 1991, *146* (1), 1–23.

187. Sun, J.S.; Yan, P.; Sun, X.B.; Lu, G.Y.; Liu, F.R.; Ye, W.Y.; Yang, J.Q. Tribological properties of nitrogen ion implanted WC–Co. Wear 1997, *213* (1–2), 131–134.

188. Zhang, T.H.; Wu, Y.G. *Ion Beam Materials Modification: Science and Applications*; Science Press: Beijing, 1999. *in Chinese*.

189. Miyagawa, Y.; Nakao, S.; Ikeyama, M.; Saitoh, K.; Miyagawa, S. Saturated thickness of nitrided layers formed by high fluence nitrogen implantation into metals. Nucl. Instrum. Methods, B. 1997, *127/128*, 765–769.

190. Yabe, K.; Nishimura, O.; Fujihana, T.; Iwaki, M. Characterization of the surface layer of various metals implanted with nitrogen. Surf. Coat. Technol. 1994, *66*, 250–254.

191. Ziegler, J.F. *Ion Implantation Science and Technology*; Academic Press, Inc.: Boston, 1988.

192. Chen, A.; Blanchard, J.; Conrad, J.R.; Fetherson, P.; Qui, X. A study of the relationship between wear rate and nitrogen concentration profile and application to plasma source ion implanted Ti_6Al_4V alloy. Wear 1993, *165*, 97–101.

193. Wei, R. Low energy, high current density ion implantation of materials at elevated temperatures for tribological applications. Surf. Coat. Technol. 1996, *83* (1–3), 218–227.

194. Wei, R.; Shogrin, B.; Wilbur, P.J.; Ozturk, O.; Williamson, D.L.; Ivanov, I.; Metin, E. The effects of low-energy-nitrogen-ion implantation on the tribological and microstructural characteristics of AISI 304 stainless steel. J. Tribol. 1994, *116* (4), 870–876.

195. Wagh, B.G.; Godbole, V.P.; Kanetkar, S.M.; Ogale, S.B. Ion beam induced atomic mixing at the W–C interface. Surf. Coat. Technol. 1994, *66*, 296–299.

196. Ido, S.; Miyama, A.; Ogata, K. Monte Carlo simulation of the mixing effect induced by ion beam implantation. Surf. Coat. Technol. 1994, *66*, 453–457.

197. Tian, X.B.; Kwok, D.T.K.; Chu, P.K.; Chan, C. Nitrogen depth profiles in plasma implanted stainless steel. Phys. Lett., A. 2002, *299* (5–6), 577–580.

198. Zhao, W.J.; Remney, G.E.; Yan, S.; Opekounov, M.S.; Le, X.Y.; Matvienko, V.M.; Han, B.X.; Xue, J.M.; Wang, Y.G. Intense pulse ion beam sources for industrial applications. Rev. Sci. Instrum. 2001, *71*, 1045–1048.

199. Rej, D.J.; Davis, H.A.; Olson, J.C.; Remnev, G.E.; Zakoutaev, A.N.; Ryzhkov, V.A.; Struts, V.K.; Isakov, I.F.; Shlov, V.A.; Nochevaya, N.A.; Yatsui, K.; Jiang, W. Materials processing with intense pulsed ion beams. J. Vac. Sci. Technol. A. 1997, *15*, 1089–1097.

200. Sharkeev, Y.P.; Didenko, A.N.; Kozlov, E.V. High dislocation density structures and hardening produced by high fluency pulsed-ion-beam implantation. Surf. Coat. Technol. 1994, *65* (1–3), 112–120.

201. Stevie, F.A.; Wilson, R.G.; Simons, D.S.; Current, M.I.; Zalm, P.C. Review of secondary-ion characterization of contamination associated with ion implantation. J. Vac. Sci. Technol., B. 1994, *12* (4), 2263–2279.

202. Ryssel, H.; Frey, L. Contamination problems in ion implantation. In *Handbook of Ion Implantation Technology*. Ziegler, J.F, Ed.; Elsevier Science Publishers B.V., 1992; 675–692.

203. Chu, P.K.; Fu, R.K.Y.; Zeng, X.C.; Kwok, D.T.K. Metallic contamination in hydrogen plasma immersion ion implantation of silicon. J. Appl. Phys. 2001, *90* (8), 3743–3749.

204. Goebel, D.M.; Adler, R.J.; Beals, D.F.; Reass, W.A. Pulser technology. In *Handbook of Plasma Immersion Ion Implantation and Deposition;* Anders, A Ed.; John Wiley and Sons, Inc.: New York, 2000; 467–513.

205. Tian, X.B. Hybrid ion implantation/nitriding and cathodic arc plasma implantation/deposition in an immersion configuration. Ph.D. Dissertation. City University of Hong Kong: Hong Kong, 2002.

206. Tian, X.B.; Chu, P.K. Investigation of low-pressure, elevated-temperature plasma immersion ion implantation of AISI 304 stainless steel. J. Vac. Sci. Technol., A. 2001, *19*, 1008–1012.

207. Fukui, Y.; Hirose, Y.; Iwaki, M. Carbon and oxygen incorporation into surface layers during titanium implantation. Mater. Res. Soc. Symp. Proc. 1988, *100*, 191–194.

208. Bell, T.; Dong, H.; Sun, Y. Realising the potential of duplex surface engineering. Tribol. Int. 1998, *31*, 127–137.

209. Chu, P.K.; Tian, X.B. Plasma surface modification: CVD and PVD. *Surface Modification and Processing: Physical and Chemical Tribological Methodologies*; Marcel Dekker, 2003. *in print*.

210. Freller, H.; Lorenz, H.P. Hybrid processes. In *Advanced Techniques for Surface Engineering*; Gissler, W., Jehn, H.A., Eds.; Brussels and Luxembourg: Netherlands, 1992.

211. Panjan, P.; Urankar, I.; Navinsek, B.; Tercelj, M.; Turk, R.; Cekada, M.; Leskovsek, V. Improvement of hot forging tools with duplex treatment. Surf. Coat. Technol. 2002, *151*, 505–509.

212. Sun, Y.; Bell, T. Combined plasma nitriding and PVD treatments. Trans. Inst. Met. Finish. 1992, *70*, 38–44.

213. Jeong, G.H.; Hwang, M.S.; Jeong, B.Y.; Kim, M.H.; Lee, C. Effects of the duty factor on the surface characteristics of the plasma nitrided and diamond-like carbon coated high-speed steels. Surf. Coat. Technol. 2000, *124*, 222–227.

214. Kanno, I.; Nomotok, K.; Nishiura, S.; Okada, T.; Katagiri, K.; Mori, H.; Iwamoto, K. Tribological properties of aluminum modified with nitrogen ion-implantation and plasma treatment. Nucl. Instrum. Methods, B. 1991, *59*, 920–924.

215. Matossian, J.N. Plasma ion implantation technology at Hughes Research Laboratories. J. Vac. Sci. Technol. B. 1994, *12*, 850–853.

216. Ensinger, W.; Usedom, K.J.; Rauschenbach, B. Char-

acteristic features of an apparatus for plasma immersion ion implantation and physical vapour deposition. Surf. Coat. Technol. 1997, *93*, 175–180.

217. Ensinger, W.; Volz, K.; Enders, B. An apparatus for insitu or sequential plasma immersion ion beam treatment in combination with RF sputter deposition or triode DC sputter deposition. Surf. Coat. Technol. 1999, *120–121*, 343–346.

218. Brown, I.G.; Anders, A.; Anders, S.; Dickinson, M.R.; Ivanov, I.C.; MacGill, R.A.; Yao, X.Y.; Yu, K.M.

Plasma synthesis of metallic and composite thin films with automatically mixed substrate bonding. Nucl. Instrum. Methods, B. 1993, *80/81*, 1281–1287.

219. Brown, I.G.; Godechot, X.; Yu, K.M. Novel metal ion surface modification technique. Appl. Phys. Lett. 1991, *58*, 1392–1394.

220. Sano, M.; Teramoto, T.; Yukimura, K.; Maruyama, T. TiN coating to three-dimensional materials by PBII using vacuum titanium arc plasma. Surf. Coat. Technol. 2000, *128/129*, 245–248.

20

Design of Thermal Spray Processes

Bernhard Wielage and Andreas Wank
Chemnitz University of Technology, Chemnitz, Germany

Johannes Wilden
Technical University Ilmenau, Ilmenau, Germany

I. INTRODUCTION

Thermal spray processes have gained acceptance and are widely used in different industrial branches. The importance of thermal spraying has risen significantly, especially in the past 20 years, because of increasing demands on components. The combination of structural and surface demands (e.g., outstanding strength combined with high corrosion or wear resistance) can often only be met economically by applying coatings. Thermally sprayed wear and/or corrosion protective coatings as well as thermal barrier coatings are examples of this strategy. Additionally, thermal spraying is applied to repair parts.

In the standard DIN EN 657, thermal spraying is defined as a process, in which the feedstock is partially or fully melted inside or outside a spraying gun and deposited on a prepared surface. Usually, the component surface is not melted. Typical coating thicknesses range from 50 μm to 2 mm, but there are also applications with thicknesses of up to 10 mm.

Thermal spray processes feature inherent advantages:

- All materials with a liquid phase or sufficient ductility below the decomposition temperature can be applied.
- A low thermal stress of the substrate is possible.
- There is possibility to deposit coatings on large or locally very limited areas.
- There is partial possibility to use equipment on-site.

However, in general, there is a weak bond strength between the substrate and the coating; the coatings contain pores. Posttreatment (e.g., by remelting) hot isostatic pressing (HIP) or shot peening helps to improve the bond strength and/or to decrease the coating porosity. Usually, exact coating thickness is achieved by grinding, which also provides a low surface roughness.

II. THERMAL SPRAY MARKET

The international thermal spray industry has grown from an estimated US$2.7 billion in 1996 to US$3.5 billion in 2000. Thermal spraying is now used in various industrial branches (e.g., aircraft engine and automotive systems, boiler components, power generation equipment, chemical process equipment, bridges, steel mills, concrete structures, biomedical applications, land-based and marine turbines, and paper).

Among the thermal spray processes, the early-developed conventional flame spraying equipment holds the largest share, although the achievable coating quality is limited in terms of coating porosity, bond strength,

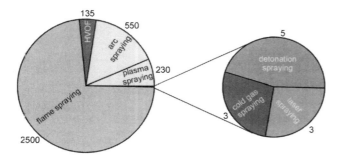

Figure 1 Thermal spray equipment in the German-speaking area.

and oxide content. However, the equipment is robust, cheap, and requires only minor education of personnel. In German-speaking areas, the share of conventional flame spraying systems still exceeds 70% (Fig. 1). Another early-developed process—arc spraying—is also widely used, which can mainly be attributed to high deposition rates and high flexibility concerning on-site application. Plasma spraying was introduced into industrial manufacturing in the early 1960s and has found many application fields because the high temperature of the plasma jet permits the processing of any material with a liquid phase. Although the share of detonation spraying equipment is vanishing these days, high-velocity oxyfuel (HVOF) processes are increasingly applied. As cold gas spraying has just been introduced into industrial manufacturing, the size of its promising market in the future is still uncertain.

The economically most interesting thermal spray work area is the sales of coating solutions. According to a study carried out in the European Union, the turnover share of coating production is more than three quarters of the complete turnover of the thermal spray market (Fig. 2). Within the turnover of coating production, the share of job shops exceeded that of the in-house production of consumer companies. Job shops are usually small-sized or medium-sized. In German-speaking areas, the annual turnover of about 50% of job shops amounts to less than US$1,000,000, which is obtained with an average of five employees. Although the turnover share of spraying feedstock takes a considerable 19%, the equipment turnover share is rather small.

III. THERMAL SPRAY MATERIALS

Thermal spray feedstock can be processed in powders, wires, or rods. Depending on the material and the

spraying process, the size of powders ranges from 5 to 125 μm. Thermoplastic granulate materials are coarser, but because of their low turnover share, these materials will be neglected further on. Generally, narrow-sized fractions are beneficial in achieving optimum coating properties, but also increase the feedstock cost.

Wire feedstock is most commonly processed with a 1.6- or 3.2-mm diameter. Especially for expensive materials, significantly smaller diameters are used. In coating large areas with corrosion-protective Al or Zn coatings, 4.8-mm-diameter wires are applied to increase the deposition rate. The diameter of spraying rods most commonly ranges from 6 to 8 mm.

Thermal spray coatings are used in a broad range of applications:

- Antifretting, abrasive, and erosion wear resistance
- Abradable and abrasive coatings
- Corrosion resistance
- High-temperature oxidation resistance
- Thermal barrier coating systems
- Electrical resistance and conductivity
- Biocompatible and bioactive coatings
- Near-net-shaped component manufacturing (free-standing bodies).

In agreement with the wide spectrum of functions that have to be met, a large material spectrum is applied as thermal spray feedstock. Different methods permit to vary the feedstock properties in wide ranges, which leads to tailored feedstock, depending on the used process and the desired coating properties. The most important manufacturing processes and materials are introduced in the following sections.

A. Powder Feedstock

In addition to pure metallic or ceramic powders, there is also a variety of composite materials. Pure metallic

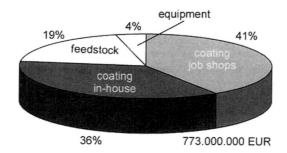

Figure 2 Thermal spray market in the European Union.

powders are usually manufactured by melt atomization (Fig. 3). Typical examples of melt-atomized powder materials are as follows:

- Al and Al12Si
- Cu, CuNi, and CuAl alloys
- Various steels (e.g., AISI 316L)
- Ni, NiCr, NiAl, and self-fluxing NiCrBSi alloys
- Cobalt-based hard alloys (e.g., stellite 6)
- MCrAlY alloys (M: Fe, Ni, and Co)
- Mo
- Ti and Ti6Al4V
- Refractory metals (e.g., Ta and W).

Aluminum alloys are used for corrosion protection and for the repair of aluminum-based or magnesium-based alloy components. Copper-based materials are applied as soft-bearing materials and cavitation-resistant coatings, or for metallizing of electrically nonconductive materials.

NiCr and NiAl are applied as bond coats because they feature good corrosion resistance, high ductility, and high bond strength. Bond coats are applied when the thermomechanical properties of the substrate and a desired coating material differ too much. For example, wear-protective oxide coatings are usually sprayed on steel components on top of a Ni5Al bond coat.

Self-fluxing NiCrBSi and cobalt hard alloys are used for combined wear and corrosion protection, whereas NiCoCrAlY alloys exhibit excellent hot gas corrosion resistance. The large difference between the solidus and liquidus temperatures of NiCrBSi alloys permits easy densification of rather porous powder flame spray

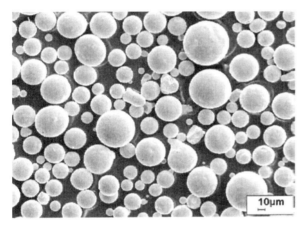

melt atomized NiCrBSi powder

agglomerated and sintered Mo powder

fused and crushed Mo powder

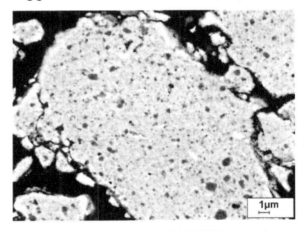

mechanically alloyed Ni-TiB$_2$ powder

Figure 3 Powder feedstock manufactured by different processes.

coatings by remelting. Molybdenum coatings are also used for wear protection, whereas titanium coatings are applied for biomedical purposes or severe corrosion protection.

Tungsten is mainly used for spraying of freestanding bodies, which are applied to components in extremely high-temperature environments. The outstanding corrosion resistance of tantalum is exploited in applications in the chemical industry.

Ceramic materials are available as fused or sintered and crushed, agglomerated (spray-dried) and sintered, or spherodized powders. In order to assure the desired purity, it can be necessary to synthesize the raw material. By coprecipitation, the optimum homogeneity of the distribution of different constituents is achieved.

Fused and crushed powders are relatively cheap and can show theoretical density. On the other hand, flow characteristics are poor because of the irregular shape of the particles (Fig. 3). Spray drying permits the manufacturing of spherical particles (Fig. 3) with accordingly good flow characteristics and with easily constituted composition. Spherodization is achieved by the melting of particles with irregular shape, or of spray-dried powders in a thermal plasma jet. When the particles leave the hot gas stream, they resolidify in a spherical shape and are collected. Depending on the used powder and the process conditions, the share of dense and hollow particles varies significantly.

Typical ceramic powders are as follows:

- Al_2O_3 and Al_2O_3–TiO_2
- Cr_2O_3
- ZrO_2–Y_2O_3, ZrO_2–CeO_2, ZrO_2–MgO, and ZrO_2–CaO
- Fused tungsten carbide (FTC; W_2C–WC), TiC, TaC, NbC, and Cr_3C_2
- Hydroxyapatite.

Al_2O_3 is used for electrical insulation and for combined corrosion and wear protection. The addition of titania slightly decreases abrasive wear resistance, but improves the grindability of manufactured coatings. Cr_2O_3 is also used for wear protection applications, especially in corrosive environments. FTC and other carbide materials are rarely applied as pure coatings, but mainly as hard phases in composite materials. For biomedical implant applications, hydroxyapatite coatings are applied to achieve a good bonding to the natural bone.

The most important composite materials these days are WC–Co(Cr) and Cr_3C_2–Ni20Cr. Usually, composite powders are produced by spray drying and sintering.

Depending on the application of the coatings, carbide size can vary from submicron size (~ 500 nm) to a medium size of 5 µm. For erosion protection, fine carbides are beneficial because the small distance between the hard phases hinders the wear out of the soft matrix. Large carbide particles have proven to be suitable for protection against severe abrasive wear.

In order to protect materials that cannot be thermally sprayed because of lack of a liquid phase, and to provide necessary cohesion inside the coating by a matrix, the chemical cladding of, for example, graphite or hexagonal boron nitride is carried out. Usually, nickel is used for cladding, but other metals can also be applied. The according coatings are used, for example, for low friction and self-lubricating bearing applications. The spraying of blends is a cheap alternative, but the homogeneity of manufactured coatings is usually poor, especially when the constituents show a strong density difference, which results in segregation. Additionally, the size of the constituents needs to be coarse in comparison to spray-dried powders to assure sufficient flowability of the blend, which results in a coarse microstructure of the coatings.

There are several further processes that permit the production of composite powders, but have not yet become valid for industrial manufacturing. In addition to self-propagating high-temperature synthesis (SHS), mechanical alloying has been used on a laboratory scale. The results show partially high potential and, recently, semicontinuously working mechanical alloying machines that permit the production of industrially valid amounts have been developed. Mechanical alloying permits the production of composites with an extremely fine distribution of hard phases in a metallic binder material (Fig. 3). Additionally, alloys with compositions that cannot be achieved by conventional metallurgy are producible.

B. Wire Feedstock

If massive wires or rods are applied, it is assured that the deposited material has been molten completely before droplets are carried from the wire or rod tip in the hot gas jet. Therefore, the incorporation of unmolten particles in the coating is limited to resolidified particles. Further advantages are the increased deposition efficiency in comparison to powder feedstock and the lower costs of massive wire feedstock. But the forming of massive wires is only possible for metallic materials with sufficient ductility, which limits the material spectrum strongly. The following materials are available as massive wires for thermal spraying:

- Aluminum and aluminum alloys
- Zinc and zinc alloys
- Copper and copper alloys (e.g., bronze and brass)
- Tin alloys
- Steels (limitations by brittleness)
- Nickel and nickel alloys
- Molybdenum
- Titanium and Ti6Al4V
- Precious metals (Au and Pt).

The application of coatings produced from wire feedstock is generally in agreement with those produced from powder feedstock. Aluminum and zinc coatings are used for corrosion protection. Copper is also used for corrosion protection and additionally for electrical works. For soft-bearing applications, copper and tin alloys are applied. Steel wires are used for the repair of steel components, or for the manufacturing of wear-protective and corrosion-protective coatings. Nickel and nickel alloy coatings are applied for corrosion and high-temperature oxidation protection and as bond coats, whereas molybdenum is used for wear protection. For biomedical purposes and protection against severe corrosion, titanium and titanium alloys are applied. The spraying of precious metals is carried out if the required coating thickness cannot be achieved economically by electroplating.

Cored wires have been well established in welding applications before their use for thermal spraying was initiated. Cored wires expand the material spectrum, which can be produced by wire feedstock, significantly. In addition to brittle alloys that do not permit the forming of wires, cermet coatings also can be sprayed by a combination of a suitable matrix material as velum and carbides, borides, or nitrides as filler.

In addition to a variety of techniques for the production of grooved cored wires, tube cored wires have been developed. Tube cored wires feature a closed velum structure, which prevents opening due to thermally induced relaxation and therefore permits a high stability of the melting behavior and a good homogeneity of the coating microstructure. High-speed charge-coupled device (CCD) camera investigations of the wire melting behavior during high-velocity combustion wire (HVCW) spraying show the benefits of the tube cored wire design. Both massive and tube cored wires show a stable melting behavior with a continuous flow of melt to the wire tip (Fig. 4). After droplets have been sheared off the tip, they are atomized to fine droplets and accelerated onto the substrate surface. In contrast, the velum of a grooved cored wire is continuously melted and droplets are carried from the coarse wire tip in the high-velocity gas jet. The filler material, however, accumulates and is torn off the wire tip as coarse particles discontinuously. The reliability of full feedstock melting made the spraying of ceramic rods an interesting alternative to powders

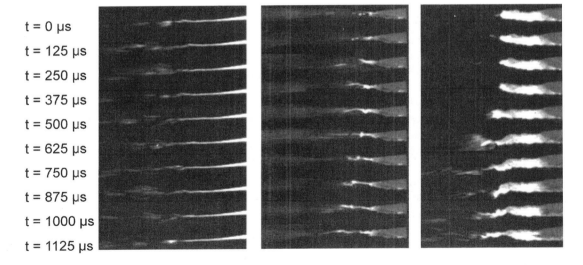

t = 0 μs
t = 125 μs
t = 250 μs
t = 375 μs
t = 500 μs
t = 625 μs
t = 750 μs
t = 875 μs
t = 1000 μs
t = 1125 μs

Figure 4 Sequences of the melting off behavior during HVCW spraying with different wire designs (left) Mo massive wire; (middle) tube cored Fe-Cr-Mn-C wire; and (right) grooved cored Ni-Al-Mo wire.

until plasma spraying was developed. In contrast to metallic materials, the price of sintered rods is significantly higher compared to powder materials. Therefore, the relevance of the according technology and feedstock vanishes. In order to permit the wire spraying of ceramic materials, flex cords, which contain ceramic powder materials in an organic or synthetic polymer velum, have been developed. As these cords are electrically nonconductive, they can only be processed by flame spraying.

C. Metallurgical Aspects

One key issue for the thermal spraying of coatings is achieving a sufficiently high bond strength. As the substrate is not melted, on one hand, mixing is avoided; on the other hand, there is no metallurgical bond. Oxide ceramic coatings sprayed directly on top of a steel substrate easily result in spallation due to the mismatch of the thermal expansion coefficient. Therefore, bond coats are applied. Ni5Al has proven to be suitable because of its high ductility and its excellent bond strength to steel substrates due to exothermic reactions at the interface. Additionally, Ni5Al protects the steel substrate against corrosive attack by chemicals, which might pass through pores or cracks of the oxide ceramic coating.

Nickel-based super alloys that form protective Al_2O_3 layers on top have proven to exhibit high hot gas corrosion and oxidation resistance. In general, oxidation resistance increases with decreasing oxide content in the coatings and with improved purity of the Al_2O_3 layer. The oxide scale grows up to a critical thickness, which causes spallation due to the thermal expansion mismatch, especially when the components undergo thermal cycling. As $ZrO_2-Y_2O_3$ thermal barrier coatings show high oxygen ion diffusivity at a working temperature of about 1000°C at the bottom side, oxidation-resistant bond coats need to be applied on nickel-based super alloy turbine blades. MCrAlY coatings have proven to qualify for this purpose.

WC-Co is of high interest for surfaces, which are subject to severe wear load. According components are produced by sintering at high pressure in an inert environment. Sintering permits comparatively low processing temperatures. Therefore, the formation of brittle mixed carbides and the degradation of wear resistance are avoided. The inert environment is necessary, as WC is oxidized at temperatures exceeding 600°C, forming volatile tungsten oxides. For the processing of WC-Co by thermal spraying, it is therefore necessary to avoid melting and long exposure of feedstock to high temperatures in oxygen-rich environments.

The excellent wear and chemical resistance of SiC, on one hand, and the low price, on the other hand, make it an attractive material for the manufacturing of protective thermal spray coatings. As SiC decomposes at temperatures exceeding 2545°C and the ductility is poor even at high temperatures, pure SiC coatings cannot be produced by conventional thermal spray processes.

The use of SiC as a hard phase in a cermet material is restricted by high reactivity, with most metals already in solid state. Attempts to spray SiC with chromium or nickel matrices resulted in low SiC and high silicide-containing coatings. Due to the brittleness of the silicides, only poor wear resistance is obtained. To achieve thermal-sprayed composite coatings with high SiC content, it is mandatory to reduce the reactivity of the metallic binder matrix and to promote the wetting of the SiC particles. Sufficiently silicon and carbon-saturated nickel or cobalt crystals permit to reduce the reactivity and to improve the wetting of the carbide particles. Adequate alloying is obtained out of the range of the primary crystal-terminated eutectic. Additionally, processes that permit particle temperatures below the melting temperature of the matrix, and high particle velocities to achieve high density and short interaction times with the hot gas jet have to be applied.

IV. MICROSTRUCTURE AND PROPERTIES OF THERMALLY SPRAYED COATINGS

As for every other technology for the production of components, thermally sprayed coatings also show a characteristic microstructure based on the principle of coating formation. A schematic of the thermal spray process is shown in Fig. 5. Different energy sources are applied to transfer heat and momentum to the feedstock, which is added in powders, wires, or rods. Heat transfer results in full or partial melting, or at least in a certain ductility increase of the feedstock, whereas the momentum transfer results in acceleration of the spray particles onto a component surface.

A. Coating Formation

Thermal spray coating formation is a stochastic process, as differences in the spray particle size and in the case of powder feedstock additionally shape differences along with the temperature and velocity distribution in the accelerating gas stream, resulting in variances of particle conditions. Each powder particle or melt droplet injected into the accelerating gas stream follows a

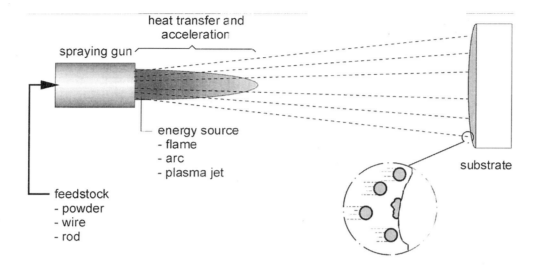

Figure 5 Schematic of the thermal spray process.

trajectory depending on its mass, density, shape, and initial velocity on its way to the substrate surface. The interaction of heat and momentum transfer determines the thermal history of the particles prior to their impact on the substrate, which, along with the spraying environment, governs the progression of reactions. Furthermore, the temperature and velocity of the impinging particles are thereby set, which, along with the substrate condition (temperature, roughness, inclination to particle trajectory, etc.), determine flattening and solidification on the substrate.

If a thermal spray coating is mainly built up of molten particles, it shows a lamellar structure due to the superposition of multiple particle impacts with specific flattening and solidification characteristics. Depending on the applied material and the process conditions, thermal spray coatings are more or less porous, can contain microcracks or macrocracks, and show heterogeneous and anisotropic properties due to microstructure anisotropy. In addition to unmolten or resolidified particles, oxidized or nitrided particles can be incorporated. Figure 6 shows a schematic of the formation of a thermal spray coating with the resulting characteristic microstructure.

For given substrate conditions and particle temperatures, the kinetic energy determines the flattening behavior. The temperature-dependent surface tension limits the spreading. Although particles impinging on polished surfaces spread radially, for rough surfaces or impingement on prior deposited particles, spreading occurs preferentially in directions where the required energy consumption due to surface tension is least. The

spreading behavior with the resulting splat morphology takes significant influence on the bond strength.

Generally, two types of splat morphologies are discerned. On one hand, splats that show a coherent spreading are observed. For high viscosity, some changes from the perfect circular boundary shape are possible. On the other hand, fingering is observed, if particles impinge with high velocity and low viscosity. The strong melt flow can also result in the formation of a corona. Solidified oxide ceramic splats often contain cracks both in the vertical and horizontal directions. Horizontal cracks are critical for the cohesive strength of the coating. Unmolten or resolidified particles usually bounce at polished substrate surfaces, whereas they

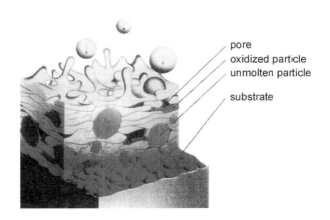

Figure 6 Formation of a thermal spray coating.

might get stuck in the profile of grit blasted substrates, however, with poor bond strength.

Heat transfer to the particles can result in changes of the spray materials composition or structure. The most important in-flight reactions and processes are as follows:

- Selective evaporation of components
- Reactions with environmental gases (oxidation, nitration, and hydration) and formation of stable or volatile reaction products or solving of gases
- Reactions of feedstock material constituents (formation of mixed phases by the reaction of metallic binder and hard phases in cermets, and metallurgical reactions of mechanically alloyed powders)
- Phase transitions of the feedstock material (e.g., γ-Al_2O_3 in coatings sprayed with pure α-Al_2O_3 powder).

Especially the oxidation of metallic spray particles on their way to the substrate takes significant influence on the coating properties. On one hand, selectively increased oxidation of corrosion-protective components such as chromium in steels reduces corrosion resistance strongly. On the other hand, oxides and solved oxygen can increase the coating hardness and thereby improve wear resistance. Formed oxides are preferentially incorporated into the coatings on the lamellae boundaries, but depending on the melt dynamics and the viscosity of the oxides formed at the particle surface, oxides can also be transported into the metallic core of the particles.

B. Adhesion of Thermally Sprayed Coatings on Substrates

As there is no metallurgical bond between thermal spray coatings and substrates, bond strength is a decisive criterion for applicability. Careful surface preparation is demanded to achieve an optimum adhesion of thermal spray coatings.

1. Coating Adhesion Mechanisms

The adhesion of thermal spray coatings on, for most instances, metallic substrates depends on a combination of the following physical and chemical mechanisms:

- Mechanical cottering
- Metallurgical reactions between spray particles and substrates, and diffusion
- Adhesion by physical or chemical adsorption.

Mechanical cottering is possible for roughened substrate surfaces. Droplets impinging on the rough surface penetrate into holes and undercuts due to the high kinetic energy and low viscosity. The effect is intensified by stress as a result of the contraction of splats during solidification.

The adhesion of thermal spray coatings can also depend on metallurgical reactions and diffusion. Both diffusion and reactions between coating and substrate result in a strong bond strength increase. These reactions do not take place across the complete contact area, but only at single microscopical "active" contact zones, which are separated by oxides and pores. For sufficiently high temperatures in the contact zone, both the formation of reaction products and interdiffusion are possible. The affinity of the involved elements and the formation enthalpy of the reaction products determine if a solid solution or the formation of new phases occurs. The density of active zones depends on the temperature and cooling rate in the interface between substrates and particles. Ceramic spray particles can react with oxides of a bond coat. Therefore, preoxidation of the bond coat can result in increased bond strength. For plasma spraying of alumina coatings on steel substrates, the formation of iron and mixed oxides is observed depending on the substrate temperature.

Although physical adsorption is based on comparatively weak van der Waals electrostatic forces, which become effective for an activated (roughened) substrate surface and the close distance between substrates and coating material atoms, chemical adsorption is based on strong covalent bonds, which depend on the affinity between the substrate and the coating material. Adsorption is strongly affected by all kinds of surface impurities.

The efficiency of the discussed mechanism and their contribution to the overall bond strength depend on the spraying process, the combination of substrates and coating materials, and the surface preparation. In addition to the interface temperature, the velocity, thermal conductivity, and specific heat of the particle; the thermal conductivity of the substrate; the residual stresses; and the properties of the substrate and particle surface take significant influence. High interface temperatures generally improve bond strength. But for the deposition of metallic coatings on metallic substrates, oxidation of the substrate surface has to be avoided.

2. Surface Pretreatment Options

Generally, all pretreatments for thermal spraying aim for clean and activated surfaces with large surface areas. For polymers and most ceramic substrates, metallurgical reactions with spray particles are not possible, which means that bonding works by mechanical cottering and physical adsorption. The accordingly demanded high surface roughness can be achieved by grit blasting. For some two-phase polymers, selective etching can also provide a high surface roughness.

Generally, rinsing with suitable solvents, especially with ultrasonic support, results in cleaner surfaces after grit blasting in comparison to brushing and vacuum cleaning, but not every component permits this procedure.

In addition to grit blasting, further pretreatment processes have been developed for metallic substrates. To activate the surface of the metallic substrate, an oxide scale removal is necessary. Etching in acids requires efficient removal of the aggressive medium prior to spraying. Vacuum plasma spraying (VPS) permits etching by sputtering. Therefore, the substrate is charged positively and the resulting transferred arc removes atoms from the surface. As the process is carried out in an inert low-pressure environment, the active surface state is maintained.

Recently, systems that activate the surface directly in front of the spray jet have been qualified for industrial application. On one hand, high-energy laser beams are used to evaporate impurities and oxide scales. This technology is attractive from the economical point of view, although laser systems are expensive and consume a lot of energy. However, the time-consuming masking and grit blasting procedure can be avoided. On the other hand, cryogenic blasting with CO_2 pellets has been tested. The activation of the surface is achieved by the flash evaporation of the pellets at the moment of impact on the substrate. The removal of substrate material takes place by a cavitation-related process. For both systems, promising preliminary results are reported.

3. Bond Strength Determination

The evaluation of the bond strength of thermal spray coatings can be carried out by quantitative or qualitative methods. The latter is based on the bending of a coated specimen with a visual inspection of the deformation behavior. Spallation of the coating due to stresses in the interface with the substrate points at bad bond strength.

The most commonly applied quantitative method to determine the bond strength of thermal spray coatings is the tensile bond strength according to the standard DIN EN 582. Specimens with a roughened surface and a diameter of 25 or 40 mm are coated with minimum coating thicknesses of 150 μm, glued to a roughened counterpart, and the joint is tested concerning the strength

perpendicular to the specimen surface. Tensile bond strength is calculated as the quotient of the maximum sustained load and the cross section of the specimen.

Depending on the coating thickness and porosity, the adhesive may penetrate down to the substrate and therefore result in the determination of an unrealistic high bond strength, whereas differences of the coating thickness cause failure of the adhesive joint at stresses significantly below the strength of the adhesive due to notch effect. Generally, two failure types are discerned: cohesive failure occurs by crack propagation through the coating material, whereas adhesive failure occurs in the interface between the coating and the substrate, or top and bond coat.

As bond strength depends on the combination of substrate and coating material as well as on residual stresses, it is necessary to spray the specimen for tensile strength tests on the real component materials under comparable conditions and with the same thickness. The strength of available adhesives is limited to about 100 MPa. Coatings produced with modern spraying equipment, especially when high particle velocities are achieved, can exceed this bond strength significantly. Therefore, a new standard is presently elaborated on.

Force transmission under working conditions can differ strongly from the perfect tensile stress in the standardized test, which results in an accordingly different failure mechanism. Therefore, tensile bond strength does not necessarily permit to conclude the performance of a thermal spray-coated component, especially for cyclical load conditions.

C. Porosity and Surface Characteristics

The porosity of thermal spray coatings depends on the coating material and the applied process. Although the density of conventional flame and arc-sprayed coatings is limited, high-energy processes such as HVOF, detonation, cold gas, and vacuum plasma spraying permit the production of coatings with nearly theoretical density. In general, ceramic coatings show higher porosity than metallic coatings, especially when coarse powder feedstock is used. Therefore, posttreatment (e.g., by application of sealers) can be necessary if corrosion-protective function is required. The use of sealers, however, results in thermal limitations of the coating.

Due to coating formation mechanisms, the surface roughness of thermal spray coatings is relatively high (5–30 μm). For components such as grippers, high roughness is beneficial; for paints and varnishes, thermal spray coatings feature excellent bond characteristics, which are exploited for flame or arc-sprayed aluminum or zinc coatings for atmospheric corrosion protection. For most applications, machining by grinding and sometimes additionally polishing are required.

D. Residual Stresses

Depending on spray conditions, thermal spray coatings can show high residual stresses due to the contraction of splats after spreading on the substrate surface during the solidification and cooling process. These tensile residual stresses decrease the bond strength and limit the admissible coating thickness. In ductile metallic coatings, residual stresses are limited by the yield strength of the coating material. In brittle metallic or ceramic coatings and in brittle components such as oxidized shares of a ductile metallic coating, residual stresses can cause crack formation and propagation. Vertical cracks in zirconia-based thermal barrier coatings can form a crack network. This segmentation can be achieved by controlled thermal deposition conditions and results in improved thermal shock resistance.

V. THERMAL SPRAY PROCESSES

One common criterion in classifying thermal spray processes is the energy source by which the feedstock is heated and accelerated (Fig. 7). The oldest process type works by direct melt atomization onto a substrate surface [1]. As this process is not applied industrially, it is neglected in the following. Additionally, processes that work by combustion of different fuels (flame spraying), by electrical discharge with direct (arc spraying) or indirect (plasma spraying) interaction with the feedstock, or by high-velocity gas streams (cold gas spraying) can be discerned. Laser spraying is also considered a thermal spray process by some authors, but because it is accompanied by melting of the substrate surface, it is not included in the following process descriptions.

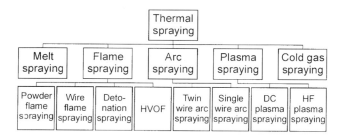

Figure 7 Classification of thermal spray processes by energy source.

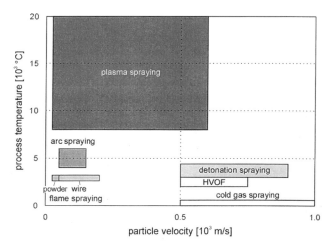

Figure 8 Process temperature and particle velocity ranges of thermal spray processes.

The different thermal spray processes feature distinct characteristics concerning the process temperature and particle velocity (Fig. 8). These characteristics determine what materials generally can be sprayed and what coating characteristics (density, oxide content, etc.) can be achieved. With increasing particle velocity, the interaction time between the energy source and the sprayed particles is reduced, which results in low oxide contents, whereas increasing process temperatures favor oxidation. On the other hand, high process temperatures are necessary to manufacture coatings of refractory brittle materials such as ceramics, as their deformation capability is too low to permit deposition without at least partial melting.

A. Flame Spraying

The energy source for the melting and acceleration of particles in conventional flame spraying processes is a gaseous oxyfuel flame. Most commonly, acetylene, propane, or hydrogen is applied as fuel gas. There is spraying equipment available for the processing of powder (Fig. 9), wire, and rod feedstock (Fig. 10).

The conventional flame spraying process with powder feedstock is mainly applied for the production of self-fluxing NiCrBSi coatings for combined wear and corrosion protection. The relatively high porosity of coatings in the assprayed state is removed by a subsequent fusion process, which also results in a high bond strength. The fusion can be carried out in furnaces, by inductive heating or by use of the spraying guns flame

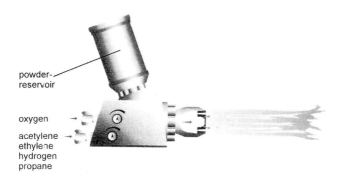

powder-
reservoir

oxygen →

acetylene →
ethylene
hydrogen
propane

Figure 9   Powder flame spraying process.

without injection of powders, which is especially suitable for on-site applications.

For the manufacturing of thermoplastic polymer
coatings, a special gun design, which prevents the overheating of the feedstock, is applied. The axially fed
granules are shrouded by a coaxial gas, which is heated
by an externally coaxial flame. Aluminium or zinc wires
are used for the production of corrosion-protective
coatings on steel structures, and molybdenum is applied
for state-of-the-art wear-protective coatings on automotive gear synchronizing disks. The main application
field for rod flame spraying is the production of wear-
protective and corrosion-protective Al_2O_3 and Al_2O_3-
TiO_2 layers and partially stabilized ZrO_2 coatings for
thermal insulation. The according rods are produced by
sintering. A cheap alternative is the application of
flexible cords filled with the according oxide powder.
Cored wires also expand the spectrum of applicable
coating materials significantly, as they permit the incorporation of hard phases in the coatings.

B.   HVOF Spraying

The HVOF process is an advancement of the conventional flame spraying process and has found broad

industrial acceptance these days due to low porosity,
low oxide content, and high bond strength of the
thereby produced coatings.

HVOF spraying features high particle velocities and
moderate particle temperatures due to the short flame–
particle interaction time. Although the particle velocity
does not exceed 50 m/sec for conventional flame spraying of powders and 200 m/sec for spraying of wire feedstock, modern HVOF spraying equipment permits
velocities up to 650 m/sec. In general, the combustion
temperature is kept low by applying propane or propylene as gaseous fuels (Fig. 11). Only a few spraying
guns for acetylene fuel are available. Due to the necessity of high gas pressures for the operation of modern
spraying guns, hydrogen and ethylene are increasingly
applied because liquefaction is easily avoided. Kerosene
permits the lowest flame temperatures, especially when
the combustion is carried out with air (HVAF) instead
of pure oxygen (HVOF). The defined addition of water
enables a further decrease of the flame temperature with
a simultaneous increase of the particle velocity. This
process (high-velocity impact fusion, HVIF) is in the
state of laboratory analysis.

The main application of HVOF processes is the
manufacturing of wear-protective and corrosion-protective cermet coatings. The most important cermet
materials are WC–Co(Cr) and Cr_3C_2–NiCr. The relatively low particle temperatures and the short interaction time of flames and particles permit to avoid the
formation of brittle mixed carbides. In optimized
HVAF-sprayed WC–Co coatings, mixed carbide formation can be kept below the detection limit of x-ray
diffraction (XRD). Further HVOF applications are hot
gas corrosion-protective MCrAlY coatings.

Due to flame powers up to 250 kW, strong heat
transfer to the substrates often necessitates cooling.
For systems with axial injection of the powder feedstock, which features particle trajectories near the
flame axis, orifices can be used to fade out the outer
jet areas. Usually, the orifice diameter is 4–6 mm smaller
than the diameter of the jet at the conventional torch

wire →

oxygen acetylene
         ethylene
         hydrogen
         propane

Figure 10   Wire flame spraying process.

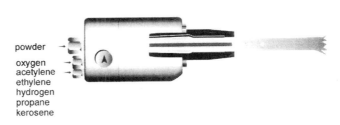

powder →

oxygen →
acetylene
ethylene
hydrogen
propane
kerosene

Figure 11   HVOF spraying process.

exit. The application of an 8-mm orifice at the exit of a 12-mm internal diameter torch reduces the heat transfer to the substrate by about 55%. In addition to the benefits concerning the necessity of a less powerful cooling system and less consumption of the cooling medium, the residual stresses of the produced coatings can be reduced.

HVOF spraying is accompanied by high noise levels up to 140 dB(A) and the expansion of the combustion gases results in a significant rebound. Therefore and because of strong heat and dust evolution, HVOF spraying is mainly carried out with robot handling systems. On-site application with manual handling is generally possible, but significant efforts for occupational safety and health are required.

In addition to systems for the processing of powders, a wire flame spraying equipment permitting increased gas and also particle velocities has been developed (HVCW). The thereby produced coatings show a significant improvement concerning surface roughness, bond strength, and porosity, but the oxide content is significantly higher than for HVOF coatings. This is due to the characteristic that wire spraying processes require complete melting of the feedstock, before droplets can be accelerated in the flame. Powders are not necessarily melted and are less oxidized due to the shorter interaction time with the flame.

C. Detonation Spraying

Detonation spraying was developed in the 1950s in the United States and is the progenitor of the HVOF processes. Detonation spraying is a discontinuous combustion process. It is characterized by a sequence of charging, in which the combustion gases—usually acetylene and oxygen—and the powder feedstock with the nitrogen carrier gas are injected into the tube, the ignition of the combustible, and the flushing of the tube with nitrogen (Fig. 12). Ignition frequency can vary from 4 to 8 Hz for early developed systems, up to 100 Hz for newly developed spraying guns, which work with

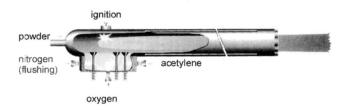

Figure 12 Detonation spraying process.

fluid dynamical control of the gas injection (high-frequency pulse detonation, HFPD).

Detonation spraying is characterized by relatively high process gas temperatures, which can be up to 4000°C, and high particle velocities of up to 900 m/sec. Oxidation of the sprayed particles is limited due to the short interaction time between the hot combustion gases and the powder particles. The high particle velocities result in high coating density and high bond strength.

Detonation spraying is, like HVOF processes, in the first place, applied for the production of wear-protective and corrosion-protective cermet coatings, but metallic or ceramic coatings are also manufactured.

On one hand, detonation spraying shows economical advantages in comparison to HVOF spraying due to lower gas consumption. On the other hand, the process produces noise levels that can exceed 140 dB(A) and therefore requires special sound-proof and explosion-proof rooms. Additionally, the comparatively high weight makes the handling of the detonation spraying guns more difficult than for HVOF guns. Therefore, mainly simple geometries, which do not require extensive movement of the gun, are coated.

D. Arc Spraying

The arc-spraying process uses an electrical current to generate an arc, which is used to melt the feedstock material. Usually, two conductive metal wires, which act as electrodes, are continuously consumed as the tips melt. A potential difference of between 15 and 50 V is applied, and the two electrically charged wires are fed on an angle, whereby the distance between their tips is gradually reduced (Fig. 13). When the distance is decreased sufficiently, the arc is struck. The arc heat melts the tips of the wires, and an atomizing gas shears off and accelerates the resulting droplets onto the substrate.

The temperature of the arc by far exceeds the melting temperatures of the sprayed materials and results in superheated particles. These extremely high particle temperatures can result in localized metallurgical interactions with the substrate surface or diffusion zones. Due to these microscopical processes, good cohesive and bond strengths can be achieved.

The atomizing gas is usually compressed air, but nitrogen or argon is also used in some cases. Due to the entrainment of environmental gases within short distances from the gun exit, the protection by inert atomizing gases is limited. Additional shrouding with inert gases or spraying in a controlled inert atmosphere, however, permits even the processing of metals with high oxygen affinity such as titanium alloys.

Figure 13 Twin wire arc spraying process.

For high-quality coatings, the droplets need to impact on the substrate under comparable conditions. Closed nozzle configurations, which apply an additional radial gas stream, permit a highly focused spray jet and are therefore beneficial to meet this requirement.

Arc spraying features outstanding energetic efficiencies and deposition rates among thermal spray processes, which makes the process economically interesting. Additionally, the robust equipment is mobile and therefore qualifies especially for on-site application. Coating quality is limited because of relatively low particle velocities of less than 150 m/sec, which result in significant porosity and oxide content due to the resulting long interaction time of the superheated droplets with the atomizing gas jet.

Arc spraying is applied for the production of corrosion-protective aluminum or zinc coatings for steel and concrete structures and for on-site repair of damaged components. Cored wires permit a significant extension of the producible material spectrum. By the addition of hard phases as filler materials, the wear resistance of the sprayed coatings can be improved significantly [2].

The single-wire arc spraying equipment is in the state of laboratory analysis. The arc is struck between the wire and the nozzle wall, which works as a nonconsumable electrode. Straightened wires serving as the anode can be processed with an extremely small divergence angle of the particle trajectories [3]. That results in deposit spots with a diameter down to 3 mm and a microstructure that does not show the typical lamellar texture.

E. Plasma Spraying

The energy source for the melting and acceleration of particles onto a substrate surface in plasma spraying processes is a thermal plasma jet. The plasma jet can be generated by DC or AC arcs, or by inductive coupling. In the earliest developed DC plasma spraying process, an arc is struck between a cylindrically shaped anode nozzle and a tungsten pin cathode (Fig. 14). Both electrodes are water-cooled to prevent immediate de-

struction. Inert gases (Ar, He, N_2, and H_2) are fed through the gap between the electrodes and pass the arc. The energy transfer of the arc to the plasma gases results in dissociation and partial ionization. The recombination of electrons and atoms or atoms under the formation of molecules results in an extremely hot gas jet with temperatures that can exceed 15,000 K at the exit, depending on the electrical power, the gas composition, and the pressure of the spraying environment. The strong temperature increase and the pressure difference between the torch inlet and the exit lead to high plasma gas velocities. These days, spraying guns with electrical power ranges between 10 and 200 kW (water-stabilized plasma spraying) are available. Usually, powder feedstock is fed radially inside the nozzle, or directly behind the nozzle exit into the thermal plasma jet. Modern systems work with computerized control of machine parameters and gun handling, which results in high reliability and reproducibility even for coatings on components with complex shape.

The extremely high temperature of the thermal plasma jet permits melting of any desired material that has a liquid phase. Therefore, the main application field of atmospheric plasma spraying is the production of wear-protective and corrosion-protective Al_2O_3–TiO_2 or Cr_2O_3 coatings, and Al_2O_3 coatings for electrical and ZrO_2-based coatings (additives: Y_2O_3, CeO_2, MgO, and CaO) for thermal insulation. Metallic and cermet coatings are also produced.

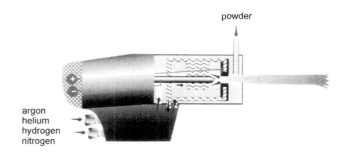

Figure 14 DC plasma spraying process.

Knowledge concerning the beneficial influence of high particle velocities on coating porosity and bond strength led to the development of spraying guns, which permit plasma gas velocities up to 3000 m/sec. The optimization of the nozzle design is still an important research subject. The spraying of wire feedstock with DC plasma torches has been a research topic since the 1980s, but has not reached industrial acceptance yet.

The plasma spraying process has also been advanced for use in controlled environments. Besides versions with high-pressure environments (controlled atmosphere plasma spraying, CAPS; $p < 300$ kPa) for the exploitation of spray particles reactions with environmental gases of low-pressure versions (VPS; low-pressure plasma spraying, LPPS; $p > 2$ kPa) have been developed. The latter are used for the production of hot gas corrosion-protective MCrAlY coatings, which feature high density and a very homogeneous oxide-free microstructure, on gas turbine blades. These coatings also work as bond coats for subsequently applied ZrO_2-based thermal insulation coatings.

For spraying of reactive materials and for less strict demands on the content of environmental gases or reaction products, the use of gas shrouds has proven to be effective. Near the torch exit, a fast-flowing, coaxial laminar stream of inert gases is utilized to prevent the entrainment of reactive environmental gas into the plasma jet. This also results in effective cooling of the substrate.

Single-cathode DC plasma spraying guns with axial powder feeding have not found industrial acceptance due to poor reliability. The Axial III (Northwest Mettech Corp.) spraying gun with three separate plasma jets, which unite in front of an axial powder injector inside the gun, permits a significant increase of the deposition rate and efficiency due to more uniform particle parameters. This is a result of the small divergence of the particle trajectories and therefore the improved homogeneity of the heat transfer conditions in comparison to conventional radial injection in single-cathode systems. The major disadvantage of this design is the high energy consumption due to the operation of three plasma jets, each with a comparable power such as that of a conventional single-cathode torch.

In another three-cathode system, the plasma power per arc is limited to about a third of the power of conventional single-cathode torches, resulting in a significant decrease of the electrical power consumption and a simultaneous increase of the electrodes' lifetime. In the Triplex torch (Sulzer Metco Corp.), three separate arcs are struck, which feature a fixed anode attachment, whereas in conventional torches, the arc moves along the nozzle surface. The fixed anode attachment results in a stable process state, which permits homogeneous heat transfer to the sprayed particles despite a radial injection in front of the three anode attachments. Additionally, the noise level drops significantly from about 120 dB(A) down to 100 dB(A). The improved heat transfer conditions permit an increased deposition rate with a simultaneous increase of the powder feed rate and an improvement of coating properties (porosity, hardness, and bond strength).

HF plasma spraying utilizes a plasma jet that is generated by inductive coupling of electrical energy into a gas stream. The main advantage of this technique is that the generating components are not in contact with the plasma gases. Therefore, even highly reactive gases and gas mixtures can be applied, and the incorporation of eroded electrode material in the coatings is ruled out. Additionally, the large plasma volume permits high deposition rates and the particles undergo homogeneous heat treatment due to the axial injection inside the torch through water-cooled probes. These days, systems with plate powers between 25 and 200 kW are available.

The temperature in the plasma jet usually does not exceed 10,000 K. The gas and therefore also the particle velocities are low (<50 m/sec) due to the large plasma volume. Gravity is commonly exploited to achieve the highest possible velocity. Recently, supersonic nozzles that permit gas velocities comparable to DC torches, whereby the cross section of the jet is reduced accordingly, have been developed for HF plasma torches.

The main disadvantages of HF torches are the poor flexibility concerning the handling of the torch due to the dependence of the particle trajectories on the angle between the torch and the gravity axis for conventional torches; and the stiff power supply design and the strong electromagnetic fields that prevent the use of electronically based substrate handling systems. Therefore, the most important application field of HF plasmas is the spherodization of powders and the analysis of materials by emission spectroscopy after complete evaporation in the jet. Additionally, simple geometries are coated.

F. Cold Gas Spraying

Cold gas spraying is a relatively new spraying process, by which coatings can be produced without significant heating of the sprayed powder. The powder particles are accelerated in a Laval-type nozzle with gas of moderate temperature (<600°C) to velocities exceeding 500 m/sec (Fig. 15). Upon impact, the kinetic energy

Figure 15 Cold gas spraying process.

is transformed to thermal energy and the particles form a well-adherent coating with nearly theoretical density by a cold welding mechanism. As the transformation from kinetic to thermal energy occurs by plastic deformation of the particles, sufficient ductility of the sprayed material is necessary. Typically, metals with cubic face-centered crystal structure and therefore high ductility such as copper, austenitic stainless steels, or nickel-based super alloys are sprayed. Zinc is also easy to process as its melting temperature is very low. Dense cermet coatings have also been produced, but the hardness is well below that of HVOF-sprayed coatings.

The cold gas spraying process requires adapted gas supply and powder feeder systems as the gas working pressure is up to 3500 kPa and flow rates of 1250 slpm are common. The gas is usually heated in a helical tube by electrical resistance heating. Most commonly, nitrogen is used as process gas, but the addition of helium or even spraying with pure helium can also be applied if materials with high demands concerning the particle velocity and heat transfer to the particles are sprayed. The high helium costs, however, will restrict its use to a few special applications.

The main advantage of cold gas spraying is the avoidance of oxidation both of the spray material and the substrate. The electrical conductivity of cold gas-sprayed copper coatings can be as high as 90% of the cast material. Coatings up to several centimeters in thickness can be produced, and even on polished glass surfaces, well-adhering conductive layers can be deposited. The small divergence of the particle trajectories permits the manufacturing of freestanding complex structures. This new technology will open up specific applications in the coming years.

G. Thermal Spray-Related Processes

The processing of conventional powder, wire, or rod feedstock results in sprayed particles with diameters exceeding 5 μm. The suspension plasma spraying process has been developed to build up coatings of nanosized droplets. Bioactive hydroxyapatite coatings have been manufactured from a mist of droplets with a

medium diameter of 20 μm by HF plasma spraying. The high temperature of the plasma jet and the long residence time of the feedstock in the hot area permit the complete evaporation of the solvent and the melting of dried powder particles. The subsequent coating formation is comparable to a conventional spraying process and a nanocrystalline structure is obtained. Another example is the production of $LaMnO_3$ perowskite coatings using a suspension of nanosized MnO_2 particles in a $LaCl_3$-saturated ethanol solution. This new technology is in the state of laboratory analysis. The thermal plasma jet cannot only be used to melt particles, but also to evaporate or to dissociate a feedstock material completely with subsequent deposition on a substrate surface. The so-called thermal plasma jet CVD (TPCVD) process has successfully been applied for the production of diamond coatings on hardmetal turning tools. Usually, methane is used as precursor gas. Methane is injected into an argon–hydrogen plasma and dissociates to carbon and hydrogen atoms due to high temperatures. In the laminar boundary layer above the substrate surface, partial recombination reactions occur before CH_x ($0 \leq x \leq 3$) radicals approach the surface and gradually form the diamond coating. Both DC and HF torches can be used, but the low gas velocity of HF plasma jets causes thick laminar boundary layers and therefore strong recombination processes resulting in low deposition rates. The strong temperature gradients of thermal plasma jets cause locally inhomogeneous deposition conditions. Therefore, high-quality diamond coatings can only be produced on small areas.

Besides diamonds, for example, SiC and superconductive $YBa_2Cu_4O_x$ coatings have been produced in laboratories. In addition to gaseous and solid state precursors, also liquid single source precursors, which feature an easy fixation of the precursor stoichiometry during the deposition process, have been used. The application of liquid single precursors is especially interesting for Si–C–N coating production, as the common gaseous precursors SiH_4 or $SiCl_4$ (evaporated) are hazardous both for machinery and personnel.

The hypersonic plasma particle deposition (HPPD) process was developed to produce nanostructured coatings and also uses gas-phase precursors. In contrast to TPCVD, the coating is not built up by the gas-phase species, but by particles formed inside the plasma jet. HPPD uses a plasma generated by a DC torch, into which gaseous reactants are injected. Subsequently, the mixture is quenched by supersonic expansion through a nozzle into a vacuum deposition chamber. The high quenching rates cause the nucleation and growth of nanosized particles from the supersaturated vapor. As the process is carried out at pressures of about 0.1 kPa, the particles are strongly accelerated and deposited on the surface of a cooled substrate. The inertial impact with velocities exceeding 1000 m/sec results in a direct consolidation of the coating without necessity for further treatment, although the latter may improve the properties of the coating.

This process has successfully been applied to produce silicon and silicon carbide coatings with deposition rates of 3600 and 1500 μm/hr, respectively [4]. Measurements of the particle size in-flight and the grain size inside the coating showed good agreement, with a mean size of about 10 nm. When the nanosized particles are focused to a narrow beam, which can, for example, be achieved by the application of a system of aerodynamic lenses, this process can also be applied for nanostructured device production. For example, SiC towers with a diameter below 100 nm and heights exceeding 1 mm have been manufactured.

VI. NEW DEVELOPMENTS AND NEW APPLICATIONS

Recently, thermal spraying has been introduced into the line production of automotive engine manufacturing for internal coating of cylinder bores [5]. For this purpose, a special gun design (which permits fast internal coating of bores with a diameter down to less than 70 mm), low heat transfer to the AlSi cast alloy engine block, and a reliable long lifetime of the gun components is required. Additionally, coatings with superior tribological characteristics in comparison to traditionally applied cast iron sleeves at reasonable costs and an integral technology considering all the necessary pretreatments and posttreatments and their interdependencies had to be developed.

The solution by Sulzer Metco Corp. is the Rota Plasma system with a specially designed plasma gun. Although the engine blocks remain stationary, the plasma gun rotates inside the cylinder bores with constant distance to the cylinder surface. Prior to the coating process, automatic grit blasting is carried out with subsequent complete removal of the erosive medium. In addition to composite powders with steel matrix and solid lubricants, low-alloyed carbon steel is applied as feedstock material. As the process is carried out in atmosphere, the particles are partially oxidized. The oxides show good lubricating characteristics during the engine operation. The combination with the coatings porosity, which results in microscopical reservoirs for the liquid lubricant, permits significant friction

reduction and improvement of wear resistance compared to cast iron. To achieve these properties, the diamond honing machining after the deposition of the coating needs to be optimized carefully.

For the so-produced engines, high reliability with long lifetime and a reduced emission of hydrocarbons due to decreased oil consumption in comparison to engines with cast iron sleeves are obtained.

Another innovative process development deals with the fast production of relatively thin area-wide coatings on large surfaces by thermal spraying [6]. For this application, a plasma spraying gun has been developed, which features typical jet diameters of 400 mm at a chamber pressure below 1 kPa with a homogeneous distribution of particles in the plasma jet. The large cross section of the plasma jet is a result of the supersonic expansion into the low-pressure environment due to a high pressure difference inside and outside the torch. At the applied chamber pressure, the collision rate of species is strongly decreased and the plasma jet may no longer be in local thermodynamic equilibrium as it can be assumed for conventional jets in plasma spraying processes.

The low density of particles in the spray jet, combined with a high-velocity relative movement of the substrate, results in the deposition of isolated splats. With successive passes, an area-wide coating is built up. The buildup in several passes results in low residual stresses of the coating.

Like TPCVD, this process is capable of filling the gap between typical thicknesses of coatings deposited by conventional CVD or PVD and thermal spray processes, and features economical advantages if large areas need to be coated because of the extremely short deposition time. A potential application for, for example, electrical purposes is the insulation by alumina, or use of conductive copper, or high-temperature ionic conductive zirconia coatings.

In addition to coatings, freestanding bodies can be thermally sprayed. Typical freestanding body materials are refractory or high-temperature-resistant metals and ceramics. Although for coating production the substrate is roughened to increase the surface, for freestanding body production, surfaces need to be rather smooth to permit separation from the substrate body. Depending on the desired material and the geometry, different techniques can be used. After spraying on graphite, the counterbody can be burned if chemical reactions are avoided, whereas bodies deposited on salts can be separated by solving in aqueous solutions. Additionally, spraying on polished substrates and separation with thermal cyclical support are possible.

The production of bulk ceramic bodies such as plates and tubes of almost any size with characteristic anisotropic physical, mechanical, and thermomechanical properties by plasma spraying is an established technology [7]. By plasma spray forming, components can be produced directly with the desired dimensions, posing an interesting alternative to conventional methods of shaping green bodies by molding and casting followed by sintering. These preferentially oxide and silicate ceramics exhibit the typical thermal spray deposit structure with 10–20% pores and laminar grains. They show low structural hardness, thermal conductivity and heat capacity, and extremely low Young's moduli. The structure advantageously provides quasi-ductile fracture behavior and outstanding thermal shock properties. For example, mullite tubes of 3 mm wall thickness survive a quench of 1200°C in 20°C water without fracture. Subsequent firing results in a slight density increase of 1–5% and a significant increase in Young's modulus. Plasma-sprayed ceramic bodies are mainly applied in hot corrosive environments of rapidly changing temperatures, where severe thermal gradients occur.

HVOF spraying is applied for the production of freestanding nickel-based super alloy bodies for supersonic engine liners. A new approach uses the vacuum plasma spraying process for the production of superplastic Ti6Al4V foils by thermal spraying. Superplastic forming demands a two-phase material with fine and nearly constant grain size at forming temperature. Ti6Al4V features a high stability of its grain structure at temperatures below the transition from the low-temperature $\alpha + \beta$ microstructure to pure β. For the production of Ti6Al4V sheets with according fine microstructure, an expensive thermomechanical treatment has to be done. The sprayed foils show extremely fine grain size and, after annealing at 800°C, a comparable microstructure to conventionally produced superplastic sheet material is achieved.

Microtechnology is a fast-growing market these days. Developments in stereo lithography and selective etching processes commonly applied for microelectronic components permit extremely small component sizes. But the choice of materials is limited. Usually, silicon or polymers are used and galvanoforming expands the applicable material spectrum by some pure or particle reinforced metals. Furthermore, particle content and distribution is not constant in microparts with high aspect ratios. The novel approach "thermal spray molding" uses thermal spraying as a forming tool to mold microstructural devices. For this purpose, the desired material is thermally sprayed on an according negative counterbody with subsequent separation.

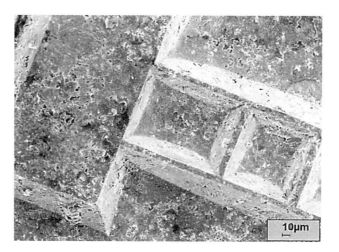

Figure 16 Thermal spray moulded molybdenum microcomponent.

There is hardly any limitation concerning the choice of materials and the small component volume requires only very short processing times.

Microcomponents with constituents smaller than 10 μm have successfully been molded by HVCW spraying of molybdenum (Fig. 16) and Ni20Cr. The surface roughness is decreased significantly in comparison to kerosene fuel HVOF spray-molded parts, as the impingement of fully molten particles on the counterbody does not cause damage. The promising technique is at the stage of transition from laboratory research to industrial application.

VII. SIMULATION OF THERMAL SPRAYING

The applicability of thermally sprayed coatings will be expanded significantly if there is success in tailoring the coatings structure. There is particular interest in the optimization of spraying processes for the production of nanocrystalline, quasi-crystalline, or amorphous coatings. For the development of such coatings, it is desirable to decrease the number of experiments to optimize the process parameters. The large quantity of influencing parameters and the boundary conditions of geometrical and spraying conditions during the coating process complicate the development of defined structures and morphologies.

Modeling of the spraying process allows a better understanding of the processes during thermal spraying. If it is possible to optimize the spraying process and the production conditions on the computer, the development costs of new coating systems will be reduced. A good agreement of the virtual spraying process with the real coating deposition is achieved by precise modeling of the particular process steps. The simulation of coating formation to estimate process parameters is an important tool to develop new coating structures with special properties.

A. Simulation Concept

The different process steps in forming a plasma-sprayed coating, which have to be regarded in the simulation, are illustrated in Fig. 17.

First, the particles are injected into the plasma. Heat and momentum are transferred to the particles. After accelerating and melting the particles, they will be partially or fully melted before their impact on the substrate. The trajectories, the velocity, and the temperature of the sprayed droplets can be calculated by the determination of temperature and velocity within the plasma or the flame, taking the plasma–particle interactions into account.

On the substrate, the particles flatten and the resulting shape depends on the particle temperature and particle velocity. Temperature-dependent properties such as viscosity and surface tension also take influence. Heat transfer processes have to be regarded in this simulation as well. As a result of the particle flattening on the substrate, temperature gradient and cooling rate in the particles can be calculated from the local temperature–time function. These parameters are the basis for the calculation of the microstructure after solidification of the sprayed particles. At least for the simulation of the coating formation, the interaction of several droplets has to be taken into account.

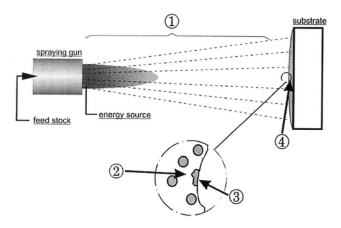

Figure 17 Process steps to be considered in simulation of thermal spray coating formation.

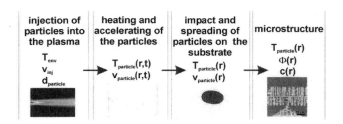

Figure 18 Sequence of simulation steps and decisive results.

The sequence of the involved simulation steps and the decisive parameters that have to be calculated are given in Fig. 18 for each process step.

B. Plasma–Particle Interaction

First, the trajectories of the particles have to be calculated. Based on the measured temperature and velocity distribution of plasma, the transfer of heat and momentum to the particles is simulated. According to Ref. 8, the following functions, with T_w as the temperature of the walls and R_{in} as the radius of the nozzle, are proven for the calculation of plasma jet temperature and velocity:

Temperature

$$T = (T_0 - T_w)[1 - (r/R_{in})^{n_T}] - T_w$$

Velocity

$$v = v_0[1 - (r/R_{in})^{n_v}]$$

The injection of particles leads to a two-phase flow of discrete particles in a continuous plasma jet. For spraying processes, particle size and particle size distribution are important. To consider these parameters, a Gaussian particle size distribution with a mean size of 40 μm and a standard divergence of 24 μm of the injected particles can be chosen, and the particle trajectories for different sizes of the particles and for different injection velocities can be calculated. It can clearly be seen that higher injection velocities lead to a broader particle jet.

Simultaneously, the resulting particle temperatures depending on the particle size are calculated. Low injection velocities lead to a more homogeneous temperature distribution within the particle jet compared to high injection velocities. In addition, particle velocities depending on the injection velocity are calculated. Particle velocities and temperatures are the starting conditions for the simulation of the impact and the spreading of the sprayed particles on the substrate.

C. Particle–Substrate Interaction

To simulate coating formation, the spreading of more than one particle has to be calculated. Depending on the calculated particle temperature and velocity, the shape of the flattened particle changes. Different starting conditions for the particles are possible. Particles can be fully melted, partially melted with a solid core, partially solidified on the particle surface, etc.

The spreading behavior of the splashed particles can be described with the Navier–Stokes equation for the dynamic, viscous, and incompressible flow [9]. The flattening of the particle on a plane substrate is then calculated by the volume of fluid (VOF) model [9,10]. The change of the particle temperature is calculated by the change of enthalpy in each individual cell. Thereby, the heat flux generated by convection, radiant heat, and thermal conduction is considered.

In general, the simulation parameters for the description of the particle flattening process on the substrate during plasma spraying are temperature-dependent and an overview on the regarded parameters is given in the following table:

Particle parameters

- Particle size
- Temperature dependent surface tension
- Temperature dependent viscosity
- Thermal conductivity of the particle
- Specific heat of the particle
- Latent heat of the particle
- Density of the particle
- Melting point
- Particle temperatures
- Particle velocities

Substrate parameters

- Thermal conductivity of the substrate
- Specific heat of the substrate
- Substrate density
- Substrate temperature

Interaction

- Pressure
- Temperature around
- Heat transfer coefficient
- Emission coefficient
- Hot gas jet

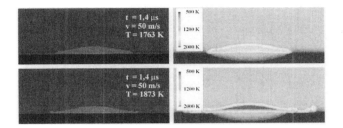

Figure 19 Spreading behavior of nickel particles depending on impact conditions.

The particle velocity significantly influences the particle shape after the spreading of a single particle. High velocities (150 m/sec) lead to an increased spreading of the particle. For sprayed droplets with a high temperature before the impact, the spreading and solidification time are longer than for lower temperatures because the temperature within the particle after the splashing is also higher (Fig. 19). In Fig. 20, the time-dependent temperature at different positions in a single particle is given. For positions near the substrate, temperature decreases faster by the stronger heat transfer from the particle to the substrate. The temperature behavior within the first particle is influenced by the second particle after the impact on the substrate. The cooling rate for the three different positions changes as follows: Cooling starts after the impact on the substrate. The release of the latent heat during the solidification of the particles leads to a reduction of the cooling rate. The

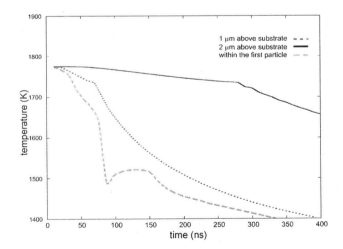

Figure 20 Local temperature evolution in a sprayed particle after impact on the substrate.

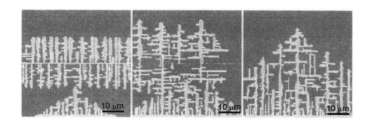

Figure 21   Dendrite growth in nickel particles after impact on the substrate depending on the particle interface temperature.

impact of the second particle leads to a decrease of the cooling rate within the first particle. After solidification, the cooling rate increases. Heat transfer to the substrate leads to temperature balance and the cooling rate decreases exponentially. The calculated temperature gradient and the cooling rate are the boundary conditions for microstructure simulation.

D.   Microstructure Simulation

The microstructure of solidified particles depends on the boundary conditions and can be calculated using the phase field model. The field function can be deducted by the Landau–Ginzburg representation of the function of the free energy F. The free energy is used to derive the kinetic equation for the phase field in a manner that the free energy decreases monotonically in time. The free energy equation is solved to calculate the position of the phase field. This leads to an equation to calculate the phase field as a function of temperatures, boundary conditions, and material parameters. For the microstructure formation of thermal spray coatings, the temperature of the interface of the subsequent flattening particles is important as the nucleation occurs on two interfaces: substrate/lower particle and the interface between the two particles. Dendrites grow from the nucleus on the substrate into the molten particle, and from the interface into the two flattened particles. The growth of the lower dendrites takes influence on the resulting microstructure and the duration of the solidification.

For different temperatures of the interface between the particles, different microstructures will be formed. The calculated microstructure for different interface temperatures is shown in Fig. 21. Low interface temperatures lead to heterogenous nucleation on the interface. The dendrites grow in different directions with a broader structure of the upper dendrites. With increasing interface temperature, the interface will be melted gradually. Less nuclei are formed and the dendrites can grow partially through the two particles. With an interface temperature above the melting point, the interface is fully liquid and the dendrites can grow through the two particles. The temperature takes influence on the dendrite spacing.

Another important aspect of the microstructure formation is that, normally, multicomponent materials and not pure metals are used for industrial applications. As a first approach to this, the solidification of a binary alloy is simulated. In Fig. 22, the composition and the structure of a solidified Ni–Al alloy are shown. Solidification starts at two nuclei set on the substrate. The starting composition is 20 at.% Al. After the solidification, a eutectic with the two phases Ni and AlNi$_3$ is formed.

E.   Experimental Verification

In experimental verification, a good agreement between simulation results and experimental particle spreading is found. Even the microstructure simulation can be validated experimentally. At first, the dendrite structure in one particle is shown in Fig. 23. It clearly can be seen that directional solidification of Ni dendrites occurred. Dendrites growing through three particles can be seen in Fig. 24.

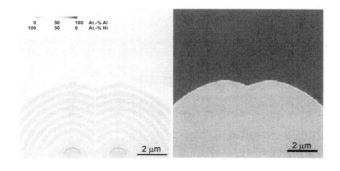

Figure 22   Composition and solidification boundary evolution in a Ni-20Al melt.

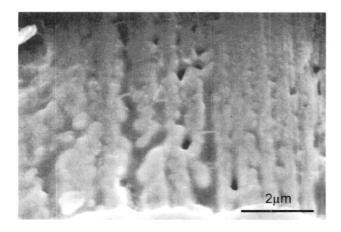

Figure 23 Dendritically solidified particle in a VPS nickel coating.

VIII. QUALITY MANAGEMENT

The importance of quality management has increased significantly worldwide in the past 10 years. Aggravated laws concerning product liability have supported this development in Europe. Additionally, the unification of the European market leads to increased competition of companies and products. For special applications with high security relevance such as turbine blade coatings, company-dependent specifications have been developed by end-product manufacturers (e.g., Pratt and Whitney). However, to date, there has been no single thermal spray company that fulfills the demands of the standard Ford 101.

Due to the high costs involved in certification, most of the German thermal spray companies and research institutes have joined the Gemeinschaft Thermisches Spritzen (GTS) Thermal Spray Community and have worked out a technology-specific standard. The GTS quality guidelines are based upon the standards DIN ISO 9000 and DIN ISO 9004, with adaptations for thermal spray surface technology. This certification standard constantly gains acceptance in different industrial branches served by the thermal spray industry. A parallel standard in other nations is not known to date.

Generally, methods and tools for on-line monitoring and control of the spray process and for nondestructive testing of the thereby produced coatings have to be developed and applied in order to make thermal spraying an unquestionable reliable coating process. Corresponding developments are introduced and discussed in the following.

A. Process Control Tools

The monitoring and control of the machine parameters such as gas flow rates or plasma current are state of the art. The large number of machine parameters is optimized prior to the coating of components. However, even the complete control of the machine parameters does not guarantee constant coating properties, as the particle parameters (temperature and velocity) at the moment of impact on the substrate, which determine the coating structure in combination with the substrate temperature, are also liable for, for example, torch degradation. Therefore, efficient process control tools have to provide characteristic data collected in the spray jet. Different systems are currently being tested concerning their industrial practicability.

The fast pyrometry-based DPV2000 (Tecnar Automation Ltd., Canada) system permits the determination of absolute particle parameters and particle size distribution. It was initially designed for process understanding, modeling, and development. Meanwhile, it has evolved to a successfully applied monitoring tool [11], but the high price and the low flexibility concerning handling inhibit its use, especially in small job shops.

Additionally, two CCD camera-based systems have been developed and are presently being introduced in series production. The SprayWatch system (Oseir Ltd., Finland) works by recording two successive pictures of different wavelengths, allowing the determination of the temperature and velocity of individual particles as well as their spatial distribution [12]. The handling of the diagnostic tool along with the spraying gun is possible,

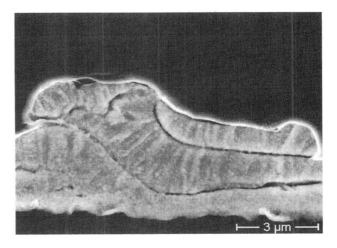

Figure 24 Dendrite growth through several particles in a VPS nickel coating.

but the flexibility is decreased due to its large size and relatively high price.

The other CCD camera-based approach, Particle Flux Imaging (PFI), is used to monitor the process state with the aim of detecting deviations from an optimized state. Images of the torch plume and the particle jet are recorded and the pictures' grey scale distribution is patterned by ellipses. Deviations in the ellipses' position and axis direction point at changes in process conditions [13]. The system permits in situ application and is comparatively cheap. However, the determination of boundary values for the ellipses' characteristics requires extensive preinvestigations.

An emission spectroscopy-based system, Flumesys (Flumesys GmbH, Germany), which like PFI aims to detect deviations from an optimized process state by location resolved data of characteristic wavelengths intensity [14]. As only a small optical setup linked to a spectrometer by optical fibers is moved along with the spraying gun, emission spectroscopy shows high flexibility and is already implemented in the production of turbine blade coatings for aviation applications and for cylinder bore coating manufacturing in the automotive industry.

Although the price of emission spectrometers has decreased significantly in recent years, an extensive use by job shops is hindered by missing strategies for the determination of intensity value boundaries with necessity of only few preexperiments. Emission spectroscopy is a highly interesting tool because it is also capable of detecting feedstock impurities, as atomic species of each element emit photons with characteristic wavelengths. For example, coinjection of copper during vacuum plasma spraying of aluminium coatings can easily be detected by monitoring characteristic wavelengths (Fig. 25).

A cost-efficient system that detects critical changes of the process conditions based on a correlation of the acoustic emission of a spraying gun is presently being evaluated concerning its applicability under industrial conditions. Finally, the acoustic emission of particles impinging on the substrate is correlated with the size, density, velocity, and viscosity of the particles. Thereby, an online control of the coating formation will be possible. So far, all systems lack a closed loop and therefore do not permit the control of the spraying process.

B. Nondestructive Testing of Thermally Sprayed Coatings

These days, mainly the destructive metallographical inspection of thermally sprayed coatings is applied. Small specimens are coated along with the real components in order to permit components intactness. This method is not very reliable, as during the spraying of a large component, critical process fluctuations resulting in local bad spots are more likely to occur in comparison to spraying on the small specimen. The same problem applies to the inspection of small parts taken out of the component at locations that are not critical for its

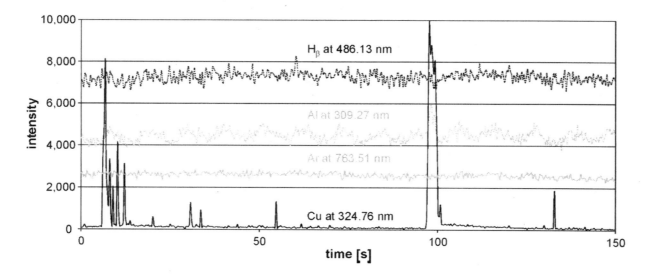

Figure 25 Emission intensity of characteristic wavelengths during vacuum plasma spraying of aluminum and co-injection of copper powder at 10 and 95 s.

use. Additionally, the thermal conditions of uncoupled specimens and a component can differ significantly.

The nondestructive testing of thermal spray coatings is still in the stage of laboratory investigations. Besides ultrasonic testing, optical holography, and eddy current testing, thermography has been applied. The latter permits a comparatively fast detection of bad spots with a dimension down to 100 μm for coatings on flat surfaces. This promising method still has to be qualified for the inspection of coatings on components with complex shape.

IX. ENVIRONMENTAL ASPECTS

These days, thermal spraying is considered as a coating technology with broad industrial acceptance, which features a high standard of environmental sustainability. Feedstock that does not contribute to the coating formation and dust involved in thermal spraying are collected in special filters and disposed of separately. For common spray materials, hazardous waste is avoided. Exemplary studies carried out for atmospheric plasma spraying show that, in addition to the evolved dust, the generation of gaseous species and fumes, and the emission of noise and radiation have to be considered.

Despite the broad application of thermal spray processes, only little information on hazards for personnel during the spraying procedure has been elaborated. Due to the generally rising ecological awareness, this deficit has been minimized. Furthermore, personal safety equipment and methods for avoidance of any kind of emission are being developed.

X. CONCLUSIONS AND PERSPECTIVES

Coating technologies have become key factors for the manufacture of innovative products and for general technological advancement. Due to the high flexibility concerning both substrate and feedstock material, and processing conditions by a variety of processes, thermal spraying has opened up versatile markets and industrial applications. Adjusted combinations of feedstock and spraying equipment permit the tailoring of coating properties in wide ranges.

The use of tools that monitor relevant data out of the gas jet–particle interaction zone will increase the acceptance of thermal spray processes further on due to the improved reliability of the coating properties.

Although in the past the focus of thermal spray applications was in the area of wear and corrosion pro-

tection or thermal insulation, these days, coatings with special physical, electrical, chemical, or electrochemical properties gain more and more interest. Computer-aided engineering will permit to decrease the necessary effort for the development of adapted spray processes and the optimization of process parameters and feedstock.

REFERENCES

1. Berndt, C.C. The origins of thermal spray literature. Proceedings of the International Thermal Spray Conference 2001; Singapore, 2001; 1351–1360.
2. Wilden, J.; Wank, A.; Schreiber, F. Wires for arc- and high velocity flame spraying—wire design, materials and coating properties. Proceedings of the International Thermal Spray Conference 2000; Montreal, Canada, 2000; 609–617.
3. Heberlein, J.; Kelkar, M.; Hussary, N.; Carlson, R. New developments in wire arc spraying. Schriftenreihe Werkstoffe und Werkstofftechnische Anwendungen: Band 4.3. Verlag Mainz, Aachen: Werkstofftechnisches Kolloquium, Chemnitz, Germany, 2000; 10–20 ISBN 3-89653-531-5.
4. Tymiak, N.; Iordanoglou, D.I.; Neumann, D.; Gidwani, A.; Di Fonzo, F.; Fan, M.H.; Rao, N.P.; Gerberich, W.W.; McMurry, P.H.; Heberlein, J.V.R.; Girshick, S.L. Hypersonic plasma particle deposition of nano-structured silicon carbide films. Proceedings of the 14th International Symposium on Plasma Chemistry; Prague, Czech Republic, 1999; 1989–1994.
5. Barbezat, G. The internal plasma spraying on powerful technology for the aerospace and automotive industries. Proceedings of the International Thermal Spray Conference 2001; Singapore, 2001; 135–139.
6. Loch, M.; Barbezat, G. Characteristics and potential applications of thermally sprayed thin film coatings. *Thermal Spray: Surface Engineering via Applied Research*; Berndt, C.C., Ed.; ASM International: Material Park, OH, USA, 2000; 1141 pp.
7. Lutz, E.H.; Steinhauser, U.; Braue, W. Microstructure and Properties of Plasma-Spray Formed Ceramics. Fall Meeting 1997; Materials Research Society: Boston, MA, USA, 1997.
8. Fincke, J.R. A Computational Examination of the Sources of Statistical Variance in Particle Parameters During Thermal Plasma Spraying. Proceedings of ITSC Montreal, 2000.
9. Pasandideh-Fard, M.; Mostaghimi, J. Droplet impact and solidification in a thermal spray process: Droplet–substrate interactions. Thermal Spray: Practical Solutions for Engineering Problems. Proceedings of NTSC '96, 1996; 637–646.
10. Fluent 4.4. User's Guide. Vols. 1–2. Fluent Inc., USA, 1997.

11. Refke, A.; Barbezat, G.; Loch, M. The benefit of an on-line diagnostic system for the optimization of plasma spray devices and parameters. Proceedings of the International Thermal Spray Conference 2001; Singapore, 2001; 765–770.

12. Vuoristo, P., et al. Optimization and monitoring of spray parameters by imaging a CCD camera based imaging thermal spray monitor. Proceedings of the International Thermal Spray Conference 2001; Singapore, 2001; 727–736.

13. Zierhut, J., et al. Verification of Particle Flux Imaging (PFI), An In Situ Diagnostic Method. Proceedings of the International Thermal Spray Conference 2001; Singapore, 2001; 787–790.

14. Aumüller, B.; Lang, A.; Dotzler, K.; Bergmann, D.H.W.; Schulte, K. Experiences with the application of a spectroscopic measurement system for thermal spray process control and optimisation. Proceedings of the United Thermal Spray Conference 1999; Düsseldorf: Germany, 1999; 2, 747–749.

21

Designing a Surface for Endurance: Coating Deposition Technologies

Joaquin Lira-Olivares
Simon Bolívar University, Caracas, Venezuela

I. INTRODUCTION

An increasing awareness on the environmental impli-
cations of any manufacturing process has given new air
to the use of coatings in the manufacturing process.
Historically, coatings have been used to protect structures (painting and galvanizing) and pieces (metallizing)
and often to reconstruct worn pieces. These activities,
tending to increase the useful life of structures, machinery, and parts, contribute to diminish scrap piles
and the exploitation of raw materials. Presently, the
consciousness on the life expectancy and the cycle of
manufactured items, the need to respond to the community of the "from-cradle-to-grave" cycle of their
products, pushed original-parts manufacturers to include in their process coating technologies. Coating a
structure, machinery, equipment, or elements of those
usually provides a practical and a less expensive way of
manufacturing high-quality and high-performance
items, especially if incompatible characteristics are required from the bulk than from its surface. For example,
an inexpensive but mechanically resistant body with a
relatively small amount of more expensive material as
coating could provide a piece with a surface that is acceptable for the envisioned application, which improves
the quality of the piece, adding priced qualities to the
resistant bulk. In many cases, if the piece takes part in
the production of new products, the quality of the
product will also improve. Coatings increase the endurance of the piece, improving its thermal resistance,

corrosion, and wear resistance, making the piece more
appealing by improving its physical appearance (ornamental). Also, coatings could make parts more perform-
ant by changing its physical properties, like reflectivity,
light absorbance, or conductivity, or by making them
acceptable of being introduced in a human body that is
biocompatible.

Usually, the part in question is mechanically finished
first and then its surface is prepared to be coated, for
instance, made more coarse, washed, cleaned, and
dried. Then coating is deposited, sometimes heat-treated to improve its adhesion to the substrate or diminish
porosity, and, finally, mechanically or chemically finished to obtain the desired surface conditions. Only in
very thick (welded or cladded) or sometimes very thin
coatings (i.e., galvanic plating) some bending or forming can be accomplished after the piece is coated
without rupturing the overlayer. In thin-coated parts,
the elasticity of the coating could endure bending
without rupture.

Choosing the coating material, the appropriate process and its thickness, is one of the hardest decisions
for the manufacturer. The concepts of quality (i.e.,
durability) as well as economic considerations are
usually determinant in this choice. Production time
and technological complexity must be therefore considered. Depending on the physical characteristics,
thickness, and its mechanical properties of the coating,
the manufacturers must consider the system that fits
better in his production chain, with a minimum of

specialization for the operators, time required for the application, initial investment, and production expenses. In sensitive parts, however, price might not be limiting, thus complex coating systems and expensive materials could be justified due to the assurance they give to the performance of the part. Examples of these are biomedical implants, like femur prosthesis's heads and artificial dental implants, where the lifetime durability of the ceramic-coated metal implant must be assured. Also, turbine blades, especially for aircraft applications and some defense equipment, justify large investments in sophisticated processes and expensive consumables. The price of production is acceptable when human life depends on it or the consumer requires high quality.

Coating technology has increased in number, sophistication, and types of materials available for coatings. Also, the diversity of applications has been expanding. So the know-how on characterization of coatings and consumables, like powders, wires, and the like, has become more sophisticated [1].

The two main factors to be aware of on a coating performance are its final surface characteristics and its adherence; however, the mechanical resistance of the coating itself seems to be a factor impinging on its durability [2]. Fatigue resistance [3] as well as shear resistance seem to be important on coating durability. In the case of sprayed particles, particle decohesion has proven to be a major source of failure [4]. The coating quality depends not only in its chemistry, but also in the morphology of the deposited material, be it atomic or molecular, drops or particles, which is influenced by the manufacturing method of these consumables [1]. Thus wire, bar, or particle characterization is imperative to assure a good coating when these are used including chemical composition; it is also important when fluids are used. Surface preparation of the substrate, in special cleanliness and finishing (roughness or coarseness), is also important for adhesion [5].

II. SURFACE TREATMENTS

Surface treatments, like shot pinning, heat treatments, cementation (carbonitruration), or ion implantation, could not be considered as coatings, even if they provide comparable surface effects. To be considered a coating, it must be always an overlayer, smaller in thickness than the substrate that has been covered. If the coating thickness is comparable with that of the substrate, it is referred either as a metal couple (cladded), layered material, or another appropriate name, but not as coating. Surface treatments with and without material depositions on the surfaces will be discussed in this section.

A. Shot Peening

Shot peening is a cold-working method accomplished by pelting the surface of a metal part with round metallic shot (steel balls) thrown at a relatively high velocity. Each shot acts as a tiny peen hammer, making a small dent in the surface of the metal and stretching the surface radially as it hits. The impact of the shot causes a plastic flow of the surface grains, extending to a depth depending upon the kinetic energy and impinging angle of impact of the shot and the physical properties of the surface (hardness and toughness). Kinetic energy transfer is the controlling process of shot peening. Depths varying from 0.05 to 0.30 in. are rather common, but values either higher or lower than this range can be practical [6].

There is a momentary rise of temperature of the surface due to the transformation of energy, possibly enough to affect the plastic flow of the surface grains. However, the effect of shot peening is considered as cold working, as the temperature reached is not high enough to produce annealing, creep, or metal flow at high temperatures [6].

The grains underneath the top layer are not deformed to their yield point and therefore retain their elasticity. The under grains are of course bonded to the plastically deformed upper grains and would tend to remain undeformed. In the equilibrium that results, the surface grains are in residual compressive stress while the inner fibers are in tension, thus providing conditions that will curtail the growth of surface cracks.

Experimental work indicates that the surface compressive stress is several times greater than the tensile stress in the interior of the section, so that when working stresses are applied that would ordinarily impose a tensile stress on the surface layer, the residual compressive stress will counteract them. Fatigue failures, which cause about 90% of the mechanical catastrophic failures in metal parts and structures, largely start at surface cracks that grow under tensile stresses; thus shot peening generally results in considerably greater fatigue strength. An equivalent result could be obtained with a tough welded protective coating, as cooling of the coating will produce compressive stresses on the undeformed substrate. Nickel alloys or manganese steel coatings, which are tougher than most carbon steels, are also used to reduce fatigue cracking on these materials.

B. Heat Treatments

Many machined, welded, and even coated pieces and structures retain stresses, mainly tensile, that could lead to failure when in use. Increasing the temperature of the surface allows the climbing of dislocations, thus the untangling of dislocation networks, which will provide a larger mean free path for dislocation gliding and thus permit more plastic deformation. These mechanisms absorb mechanical energy that otherwise would lead to dislocation alignment at obstacles near cleavage planes and subsequent crack generation.

Heat treatments are then used to "relax" the surface (or the bulk), reducing the chances of failure. Deeper or more prolonged treatments might provide enough energy for recrystallization and grain growth, thus softening the surface. These treatments are used after welding, hammering, cladding, and other machining processes when large residual stresses are feared to remain. Some coated materials are also heat-treated for the same reasons and, in some cases, for melting the coat, providing better adhesion to the surface, liquid sintering of the deposited grains, and reduction of porosity. It is often used when eutectic material is deposited in the so-called self-fluxing alloys (i.e., Ni–B–Si alloys).

C. Cementation

Cementation is a generic name for surface-treating steel (or another alloy) by exposing it to a bead of carbon- or nitrogen-rich material (could be extended to other relatively small atom sources) that upon heating will promote the penetration of these elements into the surface of a steel substrate by diffusion [7].

Fick's first law of molecular diffusion is also a proportionality between a flux and a concentration gradient. For a binary mixture of A and B [8]

$$J_{A_z} = -D_{AB}\frac{dC_a}{dz} \qquad (1)$$

where in Eq. (1), J_{A_z} is the molar flux of A by ordinary molecular diffusion relative to the molar average velocity of the mixture in the positive z direction, D_{AB} is the mutual diffusion coefficient of A in B, and dC_a/dz is the concentration gradient of A, which is negative in the

direction of ordinary molecular diffusion. If the gas, liquid, or solid mixture through molecular diffusion is isotropic, then the value of D_{AB} is independent of direction. Nonisotropic (anisotropic) materials include fibrous and laminated solids as well as single noncubic crystals. The diffusion coefficient is also referred to as the diffusivity and the mass diffusivity (to distinguish it from thermal and momentum diffusivities).

Many alternative forms of Eq. (1) are used depending on the choice of driving force or potential in the gradient. For example, we can express Eq. (1) as

$$J = -CD_{AB}\frac{dx_A}{dz} \qquad (2)$$

where, for convenience, the subscript on J has been dropped, C = total molar concentration, and x_I = mole fraction of species I [8].

The atoms in the cementation process will move mostly interstitially into the metal and produce a slow-down motion of dislocations. Cottrell atmospheres are considered the important mechanism to restrain dislocation motion when interstitial free carbon or nitrogen is present in steel.

The surface hardening with carbon is also called carburizing [9], with nitrogen, it is called nitriding, and with both, it is called carbonitruration. Other small atoms might play similar roles on steel and other metals, but in many cases, ad hoc names are given to these other diffusion effects.

There are essentially two methods in use for carburizing steel surfaces. They are pack carburizing and fluid carburizing. In both instances, low-carbon steels with 0.15% to 0.25% carbon are heated in contact with a carbonaceous material or a carbon gas [10].

In the first processes, the material is sealed into a pack with a commercially prepared carbon material or some other material such as coke charcoal or leather. The compound dissociates, releasing carbon, which diffuses below the surface of the steel to form iron carbides. In the fluid processes, the steel is heated in a retort furnace. Carbon monoxides or hydrocarbons (ethane, methane, propane, or natural gas) are introduced into the furnace. When heated, the carbon will penetrate into the sample. Liquid cyanide is another fluid carburizing process [10].

In either process, the materials are heated well above the critical temperature for a predetermined period of time. The carburizing temperature is generally above about 1750°F, which is well up into the austenite range. The length of time employed depends upon the carburizing materials, the size of the part, and the depth of penetration desired. The soaking period generally takes from 4 to 10 hr at the carburizing temperature. If the part is heated, soaked, and removed from the furnace, the case is not too thick, and the change from high to low carbon is abrupt. If the part is furnace-cooled, the penetration is deeper and more gradual. If the grain structure of the material is coarse, the penetration will be greater than if the grains are fine [11].

More details of carburizing can be found in Avner [12], Introduction to Physical Metallurgy.

Another method of carbonitrurizing is the plasma carbonitruration. The piece is exposed to a plasmogenic atmosphere rich in C and N, and the gas is plasmolyzed (ionized), usually by a spark, making the piece negative (cathodic); with respect to an anode, the ionized C and N atoms will hit the surface, warming it by transforming its kinetic energy, and penetrating it by diffusion, thanks to the concentration gradient between surface and bulk. Some high-energy ions will penetrate by ion implantation, but in any case, the final result will be a cemented surface.

Cementation is the simplest and oldest hydrometallurgical process. It has been used in extractive metallurgy to recover valuable metals as well as to remove unwanted impurities. Iron and zinc are the most common reductants in commercial applications of cementation process. There is little information in the literature about cementation on other metals [13]. Cao and Duby [13] have been investigating about thermochemical properties, electrochemical behavior, and potential use for cementation on two ferromanganese alloys, standard ferromanganese (6.8% carbon) and medium carbon ferromanganese (1.4% carbon). Cementation of cobalt on ferromanganese has also been investigated. The rate is measured by rotating-disk experiments, and the results show that cobalt cementation on ferromanganese is a first-order reaction controlled by mass transfer. The reaction is fast in the pure sulfate solution, and the efficiency of use of manganese is close to 100%.

Some of these cementation processes are substitutional in nature; that is, the atoms diffuse in substituting atoms in the crystallographic matrix of the substrate as the cementing atoms are very large to diffuse through interstitial spaces. In any case, hardening is often pursued by cementation, but some other applications might also be engineered.

A related process is the oxidation of metals like aluminum alloys and chrome steels to either passivate them or to color their surface. These processes are accelerated by heating the metal pieces in a reactive (oxygen-rich) atmosphere.

D. Ion Implantation

Ion implantation is a result of ion bombardment of a surface. The source of ions is usually a plasma. The incoming ion could hit a surface atom, and if its kinetic energy is enough to remove the atom from the lattice position, it will either produce a so-called "knock on," which at its time will hit another atom to produce a cascade effect (spique), or simply replace the lattice atom upon displacing it. If the energy of the incoming ion is very high, its cross section to collision [differential cross section for transfer of energy to nucleus [14]] is very small and will be able to penetrate deeply into the substrate. The implanted ions might substitute the atoms of the crystal lattice, changing the electron density at different energy levels. If we have, for instance, a silicon (Si) substrate (as in a chip), an atom of valence 5 (i.e., P, As, or Sb) will contribute with excess electrons, thus energetically contribute to a donor energy level, making Si an intrinsic semiconductor into n-type extrinsic semiconductor. An implanted ion of valence 3 (i.e., B, Al, and Ga) will make a vacancy contribution and a p-type semiconductor. In both cases, the energy gap is lowered to a 100th of an electron volt [15,16].

Cheng et al. [17] have been observing medium-range order in ion-implanted amorphous silicon using fluctuation electron microscopy. In fluctuation electron microscopy, variance of dark field image intensity contains the information of high-order atomic correlations, primarily in medium-range-order length scale (1–3 nm). Thermal annealing greatly reduces the order and leaves a random network. It appears that the free energy change previously observed on relaxation may therefore be associated with randomization of the network.

Zhang et al. [18] prepared silver-doped titanium oxide coatings for biomedical application by ion beam-assisted deposition in an oxygen atmosphere. X-ray photoelectron spectroscopy (XPS) was used to examine the chemical states and composition, and the critical surface tension and dispersive and polar components of the surface energy were investigated by contact angle analysis. XPS analysis confirmed the presence of TiO, Ti_2O_3, and TiO_2, and the concentrations of the different oxidation states were influenced by the deposition rate of Ti and Ag. Dispersive and polar components of the surface energy were significantly affected by the oxygen atmosphere, and samples with a higher concentration of silver exhibited a somewhat large critical surface tension.

Wei et al. [19] presented the experimental results of the formation of AlHf phase and the associated mechanical properties of the Al matrix implanted by Hf ions. The intermetallic compound Al_3Hf is very attractive, owing to its melting point (1540°C). The Al_3Hf phase has been formed by some methods. The high aluminum intermetallic compounds DO_{23}–Al_3Hf and Ll_2–Al_3Hf were directly formed by Hf ion implantation into Al matrix with an average current density of 64 μA/cm^2 using a metal vapor vacuum arc (MEVVA) ion source at a dose from 3×10^{17} to 7×10^{17} ions/cm^2. The surface layer hardness of the sample implanted by Hf ions at a dose of 7×10^{17} ions/cm^2 was approximately three times larger than that of the aluminum without any Hf ion implantation.

Iron aluminides are of considerable interest from low to intermediate temperature structural applications in which low cost, low density, and good corrosion or oxidation resistance are required [20]. However, their application is currently limited by room-temperature brittleness and low-corrosion resistance. Choe [20], in order to improve the wear and corrosion resistance of iron aluminides in acidic solutions, made a study where steel containing Mo and Cr were fabricated and nitrogen ion-implanted on the surfaces of samples with doses of 3.0×10^{17} ions/cm^2 at an energy of 150 keV. Nitrogen implantation promotes the corrosion potential, the pitting, and the repassivation potential for iron aluminides containing Mo. He considered that the corrosion problem of iron aluminides could be solved, to some extent, by the addition of Mo and Cr to iron aluminides and nitrogen implantation on the surface of iron aluminides.

Thus ion implantation could be used for electronic applications, as well as for the control of some mechanical and chemical properties, in a similar way as the coatings.

These surface treatment techniques, as was mentioned before, might provide desired changes on the surface without changing much the surface profile of the piece. Coatings contribute to the thickness of the pieces or structures.

III. COATINGS

A coating is an overlay or film of a determined material (metal, alloy, ceramic, composite, or polymer; could be thin, medium, or thick), which is deposited on the surface of another material to change the characteristics or superficial proprieties (mechanical, chemical, or physical characteristics).

A. Coating Thickness

The thickness limits for thin and thick coatings are somewhat arbitrary. Millimetric coatings are considered thick; micrometric ones are considered thin. A thickness of 10,000 Å (1 μm) is often accepted as the boundary between thin and thick films. A recent viewpoint is that a film can be considered thin or thick depending on whether it exhibits surface-like or bulk-like properties [21].

Almost all material compositions used as substrates can be coated by a large number of materials if the substrate's surface is accessible to the coating process and is duly prepared to enhance adhesion. Paint on metals, polymers, and ceramics, considered a thin polymer coating, will cover the surfaces, without much previous preparation other than cleaning. Contrasting with it, the thick welded coating of metal over metal requires machining and cleaning before coating. Historically, those materials of lower fusion temperature were used earlier (polymers and eutectic metal alloys) than those of high fusion temperature, like refractory metals and ceramics. Chemical or electrochemical processes could build up a layer of material on a substrate, depositing molecules or ions on the surface, thus protecting or ornamenting it. Diamonds, for instance, could be grown on steel by chemical reactions, leaving a wear protection coat; also, chrome oxide could be deposited or formed (stainless steel) for corrosion protection. Nickel, on the other hand, could coat a fender by galvanic means, having a strong bonding due to its growth pattern and yield not only protection, but also beauty. Evaporation, ablation, and sputtering of sources can deposit atoms, forming thin layers on a clean substrate, which are useful for electronic applications. Other processes deposit micrometric or nanometric particles, building up millimetric and even centimetric layers, from drops, small particles (micrometric), or simply wetting and drying. In must applications, one of the deciding factors of stability is the similarity between the thermal expansion coefficients of the coating and substrate to avoid bending of thin pieces and detachment from thick ones.

B. Adhesion of Coatings

It is reasonable to think that historically, metallic coatings and later polymeric (paints) were used earlier than ceramic coatings due to the dependence of the coating on temperature for its adhesion. However, caves, ceramic pots, and tombs were decorated with oxide (ceramic) layers that could be considered coatings. In all of those, adhesion to the surface seems to be a major concern for endurance.

Adhesion of the coating to the substrate requires a close contact between the materials at the interface, and interfacial contact is better obtained if at least one of the faces is in the liquid state. Thus to deposit metals or ceramics, one must reach its fusion temperature or close to it when the particle, wire, or band contacts the substrate. That energy must come from heating the material prior to deposition (molten metal, deposited by immersion), while being deposited (plasma or oxy-acetylene-sprayed coatings), or after being deposited (thermally sprayed self-fusing alloys). Otherwise, the material must be in liquid state in a suspension, like paint or like sol–gel deposits, where the material is in a colloidal suspension and is deposited and subsequently dried. Lately, the energy for melting is obtained from the impact kinetic energy of the impinging particles, as in high-velocity oxygen fuel (HVOF) processes, the novel high-velocity cold process, and exothermic reactions like aluminum oxidation [22].

Adhesion of the coating to the substrate depends on the intimate contact between the two materials, as well as the above-mentioned similarity between the thermal expansion coefficients. Once the materials are chosen, adhesion is best accomplished if two materials, coating and substrate, are at atomic or molecular distance from each other, which is obtained by atomic or molecular deposition in vacuum or by wetting. The free bonds and the surface dipoles from the two layers can then interact at the interface either weakly (van der Waals and other polar forces) or strongly (metallic, ionic, or covalent bonding). Moreover, if diffusion is possible, that is, if there is enough heat for diffusion, the two materials can interdiffuse, or even react, and new phases could appear providing better bonding as in welding. Of course, these phenomena are dependent on the chemical composition and even crystal structure of the deposited coat and that of the substrate, which will define interdiffusion and reactivity at a given temperature. Thus improving the bonding in itself defines the technique to be chosen and the deposition parameters: temperature, pressure, atmosphere, etc. once the coating material has been decided for a given substrate. The microstructure to be obtained is also decisive, as from it depends its mechanical, chemical, and thermal behavior among others. Amorphous coatings, by rapid solidification, porous coatings, small grains, martensitic structure, etc., depend on the method of deposition and subsequent heat treatment. The method of deposition might affect the substrate itself. Welding, for example, due to

the high heat input, could produce new phases, which weaken locally the substrate. Also, arch welding and, to some extent, galvanizing might produce hydrogen contamination, leading to embrittlement of the interface or deposit [23].

Vacuum techniques, for depositing atomic or molecular layers, increase greatly the costs and are usually cumbersome, thus physical vapor deposition (PVD), chemical vapor deposition (CVD), and the like are mostly used for electronics, aviation, biomedical, and other high-technology applications, where cost can be justified. These sophisticated techniques often induce epitaxial growth of the coating on the substrate, ensuring structural continuity at the interface and thus good bonding. Deposition of drops or solid particles, soften or almost molten (wet and thermal spraying), requires either heat produced by burning fuels, by electrical sparks, or by plasma or exothermic reactions, like with aluminum particles. High-velocity impacts, like those achieved by the now popular high-velocity oxygen fuel (HVOF), where the particle at the speed of sound impinges the surface, transforming its kinetic energy into heat, melt the particle at the impact and wet the surface. Thus adhesion is associated to wettability, and the force, stress, or energy of adhesion could be related to changes of surface tension and energy, but not necessarily be equated to it. Surface wettability is proportional to surface energy differences. Adhesion energy is the work per unit area required to separate the coating from the substrate. Thus it is linearly proportional to the difference between the surface energy air substrate (γ_{AS}) and air coating (γ_{AC})

$$\text{adhesion} = E + (\gamma_{AS} + \gamma_{AC}) \qquad (3)$$

E is the added energy due to mechanical anchoring and other bonding factors.

Other factors contribute to the macroscopic (mechanically measurable) adhesion force or energy as might be mechanical anchoring of the coating in the irregularities of the substrate or diffusion and subsequent formation of new interfacial phases and chemical bonding. The chemical bonding is considered stronger than the other types of bonding.

Adhesion tests have been normalized, and peeling (for thin films) (refer to ASTM C 633-79, "Standard Test Method for Monotonic Stress or Pulling") and shear test (also refer to ASTM C 633-79) are usually accepted to prove good adhesion. Other tests like bending and twisting of the overlaid rod or plate and fatigue tests by cyclic deformation of the substrate and coating are also used. However, good correlation of these tests to the adhesion observed when the part is in use requires more statistical work. In most cases, however, the tested samples fail at the coating itself, near the interface coating substrate, more than at the interface itself. In sprayed coatings, it has been shown that the apparent interface peeling is due mostly to particle decohesion [4] in the coating rather than separation of the coating from the substrate.

C. Types of Coatings

Coatings can be classified according to multiple criteria. For instance, thickness is one of them (as mentioned before) [thin solid films (angstrom or micron thickness) and thick layers like welded metallizing (millimetric or centimetric)]. Another is according to composition (metallic, ceramic, polymeric, composites, like cermets) and also according to the expected added property to the substrate, as corrosion-resistant, wear-resistant, heat-resistant, opaque, reflective, and the like.

The chemical composition of some typical coatings and their application can be seen in Section VII. In general, commercial coatings are presented with an oversimplification of its applicability. For instance, the label "corrosion-resistant" might be shown without specifying the media and conditions to which it might be resistant.

The color and brightness of the coating are also popular classification parameters, like dull coat, bright coat, etc.; terms as hard coatings, refractory, and heat-resistant coatings tend to make the classification more complex.

Usually "hard coatings" are presented as wear-resistant coatings; however, hardness is not necessarily the determining factor in the resistance of a coating to say, friction, where the wear is more dependent on the similarity or difference of the surfaces of the tribological pair [24] than to the hardness of the coating. Also, in the case of fretting corrosion and even abrasion, the resistance of the coating depends more in toughness than in hardness [25]. Cermets, like tungsten or titanium carbide in cobalt or nickel matrix, are considered hard coatings. In these materials, the carbide gives the overall hardness, and the metal gives the required toughness [26]. Erosion could be controlled sometimes by soft elastomes (rubber) instead of hard coatings.

Thermal barriers on the other hand depend on the material used (alumina, zirconia, partially stabilized zirconia—PSZ) and also on the porosity. Thus the same material, with a different application process, could yield different properties to the coating. Ceramics, by

and large, could be applied by CVD, oxyacetylene torch, or plasma torch, but with plasma, the coating will be more mechanically stable than with oxyacetylene and CVD is used for relatively thin coatings. This leads us to think that our attention, once we know about the possible nature of the finished coating, should concentrate on the coating method or process [27].

D. Time for Coating

Coating is usually done when the self-sustaining structure or piece is completed (after the manufacturing) for aesthetics (painting, plating), protection (galvanizing, CVD), or reconstruction (metallizing), during the manufacturing process, when the coating is part of the piece design (OPM: original part manufacturing), for some purposes like corrosion, wear protection, or thermal barriers, and during piece or equipment assembly, like coating of electronic devices. Some users, however, coat the original pieces of new machinery or equipment to increment their service life, and others use coatings for repair purposes. Some automobile pieces are coated after being manufactured and before assembling the vehicle, for instance, Babbitt coating on rolling bearings, to reduce fiction; also, the full body is immersed in anticorrosive fluid after the body is mechanically finished. Similarly, water and oil pipes are coated as part of the manufacturing process, and electronic parts and medical implants are also coated during manufacturing.

More and more original parts are coated during production and, also, structures are protected by coatings once assembled, like oil (petrol) rigs, bridges, etc.

IV. CHOOSING THE APPROPRIATE COATING

Obviously, the appropriate coating would depend, as we have said before, on the mechanical and chemical properties as well as other requirements that the surface should supply and also on the adherence of the surface to the substrate, which depends on chemical composition, among other things. In this chapter, we will emphasize surfacing or coating for wear resistance and corrosion resistance.

A. Wear Resistance

Many commercial products are labeled "wear-resistant"; however, wear is a very generic word. Using a somewhat classical definition, wear is the volume of material lost from the surface of a piece, either by chemical or mechanical work done on the surface; thus we could talk about chemical wear and mechanical wear, and these two subgroups are subdivided in apparently different phenomena, as follows:

Chemical wear: corrosion and oxidation [28]
Mechanical wear: adhesion (friction), abrasion, erosion, cavitation, fretting corrosion, surface fatigue [29]

However, Suh [25] considers that surface fatigue supersedes all other wear processes; thus if one controls surface fatigue, one controls wear. Fatigue is controlled by toughness in the material, thus a tough coating should suffice in most wear processes. This, in practice, is not totally true, as adhesion requires that the unlubricated tribological pairs are made of "incompatible" materials, that means, materials that do not weld to each other. Thus the coating to control friction must be either a solid lubricant (like graphite, Teflon, and molybdenum disulfide) or should not obey, if possible, the Hume Rothery rules of dilution to avoid welding between the asperities of the surfaces in contact. A table of comparison of metallic elements was published some time ago by Rabinowicz [30] and is shown below. Alloy couples must be tried ad hoc, and ceramic coatings are usually good couples used against adhesion due to their high-melting temperature, creep resistance, and low ductility. However, caution must be put on pairing ceramics with soft metals as the latter will probably wear faster.

Abrasion control is usually done with composite materials, like hard carbides, borides, and nitrides, as hard particles embedded in a tough metal or polymeric matrix [30,31]. These called cermets, in the case of metal–carbide composites, are used as machining bits, abrasion wheels, and as coatings for tractor loaders, trucks, platforms, etc. to protect them against abrasion produced by gravel and sand. A list of usual coatings and related applications is presented later. However, each practical case should be treated separately, and the coating should be chosen for each case presented. No general solutions for wear problems are available, not even for "kinds" of problems, like erosion. An erosion problem, produced by low-energy impinging particles, might be solved with an elastomeric coating that bounces away the erosive particle. However, if the particles are amorphous with sharp edges and have high kinetic energy, a hard material, with certain toughness, like zirconia partially stabilized with yttria (Y-PSZ), might be the solution. Also, a high-density cermet could be applied (i.e., Co–WC).

The real problems in the field often present several wear modes, like erosion–corrosion, fatigue–corrosion, adhesion–abrasion, and so on and not the academically identified phenomena [30].

Each manufacturer has a name for his product and some of them guard fiercely their chemical composition and manufacturing process; thus one must, in many cases, examine well the consumable product before applying it, as their commercial names might be misleading. Some well-known commercial alloys used for coatings will be mentioned below, omitting their commercial identification.

B. Abrasion, Erosion, Cavitation, and Fretting

Abrasion, erosion, cavitation, and fretting have in common surface fatigue; thus tough coatings and cermets with tough matrices are often applied.

Manganese-rich steels are among the abrasion (fatigue)-resistant alloys. For erosion applications, composites of Ni–WC, Co–WC, and the like are often used, paying attention that the interparticular distance (between the hard particles) must be smaller than the average grain size of the eroding particles. Cavitation, on the other hand, is caused by imploding bubbles, causing fatigue on the surface. Due to cyclic implosions, tough corrosion-resistant alloys are required. Fretting also requires corrosion-resistant and tough alloys, as do applications where erosion and corrosion act simultaneously (incinerator vents and extractors).

C. Nonferrous Abrasion and Erosion-Resistant Alloys

- Aluminum bronze is good for machine element work. It has twice the strength and hardness of other bronzes and has much better machining characteristics. It machines easily to an excellent finish. It is very wear-resistant. While it sprays with a coarse atomization, the resulting coatings are very dense. It is recommended for all general metallizing work with bronze. Examples are pump impellers, bronze castings, plungers, armature bushing, O.D. split motor bearings, O.D. airbrake valves, valve plugs, etc. [32].
- Tobin-type bronze, which is widely used for general metallizing work. It gives fine atomization and high spraying speeds. It is more difficult to machine than phosphor bronze.

- Phosphor bronze is recommended only for special applications. It can be machine-finished fairly well.
- Bronze coating (Cu–Sn) is recommended where a commercial-type bronze is satisfactory. It machine-finishes fairly. Among its applications are armature window bushings and sand spots in castings.
- Brass (Cu–Zn) can be used for fast spraying. It atomizes finely to give a solid dense coat. It is recommended where brass is required. Brass machine-finishes fairly good. Among its applications is sand spots in brass castings.
- Monel (Ni–Mo) is recommended for machine elements subject to corrosion, for all applications where it is hard enough. Otherwise, stainless steels are recommended. It atomizes well. Monel machine-finishes good. Examples are pump plungers, shafts, seal rings, gland casings, valve plugs, tailshafts, hydraulic pumps, etc.
- Inconel and other nickel alloys, as monel, are recommended for corrosion resistance and some abrasive wear resistance. These alloys attach well to steel surfaces and thus are among the most popular for oxyacetylene and plasma metallizing. They are usually easy to machine-finish giving a fair result. Some applications are in the dairy industry, petroleum pump impellers, and beer industry plungers. The applications are similar to monel.

D. Friction Adhesion-Resistant Alloys

As was mentioned before, in tribological pairs, especially in dry friction, we must consider both surfaces in interaction. Coating will vary in either one of the surfaces of the pair or both surfaces, but in friction–adhesion phenomena, the most important quality of the pair is that it most present in each intersecting surface member different characteristics, as to avoid interdiffusion and thus welding. That is the so-called "grabbing" in the technicians argot.

The table of Rabinowichcz helps finding those metals that are so-called "incompatible", thus apt for a tribological pair. Alloys do not appear in this table, but following Hume–Rothery rule, one could easily choose the right couple. Moreover, ceramic coatings on metals tend to simplify the choices, so do the polymeric coatings like Teflon. A version of Rabinowichcz's table, simplifying the symbols, is presented in Fig. 1.

But the best friction-control coatings for dry friction are those that avoid contact between the surfaces and do

	W	M	Cr	Co	Ni	Fe	Nb	Pt	Zr	Ti	Cu	Au	Ag	Al	Zn	Mg	Cd	Sn	Pb
In				F	G				E	G	G	E	E	B	G	E	F	E	
Pb	F	F	B	B	B	B	B	B	E	E	E	B	F	G	B	B	G	G	
Sn	G		B	G	G	F	G	B	G	B	G	B	B	F	G	F	E		
Cd			G	G	F	F		E	E	G	G	E	E	B	E	E			
Mg		G		G	F	F	F		E	F	E	E	E	E	G				
Zn		G	E	E	E	E	F	E	F	G	E	E	E	E					
Al	E	G	E	G	E	E	E	G	G	E	E	E	E						
Ag	B	F	B	B	B	B	F	E	G	E	G	E							
Au	E	G	E	G	E	E		E	E	G	E								
Cu	F	B	B	E	E	F	F	E	E	G									
Ti	E	E	E	G	E	E	E	E	E										
Zr	G	E	G	E	G	F	E	G											
Pt	E	E	E	E	E	E	E												
Nb	E	E	E	E	E	E													
Fe	E	E	E	E	E														
Ni	E	E	E	E															
Co	E	E	E																
Cr	E	E																	
M	E																		

Figure 1 Rabinowicz's table showing adhesion possibilities (high friction). E = Excellent, G = Good, F = Fair, B = Bad. (From Ref. 30.)

not adhere to either of them. Graphite, molybdenum bisulfide, Babbitt, and Teflon are among the most used ones. They are mostly either soft (low hardness), like tin–antimony, or scaly, that is, formed by particles that slide over each other, like in graphite. Therefore the solid lubricants which flow between the tribological surfaces without collapsing avoid friction and griping, but choosing the tribological pair protects against friction wear even if the lubricants collapse.

E. Corrosion-Resistant Coatings

Corrosion protection coatings must be, to start with, impermeable to the corrosive media, that is, nonporous. They must also be ductile enough to avoid cracking during service. Besides, they must not dilute and, preferably, they must become passive (corrosion-wise) in the media when they are exposed. The choice of the coating should be made following the galvanic activity table [33]. Among the most resistant materials against corrosion are the polymers and ceramics. The first are usually water-resistant and can make a good seal against corrosion. Ceramic coatings, if not porous, could also be a good anticorrosion barrier especially at high temperatures, if their brittleness does not interfere in the application. Glasses, among ceramics, are probably the most corrosion-resistant materials if they are not exposed to extreme conditions. This is due to their lack of grain boundaries (noncrystalline materials) that present a homogeneous surface toward the chemically aggressive media. For that matter, attempts have been made to surface melt metals with laser, producing a glassy surface (amorphous), without grain boundaries and thus more corrosion-resistant. Some examples of corrosion-resistant metals are presented below.

F. Corrosion-Resistant Metals

The following metals are used primarily for corrosion resistance of flat areas, but they are sometimes used for corrosion protection to various machine element parts.

- Aluminum (Al) is applied either as coating or in cladding for general corrosion resistance and, specially, atmospheric corrosion, as it readily passivates in an oxygen-rich atmosphere, forming alumina (Al_2O_3), a very resistant impermeable ceramic.
- Zinc (Zn) is applied as galvanic coating by dipping or thermally sprayed (wire spraying); for general corrosion resistance, it produces an anodic surface with respect to iron and steel, thus protecting them as a sacrificial anode, donating readily Zn ions to the electron-rich environment, avoiding corrosion of the steel piece.
- Cobalt (Co) is corrosion-resistant at higher temperatures than nickel, aluminum, and, of course, zinc. It is used as coating in the interior of engines, like valves, and usually applied as an alloy, commercially called Stellite.
- Tin (Sn), as pure block tin, is manufactured by drawing into wire for metallizing work. It resists corrosion in various acid environments. Tin sprays rapidly with fine atomization or can be deposited galvanically. It is used for pipes and tank linings and food-handling equipment [34].
- Lead (Pb) is recommended as an alloy for bearings (Babbitt), similarly to Sn and for several corrosion works where it can be kept isolated to avoid lead poisoning of environment. It can be readily atomized, thus thermally sprayed. Some applications are battery connecting straps and coating to resist acid fumes. A lead alloy containing 6% antimony is recommended for all cases where a hard lead is required and for special corrosion problems. That percentage of antimony

diminishes the melting temperature of lead, making it more fluid at higher temperatures. Lead is usually avoided for ecological reasons.

Once the material is chosen, the deposition method is fairly well determined. Carbides, for example, can be deposited by few methods (plasma, oxyacetylene, HVOF); diamond coatings are also limited (CVD) and so are polymeric coatings (painting, immersion, low-temperature spraying). Temperature, speed, flux, and transportation of the supplied material determine the coating.

The following section will discuss how to make the choice.

V. COATING PROCESSES

The coating processes are determinant of the characteristics and quality of the coating, and they limit the materials to be used, its thickness, performance, durability, and also the acceptable shapes of the parts or some structures used. Some processes permit only to coat exterior surfaces or interiors, only if the ID permits (plasma, thermal spraying, PVD). Others could be applied in intricate cavities (CVD, sol–gel). There are some processes that operate better in some geometries than others, be these plane or angular (galvanic and spraying methods). These limitations comprehend part of the versatility of the system.

In function of the interaction coating substrate, the coating processes could be classified as:

- Chemical: including CVD, sol-gel, electroless, biomimetic
- Reaction-conversion: including surface reaction processes, like diamond layers' formation on the substrate's surface
- Physical processes: particles and drop spray processes (spray pyrolysis, oxyacetylene, arch, detonation gun, HVOF, and plasma spray) and physical vapor deposition (PVD)
- Electrochemical: galvanizing
- Diffusional: welding and soldering
- Wet coating and immersion: painting sol-gel, liquid spraying and drying, immersion in liquid or fused material

A brief description of some of these methods will be included in this chapter; however, a different nomenclature will sometimes be used. CVD, PVD, thermal spraying, and other methods will be covered in other chapters in detail but will also be treated here.

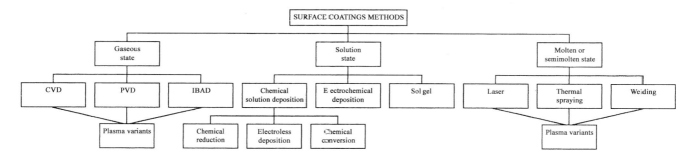

Figure 2 Surface coating methods according to deposition state.

Coating methods can also be classified according to the fabrication technique, as can be seen in Fig. 2.

VI. CHEMICAL PROCESSES

A. Chemical Vapor Deposition

Chemical vapor deposition uses a chemical process occurring between gaseous compounds when in contact with a heated material. The deposition takes place as long as the reaction produces a solid. The heating attachment must provide two temperature plateaus: the first is held at the vapor source, not far below the evaporation temperature T_E in order to have a constant vapor pressure of the source material above the crucible. The second, at the target, is higher and allows the reaction to take place. A carrier gas brings the vapor from the evaporation zone to the reaction zone. In many cases, a reactant gas such as hydrogen or oxygen is added to achieve either metal or oxide deposition. Metallic carbides, borides, and nitrides are deposited by using CH_4, BCl_3, and NH_3, respectively [36].

1. The Source Material

According to Chapman and Anderson [37], for a given layer, the choice of a source material will very much depend on the following practical considerations.

1. The use of a volatile compound will greatly simplify the technology of the gas transport. Otherwise, the pressure must be held below atmospheric pressure.

2. It must not decompose when heated; otherwise, an attempt must be made to evaporate at low temperature and low pressure (in the 0.1–10 Torr range).

3. It must react (or decompose) at a temperature consistent with usual technology, namely, not higher than 1200°C. The lower the temperature, the more versatile is the method.

From the data available in the thermodynamic literature, a choice can be made at once of the metallic compound for the source material, the reactant, and the reaction temperature.

2. The Coating Structure

CVD coatings are very varied. Metals as well as metallic compounds can be deposited. Only a suitable source material is required and, of course, a substrate where the chosen deposit would attach upon reacting.

The structure of deposits is closely related to the rate-determining step of the reaction. So choosing the conditions, which will give the desired structure with a good reproducibility, depends on the kinetics of the reactions.

In saturation conditions, the growth rate is high, forming nuclei with no time for rearrangement. The deposit is fine-grained, and the crystallites are oriented at random. The whole process is diffusion-controlled below saturation. If the feeding of vapor is slow, the growth rate is low and rearrangement on the substrate surface is possible. Large, oriented crystallites will form in these conditions: lowering the mass flux and increasing the temperature for surface diffusion allow a rearrangement of the nuclei.

Gaseous contamination of the surface influences the structure of deposits. The heterogeneous reaction starts by an adsorption of molecules; thus a strongly adsorbed gas will hinder the reaction. As the magnitude of this phenomenon varies according to the crystallite face, there is a tendency towards growth in a particular direction. This has a determining effect on the crystallization of the layer [38].

Even parts per million (ppm) levels of gaseous impurities are of importance. They are sometimes reactive gases such as oxygen and chlorine and are sometimes gaseous intermediates of the reaction. It is clear that it will be difficult to foresee what will happen as long as we cannot detect them. This is not the least important factor of irreproducibility. Thus high vacuum is important before reaction starts.

Gas adsorption effects on crystal growth should be an attractive research field, which could add a great deal to our knowledge of CVD.

3. The Coating Thickness

Thickness in some particular cases, as for thermal barriers, might be as much as 1 mm. Generally, such large values are not necessary for some needed mechanical properties, where a few hundred micrometers are sufficient (as for corrosion inhibition).

Coarseness and the lack of cohesion might be caused when larger crystals are grown; thus sharp edges and corners have a drastic influence on the quality of the coating.

4. Substrate Requirements

According to Chapman, the reactor design will vary widely due to the nature of the substrate, the size and number of parts, and the morphology of these parts. Ceramic substrates are generally heated in electric furnaces, whereas metals are heated with induced radio-frequency fields. The reaction temperature will depend on the activation energy of the reaction and the stability of the substrate; thus the maximum reaction temperature should not affect the properties of the substrate. Some metals with high melting temperature (refractory metals), mostly ceramics and graphite, can be heated to 1200–1300°C. However, copper alloys, aluminum, many glasses, and most steels must be treated at relatively low temperatures due to their low melting or creep temperature, or their low phase transformation temperature, to avoid morphological or microstructural changes. For stainless steel, that temperature usually should not surpass 400°C.

The shape and size of samples are limited by the heating possibilities. A sophisticated geometry yields nonuniform growth rates. Practically, the method is limited to samples having a rotation axis and planes of a few square centimeters in size.

The substrate treatment before deposition is rather a perplexing subject. When crystals of an electronic quality are to be grown by epitaxy from a monocrystalline (single crystal) substrate, attention must be paid to surface reparation. The substrate must be mechanically and chemically polished. Annealing is necessary to get rid of long-range damage of the crystal lattice induced by the previous abrasive treatment. With reference to gas adsorption, the atmosphere prevailing before deposition has a definite influence on the quality of the layer. This is the reason for baking the substrate in situ for 1 or 2 hr. As an example, Chapman gives getting rid of the oxygen adsorbed layer on the substrate surface by baking it in hydrogen flow.

The adsorbed gas layers are usually a barrier to CVD coating. They have a deleterious effect on the quality of the layer. For this reason, high vacuum is often employed for degassing substrates. Typical figures are 1200°C for the temperature and 10^{-5} Torr (1.3 mPa) for the residual pressure.

B. Sol–gel

A sol is a colloidal dispersion of particles in a liquid, usually aqueous, but sometimes an organosol. Sol–gel processing may be carried out by dipping or spinning and is an emergent coating technology. It involves applying the sol to a substrate, whereupon it undergoes aggregation, gelation, and final drying to form gel. Typically, the technique is used for the production of oxide ceramic films, in which case the gel is fired at above 150°C to leave the ceramic. The technique is still in the development stage, but it seems that it has considerable potential, not only as a means of producing films with controlled porosity, but also as a means of producing spheres of controlled size, which can then be used in plasma spraying processes [30].

Mullite, alumina, zirconia, and other ceramic coatings have been successfully applied by sol–gel [39] for anticorrosion and thermal barrier applications.

C. Electroless

Electroless plating processes differ from electroplating processes in that no external current source is required. Metal coatings are produced by chemical reduction with the necessary electrons supplied by a reducing agent (R.A.) present in the solution [40].

$$M^{n+} + ne - (\text{supplied by R.A.}) \xrightarrow[\text{surface}]{\text{catalytic}}$$

$$M°1(+ \text{ reaction products})$$

It can be considered as an autocatalytic method; the reduction is catalyzed by certain metals immersed in the solution and proceeds in a controlled manner on the substrate's surface. The deposit itself continues to catalyze the reduction reaction so that the deposition process becomes self-sustaining or autocatalytic. These features permit the deposition of relatively thick deposits. Thus the process is differentiated from other types of chemical reduction:

1. Simple immersion or displacement reactions in which deposition ceases when equilibrium between the coating and the solution is established (e.g., copper immersion on steel from copper sulfate solutions).
2. Homogeneous reduction where deposition occurs over all surfaces in contact with the solution (e.g., silvering-mirroring).

To prevent spontaneous reduction (decomposition), other chemicals are present; these are generally organic complexing agents and buffering agents. Other additives provide special functions as in electroplating solutions: additional stabilizers, brighteners, and stress relievers.

Bunshah gives the following as the reducing agents most widely used: sodium hypophosphite (for Ni and Ca), sodium borohydride (for Ni and Au), dimethylamineborane (or other substituted amine boranes) (for Ni, Ca, Au, Cu, and Ag), hydrazine (for Ni, Au, and Pd), and formaldehyde (for Cu).

Nickel deposits produced with hypophosphite or the boron-containing reducing agents are alloys containing the elements P or B. They are very fine polycrystalline supersaturated solid solutions or amorphous metastable alloys with hardness ranging approximately 500–650 VPN and can be precipitation-hardened, being converted to crystalline nickel and nickel phosphide (Ni_3P) or boride (Ni_3B). Maximum hardness ranging from 900 to 1100 VPN is obtained at 400°C (750°F) for 1 hr.

The phosphorus content of the deposit increases as the hypophosphite concentration increases and the pH decreases in the solution. The complexing agents in the solution influence the deposition rate (along with pH) and may also have an effect on the as-plated deposit. It appears then that the properties of the deposit may vary considerably depending on the phosphorus content which, in turn, is determined by the solution used and its operating pH.

The costs of the complexing and reducing agents used in electroless plating solutions make them noncompetitive with electroplating processes. According to Bun-

shah, the application of electroless plating is usually based on one or more of the following advantages over electroplating:

1. Deposits are very uniform without excessive buildup on corners or projections or insufficient thickness in recessed areas. Internal surfaces are also evenly coated. The uniformity is limited only by the ability of the solution to contact the surface and be replenished at the surface.
2. Deposits are usually less porous and more corrosion-resistant than electroplated deposits (of equal thickness).
3. Almost any metallic or nonmetallic, nonconducting surfaces, including polymers (plastics), ceramics, and glasses, can be plated. Those materials which are not catalytic (to the reaction) can be made catalytic by suitable sensitizing and nucleation treatments.
4. Electrical contacts are not required.
5. The deposits have unique chemical, mechanical, physical, and magnetic properties.

The disadvantages compared to electroplating include:

1. Solution instability
2. More expensive
3. Slower deposition rates
4. Frequent replacement of tanks or liners
5. Greater and more frequent control for reproducible deposits

However, electroless has been successfully industrialized and more and more electroless shops can be seen.

D. Biomimetic

Biomedical implants used in the human body as permanent or quasi-permanent replacements of bones or dental pieces must be manufactured with biocompatible materials, that is, materials which do not prompt negative body responses. Some metals have been used to undertake the structural needs of replacement of bones, dental pieces, and similar applications. Dental prosthesis, artificial femurs, and backbone spacers are probably the most common application of metals in implants. These metals are usually very expensive to produce; thus it seems more attractive to use less expensive metals coated by biocompatible materials. Also, alloys like Ti_4Al_6V are fairly inert in the human body; these do not attach promptly to a host bond unless previously coated with hydroxyapatite (HA).

HA is the mineral matter that composes our bones and is nothing more than the hydrated tricalcium phosphate $[(Ca_{10}(PO_4)_6(OH)_2]$. Plasma techniques and ionic ablation have been tried to coat implants with HA, but these techniques generate temperatures higher than 1100°C, decomposing the HA, thus producing similar to the HA, but soluble in the body fluids [41].

The so-called biomimetic method is a wet method, which means the process of coating is made in liquid media based on a process of nucleation and growth of the apatite on the substrate [42]. It seems to be one of the simplest techniques to be applied because it only requires a solution with the ionic concentration similar to those of the human plasma, the substrate to be coated, and a bioglass that is a glass composed of $CaO–SiO_2–P_2O_5$. Using the biomimetic method, coatings have been generated on varied substrates (metallic, polymeric, and ceramic), and it has been demonstrated using x-ray diffraction and Fourier transform infrared reflective spectroscopy (FTIR) that the film generated is conformed by carbonated apatite similar to that found in the bones [43] This technique is based fundamentally in immersing the substrate into the ionic solution together with granulated bioglass; the hydrated silicate provided the preferential sites for nucleation of the HA. Several compositions of glasses have been tried showing that the SiO_2 did not generate the apatite film [44]. Moreover, it has been demonstrated that at high temperatures of the media, the speed of deposition increases [45]. The growth stage of the apatite is realized in solutions except of glass because the continuous growth continues from the HA nuclei that were present.

VII. REACTION–CONVERSION PROCESSES

A. Diamond Layer Formation on the Substrate's Surface

Diamond surfacing could be obtained by different methods, among which plasma-assisted reactions are well known, including also CVD.

Diamond is the hardest known material, has the lowest coefficient of thermal expansion, is chemically inert and wear-resistant, offers low friction, has high thermal conductivity, is electrically insulating, and is optically transparent from the ultraviolet (UV) to the far infrared (IR).

World interest in diamond has been further increased by the much more recent discovery that it is possible to produce polycrystalline diamond films, or coatings, by a wide variety of chemical vapor deposition (CVD) techniques using, as process gases, nothing more exotic than a hydrocarbon gas (typically methane) in an excess of hydrogen. This CVD diamond can show mechanical, tribological, and even electronic properties comparable to those of natural diamond. There is currently much optimism that it will prove possible to scale CVD methods to the extent that they will provide an economically viable alternative to the traditional HPHT methods for producing diamond abrasives and heat sinks, while the possibility of coating large surface areas with a continuous film of diamond will open up whole new ranges of potential application for the CVD methods.

1. The CVD Diamond Process

All CVD techniques for producing diamond films require a means of activating gas-phase carbon-containing precursor molecules. This generally involves thermal (e.g., hot filament) or plasma (d.c., rf, or microwave) activation or use of a combustion flame (oxyacetylene or plasma torches). While each method differs in detail, they all share features in common. For example, the growth of diamond (rather than deposition of other less well-defined forms of carbon) normally requires that the substrate be maintained at a temperature in the range 1000–1400 K and that the precursor gas be diluted in an excess of hydrogen (typical CH_4 mixing ratio ≈ 1–2 vol.%).

The resulting films are polycrystalline with a morphology that is sensitive to the precise growth conditions (see later). Growth rates for the various deposition processes vary considerably, and it is usually found that higher growth rates can be achieved only at the expense of a corresponding loss of film quality. "Quality" here is a subjective concept. It is taken to imply some measure of factors such as the ratio of sp^3 (diamond) to sp^2-bonded (graphite) carbon in the sample, the composition (e.g., CC vs. CH bond content), and the crystallinity. In general, combustion methods deposit diamond at high rates (typically 100–1000 µm/hr, respectively), but often only over very small, localized areas and with poor process control leading to poor quality films. In contrast, the hot filament and plasma methods have much slower growth rates (0.1–10 µm/hr), but produce high-quality films. One of the great challenges facing researchers in CVD diamond technology is to increase the growth rates to economically viable rates (hundreds of micrometers per hour or even millimeters per hour) without compromising film quality. Progress is being made using microwave deposition reactors since the deposition rate has been found to scale approximately linearly with applied microwave power. Currently, the typical power rating for a microwave reactor is 5 kW approximately, but the next generation of such reactors has power ratings up to 50–80 kW. This gives a much more realistic deposition rate for the diamond, but for a much greater cost of course.

Thermodynamically, graphite, not diamond, is the stable form of solid carbon at ambient pressures and temperatures. The fact that diamond films can be formed by CVD techniques is inextricably linked to the presence of hydrogen atoms, which are generated as a result of the gas being "activated" either thermally or via electron bombardment. These H atoms are believed to play a number of crucial roles in the CVD process:

- They undergo H abstraction reactions with stable gas-phase hydrocarbon molecules, producing highly reactive carbon-containing radical species. This is important since stable hydrocarbon molecules do not react to cause diamond growth. The reactive radicals, especially methyl, CH_3, can diffuse to the substrate surface and react, forming the C–C bond necessary to propagate the diamond lattice.
- H atoms terminate the "dangling" carbon bonds on the growing diamond surface and prevent them from cross-linking, thereby reconstructing to a graphite-like surface.
- Atomic hydrogen etches both diamond and graphite, but, under typical CVD conditions, the rate of diamond growth exceeds its etch rate, while for other forms of carbon (graphite, for example), the converse is true. This is believed to be the basis for the preferential deposition of diamond rather than graphite.

One major problem that is receiving much attention is the mechanism of heteroepitaxial growth, that is, the initial stages by which diamond nucleates upon a nondiamond substrate. Several studies have shown that pre-abrasion of nondiamond substrate reduces the induction time for nucleation and increases the density of nucleation sites. Enhanced growth rates inevitably follow since formation of a continuous diamond film is essentially a process of crystallization, proceeding via nucleation, followed by 3-D growth of the various microcrystallites to the point where they eventually coalesce.

The abrasion process is usually carried out by polishing the substrate with an abrasive grit, usually diamond powder of 0.1–10 μm particle size, either mechanically or by ultrasonic agitation. Whatever the abrasion method is, however, the need to damage the surface in such a poorly defined manner prior to deposition may severely inhibit the use of CVD diamond for applications in, say, the electronics industry, where circuit geometries are frequently on a submicron scale. This worry has led to a search for more controllable methods of enhancing nucleation, such as ion bombardment. This is often performed in a microwave deposition reactor by simply adding a negative bias of a few hundred volts to the substrate and allowing the ions to damage the surface, implant into the lattice, and form a carbide interlayer.

Obviously, the crystalline morphology of a CVD diamond film is an important consideration when it comes to potential applications. A film might find use as a fine abrasive coating, but most of the envisaged uses for diamond films in optics, in thermal management applications, and as possible electronic devices require that the film surfaces be as smooth as possible. One can envisage (at least) two routes to this objective: one has to either identify growth conditions which naturally result in the formation of smooth films or to optimize ways of "polishing" away the surface roughness of the film as grown. Both concepts are presently the subject of intense research effort.

2. The Substrate

Most of the CVD diamond films reported to date have been grown on single crystal silicon wafers, but this is by no means the only possible substrate material. What are the properties required of a substrate if it is to be suitable for supporting an adherent film of CVD diamond? One requirement is obvious. The substrate must have a melting point (at the process pressure) higher than the temperature window (1000–1400 K) required for diamond growth. This precludes the use of existing CVD techniques to diamond-coat plastics or low melting metals like aluminum. It is also helpful, although not essential, that the substrate be capable of forming carbide. CVD of diamond on nondiamond substrates will usually involve initial formation of a carbide interfacial layer upon which the diamond then grows. Somewhat paradoxically, it is difficult to grow on materials with which carbon is "too reactive," i.e., many of the transition metals (e.g., iron, cobalt, etc.) with which carbon exhibits a high mutual solubility, and hence the appeal of substrates like Si, Mo, and W materials which form carbides, but only as a localized interfacial layer because of their modest mutual solubility with carbon under typical CVD process conditions. The carbide layer can be pictured as the "glue" which promotes growth of the CVD diamond and aids its adhesion by (partial) relief of stresses at the interface.

3. Present Applications and Future Prospects

A number of areas of application of CVD diamond are gradually beginning to appear.

1. Thermal management: Natural diamond has a thermal conductivity roughly four times superior to that of copper, and it is an electrical insulator. It should therefore come as a little surprise to learn that CVD diamond is now being marketed as a heat sink for laser diodes and for small microwave-integrated circuits. The natural extrapolation of this use in circuit fabrication ought to be higher speed operation since active devices mounted on diamond can be packed more tightly without overheating.

2. Cutting tools: CVD diamond is also finding applications as an abrasive and as a coating on cutting tool inserts. CVD diamond-coated drill bits, reamers, countersinks, etc. are now commercially available for machining nonferrous metals, plastics, and composite materials. Initial tests indicate that such CVD diamond-coated tools have a longer life, cut faster, and provide a better finish than conventional tungsten carbide tool bits.

3. Wear-resistant coatings: In both the previous applications, CVD diamond is performing a task that could have been fulfilled equally well by natural diamond if economics were not a consideration. However, there are many other applications at, or very close to, the marketplace where CVD diamond offers wholly new opportunities. Wear-resistant coatings are one such use. The ability to protect mechanical parts with an ultra-hard coating, in, for example, gearboxes, engines, and transmissions, may allow greatly increased lifetimes of components with reduced lubrication.

The phrase "nonferrous" is worth emphasizing here since it reminds us of one of the biggest outstanding challenges in the application of diamond film technology—whether as a wear-resistant coating or as a fine abrasive. In any application where friction is important, the diamond-coated tool bit will heat up and, in the case of ferrous materials (be it the tool substrate or the workpiece), the diamond coating will ultimately react with the iron and dissolve. Intense research efforts into suitable barrier layer materials to allow diamond coating of iron and steel machine parts are currently underway.

4. Optics: Because of its optical properties, diamond is beginning to find uses in optical components, particularly as protective coatings for infrared (IR) optics in harsh environments. Most IR windows currently in use are made from materials such as ZnS, ZnSe, and Ge, which, while having excellent IR transmission characteristics, suffer the disadvantage of being brittle and easily damaged. A thin protective barrier of CVD diamond may provide the answer, although it is more likely that future IR windows will be made from free-standing diamond films grown to a thickness of a few millimeters using improved high growth-rate techniques.

However, a major consideration when using polycrystalline CVD diamond films for optics is the flatness of the surface since roughness causes attenuation and scattering of the transmitted IR signal, with subsequent loss of image resolution, and hence the current interest in techniques for smoothing diamond films we mentioned very briefly earlier.

5. Electronic devices: The possibility of doping diamond and so changing it from being an insulator into a semiconductor opens up a whole range of potential electronic applications.

B. Oxide Coating

Chrome-rich steel could be coated by a chrome layer if it is exposed to an oxidizing atmosphere at high temperatures. Equally, Cu–Steel will form oxide and aluminum and also nickel alloys.

VIII. PHYSICAL PROCESSES

A. Thermal Spray

The thermal spray coating process is by far the most versatile modern surfacing method with regard to economics, range of materials, and scope of applications. The thermal spray process permits rapid application of high-performance materials in thicknesses from a few mils to more than 1 in. (25 mm) on parts of a variety of sizes and geometries. Thermal spray requires minimal base-metal preparation, can be applied in the field, and is a low-temperature ($>95\,^{\circ}C$ or $>200\,^{\circ}F$) method compared with techniques such as weld overlay. Typical part configurations include piston rings, journals, conveyors, shifter forks, extrusion compartments, and suspension bridges [46].

Thermal spraying reduces wear and corrosion and greatly prolongs part service life by allowing the use of a high-performance coating material over a low-cost base metal.

More than 200 coating materials with different characteristics of toughness, coefficient of friction, hardness, and other properties are available.

All thermal spray processes rely on three basic operational mechanisms:

- Heating a coating material in either wire or powder form to molten or plastic state
- Propulsion of particles of the heated material
- Impact of the material onto a workpiece whereby the particles rapidly solidify and adhere both to one another and to the substrate to form a dense, functional, protective coating

The particles bond to the substrate mechanically and, in some cases, metallurgically. Particle velocity, substrate coarseness (roughness) and cleanliness, particle size, material chemistry, particle temperature, and substrate temperature influence the bond strength of the coating material. The process was originally referred to as flame spraying, metal spraying, flame plating, or metallizing when it was limited to the oxygen–fuel (oxyfuel) wire spray method.

1. Thermal Spray Processes

Currently, five different commercially available thermal spray methods are in use

- Oxyacetylene or oxyfuel wire (OFW) spray
- Electric arc wire (EW) spray
- Oxyacetylene or oxyfuel powder (OFP) spray
- Plasma arc (PA) powder spray (transferred and nontransferred plasma)
- High-velocity oxyfuel (HVOF) powder spray

Some of these processes are complemented or aided by laser and even plasma to increase its adherence and/or compaction.

The *oxyacetylene or oxyfuel wire spray process* (also called wire flame spraying or the combustion wire process) is the oldest of the thermal spray coating methods and among the lowest in capital investment. Originally, low melting temperature materials were solely employed like zinc. The process utilizes an oxygen and mostly acetylene gas flame as heating source and coating material in wire form. Lately, solid rod feedstock has also been used. During operation, the wire is drawn into the flame by drive rolls that are powered by an adjustable air turbine or electric motor. The tip of the wire is partially melted as it enters the flame and is atomized into particles by a surrounding jet of compressed air and propelled to the workpiece.

Spray rates for this process range from 2.3 to 55 kg/hr (5 to 120 lb/hr) and are dictated by the melting point of the material and the choice of fuel gas. The wire spray gun is most commonly used as a handheld device for on-site application, although an electric motor-driven gun is recommended for fixed-mounted use in high-volume, repetitive production work. It is fully portable, thus very appropriate for coating exteriors in large structures.

Aluminum and zinc, among other metals, are often projected by the OFW process for corrosion protection of large outdoor structures, such as bridges and storage tanks, and for restoration of dimension to worn machinery components, where compatible alloys are applied (i.e., steels). It is a good choice for all-purpose spraying. Layers can be applied rapidly and at low cost. A wide variety of metal coating materials include austenitic and martensitic stainless steels, nickel aluminide, nickel chromium alloy, bronze, monel, Babbitt, aluminum, zinc, and molybdenum.

Electric arc wire spraying or spark gun is similar to the OFW and also applies coatings of selected metals in wire form, but using instead an electrical spark as energy source for melting. Two electrically charged wires with opposite polarity are fed by push–pull motors to produce a spark at the gun head. The spark created melts the wires at temperatures above 5500°C (10,000°F). A jet of compressed air atomizes the molten metal and projects it onto the surface duly prepared of the workpiece, which is mechanically turned or displaced to expose it evenly.

It is an excellent process for applications that require thick coating buildup (layer by layer) or that have large surfaces to be sprayed. The arc system can produce a relatively wide spray zone, ranging from 50 to 300 mm (2 to 12 in.), and can spray at high speeds. It is adaptable, allowing the change in parameters like voltages, distances, and fluxes, thus tailoring coating characteristics, such as toughness, hardness, or surface texture, to specific requirements.

The spark gun method produces strongly bond coating due to the high particle temperature produced. As the process uses only electricity and compressed air, it is highly portable, relatively easily transported from one site to another, and eliminates the need to stock oxygen and fuel gas supplies which is always cumbersome. Almost the same materials applied by the EAW process can be applied by the OFW process; however, consumables are hardly exchangeable.

Powder spray by oxyfuel method or oxyacetylene powder spray extends the range of available coatings' subsequent applications to include ceramics, cermets, carbides, and fusible hardfacing coatings. The powder is either fed by gravity from a canister on top of the gun, just into the flame or using pressurized gas (nitrogen, oxygen, or air), to feed the powder into the gun barrel to be carried to the nozzle, where it is melted and projected by the hot gas onto a prepared surface. High-quality and/or high spray rates require pressurized feed system. The temperatures achieved at the nozzle range from 1500°C to 2800°C, but at the workpiece, the temperature can be kept below 200°C, with proper air cooling; thus no HAZ can be expected. Probably, these powder guns are the easiest to set up and change the coating materials. Because of that, and their simple overall handling, OFP method finds the widest acceptability and is used in short-run machinery maintenance work, original parts manufacturing (i.e., oil pumps), and in the production spraying of abradable clearance-control seals for gas turbine engines [47].

Plasma arc powder could be considered the most sophisticated and versatile thermal spray method. Commercial plasma equipment could reach temperatures at the gun as high as 11,000°C (20,000°F) which are far above the melting point or even the vaporization point of any known material. However, only a small zone achieves this temperature, and decomposition of materials during spraying is minimized because of the high gas velocities produced by the plasma, which results in extremely short residence time in the small high-temperature zone. As the plasmogenic gas is usually inert (argon or nitrogen) or a reducing gas (hydrogen), the plasma process also provides a controlled atmosphere for melting and transport of the coating material, thus minimizing oxidation, and the high gas velocities produce coatings of high density due to the kinetic energy at impact.

At the plasma gun, the temperature is raised by passing a gas (plasmogenic gas) through an electric arc. The arc could be formed in the gun itself (nontransferred plasma) or between the gun and the workpiece (transferred plasma). The gas is decomposed into atoms (if molecular) and ionized by electron bombardment. The release of energy in returning the gas to its ground state results in exceedingly high temperatures. Usually, a tungsten cathode is used and a copper anode (sometimes internally cladded with tungsten for temperature endurance) makes up the nozzle of the nontransferred plasma gun. In transferred plasma, the cathode is only required, as the anode is the workpiece. Mainly the anode is cooled by a constant flow of water through internal passages in the nontransferred plasma gun. The powdered coating material, transported by

gas, is injected directly into the plasma outside the gun to avoid erosion of the electrodes, where it is melted and propelled at high velocity to the workpiece. In practice, argon is used as primary plasmogenic gas, and a small amount of a secondary gas, such as hydrogen or helium, is mixed with the primary plasma gas to increase operating voltage and thermal energy, as the molecular recombination of these bimolecular gases is highly exothermic.

Thanks to the high temperatures and high gas velocities produced by the plasma process, the resulting coatings are dense and well attached to the substrate, thus superior in mechanical and metallurgical properties to low-velocity OFW or OFP coatings. The nontransferred plasma process is particularly efficient for spraying high-quality coatings of ceramics materials, such as zirconium oxide for turbine engine combustors and chromium oxide for printing rolls. The nontransferred plasma process is also readily field-portable, and the gun-to-piece distance is not too critical (a few centimeters). It is thus used for large on-site applications such as powder utility plant boiler tubes. While the transferred plasma process is less versatile, but produces a porousless coating with a continuous interface between coating and substrate, however, the gun-to-piece distance is critical (a few millimeters), making robotics almost mandatory for this method. It is mostly applied in the reconstruction of plastic extrusion screws, metal guillotine cutting edges, and the manufacturing of pieces, where the geometry is easily followed by mechanical means.

Presently, plasma spray technology utilizes fully automatic start/stop operation and closed-loop computer control for power level, plasma gas flow, and powder flow rate. In some equipments, system problems can be diagnosed via computer. Some plasma applications require vacuum or low-pressure chambers to avoid oxidation or contamination, improving the quality of metallic or ceramic coating, reducing porosity, and increasing coating adhesion. Some, instead, are purposely applied in air or oxygen-rich atmosphere to increase porosity and avoid reduction, like zirconia partially stabilized with yttria (PSZ), for TBC applications, where porosities as high as 20% are expected [27].

The degree to which a given flame effectively melts and accelerates the powder depends on the type of coating material, the size and shape of the particles, and the residence time. Each particular coating material and gun combination has an optimum particle size. For instance, for nontransferred plasma, particle size is commonly 40 μm on the average. Particles much smaller than that will overheat, totally melt, and vaporize; much larger particles will not melt even superficially and may fall from the flame or rebound from the target or mix with the sintered particles with little or no cohesion, thus serving as crack starters.

The *high-velocity oxyfuel powder spray process* [also known as the hypervelocity oxyfuel powder spray process, the oxyfuel detonation (OFD) process, and the HVOF] represents the state of the art for thermal spray metallic coatings. Instead of heat, the HVOF process uses extremely high kinetic energy output to produce very low porosity coatings that exhibit high bond strength, fine as-sprayed surface finish, and low residual stresses.

The HVOF process operates with an oxygen–fuel mixture consisting of oxygen and either acetylene, propylene, propane, or hydrogen fuel gas, depending on coating requirements. The fuel gas is introduced into a combustion chamber, where it is thoroughly mixed with oxygen and continuously ignited by sparks. The expanded gas produced by the explosion is allowed to exhaust from the chamber through a narrow nozzle and expand to a barrel, where the powder is fed. The high-velocity gases, transporting the powder, produce uniquely characteristic multiple shock diamond patterns, which are visible in the flame, due to shock waves produced by trespassing the sound barrier. Combustion temperatures approach 2750°C (5000°F) and form a circular flame configuration. The powder particles injected into the flame axially within the barrel are uniformly heated, and powder particles are accelerated by the high-velocity gases, which typically approach a speed of 1350 m/sec (4500 ft/sec). But the interparticle cohesion and surface adhesion of the spraying coating are mainly achieved by the kinetic energy transformation into heat, upon impingement, as the residence time at high temperatures is very limited.

The low residual coating stress produced in the HVOF process allows significantly greater thickness capability than the plasma method, while providing lower porosity, lower oxide content, and higher coating adhesion. Coatings produced by the HVOF process also have much better machinability compared with other methods, and coating porosity has closely approached wrought materials, as verified by recent gas permeability testing. HVOF systems are available with closed-loop computer control and robotics capability. A new experimental process, called "cold spray," tends to fully exploit kinetic energy without any substantial heating of the powder at the gun [48].

The coating deposition rate is limited by the method used and the melting point of the coating material. Other considerations include the deposit efficiency and the target efficiency. Deposit efficiency is the quantity of coating material deposited relative to the quantity being sprayed. Target efficiency relates to the area of the spray pattern relative to the size of the part of target. Both factors influence cost.

All thermal spray coatings exhibit a degree of internal stress as a result of shrinkage from a molten state to a solid state. These stresses accumulate as the coating thickness increases and result in a shear force at the substrate interface. Ductile coating materials tend to exhibit low stress; the opposite is true for hard coating materials, such as carbides or ceramics, and also for very porous coatings. When stresses produced by rapid cooling or different thermal expansion coefficients exceed the adhesion stress, the coating can delaminate from the substrate or, more often, crack at stress concentrators near the interface (porosities, slag, and inclusions). Equipment manufacturers usually provide the practical thickness limit for each spray material.

To improve adhesion or to prevent detachment due to different expansion coefficients or interface corrosion, some applications require an intermediate coating between the functional one and the substrate, the so-called bond coatings. Bond coatings form a metallurgical bond with the substrate. They include materials that react with the substrate exothermically, such as nickel–aluminum alloys, and materials that melt at high temperatures, such as molybdenum.

These materials are frequently applied as a thin coating under a poor-bonding or high-stress topcoat to enhance the adhesion provided by grit blasting alone. Bond coat materials are typically low in internal stress, and single coats are sufficient if the physical properties meet the design specification. Selection of a particular bond coating depends on its compatibility with the spray method being used in terms of parameters such as bond strength, thermal expansion properties, related to the base metal, corrosion resistance, and oxidation resistance.

Thermal spray coatings are usually finished to dimension either by single-point machining or by grinding. Hard surface materials such as ceramics and carbides are restricted to grind finishing, typically with diamond wheels. Only the HVOF method produces coatings closest in machining and grinding behavior to wrought materials due to its very low porosity. In general, thermal spray coatings can be machined to 1 to 2 μm (40 to 80 μin.) R_a and ground to 0.25 to 0.5 μm (10 to 20 μin.) R_a. Many plasma spray ceramic coatings can be lapped down from 0.025 to 0.05 μm (1 to 2 μin.) R_a. However, both plasma and HVOF are noise pollutants and require special attention to protect the operators and the environment from noise.

B. Physical Vapor Deposition

Physical vapor deposition (PVD) is a technique that requires high vacuum to ensure a clean substrate and a large mean free path for the particles to be deposited (atoms or molecules). Then, a vacuum chamber must be used, thus limiting the number and size of the pieces to be coated. Also, the manufacturing process must be done by batches and not continuously in order to pump out the vacuum chamber after extracting the coated pieces and reloading the new batch. The electronic applications usually justify these difficulties and expenses.

PVD involves the techniques that deposit angstrom-size particles (atoms, ions, or molecules) on a substrate without chemical reactions, diffusion, or melting of the coating; thus evaporation, ion plating, ablation, and sputtering are PVD processes. These are used to deposit thin films on self-supported shapes such as printed circuits, sheets, foils, tubing, etc. The thickness of the deposits can vary from angstroms to millimeters. The application of these techniques ranges over a wide variety of applications from decorative to utilitarian, over significant segments of the engineering, chemical, nuclear, microelectronics, and related industries. Their use has been increasing at a very rapid rate since modern technology demands multiple and often conflicting sets of properties from engineering materials.

PVD technology enables depositing virtually all the inorganic materials—metals, alloys, compounds, and mixtures thereof, as well as some organic materials. The deposition rates can be varied from 10 to 750,000 Å (10^{-3} to 75 μm) per minute; the higher rates have come about in the last 20 years with the advent of electron beam-heated sources.

As mentioned before, spray processes, like oxyacetylene, plasma, HVOF, and the like, transfer particles which are drops or semimolten, sintered by pressure and heat, but usually leave porosity in the coating, which many times is undesirable; however, atomic coatings often produce epitaxially grown layers, which are without pores.

Any PVD deposit requires to synthesize the material to be deposited, usually by a transition from a condensed phase (solid or liquid) to the vapor phase or sublimation of the coating material, and, in some cases, for deposition of compounds, a reaction between the components of the compound is needed, some of which may be introduced into the chamber as a gas or vapor. The vapors or particles so produced must be transported between the source and the substrate to produce by condensation, nucleation, and growth of films, which might grow epitaxially on the substrate.

1. Advantages and Limitations of PVD

There are several advantages of PVD processes over competitive processes such as electrodeposition, CVD, and plasma spraying. They are the following [44]:

1. Extreme versatility in composition of deposit. Virtually any metal, alloy, refractory or intermetallic compound, some polymeric type materials, and their mixtures can be easily deposited. In this regard, they are superior to any other deposition process.
2. The ability to produce unusual microstructures and new crystallographic modifications, e.g., amorphous deposits.
3. The substrate temperature can be varied within very wide limits from subzero to high temperatures.
4. Ability to produce coatings or self-supported shapes at high deposition rates.
5. Deposits can have very high purity.
6. Excellent bonding to the substrate.
7. Excellent surface finish which can be equal to that of the substrate.
8. Elimination of pollutants and effluents from the process which is a very important ecological factor.

The present limitations of PVD processes are:

1. Inability to deposit polymeric materials with certain exceptions
2. Higher degree of sophistication of the processing equipment and hence a higher initial cost
3. Relatively new technology and the "mystique" associated with vacuum processes, much of which is disappearing with education and familiarity

2. Vacuum Evaporation's Theory and Mechanisms

The theory of vacuum evaporation involves thermodynamic considerations, i.e., phase transitions from which the equilibrium vapor phase pressure of materials can be derived, as well as the kinetic aspects of nucleation and growth. Both of these are of obvious importance in the evolution of the microstructure of the deposit.

The transition of solids or liquids into the gaseous state can be considered to be a macroscopic or as an atomistic phenomenon. The former is based on thermodynamics and results in an understanding of evaporation rates, source–container reactions and the accompanying effect of impurity introduction into the vapor state, changes in composition during alloy evaporation, and stability of compounds. Glang gives an excellent detailed treatment of the thermodynamic and kinetic bases of evaporation processes. He points out that the application of kinetic gas theory to interpret evaporation phenomena resulted in a specialized evaporation theory. Such well-known scientists as Hertz, Knudsen, and Langmuir were the early workers in evaporation theory. They observed deviations from ideal behavior which led to refinements in the theory to include concepts of reaction kinetics, thermodynamics, and solid state theory. From the kinetic theory of gases, the relationship between the impingement rate of gas molecules and their partial pressure p is given by:

$$\frac{dN_i}{A_w dt} = (2\pi m k T)^{-\frac{1}{2}} p \tag{4}$$

where N_i is the number of molecules striking a unit area of surface and A_w is the area of the surface.

Hertz, in 1882, first measured the evaporation rate of mercury in high vacuum and found that the evaporation rate was proportional to the difference between the equilibrium vapor pressure of mercury, p^*, at the evaporant surface and the hydrostatic pressure, p, acting on the surface, resulting from the evaporant atoms or molecules in the gas phase. Thus the evaporation rate based on the concept of the equilibrium vapor pressure (i.e., the number of atoms leaving the evaporant surface is equal to the number returning to the surface) is shown by:

$$\frac{dN_e}{A_e dt} = (2\pi m k T)^{-\frac{1}{2}} (p^* - p) \tag{5}$$

such that dN_e, the number of molecules evaporating from a surface area A_e in time dt, is equal to the impingement rate of gas molecules based on the kinetic theory of gases with the value of p^* inserted therein minus the return flux corresponding to the hydrostatic pressure p of the evaporant in the gas phase. In the above equations, m is the molecular weight, k is Boltzmann's constant, and T is the temperature in K. The maximum possible evaporation rate corresponds to the condition $p = 0$. Hertz measured the evaporation rates only about 1/10 as high as the theoretical maximum rates. The latter were subsequently measured by Knudsen in 1915. Knudsen postulated that some of the

molecules impinging on the surface were reflecting back into the gas phase rather than becoming incorporated into the liquid. As a result, there is a certain fraction $(1-\alpha_v)$ of vapor molecules which contribute to the evaporant pressure but not to the net molecular flux from the condensed phase into the vapor phase. To this end, he postulated the evaporation coefficient α_v which is defined as the ratio of the real evaporation rate in vacuum to the theoretically possible value defined by Eq. (2). This then results in the well-known Hertz–Knudsen equation [49].

$$\frac{dN_e}{A_e dt} = \alpha_v (2\pi m k T)^{-\frac{1}{2}} (p^* - p) \tag{6}$$

α_v is very dependent on the cleanliness of the evaporant surface and can range from very low values for dirty surfaces to unity for clean surfaces. In very high rate evaporation with a clean evaporant surface, it has been found that the maximum evaporation given by Eq. (2) has been exceeded by a factor of 2 to 3 for the evaporation of a light metal such as beryllium using electron beam heating. The reason for this is that the high-power input results in considerable agitation of the liquid evaporant pool resulting in a real surface area much larger than the apparent surface area.

The directionality of evaporating molecules from an evaporation source is given by the well-known cosine law. The mass deposited per unit area is:

$$\frac{dM_r(\sigma, \theta)}{dA_r} = \frac{M_e}{\pi r^2} \cos \phi \cos \theta \tag{7}$$

where M_e is the total mass evaporated.

For a point source, Eq. (4) reduces to:

$$\frac{dM_r(\sigma, \theta)}{dA_r} = \frac{M_e}{\pi r^2} \cos \phi \tag{8}$$

For a uniform deposit thickness, the point source must be located at the center of the spherical receiving surface such that r is a constant and $\cos \theta = 1$.

In high rate evaporation conditions, e.g., using a high-power electron beam heated source, the thickness distribution is steeper than with a point or small area source discussed above. This has been attributed by some authors to the existence of a virtual source of vapor located above the molten pool. On the other hand, at high power, the electron beam impact area on the surface of the molten pool is not flat but pushed down into an approximate concave spherical segment which, as Riley shows, can equally well account for the steeper thickness distribution.

IX. ELECTROCHEMICAL PROCESSES

A. Galvanizing (Electrodeposition)

Electrodeposition, also called electroplating or simply plating, has both decorative and engineering applications. The emphasis is on engineering applications and the structures and properties of deposits. It should be noted, however, that these are also of interest in decorative plating which constitutes the greatest commercial use of plated coatings, finding wide application in automotive hardware and bumpers, household articles, furniture, jewelry, and others. Since the purpose of decorative plating is to provide a durable, pleasing finish to the surfaces of manufactured articles, the corrosion characteristics of the deposits and their ability to protect the substrate are important factors. These and other deposit characteristics involved in the selection and performance of decorative coatings including hardness, wear resistance, ductility, and stress resistance are also important to engineering applications of plated coatings [50].

Engineering applications of plated coatings involve imparting special or improved properties to significant surfaces of a part or assembly and/or protecting or enhancing the function of a part in its operating environment. Other applications include salvage of mismachined or worn parts and other types of reworking as well as material savings and use of less expensive materials. Special technologies such as electroless deposition, electroforming, anodizing, thin films (for electronics), magnetic coatings, and printed circuit boards have been selected for discussion as representing specific engineering applications.

1. Mechanism of Electrodeposition

Metal deposition differs from other electrochemical processes in that a new solid phase is produced. This dynamic process complicates and introduces new factors in elucidation of the mechanisms involved in the discharge of ions at the electrode surfaces. Factors determining deposition processes include:

1. The electrical double layer (~10 Å thick) and adsorption of ions at the surface some 2–3 Å away. At any electrode immersed in an electrolyte, a double layer of charges is set up in the metal and the solution ions adjacent to the surface. At solid electrode surfaces, which are usually heterogeneous, the character and constitution of this double layer may exhibit local variations, resulting in variations in the kinetics of the deposition

process. This could affect the electrocrystallization processes involved in the overall growth process.
2. The energy and geometry of solvated ions—especially those involving complex ions. All metal ions are associated with either the solvent (water) or complexed with other solution constituents either electrostatically or by coordinated covalent bonding. Desolvation energy is required in transferring the metal ion out of solution to the growing crystal lattice.
3. Polarization effects.

Thus the condition of the metal surface to be plated is a basic determining factor in the kinetics of the deposition process and the morphology and properties of the final deposit. The presence of other inorganic ions and organic additives in the double layer or adsorbed onto the surface can greatly modify the electrocrystallization and growth process.

Based on these considerations, several deposition mechanisms have been proposed:

1. The aquo or complexed metal ion is transferred or deposited as an adion (still partially bound) to a surface site. Such sites include the plane surface, edges, corners, crevices, or holes with the plane surface providing the primary sites.
2. The adion diffuses across the surface until it meets a growing edge or step where further dehydration occurs.
3. Continued transfer or diffusion steps may occur into a kink or vacancy or coordinate with other adions, accompanied by more dehydration until it is finally fully coordinated with other ions (and electrons) and becomes part of the metal being incorporated into the lattice.

Deposition of metal ions results in depletion in the solution adjacent to the surface. These ions must be replenished if the deposition process is to be continued. This replenishment or mass transport of the ions can be accomplished in three ways: (1) ionic migration, which contributes the least; (2) convection, which is the most effective, involving substantial movement of the solution by mechanical stirring, circulation, or air agitation of the solution or moving the electrodes (parts) through the solution; and (3) diffusion of ions, producing migration in the vicinity of the electrode surface itself where convection becomes negligible. The diffusion or boundary layer is defined somewhat arbitrarily as the region where the concentrations differ by 1% or more. The diffusion layer is much thicker than the electrical double layer (approximately 15,000 to 200,000 times thicker, depending on agitation). The diffusion rate may be given as

$$R = \frac{D(C_S - C_E)}{dN} \tag{5}$$

where

R = diffusion rate (moles $cm^{-2} S^{-1}$)
D = diffusion constant ($cm^2 S^{-1}$)
C_S = solution concentration (bulk concentration)
C_E = concentration at the electrode
dN = the diffusion layer thickness (also called the Nernst thickness)

On flat, smooth electrode areas, the diffusion layer is fairly uniform; however, at rough surfaces or irregularities which have a roughness profile with dimensions about equal to the diffusion layer thickness, the diffusion layer cannot follow the surface profile, being thinner at the irregularities (valleys and crevices).

The controlling parameters for the composition, structure, and properties of galvanic coatings are: the basic electrolyte composition, including the compounds supplying the metal ions (to be deposited) and the supporting ions; and the additives, commonly called addition agents (A.A.), frequently added to plating solutions to alter desirably the character of the deposit, which are usually organic or colloidal in nature, although some are soluble inorganic compounds. When additives produce a specific effect, they are descriptively called brighteners, levelers, grain refiners, stress relievers, antipitters, etc.

Small concentrations of additives produce profound effects. In general, the effective concentration range is of the order of 10^{-4} to 10^{-2} M. The additive must be adsorbed or included in the deposit in order to exert its effect, and thus appears to be related to its role in the diffusion layer.

Presently, no generally acceptable mechanism has been devised to explain satisfactorily the brightening action of addition agents. Brightness, of course, is related to the absence of roughness on a very small scale.

Surface tension might be one of the controlling factors in attaining brightness, as it tends to minimize surface area, diminishing roughness.

The influence and effects of the operating variables depend on the solution's composition and are also interdependent, exerting an influence on the structure and properties of the deposit. They are not always predictable, and establishment of optimum ranges is usually determined empirically.

The use of ultrasonic agitation in electroplating solutions, i.e., its effect on the polarization, the diffusion layer, and the properties of deposits, has received considerable interest since the 1950s.

Ultrasonic agitation has been widely employed in degreasing, cleaning, and pickling preplating operations.

The advantage included in the use of ultrasonics in electrodeposition are higher permissible current densities resulting in higher rates of deposition, suppression of hydrogen evolution in favor of metal deposition, i.e., a shift in limiting current density, improved adhesion, less porosity, smaller stress, increased brightness, and increased hardness (especially in chromium deposits).

Ultrasonic agitation influences the grain size and appears to be the most important factor, controlling most of the other property changes.

It is not impossible to maintain a plating solution free of impurities. They are mostly provided by the substances used, impure or improperly cleaned anodes and filters, other parts introduced in the bath, decomposition of addition agents, masking materials, contaminated atmosphere, impurities in water, etc. The particles suspended in the solution may become attached to the surface, resulting in rough, nodular deposits, or leave pits if they fall off, resulting in adverse effects on the integrity and corrosion resistance of the deposit. Pitting, poor covering power, poor adhesion, and harder more brittle, dark deposits are often attributed to organic contaminants. Metallic impurities could also contribute to pitting, poor throwing power, poor adhesion, lower cathode efficiency, stress and cracking, brittleness, burning, and off-color deposits. These cations may codeposit or become entrapped in the deposit, altering its structure and properties. The distribution of the impurity in the deposit may be current-density-dependent—usually more concentrated in the low current-density areas.

With very few exceptions, all plating processes require direct current (d.c.). As the common current sources are alternating current (a.c.) sources, they must be rectified. Depending on the number of rectifying elements, the type of a.c. (single- or three-phase), and the circuitry, the output wave form can be half-wave or (usually) full wave with varying percentages of ripple, ranging from 48% to less than 4%. In most plating processes, especially from complex ion type solutions, ripple may not be too significant. However, it can be a significant factor in some plating operations, notably chromium where the ripple should be low (5–10%) since higher ripple may codeposit excessive oxides and adversely affect the deposit's structure and result in dull deposits.

Periodic reverse, or PR, under the proper conditions, produces dense fine-grained, striated, leveled, and bright deposits. It has its greatest effect and applications on deposits from cyanide solutions, notably copper, permitting smooth, heavy deposits. A typical PR cycle is 15-sec plating and 3-sec deplating; the longer the deplating (reversal) cycle, the smoother is the deposit.

The extended plating time or increased current density required by PR to deposit a given thickness led to the use of interrupted d.c. employing similar cycles. The interrupted segment's function is to permit the diffusion layer to be replenished.

The preparation of metal surfaces for plating involves the modification or replacement of interfering films to provide a surface upon which deposits can be produced with satisfactory adhesion. The type and composition of the soils present as well as the composition and metallurgical condition of the substrate determine the "preparation cycle" and the materials used.

Besides cleaning, pickling or "conditioning" the surface is required. These are acid dips which neutralize and solubilize the residual alkaline films and "micro-etch" the surface. The common acid dips are either sulfuric acid (5–15% v/v) or hydrochloric acid (5% to full strength) and are satisfactory for most alloys. Where undesirable reactions or effects may occur, the acid dip should be formulated to be compatible with the substrate composition. Finally, when etching or "activating" the surface, undesirable (from the plating viewpoint) metallurgical microconstituents are removed or rendered noninterfering, e.g., silicides in aluminum alloys, nickel or chromium in stainless steels or superalloys. These steps remove or reduce oxides and other passive conditions prevalent to some surfaces.

Alloys containing high percentages of nickel and/or chromium usually have tenacious oxide or passive films which by themselves are highly protective to corrosion; however, to get them coated, this must be destroyed with strong acids or anodic etching in strong acids using 15–25% v/v or more. Sulfuric acid are usually employed at low current densities, 2.2–5.5 A/dm^2 (20–50 A/ft^2), if metal removal is desired, or at high current densities, 10–30 A/dm^2 (100–300 A/ft^2), if smut removal or oxide alteration is desired. Both current density ranges may be employed to maximize adhesion of thick deposits.

In special cases, activation may be accomplished by cathodic treatment in acid or alkaline (cyanide) solutions. Hydrogen is deposited at the surface to reduce superficial oxide films. Solution contamination must be avoided or minimized since such contamination—

especially heavy metal ions—may be code-posited as smut.

In order to stabilize the surface, when very active materials—alloys of aluminum, magnesium, or titanium—tend to oxidize or adsorb gases readily, a necessary step involving a deposit by immersion of zinc or tin, electroless coating, or modified porous oxides is required to make the surface receptive to an adherent electrodeposit [51].

Also, electrodeposition of thin coatings from specially formulated solutions called "strikes" can be considered stabilizing since it provides a new, homogeneous, virgin surface upon which subsequent deposits are plated, but these strike solutions and plating conditions are designed to be highly inefficient electrochemically. The considerable hydrogen gas evolution assists any final cleaning, reduction of oxides, and activation of the surface, while the thin deposit covers surface defects and remaining soils (smut).

X. DIFFUSIONAL PROCESSES

A. Welding

Coating by welding has been used for a long time as a reconstructive technique for warm pieces. In most cases, steel pieces are machined to remove the damage surface, including the cracks produced by wear under the surface. The machine surface is cleaned to remove any traces of grease or other contaminants left by the mechanization, and the coating is applied using an electrode or equivalent equipment. In some OPM, when large surfaces must be coated, like in the case of overcoating the interior of pressure vessels used in nuclear reactors, the carbon steel vessel is coated with a nickel alloy using submerged arc processes with feed of a band of nickel alloy making the process much less time-consuming. Welding processes tend to overheat the surface, producing a heat-affected zone (HAZ), where deleterious effects, produced by new phases, might fragilize the piece that has been coated.

B. Soldering

Soldering or brazing is also known as a coating mechanism, however, less common than welding due to the small wear resistance of the materials that could be applied by these methods and also because alternative methods are more often used (galvanization). As lower temperatures are used in brazing, a HAZ is not probable; however, brazing materials are less wear-resistant than soldered materials, and their application is reduced to reconstruction or ornamental applications. Silver, gold, and copper alloys are often used for brazing.

XI. WET COATING AND IMMERSION PROCESSES

A. Painting

Painting is one of the oldest coating methods known to men. Its applications is usually for ornamental purposes, but painting is also applied for corrosion control. The layer could be a simple impermeable coating to avoid oxygen penetration toward the surface or an active layer that reacting with the environmental corrosive substances controls the aggressivity to the coated piece. Also, the paint might stabilize mechanically the existing oxide.

The coating adheres to the substrate by both mechanical anchoring and polar attraction (including hydrogen bonding). Therefore the surface preparation to improve the adhesion of painting should include cleaning of the substrate and providing the adequate roughness for better adherence. In the case of metals and some ceramics, drying the surface for several minutes above 100°C is advised.

Metal-rich paintings containing tin (Sn) and other metals that serve as sacrificial anodes are often used.

1. Applications of Paintings

Paintings could be applied by brush or roller cylinder, like on walls or large structures, or spraying as on car bodies, boats, and plastic or metal structures. Immersion of the whole structure or piece into the painting media is used as a protective coating for corrosion control, as on car bodies and boats.

Some of these coatings are polar in nature. Thus their attachment to the coating surface could be accelerated by an electric field. That is the case of immersion coatings and some spray coatings deposited on metal surfaces.

Spraying and immersion of liquid metals are also used as means for coating. These applications, similar to paintings, are done at high temperatures, above the melting temperature of the coating material; however, their attachment to the surface is similar in nature to that of painting.

2. Surface Preparation Standards

It is useful to present the standards for surface preparation given by the National Association of Corrosion Engineers (NACE), the Steel Structures Painting Council (SSPC), and the Swedish Standards (Sa, St).

The National Association of Corrosion Engineers (NACE) presents as standards three types:

- NACE 1 White Metal Blast Cleaning
- NACE 2 Near-White Blast Cleaning
- NACE 3 Commercial Blast Cleaning

The Steel Structures Painting Council (SSPC) presents the following standards:

- SP-1 Solvent Cleaning
- SP-2 Hand Tool Cleaning
- SP-3 Power Tool Cleaning
- SP-4 Flame Cleaning
- SP-5 White Metal Blast Cleaning
- SP-6 Commercial Blast Cleaning
- SP-7 Brush-Off Blast Cleaning
- SP-8 Pickling
- SP-9 Weathering Followed by Blast Cleaning
- SP-10 Near-White Blast Cleaning

Moreover, the Swedish Standard (St, Sa) presents the following:

- St 2 Hand Tool Cleaning
- St 3 Power Tool Cleaning
- Sa 1 Brush-Off Blast Cleaning
- Sa 2 Commercial Blast Cleaning
- Sa 2 1/2 Near-White Blast Cleaning
- Sa 3 White Metal Blast Cleaning

Some of these standards are explained below:

SSPC-SP-1

Solvent cleaning—Removal of all detrimental foreign matter such as oil, grease, dirt, soil, salts, drawing and cutting compounds, and other contaminants from steel surfaces by the use of solvents, emulsions, cleaning compounds, steam, or other similar materials and methods which involve a solvent or cleaning action.

SSPC-SP-2, St 2

Hand tool cleaning—Removal of all rust scale, mill scale, loose rust, and loose paint to the degree specified by hand wire brushing, hand sanding, hand scraping, hand chipping, or other hand impact tools or by a combination of these methods. The substrate should have a faint metallic sheen and also be free of oil, grease, dust, soil, salts, and other contaminants.

SSPC-SP-3, St 3

Power tool cleaning—Removal of all rust scale, mill scale, loose paint, and loose rust to the degree specified by power wire brushes, power impact tools, power grinders, power sanders, or by a combination of these methods. The substrate should have a pronounced metallic sheen and also be free of oil, grease, dirt, soil, salts, and other contaminants. Surface should not be buffed or polished smooth.

SSPC-SP-4

Flame cleaning—Removal of all loose scale, rust, and other detrimental foreign matter by passing high-temperature, high-velocity oxyacetylene flames over the entire surface, followed by wire brushing. Surface should also be free of oil, grease, dirt, soil, salts, and other contaminants.

SSPC-SP-5, Sa 3, NACE 1

White metal blast cleaning—Removal of all mill scale, rust, rust scale, paint, or foreign matter by the use of abrasives propelled through nozzles or by centrifugal wheels. A white metal blast cleaned surface finish is defined as a surface with a gray-white, uniform metallic color, slightly roughened to form a suitable anchor pattern for coatings. The surface, when viewed without magnification, shall be free of all oil, grease, dirt, visible mill scale, rust, corrosion products, oxides, paint, or any other foreign matter.

SSPC-SP6, Sa 2, NACE 3

Commercial blast cleaning—Removal of mill scale, rust, rust scale, paint, or foreign matter by the use of abrasives propelled through nozzles or by centrifugal wheels to the degree specified. A commercial blast cleaned surface finish is defined as one from which all oil, grease, dirt, rust scale, and foreign matter have been completely removed from the surface and all rust, mill scale, and old paint have been completely removed except for slight shadows, streaks, or discolorations caused by rust stain, mill scale oxides, or slight, tight residues of paint or coating that may remain; if the surface is pitted, slight residues of rust or paint may be found in the bottom of pits; at least two-thirds of each square inch of surface area shall be free of all visible residues, and the remainder shall be limited to the light discoloration, slight staining, or tight residues mentioned above.

SSPC-SP-7, Sa 1

Brush-off blast cleaning—Removal of loose mill scale, loose rust, and loose paint to the degree hereafter specified by the impact of abrasives propelled through nozzles or by centrifugal wheels. It is not intended that the surface shall be free of all mill scale, rust, and paint. The remaining mill scale, rust, and paint should be tight and the surface should be sufficiently abraded to provide good adhesion and bonding of paint. A brush-off blast cleaned surface finish is defined as one from which all oil, grease, dirt, rust scale, loose mill scale, loose rust, and loose paint or coatings are removed completely, but tight mill scale and tightly adhered rust, paint, and coatings are permitted to remain provided that all mill scale and rust have been exposed to the abrasive blast pattern sufficiently to expose numerous flecks of the underlying metal fairly uniformly distributed over the entire surface.

SSPC-SP-8

Pickling—Removal of all mill scale, rust, and rust scale by chemical reaction, by electrolysis, or by both. It is intended that the pickled surface shall be completely free of all scale, rust, and foreign matter. Furthermore, the surface shall be free of unreacted or harmful acid or alkali or smut.

SSPC-SP-9

Weathering followed by blast cleaning—Weathering to remove all or part of the mill scale followed by one of the blast cleaning standards.

SSPC-SP-10, Sa 2-1/2. NACE 2

Near-white blast cleaning—Removal of nearly all mill scale, rust, rust scale, paint, or foreign matter by the use of abrasives propelled through nozzles or by centrifugal wheels to the degree hereafter specified. A near-white blast cleaned surface finish is defined as one from which all oil, grease, dirt, mill scale, rust, corrosion products, oxides, paint, or other foreign matter have been completely removed from the surface except for very light shadows, very slight streaks, or slight discolorations caused by rust stain, mill scale oxides, or light, tight residues of paint or coating that may remain. At least 95% of each square inch of surface area shall be free of all visible residues, and the remainder shall be limited to the light discoloration mentioned above.

These standards for surface preparation do not differ much of the cleaning procedures mentioned for other coating processes, as seen in "Surface Preparation" below.

XII. CHOOSING THE COATING PROCESS AND SYSTEM

To give a fast assessment to the user, a comparison could be made between the application system for a given coating material and substrate and the average porosity obtained, surface covered per unit time, price of area covered, energy consumption per unit time, noise produced, material waste, heat radiation or loss, local contamination, maintainability, availability of consumables, training required for operators, and initial investment on equipment. Finally, an overall pollution assessment should be done in a comparative basis.

The thermally assisted processes used for surface modification show a wide range of temperatures (37–30,000°C), as measured at the nozzle, in the case of the plasma or oxyfuel guns, and at the tip of the ray, in the case of laser. However, the substrate does not receive such a great amount of heat, as the oxyfuel or plasma gun, for instance, are kept a few centimeters away from the surface, and cooling by air can also be implemented to reduce overheating of the piece. In the case of the laser process, the scanning motion of the ray permits to dossify the heat to obtain the required results, for instance, power melting or surface melting, without

affecting the bulk. Heat, however, dispersed in the environment, is considered a pollutant and also high-temperature equipments are high-energy consumers, increase cost, and indirectly affect earth warming and the ecology. A comparison chart can be seen in Fig. 3 [52–55].

The price differences for the coating systems are shown in Fig. 4 [56]. They involve the investment price for the equipment only, excluding the operational cost. Price increases with sophistication as shown in Fig. 4. A comparison of equipment technological complexity and sophistication, which involves more training for the operators and more on-the-job control, which gives us an idea of operational costs other than consumables, is shown in Fig. 5 [57]. In such a comparison, PVD and CVD are at the top, and galvanizing and biomimetic are at the bottom.

Direct pollution of the environment, either by gas, liquid, or powder spilling, should be also compared, as in Fig. 6 [58]. The most polluting processes are the immersion, the electrodeposition, and the electroless, which produce liquid wastes that, if not treated, contaminate the water supplies. Plasma and HVOF only pollute the atmosphere with small amounts of forbidden gases (hydrogen and some fuels). PVD is usually very clean, and so is the biomimetic process.

Comparing the investment cost of the equipments with the pollution that the equipments can produce in operation, one realizes that the most expensive equip-

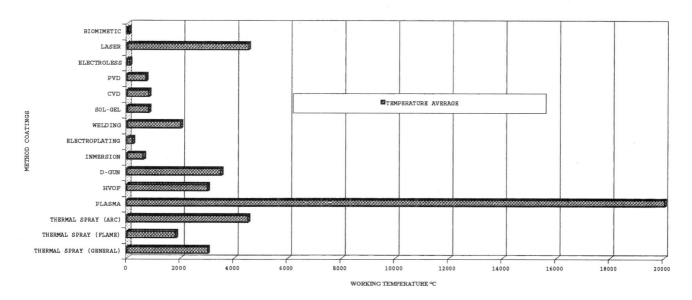

Figure 3 Comparison chart of working temperatures vs. coating method.

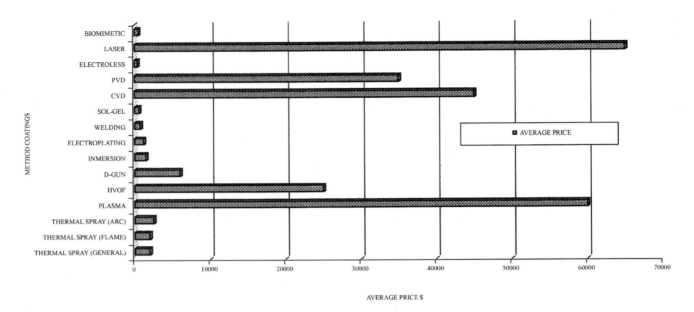

Figure 4 Comparison chart of the coating method and their prices.

ments will produce less pollution than the less expensive ones. Nevertheless, the noise pollution in equipments with high price (plasma, laser, HVOF, etc.) involves an increase in the operational cost because it is necessary to implement soundproof rooms and other isolating devices.

On the other hand, processes as D-Gun, plasma, HVOF, and PVD are not significantly contaminant other than some gas contamination.

In that realm, processes like laser, sol–gel, and bio-mimetic do not have deleterious effects, as they do not eject gases or liquids to the atmosphere during opera-

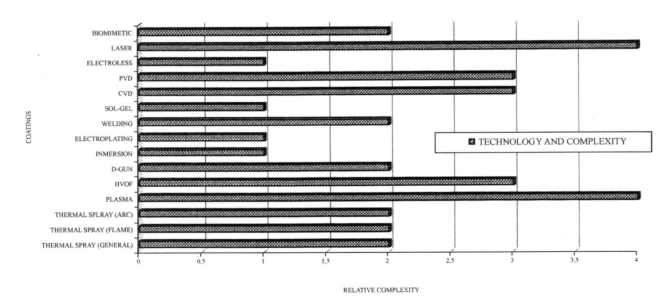

Figure 5 Comparison chart of the coating method and complexity of operation.

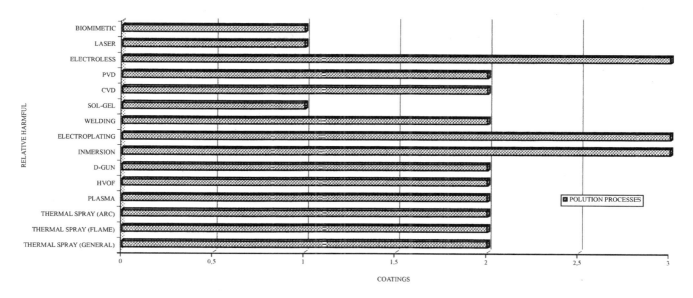

Figure 6   Comparison chart of direct pollution of environment vs. coating method.

tion; however, sol–gel and biomimetic represent a potential risk to the environment as a consequence of their liquid wastes (Fig. 7) [58].

In Fig. 5, it is shown that the largest pollutants are the immersion, the electrodeposition, and the electroless systems due to the solutions used which are highly corrosive and poisonous to plants and animals.

From Fig. 8 [59–63], it has been observed that the highest noise producers are the D-gun, plasma spray, and HVOF, while the lowest noise producers are CVD and sol–gel. However, the coatings obtained are not comparable with respect to applicability, and plasma spray could be compared to oxyacetylene spray, with the latter a lower noise producer. Energy-wise, trans-

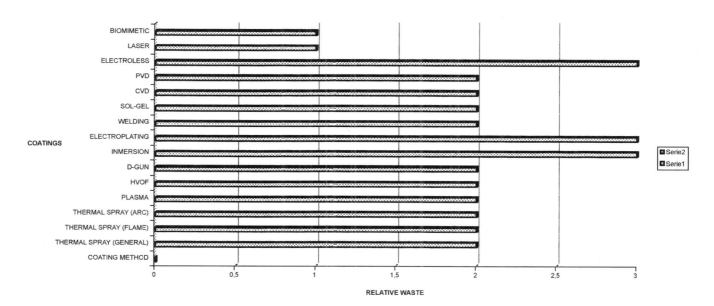

Figure 7   Comparison chart of coating method and relative amount of waste.

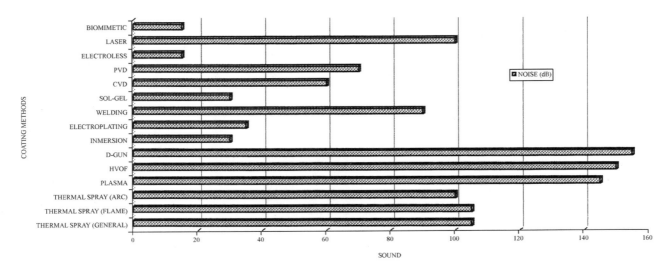

Figure 8 Comparison chart of noise production vs. coating method.

ferred plasma and nontransferred plasma as well as welding are the highest energy consumers. However, some physical deposition methods might also be high consumers due to pumping systems and power supplies, but in many cases, there are no other alternative methods.

We could summarize the parameters used to select a deposition method which depends on:

- Characterization of substrate and working conditions (shape, chemical composition, and microstructure)
- Choice of the coating material to be deposited (composition, thickness, and microstructure)
- Possible effects of process on the substrate: overheating (HAZ), deformation, and hydrogenation
- Coating speed: layers or thickness deposited by unit area and unit time
- Expected coating quality: finishing, porosity, and flaws
- Noise level (protection required for operators and environment)
- Power consumption: kilowatt per hour consumed
- Environmental pollution: minimum of wastes, gases (CO_2, monoxide, and others), hot fluids (including water), poisonous liquids, slag, ashes, and metal powders
- Transportability (if fieldwork is required): compactness and lightness of equipment
- Accessibility: easiness of approaching equipment to substrate or loading equipment

- Maintenance: availability of parts and easiness to exchange parts with a minimum of stop over time
- Versatility: should be usable in different applications and a variety of pieces and able to apply a coating in hard-to-reach segments of a piece
- Reproducibility: will provide reproducible results, without much readjusting requirements
- Investment cost: the initial investment price should be accessible to the customer and a justifiable inversion according the expected gain
- Training: should not require lengthy and costly training for the operators
- Consumables: easy to acquire in the market, reproducible quality
- Competitiveness: coated pieces should compete in price and quality with a solid piece, of same performance and coated pieces of the competition

The deposition method is then chosen according to the manufacturing needs and capabilities and, of course, the price. For example, plasma could give an excellent ceramic coating; however, for some applications, vacuum should be used, thus limiting the size of the parts to the vacuum chamber, and the positioning of the equipment in the plant must be chosen so that the noise does not bother the workers. Robotization might be required, thus increasing the investment. A similar situation is found with HVOF, which is also noisy and usually dusty due to scattering powder particles, but does not require vacuum.

A. International Standards

A relatively new constraint has been imposed to man-ufacturers in general, and those that use coatings do not escape from them: the norms ISO 9000 and ISO 14000, that is, the International Standard Organiza-tion norms, with the first referring to quality and the second referring to environment. Thus as quality covers that of the product and the working environment, the previous considerations are mandatory. By the same token, polluting techniques might prompt a reaction from the local governments, and that encompasses handling and final destiny of energy, consumables, and waste materials, including fluid, gases, powders, slag, as well as heat and noise. Some materials in themselves are considered highly contaminating and should be avoided. Among these are lead, radioactive materials, including thorium (Th), and lately, chromium (Cr), not to mention etchants like acids, cyanide, and the like.

XIII. DEPOSITION PROCEDURES

As was discussed earlier, each deposition method will have a procedure of sample preparation and coating application. However, some generic processes could be discussed at this stage, like cleaning and other accepted surface preparation methods.

A. Cleaning

It has been mentioned that cleanliness is absolutely essential in coating processes, testing, and characteriza-tion. Good practice starts with cleanliness of the coating materials. These must be free from moisture, oil, grease, dust, and other contaminants that come from normal handling and sectioning. Cleaning should be done after sectioning and before starting the mounting process. Cleanliness during polishing is essential. Care must be taken to not transfer debris from one step to subsequent steps.

Cleaning methods are generally classified as chemi-cal, mechanical, solvent, or any combination of these Chemical cleaning methods include alkaline or acid cleaning, chemical etching, and molten salt bath clean-ing. Such alkaline cleaners are nonflammable water solutions containing specially selected detergents for wetting, penetrating, emulsifying, and saponifying var-ious types of oils. Scrubbing of the sample with soap and

water using cotton swabs is a simple and routine method used to remove oil and dirt.

Hot alkaline solutions are used to remove oxides and paint films. In general, detergents can be alkaline, neutral, or acid in nature, but must be noncorrosive to the item being inspected. Acid etching is routinely used for descaling (necessary to remove oxide scale) part surfaces and for removing smeared metal.

Mechanical cleaning methods include tumbling, wet blasting. Solvent-cleaning methods include vapor degreasing which is a preferred method for removing grease-type soils from open discontinuities, solvent spraying, solvent wiping, and ultrasonic immersion. In vapor degreasing, the vapor that resides above a bath of solvent at elevated temperatures reacts chemically with the contaminants, removing them from the surface. This method is not widely used because of its chemical tox-icity, maintenance, and price. Ultrasonic cleaning is the most effective means of ensuring clean parts, but it can be a very expensive capital equipment investment. This method works by an induced ultrasonic shear wave which, upon rebound from the surface of the material, creates an implosion (collapse of a gaseous "bubble" at the surface), causing an action like cavitation erosion; a by-product of the cavitation is the cleaning of the surface. There are a variety of solvent cleaners that can be effectively utilized to dissolve organic matter as grease and oily films, waxes and sealants, paints, dry abrasive blasting, wire brushing, and high-pressure water or steam cleaning. These processes may decrease the effectiveness of the liquid penetrant examination by smearing or peening over metal surfaces and filling discontinuities open to the surface, especially for soft metals.

Drying of the specimen after cleaning can prevent many problems, such as fissuring and bubbling, due to decohesion and peeling the first, and rapid oxidation or hydrogen embrittlement, as in welded layers. Drying will generally be completed using a temperature slightly above 100°C (212°F). The substrate before coating as well as the coated surface should be duly cleaned.

B. Surface Preparation

Surface preparation may differ from surface cleaning, but we will begin by discussing cleaning. As a practical matter, a "clean surface" is one that contains no signif-icant amounts of undesirable material; thus what con-stitutes a clean surface depends on the requirements of the user. For someone studying surface crystallography, catalysts, or gas absorption, a clean surface is one that contains only a small fraction of a monolayer of foreign material (this is often called an atomically clean surface). For someone interested in joining surfaces using gross deformation in shear or high temperatures, quite a lot of contamination may be present without materially affecting the resultant bond. In some cases, such as the adhesion of deposited thin films, the presence of foreign material on the surface may be desirable or necessary. These surfaces can only be obtained through cleavage in ultrahigh vacuum (10^{-10} Torr), ion bombardment, or heating in vacuum. Adhesion is often equated with surface cleanliness; so surface treatments or the deliberate addition of foreign materials which improve adhesion may be considered to be cleaning techniques [64]. Pre-oxidized, phosphatized, and stabilized surfaces, as mentioned in galvanization, are desirable "unclean" surfaces in some cases. The reasons for cleaning may vary; it could be used to get as-received material into a uniform condition before processing or it may be necessary for processing (e.g., in solid–solid bonding or to obtain adhesion of thin films) or, as in silicon technology, for uniform diffusion. Also, cleaning may be necessary for the material to function actively such as with an electron emitter, a catalytic surface, or a gas getter. In the electronic and biomedical industries, for example, a great deal of effort was supplied to providing a "clean" environment for processing as is the case with "clean" rooms where clean only refers to particulate contamination. These rooms are provided with air conditioning, positive pressures, and "dusting-off" gates to blow off dust particles from the incoming personnel.

To design a cleaning process, we need to know the probable origins of contaminants.

Contaminants on as-received material may be deceptive since the prior history of the material is often not known. The supplier may have changed his procedures, so the type of contaminant may have changed, requiring a change in the cleaning procedure. Often, the cleaning of as-received material to put into a known condition is a good first step in the design of a cleaning process [65].

Reactions of the surface material may occur with diffused species or with contaminants in the atmosphere. As diffusion, according to Fick's law, depended on time and temperature, these parameters accelerate contamination by diffusion. Adsorption of contaminant vapors or the deposition of particulate matter from the surroundings is also time-dependent. Some process may be impossible to perform if the time between the processing steps is excessive. An example is the activation of nickel surface, which is necessary prior to the electrodeposition of gold in order to obtain an adherent film. If the activated surface is exposed to air and allowed to dry, the gold will not adhere.

If a part is left for a long time after cleaning, in a humid atmosphere, it might get too contaminated for coating.

Necessary contaminants should be recognized and cleaning steps should be designed to remove them. Unnecessary contaminants should be avoided by proper design procedures and by control over the components and cleaning materials.

The cleaning procedure may lead to an undesirable surface composition, which is produced by a high temperature, like in high vacuum baking of stainless steels which volatizes the protective chromium oxide and allows the surface to rust on exposure to the atmosphere. The surface morphology can be affected by selective etching, thermal faceting, etc., due to the substances used in the cleaning process; also, this change in morphology may affect subsequent processing.

Specific cleaning processes are designed to remove specific types of contaminants without changing the surface of interest. Examples are solvent-cleaning techniques which dissolve or emulsify contaminants without attacking the surface. Other examples are volatilization techniques where the contaminant is vaporized by heating or is reacted with a gas to form a volatile compound (e.g., the oxidation of carbon contaminants). General cleaning techniques often involve the removal of some of the surface material as well as the contaminants. Etching procedures such as sputter cleaning, electropolishing, and chemical etches fall into this category, as does mechanical abrasion. The cleaning procedure should be designed to take advantage of the strong points and weak points of each step, or the cleaning steps may defeat each other. For example, an etching step may be used to remove the surface oxide on silver, but, if it is followed by a vacuum bake, oxygen may diffuse from the bulk to the surface forming a new oxide layer. Generally, cleaning procedures should be as simple as possible.

The surface contaminants on metals may be removed by a variety of techniques, which may depend on the metals involved. As with oxide surfaces, the metal surfaces should be detergent- or solvent-cleaned to rid the surface of soluble contaminants. Ultrasonic agitation will aid in this cleaning, although cavitation may affect the surface morphology of some metals, notably aluminum [66].

Thick oxides may be removed by abrasive cleaning, etching, or fluxing. Etching may be performed by chemical etching or electropolishing, depending on the metals involved. Various fluxes may be used to dissolve or float away thick oxides. These fluxes are often molten salts. After abrasive cleaning, etching, or fluxing, care must be taken to remove residues from the surface by rinsing.

The most widely applicable cleaning technique for removing trace contamination is sputter cleaning followed by ultrahigh vacuum heating to desorb gases incorporated in the surface during the cleaning. For conductive surfaces, the applied sputtering potential may be d.c. or rf, although rf is preferred. Special sputtering configurations have been developed to sputter-clean metal surfaces in a magnetron sputtering system.

Vacuum or hydrogen firing may also be used to clean some metal surfaces. Some metal oxides have a high volatility compared with the metal. For instance, molybdenum oxide is quite volatile at high temperatures in vacuum. Gold melted on a molybdenum surface in vacuum will not wet the surface initially because of the presence of the oxide. As the oxide volatilizes, the gold will wet the molybdenum surface very well, although gold is insoluble molybdenum. Hydrogen firing will clean the oxide from some metal surfaces, but care should be taken to remove the hydrogen from the system while the metal is hot if the hydrogen is soluble in the metal at lower temperatures. A number of chemical and electrochemical cleaning techniques exist for particular metal and contaminants.

Glow discharge cleaning in an inert gas or oxygen may be used to remove hydrocarbon contamination from metal surfaces. An oxygen discharge may cause excessive oxidation of some metal which do not form passive oxides. For instance, silver will oxidize heavily in an oxygen discharge. Direct adsorption of ultraviolet radiation may cause the photodesorption of adsorbed gases. Pulsed laser bombardment has been used to clean metal surfaces in vacuum. Oxygen discharges, inert gas discharges, hydrogen discharges, and UV/O_3 cleaning have been used to clean the metal walls of ultrahigh vacuum systems.

In the semiconductor technology industry, reactive gases have been used to etch metal and dielectric surfaces to generate patterns. Recently, reactive plasma cleaning has been used to generate clean metal surfaces at a lower power input than is possible with the sputter cleaning alone.

The cleaning of gold presents a unique case. Since gold does not readily oxidize, it may be cleaned by the oxidizing treatments because it does not adhere to its oxides. These include heating in air or oxygen, plasma cleaning, chemical oxidation treatments, and UV/C_3 cleaning.

Mechanical removal may be used to clean a metal surface prior to film deposition. Grit blasting is a favor-

ite surface preparation process in thermal spraying. Wire brushing also seems to aid adhesion in some metallic coating systems.

Coddet et al. [67] describe a relatively new process, so-called Protal® process, as a substitute for degreasing and grit-blasting, comparing it with conventional processes. They describe the laser interaction with matter as absorption of the laser energy, which are photonic absorption and the inverse bremsstrahlung absorption, having as a result the excitation of electrons. The relaxation of electrons follows different paths depending on the electrical properties of the materials; insulators trap the electrons, semiconductors release the energy by radiation, and metals release energy by vibration (phonons and heat). In any case, the energy could be used for evaporation, ablation, surface fusion, and other phenomena applicable to surface preparation. It could be used for surface preparation, particularly in the case of materials sensitive to the formation of stable oxides (aluminum and titanium alloys, etc.). The objective of the laser treatment is to evaporate the grease, eliminate the oxide layer, and, in some cases, superficially melt the substrate as well as increase the surface roughness. These factors can improve the metallurgical bonding between the surface and the coating. According to Folio et al. [68], the adhesion measurements show that the new process can eliminate degreasing and grit blasting operations, obtaining similar adhesive strengths.

Folio et al. [69] assures that continuing the laser irradiation during deposition process permits the reduction of the recontamination of surfaces by dusts and condensed vapors between successive passes and the limitation of oxide layers formation, thus increasing the deposit cohesion. They conclude that if the process is performed in a single step, together with spraying, it improves the overall quality of thermal spray coatings and their global cost.

XIV. COATING CHARACTERIZATION

Coating characterization is the most important quality control activity for any industrial or laboratory metallization or protective coating. Simple observation of a coating can give some idea of possible flaws and detect major defects. Some empirical tests, such as tapping, seem to yield some useful information to any trained practitioner; however, neither of these practices guarantees or gives any valid indication of the future behavior of the coating in use. Therefore standardized methods have been developed to reduce errors and predict with some accuracy the endurance and behavior of a coating. A sequence of steps must be followed to ensure reproducibility and reliability of the tests, beginning with the cleaning of the piece or sample to be tested.

For an efficient characterization sequence, it is necessary to employ each test according to its performance. Initially, those methods that do not change significantly the original piece should be employed thereafter; the specimen could be prepared according to any special test specifications. Usually, advanced testing requires especial specimen performance, like adherence test, for which samples with standard dimensions are needed. In some cases, the specimen specifications bring several complications in its performance, especially for automatic spray processes.

All the coatings to be evaluated must be first of all assure to present a homogeneous clean surface where undesired marks, spots, cracks, and peeling are absent, and this could be made with the naked eye or with the help of a magnifying glass. Hardness testing could also be useful at this stage to determine mechanical properties of the coating. However, microcracks and pores might escape our view and they could be detected using penetrating dies or magnetic particles, among other things. Defects lying underneath the surface, like cracks produced by small coating detachments from the metal piece, are not easily detected without destroying the part. Sectioning into handy pieces is necessary for microscopic observation.

Testing the coatings will depend on its expected performance. However, the coatings used as wear protection or durability and corrosion protection are inspected for continuity (absence of flaws in the surface) and homogeneity of the surface or the cross section, using either the naked eye, magnifying lens, optical observation, or in some cases, scanning electron microscopy, and exposed to either wear or corrosion control tests when needed [1]. Indirect methods, like liquid penetrants and magnetic particles, among others, are useful to detect surface flaws.

XV. TESTING

Depending on the application, the coating must be tested using appropriate reproducible methods. If the coating has been tried already for a certain time we required to prove its reliability, few test would be needed; however, a new coatings application requires to be evaluated with more detail before it is launched into a wider-use scale.

With the evaluation of the coating, we search not only the quality of the final result, but also the quality of the process and/or the proficiency of the technicians.

It is important that the users of thermal spray, plasma spray, or other methods adopt standard tests in order to achieve consistent results. This is ever more important due to the globalization movement that requires the implementation of international norms like ISO. These are established to increment the precision and reproducibility of the final properties of the coating, given a set of parameters for a specific material and application. Thus we must consider two testing sequences, one for the evaluation of spraying coatings already in production, including their process and technical ability, and another for new developments or applications, to decide if a given material and/or process is adequate for a required application.

In the present chapter, we will concentrate on the sequence oriented to the nonexperimental applications. The tests will be presented in their performing sequence, from nondestructive tests to those that require sectioning the pieces.

A. Cleaning

One must insist that cleanliness is absolutely essential in coating testing and characterization, as it is in the coating process. Good practice starts with cleanliness of the piece or sample to be examined. This must be free from moisture, oil, grease, dust, and other contaminants that come from normal handling and sectioning. Cleaning should be done after sectioning and before the mounting process. Cleanliness during polishing is essential. Care must be taken to not transfer debris from one step to subsequent steps.

Drying of the specimen after cleaning can prevent many problems, such as fissuring and bubbling, that occur in the cold-mounting process. Drying can be done by warming the part with a heating torch, in drying ovens, by infrared lamps, forced hot air, or exposure to warm ambient temperature for long periods of time. Drying temperatures above 100°C (212°F) are most effective. Lower temperatures need longer drying times.

The substrate should be thoroughly cleaned before being coated. Coatings also need to be cleaned during metallography, as will be explained later. The next box contains a résumé of cleaning methods.

B. Nondestructive Tests

Nondestructive tests give information about the general mechanical, physical, or chemical characteristics of a coating, without deforming or damaging either the substrate or the coating. These tests are usually indirect and require interpretation based on previous experiences either with destructive tests or in the field-testing or behavior.

Note: all the tests could be made nondestructively to the piece in question, if they are ran on a test piece, witness sample, or monitoring sample that receives the same preparation than the main piece.

1. Basic Observations

Testing by visual inspection is a method of limited use, apart from meeting the requirement of aesthetic acceptability of a coated article. However, the usefulness of this method should not be ignored since it can be a rapid and comparatively cheap means of detecting failures, and it can provide much useful information in the hands of experienced inspectors.

The coated surface should be inspected with the naked eye or a low magnification lens to look for pores, cracks, and signs of spalling. If these defects or other types of discontinuity appear, they are reasons for rejection. However, if these do not appear, further tests are needed to ensure a good quality.

Gross defects such as completely uncoated areas or mechanically damaged coatings are readily detected and must be rejected, and the cause of these defects can often be seen. An uncoated area may be due to physical masking of the substrate during spraying or due to surface contamination of the substrate. Damage to coatings may often be tracked back to specific defects in handling procedures either during or after coating.

Apart from completely uncoated areas, cases where the coating thickness varies with the geometry of the component can be detected either by observation of changes of coating-surface contours relative to the shape of the part or the color and reflectivity of the surface.

Irregularities in the surface contours of a coated part might reveal defects in the substrate material or inadequate smoothing or polishing prior to coating. These substrate defects may also be revealed by blistering or debonding, although these latter defects may also be the result of inadequate precleaning or surface preparation.

Even colored, smooth, continuous surfaces are a good quality indicator.

2. Liquid Penetrant Examination

Surfaces to be coated (substrate) and the finished coatings might present undesired discontinuities, such as cracks, which are undetectable by the naked eye. Liquid penetrant examination is a nondestructive testing method for detecting discontinuities that are open to the surface, such as cracks, seams, laps, cold shuts, laminations, through leaks, or lack of fusion. The test is applicable to in-process, final, and maintenance examination. The test is useful in the examination of nonporous metallic materials, both ferrous and nonferrous, and nonmetallic materials such as densified ceramics, certain nonporous plastics, and glass. The major limitation is that liquid penetrant inspection can only detect imperfections which are open to the surface.

In comparison to magnetic particle inspection, its sensitivity is greater and it does not depend on magnetism, so it can be used on a variety of electrically conductive and nonconductive materials.

In general, penetrant inspection consists of a liquid penetrant migrating into cavities open to the surface, and once the surface has been cleaned of excess penetrant, the trapped liquid emerges from these openings. Although in some cases the amount of penetrant coming out of a surface is sufficient to be detected visually, sensitivity is vastly increased by the use of a developer, a film that enhances the liquid emerging from the flaw, as shown schematically in Fig. 1. This test includes at least five essential steps: surface preparation, penetration, removal of excess penetrant, development, and inspection.

Surface Preparation

All parts or areas must be cleaned and completely dried prior to the inspection process. The presence of any contaminants might interfere with the penetrant process.

After cleaning, it is essential that the surface of the part be thoroughly dry since liquid residues will hinder the entrance of the penetrant. Drying may be achieved by warning the part in drying ovens, with infrared lamps, forced hot air, or exposure to ambient temperature (see "Cleaning" above).

Penetration

The penetrant liquid is applied so as to form a film of penetrant over the cleaned surface. There are various ways of applying penetrants, such as dipping, brushing, flooding, or spraying. Small parts are often placed in suitable baskets and dipped into a tank of penetrant liquid. On larger parts, and those with complex geometries, penetrant can be applied effectively by brushing or spraying. Aerosol sprays are conveniently portable and suitable for on-site application. The applied film should remain on the surface long enough to allow proper penetration, as per the penetrant manufacturer's recommendations. There are two basic types of penetrants, shown as follows:

Fluorescent	Visible
Used in dark places where an ultraviolet light (commonly called a black light) is used to illuminate and show very small defects. Fluorescent indications are many times brighter than their surroundings whenviewed under black light illumination.	The visible-color contrast method, which allows inspection under white (room) light conditions, is less sensitive thanfluorescent penetrant inspection, but is widely used in industry for noncritical inspection. Red or other color lines or spots are visible where there were invisible flaws.

The cavities of interest are usually exceedingly small, often invisible to the naked eye, and for this reason, the desired sensitivity is usually the most important factor in selecting the proper type of penetrant examination system. Visible penetrant systems have only a single sensitivity, a disadvantage compared to fluorescent ones, which have different sensitivity levels, according to the size (width) of the expected imperfection, as follows: Level 1/2—very low, Level 1—low, Level 2—medium, Level 3—high, Level 4—ultrahigh. Each type of penetrant is available into three basic grades, which are selected according to the size, shape, reflectiveness of the surface and weight of workpieces, as well as the number of similar workpieces to be inspected. The three methods are broadly classified as seen in the next box.

Water-washable	Postemulsifiable (lipophilic and hydrophilic penetrants)	Solvent-removable
–The penetrant is directly water washable from the surface of the workpiece. –It can be used to process workpieces quickly and efficiently. –It is important that the washing operation is carefully controlled	–These penetrants are not directly water-washable. –They are designed to be selectively removed form the surface using a separate emulsifier. –The emulsifier combines with the excess surface penetrant to form	–In this method, excess surface penetrant can be removed by wiping until most of the penetrant has been removed. –This method is used when it is necessary to inspect only a localized area of a workpiece or to inspect a

because water-washable penetrants are susceptible to overwashing, and, as a result, the penetrant can be washed out of the discontinuities.

a water-washable mixture, which can then be rinsed from the surface.
–Proper emulsification time must be established experimentally and maintained to ensure that overemulsification does not occur, resulting in loss of discontinuity indications.

workpiece on site.
–Care should be taken to avoid the use of excess solvent, which can result in loss of discontinuity indications.

Removal of Excess Penetrant

After the required penetration time, excess penetrant must be removed from the surface. The type of penetrant used determines the removal method.

Development

A developer acts as a blotter to assist the natural seepage of the penetrant out of the openings to spread it at the edges to enhance the penetrant indication. The use of a developer is always desirable because it decreases inspection time by hastening the appearance of indications. Developer film can be applied over the surface by different modes of application such as dusting, immersing, flooding, or spraying. The choice of developer application used is determined by the size, configuration, surface condition, and numbers of parts to be inspected. Three types of developers are mentioned below.

Dry powder developers	Aqueous developers	Nonaqueous wet developers
–Used with fluorescent penetrants but should not be used with visible-dye penetrants because they do not produce a satisfactory contrast –Should be applied immediately after drying	–Used for either fluorescent or visible post-emulsifiable or solvent-removable penetrants –Supplied as a dry powder concentrate, aqueous developers have to be dispersed in water in	–Used for both fluorescent and visible penetrants. These developers yield the maximum color contrast with red visible penetrant indications and extremely brilliant fluorescent indications. –Usually supplied in a ready-to-use condition and contains particles

to ensure complete part coverage
–It is common to apply dry developers by immersing the parts in a container of the powder or in a fluidized bed.

recommended proportions. Such developers are classified as water-soluble developers and water-suspendible developers, the first having the advantage of being completely soluble and not requiring any agitation.

suspended in a mixture of volatile solvents. The solvents evaporate very rapidly at normal room temperature and do not therefore require the use of a dryer.

3. *Magnetic Particles*

The principle of magnetic particle testing can only be used with magnetic materials. An electric current applied to the test piece will generate magnetic field lines, or a flux of magnetic lines, that can attract small magnetic particles, which will position themselves following the flux lines. The magnetic flux lines are induced at right angles to the direction of the electric current producing them. At a crack on the surface of the material tested, the lines will leak. If fine iron filings are applied to the surface of the piece either as a liquid suspension or dusted as a powder, these filings will be held in the cracks by the concentration of the flux lines at that point. The process is called Magnaflux™. If the filings are made fluorescent, they can be viewed by black (ultraviolet) light very clearly. This latter process is called Magnaglow. Cracks that are perpendicular to the direction of the magnetic flux lines are more visible than those parallel to the lines.

Inspection

It should begin just after the recommended developing time. An overall inspection should be performed before the formal inspection in order to ensure that the workpiece has been properly processed, and that wet developers (aqueous and nonaqueous) are completely dry, the developer film is thin and homogeneous, and neither penetrant bleedout nor background is excessive. The more common types of flaws that can be found by penetrant inspection, together with their locations and their characteristics, are summarized in Table 1.

An experienced inspector should be able to readily determine which indications are within acceptable limits and which ones are not. A practice commonly used is to lay a flat gage of the maximum acceptable dimension of discontinuity over the indication. If the indication is not completely covered by the gage, it is not acceptable.

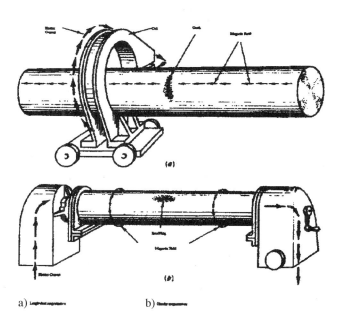

a) Longitudinal magnetization   b) Circular magnetization

There are two methods of generating flux lines:

(A) Passing a direct electric current (d.c.) through the workpiece to produce a magnetic field
(B) Introducing the piece into a coil or solenoid, so that an electric current is induced on the piece and it produces a magnetic field

Visible penetrant inspection	Fluorescent penetrant inspection
Those systems provide vivid red indications that can be seen in visible light. Visible penetrant indications can be examined in either natural or artificial light. Lighting intensity should be verified at regular intervals by the use of a suitable white light meter. The examination area must be kept free of debris, including fluorescent objects.	The inspection area should be properly darkened. Fluorescent light examinations can be performed under visible ambient light level, in which case the visible ambient light should not exceed 2-ft candles (20 lx) and under black light level control.

C. Destructive Tests

Destructive tests are those tests that require sectioning or other damaging of the part, such that it would render it unusable for its designed application.

1. Sample Preparation

Coating samples must be prepared correctly to enable them to be examined using either an optical or electronic microscope (SEM). The main aim of metallographic examinations is to examine the constituents, structure, and defects of coatings using a microscope. Because of the variations in available equipment, the wide variety of problems encountered, and the personal element in the preparation of metallographic specimens, there is little opportunity of standardization. All coatings do not respond similarly to the same procedure; some coating systems are very sensitive to variations in metallographic procedure.

The coated sample must be prepared metallographically prior to microscopic observation and for thickness measurements. This preparation, if not done correctly, can introduce defects into the samples, resulting from "pullout" or "smearing" effects, that render the subsequent analysis completely unreliable.

The following text presents those practices which experience has shown to be generally satisfactory. It does not describe all those little variations in technique, which differentiate individual metallographers and are the "tricks of the trade."

Sectioning

WHAT WOULD BE WRONG WITH SECTIONING? Problems can occur during sectioning. Care must be taken to avoid damaging coatings, debonding the coating from the substrate, or affecting its properties.

In cutting a metallographic specimen, care must be taken to avoid damaging the structure of the coating. The methods of cutting that affect most coating microstructures are sawing, abrasive cutoff wheels, and flame cutting.

Size of the specimen	Equipment and materials	Sectioning procedure
$-\frac{1}{2} \times 1$ in. (12.7–25.4 mm^2) or approximately $\frac{1}{2}$–1 in. in diameter, if the sample is circular. It should fit on	–Burning (oxyacetylene flame cutting), fracturing, hacksawing, and abrasive cutting with rotating wheels or discs	–The specimen should be held firmly by some form of gripping device to be cut or sectioned. The cutting device is

the optical microscope stage or space available in the electron microscope chamber.

–As a general rule, soft materials are cut with hard wheels, and hard materials are cut with soft wheels.
–Aluminum oxide wheels are preferred for ferrous metals, and silicon carbide wheels are preferred for nonferrous alloys.
–Abrasive cutoff wheels are essential for sectioning metals with hardness above about 35 HRC.
–Diamond cutoff wheels are more reliable to produce a clean section from a coated sample without disturbing its structure.

moved by an electrical motor at either a standard or variable speed automatically rotating the wheel or the specimen is translated to perform the cut; translation may be either manual or automatic. The specimen is usually cooled with water or cutting fluid during the cutting.

Note: sectioning the coating must begin at the outermost layer and progress inwards toward the substrate and through the substrate. Beginning at the substrate and cutting outward may damage the coating's original appearance by peeling it off or deforming the surface.

Sawing whether by hand or machine with lubrication is easy, rapid, and relatively cool. It produces a rough surface that must be removed in subsequent preparation. Using an abrasive cutoff wheel will produce a smoother surface often ready for fine grinding. This method of sectioning is normally faster than sawing.

A poor choice of cutting conditions can easily overheat the sample, producing an alteration of the microstructure. Flame cutting completely alters the structure of the metal at the flame cut edge. If it is necessary to remove the specimen from a larger piece, it should be cut sufficiently large so that it can be recut to the proper size by some less obstructive method, like the diamond wheel.

Another method of protecting coatings from sectioning damage is to first encapsulate the samples in a protective sheath or epoxy film (see "Mounting" below).

A combination of the above techniques will minimize coating damage due to sectioning: coating the specimen with a protective epoxy film, using a low-speed diamond

Table 1 Common Types, Locations, and Characteristics of Flaws or Discontinuities Revealed by Liquid Penetrant Inspection

Type	Locations	Characteristics
Relevant indications		
Shrinkage cracks	Castings (all metals)—on flat surfaces	Open
Inclusions	Castings, forgings, sheet, bar—anywhere	Tight, shallow, intermittent
Microshrinkage pores	Castings—anywhere	Spongy
Porosity	Castings, welds	Spherical
Grinding cracks	Any hard metal—ground surfaces	Tight, shallow, random
Quench cracks	Heat-treated steel	Tight to open, oxidized
Stress-corrosion cracks	Any metal	Tight to open, may show corrosion
Fatigue cracks	Any metal	Tight
Nonrelevant indications		
Burrs	Machined parts	Bleeds heavily
Nicks, dents, scratches	All parts	Visible without penetrant aids

saw, and cutting the coating beginning at the outermost layer.

After sectioning coated samples, the sample should be mounted in a suitable protective film of epoxy.

Mounting

It is not always possible to obtain coating samples of the optimum size, as samples to be examined are frequently smaller or larger than instrument sample holders or stages. Thus in the polishing of wire, strips, and other small parts, it is generally necessary to "mount" the samples because their size and shape otherwise make them hard to grip. Specimens that are too small to be handled readily during polishing (see "Sectioning" above) should be mounted to produce a surface suitable for microscopic study.

Two basic mounting procedures are used: hot and cold. Hot mounting uses considerable pressures, while cold mounting uses neither heat nor pressure. It is recommended that cold mounting materials and methods be utilized for thermally sprayed coatings to avoid damage to the specimen.

The most popular method of mounting is to embed the specimen in plastic because castable plastics or polymers such as Bakelite and epoxy fulfill these requirements and are very easy to work with, softer than most metals, and rounding of specimen edges can occur. In some cases, rounding is not important. Mounting plastics are divided into two classes: thermosetting and castable. Metallographic mounting materials should also have the ability to penetrate and fill surface-connected pores. The filling of pores with mount material will preserve their original size and shape and may minimize pullout-type damage during the grinding and polishing operation.

The final requirement for mounting materials is that their coefficient of thermal expansion be tailored to the specific materials application. For metallic coatings, epoxy-type mount materials usually adhere very well to the metal surface, whereas nonepoxy plastics such as phenolics or Bakelite do not. For the nonepoxy plastics, the difference in thermal expansion increases the tendency for the material to shrink onto the specimen during cooling from the molding temperature. Shrinkage of mount materials, in general, should be low, as coatings can be delaminated by the force generated during cooling.

Mounting procedure	Cold mounting preparation	Hot mounting preparation
—Two basic mounting procedures are used: hot and cold	—Clean the sample with soap and water or solvent and then dry it. —Adhesive is applied liberally to one end of a phenolic ring form and the surface of the ring is pressed	—Clean the sample with soap and water or solvent and then dry it. —The ram inside the central is activated by bringing a platen or pedestal base to the top The

against a flat aluminum foil. The sample is placed inside the ring form (the side to be polished against the foil) and pour the mixed epoxy (or other resin) around the sample until it is fully covered with epoxy.

–The sample is introduced into a vacuum chamber and the air and other gases are evacuated from the pores in the coating. Encapsulation is recommended for porous and brittle materials. After the epoxy is cured, the sample permanently enclosed by the ring off the aluminum foil can be pulled without damage.

specimen is placed face down (face being the surface that will be ground, polished, and viewed) on the platen.

–The thermosetting plastic is poured around the sample. The plastic is softened and maintained in this condition with pressure and heat applied for approximately 5 to 10 min and the cylinder, ram, and mount are allowed to cool. The finished mount can be ejected from the central.

vent loss of friable or loose components. The specimen is placed in a jar or container that could be evacuated, and atmospheric pressure will force the resin into the pores and holes. The resin is placed in the cup before evacuation, and the air in the sample will bubble out through the resin. Castable resins should be handled carefully because they can cause dermatitis.

Varieties of castable plastics		
Acrylic resins	Polyesters	Epoxy resins
–Consist of a powder and liquid, which will cure rapidly, over 20–30 min, to a moderate hardness. They suffer from high shrinkage porosity; some have an unpleasant odor and give off enough heat during curing to affect the coating microstructure.	–Consist of two liquids, which cure to form mounts with little heat evolution, low shrinkage, and low hardness. Curing takes 1–3 hr and the mixing ratio can be critical. These are also distinctly more expensive than acrylic resins.	–These have the best properties: transparency, moderate heat generation, negligible shrinkage, strong adhesion to the specimen, and greater hardness than other castable resins. They are expensive and take a long time, 4–8 hr to cure.

CASTABLE RESINS. The castable resins used at room temperature are generally two-component systems: a resin and a hardener, usually two liquids or a liquid and a powder. When mixed, these undergo a polymerization process, which hardens the material. The resin and hardener must be carefully measured and thoroughly mixed. Further, the mixed resin and hardener must be poured quickly while they flow easily in order to ensure penetration into pores and cavities of the coating. In order for complete filling to occur, the sample can be vacuum-impregnated, that is, gaseous spaces in the pores can be removed by placing the prepared sample (specimen in freshly poured epoxy) in a vacuum chamber. A fluorescent dye can be added at this point to ensure that the impregnated epoxy can be seen in the pores under polarized light. The mounting molds are simple cups that hold the resin until it cures. Vacuum impregnation of porous or intricate specimens will fill pores, preventing contamination and seepage, and pre-

THERMOSETTING PLASTICS. These require the use of heat (160°C) and pressure (4200 psi). The procedure consists of placing the specimen in the heated mold, face down. Resin is poured in over the specimen, the mold is closed, and pressure is applied. At the end of the cure, the pressure is relieved, the mold is opened, and the finished mount is ejected.

Varieties of thermosetting plastics		
Wood-filled phenolic resins	Di-allyphtalate resins	Filled dry epoxy resins
–Cure in 5–10 min, are relatively cheap, and can be obtained in a variety of colors and are opaque. They have a	–These are less likely to shrink and are more resistant to attach by etchants. They are more expensive than	–They adhere tightly to the sample. Commercial resins intended for metallography are usually filled with hard material,

			Porous friable areas	Bubbles in cold setting materials

tendency to pull away from the specimen leaving a space where liquid can accumulate and later seep out to stain the specimen.

phenolic resins and just as hard.

minimizing edge rounding during preparation. They are also more expensive.

What can go wrong during mounting? Defects that can occur in prepared mounts include:

Racking	Bulging	Soft mounts	Porous friable areas	Bubbles in cold setting materials
Usually occurs when the specimen is too large for the mount size or possibly the corners are too sharp. To overcome the problem, either reduce the specimen size or use a larger mount and eliminate sharp corners and burrs, wherever possible.	In phenolics, bulging is caused by insufficient curing time or insufficient pressure while the specimen is at the curing temperature.	Are due to improper mixing or curing. Soft mounts are usually caused by either an incorrect mixture of resin and hardener or by incomplete mixing when the hardener is added to the resin. To correct the problem, mix correctly.	These defects can be caused by low molding pressure, short curing times, or adding the powder to an excessively hot mold.	Are caused by stirring the hardener too vigorously with the resin, not outgassing the specimen, or not drying the specimen completely in which case all the solvents might not be removed.

Mounting time: total preparation times of less than 1 hr are now achievable.

Note: it is recommended that epoxies are mixed in a vacuum hood and steps are taken to protect eyes and skin.

In summary, when mounting specimens of thermally sprayed coatings, be careful.

- Accurately prepare and evaluate the coating sample. Rushing the procedure will likely cause difficulties later.
- Use established cold mount procedures to give superior results, but this may take longer.

Hot mounting should only be used when the coating sample is well bonded and almost fully dense.

After mounting the sample, the next steps will be grinding and polishing its surface in order to produce a flat-enough surface for microscopic inspection.

Grinding and Polishing

Grinding and polishing procedures can directly affect the results of testing and the structures that are seen under a microscope. The purpose of grinding and polishing is to produce a surface suitable for observation at both low and high magnifications.

After grinding and polishing, samples should:

1. Present an area flat enough to allow viewing the microstructure in focus (within the depth of field of the microscope), over its entire field of view without edge rounding
2. Reveal the microstructure most representative of the rest of the sample and undisturbed by preparation
3. Contain abrasive or polishing artifacts (scratches) smaller than the features of interest observable at a given magnification

GRINDING. The objective in abrasive grinding is to remove material by machining or cutting the surface away from the underlying material. Grinding consists of two steps: rough and fine.

Rough grinding (180 grit and coarser)	Fine grinding
–Used to flatten irregular or damaged cut surfaces, remove large amounts of sample material to produce a flat surface for the steps that follow. Grinding can also remove plastic mounting flash, level the mount surface, and bevel the mount edges ready for polishing.	–In fine grinding, a sample is ground on progressively finer abrasive papers using water to wash away grinding debris and to act as a lubricant. The sample should be cleaned between successive papers to prevent carryover of coarser abrasive.

a. Equipment and Materials for Grinding. Grinding operations (and polishing operations) can be carried out by hand or by automatic methods.

Hand methods consist of holding the specimen against a rotating abrasive wheel and moving the specimen in an elliptical path around the wheel against the direction of rotation. The specimen should be held firmly in contact with the wheel. Just how firm and just how fast are a matter of experience and personal preference.

Manual grinding and polishing methods should generally be avoided because the pressures and speeds are not repeatable and cannot be controlled. Every investigator has a different "touch"; therefore calibration is impossible and methods are not transferable.

Note: only automatic grinding and polishing equipment is recommended for thermal spray coatings and the like.

Some key items must be considered before beginning grinding. The first step is to select the kind of abrasive and abrasive fluid you are going to use, which depends on the coating.

b. Abrasives for Grinding. Several different types of abrasives are used in the grinding process as classified below:

Grinding stones	Abrasive papers	Diamond grinding discs
–Sometimes used for initial coarse grinding to remove damage from sectioning. –These can remove a lot of material quickly. –Damage depths using these wheels can be 170 μm or greater.	–Made with either silicon carbide or aluminum oxide abrasive particles bonded to a paper backing. Different grits of paper (fine to coarse) allow better control of process.	–The diamond is held in place by some form of coating or plating mechanism. –A problem with diamond grinding discs is that they "load" with abraded debris fairly quickly, and the loaded disc then effectively becomes a rotating press which can significantly disturb the material being ground.

Abrasive particles may be at least 2.5 times as hard as the material they are being used to cut. Silicon carbide particles have a Vickers microhardness of about 2500, aluminum oxide with approximately 2000, and synthetic diamond with hardness in excess of 8000 Vickers.

c. Fluids for Grinding. Most types of abrasive machining devices are flooded with a fluid during use. This fluid serves as a lubricant, as a coolant, and as a medium to flush abrasion debris away from the abrasion track. Water performs all of these functions well when applied in adequate doses. Materials that react with water will likely require an oil-based fluid.

d. Grinding Procedure. Grinding can be done a number of ways, ranging from rubbing the specimen on a stationary piece of abrasive paper to the use of automatic equipment. Grinding on abrasive-covered rotating disks using handheld specimens is the most common method; however, it is not advisable for coatings.

Grinding should start with the finest paper capable of flattening the specimen and removing the effects of previous steps, such as sectioning, using higher grit paper each time, until a smooth, even surface is achieved. The next paper should remove the effects of the previous paper in a short time. A sequence of papers might be 240-, 320-, 400-, and 600-grit abrasive papers (sometimes, higher grits are available). Depending on the smoothness of the sample surface, some of the coarser papers (low number) can be dismissed.

Most of the devices for automatic grinding move the specimen around a rotating wheel covered with abrasive so that the specimen follows a quasi-spiral path. In some devices, the specimen rotates on its own axis as well. The scratch pattern now consists of random arcs. Cleaning between stages is generally necessary to prevent carryover of abrasives and contamination of grinding surfaces.

After all grinding is done, the specimen must be cleaned thoroughly. Ultrasonic cleaning in a water-detergent bath is recommended. In hand operations, the hands must be washed also.

The application of a high load during the grinding of a porous coating could result in a smearing effect, the artificial filling of the voids in the coating.

e. Grinding Time. The requirement of each grinding or polishing step is to remove the volume of material, which was "damaged" by the previous step. For abrasive papers, grinding should be around 2 min per size of abrasive, changing the paper after 1 min.

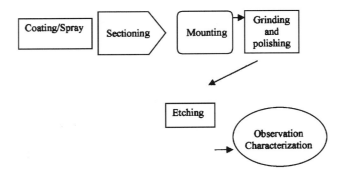

POLISHING. Polishing completes the metallographic sample preparation. The surface of the sample must be free of fine scratches and shiny for microscopic examination. The first and most important objective of polishing is to remove the abrasion-damaged layer produced by the final grinding. Each of the requirements stated previously for grinding also applies to polishing. Material removal rates are a lot lower than those for grinding. Polishing is usually distinguished from grinding by the use of loose abrasive embedded in a lubricated supporting surface. The choice of abrasive, lubricant, and surface is often specific to the coating and substrate material. The use of progressively finer-sized diamond paste as the abrasive can reduce the number of combinations. The supports most commonly used are nylon or nonwoven textiles.

For the final polish, may be a 1-μm diamond paste or, for high-precision work, the 1-μm diamond paste polish may be followed by polishing on short nap synthetic suede using an aqueous suspension of 0.05-μm gamma alumina. Other abrasives and supports are often required depending on the particular task in hand.

The final polishing is typically made using diamond suspensions or pastes or gradually decreasing grains (from 3 down to 0.25 μm) on hard surface cloths and might be finished using aluminum oxide (alumina) suspension on soft cloth. Any pores present in a final ground and polished sample should not be larger than the largest powder particle; thus if a pore of 50 μm appears when the largest powder particle was less than 30 μm, this suggests faulty preparation techniques. Careful cleaning of the specimen between stages is required to prevent contamination by coarser abrasives. Ultrasonic cleaning is recommended between each polishing step.

Abrasives	Cloths	Lubricants	Pressure
Silicon carbide, aluminum oxide, and, for a final stage, diamond abrasives either as a suspension, as an aerosol, or as a paste.	Papers, felts, woven cloths, and napped cloths.	Water-based; oil-based	The lower pressure range is adequate to polish virtually all metallographic mount types, plasma, and other thermally sprayed coatings. Ceramic materials, which easily crack during abrasion processes, can be expected to be damaged more by higher grinding pressures.

Etching

Etching is commonly done to observe microstructure in more detail. The purpose of etching is revealing the grain structure and phases present using selective etchants that attack or color-specific phases of interest. Etching is accomplished by a severe, however short, chemical, or electrochemical attack on the surface.

Most metallic specimens will not show evidence of scratches immediately after polishing and before etching. However, a significant network of scratches often appears upon etching or when viewed under polarized light, as the etchants remove the loosely attached atoms, deepening the crevices, grain borders, and other interfaces and defects, making them visible.

ETCHING PROCEDURE. After cleaning a polished surface, maintaining the sample surface free of any contaminants, and avoiding touching the finished surface, the surface is immersed for a short period of time in an etchant solution. The solutions used for each specific macroetching (see Tables 2–5) should be prepared fresh. They must present a good grade reagent but need not be chemically pure. The solution should be clean and clear, free of suspended particles. Caution must be observed in mixing, so various chemicals should be added slowly to the water or solvent while stirring. Some etchants require temperatures different from room temperature, like in an ice–brine bath surrounding the etchant container or a hot bath in some other cases. The specimen should be held on a nonreactive support. Care must be taken that the whole surface to be observed is wetted by the etchant uniformly; thus placing directly the specimen on glass rods on the bottom of the acid container could assure good etching. Agitating the solution during etching is usually needed in order to obtain a more uniform etch. The etched specimen should be immediately washed in clean running water or hot water, blow dried and protected when needed to assure an observable surface. It is important to keep in mind that light etch is better than a heavy one because overetching can often lead to misinterpretations. The tables below show some usual macroetchants to give an idea of procedures and composition.

In order to obtain more information about the etching process and macroetchants for different materials, look for the standard test methods for macroetching metals and alloys (ASTM E340-87) and microetching metals and alloys (ASTM E407-70). There is also a standard method of macroetching testing, inspection, and rating steel products, comprising bars, billets, blooms, and forgings designated as ASTM E381-79.

Table 2 Macroetchants for Aluminum and Aluminum Alloys Coatings and Substracts

Alloy	Composition		Procedure	Comments
All	NaOH	10 g	Immerse sample 5 to 15 min in solution heated to 60–70°C. Rinse in water and remove smut in strong HNO_3 solution. Rinse and repeat if necessary.	It can be used on almost all aluminum alloys; does not require fine grinding.
	H_2O	100 mL		

2. Microstructural Characterization

A reflecting optical microscope (also called a metallographic microscope) can be adequate for most of the metallographic characterization; however, a scanning electron microscope (SEM) can be very useful in many cases, especially if provided with x-ray chemical analyzer. The use of one microscope or the other will depend on the size of the samples, thickness of the coatings, and, of course, availability.

The microscopic observation of a coating has four important goals: to characterize the surface profile, coating thickness, coating substrate interface, and the microstructure of the coating and nearby substrate. Surface crevices and voids, size and morphology of deposited powder particles, grain sizes within particles, and phases and defects are all important items used to understand the structure of coatings.

Surface Profile

Carefully cutting and mounting a sample, to show the cross section of the coated material, exposes the profile of the surface, showing the roughness of the as-sprayed or finished surface. Deep crevices, microcracks, and voids, can be easily seen as hills and valleys. Deep grooves and interconnected pores opening to the surface can allow oxygen and contaminants to reach the interior of the coating—substrate interface producing chemical changes that could affect adhesion. Oxide or corrosion layers are easily detected by contrast.

Coating Thickness

One of the most important coating specifications is the coating thickness because this is often an important factor in the performance of a coating in service. There are nondestructive, semidestructive, and destructive methods for measuring coating thickness. Further information can be found in the standard guide for measuring thickness of metallic and inorganic coatings designated as ASTM B659-90.

NONDESTRUCTIVE METHODS OF MEASURING COATING THICKNESS. Nondestructive methods are mechanical, like measuring the thickness difference before and after coating with a caliper or a precision micrometer, the magnetic, eddy current, x-ray fluorescence, and beta backscatter methods, as described below:

Mechanical methods	If the piece has a simple geometry, its thickness after surface preparation can be measure with common devices like vernier calipers and micrometers. After coating, the new thickness will give, by a simple subtraction from the previous uncoated measure, the thickness of the deposited material. A variation of this method is to weigh the sample before and after thermal spraying or if the flow is constant to measure the amount of coating material used to determine the coating weight. Knowing the surface coated, the thickness can be calculated by the weight difference, which is proportional to the volume of material deposited, that is, to the area times the thickness. The density of the coating is needed to calculate the volume (density is weight divided by volume).
Magnetic methods	These methods use instruments that measure the magnetic attraction between a magnet and a coating or the substrate or both, or that

measure the magnetic flux path passing through a coating and the substrate. Coating thickness gages of this type are available commercially. There is a "Standard magnetic method to measure the thickness of nonmagnetic coatings on magnetic base metals," designated as ASTM B499-88.

Eddy-current methods
This method uses an instrument that generates a high-frequency current in a probe, inducing eddy currents near the surface of the test specimen. The magnitude of the eddy currents is a function of the relative conductivities of the coating and substrate materials and the coating thickness. Eddy-current-type coating thickness gages are available commercially. More information about this technique is given in the "Standard practice for measuring coating thickness by magnetic-field or eddy-current (electromagnetic) test methods" (ASTM E376-89).

X-ray fluorescence methods
These methods cover the use of emission and absorption x-rays for determining the thickness of metallic coatings up to about 15 μm. After exposure to x-rays, the intensity of the secondary radiation emitted by the coating or the substrate is a function of the coating thickness. More information about this method is found in ASTM B5687-91, "Standard test method for measurement of coating thickness by x-ray spectrometry."

Beta backscatter methods
The beta backscatter method employs radioisotopes that emit beta (fast electrons) radiation and a detector that measures the intensity of the beta radiation backscattered by the test specimen. Coating thickness gages of this type are available commercially. There is a "Standard test method for measurement of coating thickness by the beta backscatter method" designated as ASTM B567-91.

SEMIDESTRUCTIVE METHODS. Among the semidestructive methods, there are Coulombmetric (or coulometric) and double-beam interference microscope methods.

Coulombmetric method
Coating thickness may be determined by measuring the quantity of electricity consumed in dissolving the coating from an accurately defined area when the article is made positive in a suitable conductive liquid (electrolyte) under suitable conditions. The voltage change measured when the substrate is exposed indicates the endpoint of the dissolution. The standard test method is ASTM B504-90, "Measurement of thickness of metallic coatings by the coulometric method."

The double-beam interference microscope method
Information about this could be found at "Method for measurement of thickness of transparent or opaque coatings by double-beam interference microscope technique" (ASTM B588-88).

DESTRUCTIVE METHODS. The most common destructive methods used are the mechanical methods, gravimetric, optical microscopy, and Tooke inspector gage methods.

Mechanical method
A micrometer, or any other mechanical device that allows us to compare thickness, can be used if the coating can be partially removed or scratched, exposing the substrate or a section of the part left uncoated, by protecting it with a masking tape. Measuring directly the thickness of the coated sample and subtracting that of the uncoated region gives us the thickness of the coating. Profilometers are also used in this manner.

Gravimetric method
Weighing a sample before and after dissolving the coating without attack of the substrate or by weighing the coating after dissolving the substrate without attack of the coating gives you indirectly its thickness, as the weight is

proportional to the volume and we can measure the area coated; thus from the volume, as before, coated area times thickness, we can get the thickness. More information about this can be found in the ASTM B767-88, "Standard guide for determining mass per unit area of electrodeposited and related coatings by gravimetric and other chemical analysis procedures."

Optical microscopy method This technique is the most commonly used to measure the coating thickness. The method is useful for the direct measurement of the thickness of metallic coatings and of individual layers of composite coatings, particularly for thin layers. Knowing the magnification of your photographs, including the enlargement made, if any, allows you to measure directly from a cutting perpendicular to the surface (cross section), the thickness of the coating. Computer programs to aid in these measurements are now available (digitalized images). More information about this method is available in the "Standard test method for measurement of thickness of metallic coatings by measurement of cross section with a microscope" (ASTM B748-90).

Tooke inspector gage method This method covers the measurement of dry film thickness of coating films by microscopical observation of precision-cut angular grooves in the coating film. This technique is not recommended for excessively elastomeric or brittle films, although cooling or heating the surface can often modify the coating in the desired direction. The range of thickness measurement is 0 to 50 mils (0 to 1.3 mm). A scribe cutter of known configuration and illuminated microscope are combined as single instrument (Tooke inspector gage). The illuminated, 50-power microscope contains a reticle scaled from 0 to 100 divisions. The configuration of the tungsten carbide cutting tips shall be designated to provide a very smooth incision in the film at precise angle to the surface (see Fig. 7). For more information about this technique, there is an ASTM D4138-88, "Standard test method for measurement of dry film thickness of protective coating systems by destructive means."

Interface Characterization

Failure of a coating–substrate interface or boundary is the greatest worry of a spray technician. A dark line or a series of dark dots or spots between the coating and the substrate are a serious indication of a gap at the interface, a serious failure that results in peeling of the coating. Poor adhesion can be detected by this method (see Fig. 8). Poor preparation of the surface before deposition can leave dust, oxides, oil, or other contaminants at the interface that will hamper the sticking or adhesion of the coating to the substrate. The existence of desired roughness produced by gritting and the possibilities of an oxide film left behind in the cleaning process can also be detected. Also, the diffusion of components to or from the interface and their reaction with the media, in the case of porous coatings, can be seen due to their color contrast with the coating and substrate.

An interface observation (see Fig. 9) can be made as a preventive analysis, to check that the deposition method and surface preparation were adequate, or as a failure analysis, to determine if the case of failure of a coating was due to the interface or some other reason. Peeling of coatings is usually blamed on bad adhesion of the coating to the substrate. However, many researchers point out that these failures occur more often due to decohesion of sprayed particles than substrate–coating separation. Therefore direct observation of the interface must be done to clarify these issues.

For the observation of the interface, a cross section of the sample must be mounted in Bakelite or epoxy and polished as described in "Sample Preparation."

ADHESION. The most serious observable features are detachment of the coating from the substrate and/or high interfacial porosity.

Detachment or debonding appears as a distinct gap between the coating and the substrate. It could be partial (along some length of the interface only) or total. It must be emphasized that poor handling of the sample, especially for ceramic or composite coating materials, can also cause detachment. For this reason, care must be taken when mounting a sample in epoxy or Bakelite, as needed, before the final cut and before grinding and polishing for the microscope, as explained in "Sample Preparation." Porosity appears as very dark areas, round (spherical) in nature, and, in some cases, elongated. When a large number of pores coincide at the interface, they coalesce producing the separation of the coating.

OXIDATION/CORROSION. Bad substrate preparation, that is, poor cleaning or inadequate handling,

Table 3 Macroetchants for Iron and Steel Coatings and Substracts

Alloy	Composition		Procedure	Comments
Plain and alloy steels	$CuCl_2$ HCl (c) H_2O	45 g 180 mL 100 mL	As above.	Modified Fry's reagent. Same as for reagent no. 9 but modified by Wazau, may give more contrast; specimen can be washed in water without depositing copper.
Stainless and high-Cr steels	HCl Alcohol Picric acid	10 mL 100 mL 1 g	Immerse specimen in solution at room temperature until desired contrast is obtained. Rinse and dry.	Vilella's reagent.

can leave oxide and dirt layers that can impair adhesion of the coating to the substrate, those allowing debonding. Such layers will appear as a continuous or sometimes discontinuous lines between the substrate and the coating and are observable as a thin layer of different color. Oxides, in most cases, are brittle and, in many cases, detachment of the coating can be observed in some sections of the interface.

In coatings that have been thermally treated, that is, melted after spraying, heated by induction, or otherwise, an interlayer might appear, which could be essential for good adhesion or damaging if oxidized.

Table 4 Macroetchants for Stainless Steels and High-Temperature Alloy Coatings and Substrates

Alloy	Composition		Procedure	Comments
Stainless steels and iron-base high-temperature alloys	HCl (c)	50 mL	Immerse specimen in solution heated to 160–180°F for 30 min. Desmut by vigorous scrubbing with vegetable brush under running water. Stainless steels may be desmutted by dipping in warm 20% HNO_3 to give bright finish.	General purpose.
Iron-, cobalt-, and nickel-base high-temperature alloys.	HCl (c) HNO_3 (c) H_2O	50 mL 25 mL 25 mL	Immerse specimen in solution at room temperature for 10–30 min. Rinse and dry.	Ratio HCl + HNO_3 runs 2 + 1 to 3 + 1.
Stainless steels and high-temperature alloys.	HNO_3 HF (48%) H_2O to HNO_3 (c) HF (48%) H_2O	10 mL 3 mL 87 mL 40 mL 10 mL 50 mL	Immerse specimen in solution heated to 160–180°F until desired contrast is obtained. Rinse and dry.	Ratio HNO_3–HF varies.

Table 5 Macroetchants for Nickel and Nickel Alloy Coatings and Substracts

Alloy	Composition			Procedure	Comments
Ni	$CuSO_4$	10 g		Immerse specimen in solution at room temperature until desired contrast is obtained. Rinse and dry.	Marble's reagent, for grain structure.
	HCl	50 mL			
	H_2O	50 mL			
Alloys containing Cr, Fe, and other elements	HNO_3	50 mL		Immerse specimen in hot solution.	
	Acetic acid	50 mL		Rinse in hot water and dry.	
	Acetic acid	50 mL		Swab.	
	HNO_3	50 mL			
	Sat soln of $CuSO_4$ in H_2O	50 mL		Swab etchant.	

In order to prove the existence of compositional changes at the interface, a microprobe analysis must be done so that the new chemical species or the concentration of unexpected elements can be detected. A complete explanation of these methods is beyond the scope of this book.

DEBRIS AND OTHERS. Debris from poorly cleaned surfaces, left from atmospheric contaminants, oxides, and/or particles from the grit blasting process, can become trapped at the interface. In most cases, large amounts of debris will yield poor adhesion of the coating to substrate. Debris particles can be clearly observed with optical microscopy.

Multilayer coatings. In many applications, several layers of different compositions must be applied. The layer closest to the substrate is meant to match the expansion coefficient of the coating to that of the sub-strate, so that the differences in thermal expansion or contraction between substrate and coating are minimized. The interlayer is usually sprayed and fused to assure a good bonding. If the system is heated afterwards, either during the top layer deposition or during use, new layers of different compositions might appear, some of which could be brittle, sensitive to corrosion or oxidation, or otherwise damaging (see Fig. 10). The bond layer, when the top coating is porous, is also chosen to be corrosion-resistant. The choice of this interlayer will depend on the adhesion of the coating, and if poorly chosen, a dark continuous or segmented band at the interface will be visible showing the separation between the interlayer and coating or substrate (Fig. 9).

Layered coatings are among the latest applications of thermal-sprayed and CVD coatings. These sometimes consist of sandwiches of metallic and ceramic layers as to offer transitions between tough and hard materials to slow down cracking of a relatively thick composite layer.

When a thick coating is required, several passes of the torch will provide with slow thickening of the coating, producing in fact a multilayer coating, but of the same material. If there are no contaminants between these layers, they are stable. Oxides between the layers produced by the fusing components could be observed many times (see "Inclusions and Gross Precipitates" below), but if they are not too large, they are not dangerous.

Microstructural Characterization

Microstructural characterization is a discipline in itself. The only assurance of a good characterization is

Figure 9 Microstructure of a composite coating. Matrix in gray. Cross section (80×).

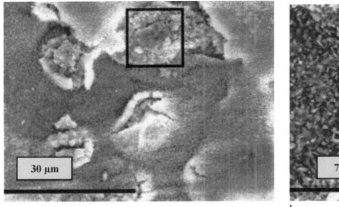

a   b

Figure 10   a) Microphotograph of dioxide Ti-Mo-Al alloy after 150 hours of continuous heating in oxygen atmosphere at 700°C. A new phase protruding through the top layer (200×). b) Detail (800×).

experience. It is advisable to compare one's samples with those of others that have been well characterized.

The microstructures of coatings differ significantly from bulk materials due to the nature of its deposition process. Frequently, powders are the starting materials for thermal spray. Powder particles usually contain several grains.

The most important feature to be observed is the shape or morphology of the as-deposited particles. If spherical particles are found, these indicate that their residence time in the spray jet was too short, so they did not melt or soften, thus arriving to the surface too hard to flatten out and cannot adhere to other particles or to the substrate. Over heating of particles can also occur if

the residence time is too long, causing splashing or particle oxidation.

The higher the temperature and/or speed of particles, the flatter the as-sprayed particles in the coating and the better the bonding. The feed powder size or the feed speed in the case of rods and wires predetermines the end size of the deposited particles.

In most applications, small particles (from 45 to 10 μm) and small grains within the particles are preferred.

GRAIN SIZES, MORPHOLOGY, AND MICROSTRUCTURE.   Most crystalline materials used are polycrystalline, that is, they are a collection of crystals or grains, which have formed independently and have different

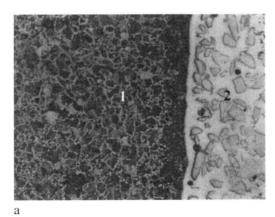
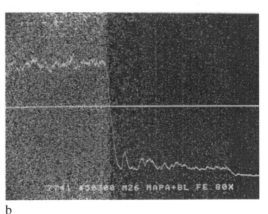

a   b

Figure 11   a) Effect of temperature after heating the deposit layer (optical microphotograph 100×). b) Profile of iron concentration from substrate to the coating, indicating diffusion (SEM 100×).

geometrical orientations of their atomic pattern. Thus grains are considered to be the smaller solid fractions or units which contain the characteristic chemical composition, phases, and crystal structure of the whole material. An as-sprayed particle might contain many grains. Within these grains, typical or characteristic defects, some are too small to be seen without a transmission electron microscope (TEM); however, they influence somewhat the behavior of the coating.

But for the purpose of quality control of a coating, more important to mechanical resistance are larger defects as pores and inclusions, and, above all, grain size and particle shape. There are some methods of estimation of average grain size, but they are not precise; however, usually comparative measurements are more important. The basic procedures for estimating grain size are (ASTM 112, "Standard Test Methods for Determining Average Grain Size"):

Comparison procedure	The borders between grains in the materials appear as a net under the microscope, similar to chicken wire or cobbles in the cement of a walkway. The grain size is the average width of these grains. It is important to measure it, as it affects the behavior of the material (hardness, resistance, and even corrosion).
Planimetric procedure	This method consists in taking a micrograph and describing a circle or a rectangle of known area on it. This area could be 5000 mm^2. It is recommended that the image contain at least 50 grains. Count the number of grains within the selected area by adding all the grains included completely within the known area plus one half the number of the grain intersected by the circumference of the area and divide by the area to get grains per square millimeter.
Intercept procedure	This method is applied particularly for structures consisting of uniform equiaxed grains. Placing a linear scale (or segment of known length) on the image field or on a photomicrograph of known dimensions, the number of grains intersecting the scale is counted. The average number of grains per unit length is thus measured, and dividing the length of the scale or segment by the number of grains, the grain size is obtained. When the grains are larger in some directions, this procedure can be used making separate size estimates in each of the three principal directions.

PHASES. A phase is a zone within a grain, which have different chemical composition or different crystalline structure from those surrounding it. These phases and their proportion in the grains, thus in the coating, control in great part the mechanical and chemical behavior of the coating. That is, the hardness, toughness, and corrosion resistance are mostly controlled by the phases present. New phases could be formed or precipitate during solidification or posterior melting of the coating. These could be large enough to be seen with a microscope. These can be better observed after chemical etching to accentuate grain borders and expose interphase boundaries or layers between coating and substrate, as mentioned before. Composition homogeneity and/or diffusion of components can be depicted with the electron microprobe (see Figs. 11 and 12). Statistical proportion of phases can also be obtained by the previously described method. The user should know which phases are acceptable in the coating application and thus search for them.

INCLUSIONS AND GROSS PRECIPITATES. Inclusions are grains that solidify before the rest of the material (matrix) does and appear optically different many times across grain borders or within grains. Precipitates are formed in the grains, or grain boundaries as new crystals, from atoms that could not dissolve in the grain.

Inclusions	Transgranular inclusions are observed, detected as estrange (different color and shape) bodies that cross over the grains. These are usually sites of crack formation and are produced by low-quality materials, thus must be avoided. Layers of oxide between the metal layers of self-fluxing alloys are easy to detect as continuous light bands, usually parallel to the surface (see Figs. 13–16). These layers are usually the sites of crack formation and could be avoided by proper heating between deposited layers (see Figs. 12 and 17). Cracks often propagate from an inclusion to another, as seen in Fig. 18.
Precipitates	Homogeneously dispersed precipitates, in general, are expected to have hardening. These cannot be seen by optical or SEM. Overheating some alloys, where precipitation hardening is expected, like aluminum alloys could produce weakening of the alloy. The aluminum coating will then reveal small darker spots like a rosary in the grain borders as evidence of overheating or "overaging," as it is called, and the coating will be softer and more prone to corrosion pitting. Microprobe analysis would detect concentrations of solute on these particles, as well as impoverishment of solute in their surrounding area.

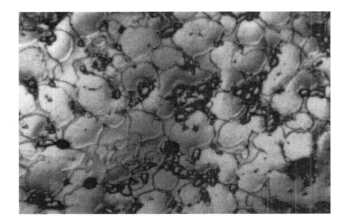

Figure 12 Two phase coating. Light matrix areas (Nickel alloy) presenting a continuous phase containing tungsten carbide segregated to the grain boundaries (dark particles).

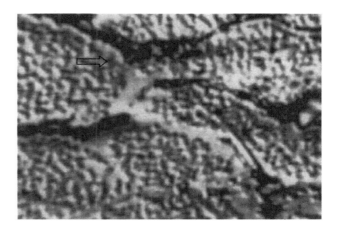

Figure 14 Detail of Fig. 13.

PORES. Pores are voids, usually found between the particles of the coating, and sometimes within the grains. Large pores between particles can be observed as dark spots; however, some small pores, which might connect to the large ones, are difficult to observe even with the SEM (see Figs. 13 and 16). The size, distribution, and morphology of the pores influence both the mechanical and chemical stability of the coating. Interconnected pores lead to corrosion (oxidation) of the interface and accumulation of pores in a zone (i.e., interface) could lead to early crack formation. Pore density can be measured and calculated, either from density experiments, porosimeters, or other devices.

Image analysis of the micrographs attained with optical or electronic microscopes are usually sufficient to measure pore density, distribution, and size. In some applications, pores are desired; thus their accepted density, size, and distribution should be measured. Among these applications are self-lubricating coatings, biomedical implants, and thermal barriers. The client or user should, in these cases, specify his requirements (see Fig. 16).

Operating Procedure

OPTICAL METALLOGRAPHY. A sample properly mounted either in an epoxy or Bakelite mount is on the stage of the optical microscope, making sure, by

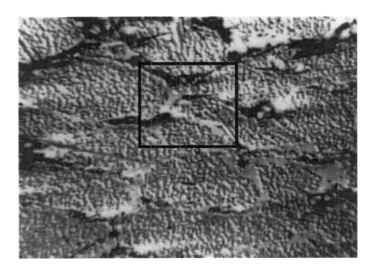

Figure 13 Glassy oxide phase between layers of a Ni-B-Si deposit.

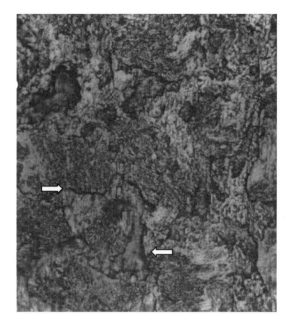

Figure 15 Crack propagating between two inclusions (see arrows).

looking carefully as it is approached to the stage, that the light beam is reflecting on it near the interface.

The sample is viewed first with the smallest objective to assure one is looking at the coating, preferably horizontal to one's eyesight. Moving the stage, the upper border can be scanned to observe the regularity of surface profile and then scan the interface to observe signs of improper adhesion. No cracks should be seen at the interface. With more magnification, the shape of the

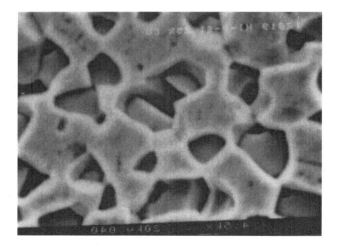

Figure 16 High porosity in as-deposit coating (Ni-B-Si), probably enhanced by etching (40 μm particles).

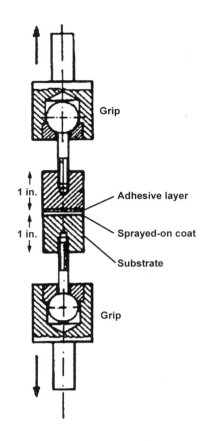

Figure 17 Pull-off adhesion test (ASTM C633-79).

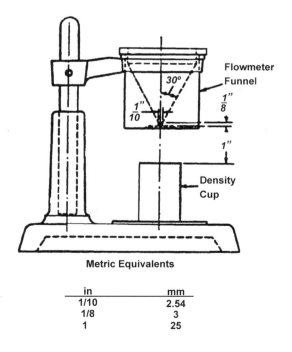

Metric Equivalents

in	mm
1/10	2.54
1/8	3
1	25

Figure 18 Flowmeter funnel (ASTM B212-89).

particles deposited can be seen. If they are mostly flatten in the direction of the surface, the coating was correctly deposited. If a large number of round particles are observed, the coating is not proper and will fail by decohesion. The same will happen if there are many pores between particles. In self-fluxing alloys, if you observe light color fringes, much larger than a particle, you probably are observing vitreous layers formed during melting of the alloy. If they appear too often (cover a large percentage of the observed surface), the coating may delaminate during performance.

Hardness Testing

Hardness can be loosely defined as a material's resistance to be penetrated by an indenter (hard-pointed implement). Hardness is one of the easiest-to-measure mechanical properties in a material, and it can be related to many other so-called "basic properties." For this reason, hardness tests are widely used as a rough guide to the mechanical strength of coatings and other materials. These tests are rapid and often "nondestructive"; therefore hardness represents an important means of quality control and in order to be nondestructive, it requires to be made on witness or monitoring sample that receives the same preparation than the main piece. There are different kinds of hardness tests which are not really equivalent; however, there are tables that allow us to transform a hardness number obtained from a method to one obtained from another method, allowing us to compare results.

Hardness tests are usually classified as (macro) hardness, measurements made with large loads (100 to 3000 kg) and large indenters, in the millimeter size (indenters), and microhardness, which uses small indenters, in the micrometer size, and small loads, not more than a few hundred to 1000 g (1 kg). These lastly mentioned tests could be considered nondestructive if one manages to perform them without cutting the piece. However, for thermal-sprayed coatings, in most cases, the piece must be cut to make samples easier to handle and subsequently prepare the smaller samples to measure the hardness at the cross section of the coating.

Microhardness Testing (Vickers and Knoop)

Wear and mechanical resistance are both associated with hardness, and phases could be qualitatively determined by hardness testing. A microhardness test produces a microindentation. Using a calibrated machine to force a diamond indenter of specific geometry, under a load of 1 to 1000 g, into the cross section of the coating and the optical measurement of the diagonal or diago-

nals of the indentation produced, enables a hardness number to be calculated.

Two types of microhardness tests will be mainly considered in this section: Knoop hardness (KH) and Vickers hardness (VH).

Knoop hardness (HN)	–Is applied on highly polished surface by a fixture attached to an optical microscope that mechanically or hydraulically pushes the indenter on the surface of the specimen. The tester uses loads of light weight to push a rhombic-based pyramidal diamond indenter into the sample. As expected, the indenter produces a rhombic scar or indentation. The long diagonal of the rhombus is measured under the microscope containing a scale in the ocular lens. The numbers obtained by dividing the applied load expressed in kilograms by the indentation mark surface, in square millimeters, are the Knoop Hardness Numbers (KHN) The KHN can be read from a table that contains the size of the diagonal in μm in the vertical and the KHN is given in a row, under the size of the indenter employed.
Vickers hardness	Differs from the Knoop in the indenter's shape, which is a highly polished, pointed square-based pyramidal diamond. Otherwise, the equipment and measurement techniques are similar.

Both tests make small indentations on the specimen, and the impression or indentation should be observed with the naked eye. The shape of the indenter and therefore of the scar are different for Knoop and Vickers tests. Other tests as Tukon tester, the shore scleroscope, and scratch hardness testing (Mohs hardness) are some of the less traditional hardness, yet useful tests.

Macrohardness Testing (Rockwell and Brinell)

In this section, the Brinell and Rockwell hardness tests will be considered. Both testing hardness machines usually contain a hydraulic press to push the indenter perpendicularly against the sample, which rests fastened to a massive platform.

In Brinell and Rockwell hardness measurements, the indentation mark is quite large and may cause damage to finished products, making this a "destructive" evaluation method in most cases.

Brinell hardness test	–The surface of the specimen is cleaned and made smooth and flat, previous to testing, using abrasive means to remove scale, oxide films, large pits, and foreign material that might affect the results.

Rockwell
hardness
test

–The specimen must be thick enough so that no bulge appears on the opposite face during penetration by the ball and should preferably be 10 times as great in thickness as the depth of the impression. For thermal-sprayed coatings, it is the thickness of the coating that counts. Otherwise, one would be measuring a composed hardness of coating and substrate. Thus soft coatings must be very thick if Brinell hardness is to be measured. Impressions should not be made within two and one-half diameters of the specimen's edge; thus BH tests are rarely applied on cross sections of the coatings, unless they are exceptionally thick. The 500-kg load should be applied for a period of at least 30 sec and the 3000-kg load for at least 15 sec.

–This test is more sensitive to irregularities than the Brinell test. A pitted surface may give erratic readings, owing to some indentations being near the edge of a previous depression. The metal around the indenter will be easily pushed to the empty space of the pit, giving a low reading. Oiled surfaces also give generally lower readings than cleaned and dry surfaces because of the reduced friction under the indenter. The thickness of the specimen should also be about 10 times larger than the depth of penetration to assure a correct reading.

To minimize the effects of surface irregularities, the initial setting (SET) of the scale is made after a minor load (10 kg) is applied to preset the indenter into the specimen (prepenetration). The scale reads from 0 to 100 in units of 0.002 mm and is graduated in a direction so that the greater the penetration, the lower the reading is.

–The load and the type of indenter (shape and material) depend on the material under study. For hard materials, a diamond (ball) indenter and 150-kg major load are ordinarily used (this is called Rockwell C scale). For softer materials, a 1/16 in. steel ball indenter and a 100-kg load are commonly used (Rockwell B scale). A number of other scales are used for specific purposes and the operators will learn them as needed. For testing surface-hardened materials, thin samples, or coatings, a superficial hardness tester uses units of 0.001 mm.

Adhesion Testing

The adhesion of any metal coating to its substrate may be seriously impaired by incorrect operation of either pretreatment or coating processes, and adhesion tests are necessary to detect such processing irregularities and to measure the bond strength of coatings.

This test measures the degree of adhesion, or bonding strength, of a coating to a substrate under a normal tension. It should not be considered to provide an intrinsic value for direct use in making calculations such as to determine if a coating will withstand specific environmental stresses. Real coating–substrate adher-

ence is determined by several parameters such as the final surface treatment, piece shape, environment, and applied stresses, including fatigue. Also, a coating in use is stressed in a much more complicated manner than is practical in a standardized test.

It is considered that "acceptable" adhesion exists if the bonding between a coating and the substrate is greater than the cohesive strength of either. For many purposes, the adhesion test has the objective of detecting any adhesion less than acceptable. Usually, any means available to attempt to separate the coating from the substrate are used. This may be prying, hammering, bending, beating, heating, sawing, grinding, pulling, scribing, chiseling, or a combination of such treatments. In most cases, the coating itself gives in before the interface coating–substrate is separated. As coatings are usually full of defects, as mentioned before, and as care was taken to ensure good bonding, the pores, inclusions, and even particle to particle bonds will be weaker than the interface bond itself. Particle separation is a common failure in ceramic coatings, when adhesion is tested, and in fatigue tests, cracks tend to be formed in the coating near the interface, but not on it.

The method or methods to be used must be specified especially when they are used for acceptance inspection. The results of tests in cases of marginal adhesion are open to interpretation; agreement shall be reached on what is acceptable.

Test conditions	Adhesion equipment
The common "pull-off" adhesion test is performed at room temperature	–A device that holds on one side the substrate and on the other side the coating is required. It must allow us to try to pull them apart by moving them apart in opposite directions. Ideally, the equipment should permit to apply increasing pull (tensile load) at a constant rate. The machine should include a load-indicating device that registers the maximum load applied before rupture occurs (see Fig. 17). The speed for loading the equipment, size, and materials used and other conditions are given in a norm (ASTM C 633-79) for adhesive and cohesive strength.
	–For these tests, two solid cylinders of equal diameter and length at least double than the diameter are machined, so that the circular face of one cylinder, made of the substrate material, is prepared and coated. The other cylinder is glued by strong glue (bonding agent) to the coated side. The coating thickness should be uniform across the cylinder's face and should not be more than 0.015 in. (0.38 mm). The equipment holds tightly the cylinders and puts a load on them tending to pull off the coating from the substrate, if the glue resists.

PROBLEMS ABOUT THIS METHOD. This test method is limited to testing flame-sprayed coatings that can be applied in thickness greater than 0.015 in. (0.38 mm). The limitation is imposed because the glue could penetrate the porosity and glue together the two test cylinders, through the coating, without testing the coating.

Unless proved satisfactorily by comparison testing, any agent requiring elevated temperature for curing should be avoided because viscosity may decrease at high temperature, allowing penetration. It is a controversial test, as it seldom simulates the operating conditions.

RESULTS OF ADHESION TEST. To calculate the rate of adhesion or cohesive strength, use the following formula:

$$\text{adhesion or cohesive strength} = \frac{\text{maximum load}}{\text{cross} - \text{section area}}$$

The adhesion or cohesive strength value measured represents the weakest part of the system, whether in the coating or in the interface. The adhesion strength of the coating gives the failure entirely at the interface between the coating and the surface. The cohesive strength of the coating gives the rupture only within the coating.

Shear Test

Another test, probably very relevant for many users, is presented by the German Standard DIN 50 161. The test consists of peeling off a coating deposited onto a solid cylinder by shearing it using an open cylinder of a slightly larger diameter than that of the coated cylinder, made of hardened steel.

Two substrate surface preparations are usually used. One consists of machining the cylinder in a spiral (screw) way and another in the more standard sand blasting.

The ultimate strength of the coating is calculated from

$$\sigma = L_{\max}/A_{\text{sample}}$$

where the maximum load to failure is $L_{\max}$ and A_{sample} is the coated area of the sample. Several sections of the cylinder could be coated, thus reducing the number of samples. Many other tests have been proposed for adhesion; however, few are standardized.

D. Raw Materials Characterization

1. Powder Characterization

The root cause of coating variability and poor coating properties may be due to a powder-feeding problem or powder composition. Manufacturers often supply powders of varying compositions due to labeling errors, or powders might be mistakenly stored in containers wrongly labeled by the user. Also, changes of raw material supplier might result in powders with slight changes in composition. Differences in grain size might yield technical difficulty at the initial stage of powder feeding affecting the quality of the coating. The choice of an acceptable powder feed system is a very important consideration for every powder and thermal spray process.

Some powder quality control tests that should be considered are powder manufacture method, particle size, particle size distribution and range (e.g., Micro-Trac or Horiba), apparent density, tap density, morphology (or shape), material composition (chemistry), Hall flow rate (flowability), "sprayability," and hazards such as toxicity, flammability, or pyrophoricity.

The raw materials for thermal spray and the parameters for processing these materials can be, and often are, different enough to affect the bonding characteristics of the finished product. It is difficult to say whether these variations are due to the raw material, the spray process conditions, or the test method. A powder qualification program is recommended and necessary to ensure high quality and repeatability.

The initial powder size distribution, chemistry, and phases may vary from time to time. There are also variations in the test results between regions of a coating containing oxide clusters, unmelted particles, and porosity compared to those areas that are free of defects.

The most important powder parameters are particle size, size distribution and range, chemical and phase composition, purity, density, homogeneity of component's distribution, shape or morphology of particles, internal porosity, and flowability.

Composition

In order to determine the chemical composition and phase changes, there are some important techniques that an operator should know, or have experienced personnel to determine powder composition routinely for the user.

Chemical composition	May be found using any of the following techniques: —Electron microprobe of the scanning electron microscope, using energy dispersive x-ray spectroscopy (EDS) or wavelength dispersive x-ray spectroscopy (WDS).

Phase analysis	—Inductively coupled plasma optical emission spectroscopy (ICP) (ASTM E1479-92) —X-ray diffraction (XRD) (ASTM E1172-87) —X-ray fluorescence (ASTM E1476-92, ASTM C982-88, and ASTM C1118-89) —Agglomerated or clad powder particles contain different phases (elements) and it might be important to know whether they are distributed homogeneously in the particle (spray-dried powders) or whether cladding covers the core (cladded powders). It is necessary to prepare a metallographical cross section of the powder particles. The 2-D distribution of the elements can be obtained using energy dispersive x-ray spectroscopy (EDS) or wavelength dispersive x-ray spectroscopy (WDS). —Phase analysis can also be carried out using x-ray diffraction (XRD) technique in order to identify the crystalline structures of a material.

Granulometry

Granulometry is a term used for both grains in the microstructure and powder particle measurements. Basically, it is important to determine the particle size and shape for its use in spraying. Either sieving, optical microscopy, scanning electron microscopy, or gravitational sedimentation and laser scattering (MicroTrac or Horiba) can determine particle size. The particle shape or morphology can be determined by optical or scanning electron microscopy.

PARTICLE SIZE. Testing procedures and measurements enable us to determine the particle size of the feedstock. Powder size distribution determination methods utilized for plasma and thermal spray powders are sieving, electrical resistance, sedimentation, centrifuge, image analysis, and light scattering. It has been found that these methods give very different particle size distributions for batches of the same material. Then, it is very important that particle size results should always be related to the technique by which the measurement was performed.

When the particle size analysis is carried out using only a few grams of powder, then it is important to pick a representative sample. The sampling can be made with the help of a special device sampler as described by ASTM B215-82. However, care has to be taken to ensure that the sampled powder particles do not agglom-

erate together. Poor preparation of the powder sample can lead to false results.

Important: particle size is dependent on the measuring technique. Therefore it is important that the technique used to measure the mean particle size or the particle size distribution always be referenced. The sieve method is perhaps the most commonly used, although some controversy exists as to its accuracy.

Sieving	—The powder is deposited on a set of standard sieves, arranged with the one of widest mesh on top and the thinnest at the bottom. —Usually, a sieve shaker mechanically operated imparts to the set of sieves a rotary motion and tapping action of uniform speed. The apparatus is mounted in a rigid base, preferably concrete. The powder that does not go through the top sieve is weighed, continuing to shake the material, until most of it has gone through the second sieve. —The amount of powder that rests on the second sieve is then removed and weighed, continuing with the same pattern. The powder grains are classified in size ranges, indicated in terms of minus and plus. —The weight of the fractions retained on each sieve and the weight of the fraction in the pan, at the bottom of the sieve pile, are expressed in percentages of the weight of the specimen (ASTM B214-92).
Electrical resistance particle counter	—This method (ASTM F662-86) is used for particle size range approximately from 0.5 to 200 µm. The instrument uses an electric current path of small dimensions which is modulated by individual particle passage through an aperture and produces individual pulses of amplitude proportional to the particle volume. —Basically, a test sample is made conductive with addition of a clean electrolyte and placed in the instrument's sample stand counting position. The suspension is forced through a restricting aperture. Each particle passing is recorded on an electronic counter according to selected size levels. It is important to keep in mind that challenge test

Gravitational sedimentation

particles soluble in water cannot be analyzed by this test method unless alcohol can be substituted for water in the filter testing.

–This technique is based on Stokes' law of fluid dynamics. The powder is classified by their rate of settling in a fluid. This method is used to determine particle size and size distribution. It can be used when particles are larger than 5 μm in diameter; for sedimentation in air, smaller particles (0.1 μm or less) can be determined by sedimentation in liquids.

–Basically, this method consists of the dispersion of a powder homogeneous suspension that settles in an electronically programmed cell that can move downward with respect to a fixed collimated x-ray beam of constant intensity. This net x-ray signal is inversely proportional to the sample concentration, and the particle diameter is related to cell position. Finally, an x–y recorder is plotted based on the cumulative mass percent vs. equivalent spherical diameter in order to produce a particle size distribution curve (ASTM B761-90 and ASTM C958-81).

Microscopy and image analysis

–This test method (ASTM E1382-91) may be used to determine the mean particle size, or the distribution of grain intercept lengths or areas, in metallic polycrystalline materials. The test methods may be applied to specimens with equiaxed or elongated grain structures with either uniform or duplex particle size distribution. Either semiautomatic or automatic image analysis devices may be utilized to perform the measurements.

–Basically, the determination of the mean particle size is based on measurement of the number of particles per unit area, the length of grain boundaries in unit area, the grain areas, and the number of grain intercept lengths. These measurements are made for large number of grains, or all of the grains in a given area, within a microscopically field and then repeated on additional fields to obtain an adequate number of measurements to achieve the desired degree of

–statistical precision. The measurement is made with semiautomatic digitizing tablet or by automatic image analysis using an image of the grain structure produced by a microscope. The test method is applicable to any type of particle structure or particle size distribution as long as the particle boundaries can be clearly delineated.

PARTICLE SHAPE OR MORPHOLOGY.   The ideal powder consists of a rather narrow-sized distribution of spherical particles that follow an ideal path to and through the thermal source. The ideal shape or morphology is spherical.

The term "morphology" refers to the shape of individual particles. Shape is commonly determined by optical or scanning electron microscopy. Particle shape is controlled by the production process. It is incorrect to assume that the powder surface will be smooth or that each particle will be chemically homogeneous and 100% dense. Particles in a powder can be angular or fragmented spherical, rounded, or composite. They can be in the shape of fibers or in an acicular needlelike form. Particles can also be shaped as platelets, rods, or "dog bones" and even as wire or shots. Particle shape affects the material transport from the powder feed hopper to the spray torch; therefore shape of the powder particles is an essential characteristic to be determined. Particle size and distribution are also important parameters to determine since this is optimized with regard to the coating quality.

The observation of the grain shape enables recognition of the manufacturing technique. To get complete information about the powder, it is necessary to observe the particles from the outside (usually SEM; for more information about SEM performance characterization, consult ASTM E986-86) as well as the inside. The latter could be made with the specimens prepared by embedding the powder in a mounting resin and successively grinding and polishing (see "Sample Preparation" above). The specimens can be observed using an optical microscope.

Morphology observation, however, might not be sufficient to determine the manufacturing method, and chemical or phase analyses might also be necessary.

APPARENT DENSITY AND FLOWABILITY.   There are various standardized techniques used to determine the apparent density and flowability of powders. Among them, there are standard test methods for apparent density measurement of metal powders using a powder flowmeter funnel (ASTM B212-89) or using the Arnold Meter (ASTM B703-88). A standard test method for flow rate of metal powders is designated as B213-90.

a. Apparent Density.   The apparent density is the weight of an amount of powder divided by its volume and multiplied by a constant [apparent density $(g/cm^3) =$ weight $(g) \times 0.04$]. It gives a measure of the easiness of flow of a powder that is transported by an airstream. The following flowmeter funnel method determines the apparent density of free-flowing metal powders that will flow unaided through the Hall flowmeter funnel. The Arnold meter method determines the apparent density of both and nonfree flowing powders and premixes.

Flowmeter funnel method	Arnold meter method	Powder flowmeter funnel
Is based on the flow of a powder volume into a contained or definite volume under controlled conditions. Therefore the weight of powder per unit volume is measured as apparent density. Consists on a standard Hall flowmeter funnel having a calibrated orifice (Fig. 18).	Is based on sliding a bushing filled with the test specimen over a steel block's hole. Before the filling, collecting, and weighting of the powder, it is possible to calculate its apparent density.	The flow rate of a metal powder gives to the operator ideas about the rate of filling of die cavities in the pressing of sintered metal parts or bearings and possible rates of powders through the nozzle during a spraying process. A vibration-free base is used to support the powder flowmeter funnel.

Hardness

In order to determine powder hardness, there are some basic methods. The relative abrasive method, invented by Friedrich Mohs, could be adapted, although it is not a standard test, but is a useful one.

Microcompression equipment could also be used to determine the powder hardness. First, a small amount of particles are placed on a sample carrier. Using a microscope, a particle of average diameter is selected. Then, an indentation is applied to the particle using a plane indenter of 50 μm Φ. Finally, a chart of applied effort vs. deformation is produced, and the point where the powder fracture happens is identified.

Quality control is needed not only to determine whether the powder is good from the manufacturers,

but also to ascertain if the appropriate powder, and not a mistaken one, is to be applied.

2. Characterization of Wires and Rods

The objective in nondestructive inspection is to detect conditions in the material that may cause an unsatisfactory end use of the product. The nondestructive inspection of rods and other semifinished products provides rapid feedback of information and can be used as an on-line or off-line system (see "Liquid Penetrant Examination"). There are some defects commonly found in wires and bars that could affect the end use of the product: pipe, porosity, inclusions, laminations, slivers, scabs, pits, blisters, cracks, seams, laps, and chevrons.

In terms of the results obtained from wire and rod spraying, a good quality control of wires and bars manufacturing are important during the spraying process. Small differences in the inlet material characteristics could generate major differences in the end results; in some instances, they could make the materials impossible to spray.

There are some important characteristics associated with equipment operation conditions that the operator has to keep in mind when the end results present problems: wire/bar composition, dimensions (diameter and length), and surface roughness.

The wire/rod composition is important in determining when the operation presents problems. Sometimes, a small difference of only 1% of an element in an alloy could make spraying impossible. In order to obtain the inlet material composition, there are some analytical techniques such as x-ray fluorescence spectroscopy and inductively coupled emission spectroscopy. More information about this technique is found in the discussion of chemical composition under "Powder Characterization" above.

The wire and bar dimensions are also important due to equipment size tolerances. The equipment manufacturer specifies these size tolerances. Therefore the operator should only use the correct size. An oversize wire/bar may stick in the gun parts, wearing them and producing defective coatings. In order to measure the wire/bar dimensions, simple measurement devices such as calipers can be used.

The inlet material surface roughness should be low to avoid problems during the spraying process. A high roughness of wires and rods could cause significant wear of key equipment parts (nozzle, drive rolls, and guides) and difficult handling of inlet material. Simple devices can be used to measure wire/bar roughness.

Appendix Comparative Characteristics of Some of the Main Coating Methods

		Deposition rate	Component size	Substrate material	Pretreatment	Posttreatment	Control of deposit thickness	Uniformity of coating	Bonding mechanism	Distortion of substrate
Gaseous state processes	PVD	Up to 0.5 per source	Limited by chamber size	Wide choice	Mechanical/ Chemical plus ion bombardment	None	Good	Good	Atomic	Low
	PAPVD	Up to 0.2		Wide choice	Mechanical/ Chemical plus ion bombardment	None	Good	Good	Atomic plus diffusion	Low
	CVD	Up to 1		Limited by deposition temperature	Mechanical/ Chemical	Substrate stress relief/ mechanical properties	Fair/ good	very good	Atomic	Can be high
	PACVD	Up to 0.5		Some restrictions	Mechanical/ Chemical plus ion bombardment	None	Fair/good	Good	Atomic plus diffusion	Low/ moderate
	Ion implantation			Some restrictions	Chemical plus ion bombardment	None	Good	Line of sight	Integral	Low
Solution processes	Sol–gel	0.1–0.5	Limited by solution bath	Wide choice	Grit blast and/or chemical clean	High-temperature calcine	Fair/ good	Fair/ good	Surfaces forces	Low
	Electroplating	0.1–0.5		Some restrictions	Chemical cleaning and etching	None/ thermal treatment	Fair/ good	Fair/ good		Low
Molten or semimolten state processes	Laser	0.1–1	May be limited by chamber size	Wide choice	Mechanical and chemical cleaning	None/ substrate stress relief	Fair/ good	Fair	Mechanical/ chemical	Low/ moderate
	Thermal spraying	0.1–1		Wide choice			Manual variable automated-good	Variable		Low/ moderate

REFERENCES

1. Lira-Olivares, J. *Thermal Spray: Testing, Practical Learning Series*; TSS ASM International: Materials Park, OH; October 2001; 41–43.
2. Mora-Márquez, J.G.; Lira-Olivares, J. *A Study Of Crack Initiation And Propagation In Nickel-Chrome Thermally Sprayed Coatings Using Acoustic Emission Techniques*; Elsevier Sequoia, Thin Solid Films: New York, 243–252.
3. Lira-Olivares, J.; Brito, M.; Mutoh, Y.; Takahashi, M. Proceedings of the 29th International Conference on Metallurgical Coatings and Thin Films, San Diego, California, USA. April 22–26. Elsevier. To be published on Thin Solid Films.
4. Lira-Olivares, J.; Grigorescu, I.C. *Friction And Wear Behavior Of Thermally Sprayed Nichrome-WC Coatings*, 14th International Conference on Metallurgical Coatings, San Diego, March 25, 1987; 183–190.
5. Lira-Olivares, J. *Thermal Spray: Testing, Practical Learning Series*; TSS ASM International: Materials Park, OH, October 2001; 3 pp.
6. Wheelabrator Corporation. *Shot Peening*; The Wheelabrator Corporation: Mishawaka, Indiana, 1965; 2 pp.
7. Flinn, R.; Trojan, P. *Engineering Materials and Their Applications*; Houghton Mifflin Company, 1975; 112–114.
8. Seader, J.D.; Henley, E.J. *Separation Process Principles*; John Wiley & Sons Inc., 1998; 92 pp.
9. Van Vlack; Lawrence, H. *Elements of Materials Science and Engineering*; Addison-Wesley Publishing Company: Mischawaka, Indiana, 1975; 451–452.
10. Pollack; Herman, W. *Materials Science and metallurgy*; Reston Publishing Company, 1977; 197 pp.
11. Seader, J.D.; Henley, E.J. *Separation Process Principles*; John Wiley & Sons Inc., 1998; 197–198.
12. Avner; Sydney, H. *Introduction to Physical metallurgy*; McGraw-Hill: New York, 1974; 317–331. Spanish version.
13. Cao, Yang; Duby, Paul. *Cobalt Cementation with Ferromanganese*; Hydrometallurgy, Elsevier: New York, November 24, 1999; 195–205.
14. Chadderton; Lewis, T. *Radiation Damage in Crystals*; John Wiley & Sons: New York, 1965; 144–145.
15. Rose, R.M.; Shepard, L.A.; Wulff, J. *Electronic Properties, 1964*; John Wiley and Sons: New York, 1968; 110–111.
16. Kittel; Charles. *Introduction to Solid State Physics*; John Wiley & Sons Inc.: New York, 1953; 351 pp.
17. Cheng, Ju-Yin; Gibson, J.M.; Jacobson, D.C. *Observations of Structural Order in Ion-Implanted Amorphous Silicon*; Materials Research Society: Warrendale, 2001; 3030–3033.
18. Zhang, F.; Wolf, G.K.; Wang, X.; Liu, X. Surface Properties of Silver Doped Titanium Oxide Films. Elsevier, Surface and Coatings Technology: New York, 2001, *148*, 65–70.
19. Wei, M.; Kun, T.; Xingtao, L.; Baixin, L. Formation of Al$_3$Hf by Ion Implantation Into Aluminum Using a Metal Vapor Vacuum Arc Ion Source. Elsevier, Surface and Coatings Technology: New York, 2001, *140*, 136–140.
20. Choe, Han-Cheol. Effects of Nitrogen Ion implantation on the Surface Characteristics of Iron Aluminides. Elsevier, Surface and Coatings Technology: New York, 2001, *148*, 77–87.
21. Bunshah, R.F. *Deposition Technologies for Films and Coatings, Developments and Applications*; Noyes Publications: Park Ridge, New Jersey, 1982; 83 pp.
22. Tomsia, P.; Loehman, R. *Reactions and Microstructure at Selected Ceramic/Metal Interfaces, Surfaces Modification Technologies VII*; The Institute of Materials: London, 1994; 327–341.
23. Colangelo, V.J.; Heiser, F.A. *Analysis of Metallurgical Failures*; John Wiley and Sons: New York, 1974; 230 pp.
24. Gee, A. W. Friction and Wear as Related to the Composition, Structure and Properties of Metals, Int. Met. Rev. 1979, (2), 57–67.
25. Suh, N.P. The delamination theory of wear. Wear 1973, *25*, 111–124.
26. Grigorescu, I.C.; Di Rauso, C.; Drira-Halouani, R.; Lavelle, B.; Giampaolo, R.; Lira-Olivares, J. Phase Characterization in Ni Alloy-hard Carbide Composites for Fused Coatings. Elsevier, Surface Coatings Technology: New York, 1995, *76*, 494–498.
27. Mutoh, Y.; Ohki, M.; Lira-Olivares, J.; Takahashi, M. *Thermal Barrier Function and Damage of Plasma-Sprayed Coatings for Corrosion Protection*. Conference on Corrosion CONCOR. Elsevier: New York, 1997; 357–366.
28. Fontana Greene, M.G. *Corrosion Engineering*; McGraw-Hill Book Company: New York, 1967; 26–27.
29. Chattopaday, R. *Surface Wear: Analysis, Treatment, and Prevention*; ASM International: Materials Park, OH; 2001; 26–27.
30. Rabinowicz, E. *Friction and Wear of Materials*; Wiley & Sons: New York, 1965; p. 65.
31. Lira-Olivares, J.; Grigorescu, I.C. Microstructure Development and Mechanical Properties of Ni Matrix/-Carbide Composite. Adv. Perform. Mater. Long Island: New York, 1997, (4), 95–103.
32. Ingham, H.S.; Shepard, A.P. *Flame Spray Handbook*; METCO INC: New York, 1964; A–54.
33. Fontana Greene, M.G. *Corrosion Engineering*; McGraw-Hill Book Company, 1967; 26–27.
34. Ingham, H.S.; Shepard, A.P. *Flame Spray Handbook*; METCO INC, Long Island: New York, 1964; A–55.
35. Holmberg, K.; Matthews, A. *Coatings Tribology, Properties, Techniques and Applications in Surface Engineering*; Elsevier: New York, 1994; 8 pp.
36. Chapman, B.; Anderson, J.C. *Science and Technology of Surface Coating*; Academic Press Inc.: New York, 1997; 149 pp.
37. Chapman, B.; Anderson, J.C. *Science and Technology of Surface Coating*; Academic Press Inc.: New York, 1997; 149–150.
38. Chapman, B.; Anderson, J.C. *Science and Technology of*

Surface Coating; Academic Press Inc.: New York, 1997; 156 pp.

39. Giampaolo, A.R.; Castell, R.; Perril, H.; Sainz, C.; Guerrero, A.; Calatroni, J.; Lira-Olivares, J. *Study of Sol–Gel Transition in Ceramic Systems by High Resolution Refractometry*; Elsevier Science Publishers LTD: New York, 300–308.

40. Bunshah, R.F. *Deposition Technologies for Films and Coatings, Developments and Applications*; Noyes Publications: Park Ridge, New Jersey, 1982; 412 pp.

41. Dasarathy, H.; Riley, C.; Coble, H.D. Analysis of Apatite Deposits on Substrates. J. Biomed. Mater. Res. 1993, *27*, 477–482.

42. Kokubo, T.; Yamamuro, T. Apatite Coating on Ceramics, Metals and Polymers utilizing a Biological Process. J. Mater. Sci. Mater. Med. 1990, *V1*, 233–238.

43. Kokubo, T.; Minoda, M.; Tanashi, M.; Yao T. Apatite Coatings on Organic Polymers by a Biomimetic Process. J. Am. Ceram. Soc. 1994, *V77* (11), 2805–2808.

44. Bunshah, R.F. *Deposition Technologies for Films and Coatings, Developments and Applications*; Noyes Publications: Park Ridge, New Jersey, 1982; 89 pp.

45. Ohtsuki, C.; Kokubo, T.; Takatsuya, K.; Yamamuro, T. Compositional Dependence of Bioactivity of glasses in the System $CaO–SiO_2–P_2O_5$: Its in Vitro Evaluation. J. Ceram. Soc. Jpn. 1991, *V99*, 2–6.

46. ASM International Handbook Committee. *ASM Handbook, Friction, Lubrication and Wear Technology*; ASM International: Materials Park, OH, 1992, *18*. 829 pp.

47. ASM International Handbook Committee. *ASM Handbook, Friction, Lubrication and Wear Technology*; ASM International: Materials Park, OH, 1992, *18*. 830 pp.

48. Chattopaday, R. *Surface Wear: Analysis, Treatment, and Prevention*; ASM International: Materials Park, OH, 2001; 179–180.

49. Bunshah, R.F. *Deposition Technologies for Films and Coatings: Developments and Applications*; Noyes Publications: Park Ridge, New Jersey, 1982; 91 pp.

50. Bunshah, R.F. *Deposition Technologies for Films and Coatings, Developments and Applications*; Noyes Publications: Park Ridge, New Jersey, 1982; 385 pp.

51. Bunshah, R.F. *Deposition Technologies for Films and Coatings, Developments and Applications*; Noyes Publications: Park Ridge, New Jersey, 1982; 402–403.

52. Saber, J.P.; Sahoo, P. *Hvof Process Using Aimen and Temperature Measurement*; 2000; 3 pp.

53. Safety and Health Fact Sheet N°20, 1998 American Welding Society, 1998.

54. Chattopaday, R. *Surface Wear: Analysis, Treatment, and Prevention*; ASM International: Materials Park, OH, 2001; 162–169.

55. Mencino, L.; Vartanian, V. *Point of Use Abatement Analysis for Advanced CVD Applications*; Future Fab Intl., Montgomery Research, Inc.: London, 2003; Vol. 14.

56. http://www.sulzermetco.com/tech/ap-condt.html; http://www.ewi.org/technologies/arcwelding/thermalspray.asp; http://www.sandia.gov/isrc/thermal-spray.html; http://www.asbindustries.com/grant.asp; http://www.harperimage.com/Directory/coating_rolls.htm; http://www.splasers.com/contact/index.html, http://lasertag.org/; http://www.p1diamond.com/man.html; http://www.finishing.com/Products/index.html.

57. http://www.finishes.org.uk/bens.htm; http://www.enla.com/applications.html; http://www.palminc.com/enschool/; http://www.ewi.org/technologies/arcwelding/thermalspray.asp; http://www.sandia.gov/isrc/thermalspray.html; http://www.asbindustries.com/grant.asp; http://www.harperimage.com/Directory/coating_rolls.htm; http://www.splasers.com/contact/index.html; http://lasertag.org/.

58. http://www.epa.gov/opptintr/dfe/projects/pwb/about.htm; http://www.svtc.org/hightech_prod/liaisons/dfe/connections1.htm; Safety and Health Fact Sheet N°20, 1998 American Welding Society.; http://www.osha-slc.gov/SLTC/laserhazards/; http://www.triumf.ca/safety/tsn/tsn_1_4/subsection3_1_2.html; http://www.tribology.dti.dk/pvd.html; http://www.pao.nrl.navy.mil/rel-99/18-99r.html; http://sprg.ss1.berkeley.edu/wind3dp/esahome.html#top.

59. Safety and Health Fact Sheet N°20, American Welding Society, 1998.

60. Princenton electronic Series VCOs. Specifications 2002.

61. Vemuri, Gautam "*Laser noise for better communications and diagnosis*" research in IUPUI school of science department of physics; IUOUI, School of Science, Department of Physics: Indianapolis, 2001.

62. http://www.tvc.nrao.edu/2002.

63. Protection of the human environment occupational and community noise, fact sheet N° 258, http://www.who.int/peh/noise/noiseindex.html.

64. Bunshah, R.F. *Deposition Technologies for Films and Coatings, Developments and Applications*; Noyes Publications: Park Ridge, New Jersey, 1982; 72 pp.

65. Bunshah, R.F. *Deposition Technologies for Films and Coatings, Developments and Applications*; Noyes Publications: Park Ridge, New Jersey, 1982; 73 pp.

66. Bunshah, R.F. *Deposition Technologies for Films and Coatings, Developments and Applications*; Noyes Publications: Park Ridge, New Jersey, 1982; 77–78.

67. Coddet, C.; Montavon, G.; Marchione T.; Freneaux, O. Surface Preparation and Thermal Spray in a Single Step: The Protal Process, Proceedings of the International Thermal Spray Conference, Nice, France, May 25–29, 1998; ASM International: Materials Park, OH, 1321–1325 pp.

68. Folio, F.; Barbezat, Ch.G.; Coddet, C.; Montavon, G.; Costil, S.; Frenneaux, O. Thermal Spray Deposition of Ceramic Coating on Aluminum and Titanium Alloys Using PROTAL® PROCESS, 196–201.

69. Folio, F.; Barbezat, Ch.G.; Coddet, C.; Montavon, G.; Costil, S.; Frenneaux, O. Thermal Spray Deposition of Ceramic Coating on Aluminum and Titanium Alloys Using PROTAL® PROCESS, 1321–1325.

22

Designing for Machining: Machinability and Machining Performance Considerations

I. S. Jawahir
University of Kentucky, Lexington, Kentucky, U S.A.

I. INTRODUCTION

Machining operations constitute a large segment of the manufacturing sector in the United States. However, a recent CIRP (the International Institution for Production Engineering Research) working paper [1] reports the survey results of a major cutting tool manufacturer as "... In the USA, the correct cutting tool is selected less than 50% of the time, the tool is used at the rated cutting speed only 58% of the time, and only 38% of the tools are used up to their full tool-life capability...." This situation urges the need for development of scientific approaches to select cutting tools and cutting conditions for *optimum economic and technological machining performance.*

The selection of cutting tools and cutting conditions represents an essential element in process planning for machining. This task is traditionally carried out on the basis of the experience of process planners with the help of data from machining handbooks and tool catalogs. Process planners continue to experience great difficulties because of the lack of data on the numerous new commercial cutting tools produced from different materials, with a range of coatings, geometry, and chip-groove configurations. Also, specific data on relevant machining performance measures such as tool-life, surface roughness, chip-form, etc. are hard to find because of the lack of predictive models for these measures. Consequently, process planners are forced to choose and recommend suboptimal cutting conditions for machin-

ing operations. It is, however, well recognized that in *designing for machining operations*, it is important to identify and utilize optimum cutting conditions and cutting tools for enhanced productivity and for savings on production costs.

This chapter presents a summary of the present knowledge on designing for machining, focusing on machinability and machining performance evaluation methods that largely affect the process of designing for machining. Section 2 of this chapter presents an overview of the current machinability evaluation methods and their limitations. Section 3 provides a systematic analysis of machining performance evaluation including analytical, numerical, and experimental/empirical methods currently prevalent for the various machining performance measures. Section 4 presents the most common machining optimization methods for use in design for machining, and Section 5 covers the influence of jigs and fixtures in designing for machining. Practical operations requiring specialized methods/devices for design for machining are then illustrated in Section 6.

II. MACHINABILITY EVALUATION IN DESIGNING FOR MACHINING

A. Machinability: Definition, Criteria, and Testing Methods

The traditionally known term *machinability* has been used to indicate the ease or difficulty with which a material can be machined to the required specifications. Historically, this term has been defined and understood as a *property of work materials*. Several dedicated and generic machinability database systems have been developed and are being used worldwide. These systems primarily depend on exhaustive experimental data and very limited analytical methods. In these systems, machinability ratings are generally expressed in terms of a reference index. Table 1 shows the typical machinability ratings of several plain carbon steel work materials [2]. It has also been shown that a considerable scatter exists in the machinability data obtained for various materials (Fig. 1). The effects of small variations in chemical composition or in the microstructure of the work material on machinability are quite often found to be greater than those from variations in hardness [3]. Also, the metallurgical effects of the work material, particularly those involving the secondary shear zone deformation, have been shown to be a prime cause for serrated chip formation and the associated tool-wear process, both contributing to machinability [4]. It has been shown that the basic machinability of a material is a function of its chemistry, structure, and compatibility with a tool material [5].

In general, several machining variables affect the machinability. This includes:

(a) Cutting conditions (cutting speed, feed, and depth of cut)
(b) Tool geometry (rake angle, inclination angle, cutting edge angle, corner radii, etc.)
(c) Tool material including coating
(d) Chip-groove geometry
(e) Cutting fluid application
(f) Rigidity of the machine tool–cutting tool system (including workholding devices)
(g) Nature of engagement (continuous, intermittent, etc.)

The most common work material properties affecting machinability are:

(a) Hardness
(b) Tensile strength
(c) Chemical composition
(d) Microstructure
(e) Degree of cold work and strain hardenability
(f) Work material's heat treatment level
(g) Shape and dimensions of the workpiece
(h) Workpiece rigidity

1. Machinability Criteria

The general criteria for machinability evaluation involves the testing of cutting tools for one or more of the following major machining performance measures:

(a) Tool-life
(b) Cutting force/Power
(c) Surface quality (e.g., surface roughness)
(d) Part accuracy
(e) Chip-form/Chip breakability

However, in some circumstances, specific machinability criteria would seem necessary. This includes:

(a) Drilling torque/Thrust
(b) Cutting temperature (at the tool tip or in the chip)
(c) Cutting ratio of the chip

2. Methods of Machinability Testing

Over the last few decades, several different tests have been developed and implemented for machinability evaluation. These tests include [2]:

(a) Tool-wear/Tool-life tests
(b) Surface finish test
(c) Cutting force test
(d) Cutting temperature test
(e) Power consumption test
(f) Cuttability test (e.g., rate of penetration of a drill under constant feed pressure)
(g) Simulated production tests (for optimum cutting conditions)

3. Test Standards

Significant efforts have been made in the past few decades to develop a standard test procedure for machinability evaluation. The most acceptable among these are the ISO Tests. ISO 3685:1993(E) [6], ISO 8688: 1989 (Part 1 and Part 2) [7,8] offer standard methods for tool-life testing in turning, face, and end milling, defining the standard tool geometry, including obstruction-type chip breaker geometry, and recommending standard cutting conditions and their limits with comprehensive tool-life criteria.

4. Specification of Tool-Life

The most significant aspect of tool-life assessment is the measure of tool-life. Various methods exist to specify tool-life. The most common among these methods are:

(a) Machine time—elapsed time of operation of machine tool
(b) Actual cutting time (most common definition of tool-life)

Table 1 Relative Machinability Ratings of Various Materials[a]

Material	BHN	MR	Material	BHN	MR
12% Chrome			A-6120	187	0.50
Stainless iron	165	0.70	A-6140	205	0.50
80B40	195	0.35	A-6145	207	0.50
81B45	179	0.60	A-6152	195	0.50
86B45	212	0.35	A-8640	170	0.55
98B40	185	0.40	A-8645	210	0.50
1020 (Castings)	134	0.60	A-8650	212	0.45
1040 (Castings)	190	0.45	A-8740	200	0.55
1330	223	0.60	A-8745	219	0.45
3140	197	0.55	A-8750	212	0.40
3250	220	0.45	AM 350	420	0.14
3312	191	0.50	AM 355	360	0.10
3340	220	0.45	AMS 6407	180	0.50
3450	197	0.45	AMS 6418	195	0.50
4130 (Castings)	175	0.35	AMS 6427	180	0.50
4130	183	0.65	B-1112	160	1.00
4140	190	0.55	B-1113	170	1.35
4140 (Leaded)	187	0.70	C-1008	155	0.55
4145	200	0.55	C-1010	150	0.55
4340 (100% Pearlitic)	221	0.45	C-1015	131	0.60
4340 (Spheroidized)	206	0.65	X-1020	148	0.65
4340 (Castings)	300	0.25	C-1025	143	0.65
4620	170	0.65	C-1030	190	0.65
4640	187	0.55	C-1040	205	0.60
4815	183	0.55	C-1045	217	0.50
5120	191	0.65	C-1050	205	0.50
6130	183	0.55	C-1095	210	0.45
6135	190	0.55	C-1117	170	0.90
6180	207	0.40	C-1118	160	0.80
8030 (Castings)	175	0.45	C-1120	160	0.80
8430 (Castings)	180	0.40	C-1137	197	0.75
8620	194	0.60	C-1141	240	0.50
8630	190	0.60	Cast Iron (soft)	160	0.60
8630 (Castings)	240	0.30	Cast Iron (medium)	195	0.40
8720	190	0.60	Cast Iron (hard)	262	0.20
9255	218	0.45	Cast Iron		
9260	221	0.45	(Chilled White)	500	0.70
9262H	255	0.25	Cast Iron		
A-286	300	0.10	(Gray Pearlitic)	190	0.70
A-3115	160	0.65	Chromaloy	293	0.50
A-3120	150	0.65	Discaloy	135	0.40
A-4023	183	0.70	E-3310	196	0.40
A-4027	189	0.70	E-4137	200	0.60
A-4032	190	0.70	E-52100	206	0.30
A-4037	200	0.65	E-6150	197	0.50
A-4042	200	0.60	E-9310	223	0.40
A-4047	209	0.55	E-9315	204	0.40
A-4150	208	0.50	H-11	190	0.45
A-4320	200	0.55	Hastelloy B (Cast)	200	0.12
A-4340	210	0.50	Hastelloy C	170	0.20
A-4820	205	0.45	Hastelloy X	197	0.09
A-5140	202	0.60	Haynes Stellite		
A-5150	207	0.50	#21 (Cast)		0.06

Table 1 Continued

Material	BHN	MR	Material	BHN	MR
Haynes Stellite #25		0.12	317	195	0.35
Haynes Stellite			403	200	0.55
#31 (Cast)		0.06	405	145	0.60
High Speed Steel 18-4-1	220	0.35	410	160	0.55
High Speed Steel 8-2-1	210	0.40	416	200	0.90
Inconel	240	0.30	418	160	0.40
Inconel X	360	0.15	420	207	0.45
Inconel 700	290	0.09	430F	147	0.65
Inconel 702	225	0.11	440C	240	0.35
Inconel 901	200	0.20	440	160	0.50
Inconel 901	300	0.15			
M-252	220	0.05	**Titanium and its alloys**		
M-308	352	0.05	A-55	160	0.30
Malleable–(Pearlitic)	185	0.90	A-70	188	0.27
Malleable (Standard			A-110	220	0.23
Mallable)	120	1.10	C-120	240	0.20
Molybdenum (Cast)	190	0.30	C-130	255	0.18
Monel–(K-R Monel)	240	0.45	C-140	285	0.15
Monel (R Monel)	208	0.45	MST	380	0.09
Monel (S Monel Cast)	300	0.25			
NE-9261	198	0.50	**Aluminum alloys cast**		
Ni-Hard	550	0.03	A-132-T		1.10
Ni-Resist	145	0.45	A-214		2.00
Nimonic 75	220	0.17	A-356-T		1.40
Nimonic 80	270	0.12	B-113		1.80
Nimonic 90	300	0.10	D-132-T		1.30
Nitrolloy (135)	200	0.45	108		1.40
Nodular Iron #1	183	0.60	112		1.80
Nodular Iron #2	200	0.50	122-T		1.40
Nodular Iron #3	230	0.40	195-T		1.90
Potomac M	200	0.45	212		1.60
Rene 41	215	0.15	218-T		2.40
Rycut 40	187	0.65	220-T		2.30
Stressproof	203	0.50	319-T		1.60
Super Triscent	180	0.40	333-T		1.30
Turbaloy	135	0.40	355-T		1.60
Udimet 500	290	0.09	750-T		1.80
V-57	375	0.08			
Vasco X4	150	0.50			
Vascojet 1000	190	0.45	**Aluminum alloys wrought**		
Waspalloy	270	0.12	2011		2.00
Tungsten Estimated	–	0.05	2014-T		1.40
			2017-T		1.40
Stainless steels			2024-T		1.50
PH 15-7 Mo	270	0.20	3003		1.80
17-4PH	388	0.28	3004		1.80
17-7PH	200	0.20	5052		1.90
301	183	0.55	5056		1.90
302	178	0.50	4032-T		1.10
303	180	0.65	6051-T		1.40
304	160	0.40	6061-T		1.90
310	160	0.30	6063-T		1.90
316	195	0.35	7075-T		1.20

Table 1 Continued

Material	BHN	MR	Material	BHN	MR
Aluminum–bronze			Leaded copper		2.40
(5% Al)		0.60	Leaded naval brass		2.10
			Leaded nickel–silver		
Other materials			(12% Ni)		1.30
Aluminum–bronze			Leaded nickel–silver		
(8% Al)		0.50	(18% Ni)		1.50
Aluminum–bronze			Leaded phosphor–bronze		
(9.25% Al)		0.50	(5% Tin)		1.50
Aluminum–bronze			Leaded silicon–bronze		1.80
(9.5% Al)		0.60	Low brass (80% Cu)		0.90
Aluminum			Low-leaded brass		1.80
silicon–bronze		1.80	Low-leaded (tube)		1.80
Architectural bronze		2.70	Manganese bronze		0.90
Beryllium–copper			Medium-leaded brass		2.10
(not heat-treated)		0.60	Muntz metal		1.20
Chromium–copper		0.60	Naval brass		0.90
Commercial bronze			Nickel		2.00
(90% Cu)		0.60	Nickel–silver (18% Ni)		0.60
Cupro-nickel		0.60	Nickel–silver (20% Ni)		0.60
Deoxidized copper		0.60	Phosphor–bronze		
Electrolytic tough-			(5% Tin)		0.60
pitch copper		0.60	Phosphor–bronze		
Extruded leaded			(8% Tin)		0.60
nickel–silver (10% Ni)		2.40	Phosphor–bronze		
Forging brass		2.40	(10% Tin)		0.60
Free-cutting brass		3.00	Red brass (35% Cu)		0.90
High-leaded brass		2.70	Selenium or		
High-leaded brass (tube)		2.40	tellurium–copper		2.70
High silicon–bronze		0.90	Special free-cutting		
Leaded commercial			phosphor-bronze		2.70
bronze		2.40	Zinc		2.00

[a] The machinability ratings, MR, in this chart were established for materials with Brinell hardness numbers BHN as listed. When a material listed is to be machined and is found to have a BHN different from that shown in the table, the ratio of the BHN in the table to the actual BHN of the workpiece is multiplied by the listed machinability rating, MR, to provide the MR of the actual workpiece. For example, a 3140 at a BHN of 197 is shown on the table to have an MR of 0.55. The 3140 to be machined has an actual BHN of 220. Therefore:

$$\frac{197}{220} \times 0.55 = 0.49 = MR \ @ \ BHN \ 220$$

The MR of the part to be machined is 0.49. On the other hand, if the 3140 to be machined has an actual hardness of 170:

$$\frac{170}{197} \times 0.55 = 0.64 = MR \ @ \ BHN$$

Source: Ref. 2.

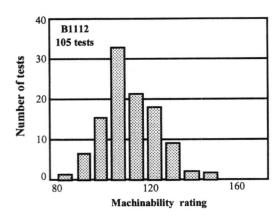 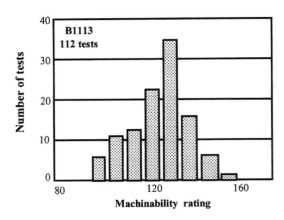

Figure 1   Machinability variations in B1112 and B1113 steels. (From Ref. 3.)

(c)   Volume of metal removed
(d)   Number of pieces machined
(e)   Equivalent cutting speed (Taylor speed)
(f)   Relative cutting speed

5.   *Tool Failure Types and Failure Criteria*

The tool failure is determined in terms of one or more of the following types of tool-wear:

(a)   Flank wear
(b)   Crater wear
(c)   Chipping
(d)   Built-up edge, thermal cracking, or deformation
(e)   Various combinations of the above

The most common criteria for tool failure include:

(a)   Complete failure
(b)   Preliminary failure (e.g., appearance on the surface finish)
(c)   Flank failure
(d)   Finish failure
(e)   Size (dimension) failure
(f)   Cutting force (power), or thrust force, or feed force failure

6.   *Major Tool-Wear Mechanisms*

Tool failure is determined in terms of one or more of the following tool-wear mechanisms:

1.   Abrasion
2.   Adhesion
3.   Diffusion
4.   Chemical reaction
5.   Oxidation

7.   *Assessment of Machinability*

In general, two broad categories of machinability tests are most commonly known: category one does not require an actual machining test and category two would require a machining test. There also exists a parallel subdivision of tests: a relative machinability test known as ranking test to compare the machinability of two or more work-tool material combinations, and an absolute test to indicate the merit of a given combination of work-tool pair. Table 2 shows a list of short-term machinability tests covering these four major types of tests for a few machining operations indicating the major test parameters [9]. Having noted the significant effect of work material microstructure, Zlatin and Field [10] showed the relative tool-life at constant cutting speed and relative cutting speed at constant tool-life for a range of microstructure types (Table 3). Field [11] and Murphy and Aylward [12] showed that tool-life is primarily dependent on the relative proportion of pearlite and ferrite and that an optimum ratio of the two exists. Subsequent works show that the minor elements (e.g., Mn, S) have a considerable effect on machinability [13–15].

Henkin and Datsko [16] developed a general machinability equation

$$V_{60} \propto \frac{B}{LH_{B}} \left(1 - \frac{A_{r}}{100}\right)^{0.5} \tag{1}$$

where B is the thermal conductivity of the material, L is a characteristic length, H_{B} is the Brinell hardness of the material, $A_{r} = \%$ reduction in area (from tensile test).

In 1958, Volvo (Sweden) developed an innovative machinability test that primarily compares the machin-

Table 2 Well-Known Short-Term Machinability Tests

Test	Type				Operation			Test parameter		
	Machining	Nonmachining	Absolute	Ranking	Turning	Drilling	Sawing	Workpiece	Tool	Cutting fluid
Chemical composition		*	*		*			*		
Microstructure		*	*		*			*		
Physical properties		*	*		*			*		
Rapid facing	*		*		*			*	*	*
Constant pressure	*		*		*	*	*	*		*
Taper turning	*			*	*			*	*	*
Variable-rate machining	*			*	*			*	*	*
Step turning	*			*	*			*	*	*
Degraded tool	*		*		*			*		
Accelerated wear	*		*		*			*		
Tapping	*		*							*
High-speed-steel tool wear rate	*			*	*			*		

Source: Ref. 9.

ability of a given bar material to that of a standard free-cutting steel in a 22-min tool-life. Results of the test provide a "B index," which defines the machinability of a given material. It was shown that this method offers a major advantage over other machinability tests in terms of its reduced requirements of work material quantity for testing [17]. In more recent times, mathematical models for the assessment of machinability have emerged [18,19].

B. Machinability Database Systems

For over several decades, machinability data have been compiled and made generally available in the form of

Table 3 Effect of Work Material Microstructure on Machinability of Steels

Type of microstructure	Brinell hardness	V_{20} cutting speed (m min^{-1}) (carbide tool)	Machinability rating	
			Relative life at constant speed (min)	Relative speed at constant tool life (m min^{-1})
10% Pearlite + 90% Ferrite	100–120	290	8	22
20% Pearlite + 80% Ferrite	120–140	260	6	20
25% Pearlite	150	—	—	15
Spheroidized	160–180	180	5	14
50% Pearlite + 50% Ferrite	150–180	—	4	11
50% Fine pearlite + 50% Network ferrite	202	—	—	10
75% Pearlite + 25% Ferrite	170–190	140	3	11
100% Pearlite	180–220	145	2	11
Tempered martensite	240	—	—	8
Tempered martensite	280–320	105	1	8
Tempered martensite	350	—	—	6
Tempered martensite	370–420	46	0·2	3

Source: Ref. 9; Based on Ref. 10.

handbooks, data sheets, charts, nomograms, etc. These are usually prepared by national organizations, leading metal cutting research institutions, cutting tool manufacturers, and universities. With the growing need for developing computer-aided process planning (CAPP) systems, during the last four decades, significant effort has been made in developing computerized machinability database systems. The use of such database systems for machining process planning is a basic requirement for the success of a CAPP system. What began as mainframe computer-based machinability databases have, over the years, transformed into highly powerful and fast personal computer-based database systems with notable systems such as those established by the General Electric Company and by the Technical University of Aachen [20–22]. Comprehensive analysis of such systems appeared in the 1970s and 1980s [23–25].

The basic objective of a machinability data system is to select cutting conditions (usually the cutting speed and feed) for a given set of characteristics such as

(a) Type of machining operation
(b) Machine tool
(c) Cutting tool
(d) Workpiece
(e) Operating parameters (other than cutting speed and feed)

Cutting tool manufacturers have developed and very effectively promoted the use of customized machining databases for the selection of cutting tools, primarily focusing on their own cutting tools. Computerized machinability data systems can be classified into two general types:

(a) Database systems; and
(b) Mathematical model systems.

The first system requires collection and storage of large quantities of data from laboratory experiments and from the shop floor experience. To collect these data, machining experiments need to be performed over a range of feasible conditions. For each set of conditions, computations are made to determine the cost of operation. These computations are generally based on the traditional machining economics concepts. The mathematical model systems attempt to predict optimum cutting conditions for an operation. These systems generally utilize the traditional Taylor-type or extended Taylor-type tool-life equations, which are overly conservative.

Notable efforts to develop a computerized databank containing several empirical relationships include the work by Venkatesh in 1986 [26] and its extension with the development of a knowledge-based expert system for machinability data selection [27].

C. Limitations of Current Machinability Database Systems

The best-known machinability data handbook produced by Metcut includes extensive data for machinability [28]. A further development is the computer package CUTDATA, which is a computerized machinability database known to contain over 90,000 machinability recommendations for over 3750 work materials and over 40 machining operations. This database has been integrated into the MetCAPP process planning system [29]. Contemporary works in Germany and Japan have produced similar systems.

Despite significant efforts being made in developing more and more machinability database systems, there are several major aspects that are not fully considered in almost all these systems. These include:

- Property and performance data for new cutting tools (material and geometry) are not available in any database system. This is particularly true for

  (a) newly developed advanced tool materials;
  (b) different types of wear-resistant tool coatings;
  (c) variations of tool geometry/tool holder applications; and
  (d) chip-grooves and chip-forming toolface configurations.

- The traditionally known machinability assessment methods are all based on one (or two, in some cases) machining performance parameters such as tool-life, cutting forces (or power consumption), surface roughness, and machining accuracy. The "Total Effect" of all major parameters affecting the machining process quality and productivity, however, remains unknown.
- The machine tool static and dynamic characteristics are generally not considered in current practice. There is a pressing need to consider *machinability as a systems property* to include the combined effects of the three functional elements (work material, cutting tool, and machine tool) into a comprehensive and total system.

D. Machinability Interrelationships

Modeling of machining processes for the integrated effects of interacting multiple machinability parameters is a complex task. A greater knowledge and understanding of the interrelated machinability parameters (cutting forces, surface finish, chip-forms/chip breakability, tool-wear, and machining accuracy) is essential for this. Some preliminary work conducted over a decade ago established the interrelationships between the surface roughness and chip-form/chip breakability (Fig. 2) [30], and between the progressive tool-wear rates and chip breakability (Fig. 3) [31]. Building on this knowledge and developing suitable analytical, numerical, or empirical models for incorporating within a computer-aided process planning system is very useful for predictive assessment. However, using predicted machinability parameters to optimize the overall machining performance is a more productive option from the eventual machining economy point of view.

E. The Cutting Tool Factor in Machinability Assessments

The research and application focus on the cutting tool, primarily from the material science viewpoint, centers around a short list of hard materials. The hard phases are incorporated in the bulk and in the coating. Table 4 shows a summary of cutting tool materials and coating processes. The ultimate cutting tool material combines the highest fracture strength (feed rate capability) and wear resistance (speed capability) at the cutting edge. Recent technological advances offer creative consolidations of hard phases in the bulk material and surface coatings to approach this goal. There are tradeoffs between wear resistance and fracture strength between tool materials, as illustrated in Fig. 4. The strength and ductility of workpiece materials determine the applied stress and the temperature field generated at the cutting edge.

1. Tool Material Requirements

The ideal requirements of a satisfactory cutting tool can easily be defined, but it is more difficult to specify a tool material that meet all these requirements over a wide range of cutting conditions. The physical and metallurgical requirements of a good cutting tool material include the following:

(a) High yield strength at cutting temperatures;
(b) High fracture toughness;
(c) High wear resistance;
(d) High fatigue resistance;
(e) High thermal capacity and thermal conductivity;
(f) Low solubility in the workplace material;
(g) High thermal shock resistance; and
(h) Good oxidation resistance.

2. Cutting Tool Classification

According to ISO [32], cutting tool materials are classified into four major groups, based roughly on their chemical compositions. These are hard metals, ceramics, boron nitride, and diamond, and they are designated by letter symbols, H, C, B, and D, respectively. These main groups and the subgroups for the most popular first two groups (i.e., H and C groups) are given in Table 5.

Table 6 shows the three most common application groups from the ISO-based classification of recommended applications where the work materials are classed into six main groups with the appropriate letter symbols—P, M, K, H, S, and N [32]. In this table, for each group, the wear resistance and cutting speed increase from bottom to top, while the feed and toughness increase from top to bottom.

The application selection table (Table 7) provides the recommended combinations of tool materials and work materials for three levels of applications: usual application, limited application, and special purpose application [33]. There exists various toughness levels within

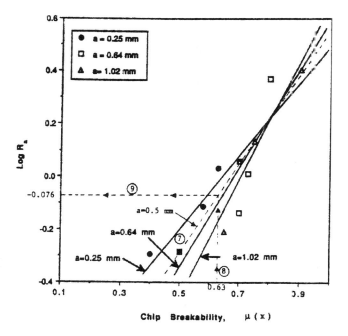

Figure 2 Surface roughness–chip breakability interrelationships. (From Ref. 30.)

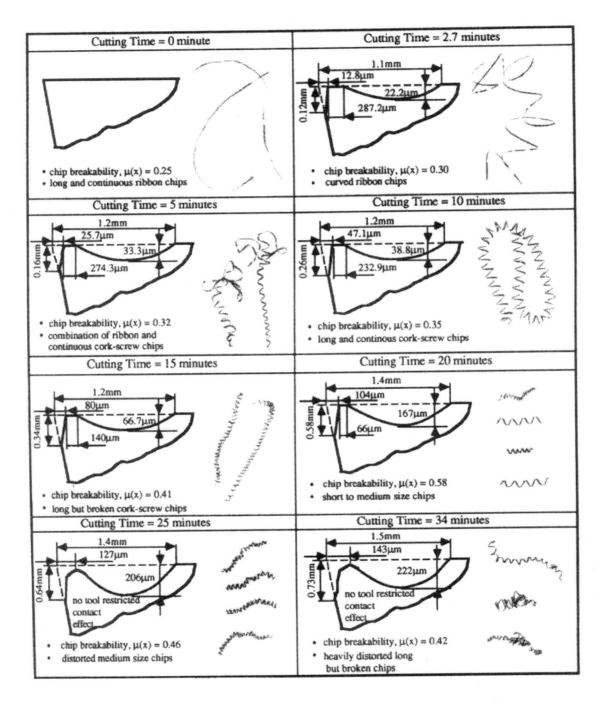

Figure 3 Tool-wear–chip breakability interrelationships. (From Ref. 31.)

each of the work material identification groups and these are recognized with numerical values: 01, 10, 20, 30, 40, etc., in an increasing order of toughness required by the machining operation—see Fig. 5. The required toughness generally increases when the feed rate increases as shown in Fig. 5.

3. Cutting Tool Geometry and Chip-Groove Selection

Having selected the cutting tool material for a given set of cutting conditions and work materials, it becomes vital to choose the type (or shape) of the cutting tool. In general, the insert shape should be selected relative to the cutting edge (or lead) angle and the accessibility or versatility required for the application. For strength and economy, the largest point (included) angle is selected. Figure 6 shows the range of standard tool shapes in varying point angles. Scale 1 shows "S" for strength (for cutting edge) and "A" for accessibility at each end, while Scale 2 shows "V" for vibration tendency and "P" for power requirement as two extreme factors for the selection of an insert shape [34]. Figure 7 shows the basic factors affecting the choice of insert shape [34].

Also, among the most significant factors considered in the selection of the tool type are the cutting tool geometry and the type of clamping of the tool insert. The geometry of the tool plays a critical role in achieving the desired surface finish producible on the machined surface and in controlling the chips (i.e., breaking the chips into small and acceptable shapes and forms). The tool nose radius affects the surface finish achievable according to the well-established (and idealized) model of surface finish generations by feed marks and the corresponding relationship. Higher feed rates and lower tool nose radius values produce rougher surfaces. However, in the finish machining range (i.e., at low feeds and depths of cut), the practically achievable surface roughness values are always larger than the theoretically estimated values [5,30]. The complex work-tool material interactions that take place at the cutting edge in finish machining has been shown to attribute to this.

Turning to the problem of chip control, it should be noted that the lack of chip control often leads to long and snarled chips, which are hazardous and do interrupt the machining operation. It is essential to produce chips broken into small and disposable sizes and shapes. The first step in breaking the chip is to effectively form the chip while it is flowing on the tool face. There are

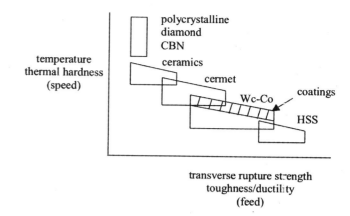

Figure 4 Tool material property trade-offs. (Courtesy of Kennametal, Inc.)

hundreds of different commercially available chip-forming tool configurations. Figure 8 shows a sample of such chip-forming tool inserts. The selection of chip formers (commonly known as chip breakers) is made based on the machining application involving a given work material and cutting conditions for an operation. Cutting tool catalogues and computer-aided databases provide some information on such selection although specifics are generally to be obtained from applications engineers and process planners. In automotive and aerospace machining, lack of chip control is still a chronic problem causing significant losses in productivity and costs. An assessment of chip former performance has been carried out in an extensive study on several different tool inserts having varying chip-groove geometries, under a wide range of cutting conditions and work materials for finish turning operations [35]. Specialized applications such as contour turning and boring operations generally require scientific studies on chip flow and effective tool geometric features for a range of cutting conditions. The significance of chip control in machining is internationally well recognized, and in 1993, the CIRP's working group on chip control produced an extensive study on chip control [36].

III. EVALUATION OF MACHINING PERFORMANCE MEASURES IN DESIGNING FOR MACHINING

Recent advances in developing and implementing CAPP systems for modern industrial environments have called for a need of predicting machining perform-

Table 4 Most Common Cutting Tool Materials and Coating Processes

Tool material	Hard phase	Binder/Matrix
Carbides	WC	Co
Cermets	TiC, TiN	Ni, Mo, Co,...
Ceramics	Al_2O_3, Si_3N_4	Glassy phases
PCD, PCBN	Diamond, CBN	Metal, ceramic

Coating Process	Coating Material
CVD	TiC, TiCN, TiN, Al_2O_3
PVD	TiN, TiCN, TiAlN
Plasma-assisted	Diamond

Table 5 ISO-based Cutting Tool Material Groups and Subgroups

Main groups of cutting materials	Letter symbol
Hardmetals	H
Ceramics	C
Boron nitride	B
Diamond	D
Groups of hardmetals	**Designation**
Uncoated hardmetals, consisting primarily of tungsten carbide	HW
Uncoated hardmetals, consisting primarily of titanium carbide or nitride	HT
Coated hardmetals	HC
Groups of ceramics	**Designation**
Oxide ceramics, consisting primarily of aluminum oxide (Al_2O_3)	CA
Mixed ceramics on the basis of aluminum oxide (Al_2O_3), with titanium carbide (TiC)	CM
Nitride ceramics	CN
Coated ceramics	CC
Whisker reinforced ceramics	CR

Source: Ref. 32.

ance. The most common technological machining performance measures are: cutting forces/power/torque, tool-wear/tool-life, chip-form/chip breakability, surface roughness/surface integrity, and part accuracy [37]. Figure 9 shows the inherent strong interactions among these performance measures and emphasizes the need for considering their integrated effect on the machining performance of a system. Table 8 (based on Ref. 37) shows the desired levels for these machining performance measures.

The term "machining performance" is discussed here from a *systems framework*, comprising three primary

Table 6 Most Common Cutting Tool Materials and Applications (ISO-Based)

Symbol	Broad categories of material to be machined	Designation	Material to be machined	Use and working conditions
		P 01	Steel, steel castings	Finish turning and boring; high cutting speeds, small chip section, accuracy of dimensions and fine finish, vibration-free operation.
		P 10	Steel, steel castings	Turning, copying, threading, and milling, high cutting speeds, small or medium chip sections.
		P 20	Steel, steel castings Malleable cast iron with long chips	Turning, copying, milling, medium cutting speeds and chip sections, planing with small chip sections.
P	Ferrous metals with long chips	P 30	Steel, steel castings Malleable cast iron with long chips	Turning, milling, planing, medium or low cutting speeds, medium or large chip sections, and machining in unfavorable conditions
		P 40	Steel, steel castings with sand inclusion and cavities	Turning, planing, slotting, low cutting speeds, large chip sections with the possibility of large cutting angles for machining in unfavorable conditions and work on automatic machines

Table 6 Continued

Symbol	Broad categories of material to be machined	Designation	Material to be machined	Use and working conditions
		P 50	Steel steel castings of medium or low tensile strength, with sand inclusion and cavities	For operations demanding very tough carbide: turning, planing, slotting, low cutting speeds, large chip sections, with the possibility of large cutting angles for machining in unfavorable conditions and work on automatic machines.
M	Ferrous metals with long or short chips and nonferrous metals	M 10	Steel, steel castings, manganese steel, gray cast iron, alloy cast iron	Turning, medium or high cutting speeds. Small or medium chip sections
		M 20	Steel, steel castings, austenitic or manganese steel, gray cast iron	Turning, milling. Medium cutting speeds and chip sections
		M 30	Steel, steel castings, austenitic steel, gray cast iron, high temperature resistant alloys	Turning, milling, planing. Medium cutting speeds, medium or large chip sections
		M 40	Mild free cutting steel, low tensile steel Nonferrous metals and light alloys	Turning, parting off, particularly on automatic machines
K	Ferrous metals with short chips, nonferrous metals, and nonmetallic materials	K 01	Very hard gray cast iron, chilled castings of over 85 Shore, high silicon aluminum alloys, hardened steel, highly abrasive plastics, hard cardboard, ceramics	Turning, finish turning, boring, milling, scraping
		K 10	Gray cast iron over 220 Brinell, malleable cast iron with short chips, hardened steel, silicon aluminum alloys, copper alloys, plastics, glass, hard rubber, hard cardboard, porcelain, stone	Turning, milling, drilling, boring, broaching, scraping
		K 20	Gray cast iron up to 220 Brinell, nonferrous metals: copper, brass, aluminum	Turning, milling, planing, boring, broaching, demanding very tough carbide
		K 30	Low hardness gray cast iron, low tensile steel, compressed wood	Turning, milling, planing, slotting, for machining in unfavorable conditions and with the possibility of large cutting angles
		K 40	Soft wood or hard wood, Nonferrous metals	Turning, milling, planing, slotting, for machining in unfavorable conditions and with the possibility of large cutting angles

Source: Ref. 32.

Table 7 ISO-Based Work Material–Cutting Tool Application Selection

Cutting tool materials		Steel (P)	High alloyed steel (M)	Cast iron (K)	Non-ferrous metals non-metallic materials (N)	Hard metallic materials (H)	Special alloys (S)
				Materials to be machined			
Hardmetals	HP	●	○				○
	HM	○	●	○		●	○
	HK		○	●	●	●	(○)
	HT	●	○	○			
	HC	●	●	●			
	HC-P	■	○				
	HC-K		○	■			
Ceramics	CA	○		●		○	
	CM	○		●		○	
	CN			●			●
	CC			●			●
	CR		●	●		●	●
Boron nitride	BN			○		●	
Diamond	DK				●		
	DP				●		

● = usual application; ○ = limited application; ■ = special purpose
Source: Ref. 32.

elements that constitute the machining system: *the machine tool, cutting tool, and work material*. As shown above, traditional approaches have only found limited applications with work material-related "machinability" as these approaches have excluded rapid technological advances in the cutting tool and machine tool industry. An example is the continuous development of advanced chip-grooves and tool coatings to improve chip control and tool-life in machining. Also, other subconstituents such as jigs and fixtures, coolants, etc. need to be considered in the assessment of the overall machining performance. Kahles [25], in his international survey on machinability data requirements for advanced machining systems, emphasized the need for reliable data in areas concerning tool-life, chip control, dimensional accuracy, surface finish, and surface integrity. Shen and van Luttervelt [38] showed the concept of a robust CAPP integrated intelligent machining system (RIMAS) with a hybrid intelligent methodology for the assessment and optimization of machining operations. The need for predictive modeling of machining perform-

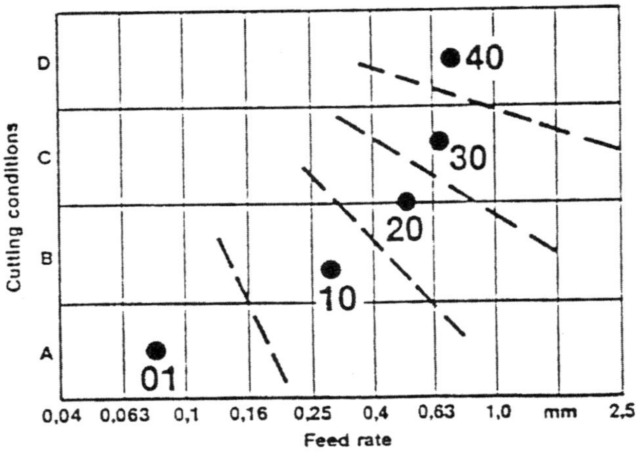

Figure 5 Tool material toughness application range. (From Ref. 33.)

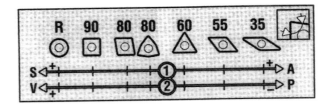

Figure 6 Tool insert shape selection guide. (From Ref. 34.)

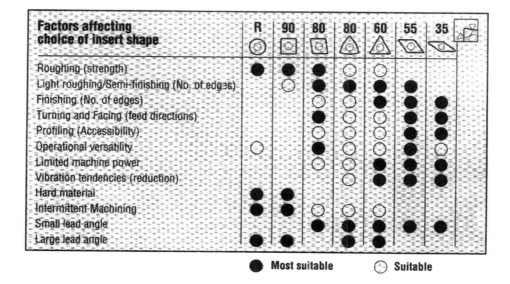

Figure 7 Factors affecting the choice of insert shape. (From Ref. 34.)

Figure 8 Sample of commercially available chip-forming tool inserts.

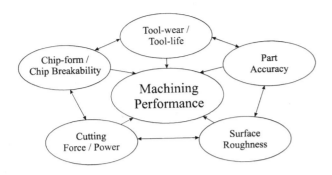

Figure 9 Machining performance interrelationships.

ance measures have been highly emphasized in recent times [39–41].

A. Predictive Modeling of Machining Performance

Modeling activities that are traditionally carried out for the prediction of machining performance measures are based on analytical (mechanics-based) techniques, computational/numerical methods [e.g., finite element model (FEM)] and artificial intelligence (AI)-based modeling using fuzzy logic techniques, neural networks, etc. Prediction of machining performance typically involves two broad phases: stage 1 involves modeling of machining variables such as stresses, strains, strain rates, temperatures, etc. and Stage 2 involves the modeling of performance measures such as those discussed in the previous section. A schematic representation of the modeling methodology is shown in Fig. 10 [42].

It can be seen that for the modeling of machining variables, typically analytical or numerical methods ar reasonably effective depending on the suitability of the input and validation from experimental work. However, from the practical modeling perspective for industrial applications, the modeling of machining performance, in the form of such measures as tool-wear/tool-life, chip-form/chip breakability, etc. throws up a serious challenge because of transitional problems in converting *machining variables to performance measures.* Hence the main theme running throughout this section of the chapter is the need for integrating different modeling methodologies to effectively predict machining performance. Usui shows the urgent need for developing predictive theories for practical machining and advocates the use of AI techniques to fill in the gaps

because of the complex nature of the machining process [43].

1. Analytical and Numerical Modeling of Cutting Forces, Chip Thickness, Chip Flow and Chip Curl

Significant effort has been made to develop analytical predictive models for machining [44–57]. The recently established universal slip-line model (SLM) for machining producing curled chip formation with restricted contact and grooved tools predicts the cutting forces, chip thickness, chip curl radius, and chip back-flow angle for a given set of cutting conditions, tool geometry, and the state of stresses [58–60]. Extensive experimental work has been conducted to validate and test this model for cutting forces, chip thickness, chip back-flow angle, and chip up-curl radius [61]. A most recent modeling effort using the universal slip-line model, in conjunction with Oxley's predictive model incorporating the effects of strains, strain rates, and temperature, offers prediction of cutting forces for 2-D machining with grooved tools [62]. Also, significant work has been conducted on finite element modeling of the machining process using grooved tools. A custom-made finite element model (FEM) has been developed for predicting stresses, strains, strain rates, and temperatures in machining with grooved tools [63,64]. In what follows, a short description of the present knowledge on SLM and FEM will be discussed, but first it is necessary to understand the strengths and weaknesses of these two methods.

Based on the rigid-plastic plane-strain material assumptions, the slip-line method does not consider the temperature effect in material deformation. It is also not easy to include the strain-hardening effect in the slip-line analysis as shown in Ref. 62. However, the slip-line analysis can give approximate solutions to chip geometry (such as chip thickness and chip curl radius) and the

Table 8 Major Machining Performance Measures and Their Desired Levels

Machining performance measure	Desired level
Cutting forces/power/torque	Minimum
Tool-wear/tool-life	Minimum/maximum
Chip form/chip breakability	Disposable/maximum
Surface roughness/surface integrity	Minimum/maximum
Part accuracy	Maximum

Source: Based on Ref. 37.

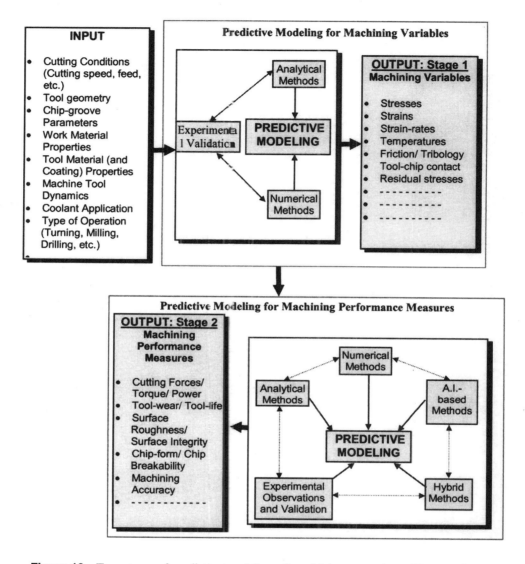

Figure 10 Two stages of predictive modeling of machining operations. (From Ref. 42.)

position of the primary shear plane (the primary shear-plane position is often indicated by the shear-plane angle) for a given state of stresses. The finite element method considers both temperature and stain-hardening effects; however, the chip thickness, chip curl radius, and shear-plane angle need to be assumed before constructing a suitable finite element mesh. Indeed, it is generally known that the most important aspect of FEM is establishing the boundary conditions to adequately describe the mechanical and thermal boundaries of the cutting region in machining. It is imperative that such boundary conditions can be easily established from the basic output parameters of the SLM. There-

fore, it is deemed promising to develop a *hybrid model* that can effectively integrate the slip-line model and the finite element model (FEM) to combine the advantages of both techniques and to overcome their shortcomings.

Slip-Line Modeling (SLM) of Chip Formation

A universal slip-line model for machining with restricted contact and grooved tools involving curled chip formation was recently developed at the University of Kentucky [58–62]. This model involves a convex upward shear plane AC and four slip-line angles $(\theta, \psi, \eta_1, \eta_2)$, as shown in Fig. 11. The tool cutting edge

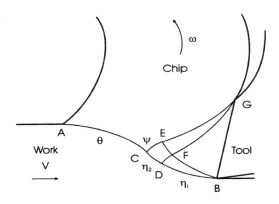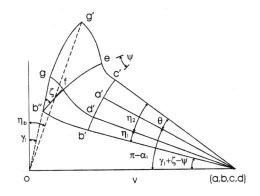

Figure 11   The universal slip-line field and hodograph for machining with restricted contact cut-away tools. (From Ref. 58.)

is still assumed to be perfectly sharp with this model. The model is universal in the sense that it incorporates six slip-line models previously develop for machining in the past five decades as special cases [58].

A FORTRAN computer program has been developed using Dewhurst and Collins' [65] matrix technique and Powell's [66] optimization algorithm. The slip-line GE (see Fig. 11) is taken as the base slip-line and its radius of curvature denoted by a column vector GE is determined by:

$$A \cdot GE = B \cdot (\rho/\omega) \cdot c \qquad (2)$$

where ρ is the magnitude of the velocity jump across the slip-line $ACDB$, ω is the angular velocity of the curled chip rotation, A and B are two matrices determined by:

$$A = I - (P_{\eta_1\eta_2}R_{\eta_1}G_\varsigma Q_{\psi\eta_1}P_{\eta_2}^* + Q_{\eta_2\eta_1}Q_{\psi\eta_2})$$
$$\times (P_{\psi\eta_2}Q_{\eta_1}^* G_\varsigma P_{\eta_2}^* + Q_{\eta_2\psi}Q_{\eta_1\eta_2}) \qquad (2a)$$

$$B = P_{\eta_1\eta_2}R_{\eta_1}G_\varsigma(P_{\eta_1\psi} + Q_{\psi\eta_1}Q_{\eta_2}^*) + Q_{\eta_2\eta_1}P_{\eta_2\psi} \qquad (2b)$$

where P, P^*, Q, Q^*, G, and R are members of a set of basic matrix operators defined by Dewhurst and Collins [65]. I is the unit matrix.

Using the universal slip-line model, some important machining parameters, such as chip thickness t_2 and chip curl radius R_u, can be determined.

$$t_2 = 2\left(\frac{V_{g'}}{\omega} - R_u\right) \qquad (3)$$

where $V_{g'}$ is the magnitude of the velocity of the chip at the moment of departing from the tool restricted contact land.

$$R_u = \frac{1}{2}\sqrt{\left(\frac{V}{\omega}\right)^2 + \left(\frac{\rho}{\omega}\right)^2 + 2\frac{V}{\omega}\frac{\rho}{\omega}\cos\alpha_1} + \frac{V_{g'}}{2\omega} \qquad (4)$$

This slip-line model has been validated using extensive experimental data covering a large range of cutting conditions [59–61]. The predicted results for cutting forces, chip thickness, chip up-curl radius, and chip back-flow angle have shown a reasonable agreement with experiments. The most recent analysis shows how the model can be extended to the modeling of machining with grooved tools, the more practical machining operation [62]. This is carried out by incorporating an additional backwall force acting on the generated chip into the model (Fig. 12). Coulomb friction conditions are assumed at the tool–chip contact point G_1 on the groove backwall. Figure 12 also shows the two components of the backwall force, i.e., frictional force F_b and the normal force N_b [62]. Good correlation is obtained between predicted and experimental resultant force—see Fig. 13.

Finite Element Modeling (FEM) of Chip Formation

Finite element techniques have been widely used in machining research [67–75]. Considering the very large plastic deformation in machining, the elastic deformation has been largely neglected and a thermo-viscoplastic finite element model is used in recent works [63,64]. The material in the machining process is approximated as the

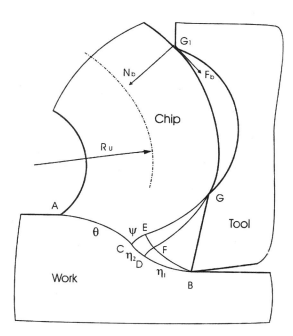

Figure 12 The universal slip-line field and hodograph for machining with restricted contact grooved tools. (From Ref. 62.)

flow of a non-Newtonian viscous fluid. Three important problems involved in the finite element analysis, i.e., constitutive material model, tool–chip friction pattern, and heat transfer, are answered as follows.

CONSTITUTIVE MATERIAL MODEL. Various types of constitutive material models have been developed and used in the numerical modeling of machining processes. For example, the material shear flow stress σ_0 is determined as [76]:

$$\sigma_0 = A\left(10^{-3}\dot{\bar{\varepsilon}}\right)^M e^{k_1 T}\left(10^{-3}\dot{\bar{\varepsilon}}\right)^m$$

$$\times \left[\int e^{-k_1 T/N}\left(10^{-3}\dot{\bar{\varepsilon}}\right)^{-m/N} d\bar{\varepsilon}\right]^N \qquad (5)$$

where A, N, M, k_1, and m are material constants determined experimentally, and $\dot{\bar{\varepsilon}}$ and T are the strain rate and temperature, respectively. Because of the very large strains, strain rates, and high temperatures involved in machining, Johnson and Cook's material law [76] is commonly used in numerical analysis. A von Mises yield criterion and an isotropic strain hardening rule are assumed. The material shear flow stress σ_0 is given by

$$\sigma_0 = \left(A + B\bar{\varepsilon}^n\right)\left(1 + C\ln\dot{\bar{\varepsilon}}^*\right)\left(1 - T^{*m}\right) \qquad (6)$$

where A, B, C, m, and n are the material parameters, and $\bar{\varepsilon}, \dot{\bar{\varepsilon}}^*$, and T^* are the effective strain, the nondimensional effective strain rate, and the nondimensioned temperature, respectively.

TOOLFACE FRICTION PATTERN. The analysis of tool–chip friction is a very complex task in machining [44,77]. In early research, Coulomb friction has been widely used to represent the tool–chip friction pattern in machining. However, extensive metal cutting experiments have shown that the Coulomb friction law does not hold in machining. Some other tool–chip friction relationships have also been developed such as [78]:

$$\tau/k = 1 - \exp(\mu\sigma/k) \qquad (7)$$

where τ and σ are the frictional and normal stresses on the tool rake face, k is the shear flow stress of the

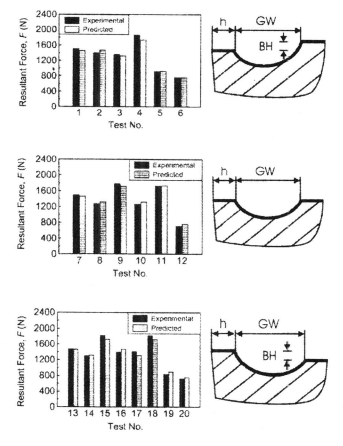

Figure 13 Comparison of predicted and experimental resultant force in machining with grooved tools. (From Ref. 62.)

strained chip on the tool rake face, and μ is the characteristic constant of friction. In recent works [63,64], the following tool–chip friction pattern was used.

$$\tau/k = f \tag{8}$$

where f is the frictional parameter. Equation (7) follows the tool–chip friction assumption used in the universal slip-line model [58].

TEMPERATURE DISTRIBUTION. The following heat conduction–convection equation [79] for determining the temperature distribution in the plastic deformation region has been employed in recent works [63,64]:

$$k_{th}\left(\frac{\partial^2 T}{\partial x^2} + \frac{\partial^2 T}{\partial y^2}\right) - \rho C_p \left(u_x \frac{\partial T}{\partial x} + u_y \frac{\partial T}{\partial y}\right) + \dot{q} = 0 \tag{9}$$

where k_{th} is the thermal conductivity, ρ is the density, C_p is the specific heat, q is the volumetric heat generation rate, and T is the temperature. Because the tool and the chip are in intimate contact along the tool–chip interface, the temperature is assumed to be continuous across the interface. Figure 14 shows a typical output from the present FEM for 2-D machining with a grooved tool.

Limitations of Numerical Modeling for Practical Machining Operations

Numerical models (typically FE-based models) are capable of predicting machining variables such as stresses, strains, temperatures, etc. for the 2-D machining process. Such type of modeling is capable of pro-

viding fundamental insight and simulation capabilities for different combinations of cutting conditions and work–tool material pairs. However, numerical modeling, on its own, runs into severe problems because of the following reasons:

1. *Difficult transition to machining performance measures*: The majority of numerical models are restricted to simplified 2-D machining processes and are very unwieldy to translate into practical machining operations such as turning, milling, etc.
2. *Need for accurate constitutive data*: The accuracy of results from the numerical model heavily depends on the constitutive data that defines the material behavior. Obtaining accurate and robust data has been a major cause for concern in the numerical modeling community.

It can be seen that numerical modeling alone is unable to provide a comprehensive solution for performance of a machining system. Hence there is a need for hybridizing the modeling approaches by combining several traditional approaches to achieve realistic solutions for machining system performance.

2. Empirical Modeling of Tool-Life

Tool-wear/tool-life investigations in machining have been among the most significant research topics during the last several decades. A large domain of specific knowledge has been acquired and many tool-wear/tool-life models have been developed through analytical modeling and experimental observations [80–82]. Early work by Colding set the pace for renewed interest in

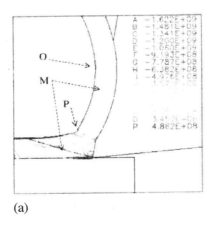
(a)

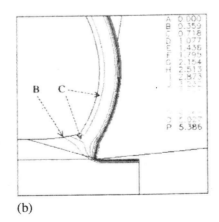
(b)

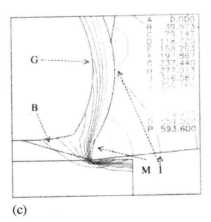
(c)

Figure 14 Typical FEM results for (a) hydrostatic stress, (b) strain, and (c) temperature distributions. (From Ref. 63.)

machining optimization for productivity and cost [83]. In his latter work [84], he expressed his initial tool-life relationship, which was established as a function of equivalent chip thickness (ECT) and cutting speed, in terms of measurable cutting temperature, by developing a new temperature–tool-life relationship. A computer-based mathematical model developed by Lindstrom [85] for tool-life estimation in an adaptive control system includes the capability for extrapolation of the cutting data field. This work was later extended to include a statistical evaluation aimed at economic optimization [86].

All presently known tool-life testing methods specified by major standards are based on the use of flat-faced cutting tool inserts and there are no effective equations for predicting tool-life in machining with grooved tools [6,87]. Significant progress has been made in developing new chip-groove geometries, tool materials, coating techniques, and in the associated development and implementation of tool-life testing methods. However, the dual role of a tool insert for providing longer tool-life and effective chip breaking has not yet been fully investigated despite obvious experimental evidence observed in machining with grooved tools, which fail too frequently because of inappropriate design and/or use of chip-grooves. The tool-life data provided by the tool manufacturers are all basically for the various tool grades with none representing the combined effects of tool grades and chip-groove configurations. Moreover, it is also a well-known practice to use a given tool-wear criterion such as flank wear and/or crater wear limits, for tool-life estimates. However, it is generally observed that the tool failure is largely a result of a number of different concurrently occurring progressive tool-wear types, such as crater wear, nose wear, flank wear, notch wear, and edge chipping [88–92].

A Summary of Currently Available Theories for Tool-Life

Among the several experimental and theoretical relationships reviewed, eight types of tool-life equations have been identified as relevant and most common in "academic" investigations, and these are summarized in Table 9. Unfortunately, these methods do not agree well with industrial production results over wide ranges of machining process parameters and work materials. Poor predictability of tool-life in current practice is largely as a result of the following reasons:

(a) The most commonly used tool-life criteria (e.g., flank wear and crater wear) often mask the influence of other types of concurrent wear types

such as the notch wear, groove backwall wear, nose wear, and edge chipping, which largely contribute to tool failure.

(b) All empirical constants for predicting tool-life are predetermined over a wide range (e.g., $n = 0.2$–0.3 for carbide tools and $n = 0.4$–0.7 for ceramic tools, etc.). Accuracy and consistency of these values are generally poor. Also, the effect of currently available different tool coatings is not yet fully considered in predictive assessments.

(c) Almost all currently existing tool-life testing methods are based on flat-faced cutting tools. The effects of varying chip breaker configurations and the related chip flow variations have not been considered as yet.

Although various tool-wear mechanisms do exist, it is generally known that the gradual (progressive) tool-wear is produced by temperature-dependent mechanisms. The extended Taylor equation is usually considered a good approximation to predict tool-life T, which is expressed in terms of cutting speed V, feed f, and depth of cut d with empirical constants C, n, m, and l as:

$$T = \frac{C}{V^{1/n} f^{1/m} d^{1/l}} \tag{10}$$

In machining with grooved tools, it has been observed that the traditional parameters for wear measurement (VB—flank wear, and KT—crater wear) are inadequate for characterizing the multiple, concurrent, and complex wear mechanisms undergone because of the interactions of the cutting conditions and the chip-groove geometry.

New Tool-Life Relationships

Jawahir et al. [90–92] have presented a new methodology for measuring the multiple tool-wear parameters in a grooved tool. Figure 15 shows these measurable grooved tool-wear parameters for a generic grooved tool. Figure 16 shows the significant variations in the tool-wear patterns and the corresponding tool failure modes in machining with various grooved tools under the same conditions [90]. It is apparent that the accurate estimation of tool-life in grooved tools is heavily dependent on the precise prediction of the tool-wear rates at different locations of a grooved tool.

Also, the recent work on tool-life includes the effects of tool coatings and chip-groove geometry, and the corresponding tool-life equation is expressed

Table 9 Historical Development of Commonly Known Tool-Life Equations

Year	Tool life equation (T: tool-life in min)	Determination of constants	Comments
1907	Taylor's basic equation: $VT^n = C$ where V is the cutting speed and T is the tool-life.	C and n are experimentally determined and currently available from many reference sources	Most widely used equation, however, C and n apply only to particular tool/workpiece combinations.
1907	Taylor's reference—speed-based equation: $\left(\dfrac{V}{V_R}\right) = \left(\dfrac{T_R}{T}\right)^n$ where V_R is the reference cutting speed for reference tool-life $T_R = 1$ min	n is experimentally determined and currently available from many reference sources	n applies only to particular tool/workpiece combinations.
1937	Temperature-based tool life equation: $\theta T^n = C_3$ where θ is the tool temperature.	• n is found between 0.01 and 0.1. • C_3 is experimentally determined.	Although the equation is not set only on an empirical basis, it is not convenient for the practical use in shopfloor environment.
1959	Colding's equation based on ECT (equivalent chip thickness): $y = K - \dfrac{(x-H)^2}{4M} - (N_0 - Lx)z$ where $x = \ln ECT$, $y = \ln V$, and $z = \ln T$	• Five constants (K, H, M, N_0, and L) are empirically determined. • Feed, depth of cut, lead angle, and cutting edge length are integrated into a single parameter ECT.	• It is claimed that the accuracy can be -50% to $+100\%$. • Tool-life prediction is inconsistent.
1973	Taylor's extended equation: $T = \dfrac{C_2}{V^p f^q d^r}$ where f is the feed and d is the depth of cut.	All constants (C_2, p, q, and r) are experimentally determined.	Gives better accuracy than Taylor's basic equation, but requires more tool-life tests
1984	Taylor's extended equation including cutting conditions and workpiece hardness: $V = \dfrac{C_5}{T^m f^y d^x (\mathrm{BHN}/200)^n}$	All constants (C_5, m, y, x, and n) are experimentally determined.	It is claimed to be a good approximation for tool-life ranges of 10–60 min
1986	Taylor's extended equation including cutting conditions and tool geometry: $T = C_4\, V^n f^m d^p r^q s^t i^u j^x$	Requires excessive tool-life tests to determine all constants (C_4, n, m, p, q, t, u, and x).	It is claimed that the data for setting up the equation are generated from both laboratory and industrial sources.
1989	Taylor's basic equation including rake angle and clearance angle: $C \propto \left[(\cot\beta - \tan\alpha)^n F(\alpha,\beta)^{1/\varepsilon}\right]^{-1}$ where $F(\alpha,\beta)$ is a function of α and β.	The influence of α and β can be theoretically determined as a part contribution to Taylor's constant C.	A complicated relationship between tool-life and rake/clearance angles.

Source: Based on Ref. 89.

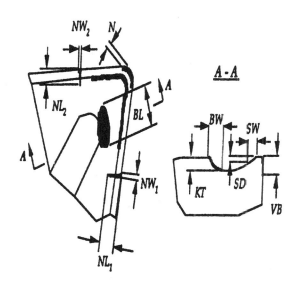

VB flank wear
BW width of groove backwall wear
BL length of groove backwall wear
KT depth of groove backwall wear
SW width of secondary face wear
SD depth of secondary face wear
N nose wear
NL_1 notch wear length on main cutting edge
NW_1 notch wear width on main cutting edge
NL_2 notch wear length on secondary cutting edge
NW_2 notch wear width on secondary cutting edge

Figure 15 Measurable tool-wear parameters in a grooved tool. (From Ref. 90.)

as [90],

$$T = T_R W_g \left(\frac{V_R}{V}\right)^{W_c/n} \tag{11}$$

where T is the tool-life, V is the cutting speed, n is Taylor's tool-life exponent, W_c is the tool coating effect factor, W_g is the chip-groove effect factor, T_R is the reference tool-life, and V_R is the reference cutting speed. The coating effect factor W_c and the chip-groove effect factor W_g are determined as

$$W_c = \frac{n}{n_c} \tag{12}$$

$$W_g = \frac{blm}{f^{n_1} d^{n_2}} \text{ and } l = \frac{d - r_\varepsilon(1 - \cos\kappa_r)}{\sin\kappa_r} + \left(\kappa_r + \sin^{-1}\frac{f}{2r_\varepsilon}\right)\frac{\pi r_\varepsilon}{180} \tag{13}$$

where n_c is the actual tool-life slope modified by the coating effect, which can be determined from the actual

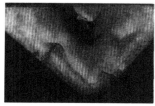

(a) CNMG 432MG

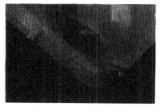

(b) CNMG 432K

(c) CNMG 432P

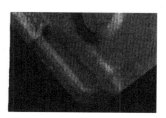

(d) CNMG 432

Figure 16 Variations in tool-wear parameters in machining with different grooved tools. Cutting Time = 2 min, V = 274 m/min, f = 0.43 mm/rev, d = 2.54 mm, Work Material = 1037 M Steel, Tool Coating = KC 850. (From Ref. 90.)

tool-life; and m is machining operation effect factor (with $m=1$ considered for turning); f is the feed; d is the depth of cut; and b, n_1, and n_2 are empirical constants. l is the engaged cutting length that can be calculated as shown above from feed f, depth of cut d, side cutting edge angle κ_r, and tool nose radius r_ε.

3. AI-Based Modeling of Chip-Form/Chip Breakability

The most common chip-form/chip breakability evaluation process is based on the "chip chart" method shown in Fig. 17. A new method for quantifying chip breakability was derived based on the fundamentals of fuzzy reasoning [93–95]. Figure 18 schematically shows the basic outline of the model. It is assumed that the following three factors determine the levels of chip breakability with the corresponding weighting factors:

- Size of chip produced (60%): defined by the dimensional features such as length and other geo-

metric parameters, e.g., diameter, or curl radius of the chip, or chip coil;
- Shape of chip produced (25%): defined by the geometric configurations of the chip such as helical form, spiral form, arc shape, etc.; and
- Difficulty/ease of chip producibility (15%): defined by the characteristics of the produced chip, for example, smoothness of the chip surface, chip color, any resulting burr formation, etc. These characteristics are expected to reflect the other performance measures such as cutting power, surface finish, tool-wear rate, etc.

It is quite obvious that all three factors are fuzzy in nature owing to the "uncertainty" in their definition levels, i.e., there are no well-defined boundaries. Chip breakability is classified into five fuzzy sets: Very Poor (VP), Poor (P), Fair (F), Good (G), and Excellent (E). In further analysis, a numerical rating system covering a range from 1 to 5 was used to represent these five levels of chip breakability. For example, chip-forms between

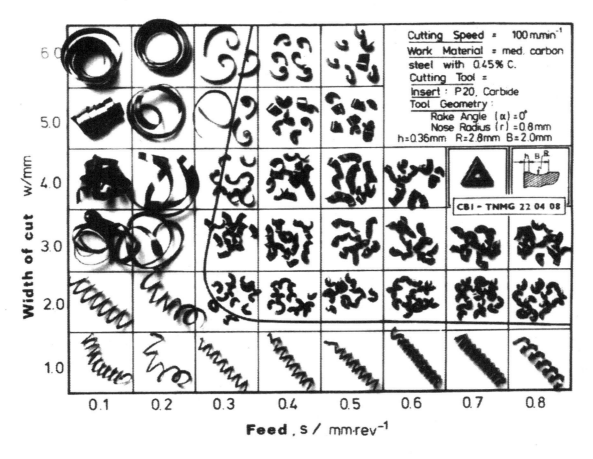

Figure 17 A typical chip chart.

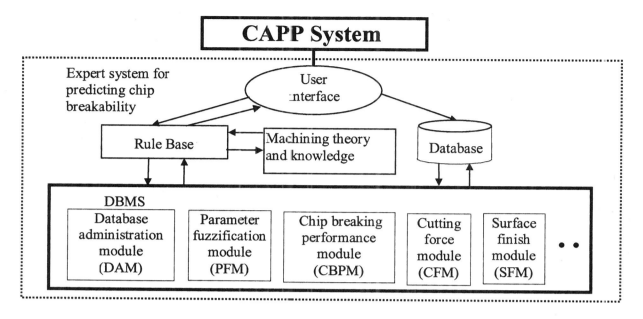

Figure 18 Modular approach used in the rule-based predictive system for chip breakability. (From Ref. 94.)

Poor (P) and Fair (F) may be assigned a value, based on their relative sizes and shapes, e.g., 2, 2.3, 2.5, 2.7, and 3. This avoids the inaccuracy caused by having to classify a chip-form that is in between F and P to a discrete value of either F or P.

The rule-based predictive system for chip breakability is developed based on the analysis of effective chip-groove profiles and is extended to include predictions for varying cutting conditions such as side-cutting edge angle, nose radius, and cutting speed with limited experiments. Suppose, a total of *n* chip-groove parameters are used as input to the rule-base system, and each parameter is fuzzified into *m* fuzzy sets. The goal is to find, based on the machining database, a rule-base that generates rules involving all possible combinations of these interacting chip-groove parameters with associated fuzzy ratings.

The rule-base is of the following form:

IF	Groove width is {GW} and Backwall height is {BH} and Backwall slope is {BS} and Rake angle is {RA}...
AND	Other conditions hold
THEN	Chip breakability is {CB}
AND	Chip breakability certainty-level is {certainty}

(14)

The results of this rule-based operation are then postprocessed using a fuzzy inference engine to obtain the final prediction. Figure 19 shows predicted chip breakability for a given cutting tool insert, based on a 2-D cross-sectional chip-groove profile obtained from a 3-D image of the cutting tool insert [94].

4. Hybrid Modeling

The discussion on modeling of machining performance in the previous section has highlighted the need for somehow integrating suitable models that use different modeling approaches to obtain practical and realistic solutions for present-day needs. An example of such an attempt at hybrid modeling is provided in the following case study, which deals with the complex problem of predicting dominant tool-wear modes in machining with grooved tools.

Case Study: Hybrid Modeling of Tool-Wear (Sensor-Based and Analytical Models)

EQUIVALENT TOOLFACE (ET) MODEL FOR PREDICTING DOMINANT MODE OF TOOL FAILURE. A new methodology for predicting the primary mode of tool failure in turning with complex grooved tools has recently been developed [96]. This method is based on the fact that the changes in cutting conditions and chip-groove geometry are reflected in the cutting forces. According to the measured, cutting forces at various stages of the tool-wear process, the geometry of the complex groove tool is approximated to a flat-faced tool and the

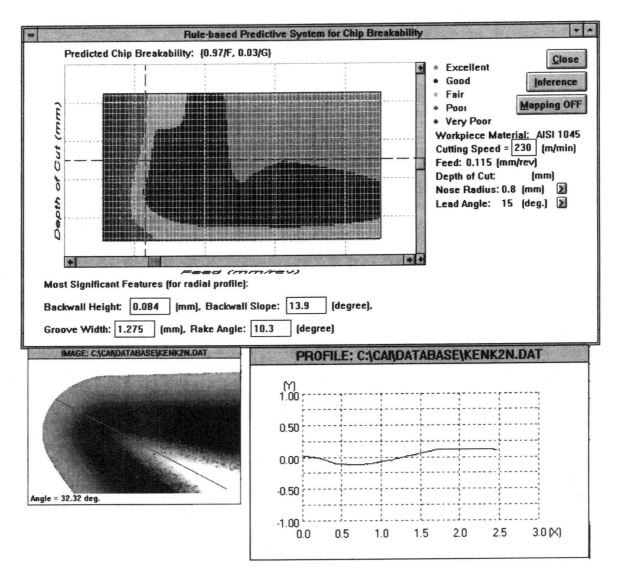

Figure 19 Predicted chip breakability for a specified grooved tool insert based on 2-D chip-groove profile obtained from a 3-D image of the tool insert. (From Ref. 94.)

geometric orientation of the grooved tool is represented by the equivalent effective rake angle and the equivalent effective inclination angle [97]. The variation of these two angles, when related to one of the four most common combinations shown in Fig. 20a, would give the most predominant regions of wear in the cutting tool. And this, in turn, can be related to the type of tool-wear according to the trends observed within the first few minutes of machining with a grooved tool. Figure 20b shows one such trend observed in machining of steel 1045 with a CNMG 432K grooved tool and the actual tool-

wear measured during the machining operation. In this case, within the first 2 min of machining, it is predicted that the tool will fail by nose wear, which was shown to be true after machining for 38 min (Fig. 20c).

INTEGRATED HYBRID MODELING. It can be seen that certain modeling approaches have distinct advantages over others: e.g., numerical predictions of temperatures, empirical predictions of tool-life and AI-based predictions of chip breakability. The challenge lies in avoiding exhausting efforts to try and find a universal

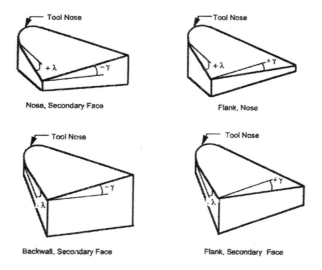

(a) Four Major Flat-faced Tool Orientations

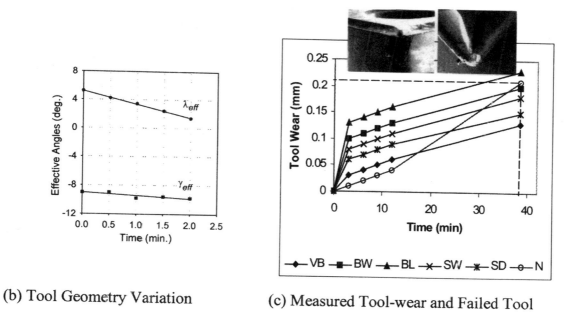

(b) Tool Geometry Variation

(c) Measured Tool-wear and Failed Tool

Figure 20 Prediction of tool failure modes. Dominant mode of tool failure: Nose wear, $f = 0.25$ mm/rev, $d = 0.6$ mm, $V = 274$ m/min, Cutting tool: CNMG 432K, Work material: Steel 1045. (From Ref. 96.)

solution for all performance measures by using just one modeling approach. The inherent disadvantages of every modeling approach, be it analytical or numerical, pose severe hurdles in adopting an individualistic methodology. The best solution would be to use the best available approach to predict different subproblems associated w'ith machining performance and then integrate these solutions to obtain a comprehensive solution for machining performance.

IV. MACHINING OPTIMIZATION IN DESIGNING FOR MACHINING

The role of optimization for machining in CAPP systems is extremely important because of the complex interactions that take place during the machining process. Traditional approaches in machining optimization have been limited to objective functions related to *cost* or *productivity* [98–101]. Although such an objective is desirable, the more critical role of optimization lies in the optimization of various *machining performance measures* for improved machining performance. The conflicting machining performance requirements, depending on specific applications, result in the need for the optimization of the machining process by the selection of the most suitable cutting conditions and tools, as well as information regarding changes that should be made to the various elements of the machining system to achieve optimized machining performance. Almost all relationships between the performance variables and process parameters employed by previous research are approximated by power functions with fixed empirical coefficients. This may be attributed to the nonavailability of quantitatively reliable machining performance models relating the machining performance measures to the process variables. The lack of technological performance data and equations as well as detailed machine tool specifications and capabilities have inhibited the widespread use of the available optimization strategies [37]. Traditional approaches in this area of performance-based optimization have not considered the functional requirement of chip breaking and the associated use of commercial complex grooved tools.

A. Hybrid Models Used in the Optimization of Machining Performance Measures

Significant progress has been made in developing predictive models for machining performance measures for turning operations at the University of Kentucky during the last 10 years. More recently, several of these analyt-

ical and empirical models, along with the selective database systems, have been integrated into hybrid predictive models for turning operations. These models are effectively used for predicting major machining performance measures, which directly depend on the cutting parameters including surface roughness, cutting force, chip breakability, tool-life, and material removal rate, in terms of cutting conditions: cutting speed, feed, and depth of cut. Because currently available metal cutting theories are unable to explicitly present all relationships between cutting conditions and machining performance, especially for complex grooved tools, an experimental database and bicubic spline data interpolation are used to supplement prediction of machining performance.

1. Surface Roughness

Turning operations give much higher measured surface roughness R_a values than predicted theoretical values in the finish turning range [102]. Therefore, a database is established by measuring R_a experimentally for a set of different cutting conditions. R_a was represented as a function of feed and depth of cut, because the effect of cutting speed was very minimal in the finish turning region.

2. Cutting Force

The cutting speed affects the cutting forces, much less significantly than the feed and depth of cut, and is hence ignored in the modeling of cutting force. The following relationship has been established for the cutting force F_c [103]:

$$F_c = F_z = C_z f^{\alpha_z} d^{\beta_z} + E_z d^{\gamma_z} \qquad (15)$$

where C_z and E_z are force constants, α_z is the feed exponent, β_z and γ_z are depth of cut exponents. The last term in the equation represents the experimentally verified cutting edge force prevalent in machining with tools having a rounded cutting edge.

3. Chip-Form/Chip Breakability

Chip-form/chip breakability is considered as a basic requirement in automated machining. In the present work, the previously established definition of chip-form/chip breakability has been used, which assumes that the size, shape, and difficulty/ease of chip producibility determine the levels of chip breakability. According to the definition, the values of chip breakability range between 0 and 1, with "0" for absolutely unbroken chips and "1" for well-broken chips [104]. Again, the effect of cutting speed on the chip breakability is ignored in the finish turning range. The predictive model for chip breakability developed from chip charts is similar to the one shown in Fig. 17, and utilizes an interactive fuzzy logic-based technique to predict chip breakability for a given set of cutting conditions, tool geometry, and work material for effective chip-groove profiles generated using analytically predicted chip flow direction as shown in Fig. 2 [93–95].

4. Tool-Life

The most commonly known extended Taylor equation shown previously (Eq. (10)) is usually considered a good approximation to predict tool-life T. However, the more recent work on tool-life including the effects of tool coating and chip-groove geometry and the corresponding tool-life equation shown in Eq. (11) would give more accurate predictions for machining with coated grooved tools. In machining with grooved tools, it has been observed that the traditional parameters for wear measurement (VB—flank wear, and KT—crater wear) are inadequate in characterizing the multiple, concurrent, and complex wear mechanisms undergone because of the interactions of the cutting conditions and the chip-groove geometry.

B. Optimization of Turning Operations

Da et al. [103–105] developed a hybrid process model that defines the relationships between the dependent performance variables (surface roughness, cutting force, chip breakability, material removal rate, and tool-life) and the independent parameters (cutting speed, feed, and depth of cut) based on the classical theories of metal cutting and a representative database of experimental results. Nonlinear optimization techniques, based on the hybrid process model, were applied to obtain the optimum cutting conditions. More recently, Da et al. [106–108] used a multiple criteria optimization method and presented a methodology for predicting the optimum cutting conditions and selection of tool inserts by considering the effects of progressive tool-wear on the machining performance.

The goal of multiple criterion optimization of machining processes is to establish trade-offs among the various conflicting machining performance measures to achieve optimum economic performance of the operation. In addition to tool-life and material removal rate, all other machining performance measures, such as surface roughness, cutting force, and chip breakability, will significantly affect the economic performance of machining, such as profit rate, production cost and so

on, directly or indirectly. The comprehensive optimization criterion proposed by Da et al. [106–108] includes the effects of all major technological machining performance measures and the effect of progressive tool-wear.

Optimization of multipass turning operations plays an important role in process planning for machining, because multipass machining operations are more widely used than single-pass machining operations in the manufacturing industry. In order to achieve overall optimal results in multipass turning operations, trade-offs are usually established not only among the various conflicting machining performance measures, but also among all passes in a given turning operation. Early work on optimization of multipass machining operations shows the economic benefits of machining [109]. Multiple constrained milling operations have been optimized by using computer-based methodologies subsequently [110,111]. More recent research highlights the use of advanced computer-aided and fuzzy logic applications for multipass machining optimization [112–114].

The optimization objective of multipass turning operations differs from that of single-pass operations. In rough turning operations, the highest possible material removal rate is desired, within the constraints of other appropriate machining performance measures. However, the surface roughness is the most important measure in comparison with all other measures in finish turning operations. To implement the importance of one or more of the machining performance measures in optimization, various weighting factors are applied to these measures in the objective function. Subsequent work includes the rated effect of progressive tool-wear [105] and a performance-based criterion for the selection of optimum cutting conditions and cutting tool selection [108] in single-pass turning. Recently, this work has been extended to cover multipass turning using genetic algorithms [115]. For multipass problems, the objective function in optimization processes is the sum of objectives of all passes. The total objective function is:

$$U(V_i, f_i, d_i, (i = 1, 2, \ldots, N))$$

$$= \sum_{i=1}^{N} \left\{ C_{R_i}\left(\frac{R'_{a_i} - R_{a_i}}{R'_{a_i}}\right) + C_{F_i}\left(\frac{F'_{c_i} - F_{c_i}}{F'_{c_i}}\right) \right.$$

$$+ C_{T_i}\left(\frac{T_i - T'_i}{T'_i}\right) + C_{M_i}\left(\frac{M_{R_i} - M'_{R_i}}{M'_{R_i}}\right)$$

$$\left. + C_{CB_i}\left(\frac{CB_i - CB'_i}{CB'_i}\right) \right\} \qquad (16)$$

where N is the number of passes in a turning operation, V_i is the cutting speed, f_i is the feed, and d_i is the depth of cut for each pass.

Constraints are represented as:

$$R_{a_i} \leq R'_{a_i}, F_{c_i} \leq F'_{c_i}, T_i \geq T'_i, M_{R_i} \geq M'_{R_i},$$
$$CB_i \geq CB'_i (i = 1, 2, \ldots, N) \qquad (17)$$

Hence the optimization problem becomes

Maximize $U(V_i, f_i, d_i, (i = 1, 2, \ldots, N))$

With respect to $V_i, f_i, d_i, (i = 1, 2, \ldots, N)$

Subject to $R_{a_i} \leq R'_{a_i}, F_{c_i} \leq F'_{c_i}, T \geq T'$,
$$M_{R_i} \geq M'_{R_i}, CB_i \geq CB'_i,$$
$$V_{\text{min}i} \leq V_i \leq V_{\text{max}i}, f_{\text{min}i} \leq f_i$$
$$\leq f_{\text{max}i}, d_{\text{min}i} \leq d_i \leq d_{\text{max}i}$$
$$(i = 1, 2, \ldots, N)$$
$$\sum d_i = d \qquad (18)$$

where d is the total depth of cut. Because the goals of optimization for different passes in multipass turning operations are different, different constraints and weighting factors of the objective function are applied to each pass.

Genetic algorithms (GA) are used in the optimization process for multipass turning operations. GA are algorithms based on the mechanics of natural selection and natural genetics, which are more robust and more likely to locate a global optimum [116]. It is because of this feature that the GAs move through the solution space starting form a group of points and not from a single point.

To apply GA in the optimization for machining, the cutting conditions are encoded as genes by binary encoding. A set of genes is combined together to form a chromosome, which is used to perform those basic mechanisms in GA such as crossover and mutation. Crossover is the operation to exchange some part of two chromosomes to generate new offspring, which is important to explore the whole search space rapidly. Mutation is applied after crossover to provide a small randomness to the new chromosome. To evaluate each individual or chromosome, the encoded cutting conditions are decoded from the chromosome and are used to predict machining performance measures. The fitness or objective function is a function needed in the optimization process and the selection of the next generation in GA. After a number of iterations of GA, optimal results of cutting conditions are obtained by comparison of values of objective functions among all individuals. Besides weighting factors and constraints, suitable

parameters of GA are required for this methodology to operate efficiently.

1. Effect of Progressive Tool-Wear

The general prediction of machining performance is based on the assumption that cutting tools are fresh in every pass. However, in the actual machining process, cutting tools are subjected to progressive tool-wear on different tool faces. The machining performance will vary significantly with the progression of the overall tool-wear, which leads to different optimum cutting conditions for a given cutting tool at different wear states. In multipass turning operations, if the same tool insert is used for all passes, the effect of tool-wear must be considered in all subsequent passes.

The reciprocal of tool-life will be defined as the wear rate R.

$$R = \frac{1}{T} = \frac{1}{kmC} V^{n_3} f^{n_1} d^{n_2} \qquad (19)$$

where n_3 and C are

$$n_3 = \frac{1}{n_c} \text{ and } C = T_R (V_R)^{\frac{1}{n_c}} \qquad (20)$$

If a tool insert has been used in N known previous operations, the tool-wear index can be defined as

$$w = \sum_{i=1}^{N} R_i t_i = \sum_{i=1}^{N} \frac{1}{kmC} V_i^{n_3} f_i^{n_1} d_i^{n_2} t_i \qquad (21)$$

where V_i, f_i, d_i, and t_i are cutting speed, feed, depth of cut, and the time interval of the i-th operation, respectively. If the cutting parameters vary continuously with time in the operation, the tool-wear index can be rewritten as,

$$w = \sum_{i=1}^{N} \left(\int_0^{t_i} R_i dt \right)$$
$$= \sum_{i=1}^{N} \left(\int_0^{t_i} \frac{1}{kmC} V_i^{n_3} f_i^{n_1} d_i^{n_2} dt \right) \qquad (22)$$

According to the definition, the tool-wear index for a new tool insert is 0. And, the tool is considered failed if the tool-wear index reaches 1. The machining performance for the tool after some known usage w can be predicted by

$$p = W_p(w) p_u(V, f, d) \qquad (23)$$

where W_p is a function of tool-wear index w independent of cutting conditions V, f, and d, which represents the

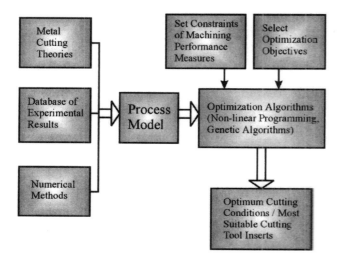

Figure 21 The new methodology developed and used for optimized machining performance and cutting tool selection.

tool-wear effect on the machining performance p. $p_u(V,f,d)$ is the machining performance predicted by assuming unworn tools.

The new methodology developed and used for machining optimization is shown in a schematic diagram in Fig. 21.

Case Study 1: Optimum Cutting Conditions for Multipass Turning Using Genetic Algorithms

The optimization results for multipass turning are shown in Table 10 [115]. Different weighting factors are defined for each turning pass. In medium turning, material removal rate is given emphasis with a weighting factor of 0.6, whereas in finish turning, surface rough-

ness is given prime importance with a weighting factor of 0.6. The other weighting factors are all assigned to be 0.1. However, all machining performance measures are considered to have the same importance in semifinish turning. Thus all weighting factors are given as 0.2 for each performance measure in semifinish turning. Different constraints for medium, semifinish, and finish turning were applied. Feasible regions of those three passes are illustrated in Fig. 22a–c. Because there is a constraint on the total depth of cut, the optimum results of the multipass turning operation are an optimum combination of cutting conditions in those feasible regions.

Case Study 2: Optimal Selection of Cutting Tool Inserts Using Nonlinear Programming

This case study illustrates the optimum selection of cutting tool inserts for specified requirements and the determination of corresponding machining conditions for optimal chip control [104]. The cutting tool will be selected from a group of three tool inserts (TNMG332-CG1, TNMG332-CG2, and TNMG331-CG1), which have very different chip-groove configurations and nose radius. The optimization results for these three tool inserts are shown in Fig. 23. We can see that there are feasible solutions for all three tools, but the objective value (chip breakability) at the optimum for TNMG331-CG1 is 5.83% larger than that for TNMG332-CG2 and 14.32% for TNMG332-CG1. Therefore the advantage for chip control is obvious by selecting TNMG331-CG1 over the other two inserts. However, the higher chip breakability is achieved with lower material removal rate and rougher surface finish. The trade-off among the various machining performance measures can be made through setting the constraints imposed on these parameters.

Table 10 Optimization Results for Three-Pass Turning

Total depth of cut (mm)		3.10			3.70		
Turning pass		Medium	Semifinish	Finish	Medium	Semifinish	Finish
Optimum cutting conditions	V (m/min)	150	200	345	150	237	334
	f (mm/rev)	0.183	0.131	0.104	0.240	0.167	0.104
	d (mm)	1.329	0.776	0.995	1.695	1.010	0.995
Predicted machining performance	R_a (μm)	1.302	1.026	0.723	2.356	1.196	0.718
	F_c (N)	686	303	305	1108	487	303
	M_R (mm³/min)	36,705	20,266	35,827	61,140	40,072	34,450
	CB	0.65	0.76	0.77	0.64	0.67	0.76
	T (min)	8.521	8.315	2.396	6.145	4.120	2.579

Source: Ref. 115.

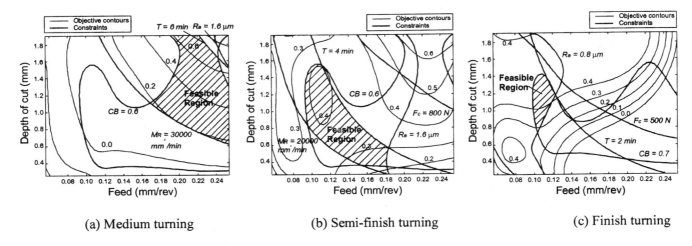

(a) Medium turning   (b) Semi-finish turning   (c) Finish turning

Figure 22   Feasible regions for optimization in multipass turning. (From Ref. 115.)

V.   JIGS AND FIXTURES IN DESIGNING FOR MACHINING

Both jigs and fixtures provide for the efficient and productive manufacture, inspection, or assembly of quality interchangeable parts by providing the means to:

- Correctly locate the workpiece with respect to the tool or gauge.
- Securely clamp and rigidly support the workpiece while the operation is being performed.
- Guide the tool (in the case of jigs).

- Position and fasten the device on a machine (in the case of most fixtures).

These features, in addition to ensuring interchangeability and accuracy of parts, also provide the following advantages in manufacturing:

- They minimize the possibility of human error.
- They permit the use of unskilled labor.
- They reduce manufacturing and inspection time.
- They eliminate the need to retool for repeat orders.

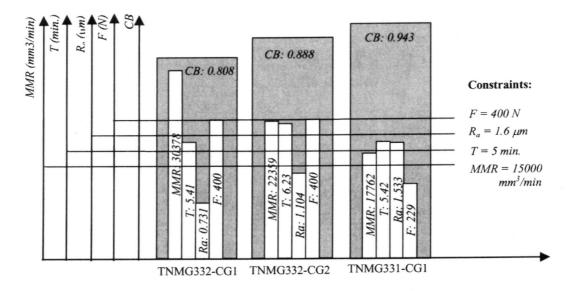

Figure 23   Optimal selection of cutting tool inserts. (From Ref. 104.)

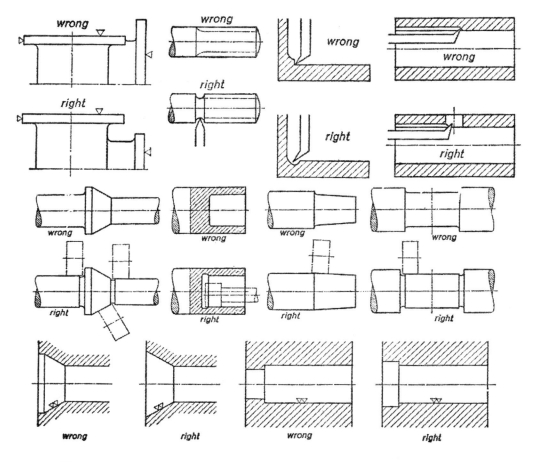

Figure 24 Designing for machining operation—tool access. (From Ref. 118.)

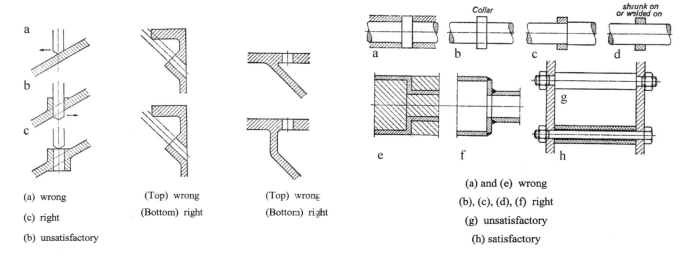

(a) wrong

(c) right

(b) unsatisfactory

(Top) wrong

(Bottom) right

(Top) wrong

(Bottom) right

(a) and (e) wrong

(b), (c), (d), (f) right

(g) unsatisfactory

(h) satisfactory

Figure 25 Designing for machining operation (drilling)— tool loading and access. (From Ref. 118.)

Figure 26 Designing for minimum machining. (From Ref. 118.)

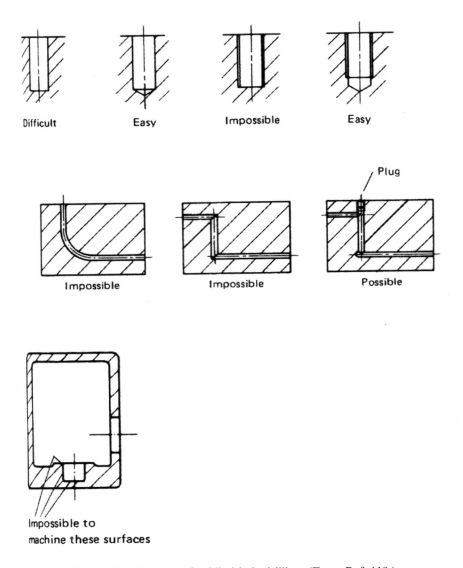

Figure 27 Designing for blind hole drilling. (From Ref. 119.)

Advances in NC and CNC machines have inevitably forced designers and manufacturers to improve the effectiveness and performance of parts production. The versatility and possibilities available using NC machines open the door to inventions in the design and development of jigs and fixtures.

The four main objectives in designing a workholding fixture for NC are accuracy, rigidity, accessibility, and speed and ease of clamping and changing the workpiece [117]. To be effective, these systems have to be able to:

* Mount onto the machining center pallet precisely and quickly with no loss in location accuracy when moved from one pallet to another.

* Guarantee quick and accurate part location, and ensure accurate repeatability with multiple part changes.
* Make clamping of parts easy.
* Facilitate one completed part per machining cycle.
* Virtually eliminate the need for single-purpose fixtures.

Jigs and fixtures are generally costly items and hence an economic justification in terms of the batch size, production frequency, and the continuity of usage must be made before the purchase of standard equipment or prior to designing and developing one to suit the specific requirement. Care must be taken in choosing a low-cost but functional design. The use of flexible fixturing and multipurpose jigs and fixtures must be considered in designing for machining processes involving several operations. Recent advances in modular tooling systems provide an excellent promise in applications.

VI. SOME PRACTICAL MACHINING CONSIDERATIONS

Designing for manufacturing by machining methods must take the following practical aspects into consideration [118,119]:

- Designing for economy (cost issue)
- Designing for clampability (jigs and fixtures issue)
- Designing for tool/equipment access and release (tool path planning issue)
- Designing for machining operation (tool access, favorable tool loading, etc.—see Figs. 24 and 25)
- Designing for Minimum Machining (Fig. 26)
- Designing for Blind Hole Drilling (Fig. 27)

VII. CONCLUDING REMARKS

The work presented in this chapter is an extended summary of the broad topic on designing for machining. The contents of this chapter are therefore aimed at providing a broad overview of the factors to be considered in designing for machining, particularly from the machinability and machining performance viewpoint. The need for developing and implementing predictive models for optimized machining performance and cutting tool selection continues to be a major, and often difficult to handle problem area requiring significant efforts by the research and application communities. However, the recognition of such need is more likely to encourage much work in this regard although the cost-effectiveness of such efforts is unlikely to materialize immediately. The academic and practical challenges in providing suitable solutions to some of the major design for machining problems in the industry are far more greater to ignore.

REFERENCES

1. Armarego, E.J.A.; Jawahir, I.S.; Ostafiev, V.A.; Venuvinod, P.K. Modeling of Machining Operations. STC-C Presentation, Working Group on Modeling of Machining Operations, Paris, France, January 1996.
2. Tool and Manufacturing Engineers Handbook, Volume 1: Machining, Society of Manufacturing Engineers, 1983.
3. Boulger, F.W. Influence of Metallurgical Properties on Metal Cutting Operations. Society of Manufacturing Engineers, 1958.
4. Sullivan, K.F.; Wright, P.K.; Smith, P.D. Metallurgical Appraisal of Instabilities Arising in Machining Metals Technology, June 1978; 181–188.
5. Shaw, M.C. *Metal Cutting Principles*; Oxford University Press Inc.: NY, 1984.
6. ISO 3685. Tool-life Testing with Single-Point Turning Tools, 1993.
7. ISO 8688-1. Tool-life Testing in Milling—Part 1: Face Milling, 1989.
8. ISO 8688-2. Tool-life Testing in Milling—Part 2: End Milling, 1989.
9. Mills, B.; Redford, A.H. Machinability of Engineering Materials. Applied Science Publishers, 1983.
10. Zlatin, N.; Field, M. Relationship of Microstructure to the Machinability of Wrought Steels and Cast Iron. In *Machining Theory and Practice*; ASM: Cleveland, OH, 1950.
11. Field, M. Relationship of Microstructure to the Machinability of Wrought Steels and Cast Iron. Int. Production Engineering Research Conference, Paper 20, Carnegie Institute of Technology, Pittsburgh, PA, September 1063.
12. Murphy, D.W.; Aylward, P.T. Measurement of Machining Performance in Steels. Metalworking of Steels, AIME, 1965; 49–82.
13. Chisholm, A.W.J.; Richardson, B.D. A Study of the Effect of Non-metallic Inclusions on the Machinability of Two Ferrous Materials. Proc. Inst. Mechanical Engineers, United Kingdom, February 1965.
14. Lorenz, G.; Evans, P.T. Improving the Machinability of Low Alloy Steels by Better Metallurgical Control. ASTME Inst. Machining and Tooling Symposium, Sydney, 1967.
15. Wilbur, W.J. The Effect of Manganese Sophie Inclusions on the Mechanics of Cutting and Tool-wear in Machining of Low Carbon Steels. Ph.D. thesis, University of Salford, England, 1970.
16. Henkin, A.; Datsko, J. The Influence of Physical Properties on Machinability. Trans. ASME, J. Eng. For Industry, November 1963; 321–327.
17. DeArdo, A.J.; Garcia, C.I.; Laible, R.M.; Eriksson, U. A Better Way to Assess Machinability. American Machinist, May 1993; 33–35.
18. Enache, S.; Strajescu, E.; Opran, C.; Minciu, C.; Zamfirache, M. Mathematical Model for the Establishment of the Machinability. Annals of the CIRP 1995, *44* (1), 79–82.
19. Ostafiev, D.; Ostafiev, V.; Kharkevich, A. Predictive Machinability Model for Global Manufacturing. Proc. CIRP Int. Symposium on Advanced Design and Manufacture in the Global Manufacturing Era, August 1997; 353–358.

20. Machinability Database System, Carboly Systems, General Electric Company, 1968.

21. Eversheim, W.; König, W.; Schwanborn, W.; Wesch, H.; Dammer, L. Computer-Aided Determination Optimization of Cutting Data, Cutting Time and Costs. Annals of the CIRP 1981, *30* (1), 409–412.

22. Konig, W.; Dammer, L. Mini computer application in a machining data bank. In *The Application of Mini and Micro Computers in Information, Documentation, and Libraries*; Keren, C. Perlmutter, L. Elsevier Science Publishers, 1983; 679–684.

23. Parsons, N.R., Ed.; *NC Machinability Data Systems*; Society of Manufacturing Engineers: Dearborn, MI, 1971.

24. Balakrishnan, P.; DeVries, M.F. A Review of Computerized Machinability Database Systems. Proc. 10th NAMRC, 1982, 348–356.

25. Kahles, J.F. Machinability data requirements for advanced machining systems. Annals of the CIRP 1987, *36* (2), 523–529.

26. Venkatesh, V.C. Computerized machinability data. Proc. AUTOMACH Conference, Australia, (MS86-464 SME Publications), 1986; 1.59–1.73.

27. Yeo, S.H.; Rahman, M.; Venkatesh, V.C. Development of an Expert System for Machinability Data Selection. J. Mech. Working Technology 1988, *17*, 51–60.

28. Machinability Data Handbook. 3rd Ed.; Metcut Associates, Inc.: Cincinnati, OH, 1980.

29. CUTDATA Computerized Machinability Database System, MetCAPP, 1997.

30. Jawahir, I.S.; Qureshi, N.U.; Arsecularatne, J.A. On the Interrelationships of Some Machinability Parameters in Finish Turning. Int. J. Machine Tools and Manufacture 1992, *32* (5), 709–723.

31. Fang, X.D.; Jawahir, I.S. The Effects of Progressive Tool-wear and Tool restricted Contact on Chip Breakability in Machining. J. Wear 1993, *160*, 243–252.

32. ISO 513. Application of carbides for Machining by Chip Removal, 1975.

33. Kunz, H.; Konig, W. Cutting Tool Materials Classification and Standardization. Annals of the CIRP 1987, *36* (2), 531–535.

34. Modern Metal Cutting—A Practical Handbook, Sandvik Coromant, 1996.

35. Ghosh, R.; Lin, M.; Jawahir, I.S.; Khetan, R.P.; Bandyopadhyay, P. Chip Breakability Assessment Using a Chip-groove Classification System in Finish Turning. Proc. ASME Int. Mech. Eng. Congress, MED-Vol. 2-1, San Francisco November 1995; 679–702.

36. Jawahir, I.S.; van Luttervelt, C.A. Recent Developments in Chip Control Research and Applications. Annals of the CIRP 1993, *42* (2), 659–693.

37. Armarego, E.J.A. Predictive Modeling of Machining Operations—A Means of Bridging the Gap Between Theory and Practice. Keynote paper, Proc. Canadian Society of Mechanical Engineers Forum, Hamilton, ON, May 1996; 18–27.

38. Shen, L.G.; van Luttervelt, C.A. A Hybrid Intelligent Methodology for Assessment and Optimization of Machining Operations in an Intelligent Machining System. CIRP International Symposium on Advanced Design and Manufacturing in the Global Manufacturing Area, Hong Kong, August 21, 1997; 21–27.

39. Jawahir, I.S.; Balaji, A.K.; Stevenson, R.; van Luttervelt, C.A. Towards Predictive Modeling and Optimization of Machining Operations. Keynote Paper, Symp. on Predictable Modeling in Metal Cutting as Means of Bridging Gap Between Theory and Practice, Manuf. Science and Engg., ASME IMECE '97, Dallas, Texas, USA, MED-Vol. 6-2, 1997; 3–12.

40. vanLuttervelt, C.A.; Childs, T.H.C.; Jawahir, I.S.; Klocke, F.; Venuvinod, P.K. Present Situation and Future Trends in Modeling of Machining Operations. Annals of the CIRP 1998, *47* (2), 587–626.

41. Jawahir, I.S.; Balaji, A.K. Predictive Modeling and Optimization of Turning Operations with Complex Grooved Tools for Curled Chip Formation and Chip Breaking. J. Machining Science and Technology, 2000, *4* (3), 399–443.

42. Jawahir, I.S.; Balaji, A.K.; Rouch, K.E.; Baker, J.E. Towards Integration of Hybrid Models for Optimized Machining Performance in Intelligent Machining Systems. Proc. Int. Manufacturing Conf. in China (IMCC 2000), Hong Kong, CD-ROM, August 2000.

43. Usui, E. Progress of "Predictive" Theories in Metal Cutting. JSME International Journal, Series III 1988, *31* (2), 303–309.

44. Oxley, P.L.B. *The Mechanics of Machining: An Analytical Approach to Assessing Machinability*; Ellis Horwood Limited: Chichester, 1989.

45. Ernst, H.; Merchant, M.E. Chip Formation, Friction and High Quality Machined Surfaces. Surface Treatment of Metals, ASM, 1941; 299 pp.

46. Merchant, M.E. Basic Mechanics of the Metal Cutting Process. Trans. ASME (J. Applied Mechanics) 1944, *66*, A168–A175.

47. Lee, E.H.; Shaffer, B.W. The Theory of Plasticity Applied to a Problem of Machining. J. Applied Mechanics 1951, *18*, 405–413.

48. Palmer, W.B.; Oxley, P.L.B. Mechanics of Metal Cutting. Proc. Inst. Mech. Engineering 1959, *173*, 623–654.

49. Shaw, M.C.; Cook, N.H.; Finnie, I. Shear Angle Relationships in Metal Cutting. Trans. ASME 1953, *75*, 273.

50. Rubenstein, C. The Application of Force Equilibrium Criteria to Orthogonal Cutting. Int. J. MTDR 1972, *12*, 121–126.

51. Usui, E.; Hirota, A. Analytical Prediction of Three Dimensional Cutting Process, Part 2: Chip Formation and Cutting Force with Conventional Single-Point Tool. Trans. ASME (J. Eng. Ind.) 1978, *100*, 229–235.

52. Dewhurst, P. On the Non-uniqueness of the Machining Process. Proc. Royal Society, London, A 1978, *350*, 587–610.

53. Armarego, E.J.A.; Whitfield, R.C. Computer Based Modeling of Popular Machining Operations for Forces and Power Prediction. Annals of the CIRP 1985, *34* (1), 65–69.

54. Shi, T.; Ramalingam, S. Modeling Chip Formation with Grooved Tools. Int. J. of Mech. Sci. 1993, *35* (9), 741.

55. Merchant, M.E. An Interpretive Look at 20th Century Research on Modeling of Machining. Inaugural Address, Proc. CIRP Int. Workshop on Modeling of Machining Operations, Atlanta, GA, May 19, 1998; 27–31.

56. Oxley, P.L.B. Development and Application of a Predictive Machining Theory. J. Machining Science and Technology 1998, *2* (2), 165–190.

57. Armarego, E.J.A. A Generic Mechanics of Cutting Approach to Predictive Technological Performance Modeling of the Wide Spectrum of Machining Operations. J. Machining Science and Technology 1998, *2* (2), 191–211.

58. Fang, N.; Jawahir, I.S.; Oxley, P.L.B. A Universal Slip-line Model with Non-unique Solutions for Machining with Curled Chip Formation and a Restricted Contact Tool. Int. Journal of Mechanical Sciences 2001, *43*, 557–580.

59. Fang, N.; Jawahir, I.S. A New Methodology for Determining the Stress State of the Plastic Region in Machining with Restricted Contact Tools. Int. Journal of Mechanical Sciences 2001, *43*, 1747–1770.

60. Fang, N.; Jawahir, I.S. Prediction and Validation of Chip Up-curl in Machining Using the Universal Slip-line Model. Trans. NAMRC 2000, *XXVIII*, 137–142.

61. Fang, N.; Jawahir, I.S. Analytical Predictions and Experimental Validation of Cutting Force Ratio, Chip Thickness, and Chip Back-flow Angle in Restricted Contact Machining Using the Universal Slip-line Model. Int. J. Machine Tools and Manufacture 2002, *42*, 681–694.

62. Fang, N.; Jawahir, I.S. An Analytical Predictive Model and Experimental Validation for Machining with Grooved Tools Incorporating the Effects of Strains, Strain-rates and Temperatures. Annals of the CIRP 2002, *51* (1) xx–xx.

63. Dillon, O.W., Jr.; Zhang, H. An Analysis of Cutting Using a Grooved Tool. Int. J. Forming Processes 2000, *3* (1–2), 115.

64. Zhang, H.; Dillon, O.W., Jr.; Jawahir, I.S. A Finite Element Analysis of 2-D Machining with a Grooved Tool. Trans. NAMRC 2001, *XXIX*, 327–334.

65. Dewhurst, P.; Collins, I.F. A Matrix Technique Constructing Slip-line Field Solutions to a Class of Plane Strain Plasticity Problems. Int. J. for Numerical Methods in Engineering 1973, *43*, 1747–1770.

66. Powell, M.J.D. A Fortran Subroutine for Solving Systems of Non-linear Algebraic Equations. In *Numerical Methods for Nonlinear Algebraic Equations*; Rabinowitz, Ed.; Gordon and Breach: London, 1970.

67. Tay, A.O. A Numerical Study of the Temperature Distribution Generated During Orthogonal Machining, Ph.D. Thesis, University of New South Wales, Sydney, Australia, 1973.

68. Klamecki, B.E. Incipient Chip Formation in Metal Cutting—A Three Dimensional Finite Element Analysis, Ph.D. Thesis, University of Illinois at Urbana–Champaign, 1973.

69. Stevenson, M.G.; Wright, P.K.; Chow, J.G. Further Development in Applying the Finite Element Method to the Calculation of Temperature Distributions in Machining and Comparisons with Experiment. ASME Journal of Engineering for Industry 1985, *105*, 149.

70. Strenkowski, J.S.; Carroll, J.T. A Finite Element Model of Orthogonal Metal Cutting. ASME Journal of Engineering for Industry 1985, *107*, 349.

71. Ueda, K.; Manabe, K. Rigid-Plastic FEM Analysis of Three-Dimensional Deformation Field in Chip Formation Process. Annals of the CIRP 1993, *42*, 35.

72. Wu, J.S.; Dillon, O.W., Jr.; Lu, W.Y. Thermo-Viscoplastic Modeling of Machining Process Using a Mixed Finite Element Method. ASME J. of Eng. for Industry 1996, *118*, 470.

73. Maekawa, K.; Maeda, M.; Kitagawa, T. Simulation Analysis of Three-dimensional Continuous Chip Formation Processes (Part 1). Int. J. Japan Soc. Prec. Eng. 1997, *31*, 39.

74. Shirakashi, T.; Obikawa, T. Recent Progress of Computational Modeling and Some Difficulties. Proceedings of the CIRP International Workshop on Modeling of Machining Operations; Jawahir, et al., Ed.; Atlanta, GA, 1998; 179 pp.

75. Tugrul, O.; Altan, T. Process Simulation Using Finite Element Method - Prediction of Cutting Forces, Tool Stresses and Temperatures in High-Speed Flat End Milling. International Journal of Machine Tools and Manufacture 2000, *40*, 713.

76. Johnson, G.R.; Cook, W.H. A Constitutive Model and Data for Metal Subjected to Large Strains, High Strain Rates and High Temperature. Proceedings of 7th International Symposium on Ballistic, The Hague, 1983; 12 pp.

77. Wallace, P.W.; Boothroyd, G. Tool-faces and Tool-Chip Friction in Orthogonal Machining. J. Mech. Eng. Sci. 1964, *6*, 74.

78. Shirakashi, T.; Usui, E. Friction Characteristics on Tool Face in Metal Machining. J. JAPE 1973, *39* (9), 966.

79. Zienkiewicz, O.C. The Finite Element Method. 3rd Ed.; McGraw-Hill: London, 1977.

80. Cook, N.H. Tool-wear and Tool-life. ASME J. of Eng. for Ind., 1973; 931–938.

81. Konig, W.; Fritsch, R.; Kammermeier, D. New Approaches to Characterizing the Performance of Coated Cutting Tools. Annals of the CIRP 1992, *41* (1), 49–54.

82. Kramer, B.M. An Analytical Approach to Tool Wear Prediction. Ph.D. Thesis, Department of Mechanical Engineering, MIT, USA, 1979.

83. Colding, B.N. A Three-Dimensional Tool-life Equation—Machining Economics. ASME J. of Eng. for Ind., 1959, 239–250.

84. Colding, B.N. A Tool-Temperature/Tool-Life Relationship Covering a Wide Range of Cutting Data. Annals of the CIRP 1991, *40* (1), 35–40.

85. Lindstrom, B. Cutting Data Field Analysis and Predictions—Part 1: Straight Taylor Slopes. Annals of CIRP 1989, *38* (1), 103–106.

86. Carisson, T.E.; Strand, F. A Statistical Model for Prediction of Tool-Life as a Basis for Economical Optimization of Cutting Process. Annals of the CIRP 1992, *41* (1), 79–82.

87. ASME. Tool Life Testing With Single-Point Turning Tools. ASME: New York, 1986.

88. Fang, X.D.; Jawahir, I.S. The effects of Progressive Tool-wear and Tool Restricted Contact on Chip Breakability in Machining. Wear 1993, *160*, 243–252.

89. Jawahir, I.S.; Ghosh, R.; Fang, X.D.; Li, P.X. An Investigation of the Effects of Chip Flow on Tool-wear in Machining with Complex Grooved Tools. Wear 1995, *184*, 145–154.

90. Jawahir, I.S.; Li, P.X.; Ghosh, R.; Exner, E.L. A New Parametric Approach for the Assessment of Comprehensive Tool-wear in Coated Grooved Tools. Annals of the CIRP 1995, *45* (1), 49–54.

91. Li, P.X.; Jawahir, I.S.; Fang, X.D.; Exner, E.L. Chipgroove Effects on Concurrently Occurring Multiple Tool-wear Parameters in Machining with Complex Grooved Tools. Trans. NAMRI 1996, *XXIV*, 33–38.

92. Jawahir, I.S.; Fang, X.D.; Li, P.X.; Ghosh, R. Method of Assessing Tool-life in Grooved Tools, U.S. Patent No: 5,689,062, November 18, 1997.

93. Lin, M.; Ghosh, R.; Jawahir, I.S. An Intelligent Technique for Predictive Assessment of Chip Breakability in Turning Operations for Use in CAPP Systems. Proc. IPMM'97, Australasia–Pacific Forum on Intelligent Processing and Manuf. Of Materials 1997, *2*, 1311–1320.

94. Lin, M.; Da, Z.J.; Jawahir, I.S. Development and Implementation of Rule-Base Algorithms in CAPP Systems for Predicting Chip Breakability in Machining. ICME 98, CIRP Int. Seminar on Intelligent Computation in Manuf. Engg., Capri (Naples), Italy; July 1–3, 1998; 517–522.

95. Jawahir, I.S.; Fei, J. A Comprehensive Evaluation of Tool Inserts for Chip Control Using Fuzzy Modeling of Machinability Parameters. Trans. NAMRI 1993, *XXI*, 205–213.

96. Jawahir, I.S.; Ghosh, R.; Li, P.X.; Balaji, A.K. Predictability of Tool Failure Modes in Turning with Complex Grooved Tools Using Equivalent Toolface (ET) Model. J. Wear 2000, *244*, 94–103.

97. Ghosh, R.; Redetzky, M.; Balaji, A.K.; Jawahir, I.S. The Equivalent Toolface (ET) Approach for Modeling Chip Curl in Machining with Grooved Tools. Proc. CSME/SCGM, Hamilton, ON, Canada, May 1996; 702–711.

98. Gilbert, W.W. Economics of Machining. Machining Theory and Practice. American Society of Metals, 1950; 465–485.

99. Okushima, K.; Hitomi, K. A Study of Economical Machining: An Analysis of the Maximum—Profit Cutting Speed. Int. J. Prod. Res. 1964, *3*, 73–xx.

100. Armarego, E.J.A.; Russell, J.K. Maximum Profit Rate as a Criterion for the Selection of Machining Conditions. Int. Jour. Of Mach. Tool Des. And Res. 1966, *6*, 1–xx.

101. Zdebrick, W.J.; DeVor, R.E. A Comprehensive Machining Cost Model and Optimization Technique. Annals of CIRP 1981, *30*(1), 405–xx.

102. Jawahir, I.S.; Qureshi, N.U.; Arsecularate, J.A. On the Interrelationships of Some Machinability Parameters in Finish Turning with Cermet Chip Forming Tool Insert. Int. J. Mach. Tools and Manuf. 1992, *32* (5), 709–723.

103. Da, Z.J.; Sadler, J.P.; Fang, X.D.; Jawahir, I.S. Optimum Machining Performance in Finish Turning with Complex Grooved Tools, Manufacturing Science and Engineering, MED-Vol. 2-1/MH-Vol. 3-1, ASME, 1995; 703–714.

104. Da, Z.J.; Jawahir, I.S. Optimal Chip Control in Turning Operations. Proceedings of the International Seminar on Improving Machine Tool Performance, San Sebastian, Spain, July 1998, II, 607–618.

105. Da, Z.J.; Sadler, J.P.; Jawahir, I.S. Predicting Optimum Cutting Conditions for Turning Operations at Varying Tool-Wear States. Transactions of NAMRI/SME 1997, *XXV*, 75–80.

106. Da, Z.J.; Sadler, J.P.; Jawahir, I.S. A New Performance-Based Criterion for Optimum Cutting Conditions and Cutting Tool Selection in Finish Turning. Trans. NAMRI/SME 1998, *XXVI*, 129–134.

107. Sadler, J.P.; Jawahir, I.S.; Da, Z.J.; Lee, S.S. Method of Predicting Optimum Machining Conditions, United States Patent No. 5,801,963, 1998.

108. Sadler, J.P.; Jawahir, I.S.; Da, Z.J.; Lee, S.S. Optimization of Machining with Progressively Worn Cutting Tools. United States Patent No. 5,903,474, 1999.

109. Ermer, D.S.; Kromodihardjo, S. Optimization of Multipas Turning with Constraints. Trans. ASME 1981, *103*, 462–468.

110. Wang, J. Constrained Optimization of Rough Milling Operations, Ph.D. Thesis. The University of Melbourne, Australia, 1993.

111. Wang, J.; Armarego, E.J.A. Computer-Aided Optimi-

zation of Multiple Constraint Single Pass Face Milling Operations. Machining Science and Technology 2001, 5 (1), 77–99.

112. Kee, P.K. Development of Computer-Aided Machining Optimisation for Multi-Pass Rough Turning Operations. Int. J. Prod. Economics 1994, 37, 215–227.

113. Mesquita, R.; Krastera; Doytchinov, S. Computer-Aided Selection of Optimum Machining Parameters in Multipass Turning. Int. J. Adv. Manuf. Technol. 1995, 10, 19–26.

114. Alberti, N.; Perrone, G. Multipass Machining Optimization by Using Fuzzy Possibilistic Programming and Genetic Algorithms. Proc. Instn. Mech. Engrs. 1999, 213 (B), 261–273.

115. Wang, X.; Da, Z.J.; Balaji, A.K.; Jawahir, I.S. Performance-based Optimal Selection of Cutting Conditions and Cutting Tools in Multi-pass Turning Operations using Genetic Algorithms. Proceedings of Intelligent Computation in Manufacturing Engineering-2, Capri, Italy, June 2000; 409–414.

116. Goldberg, D.E. Genetic Algorithms in Search, Optimization, and Machine Learning. Addison Wesley Longman, 1989.

117. William E. Boyes, Ed.; *Low Cost Jigs, Fixtures and Gages for Limited Production*, 1st Ed.; SME: Dearborn, MI, USA, 1986.

118. Matousek, R. *Engineering Design: A Systematic Approach*. Blackie and Son Ltd.: London, United Kingdom, 1972.

119. Boothroyd; Dewhurst, G.P.; Knight, W. *Product Design for Manufacture and Assembly*, 2nd Ed.; Marcel Dekker, 2002.

Index

Adhesion energy, 862
Arrhenius equation, 513
Annealing
    cooling rate, 482
    diffusion annealing, 483
    full annealing, 480
    intercritical annealing, 481
    isothermal annealing, 482
    recrystallization annealing, 481
    recrystallization temperature, 482
    spheroidizing, 482
    spheroidizing temperature, 483
Austempering, 493
Austenitizing, 477

Beer–Lambert law, 811
Biomimetic, 869
Biot number, 476, 736
Bofors equation, 483

Carbonitriding
    austenitizing temperature, 537
    carbon and nitriding diffusion, 534
    carbonitriding curves, 536
    carbon potential, 536
    control, 538
    diffusion rate, 537
    heat treatment, 538
    process, 546, 538
    role of nitrogen, 535
Carbon potential, 509
Carburizing
    atmosphere composition, 508
    atmosphere equilibrium curve, 509

[Carburizing]
    carbon potential, 509
    carbon solubility, 514
    carbon transfer coefficient (β), 512
    carburizing curves, 514
    case depth, 514, 529
    chemical absorption rate, 508
    classification, 516
    dew point, 508
    diffusion constant, 513
    diffusion rate, 512
    endo gas, 508
    fatigue strength, 528
    gas dissociation, 510
    heat treatment, 517
    intergranular oxidation, 530
    kinetics, 512
    mass transfer equation, 512
    postfinishing methods, 530
    process, 546, 859
    residual stress, 528, 531
    shot peening, 532
    steel grades, 517
    structure property, 522
    thermodynamics, 507
    times, 514
Casting, continuous (ferrous)
    air gap thickness, 275
    austenite grain growth, 278
    contamination, 263
    conventional slab casting, 254
    crack formation, 268
    direct steelmaking, 251
    heat extraction mechanism, 263

[Casting, continuous (ferrous)]
    heat transfer, 262
    heat flux, 275
    heat flux and carbon content, 269
    heat withdrawal, 284
    horizontal belt casting, 282
    indirect casting, 251
    interfacial heat transfer coefficient, 274
    instantaneous heat flux, 264
    machine characteristics, 258
    metallurgy, 258, 262
    minimill operations, 252
    minimum heat transfer coefficient, 285
    mold heat flux, 264
    mold lubricants, 266, 270
    process comparison, 286
    processing operations, 252
    process parameters, 259
    scale formation, 288
    secondary refining operations, 252
    solidification, 267, 276
    spray chamber, 266
    spray heat transfer coefficient, 267
    steel chemistry and heat flux, 264
    strand thickness, 285
    streaming velocity, 282
    strip casting, 256
    thermal capacity, 275
    thermomechanical behavior, 268
    thin slab casting, 256
    total solidification time, 285
    twin-roll casting, 269, 280
    Wood's alloy, 270
Casting, continuous (non-ferrous)
    Stepanov method, 295
Casting, sand
    alloy melting temperatures, 351
    apparent density, 373
    Brigg's logarithmic equation, 377
    buoyancy effects, 374
    bulk density, 373
    chemically bonded molding, 353
    cooling rate, 368
    dendrite arm spacing, 368
    density measurements, 373
    engineering modulus, 377
    fractgraphic correlation, 381
    graphite effect, 383
    grain development, 363
    green sand molding, 352
    hydrogen effects, 367
    liquid penetrant inspection, 374
    lost foam casting, 356
    matrix effects, 384
    metallographic property correlations, 381
    microshrinkage effects, 385

[Casting, sand]
    microstructure, 368, 370
    molding process, 350, 352
    no-bake molding, 353
    nondestructive inspection and properties, 371
    porosity on properties, 368
    procedure effects, 355
    radiation inspection, 374
    shell molding, 353
    shrinkage, 364
    shrinkage porosity, 366
    solidification, 363, 365
    true density, 373
    x-ray property correlation, 380
    ultrasonic attenuation coefficient, 377
    ultrasonic inspection, 375
Cementation, 859
Chemical vapor deposition
    compound formation, 801
    diamond process, 870
    disproportionation, 801
    metal-organic deposition, 805
    photo-assisted, 809
    photothermal, 811
    plasma-enhanced, 805
    reaction types, 800
    reactor systems, 802
    source, 867
    structure, 867
    thermodynamic criteria, 801
    thickness, 868
Coatings
    adhesion, 862
    adhesion test, 911
    characterization, 889
    corrosion resistant, 865
    deposition procedures, 887
    destructive tests, 894
    erosion resistant, 864
    friction-adhesion resistant, 865
    metallography, 908
    oxide, 872
    processes, 856
    process selection, 883, 886
    protective, 864
    shear test, 911
    testing, 890
    thickness, 861
    time, 863
    types, 863
    ultimate strength, 911
    wear resistance, 864
Corrosion resistance, 558
Cracking
    carbon equivalent, 479
    sensitivity, 479

Die casting
    casting parameters, 430
    casting segmentation design, 417
    clamping force, 413
    cooling system design, 431, 441
    cross-sectional area reduction, 422
    flow design guidelines, 412
    die characteristic line, 410
    feed system configuration, 414
    gate dimensions, 418
    gate location, 417
    gate velocity, 413, 420
    gooseneck/nozzle characteristic line, 408
    gravity fill time, 413
    heat transfer analysis, 432
    injection force, 412
    machine characteristic line, 407
    metal pressure, 407
    metal-velocity line, 411
    overflow design, 412, 425
    pressure/flow equation, 409
    processes, 401
    P-Q2 diagram, 404
    runner dimensions. 421
    runner shape, 424
    shot system instrumentation, 405
    sprue configuration, 425
    surface temperature, 435
    system design, 403, 404
    theoretical maximum flow rate, 408
    thermal conductivity, 446
    true gate velocity, 421
    vents, 425
    water line placement, 446
    water temperature, 449
Direct thermal processing, 496
Duplex processes, 573

Electrodeposition
    diffusion rate, 879
    effective concentration range, 879
    mechanism, 878
    ultrasonic agitation, 879
Electroless, 868
Extrusion
    alloy development, 201
    aluminum section design, 153
    applications, 154
    approximate temperature field, 195
    area of relief, 150
    burb cycle, 143
    circumscribed diameter, 144
    cross-sectional area, 144
    dies, 141
    die deflection, 200
    Drucker's hardening postulates, 176

[Extrusion]
    dummy block, 143
    ecodesign strategy wheel, 153
    emptying diagrams, 167, 189
    flow conditions, 162, 170
    flow patterns, 165
    flow rule, 177
    friction, 200
    frictionless extrusion, 186
    Geiringer equations, 188
    geometrical deviation, 148
    Hencky's equations, 184
    iso-residence-time curve, 167
    limits of extrudability, 201
    marker material technique, 164
    metallurgy, 168
    Mohr diagram, 179
    parameters, 137
    Prager's method, 186
    press cycle, 145
    process, 4
    quality, 146
    product development, 151
    reduction ratio, 140, 144
    runout table, 144
    section design limitations, 158
    slab model, 194
    straightness angles, 149
    strain hardening, 174
    surface defects, 150
    Tresca criterion, 173
    Valberg plot, 166
    von Mises criterion, 173
    yield stress, 171

Fatigue resistance, 558
Fick's first law of diffusion, 512, 859
Fick's second law of diffusion, 513
Forming, bulk
    adiabatic temperature rise, 8
    bulk deformation, 1
    classification, 1
    cold working, 1
    constant friction factor, 10
    Coulomb's law, 10
    defects, 12
    deformation modeling, 14
    deformation pressure, 9
    Δ parameter, 11
    drawing, 6
    effective stress, 9
    flow behavior classification, 2
Forging
    forging hammers, 4
    formability, 4
    friction, 9

[Forging]
  geometrical design, 11
  hardening, 10
  hot working, 1
  hydraulic presses, 5
  metallurgical design, 12
  Mises criterion, 10
  Mises equivalent strain, 8
  process, 3
  processing temperature, 8
  ring test, 15
  rolling, 5
  slab equilibrium, 15
  slip line method, 17
  strain, 8
  strain rate, 8
  stress, 9
  stress state classification, 9
  Tresca yield criterion, 10
  warm working, 1, 2
  workability, 12
  workability testing, 12
Forming, semisolid
  constitutive equations, 126
  deformation, 117
  die casting, 129
  extrusion, 131
  flow, 118
  flow stress, 120, 127
  forging, 134
  injection molding, 129
  joining, 117
  manufacturing, 128, 130
  morphology, 115
  mushy metal, 115, 116
  rolling, 132
  stirring and mixing, 117
  stress-strain curve, 119
  temperature ranges, 116
  viscosity, 121
  yield criterion, 122, 124
Forming, sheet metal
  anticlastic curvature, 34
  beading, 33
  bending, 33, 36
  bending allowance, 36
  blank holder pressure, 28
  bulging, 25, 33
  classification, 24
  coining, 25
  cold heading, 25
  deep drawing, 23, 25
  deformation modeling, 46
  dome forming, 25
  drawing ratio, 29
  edge rolling, 25

[Forming, sheet metal]
  elastic bending, 34
  Erichsen test, 44
  expanding, 33
  flanging, 25
  folding, 36
  formability, 39
  formability testing, 43
  friction, 37
  friction tests, 45
  geometrical design, 38
  Guerin process, 29
  Hill criterion, 38
  hydroform process, 30
  ironing, 31
  Limiting Drawing Ratio (LDR), 26
  Marform process, 30
  maximum drawing force, 27
  metallurgy and microstructure design, 38
  moment of elastic unloading, 34
  necessary moment, 34
  necking, 25
  neutral line, 35
  processing strain, 37
  processing temperature, 37
  redrawing, 28
  residual stresses, 34
  roll forming, 36
  roller flanging, 36
  rubber forming, 29
  simulative testing, 44
  spinning, 30
  spring back, 34
  strain rate, 37
  stress, 37
  stretcher leveling, 32
  stretching, 24, 32
  Swift test, 44
  yield criteria, 37

Galvanizing, 878

Hall–Petch equation, 76, 285, 480
Hardenability
  depth of hardening, 471
  equivalent bar diameter, 471
  grain size, 473
  Grossmann hardenability, 475
  H-values, 476
  hardenability bands, 475
  hardenability factors, 472
  ideal critical diameter, 472, 475
  influence of carbon, 470
  Jominy end-quench test, 473
Harris equation, 513